Symbol	Description
a_i	Action i in a decision analysis 1015
A^c	Complement of event A 147
A_2	A constant used in constructing $\bar{x}$-charts 727
A, B	Events 131
A $\cup$ B	Union of events A and B 142
A $\cap$ B	Intersection of events A and B 142
α (alpha)	Probability of rejecting H_0 if in fact H_0 is true (Type I error) 342
β (beta)	Probability of accepting H_0 if in fact H_0 is false (Type II error) 343
β_0	y-intercept in regression models 460
$\hat{\beta}_0$	Least squares estimator of β_0 464
β_1	Slope of straight-line regression model 460
$\hat{\beta}_1$	Least squares estimator of β_1 464
β_i	Coefficient of independent variable x_i in a multiple regression model 526
$\hat{\beta}_i$	Least squares estimator of β_i 528
C_t	Cyclical effect in a time series 808
CS	Cost of sampling 1057
χ^2 (chi-square)	Probability distribution of various test statistics 945
d	Test statistic for Durbin–Watson test of autocorrelation 838
D_0	Hypothesized difference between two population means 393
D_3, D_4, d_2, d_3	Constants used in constructing R-charts 744
df	Degrees of freedom for the t, χ^2, and F distributions 318
$E(n_{ij})$	Expected count in cell (i, j) of a contingency table 981
$\hat{E}(n_{ij})$	Estimated expected count in cell (i, j) of a contingency table 981
$E(x)$	Expected value of the random variable x 190
ENGS	Expected net gain of sampling in a decision analysis 1057
$EP(a_i)$	Expected payoff for action i in a decision analysis 1021
EPNS	Expected payoff for no sampling in a decision analysis 1057
EPS	Expected payoff of sampling in a decision analysis 1056
$EU(a_i)$	Expected utility for action i in a decision analysis 1030
EVSI	Expected value of sample information 1056
ϵ (epsilon)	Random error component in regression models 460
$f(x)$	Probability density function for a continuous random variable x 226
F	Test statistic used to compare two variances, to compare p population means, and to test several terms in a multiple regression 412
F_α	Value of F distribution with area α to its right 413
H	Test statistic for the Kruskal–Wallis nonparametric analysis of variance 944
H_a	Alternative (or research) hypothesis 340
H_0	Null hypothesis 340

Symbol	Description
I	Sample information in a decision analysis 1042
I_t	Value of the index I at time t 779
λ (lambda)	(1) Mean of the Poisson distribution 213 (2) Reciprocal of mean for exponential distribution 248
MAD	Mean absolute deviation of a set of forecasts 818
MSE	Estimator of variance, σ^2, in an analysis of variance 533
MST	Mean square for treatments in an analysis of variance 860
μ (mu)	Population mean 53
μ_D	Population mean of differences in a paired-difference experiment 422
μ_i	True (population) mean for ith treatment in an analysis of variance 859
μ_R	Mean of the sampling distribution of R 743
$\mu_{\bar{x}}$	Mean of the sampling distribution of $\bar{x}$ 280
n	(1) Sample size, total number of measurements in a sample 51 (2) Number of trials in a binomial experiment 198
n_D	Number of differences (or pairs) in a paired-difference experiment 423
n_i	(1) Number of observations for treatment i in an analysis of variance 860 (2) Count in ith cell in a multinomial sample 972 (3) Number of sampling units selected from stratum i in stratified random sampling 1093
n_{ij}	Observed count in cell (i, j) in a contingency table 984
$n!$	n factorial; product of first n integers $(0! = 1)$ 137
$\binom{N}{n}$	Number of possible combinations when n elements are drawn without replacement from N elements 137
N	Number of elements in the population 137
N_i	Number of sampling units in stratum i 1093
ν_1 (nu)	Numerator degrees of freedom for an F statistic 412
ν_2	Denominator degrees of freedom for an F statistic 412
O_{ij}	Outcome corresponding to action i and state of nature j in a decision analysis 1010
p	(1) Observed significance level for a hypothesis test 356 (2) Probability of Success for the binomial and geometric distributions 198
$p(x)$	Probability distribution for a discrete random variable x 185
$p_1 - p_2$	Difference between true proportions of success in two independent binomial experiments 433
p_i	Probability of outcome i in a multinomial probability distribution 972
p_{ij}	Probability associated with the (i, j) cell in a contingency table 980
$\hat{p}$	(1) Estimator of binomial probability p; sample proportion of successes 325 (2) Pooled estimator of $p_1 = p_2 = p$ in test of equality of two population proportions 435

STATISTICS

for Business and Economics

STATISTICS

For Business and Economics

SIXTH EDITION

James T. McClave
Info Tech, Inc.
University of Florida

P. George Benson
Graduate School of Management
Rutgers University

DELLEN
an imprint of
MACMILLAN COLLEGE PUBLISHING COMPANY
New York

MAXWELL MACMILLAN CANADA
Toronto

MAXWELL MACMILLAN INTERNATIONAL
New York Oxford Singapore Sydney

. . . .

On the cover: "Untitled" was executed by Los Angeles artist Charles Arnoldi. The work is watercolor on paper. Arnoldi's work may be seen at the Stremmel Gallery in Reno, Nevada, the Hampton Gallery in Venice, California, and the Charles Cowles Gallery in New York City. His work is included in the permanent collections of the Albright-Knox Museum in Buffalo, New York, the Chicago Art Institute, the Museum of Modern Art in New York City, the Metropolitan Museum, the Los Angeles County Museum, and the San Francisco Museum of Modern Art.

Printed in the United States of America.

Macmillan College Publishing Company
866 Third Avenue
New York, NY 10022

Macmillan College Publishing Company is part of the
Maxwell Communication Group of Companies.

Maxwell Macmillan Canada, Inc.
1200 Eglinton Avenue East, Suite 200
Don Mills, Ontario M3C 3Nl

Library of Congress Cataloging-in-Publication Data

McClave, James T.
Statistics for business and economics / James T. McClave, P. George Benson. —6th ed.
p. cm.
Includes indexes.
ISBN 0-02-379201-9
1. Commercial statistics. 2. Economics—Statistical methods. 3. Statistics. I. Benson, P. George, 1946– . II. Title.
HF1017.M36 1994
519.5—dc20 93-33416
CIP

Printing: 1 2 3 4 5 6 7 8 9 Year: 3 4 5 6 7

Contents

CHAPTER THREE

Probability 127

CHAPTER FOUR

Discrete Random Variables 181

CHAPTER FIVE

Continuous Random Variables 225

CHAPTER SIX

Sampling Distributions 269

CHAPTER SEVEN

CHAPTER EIGHT

CHAPTER NINE

CHAPTER TEN

Simple Linear Regression 457

CHAPTER ELEVEN

Multiple Regression 525

CHAPTER TWELVE

Introduction to Model Building 609

CHAPTER SIXTEEN

Design of Experiments and Analysis of Variance 851

CHAPTER SEVENTEEN

Nonparametric Statistics 919

CHAPTER EIGHTEEN

The Chi-Square Test and the Analysis of Contingency Tables 971

CHAPTER NINETEEN

Decision Analysis 1003

CHAPTER TWENTY

APPENDIX A

APPENDIX B

Preface

The sixth edition of *Statistics for Business and Economics* has four primary objectives: (1) increased emphasis on methods for quality improvement, (2) greater use of statistical computer output, (3) many new business applications, and (4) to accomplish these objectives without increasing the total size of the text.

Accomplishing the first objective includes adding flowcharting, cause-and-effect diagrams, and additional systems theory to Chapter 13, Methods for Quality Improvement. The authors believe that this chapter represents the most complete and authoritative treatment of quality in any introductory statistics text. The second and third objectives were achieved by adding many modern business applications, with emphasis on computer output, and the interpretation of observed significance levels (p-values) that are typically produced by statistical software.

The fourth objective is perennially (or, more accurately, triennially) our most difficult. When we poll our users, they overwhelmingly agree that the text is too large, but they cannot agree on what should be eliminated. In fact, many of our reviewers advise us that we should eliminate nothing, while at the same time lamenting the size of the book. We decided to be more aggressive in our quest for brevity than in the past, and have taken the following actions:

1. Chapter 2 (Graphical Descriptions of Data) and Chapter 3 (Numerical Descriptive Measures) have been combined, with coverage of some of the graphical methods relating to categorical (nominal and ordinal) data through examples, case studies, and exercises rather than separate sections.
2. Randomized block designs have been eliminated from the Analysis of Variance and Nonparametric Statistics chapters.
3. Moving averages have been removed from the Time Series chapters, as has the section Forecasting with Autoregressive Models.
4. The section on the Expected Value of Perfect Information has been eliminated from the Decision Analysis chapter.
5. The exercises have been updated, with many new business applications added; however, slightly more "dated" exercises were eliminated than new ones added, providing additional space-savings.
6. Cluster sampling has been eliminated from the Survey Sampling chapter.

While these decisions were most difficult, we believe that the slight reduction in size has been achieved at a relatively modest cost. As usual, we anxiously await our users' assessment.

A number of other minor changes were made to make the text more readable, including separation of the concepts of mutually exclusive and independent events into separate sections of the Probability chapter, condensing the discussion of residual analysis in the regression chapters, and several others.

We continue to make statistical inference the primary theme of the text, but we think that our treatment of inference blends well with contemporary thinking about teaching statistics. For example, if you want to emphasize processes and process improvement in teaching statistics, you might cover all of Chapter 1, followed immediately by the first five sections of Chapter 13, Methods for Quality Improvement. Then later, after covering the relevant sections of Chapters 6, 7, and 8, return to Chapter 13 and cover the last six sections on statistical process control and process diagnoses. An even fuller treatment of processes would include time series and forecasting in Chapters 14 and 15. Chapters 13, 14, and 15 are devoted entirely to the study of processes and their output.

We have maintained the features of this text that we believe make it unique among introductory statistics texts for business courses. These features, which assist the student in achieving an overview of statistics and an understanding of its relevance in the solution of business problems, are as follows:

1. **Case Studies.** (See the list of case studies on page xxi.) Many important concepts are emphasized by the inclusion of case studies, which consist of brief summaries of actual business applications of the concepts and are often drawn directly from the business literature. These case studies allow the student to see business applications of important statistical concepts immediately after introduction of the concepts. The case studies also help to answer by example the often asked questions, "Why should I study statistics? Of what relevance is statistics to business?" Finally, the case studies constantly remind the student that each concept is related to the dominant theme—statistical inference.
2. **Where We've Been . . . Where We're Going . . .** The first page of each chapter is a "unification" page. Our purpose is to allow the student to see how the chapter fits into the scheme of statistical inference. First, we briefly show how the material presented in previous chapters helps us to achieve our goal (Where We've Been). Then, we indicate what the next chapter (or chapters) contributes to the overall objective (Where We're Going). This feature allows us to point out that we are constructing the foundation block by block, with each chapter an important component in the structure of statistical inference. Furthermore, this feature provides a series of brief résumés of the material covered as well as glimpses of future topics.
3. **Many Examples and Exercises.** We believe that most students learn by doing. The text contains many worked examples to demonstrate how to solve various types of problems. We then provide the student with a large number (more than

1,300) of exercises. The answers for odd-numbered exercises are included at the end of the text. The exercises are of two types:

a. **Learning the Mechanics.** These exercises are intended to be straightforward applications of the new concepts. They are introduced in a few words and are unhampered by a barrage of background information designed to make them "practical," but which often detracts from instructional objectives. Thus, with a minimum of labor, the student can recheck his or her ability to comprehend a concept or a definition.

b. **Applying the Concepts.** The mechanical exercises described above are followed by realistic exercises that allow the student to see applications of statistics to the solution of problems encountered in business and economics. Once the mechanics are mastered, these exercises develop the student's skills at comprehending realistic problems that describe situations to which the techniques may be applied.

4. **On Your Own . . .** The chapters end with an exercise entitled **On Your Own . . . .** The intent of this exercise is to give the student some hands-on experience with a business application of the statistical concepts introduced in the chapter. In most cases, the student is required to collect, analyze, and interpret data relating to some business phenomenon.

5. **Using the Computer.** Another feature at the end of most chapters encourages the use of computers in the analysis of real data. A demographic data base, consisting of 1,000 observations on 15 variables, has been described in Appendix C and is available on diskette from the publisher. Each **Using the Computer** section provides one or more computer exercises that utilize the data in Appendix C and enhance the new material covered in the chapter.

6. **A Simple, Clear Style.** We have tried to achieve a simple and clear writing style. Subjects that are tangential to our objective have been avoided, even though some may be of academic interest to those well versed in statistics. We have not taken an encyclopedic approach in the presentation of material.

7. **An Extensive Coverage of Multiple Regression Analysis and Model Building.** This topic represents one of the most useful statistical tools for the solution of business problems. Although an entire text could be devoted to regression modeling, we feel that we have presented a coverage that is understandable, usable, and much more comprehensive than the presentations in other introductory business statistics texts. We devote three chapters to discussing the major types of inferences that can be derived from a regression analysis, showing how these results appear in computer printouts and, most important, selecting multiple regression models to be used in an analysis. Thus, the instructor has the choice of a one-chapter coverage of simple regression, a two-chapter treatment of simple and multiple regression, or a complete three-chapter coverage of simple regression, multiple regression, and model building. This extensive coverage of such useful statistical tools will provide added evidence to the student of the relevance of statistics to the solution of business problems.

8. Footnotes and Appendix A. Although the text is designed for students with a noncalculus background, footnotes explain the role of calculus in various derivations. Footnotes are also used to inform the student about some of the theory underlying certain results. Appendix A presents some useful counting rules for the instructor who wishes to place greater emphasis on probability. Consequently, we think the footnotes and Appendix A provide an opportunity for flexibility in the mathematical and theoretical level at which the material is presented.
9. Supplementary Material. Solutions manuals, a study guide, a Minitab supplement, an integrated software system, a computer-generated test system, a test bank, and a 1,000-observation demographic data base, and data on disk are available.
   a. Student's Solutions Manual (by Nancy S. Boudreau). The student's solutions manual presents detailed solutions to most odd-numbered exercises in the text. Many points are clarified and expanded to provide maximum insight into and benefits from each exercise.
   b. Instructor's Solutions Manual (by Nancy S. Boudreau). The instructor's solutions manual presents the full solutions to the even-numbered exercises contained in the text. For adopters, the manual is complimentary from the publisher.
   c. Study Guide (by Susan L. Reiland). For each chapter the study guide includes (1) a brief summary that highlights the concepts and terms introduced in the textbook; (2) section-by-section examples with detailed solutions; and (3) exercises (with answers provided at the end of the study guide) that allow the student to check mastery of the material in each section.
   d. Minitab Supplement (by Ruth K. Meyer and David D. Krueger). The Minitab computer supplement was developed to be used with Minitab Release 8.0, a general-purpose statistical computing system. The supplement, which was written especially for the student with no previous experience with computers, provides step-by-step descriptions of how to use Minitab effectively as an aid in data analysis. Each chapter begins with a list of new commands introduced in the chapter. Brief examples are then given to explain new commands, followed by examples from the text illustrating the new and previously learned commands. Where appropriate, simulation examples are included. Exercises, many of which are drawn from the text, conclude each chapter.

      A special feature of the supplement is a chapter describing a survey sampling project. The objectives of the project are to illustrate the evaluation of a questionnaire, provide a review of statistical techniques, and illustrate the use of Minitab for questionnaire evaluation.
   e. ASP statistical software diskette. New to this edition, the text includes a 3½" diskette containing the ASP program, A *Statistical Package for Business, Economics, and the Social Sciences*. ASP, from DMC Software, Inc., is a user-friendly, totally menu-driven program that contains all of the major statistical applications covered in the text, plus many more. ASP runs on any IBM-compatible PC with at least 516K of memory. With ASP, students with no

knowledge of computer programming can create and analyze data sets easily and quickly. Appendix E contains start-up procedures and a short tutorial on the use of ASP. Full documentation is provided complimentary to adopters of the text.

f. **ASP Tutorial and Student Guide** (by George Blackford). Most students have little trouble learning to use ASP without documentation. Some, however, may want to purchase the *ASP Tutorial and Student Guide*. Bookstores can order the tutorial from DMC Software, Inc., 6169 Pebbleshire Drive, Grand Blanc, MI 48439.

g. **Test Bank** (by Mark Dummeldinger). This manual provides a large number of test items utilizing real data.

h. **Dellen Test.** This unique computer-generated random test system is available to instructors without cost. Utilizing an IBM PC computer and a number of commonly used dot-matrix printers, the system will generate an almost unlimited number of quizzes, chapter tests, final examinations, and drill exercises. At the same time, the system produces an answer key and student worksheet with an answer column that exactly matches the column on the answer key.

i. **Data Base.** A demographic data set was assembled based on a systematic random sample of 1,000 U.S. zip codes. Demographic data for each zip code area selected were supplied by CACI, an international demographic and market information firm. Fifteen demographic measurements (including population, number of households, median age, median household income, variables related to the cost of housing, educational levels, the work force, and purchasing potential indexes based on the Bureau of the Census Consumer Expenditure Surveys) are presented for each zip code area.

   Some of the data are referenced in the **Using the Computer** sections. The objectives are to enable the student to analyze real data in a relatively large sample using the computer, and to gain experience using the statistical techniques and concepts on real data.

j. **Data on Disk.** We have placed all of the large exercise data sets, as well as the Demographic Data Base described in the preceding paragraphs, on a computer disk. The Appendix of the text contains an index that describes these data sets, indicates the exercise number, the page number, and the file names of the data sets. Instructors who adopt the text may obtain the disk by writing to Dellen Publishing Company, 400 Pacific Avenue, San Francisco, California 94133, or by calling 415-433-9900.

Acknowledgments

We owe thanks to the many people who assisted in reviewing and preparing this textbook. Their names are listed on pages xviii–xx. We particularly acknowledge the editorial assistance of Susan L. Reiland, the typing and assistance of Brenda Dobson, and the administrative support of Jane Oas Benson. Without these three, we never could have completed this work.

Gordon J. Alexander
University of Minnesota

Richard W. Andrews
University of Michigan

Larry M. Austin
Texas Tech University

Golam Azam
North Carolina Agricultural & Technical University

Donald W. Bartlett
University of Minnesota

Clarence Bayne
Concordia University

Carl Bedell
Philadelphia College of Textiles and Science

David M. Bergman
University of Minnesota

William H. Beyer
University of Akron

Atul Bhatia
University of Minnesota

Jim Branscome
University of Texas at Arlington

Francis J. Brewerton
Middle Tennessee State University

Daniel G. Brick
University of St. Thomas

Robert W. Brobst
University of Texas at Arlington

Michael Broida
Miami University of Ohio

Glenn J. Browne
University of Maryland, Baltimore

Edward Carlstein
University of North Carolina at Chapel Hill

John M. Charnes
University of Miami

Chih-Hsu Cheng
Ohio State University

Larry Claypool
Oklahoma State University

Edward R. Clayton
Virginia Polytechnic Institute and State University

Ronald L. Coccari
Cleveland State University

Ken Constantine
University of New Hampshire

Robert Curley
University of Central Oklahoma

Joyce Curry-Daly
California Polytechnic State University

Jim Daly
California Polytechnic State University

Jim Davis
Golden Gate University

Dileep Dhavale
University of Northern Iowa

Mark Eakin
University of Texas at Arlington

Rick L. Edgeman
Colorado State University

Carol Eger
Stanford University

Robert Elrod
Georgia State University

Douglas A. Elvers
University of North Carolina at Chapel Hill

Iris Fetta
Clemson University

Susan Flach
General Mills, Inc.

Alan E. Gelfand
University of Connecticut

Joseph Glaz
University of Connecticut

Edit Gombay
University of Alberta

Paul W. Guy
California State University, Chico

Michael E. Hanna
University of Texas at Arlington

Don Holbert
East Carolina University

James Holstein
University of Missouri, Columbia

Warren M. Holt
Southeastern Massachusetts University

Steve Hora
University of Hawaii, Hilo

Petros Ioannatos
GMI Engineering & Management Institute

Marius Janson
University of Missouri, St. Louis

Ross H. Johnson
Madison College

Timothy J. Killeen
University of Connecticut

David D. Krueger
St. Cloud State University

Richard W. Kulp
Wright-Patterson AFB, Air Force Institute of Technology

Martin Labbe
State University of New York College at New Paltz

James Lackritz
California State University at San Diego

Philip Levine
William Patterson College

Eddie M. Lewis
University of Southern Mississippi

Fred Leysieffer
Florida State University

Pi-Erh Lin
Florida State University

Robert Ling
Clemson University

Karen Lundquist
University of Minnesota

G. E. Martin
Clarkson University

Brenda Masters
Oklahoma State University

Ruth K. Meyer
St. Cloud State University

Paul I. Nelson
Kansas State University

Paula M. Oas
General Office Products

Dilek Önkal
Bilkent University, Turkey

Vijay Pisharody
University of Minnesota

P. V. Rao
University of Florida

Don Robinson
Illinois State University

Jan Saraph
St. Cloud State University

Craig W. Slinkman
University of Texas at Arlington

Robert K. Smidt
California Polytechnic State University

Donald N. Steinnes
University of Minnesota at Duluth

Virgil F. Stone
Texas A & I University

Katheryn Szabet
La Salle University

Alireza Tahai
Mississippi State University

Chipei Tseng
Northern Illinois University

Pankaj Vaish
Arthur Andersen & Company

Robert W. Van Cleave
University of Minnesota

Charles F. Warnock
Colorado State University

William J. Weida
United States Air Force Academy

T. J. Wharton
Oakland University

Kathleen M. Whitcomb
University of South Carolina

Edna White
Florida Atlantic University

Steve Wickstrom
University of Minnesota

James Willis
Louisiana State University

Douglas A. Wolfe
Ohio State University

Gary Yoshimoto
St. Cloud State University

Fike Zahroon
Moorhead State University

Christopher J. Zappe
Bucknell University

Case Studies

STATISTICS

for Business and Economics

CHAPTER ONE

What Is Statistics?

Contents

Case Studies

Where We're Going

What is statistics? Is it a field of study, a group of numbers that summarize a business operation, or—as the title of a popular book (Tanur et al., 1989) suggests—"a guide to the unknown"? We will begin to see in Chapter 1 that each of these descriptions has some applicability in understanding statistics. We will see that *descriptive statistics* focuses on developing numerical summaries that describe some business phenomenon, whereas *inferential statistics* uses numerical summaries to assist in making business decisions. Our main objective is to show how statistics can be useful to you in business decision-making, so the primary theme of this text is inferential statistics.

1.1 Statistics: What Is It?

What does statistics mean to you? Does it bring to mind batting averages, the Dow Jones Industrial Average, unemployment figures, numerical distortions of facts (lying with statistics!), or simply a college requirement you have to complete? We hope to convince you that statistics is a meaningful, useful science with a broad, almost limitless scope of application to business and economic problems. We also want to show that statistics lie only when they are misapplied. Finally, our objective is to paint a unified picture of statistics to leave you with the impression that your time was well spent studying a subject that will prove useful to you in many ways.

Statistics means "numerical descriptions" to most people. The Dow Jones Industrial Average, monthly unemployment figures, and the fraction of women executives in a particular industry are all statistical descriptions of large sets of data. Often, the purpose of calculating these numbers goes beyond the description of the particular set of data. Frequently, the data are regarded as a sample selected from some larger set of data whose characteristics we wish to estimate. For example, a sampling of unpaid accounts for a large merchandiser would allow you to calculate an estimate of the average value of unpaid accounts. This estimate could be used as an audit check on the total value of all unpaid accounts held by the merchandiser. So, the applications of statistics can be divided into two broad areas: **descriptive** and **inferential** statistics.

> **Descriptive statistics** utilizes numerical and graphical methods to look for patterns, summarize, and present the information in a set of data.
>
> **Inferential statistics** utilizes sample data to make estimates, predictions, or other generalizations about a larger set of data, frequently as an aid to decision making.

Although both descriptive and inferential statistics are discussed in the following chapters, the primary theme of the text is **inference.** Let us examine some case studies that illustrate applications of descriptive and inferential statistics in business and government.

CASE STUDY 1.1 / The Consumer Price Index

A data set of interest to virtually all Americans is the set of prices charged for goods and services in the U.S. economy. The general upward movement in this set of prices is referred to as *inflation*; the general downward movement is referred to as *deflation*. In order to *estimate* the change in prices over time, the Bureau of Labor Statistics (BLS) of the U.S. Department of Labor developed the Consumer Price Index (CPI). Each month, the BLS collects price data about a specific collection of goods and services (called a *market basket*) from 85 urban areas around the country. Statistical procedures are used to compute the CPI (a descriptive statistic) from this sample price data and other information about consumers' spending habits. By comparing the level of the CPI at different points in time, it is possible to *estimate* (make an inference about) the rate of inflation over particular time intervals and to compare the purchasing power of a dollar at different points in time.

One major use of the CPI as an index of inflation is as an indicator of the success or failure of government economic policies. A second use of the CPI is to escalate income payments. Millions of workers have *escalator clauses* in their collective bargaining contracts; these clauses call for increases in wage rates based on increases in the CPI. In addition, the incomes of Social Security beneficiaries and retired military and federal civil service employees are tied to the CPI. It has been estimated that a 1% increase in the CPI can trigger an increase of over $1 billion in income payments. Thus, it can be said that the very livelihoods of millions of Americans depend on the behavior of a statistical estimator, the CPI (U.S. Department of Labor, 1978). [*Note:* We discuss the Consumer Price Index in greater detail in Chapter 14.]

CASE STUDY 1.2 / Taste-Preference Scores for Beer

Two sets of data of interest to the marketing department of a food-products firm are (1) the set of taste-preference scores given by consumers to its product and to competitors' products when all brands are clearly labeled and (2) the taste-preference scores given by the same set of consumers when all brand labels have been removed and the consumer's only means of product identification is taste. With such information, the marketing department should be able to determine whether taste preference arose because of perceived physical differences in the products or as a result of the consumer's image of the brand. (Brand image is, of course, largely a result of a firm's marketing efforts.) Such a determination should help the firm develop marketing strategies for its product.

A study using these two types of data was conducted by Ralph Allison and Kenneth Uhl (1965) to determine whether beer drinkers could distinguish among major brands of unlabeled beer. A sample of 326 beer drinkers was randomly selected from the set of beer drinkers identified as males who drank beer at least three times a week. During the first week of the study, each of the 326 participants was given a six-pack of unlabeled beer containing three major brands and was asked to taste-rate each beer on a scale from 1 (poor) to 10 (excellent). During the second week, the same set of drinkers was given a six-pack containing six major brands. This time, however, each bottle carried its usual label. Again, the drinkers were asked to taste-rate each beer from 1 to 10. From a statistical analysis of the two sets of data yielded by the study, Allison and Uhl concluded that the 326 beer drinkers studied could not distinguish among brands by taste on an overall basis. This result enabled them to *infer* statistically that such was also the case for beer drinkers in general. Their results also indicated that brand labels and their associations did significantly influence the tasters' evaluations. These findings suggest that physical differences in the products have less to do with their success or failure in the marketplace than the image of the brand in the consumers' minds. As to the benefits of such a study, Allison and Uhl note, "to the extent that product images, and their changes, are believed to be a result of advertising . . . the ability of firms' advertising programs to influence product images can be more thoroughly examined."

CASE STUDY 1.3 / Monitoring the Unemployment Rate

The employment status (employed or unemployed) of each individual in the U.S. work force is a set of data that is of interest to economists, businesspeople, and sociologists. These data provide information on the social and economic health of our society. To obtain information about the employment status of the work force, the U.S. Bureau of the Census conducts what is known as the *Current Population Survey*. Each month approximately 1,500 interviewers visit about 59,000 of the 91.9 million households in the United States and question the occupants over 14 years of age about their employment status. Their responses enable the Bureau of the Census to *estimate* the percentage of people in the labor force who are unemployed (the *unemployment rate*). Thus, a *statistical estimator* serves as a monthly indicator of the nation's economic welfare.

Perhaps you are wondering how a reliable estimate of this percentage can be obtained from a sample that includes only about .1% of the households in the United States. The answer lies in the method used to select the sample of households. The method was designed to enable the Bureau of the Census to control the precision of its estimate while obtaining a sample that is representative of the set of all households in the country. That reliable estimates of nationwide characteristics can be obtained from relatively small sample sizes is an illustration of the power of statistics (U.S. Department of Commerce, 1978). [*Note:* We discuss sampling methods in detail in Chapter 20.]

CASE STUDY 1.4 / Auditing Parts and Equipment for Airline Maintenance

The United Airlines Maintenance Base in San Francisco is responsible for the maintenance and overhaul of all United Airlines aircraft. Its storeroom receives, stores, and distributes all the parts needed for maintenance of the aircraft. To control the stock of spare parts and to determine the value of parts on hand, *inventory counts* (a descriptive statistic) of the number of each item in stock are taken. It is the responsibility of the Auditing Division of United Airlines to verify the accuracy of the inventory counts. Rather than verifying the accuracy of the counts by recounting all of the inventory item groups, the accountants sample a small number of these groups and recount them. If they find a large number of discrepancies between the original counts and their sample counts, they *infer* that many of the rest of the item counts (those not sampled and recounted) are also in error. They conclude that the original inventory counts are unacceptable and must be recounted. On the other hand, if they find only a small number of discrepancies, they *infer* that most of the item counts not rechecked are accurate, and conclude that the original inventory counts are satisfactory.

Before this inferential statistical procedure was implemented, the Auditing Division verified inventory counts by recounting all the items in stock. The inferential procedure enables the accountants to maintain the quality of their verifications with a substantial reduction in work-hours (Hunz, 1956).

CASE STUDY 1.5 / The Decennial Census of the United States

The following description is quoted from the U.S. Bureau of the Census, *Statistical Abstract of the United States: 1992*:

The U.S. Constitution provides for a census of the population every 10 years, primarily to establish a basis for apportionment of members of the House of Representatives among the States. For over a century after the first census in 1790, the census organization was a temporary one, created only for each decennial census. In 1902, the Bureau of the Census was established as a permanent Federal agency, responsible for enumerating the population and also for compiling statistics on other subjects.

The census of the population is a complete count. That is, an attempt is made to account for every person, for each person's residence, and for other characteristics (sex, age, family relationships, etc.). Since the 1940 census, in addition to the complete count information, some data have been obtained from representative samples of the population. In the 1990 census, variable sampling rates were employed. For most of the country, one in every six households (about 17%) received the long form or sample questionnaire; in governmental units estimated to have fewer than 2500 inhabitants, every other household (50%) received the sample questionnaire to enhance the reliability of sample data for small areas. Exact agreement is not to be expected between sample data and the complete census count.

Census statistics regarding total numbers of people in various age groups are examples of numerical *descriptions* that require no statistical inference, since they are (purportedly) complete counts. However, income data collected by the Bureau of the Census from "representative samples" might be used to make *inferences* about the incomes of *all* persons. You will learn that the reliability of statistical inferences is dependent on the sampling procedure, characteristics of the data, and the methodology employed to make the inferences.

Why study statistics in a business program? The quantification of business research and business operations (quality control, statistical auditing, forecasting, etc.) has been truly astounding over the past several decades. Econometric modeling, market surveys, and the creation of indexes such as the Consumer Price Index all represent relatively recent attempts to quantify economic behavior. It is extremely important that today's business graduate understand the methods and language of statistics, since the alternative is to be swamped by a flood of numbers that are more confusing than enlightening to the untutored mind. The business student should develop a discerning sense of rational thought that will distill the information contained in these numbers so it can be used to make intelligent decisions, inferences, and generalizations. We believe that the study of statistics is essential to the ability to operate effectively in the modern business environment.

1.2 The Elements of Statistics

Statistical methods are particularly useful for studying, analyzing, and learning about **populations**.

> **Definition 1.1**
>
> A **population** is a set of existing units (usually people, objects, transactions, or events).

Examples of populations include (1) all employed workers in the United States, (2) all registered voters in California, (3) everyone who has purchased a particular brand of cellular telephone, (4) all the cars produced last year by a particular assembly line, (5) the current stock of spare parts at United Airlines' maintenance facility, (6) all sales made at the drive-in window of a fast-food restaurant during a given year, and (7) the set of all accidents occurring on a particular stretch of interstate highway during a holiday period. Notice that the first three population examples (1–3) are sets (groups) of people; the next two (4–5) are sets of objects; the next (6) is a set of transactions; and the last (7) is a set of events.

In studying a population, we focus on one or more characteristics or properties of the units in the population. For example, we may be interested in the age, income, or the number of years of education of the people currently unemployed in the United States. We call such characteristics **variables**.

> **Definition 1.2**
>
> A **variable** is a characteristic or property of an individual population unit. The name *variable* is derived from the fact that any particular characteristic may *vary* among the units in a population.

In studying a particular variable, it is helpful—as we will see in forthcoming chapters—to be able to obtain a numerical representation for the variable. Thus, when numerical representations are not readily available, the process of measurement plays an important supporting role in statistical studies. **Measurement** is the process by which numbers are assigned to variables of individual population units. Measurement may entail asking a consumer to rate the taste of a product on a scale from 1 to 10 or simply asking a worker how old she is. Frequently, however, it involves the use of instruments such as stopwatches, scales, calipers, etc. We discuss measurement in more detail in the next chapter.

If the population you wish to study is small in size, then it is feasible to measure a variable for every unit in the population. For example, if you are measuring the starting salary for all University of Michigan MBA graduates in 1993, it is at least feasible to obtain every salary. When we measure a variable for every unit of a population, it is called a **census** of the population. However, the populations of interest in business problems are typically much larger, involving perhaps many thousands of units. Some examples of large populations were given after Definition 1.1; others are all invoices produced in the last year by a Fortune 500 company, all potential buyers of a new facsimile machine, and all stockholders of a firm listed on the New York Stock Exchange. In studying such populations it would typically be too time-consuming or too costly to conduct a census. A more reasonable alternative would be to select and study a subset (a portion) of the units in the population.

Thus, for example, instead of examining all 15,472 invoices produced by a company during a given year, an auditor may select and examine a **sample** of only 100 invoices. If he is interested in the variable *dollar value*, then he would record (measure) the dollar value of each invoice.

Definition 1.3

A **sample** is a subset of the units of a population.

The method of selecting the sample is called the sampling procedure, or sampling plan. One very important sampling procedure is **random sampling**, one that assures that every subset of units in the population has the same chance of being included in the sample. Thus, if an auditor samples 100 of the 15,472 invoices in the population so that every invoice (and subset of invoices) has an equal chance of being included in the sample, he has devised a random sample. Random sampling is discussed in Chapter 3, and various other sampling procedures are discussed in Chapter 20.

After selecting the sample and measuring the variable(s) of interest for every sampled unit, the information contained in the sample is used to make *inferences* about the population.

Definition 1.4

A **statistical inference** is an estimate, prediction, or other generalization about a population based on information contained in a sample.

That is, *we use the information contained in the smaller sample to learn about the larger population.** Thus, from an examination of the sample of 100 invoices, the auditor may estimate the total number of invoices containing errors in the population

*The terms *population* and *sample* are often used to refer to the sets of measurements themselves, in addition to the units on which the measurements are made. For applications in which a single variable of interest is being measured, this will cause little confusion. When the terminology is potentially ambiguous, the measurements will be referred to as *population data sets* and *sample data sets*, respectively.

of 15,472 invoices or predict the total number of invoices with errors in next year's population of invoices.

Managers use statistical inferences to guide and support their decisions. For example, the auditor's inference about the quality of the firm's invoices can be used in deciding whether to modify the firm's billing operations. The decision to market a new product may hinge on an inference about the willingness of a population of consumers to buy the product—an inference based on the results of testing the product on a sample of consumers.

The preceding definitions identify four of the five elements of an inferential statistical problem: a population, one or more variables of interest, a sample, and an inference. The fifth—and perhaps most important—is a measure of reliability for the inference. This is the topic of Section 1.3.

EXAMPLE 1.1

A large paint retailer has had numerous complaints from customers about underfilled paint cans. As a result, the retailer has begun inspecting incoming shipments of paint from suppliers. Shipments with underfill problems will be returned to the supplier. A recent shipment contained 2,440 gallon-size cans. The retailer randomly selected 100 cans and weighed each on a scale capable of measuring weight to four decimal places. Properly filled cans weigh 10 pounds.

a. Describe the population.

b. Describe the variable of interest.

c. Describe the sample.

d. Describe the inference.

Solution

a. The population is the set of units of interest to the retailer, which is the shipment of 2,440 cans of paint.

b. The weight of the paint cans is the variable the retailer wishes to evaluate.

c. The sample must be a subset of the population. In this case, it is the 100 cans of paint selected by the retailer. If the 100 cans represent a random sample, then the sampling procedure used must be such that each of the 2,440 cans had an equal chance of being included in the sample.

d. The inference of interest involves the *generalization* of the information contained in the weights of the sample of paint cans to the population of paint cans. In particular, the retailer wants to learn about the extent of the underfill problem (if any) in the population. This might be accomplished by finding the average* weight of the cans in the sample and using it to estimate the average weight of the cans in the population.

*Although we will not formally define the term *average* until Chapter 2, *typical* or *middle* can be substituted here without confusion.

EXAMPLE 1.2

Cola wars is the popular media term for the intense competition between the marketing campaigns of Coca-Cola and Pepsi. The campaigns have featured movie and television stars, rock videos, athletic endorsements, and claims of consumer preference based on taste tests. Suppose a particular Pepsi bottler gives 1,000 cola consumers in the bottler's marketing region a "blind" taste test (i.e., a taste test in which the two brand names are disguised). Each consumer is asked to state a preference for brand A or brand B.

a. Describe the population.
b. Describe the variable of interest.
c. Describe the sample.
d. Describe the inference.

Solution

a. The population of interest to the Pepsi bottler is the collection or set of all cola consumers in the marketing region.
b. The characteristic of each cola consumer that the bottler wishes to study is the consumer's cola preference as revealed under the conditions of a blind taste test. Thus, cola preference is the variable of interest.
c. The sample is the group of 1,000 cola consumers from the bottler's marketing region.
d. The inference of interest is the *generalization* of the cola preferences of the 1,000 sampled consumers to the population of all cola consumers in the bottler's marketing region. In particular, the preferences of the consumers in the sample can be used to *estimate* the percentage of all cola consumers in the region who prefer each brand.

1.3 Statistics: Witchcraft or Science?

The primary objective of statistics is inference. In the previous section, we described inference as making generalizations about populations based on information contained in a sample. But making the inference is only part of the story. We also need to know how good the inference is. The only way we could be reasonably certain that an inference about a population is correct would be to include the entire population in our sample. But, due to resource constraints (i.e., insufficient time or money), this is generally not an option. In basing inferences on only a portion of the population (a sample), we introduce an element of uncertainty into our inferences. In general, the smaller the sample size, the less certain we are about the inference. Thus, an inference based on a sample of size 5 is (usually) less reliable than an inference based on a sample of size 100. Consequently, whenever possible, it is important to determine and report the **reliability** of each inference made; this is the fifth element of a statistical problem.

The measure of reliability that accompanies an inference separates the science of statistics from the art of fortune-telling. A palm reader, like a statistician, may examine a sample (your hand) and make inferences about the population (your life). However, unlike statistical inferences, no measure of reliability can be attached to the palm reader's inferences.

Suppose, as in Example 1.1, we are interested in estimating the average weight of a population of paint cans from the average weight of a sample of cans. Using statistical methods, we can determine a *bound on the estimation error*. This bound is simply a number that our estimation error (the difference between the average weight of the sample and the average weight of the population of cans) is not likely to exceed. We will see in later chapters that this bound is a measure of the uncertainty of our inference. The reliability of statistical inferences is discussed throughout this text. For now, we simply want you to realize that an inference is incomplete without a measure of its reliability.

We conclude this section with a summary of the elements of inferential statistical problems and an example to illustrate a measure of reliability.

Five Elements of Inferential Statistical Problems

1. The population of interest
2. One or more variables (characteristics of the population units) that are to be investigated
3. The sample of population units
4. The inference about the population based on information contained in the sample
5. A measure of reliability for the inference

EXAMPLE 1.3

Refer to Example 1.2, in which 1,000 consumers indicated their cola preferences in a taste test. Describe how the reliability of an inference concerning the preferences of all cola consumers in the Pepsi bottler's marketing region could be measured.

Solution

When the preferences of 1,000 consumers are used to estimate the preferences of all consumers in the region, the estimate will not exactly mirror the preferences of the population. For example, if the taste test shows that 56% of the 1,000 consumers preferred Pepsi, it does not follow (nor is it likely) that exactly 56% of all cola drinkers in the region prefer Pepsi. Nevertheless, we may be able to use sound statistical reasoning (which is presented later in the text) to ensure that the sampling procedure used will generate estimates that are almost certainly within a specified limit of the true percentage of all consumers who prefer Pepsi. For example, such reasoning might assure us that the estimate of the preference for Pepsi is almost certainly within 5% of the actual population preference. The implication is that the actual preference for

Pepsi is between 51% [i.e., (56 − 5)%] and 61% [i.e., (56 + 5)%]. This interval represents a measure of reliability for the inference.

1.4 Processes (Optional)

Sections 1.2 and 1.3 focused on the use of statistical methods to analyze and learn about populations, which are sets of *existing* units. Statistical methods are equally useful for analyzing and making inferences about **processes**.

> **Definition 1.5**
>
> A **process** is a series of actions or operations that transforms inputs to outputs. A process produces or generates output over time.

The most obvious processes that are of interest to businesses are production or manufacturing processes. A manufacturing process uses a series of operations performed by people and machines to convert inputs, such as raw materials and parts, to finished products (the outputs). Examples include the process used to produce the paper on which these words are printed, automobile assembly lines, and oil refineries.

Figure 1.1 presents a general description of a process and its inputs and outputs. In the context of manufacturing, the process in the figure (i.e., the transformation process) could be a depiction of the overall production process or it could be a depiction of one of the many processes (sometimes called subprocesses) that exist within an overall production process. Thus, the output shown could be finished goods that will be shipped to an external customer or merely the output of one of the steps or subprocesses of the overall process. In the latter case, the output becomes input for the next subprocess. For example, Figure 1.1 could represent the overall automobile assembly process, with its output being fully assembled cars ready for shipment to dealers. Or, it could depict the windshield-assembly subprocess, with its output of partially assembled cars with windshields ready for "shipment" to the next subprocess in the assembly line.

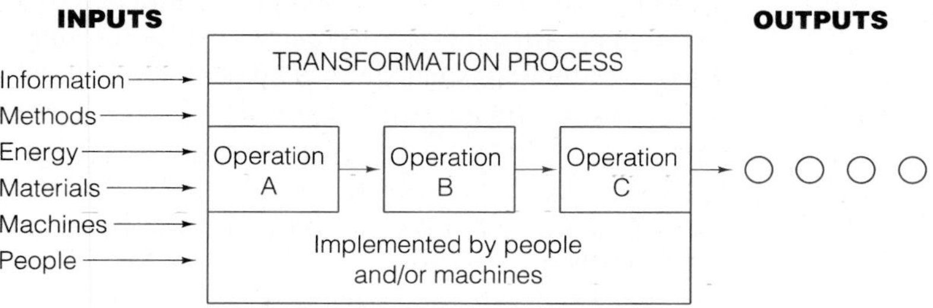

FIGURE 1.1 ► Graphical depiction of a manufacturing process

Besides physical products and services, businesses and other organizations generate streams of numerical data over time that are used to evaluate the performance of the organization. Examples include weekly sales figures, quarterly earnings, and yearly profits. The U.S. economy (a complex organization) can be thought of as generating streams of data that include the Gross National Product (GNP), stock prices, and the Consumer Price Index. Statisticians and other analysts conceptualize these data streams as being generated by processes. Typically, however, the series of operations or actions that cause particular data to be realized are either unknown or so complex (or both) that the processes are treated as **black boxes**.

Definition 1.6

A process whose operations or actions are unknown or unspecified is called a **black box**.

Frequently, when a process is treated as a black box, its inputs are not specified either. The entire focus is on the output of the process. A black box process is illustrated in Figure 1.2.

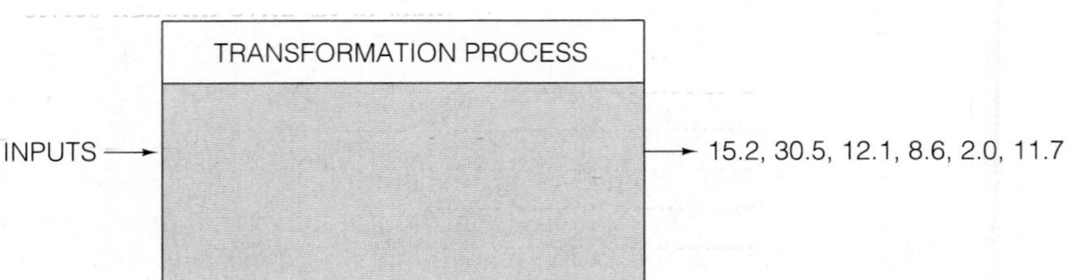

FIGURE 1.2 ► **A black box process with numerical output**

In studying a process, we generally focus on one or more characteristics, or properties, of the output. For example, we may be interested in the weight or the length of the units produced or even the time it takes to produce each unit. As with characteristics of population units, we call these characteristics **variables**. In studying processes whose output is already in numerical form (i.e., a stream of numbers), the characteristic, or property, represented by the numbers (e.g., sales, GNP, or stock prices) is typically the variable of interest. If the output is not numeric, we use **measurement processes** to assign numerical values to variables.* For example, if in the automobile assembly process the weight of the fully assembled automobile is the variable of interest, a measurement process involving a large scale will be used to assign a numerical value to each automobile.

*A process whose output is already in numerical form necessarily includes a measurement process as one of its subprocesses.

As with populations, we use sample data to analyze and make inferences (estimates, predictions, or other generalizations) about processes. But the concept of a sample is defined differently when dealing with processes. Recall that a population is a set of existing units and that a sample is a subset of those units. In the case of processes, however, the concept of a set of existing units is not relevant or appropriate. Processes generate or create their output *over time*—one unit after another. For example, a particular automobile assembly line produces a completed vehicle every 4 minutes. We define a sample from a process in the box.

Definition 1.7

Any set of output (objects or numbers) produced by a process is called a **sample**.

Thus, the next 10 cars turned out by the assembly line constitute a sample from the process, as do the next 100 cars or every fifth car produced today.

EXAMPLE 1.4

A particular fast-food restaurant chain has 6,289 outlets with drive-through windows. To attract more customers to its drive-through services, the company is considering offering a 50% discount on the price of an order that a customer must wait more than a specified number of minutes to receive. To help determine what the time limit should be, the company decided to estimate the average waiting time at a particular drive-through window in Dallas, Texas. For 7 consecutive days, the worker taking customers' orders recorded the time that every order was placed. The worker who handed the order to the customer recorded the time of delivery. In both cases, workers used synchronized digital clocks that reported the time to the nearest second. At the end of the 7-day period, 2,109 orders had been timed.

a. Describe the process of interest at the Dallas restaurant.
b. Describe the variable of interest.
c. Describe the sample.
d. Describe the inference of interest.
e. Describe how the reliability of the inference could be measured.

Solution

a. The process of interest is the drive-through window at a particular fast-food restaurant in Dallas, Texas. It is a process because it "produces," or "generates," meals over time. That is, it services customers over time.
b. The variable the company monitored is customer waiting time, the length of time a customer waits to receive a meal after placing an order. Since the study is focusing only on the output of the process (the time to produce the output) and not the internal operations of the process (the tasks required to produce a meal for a customer), the process is being treated as a black box.

c. The sampling plan was to monitor every order over a particular 7-day period. The sample is the 2,109 orders that were processed during the 7-day period.

d. The company's immediate interest is in learning about the drive-through window in Dallas. They plan to do this by using the waiting times from the sample to make a statistical inference about the drive-through process. In particular, they might use the average waiting time for the sample to estimate the average waiting time at the Dallas facility.

e. As for inferences about populations, measures of reliability can be developed for inferences about processes. The reliability of the estimate of the average waiting time for the Dallas restaurant could be measured by a bound on the error of estimation. That is, we might find that the average waiting time is 4.2 minutes, with a bound on the error of estimation of .5 minute. The implication would be that we could be reasonably certain that the true average waiting time for the Dallas process is between 3.7 and 4.7 minutes.

Notice that there is also a population described in this example: the company's 6,289 existing outlets with drive-through facilities. In the final analysis, the company will use what it learns about the process in Dallas and, perhaps, similar studies at other locations to make an inference about the waiting times in its populations of outlets.

Note that output already generated by a process can be viewed as a population. Suppose a soft-drink canning process produced 2,000 twelve-packs yesterday, all of which were stored in a warehouse. If we were interested in learning something about those 2,000 packages—such as the percentage with defective cardboard packaging—we could treat the 2,000 packages as a population. We might draw a sample from the population in the warehouse, measure the variable of interest, and use the sample data to make a statistical inference about the 2,000 packages, as described in Sections 1.2 and 1.3.

CASE STUDY 1.6 / Quality Improvement: U.S. Firms Respond to the Challenge from Japan

Over the last 2 decades, U.S. firms have been seriously challenged by products of superior quality from overseas, particularly from Japan. Japan currently produces 26% of the cars sold in the United States, and some predict this figure will climb to 40% within a decade. Only one U.S. firm still manufactures televisions; the rest are made in Japan.

To meet this competitive challenge, more and more U.S. firms—both manufacturing and service firms—have begun quality-improvement initiatives of their own. Many of these firms now stress the management of quality in all phases and aspects of their business, from the design of their products to production, distribution, sales, and service.

Broadly speaking, quality-improvement programs are concerned with (1) finding out what the customer wants, (2) translating those wants into a product design, and (3) producing and delivering a product or service that meets or exceeds the specifications of the product design. In all these areas, but particularly in the third, *improvement of quality requires improvement of processes*—including production processes, distribution processes, and service processes.

But what does it mean to say that a process has been improved? Generally speaking, it means that the customer of the process (i.e., the user of the output) indicates a greater satisfaction with the output. Frequently, such increases in satisfaction require a reduction in the variation of one or more process variables. That is, a reduction in the variation of the output stream of the process is needed.

But how can process variation be monitored and reduced? In the mid-1920s, Walter Shewhart of the Bell Telephone Laboratories made perhaps the most significant breakthrough of this century for the improvement of processes. He recognized that variation in process output was inevitable. No two parts produced by a given machine are the same; no two transactions performed by a given bank teller are the same. He also recognized that variation could be understood, monitored, and controlled using statistical methods. He developed a simple graphical technique—called a **control chart**—for determining whether product variation is within acceptable limits. This method provides guidance for when to adjust or change a production process and when to leave it alone. It can be used at the end of the production process or, most significantly, at different points within the process. We discuss control charts and other tools for improving processes in Chapter 13.

In recent years, largely as a result of the Japanese challenge to the supremacy of U.S. products, control charts and other statistical tools have gained widespread use in the United States. As evidence for the claim that U.S. firms are responding well to Japan's competitive challenge, consider this: The most prestigious quality improvement prize in the world that a firm can win is the Deming Prize. It is awarded by the Japanese. In 1989 it was won for the first time by an American company—Florida Power and Light Company.

In this section we have presented a brief introduction to processes and the use of statistical methods to analyze and learn about processes. In Chapters 13, 14, and 15 we present an in-depth treatment of these subjects. If you would like further amplification now of the ideas presented in this section, we suggest that you read Section 13.4 "Systems and Systems Thinking."

1.5 The Role of Statistics in Managerial Decision-Making

Managers frequently rely on input from statistical analyses to help them make decisions. The role statistics can play in managerial decision-making is indicated in the flow diagram in Figure 1.3. Every managerial decision-making problem begins with a real-world problem. This problem is then formulated in managerial terms and framed as a managerial question. The next sequence of steps (proceeding counterclockwise around the flow diagram) identifies the role that statistics can play in this process. The managerial question is translated into a statistical question, the sample data are collected and analyzed, and the statistical question is answered. The next step in the process is using the answer to the statistical question to reach an answer to the managerial question. The answer to the managerial question may suggest a reformulation of the original managerial problem, suggest a new managerial question, or lead to the solution of the managerial problem.

FIGURE 1.3 ▶
Flow diagram showing the role of statistics in managerial decision-making
***Source:* Chervany, Benson, and Iyer (1980)**

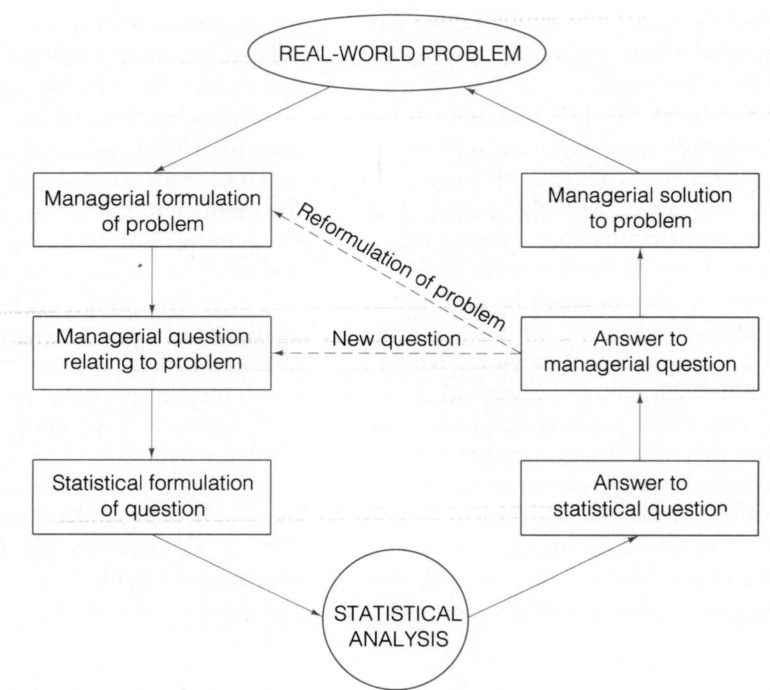

One of the most difficult steps in the decision-making process—one that requires a cooperative effort among managers and statisticians—is the translation of the managerial question into statistical terms (for example, into a question about a population). This statistical question must be formulated so that, when answered, it will provide the key to the answer to the managerial question. Thus, as in the game of chess, you must formulate the statistical question with the end result, the solution to the managerial question, in mind.

Exercises 1.1 – 1.18

Note: Starred (*) exercises refer to the optional section.

Learning the Mechanics

1.1 Explain the difference between descriptive and inferential statistics.

1.2 List and define the five elements of an inferential statistical problem.

1.3 Explain how populations and variables differ.

1.4 Explain how populations and samples differ.

1.5 Why would a statistician consider an inference incomplete without an accompanying measure of its reliability?

***1.6** Explain the difference between a population and a process.

1.7 Consider the set of all students enrolled in your statistics course this term. Suppose you are interested in learning about the current grade point averages (GPAs) of this group.

a. Define the population and variable of interest.

b. Suppose you determine the GPA of every member of the class. What would this represent, a census or a sample?

c. Suppose you determine the GPA of 10 members of the class. Would this represent a census or a sample?

d. If you determine the GPA of every member of the class and then calculate the average, how much reliability does this have as an "estimate" of the class average GPA?

e. If you determine the GPA of ten members of the class and then calculate the average, will the number you get necessarily be the same as the average GPA for the whole class? On what factors would you expect the reliability of the estimate to depend?

1.8 Refer to Exercise 1.7. What must be true in order for the sample of 10 students you select from your class to be considered a random sample?

Applying the Concepts

1.9 Pollsters regularly conduct opinion polls to determine the popularity rating of the current president. Suppose a poll is to be conducted tomorrow in which 2,000 individuals will be asked whether the president is doing a good or poor job.

a. What is the relevant population?

b. What is the variable of interest? Is it numerical or nonnumerical?

c. What is the sample?

d. What is the inference of interest to the pollster?

1.10 Refer to Exercise 1.9. Suppose the poll is conducted as described therein, except that each of the 2,000 individuals polled is asked to rate the job performance of the president on a scale from 0 to 100. How do your answers to parts **a–d** change, if at all?

1.11 *Fortune* publishes a list of the 500 industrial corporations in the United States with the largest annual sales volume ("The Fortune 500," 1992). Throughout the year, each of these companies is then referred to as a *Fortune 500 company*. Assume a researcher with limited time and money wants to predict the number of new employees these 500 companies anticipate hiring next year. She selects 50 of the 500 companies, and asks a representative of each company how many new employees the company is likely to hire in the coming year.

a. Identify the population from which the sample was drawn.

b. What is the variable of interest?

c. Identify the sample.

d. What could we infer about the population from the sample data?

e. When answering part **d**, are we employing descriptive or inferential statistics? Explain.

f. *Fortune* lists the net profit of each of the 500 companies. If we compute the average of these 500 net profits to learn more about the Fortune 500 companies, are we employing descriptive or inferential statistics? Explain.

*1.12 Coca-Cola and Schweppes Beverages Limited (CCSB), which was formed in 1987, is 49% owned by the Coca-Cola Company. David Nellist (1992) reports that CCSBs new Wakefield plant can produce 4,000 cans of soft drink per minute. The automated process consists of measuring and dispensing the raw ingredients into storage vessels to create the syrup, and then injecting the syrup, along with carbon dioxide, into the beverage cans. Suppose that to monitor the subprocess that adds carbon dioxide to the cans, five filled cans are pulled off the line every 15 minutes and the amount of carbon dioxide in each of these five is measured to determine whether the amounts are within prescribed limits.

a. Describe the process studied.
b. Describe the variable of interest.
c. Describe the sample.
d. Describe the inference of interest.
e. *Brix* is a unit for measuring sugar concentration. If a technician is assigned the task of estimating the average brix level of all 240,000 cans of beverage stored in a warehouse near Wakefield, will the technician be examining a process or a population? Explain.

1.13 An insurance company would like to determine the proportion of all medical doctors who have been involved in one or more malpractice suits. The company selects 500 doctors at random from a professional directory and determines the number in the sample who have ever been involved in a malpractice suit.

a. Identify the population and variable of interest to the insurance company.
b. Describe the sample and identify the type of inference the insurance company wishes to make.
c. What is meant by the phrase *at random* in the context of this exercise?

1.14 *Corporate merger* is a means through which one firm (the bidder) acquires control of the assets of another firm (the target). Carol E. Eger (1982) identified a total of 497 mergers between firms listed on the New York Stock Exchange that resulted in the delisting of the acquired (target) firm's stock during the period 1958–1980. She sampled 38 of these mergers and evaluated the effects of the merger on the value of the holdings of the bidder firm's bondholders. In particular, she wanted to learn whether the value of the holdings increased or decreased as a result of the merger.

a. Identify the population studied.
b. Identify the variable of interest.
c. Describe the sample.
d. Discuss types of inferences we might wish to make about the population.

1.15 *Job-sharing* is an innovative employment alternative that originated in Sweden and is becoming very popular in the United States. Firms that offer job-sharing plans allow two or more persons to work part-time, sharing one full-time job. For example, two job-sharers might alternate work weeks, with one working while the other is off. Job-sharers never work at the same time and may not even know each other. Job-sharing is particularly attractive to working mothers and to people who frequently lose their jobs due to fluctuations in the economy ("Your Job in the 1980's," 1980). To evaluate employers' satisfaction with job-sharing plans, a government agency contacted 100 firms that offer job-sharing. Each firm's director of personnel was asked whether the firm was satisfied with the productivity of workers with shared jobs.

a. Identify the population from which the sample was selected.
b. Identify the variable measured.
c. Identify the sample selected.
d. What type of inference is of interest to the government agency?

1.16 Myron Gable and Martin T. Topol (1988) sampled 218 department-store executives to study the relationship between job satisfaction and the degree of Machiavellian orientation. Briefly, the Machiavellian orientation is one in which the executive exerts very strong control—even to the point of deception and cruelty—over the employees supervised. The authors administered a questionnaire to each of the sampled executives and obtained both a job satisfaction score and a Machiavellian rating. They concluded that those with higher satisfaction scores are likely to have a lower "Mach" rating.

a. What is the population from which the sample was selected?
b. What variables were measured by the authors?
c. Identify the sample.
d. What inference was made by the authors?

1.17 Manufacturers of consumer goods rely on the information provided by consumer preference surveys to guide both the design and the marketing of new products. In the winter of 1986, Onan Corporation, a manufacturer of built-in generators for recreational vehicles (RVs), was considering developing and marketing a portable generator. Such a product could potentially be marketed both to RV owners and RV manufacturers. To determine RV owners' preferences with respect to the features of a portable generator (e.g., size and manual or electric start), 3,000 questionnaires were mailed to RV owners in the continental United States. One thousand fifty-two (1,052) responses were received by Onan.*

a. Identify the population, the variables, the sample, and the inferences of interest to Onan.
b. Chapters 6–9 indicate that the reliability of an inference is related to the size of the sample used. In addition to sample size, what other factors might affect the reliability of inferences based on the responses to a mailed questionnaire?

1.18 The Wallace Company of Houston is a distributor of pipes, valves, and fittings to the refining, chemical, and petrochemical industries. The company was one of four winners of the Malcolm Baldrige National Quality Award in 1990. Don Nichols (1991) explains that one of the steps the company takes to monitor the quality of its distribution process is to send out a survey twice a year to a subset of its current customers, asking the customers to rate the speed of deliveries, the accuracy of invoices, and the quality of the packaging of the products they have received from Wallace.

a. Describe the process studied.
b. Describe the variables of interest.
c. Describe the sample.
d. Describe the inferences of interest.
e. What are some of the factors that are likely to affect the reliability of the inferences?

On Your Own

If you could start your own business right now, what kind would it be? Identify a large population and a characteristic of the units in the population (i.e., a variable) that would be of interest to you and your firm. How would you measure the variable of interest for a sample of units drawn from the population? Is the data set you identified a sample data set or a population data set? How could you use this data set in the operation of your business?

*Information by personal communication with Thomas J. Roess, Manager, Market Analysis, Onan Corporation, Fridley, Minnesota.

References

Allison, R. I. and Uhl, K. P. "Influence of beer brand identification on taste perception." *Journal of Marketing Research*, Aug. 1965, pp. 36–39.

Careers in Statistics. Washington, D.C.: American Statistical Association and the Institute of Mathematical Statistics, 1974.

Chervany, N. L., Benson, P. G., and Iyer, R. K. "The planning stage in statistical reasoning." *The American Statistician*, Nov. 1980, pp. 222–226.

Eger, C. E. "Corporate mergers: An analytical analysis of the role of risky debt." Unpublished Ph.D. dissertation. University of Minnesota, 1982.

"The Fortune 500." *Fortune*, Apr. 20, 1992, pp. 220–239.

Gable, M. and Topol, M. T. "Machiavellianism and the department-store executive." *Journal of Retailing*, Spring 1988, pp. 68–84.

Hunz, E. "Application of statistical sampling to inventory audits." *The Internal Auditor*, 1956, *13*, p. 38.

Nellist, David. "Quality teamwork at Wakefield." *Industrial Management and Data Systems*, 1992, Vol. 92, No. 2, pp. 21–23.

Nichols, Don. "Quality wins." *Small Business Reports*, May 1991, pp. 26–35.

Tanur, J. M., Mosteller, F., Kruskal, W. H., Link, R. F., Pieters, R. S., and Rising, G. R. *Statistics: A Guide to the Unknown*. San Francisco: Holden-Day, 1989.

U.S. Bureau of the Census. *Statistical Abstract of the United States: 1989*. Washington, D.C.: U.S. Government Printing Office, 1989.

U.S. Department of Commerce. *An Error Profile: Employment as Measured by the Current Population Survey*. Statistical Policy Working Paper 3. Washington, D.C.: U.S. Government Printing Office, 1978.

U.S. Department of Labor. *The Consumer Price Index: Concepts and Content over the Years*. Bureau of Labor Statistics, Report 517. Washington, D.C.: U.S. Government Printing Office, May 1978.

Willis, R. E. and Chervany, N. L. *Statistical Analysis and Modeling for Management Decision-making*. Belmont, Calif.: Wadsworth, 1974, Chapter 1.

"Your job in the 1980's." *Consumer's Digest*, Nov.–Dec. 1980, pp. 32–36.

CHAPTER TWO

Methods for Describing Sets of Data

Contents

Case Studies

Where We've Been

In Chapter 1, we examined some examples of the use of statistics in business. We discussed the role that statistics plays in supporting managerial decision-making. We introduced you to descriptive and inferential statistics and to the five elements of inferential statistics: a population, one or more variables, a sample, an inference, and a measure of reliability for the inference. We described the primary goal of inferential statistics as using sample data to make inferences (estimates, predictions, or other generalizations) about the population from which the sample was drawn.

Where We're Going

Before we make an inference, we must be able to describe and extract information from the sample data. This can be accomplished through both graphical and numerical methods.

Before we can use the information in a sample to make inferences about a population, we must be able to extract the relevant information from the sample. That is, we need methods to summarize and describe the sample measurements. For example, if we look at last year's sales for 100 randomly selected companies, we are unlikely to extract much information by looking at the set of 100 sales figures. We would get a clearer picture of the data by calculating the average sales for all 100 companies, by determining the highest and lowest company sales, by drawing a graph that shows the spread of the 100 sales figures, or, in general, by using some technique that will extract and summarize relevant information from the data and, at the same time, allow us to obtain a clearer understanding of the sample.

In this chapter, we first define four different types of data and then present some graphical and numerical methods for describing data of each type. You will see that graphical methods for describing data are intuitively appealing and can be used to describe either a sample or a population. However, numerical methods for describing data are the keys that unlock the door to population inference-making.

2.1 Types of Data

In Chapter 1, you learned that statistics, both descriptive and inferential, is concerned with measurements of one or more variables of a sample of units drawn from a population. These measurements are referred to as **data**. We will generally classify data as one of four types: **nominal**, **ordinal**, **interval**, or **ratio**.

Definition 2.1

Nominal data are measurements that simply classify the units of the sample (or population) into categories.

Nominal data (also referred to as **categorical data**) are labels or names that identify the category to which each unit belongs. The following are examples of nominal data:

1. The political party affiliation of each individual in a sample of 50 business executives
2. The gender of each individual in a sample of seven applicants for a computer programming job
3. The state in which each of a sample of 100 U.S. firms had its highest sales revenue in 1993.

Note that in each case—political party, gender, and state—the measurement is no more than a categorization of each sample unit. Nominal data are often reported as nonnumerical labels, such as Democrat, woman, and Ohio. Even if the labels are converted to numbers, as they often are for ease of computer entry and analysis, the numerical values are simply codes. They cannot be meaningfully added, subtracted, multiplied, or divided. For example, we might code Democrat = 1, Republican = 2, and Other = 3. These are simply numerical codes for each of the categories into which units may fall and have no further significance.

Definition 2.2

Ordinal data are measurements that enable the units of the sample (or population) to be ordered with respect to the variable of interest.

Ordinal data are measurements that indicate the *relative* amount of a property possessed by the units. The following are examples of ordinal data:

1. The size of car rented by each individual in a sample of 30 business travelers: compact, subcompact, midsize, or full-size
2. A taste-tester's ranking of four brands of barbecue sauce
3. A supervisor's annual ranking of the performance of her 10 employees using a scale of 1 (worst performance) to 10 (best performance)

Note that in each case—size, flavor preference, and performance ranking—more than a categorization of units is involved. In addition to providing a categorization, the measurements actually rank the units. For example, we know that a midsize car is larger than a subcompact and that an employee with a performance ranking of 9 performed better, in the opinion of the supervisor, than one with a ranking of 7. We also know that a taster preferred brand C barbecue sauce to brand A if he gives brand C a higher flavor-preference ranking than brand A.

Ordinal data are said to represent a "higher" level of measurement than nominal data because ordinal data contain all the information of nominal data (i.e., category labels that differentiate units) *plus* an ordering of the units. As with nominal measurements, the distance between ordinal measurements is not meaningful. For example, we do not know whether the difference in size between a full-size and a midsize car is the same as the difference between the midsize and the subcompact. Nor do we know whether the extent of flavor preference between barbecue sauce brands C and A is the same as that between brands A and B if a taster reports her flavor preference as C > A > B > D, where > means "more flavorful than."

As with nominal data, ordinal data can be reported with or without numbers. For example, the automobile sizes are nonnumerically labeled, whereas the supervisor's performance rankings are numerical. Even if numbers are used, we must again be careful: They simply provide an ordering or ranking of the units in the sample or population. The arithmetic operations of addition, subtraction, multiplication, and division are not meaningful for ordinal data.

Definition 2.3

Interval data are measurements that enable the determination of how much more or less of the measured characteristic is possessed by one unit of the sample (or population) than another.

Interval data are always numerical, and the numbers assigned to two units can be subtracted to determine the *difference* between the units with respect to the variable measured. The following are examples of interval data:

1. The temperature (in degrees Fahrenheit) at which each of a sample of 20 pieces of heat-resistant plastic begins to melt
2. The scores of a sample of 150 MBA applicants on the GMAT, a standardized business graduate school entrance exam administered nationwide
3. The time at which the 5 P.M. Washington–to–New York air shuttle arrives at LaGuardia on each of a sample of 30 weekdays

Note that in each case—temperature, score, and arrival time—more than a ranking is involved. The difference between the numerical values assigned to the units is meaningful. For example, the difference between scores of 600 and 580 on the GMAT is the same as that between scores of 520 and 500. Also, the morning shuttle due at 9 A.M. but arriving at 9:20 A.M. is just as late as the afternoon shuttle due at 5:30 P.M. and arriving at 5:50 P.M. Note in each case that the difference is the key, not the numerical measurement itself.

Interval data represent a higher level of measurement than ordinal data, because in addition to ranking the units, interval data reflect the difference between the units with respect to the variable measured. Although adding or subtracting interval data is valid, multiplying or dividing them is not. This is because the zero point (the origin, or 0) does not indicate an absence of the characteristic of interest. For example, the origin on the temperature scale differs for the Fahrenheit and Celsius scales and does not indicate an absence of heat on either scale. Temperatures lower than 0° (e.g., −10°C and −10°F) indicate that less heat is present, so 0° does not mean "no heat." The result is that we cannot say that a temperature of 100°F indicates twice the heat of 50°F. Similarly, since GMAT scores range from 200 to 800, a zero score is not even possible, and thus has no meaning. The result is that a score of 600 cannot be interpreted as being 50% higher than a score of 400.

Most numerical business data are measured on scales for which the origin is meaningful. Thus, most numerical measurements encountered in business are *ratio data*.

Definition 2.4

Ratio data are measurements that enable the determination of how many times as much of the measured characteristic is possessed by one unit of the sample (or population) than another.

Ratio data are always numerical, and the ratio between the numbers assigned to two units can be interpreted as the multiple by which the units differ. The following are examples of ratio data:

1. The sales revenue for each firm in a sample of 100 U.S. firms
2. The unemployment rate (reported as a percentage) in the United States for each of the past 60 months
3. The number of female executives employed in each of a sample of 50 manufacturing companies

Note that in each case—dollars of revenue, percentage unemployed, and count of female executives—the scale measures the absolute amount of the characteristic possessed by the unit. The result is that the ratio of measurements between units is meaningful. That is, a company with revenue of $100 million has twice the revenue of a company with $50 million in revenue. Similarly, an unemployment rate of 8% means twice as many unemployed as with a rate of 4%. And a company with 30 female executives has 1.5 times as many as one with 20 female executives.

Ratio data represent the highest level of measurement. The numbers can be used to categorize, rank, differentiate, and measure multiples of one unit with respect to another. All arithmetic operations performed on ratio data are meaningful.

The key to differentiating interval and ratio data is that for ratio data the zero point, or origin, denotes an absence of the characteristic being measured. For example, zero revenue, zero unemployment, and zero female executives mean *absence* of income, unemployment, and female executives, respectively. Most measurement scales utilized in business yield ratio data: measures of monetary value, distance, weight, height, percentages, and numerical counts all usually generate ratio data.

The four types of data are often combined into two classes that are sufficient for most statistical applications. Nominal and ordinal data are often referred to as **qualitative** data, whereas interval and ratio data are called **quantitative** data.

The properties of the four types of data are summarized in the box. As you would expect, the methods for describing and reporting data depend on the type of data analyzed. We devote the remainder of this chapter to graphical methods for describing qualitative and quantitative data.

Types of Data

Nominal	Classification of sample (or population) units into categories Often uses labels rather than numbers
Ordinal	Rank-orders the sample (or population) units May be verbal labels or numbers
Interval	Enables comparison of sample (or population) units according to differences between values Always numerical, but the zero point on the scale does not indicate an absence of the measured characteristic
Ratio	Enables comparison of sample (or population) units according to multiples of the values Always numerical, and the zero point on the scale denotes an absence of the measured characteristic
Qualitative	Includes nominal and ordinal data types
Quantitative	Includes interval and ratio data types

Exercises 2.1–2.8

Learning the Mechanics

2.1 **a.** Explain the difference between nominal and ordinal data.
b. Explain the difference between interval and ratio data.
c. Explain the difference between qualitative and quantitative data.

2.2 Each of the following descriptions of data defines one of the following types: nominal, ordinal, interval, ratio. Match the correct type to each description.
a. Data that enable the units of the sample to be compared by the differences between their numerical values
b. Data that enable the units of the sample to be classified into categories
c. Data that enable the units of the sample to be rank-ordered
d. Data that enable the units of the sample to be compared by computing the ratios of the numerical values

2.3 Suppose you are provided a data set that classifies each sample unit into one of four categories: A, B, C, or D. You plan to create a computer database consisting of these data, and you decide to code the data as A = 1, B = 2, C = 3, and D = 4 for entering them into the computer. Are the data consisting of the classifications A, B, C, and D qualitative or quantitative? After the data are entered as 1, 2, 3, or 4, are they qualitative or quantitative? Explain your answers.

Applying the Concepts

2.4 A food-products company is considering marketing a new snack food. To see how consumers react to the product, the company conducted a taste-test using a sample of 100 shoppers at a suburban shopping mall. The shoppers were asked to taste the snack food and then fill out a short questionnaire that requested the following information:

a. What is your age?

b. Are you the person who typically does the food shopping for your household?

c. How many people are in your family?

d. How would you rate the taste of the snack food on a scale of 1 to 10, where 1 is least tasty?

e. Would you purchase this snack food if it were available on the market?

f. If you answered yes to question **e**, how often would you purchase it?

Each of these questions defines a variable of interest to the company. Classify the data generated for each variable as nominal, ordinal, interval, or ratio. Justify your classification.

2.5 Classify the following examples of data as nominal, ordinal, interval, or ratio. Justify your classification.

a. Ten college freshmen were asked to indicate the brand of jeans they prefer.

b. Fifteen television cable companies were asked how many hours of sports programming they carry in a typical week.

c. Fifty executives were asked what percentage of their workday is spent in meetings.

d. The number of long-distance phone calls made from each of 100 public telephone booths on a particular day was recorded.

e. The Scholastic Aptitude Test (SAT) scores of 250 incoming freshmen to a small college were compiled.

2.6 Classify the following examples of data as either qualitative or quantitative:

a. The brand of calculator purchased by each of 20 business statistics students

b. The list price of calculators purchased by each of 20 business statistics students

c. The number of automobiles purchased during the past 5 years by each household in a sample of 50 randomly selected households

d. The month indicated by each of 41 randomly selected business firms as the month during which it had the highest sales

e. The depth of tread remaining on each of 137 randomly selected automobile tires after 20,000 miles of wear

2.7 Windows is a computer software product made by Microsoft Corporation. In designing Windows Version 3.1, Microsoft telephoned 60,000 users of Windows 3.0 (an older version) and asked them how the product could be improved (Roberts, 1992). Assume customers were asked the following questions:

I. Are you the most frequent user of Windows 3.0 in your household?

II. What is your age?

III. How would you rate the helpfulness of the Tutorial instructions that accompany Windows 3.0, on a scale of 1 to 10, where 1 is not helpful?

IV. When using a printer with Windows 3.0, do you most frequently use a dot-matrix printer or another type of printer?

V. If the speed of Windows 3.0 could be changed, which one of the following would you prefer: slower, unchanged, faster?

VI. How many people in your household have used Windows 3.0 at least once?

Each of these questions defines a variable of interest to the company. Classify the data generated for each variable as nominal, ordinal, interval, or ratio. Justify your classification.

2.8 Classify the examples of data in parts **a–d** as either qualitative or quantitative:

a. The number of corporate mergers during each of the last 15 years

b. The change in the Consumer Price Index during each of the last 6 months

c. The length of time before each of 30 dry-cell batteries goes dead

d. The American automobile manufacturer that each of 25 service station mechanics indicated as producing the most reliable cars

e. Classify each of the preceding as nominal, ordinal, interval or ratio data.

2.2 Graphical Methods for Describing Quantitative Data: Histograms and Stem-and-Leaf Displays

Recall from Section 2.1 that quantitative data sets consist of either interval or ratio data. Most business data are quantitative, so that methods for summarizing quantitative data are especially important.

For example, suppose a financial analyst is interested in the amount of resources spent by computer hardware and software companies on research and development (R&D). She samples 50 of these high-technology firms and calculates the amount each spent last year on R&D as a percentage of their total revenues. The results are given in Table 2.1. As numerical measurements made on the sample of 50 units (the firms),

TABLE 2.1 Percentage of Revenues Spent on Research and Development

Company	Percentage	Company	Percentage	Company	Percentage	Company	Percentage
1	13.5	14	9.5	27	8.2	39	6.5
2	8.4	15	8.1	28	6.9	40	7.5
3	10.5	16	13.5	29	7.2	41	7.1
4	9.0	17	9.9	30	8.2	42	13.2
5	9.2	18	6.9	31	9.6	43	7.7
6	9.7	19	7.5	32	7.2	44	5.9
7	6.6	20	11.1	33	8.8	45	5.2
8	10.6	21	8.2	34	11.3	46	5.6
9	10.1	22	8.0	35	8.5	47	11.7
10	7.1	23	7.7	36	9.4	48	6.0
11	8.0	24	7.4	37	10.5	49	7.8
12	7.9	25	6.5	38	6.9	50	6.5
13	6.8	26	9.5				

these percentages represent quantitative data. The analyst's initial objective is to describe these data in order to extract relevant information.

A **relative frequency histogram** for these 50 R&D percentages is shown in Figure 2.1. The horizontal axis of Figure 2.1, which gives the percentage spent on R&D for each company, is divided into intervals commencing with the interval from 5.15 to 6.25 and proceeding in intervals of equal size to 12.85 to 13.95 percent. The vertical axis gives the proportion (or **relative frequency**) of the 50 percentages that fall in each interval. Thus, you can see that the bulk of the companies spend between 6.25% and 10.65% of their revenues on research and development, while only .06 of the companies spend more than 12%. Many other summary statements can be made by further study of the histogram.

FIGURE 2.1 ► **Relative frequency histogram for the 50 computer companies' R&D percentages**

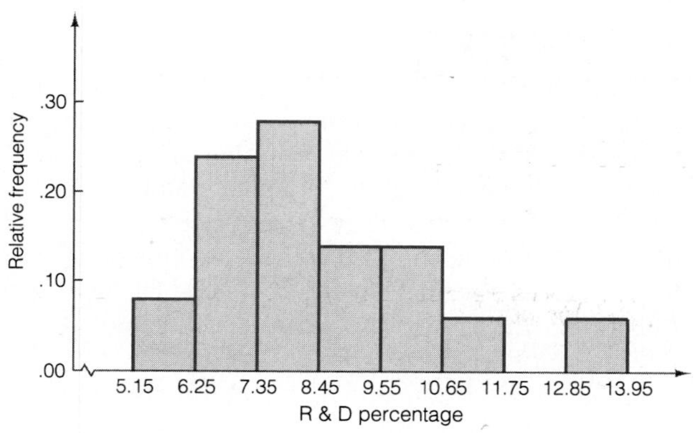

Another graphic representation of these same data, a **stem-and-leaf** display, is shown below. In these displays the **stem** is the portion of the observation to the left of the decimal point, whereas the remaining portion to the right of the decimal point is the **leaf**.

The stems and leaves for the R&D percentages 7.4, 10.5, and 13.2 are shown here:

Stem	*Leaf*
7	4

Stem	*Leaf*
10	5

Stem	*Leaf*
13	2

The stem-and-leaf display for all 50 R&D percentages is shown in Figure 2.2 (page 30). Note that the leaves corresponding to each stem are arranged in ascending order, and a key is included with the display to specify the units of the leaf (and, by implication, the units of the stem).

FIGURE 2.2 ►
Stem-and-leaf display for 50 computer companies' research and development percentages

Stem	*Leaf*
5	2 6 9
6	0 5 5 5 6 8 9 9 9
7	1 1 2 2 4 5 5 7 7 8 9
8	0 0 1 2 2 2 4 5 8
9	0 2 4 5 5 6 7 9
10	1 5 5 6
11	1 3 7
12	
13	2 5 5

Key: Leaf units are tenths.

Note that although the stem 12 has no leaves (meaning that none of the 50 observations fell in the range from 12.0 to 12.9), we include the 12 stem in the display so that this fact is visually obvious. Note also that the decimal point is not included in the display. When there is no confusion caused by its omission, we can usually obtain a less cluttered graphical description without it.

Several descriptive facts about these data are easily seen in the stem-and-leaf display. Most of the sampled computer companies (37 of 50) spent between 6.0% and 9.9% of their revenues on R&D, and 11 of them spent between 7.0% and 7.9%. Relative to the rest of the sampled companies, three spent a high percentage of revenues on R&D—in excess of 13%.

Both the histogram and stem-and-leaf displays provide useful graphic descriptions of quantitative data. Since most statistical software packages can be used to construct these displays, we will focus on their interpretation rather than their construction.

Histograms can be used to display either the **frequency** or **relative frequency** of the measurements falling into specified intervals (called **measurement classes**). The frequency is just a count of the number of measurements in a class, while the relative frequency is the proportion, or fraction, of measurements in the class. The measurement classes, frequencies, and relative frequencies for the R&D data are shown in Table 2.2.

By looking at a histogram (say, the relative frequency histogram in Figure 2.1), you can see two important facts. First, note the total area under the histogram, and then note the proportion of the total area that falls over a particular interval of the horizontal axis. You will see that the proportion of the total area that falls above an interval is equal to the relative frequency of the measurements that fall in the interval.*

*Some histograms are constructed with all class intervals of equal width except the first and last, which are open-ended. The proportionality between area and relative frequency will not hold for such histograms. We will restrict our attention to histograms that have equal-sized class intervals, because later we will want to establish a correspondence between relative frequency histograms and probability distributions.

TABLE 2.2 Measurement Classes, Frequencies, and Relative Frequencies for the R&D Percentage Data

Class	Measurement Class	Class Frequency	Class Relative Frequency
1	5.15– 6.25	4	$4/50 = .08$
2	6.25– 7.35	12	$12/50 = .24$
3	7.35– 8.45	14	$14/50 = .28$
4	8.45– 9.55	7	$7/50 = .14$
5	9.55–10.65	7	$7/50 = .14$
6	10.65–11.75	3	$3/50 = .06$
7	11.75–12.85	0	$0/50 = .00$
8	12.85–13.95	3	$3/50 = .06$
			1.00

For example, the relative frequency for the class interval 5.15–6.25 is .08. Consequently, the rectangle above that interval contains 8% of the total area under the histogram.

Second, you can imagine the appearance of the relative frequency histogram for a very large set of data (say, a population). As the number of measurements in a data set is increased, you can obtain a better description of the data by decreasing the width of the class intervals. When the class intervals become small enough, a relative frequency histogram will (for all practical purposes) appear as a smooth curve (see Figure 2.3).

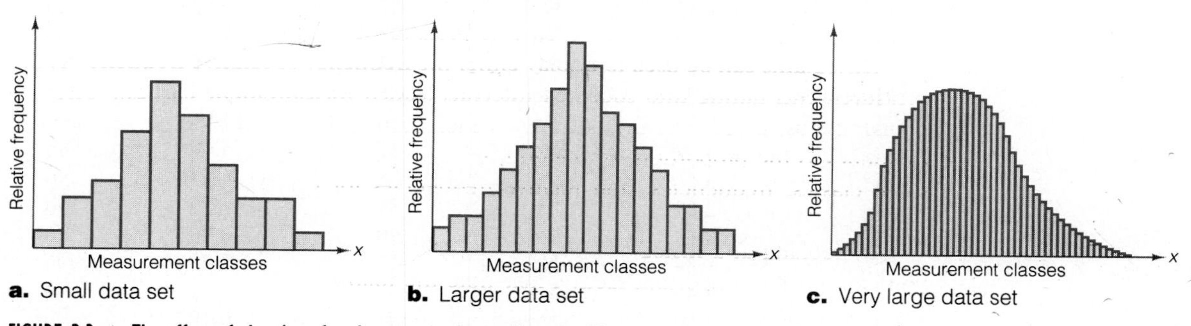

FIGURE 2.3 ▲ The effect of the size of a data set on the outline of a histogram

While histograms provide good visual descriptions, particularly of very large data sets, individual measurements cannot be identified by looking at a histogram. In contrast, each of the original measurements is "visible" in a stem-and-leaf display. The stem-and-leaf display arranges the data in ascending order, which enables easy location of the individual measurements. For example, in Figure 2.2 we can easily see that three of the R&D measurements are equal to 8.2%, whereas that fact is not evident by inspection of the histogram in Figure 2.1. However, stem-and-leaf displays can become unwieldy for very large data sets. A very large number of stems and leaves causes the vertical and horizontal dimensions of the display to become cumbersome, so that the usefulness of the visual display is diminished.

EXAMPLE 2.1

A manufacturer of industrial wheels suspects that profitable orders are being lost because of the long time the firm takes to develop price quotes for potential customers. To investigate this possibility, 50 requests for price quotes were randomly selected from the set of all quotes made last year, and the processing time was determined for each quote. The processing times are displayed in Table 2.3, and each quote was classified according to whether the order was "lost" or not (i.e., whether or not the customer placed an order after receiving a price quote).

TABLE 2.3 Price Quote Processing Times (Days)

Request Number	Processing Time	Lost?	Request Number	Processing Time	Lost?
1	2.36	No	26	3.34	No
2	5.73	No	27	6.00	No
3	6.60	No	28	5.92	No
4	10.05	Yes	29	7.28	Yes
5	5.13	No	30	1.25	No
6	1.88	No	31	4.01	No
7	2.52	No	32	7.59	No
8	2.00	No	33	13.42	Yes
9	4.69	No	34	3.24	No
10	1.91	No	35	3.37	No
11	6.75	Yes	36	14.06	Yes
12	3.92	No	37	5.10	No
13	3.46	No	38	6.44	No
14	2.64	No	39	7.76	No
15	3.63	No	40	4.40	No
16	3.44	No	41	5.48	No
17	9.49	Yes	42	7.51	No
18	4.90	No	43	6.18	No
19	7.45	No	44	8.22	Yes
20	20.23	Yes	45	4.37	No
21	3.91	No	46	2.93	No
22	1.70	No	47	9.95	Yes
23	16.29	Yes	48	4.46	No
24	5.52	No	49	14.32	Yes
25	1.44	No	50	9.01	No

a. Use a statistical software package to create a frequency histogram for these data. Then shade the area under the histogram that corresponds to lost orders.
b. Use a statistical software package to create a stem-and-leaf display for these data. Then shade each leaf of the display that corresponds to a lost order.
c. Compare and interpret the two graphic displays of these data.

Solution

a. We used SAS to generate the relative frequency histogram in Figure 2.4.

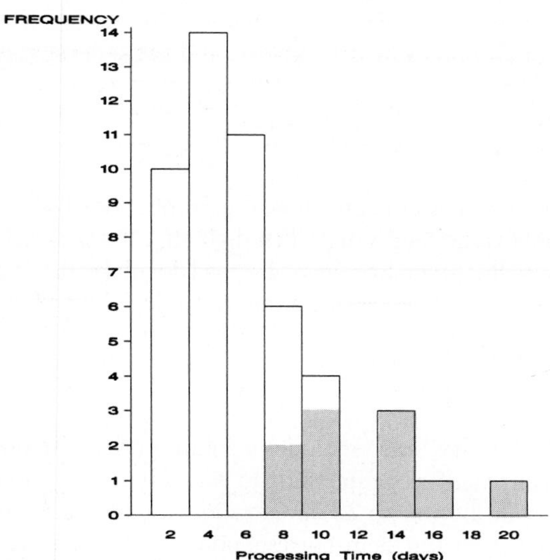

FIGURE 2.4 ► **Frequency histogram for the quote processing time data**

SAS, like most statistical software, offers the user the choice of accepting default class intervals and interval widths, or the user can make his or her own selections. After some experimenting with various numbers of class intervals and interval widths, we used 10 intervals. SAS then created intervals of width 2 days, beginning at 1 day, just below the smallest measurement of 1.25 days, and ending with 21 days, just above the largest measurement of 20.2 days. Note that SAS labels the midpoint of each bar, rather than its endpoints. Thus, the bar labeled "2" represents measurements from 1.00 to 2.99, the bar labeled "4" represents measurements from 3.00 to 4.99, etc. This histogram clearly shows the clustering of the measurements in the lower end of the distribution (between approximately 1 and 7 days), and the relatively few measurements in the upper end of the distribution (greater than 12 days). The shading of the area of the frequency histogram corresponding to lost orders clearly indicates that they lie in the upper tail of the distribution.

b. We used the statistical package SPSS[R] to generate the stem-and-leaf display in Figure 2.5 (page 34).

FIGURE 2.5 ▶
Stem-and-leaf display for the quote processing time data

```
Frequency      Stem &  Leaf

     5.00         1 .  24789
     5.00         2 .  03569
     8.00         3 .  23344699
     6.00         4 .  034469
     6.00         5 .  114579
     5.00         6 .  01467
     5.00         7 .  24557
     1.00         8 .  2
     3.00         9 .  049
     1.00        10 .  0
      .00        11 .
      .00        12 .
     1.00        13 .  4
     4.00 Extremes     (14.1), (14.3), (16.3), (20.2)

Stem width:        1.00
Each leaf:           1 case(s)
```

Note that the stem consists of the number of whole days (units and ten digits), and the leaf is the tenths digit (first digit after the decimal) of each measurement.* The hundredths digit has been dropped to make the display more visually effective. Thus, the first processing time in Table 2.3, 2.36 days, is partitioned as follows:

Stem	*Leaf*
2	3

Note that SPSS also includes a column titled "Frequency" showing the number of measurements corresponding to each stem. Also, note that instead of extending the stems all the way to 20 days to show the largest measurement, SPSS truncates the display after the stem corresponding to 13 days, labels the largest four measurements as "Extremes," and simply lists them horizontally in the last row of the display. Extreme observations that are detached from the remainder of the data are called **outliers**, and they usually receive special attention in statistical analyses. Although outliers may represent legitimate measurements, they are frequently mistakes: incorrectly recorded, miscoded during data entry, or taken from a population different from the one from which the rest of the sample was selected. Stem-and-leaf displays are useful for identifying outliers.

c. As is usually the case for data sets that are not too large (say, fewer than 100 measurements), the stem-and-leaf display provides more detail than the histogram without being unwieldy. For the processing time data, note that the stem-and-leaf display in Figure 2.5 clearly indicates not only that the lost orders are associated with high processing times (as does the histogram in Figure 2.4), but also exactly which of the times correspond to lost orders. Histograms are most useful for displaying very large data sets, when the overall shape of the distribution of measure-

*In the examples in this section, the stem was formed from the digits to the left of the decimal. This is not always the case. For example, in the following data set the stems could be the tenths digit and the leaves the hundredths digit: .12, .15, .22, .25, .28, .33.

ments is more important than the identification of individual measurements. Nevertheless, the message of both graphical displays is clear: establishing processing time limits may well result in fewer lost orders.

While stem-and-leaf displays may generally provide more information than histograms for quantitative (interval or ratio) data, they are not useful for qualitative (nominal or ordinal) data. However, histograms can be used to graph the **frequency** or **relative frequency** corresponding to each classification of the qualitative variable. Histograms for qualitative data are often called **bar charts**, and their usefulness is demonstrated in the Case Study 2.1.

CASE STUDY 2.1 / Pareto Analysis

Vilfredo Pareto (1843–1923), an Italian economist, discovered that approximately 80% of the wealth of a country lies with approximately 20% of the people. Others have noted similar findings in other areas: 80% of sales are attributable to 20% of the customers; 80% of customer complaints result from 20% of the components of a product; 80% of defective items produced by a process result from 20% of the types of errors that are made in production (Kane, 1989). These examples illustrate the idea of "the vital few and the trivial many," the **Pareto principle**. As applied to the last example, a "vital few" errors account for most of the defectives produced. The remaining defectives are due to many different errors, the "trivial many."

In general, **Pareto analysis** involves the categorization of items and the determination of which categories contain the most observations. These are the "vital few" categories. Pareto analysis is used in industry today as a problem-identification tool. Managers and workers use it to identify the most important problems or causes of problems that plague them. Knowledge of the "vital few" problems permits management to set priorities and focus their problem-solving efforts.

The primary tool of Pareto analysis is the **Pareto diagram**. The Pareto diagram is simply a frequency or relative frequency histogram, or bar chart, with the bars arranged in descending order of height from left to right across the horizontal axis. That is, the tallest bar is positioned at the left and the shortest is at the far right. This arrangement locates the most important categories—those with the largest frequencies—at the left of the chart. Since the data are qualitative, there is no inherent numerical order: They can be rearranged to make the display more useful.

Consider the following example from the automobile industry (adapted from Kane, 1989). All cars produced on a particular day were inspected for defects. The defects were categorized by type as follows: body, accessories, electrical, transmission, and engine. The resulting Pareto diagram for these qualitative data is shown in Figure 2.6(a) on page 36. The diagram reveals that most of the defects were found on the bodies of the cars or in their accessories (radio, wipers, etc.).

Sufficient data were collected when the cars were inspected to take the Pareto analysis one step farther. All 70 body defects were further classified as to whether they were paint defects, dents, upholstery defects, windshield defects, or chrome defects. All 50 accessory defects were further classified as to whether they were defects in the air conditioning (A/C) system, the radio, the power steering, the cruise control, or the windshield (W/S) wipers. Two more Pareto diagrams were constructed from these data. They are shown in panels (b) and (c) of Figure 2.6. This decomposition of the original Pareto diagram is called **exploding the Pareto diagram**.

It can be seen that paint defects and dents were the predominant types of body defects and that the accessory with the most problems is the air conditioning system. These are the "vital few" types of defects. Their identification permitted management to target them for special attention by managers, engineers, and assembly-line workers.

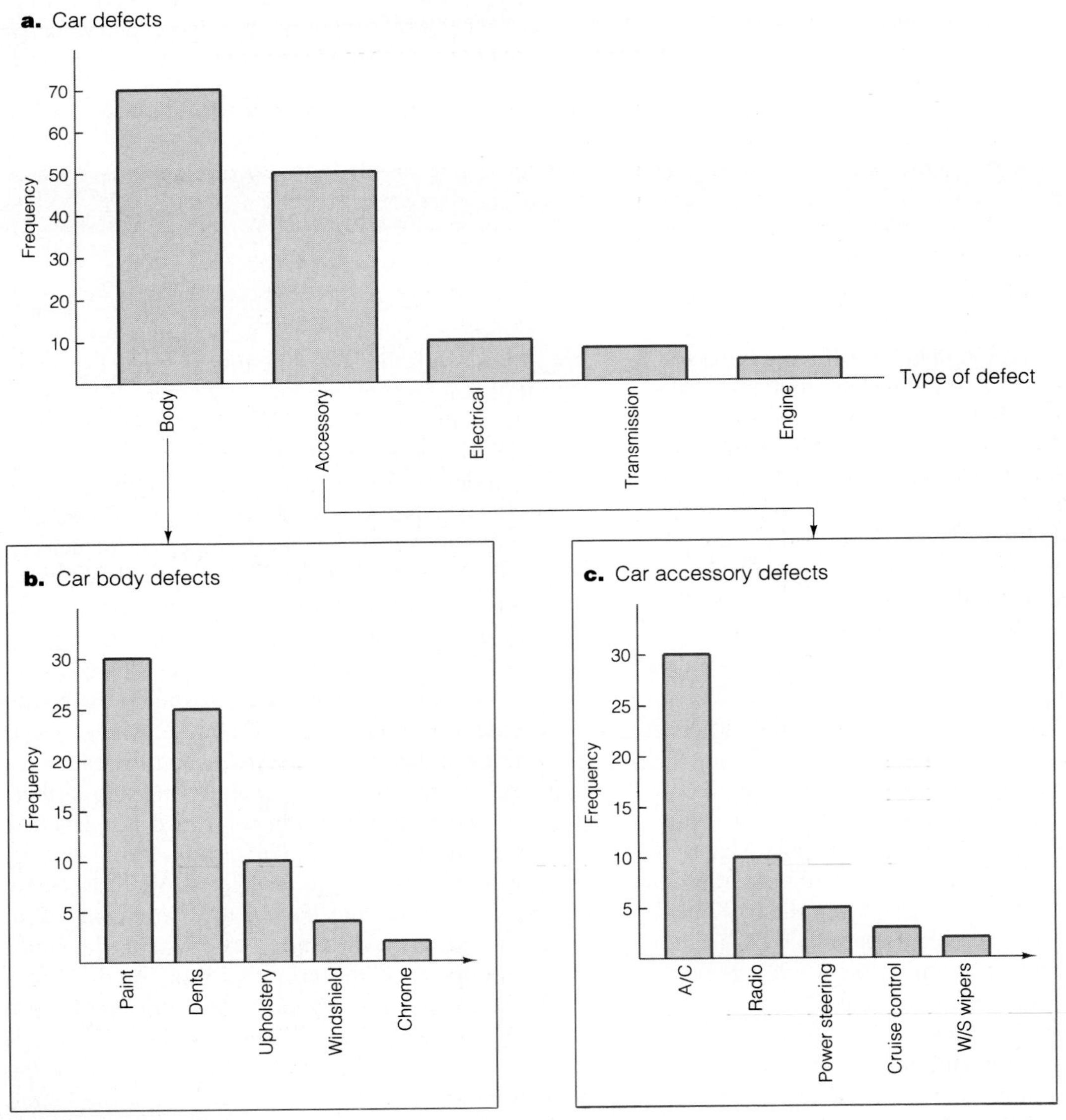

FIGURE 2.6 ▲ **Pareto diagrams**

Most statistical software packages can be used to generate histograms and stem-and-leaf displays. Both are useful tools for graphically describing data sets. We recommend generating and comparing both displays when feasible. The histogram will generally be more useful for very large data sets or for qualitative data, whereas the stem-and-leaf display provides useful detail for smaller quantitative data sets.

CASE STUDY 2.2 / Appraising the Market Value of an Asset

The *market value* of an asset is the price negotiated by a willing buyer and a willing seller of the asset, each acting rationally in his or her own self-interest. The *book value* of an asset is the value of the asset as shown in its owner's accounting records. Generally speaking, it is the amount the owner paid for the asset, less any depreciation expense (Davidson, Stickney, and Weil, 1979).

Robert R. Sterling and Raymond Radosevich (1969) examined the hypothesis that accountants generally agree on the book value of a depreciable asset but do not agree on its current market value. A questionnaire was prepared in which the installment purchase of a depreciable asset was described and the respondent was asked to determine the market value of the asset. The questionnaire also contained a series of questions relating to the book value of the asset. These questions enabled Sterling and Radosevich to calculate a book value for the asset for each of the respondents. The questionnaire was mailed to 500 randomly selected certified public accountants (CPAs) in the United States; 114 and 99 usable book value and market value responses, respectively, were returned.

The frequency distributions of book values and market values obtained from the returned questionnaires appear in Figure 2.7. In both histograms, the intervals from \$150 to \$200 and \$600 to \$650 include all responses less than \$200 and greater than \$600, respectively. The histograms suggest disagreement among the CPAs as to both the book value and the market value of the asset. Thus, Sterling and Radosevich rejected the hypothesis that accountants tend to agree on book values and to disagree on market values.

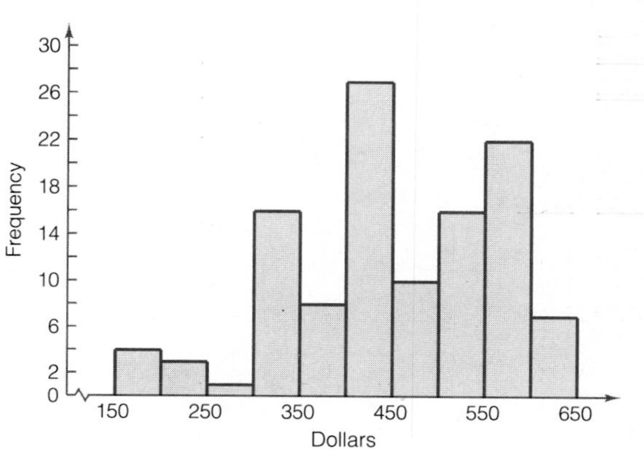

a. Book values

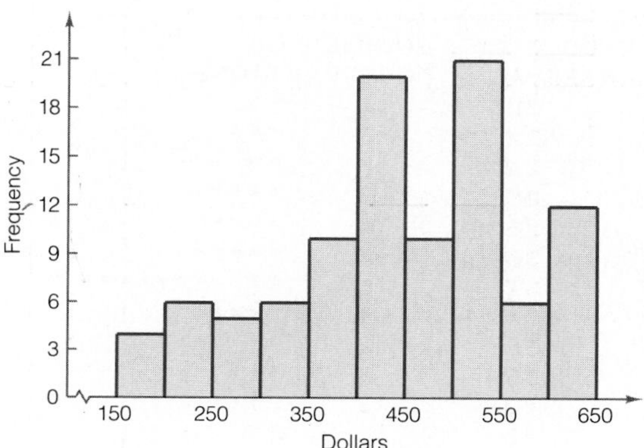

b. Market values

FIGURE 2.7 ▲ Frequency histograms for book and market values as assessed by CPAs

Note that decisions based on a visual comparison of histograms (or other graphical descriptions) are risky because they are subject to an unknown probability of error. For example, we might wonder whether disagreement among the CPAs really exists or whether the difference we see in the histogram is due to random variation that would be present from sample to sample. We begin to answer questions of this type in Chapter 6.

Exercises 2.9–2.22

Note: *Starred exercises (*) require the use of a computer.*

Learning the Mechanics

2.9 Graph the relative frequency histogram for the 500 measurements summarized in the accompanying relative frequency table.

Measurement Class	*Relative Frequency*
.5– 2.5	.10
2.5– 4.5	.15
4.5– 6.5	.25
6.5– 8.5	.20
8.5–10.5	.05
10.5–12.5	.10
12.5–14.5	.10
14.6–16.5	.05

2.10 Refer to Exercise 2.9. Calculate the number of the 500 measurements falling into each of the measurement classes. Then graph a frequency histogram for these data.

2.11 The statistical package Minitab was used to generate the following histogram:

```
MIDDLE OF    NUMBER OF
INTERVAL     OBSERVATIONS
    20           1    *
    22           3    ***
    24           2    **
    26           3    ***
    28           4    ****
    30           7    *******
    32          11    ***********
    34           6    ******
    36           2    **
    38           3    ***
    40           3    ***
    42           2    **
    44           1    *
    46           1    *
```

a. Is this a frequency histogram or relative frequency histogram? Explain.
b. How many measurement classes were used in the construction of this histogram?
c. How many measurements are there in the data set described by this histogram?

2.12 SAS was used to generate the stem-and-leaf display below. Note that SAS arranges the stems in descending order. Also, the instruction to "Multiply Stem.Leaf by 10**+0.1" indicates that each number should be multiplied by 10. For example, the top number in the display, 5.1, represents an observation of 10(5.1) = 51.

```
Stem Leaf                    #
   5 1                       1
   4 457                     3
   3 00036                   5
   2 1134599                 7
   1 2248                    4
   0 012                     3
     ----+----+----+----+
  Multiply Stem.Leaf by 10**+01
```

a. How many observations were in the original data set?
b. In the bottom row of the stem-and-leaf display, identify the stem, the leaves, and the numbers in the original data set represented by this stem and its leaves.

2.13 Bonds can be issued by the federal government, state and local governments, and U.S. corporations. A *mortgage bond* is a promissory note in which the issuing company pledges certain real assets as security in exchange for a specified amount of money. A *debenture* is an unsecured promissory note, backed only by the general credit of the issuer. The bond price of either a mortgage bond or debenture is negotiated between the asked price (the lowest price anyone will accept) and the bid price (the highest price anyone wants to pay). (See *How the Bond Market Works*, 1988.) The accompanying table contains the bid prices on December 31, 1991, for a sample of 30 publicly traded debenture bonds issued by utility companies.

Utility Company	*Bid Price*	*Utility Company*	*Bid Price*
Gulf States Utilities	108⅜	Indiana & Michigan Electric	98¼
Northern Natural Gas	101⅛	Toledo Edison Co.	105⅛
Indiana Gas	101	Dayton Power and Light	102¼
Appalachian Power	97⅞	Atlantic City Electric	94¼
Empire Gas Corp.	57½	Long Island Lighting	105⅛
Wisconsin Electric Power	100⅜	Portland General Electric	108
Pennsylvania Electric	97⅞	Boston Gas	105⅛
Commonwealth Edison	88	Duquesne Light Co.	69¾
El Paso Natural Gas	106⅜	General Electric Co.	99⅜
Montana Power Co.	100¼	Ohio Power Co.	94⅜
Elizabethtown Water	97⅞	Texas Power and Light	100½
Cascade Natural Gas	109¾	Columbia Gas System	78
Tennessee Gas Pipeline	102⅝	Central Power and Light	108⅛
Western Electric	99¾	Boston Edison	113⅛
Dallas Power and Light	108⅛	Philadelphia Electric	102⅝

Source: *Bond Guide* (a publication of the Standard & Poor Corporation), January 1992.

a. A frequency histogram was generated using a statistical software package. (See the accompanying figure.) Note that the software labels the midpoint of each measurement class rather than the two endpoints, and plots the bars horizontally rather than vertically. Interpret the histogram.

```
price        N =          30

Midpoint   Count
      60       1  *
      65       0
      70       1  *
      75       0
      80       1  *
      85       0
      90       1  *
      95       2  **
     100      12  ************
     105       6  ******
     110       5  *****
     115       1  *
                 +---------+---------+---------+---------+---------+
                0.0       3.0       6.0       9.0      12.0      15.0
```

b. Use the histogram to determine the number of bonds in the sample that had a bid price greater than \$97.50. What proportion of the total number of bonds is this group?

c. Identify the area under the histogram of part **a** that corresponds to this proportion.

Applying the Concepts

2.14 Production processes may be classified as *make-to-stock processes* or *make-to-order* processes. Make-to-stock processes are designed to produce a standardized product that can be sold to customers from the firm's inventory. Make-to-order processes are designed to produce products according to customer specifications. The McDonald's and Burger King fast-food chains are classic examples of these two types of processes. McDonald's produces and stocks standardized hamburgers; Burger King—whose slogan is "Your way, right away"—makes hamburgers according to the ingredients specified by the customer (Schroeder, 1993). In general, performance of make-to-order processes is measured by delivery time—the time from receipt of an order until the product is delivered to the customer. The following data set is a sample of delivery times (in days) for a particular make-to-order firm last year. The delivery times marked by an asterisk are associated with customers who subsequently placed additional orders with the firm.

50*	64*	56*	43*	64*
82*	65*	49*	32*	63*
44*	71	54*	51*	102
49*	73*	50*	39*	86
33*	95	59*	51*	68

The Minitab stem-and-leaf display of these data is shown here.

```
Stem-and-leaf of Time        N  = 25
Leaf Unit = 1.0

     3     3 239
     7     4 3499
   (7)     5 0011469
    11     6 34458
     6     7 13
     4     8 26
     2     9 5
     1    10 2
```

a. Circle the individual leaves that are associated with customers who did not place a subsequent order.
b. Concerned that they are losing potential repeat customers because of long delivery times, the management would like to establish a guideline for the maximum tolerable delivery time. Using the stem-and-leaf display, suggest a guideline. Explain your reasoning.

2.15 In order to better understand the interactions that take place between salespeople and customers, Ronald P. Willett and Allan L. Pennington (1966) monitored the interactions of appliance salespeople and customers on the floor of a large department store. Part of their research involved observing the length of time customers and salespeople interacted prior to the close of the sale or the departure of the customer. The data below, adapted from the article, are the lengths of time (in minutes) from the first customer–salesperson contact to the close of the sale or the customer's departure for 132 customers who completed their appliance purchase either at the time they were observed or within the following 2 weeks. Instances where a purchase was made by the customer at the time he or she was observed are denoted with an asterisk.

1.0*	3.2	49.1*	3.3	5.4*	6.0*	7.9*	39.9	1.1	25.4*	14.9
7.4*	12.4	10.9*	20.1*	30.5	27.2*	.9	50.1	48.6	10.0*	40.0
41.3*	26.2	118.4	12.0*	30.5	7.0*	66.1*	105.2*	47.6*	10.1*	12.2*
15.1*	33.3	8.8	30.0*	26.4*	8.4	1.7	12.5	23.0	11.1	21.9
6.1	.7	1.5	13.0	9.0*	7.6	10.9	13.5	8.0*	35.1	41.6
22.3	44.4*	18.4	8.9*	16.9*	34.6*	16.2	98.2	11.0	43.1	31.8
.4*	32.2	37.0	18.0*	14.2	39.2*	8.1*	4.5*	69.1	24.8	15.0*
17.4*	28.7	15.0*	14.2*	20.6*	27.7	7.9	18.7*	8.4	15.9	38.2
11.0*	7.8*	15.0	6.0	77.1	26.0	7.0*	7.4*	3.0	11.1	35.0*
35.4	25.6*	1.9*	40.1	.8	19.2*	42.3	15.5	13.3*	81.0	20.1*
12.8*	14.9	38.1	9.7	17.7*	7.7	42.1	4.1*	17.6*	5.1*	30.0
8.1*	25.1*	29.2	12.3	15.9	60.2*	27.7	10.3	14.0*	30.0*	10.5

SAS was used to construct a relative frequency histogram for each of the following data sets (see page 42):
(1) The complete set of 132 times
(2) The set of times associated with customers who made appliance purchases at the time they were being observed
(3) The set of times associated with customers who made the appliance purchases at a later date

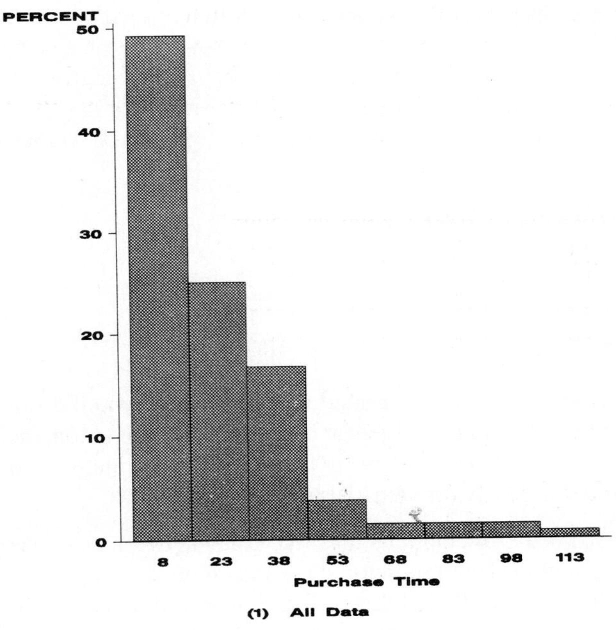

(1) All Data

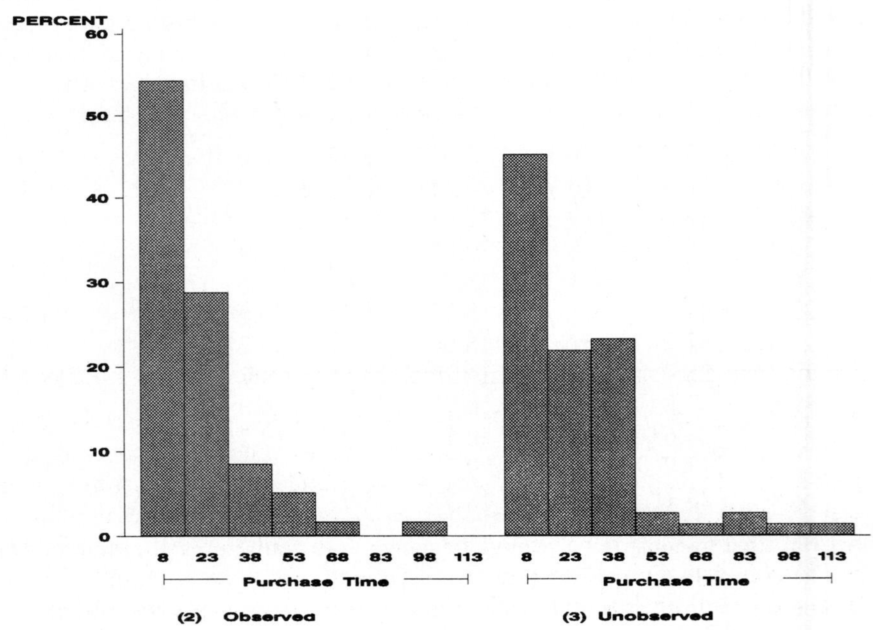

(2) Observed (3) Unobserved

a. Interpret the three histograms. Describe any differences you detect between the histograms of parts (2) and (3).

b. Suggest possible explanations for the differences you noted in part **a**.

2.16 In a manufacturing plant a *work center* is a specific production facility that consists of one or more people and/or machines and is treated as one unit for the purposes of capacity requirements planning and job scheduling. If jobs arrive at a particular work center at a faster rate than they depart, the work center impedes the overall production process and is referred to as a *bottleneck* (Fogarty, Blackstone, and Hoffmann, 1991). The data in the table were collected by an operations manager for use in investigating a potential bottleneck work center.

Number of Items Arriving at Work Center per Hour

155	115	156	156	109	127
150	159	163	148	135	119
172	143	159	140	127	115
166	148	175	122	99	106
151	161	138	171	123	135
148	129	135	125	107	152
140	152	139	111	137	161

The stem-and-leaf displays for the two sets of data are shown below:

Arrivals Stem	Leaf	*Departures* Stem	Leaf
		9	9
		10	6 7 9
11	5	11	1 5 9
12	9	12	2 3 5 7 7
13	5 8 9	13	5 5 7
14	0 3 8 8	14	0 8
15	0 1 2 5 6 9 9	15	2 6
16	1 3 6	16	1
17	2 5	17	1

Do the stem-and-leaf displays suggest that the work center may be a bottleneck? Explain.

***2.17** The ability to fill a customer's order on time depends to a great extent on being able to estimate how long it will take to produce the product in question. In most production processes, the time required to complete a particular task will be shorter each time the task is undertaken. Furthermore, it has been observed that in most cases the task time will decrease at a decreasing rate the more times the task is undertaken. Thus, in order to estimate how long it will take to produce a particular product, a manufacturer may want to study the relationship between production time per unit and the number of units that have been produced. The line or curve characterizing this relationship is called a *learning curve* (Adler and Clark, 1991). Twenty-five employees, all of whom were performing the same production task for the tenth time, were observed. Each person's task-completion time (in minutes) was recorded. The same 25 employees were observed again the 30th time they performed the same task and the 50th time they performed the task. The resulting completion times are shown in the table.

Tenth Performance		Thirtieth Performance		Fiftieth Performance	
15	19	16	11	10	8
21	20	10	10	5	10
30	22	12	13	7	8
17	20	9	12	9	7
18	19	7	8	8	8
22	18	11	20	11	6
33	17	8	7	12	5
41	16	9	6	9	6
10	20	5	9	7	4
14	22	15	10	6	15
18	19	10	10	8	7
25	24	11	11	14	20
23		9		9	

a. Use a statistical software package to construct a frequency histogram for each of the three data sets.
b. Compare the histograms. Does it appear that the relationship between task completion time and the number of times the task is performed is in agreement with the observations noted above about production processes in general? Explain.

***2.18** When two firms announce plans to merge, it frequently happens that within a few weeks one firm or the other becomes dissatisfied with the consequences of merging and the merger is canceled. Dodd (1980) reported that of 151 merger announcements that he identified, 80 were canceled. Thus, at the time a proposed merger is announced, there exists a great amount of uncertainty concerning whether the merger will take place. This uncertainty may persist for a considerable period of time and it may be many months after the announcement that the merger actually occurs. In her study of 38 mergers that were consummated, Eger (1982) reported the number of *trading days* (days the New York Stock Exchange is open for business) between the merger announcement (defined as the first mention of the potential merger in the *Wall Street Journal*) and the effective date of the merger. These data are listed next:

74	45	55	74	64	97	65	82	92	116
140	62	92	78	45	93	94	57	123	128
92	73	173	116	35	124	64	84	255	277
123	80	143	112	76	214	64	86		

a. Use a statistical software package to construct a stem-and-leaf display for these data.
b. Summarize the information reflected in your stem-and-leaf display concerning the number of trading days between announcement and the effective merger date for this sample of mergers.

2.19 Typically, the more attractive a corporate common stock is to an investor, the higher the stock's price–earnings (P/E) ratio. For example, if investors expect the stock's future earnings per share to increase, the price of the stock will be bid up and a high P/E ratio will result. Thus, the level of a stock's P/E ratio is a function of both the current financial performance of the firm and an investor's expectation of future performance (Spiro, 1982). The table contains the 1986 P/E ratios for samples of firms from the electronics industry and the auto parts industry.

Auto Parts Firm	P/E Ratio	Electronics Firm	P/E Ratio
Lear Siegler	9	AMP Inc.	22
Genuine Parts	15	Raytheon	12
Federal–Mogul	12	Intel	85
PPG Industries	13	Avnet	24
Borg–Warner	13	Perkin Elmer	14
Hoover Universal	12	TRW Inc.	15
Libbey–Owens–Ford	8	Motorola	39
Dana	8	Hewlett–Packard	20
Champion Spark Plug	15	Honeywell	12
Dayco	10	American District	27
Sheller–Globe	7	Corning Glass Works	27
Arvin Industries	10	EG&G	19
Allen Group	13	Varian Associates	14
Eaton	9	M/A-Com, Inc.	14
Cummins Engine	10	Harris Corp.	19
Barnes Group	11	Texas Instruments	10
Echlin	14	IT&T	12
Johnson Controls	9	North American Philips	11
Rockwell Int.	9	GTE	8
Snap-on-Tools	14	Tektronix	16

Source: Stock reports (OTC, NYSE, American), Standard & Poor's 1986.

a. Construct a stem-and-leaf display for each of these data sets.

b. What do your stem-and-leaf displays suggest about the level of the P/E ratios of firms in the electronics industry as compared to firms in the auto parts industry? Explain.

2.20 Consider the accompanying **bar chart** (a histogram for qualitative data), which shows 1985 cigarette sales (in billions of cigarettes) by company.

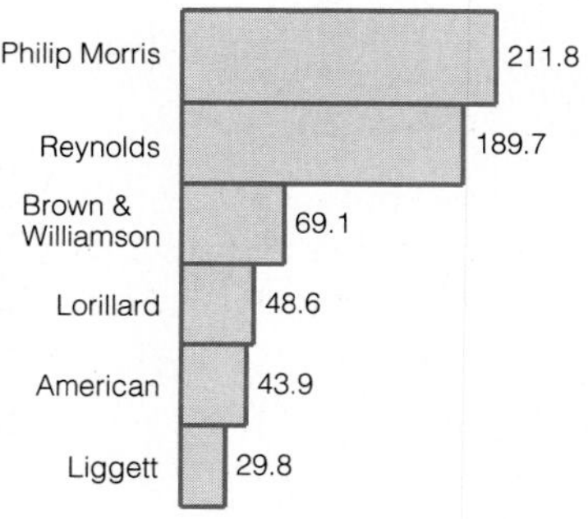

a. In general, what is described by the bar chart?
b. Which company sold the most cigarettes in 1985? Approximately how many cigarettes did the company sell?
c. Convert the bar chart to a relative frequency bar chart. Describe any problems you encounter in making the conversion.

2.21 A large midwestern city conducted a study of commuter traffic patterns and modes of transportation in 1990 and compared the results with those of a similar 1970 study ("In Twin Cities, Free(way)'s a Crowd," 1992). The comparisons will be used by the Metropolitan Council when making decisions about road improvements and mass transit. Some of the general findings were: Due to suburban job growth, most work trips are from suburb to suburb; the average commuting time has increased only 1 minute over the average in 1970. The accompanying table shows additional study findings.

	1970	*1990*
Drive alone	3,720	8,500
Carpool	1,680	1,000
Use mass transit	360	400
Other	240	100

Data: Metropolitan Council, Minneapolis, MN.

a. Construct two relative frequency bar charts for these data, one for 1970 and one for 1990.
b. Combine the bar charts you constructed in part **a** by plotting the eight relative frequencies on the same bar chart. You can do this by drawing two bars side by side for each category listed on the horizontal axis of your chart. Such a chart facilitates comparison of the two data sets.
c. *Stacking* is the combining of all the bar charts for any one time period into a single bar, by drawing one on top of the other and distinguishing one from another by the use of colors or patterns. Stack the relative frequencies of the four categories for 1970. Do the same for 1990.
d. Describe the changes in commuter behavior revealed by these data.

2.22 A research company conducted a written customer survey for a computer manufacturer to evaluate the services provided by its dealerships. Customers who had purchased personal computers were asked which one of the following four items they were least satisfied with: service, technical assistance, training, or the salesperson. Within the indicated category, each customer was asked to select the one aspect with which they were least satisfied. The accompanying table describes their responses.

	No. of Customers	
Service	10,002	
Timely completion of maintenance/service		6,040
Availability of service, repairs, and maintenance support		1,518
Responsiveness of dealer to your service needs		1,426
Quality of maintenance/service work completed		1,018
Technical Assistance	8,555	
Ability of staff to answer technical questions about installation, operations, and applications of the product		6,001
Availability of staff to assist with set-up or operation of the product		2,099
Courtesy of staff providing technical support		455

	No. of Customers	
Training	1,202	
Quality of training provided by the dealer		820
Ability of dealer to provide training to meet your specific needs and level of expertise		212
Availability of individualized training or classroom sessions to assist you in the use of your product		90
Courtesy of the person who provided training		80
Salesperson	4,100	
Salesperson's ability to clearly explain the features and benefits of the product		1,515
Salesperson's understanding of your needs		765
Salesperson's ability to provide solutions to your needs		556
Ease of contacting sales staff		451
Convenience of delivery schedule		410
Condition of product on delivery		300
Courtesy of your salesperson		103

a. How many customers responded to the survey?
b. Construct a Pareto diagram for the four categories. According to your diagram, which category represents the greatest opportunity for increasing customer satisfaction? Explain.
c. For the category you identified in part **b**, explode the Pareto diagram and identify more specifically (than in part **b**) how customer satisfaction can be increased.

2.3 Graphical Methods for Describing Quantitative Data Produced Over Time: The Time Series Plot (Optional)

Each of the previous sections has been concerned with describing the information contained in a sample or population of data. Often these data are viewed as having been produced at essentially the same point in time. Thus, time has not been a factor in any of the graphical methods described so far.

Data of interest to managers are often produced and monitored over time. Examples include the daily closing price of their company's common stock, the company's weekly sales volume and quarterly profits, and characteristics—such as weight and length—of products produced by the company.

Definition 2.5

Data that are produced and monitored over time are called **time series** data.

Recall from Section 1.4 that a process is a series of actions or operations that generates output over time. Accordingly, measurements taken of a sequence of units produced by a process—such as a production process—are time series data. In general, any sequence of numbers produced over time can be thought of as being generated by a process.

When measurements are made over time, it is important to record both the numerical value and the time or the time period associated with each measurement. With this information a **time series plot**—sometimes called a **run chart**—can be constructed to describe the time series data and to learn about the process that generated the data. A time series plot is a graph of the measurements (on the vertical axis) plotted against time or against the order in which the measurements were made (on the horizontal axis). The plotted points are usually connected by straight lines to make it easier to see the changes and movement in the measurements over time. For example, Figure 2.8 is a time series plot of a particular company's monthly sales (number of units sold per month). And Figure 2.9 is a time series plot of the weights of 30 one-gallon paint cans that were consecutively filled by the same filling head. Notice that the weights are plotted against the order in which the cans were filled rather than some unit of time. When monitoring production processes, it is often more convenient to record the order rather than the exact time at which each measurement was made.

Time series plots reveal the movement (trend) and changes (variation) in the variable being monitored. Notice how sales trend upward in the summer and how the variation in the weights of the paint cans increases over time. This kind of information would not be revealed by stem-and-leaf displays or histograms, as the following case study illustrates.

FIGURE 2.8 ▶
Time series plot of company sales

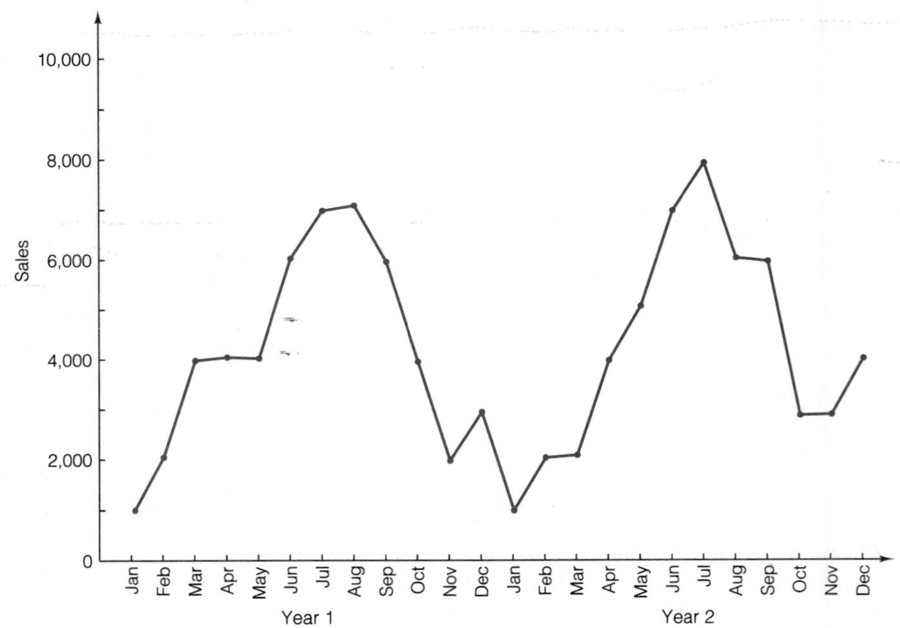

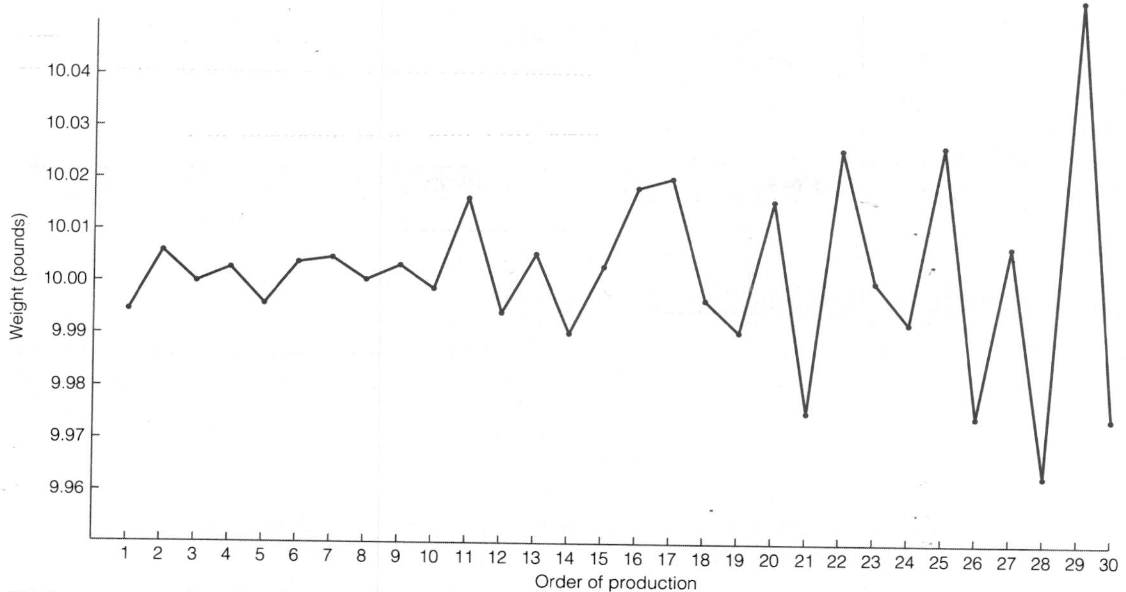

FIGURE 2.9 ▲ **Time series plot of paint can weights**

CASE STUDY 2.3 / Deming Warns Against Knee-Jerk Use of Histograms

W. Edwards Deming is one of America's most famous statisticians. He is best known for the role he played after World War II in teaching the Japanese how to improve the quality of their products by monitoring and continually improving their production processes. In his book *Out of the Crisis* (1986), Deming warns against the knee-jerk (i.e., automatic) use of histograms to display and extract information from data. As evidence he offers the following example.

Fifty camera springs were tested in the order in which they were produced. The elongation of each spring was measured under the pull of 20 grams. Both a time series plot and a histogram were constructed from the measurements. They are shown in Figure 2.10, which has been reproduced from Deming's book. If you had to predict the elongation measurement of the next spring to be produced (i.e., spring 51) and could use only one of the two plots to guide your prediction, which would you use? Why?

Only the time series plot describes the behavior *over time* of the process that produces the springs. The fact that the elongation measurements are decreasing

FIGURE 2.10 ► **Deming's time series plot and histogram**

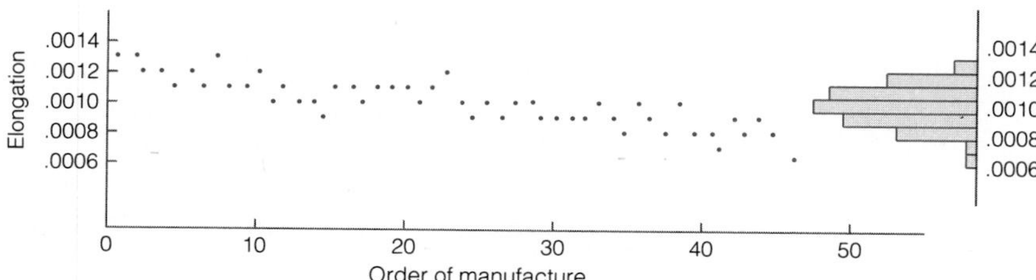

over time can only be gleaned from the time series plot. Because the histogram does not reflect the order in which the springs were produced, it in effect represents all observations as having been produced simultaneously. Using the histogram to predict the elongation of the 51st spring would very likely lead to an overestimate.

The lesson from Deming's example is this: For displaying and analyzing data that have been generated over time by a process, the primary graphical tool is the time series plot, not the histogram.

We cover many other aspects of the statistical analysis of time series data in Chapters 14 and 15.

2.4 Numerical Methods for Measuring Central Tendency

Now that we have presented some graphic techniques for summarizing and describing data sets, we turn to numerical methods for accomplishing this objective. When we speak of a data set, we refer to either a sample or a population. If statistical inference is our goal, we will wish ultimately to use sample numerical descriptive measures to make inferences about the corresponding measures for a population.

As you will see, there are a large number of numerical methods available to describe data sets. Most of these methods measure one of two data characteristics:

1. The **central tendency** of the set of measurements; i.e., the tendency of the data to cluster or to center about certain numerical values.
2. The **variability** of the set of measurements; i.e., the spread of the data.

In this section, we concentrate on measures of central tendency. In the next section, we discuss measures of variability.

The most popular and best understood measure of central tendency for a quantitative data set is the **arithmetic mean** (or simply the **mean**) of a data set.

Definition 2.6

The **mean** of a set of quantitative data is equal to the sum of the measurements divided by the number of measurements contained in the data set.

In everyday terms, the mean is the average value of the data set.

We will denote the measurements of a data set as follows:

$$x_1, x_2, x_3, \ldots, x_n$$

where x_1 is the first measurement in the data set, x_2 is the second measurement in the data set, x_3 is the third measurement in the data set, . . . , and x_n is the nth (and last) measurement in the data set. Thus, if we have five measurements in a set of data, we will write x_1, x_2, x_3, x_4, x_5 to represent the measurements. If the actual numbers are 5, 3, 8, 5, and 4, we have $x_1 = 5$, $x_2 = 3$, $x_3 = 8$, $x_4 = 5$, and $x_5 = 4$.

To calculate the mean of a set of measurements, we must sum them and divide by n, the number of measurements in the set. The sum of measurements $x_1, x_2, \ldots, x_n$ is

$$x_1 + x_2 + \cdots + x_n$$

To shorten the notation, we will write this sum as

$$x_1 + x_2 + \cdots + x_n = \sum_{i=1}^{n} x_i$$

where $\sum$ is the symbol for the summation. Verbally translate $\sum_{i=1}^{n} x_i$ as follows: "The sum of the measurements, whose typical member is x_i, beginning with the member x_1 and ending with the member x_n."

Finally, we will denote the mean of a sample of measurements by $\bar{x}$ (read "x-bar"), and represent the formula for its calculation as follows:

$$\bar{x} = \frac{\sum_{i=1}^{n} x_i}{n}$$

EXAMPLE 2.2

Calculate the mean of the following five sample measurements: 5, 3, 8, 5, 6.

Solution

Using the definition of sample mean and the shorthand notation, we find

$$\bar{x} = \frac{\sum_{i=1}^{5} x_i}{5} = \frac{5 + 3 + 8 + 5 + 6}{5} = \frac{27}{5} = 5.4$$

Thus, the mean of this sample is 5.4.*

*In the examples given here, $\bar{x}$ is sometimes rounded to the nearest tenth, sometimes the nearest hundredth, sometimes the nearest thousandth. There is no specific rule for rounding when calculating $\bar{x}$ because $\bar{x}$ is specifically defined to be the sum of all measurements divided by n; i.e., it is a specific fraction. When $\bar{x}$ is used for descriptive purposes, it is often convenient to round the calculated value of $\bar{x}$ to the number of significant figures used for the original measurements. When $\bar{x}$ is to be used in other calculations, however, it may be necessary to retain more significant figures.

EXAMPLE 2.3

Refer to Table 2.4. Calculate the mean of the percentages of revenues spent by the 50 companies on research and development.

TABLE 2.4 Percentages of Revenues Spent on Research and Development

Company	Percentage	Company	Percentage	Company	Percentage	Company	Percentage
1	13.5	14	9.5	27	8.2	39	6.5
2	8.4	15	8.1	28	6.9	40	7.5
3	10.5	16	13.5	29	7.2	41	7.1
4	9.0	17	9.9	30	8.2	42	13.2
5	9.2	18	6.9	31	9.6	43	7.7
6	9.7	19	7.5	32	7.2	44	5.9
7	6.6	20	11.1	33	8.8	45	5.2
8	10.6	21	8.2	34	11.3	46	5.6
9	10.1	22	8.0	35	8.5	47	11.7
10	7.1	23	7.7	36	9.4	48	6.0
11	8.0	24	7.4	37	10.5	49	7.8
12	7.9	25	6.5	38	6.9	50	6.5
13	6.8	26	9.5				

Solution

Using the data in Table 2.4, we have

$$\bar{x} = \frac{\sum_{i=1}^{50} x_i}{50} = \frac{13.5 + 8.4 + \cdots + 6.5}{50} = \frac{424.6}{50} = 8.49$$

The average expenditure on research and development for the 50 companies is 8.49% of revenues. Glancing at the relative frequency histogram for these data (Figure 2.11), we note that the mean falls in the middle of this data set.

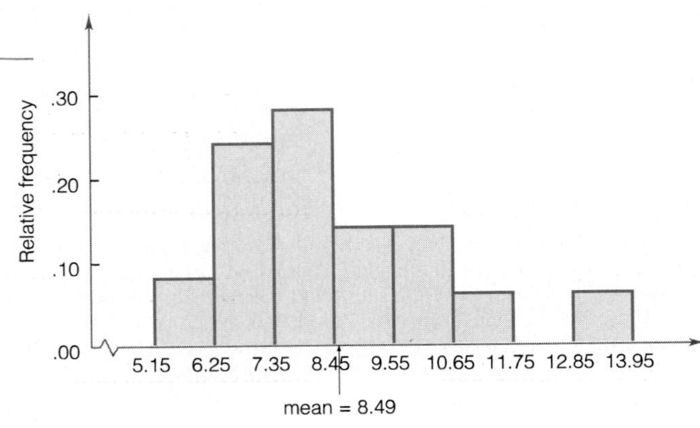

FIGURE 2.11 ► Relative frequency histogram for the 50 computer companies' R&D percentages: The mean

The sample mean will play an important role in accomplishing our objective of making inferences about populations based on sample information. For this reason, it is important to use a different symbol when we want to discuss the **mean of a population**—i.e., the mean of the set of measurements on every unit in the population. We use the Greek letter μ (mu) for the population mean. We will adopt a general policy of using Greek letters to represent population numerical descriptive measures and Roman letters to represent corresponding descriptive measures for the sample.

$\bar{x}$ = Sample mean μ = Population mean

The sample mean, $\bar{x}$, will often be used to estimate (make an inference about) the population mean, μ. For example, the percentages of revenues spent on research and development by the population consisting of *all* U.S. companies has a mean equal to some value, μ. Our sample of 50 companies yielded percentages with a mean of $\bar{x} = 8.49$. If, as is usually the case, we did not have access to the measurements for the entire population, we could use $\bar{x}$ as an estimator or approximator for μ. Then we would need to know something about the reliability of our inference. That is, we would need to know how accurately we might expect $\bar{x}$ to estimate μ. In Chapter 7, we will find that this accuracy depends on two factors:

1. *The size of the sample.* The larger the sample, the more accurate the estimate will tend to be.
2. *The variability, or spread, of the data.* All other factors remaining constant, the more variable the data, the less accurate the estimate.

In summary, the mean provides a valuable measure of the central tendency for a set of measurements. It is a very common tool in business and economic research, and therefore the mean will be the focus of much of our discussion of inferential statistics.

CASE STUDY 2.4 / Hotels: A Rational Method for Overbooking

The most outstanding characteristic of the general hotel reservation system is the option of the prospective guest, without penalty, to change or cancel his reservation or even to "no-show" (fail to arrive without notice). Overbooking (taking reservations in excess of the hotel capacity) is practiced widely throughout the industry as a compensating economic measure. This has motivated our research into the problem of determining policies for overbooking which are based on some set of rational criteria.

So said Marvin Rothstein (1974) in an article that appeared in *Decision Sciences*, a journal published by the Decision Sciences Institute. In this paper Rothstein introduces a method for scientifically determining hotel booking policies and applies it to the booking problems of the 133-room Sheraton Pocono Inn at Stroudsburg, Pennsylvania.

From the Sheraton Pocono Inn's records, the number of reservations, walk-ins (people without reservations who expect to be accommodated), cancellations, and no-shows were tabulated for each day during the period August 1–28, 1971. The inn's records for this period included approximately 3,100 guest histories. From the tabulated data, the mean or average number of room reservations per day for each of the seven days of the week was computed, as shown in Table 2.5. In applying his booking policy decision method to the Sheraton's data, Rothstein used the means listed in Table 2.5 to help portray the inn's demand for rooms.

TABLE 2.5 Mean Number of Room Reservations, August 1–28, 1971, 133 Rooms

Sunday	Monday	Tuesday	Wednesday	Thursday	Friday	Saturday
138	126	149	160	150	150	169

The mean number of Saturday reservations during the period August 1–28, 1971, is 169. This may be interpreted as an estimate of μ, the mean number of rooms demanded via reservations (walk-ins also contribute to the demand for rooms) on a Saturday during 1971. If the reservation data for all Saturdays during 1971 had been tabulated, μ could have been computed. But, since only the August data are available, they were used to estimate μ. Can you think of some problems associated with using August's data to estimate the mean for the entire year?

Another very important measure of central tendency is the **median** of a set of measurements:

Definition 2.7

The **median** of a data set is the middle number when the measurements are arranged in ascending (or descending) order.

The median is of most value in describing large data sets. If the data set is characterized by a relative frequency histogram (see Figure 2.12), the median is the point on the x-axis such that half the area under the histogram lies above the median and half lies below. [*Note:* In Section 2.2, we observed that the relative frequency associated with a particular interval on the x-axis is proportional to the area under the histogram that lies above the interval.]

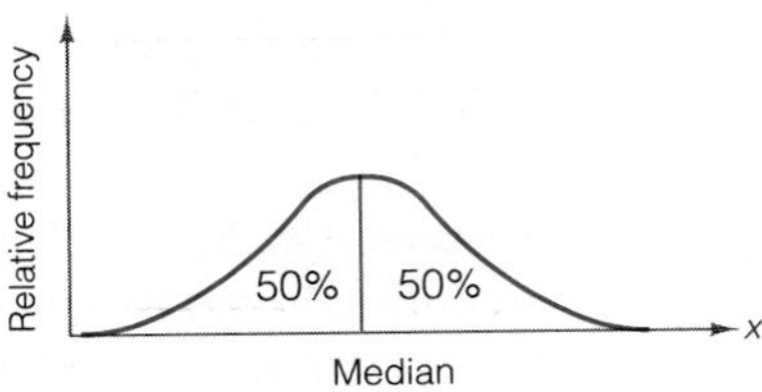

FIGURE 2.12 ▶ Location of the median

For a small—or even a large but finite—number of measurements, there may be many numbers that satisfy the property indicated in Figure 2.12. For this reason, we will arbitrarily calculate the median of a data set as follows:

Calculating the Median

Arrange the n measurements from the smallest to the largest.

1. If n is odd, the median is the middle number.
2. If n is even, the median is the mean (average) of the middle two numbers.

EXAMPLE 2.4

Consider the following sample of $n = 7$ measurements: 5, 7, 4, 5, 20, 6, 2.

a. Calculate the median of this sample.

b. Eliminate the last measurement (the 2), and calculate the median of the remaining $n = 6$ measurements.

Solution

a. The seven measurements in the sample are first ranked in ascending order:

2, 4, 5, 5, 6, 7, 20

Since the number of measurements is odd, the median is the middle measurement. Thus, the median of this sample is 5.

b. After removing the 2 from the set of measurements, we rank the sample measurements in ascending order as follows:

4, 5, 5, 6, 7, 20

Now the number of measurements is even, so we average the middle two measurements. The median is $(5 + 6)/2 = 5.5$.

In certain situations, the median may be a better measure of central tendency than the mean. In particular, the median is less sensitive than the mean to extremely large or small measurements. To illustrate, note that all but one of the measurements in Example 2.4 center about $x = 5$. The single large measurement, $x = 20$, does not affect the value of the median, 5, but it causes the mean, $\bar{x} = 7$, to lie to the right of most of the measurements.

As another example, if you were interested in computing a measure of central tendency of the incomes of a company's employees, the mean might be misleading. If all blue- and white-collar employees' incomes are included in the data set, the high incomes of a few executives will influence the mean more than the median. Thus, the median will often provide a more accurate picture of the typical income for an employee. Similarly, the median yearly sales for a set of companies would locate the middle of the sales data. However, the very large yearly sales of a few companies would

TABLE 2.6 Percentages of Revenues Spent on Research and Development, in Ascending Order

5.2	7.1	8.2	9.9
5.6	7.2	8.2	10.1
5.9	7.2	8.2	10.5
6.0	7.4	8.4	10.5
6.5	7.5	8.5	10.6
6.5	7.5	8.8	11.1
6.5	7.7	9.0	11.3
6.6	7.7	9.2	11.7
6.8	7.8	9.4	13.2
6.9	7.9	9.5	13.5
6.9	8.0	9.5	13.5
6.9	8.0	9.6	
7.1	8.1	9.7	

greatly influence the mean, making it deceptively large. That is, the mean could exceed a vast majority of the sample measurements, making it a misleading measure of central tendency.

For an example using more measurements, we have arranged the 50 R&D percentages in ascending order in Table 2.6. Since the number of measurements is even, the median equals the mean of the middle two numbers (shaded)—that is, the mean of the 25th and 26th numbers in the ordered list:

$$\frac{8.0 + 8.1}{2} = 8.05$$

Note that the median is smaller than the mean (8.49) for these data. This fact indicates that the data are **skewed** to the right—i.e., there are more extreme measurements in the right tail of the distribution than in the left tail. This affects the mean more than the median, because the extreme values (large or small) are used explicitly in the calculation of the mean. On the other hand, the median is not affected directly by extreme measurements, since the middle measurement(s) is (are) the only one(s) explicitly used to calculate the median. Consequently, if measurements are pulled toward one end of the distribution, the mean will shift toward that tail more than the median. The skewness of the R&D data set is evident in Figure 2.13, where we show the median and mean of the R&D percentages.

FIGURE 2.13 ► **Relative frequency histogram for the R&D percentages: Mean and median**

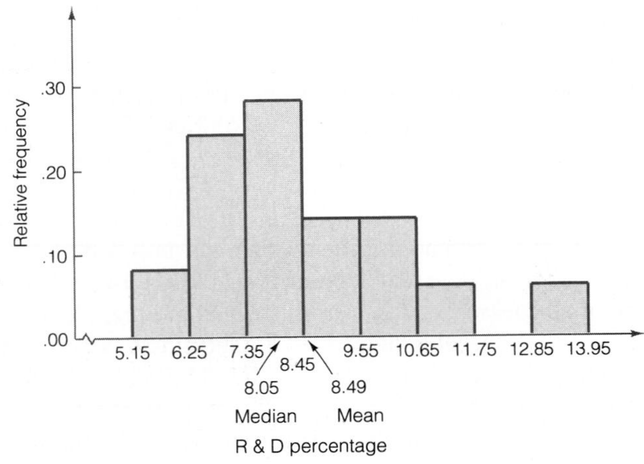

A comparison of the mean and median gives us a general method for detecting skewness in data sets, as shown in the box.

Comparing the Mean and the Median

1. If the median is less than the mean, the data set is skewed to the right:

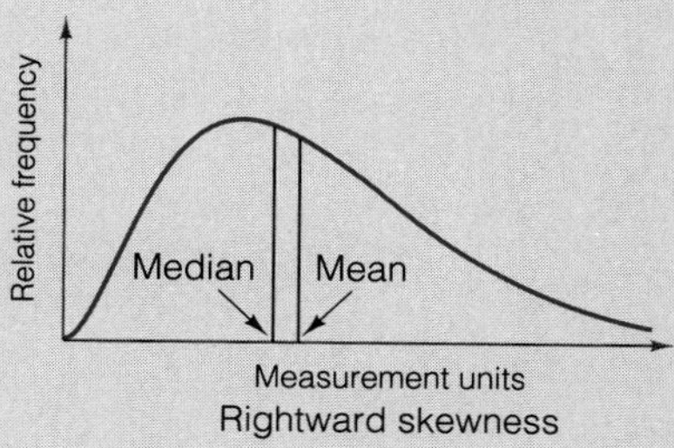

Rightward skewness

2. The median will equal the mean when the data set is symmetric:

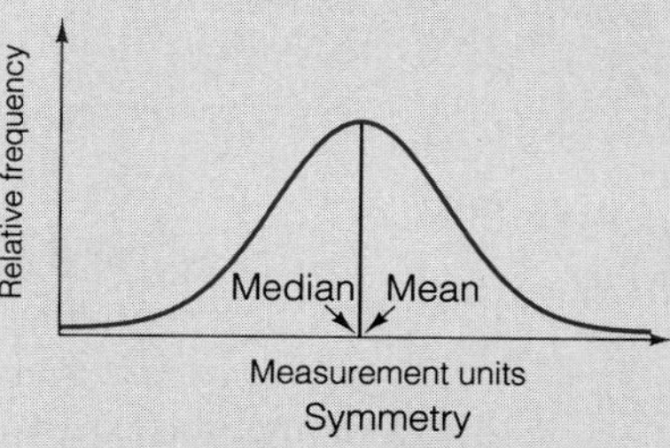

Symmetry

3. If the median is greater than the mean, the data set is skewed to the left:

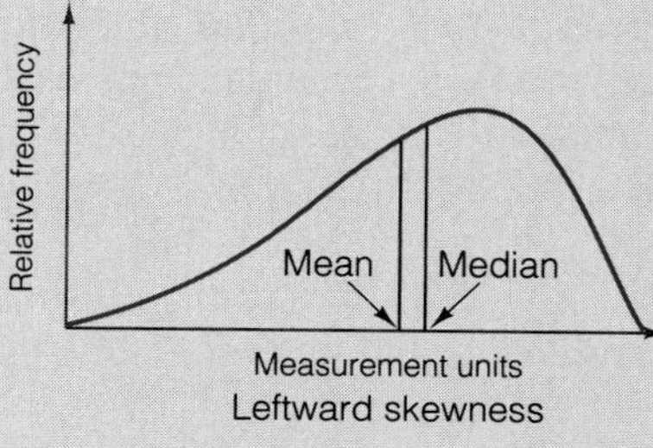

Leftward skewness

CASE STUDY 2.5 / The Delphi Technique for Obtaining a Consensus of Opinion

George T. Milkovich, Anthony J. Annoni, and Thomas A. Mahoney (1972) explain the delphi technique as follows:

The delphi technique, a set of procedures originally developed by the Rand Corporation in the late 1940's, is designed to obtain the most reliable consensus of opinion of a group of experts. Essentially, the delphi is a series of intensive interrogations of each individual expert (by a series of questionnaires) concerning some primary question interspersed with controlled feedback. The procedures are designed to avoid direct confrontation of the experts with one another.

The interaction among the experts is accomplished through an intermediary who gathers the data requests of the experts and summarizes them along with the experts' answers to the primary question. This mode of controlled interaction among the experts is a deliberate attempt to avoid the disadvantages associated with more conventional use of experts such as in round table discussions or direct confrontation of opposing views. The developers of the delphi argue the procedures are more conducive to independent thought and allow more gradual formulation to a considered opinion.

This article presents a study of the usefulness of the delphi procedure in projecting labor requirements in a low profit margin national retail firm. Seven company executives formed the panel of experts. Five questionnaires submitted at approximately 8-day intervals were used to interrogate the seven experts. On questionnaires 2–5, they were each asked the primary question, "How many buyers will the firm need 1 year from now?" Their individual responses along with the median response of the group for each questionnaire appear in Table 2.7.

TABLE 2.7 Projected Demand for Buyers

Questionnaire	Experts							Median
	A	B	C	D	E	F	G	
2	55	35	33	35	55	33	32	35
3	45	35	41	35	41	34	32	35
4	45	38	41	35	41	34	34	38
5	45	38	41	35	45	34	34	38

Note that the median response increased from 35 on questionnaire 2 to 38 on questionnaires 4 and 5. This increase indicates an upward shift in the distribution of the experts' forecasts as to the number of buyers the firm would need a year from now. If a single number is needed to represent the group's forecast, the median of the last questionnaire—in this case 38—is frequently used.

One conclusion of the study was that the delphi technique provided a better forecast of the actual number of buyers (37) needed by the firm 1 year later than did other more conventional forecasting techniques.

A third measure of central tendency is the **mode** of a set of measurements.

Definition 2.8

The **mode** is the measurement that occurs with greatest frequency in the data set.

Because it emphasizes data concentration, the mode is often used with large data sets to locate the region in which much of the data is concentrated. A retailer of men's clothing would be interested in the modal neck size and sleeve length of potential customers. The modal income class of the laborers in the United States is of interest to the Labor Department. Thus, the mode provides a useful measure of central tendency for many business applications.

Unless a quantitative data set is very large, its mode may not be very meaningful. For example, consider the percentage of revenues spent on research and development (R&D) by 50 companies. These data, first presented in Table 2.1, were analyzed in Section 2.2. A reexamination of the data reveals that three of the measurements are repeated three times: 6.5%, 6.9%, and 8.2%. Thus, there are three modes in the sample, and none is particularly useful as a measure of central tendency.

We can calculate a more meaningful measure by first constructing a relative frequency histogram for the data. The measurement class containing the most measurements is called the **modal class**, and the mode is taken to be the midpoint of this interval. For the R&D data, the modal class is the one corresponding to the interval 7.35 to 8.45, as shown by the shaded rectangle in Figure 2.14. The mode is taken to be the midpoint of this interval, that is, $(7.35 + 8.45)/2 = 7.90$. This modal class (and the mode itself) identifies the area in which the data are most concentrated, and in that sense is a measure of central tendency. However, for most applications involving quantitative data, the mean and median provide more descriptive information than the mode.

FIGURE 2.14 ▶ **Relative frequency histogram for the computer companies' R&D percentages: The modal class**

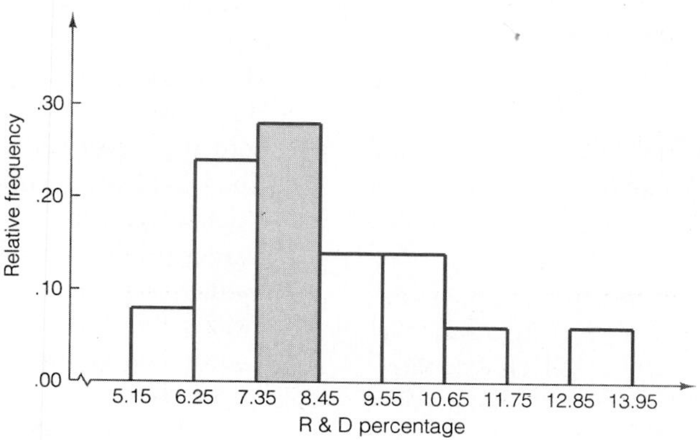

Exercises 2.23–2.34

Learning the Mechanics

2.23 Calculate the mean, median, and mode of the following data: 18, 10, 15, 13, 17, 15, 12, 15, 18, 16, 11.

2.24 Calculate the mean and median of the following grade-point averages: 3.2, 2.5, 2.1, 3.7, 2.8, 2.0.

2.25 Explain the difference between the calculation of the median for an odd and an even number of measurements. Construct one data set consisting of 5 measurements and another consisting of 6 measurements for which the medians are equal.

2.26 Explain how the relationship between the mean and median provides information about the symmetry or skewness of the data's distribution.

2.27 Refer to Case Study 2.5. Practitioners of the delphi technique frequently use the median of the set of responses given by the panel of experts on the last questionnaire to describe "the opinion of the experts." Why do you suppose the median response is used rather than the mean response?

Applying the Concepts

2.28 According to *Consumers' Digest* (Nov.–Dec. 1980), in 1980 sugar was the leading food additive in the U.S. food supply. Sugar may be listed more than once on a product's ingredient list since it goes by different names depending on its source (e.g., sucrose, corn sweetener, fructose, and dextrose). Thus, when you read a product's label you may have to total up the sugar in the product to see how much sweetener it contains. The accompanying table gives a list of candy bars and the percentage of sugar they contain relative to their weight.

Brand	*Percentage of Sugar by Weight*	Brand	*Percentage of Sugar by Weight*
Baby Ruth	23.7	Power House	30.6
Butterfinger	29.5	Bit-O-Honey	23.5
Mr. Goodbar	34.2	Chunky	38.4
Milk Duds	36.0	Milk Chocolate Covered Raisinettes	24.7
Mello Mint	79.6	Oh Henry!	31.2
M & M Plain Chocolate Candies	52.2	Borden Cracker Jack	14.7
Mars Chocolate Almond	36.4	Good & Plenty Licorice	28.2
Milky Way	26.8	Nestle's Crunch	43.5
Marathon	36.7	Planter Jumbo Block Peanut Candy	21.5
Snickers	28.0	Switzer Licorice	8.4
3 Musketeers	36.1	Switzer Red Licorice	2.8
Junior Mints	45.3	Tootsie Pop Drops	54.1
Pom Poms	29.5	Tootsie Roll	21.1
Sugar Babies	41.0	Fancy Fruit Lifesavers	77.6
Sugar Daddy	22.0	Spear-O-Mint Lifesavers	67.6
Almond Joy	20.0		

Source: *National Confectioners Association Brand Name Guide to Sugar* (Nelson Hall Paperback).
Secondary Source: *Consumers' Digest*, Nov.–Dec. 1980, p. 11.

a. Calculate the mean percentage of sugar per bar for the candy bars listed.
b. Find the median for the data set.
c. What do the mean and median indicate about the skewness of the data set?
d. Construct a relative frequency histogram for the data set. Indicate the location of the mean, median, and modal class of the data set on your histogram.

2.29 During the 1980s the pharmaceutical industry placed an increased emphasis on producing revolutionary new products. As a result, R&D costs increased and companies took a greater interest in R&D management. The table lists the 1984 R&D expenditures (in millions of dollars) of the world's largest pharmaceutical manufacturers.

Company	*R&D Expenditures*	*Company*	*R&D Expenditures*
Abbott	$110	Pfizer	159
American Home	90	Rhone–Poulenc	110
Bayer	200	Sandoz	181
Boehringer Ingelheim	176	Schering–Plough	129
Bristol–Myers	162	Smith Kline–Beckman	158
Ciba–Geigy	230	Squibb	114
Hoechst	274	Takeda	125
Hoffmann–LaRoche	363	Upjohn	200
Johnson & Johnson	187	Warner–Lambert	162
Merck	290		

Source: *Business Quarterly*, Fall 1985, p. 81.

a. Calculate the mean and median for this data set.
b. What do the mean and median indicate about the skewness of this data set?
c. Will the median of a data set always be equal to an actual value in the data set, as was the case in part **a**?

2.30 The table lists the mean age for each team in the National Basketball Association (NBA), along with the number of players on each team at the start of the 1982–1983 season. What was the population mean age at the start of the 1982–1983 season? [*Hint:* $\sum_{i=1}^{n} x_i = n\bar{x}$]

Team	*Number of Players*	*Mean Age*	*Team*	*Number of Players*	*Mean Age*
1. Indiana	12	24.720	13. San Antonio	12	26.25
2. Portland	12	24.724	14. Los Angeles	13	26.42
3. Golden State	15	24.96	15. New York	14	26.45
4. Dallas	13	25.17	16. Cleveland	14	26.48
5. New Jersey	13	25.26	17. San Diego	13	26.71
6. Kansas City	12	25.27	18. Boston	12	26.94
7. Chicago	12	25.32	19. Seattle	11	27.09
8. Detroit	13	25.65	20. Atlanta	12	27.12
9. Philadelphia	12	25.96	21. Denver	11	27.59
10. Phoenix	13	26.10	22. Houston	12	29.21
11. Washington	13	26.10	23. Milwaukee	13	29.72
12. Utah	11	26.13			

Source: *Basketball Weekly*, Jan. 3, 1983, Vol. 16, No. 5, pp. 10–11.

2.31 According to the U.S. Energy Information Association, the average price of regular unleaded gasoline in the United States as of March 1989 was 14.4 cents per gallon cheaper than the average price of premium gas. The table lists the average price excluding excise tax (in cents per gallon) of regular unleaded gas in each of a sample of 20 states.

State	*Price*	*State*	*Price*
Alaska	92.5	Nevada	64.9
Arkansas	63.3	New Hampshire	74.7
Connecticut	76.8	New York	70.6
Delaware	68.5	North Dakota	72.4
Louisiana	67.6	Oklahoma	64.6
Maine	76.8	Oregon	70.8
Massachusetts	74.9	Pennsylvania	65.5
Michigan	63.9	Texas	64.6
Missouri	63.4	Wisconsin	65.7
Montana	69.0	Wyoming	69.7

Data: U.S. Energy Information Association, *Petroleum Marketing Monthly*, March 1989.
Source: *Statistical Abstract of the United States*, 1989.

a. Calculate the mean, median, and mode of this data set.
b. Eliminate the highest price from the data set and repeat part **a**. What effect does dropping this measurement have on the measures of central tendency calculated in part **a**?
c. Arrange the 20 prices in order from lowest to highest. Next, eliminate the lowest two prices and the highest two prices from the data set and calculate the mean of the remaining prices. The result is called an **80% trimmed mean**, since it is calculated using the central 80% of the values in the data set. An advantage of the trimmed mean is that it is not as sensitive as the arithmetic mean to extreme observations in the data set.

2.32 At the end of 1982, McDonald's had 7,300 restaurants and total sales for the year of $7.8 billion. Burger King had 3,400 restaurants and total sales of $2.4 billion. McDonald's opens new restaurants at the rate of 500 per year and Burger King at the rate of 200 per year (*Minneapolis Tribune*, July 3, 1983).
a. Calculate the average sales per restaurant for McDonald's in 1982 and compare it with the average sales per restaurant for Burger King in 1982.
b. On average, how many restaurants did McDonald's open per month? Per week? Per day?

2.33 In 1985, U.S. consumers redeemed 6.49 billion manufacturers' coupons worth a total of $2.24 billion (McCullough, 1986). Find the mean value per coupon.

2.34 In 1972, Kroger Corporation, one of the largest supermarket chains in the United States, made a major strategic decision. Instead of continuing to hold prices high and trying to attract customers with weekend specials and heavy advertising, Kroger decided to emphasize price competition all week long. Its new strategy involved selling "brand-name groceries for less, on average, than its competitors were doing, and to advertise this fact strenuously" ("Keeping Up," 1979). For the situation described, explain what is meant by "on average" in the preceding quote.

2.5 Numerical Methods for Measuring Variability

Measures of central tendency provide only a partial description of a quantitative data set. Our information is incomplete without a measure of the **variability**, or **spread**, of the data set. Note that in describing a data set, we refer to either the sample or the population. Ultimately (in Chapter 7) we will use the sample numerical descriptive measures to make inferences about the corresponding descriptive measures for the population from which the sample was selected.

If you examine the two histograms in Figure 2.15, you will notice that both hypothetical data sets are symmetric, with equal modes, medians, and means. However, in data set 1 in Figure 2.15(a), the measurements occur with almost equal frequency in the measurement classes, whereas in data set 2 in Figure 2.15(b), most of the measurements are clustered about the center. For this reason, a measure of variability is needed, along with a measure of central tendency, to describe a data set.

FIGURE 2.15 ►
Two hypothetical data sets

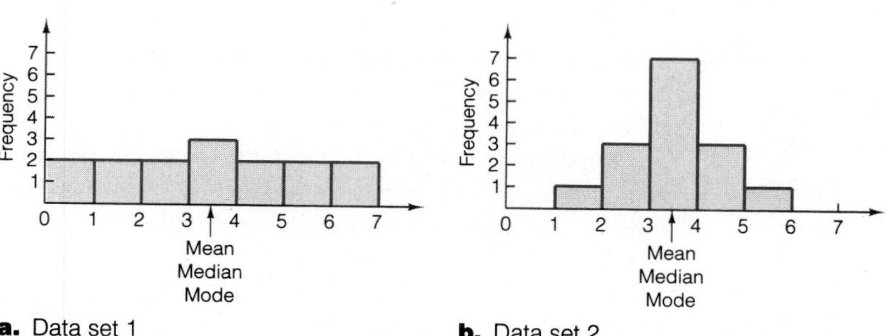

Perhaps the simplest measure of the variability of a quantitative data set is its **range**.

> **Definition 2.9**
>
> The **range** of a data set is equal to the largest measurement minus the smallest measurement.

The range measures the spread of the data by measuring the distance between the smallest and largest measurements. For example, stock A may vary in price during a given year from $32 to $36, whereas stock B may vary from $10 to $58, as shown in

Figure 2.16. The range in price of stock A is \$36 − \$32 = \$4, while that for stock B is \$58 − \$10 = \$48. A comparison of ranges tells us that the price of stock B was much more variable than the price of stock A.

FIGURE 2.16 ▶ **Ranges of stock prices for two companies**

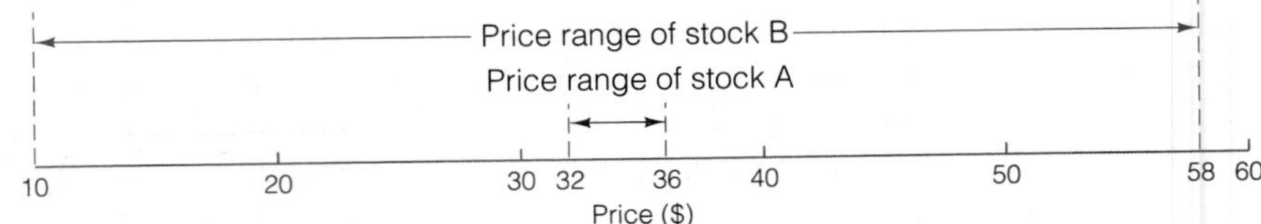

The range is not always a satisfactory measure of variability. For example, suppose we are comparing the profit margin (as a percentage of the total bid price) per construction job for 100 construction jobs for each of two cost estimators working for a large construction company. We find that the profit margins range from −10% (loss) to +40% (profit) for both cost estimators and therefore that the ranges for the two data sets, 40% − (−10%) = 50%, are equal. Because of this, we might be inclined to conclude that there is little or no difference in the performance of the two estimators.

But, suppose the histograms for the two sets of 100 profit margin measurements appear as shown in Figure 2.17. Although the ranges are equal and all central tendency measures are the same for these two symmetric data sets, there is an obvious difference between the two sets of measurements. The difference is that estimator B's profit margins tend to be more stable—i.e., to pile up or to cluster about the center of the data set. In contrast, estimator A's profit margins are more spread out over the range, indicating a higher incidence of some high profit margins, but also a greater risk of losses. Thus, even though the ranges are equal, the profit margin record of estimator A is more variable than that of estimator B, indicating a distinct difference in their cost estimating characteristics. We therefore need to develop more informative numerical measures of variability than the range. In particular, we need a measure that takes into consideration the magnitude of all measurements, not just the largest and smallest.

FIGURE 2.17 ▶ **Profit margin histograms for two cost estimators**

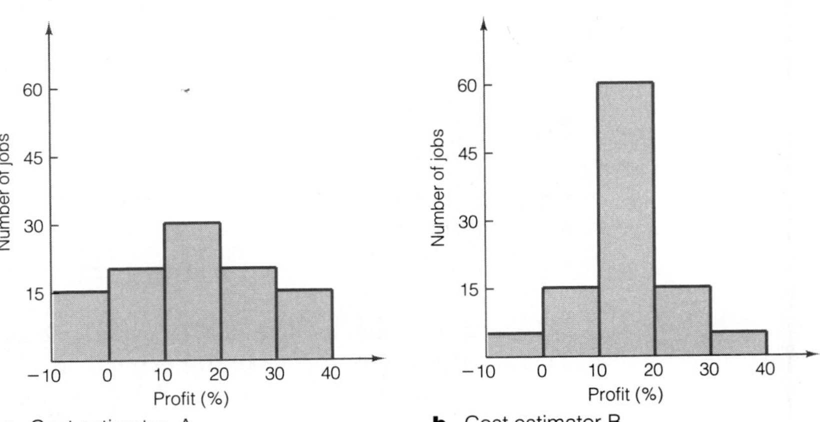

CASE STUDY 2.6 / More on the Delphi Technique

You will recall from Case Study 2.5 that the delphi technique is a set of procedures that may be used to obtain a consensus opinion from a group of experts through a series of questionnaires. Case Study 2.5 illustrated the use of the median as a measure of central tendency for the distribution of expert opinions elicited by the questionnaires. As a measure of variability of the data (i.e., the opinions), Milkovich et al. (1972) used the range. Table 2.7, showing the experts' opinions, is repeated here as Table 2.8, with the addition of the range of the distribution of opinions of each questionnaire in the right-hand column.

The range of 23 on questionnaire 2 indicates that at that time the experts' opinions were widely dispersed. The decrease in the range to 11 following questionnaire 4 indicates that as the experts received more information about the firm's needs and learned about one another's opinions, the variability in the distribution of their opinions decreased. Milkovich et al. noted that the decrease in the range was an indication that experts' opinions were converging.

TABLE 2.8 Projected Demand for Buyers

Questionnaire	Experts							Median	Range
	A	B	C	D	E	F	G		
2	55	35	33	35	55	33	32	35	23
3	45	35	41	35	41	34	32	35	13
4	45	38	41	35	41	34	34	38	11
5	45	38	41	35	45	34	34	38	11

Let us see if we can find a measure of data variability that is more sensitive than the range. Recall that we represent the n measurements in a sample by the symbols x_1, $x_2, \ldots, x_n$, and we represent their mean by $\bar{x}$. What would be the interpretation of $x_1 - \bar{x}$? It is the distance, or **deviation**, between the first sample measurement, x_1, and the sample mean, $\bar{x}$. If we were to calculate this distance for *every* measurement in the sample, we would create a set of distances from the mean:

$$x_1 - \bar{x}, \quad x_2 - \bar{x}, \quad x_3 - \bar{x}, \quad \ldots, \quad x_n - \bar{x}$$

What information do these distance contain? If they tend to be large, the interpretation is that the data are spread out or highly variable. If the distances are mostly small, the data are clustered around the mean $\bar{x}$ and therefore do not exhibit much variability. As a simple example, consider the two samples in Table 2.9 (page 66), which have five measurements (we have ordered the numbers for convenience). You will note that both samples have a mean of 3. However, a glance at the distances shows that sample 1 has greater variability—i.e., more large distances from $\bar{x}$—than sample

2, which is clustered around $\bar{x}$. You can see this clearly by looking at these distances in Figure 2.18. Thus, the distances provide information about the variability of the sample measurements.

TABLE 2.9

	Sample 1	Sample 2
Measurements	1, 2, 3, 4, 5	2, 3, 3, 3, 4
Mean	$\bar{x} = \frac{1 + 2 + 3 + 4 + 5}{5} = \frac{15}{5} = 3$	$\bar{x} = \frac{2 + 3 + 3 + 3 + 4}{5} = \frac{15}{5} = 3$
Distances from $\bar{x}$	1 − 3, 2 − 3, 3 − 3, 4 − 3, 5 − 3 or −2, −1, 0, 1, 2	2 − 3, 3 − 3, 3 − 3, 3 − 3, 4 − 3 or −1, 0, 0, 0, 1

FIGURE 2.18 ► **Distances from the mean for two data sets**

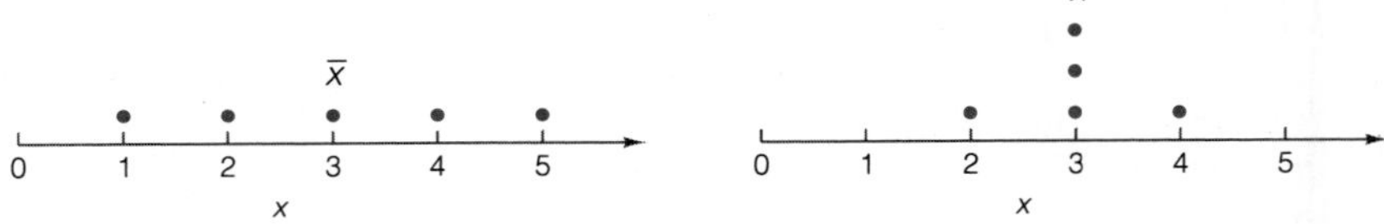

a. Sample 1 **b.** Sample 2

The next step is to condense the information on distances from $\bar{x}$ into a single numerical measure of variability. Simply averaging the distances from $\bar{x}$ will not help. For example, in samples 1 and 2 the negative and positive distances cancel, so that the average distance is 0. Since this is true for any data set—i.e., the sum of the deviations, $\sum_{i=1}^{n} (x_i - \bar{x})$, is always 0—we gain no information by averaging the distances from $\bar{x}$.

There are two methods for dealing with the fact that positive and negative distances from the mean cancel. The first is to treat all the distances as though they were positive, ignoring the sign of the negative distances. We will not pursue this line of thought because the resulting measure of variability (the mean of the absolute values of the distances) presents analytical difficulties beyond the scope of this text. A second method of eliminating the minus signs associated with the distances is to square them. The quantity we can calculate from the squared distances will provide a meaningful description of the variability of a data set.

To use the squared distances calculated from a data set, we first calculate the **sample variance**.

Definition 2.10

The **sample variance** for a sample of n measurements is equal to the sum of the squared distances from the mean divided by $(n - 1)$. In symbols, using s^2 to represent the sample variance,

$$s^2 = \frac{\sum_{i=1}^{n} (x_i - \bar{x})^2}{n - 1}$$

Referring to the two samples in Table 2.9, you can calculate the variance for sample 1 as follows:

$$s^2 = \frac{(1 - 3)^2 + (2 - 3)^2 + (3 - 3)^2 + (4 - 3)^2 + (5 - 3)^2}{5 - 1}$$

$$= \frac{4 + 1 + 0 + 1 + 4}{4} = 2.5$$

The second step in finding a meaningful measure of data variability is to calculate the **standard deviation** of the data set.

Definition 2.11

The **sample standard deviation,** s, is defined as the positive square root of the sample variance, s^2. Thus,

$$s = \sqrt{s^2} = \sqrt{\frac{\sum_{i=1}^{n} (x_i - \bar{x})^2}{n - 1}}$$

The **population variance**, denoted by the symbol σ^2 (sigma squared), is the average of the squared distances of the measurements of *all* units in the population from the mean, μ. σ (sigma) is the square root of this quantity. Since we never really compute σ^2 or σ from the population (the object of sampling is to avoid this costly procedure), we simply denote these two quantities by their respective symbols.

s^2 = Sample variance	s = Sample standard deviation
σ^2 = Population variance	σ = Population standard deviation

In contrast to the variance, the standard deviation is expressed in the original units of measurement. For example, if the original measurements are in dollars, the variance is expressed in the peculiar units "dollars squared," but the standard deviation is expressed in dollars.

You may wonder why we use the divisor $(n - 1)$ instead of n when calculating the sample variance. Although the use of n may seem logical, since then the sample variance would be the average squared distance from the mean, the use of n tends to underestimate the population variance, σ^2. The use of $(n - 1)$ in the denominator provides the appropriate correction for this tendency.* Since the primary use of sample statistics like s^2 is to estimate population parameters like σ^2, $(n - 1)$ is preferred to n when defining the sample variance.

EXAMPLE 2.5

Calculate the standard deviation of the following sample: 2, 3, 3, 3, 4.

Solution

For this set of data, $\bar{x} = 3$. Then,

$$s = \sqrt{\frac{(2-3)^2 + (3-3)^2 + (3-3)^2 + (3-3)^2 + (4-3)^2}{5-1}}$$

$$= \sqrt{\frac{2}{4}} = \sqrt{.5} = .71$$

As the number of measurements in the sample becomes larger, the sample variance becomes more difficult to calculate. We must calculate the distance between each measurement and the mean, square it, sum the squared distances, and finally divide by $(n - 1)$. Fortunately, as we show in Example 2.8, we can get around this difficulty by using a statistical software package (or most handheld calculators) to compute s^2 and s. However, if you must calculate it by hand, there is a shortcut formula for computing the sample variance.

*_Appropriate_ here means that s^2 with a divisor of $(n - 1)$ is an **unbiased** estimator of σ^2. We define and discuss unbiasedness of estimators in Chapter 6.

Shortcut Formula for Sample Variance

$$s^2 = \frac{\begin{pmatrix}\text{Sum of squares of}\\ \text{sample measurements}\end{pmatrix} - \dfrac{\begin{pmatrix}\text{Sum of sample}\\ \text{measurements}\end{pmatrix}^2}{n}}{n-1} = \frac{\sum_{i=1}^{n} x_i^2 - \dfrac{\left(\sum_{i=1}^{n} x_i\right)^2}{n}}{n-1}$$

Note that the formula requires only the sum of the sample measurements, $\sum_{i=1}^{n} x_i$, and the sum of the squares of the sample measurements, $\sum_{i=1}^{n} x_i^2$. Be careful when you calculate these two sums. Rounding the values of x^2 that appear in $\sum_{i=1}^{n} x_i^2$ or rounding the quantity $\left(\sum_{i=1}^{n} x_i\right)^2 / n$ can lead to substantial errors in the calculation of s^2.

EXAMPLE 2.6

Use the shortcut formula to compute the variances of these two samples of five measurements each:

Sample 1: 1, 2, 3, 4, 5 *Sample 2*: 2, 3, 3, 3, 4

Solution

We first work with sample 1. The two quantities needed are

$$\sum_{i=1}^{5} x_i = 1 + 2 + 3 + 4 + 5 = 15$$

and

$$\sum_{i=1}^{5} x_i^2 = 1^2 + 2^2 + 3^2 + 4^2 + 5^2 = 1 + 4 + 9 + 16 + 25 = 55$$

Then the sample variance for sample 1 is

$$s^2 = \frac{\sum_{i=1}^{5} x_i^2 - \dfrac{\left(\sum_{i=1}^{5} x_i\right)^2}{5}}{5-1} = \frac{55 - \dfrac{(15)^2}{5}}{4} = \frac{55 - 45}{4} = \frac{10}{4} = 2.5$$

Similarly, for sample 2 we get

$$\sum_{i=1}^{5} x_i = 2 + 3 + 3 + 3 + 4 = 15$$

and

$$\sum_{i=1}^{5} x_i^2 = 2^2 + 3^2 + 3^2 + 3^2 + 4^2 = 4 + 9 + 9 + 9 + 16 = 47$$

Then the variance for sample 2 is

$$s^2 = \frac{\sum_{i=1}^{5} x_i^2 - \frac{\left(\sum_{i=1}^{5} x_i\right)^2}{5}}{5 - 1} = \frac{47 - \frac{(15)^2}{5}}{4} = \frac{47 - 45}{4} = \frac{2}{4} = .5$$

EXAMPLE 2.7

The 50 companies' percentages of revenues spent on R&D are repeated here. Calculate the sample variance, s^2, and the standard deviation, s, for these measurements.

13.5	9.5	8.2	6.5	8.4	8.1	6.9	7.5	10.5	13.5
7.2	7.1	9.0	9.9	8.2	13.2	9.2	6.9	9.6	7.7
9.7	7.5	7.2	5.9	6.6	11.1	8.8	5.2	10.6	8.2
11.3	5.6	10.1	8.0	8.5	11.7	7.1	7.7	9.4	6.0
8.0	7.4	10.5	7.8	7.9	6.5	6.9	6.5	6.8	9.5

Solution

The calculation of the sample variance, s^2, would be very tedious for this sample if we tried to use the formula

$$s^2 = \frac{\sum_{i=1}^{50} (x_i - \bar{x})^2}{50 - 1}$$

because it would be necessary to compute all 50 squared distances from the mean. However, for the shortcut formula we need compute only

$$\sum_{i=1}^{50} x_i = 13.5 + 8.4 + \cdots + 6.5 = 424.6$$

and

$$\sum_{i=1}^{50} x_i^2 = (13.5)^2 + (8.4)^2 + \cdots + (6.5)^2 = 3{,}797.92$$

Then

$$s^2 = \frac{\sum_{i=1}^{50} x_i^2 - \frac{\left(\sum_{i=1}^{50} x_i\right)^2}{50}}{50 - 1} = \frac{3{,}797.92 - \frac{(424.6)^2}{50}}{49} = 3.9228$$

The standard deviation is

$$s = \sqrt{s^2} = \sqrt{3.9228} = 1.98$$

Notice that we retained all the decimal places in the calculation of the sum of squares of the measurements. This was done to reduce the rounding error in the calculations, even though the original data were accurate to only one decimal place.*

EXAMPLE 2.8

Use a statistical software package to compute the median, mean, variance, and standard deviation of the R&D data given in Example 2.7.

Solution

The SAS/PC printout is shown in Figure 2.19. The median, mean, variance, and standard deviation are shaded on the printout. Although the SAS procedure (PROC

FIGURE 2.19 ▶ SAS/PC printout for mean, median, variance, and standard deviation

```
                    Moments

N                  50   Sum Wgts          50
Mean            8.492   Sum            424.6
Std Dev      1.980604   Variance    3.922792
Skewness     .8546013   Kurtosis    .4192877
USS           3797.92   CSS         192.2168
CV           23.32317   Std Mean    .2800997
T:Mean=0     30.31778   Prob>|T|      0.0001
Sgn Rank        637.5   Prob>|S|      0.0001
Num ^= 0           50
W:Normal     .9328984   Prob<W         0.009

              Quantiles(Def=5)

 100% Max        13.5          99%      13.5
  75% Q3          9.6          95%      13.2
  50% Med        8.05          90%      11.2
  25% Q1          7.1          10%       6.5
   0% Min         5.2           5%       5.9
                                1%       5.2
```

*The accuracy of the original data has nothing to do with the degree of accuracy used in computing s^2 and s. You should retain twice as many decimal places in s^2 as you want in s. For example, if you want to calculate s to the nearest hundredth, you should calculate s^2 to the nearest ten-thousandth.

UNIVARIATE) generates many other descriptive statistics, some of which we discuss later, we can easily pick out those of interest to us. Note that many decimal places are carried by the program, but when we round to the same number of decimal places used in the previous examples, we find

$$\text{Median} = 8.05 \qquad s^2 = 3.9228$$

$$\bar{x} = 8.49 \qquad s = 1.98$$

The answers are identical to those obtained by hand calculation.

You know that the standard deviation measures the variability of a set of data and you know how to calculate it. But how can we interpret and use the standard deviation? This is the topic of Section 2.6.

Exercises 2.35–2.46

Learning the Mechanics

2.35 The range, variance, and standard deviation provide information about the variation in a data set.
a. Describe the information each conveys.
b. Discuss the advantages and disadvantages of using each to measure the variability of a data set.

2.36 Describe the sample variance using words rather than a formula. Do the same with the population variance.

2.37 Given the following information about two data sets, compute $\bar{x}$, the sample variance, and the standard deviation for each:

a. $n = 25, \quad \sum_{i=1}^{n} x_i^2 = 1{,}000, \quad \sum_{i=1}^{n} x_i = 50$ b. $n = 80, \quad \sum_{i=1}^{n} x_i^2 = 270, \quad \sum_{i=1}^{n} x_i = 100$

2.38 For each of the following data sets, compute $\sum_{i=1}^{n} x_i$, $\sum_{i=1}^{n} x_i^2$, and $\left(\sum_{i=1}^{n} x_i\right)^2$:

a. 5, 9, 6, 3, 7 b. 3, 1, 4, 3, 0, −2 c. 90, 12, 40, 15
d. −1, 4, 1, 0, 5 e. 1, 0, 0, 1, 0, 10

2.39 Compute $\bar{x}$, s^2, and s for each of the following data sets:
a. 10, 1, 0, 0, 20 b. 5, 9, −1, 100

2.40 Compute $\bar{x}$, s^2, and s for each of the following data sets. If appropriate, specify the units in which your answer is expressed.
a. 3, 1, 10, 10, 4 b. 8 feet, 10 feet, 32 feet, 5 feet c. −1, −4, −3, 1, −4, −4
d. 1/5 ounce, 1/5 ounce, 1/5 ounce, 2/5 ounce, 1/5 ounce, 4/5 ounce

2.41 Using only integers between 0 and 10, construct two data sets with at least ten observations each that have the same mean but different variances. Construct dot diagrams for each of your data sets (see Figure 2.18), and mark the mean of each data set on its dot diagram.

2.42 Using only integers between 0 and 10, construct two data sets with at least ten observations each that have the same range but different means. Construct a dot diagram for each of your data sets (see Figure 2.18), and mark the mean of each data set on its dot diagram.

2.43 Can the variance of a data set ever be negative? Explain. Can the variance ever be smaller than the standard deviation? Explain.

Applying the Concepts

2.44 The Consumer Price Index (CPI) measures the price change of a constant market basket of goods and services. The Bureau of Labor Statistics publishes a national CPI (called the U.S. City Average Index) as well as separate indexes for each of 28 different cities in the United States. The national index and some of the city indexes are published monthly; the remainder of the city indexes are published semiannually. The CPI is used in cost-of-living escalator clauses of many labor contracts to adjust wages for inflation (U.S. Department of Labor, 1978). For example, in the printing industry of Minneapolis and St. Paul, hourly wages are adjusted every 6 months (based on October and April values of the CPI) by 4¢ for every point change in the Minneapolis/St. Paul CPI.

The table lists the published values of the U.S. City Average Index and the Chicago Index during 1991 and 1992. The sums and sums of squares for the U.S. City Average Index are $\Sigma x = 3{,}318.1$ and $\Sigma x^2 = 458{,}877.49$, respectively. For the Chicago Index, these quantities are $\Sigma x = 3{,}337.2$ and $\Sigma x^2 = 464{,}169.38$.

Month	*U.S. City Average Index*	*Chicago*	*Month*	*U.S. City Average Index*	*Chicago*
January 1991	134.6	135.1	January 1992	138.1	138.9
February	134.8	135.5	February	138.6	139.2
March	135.0	136.2	March	139.3	139.7
April	135.2	136.1	April	139.5	139.8
May	135.6	136.8	May	139.7	140.5
June	136.0	137.3	June	140.2	141.2
July	136.2	137.3	July	140.5	141.4
August	136.6	137.6	August	140.9	141.9
September	137.2	138.3	September	141.3	142.7
October	137.4	138.0	October	141.8	142.1
November	137.8	138.0	November	142.0	142.4
December	137.9	138.3	December	141.9	142.9

Source: U.S. Department of Labor: Labor Statistics, 1991–1993.
Secondary Source: CPI Chicago, pp. 91–93.

a. Calculate the mean values for the U.S. City Average Index and the Chicago Index.
b. Find the ranges of the U.S. City Average Index and the Chicago Index.
c. The standard deviation of the U.S. City Average Index over the 24 months described in the table is 2.43. Calculate the standard deviation for the Chicago Index over the time period described in the table.

d. Which index displays greater variation about its mean over the time period in question? Justify your response.

2.45 To set an appropriate price for a product, it is necessary to be able to estimate its cost of production. One element of the cost is based on the length of time it takes workers to produce the product. The most widely used technique for making such measurements is the **time study**. In a time study, the task to be studied is divided into measurable parts and each is timed with a stopwatch or filmed for later analysis. For each worker, this process is repeated many times for each subtask. Then the average and standard deviation of the time required to complete each subtask are computed for each worker. A worker's overall time to complete the task under study is then determined by adding his or her subtask-time averages (Chase and Aquilano, 1977). The data (in minutes) given in the table are the result of a time study of a production operation involving two subtasks.

Repetition	*Worker A*		*Worker B*	
	Subtask 1	*Subtask 2*	*Subtask 1*	*Subtask 2*
1	30	2	31	7
2	28	4	30	2
3	31	3	32	6
4	38	3	30	5
5	25	2	29	4
6	29	4	30	1
7	30	3	31	4

a. Find the overall time it took each worker to complete the manufacturing operation under study.
b. For each worker, find the standard deviation of the seven times for subtask 1.
c. In the context of this problem, what are the standard deviations you computed in part **b** measuring?
d. Repeat part **b** for subtask 2.
e. If you could choose workers similar to A or workers similar to B to perform subtasks 1 and 2, which type would you assign to each subtask? Explain your decisions on the basis of your answers to parts **a–d**.

2.46 The table lists the 1989 profits (in millions of dollars) for a sample of seven airlines.

Airline	*Profit*
Continental	3.06
Eastern	−852.32
Northwest	355.25
Pan Am	−414.73
Delta	473.17
TWA	−298.55
United	358.09

Source: *Air Transport* (Washington, D.C.: Air Transport Association of America, 1990).

a. Calculate the range, variance, and standard deviation of the data set.
b. Specify the units in which each of your answers to part **a** is expressed.
c. Suppose Eastern Airlines had a profit of $0 instead of a loss of $852.32 million. Would the range of the data set increase or decrease? Why? Would the standard deviation of the data set increase or decrease? Why?

2.6 Interpreting the Standard Deviation

As we have seen, if we are comparing the variability of two samples selected from a population, the sample with the larger standard deviation is the more variable of the two. Thus, we know how to interpret the standard deviation on a relative or comparative basis, but we have not explained how it provides a measure of variability for a single sample.

One way to interpret the standard deviation as a measure of variability of a data set would be to answer questions such as the following: How many measurements are within 1 standard deviation of the mean? How many measurements are within 2 standard deviations? For a specific data set, we can answer the questions by counting the number of measurements in each of the intervals. However, if we are interested in obtaining a general answer to these questions, the problem is more difficult.

In Table 2.10, we present two sets of guidelines to help answer the questions of how many measurements fall within 1, 2, and 3 standard deviations of the mean. The first set, which applies to any sample, is derived from a theorem proved by the Russian mathematician, Chebyshev. The second set, the Empirical Rule, is based on empirical evidence that has accumulated over time and applies to samples that possess mound-shaped frequency distributions—those that are approximately symmetric, with a clustering of measurements about the midpoint of the distribution (the mean, median, and mode should all be about the same) and that tail off as we move away from the center of the histogram. Thus, the histogram will have the appearance of a mound or bell, as shown in Figure 2.20 on page 76. The percentages given for the various intervals (particularly the interval $\bar{x} - 2s$ to $\bar{x} + 2s$) in Table 2.10 provide remarkably good approximations even when the distribution of the data is slightly skewed or asymmetric.*

TABLE 2.10 Aids to the Interpretation and Use of a Standard Deviation

1. A rule (from Chebyshev's theorem) that applies to any sample of measurements, regardless of the shape of the frequency distribution:
   a. It is possible that none of the measurements will fall within the interval $\bar{x} \pm s$ or $(\bar{x} - s, \bar{x} + s)$—i.e., within 1 standard deviation of the mean.
   b. At least 3/4 of the measurements will fall within $(\bar{x} - 2s, \bar{x} + 2s)$—i.e., within 2 standard deviations of the mean.
   c. At least 8/9 of the measurements will fall within $(\bar{x} - 3s, \bar{x} + 3s)$—i.e., within 3 standard deviations of the mean.
   d. Generally, at least $(1 - 1/k^2)$ of the measurements will fall within $(\bar{x} - ks, \bar{x} + ks)$—i.e., within k standard deviations of the mean, where k is any number greater than 1.

**Note to the instructor:* It is our intention to imply that the Empirical Rule of Table 2.10 not only applies to normal distributions of data, but also applies very well to mound-shaped distributions of large data sets and to distributions that are moderately skewed.

2. A rule of thumb called the Empirical Rule applies to samples with frequency distributions that are mound-shaped:
   a. Approximately 68% of the measurements will fall within the interval $\bar{x} \pm s$ or $(\bar{x} - s, \bar{x} + s)$—i.e., within 1 standard deviation of the mean.
   b. Approximately 95% of the measurements will fall within $(\bar{x} - 2s, \bar{x} + 2s)$—i.e., within 2 standard deviations of the mean.
   c. Essentially all the measurements will fall within $(\bar{x} - 3s, \bar{x} + 3s)$—i.e., within 3 standard deviations of the mean.

FIGURE 2.20 ▶ Histogram of a mound-shaped sample

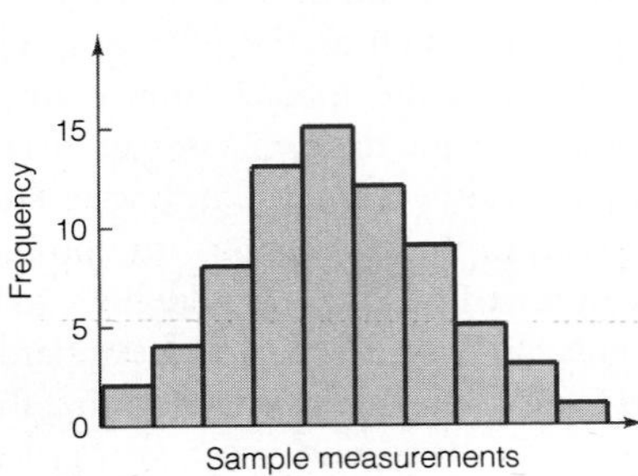

EXAMPLE 2.9

The 50 companies' percentages of revenues spent on R&D are repeated here:

13.5	9.5	8.2	6.5	8.4	8.1	6.9	7.5	10.5	13.5
7.2	7.1	9.0	9.9	8.2	13.2	9.2	6.9	9.6	7.7
9.7	7.5	7.2	5.9	6.6	11.1	8.8	5.2	10.6	8.2
11.3	5.6	10.1	8.0	8.5	11.7	7.1	7.7	9.4	6.0
8.0	7.4	10.5	7.8	7.9	6.5	6.9	6.5	6.8	9.5

We have previously shown that the mean and standard deviation of these data are 8.49 and 1.98, respectively. Calculate the fraction of these measurements that lie within the intervals $\bar{x} \pm s$, $\bar{x} \pm 2s$, and $\bar{x} \pm 3s$, and compare the results with those in Table 2.10.

Solution

We first form the interval

$$(\bar{x} - s, \bar{x} + s) = (8.49 - 1.98, 8.49 + 1.98) = (6.51, 10.47)$$

A check of the measurements reveals that 34 of the 50 measurements, or 68%, are within 1 standard deviation of the mean.

The interval

$$(\bar{x} - 2s, \bar{x} + 2s) = (8.49 - 3.96, 8.49 + 3.96) = (4.53, 12.45)$$

contains 47 of the 50 measurements, or 94%.

The 3 standard deviation interval around $\bar{x}$,

$$(\bar{x} - 3s, \bar{x} + 3s) = (8.49 - 5.94, 8.49 + 5.94) = (2.55, 14.43)$$

contains all the measurements.

In spite of the fact that the distribution of these data is skewed to the right (see Figure 2.13), the percentages within 1, 2, and 3 standard deviations (68%, 94%, and 100%) agree very well with the approximations of 68%, 95%, and 100% given by the Empirical Rule. You will find that unless the distribution is extremely skewed, the mound-shaped approximations will be reasonably accurate. Of course, no matter what the shape of the distribution, Chebyshev's theorem from Table 2.10 assures that at least 75% and at least 89% (8/9) of the measurements will lie within 2 and 3 standard deviations of the mean, respectively.

EXAMPLE 2.10

Chebyshev's theorem and the Empirical Rule (Table 2.10) are useful as a check on the calculation of the standard deviation. For example, suppose we calculated the standard deviation of the R&D percentages to be 3.92. Are there any clues in the data that enable us to judge whether this number is reasonable?

Solution

The range of the R&D percentages is $13.5 - 5.2 = 8.3$. From the Empirical Rule we know that most of the measurements (approximately 95% if the distribution is not extremely skewed) will be within 2 standard deviations of the mean. And from Chebyshev's theorem, regardless of the shape of the distribution, almost all of the measurements (at least 8/9) will fall within 3 standard deviations of the mean. Consequently, we would expect the range of measurements to be between 4 (± 2) and 6 (± 3) standard deviations in length (see Figure 2.21). For the R&D data, this means that the standard deviation s should fall between

$$\frac{\text{Range}}{6} = \frac{8.3}{6} = 1.38 \quad \text{and} \quad \frac{\text{Range}}{4} = \frac{8.3}{4} = 2.08$$

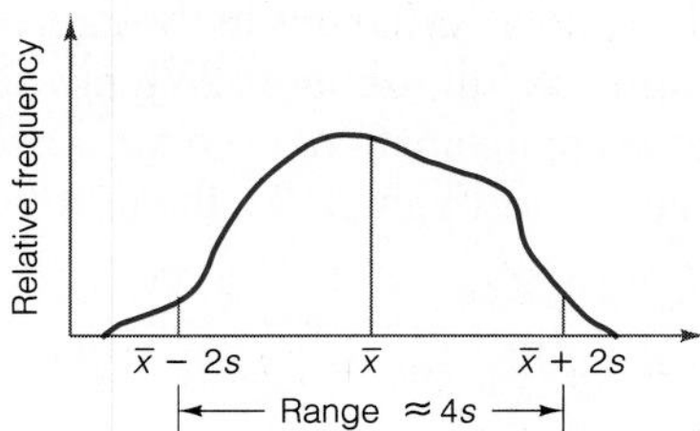

FIGURE 2.21 ► **The relation between the range and the standard deviation**

Thus, we would have reason to believe that a calculated standard deviation of 3.92 for these data is too large, since it far exceeds one-fourth the range, 2.08. A check of our work reveals that 3.92 is the variance s^2, not the standard deviation s (see Example 2.7). We "forgot" to take the square root, a common error. The correct value of 1.98 is between one-sixth and one-fourth the range.

In examples and exercises we will sometimes use $s \approx$ range/4 to obtain a crude, and usually conservatively large, approximation for s. However, we stress that this is no substitute for calculating the exact value of s when possible.

Beginning with the following example, we will use the concepts in Chebyshev's theorem and the Empirical Rule to build the foundation for statistical inference-making.

EXAMPLE 2.11

A manufacturer of automobile batteries claims that the average length of useful life for its grade A battery is 60 months. However, the guarantee on this brand is for only 36 months. Assume that the standard deviation of the lifelengths is known to be 10 months and that the frequency distribution of the lifelengths is known to be mound-shaped.

a. Approximately what percentage of the manufacturer's grade A batteries will last more than 50 months, assuming the manufacturer's claim about the mean life-length is true?

b. Approximately what percentage of the manufacturer's batteries will last less than 40 months, assuming that the manufacturer's claim about the mean lifelength is true?

c. Suppose that your grade A battery lasts 37 months. What would you infer about the manufacturer's claim that the mean lifelength is 60 months?

Solution

Assuming that the distribution of lifelengths is approximately mound-shaped, with a mean of 60 months and a standard deviation of 10 months, it would appear as shown in Figure 2.22. Note that we can take advantage of the fact that mound-shaped distributions are (approximately) symmetric about the mean, so that the percentages given by the Empirical Rule can be split equally between the halves of the distribution on each side of the mean. The approximations given in Figure 2.22 are more dependent on the assumption of a mound-shaped distribution than those given by the Empirical Rule because the approximations in the figure depend on the (approximate) symmetry of the mound-shaped distribution. We saw in Example 2.9 that the Empirical Rule can yield good approximations even for skewed distributions. This will *not* be true of the approximations in Figure 2.22; the distribution must be mound-shaped and (approximately) symmetric.

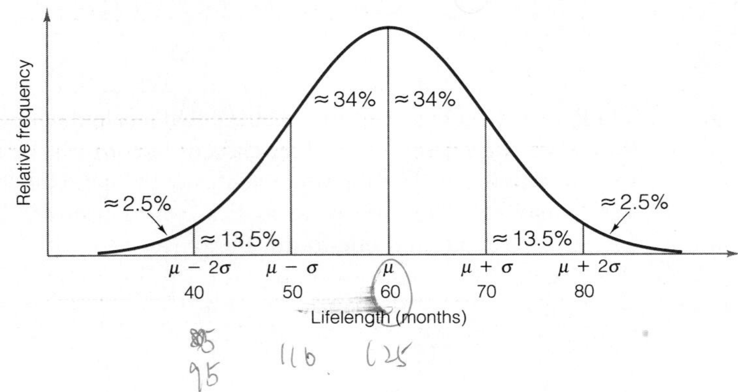

FIGURE 2.22 ► Grade A battery lifelength distribution; manufacturer's claim assumed true

For example, since approximately 68% of the measurements will fall within 1 standard deviation on both sides of the mean, the distribution's symmetry implies that approximately ½ (68%) = 34% of the measurements will fall between the mean and 1 standard deviation on each side. This concept is demonstrated in Figure 2.22. Note that the 2.5% of the measurements beyond 2 standard deviations in each direction from the mean follows from the fact that, if approximately 95% of the measurements fall within 2 standard deviations, then about 5% fall outside 2 standard deviations. If the distribution is approximately symmetric, then about 2.5% fall beyond 2 standard deviations on each side of the mean.

a. Using Figure 2.22, it is easy to see that the percentage of batteries lasting more than 50 months is, approximately, 34% (between 50 and 60 months) plus 50% (greater than 60 months). Thus, approximately (50 + 34)% = 84% of the batteries should last longer than 50 months.

b. The percentage of batteries that last less than 40 months can also be determined from Figure 2.22. Since 40 is 2 standard deviations below the claimed mean of 60 months, approximately 2.5% of the batteries should fail prior to 40 months, assuming the manufacturer's claim is true.

c. If you are so unfortunate that your grade A battery fails at 37 months, one or two inferences can be made: Either your battery was one of the approximately 2.5% that fail prior to 40 months, or the manufacturer's claim is not true. Because the chances are so small that a battery fails before 40 months if the claim is true, you would have good reason to have serious doubts about the manufacturer's claim. A mean smaller than 60 months and/or a standard deviation greater than 10 months would increase the likelihood of a battery's failure prior to 40 months.*

Example 2.11 is our initial demonstration of the statistical inference-making process. At this point you should realize that we will use sample information (in Example 2.11 your battery's failure at 37 months) to make inferences about the population (in Example 2.11, the manufacturer's claim about the mean lifelength for the population of all batteries). We will build on this foundation as we proceed.

CASE STUDY 2.7 / Becoming More Sensitive to Customer Needs

The degree of sensitization on the part of a firm to the needs and wants of its consumers is frequently an important factor in determining the firm's overall success. Namias (1964) presents a procedure for achieving such sensitivity. The procedure uses the rate of consumer complaints about a product to determine when and

*The assumption that the distribution is mound-shaped and symmetric may also be incorrect. However, if the distribution were skewed to the right, as lifelength distributions often tend to be, the percentage of measurements more than 2 standard deviations *below* the mean would be even less than 2.5%.

when not to conduct a search for specific causes of consumer complaints. For simplification, we will discuss Namias's paper as if the procedure described used the number of complaints per 10,000 units of a product sold to determine when and when not to conduct a search for specific causes of consumer complaints. The details of the procedure are discussed in Case Study 2.8.

Namias's procedure, given our simplification, makes use of the **distribution** of the number of consumer complaints received about a product per 10,000 units of the product sold. To visualize such a distribution, imagine that a company produces its product in lots of 10,000 units and keeps track of the number of complaints received about items in each lot. The company's complaint records will show a series of numbers, perhaps 100, 96, 145, 201, etc., each of which is the number of complaints received about a particular lot of 10,000 units. This series of numbers is a quantitative data set from which a relative frequency histogram can be drawn. The histogram constructed from this data set is a representation of the distribution of interest—i.e., the distribution of the number of consumer complaints received about a product per 10,000 units of the product sold. It is assumed that this distribution remains stable over time. The variance and standard deviation of this distribution are measures of the variation in the number of consumer complaints received. Namias determined that this distribution was mound-shaped. Accordingly, it can be said that approximately 95% of the time the number of complaints about a product will be within 2 standard deviations of the mean number of complaints. It is upon this fact, as we shall see in Case Study 2.8, that Namias's procedure for determining when it would be worthwhile to conduct a search for specific causes of consumer complaints is founded.

If it could not have been determined that the distribution of the number of complaints was mound-shaped, it would have been necessary for Namias to use Chebyshev's theorem (Table 2.10). In that case, it could have been said only that at least 75% of the time the number of complaints about a product will be within 2 standard deviations of the mean number of complaints.

Exercises 2.47–2.61

Learning the Mechanics

2.47 To what kind of data sets can be Chebyshev's theorem be applied? The Empirical Rule?

2.48 The output from a statistical computer program indicates that the mean and standard deviation of a data set consisting of 200 measurements are $1,500 and $300, respectively.

a. What are the units of measurement of the variable of interest? Based on the units, are these data nominal, ordinal, interval, or ratio?

b. What can be said about the number of measurements between $900 and $2,100? Between $600 and $2,400? Between $1,200 and $1,800? Between $1,500 and $2,100?

2.49 For any set of data, what can be said about the percentage of the measurements contained in each of the following intervals?

a. $\bar{x} - s$ to $\bar{x} + s$ **b.** $\bar{x} - 2s$ to $\bar{x} + 2s$ **c.** $\bar{x} - 3s$ to $\bar{x} + 3s$

2.50 As a result of government and consumer pressure, automobile manufacturers in the United States are deeply involved in research to improve their products' gasoline mileage. One manufacturer, hoping to achieve 40 miles per gallon on one of its compact models, measured the mileage obtained by 36 test versions of the model with the following results (rounded to the nearest mile for convenience):

43	35	41	42	42	38	40	41	41
40	40	41	42	36	43	40	38	40
38	45	39	41	42	37	40	40	44
39	40	37	39	41	39	41	37	40

The mean and standard deviation of these data are 40.1 and 2.2, respectively.

a. What are the units in which the mean and standard deviation are expressed?

b. If the manufacturer would be satisfied with a (population) mean of 40 miles per gallon, how would it react to the above test data?

c. Use the information in Table 2.10 to check the reasonableness of the calculated standard deviation, $s = 2.2$.

d. Construct a relative frequency histogram of the data set. Is the data set mound-shaped?

e. What percentage of the measurements would you expect to find within the intervals $\bar{x} \pm s$, $\bar{x} \pm 2s$, and $\bar{x} \pm 3s$?

f. Count the number of measurements that actually fall within the intervals of part **e**. Express each interval count as a percentage of the total number of measurements. Compare these results with your answers to part **e**.

2.51 Given a data set with a largest value of 760 and a smallest value of 135, what would you estimate the standard deviation to be? Explain the logic behind the procedure you used to estimate the standard deviation. Suppose the standard deviation is reported to be 25. Is this feasible? Explain.

Applying the Concepts

2.52 A manufacturer of video cassette recorders is disturbed because retailers were complaining that they were not receiving shipments of recorders as fast as they had been promised. The manufacturer decided to run a check on the distribution network. Each of the 50 warehouses owned by the manufacturer throughout the country had been instructed to maintain at least 200 recorders in stock at all times so that a supply would always be readily available for retailers. The manufacturer checked the inventories of 20 of these warehouses and obtained the following numbers of video cassette recorders in stock:

40	10	44	142	14	301	175	0	38	202
220	32	400	78	16	99	0	176	5	86

The mean and standard deviation are 103.9 and 111.3, respectively.

a. Calculate the median. Is the distribution of measurements skewed or symmetric? Explain.

b. Using your answer to part **a**, would you advise using the Empirical Rule to describe these data? Why?

c. According to Chebyshev's theorem, what percentage of the measurements would you expect to find outside the intervals $\bar{x} \pm s$, $\bar{x} \pm 2s$, and $\bar{x} \pm 3s$?

d. Check the numbers of measurements that actually fall outside the intervals specified in part **c**. Express each count as a percentage of the total number of measurements. Compare these results with your answers to part **c**.

2.53 Twenty-five mergers were sampled from the population of mergers that occurred between firms (excluding railroads) listed on the New York Stock Exchange during the period 1958–1980. For each merger, the ratio of the target firm's sales for the preceding year to the bidder firm's sales was calculated. (See Exercise 1.14 for definitions of target and bidder firms.) The 25 sales ratios are listed below (Eger, 1982, p. 133):

.16	.07	.32	.05	.79
.30	.18	.04	.04	.05
.14	.08	.14	.29	.02
.14	.10	.34	.10	.02
.03	.15	.09	.06	.02

For these data, $\Sigma x = 3.72$ and $\Sigma x^2 = 1.2088$.

a. What is the mean sales ratio, $\bar{x}$, for this sample of mergers? Also, find s^2 and s.

b. According to Chebyshev's theorem, what percentage of the measurements would you expect to find in the intervals $\bar{x} \pm .75s$, $\bar{x} \pm 2.5s$, and $\bar{x} \pm 4s$?

c. What percentage of measurements actually fall in the intervals of part **b**? Compare these results with the results of part **b**.

d. What percentage of the mergers involved a target firm with larger sales than the bidder firm?

2.54 Tests have demonstrated that the shelf life of cake mix A has a mound-shaped distribution with a mean of 275 days and a standard deviation of 55 days. Mix B has a shelf life whose distribution is also mound-shaped, but with a mean of 286 days and a standard deviation of 22 days.

a. What percentage of boxes of cake mix A remain fresh for 330 days or more?

b. What percentage of boxes of mix B remain fresh for 330 days or more?

c. Which cake mix is more likely to remain fresh for more than 330 days? Explain.

d. Now assume the shapes of the two shelf-life distributions are unknown. Answer parts **a–c** by applying Chebyshev's theorem.

2.55 A company that bottles sparkling water has determined that it lost an average of 30.4 cases per week last year due to breakage in transit. The standard deviation of the number of cases lost per week was 3.8 cases. With only this information, what can you say about the number of weeks last year that the company lost more than 38 cases due to breakage in transit? Justify your answer.

2.56 A chemical company produces a substance composed of 98% cracked corn particles and 2% zinc phosphide for use in controlling rat populations in sugarcane fields. Production must be carefully controlled to maintain the zinc phosphide at 2% because too much zinc phosphide will damage the sugarcane and too little will be ineffective in controlling the rat population. Records from past production indicate that the distribution of the actual percentage of zinc phosphide present in the substance is approximately mound-shaped, with a mean of 2.0% and a standard deviation of .08%. If the production line is operating correctly and a batch is chosen at random from a day's production, what is the approximate probability that it will contain less than 1.84% zinc phosphide?

2.57 For 50 randomly selected days, the number of vehicles that used a certain road was ascertained by a city engineer. The mean was 385; the standard deviation was 15. Suppose you are interested in the proportion of days that there were between 340 and 430 vehicles using the road. What does Chebyshev's theorem tell you about this proportion?

2.58 A boat dealer has determined that the frequency distribution of the number of outboard motor sales per month over the last five years is mound-shaped, with a sample mean of 30 and a sample variance of 4.

Approximately what percentage of the recorded monthly sales figures of the past five years would be expected to be greater than 34? Less than 26? Greater than 36?

2.59 Solar energy is considered by many to be the energy of the future. A recent survey was taken to compare the cost of solar energy with the cost of gas or electric energy. Results of the survey revealed that the average monthly utility bill of a three-bedroom house using gas or electric energy was $125 and the standard deviation was $15.

a. If nothing is known about the distribution of utility bills, what can you say about the fraction of all three-bedroom homes with gas or electric energy that have bills between $80 and $170?

b. If it is reasonable to assume that the distribution of utility bills is mound-shaped, approximately what proportion of three-bedroom homes would have monthly bills less than $110?

c. Suppose that three houses with solar energy units had the following utility bills: $78, $92, $87. Does this suggest that solar energy units might result in lower utility bills? Explain. [*Note:* We present a statistical method for testing this conjecture in Chapter 7.]

2.60 When it is working properly, a machine that fills 25-pound bags of flour dispenses an average of 25 pounds per fill; the standard deviation of the amount of fill is .1 pound. To monitor the performance of the machine, an inspector weighs the contents of a bag coming off the machine's conveyor belt every half-hour during the day. If the contents of two consecutive bags fall more than 2 standard deviations from the mean (using the mean and standard deviation given above), the filling process is said to be out of control and the machine is shut down briefly for adjustments. The data given in the table are the weights measured by the inspector yesterday. Assume the machine is never shut down for more than 15 minutes at a time. At what times yesterday was the process shut down for adjustment? Justify your answer.

Time	*Weight (pounds)*	*Time*	*Weight (pounds)*
8:00 A.M.	25.10	12:30 P.M.	25.06
8:30	25.15	1:00	24.95
9:00	24.81	1:30	24.80
9:30	24.75	2:00	24.95
10:00	25.00	2:30	25.21
10:30	25.05	3:00	24.90
11:00	25.23	3:30	24.71
11:30	25.25	4:00	25.31
12:00	25.01	4:30	25.15
		5:00	25.20

2.61 A buyer for a lumber company must determine whether to buy a piece of land containing 5,000 pine trees. If 1,000 of the trees are at least 40 feet tall, he will purchase the land; otherwise, he will not. The owner of the land reports that the distribution of the heights of the trees has a mean of 30 feet, and a standard deviation of 3 feet. Based on this information, what should the buyer decide?

2.7 Calculating a Mean and Standard Deviation from Grouped Data (Optional)

If your data have been grouped in classes of equal width and arranged in a frequency table, you can use the following formulas to calculate $\bar{x}$, s^2, and s:

Formulas for Calculating a Mean and Standard Deviation from Grouped Data

$$\bar{x} = \frac{\sum_{i=1}^{k} x_i f_i}{n}$$

$$s^2 = \frac{\sum_{i=1}^{k} (x_i - \bar{x})^2 f_i}{n - 1} \qquad s = \sqrt{s^2}$$

Shortcut formula: $$s^2 = \frac{\sum_{i=1}^{k} x_i^2 f_i - \frac{\left(\sum_{i=1}^{k} x_i f_i\right)^2}{n}}{n - 1}$$

where x_i = Midpoint of the ith class

f_i = Frequency of the ith class

k = Number of classes

EXAMPLE 2.12

Compute the mean and standard deviation for the companies' R&D percentages based on the groupings used to construct the relative frequency histogram of Figure 2.13.

Solution

The frequency table is repeated in Table 2.11, with a column showing the class midpoints added.

Substituting the class midpoints and frequencies into the formula for the mean of grouped data, we obtain

$$\bar{x} = \frac{\sum_{i=1}^{k} x_i f_i}{n} = \frac{(5.70)(4) + (6.80)(12) + \cdots + (13.40)(3)}{50}$$

$$= \frac{422.5}{50} = 8.45$$

TABLE 2.11 Measurement Classes, Frequencies, and Class Midpoints for the R&D Percentage Data

Class	Measurement Class	Class Midpoint	Class Frequency
1	5.15– 6.25	5.70	4
2	6.25– 7.35	6.80	12
3	7.35– 8.45	7.90	14
4	8.45– 9.55	9.00	7
5	9.55–10.65	10.10	7
6	10.65–11.75	11.20	3
7	11.75–12.85	12.30	0
8	12.85–13.95	13.40	3

Next, using the shortcut formula for calculating the sample variance for grouped data, we obtain

$$s^2 = \frac{\sum_{i=1}^{k} x_i^2 f_i - \frac{\left(\sum_{i=1}^{k} x_i f_i\right)^2}{n}}{n-1}$$

$$= \frac{(5.70)^2(4) + (6.80)^2(12) + \cdots + (13.40)^2(3) - \frac{(422.5)^2}{50}}{50-1}$$

$$= \frac{3{,}754.65 - 3{,}570.125}{49} = 3.7658$$

$$s = \sqrt{3.7658} = 1.94$$

You will note that the sample mean and standard deviation for the grouped data, 8.45 and 1.94, agree well with the exact mean and standard deviation for the 50 measurements, $\bar{x} = 8.49$ and $s = 1.98$. As long as a reasonable number of class intervals is used, the approximations that are based on grouped data will usually be good.

2.8 Measures of Relative Standing

As we have seen, numerical measures of central tendency and variability describe the general nature of a data set (either a sample or a population). We may also be interested in describing the relative location of a particular measurement within a data set.

Descriptive measures of the relationship of a measurement to the rest of the data are called **measures of relative standing**.

One measure of the relative standing of a particular measurement is its **percentile ranking**:

Definition 2.12

Let $x_1, x_2, \ldots, x_n$ be a set of n measurements arranged in increasing (or decreasing) order. The **pth percentile** is a number x such that $p\%$ of the measurements fall below the pth percentile and $(100 - p)\%$ fall above it.

For example, if oil company A reports that its yearly sales are in the 90th percentile of all companies in the industry, the implication is that 90% of all oil companies have yearly sales less than company A's, and only 10% have yearly sales exceeding company A's. This is demonstrated in Figure 2.23.

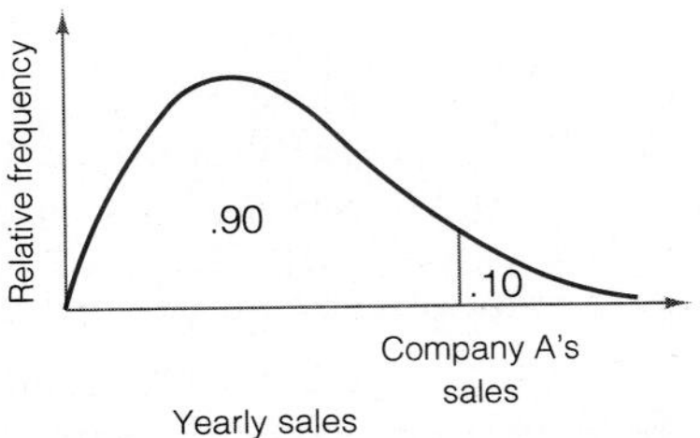

FIGURE 2.23 ► Relative frequency distribution for yearly sales of oil companies

Another measure of relative standing in popular use is the **z-score**. As you can see in Definition 2.13, the z-score makes use of the mean and standard deviation of the data set in order to specify the location of a measurement:

Definition 2.13

The **sample z-score** for a measurement x is

$$z = \frac{x - \bar{x}}{s}$$

The **population z-score** for a measurement x is

$$z = \frac{x - \mu}{\sigma}$$

Note that the z-score is calculated by subtracting $\bar{x}$ (or μ) from the measurement x and then dividing the result by s (or σ). The result, the z-score, represents the distance between a given measurement, x, and the mean, expressed in standard deviations.

EXAMPLE 2.13

Suppose 200 steelworkers are selected, and the annual income of each is determined. The mean and standard deviation are $\bar{x} = \$24{,}000$ and $s = \$2{,}000$. Suppose Joe Smith's annual income is \$22,000. What is his sample z-score?

FIGURE 2.24 ▶ **Annual income of steelworkers**

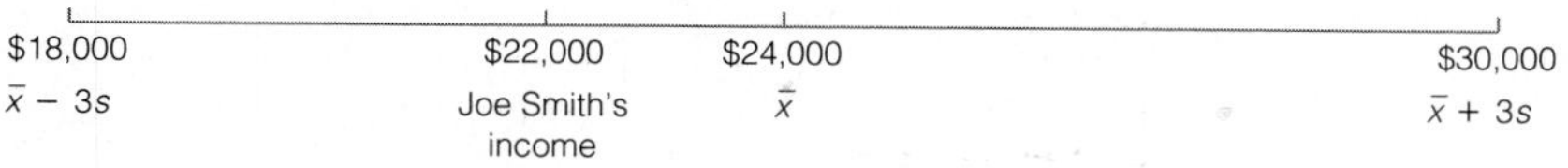

Solution

Joe Smith's annual income lies below the mean income of the 200 steelworkers (see Figure 2.24). We compute

$$z = \frac{x - \bar{x}}{s} = \frac{\$22{,}000 - \$24{,}000}{\$2{,}000} = -1.0$$

which tells us that Joe Smith's annual income is 1.0 standard deviation *below* the sample mean, or, in short, his sample z-score is -1.0.

The numerical value of the z-score reflects the relative standing of the measurement. A large positive z-score implies that the measurement is larger than almost all other measurements, whereas a large negative z-score indicates that the measurement is smaller than almost every other measurement. If a z-score is 0 or near 0, the measurement is located at or near the mean of the sample or population.

We can be more specific if we know that the frequency distribution of the measurements is mound-shaped. In this case, the following interpretation of the z-scores can be given:

Interpretation of z-Scores for Mound-Shaped Distributions of Data

1. Approximately 68% of the measurements will have a z-score between -1 and 1.
2. Approximately 95% of the measurements will have a z-score between -2 and 2.
3. All or almost all the measurements will have a z-score between -3 and 3.

Note that this interpretation of z-scores is identical to that given in Table 2.10 for samples that exhibit mound-shaped frequency distributions. The statement that a measurement falls in the interval $(\mu - \sigma, \mu + \sigma)$ is identical to the statement that a measurement has a population z-score between -1 and 1, since all measurements between $(\mu - \sigma)$ and $(\mu + \sigma)$ are within 1 standard deviation of μ (see Figure 2.25 on page 88).

FIGURE 2.25 ► Population *z*-scores for a mound-shaped distribution

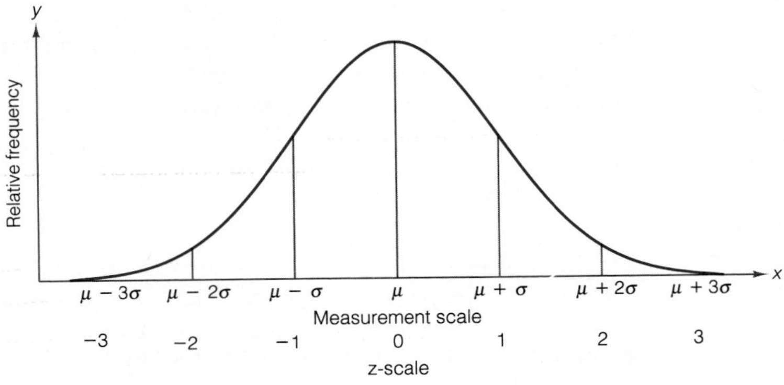

EXAMPLE 2.14

Suppose a female bank employee believes her salary is low as a result of sex discrimination. To substantiate her belief, she collects information on the salaries of her male counterparts in the banking business. She finds that their salaries have a mean of \$34,000 and a standard deviation of \$2,000. Her salary is \$27,000. Does this information support her claim of sex discrimination?

Solution

The analysis might proceed as follows. First, we calculate the *z*-score for the woman's salary with respect to those of her male counterparts. Thus,

$$z = \frac{\$27,000 - \$34,000}{\$2,000} = -3.5$$

The implication is that the woman's salary is 3.5 standard deviations *below* the mean of the male salary distribution. Furthermore, if a check of the male salary data shows that the frequency distribution is mound-shaped, we can infer that very few salaries in this distribution should have a *z*-score less than -3, as shown in Figure 2.26. Therefore, a *z*-score of -3.5 represents either a measurement from a distribution different from the male salary distribution or a very unusual (highly improbable) measurement for the male salary distribution.

FIGURE 2.26 ► Male salary distribution

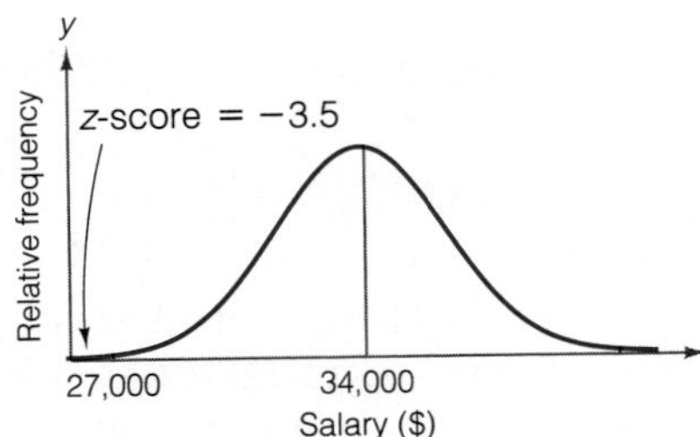

Which of the two situations do you think prevails? Do you think the woman's salary is simply an unusually low one in the distribution of salaries, or do you think her claim of salary discrimination is justified? Most people would probably conclude that her salary does not come from the male salary distribution. However, the careful investigator should require more information before inferring sex discrimination as the cause. We would want to know more about the data collection technique the woman used, and more about her competence at her job. Also, perhaps other factors such as the length of employment should be considered in the analysis.

The method of Example 2.14 exemplifies an approach to statistical inference that might be called the **rare event approach**. An experimenter hypothesizes a specific frequency distribution to describe a population of measurements. Then a sample of measurements is drawn from the population. If the experimenter finds it unlikely that the sample came from the hypothesized distribution, the hypothesis is concluded to be false. Thus, in Example 2.14 the woman believes her salary reflects sex discrimination. She hypothesizes that her salary should be just another measurement in the distribution of her male counterparts' salaries if no discrimination exists. However, it is so unlikely that the sample (in this case, her salary) came from the male frequency distribution that she rejects that hypothesis, concluding that the distribution from which her salary was drawn is different from the distribution for the men.

This rare event approach to inference-making is discussed further in later chapters. Proper application of the approach requires a knowledge of probability, the subject of our next chapter.

CASE STUDY 2.8 / Deciding When to Respond to Consumer Complaints

Now we will finish the discussion begun in Case Study 2.7 of a method proposed by Namias (1964) to determine when and when not to conduct a search for specific causes of consumer complaints.

The rate of consumer complaints about a product may change or vary merely as a result of chance or fate, or perhaps because of some specific cause, such as a decline in the quality of the product. Concerning the former, Namias says:

In any operation or production process, variability in the output or product will occur, and no two operational results may be expected to be exactly alike. Complete constancy of consumer rates of complaint is not possible, for the vagaries of fate and chance operate even within the most rigid framework of quality or operation control.

Namias provides a decision rule with which to determine when the observed variation in the rate of consumer complaints is due to chance and when it is due to specific causes. If the observed rate is 2 standard deviations or less away from the mean rate of complaint, it is attributed to chance. If the observed rate is farther than 2 standard deviations above the mean rate, it is attributed to a specific problem in the production or distribution of the product. The reasoning is that if there are no problems with the production and distribution of the product, 95% of the time the rate of complaints should be within 2 standard deviations of the mean rate. If the production and distribution processes were operating normally, it would be very unlikely for a rate higher than 2 standard deviations above the mean to occur. Instead, it is more likely that the high complaint rate is caused by abnormal operation of the production and/or distribution process; that is, something specific is wrong with the process.

Namias recommends searching for the cause (or causes) only if the observed variation in the rate of complaints is determined by the rule to be the result of a specific cause (or causes). The degree of variability due to chance must be tolerated. Namias says,

> *As long as the results exhibit chance variability, the causes are common, and there is no need to attempt to improve the product by making specific changes. Indeed this may only create more variability, not less, and it may inject trouble where none existed, with waste of time and money. . . . On the other hand, time and money are again wasted through failure to recognize specific conditions when they arise. It is therefore economical to look for a specific cause when there is more variability than is expected on the basis of chance alone.*

Namias collected data from the records of a beverage company for a two-week period to demonstrate the effectiveness of the rule. Consumer complaints concerned chipped bottles that looked dangerous. For one of the firm's brands the mean complaint rate was determined to be 26.01 and the rate 2 standard deviations above the mean was determined to be 48.78 complaints per 10,000 bottles sold. The complaint rate observed during the two weeks under study was 93.12 complaints per 10,000 bottles sold. Since 93.12 is many more than 2 standard deviations above the mean rate, it was concluded that the high rate of complaints must have been caused by some specific problem in the production or distribution of the particular brand of beverage and that a search for the problem would probably be worthwhile. The problem was traced to rough handling of the bottled beverage in the warehouse by newly hired workers. As a result, a training program for new workers was instituted.

In Chapter 13, "Methods of Quality Improvement," we discuss in detail methods for monitoring the variation of a production process and deciding when to take action to improve the process.

Exercises 2.62–2.74

Learning the Mechanics

2.62 What is the 50th percentile of a quantitative data set? What is another name for the 50th percentile?

2.63 In each of the following compute the z-score for the x value, and note whether your result is a sample z-score or a population z-score:

a. $x = 31, \quad s = 7, \quad \bar{x} = 24$ b. $x = 95, \quad s = 4, \quad \bar{x} = 101$
c. $x = 5, \quad \mu = 2, \quad \sigma = 1.7$ d. $\mu = 17, \quad \sigma = 5, \quad x = 14$

2.64 Compare the z-scores to determine which of the following x values lie the greatest distance above the mean and the greatest distance below the mean.

a. $x = 100, \quad \mu = 50, \quad \sigma = 25$ b. $x = 1, \quad \mu = 4, \quad \sigma = 1$
c. $x = 0, \quad \mu = 200, \quad \sigma = 100$ d. $x = 10, \quad \mu = 5, \quad \sigma = 3$

2.65 At the University of Statistics (US), the students are given z-scores at the end of each semester rather than the traditional GPAs. The mean and standard deviation of all students' cumulative GPAs at US, on which the z-scores are based, are 2.7 and .5, respectively.

a. Translate each of the following z-scores to a corresponding GPA: $z = 2.0$, $z = -1.0$, $z = .5$, $z = -2.5$.
b. Students with z-scores below -1.6 are put on probation at US. What is the corresponding probationary GPA?

c. The president of US wishes to graduate the top 16% of the students with cum laude honors and the top 2.5% with summa cum laude honors. Where should the limits be set in terms of z-scores? In terms of GPAs? What assumption, if any, did you make about the distribution of the GPAs at US?

2.66 Suppose that 40 and 90 are two elements of a population data set and that their z-scores are -2 and 3, respectively. Using only this information, is it possible to determine the population's mean and standard deviation? If so, find them. If not, explain why it is not possible.

Applying the Concepts

2.67 In 1987 the United States imported merchandise valued at \$406 billion and exported merchandise worth \$253 billion. The difference between these two quantities (exports minus imports) is referred to as the *merchandise trade balance*. Since more goods were imported than exported in 1987, the merchandise trade balance was a *negative* \$153 billion. The accompanying table lists the United States exports to and imports from a sample of ten countries in 1987 (in millions of dollars).

Country	*Exports*	*Imports*
Brazil	4,040	7,865
Egypt	2,210	465
France	7,943	10,730
Italy	5,530	11,040
Japan	28,249	84,575
Mexico	14,582	20,271
Panama	743	356
Soviet Union	1,480	425
Sweden	1,894	4,758
Turkey	1,483	821

Source: *Statistical Abstract of the United States: 1989*, pp. 788–791.

a. Calculate the U.S. merchandise trade balance with each of the ten countries. Express your answers in billions of dollars.

b. Use a z-score to identify the relative position of the U.S. trade balance with Japan within the data set you developed in part **a**. Do the same for the trade balance with the Soviet Union. Write a sentence or two that describes the relative positions of these two trade balances.

2.68 The accompanying table lists the unemployment rate in 1987 for a sample of nine countries.

Country	*Percent Unemployed*	*Country*	*Percent Unemployed*
Australia	8.1	Italy	7.9
Canada	8.9	Japan	2.9
France	11.1	Sweden	1.9
Germany	6.9	United States	6.2
Great Britain	10.3		

Source: *Statistical Abstract of the United States: 1989*, p. 829.

The mean and standard deviation of the nine countries' unemployment rates are 7.1 and 3.1, respectively.

a. Calculate the z-scores of the unemployment rates of the United States, Australia, and Japan.

b. Describe the information conveyed by the sign (positive or negative) of the z-scores you calculated in part **a**.

2.69 In *Fortune*'s 1990 ranking of the 500 largest industrial corporations in the United States, Control Data Corporation ranked 153rd in terms of 1989 sales. In 1985, it ranked 71st. Use percentiles to describe Control Data Corporation's position in each year's sales distribution.

2.70 A parking lot owner's accountant determined the owner's receipts for each of 100 randomly chosen days from the past year. The mean and standard deviation for the 100 days were \$360 and \$25, respectively. Yesterday's receipts amounted to \$370.

a. Find the sample z-score for yesterday's receipts.

b. How many standard deviations away from the mean is the value of yesterday's receipts?

c. Would you consider yesterday's receipts to be unusually high? Why or why not?

2.71 It is known that the frequency distribution of the number of videocassette recorders (VCRs) sold each week by a large department store in Atlanta is mound-shaped, with a mean of 35 and a variance of 9.

a. Approximately what percentage of the measurements in the frequency distribution should fall between 32 and 38? Between 26 and 44?

b. If the z-score for last week's sales was -1.33, how many VCRs did the store sell last week?

c. If it is known that the number of VCRs sold each week by a rival department store has a mound-shaped frequency distribution with a mean of 35 and a standard deviation of 2, for which store is it more likely that more than 41 VCRs will be sold in a week? Why?

2.72 One of the ways the federal government raises money is through the sale of securities such as **Treasury bonds, Treasury bills (T-bills)**, and **U.S. savings bonds**. Treasury bonds and bills are marketable (i.e., they can be traded in the securities market) long-term and short-term notes, respectively. U.S. savings bonds are non-marketable notes; they can be purchased and redeemed only from the U.S. Treasury. On June 30, 1983, the interest rate on 3-month T-bills was 8.75%. Within the next week, the *Wall Street Journal* sampled 17 economists and asked them to forecast the interest rate of 3-month T-bills on September 30, 1983. (T-bills are offered for sale weekly by the government, and their interest rates typically vary with each offering.) The forecasts obtained are listed in the table.

Economist	*Interest Rate Forecast* (%)	*Economist*	*Interest Rate Forecast* (%)
Alan Greenspan	8.70	Robert Parry	8.50
Timothy Howard	8.75	John Paulus	9.50
Lacy H. Hunt	9.35	Norman Robertson	8.50
Edward Hyman	7.80	Francis Schott	8.50
David Jones	9.25	Stuart Schweitzer	9.00
Irwin Kellner	8.25	Allen Sinai	9.15
Alan Lerner	9.25	Thomas Thompson	9.25
Donald Maude	7.70	John Wilson	10.00
Anne Parker Mills	8.50		

Source: *Wall Street Journal*, July 5, 1983, p. 2. Reprinted by permission.

The mean and standard deviation of the 17 forecasts are 8.82% and .61%, respectively.

a. Calculate the *z*-scores of Alan Greenspan's forecast and John Wilson's forecast. What do the *z*-scores tell you about their forecasts relative to the forecasts of the other economists?

b. Write a sentence or two that summarizes the 17 forecasts. In your summary, use a measure of central tendency and a measure of variability.

2.73 The mean and standard deviation of the gross weekly income distribution of a local firm's 120 employees were determined to be \$170 and \$10, respectively.

a. Approximately what percentage of the employees would be expected to have incomes over \$190 per week? Under \$160 per week? Over \$200 per week?

b. If you were employed by this firm and your weekly income was \$185, what would your *z*-score be, and how many standard deviations would your salary be away from the mean salary?

2.74 Refer to Exercise 2.73. Suppose it is known that the weekly gross income distribution is mound-shaped.

a. Approximately what percentage of the employees would be expected to have incomes over \$190 per week? Under \$160 per week? Over \$200 per week?

b. If you and a friend both worked at the firm and your income was \$160 per week and hers was \$195 per week, how many standard deviations apart are your incomes?

c. If you randomly chose an employee of this firm, is it more likely that his or her gross income is over \$190 per week or under \$145 per week? Why?

2.9 Box Plots: Graphical Descriptions Based on Quartiles (Optional)

The **box plot**, a relatively recent introduction to the methodology of descriptive measures, is based on the **quartiles** of a data set. Quartiles are values that partition the data set into four groups, each containing 25% of the measurements. The lower quartile Q_L is the 25th percentile, the middle quartile is the median m (the 50th percentile), and the upper quartile Q_U is the 75th percentile (see Figure 2.27).

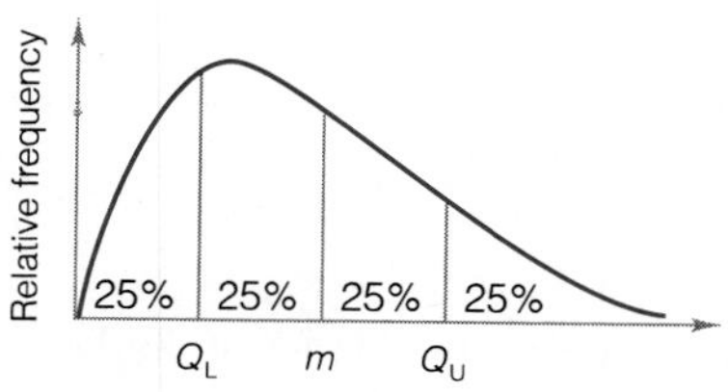

FIGURE 2.27 ▶ **The quartiles for a data set**

Definition 2.14

The **lower quartile** is the 25th percentile of a data set, the **middle quartile** is the median, and the **upper quartile** is the 75th percentile.

A box plot is based on the **interquartile range (IQR)**, the distance between the lower and upper quartiles:

$$IQR = Q_U - Q_L$$

Definition 2.15

The **interquartile range** is the distance between the lower and upper quartiles:

$$IQR = Q_U - Q_L$$

The box plot for the 50 companies' percentages of revenues spent on R&D (Table 2.6) is given in Figure 2.28. The box plot shown there was generated by the Minitab statistical software package for personal computers.* Note that a rectangle (the *box*) is drawn, with the ends of the rectangle (the *hinges*, represented by the "I's" at the ends of the box drawn by the Minitab program) drawn at the quartiles Q_L and Q_U. By definition, then, the "middle" 50% of the observations—those between Q_L and Q_U—fall inside the box. For the R&D data, these quartiles appear to be at (approximately) 7.0 and 9.5 Thus,

$$IQR = 9.5 - 7.0 = 2.5 \quad \text{(approximately)}$$

The median is shown at about 8.0 by a + sign within the box.

FIGURE 2.28 ▶ **Minitab box plot for R&D percentages**

```
                -------------------
------------I          +          I------------------------   *
                -------------------

------+---------+---------+---------+---------+---------+
     6.0       7.5       9.0      10.5      12.0      13.5
```

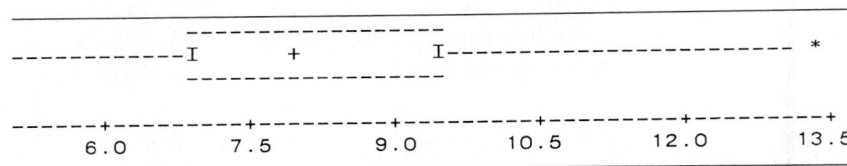

To guide the construction of the "tails" of the box plot, two sets of limits, called *inner fences* and *outer fences*, are used. Neither set of fences actually appears on the box plot. Inner fences are located at a distance of 1.5(IQR) from the hinges. Emanating from each hinge of the box are dashed lines called the *whiskers*. The two whiskers will

*Although box plots can be generated by hand, the amount of detail required makes them particularly well-suited for computer generation. We will use computer software to generate the box plots in this section.

extend to the most extreme observation inside the inner fences. For example, the inner fence on the lower side of the R&D percentage plot is (approximately):

$$\begin{aligned} \text{Lower inner fence} &= \text{Lower hinge} - 1.5(\text{IQR}) \\ &\approx 7.0 - 1.5(2.5) \\ &= 7.0 - 3.75 = 3.25 \end{aligned}$$

The smallest measurement in the data set is 5.2, which is well inside this inner fence. Thus, the lower whisker extends to 5.2. On the upper end, the inner fence is at about $(9.5 + 3.75) = 13.25$. The largest measurement inside this fence is the third largest measurement, 13.2. Note that the longer upper whisker reveals the rightward skewness of the R&D distribution.

Values that are beyond the inner fences receive special attention, because they are extreme values that represent relatively rare occurrences. In fact, for mound-shaped distributions, fewer than 1% of the observations are expected to fall outside the inner fences. As discussed above, none of the R&D measurements fall outside the lower inner fence. However, the two measurements at 13.5 fall outside the upper inner fence which is located at 13.25. These measurements are represented by asterisks (*), and they further emphasize the rightward skewness of the distribution. Note that the box plot does not reveal that there are *two* measurements at 13.5, since only a single symbol is used to represent both observations at that point.

The other pair of imaginary fences, the outer fences, are defined at a distance 3(IQR) from each end of the box. Measurements that fall beyond the outer fences are represented by zeros (0) and are very extreme measurements that require special attention and analysis. Less than one-hundredth of 1% (.01%, or .0001) of the measurements from mound-shaped distributions are expected to fall beyond the outer fences. Since no measurement in the R&D box plot (Figure 2.28) is represented by a 0, we know that none of the measurements fall outside the outer fences.

Generally, any measurements that fall beyond the inner fences, and certainly any that fall beyond the outer fences, are considered potential **outliers**. Outliers are extreme measurements that stand out from the rest of the sample and may be faulty—incorrectly recorded observations or members of a different population from the rest of the sample. At the least, they are very unusual measurements from the same population. For example, the two R&D measurements at 13.5 may be considered outliers, because they exceed the inner fence ("*" representation in Figure 2.28). When we analyze these measurements, we find they are correctly recorded. However, it turns out that both represent R&D expenditures of relatively young and fast-growing companies. Thus, the outlier analysis may have revealed important factors that relate to the R&D expenditures of high-tech companies: their age and rate of growth. Outlier analysis often reveals useful information of this kind, and therefore plays an important role in the statistical inference-making process.

The elements (and nomenclature) of box plots are summarized in the next box. Some aids to the interpretation of box plots are also given.

Elements of a Box Plot

1. A rectangle (the **box**) is drawn with the ends (the **hinges**) drawn at the lower and upper quartiles (Q_L and Q_U). The median of the data is shown in the box, usually by a "+".
2. The points at distances 1.5(IQR) from each hinge mark the **inner fences** of the data set. Horizontal lines (the **whiskers**) are drawn from each hinge to the most extreme measurement inside the inner fence.
3. A second pair of fences, the **outer fences**, exist at a distance of 3 interquartile ranges, 3(IQR), from the hinges. One symbol (usually "*") is used to represent measurements falling between the inner and outer fences, and another (usually "0") is used to represent measurements beyond the outer fences.
4. The symbols used to represent the median and the extreme data points (those beyond the fences) will vary depending on the software you use to construct the box plot. (You may use your own symbols if you are constructing a box plot by hand.) You should consult the program's documentation to determine exactly which symbols are used.

Aids to the Interpretation of Box Plots

1. Examine the length of the box. The IQR is a measure of the sample's variability, and is especially useful for the comparison of two samples (see Example 2.16).
2. Visually compare the lengths of the whiskers. If one is clearly longer, the distribution of the data is probably skewed in the direction of the longer whisker.
3. Analyze any measurements that lie beyond the fences. Fewer than 5% should fall beyond the inner fences, even for very skewed distributions. Measurements beyond the outer fences are probably **outliers**, with one of the following explanations:
   a. The measurement is incorrect. It may have been observed, recorded, or entered into the computer incorrectly.
   b. The measurement belongs to a population different from that from which the rest of the sample was drawn (see Example 2.16).
   c. The measurement may be correct and from the same population as the rest, but represents a rare event. Generally, we accept this explanation only after carefully ruling out all others.

EXAMPLE 2.15

In Example 2.1 we analyzed 50 processing times for the development of price quotes by the manufacturer of industrial wheels. The intent was to determine whether the success or failure in obtaining the order was related to the amount of time to process the price quotes. Each quote that corresponds to "lost" business was so classified. The data are repeated in Table 2.12. Use a statistical software package to draw a box plot for these data.

TABLE 2.12 Price Quote Processing Times (Days)

Request Number	Processing Time	Lost?	Request Number	Processing Time	Lost?
1	2.36	No	26	3.34	No
2	5.73	No	27	6.00	No
3	6.60	No	28	5.92	No
4	10.05	Yes	29	7.28	Yes
5	5.13	No	30	1.25	No
6	1.88	No	31	4.01	No
7	2.52	No	32	7.59	No
8	2.00	No	33	13.42	Yes
9	4.69	No	34	3.24	No
10	1.91	No	35	3.37	No
11	6.75	Yes	36	14.06	Yes
12	3.92	No	37	5.10	No
13	3.46	No	38	6.44	No
14	2.64	No	39	7.76	No
15	3.63	No	40	4.40	No
16	3.44	No	41	5.48	No
17	9.49	Yes	42	7.51	No
18	4.90	No	43	6.18	No
19	7.45	No	44	8.22	Yes
20	20.23	Yes	45	4.37	No
21	3.91	No	46	2.93	No
22	1.70	No	47	9.95	Yes
23	16.29	Yes	48	4.46	No
24	5.52	No	49	14.32	Yes
25	1.44	No	50	9.01	No

Solution

The SAS/PC box plot printout for these data is shown in Figure 2.29 (page 98). Note that the SAS program draws the box plot vertically, and shows a scale for the processing times on the left. SAS uses a horizontal dashed line in the box to represent the median, and a plus (+) sign to represent the mean. (SAS shows the mean in box plots, unlike many other statistical programs.) Also, note that SAS uses the symbol "0" to represent measurements between the inner and outer fences and "*" to represent observations beyond the outer fences.

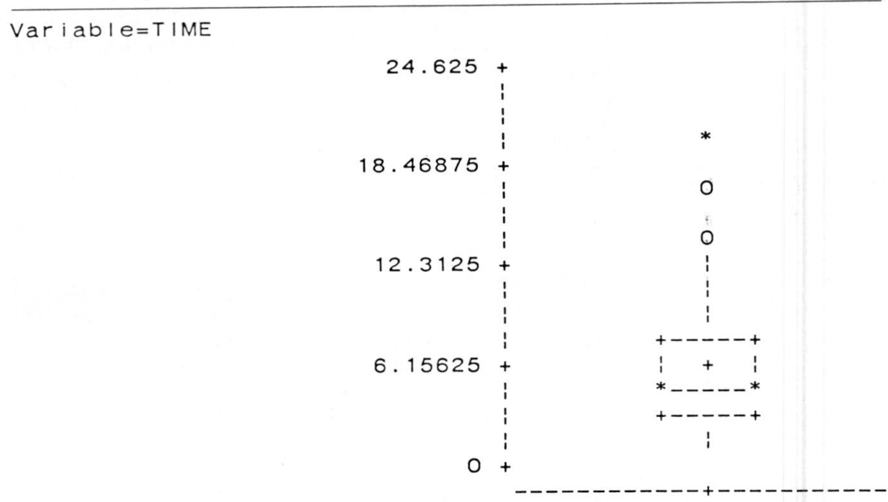

FIGURE 2.29 ▶ SAS box plot for processing time data

Note that the upper whisker is longer than the lower whisker and that the mean lies above the median; these characteristics reveal the rightward skewness of the data. However, the most important feature of the data is made very obvious by the box plot: There are at least two measurements between the inner and outer fences (in fact, there are three, but two are almost equal and are represented by the same "0") and at least one beyond the outer fence, all on the upper end of the distribution. Thus, the distribution is extremely skewed to the right, and several measurements need special attention in our analysis. We offer an explanation for the outliers in the following example.

EXAMPLE 2.16

The box plot for the 50 processing times (Figure 2.29) does not explicitly reveal the differences, if any, between the set of times corresponding to the success and the set of times corresponding to the failure to obtain the business. Box plots corresponding to the 39 "won" and 11 "lost" bids were generated using the SAS/PC program, and are shown in Figure 2.30. Interpret them.

Solution

The division of the data set into two parts, corresponding to won and lost bids, eliminates any observations that are beyond inner or outer fences. Furthermore, the skewness in the distributions has been reduced, as evidenced by the facts that the upper whiskers are only slightly longer than the lower, and that the means are closer to the medians than for the combined sample. The box plots also reveal that the processing times corresponding to the lost bids tend to exceed those of the won bids. A plausible explanation for the outliers in the combined box plot (Figure 2.29) is that they are from a different population than the bulk of the times. In other words, there are two populations represented by the sample of processing times—one corresponding to lost bids, and the other to won bids.

FIGURE 2.30 ▶
Box plots of processing time data: Won and lost bids

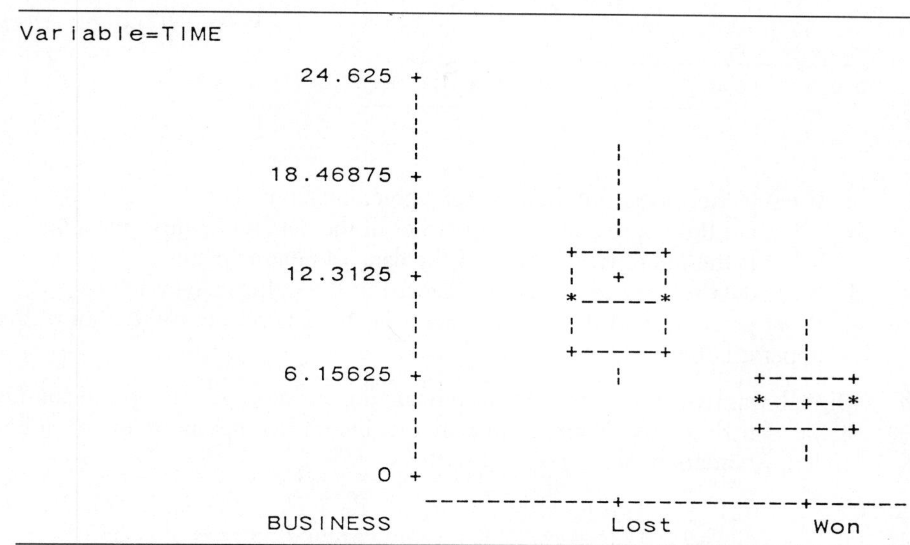

The box plots lend support to the conclusion that the price quote processing time and the success of acquiring the business are related. However, whether the visual differences between the box plots generalize to inferences about the populations corresponding to these two samples is a matter for inferential statistics, not graphical descriptions. We will discuss how to use samples to compare two populations using inferential statistics in Chapter 9.

Exercises 2.75–2.84

Note: Starred (*) exercises require the use of a computer.

Learning the Mechanics

2.75 Define the 25th, 50th, and 75th percentiles of a data set. Explain how they provide a description of the data.

2.76 Suppose a data set consisting of exam scores has a lower quartile $Q_L = 60$, a median $m = 75$, and an upper quartile $Q_U = 85$. The scores on the exam ranged from 18 to 100. Without having the actual scores available to you, construct as much of the box plot as possible.

2.77 Minitab was used to generate the following box plot:

```
                        ----------
 * *     -------------I    +    I--------
                        ----------

 +---------+---------+---------+---------+
0.0      15.0      30.0      45.0      60.0
```

a. What is the median of the data set (approximately)?
b. What are the upper and lower quartiles of the data set (approximately)?
c. What is the interquartile range of the data set (approximately)?
d. Is the data set skewed to the left, skewed to the right, or symmetric?
e. What percentage of the measurements in the data set lie to the right of the median? To the left of the upper quartile?

2.78 Minitab was used to generate the accompanying box plots. Compare and contrast the frequency distributions of the two data sets. Your answer should include comparisons of the following characteristics: central tendency, variation, skewness, and outliers.

```
      --------
-----I +    I--------                    *            O O
      --------

--------+---------+---------+---------+---------+
       4.0       8.0      12.0      16.0      20.0
```

```
               --------------------
   ----------I         +          I----------
               --------------------

   +---------+---------+---------+---------+
 -20.0     -10.0      0.0      10.0      20.0
```

***2.79** Use a statistical software package to construct a box plot for the following set of sample measurements:

```
1.11   1.39   1.66   1.33   1.30
1.72   1.36   1.26   1.35   1.46
1.55   1.24   1.65   1.40   1.50
1.31   1.41   1.24   1.38   1.28
2.00   1.12   1.25   1.49   2.10
1.86   1.55   1.31   1.14   1.82
```

***2.80** Consider the following two sample data sets:

Sample A			*Sample B*		
121	171	158	171	152	170
173	184	163	168	169	171
157	85	145	190	183	185
165	172	196	140	173	206
170	159	172	172	174	169
161	187	100	199	151	180
142	166	171	167	170	188

a. Use a statistical software package to construct a box plot for each data set.

b. Using the information reflected in your box plots, describe the similarities and differences between the two data sets.

c. Identify any outliers that may exist in the two data sets.

Applying the Concepts

2.81 The table contains the top salary offer (in thousands of dollars) received by each member of a sample of 50 undergraduate business majors (excluding accounting majors) who graduated from the Carlson School of Management at the University of Minnesota in 1991–1992.

Salary Offers to 1991–1992 Graduates

23.7	26.1	26.6	27.9	40.0	22.8	26.0	26.5	27.5	37.2
22.0	25.9	26.5	27.5	36.1	21.6	25.7	24.7	27.3	34.8
20.9	25.6	24.6	27.2	32.8	20.7	25.4	25.6	26.9	29.1
19.8	25.3	26.4	26.9	28.9	18.7	24.8	26.4	26.8	28.3
17.5	24.2	26.3	26.7	28.2	14.0	23.9	26.1	26.7	28.1

Source: Placement Office, Carlson School of Management, University of Minnesota.

a. The mean and standard deviation are 26.2 and 4.56, respectively. Find and interpret the z-score associated with the highest salary offer, the lowest salary offer, and the mean salary offer. Would you consider the highest offer to be unusually high? Why or why not?

***b.** Use a statistical software package to construct a box plot for this data set. Which salary offers (if any) are potentially faulty observations? Explain.

***2.82** A firm's earnings per share (E/S) of common stock is a measure used by investors to monitor the financial performance of a firm. Thirty firms were sampled from *Fortune*'s 1990 listing of the 500 largest corporations in the United States, and their earnings per share for 1989 are recorded in the table at the top of page 102.

a. Use a statistical software package to construct a box plot for this data set. Identify any outliers that may exist in this data set.

b. For each outlier identified in part **a**, determine how many standard deviations it lies from the mean of the E/S data set.

Firm	E/S	Firm	E/S
Illinois Tool Works	$3.06	Dow Jones	$ 3.15
Sara Lee	3.50	United Brands	1.70
Reynolds Metals	9.20	Washington Post	15.50
Scott Paper	5.11	Avon Products	.34
Phelps Dodge	7.59	Valhi	.91
Westmoreland Coal	1.39	American Cyanamid	3.12
Avery International	1.96	EG&G	2.40
Warner–Lambert	6.10	Asarco	5.50
Borden	−.41	Snap-on Tools	2.55
General Electric	4.36	McCormick	3.09
Cooper Industries	2.51	Exxon	2.74
Lockheed	.03	Georgia–Pacific	7.42
Kellogg	3.85	Crown Cork & Seal	3.58
FMC	3.79	DuPont (E.I.) de Nemours	3.53
Oxford Industries	.99	Molex	2.28

Source: *Fortune*, Apr. 23, 1990, pp. 337–396.

***2.83** A manufacturer of minicomputer systems is interested in improving its customer support services. As a first step, its marketing department has been charged with the responsibility of summarizing the extent of customer problems in terms of system down time. The 40 most recent customers were surveyed to determine the amount of down time (in hours) they had experienced during the previous month. These data are listed in the table.

Customer Number	*Down Time*	*Customer Number*	*Down Time*	*Customer Number*	*Down Time*	*Customer Number*	*Down Time*
230	12	240	24	250	4	260	34
231	16	241	15	251	10	261	26
232	5	242	13	252	15	262	17
233	16	243	8	253	7	263	11
234	21	244	2	254	20	264	64
235	29	245	11	255	9	265	19
236	38	246	22	256	22	266	18
237	14	247	17	257	18	267	24
238	47	248	31	258	28	268	49
239	0	249	10	259	19	269	50

a. Use a statistical software package to construct a box plot for these data. Use the information reflected in the box plot to describe the frequency distribution of the data set. Your description should address central tendency, variation, and skewness.

b. Use your box plot to determine which customers are having unusually lengthy down times.

c. Find and interpret the *z*-scores associated with the customers you identified in part **b**.

2.84 The accompanying Minitab-generated box plots describe the U.S. Environmental Protection Agency's 1986 automobile mileage estimates for all models manufactured by Ford and Honda.

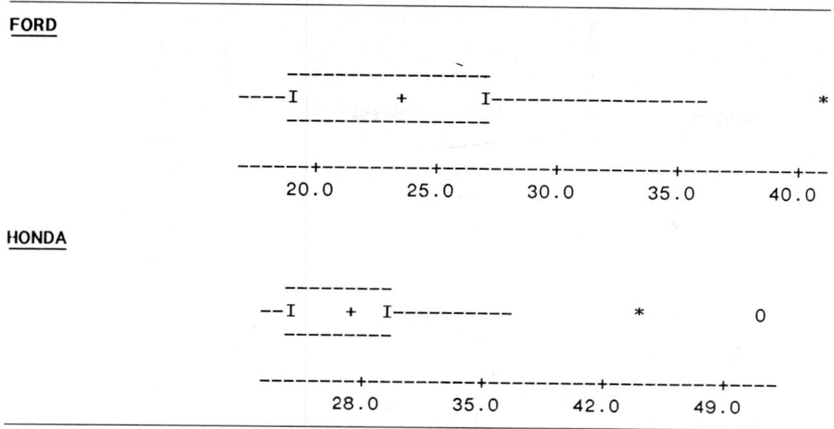

a. Which manufacturer has the higher median mileage estimate?
b. Which manufacturer's mileage estimates have the greater range?
c. Which manufacturer's mileage estimates have the greater interquartile range?
d. Which manufacturer has the model with the highest mileage estimate? Approximately what is that mileage?

2.10 Distorting the Truth with Descriptive Techniques

While it may be true in telling a story that a picture is worth a thousand words, it is also true that pictures can be used to convey a colored and distorted message to the viewer. So the old adage "Let the buyer (reader) beware" applies. Examine relative frequency histograms, time series plots, and, in general, all graphical descriptions with care. In this section, we mention a few of the pitfalls to watch for when analyzing a chart or graph.

One common way to change the impression conveyed by a graph is to change the scale on the vertical axis, the horizontal axis, or both. For example, if you want to show that the change in firm A's market share over time is moderate, you could pack in a large number of units per inch on the vertical axis. That is, make the distance between successive units on the vertical scale small, as shown in Figure 2.31. You can see that the change in the firm's market share over time appears to be minimal.

FIGURE 2.31 ▶ Firm A's market share, 1980–1990: Packed vertical axis

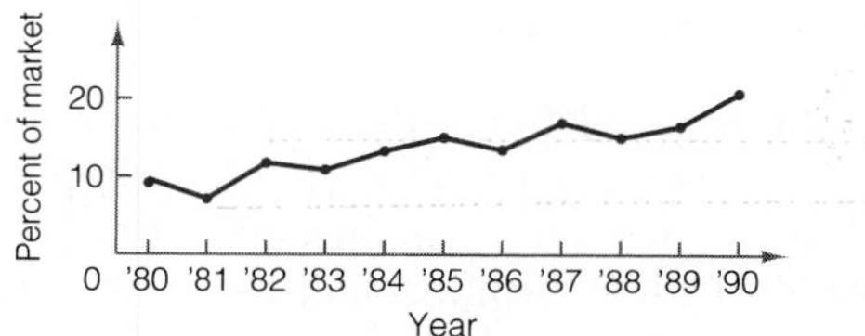

To make the changes in firm A's market share appear large, you could increase the distance between successive units on the vertical axis. That is, you stretch the vertical axis by graphing only a few units per inch, as shown in Figure 2.32. The telltale sign of stretching is a long vertical axis, but this is often hidden by starting the vertical axis at some point above 0, as shown in Figure 2.33(a). Or, the same effect can be achieved by using a broken line called a *scale break* for the vertical axis, as shown in Figure 2.33(b).

FIGURE 2.32 ► **Firm A's market share, 1980–1990: Stretched vertical axis**

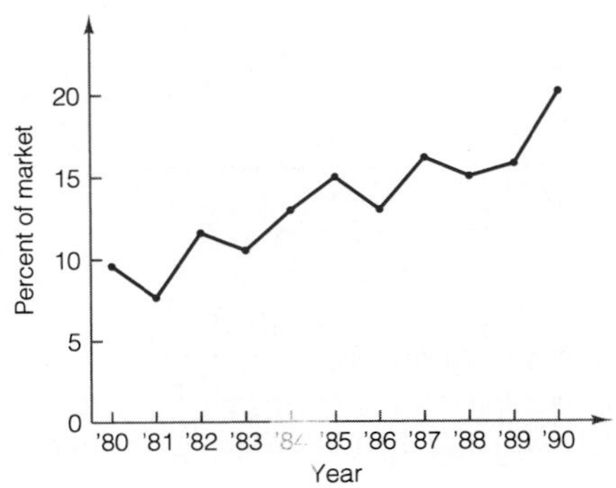

FIGURE 2.33 ► **Daily stock sales on the New York Stock Exchange from May to July 1979**

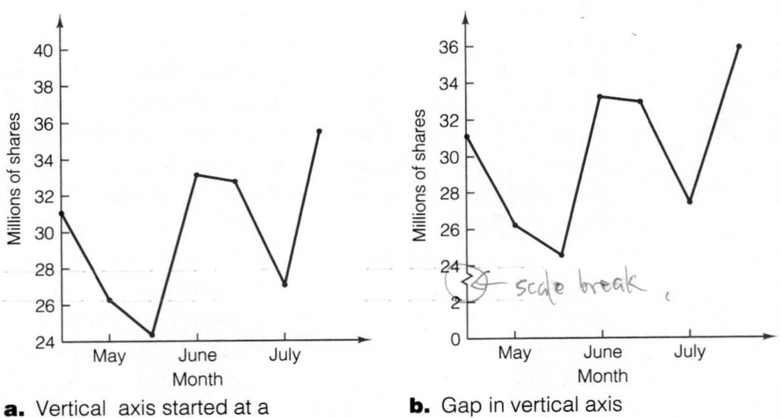

a. Vertical axis started at a point greater than 0

b. Gap in vertical axis

Stretching the horizontal axis (increasing the distance between successive units) may also lead you to incorrect conclusions. For example, Figure 2.34(a) depicts rental income in the United States from the first quarter of 1978 to the first quarter of 1980. If you increase the length of the horizontal axis, as in Figure 2.34(b), the change in the rental income over time seems to be less pronounced.

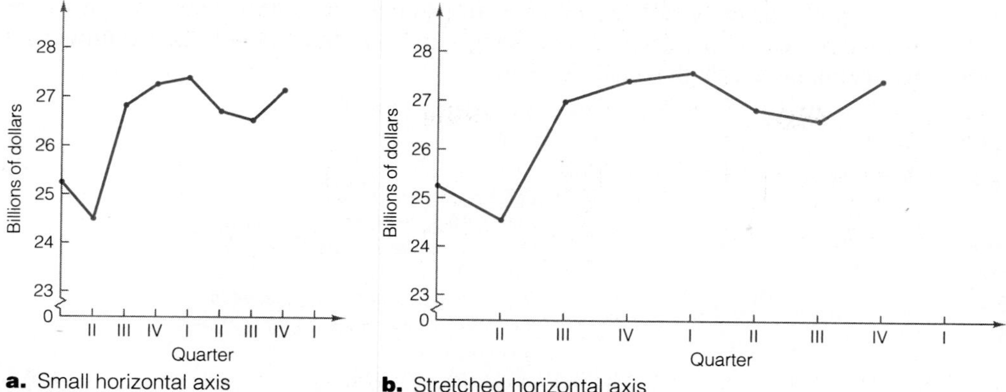

a. Small horizontal axis **b.** Stretched horizontal axis

FIGURE 2.34 ▲ **Rental income from the first quarter of 1978 to the first quarter of 1980**

The changes in categories indicated by a bar chart can also be emphasized or deemphasized by stretching or shrinking the vertical axis. Another method of achieving visual distortion with bar charts is by making the width of the bars proportional to their height. For example, look at the bar chart in Figure 2.35(a), which depicts the percentage of a year's total automobile sales attributable to each of the four major manufacturers. Now suppose we make the width as well as the height grow as the market share grows. This is shown in Figure 2.35(b). The reader may tend to equate the *area* of the bars with the relative market share of each manufacturer. In fact, the true relative market share is proportional only to the height of the bars.

FIGURE 2.35 ► **Relative share of the automobile market for each of four major manufacturers**

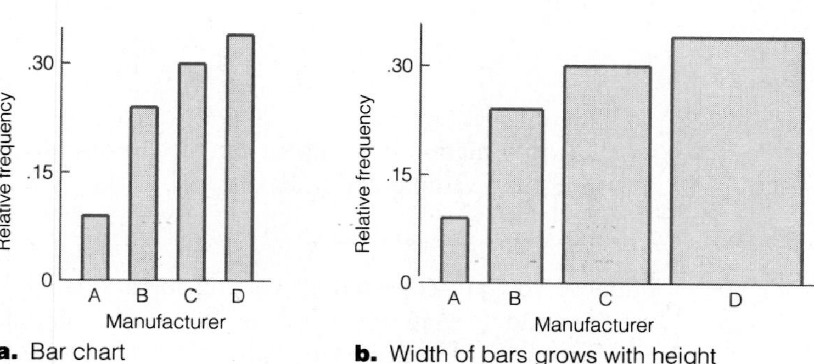

a. Bar chart **b.** Width of bars grows with height

Sometimes, as noted by Selazny (1975), we do not need to manipulate the graph to distort the impression it creates. Modifying the verbal description that accompanies the graph can change the interpretation that will be made by the viewer. Figure 2.36 (page 106) provides a good illustration of this ploy.

We have presented only a few of the ways that graphs can be used to convey misleading pictures of business phenomena. However, the lesson is clear. Examine all graphical descriptions of data with care. In particular, check the axes and the size of the units on each axis. Ignore visual changes and concentrate on the actual numerical changes indicated by the graph or chart.

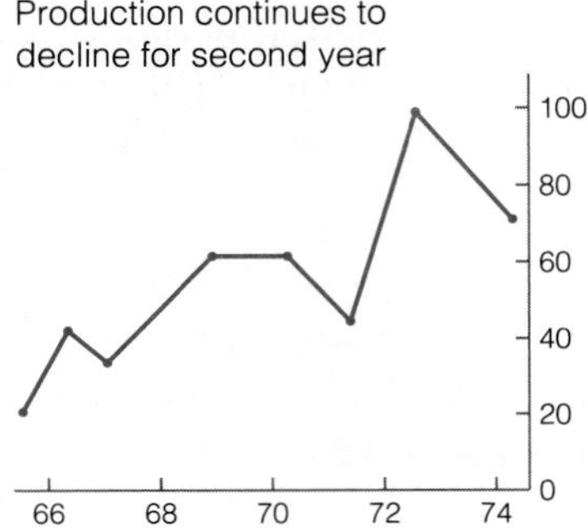

For our production, we need not even change the chart, so we can't be accused of fudging the data. Here we'll simply change the title so that for the Senate subcommittee, we'll indicate that we're not doing as well as in the past...

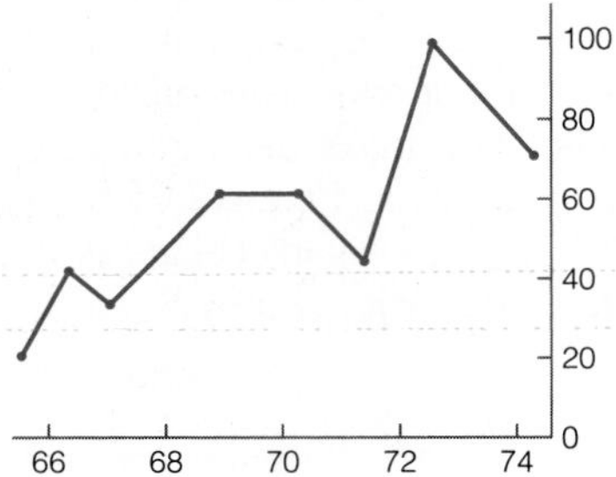

whereas for the general public, we'll tell them that we're still in the prime years.

FIGURE 2.36 ► Changing the verbal description to change a viewer's interpretation. Source: Reprinted by permission of the publisher, from "Grappling with Graphics," by Gene Selazny, *Management Review*, Oct. 1975, p. 7. © 1975 American Management Association, New York. All rights reserved.

The information in a data set can also be distorted by using numerical descriptive measures, as Example 2.17 indicates.

EXAMPLE 2.17

Suppose you are considering working for a small firm that presently has a senior member and three junior members. You inquire about the salary you could expect to earn if you join the firm. Unfortunately, you receive two answers:

Answer A: The senior member tells you that an "average employee" earns \$57,500.

Answer B: One of the junior members later tells you that an "average employee" earns \$45,000.

Which answer can you believe? The confusion exists because the phrase "average employee" has not been clearly defined. Suppose the four salaries paid are \$45,000 for each of the three junior members and \$95,000 for the senior member. Thus,

$$\bar{x} = \frac{3(\$45{,}000) + \$95{,}000}{4} = \frac{\$230{,}000}{4} = \$57{,}500$$

$$\text{Median} = \$45{,}000$$

You can now see how the two answers were obtained. The senior member reported the mean of the four salaries, and the junior member reported the median. The information you received was distorted because neither person stated which measure of central tendency was being used.

Another distortion of information in a sample occurs when *only* a measure of central tendency is reported. Both a measure of central tendency and a measure of variability are needed to obtain an accurate mental image of a data set.

Suppose you want to buy a new car and are trying to decide which of two models to purchase. Since energy and economy are both important issues, you decide to purchase model A because its EPA mileage rating is 32 miles per gallon in the city, whereas the mileage rating for model B is only 30 miles per gallon in the city.

However, you may have acted too quickly. How much variability is associated with the ratings? As an extreme example, suppose that further investigation reveals that the standard deviation for model A mileages is 5 miles per gallon, whereas that for model B is only 1 mile per gallon. If the mileages form a mound-shaped distribution, they might appear as shown in Figure 2.37. Note that the larger amount of variability associated with model A implies that more risk is involved in purchasing model A. That is, the particular car you purchase is more likely to have a mileage rating that will greatly differ from the EPA rating of 32 miles per gallon if you purchase model A, while a model B car is not likely to vary from the 30 miles per gallon rating by more than 2 miles per gallon.

FIGURE 2.37 ► **Mileage distributions for two car models**

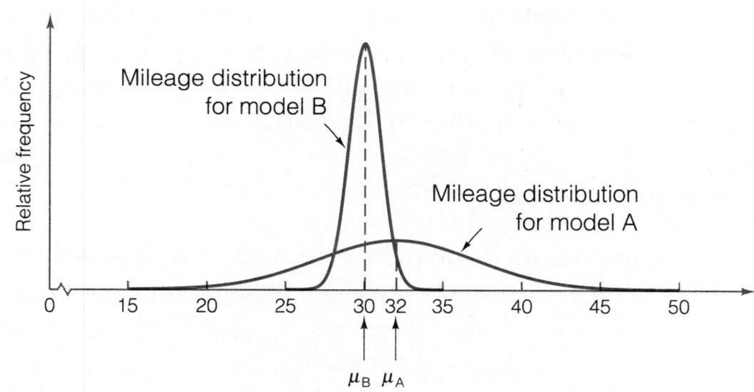

Summary

Data may be classified as one of four types: **nominal, ordinal, interval**, and **ratio**. Nominal data simply classify the sample or population units, whereas ordinal data enable the units to be ranked. Nominal and ordinal data are often referred to collectively as **qualitative**. Interval data are numbers that enable the comparison of sample or population units by calculation of their numerical differences, whereas ratio data enable the comparison using multiples, or ratios, of the numerical values. Interval and ratio data are often referred to collectively as **quantitative**. Since we want to use sample data to make inferences about the population from which it is drawn, it is important for us to be able to describe the data. **Graphic methods** are important and useful tools for describing data sets. Our ultimate goal, however, is to use the sample to make inferences about the population. We are wary of using graphic techniques to accomplish this goal, since they do not lend themselves to a measure of the reliability for an inference. We therefore developed **numerical measures** to describe a data set.

These numerical methods for describing **quantitative** data sets can be grouped as follows:

1. Measures of central tendency
2. Measures of variability

The measures of central tendency we presented were the **mean**, **median**, and **mode**. The relationship between the mean and median provides information about the **skewness** of the frequency distribution. For making inferences about the population, the sample mean will usually be preferred to the other measures of central tendency. The **range**, **variance**, and **standard deviation** all represent numerical measures of variability. Of these, the variance and standard deviation are in most common use, especially when the ultimate objective is to make inferences about a population.

The mean and standard deviation may be used to make statements about the fraction of measurements in a given interval. For example, we know that at least 75% of the measurements in a data set lie within 2 standard deviations of the mean. If the frequency distribution of the data set is mound-shaped, approximately 95% of the measurements will lie within 2 standard deviations of the mean.

Measures of relative standing provide still another dimension on which to describe a data set. The objective of these measures is to describe the location of a specific measurement relative to the rest of the data set. By doing so, you can construct a mental image of the relative frequency distribution. **Percentiles** and ***z*-scores** are important examples of measures of relative standing.

The **rare event** concept of statistical inference means that if the chance that a particular sample came from a hypothesized population is very small, we can conclude either that the sample is extremely rare or that the hypothesized population is not the one from which the sample was drawn. The more unlikely it is that the sample came from the hypothesized population, the more strongly we favor the conclusion that the hypothesized population is not the true one. We need to be able to assess accurately the rarity of a sample, and this requires knowledge of probability, the subject of our next chapter.

Finally, we gave some examples that demonstrated how descriptive statistics may be used to distort the truth. You should be very critical when interpreting graphic or numerical descriptions of data sets.

Supplementary Exercises 2.85–2.126

Note: Starred (*) exercises require the use of a computer. Double-starred exercises (**) refer to optional sections in this chapter.

2.85 Classify the following data as one of four types: nominal, ordinal, interval, or ratio.

a. The length of time it takes each of 15 telephone installers to hook up a wall phone
b. The style of music preferred by each of 30 randomly selected radio listeners
c. The arrival time of the 5 P.M. train from New York to Newark
d. A sample of 100 customers in a fast-food restaurant asked to rate their hamburger on the following scale: poor, fair, good, excellent
e. Classify each of the data sets in parts **a–d** as qualitative or quantitative.

2.86 Discuss the conditions under which the median is preferred to the mean as a measure of central tendency.

2.87 Compute $\sum_{i=1}^{n} x_i^2$, $\sum_{i=1}^{n} x_i$, and $\left(\sum_{i=1}^{n} x_i\right)^2$ for each of the following data sets:

a. 11, 1, 2, 8, 7 **b.** 15, 15, 2, 6, 12 **c.** −1, 2, 0, −4, −8, 13 **d.** 100, 0, 0, 2

2.88 Compute s^2 and s for each of the data sets in Exercise 2.87.

****2.89** A time series plot similar to the one shown here appeared in a recent advertisement for a well-known golf magazine. One person might interpret the plot's message as the longer you subscribe to the magazine, the better golfer you should become. Another person might interpret it as indicating that if you subscribe for 3 years, your game should improve dramatically.

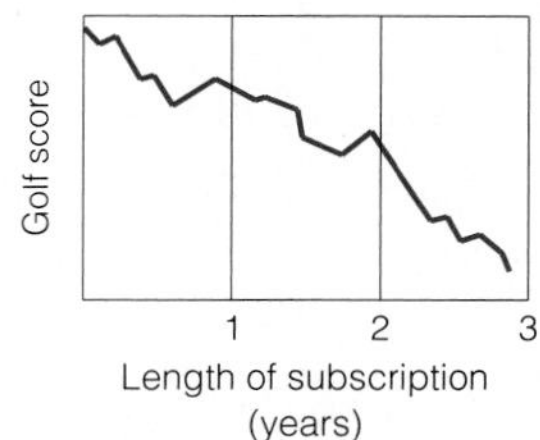

a. Explain why the plot can be interpreted in more than one way.
b. How could the plot be altered to rectify the current distortion?

2.90 Explain why we generally prefer the standard deviation to the range as a measure of variability for quantitative data.

2.91 Construct a relative frequency histogram for the data summarized in the accompanying table.

Measurement Class	Relative Frequency	Measurement Class	Relative Frequency
.00– .75	.02	5.25–6.00	.15
.75–1.50	.01	6.00–6.75	.12
1.50–2.25	.03	6.75–7.50	.09
2.25–3.00	.05	7.50–8.25	.05
3.00–3.75	.10	8.25–9.00	.04
3.75–4.50	.14	9.00–9.75	.01
4.50–5.25	.19		

2.92 Compute s^2 for data sets with the following characteristics:

a. $\sum_{i=1}^{n} x_i^2 = 246, \quad \sum_{i=1}^{n} x_i = 63, \quad n = 22$ **b.** $\sum_{i=1}^{n} x_i^2 = 666, \quad \sum_{i=1}^{n} x_i = 106, \quad n = 25$

c. $\sum_{i=1}^{n} x_i^2 = 76, \quad \sum_{i=1}^{n} x_i = 11, \quad n = 7$

2.93 Climatologists around the world use a 30-year mean as a standard of "normal" temperature or precipitation for a region. These data are regularly used by the tourist, transportation, and construction industries, as well as by farmers. Normals are computed every 10 years. The normals used from 1990–1999 are based on 1961–1990 National Weather Service data ("A Plain Old '30-Year Mean' is Forecaster's Yardstick," 1990). The table contains 1961–1990 temperature data in degrees Fahrenheit for Minnesota.

Year	Mean Daily January Temperature	Year	Mean Daily January Temperature
1961	12.0	1976	11.6
1962	7.1	1977	.3
1963	2.9	1978	5.5
1964	20.0	1979	3.2
1965	10.0	1980	15.3
1966	3.3	1981	18.0
1967	14.6	1982	2.3
1968	14.3	1983	19.6
1969	9.4	1984	12.0
1970	5.6	1985	10.1
1971	6.5	1986	17.5
1972	5.5	1987	21.2
1973	17.4	1988	10.4
1974	11.9	1989	21.2
1975	14.5	1990	26.3

a. What is the "normal" Minnesota temperature for January?

b. What is the standard deviation of the 30 temperatures? What information about the temperatures is conveyed by the standard deviation?

c. According to Chebyshev's theorem, what percentage of the 30 temperatures would be expected to fall within 2 standard deviations of the normal temperature? How many actually fell in that interval?

***2.94** If it is not examined carefully, the graphical description of U.S. peanut production shown here can be misleading.

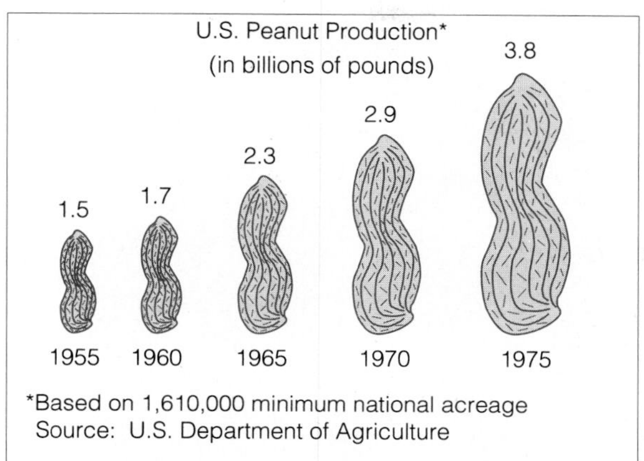

a. Explain why the graph may mislead some readers.

b. Construct an undistorted graph of U.S. peanut production for the given years.

2.95 In some locations, radiation levels in homes are measured at well above normal background levels in the environment. As a result many architects and builders are making design changes to assure adequate air exchange so that radioactive radon gas will not be trapped in homes. In one such location, 50 homes' radon levels were measured, and the mean level was 10 ppb (parts per billion), the median was 8 ppb, and the standard deviation was 3 ppb. Background levels in this location are at about 4 ppb.

a. Based on these results, is the distribution of the 50 homes' radon levels symmetric, skewed to the left, or skewed to the right? Why?

b. Use both Chebyshev's theorem and the Empirical Rule to describe the distribution of radon levels. Which do you think is most appropriate in this case? Why?

c. Use the results from part **b** to approximate the number of homes in this sample with radioactive radon gas levels above the background level.

d. Suppose another home is measured at a location 10 miles from the one sampled, and has a level of 20 ppb. What is the *z*-score for this measurement relative to the 50 homes sampled in the other location? Is it likely that this new measurement comes from the same distribution of levels as the other 50? Why? How would you confirm your conclusion?

2.96 As part of a study of property values in Minneapolis, the Robinson Appraisal Company sampled 79 apartment buildings and determined, among other things, each building's size, age, and condition as of January 1982. The following data are the ages (in years) of the buildings:

82	67	18	82	21	65	79	13	70	74	65	82	22	21	57	36
62	72	50	69	58	66	82	18	55	23	82	70	82	24	82	10
59	56	19	32	82	13	71	63	19	70	67	15	67	64	23	56
51	70	50	21	18	22	66	69	82	13	73	57	55	14	69	82
18	53	64	82	19	71	72	75	21	79	76	54	70	3	71	

Consider the following computer-generated stem-and-leaf display for these data:

```
Stem Leaf                              #
   8 22222222222                      11
   7 5699                              4
   7 000001112234                     12
   6 5566777999                       10
   6 2344                              4
   5 55667789                          8
   5 00134                             5
   4
   4
   3 6                                 1
   3 2                                 1
   2
   2 111122334                         9
   1 58888999                          8
   1 03334                             5
   0
   0 3                                 1
     ----+----+----+----+
 Multiply Stem.Leaf by 10**+1
```

a. Use the display to determine the age that approximately 50% of the buildings exceed.
b. How many buildings are more than 75 years old? Less than 10 years old?

2.97 Various state and national automobile associations regularly survey gasoline stations to determine the current retail price of gasoline. Suppose one such national association decides to survey 200 stations in the United States and intends to determine the price of regular unleaded gasoline at each station.
a. Identify the population of interest.
b. Identify the sample.
c. Identify the variable of interest.
d. In the context of this problem, define the following numerical descriptive measures: μ, σ, $\bar{x}$, s.
e. Suppose the sample of 200 stations is selected, and the mean and standard deviation of their regular unleaded prices (per gallon) are \$1.39 and \$.12, respectively. Interpret these descriptive statistics and describe the probable distribution of the 200 prices at the time of the survey.
f. One station in the southeast priced unleaded gasoline at \$1.09 per gallon at the time of the survey. Describe the relative standing of this price in the national price distribution as indicated by the sample.

***2.98** Compute the mean and variance of the following data sets:

a.

Class	*Class Frequency*
1– 5	2
6–10	5
11–15	12
16–20	6

b.

Class	*Class Frequency*
.25– .50	0
.50– .75	5
.75–1.00	12
1.00–1.25	8
1.25–1.50	6
1.50–1.75	2

2.99 In experimenting with a new technique for imprinting paper napkins with designs, names, etc., a paper products company discovered that four different results were possible:

(A) Imprint successful
(B) Imprint smeared
(C) Imprint off-center to the left
(D) Imprint off-center to the right

To test the reliability of the technique, the company imprinted 1,000 napkins and obtained the results shown in the graph.

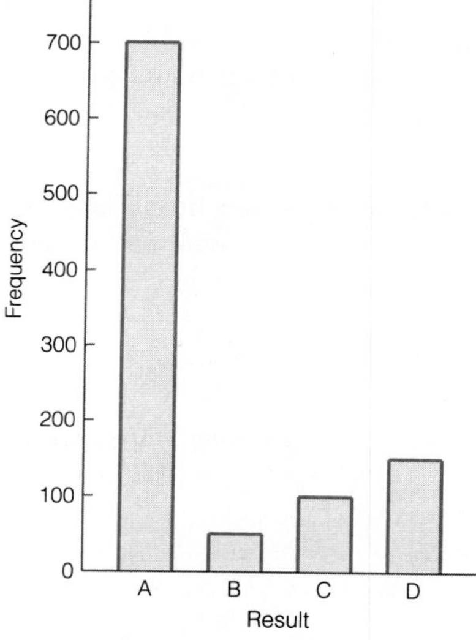

a. What type of graphical tool is the figure?
b. What information does the graph convey to you?
c. From the information provided by the graph, how might you numerically describe the reliability of the imprinting technique?

2.100 The data in the table describe the distribution of rent amounts paid by U.S. apartment dwellers in 1987. Use a relative frequency histogram to describe the data.

Rent (in Dollars)	*Percentage of Renters in Rent Class*
Less than 350	7%
350–399	12%
400–449	12%
450–499	15%
500–549	11%
550 or more	43%

Source: U.S. Bureau of the Census, *Statistical Abstract of the United States: 1988.*

2.101 Under what circumstances might the standard deviation be preferred to the variance as a measure of the variation in a data set?

2.102 In reference to a particular measurement, what is a measure of relative standing?

2.103 How does a z-score locate a measurement within a set of measurements? Explain.

2.104 In 1985, consumers spent an estimated average of $11,883 for a new car, compared with $5,414 in 1976 (*Minneapolis Star and Tribune*, Mar. 15, 1985). To be able to determine the 1985 average *exactly*, what information is needed?

2.105 Explain why the mode is often an unacceptable measure of central tendency for quantitative data sets.

2.106 A quality control inspector is interested in determining whether a metal lathe used to produce machine bearings is properly adjusted. He plans to do so by using 30 bearings selected randomly from the last 2,000 bearings produced by the machine to estimate the average diameter of bearings being produced. Define and explain the meanings of the following symbols *in the context of this problem*:

a. μ **b.** $\bar{x}$ **c.** σ **d.** s **e.** $\mu + 2\sigma$

2.107 A manufacturer of industrial wheels is losing many profitable orders because of the long time it takes the firm's marketing, engineering, and accounting departments to develop price quotes for potential customers. To remedy this problem the firm's management would like to set guidelines for the length of time each

Price Quote Processing Times (in Days)

Request Number	*Marketing*	*Engineering*	*Accounting*	*Lost?*
1	7.0	6.2	.1	No
2	.4	5.2	.1	No
3	2.4	4.6	.6	No
4	6.2	13.0	.8	Yes
5	4.7	.9	.5	No
6	1.3	.4	.1	No
7	7.3	6.1	.1	No
8	5.6	3.6	3.8	No
9	5.5	9.6	.5	No
10	5.3	4.8	.8	No
11	6.0	2.6	.1	No
12	2.6	11.3	1.0	No
13	2.0	.6	.8	No
14	.4	12.2	1.0	No
15	8.7	2.2	3.7	No
16	4.7	9.6	.1	No
17	6.9	12.3	.2	Yes
18	.2	4.2	.3	No
19	5.5	3.5	.4	No
20	2.9	5.3	22.0	No
21	5.9	7.3	1.7	No
22	6.2	4.4	.1	No
23	4.1	2.1	30.0	Yes
24	5.8	.6	.1	No
25	5.0	3.1	2.3	No
26	.6	2.2	.5	No
27	6.0	1.8	.2	No
28	5.8	.6	.5	No
29	7.8	7.2	2.2	Yes
30	3.2	6.9	.1	No
31	11.0	1.7	3.3	No
32	6.2	1.3	2.0	No
33	6.9	6.0	10.5	Yes
34	5.4	.4	8.4	No
35	6.0	7.9	.4	No
36	4.0	1.8	18.2	Yes
37	4.5	1.3	.3	No
38	2.2	4.8	.4	No
39	3.5	7.2	7.0	Yes
40	.1	.9	14.4	No
41	2.9	7.7	5.8	No
42	5.4	3.8	.3	No
43	6.7	1.3	.1	No
44	2.0	6.3	9.9	Yes
45	.1	12.0	3.2	No
46	6.4	1.3	6.2	No
47	4.0	2.4	13.5	Yes
48	10.0	5.3	.1	No
49	8.0	14.4	1.9	Yes
50	7.0	10.0	2.0	No

department should spend developing price quotes. To help develop these guidelines, 50 requests for price quotes were randomly selected from the set of all price quotes made last year; the processing time was determined for each price quote for each department. These times are displayed in the table. The price quotes are also classified by whether they were "lost" (i.e., whether or not the customer placed an order after receiving the price quote).

a. Stem-and-leaf displays for each of the departments and for the total processing time were produced using Minitab. Note that very high processing times that are "disconnected" from the other times are shown in a list under the heading of "HI" at the bottom of the display. Also, note that the units of the leaves for the total processing time are units (1.0), while the leaf units for each of the three components are tenths (.1). Shade the leaves that correspond to "lost" orders in each of the displays, and interpret each of the displays.

```
Stem-and-leaf of Mkt
Leaf Unit = 0.10

   6     0 112446
   7     1 3
  14     2 0024699
  16     3 25
  22     4 001577
 (10)    5 0344556889
  18     6 0002224799
   8     7 0038
   4     8 07
   2     9
   2    10 0
   1    11 0
```

```
Stem-and-leaf of Engr
Leaf Unit = 0.10

   7     0 4466699
  14     1 3333788
  19     2 12246
  23     3 1568
  (5)    4 24688
  22     5 233
  19     6 01239
  14     7 22379
   9     8
   9     9 66
   7    10 0
   6    11 3
   5    12 023
   2    13 0
   1    14 4
```

```
Stem-and-leaf of Accnt
Leaf Unit = 0.10

  19     0 1111111111122333444
  (8)    0 55556888
  23     1 00
  21     1 79
  19     2 0023
  15     2
  15     3 23
  13     3 78
  11     4
  11     4
  11     5
  11     5 8
  10     6 2
   9     6
   9     7 0
   8     7
   8     8 4

     HI   99, 105, 135, 144,
          182, 220, 300
```

```
Stem-and-leaf of Total
Leaf Unit = 1.0

   1     0 1
   3     0 33
   5     0 45
  11     0 666677
  17     0 888999
  21     1 0000
  (5)    1 33333
  24     1 4444445555
  14     1 6677
  10     1 8999
   6     2 0
   5     2 3
   4     2 44

     HI   30, 36
```

b. Using your results from part **a**, develop "maximum processing time" guidelines for each department that, if followed, will help the firm reduce the number of lost orders.

2.108 Refer to Exercise 2.107. The means and standard deviations for the processing times are given here.

	Marketing	*Engineering*	*Accounting*	*Total*
$\bar{x}$	4.77	5.04	3.65	13.46
s	2.58	3.84	6.26	6.82

a. Calculate the z-score corresponding to the maximum processing time guideline you developed in Exercise 2.107 for each department, and for the total processing time.
b. Calculate the maximum processing time corresponding to a z-score of 3 for each of the departments. What percentage of the orders exceed these guidelines? How does this agree with Chebyshev's theorem and the Empirical Rule?
c. Repeat part **b** using a z-score of 2.
d. Compare the percentage of "lost" quotes with corresponding times that exceed at least one of the guidelines in part **b** to the same percentage using the guidelines in part **c**. Which set of guidelines would you recommend be adopted? Why?

2.109 What is Pareto analysis?

2.110 A company has roughly the same number of people in each of five departments: Production, Sales, R&D, Maintenance, and Administration. The following table lists the number and type of major injuries that occurred in each department last year.

Type of Injury	*Department*	*Number of Injuries*	*Type of Injury*	*Department*	*Number of Injuries*
Burn	Production	3	Cuts	Production	4
	Maintenance	6		Sales	1
Back strain	Production	2		R&D	1
	Sales	1		Maintenance	10
	R&D	1	Broken arm	Production	2
	Maintenance	5		Maintenance	2
	Administration	2	Broken leg	Sales	1
Eye damage	Production	1		Maintenance	1
	Maintenance	2	Broken finger	Administration	1
	Administration	1	Concussion	Maintenance	3
Deafness	Production	1		Administration	1
			Hearing loss	Maintenance	2

a. Construct a Pareto diagram to identify which department or departments have the worst safety record.
b. Explode the Pareto diagram of part **a** to identify the most prevalent type of injury in the department with the worst safety record.

2.111 One hundred management trainees were given an examination in basic accounting. Their test scores were found to have a mean and variance of 75 and 36, respectively.

a. Make a statement about the percentage of the test scores that would be expected to fall between 69 and 81.

b. If a grade of 63 was required to pass the test, make a statement about the percentage of the trainees who would be expected to fail.

2.112 Redo Exercise 2.111 assuming that the distribution of test scores was determined to be mound-shaped.

2.113 A national chain of automobile oil-change franchises claims that "your hood will be open for less than 12 minutes when we service your car." To check their claim, an undercover consumer reporter from a local television station monitored the "hood time" of 25 consecutive customers at one of the chain's franchises. The resulting data follow. Construct a time series plot for these data and describe in words what it reveals.

Customer Number	*Hood Open (Minutes)*	*Customer Number*	*Hood Open (Minutes)*
1	11.50	14	12.50
2	13.50	15	13.75
3	12.25	16	12.00
4	15.00	17	11.50
5	14.50	18	14.25
6	13.75	19	15.50
7	14.00	20	13.00
8	11.00	21	18.25
9	12.75	22	11.75
10	11.50	23	12.50
11	11.00	24	11.25
12	13.00	25	14.75
13	16.25		

2.114 The vice president in charge of sales for the conglomerate you work for has asked you to evaluate the sales records of two of the firm's divisions. You note that the range of monthly sales for division A over the last 2 years is \$50,000 and the range for division B is only \$30,000. You compute each division's mean monthly sales for the same time period and discover that both divisions have a mean of \$110,000. Assume that is all the information you have about the division's sales records. Would you be willing to say which of the divisions has a more consistent sales record? Why or why not?

2.115 Refer to Exercise 2.114.

a. Estimate the standard deviation of the monthly sales distribution of division B.

b. Based on your estimate in part **a**, would you say it is more likely that division B's sales next month will be over \$120,000 or under \$90,000?

c. Is it possible for division B's sales next month to be over \$160,000? Explain.

2.116 The Age Discrimination in Employment Act mandates that workers 40 years of age or older be treated without regard to age in all phases of employment (hiring, promotions, firing, etc.). Age discrimination cases are of two types: *disparate treatment* and *disparate impact*. In the former, the issue is whether workers have been intentionally discriminated against. In the latter, the issue is whether employment practices adversely affect the protected class (i.e., workers 40 and over) even though no such effect was intended by the employer (Zabell, 1989). During the recession of the early 1990s, a small computer manufacturer laid off 10 of its 20 software engineers. The ages of all the engineers at the time of the lay-off are shown in the table.

Not Laid off:	34	55	42	38	42	32	40	40	46	29
Laid off:	52	35	40	41	40	39	40	64	47	44

a. Find the standard deviation of the 20 ages. Describe how the standard deviation can be used to characterize the set of ages.
b. If a software engineer were selected at random from these 20 to be laid off, what is the probability that the engineer would be 40 or older? [*Note:* We will formally define "probability" in the next chapter.]
c. Find the median age of all 20 engineers and the median age of the 10 who were not laid off.
d. Given your answers to parts **b** and **c** (and any other evidence you care to use), does it appear that the company may be vulnerable to a disparate impact claim? Explain.

2.117 Suppose you used the following formula as a measure of the variability (V) of a data set:

$$V = \frac{\sum_{i=1}^{n} (x_i - \bar{x})}{n}$$

What information can be learned about the variability of a data set using this formula? Using the data in part **a** of Exercise 2.87, find V.

2.118 Many firms use on-the-job training to teach their employees computer programming. Suppose you work in the personnel department of a firm that just finished training a group of its employees to program, and you have been requested to review the performance of one of the trainees on the final test that was given to all trainees. The mean and standard deviation of the test scores are 80 and 5, respectively, and the distribution of scores is mound-shaped.
a. The employee in question scored 65 on the final test. Compute the employee's z-score.
b. Approximately what percentage of the trainees will have z-scores equal to or less than the employee of part **a**?
c. If a trainee were randomly selected from those who had taken the final test, is it more likely that he or she would score 90 or above, or 65 or below?

***2.119** A company that bags and sells wild rice has received numerous complaints from its customers about underfilled bags. Most complaints concern their 5-pound bag. The operations manager suspects that the problem may be due to differences between fill-machine operators. To investigate her suspicion, the manager weighed the last 10 bags filled by machine operator 1 before operator 2 took over at 3:00 P.M. She also weighed the first 10 bags produced by operator 2. The data follow.

Bag Number	*Operator Number*	*Bag Weight (Pounds)*	*Bag Number*	*Operator Number*	*Bag Weight (Pounds)*
1	1	5.00	11	2	5.35
2	1	5.10	12	2	4.70
3	1	4.90	13	2	5.05
4	1	5.05	14	2	5.30
5	1	5.05	15	2	4.50
6	1	4.95	16	2	5.40
7	1	5.00	17	2	5.20
8	1	4.90	18	2	5.00
9	1	5.10	19	2	5.20
10	1	5.00	20	2	4.60

a. Construct a time series plot for these data.

b. What do these data reveal about differences in the performances of the two operators?
c. Could such differences account for the customer complaints? Explain.
d. Describe any other filling problems revealed by the plot.
e. Suggest a better plan for selecting the bags to be weighed.

2.120 The advertising expenditures (in thousands of dollars) by media for a recent year are described for the 16 top-selling brandies and cordials by the following Minitab-generated box plots.

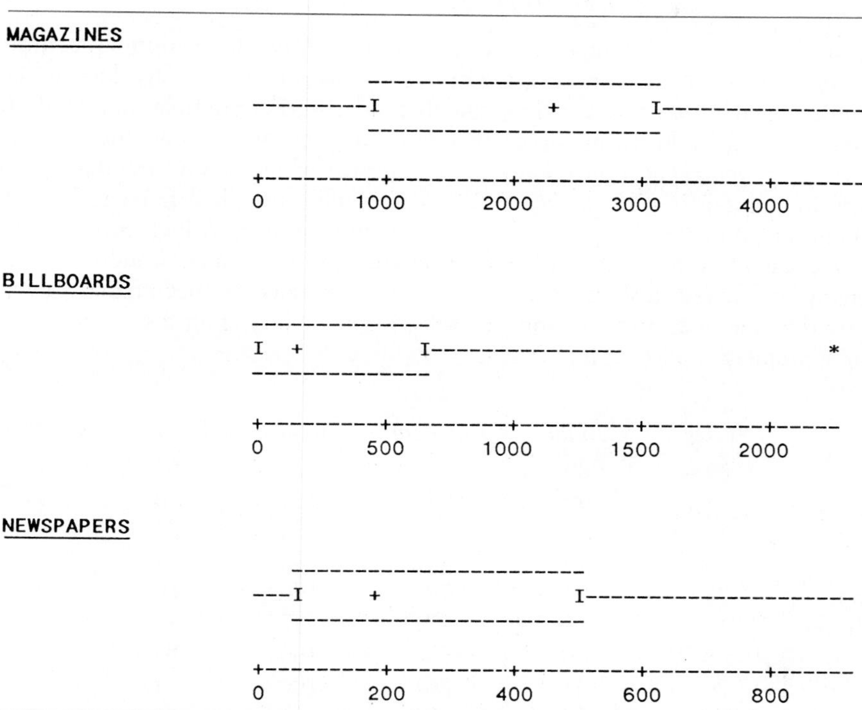

a. For which medium are median expenditures the highest?
b. For which medium is the range of expenditures the largest? The smallest?
c. For which medium is the interquartile range the largest? The smallest?
d. Describe the shape of each data set's frequency distribution.
e. Do any of the media receive advertising expenditures from all 16 brands of brandies and cordials? Explain.

2.121 Chebyshev's theorem (presented in Table 2.10) states that at least $(1 - 1/k^2)$ of a set of measurements will lie within k standard deviations of the mean of the data set. Use Chebyshev's theorem to state the fraction of a set of measurements that will lie:
a. Within 2 standard deviations of the mean μ ($k = 2$).
b. Within 3 standard deviations of the mean.
c. Within 1.5 standard deviations of the mean.
d. More than 2.75 standard deviations of the mean.

**2.122 Calculate the mean and standard deviation of the following data sets:

a.

Class	Class Frequency
10–19	15
20–29	12
30–39	8
40–49	5
50–59	2

b.

Class	Class Frequency
\$ −99 to \$ −50	20
−49 to 0	55
1 to 50	102
51 to 100	63
101 to 150	18

2.123 Economic theory suggests that the vigor of competition in an industry is related to the number of firms in the industry. As a measure of competitiveness, however, the number of firms in an industry does not take into consideration the extent to which a few firms may dominate that industry. For example, in an industry with 100 firms, each may produce 1% of industry output, or three firms may dominate, producing 75% compared to the other 97 firms' 25%. The *market concentration ratio* is a measure of competitiveness that reflects such inequalities among firms in an industry. The market concentration ratio is usually defined as the percentage of total industry sales contributed by the largest few firms (usually three or four). A high concentration ratio indicates an industry dominated by a few firms. A low concentration ratio indicates an industry with much competition among many firms. The table contains 1970 three-firm market concentration ratios for 12 industries in each of six different countries. Answer the following questions using the methods from this chapter that you deem appropriate.

Industry	*United States*	*Canada*	*United Kingdom*	*Sweden*	*France*	*West Germany*
Brewing	39	89	47	70	63	17
Cigarettes	68	90	94	100	100	94
Fabric weaving	30	67	28	50	23	16
Paints	26	40	40	92	14	32
Petroleum refining	25	64	79	100	60	47
Shoes (except rubber)	17	18	17	37	13	20
Glass bottles	65	100	73	100	84	93
Cement	20	65	86	100	81	54
Ordinary steel	42	80	39	63	84	56
Antifriction bearings	43	89	82	100	80	90
Refrigerators	64	75	65	89	100	72
Storage batteries	54	73	75	100	94	82

Source: F. M. Scherer, Alan Beckenstein, Erich Kaufer, and R. D. Murphy. *The Economics of Multiplant Operation: An International Comparisons Study* (Cambridge, Mass.: Harvard University Press, 1975). Reprinted by permission.

a. For each of the six countries, characterize the magnitude and variability of the sample of 12 market concentration ratios.

b. For each of the 12 industries, characterize the magnitude and variability of the sample of six market concentration ratios.

c. Which nation has on average the most competition within its industries? The least competition?

d. Which are the three most competitive industries? The three least competitive industries?

2.124 The **pie chart** is an alternative graphic technique for describing qualitative data sets. Each category of the data set is assigned one slice of the pie, the size of which is proportional to the relative frequency of the category.

The sequence of pie charts portrays the evolution of the structure of the top 500 firms in the United States and the top 200 firms in the United Kingdom over the period 1950–1980. Describe the trends that are revealed by these pie charts.

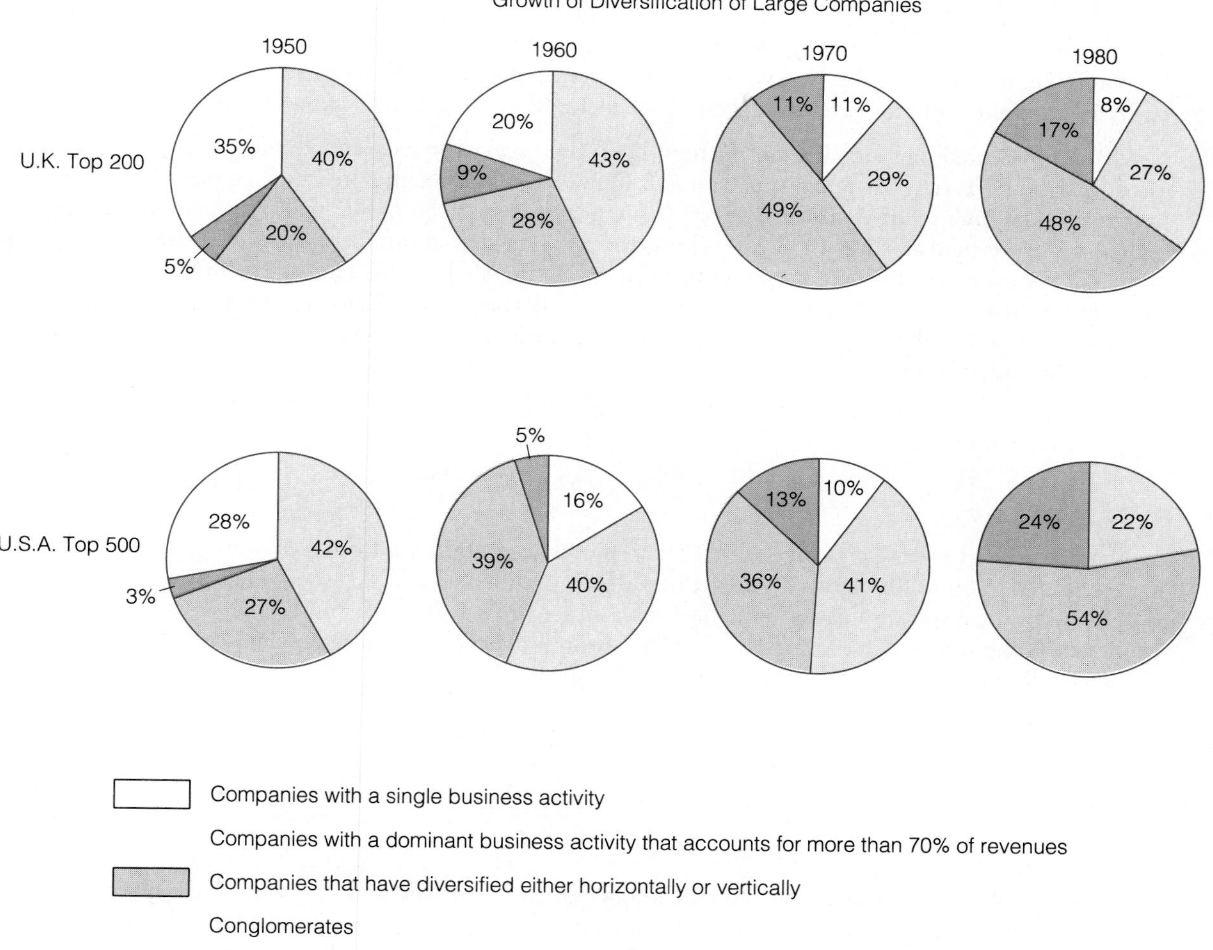

Source: Adapted from *Long Range Planning*. Vol 19, No 1, 1986, pp. 52-60. Reprinted with permission. Copyright 1986, Pergamon Press, Ltd.

***2.125** *Consumer Reports* is a magazine that contains ratings and reports for consumers on goods, services, health, and personal finances. It is published by Consumers Union, a nonprofit organization established in 1936. Consumers Union reported on the testing of 46 brands of toothpaste ("A Guide to Good Dental Care," 1992). Each was rated on: package design, flavor, cleaning ability, fluoride content, and cost per month (a cost estimate based on brushing with ½ inch of toothpaste twice daily). The data on page 122 are the costs per month for the 46 brands. Costs marked by an asterisk represent those brands that carry the American Dental Association (ADA) seal verifying effective decay prevention.

.58	.66	1.02	1.11	1.77	1.40	.73*	.53*	.57*	1.34
1.29	.89*	.49	.53*	.52	3.90	4.73	1.26	.71*	.55*
.59*	.97	.44*	.74*	.51*	.68*	.67	1.22	.39	.55
.62	.66*	1.07	.64	1.32*	1.77*	.80*	.79	.89*	.64
.81*	.79*	.44*	1.09	1.04	1.12				

a. Use a statistical software package to construct a stem-and-leaf display for the data.
b. Circle the individual leaves that represent those brands that carry the ADA seal.
c. What does the pattern of circles suggest about the costs of those brands approved by the ADA?

***2.126** Newly manufactured tires are rated for traction, heat, and tread wear according to the Uniform Tire Quality Grading System. Traction is rated using and A-B-C system to indicate the tire's braking performance on wet pavement. The tire's ability to dissipate heat is also measured using an A-B-C scale (with A representing the highest rating for both traction and heat). "Tread wear is graded on a number system with the rating for any particular tire expressed as a comparison of that tire with a tire rated at 100. Thus, if the tire tested wore twice as long as the baseline tire, it would carry a rating of 200" (Brand, 1991). Forty new tires, all of the same type and manufactured by the same company, were tested and each tire's tread-wear life in months was recorded. The data are listed below.

48	46	45	48	51	53	49	46	48	47
50	49	43	52	52	50	43	44	43	52
44	47	51	52	45	49	48	43	53	50
53	51	44	43	45	47	53	44	46	53

a. If a tire with a tread-wear rating of 100 lasts 48 months, compute the tread-wear ratings of each of the 40 tires tested. Round the ratings to the nearest whole number.
b. Use a statistical software package to construct a histogram for these 40 ratings.
c. Based on the shape of the histogram, is there reason for concern about the quality of the manufacturing process that is producing these particular tires? Explain.

On Your Own

A number of sources of business data are listed on the following page. Use these and any others you or your instructor can find to select two real business-oriented data sets, one qualitative and one quantitative. Describe both data sets using the graphic and numerical descriptive tools introduced in this chapter. Count the number of observations in your quantitative data set within 1, 2, and 3 standard deviations of the mean of the data and compare the counts with those you would expect using either Chebyshev's theorem or the Empirical Rule.

Be sure to select data of interest to you, since we will be referring to these data sets in **On Your Own** sections of later chapters.

Suggested Secondary Data Sources:*

Annual Reports to Shareholders.

Board of Governors of the Federal Reserve System. *Federal Reserve Bulletin* (monthly).

Business Week (magazine).

Dun & Bradstreet's *Million Dollar Directory* (yearly).

Economic Indicators. Council of Economic Advisors.

Forbes (magazine).

Fortune (magazine).

Handbook of Basic Economic Statistics. Economic Statistics Bureau of Washington, D.C.

Moody's Manuals (set of 8).

Simmons Study of Media and Markets. Simmons Market Research Bureau, Inc.

Sourcebook of Zip Code Demographics.

State Demographics: Population Profiles of the 50 States. Dow Jones-Irwin.

Survey of Buying Power Demographics USA.

Thomas Register of America Manufacturers.

U.S. Bureau of the Census. *Census of Manufacturers* (every 5 years).

U.S. Bureau of the Census. *County Census Patterns* (yearly).

U.S. Bureau of the Census. *Statistical Abstract of the United States* (yearly).

U.S. Department of Commerce, Office of Business Economics, *Business Statistics*.

U.S. Bureau of Commerce, Office of Business Economics. *Survey of Current Business* (monthly).

U.S. Department of Health and Human Services. *Vital Statistics of the United States*.

U.S. Department of Labor. *Monthly Labor Review*.

U.S. Department of Labor, Bureau of Labor Statistics. *Employment and Earnings*.

U.S. Department of Labor, Bureau of Labor Statistics. *National Survey of Professional Administrative, Technical, and Clerical Pay*.

Value Line Investment Survey, Value Line, Inc.

Wall Street Journal (daily).

Your state's statistical abstract.

Guides to Finding Secondary Data*

American Statistical Index: A Comprehensive Guide to the Statistical Publications of the United States Government.

Business Periodical Index.

Business Rankings and Salaries Index. Detroit: Gale Research Co.

Encyclopedia of Business Information Sources. Detroit: Gale Research Co.

Federal Statistical Directory: The Guide to Personnel and Data Sources.

Guide to Bureau of Economic Analysis.

Guide to Special Issues and Indexes of Periodicals.

Houser and Leonard. *Government Statistics for Business Use*.

New York Times Index.

Predicasts F&S Indexes.

Sourcebook of Global Statistics.

Statistical Reference Index: A Selective Guide to American Statistical Publications from Sources Other Than the U.S. Government. Congressional Information Services, Inc.

Statistics Sources, edited by Paul Wasserman. Detroit: Gale Research Co.

U.S. Bureau of the Census. *Directory of Non-federal Statistics for States and Local Areas*.

Wall Street Journal Index.

Your local Chamber of Commerce.

**Primary data* are data you (or someone in your organization) collect for the study at hand. *Secondary data* are data collected by someone outside the organization for purposes other than the study at hand.

Many of these sources and guides are now available on computer media, including CD-ROMs and online data services.

Using the Computer

We have supplied a set of data in Appendix C (also available on diskette from the publisher) that will be used as a source for the **Using the Computer** exercises at the end of most chapters. A complete description of the data can be found in Appendix C. Briefly, the data set includes such variables as population size, number of households, average household size, average income, percentage of college graduates, percentage of women in the work force, and purchasing-potential indexes for groceries, sporting goods, and home improvements for each of 1,000 U.S. zip codes.

a. Consider the percentage of women in the work force. Use a statistical software package to generate a stem-and-leaf display and a relative frequency histogram for these 1,000 percentages. Compare the graphical descriptions you obtain, and discuss what each reveals about the distribution of the percentage of women in the work force in the sample of 1,000 zip codes.

b. Repeat part **a** for one of the census regions, perhaps the one in which you currently reside. Compare the regional distribution to the national distribution you obtained in part **a**.

c. Finally, repeat part **a** for the zip codes corresponding to one state, perhaps the one in which you currently reside. Compare the state distribution to the national and regional distributions of parts **a** and **b**.

d. Now compute the mean and standard deviation of the same data set over the 1,000 zip codes and over the zip codes in the region you selected. Use the computer to count the number of zip codes' percentages within the intervals $\bar{x} \pm s$, $\bar{x} \pm 2s$, and $\bar{x} \pm 3s$. Compare the results with those given by Chebyshev's theorem and the Empirical Rule (Table 2.10).

References

Adler, Paul S. and Clark, Kim B. "Behind the learning curve: A sketch of the learning process." *Management Science*, Mar. 1991, p. 267.

"A guide to good dental care." *Consumer Reports*, Consumers Union, Sept. 1992, p. 601.

"A plain old '30-year mean' is forecaster's yardstick." *Star Tribune* (Minneapolis). Sept. 2, 1990.

Brand, Paul. "Tire's initial grade loses importance as tread depth starts to wear down." *Star Tribune* (Minneapolis). Sept. 22, 1991.

Chase, R. B. and Aquilano, N. J. *Production and Operations Management*, rev. ed. Homewood, Ill.: Richard D. Irwin, 1977.

Davidson, S., Stickney, C. P., and Weil, R. L. *Financial Accounting*, 2d ed. Chicago: Dryden Press, 1979.

Deming, W. E. *Out of the Crisis*. Cambridge, Mass. M.I.T. Center for Advanced Engineering Study, 1986.

Dodd, P. "Merger proposals, management discretion, and stockholder wealth." *Journal of Financial Economics*, 1980, 8, pp. 105–137.

Eger, C. E. "Corporate mergers: An empirical analysis of the role of risky debt." Unpublished Ph.D. dissertation. School of Management, University of Minnesota, 1982.

Fogarty, D. W., Blackstone, J. H., Jr., and Hoffmann, T. R. *Production and Inventory Management*. Cincinnati, Ohio: South-Western, 1991.

"The *Fortune* 500." *Fortune*, Apr. 23, 1990, pp. 337–396.

Gitlow, H., Gitlow, S., Oppenheim, A., and Oppenheim, R. *Tools and Methods for the Improvement of Quality*. Homewood, Ill.: Irwin, 1989.

How the Bond Market Works. The New York Institute of Finance. New York: 1988.

Huff, D. *How to Lie with Statistics*. New York: Norton, 1954.

Ishikawa, K. *Guide to Quality Control*, 2d ed. White Plains, N.Y.: Kraus International Publications, 1982.

Juran, J. M. *Juran on Planning for Quality*. New York: The Free Press, 1988.

Kane, V. E. *Defect Prevention*. New York: Marcel Dekker, Inc., 1989.

"Keeping up." *Fortune*, Aug. 13, 1979, p. 100.

McCullough, B. "Should you be a coupon clipper?" *Minneapolis Star and Tribune*, Mar. 17, 1986.

Mendenhall, W. *Introduction to Probability and Statistics*, 8th ed. Boston: Duxbury, 1991. Chapter 3.

Milkovich, G. T., Annoni, A. J., and Mahoney, T. A. "The use of the delphi procedures in manpower forecasting." *Management Science*, Dec. 1972, *19*, part 1, pp. 381–388.

Namias, J. "A method to detect specific causes of consumer complaints." *Journal of Marketing Research*, Aug. 1964, pp. 63–68.

Neter, J., Wasserman, W., and Whitmore, G. A. *Applied Statistics*, 2d ed. Boston: Allyn & Bacon, 1982. Chapter 3.

Postlewaite, S. "Salad bars sprout as fast-food battle turns from the burger." *Minneapolis Tribune*, July 3, 1983.

Roberts, Jonathan. "Microsoft speaks: Windows Version 3.1 succeeds by listening to users." *Cue News*. Egghead Software, June 1992, p. 3.

Rothstein, M. "Hotel overbooking as a Markovian sequential decision process." *Decision Sciences*, July 1974, *5*, pp. 389–405.

Schroeder, R. G. *Operations Management*, 4th ed. New York: McGraw-Hill, 1993.

Selazny, G. "Grappling with graphics." *Management Review*, Oct. 1975, p. 7.

Spiro, H. T. *Finance for the Nonfinancial Manager*. New York: Wiley, 1982. Chapter 17.

Sterling, R. R. and Radosevich, R. "A valuation experiment." *Journal of Accounting Research*, Spring 1969, pp. 90–95.

Willett, R. P. and Pennington, A. L. "Customer and salesman: The anatomy of choice and influence in a retail setting." *Science, Technology and Marketing*, Proceedings of the 1966 Fall Conference of the American Marketing Association, Raymond M. Haas (ed.), 1966, pp. 598–616.

U.S. Department of Labor. *The Consumer Price Index: Concepts and Content over the Years*. Bureau of Labor Statistics, Report 517, May 1978.

Zabell, S. L. "Statistical proof of employment discrimination." *Statistics: A Guide to the Unknown*, 3d ed. Pacific Grove, Calif.: Wadsworth, 1989.

CHAPTER THREE

Probability

Contents

Case Studies

Where We've Been

In Chapter 1, we identified inference from a sample to a population as one of the major goals of statistics. In Chapter 2, we learned how to describe a set of measurements using graphical and numerical descriptive methods.

Where We're Going

We now begin to consider the problem of making an inference. What permits us to make the inferential jump from sample to population and then to give a measure of reliability for the inference? As you will see, the answer is *probability*. This chapter is devoted to probability—what it is and some of the basic concepts of the theory that surrounds it.

You will recall that statistics is concerned with inferences about a population based on sample information. Understanding how this will be accomplished is easier if you understand the relationship between population and sample. This understanding is enhanced by reversing the statistical procedure of making inferences from sample to population. In this chapter we assume the population *known* and calculate the chances of obtaining various samples from the population. Thus, probability is the "reverse" of statistics: In probability we use the population information to infer the probable nature of the sample.

Probability plays an important role in inference-making. To illustrate, suppose you have an opportunity to invest in an oil exploration company. Past records show that for ten out of ten previous oil drillings (a sample of the company's experiences), all ten resulted in dry wells. What do you conclude? Do you think the chances are better than 50–50 that the company will hit a producing well? Should you invest in this company? We think your answer to these questions will be an emphatic no. If the company's exploratory prowess is sufficient to hit a producing well 50% of the time, a record of ten dry wells out of ten drilled is an event that is just too *improbable*. Do you agree?

As another illustration, suppose you are playing poker with what your opponents assure you is a well-shuffled deck of cards. In three consecutive five-card hands, the person on your right is dealt four aces. Based on this sample of three deals, do you think the cards are being adequately shuffled? Again, we think your answer will be no and that you will reach this conclusion because dealing three hands of four aces is just too improbable, assuming that the cards were properly shuffled.

Note that the conclusion concerning the potential success of the oil drilling company and the conclusion concerning the card shuffling were both based on probabilities—namely, the probabilities of certain sample results. Both situations were contrived so you could easily conclude that the probabilities of the sample results were small. Unfortunately, the probabilities of many observed sample results are not so easy to evaluate. For these cases, we will need the assistance of a theory of probability.

3.1 Events, Sample Spaces, and Probability

Most sets of data that are of interest to the business community are generated by some **experiment**.

Definition 3.1

An **experiment** is an act or process that leads to a single outcome that cannot be predicted with certainty.

Our definition of *experiment* is broader than that used in the physical sciences, where we might picture test tubes, microscopes, and other equipment. Examples of statistical experiments in business are recording whether a customer prefers one of two brands of coffee (say, brand A or brand B), measuring the change in the Dow Jones Industrial Average from one day to the next, recording the weekly sales of a business firm, and counting the number of errors on a page of an accountant's ledger.

A "single outcome" of an experiment is called a **simple event**.

Definition 3.2

A **simple event** is an outcome of an experiment that cannot be decomposed into a simpler outcome.

In Table 3.1 we present three examples of experiments and their simple events. We begin with simple coin and dice examples because they are most likely to be familiar to you. Experiment **a** in Table 3.1 is to toss a coin and observe the up face. You will undoubtedly agree that the most basic possible outcomes of this experiment are Observe a head and Observe a tail. Experiment **b** in Table 3.1 is to toss a die and observe the up face. Notice that the simple events Observe a 1, Observe a 2, etc., cannot be decomposed into simpler outcomes. However, the outcome Observe an even number can be decomposed into Observe a 2, Observe a 4, and Observe a 6. Thus, Observe an even number is not a simple event. The reasoning is similar for experiment **c** in Table 3.1.

TABLE 3.1 Experiments and Their Simple Events

a.	Experiment:	Toss a coin and observe the up face.
	Simple events:	1. Observe a head
		2. Observe a tail
b.	Experiment:	Toss a die and observe the up face.
	Simple events:	1. Observe a 1
		2. Observe a 2
		3. Observe a 3
		4. Observe a 4
		5. Observe a 5
		6. Observe a 6
c.	Experiment:	Toss two coins and observe the up faces.
	Simple events:	1. Observe H_1, H_2
		2. Observe H_1, T_2
		3. Observe T_1, H_2
		4. Observe T_1, T_2

(where H_1 means head on coin 1, H_2 means head on coin 2, etc.)

Outcomes such as Observe an even number, which can be decomposed into simpler outcomes (Observe a 2, Observe a 4, Observe a 6), are called **events**.

> **Definition 3.3**
>
> An **event** is a collection of one or more simple events.

Thus, in experiment **b** of Table 3.1, three examples of events are Observe an even number, Observe a number less than 4, and Observe a 6. Note that the last event, Observe a 6, cannot be decomposed into a simpler outcome and so is also a simple event. If the experiment is counting the number of errors on a page of an accountant's ledger, three examples of events are Observe no errors, Observe fewer than five errors, and Observe more than ten errors. Only the first, Observe no errors, is also a simple event. Our goal is to be able to calculate the probability that a particular event will occur when an experiment is performed.

The first step in achieving this goal is to note that simple events have an important property. If the experiment is conducted once, you will observe one and only one simple event. For example, if the experiment is to toss a coin and observe the up face, you cannot Observe a tail and Observe a head on the same toss. Or, if you toss a die and Observe a 2, you cannot Observe a 6 on the same toss. That is, for a single performance of an experiment, one and only one of the simple events will occur. To see that this property does not hold for all events, consider the event Observe an even number. It is possible to Observe an even number and Observe a 2 on the same toss of a die.

The collection of all the simple events of an experiment is called the **sample space**.

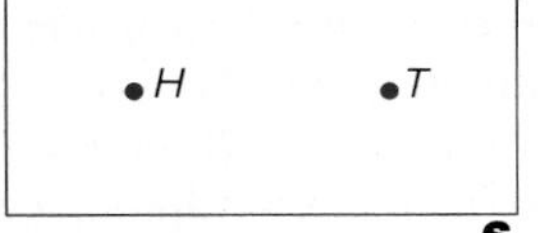

a. Experiment: Observe the up face on a coin

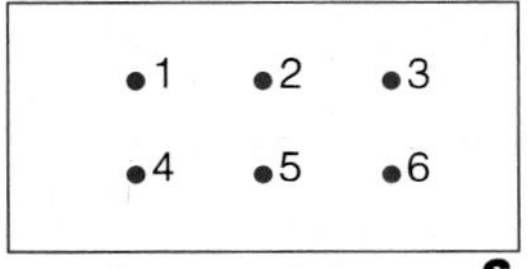

b. Experiment: Observe the up face on a die

•H_1, H_2 •H_1, T_2

•T_1, H_2 •T_1, T_2

S

c. Experiment: Observe the up faces on two coins

FIGURE 3.1 ▲
Venn diagrams for the three experiments from Table 3.1

> **Definition 3.4**
>
> The **sample space** of an experiment is the collection of all its simple events.

For example, there are six simple events associated with experiment **b** in Table 3.1. These six simple events comprise the sample space for the experiment. Similarly, for experiment **c** in Table 3.1, there are four simple events in the sample space.

A graphical method called a **Venn diagram** is useful for presenting a sample space and its simple events. The sample space is shown as a closed figure, labeled S. This figure contains a set of points, called **sample points**, with each point representing a simple event. Figure 3.1 shows the Venn diagram for each of the three experiments in Table 3.1. Note that the number of sample points in a sample space S is equal to the number of simple events associated with the respective experiment: two for experiment **a**, six for experiment **b**, and four for experiment **c**.

Now that we have defined the terms *simple event* and *sample space*, we are prepared to define the **probabilities of simple events**. The probability of a simple event is a number that indicates the likelihood that the event will occur when the experiment is performed. This number is usually taken to be the relative frequency of the occur-

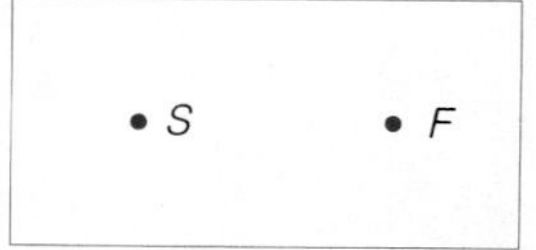

FIGURE 3.2 ▲
Experiment: Invest in a business venture and observe whether it succeeds (*S*) or fails (*F*)

rence of a simple event in a very long series of repetitions of the experiment. Or, when this information is not available, we select the number based on experience. For example, if we are assigning probabilities to the two simple events in the coin-toss experiment (Observe a head and Observe a tail), we might reason that if we toss a balanced coin a very large number of times, the simple events Observe a head and Observe a tail will occur with the same relative frequency of .5. Thus, the probability of each simple event is .5.

For some experiments we may assign probabilities to the simple events based on general information about the experiment. For example, if the experiment is to invest in a business venture and to observe whether it succeeds or fails, the sample space would appear as in Figure 3.2.

We are unlikely to be able to assign probabilities to the simple events of this experiment based on a long series of repetitions, since unique factors govern each performance of this kind of experiment. Instead, we may consider factors such as the personnel managing the venture, the general state of the economy at the time, the rate of success of similar ventures, and any other information deemed pertinent. If we finally decide that the venture has an 80% chance of succeeding, we assign a probability of .8 to the simple event Success. This probability can be interpreted as a measure of our degree of belief in the outcome of the business venture.

Such subjective probabilities should be based on expert information and must be carefully assessed. Otherwise, we run the risk of being misled by uninformed and/or biased probability statements. That is, we may be misled on any decisions based on these probabilities or based on any calculations in which they appear. We discuss the assessment of subjective probabilities in more detail in Case Study 19.1.*

No matter how you assign the probabilities to simple events, the probabilities assigned must obey two rules:

1. All simple event probabilities must lie between 0 and 1, inclusive.
2. The probabilities of all the simple events in the sample space must sum to 1.

Recall that an event is a collection of one or more simple events. Let us examine this statement in greater detail. Consider the die-tossing experiment and the event Observe an even number. This event will occur if and only if one of the three simple events, Observe a 2, Observe a 4, or Observe a 6, occurs. Consequently, you can think of the event Observe an even number as the collection of the three simple events, Observe a 2, Observe a 4, and Observe a 6. This event, which we will denote by the symbol A, can be represented in a Venn diagram by a closed figure inside the sample space S. The closed figure A will contain the simple events that constitute event A, as shown in Figure 3.3.

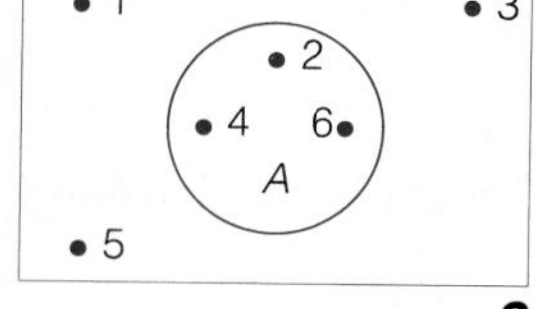

FIGURE 3.3 ▲
Die-toss experiment with event *A*: Observe an even number

How do you decide which simple events belong to the set associated with an event A? Test each simple event in the sample space S. If event A occurs when a particular simple event occurs, then that simple event is in the event A. For example, in the die-toss experiment, the event Observe an even number (event A) will occur if the

*For a text that deals in detail with subjective probabilities, see Winkler (1972) or Lindley (1985).

Bloom County Probabilities

The issue of whether probability should be defined as the relative frequency in a long series of repetitions of an experiment or as a subjective measure of belief is one that has been debated for many years by probabilists, statisticians, and even philosophers. A considerably lighter side of this debate is illustrated in the accompanying Bloom County comic strip.

simple event Observe a 2 occurs. By the same reasoning, the simple events Observe a 4 and Observe a 6 are in event A.

Now return to our original objective—finding the probability of an event. Consider the problem of finding the probability of observing an even number (event A) in the single toss of a die. You will recall that A will occur if one of the three simple events, toss a 2, 4, or 6, occurs. Since two or more simple events cannot occur at the same time, we can easily calculate the probability of event A by summing the probabilities of the three simple events. If we assume the die is fair (i.e., balanced), then each simple event is equally likely to occur. Accordingly, we would attach a probability equal to $1/6$ to each of the simple events, so the probability of observing an even number (event A), denoted by the symbol $P(\text{A})$, would be

$$\begin{aligned} P(\text{A}) &= P(\text{Observe a 2}) + P(\text{Observe a 4}) + P(\text{Observe a 6}) \\ &= 1/6 + 1/6 + 1/6 = 1/2 \end{aligned}$$

The previous example leads us to a general procedure for finding the probability of an event A.

> The probability of an event A is calculated by summing the probabilities of the simple events in A.

Thus, we can summarize the steps for calculating the probability of any event.

> **Steps for Calculating Probabilities of Events**
>
> 1. Define the experiment.
> 2. List the simple events.
> 3. Assign probabilities to the simple events.
> 4. Determine the collection of simple events contained in the event of interest.
> 5. Sum the simple event probabilities to obtain the event probability.

EXAMPLE 3.1

Consider the experiment of tossing two coins and observing the up faces. Assume both coins are fair.

a. List the simple events and assign them reasonable probabilities.
b. Consider the events

A: {Observe exactly one head} B: {Observe at least one head}

and calculate $P(A)$ and $P(B)$.

Solution

a. The simple events are

$$H_1, H_2 \qquad H_1, T_2 \qquad T_1, H_2 \qquad T_1, T_2$$

where H_1 denotes Observe a head on coin 1, H_2 denotes Observe a head on coin 2, etc. If both coins are fair, we can again use the concept of relative frequency in a long series of experimental repetitions to conclude that each simple event should be assigned a probability of 1/4.

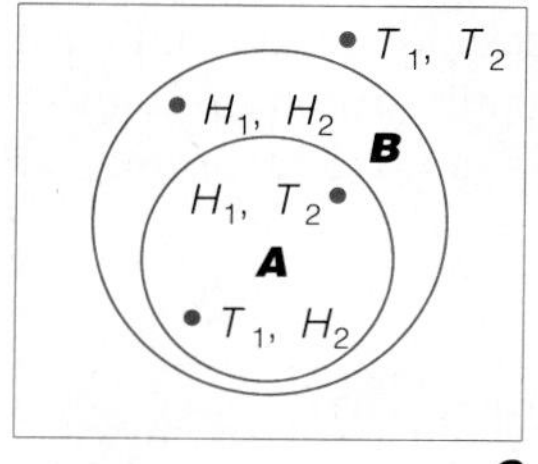

FIGURE 3.4 ▲
Venn diagram of the two-coin toss

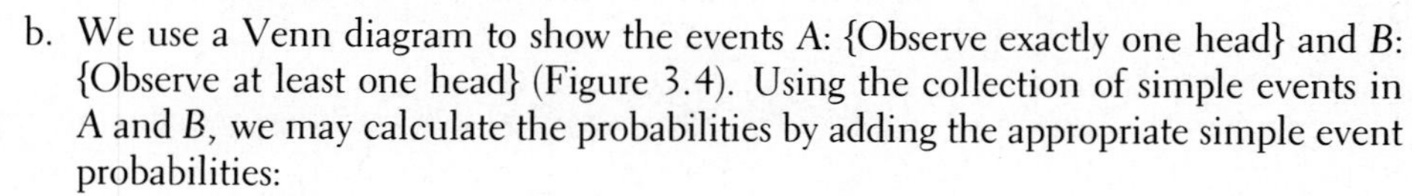

b. We use a Venn diagram to show the events A: {Observe exactly one head} and B: {Observe at least one head} (Figure 3.4). Using the collection of simple events in A and B, we may calculate the probabilities by adding the appropriate simple event probabilities:

$$P(A) = P(H_1, T_2) + P(T_1, H_2) = 1/4 + 1/4 = 1/2$$
$$P(B) = P(H_1, H_2) + P(H_1, T_2) + P(T_1, H_2) = 1/4 + 1/4 + 1/4 = 3/4$$

EXAMPLE 3.2

A retail computer store owner sells two basic types of microcomputers: IBM personal computers (IBM PCs) and IBM compatibles (PCs that run all or most of the same software as an IBM PC, but that are not manufactured by IBM). One problem facing the owner is deciding how many of each type of PC to stock. One important factor affecting the solution is the proportion of customers who purchase each type of PC. Show how this problem might be formulated in the framework of an experiment, with simple events and a sample space. Indicate how probabilities might be assigned to the simple events.

Solution

Using the term *customer* to refer to a person who purchases one of the two types of PCs, we can define the experiment as the entrance of a customer and the observation of which type of PC is purchased. There are two simple events in the sample space corresponding to this experiment:

Experiment: Observe the type of PC purchased by a customer.

Simple events: 1. *I*: {The customer purchases an IBM PC}
2. *C*: {The customer purchases an IBM compatible}

The difference between this experiment and the coin-toss experiment becomes apparent when we attempt to assign probabilities to the two simple events. What probability should we assign to the simple event *I*? If you answer .5, you are assuming that the events *I* and *C* should occur with equal likelihood, just as the simple events Heads and Tails in the coin-toss experiment. The assignment of simple event probabilities for the PC purchase experiment is not so easy. Suppose a check of the store's records indicates that 80% of its customers purchase IBM PCs. Then it might be reasonable to assign the probability of the simple event *I* as .8 and that of the simple event *C* as .2. The important points are that simple events are not always equally likely, and that the probabilities of simple events are not always easy to assign, particularly for experiments that represent real applications (as opposed to coin- and die-toss experiments).

EXAMPLE 3.3

A poll of "computer-familiar" adults who do not own a home computer was conducted by *USA Today* (Sept. 25, 1985). Each adult was asked to identify which of 10 electronic appliances was his or her highest-priority purchase, if any. The results are summarized in Table 3.2.

TABLE 3.2 Electronic Appliances of Highest Priority to Purchase by Computer-Familiar Adults

Electronic Appliance	Percent Response[a]	Electronic Appliance	Percent Response[a]
Home computer (HC)	24	Video cassette recorder (VCR)	17
Microwave oven (MO)	13	Video camera (VC)	9
Compact disc player (CDP)	4	Big-screen TV (BSTV)	7
Phone-answering machine (PAM)	6	Movie camera (MC)	3
Car telephone (CT)	5	None (N)	9
Programmable phone (PP)	3		

[a]Response percentages in the *USA Today* article did not add to 100% due to rounding. We have added 1% to the two smallest responses to facilitate the solution to this example.

Source: "Buying a computer is in our budget," *USA Today*, Sept. 25, 1985. Copyright 1985, USA Today. Excerpted with permission.

a. Define the experiment that generated the data in Table 3.2 and list the simple events.

b. Assign probabilities to the simple events.

c. What is the probability that a telephonic appliance is of highest priority?

d. What is the probability that a video appliance is of highest priority?

Solution

a. The experiment is the act of polling a computer-familiar adult. The simple events, the simplest outcomes of the experiment, are the 11 response categories listed in Table 3.2. They are shown in the Venn diagram in Figure 3.5.

FIGURE 3.5 ▶
Venn diagram for electronic appliance poll

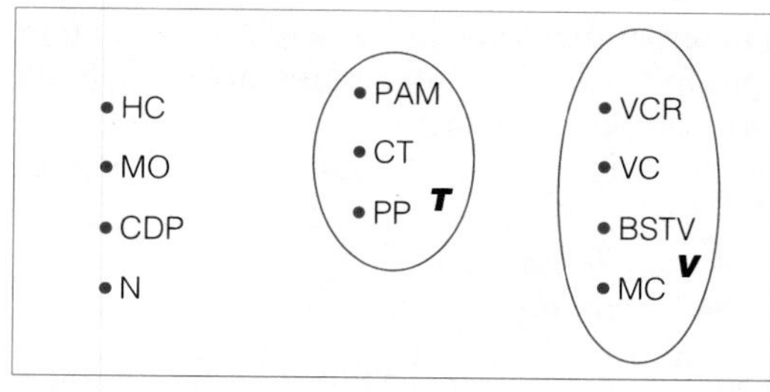

b. In Example 3.1, the simple events were assigned equal probabilities. If we were to assign equal probabilities in this case, each of the response categories would be assigned a probability of 1/11, or .09. However, you can see by examining Table 3.2 that equal probabilities are not reasonable in this case, because the response percentages were not the same in the 11 classifications. Instead, we assign a probability equal to the actual response percentage in each class, as shown in Table 3.3.

TABLE 3.3 Simple Event Probabilities for Electronic Appliance Poll

Simple Event	Probability
HC	.24
MO	.13
CDP	.04
PAM	.06
CT	.05
PP	.03
VCR	.17
VC	.09
BSTV	.07
MC	.03
N	.09

c. The event T that a telephonic appliance is the highest priority is not a simple event, because it consists of more than one of the response classifications (the simple events). In fact, as shown in Figure 3.5, T consists of three simple events. The probability of T is defined to be the sum of the probabilities of the simple events in T:

$$P(T) = P(\text{PAM}) + P(\text{CT}) + P(\text{PP}) = .06 + .05 + .03 = .14$$

d. The event V that a video appliance is identified as the highest priority purchase consists of four simple events, and the probability of V is the sum of the corresponding simple event probabilities:

$$P(\text{V}) = P(\text{VCR}) + P(\text{VC}) + P(\text{BSTV}) + P(\text{MC})$$
$$= .17 + .09 + .07 + .03 = .36$$

EXAMPLE 3.4

You have the capital to invest in two of four ventures, each of which requires approximately the same amount of investment capital. Unknown to you, two of the investments will eventually fail and two will be successful. You research the four ventures because you think that your research will increase your probability of a successful choice over a purely random selection, and you eventually decide on two. What is the lower limit of your probability of selecting the two best out of four? That is, if you used none of the information generated by your research, and selected two ventures at random, what is the probability that you would select the two successful ventures? At least one?

Solution

Denote the two successful enterprises as S_1 and S_2 and the two failing enterprises as F_1 and F_2. The experiment involves a random selection of two out of the four ventures, and each possible pair of ventures represents a simple event. The six simple events that make up the sample space are

1. S_1, S_2 2. S_1, F_1
3. S_1, F_2 4. S_2, F_1
5. S_2, F_2 6. F_1, F_2

The next step is to assign probabilities to the simple events. If we assume that the choice of any one pair is as likely as any other, then the probability of each simple event is 1/6. Now check to see which simple events result in the choice of two successful ventures. Only one such simple event exists—namely, S_1, S_2. Therefore, the probability of choosing two successful ventures out of the four is

$$P(S_1, S_2) = 1/6$$

The event of selecting at least one of the two successful ventures includes all the simple events except F_1, F_2.

$$\begin{aligned} &P(\text{Select at least one success}) \\ &\quad = P(S_1, S_2) + P(S_1, F_1) + P(S_1, F_2) + P(S_2, F_1) + P(S_2, F_2) \\ &\quad = 1/6 + 1/6 + 1/6 + 1/6 + 1/6 = 5/6 \end{aligned}$$

Therefore, the worst that you could do in selecting two ventures out of four may not be too bad. With a random selection, the probability of selecting two successful ventures will be at least 1/6 and the probability of selecting at least one successful venture out of two is at least 5/6.

The preceding examples have one thing in common: The number of simple events in each of the sample spaces was small; hence, the simple events were easy to identify and list. How can we manage this when the simple events run into the thousands or millions? For example, suppose you wish to select five business ventures from a group of 1,000. Then each different group of five ventures would represent a simple event. How can you determine the number of simple events associated with this experiment?

One method of determining the number of simple events for a complex experiment is to develop a counting system. Start by examining a simple version of the experiment. For example, see if you can develop a system for counting the number of ways to select two people from a total of four (this is exactly what was done in Example 3.4). If the ventures are represented by the symbols V_1, V_2, V_3, and V_4, the simple events could be listed in the following pattern:

V_1, V_2 $\quad$ V_2, V_3 $\quad$ V_3, V_4
V_1, V_3 $\quad$ V_2, V_4
V_1, V_4

Note the pattern and now try a more complex situation—say, sampling three ventures out of five. List the simple events and observe the pattern. Finally, see if you can deduce the pattern for the general case. Perhaps you can program a computer to produce the matching and counting for the number of samples of 5 selected from a total of 1,000.

A second method of determining the number of simple events for an experiment is to use **combinatorial mathematics**. This branch of mathematics is concerned with developing counting rules for given situations. For example, there is a simple rule for finding the number of different samples of five ventures selected from 1,000. This rule is given by the formula

$$\binom{N}{n} = \frac{N!}{n!(N - n)!}$$

where N is the number of elements in the population; n is the number of elements in the sample; and the factorial symbol (!) means that, say,

$$n! = n(n - 1)(n - 2) \cdot \cdots \cdot 3 \cdot 2 \cdot 1$$

Thus, $5! = 5 \cdot 4 \cdot 3 \cdot 2 \cdot 1$. (The quantity 0! is defined to be equal to 1.)

EXAMPLE 3.5

Refer to Example 3.4, in which we selected two ventures from four in which to invest. Use the combinatorial counting rule to determine how many different selections can be made.

Solution

For this example, $N = 4$, $n = 2$, and

$$\binom{4}{2} = \frac{4!}{2!2!} = \frac{4 \cdot 3 \cdot 2 \cdot 1}{(2 \cdot 1)(2 \cdot 1)} = 6$$

You can see that this agrees with the number of simple events obtained in Example 3.4.

EXAMPLE 3.6

Suppose you plan to invest equal amounts of money in each of five business ventures. If you have 20 ventures from which to make the selection, how many different samples of five ventures can be selected from the 20?

Solution

For this example, $N = 20$ and $n = 5$. Then the number of different samples of 5 that can be selected from the 20 ventures is

$$\binom{20}{5} = \frac{20!}{5!(20-5)!} = \frac{20!}{5!15!}$$

$$= \frac{20 \cdot 19 \cdot 18 \cdot \cdots \cdot 3 \cdot 2 \cdot 1}{(5 \cdot 4 \cdot 3 \cdot 2 \cdot 1)(15 \cdot 14 \cdot 13 \cdot \cdots \cdot 3 \cdot 2 \cdot 1)} = 15{,}504$$

The symbol $\binom{N}{n}$, meaning the **number of combinations of N elements taken n at a time**, is just one of a large number of counting rules that have been developed by combinatorial mathematicians. This counting rule applies to situations in which the experiment calls for selecting n elements from a total of N elements, without replacing each element before the next is selected. If you are interested in learning other methods for counting simple events for various types of experiments, you will find a few of the basic counting rules in Appendix A. Others can be found in the references listed at the end of this chapter.

Exercises 3.1–3.15

Learning the Mechanics

3.1 What is the difference between a *simple event* and an *event*?

3.2 The diagram describes the sample space of a particular experiment and events A and B.

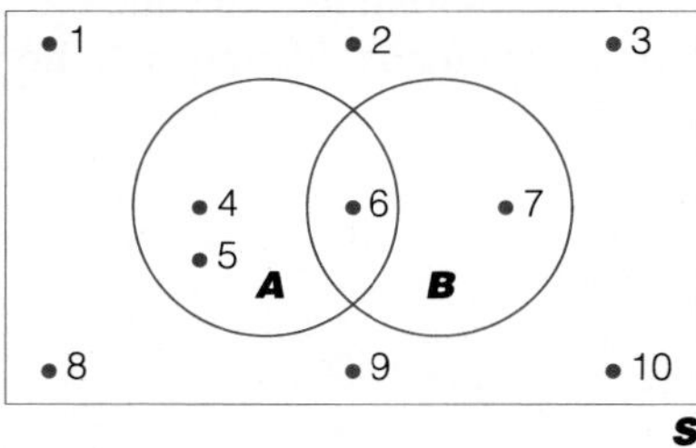

a. What is this type of diagram called?
b. Suppose the simple events are equally likely. Find $P(A)$ and $P(B)$.

c. Suppose $P(1) = P(2) = P(3) = P(4) = P(5) = 1/20$ and $P(6) = P(7) = P(8) = P(9) = P(10) = 3/20$. Find $P(A)$ and $P(B)$.

3.3 An experiment results in one of the following simple events: E_1, E_2, E_3, E_4, and E_5.
a. Find $P(E_3)$ if $P(E_1) = .1$, $P(E_2) = .2$, $P(E_4) = .3$, and $P(E_5) = .1$.
b. Find $P(E_3)$ if $P(E_1) = P(E_3)$, $P(E_2) = .1$, $P(E_4) = .2$, and $P(E_5) = .2$.
c. Find $P(E_3)$ if $P(E_1) = P(E_2) = P(E_4) = P(E_5) = .1$.

3.4 Compute each of the following:

a. $\binom{9}{4}$ b. $\binom{7}{2}$ c. $\binom{4}{4}$ d. $\binom{5}{0}$ e. $\binom{6}{5}$

Applying the Concepts

3.5 The population of the amounts of 500 automobile loans made by a bank last year can be described as shown in the table. For the purpose of checking the accuracy of the bank's records, an auditor will randomly select one of these loans for inspection (i.e., each loan has an equal probability of being selected).

Amount of Loan ($)			
Under 2,000	*2,000–4,999*	*5,000–7,999*	*8,000 or more*
35	73	300	92

a. List the simple events in this experiment.
b. What is the probability that the loan selected will be for $8,000 or more?
c. What is the probability that the loan will be for less than $5,000?

3.6 Communications products (telephones, fax machines, etc.) can be designed to operate on either an analog or a digital system. Because of improved accuracy, it is likely that a digital signal will soon replace the current analog signal used in telephone lines. The result will be a flood of new digital products for consumers to choose from (Kozlov, 1992). Suppose a particular firm plans to produce a new fax machine in both analog and digital forms. Concerned with whether the products will succeed or fail in the marketplace, a market analysis is conducted that results in the simple events and associated probabilities of occurrence listed in the table (S_a: analog succeeds, F_a: analog fails, etc.). Find the probability of each of the following events:

A: {Both new products are successful}
B: {The analog design is successful}
C: {The digital design is successful}
D: {At least one of the two products is successful}

Simple Events	*Probabilities*
S_aS_d	.31
S_aF_d	.10
F_aS_d	.50
F_aF_d	.09

3.7 Of six cars produced at a particular factory between 8 and 10 A.M. last Monday morning, three are known to be "lemons." Three of the six cars were shipped to dealer A and the other three to dealer B. Just by chance, dealer A received all three lemons. What is the probability of this event occurring if, in fact, the three cars shipped to dealer A were selected at random from the six produced?

3.8 A buyer for a large metropolitan department store must choose two firms from the four available to supply the store's fall line of men's pants. The buyer has not dealt with any of the four firms before and considers their products equally suitable. Unknown to the buyer, two of the four firms are having serious financial problems that may result in their not being able to deliver the pants as soon as promised. The firms are identified as G_1 and G_2 (firms in good financial condition) and P_1 and P_2 (firms in poor financial condition).

a. List each simple event (i.e., list each possible pair of firms that could be selected by the buyer).

b. Assume the probability of selecting a particular pair from among the four firms is the same for each pair. What is that probability?

c. Find the probability of each of the following events:

A: {Buyer selects two firms in poor financial condition}

B: {Buyer selects at least one firm in poor financial condition}

3.9 Simulate the experiment in Exercise 3.8 by marking four poker chips (or cards), one corresponding to each of the four firms. Mix the chips, randomly draw two, and record the results. Replace the chips. Now repeat the experiment a large number of times (at least 100).

a. Calculate the proportion of times event A occurs. How does this proportion compare with $P(A)$? Should the proportion equal $P(A)$? Explain.

b. Calculate the proportion of times event B occurs and compare it with $P(B)$.

3.10 The Value Line Survey, a service for common stock investors, provides its subscribers with up-to-date evaluations of the prospects and risks associated with the purchase of a large number of common stocks. Each stock is ranked 1 (highest) to 5 (lowest) according to Value Line's estimate of the stock's potential for price appreciation during the next 12 months.

Suppose you plan to purchase stock in three electrical utility companies from among seven that possess rankings of 2 for price appreciation. Unknown to you, two of the companies will experience serious difficulties with their nuclear facilities during the coming year. If you randomly select the three companies from among the seven, what is the probability that you select:

a. None of the companies with prospective nuclear difficulties?

b. One of the companies with prospective nuclear difficulties?

c. Both of the companies with prospective nuclear difficulties?

3.11 Approximately 77 million Visa and 60 million Mastercard credit cards had been issued in the United States by the end of 1984. Both were issued by thousands of banks, including Citicorp. Citicorp also issued competing cards of its own: Diners Club, Carte Blanche, and Choice are wholly owned by Citicorp. The accompanying table describes the population of credit cards issued by Citicorp. One Citicorp credit card customer is to be selected at random and the type of credit card will be recorded. (Assume that each customer has only one Citicorp-issued card.)

Credit Card	*Number Issued (in millions)*	*Credit Card*	*Number Issued (in millions)*
Visa and Mastercard	6.0	Carte Blanche	.3
Diners Club	2.2	Choice	1.0

Source: *Fortune*, Feb. 4, 1985, p. 21.

a. List the simple events in this experiment.
b. Find the probability of each simple event.
c. What is the probability that the customer selected uses one of Citicorp's wholly owned credit cards?

3.12 You are a lawyer for a client who has committed a felony, and there are seven judges who could hear your motion to set bail. Four judges are strict, and the other three are lenient. As you walk into the courtroom, judge A (a strict judge) is leaving to go home.
a. What is the probability of drawing a lenient judge for your client?
b. What is the probability of drawing a lenient judge for your client if the probability of getting judge B (a strict judge) is .3, the probability of getting judge C (a strict judge) is .4, and the probabilities of getting any of the other four judges are equally likely?
c. Suppose you know that judge D (a lenient judge) never follows judge A. What is the probability of drawing a lenient judge for your client if the probabilities in part **b** are valid?

3.13 *Sustainable development* or *sustainable farming* means "finding ways to live and work the Earth without jeopardizing the future" (Schmickle, 1992). Studies were concluded in five midwestern states to develop a profile of a sustainable farmer. Study results revealed that farmers can be classified along a sustainability scale, depending on whether they are likely or unlikely to engage in the following practices: (1) Raise a broad mix of crops; (2) Raise livestock; (3) Use chemicals sparingly; (4) Use techniques for regenerating the soil, such as crop rotation.
a. List the different sets of classifications that are possible.
b. Suppose you are planning to interview farmers across the country to determine the frequency with which they fall into the classification sets you listed for part **a**. Since no information is yet available, assume initially that there is an equal chance of a farmer falling into any single classification set. Using that assumption, what is the probability that a farmer will be classified as unlikely on all four criteria (i.e., classified as a non-sustainable farmer)?
c. Using the same assumption as in part **b**, what is the probability that a farmer will be classified as likely on at least three of the criteria (i.e., classified as a near-sustainable farmer)?

3.14 J. D. Power, a market-research firm in Agoura Hills, California, annually surveys thousands of U.S. motor vehicle owners to determine which automobiles and trucks the consumers believe to be the best based on such criteria as repair frequency, mileage, and dealer etiquette. The top three automobiles for 1991 were (in no particular order): Infiniti, Saturn sedan, and Lexus ("GM's Saturn Takes 3rd in New-Car Owner Poll," 1992).
a. List all possible sets of rankings for these three automobiles.
b. Assuming that each set of rankings in part **a** is equally likely, what is the probability that consumers ranked Saturn first? That consumers ranked Saturn last? That consumers ranked Lexus first and Infiniti second (which is, in fact, what they did)?

3.15 Probabilities are often expressed in terms of **odds**, especially in gambling settings. For example, handicappers for horse races express their belief about the probability of each horse winning a race in terms of odds. If the probability of event E is $P(E)$, then the **odds in favor of E** are $P(E)$ to $[1 - P(E)]$. Thus, if a handicapper assesses a probability of .25 that Snow Chief will win the Belmont Stakes, the odds in favor of Snow Chief are $25/100$ to $75/100$, or 1 to 3. It follows that the **odds against E** are $[1 - P(E)]$ to $P(E)$, or 3 to 1 against a win by Snow Chief. In general, if the odds in favor of event E are a to b, then $P(E) = a/(a + b)$.
a. A second handicapper assesses the probability of a win by Snow Chief to be $1/3$. According to the second handicapper, what are the odds in favor of a Snow Chief win?
b. A third handicapper assesses the odds in favor of Snow Chief to be 1 to 1. According to the third handicapper, what is the probability of a Snow Chief win?

c. A fourth handicapper assesses the odds against Snow Chief winning to be 3 to 2. Find this handicapper's assessment of the probability that Snow Chief will win.

3.2 Unions and Intersections

An event can often be viewed as a composition of two or more other events. Such an event, called a **compound event**, can be formed (composed) in two ways:

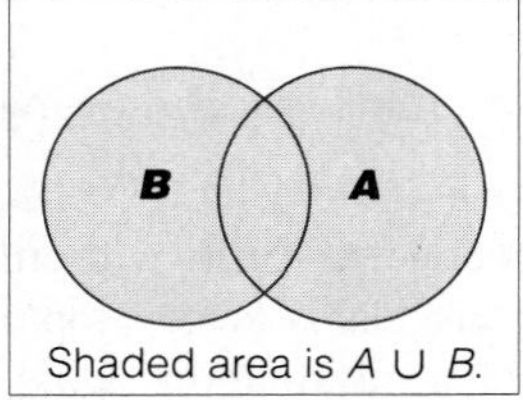

Shaded area is $A \cup B$.

Definition 3.5

The **union** of two events A and B is the event that occurs if either A or B or both occur on a single performance of the experiment. We will denote the union of events A and B by the symbol $A \cup B$.

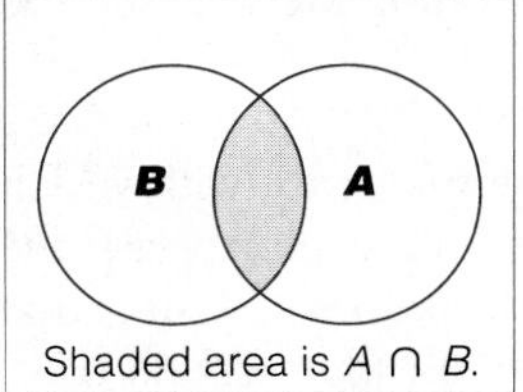

Shaded area is $A \cap B$.

Definition 3.6

The **intersection** of two events A and B is the event that occurs if both A and B occur on a single performance of the experiment. We will write $A \cap B$ for the intersection of events A and B.

EXAMPLE 3.7

Consider the die-toss experiment. Define the following events:

A: {Toss an even number}
B: {Toss a number less than or equal to 3}

a. Describe $A \cup B$ for this experiment.
b. Describe $A \cap B$ for this experiment.
c. Calculate $P(A \cup B)$ and $P(A \cap B)$ assuming the die is fair.

Solution

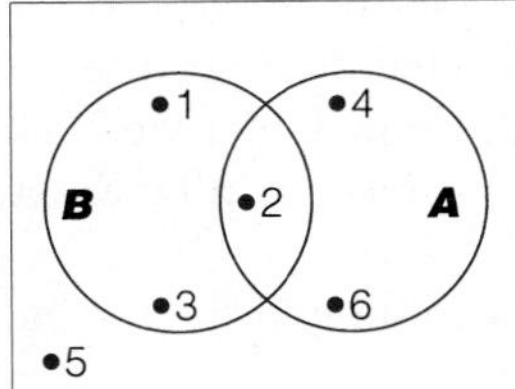

a. The **union** of A and B is the event that occurs if we observe an even number, a number less than or equal to 3, or both on a single throw of the die. Consequently, the simple events in the event $A \cup B$ are those for which A occurs, B occurs, or both A and B occur. Testing the simple events in the entire sample space, we find that the collection of simple events in the union of A and B is

$$A \cup B = \{1, 2, 3, 4, 6\}$$

b. The **intersection** of A and B is the event that occurs if we observe *both* an even number and a number less than or equal to 3 on a single throw of the die. Testing the simple events to see which imply the occurrence of *both* events A and B, we see that the intersection contains only one simple event:

$$A \cap B = \{2\}$$

In other words, the intersection of A and B is the simple event Observe a 2.

c. Recalling that the probability of an event is the sum of the probabilities of the simple events of which the event is composed, we have

$$P(A \cup B) = P(1) + P(2) + P(3) + P(4) + P(6)$$
$$= 1/6 + 1/6 + 1/6 + 1/6 + 1/6 = 5/6$$

and

$$P(A \cap B) = P(2) = 1/6$$

EXAMPLE 3.8

Many firms undertake direct marketing campaigns to promote their products. The campaigns typically involve mailing information to millions of households; the responses to the mailings are carefully monitored to determine the demographic characteristics of respondents. By studying tendencies to respond, the firm can better target future mailings to those segments of the population most likely to purchase the products.

Suppose a distributor of mail-order tools is analyzing the results of a recent mailing. The probability of response is believed to be related to income and age of the head of the household. The percentages of the total number of respondents to the mailing are given by income and age classification in Table 3.4.

TABLE 3.4 Percentage of Respondents in Age–Income Classes

		Income		
		<\$25,000	\$25,000–\$50,000	>\$50,000
Age	<30 yrs.	5%	12%	10%
	30–50 yrs.	14%	22%	16%
	>50 yrs.	8%	10%	3%

Define the following events:

A: {A respondent's income is more than \$50,000}

B: {A respondent's age is 30 years or more}

a. Find $P(A)$ and $P(B)$. b. Find $P(A \cup B)$. c. Find $P(A \cap B)$.

Solution

Following the steps for calculating probabilities of events (given in the box on page 133), we first note that the objective is to characterize the income and age distribution of respondents to the mailing. To accomplish this, we define the experiment as selecting a respondent from the collection of all respondents, and observing which income and age class he or she occupies. The simple events are the nine different age–income classes:

E_1: {<30 yrs., <$25,000}
E_2: {30–50 yrs., <$25,000}
⋮ ⋮
E_9: {>50 yrs., >$50,000}

Next, we assign probabilities to the simple events. If we blindly select one of the respondents, the probability that he or she will occupy a particular age–income class is just the proportion, or relative frequency, of respondents in that class. These proportions are given (as percentages) in Table 3.4. Thus,

$P(E_1)$ = Relative frequency of respondents in age–income class (<30 yrs., <$25,000)
= .05
$P(E_2) = .14$

and so forth. You may verify that the simple event probabilities add to 1.

a. To find $P(A)$ we first determine the collection of simple events contained in event A. Since A is defined as {>$50,000}, we see from Table 3.4 that A contains the three simple events represented by the last column of the table. In words, the event A consists of the income class (>$50,000) and all three age classes within that income class. The probability of A is the sum of the probabilities of the simple events in A:

$$P(A) = .10 + .16 + .03 = .29$$

Similarly, B consists of the six simple events in the second and third rows of Table 3.4:

$$P(B) = .14 + .22 + .16 + .08 + .10 + .03 = .73$$

b. The union of events A and B, $A \cup B$, consists of all simple events in *either* A *or* B *or both* A *and* B. That is, the union of A and B consists of all respondents whose income exceeds $50,000 *or* whose age is 30 or more. In Table 3.4 this is any simple event found in the third column *or* the last two rows. Thus,

$$P(A \cup B) = .10 + .14 + .22 + .16 + .08 + .10 + .03 = .83$$

c. The intersection of events A and B, $A \cap B$, consists of all simple events in *both* A *and* B. That is, the intersection of A and B consists of all respondents whose income exceeds $50,000 *and* whose age is 30 or more. In Table 3.4 this is any simple event found in the third column *and* the last two rows (i.e., in the last two rows of the third column). Thus,

$$P(A \cap B) = .16 + .03 = .19$$

3.3 The Additive Rule and Mutually Exclusive Events

In the previous section, we showed how to determine which simple events are contained in a union. Then we showed that the probability of the union can be calculated by adding the probabilities of the simple events in the union. It is also possible to obtain the probability of a union of two events by using the additive rule.

The union of two events will often contain many simple events, since the union occurs if either one or both of the events occur. By studying the Venn diagram in Figure 3.6, you can see that the probability of the union of two events A and B can be obtained by summing $P(A)$ and $P(B)$ and subtracting the probability corresponding to $A \cap B$. Therefore, the formula for calculating the probability of the union of two events is as given in the box.

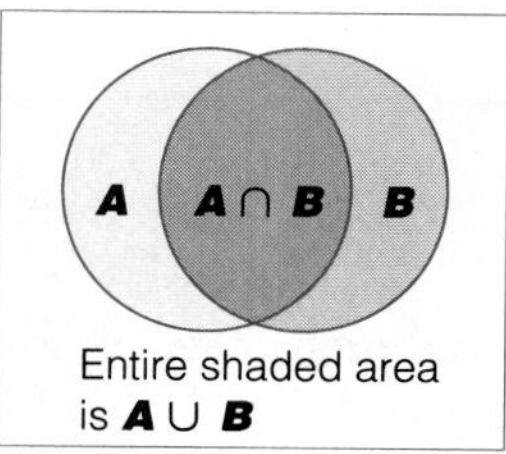

FIGURE 3.6 ▲
Venn diagram of union

Additive Rule of Probability

The probability of the union of events A and B is the sum of the probabilities of events A and B minus the probability of the intersection of events A and B:

$$P(A \cup B) = P(A) + P(B) - P(A \cap B)$$

Note that we must subtract the probability of the intersection because, when we add the probabilities of A and B, the intersection probability is counted twice.

EXAMPLE 3.9

Consider the die-toss experiment. Define the events

A: {Observe an even number}

B: {Observe a number less than or equal to 3}

Assuming the die is fair, calculate the probability of the union of A and B by using the additive rule of probability.

Solution

The formula for the probability of a union requires that we calculate the following:

$$P(A) = P(2) + P(4) + P(6) = \tfrac{1}{6} + \tfrac{1}{6} + \tfrac{1}{6} = \tfrac{3}{6}$$

$$P(B) = P(1) + P(2) + P(3) = \tfrac{1}{6} + \tfrac{1}{6} + \tfrac{1}{6} = \tfrac{3}{6}$$

$$P(A \cap B) = P(2) = \tfrac{1}{6}$$

Now we can calculate the probability of $A \cup B$:

$$P(A \cup B) = P(A) + P(B) - P(A \cap B) = 3/6 + 3/6 - 1/6 = 5/6$$

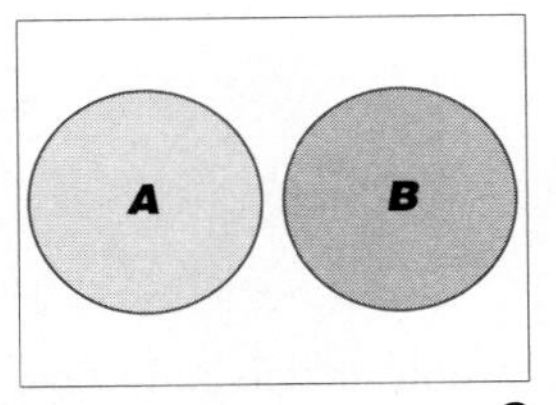

FIGURE 3.7 ▲
Venn diagram of mutually exclusive events

If two events A and B do not intersect—i.e., when $A \cap B$ contains no simple events—we call the events A and B **mutually exclusive** events:

> **Definition 3.7**
>
> Events A and B are **mutually exclusive** if $A \cap B$ contains no simple events.

Figure 3.7 shows a Venn diagram of two mutually exclusive events. The events A and B have no simple events in common; i.e., A and B cannot occur simultaneously, and $P(A \cap B) = 0$. Thus, we have the following important relationship:

> **If two events *A* and *B* are mutually exclusive**, the probability of the union of A and B equals the sum of the probabilities of A and B:
>
> $$P(A \cup B) = P(A) + P(B)$$

EXAMPLE 3.10

Consider the experiment of tossing two balanced coins. Find the probability of observing *at least* one head.

Solution

Define the events

A: {Observe at least one head}
B: {Observe exactly one head}
C: {Observe exactly two heads}

Note that

$$A = B \cup C$$

and that $B \cap C$ contains no simple events (see Figure 3.8). Thus, B and C are mutually exclusive, so that

$$P(A) = P(B \cup C) = P(B) + P(C) = 1/2 + 1/4 = 3/4$$

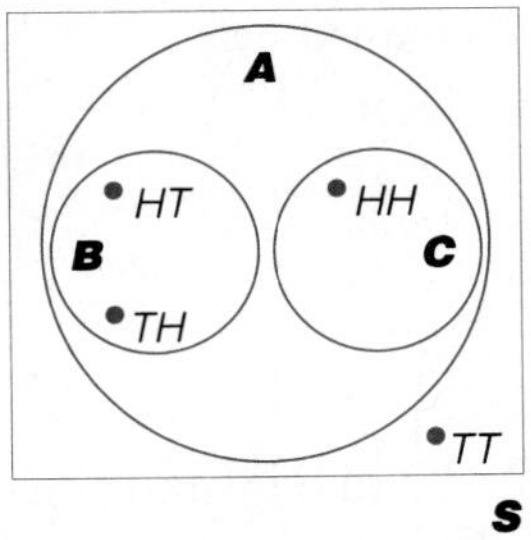

FIGURE 3.8 ▲
Venn diagram for coin toss experiment

Although Example 3.10 is very simple, the concept of writing events with verbal descriptions that include the phrases "at least" or "at most" as unions of mutually exclusive events is a very useful one. This enables us to find the probability of the event by adding the probabilities of the mutually exclusive events.

3.4 Complementary Events

A very useful concept in the calculation of event probabilities is the notion of **complementary events**:

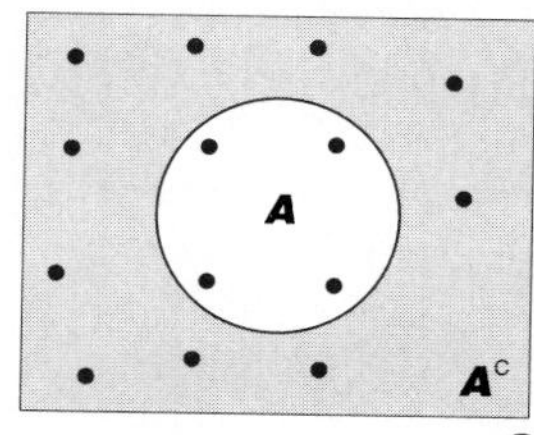

FIGURE 3.9 ▲ Venn diagram of complementary events

> **Definition 3.8**
>
> The **complement** of any event A is the event that A does not occur. We will denote the complement of A by A^c.

Since an event A is a collection of simple events, the simple events included in A^c are just those that are not in A. Figure 3.9 demonstrates this. You will note from the figure that all simple events in S are included in either A or A^c, and that *no* simple event is in both A and A^c. This leads us to conclude that the probabilities of an event and its complement *must sum to 1*:

> The sum of the probabilities of complementary events equals 1; i.e.,
>
> $$P(A) + P(A^c) = 1$$

In many probability problems, it will be easier to calculate the probability of the complement of the event of interest than the event itself. Then, since

$$P(A) + P(A^c) = 1$$

we can calculate $P(A)$ by using the relationship

$$P(A) = 1 - P(A^c)$$

EXAMPLE 3.11

Consider the experiment of tossing two fair coins. Calculate the probability of event A: {Observe at least one head} by using the complementary relationship.

Solution

We know that the event A: {Observe at least one head} consists of the simple events

$$A = \{H_1, H_2;\quad H_1, T_2;\quad T_1, H_2\}$$

The complement of A is defined as the event that occurs when A does not occur. Therefore,

$$A^c = \{T_1, T_2\}$$

FIGURE 3.10 ► Complementary events in the toss of two coins

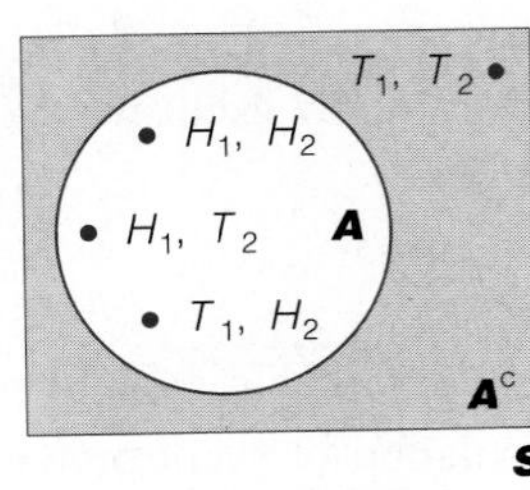

This complementary relationship is shown in Figure 3.10. Assuming the coins are balanced,

$$P(A^c) = P(T_1, T_2) = 1/4 \quad \text{and} \quad P(A) = 1 - P(A^c) = 1 - 1/4 = 3/4$$

Exercises 3.16–3.31

Learning the Mechanics

3.16 Consider the Venn diagram shown, where $P(E_1) = .13$, $P(E_2) = .05$, $P(E_3) = P(E_4) = .2$, $P(E_5) = .06$, $P(E_6) = .3$, and $P(E_7) = .06$. Find each of the following probabilities.

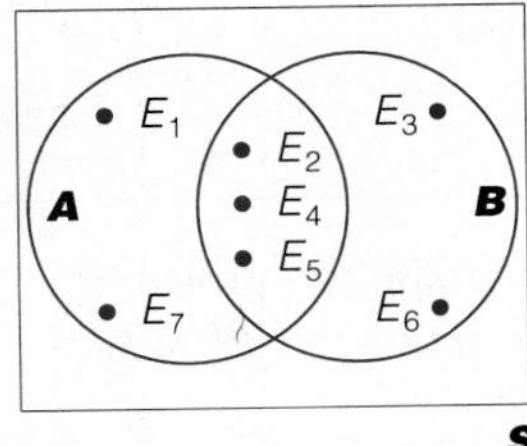

a. $P(A^c)$ b. $P(B^c)$ c. $P(A^c \cap B)$
d. $P(A \cup B)$ e. $P(A \cap B)$ f. $P(A^c \cup B^c)$

3.17 What are mutually exclusive events? Give a verbal description, and then draw a Venn diagram.

3.18 Two dice are tossed and the following events are defined:

A: {The total number of dots on the upper faces of the two dice is equal to 5}
B: {At least one of the two dice has three dots on the upper face}

a. Identify the simple events in each of the following events: A, B, $A \cap B$, and $A \cup B$.
b. Find $P(A)$ and $P(B)$ by summing the probabilities of the appropriate simple events.
c. Find $P(A \cap B)$ and $P(A \cup B)$ by summing the probabilities of the appropriate simple events.
d. Find $P(A \cup B)$ using the additive rule. Compare your answer to that for the same event in part **c**.
e. Are A and B mutually exclusive events? Why?

3.19 Three coins are tossed and the following events are defined:

A: {Observe at least one head}

B: {Observe exactly two heads}

C: {Observe exactly two tails}

Assume that the coins are balanced. Calculate the following probabilities by summing the probabilities of the appropriate simple events:

a. $P(A \cup B)$ b. $P(A \cap B)$ c. $P(A \cap C)$ d. $P(B \cap C)$

e. Are events B and C mutually exclusive?

f. Find $P(B \cup C)$.

3.20 The accompanying table describes the adult population of a small suburb of a large southern city.

		Income		
		Under $20,000	$20,000–$50,000	Over $50,000
Age	Under 25	950	1,000	50
	25–45	450	2,050	1,500
	Over 45	50	950	1,000

A marketing research firm plans to randomly select one adult from this suburb to evaluate a new food product. For this experiment the nine age–income categories are the simple events. Consider the following events:

A: {Person is under 25}

B: {Person is between 25 and 45}

C: {Person is over 45}

D: {Person has income under $20,000}

E: {Person has income of $20,000–$50,000}

F: {Person has income over $50,000}

Convert the frequencies in the table to relative frequencies and use them to calculate the following probabilities:

a. $P(B)$ b. $P(F)$ c. $P(C \cap F)$ d. $P(B \cup C)$ e. $P(A^c)$ f. $P(A^c \cap F)$

g. Consider each pair of events (A and B, A and C, etc.) and list the pairs of events that are mutually exclusive. Justify your choices.

3.21 Refer to Exercise 3.20 and use the same event definitions to solve the following.

a. Write the event that the person selected is under 25 with an income over $50,000 as an intersection of two of the events defined in Exercise 3.20.

b. Write the event that a person of age 25 or over is selected as the union of two mutually exclusive events.

c. Write the event that a person of age 25 or over is selected as the complement of an event.

3.22 Three fair coins are tossed, and we wish to find the probability of the event A: {Observe at least one head}.
  a. Express A as the union of three mutually exclusive events. Find the probability of A using this expression.
  b. Express A as the complement of an event. Find the probability of A using this expression.

Applying the Concepts

3.23 A state energy agency mailed questionnaires on energy conservation to 1,000 homeowners in the state capital. Five hundred questionnaires were returned. Suppose an experiment consists of randomly selecting and reviewing one of the returned questionnaires. Consider the events:

A: {The home is constructed of brick}

B: {The home is more than 30 years old}

C: {The home is heated with oil}

Describe each of the following events in terms of unions, intersections, and complements (i.e., $A \cup B$, $A \cap B$, A^c, etc.):
  a. The home is more than 30 years old and is heated with oil.
  b. The home is not constructed of brick.
  c. The home is heated with oil or is more than 30 years old.
  d. The home is constructed of brick and is not heated with oil.

3.24 Identifying managerial prospects who are both talented and motivated is difficult. A personnel manager constructed the table shown here to define nine combinations of talent–motivation levels. The numbers in the table are the manager's assessments of the probabilities that a managerial prospect will be classified in the respective categories.

		Talent		
		High	Medium	Low
Motivation	High	.05	.16	.05
	Medium	.19	.32	.05
	Low	.11	.05	.02

Suppose the personnel manager has decided to hire a new manager. Define the following events:

A: {Prospect places in the high motivation category}

B: {Prospect places in the high talent category}

C: {Prospect rates medium or better in both categories}

D: {Prospect rates low in at least one of the categories}

E: {Prospect places high in both categories}

  a. Does the sum of the probabilities in the table equal 1?
  b. Find the probability of each event defined above.
  c. Find $P(A \cup B)$, $P(A \cap B)$, and $P(A \cup C)$.
  d. Find $P(A^c)$ and explain what this means from a practical point of view.

e. Consider each pair of events (A and B, A and C, etc.) and list the pairs of events that are mutually exclusive. Justify your choices.

3.25 After completing an inventory of three warehouses, a manufacturer of golf club shafts described its stock of 20,125 shafts with the percentages given in the table. Suppose a shaft is selected at random from the 20,125 currently in stock and the warehouse number and type of shaft are observed.

		Type of Shaft		
		Regular	Stiff	Extra Stiff
Warehouse	1	41%	6%	0%
	2	10%	15%	4%
	3	11%	7%	6%

a. List all the simple events for this experiment.
b. What is the set of all simple events called?
c. Let C be the event that the shaft selected is from warehouse 3. Find P(C) by summing the probabilities of the simple events in C.
d. Let F be the event that the shaft chosen is an extra stiff type. Find P(F).
e. Let A be the event that the shaft selected is from warehouse 1. Find P(A).
f. Let D be the event that the shaft selected is a regular type. Find P(D).
g. Let E be the event that the shaft selected is a stiff type. Find P(E).

3.26 Refer to Exercise 3.25. Define the characteristics of a golf club shaft portrayed by the following events, and then find the probability of each. For each union, use the additive rule to find the probability. Also, determine whether the events are mutually exclusive.
a. $A \cap F$ b. $C \cup E$ c. $C \cap D$ d. $A \cup F$ e. $A \cup D$

3.27 Refer to Exercise 3.5, in which a bank's loan records were being audited. Suppose each of the 500 automobile loans made by the bank last year is now classified according to two characteristics: amount of loan and length of loan. As before, an auditor is planning to choose one loan at random for inspection.

		Amount of Loan ($)			
		Under 2,000	2,000–4,999	5,000–7,999	8,000 or more
Length of Loan (Months)	12	30	4	0	0
	24	5	18	2	0
	36	0	20	89	4
	42	0	31	95	37
	48	0	0	114	51

a. List the simple events in this experiment.
b. What is the probability that the loan selected will be for $8,000 or more? Does your answer agree with your answer to part **b** in Exercise 3.5?
c. What is the probability that the loan selected is a 3-year loan for more than $7,999?

d. What is the probability that the loan selected is a 3- or 4-year loan?
e. What is the probability that the loan selected is a 42-month loan for $2,000 or more?

3.28 The long-run success of a business depends on its ability to market products with superior characteristics that maximize consumer satisfaction and that give the firm a competitive advantage (Bagozzi, 1986). Ten new products have been developed by a food products firm. Market research has indicated that the 10 products have the characteristics described by the Venn diagram shown here.

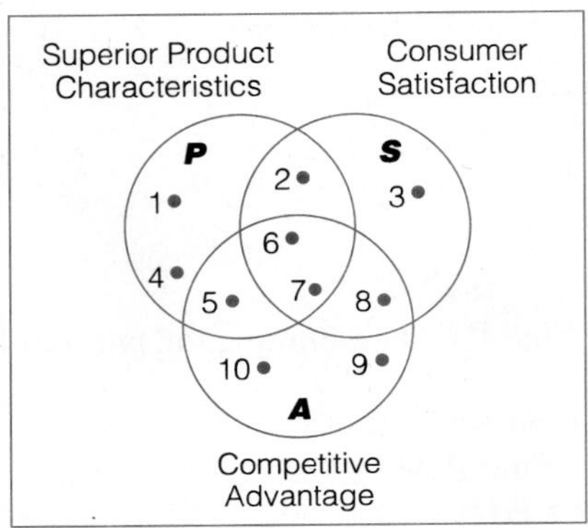

a. Write the event that a product possesses all the desired characteristics as an intersection of the events defined in the Venn diagram. Which products are contained in this intersection?
b. If one of the 10 products were selected at random to be marketed, what is the probability that it would possess all the desired characteristics?
c. Write the event that the randomly selected product would give the firm a competitive advantage or would satisfy consumers as a union of the events defined in the Venn diagram. Find the probability of this union.
d. Write the event that the randomly selected product would possess superior product characteristics and satisfy consumers. Find the probability of this intersection.

3.29 Whether purchases are made by cash or credit card is of concern to merchandisers because they must pay a certain percentage of the sale value to the credit agency. To better understand the relationship between type of purchase (credit or cash) and type of merchandise, a department store analyzed 10,000 sales and placed them in the categories shown in the table. Suppose a single sale is selected at random from the 10,000 and the following events are defined:

A: {Sale was paid by credit card}
B: {Merchandise purchased was women's wear}
C: {Merchandise purchased was men's wear}
D: {Merchandise purchased was sportswear}

		Type of Merchandise			
		Women's Wear	Men's Wear	Sportswear	Household
Type of Purchase	Cash	6%	9%	3%	7%
	Credit Card	41%	9%	22%	3%

Describe the characteristics of a sale implied by the following events, and find the probability of each.

a. $A \cup B$ b. $B \cup C$ c. $B \cap A$ d. $D \cap A$

e. Which pair of events are mutually exclusive? Why?

3.30 Refer to Exercise 3.29. The following events are defined:

A: {Sale was paid by credit card}

B: {Merchandise purchased was women's wear}

a. Describe the events A^c and B^c. b. Find $P(A^c)$. c. Find $P(B^c)$.

d. Find the probability that the sale was in neither men's wear nor women's wear.

e. Find the probability that the sale was *not* a credit card purchase in the sportswear department.

3.31 The types of occupations of the 117,342,000 employed workers (age 16 years and older) in the United States in 1989 are described in the table, and their relative frequencies are listed. A worker is to be selected at random from this population and his or her occupation is to be determined. (Assume that each worker in the population has only one occupation.)

Occupation	*Relative Frequency*	
Male worker	.548	
Managerial/professional		.142
Technical/sales/administrative		.108
Service		.052
Precision production, craft, and repair		.108
Operators/fabricators		.114
Farming, forestry, and fishing		.024
Female workers	.452	
Managerial/professional		.117
Technical/sales/administrative		.200
Service		.080
Precision production, craft, and repair		.010
Operators/fabricators		.040
Farming, forestry, and fishing		.005

Source: *Statistical Abstract of the United States: 1991*, p. 395.

a. What is the probability that the worker will be a male service worker?
b. What is the probability that the worker will be a manager or a professional?
c. What is the probability that the worker will be a female professional or a female operator/fabricator?
d. What is the probability that the worker will not be in a technical/sales/administrative occupation?

3.5 Conditional Probability

The probabilities we assign to the simple events of an experiment are measures of our belief that they will occur when the experiment is performed. When we assign these probabilities, we should make no assumptions other than those contained in or implied by the definition of the experiment. However, at times we will want to make assumptions other than those implied by the experimental description, and these extra assumptions may alter the probabilities we assign to the simple events of an experiment.

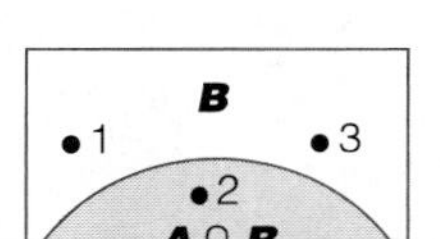

FIGURE 3.11 ▲ **Reduced sample space for the die-toss experiment—given that event B has occurred**

For example, we have shown that the probability of observing an even number (event A) on a toss of a fair die is ½. However, suppose you are given the information that on a particular throw of the die the result was a number less than or equal to 3 (event B). Would you still believe that the probability of observing an even number on that throw of the die is equal to ½? If you reason that making the assumption that B has occurred reduces the sample space from six simple events to three simple events (namely, those contained in event B), the reduced sample space is as shown in Figure 3.11.

Since the reduced sample space contains only three simple events (Observe a 1, Observe a 2, Observe a 3), each is assigned a new probability, called a **conditional probability**. Because the simple events for the die-toss experiment are equally likely, each of the three simple events in the reduced sample space is assigned a conditional probability of ⅓. Since the only even number of the three numbers in the reduced sample space B is the number 2 and since the die is fair, we conclude that the probability that A occurs *given that B occurs* is one in three, or ⅓. We will use the symbol $P(A \mid B)$ to represent the probability of event A given that event B occurs. For the die-toss example,

$$P(A \mid B) = 1/3$$

To get the probability of event A given that event B occurs, we proceed as follows: We divide the probability of the part of A that falls within the reduced sample space B—namely, $P(A \cap B)$—by the total probability of the reduced sample space—namely, $P(B)$. Thus, for the die-toss example with event A: {Observe an even number} and event B: {Observe a number less than or equal to 3}, we find

$$P(A \mid B) = \frac{P(A \cap B)}{P(B)} = \frac{P(2)}{P(1) + P(2) + P(3)} = \frac{1/6}{3/6} = 1/3$$

This formula for $P(A \mid B)$ is true in general:

> To find the **conditional probability that event A occurs given that event B occurs,** divide the probability that *both* A and B occur by the probability that B occurs; that is,
>
> $$P(A \mid B) = \frac{P(A \cap B)}{P(B)}$$

This formula adjusts the probability of $A \cap B$ from its original value in the complete sample space S to a conditional probability in the reduced sample space B. If the simple events in the complete sample space are equally likely, then the formula will assign equal probabilities to the simple events in the reduced sample space, as in the die-toss experiment. If, on the other hand, the simple events have unequal probabilities, the formula will assign conditional probabilities proportional to the probabilities in the complete sample space.

EXAMPLE 3.12

Suppose you are interested in the probability of the sale of a large piece of earth-moving equipment. A single prospect is contacted. Let F be the event that the buyer has sufficient money (or credit) to buy the product and let F^c denote the complement of F (the event that the prospect does not have the financial capability to buy the product). Similarly, let B be the event that the buyer wishes to buy the product and let B^c be the complement of that event. Then the four simple events associated with the experiment are shown in Figure 3.12, and their probabilities are given in Table 3.5.

TABLE 3.5 Probabilities of Customer Desire to Buy and Ability to Finance

		Desire	
		To Buy, B	Not to Buy, B^c
Able to Finance	Yes, F	.2	.1
	No, F^c	.4	.3

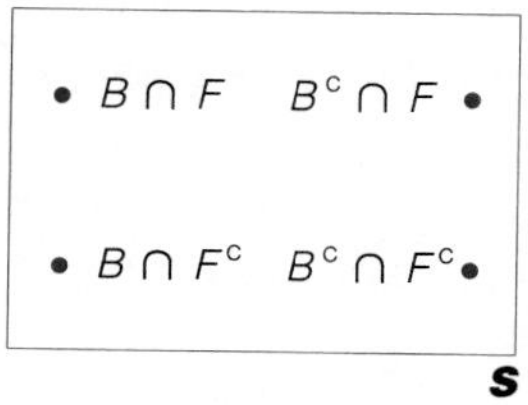

• $B \cap F$	$B^c \cap F$ •
• $B \cap F^c$	$B^c \cap F^c$ •

S

FIGURE 3.12 ▲
Sample space for contacting a sales prospect

Find the probability that a single prospect will buy, given that the prospect is able to finance the purchase.

Solution

Suppose you consider the large collection of prospects for the sale of your product and randomly select one person from this collection. What is the probability that the person selected will buy the product? In order to buy the product, the customer must be financially able and have the desire to buy, so this probability would correspond to

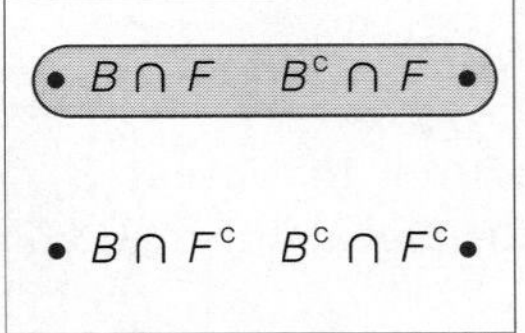

FIGURE 3.13 ▲
Subspace (shaded) containing sample points implying a financially able prospect

the entry in Table 3.5 below B and next to F, or $P(B \cap F) = .2$. This is called the **unconditional probability** of the event $B \cap F$.

In contrast, suppose you know that the prospect selected has the financial capability for purchasing the product. Now you are seeking the probability that the customer will buy given (the condition) that the customer has the financial ability to pay. This probability, the **conditional probability** of B given that F has occurred and denoted by the symbol $P(B \mid F)$, would be determined by considering only the simple events in the reduced sample space containing the simple events $B \cap F$ and $B^c \cap F$—i.e., simple events that imply the prospect is financially able to buy. (This subspace is shaded in Figure 3.13.) From our definition of conditional probability,

$$P(B \mid F) = \frac{P(B \cap F)}{P(F)}$$

where $P(F)$ is the sum of the probabilities of the two simple events corresponding to $B \cap F$ and $B^c \cap F$ (given in Table 3.5). Then

$$P(F) = P(B \cap F) + P(B^c \cap F) = .2 + .1 = .3$$

and the conditional probability that a prospect buys, given that the prospect is financially able, is

$$P(B \mid F) = \frac{P(B \cap F)}{P(F)} = \frac{.2}{.3} = .667$$

As we would expect, the probability that the prospect will buy, given that he or she is financially able, is higher than the unconditional probability of selecting a prospect who will buy.

Note in Example 3.12, that the conditional probability formula assigns a probability to the event $(B \cap F)$ in the reduced sample space that is proportional to the probability of the event in the complete sample space. To see this, note that the two simple events in the reduced sample space, $(B \cap F)$ and $(B^c \cap F)$, have probabilities of .2 and .1, respectively, in the complete sample space S. The formula assigns conditional probabilities ⅔ and ⅓ (use the formula to check the second one) to these events in the reduced sample space F, so that the conditional probabilities retain the 2 to 1 proportionality of the original simple event probabilities.

EXAMPLE 3.13

The investigation of consumer product complaints by the Federal Trade Commission has generated much interest by manufacturers in the quality of their products. A manufacturer of an electromechanical kitchen aid conducted an analysis of a large number of consumer complaints and found that they fell into the six categories shown in Table 3.6. If a consumer complaint is received, what is the probability that the cause of the complaint was product appearance, given that the complaint originated prior to the end of the guarantee period?

TABLE 3.6 Distribution of Product Complaints

	Reason for Complaint		
	Electrical	Mechanical	Appearance
During Guarantee Period	18%	13%	32%
After Guarantee Period	12%	22%	3%

Solution

Let A represent the event that the cause of a particular complaint was product appearance and let B represent the event that the complaint occurred during the guarantee period. Checking Table 3.6, you can see that $(18 + 13 + 32)\% = 63\%$ of the complaints occurred during the guarantee time. Hence, $P(B) = .63$. The percentage of complaints that were caused by appearance, A, *and* occurred during the guarantee period, B, is 32%. Therefore, $P(A \cap B) = .32$. Using these probability values, we can calculate the conditional probability $P(A \mid B)$ that the cause of a complaint is appearance given that the complaint occurred prior to the termination of the guarantee time:

$$P(A \mid B) = \frac{P(A \cap B)}{P(B)} = \frac{.32}{.63} = .51$$

Consequently, you can see that slightly more than half of the complaints that occurred during the guarantee period were due to scratches, dents, or other imperfections in the surface of the kitchen devices.

You will see in later chapters that conditional probability plays a key role in many applications of statistics. For example, we may be interested in the probability that a particular stock gains 10% during the next year. We may assess this probability using information such as the past performance of the stock or the general state of the economy at present. However, our probability may change drastically if we assume that the Gross Domestic Product (GDP) will increase by 10% in the next year. We would then be assessing the *conditional probability* that our stock gains 10% in the next year given that the GDP gains 10% in the same year. Thus, the probability of any event that is calculated or assessed based on an assumption that some other event occurs concurrently is a conditional probability.

CASE STUDY 3.1 / Purchase Patterns and the Conditional Probability of Purchasing

In his doctoral dissertation, Alfred A. Kuehn (1958) examined sequential purchase data to gain some insight into consumer brand switching. He analyzed the frozen orange juice purchases of approximately 600 Chicago families during 1950–1952. The data were collected by the *Chicago Tribune* Consumer Panel. Kuehn was interested in determining the influence of a consumer's last four orange juice purchases on the next purchase. Thus, sequences of five purchases were analyzed.

Table 3.7 contains a summary of the data collected for Snow Crop brand orange juice and part of Kuehn's analysis of the data. In the column labeled "Previous Purchase Pattern" an S stands for the purchase of Snow Crop by a consumer and an O stands for the purchase of a brand other than Snow Crop. Thus, for example, SSSO is used to represent the purchase of Snow Crop three times in a row followed by the purchase of some other brand of frozen orange juice. The column labeled "Sample Size" lists the number of occurrences of the purchase sequences in the first column. The column labeled "Frequency" lists the number of times the associated purchase sequence in the first column led to the next purchase (i.e., the fifth purchase in the sequence) being Snow Crop.

The column labeled "Observed Approximate Probability of Purchase" contains the relative frequency with which each sequence of the first column led to the next purchase being Snow Crop. These relative frequencies, which give approximate probabilities, are computed for each sequence of the first column by dividing the frequency of the sequence by the sample size of the sequence. Notice that these approximate probabilities are really conditional probabilities. For the sequences of five purchases analyzed, each of the entries in the fourth column is the approximate probability that the next purchase is Snow Crop, given that the previous four purchases were as noted in the first column. For example, .806 is the approximate probability that the next purchase will be Snow Crop given that the previous four purchases were also Snow Crop.

An examination of the approximate probabilities in the fourth column indicates that both the most recent brand purchased and the number of times a brand is purchased have an effect on the next brand purchased. It appears that the influence on the next brand of orange juice purchased by the second most recent

TABLE 3.7 Observed Approximate Probability of Purchasing Snow Crop, Given the Four Previous Brand Purchases

Previous Purchase Pattern S = Snow Crop, O = Other Brand	Sample Size	Frequency	Observed Approximate Probability of Purchase
SSSS	1,047	844	.806
OSSS	277	191	.690
SOSS	206	137	.665
SSOS	222	132	.595
SSSO	296	144	.486
OOSS	248	137	.552
SOOS	138	78	.565
OSOS	149	74	.497
SOSO	163	66	.405
OSSO	181	75	.414
SSOO	256	78	.305
OOOS	500	165	.330
OOSO	404	77	.191
OSOO	433	56	.129
SOOO	557	86	.154
OOOO	8,442	405	.048

purchase is not so strong as the most recent purchase but is stronger than the third most recent purchase. In general, it appears that the probability of a particular consumer purchasing Snow Crop the next time he or she buys orange juice is inversely related to the number of consecutive purchases of another brand he or she made since last purchasing Snow Crop and is directly proportional to the number of Snow Crop purchases among the four purchases.

Kuehn, of course, goes on to conduct a more formal statistical analysis of these data, which we will not pursue here. We simply want you to see that probability is a basic tool for making inferences about populations using sample data.

3.6 The Multiplicative Rule and Independent Events

The probability of an intersection of two events can be calculated using the multiplicative rule, which employs the conditional probabilities we defined in the previous section. Actually, we have already developed the formula in another context. You will recall that the formula for calculating the conditional probability of B given A is

$$P(B \mid A) = \frac{P(A \cap B)}{P(A)}$$

If we multiply both sides of this equation by $P(A)$, we get a formula for the probability of the intersection of events A and B:

Multiplicative Rule of Probability

$$P(A \cap B) = P(A)P(B \mid A) \quad \text{or, equivalently,} \quad P(A \cap B) = P(B)P(A \mid B)$$

The second expression in the box is obtained by multiplying both sides of the equation $P(A \mid B) = P(A \cap B)/P(B)$ by $P(B)$.

Before working an example, we emphasize that the intersection often contains only a few simple events, in which case the probability is easy to calculate by summing the appropriate simple event probabilities.

The formula for calculating intersection probabilities plays a very important role in an area of statistics known as **Bayesian statistics**. (More complete discussions of Bayesian statistics are contained in Chapter 19 and in the references at the end of this chapter.)

EXAMPLE 3.14

Suppose an investment firm is interested in the following events:

A: {Gross Domestic Product gains 10% next year}

B: {Common stock in XYZ Corporation gains 10% next year}

The firm has assigned the following probabilities on the basis of available information:

$P(B \mid A) = .8 \qquad P(A) = .3$

That is, the investment company believes the probability is .8 that XYZ common stock will gain 10% in the next year *assuming that* the GDP gains 10% in the same time period. In addition, the company believes the probability is only .3 that the GDP will gain 10% in the next year. Use the formula for calculating the probability of an intersection to determine the probability that XYZ common stock *and* the GDP gain 10% in the next year.

Solution

We want to calculate $P(A \cap B)$. The formula is

$$P(A \cap B) = P(A)P(B \mid A) = (.3)(.8) = .24$$

Thus, according to this investment firm, the probability is .24 that both XYZ common stock and the GDP will gain 10% in the next year.

In the previous section, we showed that the probability of an event A may be substantially altered by the knowledge that the event B has occurred. However, this will not always be the case. In some instances the assumption that event B has occurred will not alter the probability of event A at all. When this is true, we call events A and B **independent**.

Definition 3.9

Events A and B are **independent** if the assumption that B has occurred does not alter the probability that A occurs; i.e., events A and B are independent if

$$P(A \mid B) = P(A)$$

Equivalently, events A and B are **independent** if

$$P(B \mid A) = P(B)$$

Events that are not independent are said to be **dependent**.

EXAMPLE 3.15

Suppose that we decide to change the definition of event B in the die-toss experiment to {Observe a number less than or equal to 4} but we let event A remain {Observe an even number}. Are events A and B independent (assuming a fair die)?

Solution

The Venn diagram for this experiment is shown in Figure 3.14. We first calculate

$$P(A) = \tfrac{1}{2}$$

$$P(B) = P(1) + P(2) + P(3) + P(4) = \tfrac{4}{6} = \tfrac{2}{3}$$

$$P(A \cap B) = P(2) + P(4) = \tfrac{2}{6} = \tfrac{1}{3}$$

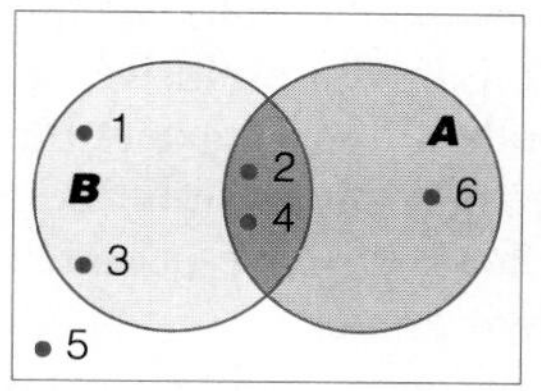

FIGURE 3.14 ▲
Die-toss experiment, Example 3.15

Now assuming B has occurred, the conditional probability of A is

$$P(A \mid B) = \frac{P(A \cap B)}{P(B)} = \frac{1/3}{2/3} = 1/2 = P(A)$$

Thus, the assumption of the occurrence of event B does not alter the probability of observing an even number—it remains ½. Therefore, the events A and B are independent. Note that if we calculate the conditional probability of B given A, our conclusion is the same:

$$P(B \mid A) = \frac{P(A \cap B)}{P(A)} = \frac{1/3}{1/2} = 2/3 = P(B)$$

EXAMPLE 3.16

Refer to the consumer product complaint study in Example 3.13. The percentages of complaints of various types in the pre- and post-guarantee periods are shown in Table 3.6. Define the following events:

A: {Cause of complaint is product appearance}
B: {Complaint occurred during the guarantee term}

Are A and B independent events?

Solution

Events A and B are independent if $P(A \mid B) = P(A)$. We calculated $P(A \mid B)$ in Example 3.13 to be .51, and from Table 3.6 (page 157) we can see that

$$P(A) = .32 + .03 = .35$$

Therefore, $P(A \mid B) \neq P(A)$, and A and B are not independent events.

To gain an intuitive understanding of independence, think of situations in which the occurrence of one event does not alter the probability that a second event will occur. For example, suppose two small companies are being monitored by a financier for possible investment. If the businesses are in different industries and they are otherwise unrelated, then the success or failure of one company may be *independent* of the success or failure of the other. That is, the event that company A fails may not alter the probability that company B will fail.

As a second example, consider an election poll in which 1,000 registered voters are asked to state their preference between two candidates. One objective of the pollsters is to select a sample of voters so that the responses will be independent. That is, the sample is selected so that the event that one polled voter prefers candidate A does not alter the probability that a second polled voter prefers candidate A.

We will make three final points about independence. The first is that the property of independence, unlike the mutually exclusive property, cannot be shown on or gleaned from a Venn diagram. In general, the only way to check for independence is by performing the calculations of the probabilities in the definition.

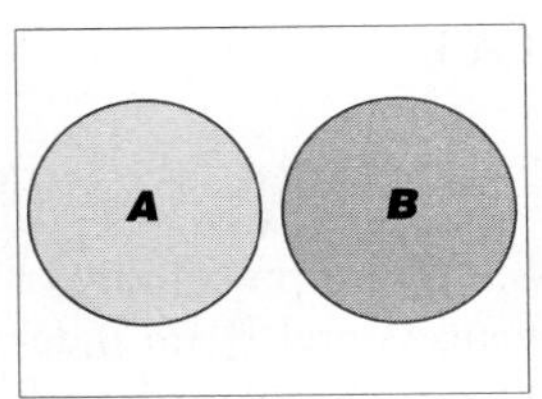

FIGURE 3.15 ▲
Mutually exclusive events are dependent.

The second point concerns the relationship between the mutually exclusive and independence properties. Suppose that events A and B are mutually exclusive, as shown in Figure 3.15. Are these events independent or dependent? That is, does the assumption that B occurs alter the probability of the occurrence of A? It certainly does, because if we assume that B has occurred, it is impossible for A to have occurred simultaneously. Thus, mutually exclusive events are dependent events.

The third point is that the probability of the intersection of independent events is very easy to calculate. Referring to the formula for calculating the probability of an intersection, we find

$$P(A \cap B) = P(B)P(A \mid B)$$

Thus, since $P(A \mid B) = P(A)$ when A and B are independent, we have the following useful rule:

> **If events A and B are independent**, the probability of the intersection of A and B equals the product of the probabilities of A and B:
>
> $$P(A \cap B) = P(A)P(B)$$
>
> The converse is also true: If $P(A \cap B) = P(A)P(B)$, then events A and B are independent.

In the die-toss experiment, we showed in Example 3.15 that the events A: {Observe an even number} and B: {Observe a number less than or equal to 4} are independent if the die is fair. Thus,

$$P(A \cap B) = P(A)P(B) = (1/2)(2/3) = 1/3$$

This agrees with the result

$$P(A \cap B) = P(2) + P(4) = 2/6 = 1/3$$

that we obtained in the example.

EXAMPLE 3.17

Almost every retail business has the problem of determining how much inventory to purchase. Insufficient inventory may result in lost business, and excess inventory will have a detrimental effect on profits. Suppose a retail computer store owner is planning to place an order for personal computers (PCs). She is trying to decide how many IBM PCs and how many IBM compatibles (personal computers that run all or most of the same software as the IBM PC but that are not manufactured by IBM) to order. The owner's records indicate that 80% of the previous PC customers purchased IBM PCs, and 20% purchased compatibles.

a. What is the probability that the next two customers will purchase compatibles?

b. What is the probability that the next 10 customers will purchase compatibles?

Solution

a. Let C_1 represent the event that customer 1 will purchase a compatible, and C_2 the event that customer 2 will purchase a compatible. The event that *both* customers purchase compatibles is the intersection of these events, $C_1 \cap C_2$. From past records, the store owner could reasonably conclude that the probability of C_1 is equal to .2 (based on the fact that 20% of past customers have purchased compatibles), and the same reasoning would apply to C_2. However, in order to compute the probability of the intersection $C_1 \cap C_2$, more information is needed. Either the records must be examined for the occurrence of consecutive purchases of compatibles, or some assumption must be made to enable the calculation of $P(C_1 \cap C_2)$ from the multiplicative rule. It seems reasonable to make the assumption that the two events are independent, since the decision of the first customer is not likely to affect that of the second customer. Assuming independence, we have

$$P(C_1 \cap C_2) = P(C_1)P(C_2) = (.2)(.2) = .04$$

b. To see how to compute the probability that 10 consecutive purchases will be of compatibles, first consider the event that three consecutive customers purchase compatibles. If C_3 represents the event that the third customer purchases a compatible, then we want to compute the probability of the intersection of $C_1 \cap C_2$ with C_3. Again assuming independence of the purchasing decisions, we have

$$P(C_1 \cap C_2 \cap C_3) = P(C_1 \cap C_2)P(C_3) = (.2)^2(.2) = .008$$

Similar reasoning leads to the conclusion that the intersection of 10 such events can be calculated as follows:

$$\begin{aligned} P(C_1 \cap C_2 \cap \cdots \cap C_{10}) &= P(C_1)P(C_2) \cdot \cdots \cdot P(C_{10}) \\ &= (.2)^{10} \\ &= .0000001024 \end{aligned}$$

Thus, the probability that 10 consecutive customers purchase IBM compatibles is about 1 in 10 million, assuming that the probability of each customer's purchase of a compatible is .2, and that the purchase decisions are independent.

Exercises 3.32–3.47

Learning the Mechanics

3.32 An experiment results in one of three mutually exclusive events, A, B, or C. It is known that $P(A) = .30$, $P(B) = .55$, and $P(C) = .15$. Find each of the following probabilities:

a. $P(A \cup B)$ b. $P(A \cap C)$ c. $P(A \mid B)$ d. $P(B \cup C)$

e. Are B and C independent events? Explain.

3.33 Consider the experiment depicted by the Venn diagram on page 164, with the sample space S containing five simple events. The simple events are assigned the following probabilities: $P(E_1) = .20$, $P(E_2) = .30$, $P(E_3) = .30$, $P(E_4) = .10$, $P(E_5) = .10$.

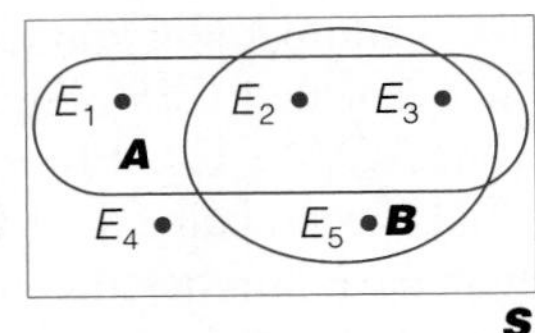

a. Calculate $P(A)$, $P(B)$, and $P(A \cap B)$.
b. Suppose we know that event A has occurred, so that the reduced sample space consists of the three simple events in A—namely, E_1, E_2, and E_3. Use the formula for conditional probability to adjust the probabilities of these three simple events for the knowledge that A has occurred [i.e., $P(E_i \mid A)$]. Verify that the conditional probabilities are in the same proportion to one another as the original simple event probabilities.
c. Calculate the conditional probability $P(B \mid A)$ in two ways: (1) Add the adjusted (conditional) probabilities of the simple events in the intersection $A \cap B$, since these represent the event that B occurs given that A has occurred; (2) Use the formula for conditional probability:

$$P(B \mid A) = \frac{P(A \cap B)}{P(A)}$$

Verify that the two methods yield the same result.
d. Are events A and B independent? Why or why not?

3.34 An experiment results in one of five simple events, with the following probabilities: $P(E_1) = .22$, $P(E_2) = .31$, $P(E_3) = .15$, $P(E_4) = .22$, and $P(E_5) = .10$. The following events have been defined:

$$A = \{E_1, E_3\} \qquad B = \{E_2, E_3, E_4\} \qquad C = \{E_1, E_5\}$$

Find each of the following probabilities:
a. $P(A)$ b. $P(B)$ c. $P(A \cap B)$ d. $P(A \mid B)$ e. $P(B \cap C)$ f. $P(C \mid B)$

3.35 Refer to Exercise 3.34. Which of the following pairs of events are independent? Explain.
a. A and B b. A and C c. B and C

3.36 Three coins are tossed and the following events are defined:

A: {Observe exactly two heads}
B: {Observe at least one head}
C: {Observe at most one head}
D: {Observe at least two tails}

Use the formulas of this section to calculate the following:
a. $P(B \cap C)$ b. $P(B \cup C)$
c. $P(A \cap D)$ d. $P(A \cup D)$
e. $P(B^c \cap C)$ f. $P(A \cap C)$
g. Determine which pairs of events, if any, are independent.

3.37 Defend or refute each of the following statements:
a. Dependent events are always mutually exclusive.
b. Mutually exclusive events are always dependent.

c. Independent events are always mutually exclusive.
d. Mutually exclusive events are always independent.

3.38 If $P(R) = \frac{1}{3}$, $P(S) = \frac{1}{3}$, and events R and S are mutually exclusive, find $P(R \mid S)$ and $P(S \mid R)$.

3.39 Two dice are tossed and the following events are defined:

A: {Sum of the numbers showing is an odd number}
B: {Sum of the numbers showing is 8, 9, 10, or 11}

3.40 Two hundred shoppers at a large suburban mall were asked two questions: (1) Did you see a television ad for the sale at department store X during the past 2 weeks? (2) Did you shop at department store X during the past 2 weeks? The responses to these questions are summarized in the table. One of the 200 shoppers questioned is to be chosen at random.

	Shopped at X	*Did Not Shop at X*
Saw ad	100	25
Did not see ad	25	50

a. What is the probability that the person selected saw the ad?
b. What is the probability that the person selected saw the ad and shopped at store X?
c. Find the conditional probability that the person shopped at store X given that the person saw the ad.
d. What is the probability that the person selected shopped at store X?
e. Use your answers to parts **a**, **b**, and **d** to check the independence of the events Saw ad and Shopped at X.
f. Are the two events {Did not see ad} and {Did not shop at X} mutually exclusive? Explain.

Applying the Concepts

3.41 Businesses that offer credit to their customers are inevitably faced with the task of collecting unpaid bills. Richard L. Peterson (1986) conducted a study of collection remedies used by creditors. As part of the study he asked samples of creditors in four states about how they deal with past-due bills. Their responses are tallied in the table. "Tough actions" included filing a legal action, turning the debt over to a third party such as an attorney or collection agency, garnishing wages, and repossessing secured property. Suppose one of the creditors questioned is selected at random.

	Wisconsin	*Illinois*	*Arkansas*	*Louisiana*
Take tough action early	0	1	5	1
Take tough action late	37	23	22	21
Never take tough action	9	11	6	15

a. What is the probability that the creditor is from Wisconsin or Louisiana?
b. What is the probability that the creditor is not from Wisconsin or Louisiana?
c. What is the probability that the creditor never takes tough action?
d. What is the probability that the creditor is from Arkansas and never takes tough action?
e. What is the probability that the creditor never takes tough action, given that the creditor is from Arkansas?
f. If the creditor takes tough action early, what is the probability that the creditor is from Arkansas or Louisiana?
g. What is the probability that a creditor from Arkansas never takes tough action?

3.42 A particular automatic sprinkler system for high-rise apartment buildings, office buildings, and hotels has two different types of activation devices for each sprinkler head. One type has a reliability of .91 (i.e., the probability that it will activate the sprinkler when it should is .91). The other type, which operates independently of the first type, has a reliability of .87. Suppose a serious fire starts near a particular sprinkler head.

a. What is the probability that the sprinkler head will be activated?

b. What is the probability that the sprinkler head will not be activated?

c. What is the probability that both activation devices will work properly?

d. What is the probability that only the device with reliability .91 will work properly?

3.43 A soft-drink bottler has two quality control inspectors independently check each case of soft drinks for chipped or cracked bottles before the cases leave the bottling plant. Having observed the work of the two trusted inspectors over several years, the bottler has determined that the probability of a defective case getting by the first inspector is .05 and the probability of a defective case getting by the second inspector is .10. What is the probability that a defective case gets by both inspectors?

3.44 The table describes the 67.1 million U.S. long-form federal tax returns filed with the Internal Revenue Service (IRS) in 1989 and the percentage of those returns that were audited by the IRS.

Income	*Number of Tax Filers (Millions)*	*Percentage Audited*
Under $25,000	29.1	.5
$25,000–$49,999	25.8	1.0
$50,000–$99,999	10.0	1.0
$100,000 or more	2.2	5.5

Source: *Statistical Abstract of the United States: 1991*, p. 324.

a. If a tax filer is randomly selected from this population of tax filers (i.e., each tax filer has an equal probability of being selected), what is the probability that the tax filer was audited?

b. If a tax filer is randomly selected from this population of tax filers, what is the probability that the tax filer had an income of $25,000–$49,999 in 1989 *and* was audited? What is the probability that the tax filer had an income of $50,000 or more in 1989 *or* was not audited?

3.45 Even with strong advertising programs, new products are often unsuccessful. A company that produces a variety of household items found that only 18% of the new products it introduced over the last 10 years have become profitable. When two new products were introduced during the same year, only 5% of the time did both products become profitable. Suppose the company plans to introduce two new products, A and B, next year. If the percentages just cited define the probabilities of success, what is the probability that:

a. Product A will become profitable?

b. Product B will not become profitable?

c. At least one of the two products will become profitable?

d. Neither of the two products will become profitable?

e. Either product A or product B (but not both) will become profitable?

3.46 Refer to Exercise 3.45.

a. If product B is profitable, what is the probability that product A becomes profitable?

b. If at least one of the products will be profitable, what is the probability that the profitable product is A?

3.47 **Total quality management (TQM)** is a management philosophy and system of management techniques to improve product and service quality and worker productivity. TQM involves such techniques as teamwork, empowerment of workers, improved communication with customers, evaluation of work processes, and statistical analysis of processes and their output (Benson, 1992). One hundred U.S. companies were surveyed and it was found that 30 had implemented TQM. Among the 100 companies surveyed, 60 reported an increase in sales last year. Of those 60, 20 had implemented TQM. Suppose one of the 100 surveyed companies is to be selected at random for additional analysis.

a. What is the probability that a firm that implemented TQM is selected? That a firm whose sales increased is selected?

b. Are the two events {TQM implemented} and {Sales increased} independent or dependent? Explain.

c. Suppose that instead of 20 TQM-implementers among the 60 firms reporting sales increases, there were 18. Now are the events {TQM implemented} and {Sales increased} independent or dependent? Explain.

3.7 Random Sampling

How a sample is selected from a population is of vital importance in statistical inference because the probability of an observed sample will be used to infer the characteristics of the sampled population. To illustrate, suppose you deal yourself four cards from a deck of 52 cards, and all four cards are aces. Do you conclude that your deck is a ordinary bridge deck, containing only four aces, or do you conclude that the deck is stacked with more than four aces? It depends on how the cards were drawn. If the four aces were always placed on the top of a standard bridge deck, drawing four aces would not be unusual—it would be certain. On the other hand, if the cards were thoroughly mixed, drawing four aces in a sample of four cards would be highly improbable. The point, of course, is that, in order to use the observed sample of four cards to make inferences about the population (the deck of 52 cards), you need to know how the sample was selected from the deck.

One of the simplest and most frequently used sampling procedures (implied in the previous examples and exercises) produces what is known as a **random sample**.

Definition 3.10

If n elements are selected from a population in such a way that every possible combination of n elements in the population has an equal probability of being selected, the n elements are said to be a **random sample**.*

*Strictly speaking, this is a **simple random sample**. There are many different types of random samples. The simple random sample is the most frequently employed. We discuss other types of sampling procedures in Chapter 20.

If a population is not too large and the elements can be marked on slips of paper or poker chips, you can physically mix the slips of paper or chips and remove n elements from the total. Then the elements that appear on the slips or chips selected would indicate the population elements to be included in the sample. Such a procedure would not guarantee a random sample because it is often difficult to achieve a thorough mix, but it provides a reasonably good approximation to random sampling.

Many samplers use a table of random numbers (see Table I in Appendix B) or a statistical software package to generate a random sample. Random-number tables are constructed in such a way that every digit occurs with (approximately) equal probability. To use a table of random numbers, we number the N elements in the population from 1 to N. Then we turn to Table I and haphazardly select a number in the table. Proceeding from this number across the row or down the column (either will do), remove and record n numbers from the table. Use only the necessary number of digits in each random number to identify the element to be included in the sample. If, in the course of recording the n numbers from the table, you obtain a number that has already been selected, simply discard the duplicate, and select a replacement at the end of the sequence. Thus, you may have to select more than n numbers to obtain a random sample of n unique numbers.

We illustrate this procedure with an example.

EXAMPLE 3.18

Suppose you wish to randomly sample five households (we will keep the number in the sample small to simplify our example) from a population of 100,000 households. Use Table I in Appendix B to select a random sample.

TABLE 3.8 Reproduction of Part of Table I of Appendix B: Random Numbers

Row \ Column	1	2	3	4	5	6
1	10480	15011	01536	02011	81647	91646
2	22368	46573	25595	85393	30995	89198
3	24130	48360	22527	97265	76393	64809
4	42167	93093	06243	61680	07856	16376
5	37570	39975	81837	16656	06121	91782
6	77921	06907	11008	42751	27756	53498
7	99562	72905	56420	69994	98872	31016
8	96301	91977	05463	07972	18876	20922
9	89579	14342	63661	10281	17453	18103
10	85475	36857	53342	53988	53060	59533
11	28918	69578	88231	33276	70997	79936
12	63553	40961	48235	03427	49626	69445
13	09429	93969	52636	92737	88974	33488
14	10365	61129	87529	85689	48237	52267
15	07119	97336	71048	08178	77233	13916

Solution

First, number the households in the population from 1 to 100,000. Then, turn to a page of Table I, say, the first page. A reproduction of part of the first page of Table I is shown in Table 3.8. Now, commence with the random number that appears in the third row, second column. This number is 48360. Proceed down the second column to obtain the remaining four random numbers. The five selected random numbers are shaded in Table 3.8. Using the first five digits to represent the households from 1 to 99,999 and the number 00000 to represent household 100,000, you can see that the households numbered

48,360 93,093 39,975 6,907 72,905

should be included in your sample.

CASE STUDY 3.2 / The 1970 Draft Lottery

From 1948 through the early years of the Vietnam War, the Selective Service System drafted men into the military service by age—oldest first, starting with 25-year-olds. A network of local draft boards was used to implement the selection process. Then, on the evening of December 1, 1969, the Selective Service System conducted a lottery to determine the order of selection for 1970 in an attempt to overcome what many believed were inequities in the system. (Such lotteries had been used during World Wars I and II, but it had been 27 years since the last one.)

The objective of the lottery was to randomly order the induction sequence of men between the ages of 19 and 26. To do this, the 366 possible days in a year were written on slips of paper and placed in egg-shaped capsules that were stored in monthly lots. The monthly lots were placed one by one into a wooden box that was ". . . turned end over end several times to mix the numbers" ("Random or not? Judge studies lottery protest," 1970). The capsules were then dumped into a large glass bowl and drawn one by one to obtain the order of induction. All men born on the first day drawn would be inducted first; those born on the second day drawn would be drafted next, etc. Thus, the lottery assigned a rank to each of the 366 birthdays. The results of the lottery are shown in Table 3.9 on page 170.

For a random sequence of numbers to be generated with this procedure, it is necessary for each (remaining) capsule in the bowl to have an equal probability of being selected on each draw. That is, by means of thorough mixing, each capsule must have an equal opportunity to come to rest precisely where the sampler's hand closes within the bowl. Although a mixing procedure with this property is almost impossible to achieve, the ideal can be closely approximated. Unfortunately, this was apparently not the case in the 1970 lottery. Even though the sequence of dates in Table 3.9 may appear to be random, there is ample statistical evidence to indicate a nonrandom selection of induction dates.*

To obtain an understanding of the problem with the 1970 lottery, we calculated the median rank for each month and plotted the medians in Figure 3.16(a) on page 171. If the sequence of ranks were randomly generated, there should be no relationship between the size of the median ranks and the months of the year. The medians should vary randomly above and below a horizontal line with intercept 183.5 (the median of the integers 1 through 366). However, Figure 3.16(a) reveals a general downward trend in the medians. Men born later in the year we more likely to be drafted before men born early in the year. Furthermore, since not all men between the ages of 19 and 25 would be

*We discuss procedures for detecting nonrandomness in Chapter 13.

TABLE 3.9 1970 Draft Lottery Results

1 Sept. 14	62 April 21	123 Dec. 28	184 Sept. 8	245 Aug. 26	306 Jan. 7
2 April 24	63 Sept. 20	124 April 13	185 Nov. 20	246 Sept. 18	307 Aug. 13
3 Dec. 30	64 June 27	125 Oct. 2	186 Jan. 21	247 June 22	308 May 28
4 Feb. 14	65 May 10	126 Nov. 13	187 July 20	248 July 11	309 Nov. 26
5 Oct. 18	66 Nov. 12	127 Nov. 14	188 July 5	249 June 1	310 Nov. 5
6 Sept. 6	67 July 25	128 Dec. 18	189 Feb. 17	250 May 21	311 Aug. 19
7 Oct. 26	68 Feb. 12	129 Dec. 1	190 July 18	251 Jan. 3	312 April 8
8 Sept. 7	69 June 13	130 May 15	191 April 29	252 April 23	313 May 31
9 Nov. 22	70 Dec. 21	131 Nov. 15	192 Oct. 20	253 April 6	314 Dec. 12
10 Dec. 6	71 Sept. 10	132 Nov. 25	193 July 31	254 Oct. 16	315 Sept. 30
11 Aug. 31	72 Oct. 12	133 May 12	194 Jan. 9	255 Sept. 17	316 April 22
12 Dec. 7	73 June 17	134 June 11	195 Sept. 24	256 Mar. 23	317 Mar. 9
13 July 8	74 April 27	135 Dec. 20	196 Oct. 24	257 Sept. 28	318 Jan. 13
14 April 11	75 May 19	136 Mar. 11	197 May 9	258 Mar. 24	319 May 23
15 July 12	76 Nov. 6	137 June 25	198 Aug. 14	259 Mar. 13	320 Dec. 15
16 Dec. 29	77 Jan. 28	138 Oct. 13	199 Jan. 8	260 April 17	321 May 8
17 Jan. 15	78 Dec. 27	139 Mar. 6	200 Mar. 19	261 Aug. 3	322 July 15
18 Sept. 26	79 Oct. 31	140 Jan. 18	201 Oct. 23	262 April 28	323 Mar. 10
19 Nov. 1	80 Nov. 9	141 Aug. 18	202 Oct. 4	263 Sept. 9	324 Aug. 11
20 June 4	81 April 4	142 Aug. 12	203 Nov. 19	264 Oct. 27	325 Jan. 10
21 Aug. 10	82 Sept. 5	143 Nov. 17	204 Sept. 21	265 Mar. 22	326 May 22
22 June 26	83 April 3	144 Feb. 2	205 Feb. 27	266 Nov. 4	327 July 6
23 July 24	84 Dec. 25	145 Aug. 4	206 June 10	267 Mar. 3	328 Dec. 2
24 Oct. 5	85 June 7	146 Nov. 18	207 Sept. 16	268 Mar. 27	329 Jan. 11
25 Feb. 19	86 Feb. 1	147 April 7	208 April 30	269 April 5	330 May 1
26 Dec. 14	87 Oct. 6	148 April 16	209 June 30	270 July 29	331 July 14
27 July 21	88 July 28	149 Sept. 25	210 Feb. 4	271 April 2	332 Mar. 18
28 June 5	89 Feb. 15	150 Feb. 11	211 Jan. 31	272 June 12	333 Aug. 30
29 Mar. 2	90 April 18	151 Sept. 29	212 Feb. 16	273 April 15	334 Mar. 21
30 Mar. 31	91 Feb. 7	152 Feb. 13	213 Mar. 8	274 June 16	335 June 9
31 May 24	92 Jan. 26	153 July 22	214 Feb. 5	275 Mar. 4	336 April 19
32 April 1	93 July 1	154 Aug. 17	215 Jan. 4	276 May 4	337 Jan. 22
33 Mar. 17	94 Oct. 28	155 May 6	216 Feb. 10	277 July 9	338 Feb. 9
34 Nov. 2	95 Dec. 24	156 Nov. 21	217 Mar. 30	278 May 18	339 Aug. 22
35 May 7	96 Dec. 16	157 Dec. 3	218 April 10	279 July 4	340 April 26
36 Aug. 24	97 Nov. 8	158 Sept. 11	219 April 9	280 Jan. 20	341 June 18
37 May 11	98 July 17	159 Jan. 2	220 Oct. 10	281 Nov. 28	342 Oct. 9
38 Oct. 30	99 Nov. 29	160 Sept. 22	221 Jan. 12	282 Nov. 10	343 Mar. 25
39 Dec. 11	100 Dec. 31	161 Sept. 2	222 Jan. 28	283 Oct. 8	334 Aug. 20
40 May 3	101 Jan. 5	162 Dec. 23	223 Mar. 28	284 July 10	345 April 20
41 Dec. 10	102 Aug. 15	163 Dec. 13	224 Jan. 6	285 Feb. 29	346 April 12
42 July 13	103 May 30	164 Jan. 30	225 Sept. 1	286 Aug. 25	347 Feb. 6
43 Dec. 9	104 June 19	165 Dec. 4	226 May 29	287 July 30	348 Nov. 3
44 Aug. 16	105 Dec. 8	166 Mar. 16	227 July 19	288 Oct. 17	349 Jan. 29
45 Aug. 2	106 Aug. 9	167 Aug. 28	228 June 2	289 July 27	350 July 2
46 Nov. 11	107 Nov. 16	168 Aug. 7	229 Oct. 29	290 Feb. 22	351 April 25
47 Nov. 27	108 Mar. 1	169 Mar. 15	230 Nov. 24	291 Aug. 21	352 Aug. 27
48 Aug. 8	109 June 23	170 Mar. 26	231 April 14	292 Feb. 18	353 June 29
49 Sept. 3	110 June 6	171 Oct. 15	232 Sept. 4	293 Mar. 5	354 Mar. 14
50 July 7	111 Aug. 1	172 July 23	233 Sept. 27	294 Oct. 14	355 Jan. 27
51 Nov. 7	112 May 17	173 Dec. 26	234 Oct. 7	295 May 13	356 June 14
52 Jan. 25	113 Sept. 15	174 Nov. 30	235 Jan. 17	296 May 27	357 May 26
53 Dec. 22	114 Aug. 6	175 Sept. 13	236 Feb. 24	297 Feb. 3	358 June 24
54 Aug. 5	115 July 3	176 Oct. 25	237 Oct. 11	298 May 2	359 Oct. 1
55 May 16	116 Aug. 23	177 Sept. 19	238 Jan. 14	299 Feb. 28	360 June 20
56 Dec. 5	117 Oct. 22	178 May 14	239 Mar. 20	300 Mar. 12	361 May 25
57 Feb. 23	118 Jan. 23	179 Feb. 25	240 Dec. 19	301 June 3	362 Mar. 29
58 Jan. 19	119 Sept. 23	180 June 15	241 Oct. 19	302 Feb. 20	363 Feb. 21
59 Jan. 24	120 July 16	181 Feb. 8	242 Sept. 12	303 July 26	364 May 5
60 June 21	121 Jan. 16	182 Nov. 23	243 Oct. 21	304 Dec. 17	365 Feb. 26
61 Aug. 29	122 Mar. 7	183 May 20	244 Oct. 3	305 Jan. 1	366 June 8

FIGURE 3.16 ▶
Median plots for lottery results: 1970 and 1971

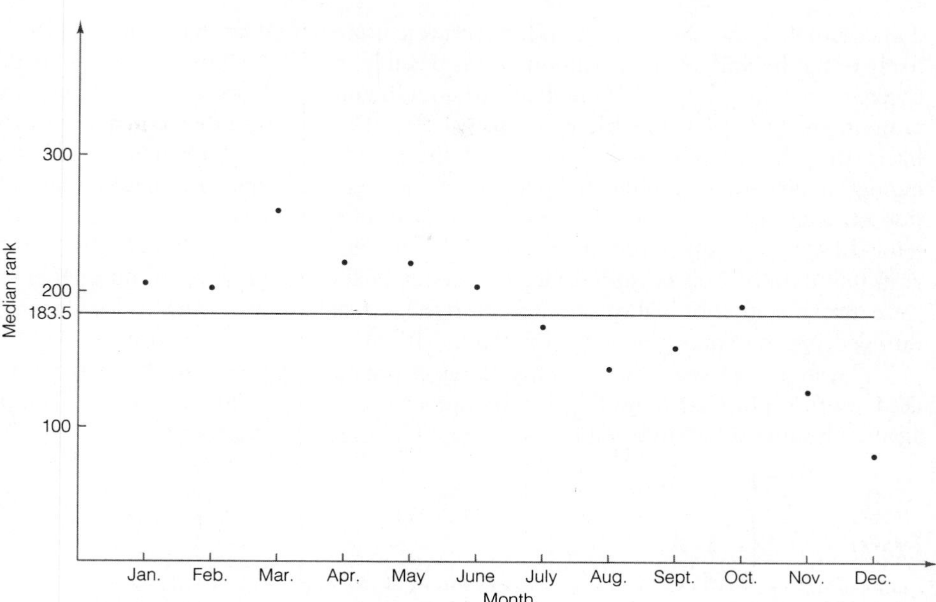

a. 1970 lottery

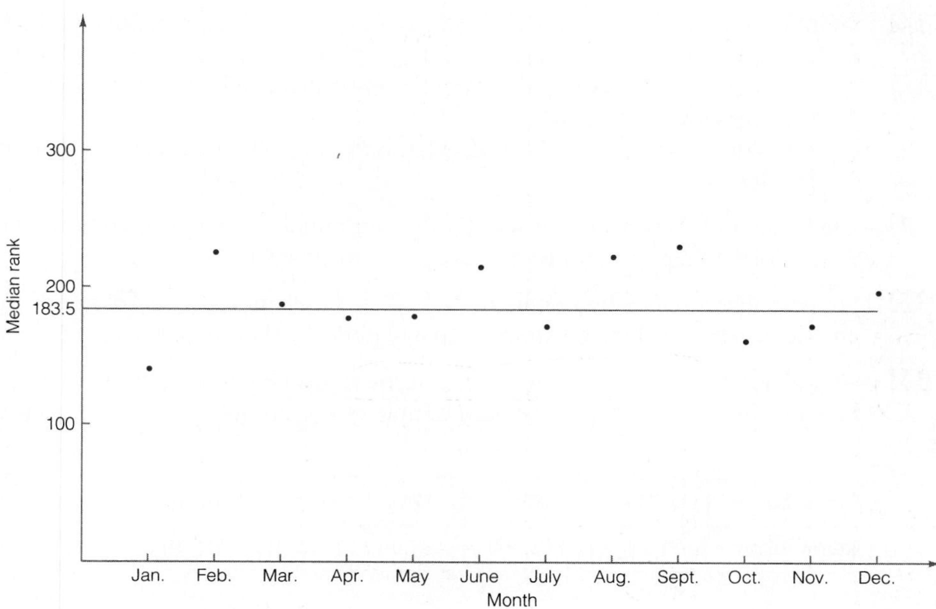

b. 1971 lottery

drafted in 1970, men born earlier in the year were more likely not to be drafted at all. Although it is possible to observe such a sequence of medians when random sampling is employed, it is highly unlikely. It is more likely that the capsules were not mixed thoroughly enough to give every capsule an equal chance of selection on each draw. The graph indicates that capsules tended to be drawn in monthly groups, and thus suggests that monthly lots of capsules in the wooden boxes were not mixed thoroughly enough after they were dumped into the large glass bowl (Williams, 1978).

The following year, the Selective Service System used more sophisticated mixing techniques to guard against the monthly clustering of selections. The median plot for the 1971 lottery is shown in Figure 3.16(b). [*Note:* The 1971 lottery involved only men born in 1951. Thus, only 365 birthdays were ranked and the sequence of monthly ranked medians should be compared to 183 instead of 183.5.] Note that no apparent trend remains; the new mixing technique appears to have been successful.

This case study emphasizes the value of using a random-number table or a random number generator of a statistical software package in the selection of a random sample. Most important, it points out the problems that may be encountered when attempting to acquire a random sample by a mechanical selection process.

Exercises 3.48–3.54

Learning the Mechanics

3.48 Suppose you wish to draw a sample of $n = 2$ elements from a population that contains $N = 10$ elements.

a. List all possible different samples of $n = 2$ elements, and also use combinatorial mathematics (see Section 3.1) to count the number of different samples that can be selected. How many different samples of $n = 2$ elements can be selected?

b. If random sampling is employed, what is the probability of drawing a particular pair of elements in your sample?

3.49 Use Table I of Appendix B to select a random sample of size $n = 20$ from among the digits 0, 1, 2, . . . , 9. (Digits may repeat themselves.) Explain your procedure.

3.50 Suppose that a population contains $N = 200{,}000$ elements. Use Table I of Appendix B to select a random sample of $n = 10$ elements from the population. Explain how you selected your sample.

3.51 Consider the population of new savings accounts opened in one business day at a bank, as shown in the table. Suppose you wish to draw a random sample of two accounts from this population.

Account Number	0001	0002	0003	0004	0005
Account Balance	\$1,000	\$12,500	\$850	\$1,000	\$3,450

a. List all possible different pairs of accounts that could be obtained.

b. What is the probability of selecting accounts 0001 and 0004?

c. What is the probability of selecting two accounts that each have a balance of \$1,000? That each have a balance other than \$1,000?

Applying the Concepts

3.52 To ascertain the effectiveness of their advertising campaigns, firms frequently conduct telephone interviews with consumers. Samples of telephone numbers may be randomly or systematically selected from telephone directories, or a more recent innovation called **random-digit dialing** may be employed. This approach involves using a random-number generator to create the sample of phone numbers to be called. Use the random-number table (Table I of Appendix B) to generate a sample of 10 seven-digit telephone numbers from area code 215. Describe the procedure you used.

3.53 When a company sells shares of stock to investors, the transaction is said to take place in the *primary market*. To enable investors to resell the stock when they wish, *secondary markets* called *stock exchanges* were created. Stock exchange transactions involve buyers and sellers exchanging cash for shares of stock, with none of the proceeds going to the companies that issued the shares (Greenleaf, Foster, and Prinsky, 1982). The results of the previous business day's transactions for stocks traded on the New York Stock Exchange (NYSE) and five regional exchanges—the Midwest, Pacific, Philadelphia, Boston, and Cincinnati stock exchanges—are summarized each business day in the NYSE–Composite Transactions table in the *Wall Street Journal*.

a. Examine the NYSE–Composite Transactions table in a recent issue of the *Wall Street Journal* and explain how to draw a random sample of stocks from the table.

b. Use the procedure you described in part **a** to draw a random sample of 20 stocks from a recent NYSE–Composite Transactions table. For each stock in the sample, list its name (i.e., the abbreviation given in the table), its sales volume, and its closing price.

3.54 The total labor force has increased annually, from 69.6 million in 1960 to 106.9 million in 1980 to 124.8 million in 1990. From 1960 to 1980, the percentage of the labor force age 16–34 steadily increased, and the percentage of the labor force age 35–64 decreased. However, beginning around 1980, with the aging of the baby-boom generation, these percentages began moving in the opposite directions.

Age	*Percent in the Labor Force*		
	1960	*1980*	*1990*
16–34	37.3	51.0	45.7
35–64	58.1	46.1	51.4
65+	4.6	2.9	2.8

Source: *Statistical Abstract of the United States: 1991*, p. 388.

a. If the year were 1990 and one laborer were randomly selected from the labor force, what is the probability that a particular laborer would be selected? What is the probability that the laborer would be between 16 and 34 years of age?

b. Repeat part **a** for the year 1960, and compare the answers.

c. If two laborers were randomly selected from the 1990 labor force, what is the probability of selecting a particular pair of laborers?

Summary

We have developed some of the basic tools of probability that enable us to determine the probabilities of various sample outcomes, given a specific population structure. Although many of the examples we presented were of no practical importance, they accomplished their purpose if you now understand the basic concepts and definitions of probability.

The basic understanding of probability presented in this chapter includes the following concepts: **Experiments** are the basis for the generation of data. The most basic outcomes of experiments, called **simple events**, cannot be predicted with certainty. Therefore, we assign **probabilities** to the simple events, such that they obey two rules: (1) all probabilities must be between 0 and 1, and (2) the simple event probabilities must sum to 1. The collection of all the simple events is called the **sample space** of the experiment.

Unions and **intersections** are useful combinations of events. **Unions** are inclusive: either or both events occur. **Intersections** are limiting: both events occur. **Complements** are exclusive: the event does not occur. The **additive rule** is useful for calculating the probability of unions, and is particularly simple when the events have no intersection, in which case they are said to be **mutually exclusive**.

Conditional probability applies when we have some knowledge about what has occurred, reducing the size of the sample space. The **multiplicative rule** is useful for relating the probability of an intersection to conditional probability. When an intersection probability is equal to the product of the two (unconditional) event probabilities, we say the events are **independent**.

Our primary reason for studying the basics of probability is to enable the quantitative evaluation of alternative explanations (or models) of real data. To be able to rule out some explanations as "unlikely" and to accept others as "probable," we must first be able to quantify "unlikely" and "probable," and the theory of probability will provide the foundation for this endeavor.

In the next several chapters, we will present probability models that can be used to solve practical business problems. You will see that for most applications, we will need to make inferences about unknown aspects of these probability models; i.e., we will need to apply inferential statistics to the problem.

Supplementary Exercises 3.55–3.79

3.55 What are the two rules that probabilities assigned to simple events must obey?

3.56 Are mutually exclusive events also dependent events? Explain.

3.57 Given that $P(A \cap B) = .4$ and $P(A \mid B) = .8$, find $P(B)$.

3.58 Which of the following pairs of events are mutually exclusive? Justify your response.

a. The Dow Jones Industrial Average increases on Monday.
A large New York bank decreases its prime interest rate on Monday.

b. The next sale by a PC retailer is an IBM microcomputer.
The next sale by a PC retailer is an Apple microcomputer.

c. You reinvest all your dividend income for 1990 in a limited partnership.
You reinvest all your dividend income for 1990 in a money market fund.

3.59 A manufacturer of electronic digital watches claims that the probability of its watch running more than 1 minute slow or 1 minute fast after 1 year of use is .05. A consumer protection agency has purchased four of the manufacturer's watches with the intention of testing the claim.

a. Assuming that the manufacturer's claim is correct, what is the probability that all four of the watches are as accurate as claimed?

b. Assuming that the manufacturer's claim is correct, what is the probability that exactly two of the four watches fail to meet the claim?

c. Suppose that three of the four tested watches failed to meet the claim. What inference can be made about the manufacturer's claim? Explain.

d. Suppose that all four tested watches failed to meet the claim. Is it necessarily true that the manufacturer's claim is false? Explain.

3.60 The state legislature has appropriated $1 million to be distributed in the form of grants to individuals and organizations engaged in the research and development of alternative energy sources. You have been hired by the state's energy agency to assemble a panel of five energy experts whose task it will be to determine which individuals and organizations should receive the grant money. You have identified 11 equally qualified individuals who are willing to serve on the panel. How many different panels of five experts could be formed from these 11 individuals?

3.61 A research and development company surveyed all 200 of its employees over the age of 60 and obtained the information given in the table. One of these 200 employees is selected at random.

	Under 20 Years with Company		Over 20 Years with Company	
	Technical Staff	Nontechnical Staff	Technical Staff	Nontechnical Staff
Plan to Retire at Age 65	31	5	45	12
Plan to Retire at Age 68	59	25	15	8

a. What is the probability that the person selected is on the technical staff?

b. If the person selected has over 20 years of service with the company, what is the probability that the person plans to retire at age 68?

c. If the person selected is on the technical staff, what is the probability that the person has been with the company less than 20 years?

d. What is the probability that the person selected has over 20 years with the company, is on the nontechnical staff, and plans to retire at age 65?

3.62 Refer to Exercise 3.61.

a. Consider the events A: {Plan to retire at age 68} and B: {On the technical staff}. Are events A and B independent? Explain.

b. Consider the event D: {Plan to retire at age 68 *and* on the technical staff}. Describe the complement of event D.

c. Consider the event E: {On the nontechnical staff}. Are events B and E mutually exclusive? Explain.

3.63 Two marketing research companies, Richard Saunders International and Marketing Intelligence Service, joined forces to create a consumer preference poll called Acupoll. Acupoll is used to predict whether newly developed products will succeed if they are brought to market. The reliability of the Acupoll has been described as follows: The probability that Acupoll predicts the success of a particular product, given that later the product actually is successful, is .89 (Hall, 1992). A company is considering the introduction of a new product and assesses the product's probability of success to be .90. If this company were to have its product evaluated through Acupoll, what is the probability that Acupoll predicts success for the product and the product actually turns out to be successful?

3.64 The performance of quality inspectors affects both the quality and the cost of outgoing products. A product that passes inspection is assumed to meet quality standards; a product that fails inspection may be reworked, scrapped, or reinspected. Quality engineers at Westinghouse Electric Corporation evaluated inspectors' performances in judging the quality of solder joints by comparing each inspector's classifications of a set of 153 joints with the consensus evaluation of a panel of experts (Meagher and Scazzero, 1985). Suppose the results for a particular inspector are as shown in the table, and that one of the 153 solder joints is to be selected at random.

		Inspector's Judgment	
		Joint Acceptable	Joint Rejectable
Committee's Judgment	Joint Acceptable	101	10
	Joint Rejectable	23	19

a. What is the probability that the inspector judges the joint to be acceptable? That the committee judges the joint to be acceptable?
b. What is the probability that both the inspector and the committee judge the joint to be acceptable? That neither judges the joint to be acceptable?
c. What is the probability that the inspector and the committee disagree? Agree?

3.65 A local country club has a membership of 600 and operates facilities that include an 18-hole championship golf course and 12 tennis courts. Before deciding whether to accept new members, the club president would like to know how many members regularly use each facility. A survey of the membership indicates that 70% regularly use the golf course, 50% regularly use the tennis courts, and 5% use neither of these facilities regularly.
a. Construct a Venn diagram to describe the results of the survey.
b. If one club member is chosen at random, what is the probability that the member uses either the golf course or the tennis courts or both?
c. If one member is chosen at random, what is the probability that the member uses both the golf and the tennis facilities?
d. A member is chosen at random from among those known to use the tennis courts regularly. What is the probability that the member also uses the golf course regularly?

3.66 Insurance companies use *mortality tables* to help them determine how large a premium to charge a particular individual for a particular life insurance policy. The accompanying table shows the probability of survival to age 65 for persons of the specified ages.

Age	Probability of Survival to Age 65	Age	Probability of Survival to Age 65
0	.72	40	.77
10	.74	45	.79
20	.74	50	.81
30	.75	55	.85
35	.76	60	.90

a. For a person 20 years old, what is the probability that he or she will die before age 65?
b. Describe in words the trend indicated by the increasing probabilities in the second and fourth columns.

3.67 Explain why the following statement is or is not valid: If an individual is chosen at random from all U.S. citizens living in the 50 states, the probability that this individual lives in New Hampshire is 1/50.

3.68 A manufacturer of 35-mm cameras knows that a shipment of 30 cameras sent to a large discount store contains six defective cameras. The manufacturer also knows that the store will choose two of the cameras at random, test them, and accept the shipment if neither is defective.
a. What is the probability that the first camera chosen by the store will be defective?
b. Given that the first camera chosen passed inspection, what is the probability that the second camera chosen will fail inspection?
c. What is the probability that the shipment will be accepted?

3.69 The accompanying figure is a schematic representation of a system comprised of three components connected *in series*. The system functions properly only if all three components operate properly. The components could be mechanical or electrical; they could be work stations in an assembly process; or could represent the functions of three different departments in an organization. The probability of failure for each of the components follows: #1 = .12, #2 = .09, #3 = .11.

A System Comprised of Three Components in Series

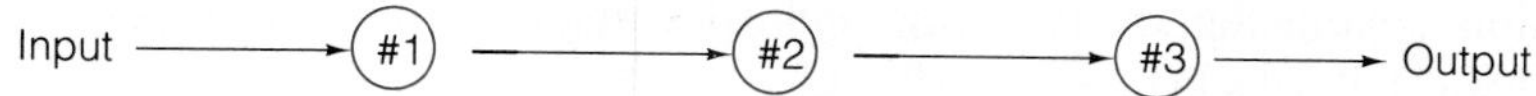

a. Find the probability that the system operates properly.
b. What is the probability that at least one of the components will fail and, therefore, that the system will fail?

3.70 To accompanying figure is a representation of a system comprised of two subsystems that are said to operate *in parallel*. Each subsystem has two components that operate in series (refer to Exercise 3.69). The system will operate properly as long as at least one of the subsystems functions properly. The probability of failure for each component in the system is .1. Assume that the components operate independently of each other.

A System Comprised of Two Parallel Subsystems

Subsystem A
#1
#2
Input
Output
Subsystem B
#3
#4

a. Find the probability that the system operates properly.
b. Find the probability that exactly one subsystem fails.
c. Find the probability that the system fails to operate properly.
d. How many parallel subsystems like the two shown here would be required to guarantee that the system would operate properly more than 99% of the time?

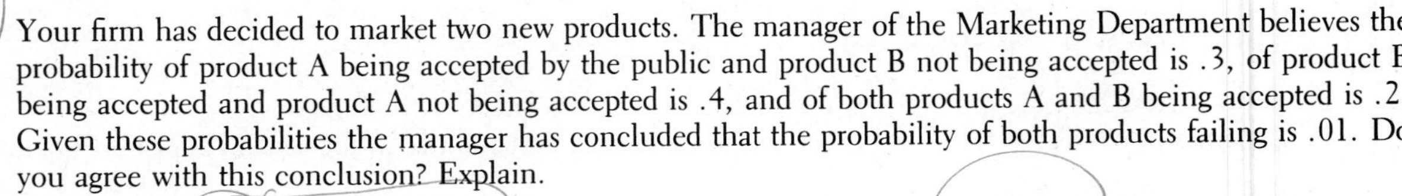

3.71 Your firm has decided to market two new products. The manager of the Marketing Department believes the probability of product A being accepted by the public and product B not being accepted is .3, of product B being accepted and product A not being accepted is .4, and of both products A and B being accepted is .2. Given these probabilities the manager has concluded that the probability of both products failing is .01. Do you agree with this conclusion? Explain.

3.72 Suppose only two daily newspapers are available in your town—a local paper and one from a nearby city—and that 1,000 people in town subscribe to a daily paper. Assume that 65% of the people in town who subscribe to a daily newspaper subscribe to the local paper and 40% of those who subscribe to a daily paper subscribe to the city paper.
a. Use a Venn diagram to describe the population of newspaper subscribers.
b. If one of the 1,000 subscribers is chosen at random, what is the probability that he or she subscribes to both newspapers?

3.73 Six people apply for two identical positions in a company. Four are minority applicants and the remainder are nonminority. Define the following events:

A: {Both persons selected for the positions are nonminority candidates}
B: {Both persons selected for the positions are minority candidates}
C: {At least one of the persons selected is a minority candidate}

If all the applicants are equally qualified and the choice is therefore a random selection of two applicants from the six available, find the following:
a. $P(A)$ b. $P(B)$ c. $P(C)$ d. $P(B \mid C)$
e. For the purpose of identification, assume that the minority candidates are numbered 1, 2, 3, and 4. Define the event D: {Minority candidate 1 is selected}. Find $P(D \mid C)$.

3.74 Suppose there are 500 applicants for five equivalent positions at a factory and the company is able to narrow the field to 30 equally qualified applicants. Seven of the finalists are minority candidates. Assume that the five who are chosen are selected at random from this final group of 30.
a. What is the probability that none of the minority candidates is hired?
b. What is the probability that no more than one minority candidate is hired?

3.75 The probability that a microcomputer salesperson sells a computer to a prospective customer on the first visit to the customer is .4. If the salesperson fails to make the sale on the first visit, the probability that the sale will be made on the second visit is .65. The salesperson never visits a prospective customer more than twice. What is the probability that the salesperson will make a sale to a particular customer?

3.76 A credit counselor claims that the probability that at least two local firms go bankrupt next year is .15, and the probability that exactly two local firms go bankrupt is .20. Can this statement be true? Explain.

3.77 Use a Venn diagram to show that

$$P(A \cap B^c) = P(A) - P(A \cap B)$$

3.78 Use your intuitive understanding of independence to form an opinion about whether each of the following scenarios represents independent events.

  a. The results of consecutive tosses of a coin
  b. The opinions of randomly selected individuals in a pre-election poll
  c. A major-league baseball player's results in two consecutive at-bats
  d. The amount of gain or loss associated with investments in different stocks that are bought on the same day, and sold on the same day 1 month later
  e. The amount of gain or loss associated with investments in different stocks that are bought and sold in different time periods, 5 years apart
  f. The prices bid by two different development firms in response to a building construction proposal issued by a university.

3.79 A fair coin is flipped 20 times and 20 heads are observed. In such cases it is often said that a tail is due on the next flip. Is this statement true or false? Explain.

On Your Own

Obtain a standard bridge deck of 52 cards and think of the cards as the 52 items your firm produces each day. Let the four aces and four kings in the deck represent defective items.

a. If one item is randomly sampled from a day's production, what is the probability of its being defective?
b. Shuffle the cards, draw one, and record whether it is a defective item. Then replace the card and repeat the process. After each draw, recalculate the proportion of the draws that have resulted in a defective item. Construct a graph with the proportion of defectives on the y-axis and the number of draws on the x-axis. Notice how the proportion defective stabilizes as the number of draws increases.
c. Draw a horizontal line on the graph in part **b** at a height equal to the probability you calculated in part **a**. Compare the calculated proportion of defectives to this probability. As the number of draws is increased, does the calculated proportion of defectives more closely approach the actual probability of drawing a defective?

Using the Computer

Suppose a large bank is planning a national mailing to market a major credit card. However, the bank will first test its marketing materials by mailing to a single zip code.

a. If one of the 1,000 zip codes in Appendix C were to be selected randomly for a test mailing, what is the probability that the selected zone is one for which the average income exceeds $35,000?
b. Suppose the zip code were to be selected from the Northeast census region (among those in Appendix C). What is the probability that the selected zone is one for which the average income exceeds $35,000?
c. Are the events described in parts **a** and **b** independent? Why or why not? What are the practical implications of the independence, or dependence, of the events?

References

Bagozzi, R. P. *Principles of Marketing Management*. Chicago: SRA, 1986, p. 215.

Benson, P. George. "Process thinking: The quality catalyst." *Minnesota Management Review*, Carlson School of Management, University of Minnesota, Minneapolis, Fall 1992.

Feller, W. *An Introduction to Probability Theory and Its Applications*, 3d ed. Vol. 1. New York: Wiley, 1968. Chapters 1, 4, and 5.

"GM's Saturn takes 3rd in new-car owner poll." *Star Tribune* (Minneapolis). June 30, 1992.

Greenleaf, J., Foster, R., and Prinsky, R. "Understanding financial data in the *Wall Street Journal*." *Wall Street Journal*, Special Education Edition, 1982, p. 19.

Hall, Trish. "Pollsters think they have trick for product picks." *Star Tribune* (Minneapolis). Dec. 16, 1992.

Kozlov, Alex. "Business products for the technophobic and parsimonious." *Newsweek*, Nov. 16, 1992, p. 7.

Kuehn, A. A. "An analysis of the dynamics of consumer behavior and its implications for marketing management." Unpublished doctoral dissertation, Graduate School of Industrial Administration, Carnegie Institute of Technology, 1958.

Lindley, D. V. *Making Decisions*, 2d ed. London: Wiley, 1985.

Meagher, J. J. and Scazzero, J. A. "Measuring inspector variability." *39th Annual Quality Congress Transactions*, American Society for Quality Control, May 1985, pp. 75–81.

Parzen, E. *Modern Probability Theory and Its Applications*. New York: Wiley, 1960. Chapters 1 and 2.

Peterson, R. L. "Creditors' use of collection remedies." *Journal of Financial Research*, Vol. 9 , No. 1, Spring 1986, pp. 71–86.

Press, S. J. *Bayesian Statistics*. New York: Wiley, 1989.

"Random or not? Judge studies lottery protest." *The National Observer*, Jan. 12, 1970, p. 2.

Schmickle, Sharon. " 'Sustainable' farming viable, study suggests." *Star Tribune* (Minneapolis). June 20, 1992.

Williams, B. *A Sampler on Sampling*. New York: Wiley, 1978. pp. 5–8.

Winkler, R. L. *An Introduction to Bayesian Inference and Decision*. New York: Holt, Rinehart and Winston, 1972. Chapter 2.

Winkler, R. L. and Hays, W. L. *Statistics: Probability, Inference, and Decision*, 2d ed. New York: Holt, Rinehart and Winston, 1975. Chapters 1 and 2.

CHAPTER FOUR

Discrete Random Variables

Contents

Case Studies

Where We've Been

By illustration, we indicated in Chapter 3 how probability would be used to make an inference about a population from information contained in a sample. We also noted that probability would be used to measure the reliability of the inference.

Where We're Going

Most events in Chapter 3 were described in words or were denoted by capital letters. In real life, most sample observations are numerical—in other words, they are quantitative data. In this chapter, we will learn that data can be characterized as observed values of random variables. We will study two important random variables and will learn how to find the probabilities of specific numerical outcomes.

4

You may have noticed that most of the examples of experiments given in Chapter 3 generated quantitative (numerical) data. This is frequently true; observations on many types of phenomena are numerical measurements. The Consumer Price Index, unemployment rate, number of sales made in a week, and yearly profit of a company are all examples of numerical measurements of some business phenomenon. Thus, most experiments have simple events that correspond to values of some numerical variable.

Definition 4.1

A **random variable** is a rule that assigns one (and only one) numerical value to each simple event of an experiment.*

The term **random variable** is more meaningful than the simpler term *variable* because the adjective *random* indicates that the experiment may result in one of the several possible values of the variable, according to the *random* outcome of the experiment. For example, if the experiment is to count the number of customers who use the drive-up window of a bank each day, the random variable (the number of customers) will vary from day to day, partly because of the random phenomena that influence whether customers use the drive-up window. Thus, the possible values of this random variable range from zero to the maximum number of customers the window could possibly serve in a day.

We define two different types of random variables, *discrete* and *continuous*, in Section 4.1. Then we spend the remainder of the chapter discussing specific types of discrete random variables and the aspects that make them important in business applications. We discuss continuous random variables in Chapter 5.

4.1 Two Types of Random Variables

Recall that the simple event probabilities corresponding to an experiment must sum to 1. Assigning one unit of probability to the simple events in a sample space, and consequently to the values of a random variable, is not always as easy as the examples in Chapter 3 may lead you to believe. If the number of simple events is finite, the job is easy. If the number of simple events is infinite but you can list them in order (we call this **countable**), the task is still not too difficult. But if the simple events are numerical and correspond to the infinitely large number of points contained in an interval, the task is impossible. Why? Because you cannot assign a small portion of probability to each of the simple events in this infinitely large set—the sum of the probabilities will exceed 1 (in fact, the sum will be infinitely large). The consequences of this mathe-

*By *experiment*, we mean an experiment that yields random outcomes (as defined in Chapter 3).

matical fact are important. We will have to use two different probability models, depending on whether the number of simple events in a sample space (or equivalently, the values that a random variable can assume) are countable or they correspond to the infinitely large number of points contained in one or more intervals.

Definition 4.2

Random variables that can assume a *countable* number of values are called **discrete**.

Definition 4.3

Random variables that can assume values corresponding to any of the points contained in one or more intervals are called **continuous**.

Examples of **discrete** random variables are:

1. The number of sales made by a salesperson in a given week: $x = 0, 1, 2, \ldots$.
2. The number of people in a sample of 500 who favor a particular product over all competitors: $x = 0, 1, 2, \ldots, 499, 500$.
3. The number of bids received in a bond offering: $x = 0, 1, 2, \ldots$.
4. The number of errors on a page of an accountant's ledger: $x = 0, 1, 2, \ldots$.
5. The number of customers waiting to be served in a restaurant at a particular time: $x = 0, 1, 2, \ldots$.

Note that each of the examples of discrete random variables begins with the words "the number of." This is very common because the discrete random variables most frequently observed are counts.

Examples of **continuous** random variables are:

1. The length of time between arrivals at a hospital clinic: $0 \leq x < \infty$ (infinity).
2. For a new apartment complex, the length of time from completion until a specified number of apartments are rented: $0 \leq x < \infty$.
3. The amount of carbonated beverage loaded into a 12-ounce can in a can filling operation: $0 \leq x \leq 12$.
4. The depth at which a successful oil drilling venture first strikes oil.
5. The weight of a food item bought in a supermarket.

In Section 4.2 we discuss how to find the probability distribution for a discrete random variable. Then, several types of discrete random variables that play important roles in business decisions will be presented in subsequent sections. Probability distributions for some useful types of continuous random variables will be the subject of Chapter 5.

Exercises 4.1–4.10

Applying the Concepts

4.1 What is a random variable?

4.2 How do discrete and continuous random variables differ?

4.3 Which of the following describe continuous random variables, and which describe discrete random variables? Justify your answers.
- **a.** The amount of water flowing through the Hoover Dam in a year
- **b.** The number of people who fly American Airlines each day
- **c.** The length of time it takes to assemble one Ford Thunderbird
- **d.** The number of patients that are admitted per week in a particular hospital

4.4 Which of the following describe continuous random variables, and which describe discrete random variables?
- **a.** The number of newspapers sold by the *New York Times* each month
- **b.** The amount of ink used in printing a Sunday edition of the *New York Times*
- **c.** The actual number of ounces in a 1-gallon bottle of laundry detergent
- **d.** The number of defective parts in a shipment of nuts and bolts
- **e.** The number of people collecting unemployment insurance each month

4.5 Give two examples of a business-oriented discrete random variable. Do the same for a continuous random variable.

4.6 Give an example of a discrete random variable that would be of interest to a banker.

4.7 Give an example of a continuous random variable that would be of interest to an economist.

4.8 Give an example of a discrete random variable that would be of interest to the manager of a hotel.

4.9 Give two examples of discrete random variables that would be of interest to the manager of a clothing store.

4.10 Give an example of a continuous random variable that would be of interest to a stockbroker.

4.2 Probability Distributions for Discrete Random Variables

Since a random variable assigns a numerical value to each of the simple events associated with an experiment, a complete description of a random variable requires that we specify its **probability distribution**. Note that each simple event is assigned one and only one value of the random variable, and hence, the values of the random variable represent mutually exclusive events.

Definition 4.4

The **probability distribution** of a discrete random variable is a graph, table, or formula that specifies the probability associated with each possible value the random variable can assume.

To illustrate, consider Example 4.1.

EXAMPLE 4.1

Recall the experiment of tossing two coins (Chapter 3), and let x be the number of heads observed. Find the probability distribution for the random variable x, assuming the two coins are fair.

Solution

Recall from Chapter 3 that the sample space and simple events for this experiment are as shown in Figure 4.1, and the probability associated with each of the four simple events is $\frac{1}{4}$. The random variable x can assume values 0, 1, 2. Then, identifying the probabilities of the simple events associated with each of these values of x, we have

$$P(x = 0) = P(T_1, T_2) = \tfrac{1}{4}$$
$$P(x = 1) = P(T_1, H_2) + P(H_1, T_2) = \tfrac{1}{4} + \tfrac{1}{4} = \tfrac{1}{2}$$
$$P(x = 2) = P(H_1, H_2) = \tfrac{1}{4}$$

We denote the probability of the discrete random variable x by the symbol $p(x)$. Then, for this example, $p(0) = \frac{1}{4}$, $p(1) = \frac{1}{2}$, and $p(2) = \frac{1}{4}$. Table 4.1 shows the probability distribution of x in tabular form, and Figure 4.2 on page 186 shows it in two alternative graphical forms. Figure 4.2(a) shows the probabilities concentrated at the points, $x =$ 0, 1, and 2. The heights of the vertical line segments give the probabilities that correspond to each of these values of x. The probability distribution in Figure 4.2(b) is shown as a histogram with one class corresponding to each of the three values of x. Although the probabilities for discrete random variables are, in fact, concentrated at specific points, the probability histogram will be a convenient way of viewing the probability distribution for a discrete random variable when we attempt to approximate certain probabilities in Section 5.5.

FIGURE 4.1 ► Venn diagram for the two-coin-toss experiment

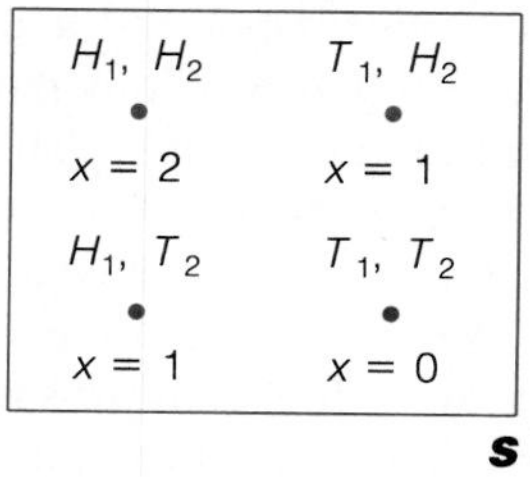

H_1, H_2	T_1, H_2
$x = 2$	$x = 1$
H_1, T_2	T_1, T_2
$x = 1$	$x = 0$

S

TABLE 4.1 Probability Distribution: Tabular Form

x	0	1	2
$p(x)$	$\frac{1}{4}$	$\frac{1}{2}$	$\frac{1}{4}$

FIGURE 4.2 ▶
Probability distribution: Graphical forms

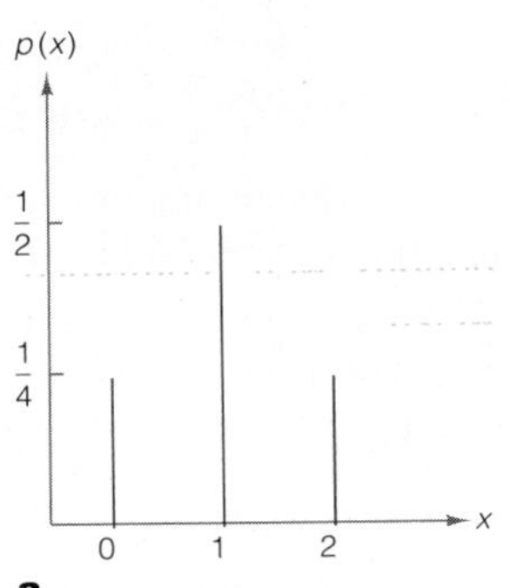

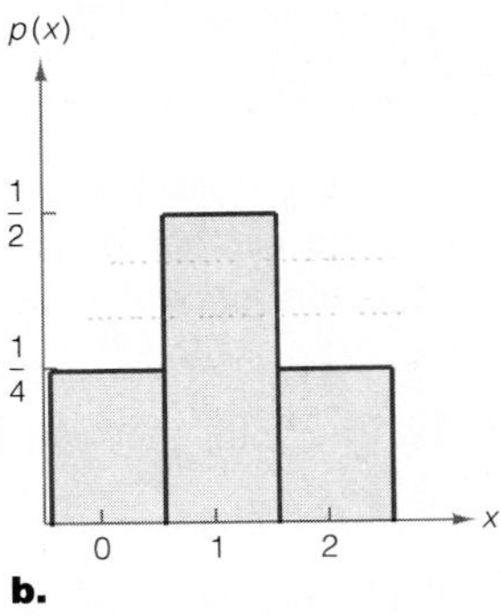

We could also present the probability distribution for x as a formula, but this would unnecessarily complicate a very simple example. We will give the formulas for the probability distributions of some discrete random variables later in this chapter.

Two requirements must be satisfied by all probability distributions for discrete random variables:

> **Requirements for the Probability Distribution of a Discrete Random Variable x**
>
> 1. $p(x) \geq 0$ for all values of x
> 2. $\sum_{\text{All } x} p(x) = 1$

Example 4.1 illustrates how the probability distribution for a discrete random variable can be derived, but for many practical examples, the task is much more difficult. Fortunately, experiments and associated discrete random variables with identical characteristics are found in many different areas of business. That is, although the data may be collected in an area of accounting, marketing, economics, or management, for all practical purposes the data may represent observed values of the same type of random variable. This fact simplifies the problem for the business analyst. All that must be done is to define the nature of the experiment, define the type of random variable, and specify the probability distribution of the random variable.

In Sections 4.4 and 4.5 we describe two important types of discrete random variables, give their probability distributions, and explain where and how they can be applied in business. (Mathematical derivations of the probability distributions will be omitted, but these details can be found in the references at the end of the chapter.)

But first, in Section 4.3, we discuss some descriptive measures of these sometimes complex probability distributions. Since probability distributions are analogous to the relative frequency distributions of Chapter 2, it should be no surprise that the mean and standard deviation are useful descriptive measures.

CASE STUDY 4.1 / Assessing the Effects of the Deadly Dutch Elm Disease

Since 1930 when the Dutch elm disease fungus was first discovered in the U.S. on the east coast, it has spread westward destroying elm trees all across the continent. It has been called ". . . the most destructive and widespread plague of trees of our time" (Webster, 1978). The fungus is spread primarily by English bark beetles which breed in diseased elm trees. Because of a lack of an effective chemical treatment, the major tactic in combating Dutch elm disease is a sanitation program which involves the removal of the diseased elm trees before they become a breeding ground for the fungus-carrying beetle (Chervany et al., 1980).

At the start of 1977, the city of Minneapolis had approximately 200,000 elms and had not lost more than 7,200 elms in any single year to Dutch elm disease. Since a major reforestation program had recently been inaugurated, this loss rate was not overly alarming. In 1977, however, Minneapolis lost 31,475 elms, approximately 16% of the existing elm population.

Prior to the 1978 growing season the Minneapolis Park and Recreation Board (Park Board), the organization responsible for managing the city's Dutch Elm Disease Sanitation Program, hired a team of management scientists to help in the design and operation of a sanitation program capable of dealing with the increased disease incidence. An integral part of the program proposed by the team involved forecasting the number of elms to be lost in the coming growing season. This information would enable the Park Board to plan staffing, evaluate potential bottlenecks in the elm removal operation, and inform the citizenry of the extent of the damage to the urban forest expected during the next year.

Because of the scarcity of historical data, the forecast was developed from the opinions of four disease experts. Each expert developed a discrete probability distribution to represent his beliefs regarding the number of elms that would be infected by the disease during the 1978 growing season. These four probability distributions were combined (by averaging using equal weights) to obtain a single discrete probability distribution reflecting the beliefs of all the experts regarding the number of elms to be infected. A slightly modified version of this probability distribution appears in the table.

Number of Elms that Would be Infected in 1978

x	$p(x)$
5,000– 9,999	.02
10,000–14,999	.07
15,000–19,999	.16
20,000–24,999	.17
25,000–29,999	.12
30,000–34,999	.13
35,000–39,999	.11
40,000–44,999	.08
45,000–49,999	.07
50,000 or more	.07

This distribution was used to make probabilistic forecasts of the number of trees that would be infected in 1978. It indicated that, according to the combined opinions of the disease experts, it was most likely that the number of trees to be infected would be between 20,000 and 24,999. Furthermore, probability statements like the following were made about the number of trees to be infected, x:

$$P(15{,}000 \le x \le 29{,}999) = .45$$
$$P(15{,}000 \le x \le 39{,}999) = .69$$
$$P(10{,}000 \le x \le 49{,}999) = .91$$
$$P(x < 15{,}000) = .09$$
$$P(x \ge 40{,}000) = .22$$

It is interesting to note that the actual number of losses to Dutch elm disease in 1978 was 20,817 trees, which is within the modal interval of the forecast distribution (Chervany et al., 1980).

Exercises 4.11–4.17

Learning the Mechanics

4.11 Three coins are tossed. Let x equal the number of heads observed.

a. Identify the simple events associated with this experiment and assign a value of x to each simple event, assuming the coins are fair.

b. Calculate $p(x)$ for each value of x.

c. Display the probability distribution of x in graphical form.

4.12 A die is tossed. Let x be the number of spots observed on the upturned face of the die.

a. Find the probability distribution of x and display it in tabular form.

b. Display the probability distribution of x in graphical form.

4.13 Explain why each of the following is or is not a valid probability distribution for a random variable x.

a.

x	0	1	2	3
$p(x)$	.1	.3	.3	.2

b.

x	−2	−1	0
$p(x)$	.25	.50	.25

c.

x	4	9	20
$p(x)$	−.3	.4	.3

d.

x	2	3	5	6
$p(x)$	.15	.15	.45	.35

4.14 The random variable x has the following discrete probability distribution:

x	1	3	5	7	9
$p(x)$	.1	.2	.4	.2	.1

a. Find $P(x \leq 3)$. b. Find $P(x < 3)$. c. Find $P(x = 7)$.
d. Find $P(x \geq 5)$. e. Find $P(x > 2)$. f. Find $P(3 \leq x \leq 9)$.

Applying the Concepts

4.15 The following probability distribution characterizes a marketing analyst's beliefs concerning the probabilities associated with the number, x, of sales that a company might expect per month for a new super-computer:

x	0	1	2	3	4	5	6	7	8
$p(x)$	.02	.08	.15	.19	.24	.17	.10	.04	.01

a. Display the probability distribution for x in graphical form.
b. Based on the probability distribution for x, what is the probability that the company will sell more than three computers per month? More than four?

4.16 Experience has shown that a builder of custom houses makes a profit on 95% of his contracts. Assume the event that the builder makes a profit on any one job is independent of whether he makes a profit on any other. The builder typically contracts to build three houses per month. Let x equal the number of houses per month that result in a profit.
a. Find $p(x)$. b. Graph $p(x)$. c. Find the probablity that $x \geq 2$.

4.17 The state of Minnesota has a "Pick Three" lottery game. The object of the game is to pick three numbers between 0 and 9, inclusive, that match three numbers selected in an official daily drawing. In the drawing, one ball is randomly selected from each of three bins containing the numbers 0–9. To win, the order of the numbers drawn must match the order in which you selected the numbers.
a. Suppose you picked the numbers 1, 3, and 7. What is the probability that none of your numbers will match those of the official drawing? That all three of your numbers will match?
b. What is the probability that you will match one number? Two numbers?
c. Will your answers to parts **a** and **b** change if it is not required that the order of the numbers drawn match the order in which you selected the numbers? Explain.

4.3 Expected Values of Discrete Random Variables

If a discrete random variable, x, were observed a very large number of times and if the data generated were arranged in a relative frequency distribution, the relative frequency distribution would be indistinguishable from the probability distribution for the random variable. Thus, the probability distribution for a random variable can be viewed as a theoretical model for the relative frequency distribution of a population. To the extent that the two distributions are equivalent (and we will assume they are), the probability distribution for x possesses a mean μ and a variance σ^2 that are identical to the corresponding descriptive measures for the population. The purpose of this section is to explain how you can find the mean value—or *expected value*, as it is called—for a random variable. We will illustrate the procedure with an example.

Examine the probability distribution for x (the number of heads observed in the toss of two fair coins) in Figure 4.3. Try to locate the mean of the distribution intuitively. We may reason as follows that the mean μ of this distribution is equal to 1: In a large number of experiments, ¼ should result in $x = 0$, ½ in $x = 1$, and ¼ in $x = 2$ heads. For example, if you were to perform the experiment 1,000 times, approximately 250 should result in $x = 0$, approximately 500 in $x = 1$, and approximately 250 in $x = 2$. Therefore, the average number of heads is

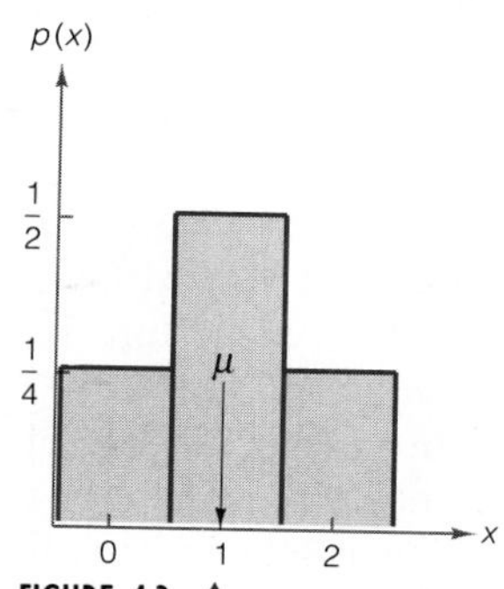

FIGURE 4.3 ▲ **Probability distribution for a two-coin toss**

$$\mu = \frac{250(0) + 500(1) + 250(2)}{1{,}000} = 0(1/4) + 1(1/2) + 2(1/4) = 0 + 1/2 + 1/2 = 1$$

Note that to obtain the mean of the random variable x, we multiply each possible value of x by its probability $p(x)$, and then we sum this product over all possible values of x. The **mean of x** is also referred to as the **expected value of x**, denoted by $E(x)$.

Definition 4.5

The **mean**, or **expected value**, of a discrete random variable x is

$$\mu = E(x) = \sum_{\text{All } x} xp(x)$$

The term *expected* is a mathematical term and should not be interpreted as it is typically used. Specifically, it is possible that a random variable might never equal its *expected* value. Rather, the expected value is the mean of the probability distribution, a measure of its central tendency. You can think of μ as the mean value of x in a *very large* (actually, infinite) number of repetitions of the experiment.

EXAMPLE 4.2

Suppose you work for an insurance company and you sell a \$10,000 whole-life insurance policy at an annual premium of \$290. Actuarial tables show that the probability of death during the next year for a person of your customer's age, sex, health, etc., is .001. What is the expected gain (amount of money made by the company) for a policy of this type?

Solution

The experiment is to observe whether the customer survives the upcoming year. The probabilities associated with the two simple events, Live and Die, are .999 and .001, respectively. The random variable you are interested in is the gain, x, which can assume the following values:

Gain x	*Simple Event*	*Probability*
\$290	Customer lives	.999
\$290–\$10,000	Customer dies	.001

If the customer lives, the company gains the \$290 premium as profit. If the customer dies, the gain is negative because the company must pay \$10,000, for a net "gain" of \$(290 − 10,000). The expected gain is therefore

$$\mu = E(x) = \sum_{\text{All } x} xp(x) = (290)(.999) + (290 - 10{,}000)(.001) = \$280$$

In other words, if the company were to sell a very large number of 1-year \$10,000 policies to customers possessing the characteristics just described, it would (on the average) net \$280 per sale in the next year.

This example illustrates that the expected value of a random variable x need not be a possible value of x. That is, the expected value is \$280, but x will equal either \$290 or −\$9,710 each time the experiment is performed (a policy is sold and a year elapses). The expected value is the mean, a measure of central tendency—but not necessarily a possible value of x.

We found in Chapter 2 that the mean and other measures of central tendency tell only part of the story about a set of data. The same is true about probability distributions. We need to measure variability as well. Since a probability distribution can be viewed as a representation of a population, we will use the population variance to measure its variability.

The **population variance**, σ^2, is defined as the average squared distance of x from the population mean, μ. Since x is a random variable, the squared distance, $(x - \mu)^2$, is also a random variable. Applying the same logic used to find the mean value of x, we find the mean value of $(x - \mu)^2$ by multiplying all possible values of $(x - \mu)^2$ by $p(x)$ and then summing over all possible x values. This quantity,

$$E[(x - \mu)^2] = \sum_{\text{All } x} (x - \mu)^2 p(x)$$

is also called the **expected value of the squared distance from the mean**; i.e., $\sigma^2 = E[(x - \mu)^2]$.* The **standard deviation of x** is defined as the square root of the variance.

Definition 4.6

The **variance** of a discrete random variable x is

$$\sigma^2 = E[(x - \mu)^2] = \sum_{\text{All } x} (x - \mu)^2 p(x)$$

EXAMPLE 4.3

Suppose you invest a fixed sum of money in each of five business ventures. Assume you know that 70% of such ventures are successful, the outcomes of the ventures are independent of one another, and the probability distribution for the number, x, of successful ventures out of five is:

x	0	1	2	3	4	5
$p(x)$	.002	.029	.132	.309	.360	.168

a. Find $\mu = E(x)$.
b. Find $\sigma = \sqrt{E[(x - \mu)^2]}$.

*It can be shown that $E[(x - \mu)^2] = E(x^2) - \mu^2$, where $E(x^2) = \sum_{\text{All } x} x^2 p(x)$. Note the similarity between this expression and the shortcut formula $\sum_{i=1}^{n} (x_i - \bar{x})^2 = \sum_{i=1}^{n} x_i^2 - \frac{(\sum x_i)^2}{n}$ given in Chapter 2.

c. Graph $p(x)$. Locate μ and the interval $\mu \pm 2\sigma$ on the graph. Explain how μ and σ can be used to describe $p(x)$.

Solution

a. Applying the formula, we obtain

$$\mu = E(x) = \sum_{\text{All } x} xp(x)$$
$$= 0(.002) + 1(.029) + 2(.132) + 3(.309) + 4(.360) + 5(.168)$$
$$= 3.50$$

b. Now we calculate the variance of x:

$$\sigma^2 = E[(x - \mu)^2] = \sum_{\text{All } x} (x - \mu)^2 p(x)$$
$$= (0 - 3.5)^2(.002) + (1 - 3.5)^2(.029) + (2 - 3.5)^2(.132) + (3 - 3.5)^2(.309) + (4 - 3.5)^2(.360) + (5 - 3.5)^2(.168)$$
$$= 1.05$$

Thus, the standard deviation is

$$\sigma = \sqrt{\sigma^2} = \sqrt{1.05} = 1.02$$

c. The graph of $p(x)$ is shown in Figure 4.4. Note that the mean μ and the interval $\mu \pm 2\sigma$ are shown on the graph. We can use μ and σ to describe the probability distribution in the same way that we used $\bar{x}$ and s to describe a relative frequency distribution in Chapter 2. Note in particular that $\mu = 3.5$ locates the probability distribution along the x-axis. If the investment is made in the five ventures, we expect to obtain a number x of successes near 3.5. Similarly, $\sigma = 1.02$ measures the spread of the probability distribution. Since this distribution can be

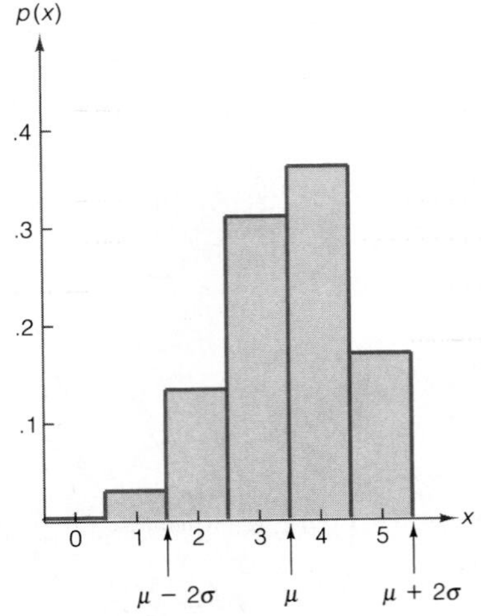

FIGURE 4.4 ▶ Graph of $p(x)$ for Example 4.3

interpreted as a relative frequency distribution that is moderately mound-shaped (see Figure 4.4), we expect (see Table 2.10 on page 75) at least 75% and, more likely, near 95% of observed x values to fall in the interval $\mu \pm 2\sigma$—i.e., between 1.46 and 5.54. Compare this with the actual probability that x falls in the interval $\mu \pm 2\sigma$. From Figure 4.4 you can see that this probability includes the sum of $p(x)$ for all values of x except $p(0) = .002$ and $p(1) = .029$. Therefore, 96.9% of the probability distribution lies within 2 standard deviations of the mean. This percentage is consistent with Table 2.10.

CASE STUDY 4.2 / Portfolio Selection

Investors—be they large corporations, banks, pension funds, mutual funds, or individuals—seldom hold a single financial asset; rather, they hold portfolios of financial assets. Thus, the investor should be less concerned with the rate of return achieved by, say, a particular stock in the portfolio* than in the overall rate of return of the portfolio. Since the future rate of return of a portfolio is uncertain, a probability distribution can be used to characterize a portfolio's future rate of return. Two examples of such distributions are shown in Figure 4.5. Alternatively, we might use just the mean and standard deviation of the probability distribution to characterize the future rate of return. Notice that portfolios A and B have the same mean but that A has the greater standard deviation. As a result, A has the higher probability of yielding a negative rate of return. Accordingly, it should not be surprising that the standard deviation of a portfolio's rate of return distribution is frequently used as a measure of the risk associated with the portfolio—the higher the standard deviation, the

FIGURE 4.5 ▶
Rate of return distributions

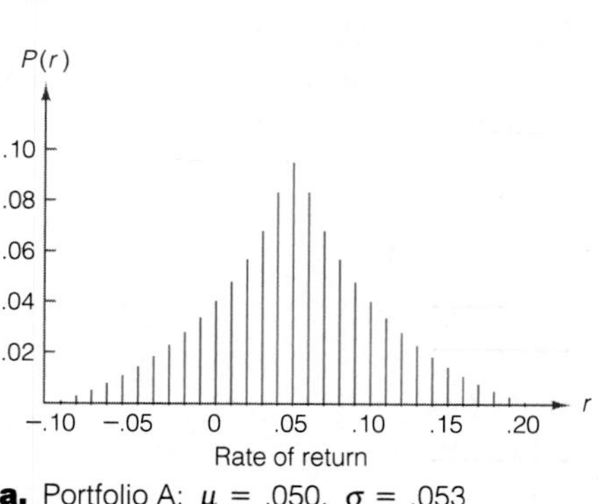

a. Portfolio A: $\mu = .050$, $\sigma = .053$

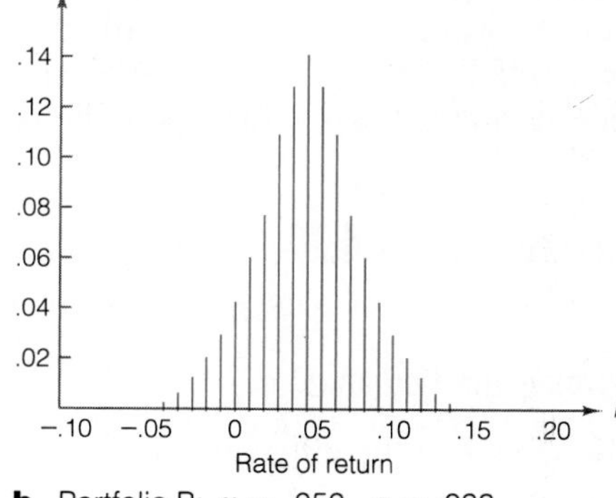

b. Portfolio B: $\mu = .050$, $\sigma = .033$

*If no dividends have been declared on the day in question, a stock's rate of return on day t is defined as $(P_t - P_{t-1})/P_{t-1}$, where P_t is the closing price of the stock on day t and P_{t-1} is its closing price the day before.

greater the risk of the portfolio (i.e., the greater the uncertainty of the portfolio's rate of return) and vice versa.

Typically, an investor can choose from among many different assets to form a portfolio. Or put another way, there are many different portfolios for the investor to choose among. But which portfolio should the investor select? This problem was addressed by Harry M. Markowitz (1952) in a classic article published in the *Journal of Finance*. Characterizing portfolios by the mean and standard deviation of their rates of return, Markowitz proposed a two-step procedure for choosing among portfolios. First, the set of all possible portfolios—the *feasible set*—must be reduced to an *efficient set* of portfolios. An efficient portfolio is one that provides the highest possible mean rate of return for any given degree of risk (i.e., any given standard deviation) or the lowest possible degree of risk for any given mean rate of return. Second, from the efficient set, the investor should choose the portfolio that best suits his or her needs.

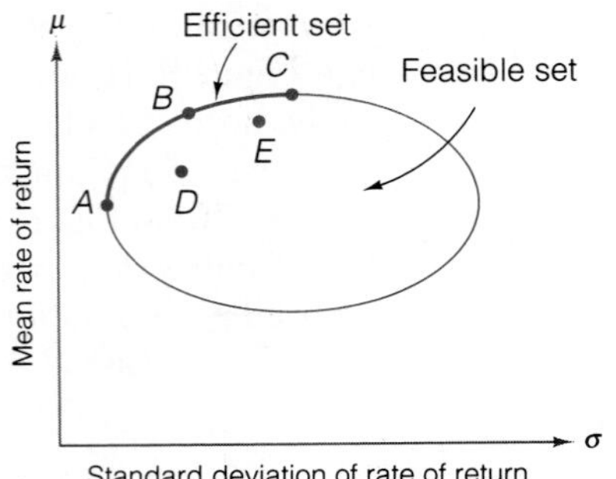

FIGURE 4.6 ▲
Efficient set of portfolios

The graph in Figure 4.6 shows the mean and standard deviation of the rate of return for various portfolios. Portfolios identified by a value of μ and σ that fall within the ellipse represent the feasible set of portfolios for a particular investor. The efficient set of portfolios is denoted by the boundary line *ABC* and is sometimes called the *efficient frontier*. Portfolios to the left of *ABC* are not obtainable because they fall outside the feasible set. Portfolios to the right of *ABC* are not efficient because there always exists a portfolio on the efficient frontier that could provide (1) a higher mean return for a given level of standard deviation of returns (compare points *B* and *D*), or (2) a lower standard deviation (lower degree of risk) for a given mean return (compare points *B* and *E*). For details on how to determine the efficient set of portfolios for an investor, see Elton and Gruber (1981) or Alexander and Francis (1986).

This case study illustrates that the probability distribution for the rate of return of a portfolio of financial assets provides a useful way to characterize the likelihood that the portfolio will yield an overall gain or loss. It also demonstrates that the process of selecting a portfolio of financial assets can be improved by comparing the means and standard deviations of the rates of return of alternative portfolios.

Exercises 4.18–4.29

Learning the Mechanics

4.18 Describe the differences in the meanings of the symbols μ, $\bar{x}$, and $E(x)$.

4.19 Consider the following probability distribution for the random variable x:

x	0	1	2	5	10
$p(x)$	.05	.25	.30	.20	.20

a. Find $E(x)$. b. Graph $p(x)$ and locate $E(x)$ on the graph.
c. Interpret the value you obtained for $E(x)$.

4.20 Consider the following probability distribution for the random variable x:

x	-3	-1	0	1	5
$p(x)$	.10	.20	.40	.20	.10

a. Find μ. b. Graph $p(x)$ and locate μ on the graph.
c. Interpret the value you obtained for μ.
d. In this case, can the random variable x ever assume the value μ? Explain.
e. In general, can a random variable ever assume a value equal to its expected value? Explain.

4.21 Consider the following probability distribution for the random variable x:

x	30	40	50	60	70	80
$p(x)$	.05	.20	.10	.25	.15	.25

a. Find μ, σ^2, and σ. b. Graph $p(x)$.
c. Locate μ and the interval $\mu \pm 2\sigma$ on your graph. What is the probability that x will fall within the interval $\mu \pm 2\sigma$?

4.22 Consider the following probability distribution for the random variable x:

x	-3	-1	0	1	5
$p(x)$	.05	.25	.45	.15	.10

a. Find μ, σ^2, and σ. b. Graph $p(x)$.
c. Locate μ and the interval $\mu \pm \sigma$ on your graph. What is the probability that x will fall within the interval $\mu \pm \sigma$?
d. Locate the interval $\mu \pm 3\sigma$ on your graph. What is the probability that x falls within this interval?

Applying the Concepts

4.23 **Economic risk** is sometimes measured by computing the variance or standard deviation of the probability distribution that describes the potential gains or losses of the firm. This follows from the fact that the greater the variation in potential outcomes, the greater the uncertainty faced by the firm; the smaller the variation, the more predictable the firm's gains or losses (Williams and Heins, 1986). The two discrete probability distributions given in the table were developed from historical data. They describe the potential total physical damage losses next year to the fleets of delivery trucks of two different firms.

Firm A		*Firm B*	
Loss next year	*Probability*	*Loss next year*	*Probability*
$ 0	.01	$ 0	.00
500	.01	200	.01
1,000	.01	700	.02
1,500	.02	1,200	.02
2,000	.35	1,700	.15
2,500	.30	2,200	.30
3,000	.25	2,700	.30
3,500	.02	3,200	.15
4,000	.01	3,700	.02
4,500	.01	4,200	.02
5,000	.01	4,700	.01

a. Verify that both firms have the same expected total physical damage loss.
b. Compute the standard deviation of each probability distribution, and determine which firm faces the greater risk of physical damage to its fleet next year.

4.24 In 1965 the U.S. Weather Bureau (now called the National Weather Service) initiated a nationwide program in which precipitation probabilities were included in all public weather forecasts. This was the first time in any field of application that probabilities were issued on such a large scale. The program continues today. Precipitation forecasts indicate the likelihood of measurable precipitation ($\geq .01$ inch) at a specific point (the official rain gauge) during a given time period (Murphy and Winkler, 1984). Suppose that if a measurable amount of rain falls during the next 24 hours, a river will reach flood stage and a business will incur damages of $300,000. The National Weather Service has indicated that there is a 30% chance of a measurable amount of rain during the next 24 hours.
a. Construct the probability distribution that describes the potential flood damages.
b. Find the firm's expected loss due to flood damage.

4.25 A stock market analyst believes that the probability of stock ABC increasing in price by the close of business tomorrow is .6, the probability of it decreasing in price is .2, and the probability of tomorrow's price remaining the same as today's is .2. Assume that when stock ABC's price changes, it does so by exactly $2.
a. Based on the analyst's assumptions, what is the expected change in price of ABC at the close of business tomorrow?
b. Can the change in ABC's price at the close of business tomorrow actually equal its expected value? Explain.

4.26 Banks and finance companies compete for personal loan customers. For purposes of product design and marketing, it is important for these financial institutions to understand the similarities and differences of the customers they attract. To further such an understanding, Robert W. Johnson and A. Charlene Sullivan (1981) sampled and questioned 488 bank borrowers and 87 finance company borrowers. One of the characteristics they investigated was the credit-worthiness of borrowers. They evaluated each borrower using a credit-scoring system of a large commercial bank. (Credit score is inversely related to credit risk.) The table summarizes the scores they obtained. The relative frequencies can be thought of as the approximate probabilities of a randomly selected borrower having the associated credit score.
a. Find the expected credit score for bank borrowers. Find the expected credit score for finance company borrowers. Interpret both these values in the context of the problem.

Credit Score	Relative Frequency		Credit Score	Relative Frequency	
	Bank Customers	Finance Company Customers		Bank Customers	Finance Company Customers
210	.109	.000	160	.102	.069
200	.117	.023	150	.102	.161
190	.109	.034	140	.074	.172
180	.113	.034	130	.035	.105
170	.219	.184	120	.020	.218

b. Find the standard deviation for the credit scores of bank borrowers. For finance company borrowers.
c. Using the probability distributions in the table and your results from parts **a** and **b**, compare the credit-worthiness of bank borrowers and finance company borrowers.

4.27 A patient complaining of severe stomach pains checked into a local hospital. After a series of tests, the doctors narrowed their diagnosis to four possible ailments. They believe there is a 40% chance that the patient has hepatitis; a 10% chance that she has cirrhosis; a 45% chance of gallstones; and a 5% chance of cancer of the pancreas. The doctors feel certain that the patient has only one of the diseases, but will not know which disease until further tests are performed. The cost associated with treating each disease is given in the table:

Disease	Hepatitis	Cirrhosis	Gallstones	Pancreatic Cancer
Cost	\$700	\$1,110	\$3,320	\$16,450

a. Construct the probability distribution for the cost of treating the patient.
b. Calculate the mean of the probability distribution you constructed in part **a**. What does this number represent?
c. Further testing reveals that the patient has either hepatitis or cirrhosis. Given this information, construct the new probability distribution for the cost of treating the patient.
d. Calculate the mean of the probability distribution you constructed in part **c**. What does this number represent?

4.28 Suppose you own a company that bonds financial managers. Based on past experience, you assess the probability that you will have to forfeit any particular bond to be .001. How much should you charge for a \$1 million bond in order to break even on all such bonds?

4.29 One of the primary responsibilities of the personnel department of a firm is to maintain accurate and up-to-date professional and personal profiles of current employees and potential employees. Such information is utilized to match people to project assignments within the firm (Misshauk, 1979). A company is interested in hiring a person with an MBA degree and at least 2 years experience in a marketing department of a computer products firm. The company's personnel department has determined that it will cost the company \$1,000 per job candidate to collect the required background information and to interview the candidate. As a result, the company will hire the first qualified person it finds and will interview no more than three candidates. The company has received job applications from four persons who appear to be qualified but, unknown to the company, only one actually possesses the required background. Candidates to be interviewed will be randomly selected from the pool of four applicants.

a. Construct the probability distribution for the total cost to the firm of the interviewing strategy.

b. What is the probability that the firm's interviewing strategy will result in none of the four applicants being hired?

c. Calculate the mean of the probability distribution you constructed in part **a**.

d. What is the expected total cost of the interviewing strategy?

4.4 The Binomial Random Variable

A common source of business data is an opinion or preference survey. Many of these surveys result in dichotomous responses—i.e., responses that admit one of two possible alternatives, such as Yes–No. The number of Yes responses (or No responses) will usually have a **binomial probability distribution**. For example, suppose a random sample of current customers is selected from a firm's data base to evaluate a new product. The number of customers in the sample who prefer the product to its competition is a random variable that has a binomial probability distribution.

All experiments that have the characteristics of the coin-tossing experiments of Chapter 3 and the preceding sections of this chapter yield **binomial random variables**. Imagine an experiment that is equivalent to tossing a coin n times. You are interested in observing the number of heads, x, in the n tosses. For such an experiment, x is a binomial random variable. In general, to decide whether a discrete random variable has a binomial probability distribution, check it against the characteristics listed in the box.

Characteristics of a Binomial Random Variable

1. The experiment consists of n identical trials.
2. There are only two possible outcomes on each trial. We denote one outcome by S (for Success) and the other by F (for Failure).
3. The probability of S remains the same from trial to trial. This probability is denoted by p, and the probability of F is denoted by q. Note that $p + q = 1$.
4. The trials are independent.
5. The binomial random variable x is the number of S's in n trials.

EXAMPLE 4.4

For each of the following examples, decide whether x is a binomial random variable:

a. You randomly select three bonds out of a possible ten for an investment portfolio. Unknown to you, eight of the ten will maintain their present value, and the other two will lose value due to a change in their ratings. Let x be the number of the three bonds you select that lose value.

b. Before marketing a new product on a large scale, many companies conduct a consumer preference survey to determine whether the product is likely to be successful. Suppose a company develops a new diet soda and then conducts a taste-preference survey with 100 randomly chosen consumers stating their preference among the new soda and the two leading sellers. Let x be the number of the 100 who choose the new brand over the two others.

c. Some surveys are conducted using a method of sampling other than simple random sampling (defined in Chapter 3). For example, suppose a television cable company is trying to decide whether to establish a branch in a particular city. The company plans to conduct a survey to determine the fraction of households in the city that would use the cable television service. The sampling method is to choose a city block at random and then to survey every household on that block. This sampling technique is called **cluster sampling.** Suppose 10 blocks are sampled in this manner, producing a total of 124 household responses. Let x be the number of the 124 households that would use the cable television service.

Solution

a. In checking the binomial characteristics, a problem arises with independence (characteristic 4 in the box). Suppose the first bond you picked was one of the two that will lose value. This reduces the chance that the second bond you pick will lose value, since now only one of the nine remaining bonds are in that category. Thus, the choices you make are dependent, and therefore x, the number of the three bonds you select that lose value, is *not* a binomial random variable.

b. Surveys that produce dichotomous responses and use random sampling techniques are classic examples of binomial experiments. In our example, each randomly selected consumer either states a preference for the new diet soda or does not. The sample of 100 consumers is a very small proportion of the totality of potential consumers, so the response of one would be, for all practical purposes, independent of another. Thus, x is a binomial random variable.

c. This example is a survey with dichotomous responses (Yes or No to the cable service), but the sampling method is not simple random sampling. Again, the binomial characteristic of independent trials would probably not be satisfied. The responses of households within a particular block would almost surely be dependent, since households within a block tend to be similar with respect to income, level of education, and general interests. Thus, the binomial model would not be satisfactory for x if the cluster sampling technique were used.

To see how to compute probabilities for binomial random variables, consider the following example.

EXAMPLE 4.5

A retail computer store sells IBM personal computers (PCs) and compatibles. Assume that 80% of the PCs that the store sells are IBM PCs, and 20% are compatibles.

a. Use the steps given in Chapter 3 (box on page 133) to find the probability that all of the next four PC purchases are compatibles.

b. Find the probability that three of the next four PC purchases are compatibles.

c. Let x represent the number of the next four PC purchases that are compatibles. Explain why x is a binomial random variable.

d. Use the answers to parts **a** and **b** to derive a formula for $p(x)$, the probability distribution of the binomial random variable, x.

Solution

a. 1. The first step is to define the experiment. Here we are interested in observing the type of PC purchased by each of the next four (buying) customers: IBM (I) or compatible (C).

2. Next, we list the simple events associated with the experiment. Each simple event consists of the purchase decisions made by the four customers. For example, *IIII* represents the simple event that all four purchase IBM PCs, and *CIII* represents the simple event that customer 1 purchases a compatible, while customers 2, 3, and 4 purchase IBM PCs. The 16 simple events are listed in Table 4.2.

TABLE 4.2 Simple Events for PC Experiment of Example 4.5

IIII	*CIII*	*CCII*	*ICCC*	*CCCC*
	ICII	*CICI*	*CICC*	
	IICI	*CIIC*	*CCIC*	
	IIIC	*ICCI*	*CCCI*	
		ICIC		
		IICC		

3. We now assign probabilities to the simple events. Note that each simple event can be viewed as the intersection of four customers' decisions and, assuming the decisions are made independently, the probability of each simple event can be obtained using the multiplicative rule, as follows:

$$\begin{aligned} P(IIII) &= P[(\text{customer 1 chooses IBM}) \cap (\text{customer 2 chooses IBM}) \\ &\qquad \cap (\text{customer 3 chooses IBM}) \cap (\text{customer 4 chooses IBM})] \\ &= P(\text{customer 1 chooses IBM}) \times P(\text{customer 2 chooses IBM}) \\ &\qquad \times P(\text{customer 3 chooses IBM}) \times P(\text{customer 4 chooses IBM}) \\ &= (.8)(.8)(.8)(.8) = (.8)^4 \\ &= .4096 \end{aligned}$$

All other simple event probabilities are calculated using similar reasoning. For example,

$$P(CIII) = (.2)(.8)(.8)(.8) = (.2)(.8)^3 = .1024$$

You can check that this reasoning results in simple event probabilities that add to 1 over the 16 simple events in the sample space.

4. Finally, we add the appropriate simple event probabilities to obtain the desired event probability. The event of interest is that all four customers purchase compatibles. In Table 4.2 we find only one simple event, *CCCC*, contained in this event. All other simple events imply that at least one IBM is purchased. Thus,

$$P(\text{All four purchase compatibles}) = P(CCCC) = (.2)^4 = .0016$$

That is, the probability is only 16 in 10,000 that all four customers purchase compatibles.

b. The event that three of the next four buyers purchase compatibles consists of the four simple events in the fourth column of Table 4.2: *ICCC*, *CICC*, *CCIC*, and *CCCI*. To obtain the event probability we add the simple event probabilities:

$$\begin{aligned}&P(3 \text{ of next 4 customers purchase compatibles})\\&\quad = P(ICCC) + P(CICC) + P(CCIC) + P(CCCI)\\&\quad = (.2)^3(.8) + (.2)^3(.8) + (.2)^3(.8) + (.2)^3(.8)\\&\quad = 4(.2)^3(.8) = .0256\end{aligned}$$

Note that each of the four simple event probabilities is the same, because each simple event consists of three *C*s and one *I*; the order does not affect the probability because the customers' decisions are (assumed) independent.

c. We can characterize the experiment as consisting of four identical trials—the four customers' purchase decisions. There are two possible outcomes to each trial, *I* or *C*, and the probability of *C*, $p = .2$, is the same for each trial. Finally, we are assuming that each customer's purchase decision is independent of all others, so that the four trials are independent. Then it follows that x, the number of the next four purchases that are compatibles, is a binomial random variable.

d. The event probabilities in parts **a** and **b** provide insight into the formula for the probability distribution $p(x)$. First, consider the event that three purchases are compatibles (part **b**). We found that

$$P(x = 3) = 4(.2)^3(.8)$$

Recall that this probability is the sum of four simple event probabilities, each equal to $(.2)^3(.8)$. This line of reasoning leads to the following formula:

$$\begin{aligned}P(x = 3) &= (\text{Number of simple events for which } x = 3)\\&\quad \times (.2)^{\text{Number of compatibles purchased}}\\&\quad \times (.8)^{\text{Number of IBMs purchased}}\\&= 4(.2)^3(.8)^1\end{aligned}$$

In general, we can use combinatorial mathematics to count the number of simple events. For example,

Number of simple events for which $x = 3$

$= $ Number of different ways of selecting 3 of the 4 trials for C purchases

$$= \binom{4}{3} = \frac{4!}{3!(4-3)!} = \frac{4 \cdot 3 \cdot 2 \cdot 1}{(3 \cdot 2 \cdot 1) \cdot 1} = 4$$

The formula that works for any value of x can be deduced as follows:

$$P(x = 3) = \binom{4}{3}(.2)^3(.8)^1 = \binom{4}{x}(.2)^x(.8)^{4-x}$$

The component $\binom{4}{x}$ counts the number of simple events with x compatibles, and the component $(.2)^x(.8)^{4-x}$ is the probability associated with each simple event having x compatibles.

For the general binomial experiment, with n trials and probability of Success p on each trial, the probability of x Successes is

$$p(x) = \binom{n}{x} \qquad p^x(1-p)^{n-x}$$

↑ No. of simple events with x S's

↑ Probability of x S's and $(n - x)$ F's in any simple event

In theory, you could always resort to first principles to calculate binomial probabilities: list the simple events and sum their probabilities. However, as the number of trials (n) increases, the number of simple events grows very rapidly (the number of simple events is 2^n). Thus, we prefer the formula for calculating binomial probabilities, since its use avoids listing simple events.

The binomial probability distribution is summarized in the box.

The Binomial Probability Distribution

$$p(x) = \binom{n}{x} p^x q^{n-x} \qquad (x = 0, 1, 2, \ldots, n)$$

where p = Probability of a success on a single trial

$q = 1 - p$

n = Number of trials

x = Number of successes in n trials

$$\binom{n}{x} = \frac{n!}{x!(n-x)!}$$

As noted in Chapter 3, the symbol 5! means $5 \cdot 4 \cdot 3 \cdot 2 \cdot 1 = 120$. Similarly, $n! = n(n-1)(n-2) \cdot \cdots \cdot 3 \cdot 2 \cdot 1$; remember, $0! = 1$.

The mean, variance, and standard deviation for the binomial random variable x are shown in the box.

Mean, Variance, and Standard Deviation for a Binomial Random Variable

Mean: $\mu = np$

Variance: $\sigma^2 = npq$

Standard deviation: $\sigma = \sqrt{npq}$

As we demonstrated in Chapter 2, the mean and standard deviation provide measures of the central tendency and variability, respectively, of a distribution. Thus, we can use μ and σ to obtain a rough visualization of the probability distribution for x when the calculation of the probabilities is too tedious. To illustrate the use of the binomial probability distribution, consider Example 4.6.

EXAMPLE 4.6

A machine that produces stampings for automobile engines is malfunctioning and producing 10% defectives. The defective and nondefective stampings proceed from the machine in a random manner. If the next five stampings are tested, find the probability that three of them are defective.

Solution

Let x equal the number of defectives in $n = 5$ trials. Then x is a binomial random variable with p, the probability that a single stamping will be defective, equal to .1, and $q = 1 - p = 1 - .1 = .9$. The probability distribution for x is given by the expression

$$\begin{aligned} p(x) &= \binom{n}{x} p^x q^{n-x} = \binom{5}{x} (.1)^x (.9)^{5-x} \\ &= \frac{5!}{x!(5-x)!} (.1)^x (.9)^{5-x} \qquad (x = 0, 1, 2, 3, 4, 5) \end{aligned}$$

To find the probability of observing $x = 3$ defectives in a sample of $n = 5$, substitute $x = 3$ into the formula for $p(x)$ to obtain

$$\begin{aligned} p(3) &= \frac{5!}{3!(5-3)!} (.1)^3 (.9)^{5-3} = \frac{5!}{3!2!} (.1)^3 (.9)^2 \\ &= \frac{5 \cdot 4 \cdot 3 \cdot 2 \cdot 1}{(3 \cdot 2 \cdot 1)(2 \cdot 1)} (.1)^3 (.9)^2 = 10(.1)^3 (.9)^2 \\ &= .0081 \end{aligned}$$

Note that the binomial formula tells us that there are 10 simple events having 3 defectives (check this by listing them), each with probability $(.1)^3(.9)^2$.

EXAMPLE 4.7

Refer to Example 4.6 and find the values of $p(0)$, $p(1)$, $p(2)$, $p(4)$, and $p(5)$. Graph $p(x)$. Calculate the mean μ and standard deviation σ. Locate μ and the interval $\mu - 2\sigma$ to $\mu + 2\sigma$ on the graph. If the experiment were to be repeated many times, what proportion of the x observations would fall within the interval $\mu - 2\sigma$ to $\mu + 2\sigma$?

Solution

Again, $n = 5$, $p = .1$, and $q = .9$. Then, substituting into the formula for $p(x)$:

$$p(0) = \frac{5!}{0!(5-0)!}(.1)^0(.9)^{5-0} = \frac{5 \cdot 4 \cdot 3 \cdot 2 \cdot 1}{(1)(5 \cdot 4 \cdot 3 \cdot 2 \cdot 1)}(1)(.9)^5 = .59049$$

$$p(1) = \frac{5!}{1!(5-1)!}(.1)^1(.9)^{5-1} = 5(.1)(.9)^4 = .32805$$

$$p(2) = \frac{5!}{2!(5-2)!}(.1)^2(.9)^{5-2} = (10)(.1)^2(.9)^3 = .07290$$

$$p(4) = \frac{5!}{4!(5-4)!}(.1)^4(.9)^{5-4} = 5(.1)^4(.9) = .00045$$

$$p(5) = \frac{5!}{5!(5-5)!}(.1)^5(.9)^{5-5} = (.1)^5 = .00001$$

The graph of $p(x)$ is shown as a probability histogram in Figure 4.7. [$p(3)$ is taken from Example 4.6 to be .0081.]

To calculate the values of μ and σ, substitute $n = 5$ and $p = .1$ into the following formulas:

$$\mu = np = (5)(.1) = .5 \qquad \sigma = \sqrt{npq} = \sqrt{(5)(.1)(.9)} = \sqrt{.45} = .67$$

To find the interval $\mu - 2\sigma$ to $\mu + 2\sigma$, we calculate

$$\mu - 2\sigma = .5 - 2(.67) = -.84 \qquad \mu + 2\sigma = .5 + 2(.67) = 1.84$$

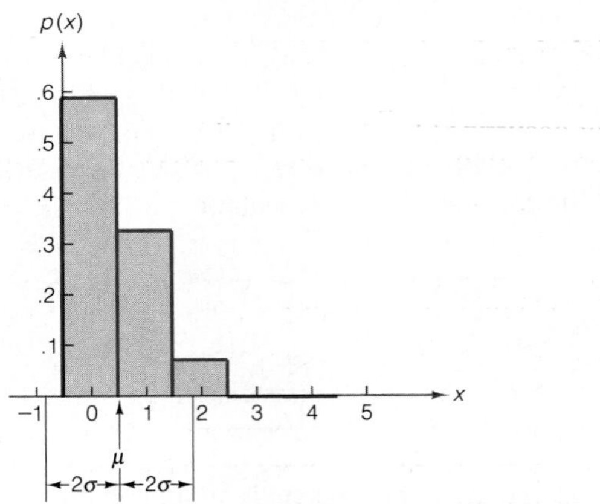

FIGURE 4.7 ▶ The binomial distribution: $n = 5$, $p = .1$

If the experiment were to be repeated a large number of times, what proportion of the x observations would fall within the interval $\mu - 2\sigma$ to $\mu + 2\sigma$? You can see from

Figure 4.7 that all observations equal to 0 or 1 will fall within the interval. The probabilities corresponding to these values are .5905 and .3280, respectively. Consequently, you would expect .5905 + .3280 = .9185, or approximately 91.9%, of the observations to fall within the interval $\mu - 2\sigma$ or $\mu + 2\sigma$. This again emphasizes that for most probability distributions, observations rarely fall more than 2 standard deviations from μ.

CASE STUDY 4.3 / The Space Shuttle *Challenger:* Catastrophe in Space

On January 28, 1986, at 11:39.13 A.M., while traveling at Mach 1.92 at an altitude of 46,000 feet, the space shuttle *Challenger* was totally enveloped in an explosive burn that destroyed the shuttle and resulted in the deaths of all seven astronauts aboard. What happened? What was the cause of this catastrophe? This was the 25th shuttle mission. Each of the preceding 24 missions had been successful.

The report of the Presidential Commission assigned to investigate the accident concluded that the explosion was caused by the failure of the O-ring seal in the joint between the two lower segments of the right solid rocket booster. The seal is supposed to prevent superhot gases from leaking through the joint during the propellant burn of the booster rocket. The failure of the seal permitted a jet of white-hot gases to escape and to ignite the liquid fuel of the external fuel tank. The fuel tank fireburst destroyed the *Challenger*.

What were the chances of this event occurring? In a 1985 report, the National Aeronautics and Space Administration (NASA) claimed that the probability of such a failure was about 1/60,000, or about once in every 60,000 flights. But a 1983 risk-assessment study conducted for the Air Force assessed the probability of a shuttle catastrophe due to booster rocket "burn-through" to be 1/35, or about once in every 35 missions.

If it is assumed that (1) p, the probability of shuttle catastrophe due to booster failure, remains the same from mission to mission, and (2) the performance of the booster rockets on one mission is independent of the performance of the boosters on other missions, then the number, x, of shuttle catastrophes due to booster failure in n missions can be treated as a binomial random variable. Accordingly, the probability that no disasters would have occurred during 25 missions is

$$P(x = 0) = \binom{25}{0} p^0(1 - p)^{25-0}$$

$$= \frac{25!}{0!25!} p^0(1 - p)^{25} = (1 - p)^{25}$$

If we use NASA's probability of shuttle catastrophe ($p = 1/60{,}000 = .0000167$), the probability of no catastrophes in 25 missions is approximately .9996. If we use the probability of catastrophe from the study prepared for the Air Force ($p = 1/35 = .02857$), the probability of no catastrophes in 25 missions is approximately .4845. Or, if we consider the complementary event that at least one catastrophe occurs in 25 missions, the chances are .0004, or about 4 in 10,000, given NASA's assumptions. On the other hand, the probability of at least one catastrophe under the Air Force's assumptions is .5155, or slightly more than 50–50. Given the events of January 28, 1986, which risk assessment—NASA's or the Air Force's—appears to be more appropriate? The probability of one or more disasters in 25 missions is so remote using NASA's assessment that it casts serious doubt on the risk assessment practices used by NASA prior to the *Challenger*'s fatal mission (McKean, 1986; Biddle, 1986; Robinson, 1986; "'83 Report Put Booster Accident as Most Likely," 1986).

Using Binomial Tables

Calculating binomial probabilities becomes tedious when n is large. For some values of n and p the binomial probabilities have been tabulated in Table II of Appendix B. Part of Table II is shown in Table 4.3; a graph of the binomial probability distribution for $n = 10$ and $p = .10$ is shown in Figure 4.8. Table II actually contains a total of nine tables, labeled (**a**) through (**i**), one each corresponding to $n = 5, 6, 7, 8, 9, 10, 15, 20$, and 25. In each of these tables the columns correspond to values of p, and the rows correspond to values of the random variable x. The entries in the table represent **cumulative** binomial probabilities. Thus, for example, the entry in the column corresponding to $p = .10$ and the row corresponding to $x = 2$ is .930 (shaded), and its interpretation is

$$P(x \leq 2) = P(x = 0) + P(x = 1) + P(x = 2) = .930$$

TABLE 4.3 Reproduction of Part of Table II of Appendix B: Cumulative Binomial Probabilities for $n = 10$

k \ p	.01	.05	.10	.20	.30	.40	.50	.60	.70	.80	.90	.95	.99
0	.904	.599	.349	.107	.028	.006	.001	.000	.000	.000	.000	.000	.000
1	.996	.914	.736	.376	.149	.046	.011	.002	.000	.000	.000	.000	.000
2	1.000	.988	.930	.678	.383	.167	.055	.012	.002	.000	.000	.000	.000
3	1.000	.999	.987	.879	.650	.382	.172	.055	.011	.001	.000	.000	.000
4	1.000	1.000	.998	.967	.850	.633	.377	.166	.047	.006	.000	.000	.000
5	1.000	1.000	1.000	.994	.953	.834	.623	.367	.150	.033	.002	.000	.000
6	1.000	1.000	1.000	.999	.989	.945	.828	.618	.350	.121	.013	.001	.000
7	1.000	1.000	1.000	1.000	.998	.988	.945	.833	.617	.322	.070	.012	.000
8	1.000	1.000	1.000	1.000	1.000	.998	.989	.954	.851	.624	.264	.086	.004
9	1.000	1.000	1.000	1.000	1.000	1.000	.999	.994	.972	.893	.651	.401	.096

This probability is also shaded in the graphical representation of the binomial distribution with $n = 10$ and $p = .10$ in Figure 4.8.

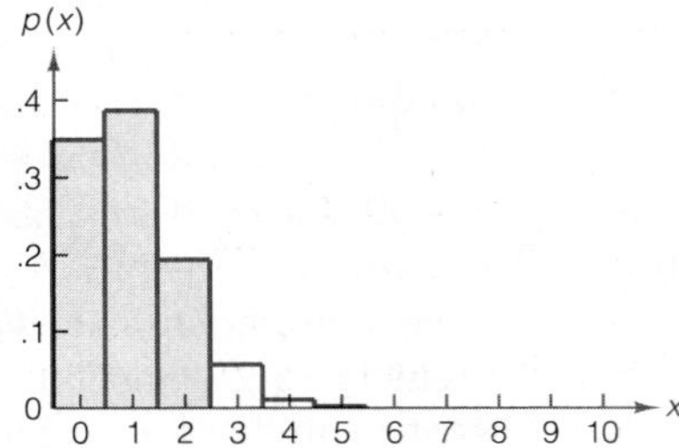

FIGURE 4.8 ▶ Binomial probability distribution for $n = 10$ and $p = .10$; $P(x \leq 2)$ shaded

You can also use Table II to find the probability that x equals a specific value. For example, suppose you want to find the probability that $x = 2$ in the binomial distribution with $n = 10$ and $p = .10$. This is found by subtraction as follows:

$$\begin{aligned} P(x = 2) &= [P(x = 0) + P(x = 1) + P(x = 2)] - [P(x = 0) + P(x = 1)] \\ &= P(x \leq 2) - P(x \leq 1) \\ &= .930 - .736 = .194 \end{aligned}$$

The probability that a binomial random variable exceeds a specified value can be found using Table II and the notion of complementary events. For example, to find the probability that x exceeds 2 when $n = 10$ and $p = .10$, we use

$$P(x > 2) = 1 - P(x \leq 2) = 1 - .930 = .070$$

Note that this probability is represented by the unshaded portion of the graph in Figure 4.8.

All probabilities in Table II are rounded to three decimal places. Thus, although none of the binomial probabilities in the table is exactly zero, some are small enough (less than .0005) to round to .000. For example, using the formula to find $P(x = 0)$ when $n = 10$ and $p = .6$, we obtain

$$P(x = 0) = \binom{10}{0}(.6)^0(.4)^{10-0} = .4^{10} = .00010486$$

but this is rounded to .000 in Table II of Appendix B (see Table 4.3).

Similarly, none of the table entries is exactly 1.0, but when the cumulative probabilities exceed .9995, they are rounded to 1.000. The row corresponding to the largest possible value for x, $x = n$, is omitted, because all the cumulative probabilities in that row are equal to 1.0 (exactly). For example, in Table 4.3 with $n = 10$, $P(x \leq 10) = 1.0$, no matter what the value of p.

The following example further illustrates the use of Table II.

EXAMPLE 4.8

Suppose a poll of 20 employees is taken in a large company. The purpose is to determine x, the number who favor unionization. Suppose that 60% of all the company's employees favor unionization.

a. Find the mean and standard deviation of x.
b. Use Table II of Appendix B to find the probability that $x < 10$.
c. Use Table II to find the probability that $x > 12$.
d. Use Table II to find the probability that $x = 11$.

Solution

a. The number of employees polled is presumably small compared with the total number of employees in this company. Thus, we may treat x, the number of the 20 who favor unionization, as a binomial random variable. The value of p is the fraction of the total employees who favor unionization; i.e., $p = .6$. Therefore, we calculate the mean and variance:

$$\mu = np = 20(.6) = 12 \qquad \sigma^2 = npq = 20(.6)(.4) = 4.8$$

The standard deviation is then

$$\sigma = \sqrt{4.8} = 2.19$$

b. The tabulated value is

$$P(x \leq 9) = .128$$

c. To find the probability

$$P(x > 12) = \sum_{x=13}^{20} p(x)$$

we use the fact that for all probability distributions, $\sum_{\text{All } x} p(x) = 1$. Therefore,

$$P(x > 12) = 1 - P(x \leq 12) = 1 - \sum_{x=0}^{12} p(x)$$

Consulting Table II, we find the entry in row $k = 12$, column $p = .6$ to be .584. Thus,

$$P(x > 12) = 1 - .584 = .416$$

d. To find the probability that exactly 11 employees favor unionization, recall that the entries in Table II are cumulative probabilities and use the relationship

$$\begin{aligned} P(x = 11) &= [p(0) + p(1) + \cdots + p(10) + p(11)] \\ &\quad - [p(0) + p(1) + \cdots + p(9) + p(10)] \\ &= P(x \leq 11) - P(x \leq 10) \end{aligned}$$

Then

$$P(x = 11) = .404 - .245 = .159$$

The probability distribution for x in this example is shown in Figure 4.9. Note that the interval $\mu \pm 2\sigma$ is (7.6, 16.4).

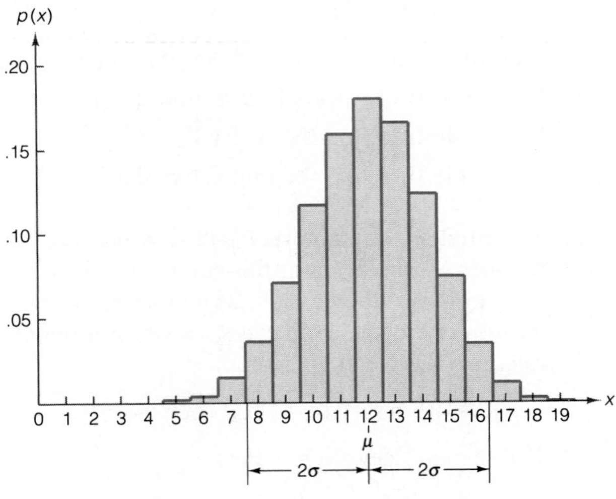

FIGURE 4.9 ► **The binomial probability distribution for x in Example 4.8: $n = 20$, $p = .6$**

CASE STUDY 4.4 / Evaluating Customer Response to a New Sales Program

Arthur A. Brown, Frank T. Hulswit, and John D. Kettelle (1956) were asked by a large firm to study, and perhaps determine reasons for and solutions to, its lack of growth over the prior 5 years. In their article, Brown et al. refer to the firm as "Penstock Press, a large commercial printing company."

The primary concern of the study was Penstock's sales operations. Accordingly, Brown et al. conducted an experiment to study the sales effectiveness of Penstock's salespeople. The salespeople were instructed to increase their sales efforts toward all of Penstock's customers, but in particular toward 60 of the larger customers, for a 4-month experimental period. At the end of the 4-month period it was determined that the probability of a customer making a genuinely positive response to the increased sales effort merely by chance was .25. Of Penstock's 60 large customers, it was noted that 24 made what appeared to be genuinely positive responses. But before concluding that the increased sales effort toward the 60 large customers had paid off, Brown et al. felt it was important to determine how likely it would be for 24 or more of the 60 customers to make positive responses merely by chance. Assuming that the probability of a positive response occurring by chance (.25) is the same for each of the 60 customers and that the response of one customer does not affect that of another, the number of positive responses observed has a binomial probability distribution. Accordingly, the probability of 24 or more positive responses from the 60 customers can be determined as follows:

$$P(x \geq 24) = \sum_{x=24}^{60} \binom{60}{x} .25^x (.75)^{60-x} = .004$$

(Due to the large value of n, the number of customers in the experiment, it would be unrealistic to try to compute the above probability by hand. If you had access to binomial tables for $n = 60$ and $p = .25$, the probability could be found using the tables. Otherwise, it would be necessary to use a computer or an approximation such as the one we will discuss in Section 5.5.) The fact that the probability of observing 24 or more genuinely positive responses from the 60 customers merely by chance is only .004 indicates that in actually observing 24 such responses either a very rare event has occurred or the responses were in fact genuine and the increased sales effort did influence the increase in sales.

Brown and his colleagues concluded that the number of positive responses could not be explained by chance, but that they in fact "implied deliberate continuing business from the customers," i.e., genuine responses. The authors noted that this conclusion was supported by Penstock's salespeople.

Exercises 4.30–4.45

Learning the Mechanics

4.30 Compute the following:

a. $\dfrac{4!}{2!(4-2)!}$ b. $\binom{6}{4}$ c. $\binom{5}{0}$ d. $\binom{3}{3}$ e. $\binom{8}{1}$

4.31 Consider the following probability distribution:

$$p(x) = \binom{6}{x}(.4)^x(.6)^{6-x} \qquad (x = 0, 1, 2, \ldots, 6)$$

a. Is x a discrete or a continuous random variable? Explain.

b. What is the name of this probability distribution?
c. Graph the probability distribution.
d. Find the mean and standard deviation of x.
e. Show the mean and the 2-standard-deviation interval on each side of the mean on the graph you drew in part **c**.

4.32 Suppose x is a binomial random variable with $n = 4$ and $p = .7$.
a. Use the formula for the binomial probability distribution to calculate the values of $p(x)$ for $x = 0, 1, 2, 3, 4$.
b. Graph $p(x)$.

4.33 Given that x is a binomial random variable, compute $p(x)$ for each of the following cases:
a. $n = 8, \quad x = 5, \quad p = .5$ b. $n = 7, \quad x = 2, \quad q = .3$ c. $n = 4, \quad x = 4, \quad p = .6$

4.34 Suppose x is a binomial random variable with $n = 13$ and $p = .6$.
a. Display $p(x)$ in tabular form. b. Compute the mean and variance of x.
c. Graph $p(x)$ and locate $E(x)$ and the interval $\mu \pm 2\sigma$ on the graph.
d. What is the probability that x falls within the interval $\mu \pm 2\sigma$?

4.35 Use the results of Exercise 4.34 to find the following probabilities:
a. $P(x \leq 5)$ b. $P(x \geq 3)$ c. $P(x < 7)$

4.36 Given that x is a binomial random variable with $n = 15$ and $p = .7$, use Table II of Appendix B to find the following probabilities:
a. $P(x \geq 11)$ b. $P(x < 7)$ c. $P(x > 7)$ d. $P(x = 9)$ e. $P(x \leq 5)$ f. $P(x > 4)$

4.37 The binomial probability distribution is a family of probability distributions, with each individual distribution depending on the values of n and p. Assume that x is a binomial random variable with $n = 6$.
a. Determine the value of p such that the probability distribution of x is symmetric.
b. Determine a value of p such that the probability distribution of x is skewed to the right.
c. Determine a value of p such that the probability distribution of x is skewed to the left.
d. Graph each of the binomial distributions you obtained in parts **a**, **b**, and **c**. Locate the mean for each distribution on its graph.
e. In general, for what values of p will a binomial distribution be symmetric? Skewed to the right? Skewed to the left?

Applying the Concepts

4.38 Your firm's accountant believes that 10% of the company's invoices contain arithmetic errors. To check this theory, the accountant randomly samples 25 invoices and finds that seven contain errors. What is the probability that of the 25 invoices written, seven or more would contain errors if the accountant's theory was valid? What assumptions do you have to make to solve this problem using the methodology of this chapter?

4.39 According to the Internal Revenue Service (IRS), the chances of your tax return being audited are about 6 in 1,000 if your income is less than \$25,000; they increase to about 14 in 1,000 if your income is \$25,000 or more; they increase to about 46 in 1,000 if your income is \$100,000 or more (*Statistical Abstract of the United States: 1991*, p. 324).
a. What is the probability that a taxpayer with income less than \$25,000 will be audited by the IRS? With income \$25,000 or more? With income \$100,000 or more?

b. If five taxpayers with incomes under \$25,000 are randomly selected, what is the probability that exactly one will be audited? That more than one will be audited?

c. Repeat part **b** assuming that five taxpayers with incomes of \$25,000 or more are randomly selected.

d. If two taxpayers with incomes under \$25,000 are randomly selected and two with incomes more than \$100,000 are randomly selected, what is the probability that none of these taxpayers will be audited by the IRS?

e. What assumptions did you have to make in order to answer these questions using the methodology presented in this section?

4.40 A problem of considerable economic impact on the economy is the burgeoning cost of Medicare and other public-funded medical services. One aspect of this problem concerns the high percentage of people seeking medical treatment who, in fact, have no physical basis for their ailments. One conservative estimate is that the percentage of people who seek medical assistance and who have no real physical ailment is 10%, and some doctors believe that it may be as high as 40%. Suppose we were to randomly sample the records of a doctor and found that five of 15 patients seeking medical assistance were physically healthy.

a. What is the probability of observing five or more physically healthy patients in a sample of 15 if the proportion, p, that the doctor normally sees is 10%?

b. What is the probability of observing five or more physically healthy patients in a sample of 15 if the proportion, p, that the doctor normally sees is 40%?

c. Why might your answer to part **a** make you believe that p is larger than .10?

4.41 According to the U.S. Golf Association (USGA), "The weight of the [golf] ball shall not be greater than 1.620 ounces avoirdupois (45.93 grams). The diameter of the ball shall not be less than 1.680 inches. The velocity of the ball shall be not greater than 250 feet per second" (USGA, 1993). The USGA periodically checks the specifications of golf balls sold in the United States by randomly sampling balls from pro shops around the country. Two dozen of each kind are sampled, and if more than three do not meet size and/or velocity requirements, that kind of ball is removed from the USGA's approved-ball list (*Golf World*, Sept. 10, 1982).

a. What assumptions must be made and what information must be known in order to use the binomial probability distribution to calculate the probability that the USGA will remove a particular kind of golf ball from its approved-ball list?

b. Suppose 10% of all balls produced by a particular manufacturer are less than 1.680 inches in diameter, and assume that the number of such balls, x, in a sample of two dozen balls can be adequately characterized by a binomial probability distribution. Find the mean and standard deviation of the binomial distribution.

c. Refer to part **b**. If x has a binomial distribution, then so does the number, y, of balls in the sample that meet the USGA's minimum diameter. [*Note:* $x + y = 24$.] Describe the distribution of y. In particular, what are p, q, and n? Also, find $E(y)$ and the standard deviation of y.

4.42 Suppose you have purchased 50,000 electrical switches and have been guaranteed by the supplier that the shipment will contain no more than .1% defectives. To check the shipment, you randomly sample 500 switches, test them, and find that four are defective. Assuming the supplier's claim is true, compute μ and σ for the number of defectives in a sample of 500 switches. If the supplier's claim is true, is it likely that you would have found four defective switches in the sample? Based on this sample, what inference would you make concerning the supplier's guarantee?

4.43 Many firms utilize sampling plans to control the quality of manufactured items ready for shipment or the quality of incoming items (parts, raw materials, etc.) that have been purchased. To illustrate the use of a sampling plan, suppose you are shipping electrical fuses in lots, each containing 5,000 fuses. The plan specifies that you will randomly sample 25 fuses from each lot and accept (and ship) the lot if the number of defective fuses, x, in the sample is less than 3. If $x \geq 3$, you will reject the lot. Find the probability of accepting a lot ($x = 0$, 1, or 2) if the actual fraction defective in the lot is:

a. 0 **b.** .01 **c.** .10 **d.** .30 **e.** .50 **f.** .80 **g.** .95 **h.** 1

Construct a graph showing $P(A)$, the probability of lot acceptance, as a function of the lot fraction defective, p. This graph is called the **operating characteristic curve** for the sampling plan.

4.44 Refer to Exercise 4.43. Suppose the sampling plan called for sampling $n = 25$ fuses and accepting a lot of $x \leq 3$. Calculate the quantities specified in Exercise 4.43, and construct the operating characteristic curve for this sampling plan. Compare this curve with the curve obtained in Exercise 4.43. (Note how the curve characterizes the ability of the plan to screen bad lots from shipment.)

4.45 A local newspaper claims that 65% of the items advertised in its classified advertisement section are sold within 1 week of the first appearance of the ad. To check the validity of its claim, the newspaper randomly selected $n = 800$ advertisements from last year's classified advertisements and contacted the people who placed the ads. They found that $x = 472$ of the 800 items sold within 1 week.

a. Compute μ and σ for the random variable x, the number of items sold within 1 week.

b. Based on a sample of 800, is it likely that you would observe $x \leq 472$ if the newspaper's claim were true? Explain.

c. Do the results of the newspaper's survey support the newspaper's claim? Explain.

4.5 The Poisson Random Variable (Optional)

A type of probability distribution useful in describing the number of events that will occur in a specific period of time or in a specific area or volume is the **Poisson distribution** (named after the 18th-century physicist and mathematician, Siméon Poisson). The following are typical examples of random variables for which the Poisson probability distribution provides a good model:

1. The number of industrial accidents in a given manufacturing plant per month
2. The number of noticeable surface defects (scratches, dents, etc.) found by quality inspectors on a new automobile (or any manufactured product)
3. The parts per million of some toxicant found in water or air emission from a manufacturing plant (a random variable of great interest to both the business community and the Environmental Protection Agency)
4. The number of arithmetic errors per 100 invoices (or per 1,000 invoices, etc.) in the accounting records of a company
5. The number of customer arrivals per unit time at a service counter (a service station, a hospital clinic, a supermarket checkout counter, etc.)
6. The number of death claims per day received by an insurance company
7. The number of breakdowns of an electronic computer per month

Characteristics of a Poisson Random Variable

1. The experiment consists of counting the number of times a particular event occurs during a given unit of time or in a given area or volume (or weight, distance, or any other unit of measurement).
2. The probability that an event occurs in a given unit of time, area, or volume is the same for all the units.
3. The number of events that occur in one unit of time, area, or volume is independent of the number that occur in other units.
4. The mean (or expected) number of events in each unit will be denoted by the Greek letter lambda, λ.

The characteristics of the Poisson random variable described in the box are usually difficult to verify for practical examples. The previous examples satisfy them well enough that the Poisson distribution provides a good model. As with all probability models, the real test of the adequacy of the Poisson model is whether it provides a reasonable approximation to reality—that is, whether empirical data support it.

The probability distribution, mean, and variance for a Poisson random variable are shown in the next box.

The calculation of Poisson probabilities is made easier by the use of Table III in Appendix B. There the cumulative probabilities $P(x \leq k)$ for various values of λ are given. The use of Table III is illustrated in Example 4.9.

Probability Distribution, Mean, and Variance for a Poisson Random Variable*

$$p(x) = \frac{\lambda^x e^{-\lambda}}{x!} \quad (x = 0, 1, 2, \ldots)$$

$$\mu = \lambda \qquad \sigma^2 = \lambda$$

where λ = Mean number of events during a given time period, or over a specific area or volume

$e = 2.71828\ldots$

EXAMPLE 4.9

Suppose the number, x, of a company's employees who are absent on Mondays has (approximately) a Poisson probability distribution. Furthermore, assume that the average number of Monday absentees is 2.6.

*The Poisson probability distribution also provides a good approximation to a binomial probability distribution with mean $\lambda = np$ when n is large and p is small (say, $np \leq 7$).

a. Find the mean and standard deviation of x, the number of employees absent on Monday.
b. Use Table III of Appendix B to find the probability that fewer than two employees are absent on a given Monday.
c. Use Table III to find the probability that more than five employees are absent on a given Monday.
d. Use Table III to find the probability that exactly five employees are absent on a given Monday.

Solution

a. The mean and variance of a Poisson random variable are both equal to λ. Thus, for this example,

$$\mu = \lambda = 2.6 \qquad \sigma^2 = \lambda = 2.6$$

Then the standard deviation of x is

$$\sigma = \sqrt{2.6} = 1.61$$

Remember that the mean measures the central tendency of the distribution, and does not necessarily equal a possible value of x. In this example, the mean is 2.6 absences, and of course there cannot be 2.6 absences on a given Monday. Similarly, the standard deviation of 1.61 measures the variability of the number of Monday absences. Perhaps a more helpful measure is the interval $\mu \pm 2\sigma$, which in this case stretches from $-.62$ to 5.82. We expect the number of absences to fall in this interval most of the time—with at least 75% relative frequency, and probably with more than 90% relative frequency. The mean and the 2-standard-deviation interval around it are shown in Figure 4.10.

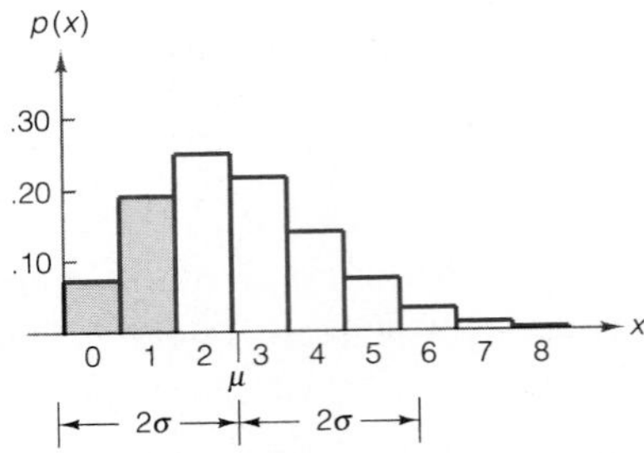

FIGURE 4.10 ▶ Probability distribution for number of Monday absences

b. A partial reproduction of Table III is shown in Table 4.4. The rows of the table correspond to different values of λ, and the columns correspond to different values of the Poisson random variable x. The entries in the table are cumulative probabilities (much like the binomial probabilities in Table II).

To find the probability that fewer than two employees are absent on a given Monday, we first note that

$$P(x < 2) = P(x \leq 1)$$

This probability is a cumulative probability and is therefore the entry in Table III in the row corresponding to $\lambda = 2.6$ and the column corresponding to $x = 1$. The entry is .267, shown shaded in Table 4.4. This probability corresponds to the shaded area in Figure 4.10, and may be interpreted as meaning that there is a 26.7% chance that fewer than two employees will be absent on a given Monday.

TABLE 4.4 Reproduction of Part of Table III of Appendix B: Poisson Probabilities

λ \ x	0	1	2	3	4	5	6	7	8	9
2.2	.111	.335	.623	.819	.928	.975	.993	.998	1.000	
2.4	.091	.308	.570	.779	.904	.964	.988	.997	.999	1.000
2.6	.074	.267	.518	.736	.877	.951	.983	.995	.999	1.000
2.8	.061	.231	.469	.692	.848	.935	.976	.992	.998	.999
3.0	.050	.199	.423	.647	.815	.916	.966	.988	.996	.999
3.2	.041	.171	.380	.603	.781	.895	.955	.983	.994	.998
3.4	.033	.147	.340	.558	.744	.871	.942	.977	.992	.997
3.6	.027	.126	.303	.515	.706	.844	.927	.969	.988	.996
3.8	.022	.107	.269	.473	.668	.816	.909	.960	.984	.994
4.0	.018	.092	.238	.433	.629	.785	.889	.949	.979	.992
4.2	.015	.078	.210	.395	.590	.753	.867	.936	.972	.989
4.4	.012	.066	.185	.359	.551	.720	.844	.921	.964	.985
4.6	.010	.056	.163	.326	.513	.686	.818	.905	.955	.980
4.8	.008	.048	.143	.294	.476	.651	.791	.887	.944	.975
5.0	.007	.040	.125	.265	.440	.616	.762	.867	.932	.968
5.2	.006	.034	.109	.238	.406	.581	.732	.845	.918	.960
5.4	.005	.029	.095	.213	.373	.546	.702	.822	.903	.951
5.6	.004	.024	.082	.191	.342	.512	.670	.797	.886	.941
5.8	.003	.021	.072	.170	.313	.478	.638	.771	.867	.929
6.0	.002	.017	.062	.151	.285	.446	.606	.744	.847	.916

c. To find the probability that more than five employees are absent on a given Monday, we consider the complementary event:

$$P(x > 5) = 1 - P(x \leq 5) = 1 - .951 = .049$$

where .951 is the entry in Table III corresponding to $\lambda = 2.6$ and $x = 5$ (see Table 4.4). Note from Figure 4.10 that this is the area in the interval $\mu \pm 2\sigma$, or $-.62$ to 5.82. Then the number of absences should exceed 5 or, equivalently, should be more than 2 standard deviations from the mean, on only about 4.9% of all Mondays. Note that this percentage agrees remarkably well with that given by the Empirical Rule for mound-shaped distributions, which tells us to expect approximately 5% of the measurements (observed values of the random variable) to lie further than 2 standard deviations from the mean.

d. To use Table III to find the probability that *exactly* five employees are absent on a given Monday, we must write the probability as the difference between two cumulative probabilities:

$$P(x = 5) = P(x \leq 5) - P(x \leq 4) = .951 - .877 = .074$$

Note that probabilities in Table III are all rounded to three decimal places. Thus, although in theory a Poisson random variable can assume infinitely large values, the values of x in Table III are extended only until the cumulative probability is 1.000. This does not mean that x *cannot* assume larger values, but only that the likelihood is less than .001 (in fact, less than .0005) that it will do so.

Finally, you may need to calculate Poisson probabilities for values of λ not found in Table III. You may be able to obtain an adequate approximation by interpolation, but if not, consult more extensive tables for the Poisson distribution.

Exercises 4.46–4.58

Learning the Mechanics

4.46 Consider the following probability distribution:

$$p(x) = \frac{2^x e^{-2}}{x!} \qquad (x = 0, 1, 2, \ldots)$$

a. Is x a discrete or continuous random variable? Explain.
b. What is the name of this probability distribution?
c. Graph the probability distribution.
d. Find the mean and standard deviation of x.
e. Find the mean and standard deviation of the probability distribution.

4.47 Given that x is a random variable for which a Poisson probability distribution provides a good approximation, use Table III of Appendix B to compute the following:
a. $P(x \leq 3)$ when $\lambda = 5$
b. $P(x \leq 3)$ when $\lambda = 3$
c. $P(x \leq 3)$ when $\lambda = 1$
d. What happens to the probability of the event $\{x \leq 3\}$ as λ decreases from 5 to 1? Is this intuitively reasonable?

4.48 Assume that x is a random variable having a Poisson probability distribution with a mean of 3.2. Use Table III of Appendix B to find the following probabilities:
a. $P(x \geq 4)$ b. $P(x > 2)$ c. $P(x = 3)$
d. $P(x < 7)$ e. $P(x \leq 4)$ f. $P(2 \leq x \leq 6)$

4.49 Given that x is a random variable for which a Poisson probability distribution with $\lambda = 5$ provides a good characterization:

a. Graph $p(x)$ for $x = 0, 1, 2, \ldots, 13, 14, 15$.
b. Find μ and σ for x, and locate μ and the interval $\mu \pm 2\sigma$ on the graph.
c. What is the probability that x will fall within the interval $\mu \pm 2\sigma$?

Applying the Concepts

4.50 The Federal Deposit Insurance Corporation (FDIC), established in 1933, insures deposits of up to $100,000 in banks that are members of the Federal Reserve System (and others that voluntarily join the insurance fund) against losses due to bank failure or theft. From 1981 through 1990 the average number of bank failures per year among insured banks was approximately 124.5 (*Statistical Abstract of the United States: 1991*, p. 502). Assume that x, the number of bank failures per year among insured banks, can be adequately characterized by a Poisson probability distribution with mean 124.5.

a. Find the expected value and standard deviation of x.
b. In 1988, 221 insured banks failed. How far (in standard deviations) does $x = 221$ lie above the mean of the Poisson distribution? That is, find the z-score for $x = 221$.
c. In 1986, 145 insured banks failed. Indicate how to calculate $P(x \leq 145)$. Do not actually perform the calculation.
d. Discuss conditions that would make the Poisson assumption plausible.

4.51 The Environmental Protection Agency (EPA), established in 1970 as part of the executive branch of the federal government, issues pollution standards that vitally affect the safety of consumers and the operations of industry (*The United States Government Manual 1985–1986*). For example, the EPA states that manufacturers of vinyl chloride and similar compounds must limit the amount of these chemicals in plant air emissions to no more than 10 parts per million. Suppose the mean emission of vinyl chloride for a particular plant is 4 parts per million. Assume that the number of parts per million of vinyl chloride in air samples, x, follows a Poisson probability distribution.

a. What is the standard deviation of x for the plant?
b. Is it likely that a sample of air from the plant would yield a value of x that would exceed the EPA limit? Explain.
c. Discuss conditions that would make the Poisson assumption plausible.

4.52 U.S. airlines fly aproximately 26 billion passenger-miles per month and average about 11.8 fatalities per month (*Statistical Abstract of the United States: 1991*, pp. 627, 629). Assume the probability distribution for x, the number of fatalities per month, can be approximated by a Poisson probability distribution.

a. What is the probability that no fatalities will occur during any given month? [*Hint:* Either use Table III of Appendix B and interpolate to approximate the probability, or use a calculator or computer to calculate the probability exactly.]
b. Find $E(x)$ and the standard deviation of x.
c. Use your answers to part **b** to describe the probability that as many as 20 fatalities will occur in any given month.
d. Discuss conditions that would make the Poisson assumption plausible.

4.53 The safety supervisor at a large manufacturing plant believes the expected number of industrial accidents per month to be 3.4. What is the probability of exactly two accidents occurring next month? Three or more? What assumptions do you need to make to solve this problem using the methodology of this chapter?

4.54 As a check on the quality of the wooden doors produced by a company, its owner requested that each door undergo inspection for defects before leaving the plant. The plant's quality control inspector found that 1 square foot of door surface contains, on the average, .5 minor flaw. Subsequently, 1 square foot of each door's surface was examined for flaws. The owner decided to have all doors reworked that were found to have two or more minor flaws in the square foot of surface that was inspected. What is the probability that a door will fail inspection and be sent back for reworking? What is the probability that a door will pass inspection?

4.55 A large manufacturing plant has 3,200 incandescent light bulbs illuminating the manufacturing floor. If the rate at which the bulbs fail follows a Poisson distribution with a mean of three bulbs per hour, what is the probability that exactly three light bulbs fail in an hour? What is the probability that no bulbs fail in an hour? That no bulbs fail in an 8-hour shift? What assumption is required to calculate the last probability?

4.56 The random variable x, the number of people who arrive at a cashier's counter in a bank during a specified period of time, often possesses (approximately) a Poisson probability distribution. If the mean arrival rate, λ, is known, the Poisson probability distribution can be used to aid in the design of the customer service facility. Suppose you estimate that the mean number of arrivals per minute for cashier service at a bank is one person per minute. What is the probability that in a given minute, the number of arrivals will equal three or more? Can you tell the bank manager that the number of arrivals will rarely exceed three per minute?

4.57 The probability that a health insurance company must pay a major medical claim for a policy is .001. If a group of 1,000 policyholders represents a random sample of all possible policyholders, what is the probability that the insurance company will have to pay at least one major medical claim in this sample?

4.58 The Department of Commerce in a particular state has determined that the number of small businesses that declare bankruptcy per month has approximately a Poisson distribution with a mean equal to 6.4.

a. Find the probability of at least five bankruptcies occurring next month.

b. Find the probability of exactly four bankruptcies occurring next month.

Summary

Random variables are rules that assign numerical values to the outcomes of an experiment. **Discrete random variables** have countable numbers of possible values, and **continuous random variables** can assume an uncountable number of values corresponding to the points in one or more intervals. For purposes of distinguishing the two, the values of a discrete random variable can be listed, whereas those of a continuous random variable cannot.

The **probability distribution** of a discrete random variable specifies the probability associated with each possible value the random variable can assume. The mean and standard deviation of the probability distribution provide numerical descriptive measures of the distribution. Many applications of statistics involve the estimation of the mean and standard deviation of a probability distribution based on sample data.

The **binomial** and **Poisson** probability distributions describe two specific types of discrete random variables that have many applications in business. The characteristics of these important random variables were presented in this chapter, along with many examples of business data for which each would be an appropriate model. The formulas for the probability distributions, means, and variances were also presented, along with tables to assist you in calculating probabilities associated with each of these distributions.

Supplementary Exercises 4.59–4.76

Note: Starred (*) exercises refer to optional section in this chapter.

4.59 Given that x is a binomial random variable, compute $p(x)$ for each of the following cases:

a. $n = 7, \quad x = 3, \quad p = .5$
b. $n = 4, \quad x = 3, \quad p = .8$
c. $n = 15, \quad x = 1, \quad p = .1$

4.60 For each of the following examples, decide whether x is a binomial random variable, and explain your decision:

a. Of five applicants for a job in the accounting department, two will be selected. Although all applicants appear to be equally qualified, only three have the ability to fulfill the expectations of the company. Suppose that the two selections are made at random from the five applicants, and let x be the number of qualified applicants selected.
b. A software developer establishes a support hotline for customers to call in with questions regarding use of the software. Let x represent the number of calls received on the support hotline during a specified workday.
c. Florida is one of a minority of states with no state income tax. A poll of 1,000 registered voters is conducted to determine how many would favor a state income tax in light of the state's current fiscal condition. Let x be the number in the sample who would favor the tax.

4.61 Given that x is a Poisson random variable, compute $p(x)$ for each of the following cases:

a. $\lambda = 14, \quad x = 11$ b. $\lambda = 11, \quad x = 14$ c. $\lambda = 13, \quad x = 7$

4.62 Given that x is a random variable for which a Poisson probability distribution with $\lambda = 4$ provides a good characterization, compute the following:

a. $P(x = 0)$ b. $P(x = 3)$ c. $P(x = 1)$ d. $P(x = 5)$ e. $P(x \leq 2)$ f. $P(x \geq 2)$

4.63 Which of the following describe discrete random variables, and which describe continuous random variables?

a. The number of damaged inventory items
b. The average monthly sales revenue generated by a salesperson over the past year
c. The number of square feet of warehouse space a company rents
d. The length of time a firm must wait before its copying machine is fixed

4.64 Variables, x, y, and z are three discrete random variables with the same mean and the same range. Their probability distributions are shown in the figures on the following page. Which has the largest variance? The next largest? The smallest? Explain.

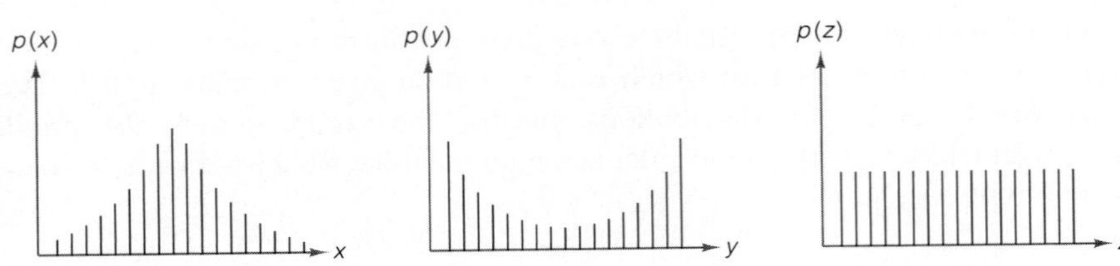

4.65 As part of a study of how pricing decisions are made by fabricare firms (the dry cleaning and laundry industry), Joe F. Goetz, Jr. (1985) contacted and questioned 103 fabricare firms. The accompanying table describes the sales volumes of these firms. Assume that the sales volume, x, for each of the six intervals can be adequately approximated by the midpoint of the interval. These midpoints are shown in the second column of the table. One of the 103 firms is to be randomly selected for more intensive questioning. In this situation the relative frequencies in the table can be interpreted as probabilities that describe the approximate likelihood of the sampled firm having sales volume x.

Sales Volume	x	*Relative Frequency*
Over \$0 to \$75,000	\$ 37,500	.14
Over \$75,000 to \$150,000	112,500	.30
Over \$150,000 to \$300,000	225,000	.20
Over \$300,000 to \$500,000	400,000	.17
Over \$500,000 to \$1,000,000	750,000	.10
Over \$1,000,000 to \$5,000,000	3,000,000	.09

a. Find $E(x)$. **b.** Find $E[(x - \mu)^2]$.
c. Interpret the values you obtained in parts **a** and **b** in the context of the problem.

4.66 The Environmental Protection Agency (EPA) tested 842 in-use automobiles to determine whether there were differences between the cars' actual gas mileage and the mileage projected in the EPA's mileage guide. For highway driving, they found that only 68% of the cars tested had fuel economies within 2 miles per gallon of the mileage guide's projection (*Environmental News*, 1978). Assume this figure holds for the population of cars currently in use for which the EPA has determined projected gas mileages. Suppose the EPA is planning to select 20 cars at random from this population and test their fuel economies to determine how many are within 2 miles per gallon of their EPA projections.

a. Decide whether the following statement is true or false and explain: The number of cars in the sample of 20 that have mileages within 2 miles per gallon of their EPA projections is not a binomial random variable but, for convenience, could be treated as a binomial random variable.

b. What is the probability (approximately) that fewer than 10 of the 20 cars selected will be within 2 miles per gallon of their EPA projections? (For convenience, use 70% as an approximation to 68%.)

4.67 The owner of construction company A makes bids on jobs so that, if it is awarded the job, company A will make a \$10,000 profit. The owner of construction company B makes bids on jobs so that, if it is awarded the job, company B will make a \$15,000 profit. Each company describes the probability distribution of the number of jobs the company is awarded per year as shown in the table.

Company A		*Company B*	
x	$p(x)$	x	$p(x)$
2	.05	2	.15
3	.15	3	.30
4	.20	4	.30
5	.35	5	.20
6	.25	6	.05

a. Find the expected number of jobs each will be awarded in a year.
b. What is the expected profit for each company?
c. Find the variance and standard deviation of the distribution of the number of jobs awarded per year for each company.
d. Graph $p(x)$ for both companies A and B. For each company, what proportion of the time will x fall in the interval $\mu \pm 2\sigma$?

4.68 An experiment is to be conducted to determine whether an acclaimed stock market analyst has extrasensory perception (ESP). Five different cards are shuffled and one is chosen at random. The analyst will try to identify which card was drawn without seeing it. The experiment is repeated 20 times and x, the number of correct decisions, is recorded. (Assume that the 20 trials are independent.)
a. If the stock market analyst is guessing (i.e., if the analyst does not possess ESP), what is the value of p, the probability of a correct decision on each trial?
b. If the analyst is guessing, what is the probability of a correct decision on at least half the trials?
c. Suppose the analyst makes the correct decision on 10 of the 20 trials. Do you think this performance is good enough to suggest that the analyst possesses ESP? Support or refute the proposition that the analyst possesses ESP using your results from part **b**.

4.69 A manufacturer considers a production lot unacceptable if 10% or more of the units in the lot are defective. In such cases the company wants to scrap (not ship) the entire lot. A company quality control inspector has proposed the following criterion for determining whether to reject a lot: In a sample of 10 units from a lot, if two or more are defective, reject the entire lot. If the lot currently under examination is 11% defective, what is the probability that this decision rule will lead the quality control inspector to the correct decision?

4.70 When the price of grain is low, many farmers participate in government-financed, on-farm storage programs rather than selling their grain. But storage invites insect infestations, and grain elevators penalize farmers who sell them insect-infested grain. In 1982, the U.S. Grain Marketing Research Laboratory estimated that 80% of the storage bins of corn in the country were infested with insects, and it has been estimated that the economic loss to farmers in the state of Minnesota alone is $12.6 million annually (*Minneapolis Tribune*, Aug. 8, 1982). Suppose 20 storage bins of corn are randomly selected and examined for insect infestation.
a. What is the probability (approximately) that less than one-half of the bins are infested?
b. What assumptions did you make in answering part **a**?
c. Why is your answer to part **a** an approximation?
d. Would you be surprised if all 20 of the bins were infested? Explain.

4.71 A wholesale office equipment outlet claims that on an average it sells 2.4 typewriters per day. If it has only five typewriters in stock at the close of business today and does not expect to receive a shipment of new typewriters until some time after the close of business tomorrow, what is the probability that the outlet's current supply of typewriters will not be sufficient to meet tomorrow's demand?

4.72 If the probability of a customer responding to one of your marketing department's mail questionnaires is .6, what is the probability that, of 20 questionnaires mailed, more than 15 will be returned?

4.73 A sales manager has determined that a salesperson makes a sale to 70% of the retailers visited.

a. Suppose the salesperson visits five retailers today and 20 tomorrow. What is the probability that she makes exactly four sales today *and* more than 10 tomorrow?

b. If the salesperson visits four retailers today and five tomorrow, what is the probability that in these 2 days she will make exactly two sales?

4.74 A small life insurance company has determined that on the average it receives five death claims per day. What is the probability that the company will receive three claims or less on a particular day? Exactly five claims? What assumptions must you make to find these probabilities?

4.75 In recent years, the use of the telephone as a data collection instrument for public opinion polls has been steadily increasing. However, one of the major factors bearing on the extent to which the telephone will become an acceptable data collection tool in the future is the *refusal rate*—i.e., the percentage of the eligible people actually contacted who refuse to take part in the poll. Suppose that past records indicate a refusal rate of 20% in a large city. A poll of 25 city residents is to be taken, and x is the number of residents contacted by telephone who refuse to take part in the poll.

a. Find the mean and variance of x. **b.** Find $P(x \leq 5)$. **c.** Find $P(x > 10)$.

4.76 [*Warning: This exercise is realistic and the calculations involved are tedious.*] According to the Orlando, Florida, *Sentinel Star* (June 1, 1977), the Red Lobster Inns of America decided to take the state of Florida to court over sales taxes. The dispute concerned the 4% sales tax levied on most purchases in the state and focused mainly on the state's "bracket tax collection system." According to the bracket system, a merchant collected 1¢ sales tax for sales between 10¢ and 25¢, 2¢ for sales between 26¢ and 50¢, 3¢ for sales between 51¢ and 75¢, and 4¢ for sales between 76¢ and 99¢. Red Lobster contended that if this system were followed, merchants would always collect more than 4%. That is, if a sale were made for $10.41, 4% would be collected on the $10, but more than 4% would be collected on the 41¢. In fact, 4.878% (2 cents/41 cents) sales tax would be collected on the 41¢. Let $x = 0, 1, 2, \ldots, 99$ be the number of cents (exceeding whole dollars) involved in a sale. Suppose that x has a probability distribution $p(x) = .01$ for $x = 0, 1, 2, \ldots, 99$ (an assumption that might be fairly accurate for restaurant sales). Find the expected value of the percentage of tax paid on the cents portion of a sale. [*Hint:* You can find the percentage tax—call it y—for each value of x. You also know the probabilities associated with each value of x.]

On Your Own

To control the quality of incoming or outgoing large lots of manufactured items, manufacturers use lot acceptance sampling plans. For example, if each lot consists of 1,000 items, the plan will call for the selection of a random sample of n items (n is usually small) from each lot. The items in each sample are carefully inspected, the number of defectives recorded, and the lot considered to be of acceptable quality if the number of defectives is less than or equal to some specified number, a. The number a is called the **acceptance number** for the plan.

To illustrate, simulate sampling from a large lot of items that contains 10% defectives. Place 10 poker chips (or marbles, etc.) in a bowl and mark one of the 10 as defective. Randomly select a sample of five items from a lot by selecting a chip from the 10, replacing it, and repeating the process four more times. Count the number of times, x, that you observe the defective chip. This process is equivalent to selecting a random sample of $n = 5$ items from a large lot containing 10% defectives.

If you choose $a = 1$ as the acceptance number for the plan, then you will accept only lots for which $x \leq 1$.

By choosing n and a, you change the ability of a sampling plan to screen out bad lots. To investigate the properties of a sampling plan with $n = 5$ and $a = 1$, simulate the process of sampling from 100 lots.

a. Collect the 100 values of x obtained from the simulation, and construct a relative frequency histogram for x. Note that this histogram is an approximation to $p(x)$. Estimate the proportion of lots that will be accepted by the plan by dividing the number of lots accepted by 100.

b. Calculate the exact values of $p(x)$ for $n = 5$ and $p = .1$, and compare these with the results of part **a**.

Using the Computer

Calculate the mean percentage of college graduates over all the zip codes for one of the census regions listed in Appendix C. Round the mean to the nearest integer, and use the result as an estimate of the percentage of college graduates among all people at least 25 years old in the region.

a. Suppose that 20 individuals from the region respond to an advertisement by an employment agency. If these people represent a random sample of all individuals at least 25 years of age in the region, what is the probability that more than half of them have college degrees?

b. If 2,000 individuals respond, find the mean μ and standard deviation σ of the number of them who have college degrees. Calculate $\mu \pm 2\sigma$ and $\mu \pm 3\sigma$, and use Chebyshev's theorem and the Empirical Rule to estimate the probability that the number of respondents with college degrees falls in each of the intervals. Based on your answers, assess the likelihood that more than half the applicants will have college degrees.

c. What are the potential problems with using the mean percentage of college graduates as an estimate of the percentage for the region? How could the estimate be improved?

References

Alexander, G. J. and Francis, J. C. *Portfolio Analysis*, Englewood Cliffs, N.J.: Prentice-Hall, 1986.

Biddle, W. "What destroyed *Challenger*?" *Discover*, Apr. 1986, pp. 40–47.

Brown, A. A., Hulswit, F. T., and Kettelle, J. D. "A study of sales operations." *Operations Research*, June 1956, *4*, pp. 296–308.

Chervany, N. L., Anderson, J. C., Benson, P. G., and Hill, A. V. "A management science approach to a Dutch elm disease sanitation program." *Interfaces*, Apr. 1980, *10*, p. 108.

"'83 report put booster accident as most likely." *Minneapolis Star and Tribune*, Feb. 11, 1986.

Elton, E. J. and Gruber, M. J. *Modern Portfolio Theory and Investment Analysis*. New York: Wiley, 1981.

Environmental News. Environmental Protection Agency. New England Regional Office, Boston, Jan. 1978, pp. 11–12.

Goetz, J. F., Jr. "The pricing decision: A service industry's experience." *Journal of Small Business Management*, Apr. 1985, *23*, pp. 61–67.

Hogg, R. V. and Craig, A. T. *Introduction to Mathematical Statistics*, 4th ed. New York: Macmillan, 1978.

Johnson, R. W. and Sullivan, A. C. "Segmentation of the consumer loan market." *Journal of Retail Banking*, Sept. 1981, 3, pp. 1–7.

Markowitz, H. M. "Portfolio selection." *Journal of Finance*, 6, Mar. 1952.

McKean, K. "They fly in the face of danger." *Discover*, Apr. 1986, pp. 48–58.

Mendenhall, W. *Introduction to Probability and Statistics*, 8th ed. Boston: Duxbury, 1991. Chapters 5 and 6.

Misshauk, M. J. *Management Theory and Practice*. Boston: Little, Brown, 1979. Chapter 13.

Murphy, A. H. and Winkler, R. L. "Probability forecasting in meteorology." *Journal of the American Statistical Association*, Sept. 1984, 79, pp. 489–500.

Parzen, E. *Modern Probability Theory and Its Applications*. New York: Wiley, 1960. Chapters 3, 4, 6, and 7.

Robinson, W. V. "NASA blamed for shuttle disaster." *Boston Globe*, June 10, 1986.

U.S. Golf Association. *The Rules of Golf*. 1993, p. 110.

The United States Government Manual 1985–1986. Office of the Federal Register, revised July 1, 1985, pp. 485–487.

Webster, A. H. "Straight answers about Dutch elm disease." *Flower and Garden*, May 1978, pp. 28–33, 46–47.

Williams, C. A., Jr., and Heins, R. M. *Risk Management and Insurance*, 6th ed. New York: McGraw-Hill, 1986.

Willis, R. E. and Chervany, N. L. *Statistical Analysis and Modeling for Management Decision-Making*. Belmont, Calif.: Wadsworth, 1974. Chapter 5.

CHAPTER FIVE

Continuous Random Variables

Contents

Case Studies

Where We've Been

Because sample data represent observed values of random variables, we needed to determine the probabilities associated with specific random variables. The probability theory of Chapter 3 provided the means for determining the probabilities associated with discrete random variables. Finding and describing this set of probabilities—the probability distribution for a discrete random variable—was the subject of Chapter 4.

Where We're Going

Since business data are derived from observations on continuous as well as discrete random variables, we need to know probability distributions associated with continuous random variables and also how to use the mean and standard deviation to describe these distributions. Chapter 5 addresses this problem and, in particular, introduces the normal probability distribution. As you will see, the normal probability distribution is one of the most useful distributions in statistics.

5

In this chapter, we consider continuous random variables. Recall that a continuous random variable is one that can assume any value within some interval or intervals. For example, the length of time between a consumer's purchase of new automobiles, the thickness of sheets of steel produced in a rolling mill, and the actual weight of the contents of 1-gallon cans of paint are all continuous random variables.

The methodology we use to describe continuous random variables is necessarily somewhat different from that used to describe discrete random variables. We first discuss the general form of continuous probability distributions, and then we present three specific types that are used in making business decisions: (1) The uniform distribution is simple but useful, and serves as a good introduction to continuous probability distributions. (2) The normal probability distribution, which plays a basic and important role in both the theory and application of statistics, is essential to the study of most of the subsequent chapters. (3) The exponential probability distribution has some specialized application to business problems, and we present it in an optional section.

5.1 Continuous Probability Distributions

The graphical form of the probability distribution for a continuous random variable, x, will be a smooth curve that might appear as shown in Figure 5.1. This curve, a function of x, is denoted by the symbol $f(x)$ and is variously called a **probability density function**, a **frequency function**, or a **probability distribution**.

The areas under a probability distribution correspond to probabilities for x. For example, the area A between the two points a and b, as shown in Figure 5.1, is the probability that x assumes a value between a and b ($a < x < b$). Because areas over intervals represent probabilities, it follows that the total area under a probability distribution, the probability assigned to all values of x, should equal 1. Note that probability distributions for continuous random variables will have different shapes depending on the relative frequency distributions of the real data that the probability distribution is supposed to model. Note too that the events $\{a < x < b\}$ and $\{a \leq x \leq b\}$ are equivalent for continuous random variables, because no probability is assigned to individual points. In terms of the probability distribution $f(x)$, no area is added by the addition of a single point (or two points) to the interval.

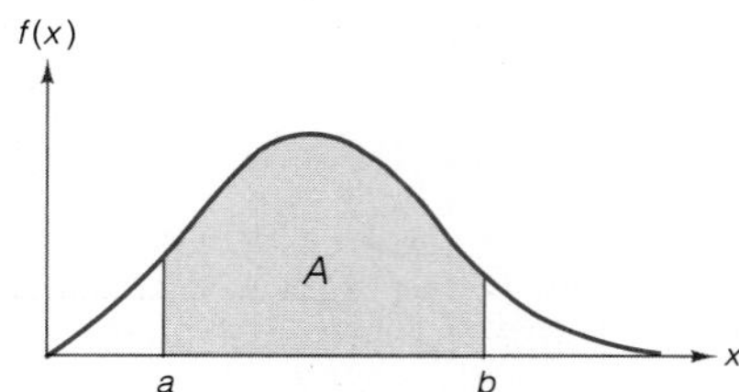

FIGURE 5.1 ▶ A probability distribution $f(x)$ for a continuous random variable x

The areas under most probability distributions are obtained by the use of calculus* or numerical methods. Because this is often a difficult procedure, we give the areas for some of the most common probability distributions in tabular form in Appendix B. Then, to find the area between two values of x, say $x = a$ and $x = b$, you can simply consult the appropriate table.

For each of the continuous random variables presented in this chapter, we will give the formula for the probability distribution along with its mean and standard deviation. These two numbers, μ and σ, will enable you to make some approximate probability statements about a random variable even when you do not have access to a table of areas under the probability distribution.

5.2 The Uniform Distribution

Perhaps the simplest of all the continuous probability distributions is the **uniform distribution**. The frequency function has a rectangular shape, as shown in Figure 5.2. Note that the possible values of x consist of all points on the real line between point c and point d. The height of $f(x)$ is constant in that interval and equals $1/(d - c)$. Therefore, the total area under $f(x)$ is given by

$$\text{Total area of rectangle} = (\text{Base})(\text{Height}) = (d - c)\left(\frac{1}{d - c}\right) = 1$$

The uniform probability distribution provides a model for continuous random variables that are *evenly distributed* over a certain interval. That is, a uniform random variable is one that is just as likely to assume a value in one interval as it is to assume a value in any other interval of equal size. There is no clustering of values around any value; instead, there is an even spread over the entire region of possible values.

FIGURE 5.2 ► The uniform probability distribution

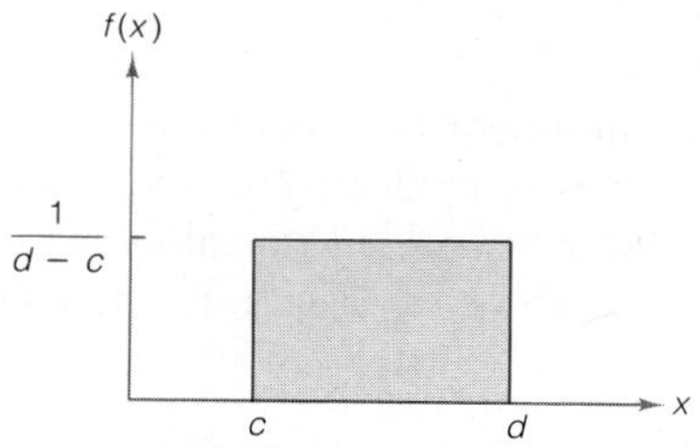

*Students with knowledge of calculus should note that the probability that x assumes a value in the interval $a < x < b$ is $P(a < x < b) = \int_a^b f(x)\,dx$, assuming the integral exists. Similar to the requirements for a discrete probability distribution, we require $f(x) \geq 0$ and $\int_{-\infty}^{\infty} f(x)\,dx = 1$.

The uniform distribution is sometimes referred to as the **randomness distribution**, since one way of generating a uniform random variable is to perform an experiment in which a point is *randomly* selected on the horizontal axis between the points c and d. If we were to repeat this experiment infinitely often and construct a relative frequency histogram for the outcomes, we would create a uniform probability distribution like that shown in Figure 5.2. The random selection of points on a line can also be used to generate random numbers such as those in a random number table (e.g., Table I of Appendix B). Recall that random numbers are selected in such a way that every digit has an equal probability of selection. Therefore, random numbers are realizations of a uniform random variable. (Random numbers were used to draw random samples in Section 3.7.)

The formulas for the uniform probability distribution and its mean and standard deviation are shown in the box.

Probability Distribution, Mean, and Standard Deviation of a Uniform Random Variable *x*

$$f(x) = \frac{1}{d - c} \quad (c < x < d)$$

$$\mu = \frac{c + d}{2} \qquad \sigma = \frac{d - c}{\sqrt{12}}$$

Suppose the interval $a < x < b$ lies within the domain of x—i.e., it falls within the larger interval $c < x < d$. Then the probability that x assumes a value within the interval $a < x < b$ is the area of the rectangle over the interval—namely,*

$$\frac{b - a}{d - c}$$

EXAMPLE 5.1

Suppose the research department of a steel manufacturer believes that one of the company's rolling machines is producing sheets of steel of varying thickness. The thickness is a uniform random variable with values between 150 and 200 millimeters. Any sheets less than 160 millimeters thick must be scrapped because they are unacceptable to buyers.

a. Calculate the mean and standard deviation of x, the thickness of the sheets produced by this machine. Then graph the probability distribution and show the mean

*The student who has knowledge of calculus should note that
$P(a < x < b) = \int_a^b f(x)\,dx = \int_a^b 1/(d - c)\,dx = (b - a)/(d - c)$

on the horizontal axis. Also show 1- and 2-standard-deviation intervals around the mean.

b. Calculate the fraction of steel sheets produced by this machine that have to be scrapped.

Solution

a. To calculate the mean and standard deviation for x, we substitute 150 and 200 millimeters for c and d, respectively, in the formulas. Thus,

$$\mu = \frac{c + d}{2} = \frac{150 + 200}{2} = 175 \text{ millimeters}$$

and

$$\sigma = \frac{d - c}{\sqrt{12}} = \frac{200 - 150}{\sqrt{12}} = \frac{50}{3.464} = 14.43$$

The uniform probability distribution is

$$f(x) = \frac{1}{d - c} = \frac{1}{200 - 150} = \frac{1}{50}$$

The graph of this function is shown in Figure 5.3. The mean and the 1- and 2-standard-deviation intervals around the mean are shown on the horizontal axis.

FIGURE 5.3 ▶
Distribution for x in Example 5.1

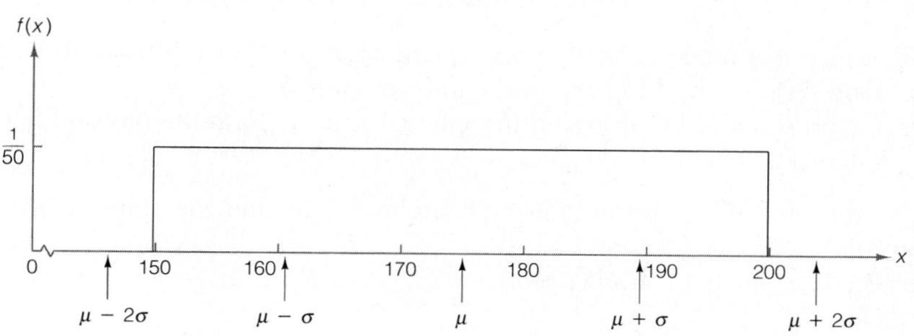

b. To find the fraction of steel sheets produced by the machine that have to be scrapped, we must find the probability that x, the thickness, is less than 160 millimeters. As indicated in Figure 5.4 on page 230, we need to calculate the area under the frequency function $f(x)$ between the points $x = 150$ and $x = 160$. This is the area of a rectangle with base $160 - 150 = 10$ and height $1/50$. The fraction that has to be scrapped is then

$$P(x < 160) = (\text{Base})(\text{Height}) = (10)\left(\frac{1}{50}\right) = \frac{1}{5}$$

That is, 20% of all the sheets made by this machine must be scrapped.

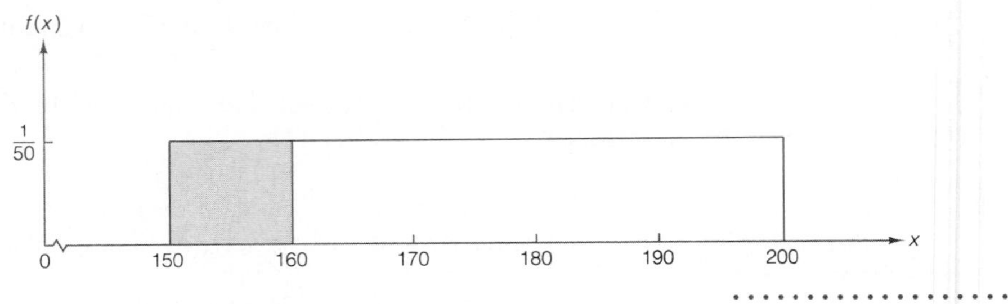

FIGURE 5.4 ▶ Probability that sheet thickness, x, is between 150 and 160 millimeters

Exercises 5.1–5.11

Learning the Mechanics

5.1 Suppose x is a random variable best described by a uniform probability distribution with $c = 20$ and $d = 45$.
a. Find $f(x)$. b. Find the mean and variance of x.
c. Graph $f(x)$ and locate μ and the interval $\mu \pm 2\sigma$ on the graph. Find the probability that x assumes a value within the interval $\mu \pm 2\sigma$.

5.2 Refer to Exercise 5.1. Find the following probabilities:
a. $P(20 \leq x \leq 30)$ b. $P(20 < x < 30)$ c. $P(x \geq 30)$ d. $P(x \geq 45)$
e. $P(x \leq 40)$ f. $P(x < 40)$ g. $P(15 \leq x \leq 35)$ h. $P(21.5 \leq x \leq 31.5)$

5.3 Suppose x is a random variable best described by a uniform probability distribution with $c = 3$ and $d = 7$.
a. Find $f(x)$. b. Find the mean and variance of x.
c. Graph $f(x)$ and locate μ and the interval $\mu \pm \sigma$. Find the probability that x assumes a value within the interval $\mu \pm \sigma$.

5.4 Use the probability distribution of Exercise 5.3 to find the value of a that makes each of the following probability statements true:
a. $P(x \geq a) = .6$ b. $P(x \leq a) = .25$ c. $P(x \leq a) = 1$ d. $P(4 \leq x \leq a) = .5$
e. $P(x > a) = .2$

5.5 The random variable x is best described by a uniform probability distribution with $c = 100$ and $d = 200$. Find the probability that x assumes a value:
a. More than 2 standard deviations from μ. b. Less than 3 standard deviations from μ.
c. Within 2 standard deviations of μ.

5.6 The random variable x is best described by a uniform probability distribution with mean 10 and standard deviation 1. Find c, d, and $f(x)$. Graph the probability distribution.

Applying the Concepts

5.7 As we noted in this section, random numbers are values of a uniform random variable. Construct a relative frequency histogram for the data set listed below. (It was created by the random-number generator of the Minitab computer package.) Except for the expected variation in relative frequencies among the class intervals, does your histogram suggest that the data are observations on a uniform random variable with $c = 0$ and $d = 100$? Explain.

38.8759	98.0716	64.5788	60.8422	.8413
88.3734	31.8792	32.9847	.7434	93.3017
12.4337	11.7828	87.4506	94.1727	23.0892
47.0121	43.3629	50.7119	88.2612	69.2875
62.6626	55.6267	78.3936	28.6777	71.6829
44.0466	57.8870	71.8318	28.9622	23.0278
35.6438	38.6584	46.7404	11.2159	96.1009
95.3660	21.5478	87.7819	12.0605	75.1015

5.8 A bus is scheduled to stop at a certain bus stop every half hour on the hour and half hour. At end of the day, buses still stop about every 30 minutes, but due to delays earlier in the day, they are equally likely to stop at any time during any given half hour. If you arrive at a bus stop at the end of the day, what is the probability that you will have to wait more than 20 minutes for the bus (no matter when you show up)? How long do you expect to wait for the bus?

5.9 Probability distributions and event probabilities can be used to express uncertainty about future events. For example, security analysts use probability distributions to forecast stock prices and quarterly earnings; weather forecasters use event probabilities to forecast precipitation, temperature ranges, and the location of a hurricane's landfall. Such forecasts are known as **probability forecasts** (Murphy and Winkler, 1984). Probability forecasts can be derived from sample data, expert judgment, or a combination of both. When a forecaster possesses relatively little information about the future value of a random variable, the uniform distribution is frequently used to make a probability forecast. Suppose a security analyst believes that the closing price of a particular stock 3 months from today will be between \$30 and \$40, but has no idea where within the interval the price will be.

a. Graph the uniform distribution the security analyst should use as a probability forecast.

b. According to the analyst's forecast, what is the probability that the stock will close higher than \$35? Higher than \$38? Between \$34 and \$36, inclusive? Higher than \$50?

5.10 The **reliability** of a piece of equipment is frequently defined to be the probability, p, that the equipment performs its intended function successfully for a given period of time under specific conditions (Martz and Waller, 1982). Because p varies from one point in time to another, some reliability analysts treat p as if it were a random variable. Suppose an analyst characterizes the uncertainty about the reliability of a particular robotic device used in an automobile assembly line using the following distribution:

$$f(p) = \begin{cases} 1 & 0 \leq p \leq 1 \\ 0 & \text{otherwise} \end{cases}$$

a. Graph the analyst's probability distribution for p. **b.** Find the mean and variance of p.

c. According to the analyst's probability distribution for p, what is the probability that p is greater than .95? Less than .95?

d. Suppose the analyst receives the additional information that p is definitely between .90 and .95, but that there is complete uncertainty about where it lies between these values. Describe the probability distribution the analyst should now use to describe p.

5.11 A tool and die machine shop produces extremely high-tolerance spindles. The spindles are 18-inch slender rods used in a variety of military equipment. A piece of equipment used in the manufacture of the spindles malfunctions on occasion and places a single gouge somewhere on the spindle. However, if the spindle can be cut so that it has 14 consecutive inches without a gouge, then the spindle can be salvaged for other purposes. Assuming that the location of the gouge along the spindle is best described by a uniform distribution, what is the probability that a defective spindle can be salvaged?

5.3 The Normal Distribution

One of the most useful and frequently encountered continuous random variables has a **bell-shaped** probability distribution, as shown in Figure 5.5. It is known as a **normal random variable**, and its probability distribution is called a **normal distribution**.

FIGURE 5.5 ▶
A normal probability distribution

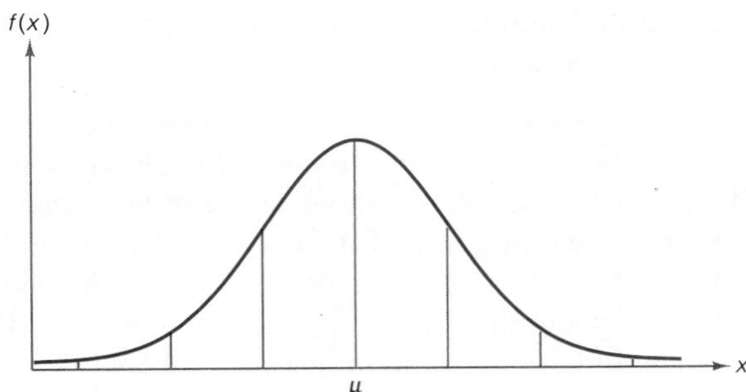

You will see during the remainder of this text that the normal distribution plays a very important role in the science of statistical inference. Many business phenomena generate random variables with probability distributions that are very well approximated by a normal distribution. For example, it has been demonstrated that the monthly rate of return (defined in the footnote on page 193) for a particular stock is approximately a normal random variable, and the probability distribution for the weekly sales of a corporation might be approximated by a normal probability distribution. The normal distribution might also provide an accurate model for the distribution of the scores on an unemployment aptitude test. You can determine the adequacy of the normal approximation to an existing population of data by comparing the relative frequency distribution of a large sample of the data to the normal probability distribution.

The normal distribution is perfectly symmetric about its mean, μ, as can be seen in the examples in Figure 5.6. Its spread is determined by the value of its standard deviation, σ.

The formula for the normal probability distribution is shown in the next box. When plotted, this formula yields a curve like that shown in Figure 5.5. Note that the mean μ and the variance σ^2 appear in this formula, so that no separate formulas for μ and σ^2 are necessary. To graph the normal curve we will have to know the numerical values of μ and σ.

FIGURE 5.6 ▶
Several normal distributions, with different means and standard deviations

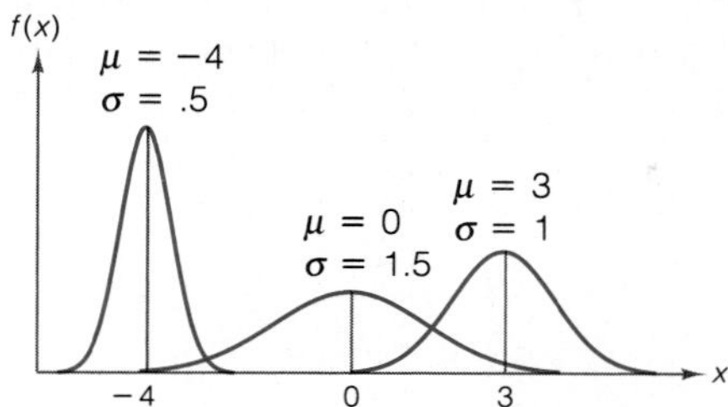

Probability Distribution for a Normal Variable x

$$f(x) = \frac{1}{\sigma\sqrt{2\pi}} e^{-(1/2)[(x-\mu)/\sigma]^2}$$

where μ = Mean of the normal random variable, x
σ^2 = Variance of the normal random variable, x
π = 3.1416. . .
e = 2.71828. . .

Computing the area over intervals under the normal probability distribution is a difficult task.* Consequently, we will use the computed areas listed in Table IV of Appendix B (and inside the front cover). Although there is an infinitely large number of normal curves—one for each pair of values for μ and σ—we have formed a single table that will apply to any normal curve. This was done by constructing the table of areas as a function of the z-score (presented in Section 2.8). The population z-score for a measurement was defined as the *distance* between the measurement and the population mean, divided by the population standard deviation. Thus, the z-score gives the distance between a measurement and the mean in units equal to the standard deviation. In symbolic form, the z-score for the measurement x is

$$z = \frac{x - \mu}{\sigma}$$

*The student with knowledge of calculus should note that there is not a closed-form expression for $P(a < x < b) = \int_a^b f(x)\,dx$ for the normal probability distribution. However, the value of this definite integral can be obtained to any desired degree of accuracy by numerical approximation procedures. The areas in Table IV of Appendix B were obtained by using such a procedure.

To illustrate the use of Table IV, suppose we know that the length of time between charges of a pocket calculator has a normal distribution, with a mean of 50 hours and a standard deviation of 15 hours. If we were to observe the length of time that elapses before the need for the next charge, what is the probability that this measurement would assume a value between 50 hours and 70 hours? This probability is the area under the normal probability distribution between 50 and 70, as shown in the shaded area, A, of Figure 5.7.

FIGURE 5.7 ▶ Normal distribution: $\mu = 50$, $\sigma = 15$

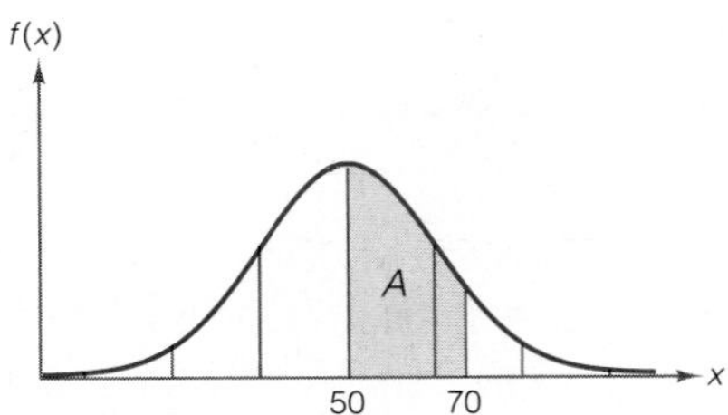

The first step in finding the area A is to calculate the z-score corresponding to the measurement 70. We calculate

$$\begin{aligned} z &= \frac{x - \mu}{\sigma} \\ &= \frac{70 - 50}{15} \\ &= \frac{20}{15} = 1.33 \end{aligned}$$

Thus, the measurement 70 is 1.33 standard deviations above the mean, 50. The second step is to refer to Table IV (a partial reproduction of this table is shown in Table 5.1). Note that z-scores are listed in the left-hand column of the table. To find the area corresponding to a z-score of 1.33, we first locate the value 1.3 in the left-hand column. Since this column lists z values to one decimal place only, we refer to the top row of the table to get the second decimal place, .03. Finally, we locate the number where the row labeled $z = 1.3$ and the column labeled .03 meet. This number represents the area between the mean, μ, and the measurement that has a z-score of 1.33:

$$A = .4082$$

Thus, the probability that the calculator operates between 50 and 70 hours before needing a charge is .4082.

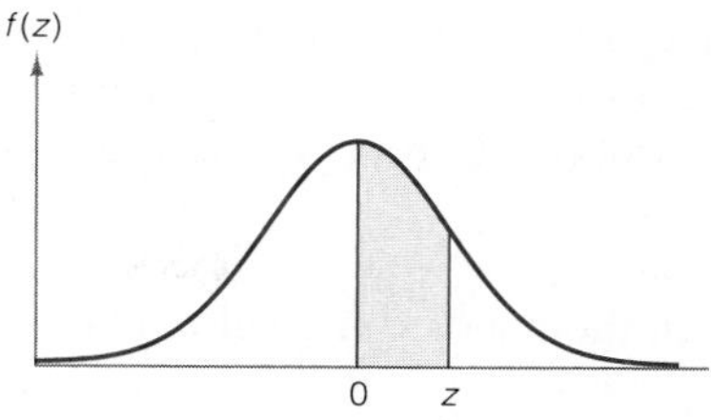

TABLE 5.1 Reproduction of Part of Table IV of Appendix B (and inside front cover): Normal Curve Areas

z	.00	.01	.02	.03	.04	.05	.06	.07	.08	.09
.0	.0000	.0040	.0080	.0120	.0160	.0199	.0239	.0279	.0319	.0359
.1	.0398	.0438	.0478	.0517	.0557	.0596	.0636	.0675	.0714	.0753
.2	.0793	.0832	.0871	.0910	.0948	.0987	.1026	.1064	.1103	.1141
.3	.1179	.1217	.1255	.1293	.1331	.1368	.1406	.1443	.1480	.1517
.4	.1554	.1591	.1628	.1664	.1700	.1736	.1772	.1808	.1844	.1879
.5	.1915	.1950	.1985	.2019	.2054	.2088	.2123	.2157	.2190	.2224
.6	.2257	.2291	.2324	.2357	.2389	.2422	.2454	.2486	.2517	.2549
.7	.2580	.2611	.2642	.2673	.2704	.2734	.2764	.2794	.2823	.2852
.8	.2881	.2910	.2939	.2967	.2995	.3023	.3051	.3078	.3106	.3133
.9	.3159	.3186	.3212	.3238	.3264	.3289	.3315	.3340	.3365	.3389
1.0	.3413	.3438	.3461	.3485	.3508	.3531	.3554	.3577	.3599	.3621
1.1	.3643	.3665	.3686	.3708	.3729	.3749	.3770	.3790	.3810	.3830
1.2	.3849	.3869	.3888	.3907	.3925	.3944	.3962	.3980	.3997	.4015
1.3	.4032	.4049	.4066	.4082	.4099	.4115	.4131	.4147	.4162	.4177
1.4	.4192	.4207	.4222	.4236	.4251	.4265	.4279	.4292	.4306	.4319
1.5	.4332	.4345	.4357	.4370	.4382	.4394	.4406	.4418	.4429	.4441

The use of the z-score simplifies the calculation of normal probabilities because if x is normally distributed with any mean and standard deviation, z is *always* a normal random variable with a mean of 0 and a standard deviation of 1. For this reason z is often referred to as the **standard normal random variable**.

Definition 5.1

The **standard normal random variable** z is defined by the formula

$$z = \frac{x - \mu}{\sigma}$$

where x is a normal random variable with mean μ and standard deviation σ. The standard normal random variable z is normally distributed with mean 0 and standard deviation 1, and can be described as the number of standard deviations between x and μ.

Since we will convert all normal random variables to standard normal in order to find probabilities based on Table IV of Appendix B, it is important that you learn to use Table IV well. The following examples illustrate its use.

EXAMPLE 5.2

Find the probability that the standard normal random variable z falls between -1.33 and $+1.33$.

Solution

The standard normal distribution is shown in Figure 5.8. Since all probabilities associated with standard normal random variables can be depicted as areas under the standard normal curve, you should always draw the curve and then equate the desired probability to an area.

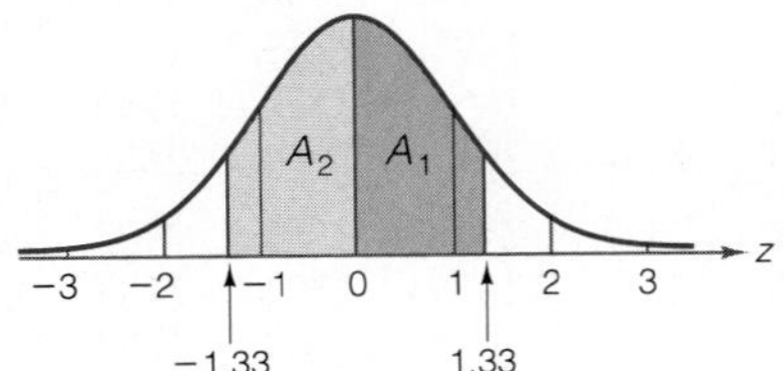

FIGURE 5.8 ► A distribution of z-scores (a standard normal distribution)

In this example we want to find the probability that z falls between -1.33 and $+1.33$, which is equivalent to the area between -1.33 and $+1.33$, shown shaded in Figure 5.8. Table IV of Appendix B provides the area between $z = 0$ and any value of z looked up, so that if we look up $z = 1.33$ we find that the area between $z = 0$ and $z = 1.33$ is .4082. This is the area labeled A_1 in Figure 5.8. To find the area A_2 between $z = 0$ and $z = -1.33$, we note that the symmetry of the normal distribution implies that the area between $z = 0$ and any point to the left is equal to the area between $z = 0$ and the point equidistant to the right. Thus, in this example the area between $z = 0$ and $z = -1.33$ is equal to the area between $z = 0$ and $z = +1.33$. That is,

$$A_1 = A_2 = .4082$$

The probability that z falls between -1.33 and $+1.33$ is the sum of the areas A_1 and A_2. Using probabilistic notation, we summarize as follows:

$$\begin{aligned} P(-1.33 < z < +1.33) &= P(-1.33 < z < 0) + P(0 \leq z < 1.33) \\ &= A_1 + A_2 = .4082 + .4082 \\ &= .8164 \end{aligned}$$

Remember that $<$ and $\leq$ are equivalent in events involving z because the inclusion (or exclusion) of a single point does not alter the probability of an event involving a continuous random variable.

EXAMPLE 5.3

Find the probability that a standard normal random variable exceeds 1.64, i.e., find $P(z > 1.64)$.

Solution

The area under the standard normal distribution to the right of 1.64 is the shaded area labeled A_1 in Figure 5.9. This area represents the desired probability that z exceeds 1.64. However, when we look up $z = 1.64$ in Table IV of Appendix B, we must remember that the probability given in the table corresponds to the area between $z = 0$ and $z = 1.64$ (the area labeled A_2 in Figure 5.9). From Table IV we find that $A_2 = .4495$. To find the area A_1 to the right of 1.64, we make use of two facts:

1. The standard normal distribution is symmetric about its mean, $z = 0$.
2. The total area under the standard normal probability distribution equals 1.

FIGURE 5.9 ► Standard normal distribution: $\mu = 0, \sigma = 1$

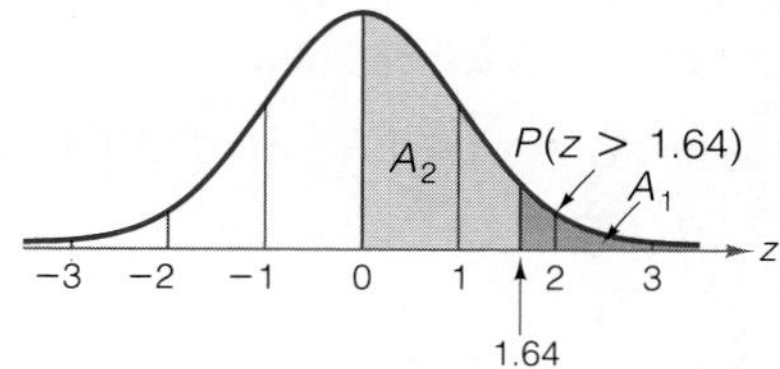

Taken together, these two facts imply that the areas on either side of the mean $z = 0$ equal .5; thus, the area to the right of $z = 0$ in Figure 5.9 is $A_1 + A_2 = .5$. Then

$$P(z > 1.64) = A_1 = .5 - A_2 = .5 - .4495 = .0505$$

To attach some practical significance to this probability, note that the implication is that the chance of a standard normal random variable exceeding 1.64 is approximately .05. Or, since z represents the number of standard deviations between *any* normal random variable and its mean, a normal random variable will exceed its mean by more than 1.64 standard deviations only about 5% of the time.

EXAMPLE 5.4

Find the probability that a normal random variable lies to the right of a point −.74 standard deviation from its mean.

Solution

We must first interpret the event in terms of a standard normal random variable, so that we can use Table IV of Appendix B to find the event's probability. Since the standard normal random variable z is simply the number of standard deviations between a normal random variable and its mean, the event that a normal random variable lies to the right of a point −.74 standard deviation from the mean is equivalent to the event that the standard normal random variable z exceeds −.74. The event is shown as the shaded area in Figure 5.10 (page 238), and we want to find $P(z > -.74)$.

FIGURE 5.10 ▶
Standard normal distribution: $\mu = 0, \sigma = 1$

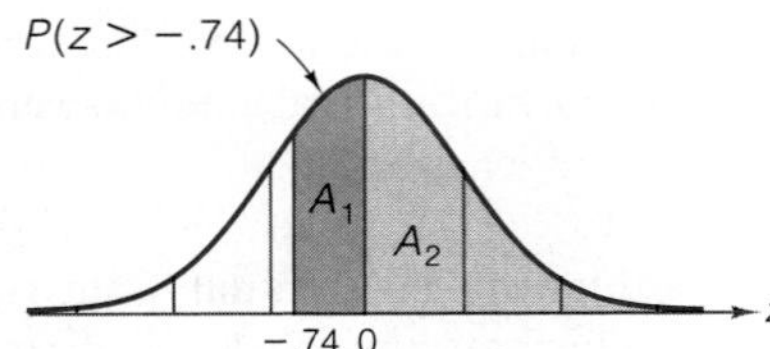

We divide the shaded area into two parts: the area A_1 between $z = -.74$ and $z = 0$, and the area A_2 to the right of $z = 0$. We must always make such a division when the desired area lies on both sides of the mean ($z = 0$), because Table IV contains areas between $z = 0$ and the point you look up. To find A_1, we remember that the sign of z is unimportant when determining the area, because the standard normal distribution is symmetric about its mean. We look up $z = .74$ in Table IV to find that $A_1 = .2704$. The symmetry also implies that half the distribution lies on each side of the mean, so the area A_2 to the right of $z = 0$ is .5. Then

$$P(z > -.74) = A_1 + A_2 = .2704 + .5 = .7704$$

EXAMPLE 5.5

Find the probability that a normal random variable lies more than 1.96 standard deviations from its mean *in either direction*.

Solution

The event that a normal random variable lies beyond 1.96 standard deviations in either direction from its mean is equivalent to the event that the standard normal random variable z exceeds 1.96 in absolute value. That is, we want to find

$$P(z > |1.96|) = P(z < -1.96 \text{ or } z > 1.96)$$

This probability is the shaded area in Figure 5.11. Note that the total shaded area is the sum of two areas, A_1 and A_2—areas that are equal because of the symmetry of the normal distribution.

FIGURE 5.11 ▶
Standard normal distribution: $\mu = 0, \sigma = 1$

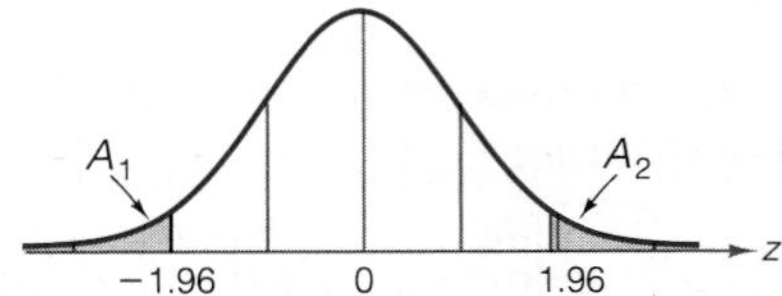

We look up $z = 1.96$ and find the area between $z = 0$ and $z = 1.96$ to be .4750. Then the area to the right of 1.96 is $.5 - .4750 = .0250$, so that

$$P(z > |1.96|) = .0250 + .0250 = .05$$

The implication is that any normal random variable lies more than 1.96 standard deviations from its mean 5% of the time. Recall (Chapter 2) that the Empirical Rule tells us that about 5% of the measurements in mound-shaped distributions will lie beyond 2 standard deviations from the mean; the normal distribution, which is certainly mound-shaped, has 5% of its area beyond 1.96 standard deviations. In fact, the normal distribution provides the model on which the Empirical Rule is based, along with much "empirical" experience with real data that often approximately obey the rule, whether they are drawn from a normal distribution or not.

EXAMPLE 5.6

Assume that the length of time, x, between charges of a pocket calculator is normally distributed with a mean of 50 hours and a standard deviation of 15 hours. Find the probability that the calculator will last between 30 and 70 hours between charges.

Solution

The normal distribution with mean $\mu = 50$ and $\sigma = 15$ is shown in Figure 5.12; the desired probability that the calculator lasts between 30 and 70 hours is shaded. In order to find the probability, we must first convert the distribution to standard normal, which we do by calculating the z-score:

$$z = \frac{x - \mu}{\sigma}$$

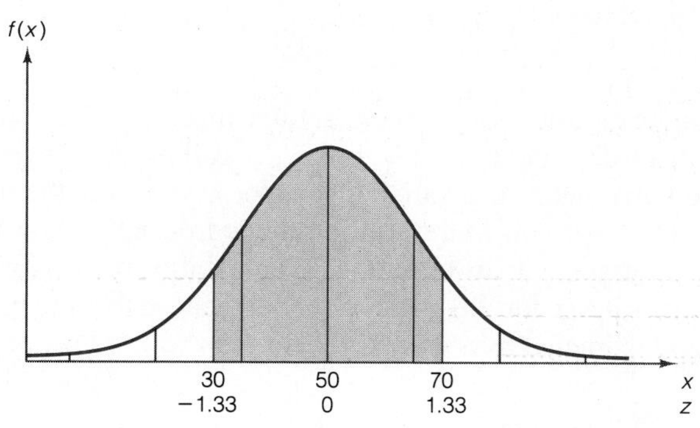

FIGURE 5.12 ► Normal probability distribution: $\mu = 50, \sigma = 15$

The z-scores corresponding to the important values of x are shown just under the x values on the horizontal axis in Figure 5.12. Note that $z = 0$ corresponds to the mean of $\mu = 50$ hours, while the x values 30 and 70 yield z-scores of -1.33 and $+1.33$, respectively. Thus, the event that the calculator lasts between 30 and 70 hours is equivalent to the event that a standard normal random variable lies between -1.33 and $+1.33$. We found this probability in Example 5.2 (see Figure 5.8) by doubling the area corresponding to $z = 1.33$ in Table IV. That is,

$$P(30 \leq x \leq 70) = P(-1.33 \leq z \leq 1.33) = 2(.4082) = .8164$$

The steps to follow when calculating a probability corresponding to a normal random variable are summarized in the box.

Steps for Finding a Probability Corresponding to a Normal Random Variable

1. Sketch the normal distribution and indicate the mean of the random variable x. Then shade the area corresponding to the probability you want to find.
2. Convert the boundaries of the shaded area from x values to standard normal random variable z values using the formula

$$z = \frac{x - \mu}{\sigma}$$

   Show the z values under the corresponding x values on your sketch.
3. Use Table IV of Appendix B (and inside the front cover) to find the areas corresponding to the z values. If necessary, use the symmetry of the normal distribution to find areas corresponding to negative z values and the fact that the total area on each side of the mean equals .5 to convert the areas from Table IV to the probabilities of the event you have shaded.

EXAMPLE 5.7

Suppose an automobile manufacturer introduces a new model that has an advertised mean in-city mileage of 27 miles per gallon. Although such advertisements seldom report any measure of variability, suppose you write the manufacturer for the details of the tests, and you find that the standard deviation is 3 miles per gallon. This information leads you to formulate a probability model for the random variable x, the in-city mileage for this car model. You believe that the probability distribution of x can be approximated by a normal distribution with a mean of 27 and a standard deviation of .3.

a. If you were to buy this model of automobile, what is the probability you would purchase one that averages less than 20 miles per gallon for in-city driving?
b. Suppose you purchase one of these new models, and it does get less than 20 miles per gallon for in-city driving. Should you conclude that your probability model is incorrect?

Solution

a. The probability model proposed for x, the in-city mileage, is shown in Figure 5.13. We are interested in finding the area, A, to the left of 20, since this area corresponds to the probability that a measurement chosen from this distribution falls below 20. Or in other words, if this model is correct, the area A represents the fraction of cars that can be expected to get less than 20 miles per gallon for in-city driving. To find A, we first calculate the z value corresponding to $x = 20$. That is,

FIGURE 5.13 ► **Normal probability distribution for x in Example 5.7: $\mu = 27$ miles per gallon, $\sigma = 3$ miles per gallon**

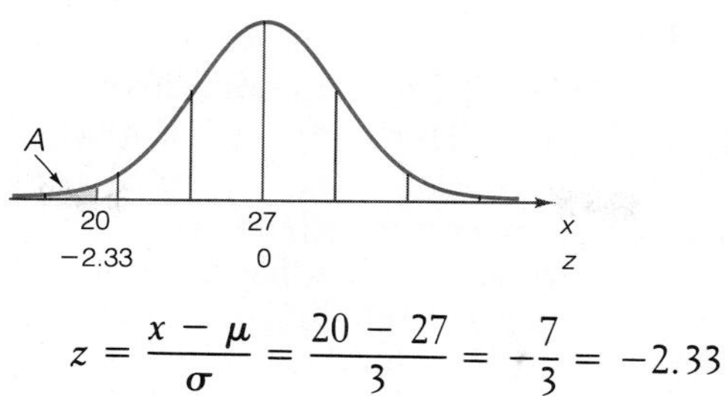

$$z = \frac{x - \mu}{\sigma} = \frac{20 - 27}{3} = -\frac{7}{3} = -2.33$$

This z value is shown in Figure 5.13 under the corresponding x value.

Because of the symmetry of the normal distribution about its mean and because Table IV provides only areas to the right of the mean, we look up 2.33 in Table IV and find that the corresponding area is .4901. This is the area between $z = 0$ and $z = -2.33$, so we find

$$A = .5 - .4901 = .0099 \approx .01$$

According to this probability model, you should have only about a 1% chance of purchasing a car of this make with an in-city mileage under 20 miles per gallon.

b. Now you are asked to make an inference based on a sample—the car you purchased. You are getting less than 20 miles per gallon for in-city driving. What do you infer? We think you will agree that one of two possibilities is true:

The probability model is correct, and you simply were unfortunate to have purchased one of the cars in the 1% that get less than 20 miles per gallon in the city.

The probability model is incorrect. Perhaps the assumption of a normal distribution is unwarranted, or the mean of 27 is an overestimate, or the standard deviation of 3 is an underestimate, or some combination of these errors was made. At any rate, the form of the actual probability model certainly merits further investigation.

You have no way of knowing with certainty which possibility is the correct one, but the evidence points to the second one. We are again relying on the rare event approach to statistical inference that we introduced earlier. The basic idea is that the sample (one measurement in this case) was so unlikely to have been drawn from the proposed probability model that it casts serious doubt on the model. We would be inclined to believe that the model is somehow in error.

Occasionally you will be given a probability and want to find the values of the normal random variable that correspond to the probability. For example, suppose the average daily production of a manufacturer is known to be normally distributed, and management wants to pay an incentive bonus to the crew when their level of production exceeds the 90th percentile of the daily production distribution. To determine the production level at which bonuses will be paid, you will need to be able to use the normal distribution probability table in reverse, as demonstrated in the following example.

EXAMPLE 5.8

Find the value of z, call it z_0, in the standard normal distribution that will be exceeded only 10% of the time. That is, find z_0 such that $P(z \geq z_0) = .10$.

Solution

In this case we are given a probability, or an area, and asked to find the value of the standard normal random variable that corresponds to the area. Specifically, we want to find the value z_0 such that only 10% of the standard normal distribution exceeds z_0 (see Figure 5.14).

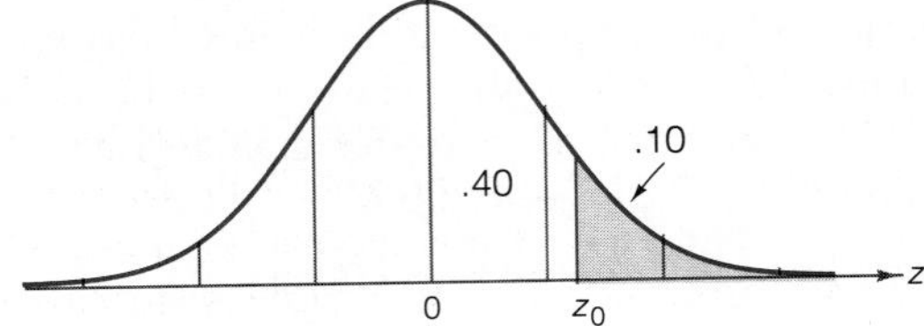

FIGURE 5.14 ► Standard normal distribution for Example 5.8

We know that the total area to the right of the mean $z = 0$ is .5, which implies that z_0 must lie to the right of (above) 0. To pinpoint the value, we use the fact that the area to the right of z_0 is .10, which implies that the area between $z = 0$ and z_0 is $.5 - .1 = .4$. But areas between $z = 0$ and some other z value are exactly the types given in Table IV of Appendix B. Therefore, we look up the *area* .4000 in the body of Table IV, and find that the corresponding z value is (to the closest approximation) $z_0 = 1.28$. Thus, the point 1.28 standard deviations above the mean is the 90th percentile of a normal distribution.

EXAMPLE 5.9

Find the value of z_0 such that 95% of the standard normal z values lie between $-z_0$ and $+z_0$, i.e.,

$$P(-z_0 \leq z \leq z_0) = .95$$

Solution

Here we wish to move an equal distance z_0 in the positive and negative direction from the mean $z = 0$ until 95% of the standard normal distribution is enclosed. This means that the area on each side of the mean will be equal to ½(.95) = .475, as shown in Figure 5.15. Since the area between $z = 0$ and z_0 is .475, we look up .475 in the body

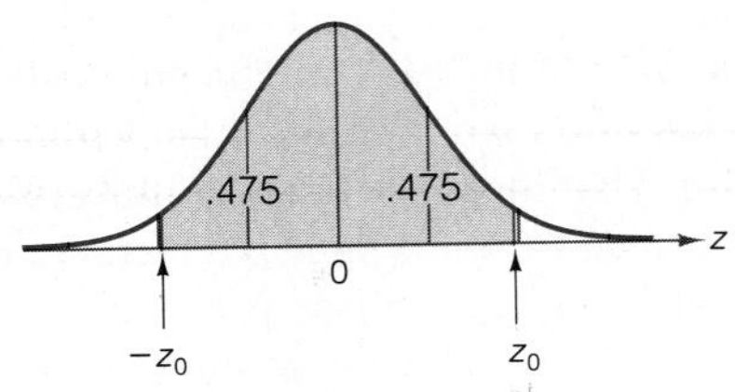

FIGURE 5.15 ► Standard normal distribution: $\mu = 0, \sigma = 1$

of Table IV and find the corresponding z value $z_0 = 1.96$. Thus, as we found in the reverse order in Example 5.5, 95% of a normal distribution lies between plus and minus 1.96 standard deviations of the mean.

Now that you have learned to use Table IV to find a standard normal z value that corresponds to a specified probability, we will demonstrate a practical application in Example 5.10.

EXAMPLE 5.10

Suppose a paint manufacturer has a daily production, x, that is normally distributed with a mean of 100,000 gallons and a standard deviation of 10,000 gallons. Management wants to create an incentive bonus for the production crew when the daily production exceeds the 90th percentile of the distribution, in hopes that the crew will, in turn, become more productive. At what level of production should management pay the incentive bonus?

Solution

First, we need to find the 90th percentile in the standard normal distribution. We did this in Example 5.8 (see Figure 5.14) and found that $z = 1.28$ is the standard normal value that will be exceeded only 10% of the time.

Next, we want to convert the standard normal value to an x value, a production level in this example. We know that the production level at which the incentive bonus will be paid is $z = 1.28$ standard deviations above the mean level. To determine the production level, remember that

$$z = \frac{x - \mu}{\sigma}$$

If we solve this equation for x, we find

$$x = \mu + z\sigma$$

In this example, $\mu = 100{,}000$, $z = 1.28$, and $\sigma = 10{,}000$ so

$$\begin{aligned} x &= 100{,}000 + 1.28(10{,}000) = 100{,}000 + 12{,}800 \\ &= 112{,}800 \end{aligned}$$

Thus, the 90th percentile of the production distribution is 112,800 gallons. Management should pay an incentive bonus when a day's production exceeds this level if its objective is to pay only when the production is in the top 10% of the current production distribution.

CASE STUDY 5.1 / Evaluating an Investment's Risk

Frederick S. Hillier (1963) described several ways a business firm can easily, but effectively, evaluate risky investment projects. He noted that, up to the time of his writing,

Such procedures as have been suggested for dealing with risk have tended to be either quite simplified or somewhat theoretical. Thus, these procedures have tended to provide management with only a portion of the information required for a sound decision, or they have assumed the availability of information which is almost impossible to obtain.

In one of his approaches to handling risk, Hillier assumes that the cash flow from an investment to the firm in the *i*th future year after the investment is made is normally distributed, and shows that the present worth, *P*, of the proposed investment is therefore normally distributed with mean μ_P and variance σ_P^2. He points out that by describing *P* with a probability distribution as he has done, management is provided with information about *P* and, as a result, with some basis upon which to evaluate the risk of the investment decision.

Hillier provides an example of how management can evaluate the risk of an investment by assuming its present worth is normally distributed:

Suppose that, on the basis of the forecasts regarding prospective cash flow from a proposed investment of \$10,000, it is determined that $\mu_P =$ \$1,000 and $\sigma_P =$ \$2,000. Ordinarily, the current procedure would be to approve the investment since $\mu_P > 0$. However, with additional information available ($\sigma_P =$ \$2,000) regarding the considerable risk of the investment, the executive can analyze the situation further. Using widely available tables for the normal distribution, he could note that the probability that $P < 0$, so that the investment won't pay, is 0.31. Furthermore, the probability is 0.16, 0.023, and 0.0013, respectively, that the investment will lose the present worth equivalent of at least \$1,000, \$3,000, and \$5,000, respectively. Considering the financial status of the firm, the executive can use this and similar information to make his decision. Suppose, instead, that the executive is attempting to choose between this investment and a second investment with $\mu_P =$ \$500 and $\sigma_P =$ \$500. By conducting a similar analysis for the second investment, the executive can decide whether the greater expected earnings of the first investment justifies the greater risk. A useful technique for making this comparison is to superimpose the drawing of the probability distribution of P for the second investment upon the corresponding drawing for the first investment. This same approach generalizes to the comparison of more than two investments.

Exercises 5.12–5.33

Learning the Mechanics

5.12 Use Table IV in Appendix B (and inside front cover) to calculate the area under the standard normal distribution between the following pairs of *z*-scores:

a. $z = 0$ and $z = 2$ **b.** $z = 0$ and $z = 3.0$ **c.** $z = 0$ and $z = 1.5$ **d.** $z = 0$ and $z = .80$

5.13 Repeat Exercise 5.12 for each of the following pairs of standard normal z values:
a. $z = -1.33$ and $z = -.33$ b. $z = -2$ and $z = 2$ c. $z = -2.33$ and $z = -1.33$
d. $z = -3$ and $z = 3$ e. $z = -2.25$ and $z = 0$ f. $z = 0.0$ and $z = 2.25$

5.14 Find each of the following probabilities for the standard normal random variable z:
a. $P(z \leq -1)$ b. $P(z \geq 2.32)$ c. $P(z \leq -2.58)$
d. $P(z \geq 2)$ e. $P(z = 0)$ f. $P(z \leq -1.645)$

5.15 Find each of the following probabilities for the standard normal random variable z:
a. $P(z = 1)$ b. $P(z \leq 1)$ c. $P(z < 1)$ d. $P(z > 1)$

5.16 Find each of the following probabilities for the standard normal random variable z:
a. $P(-1 \leq z \leq 1)$ b. $P(-1.96 \leq z \leq 1.96)$
c. $P(-1.645 \leq z \leq 1.645)$ d. $P(-2 \leq z \leq 2)$

5.17 Find each of the following probabilities for the standard normal random variable z, and compare your answers to those of Exercise 5.16.
a. $P(-1 \leq z < 1)$ b. $P(-1.96 < z < 1.96)$
c. $P(-1.645 < z \leq 1.645)$ d. $P(-2 < z < 2)$

5.18 Find a value of the standard normal random variable z, call it z_0, such that:
a. $P(z \geq z_0) = .05$ b. $P(z \geq z_0) = .025$ c. $P(z \leq z_0) = .025$
d. $P(z \geq z_0) = .10$ e. $P(z > z_0) = .10$

5.19 Find a value of the standard normal random variable z, call it z_0, such that:
a. $P(z \leq z_0) = .2090$ b. $P(z \leq z_0) = .7090$ c. $P(-z_0 \leq z < z_0) = .8472$
d. $P(-z_0 \leq z \leq z_0) = .1664$ e. $P(z_0 \leq z \leq 0) = .4798$ f. $P(-1 < z < z_0) = .5328$

5.20 Suppose the random variable x is best described by a normal distribution with $\mu = 30$ and $\sigma = 4$. Find the value of the standard normal random variable z that corresponds to each of the following x values:
a. $x = 20$ b. $x = 30$ c. $x = 27.5$ d. $x = 15$ e. $x = 35$ f. $x = 25$

5.21 Refer to Exercise 5.20. How many standard deviations away from the mean of x are each of the following x values?
a. $x = 25$ b. $x = 37.5$ c. $x = 30$ d. $x = 36$

5.22 Suppose the continuous random variable x has a normal probability distribution with mean 120 and variance 36. Draw a rough sketch (i.e., a graph) of the frequency function of x. Locate μ and the interval $\mu \pm 2\sigma$ on the graph. Find the following probabilities:
a. $P(\mu - 2\sigma \leq x \leq \mu + 2\sigma)$ b. $P(x \geq 128)$ c. $P(x \leq 108)$
d. $P(112 \leq x \leq 130)$ e. $P(114 \leq x \leq 116)$ f. $P(115 \leq x \leq 128)$

5.23 The random variable x has a normal distribution with $\mu = 1{,}000$ and $\sigma = 10$.
a. Find the probability that x assumes a value more than 2 standard deviations from its mean. More than 3 standard deviations from μ.
b. Find the probability that x assumes a value within 1 standard deviation of its mean. Within 2 standard deviations of μ.
c. Find the value of x that represents the 80th percentile of this distribution. The 10th percentile.

5.24 The random variable x has a normal distribution with mean 50 and variance 9. Find the value of x, call it x_0, such that:

a. $P(x \leq x_0) = .8413$ b. $P(x > x_0) = .025$
c. $P(x > x_0) = .95$ d. $P(41 \leq x < x_0) = .8630$

5.25 The random variable x has a normal distribution with standard deviation 25. It is known that the probability that x exceeds 150 is .90. Find the mean μ of the probability distribution.

Applying the Concepts

5.26 The tread life of a particular brand of tire is a random variable best described by a normal distribution with a mean of 60,000 miles and a standard deviation of 8,300 miles. If the manufacturer warrants the tires for the first 45,000 miles, what proportion of the tires will need to be replaced under warranty? What if the warranty is for the first 40,000 miles?

5.27 A company that sells annuities must base the annual payout on the probability distribution of the length of life of the participants in the plan. Suppose the probability distribution of the lifetimes of the participants in the plan is approximately a normal distribution with $\mu = 68$ years and $\sigma = 3.5$ years.

a. What proportion of the plan participants would receive payments beyond age 70?
b. Beyond age 75?
c. Complete the following statement: Only 15% of plan participants will receive payment beyond age ______.

5.28 **Personnel tests** are designed to test a job applicant's cognitive and/or physical abilities. An IQ test is an example of the former; a speed test involving the arrangement of pegs on a peg board is an example of the latter. (Dessler, 1986). A particular dexterity test is administered nationwide by a private testing service. It is known that for all tests administered last year the distribution of scores was approximately normal with mean 75 and standard deviation 7.5.

a. A particular employer requires job candidates to score at least 80 on the dexterity test. Approximately what percentage of the test scores during the past year exceeded 80?
b. The testing service reported to a particular employer that the score of one of its job candidates fell at the 98th percentile of the distribution (i.e., approximately 98% of the scores were lower than the candidate's, and only 2% were higher). What was the candidate's score?

5.29 Ideally, a worker seeking a new job in a particular industry should acquire information about wage rates offered by all firms in the industry. However, workers may not find it worthwhile to search until they find the highest available wage rate. The result is that managers may not have to pay top dollar to attract workers. These factors help explain the existing disparity in wage rates among firms (Blair and Kenny, 1982). Suppose the distribution of wage rates nationwide that would be offered to a particular skilled worker can be approximated by a normal distribution with $\mu = \$10.50$ per hour and $\sigma = \$1.25$ per hour. In addition, assume that the worker is offered \$12.00 per hour by the first firm contacted.

a. Suppose the worker were to undertake a nationwide job search. What proportion of the wage rates that would be offered to the worker would be greater than \$12.00 per hour?
b. If the worker were to complete a nationwide job search and then randomly select one of the many job offers received, what is the probability that the wage rate would be more than \$10.00 per hour?
c. The **median**, call it m, of a continuous random variable x is the value such that $P(x \geq m) = P(x \leq m) = .5$. That is, the median is the value m such that half the area under the probability distribution lies above m and half lies below it. Find the median of the random variable corresponding to the wage rate and compare it to the mean wage rate.

5.30 An important quality characteristic for soft-drink bottlers is the amount of soft drink injected into each bottle. This volume is determined (approximately) by measuring the height of the soft drink in the neck of the bottle and comparing it to a scale that converts the height measurement to a volume measurement (Montgomery, 1991). In a particular filling process, the number of ounces injected into 8-ounce bottles is approximately normally distributed with mean 8.00 ounces and standard deviation .05 ounce. Bottles that contain less than 7.9 ounces do not meet the bottler's quality standard and are sold at a substantial discount.

a. If 20,000 bottles are filled, approximately how many will fail to meet the quality standard?

b. Suppose that, due to the failure of one of the filling system's components, the mean of the filling process shifts to 7.95 ounces. (Assume that the standard deviation remains .05 ounce.) If 20,000 bottles are filled, approximately how many will fail to meet the quality standard?

c. Suppose that a different component fails and, although the mean of the filling process remains 8.00 ounces, the standard deviation increases to .1 ounce. If 20,000 bottles are filled, approximately how many will fail to meet the quality standard?

5.31 The **monthly rate of return** of a stock is a measure investors frequently use for evaluating the behavior of a stock over time. A stock's monthly rate of return generally reflects the amount of money an investor makes (or loses if the return is negative) for every dollar invested in the stock in a given month. Thus, stocks with high average monthly rates of return typically offer more lucrative investment opportunities than stocks with low average monthly rates of return. Eugene Fama (1976) has demonstrated that the probability distribution for the monthly rate of return of a stock can be approximated by a normal probability distribution. Suppose the monthly rates of return to stock ABC are normally distributed with mean .05 and standard deviation .03, and the monthly rates of return to stock XYZ are normally distributed with mean .07 and standard deviation .05. Assume that you have $100 invested in each stock.

a. Over the long run, which stock will yield the higher average monthly rate of return? Why?

b. Suppose you plan to hold each stock for only 1 month. What is the expected value of each investment at the end of 1 month?

c. Which stock offers greater protection against incurring a loss on your investment next month? Why?

5.32 Do security analysts do a good job of forecasting corporate earnings growth and advising their clientele? David Dreman, a *Forbes* columnist, addresses this question in an article titled "Astrology Might Be Better" (*Forbes*, Mar. 26, 1984). The basis of Dreman's article is a study by Professors Michael Sandretto of Harvard and Sudhir Milkrishnamurthi of the Massachusetts Institute of Technology. The study surveys security analysts' forecasts of annual earnings for the (then) current year for more than 769 companies with five or more forecasts per company per year. The average forecast error for this large number of forecasts was 31.3%. To apply this information to a practical situation, suppose the population of analysts' forecast errors is normally distributed with a mean of 31.3% and a standard deviation of 10%.

a. If you obtain a security analyst's forecast for a particular company, what is the probability that it will be in error by more than 50%?

b. If three analysts make the forecast, what is the probability that at least one of the analysts will err by more than 50%?

5.33 A machine used to regulate the amount of dye dispensed for mixing shades of paint can be set so that it discharges an average of μ milliliters of dye per can of paint. The amount of dye discharged is known to have a normal distribution with variance equal to .160. If more than 6 milliliters of dye are discharged when making a particular shade of blue paint, the shade is unacceptable. Determine the setting for μ so that no more than 1% of the cans of paint will be unacceptable.

5.4 The Exponential Distribution (Optional)

Another important probability distribution that is useful for describing business data is the **exponential probability distribution**. Two business phenomena with frequency functions that might be well approximated by the exponential distribution are the length of time between arrivals at a fast-food drive-through restaurant and the length of time between the filing of claims in a small insurance office. Note that in each of these examples, the measurements are the lengths of time between certain events. For this reason, the exponential distribution is sometimes called the **waiting time distribution**.

The formula for the exponential probability distribution is shown in the box, along with its mean and standard deviation.

Probability Distribution, Mean, and Standard Deviation for an Exponential Random Variable *x*

$$f(x) = \lambda e^{-\lambda x} \quad (x > 0)$$

$$\mu = \frac{1}{\lambda} \qquad \sigma = \frac{1}{\lambda}$$

Unlike the normal distribution, which has a shape and location determined by the values of the two quantities μ and σ, the shape of the exponential distribution is governed by a single quantity, λ. Further, it is a probability distribution with the property that its mean equals its standard deviation. Exponential distributions corresponding to $\lambda = .5$, 1, and 2 are shown in Figure 5.16.

FIGURE 5.16 ▶
Exponential distributions

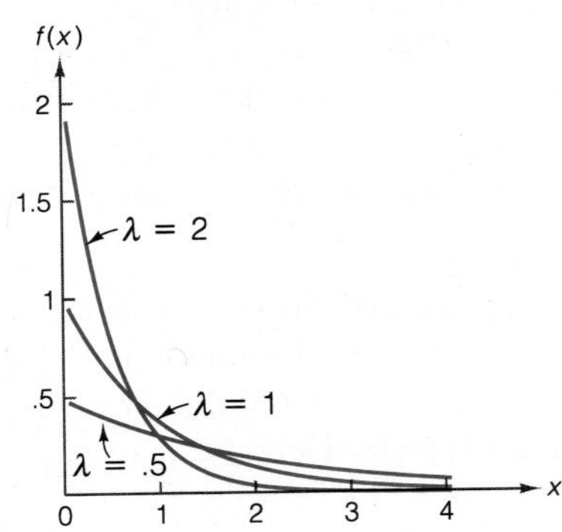

To calculate probabilities for exponential random variables, we need to be able to find areas under the exponential probability distribution. Suppose we want to find the area, A, to the right of some number, a, as shown in Figure 5.17. This area can be calculated by using the following formula:

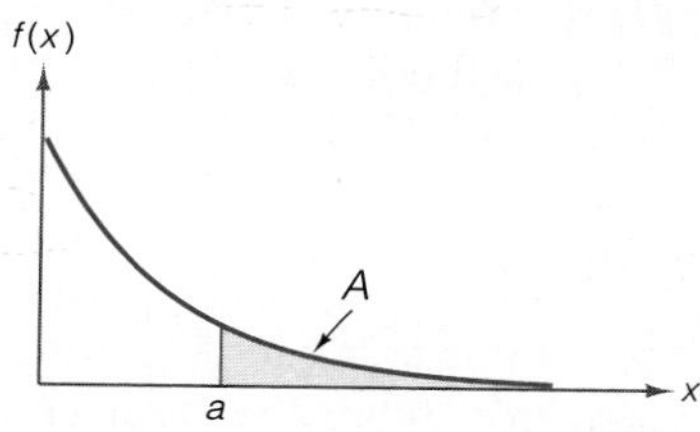

FIGURE 5.17 ▶
The area, *A*, to the right of a number, *a*, for an exponential distribution

Finding the Area, *A*, to the Right of a Number, *a*, for an Exponential Distribution*

$$A = P(x \geq a) = e^{-\lambda a}$$

Use Table V in Appendix B to find the value of $e^{-\lambda a}$ after substituting the appropriate numerical values for λ and a.

EXAMPLE 5.11

Suppose the length of time (in days) between sales for an automobile salesperson is modeled as an exponential random variable with $\lambda = .5$. What is the probability that the salesperson goes more than 5 days without a sale?

Solution

The probability we want is the area, A, to the right of $a = 5$ in Figure 5.18. To find this probability, use the formula given for area:

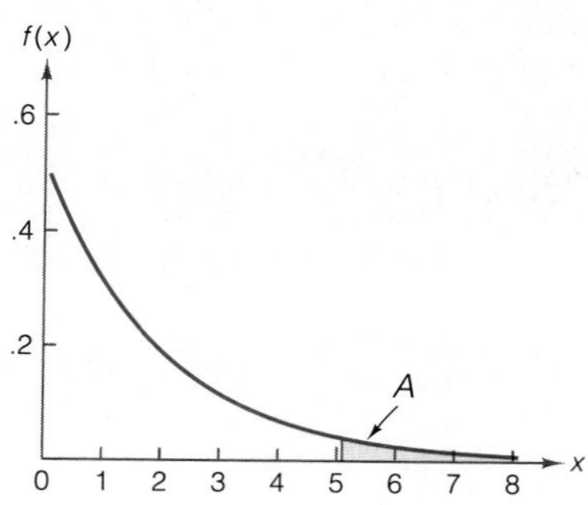

FIGURE 5.18 ▶
Exponential distribution for Example 5.11: $\lambda = .5$

*This area is calculated by integration:

$$\int_a^{\infty} \lambda e^{-\lambda x}\, dx = -e^{-\lambda x} \Big|_a^{\infty} = e^{-\lambda a}$$

$$A = e^{-\lambda a} = e^{-(.5)(5)} = e^{-2.5}$$

Referring to Table V, we find

$$A = e^{-2.5} = .082085$$

That is, our model indicates that the automobile salesperson has a probability of about .08 of going more than 5 days without a sale.

EXAMPLE 5.12

A microwave oven manufacturer is trying to determine the length of warranty period it should attach to its magnetron tube, the most critical component in the oven. Preliminary testing has shown that the length of life (in years), x, of a magnetron tube has an exponential probability distribution with $\lambda = .16$.

a. Find the mean and standard deviation of x.

b. If a warranty period of 5 years is attached to the magnetron tube, what fraction of tubes must the manufacturer plan to replace (assuming the exponential model with $\lambda = .16$ is correct)?

c. Find the probability that the length of life of a magnetron tube will fall within the interval $\mu \pm 2\sigma$.

Solution

a. Using the formulas for the mean and standard deviation for an exponential random variable, we find

$$\mu = \frac{1}{\lambda} = \frac{1}{.16} = 6.25 \text{ years}$$

Also, since $\mu = \sigma$, $\sigma = 6.25$ years.

b. To find the fraction of tubes that will have to be replaced before the 5-year warranty period expires, we need to find the area between 0 and 5 under the distribution. This area, A, is shown in Figure 5.19. To find the required probability, we recall the formula

$$P(x > a) = e^{-\lambda a}$$

Using this formula and Table V of Appendix B, we find

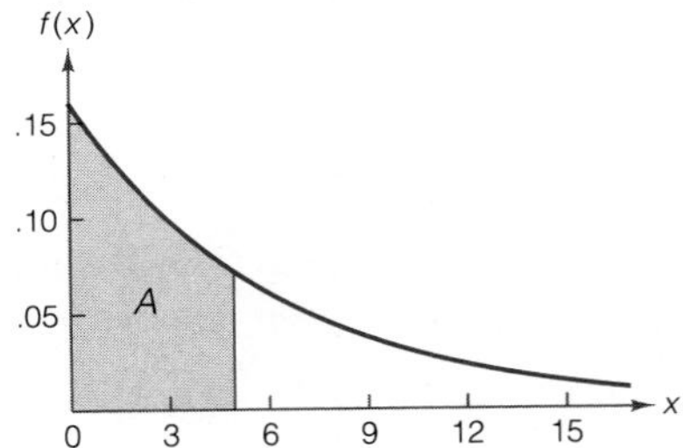

FIGURE 5.19 ► Exponential distribution for Example 5.12: $\lambda = .16$

$$P(x > 5) = e^{-\lambda(5)} = e^{-(.16)(5)} = e^{-.80} = .449329$$

To find the area A, we use the complementary relationship:

$$P(x \le 5) = 1 - P(x > 5) = 1 - .449329 = .550671$$

So, approximately 55% of the magnetron tubes will have to be replaced during the 5-year warranty period.

c. We would expect the probability that the life of a magnetron tube, x, falls within the interval $\mu \pm 2\sigma$ to be quite large (near .95 if we think the Empirical Rule in Table 2.10 might apply). A graph of the exponential distribution showing the interval from $\mu - 2\sigma$ to $\mu + 2\sigma$ is shown in Figure 5.20. Since the point $\mu - 2\sigma$ lies below $x = 0$, we need to find only the area between $x = 0$ and $a = \mu + 2\sigma = 6.25 + 2(6.25) = 18.75$. This area, A, which is shaded in Figure 5.20, is

$$\begin{aligned} A &= 1 - P(x > 18.75) \\ &= 1 - e^{-\lambda(18.75)} = 1 - e^{-(.16)(18.75)} = 1 - e^{-3} \end{aligned}$$

FIGURE 5.20 ► **Exponential distribution for Example 5.12: $\lambda = .16$**

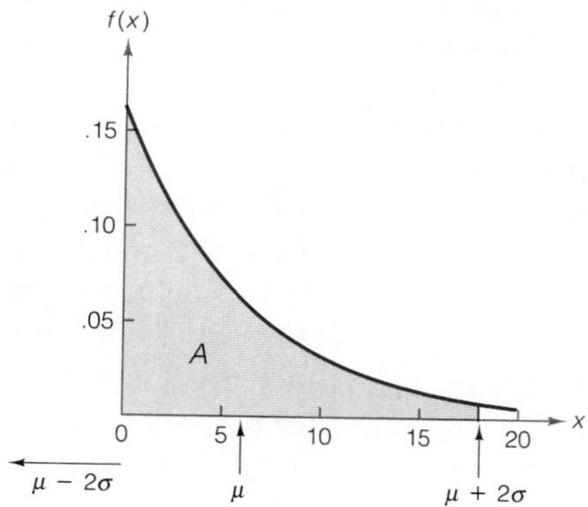

Checking Table V of Appendix B for the value of e^{-3}, we find $e^{-3} = .049787$. Therefore, the probability that the life, x, of a magnetron tube falls within the interval $\mu \pm 2\sigma$ is

$$A = 1 - e^{-3} = 1 - .049787 = .950213$$

You can see that this probability agrees very well with the Empirical Rule even though this probability distribution is not mound-shaped (it is strongly skewed to the right).

CASE STUDY 5.2 / Queueing Theory

The formation of waiting lines, or *queues*, is a phenomenon that occurs whenever the demand for a service exceeds its supply. We see this daily at bank-teller windows, supermarket checkout counters, traffic lights, etc. If long queues develop, it may be an indication that not enough service is being provided. If no queues develop, it may be an indication that too much service is being provided. Either situation can prove costly to the service provider. To assist in planning service capacity, an area of study known as **queueing theory** has been employed to model the characteristics of waiting lines.

In the basic structure assumed by most queueing models, *customers* seeking service are generated over time by an *input source*. These customers enter the *queueing system*, join a queue, and await service. Members of the queue are selected for service by some rule (for example, first-come, first-served or random sampling) known as the *service discipline*. A chosen customer is served by the service *mechanism* and then leaves the queueing system. This process is illustrated in Figure 5.21.

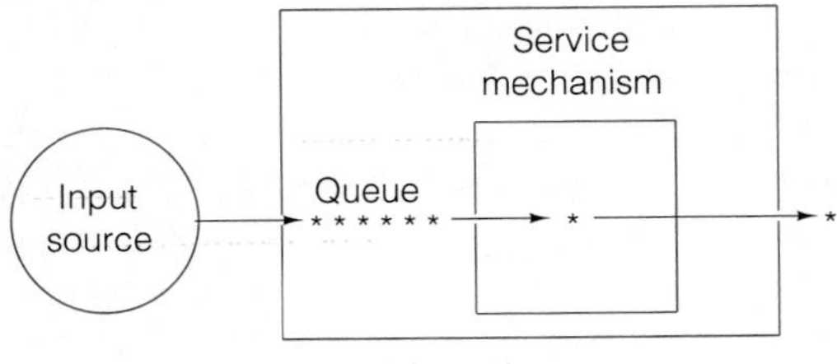

FIGURE 5.21 ▲ Basic queueing system

To complete the basic queueing model, certain assumptions must be made to model probabilistically the arrivals to and departures from the queueing system. It has been found that the *interarrival time* (the time between arrivals) of many real queues can be reasonably approximated by an exponential probability distribution. Furthermore, when the specific service required differs among individual customers, the exponential distribution has proved to provide an adequate approximation to the time required to service a customer (i.e., the time that elapses between when service begins and when it ends). Thus, the exponential distribution can be used to describe both the input source and the service mechanism. A more detailed description of the basic queueing theory model and its various assumptions can be found in Hillier and Lieberman (1990).

After developing a queueing model, we can answer such questions as: (1) What is the expected number of customers in the queueing system? (2) What is the expected length of the queue? (3) What is the expected waiting time in the system for an individual customer? (4) What is the expected waiting time in the queue for an individual customer? (5) Are the rates of arrival and departure such that the queue will continue to grow without bound?

W. Blaker Bolling (1972) describes how queueing theory was used to model the emergency room of the Richmond Memorial Hospital in Richmond, Virginia. The input source was the local population of Richmond (including visitors). The service mechanism consisted of eight fully staffed treatment tables. Thus, since it was possible to service eight patients simultaneously, the system was more complex than that described in Figure 5.21. In general, the service discipline was first-come, first-served, unless a serious emergency occurred. However, in modeling the system, a first-come, first-served discipline was assumed for simplicity. Historical data indicated that both interarrival times and service times could be adequately modeled with exponential distributions.

One use of the model was to project the capacity (number of staffed tables) needed to prevent the queue from growing without bound. For example, for the period between 8 and 9 P.M. in August 1972, the interarrival-time distribution was projected to be exponential with mean 5.96 minutes (.0993 hour), and the service-time distribution was projected to be exponential with mean 58 minutes (.9666 hour). Accordingly, the arrival rate was projected to be 10.07 per hour (i.e., $\lambda = 10.07$ for the interarrival-time distribution), and the service rate was projected to be 1.0345 patients per table per hour (i.e., $\lambda = 1.0345$ for the service-time distribution). Since there were just eight service tables, arrivals would exceed departures and, theoretically, the waiting room queue could grow without bound as long as rates of arrival and departure remained unchanged. It was determined through further analysis with the queueing model that it would be necessary to staff at least two more treatment tables to prevent this situation from occurring.

CASE STUDY 5.3 / Assessing the Reliability of Computer Software

In a discussion about the reliability of computer software, G. J. Schick (1974) says the following:

Custom software . . . is expensive to develop and requires extensive testing—the goal being to certify that the software is in error-free condition, ready to support the mission for which it was designed. Similar economies should also be expected from an integrated statistical software test program. Traditionally, there are never enough time or resources to test all possible branches and data combinations in a computer program of reasonable size.

Current practice is to design and develop a software system and then to test it to detect errors, until the amount of time and expense required to discover remaining errors is too great to justify further testing. . . . In principle, few large real-time computer programs ever have been tested completely and unequivocally in the sense that every logical data path has been successfully executed under every logical combination for the data at hand for all possible options. One management objective would be to test every logical path in the computer program at least once with some kind of numerical check. At the present state of the art, such a degree of testing is neither feasible nor realistic. In practice, the contractor must be willing to release and the customer willing to accept a level of risk associated with a program that has been less than completely checked.

In finding and correcting errors in a computer program (*debugging*) and determining the program's reliability, Schick and others have noted the importance of the distribution of the time until the next program error is found. If this distribution is assumed to be exponential, with

$$f(x) = \lambda e^{-\lambda x} \quad (x > 0, \quad \lambda > 0)$$

then, as Schick points out, its mean, $1/\lambda$, would be the average time required to find the next error.

In his article, Schick describes a method relevant to software reliability for estimating the parameter, λ, of the exponential distribution. Using computer debugging data supplied by the U.S. Navy, Schick demonstrates how this estimation procedure and the exponential distribution can be used to estimate the reliability of a computer program. [*Note:* The model used by Schick to represent the distribution for the time until the next error is based on the exponential distribution, but it is slightly more complicated because he assumes that λ varies. For our purposes, however, nothing is lost by assuming the distribution to be exponential.

After 26 program errors were found, Schick estimated λ to be .042. Accordingly, $1/\lambda = 23.8$ days. This means that the average time it would take to find the next (27th) error would be about 24 days. Thus, the probability of it taking, say, 60 or more days to find the next error is

$$P(x \geq 60) = e^{-(.042)(60)} = .08046$$

Over the next 290 days, five more errors were detected. Since this is a rate of about one error every 60 days and since $P(x \geq 60) \approx .08$, it seems unlikely that an exponential distribution with $\lambda = .042$ is an appropriate representation of the distribution for the time until the next error. Based on the number of new errors found and the length of time it took to find them, Schick reestimated λ. He found $\lambda = .0036$. Thus, $1/\lambda = 278$ days, meaning that on average the next error (32nd) would not occur for 278 days. At this point, the length of time and, therefore, the cost required to find any remaining program errors may be prohibitive. Debugging should probably be discontinued.

Exercises 5.34–5.45

Learning the Mechanics

5.34 The random variables x and y have exponential distributions with $\lambda = 3$ and $\lambda = .75$, respectively. Using Table V in Appendix B, carefully plot both distributions on the same set of axes.

5.35 Use Table V in Appendix B to determine the value of $e^{-\lambda a}$ for each of the following cases:
a. $\lambda = 1, \quad a = 1$ b. $\lambda = 1, \quad a = 2.5$ c. $\lambda = 2.5, \quad a = 3$ d. $\lambda = 5, \quad a = .3$

5.36 Suppose x has an exponential distribution with $\lambda = 3$. Find the following probabilities:
a. $P(x > 2)$ b. $P(x > 1.5)$ c. $P(x > 3)$ d. $P(x > .45)$

5.37 Suppose x has an exponential distribution with $\lambda = 2.5$. Find the following probabilities:
a. $P(x \leq 3)$ b. $P(x \leq 4)$ c. $P(x \leq 1.6)$ d. $P(x \leq .4)$

5.38 Suppose the random variable x is best approximated by an exponential probability distribution with $\lambda = 2$. Find the mean and variance of x. Find the probability that x will assume a value within the interval $\mu \pm 2\sigma$.

5.39 The random variable x can be adequately approximated by an exponential probability distribution with $\lambda = 1$. Find the probability that x assumes a value:
a. More than 3 standard deviations from μ. b. Less than 2 standard deviations from μ.
c. Within .5 standard deviation of μ.

Applying the Concepts

5.40 A taxi service based at an airport can be characterized as a transportation system with one source terminal and a fleet of vehicles that take passengers from the terminal to different destinations. Each vehicle returns to the terminal after some random trip time and makes another trip. To improve the vehicle-dispatching decisions involved in such a system (e.g., how many passengers should be allocated to a waiting taxi?), Sims and Templeton (1985) used queueing theory (see Case Study 5.2) to model and to evaluate the system. In their model, they assumed travel times of successive trips are independent exponential random variables. Assume $\lambda = .05$.
a. What is the mean trip time for the taxi service?
b. What is the probability that a particular trip will take more than 30 minutes?
c. Two taxis have just been dispatched. What is the probability that both will be gone for more than 30 minutes? That at least one of the taxis will return within 30 minutes?

5.41 The shelf-life of a product is a random variable that is related to consumer acceptance and, ultimately, to sales and profit. Suppose the shelf-life of bread is best approximated by an exponential distribution with mean equal to 2 days. What fraction of the loaves stocked today would you expect to still be saleable (i.e., not stale) 3 days from now?

5.42 A catalog company that receives the majority of its orders by telephone conducted a study to determine how long customers were willing to wait on hold before ordering a product. The length of time was found to be a random variable best approximated by an exponential distribution with a mean equal to 2.8 minutes. What proportion of customers having to hold more than 3 minutes will hang up before placing an order? More than 2 minutes? More than 1 minute?

5.43 As discussed in Case Study 5.2, Bolling (1972) used an exponential distribution with mean 58 minutes to model the service-time distribution of each of the eight treatment tables in the emergency room of Richmond Memorial Hospital.

a. Using this distribution, find the probability that it will take more than 58 minutes to treat a patient in the emergency room. More than 1.5 hours.

b. What is the probability that each of the next three patients will require more than 58 minutes for treatment?

c. Recall from part **c** of Exercise 5.29 that the median, m, of a continuous random variable x is the value such that $P(x \geq m) = P(x \leq m) = .5$. Is the median time required to treat a patient more or less than 58 minutes?

d. Using Table V in Appendix B, approximate the median of the service-time distribution.

5.44 In Case Study 5.3, the exponential distribution was used to help evaluate the reliability of a computer program. After 26 program errors were found, the time (in days) required to find the next error was determined to have an exponential distribution with $\lambda = .042$.

a. Graph this exponential distribution, and locate its mean and the interval $\mu \pm \sigma$ on your graph.

b. What is the mean time required to find the 27th program error?

c. What is the probability that it will take less than 30 days to find the 27th error?

d. Find the probability that the time required to find the 27th error is within the interval $\mu \pm \sigma$.

e. Find the probability that the time required to find the 27th error is within the interval $\mu \pm 2\sigma$. How does your answer compare with the approximate probability provided by the Empirical Rule of Chapter 2?

5.45 **Product reliability** has been defined as the probability that a product will perform its intended function satisfactorily for its intended life when operating under specified conditions (Lamberson, 1985). The **reliability function**, $R(x)$, for a product indicates the probability of the product's life exceeding x time periods. When the time until failure of a product can be adequately modeled by an exponential distribution, the product's reliability function is $R(x) = e^{-\lambda x}$. Suppose that the time to failure (in years) of a particular product is modeled by an exponential distribution with $\lambda = .5$.

a. What is the product's reliability function?

b. What is the probability that the product will perform satisfactorily for at least 4 years?

c. What is the probability that a particular product will survive longer than the mean life of the product?

d. If λ changes, will the probability that you calculated in part **c** change? Explain.

e. If 10,000 units of the product are sold, approximately how many will perform satisfactorily for more than 5 years? About how many will fail within 1 year?

f. How long should the length of the warranty period be for the product if the manufacturer wants to replace no more than 5% of the units sold while under warranty?

5.5 Approximating a Binomial Distribution with a Normal Distribution

When a binomial random variable can assume a large number of values, the calculation of its probabilities may become very tedious. To contend with this problem, we provide tables in Appendix B to give the probabilities for some values of n and p, but these tables are by necessity incomplete. For example, the binomial table (Table II) can be used only for $n = 5, 6, 7, 8, 9, 10, 15, 20$, or 25. To get around this limitation, we seek approximation procedures for calculating the probabilities associated with binomial random variables.

When n is large, a normal probability distribution can provide a good approximation to the probability distribution of a binomial random variable. To show how the approximation works, we refer to Example 4.8, in which we used the binomial distribution to model the number x of 20 employees who favor unionization. We assumed that 60% of all the company's employees favored unionization. The mean and standard deviation of x were found to be $\mu = 12$ and $\sigma = 2.19$. The binomial distribution for $n = 20$ and $p = .6$ is shown in Figure 5.22 and the approximating normal distribution with mean $\mu = 12$ and standard deviation $\sigma = 2.19$ is superimposed.

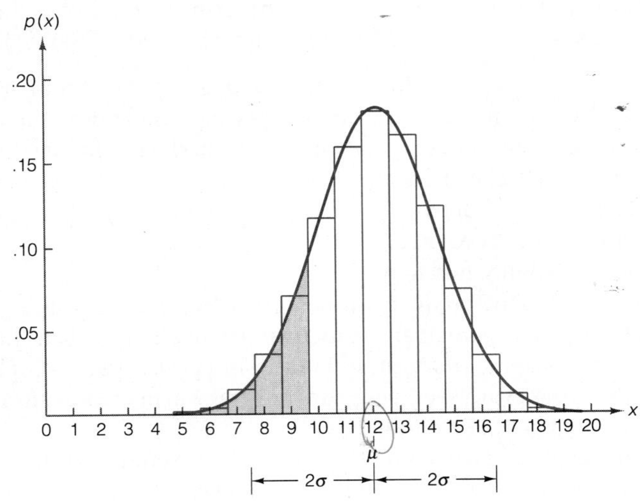

FIGURE 5.22 ▶ Binomial distribution for $n = 20$, $p = .6$ and normal distribution with $\mu = 12$, $\sigma = 2.19$

As part of Example 4.8, we used Table II to find the probability that $x < 10$. This probability, which is shaded in Figure 5.22, was found to equal .128. To find the normal approximation, we first note that $P(0 \leq x < 10)$ corresponds to the area to the left of $x = 9.5$ on the normal curve in the figure. We use $x = 9.5$ rather than $x = 9$ or $x = 10$ so that all the binomial probability corresponding to $x = 9$ is included in the approximating normal curve area, but none of that corresponding to $x = 10$ is included. Because we are approximating a discrete distribution (the binomial) with a continuous distribution (the normal), we call the use of 9.5 (instead of 9 or 10) a **correction for continuity**. That is, we are correcting the discrete distribution so that it can be approximated by the continuous one. The use of the correction for continuity leads to the calculation of the following standard normal z value:

$$z = \frac{x - \mu}{\sigma} = \frac{9.5 - 12}{2.19} = -1.14$$

Using Table IV of Appendix B, we find the area between $z = 0$ and $z = -1.14$ to be .3729. Then the probability that x is less than 10 is approximated by the area under the normal distribution to the left of 9.5, shown shaded in Figure 5.22. That is,

$$\begin{aligned} P(x \leq 9) &\approx P(z \leq -1.14) \\ &= .5 - P(-1.14 < z \leq 0) \\ &= .5 - .3729 = .1271 \end{aligned}$$

The approximation differs only slightly from the exact binomial probability, .128. Of course, when tables of exact binomial probabilities are available, we will use the exact value rather than a normal approximation.

Use of the normal distribution will not always provide a good approximation for binomial probabilities. The following is a useful rule of thumb to determine when n is large enough for the approximation to be effective: the interval $\mu \pm 3\sigma$ should lie within the range of the binomial random variable x (i.e., 0 to n) in order for the normal approximation to be adequate. The rule works well because almost all of the normal distribution falls within 3 standard deviations of the mean, so if this interval is contained within the range of x values, there is "room" for the normal approximation to work.

As shown in Figure 5.23(a) (page 258) for the example above with $n = 20$ and $p = .6$, the interval $\mu \pm 3\sigma = 12 \pm 3(2.19) = (5.43, 18.57)$ lies within the range 0 to 20. However, if we were to try to use the normal approximation with $n = 10$ and $p = .1$, the interval $\mu \pm 3\sigma$ is $1 \pm 3(.95)$, or $(-1.85, 3.85)$. As shown in Figure 5.23(b), this interval is not contained within the range of x, since $x = 0$ is a lower bound for the binomial distribution. Note in Figure 5.23(b) that the normal distribution will not "fit" in the range of x, and therefore will not provide a good approximation to the binomial probabilities.

FIGURE 5.23 ▶
Rule of thumb for normal approximation to binomial probabilities

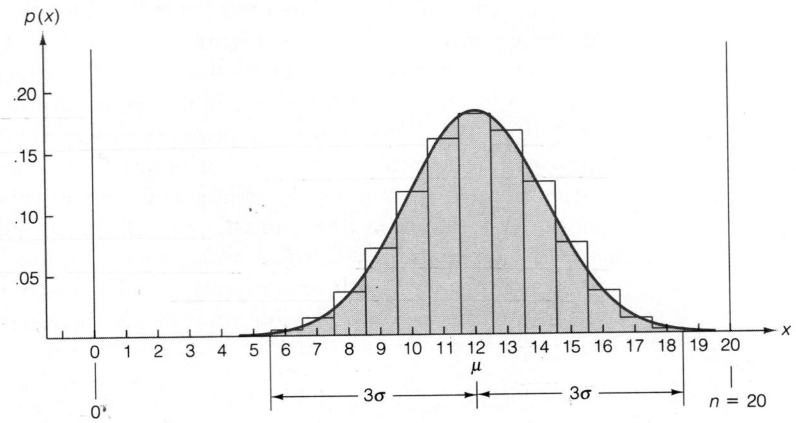

a. $n = 20$, $p = .6$: Normal approximation is good

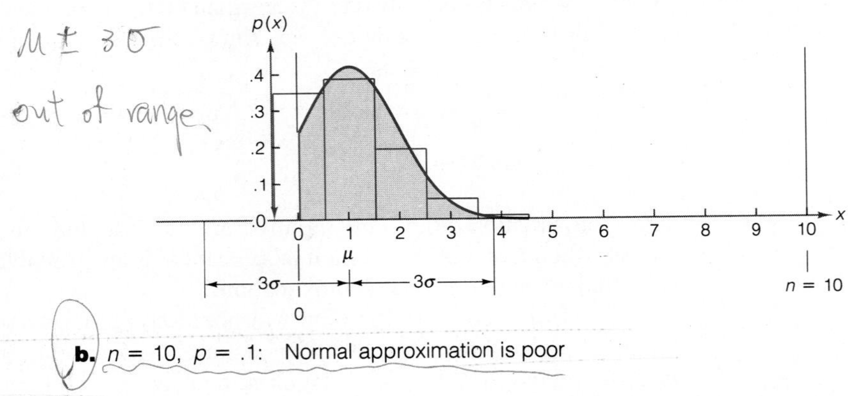

b. $n = 10$, $p = .1$: Normal approximation is poor

FIGURE 5.24 ▶
Approximating binomial probabilities by normal probabilities

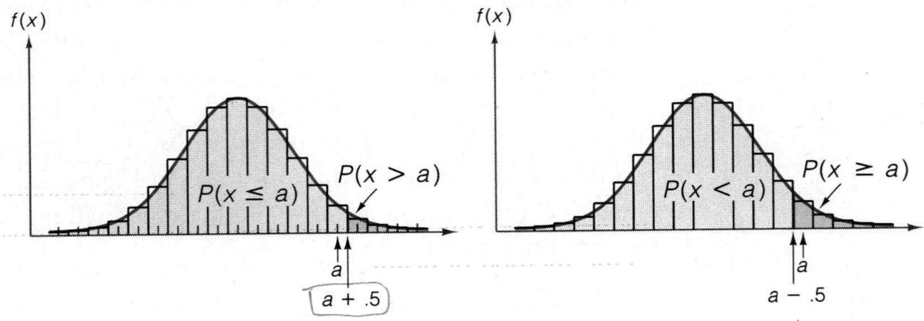

a. $P(x \leq a)$ or $P(x > a)$

b. $P(x < a)$ or $P(x \geq a)$

The steps for approximating a binomial probability by a normal probability are given in the box.

Using a Normal Distribution to Approximate Binomial Probabilities

1. After you have determined n and p for the binomial distribution, calculate the interval

$$\mu \pm 3\sigma = np \pm 3\sqrt{npq}$$

If the interval lies in the range 0 to n, the normal distribution will provide a reasonable approximation to the probabilities of most binomial events.

2. If the binomial probability to be approximated is of the form $P(x \leq a)$ or $P(x > a)$, the correction for continuity is $(a + .5)$, and the approximating standard normal z value is

$$z = \frac{(a + .5) - \mu}{\sigma}$$

See Figure 5.24(a).

3. If the binomial probability to be approximated is of the form $P(x \geq a)$ or $P(x < a)$, the correction for continuity is $(a - .5)$, and the approximating standard normal z value is

$$z = \frac{(a - .5) - \mu}{\sigma}$$

See Figure 5.24(b).

4. If the binomial probability to be approximated is an interval, such as $P(a \leq x < b)$, treat the ends of the interval separately, calculating two distinct z values according to either step 2 or step 3, whichever is appropriate.
5. Sketch the approximating normal distribution and shade the area corresponding to the probability of the event of interest, as in Figure 5.24. Using Table IV of Appendix B and the z value(s) you calculated in steps 2–4, find the shaded area. This is the approximate probability of the binomial event.

EXAMPLE 5.13

The pocket calculator has become relatively inexpensive because its solid-state circuitry is stamped by machine, thus making mass production feasible. One problem with anything that is mass-produced is quality. The production process must be monitored or audited to be sure the output of the process conforms to requirements.

One method of dealing with this problem is **lot acceptance sampling**, in which items being produced are sampled at various stages of the production process and are carefully inspected. The lot of items from which the sample is drawn is then accepted or rejected, based on the number of defectives in the sample. Lots that are accepted may be sent forward for further processing or may be shipped to customers; lots that are rejected may be reworked or scrapped. For example, suppose a manufacturer of calculators chooses 200 stamped circuits from the day's production and determines x, the number of defective circuits in the sample. Suppose that up to a 6% rate of defectives is considered acceptable for the process.

a. Find the mean and standard deviation of x, assuming the defective rate is 6%.
b. Use the normal approximation to determine the probability that 20 or more defectives are observed in the sample of 200 circuits—i.e., find the approximate probability that $x \geq 20$.

Solution

a. The random variable x is binomial, with $n = 200$ and the fraction defective $p = .06$. Thus,

$$\mu = np = 200(.06) = 12$$
$$\sigma = \sqrt{npq} = \sqrt{200(.06)(.94)} = \sqrt{11.28} = 3.36$$

b. We first note that

$$\mu \pm 3\sigma = 12 \pm 3(3.36) = 12 \pm 10.08 = (1.92, 22.08)$$

lies completely within the range from 0 to 200, so a normal probability distribution should provide an adequate approximation to this binomial distribution.

To find the approximating area corresponding to $x \geq 20$, refer to Figure 5.25. Note that we want to include all of the binomial probability histogram from 20 to 200, inclusive. But in order to include the entire rectangle corresponding to $x = 20$,

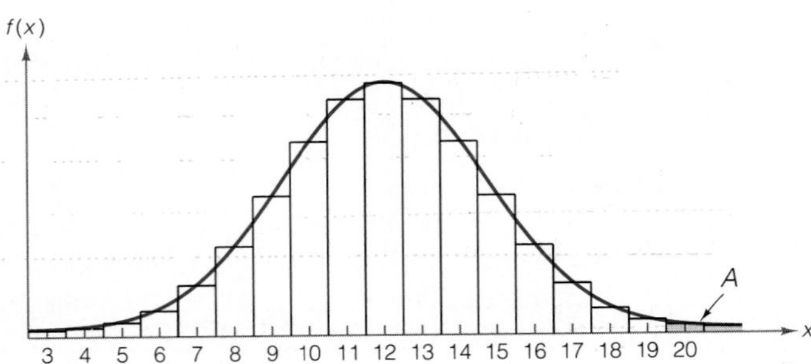

FIGURE 5.25 ▶ Normal approximation to the binomial distribution with $n = 200$, $p = .06$

we must begin the approximating area at $20 - .5 = 19.5$. In other words, since the event is of the form $x \geq a$ with $a = 20$, the correction for continuity is $a - .5 = 20 - .5 = 19.5$. Thus, the z value is

$$z = \frac{(a - .5) - \mu}{\sigma} = \frac{19.5 - 12}{3.36} = \frac{7.5}{3.36} = 2.23$$

Referring to Table IV of Appendix B, we observe that the area to the right of the mean corresponding to $z = 2.23$ (see Figure 5.26) is .4871. So, the area A is

$$A = .5 - .4871 = .0129$$

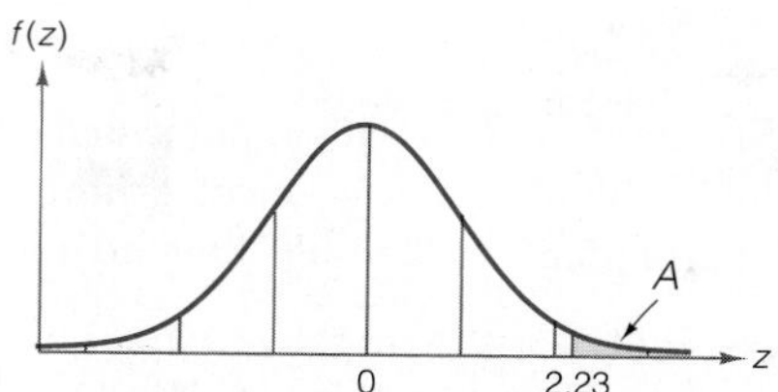

FIGURE 5.26 ▶ Standard normal distribution

Thus, the normal approximation to the binomial probability is

$$P(x \geq 20) \approx .0129$$

In other words, the probability is extremely small that 20 or more defectives would be observed in a sample of 200 circuits, *if in fact the true defective rate is 6%*. If the manufacturer were to observe $x \geq 20$, the likely reason is that the process is producing more than the acceptable 6% defectives.

Exercises 5.46–5.56

Learning the Mechanics

5.46 Why might you want to use a normal distribution to approximate a binomial distribution?

5.47 Under what circumstances is it appropriate to approximate a binomial distribution with a normal distribution?

5.48 Assume that x is a binomial random variable with n and p as specified in parts **a–f**. For which cases would it be appropriate to use a normal distribution to approximate the binomial distribution?

a. $n = 100, \quad p = .05$ **b.** $n = 100, \quad p = .2$ **c.** $n = 25, \quad p = .5$
d. $n = 1{,}000, \quad p = .01$ **e.** $n = 60, \quad p = .95$ **f.** $n = 1{,}000, \quad p = .95$

5.49 Suppose that x is a binomial random variable with $p = .7$ and $n = 25$.

a. Would it be appropriate to approximate the probability distribution of x with a normal distribution? Explain.
b. Assuming that a normal distribution provides an adequate approximation to the distribution of x, what are the mean and variance of the approximating normal distribution?
c. Use Table II of Appendix B to find the exact value of $P(x \geq 15)$.
d. Use the normal approximation to find $P(x \geq 15)$.

5.50 Assume that x is a binomial random variable with $n = 100$ and $p = .45$. Use a normal approximation to find the following:

a. $P(x \leq 45)$ **b.** $P(40 \leq x \leq 50)$ **c.** $P(x \geq 38)$

5.51 Assume that x is a binomial random variable with $n = 1{,}000$ and $p = .50$. Find each of the following probabilities:

a. $P(x > 500)$ **b.** $P(490 \leq x < 500)$ **c.** $P(x > 1{,}000)$

Applying the Concepts

5.52 The *Statistical Abstract of the United States: 1992* (p. 46) reports that 25% of the country's 94,312,000 households are inhabited by one person. If 1,000 randomly selected homes are to participate in a survey to determine television ratings, find the approximate probability that no more than 250 of these homes are inhabited by one person.

5.53 In 1982, General Motors Corporation's auto assembly plant in Fremont, California, was shut down and turned over to Toyota Motor Corporation as part of a joint venture with the Japanese firm. Before the shutdown, 5,000 workers produced 240,000 cars a year and had an absentee rate of 20%. After the Japanese took over and introduced their distinctive management style, the same number of cars were produced by only 2,500 workers and the absentee rate dropped to 2% ("The Difference Japanese Management Makes," *Business Week*, July 14, 1986, p. 47).

a. With an absentee rate of 20%, what is the probability that at least 90% of a random sample of 50 workers will be on the job on a particular day?

b. What assumption must we make in order to use the normal distribution to approximate the probability required in part **a**?

c. If the absentee rate were 2%, should the probability of the event in part **b** be approximated using a normal distribution? Explain.

5.54 In Case Study 4.3, the number of shuttle catastrophes due to booster failure in n missions was treated as a binomial random variable. Using the binomial distribution and the probability of catastrophe determined by the Air Force's risk assessment study (1/35), we determined the probability of at least 1 shuttle catastrophe in 25 missions to be .5155.

a. Based on the guidelines presented in this section, would it have been advisable to approximate this probability using the normal approximation to the binomial distribution? Explain.

b. Regardless of your answer to part **a**, use the normal distribution to approximate the binomial probability. Comment on the difference between the exact and approximate probabilities.

c. Refer to part **a**. Would the normal approximation be advisable if $n = 100$? $n = 500$? $n = 1{,}000$?

d. Approximate the probability that more than 25 catastrophes occur in 1,000 flights, assuming that the probability of a catastrophe in any given flight remains 1/35.

5.55 In May 1983, after an extensive investigation by the Consumer Product Safety Commission, Honeywell agreed to recall 770,000 potentially defective smoke detectors. The commission suggested that about 40% of the Honeywell detectors were defective. However, Honeywell found only four defectives in a random sample of 2,000 detectors and claimed the recall was not justified (Gross, 1983). Let x be the number of defective smoke detectors found in a random sample of 2,000 detectors.

a. What assumptions must be made in order to characterize x as a binomial random variable? Do these assumptions appear to be satisfied?

b. Assume that the conditions of part **a** hold. Determine the approximate probability of finding four or fewer defective smoke detectors in a random sample of 2,000 if, in fact, 40% of all detectors are defective.

c. Assume that Honeywell's sample data have been reported accurately. Is it *likely* that 40% of their detectors are defective? Explain.

d. Refer to part **c**. Is it *possible* that 40% of Honeywell's detectors are defective? Explain.

5.56 A computer disk manufacturer claims that 99.4% of its disks are defect-free. A large software company that buys and uses large numbers of the disks wants to verify this claim, and selects 1,600 disks to be tested. The tests reveal 12 defective disks. Assuming that the disk manufacturer's claim is correct, what is the probability of finding 12 or more defective disks in a sample of 1,600? Does your answer cast doubt on the manufacturer's claim? Explain.

Summary

Many **continuous random variables** in business applications have probability distributions that are well approximated by the **normal, uniform**, or **exponential probability distributions**. In this chapter we showed the graphical shape of each probability distribution, gave its mean and variance, and pointed out some practical applications of each probability model. In addition, we showed that the normal probability distribution provides a good approximation for the binomial distribution when n is sufficiently large.

Supplementary Exercises 5.57–5.75

Note: Starred (*) exercises refer to the optional section in this chapter.

5.57 Assume that x is a random variable best described by a uniform distribution with $c = 10$ and $d = 90$.
a. Find $f(x)$. b. Find the mean and standard deviation of x.
c. Graph the probability distribution for x, and locate its mean and the interval $\mu \pm 2\sigma$ on the graph.
d. Find $P(x \leq 60)$. e. Find $P(x \geq 90)$. f. Find $P(x \leq 80)$.
g. Find $P(\mu - \sigma \leq x \leq \mu + \sigma)$. h. Find $P(x > 75)$.

5.58 Use Table IV of Appendix B to calculate the area under the standard normal distribution between the following pairs of z-scores:
a. -1.96 and 1.96 b. -1.645 and 1.645 c. -3 and 3
d. -2.2 and 1.2 e. -2.6 and -1.6 f. -2.3 and 1.2

5.59 Use Table IV of Appendix B to find the following probabilities:
a. $P(z \geq .44)$ b. $P(z \leq -1.33)$ c. $P(-1.64 \leq z \leq 1.96)$ d. $P(1.64 \leq z \leq 1.96)$

5.60 The random variable x has a normal distribution with $\mu = 75$ and $\sigma = 10$. Find the following probabilities:
a. $P(x \leq 80)$ b. $P(x \geq 85)$ c. $P(70 \leq x \leq 75)$
d. $P(x > 80)$ e. $P(x = 78)$ f. $P(x \leq 110)$

5.61 Find a value of z, call it z_0, such that:
a. $P(z \geq z_0) = .5517$
b. $P(z \leq z_0) = .5080$
c. $P(z \geq z_0) = .1492$
d. $P(z_0 \leq z \leq .59) = .4773$

5.62 Suppose x is a normal random variable with mean 40 and standard deviation 6. Find the value of x, call it x_0, such that:

a. $P(\mu \leq x < x_0) = .40$ b. $P(x \geq x_0) = .10$ c. $P(x_0 \leq x < \mu) = .45$
d. $P(x < x_0) = .05$ e. $P(x > x_0) = .40$ f. $P(x \leq x_0) = .40$

***5.63** Suppose x has an exponential distribution with $\lambda = .3$. Find the following probabilities:

a. $P(x \leq 2)$ b. $P(x > 3)$ c. $P(x = 1)$ d. $P(x \leq 7)$ e. $P(4 \leq x \leq 12)$ f. $P(x = 2.5)$

5.64 Assume that x is a binomial random variable with $n = 50$ and $p = .6$. Find approximate values for the following probabilities:

a. $P(x \leq 35)$ b. $P(25 \leq x \leq 40)$ c. $P(x \geq 20)$ d. $P(40 \leq x \leq 50)$

5.65 The metropolitan airport commission is considering the establishment of limitations on noise pollution around a local airport. At the present time, the noise level per jet takeoff in one neighborhood near the airport is approximately normally distributed with a mean of 100 decibels and a standard deviation of 6 decibels.

a. What is the probability that a randomly selected jet will generate a noise level greater than 108 decibels in this neighborhood?
b. What is the probability that a randomly selected jet will generate a noise level of exactly 100 decibels?
c. Suppose a regulation is passed that requires jet noise in this neighborhood to be lower than 105 decibels 95% of the time. Assuming the standard deviation of the noise distribution remains the same, how much will the mean level of noise have to be lowered to comply with the regulation?

5.66 Suppose the present value of a risky investment is approximately normally distributed with mean \$10,000 and standard deviation \$4,000. What is the probability that the present value of the investment is less than \$1,000? Greater than \$20,000?

5.67 The **tolerance limits** for a particular quality characteristic (e.g., length, weight, or strength) of a product are the minimum and/or maximum values at which the product will operate properly. Tolerance limits are set by the engineering design function of the manufacturing operation (Juran and Gryna, 1980). The tensile strength of a particular metal part can be characterized as being normally distributed with a mean of 25 pounds and a standard deviation of 2 pounds. The upper and lower tolerance limits for the part are 30 pounds and 21 pounds, respectively. A part that falls within the tolerance limits results in a profit of \$10. A part that falls below the lower tolerance limit costs the company \$2; a part that falls above the upper tolerance limit costs the company \$1. Find the company's expected profit per metal part produced.

5.68 E. Brewer and P. Kaeser (1963) conducted a study of the factors that affect the level of production of workers paid on a piecework basis (i.e., paid according to the number of items they produce or process). Their study involved observing the performance of 36 quality control inspectors at a paper mill in England over an 8-week period. The inspectors were responsible for detecting and sorting out paper with defects such as holes, spots, creases, and rust marks. Part of the study entailed computing and analyzing the average hourly earnings for each inspector for each week of the 8-week observation period. Brewer and Kaeser constructed a frequency histogram of the average earnings data and noted that the histogram could be approximated by a normal distribution.

a. How many observations are there in the average earnings data set?
b. Suppose the average-earnings histogram can be approximated by a normal distribution with $\mu = \$7.65$ and $\sigma = \$1.25$. Approximately what proportion of the weekly average earnings are over \$8.50 per hour?

c. Using the normal distribution of part **b**, is it possible to determine approximately how many inspectors averaged over $8.50 per hour for the 8-week period? If so, how many? If not, why not?

***5.69** Based on sample data collected in the Denver area, Nicholas Kiefer (1985) found that in some cases the exponential distribution is an adequate approximation for the distribution of the time (in weeks) an individual is unemployed. In particular, he found the exponential distribution to be appropriate for white and black workers, but not for Hispanics. Use $\lambda = .075$ to answer the following questions.

a. What is the mean time workers are unemployed according to the exponential distribution?
b. What is the probability that a white worker who just lost her job will be unemployed for at least 2 weeks? More than 6 weeks?
c. What is the probability that an unemployed worker will find a new job within 12 weeks?

5.70 It is quite common for the standard deviation of a random variable to increase proportionally as the mean increases. When this occurs, the **coefficient of variation,**

$$CV = \frac{\sigma}{\mu}$$

the ratio of σ to μ, is the *proportionality constant.* To illustrate, the error (in dollars) in assessing the value of a house increases as the house increases in value. Suppose that long experience with assessors in your part of the country has shown that the coefficient of variation is .08 and that the probability distribution of assessed valuations on the same house by many different assessors is approximately normal with a mean we will call the *true value* of the house. Suppose the true value of your house is $50,000, and it is being assessed for taxation purposes. What is the probability that the assessor will assess your house in excess of $55,000?

5.71 As noted in Exercise 5.70, it is sometimes true that the larger the mean of a random variable, the larger will be its standard deviation. A measure of the relative variability of different data sets can be obtained by computing the coefficient of variation ($s/\bar{x}$) for each. The data sets in the table reflect the numbers of checks (in thousands) processed per week by three different banks over the last 6 weeks:

Bank 1		*Bank 2*		*Bank 3*	
20	25	60	65	100	110
10	31	63	70	92	105
18	12	58	54	81	129

Rank each of these data sets in terms of variability using the:

a. Range b. Standard deviation c. Coefficient of variation

***5.72** Assume that the length of the active life of baking yeast has an exponential distribution with a mean equal to 6 months. If the expiration date marked on a package of yeast is based on a life of 6 months, what is the probability that a package of the yeast will lose its potency before its expiration date?

5.73 A company has a lump-sum incentive plan for salespeople that is dependent on their level of sales. If they sell less than $100,000 per year, they receive a $1,000 bonus; from $100,000 to $200,000, they receive $5,000; and above $200,000, they receive $10,000. If the annual sales per salesperson has approximately a normal distribution with $\mu = \$180{,}000$ and $\sigma = \$50{,}000$:

a. Find p_1, the proportion of salespeople who receive a $1,000 bonus.
b. Find p_2, the proportion of salespeople who receive a $5,000 bonus.
c. Find p_3, the proportion of salespeople who receive a $10,000 bonus.

d. What is the mean value of the bonus payout for the company? [*Hint:* Review the definition for the expected value of a random variable in Chapter 4.]

5.74 Contrary to our intuition, very reliable decisions concerning the proportion of a large group of consumers who favor a particular product or a particular social issue can be based on relatively small samples. For example, suppose the target population of consumers contains 50,000,000 people and we want to decide whether the proportion of consumers, p, in the population who favor some product (or issue) is as large as some value, say .2. Suppose you randomly select a sample as small as 1,600 from the 50,000,000 and you observe the number, x, of consumers in the sample who favor the new product. Assuming that $p = .2$, find the mean and standard deviation of x. Suppose that 400 (or 25%) of the sample of 1,600 consumers favor the new product. Why might this sample result lead you to conclude that p (the proportion of consumers who favor the product in the population of 50,000,000) is larger than .2? [*Hint:* Compare the observed value of x with the values of μ and σ calculated on the assumption that $p = .2$.]

5.75 To help highway planners anticipate the need for road repairs and design future construction projects, data are collected on the volume and weight of truck traffic on specific roadways. Recently, equipment has been developed that can be built into road surfaces to measure traffic volumes and to weigh trucks without requiring them to stop at roadside weigh stations. As with any measuring device, however, the "weigh-in-motion" equipment does not always record truck weights accurately. In an experiment performed by the Minnesota Department of Transportation involving repeated weighing of a 27,907-pound truck, it was found that the weights recorded by the weigh-in-motion equipment were approximately normally distributed with mean 27,315 and a standard deviation of 628 pounds (Dahlin, 1982; Wright, Owen, and Pena, 1983). It follows that the difference between the actual weight and recorded weight, the error of measurement, is normally distributed with mean 592 pounds and standard deviation 628 pounds.

a. What is the probability that the weigh-in-motion equipment understates the actual weight of the truck?

b. If a 27,907-pound truck were driven over the weigh-in-motion equipment 100 times, approximately how many times would the equipment overstate the truck's weight?

c. What is the probability that the error in the weight recorded by the weigh-in-motion equipment for a 27,907-pound truck exceeds 400 pounds?

d. It is possible to adjust (or *calibrate*) the weigh-in-motion equipment to control the mean error of measurement. At what level should the mean error be set so the equipment will understate the weight of a 27,907-pound truck 50% of the time? Only 40% of the time?

On Your Own

For large values of n the computational effort involved in working with the binomial probability distribution is considerable. Fortunately, in many instances the normal distribution provides a good approximation to the binomial distribution. This exercise was designed to demonstrate how well the normal distribution approximates the binomial distribution.

a. Suppose the random variable x has a binomial probability distribution with $n = 10$ and $p = .5$. Using the binomial distribution, find the probability that x takes on a value in each of the following intervals: $\mu \pm \sigma$, $\mu \pm 2\sigma$, and $\mu \pm 3\sigma$.

b. Approximate the probabilities requested in part **a** using a normal approximation to the given binomial distribution.
c. Determine the magnitude of the difference between each of the three probabilities as determined by the binomial distribution and by the normal approximation.

Using the Computer

Refer to **Using the Computer** in Chapter 4. Again use the mean percentage of college graduates as an estimate of the percentage of college graduates among all people at least 25 years old in the region you selected.

a. Use the normal approximation to the binomial to estimate the probability that fewer than 10% of 20 applicants (still assuming they represent a random sample) have a college education. Compare your answer with the exact binomial probability of the same event.
b. Use the normal approximation to the binomial to estimate the probability that fewer than 10% of 200 applicants have a college education. If your statistical software package has a function that calculates exact binomial probabilities, use it to calculate the same probability you approximated, and compare the results.

References

Blair, R. D. and Kenny, L. W. *Microeconomics for Managerial Decision Making*. New York: McGraw-Hill, 1982. Chapter 10.

Bolling, W. B. "Queuing model of a hospital emergency room." *Industrial Engineering*, Sept. 1972, pp. 26–31.

Brewer, E. and Kaeser, P. "A comparative analysis of incentive plans." *Journal of Industrial Relations*, July 1963, *11*, pp. 183–198.

Dahlin, C. *Minnesota's Experience with Weighing Trucks in Motion*. St. Paul: Minnesota Department of Transportation, 1982.

Dessler, G. "Personnel tests gain in popularity." *St. Paul Pioneer Press and Dispatch*, Mar. 17, 1986.

Fama, E. F. *Foundations of Finance*. New York: Basic Books, 1976. Chapter 1.

Gross, S. "Honeywell smoke detectors recalled after 18-month probe." *Minneapolis Star and Tribune*, May 25, 1983.

Hillier, F. S. "The derivation of probabilistic information for the evaluation of risky investments." *Management Science*, Apr. 1963, *9*, pp. 443–457.

Hillier, F. S. and Lieberman, G. J. *Introduction to Operations Research*, 5th ed. New York: McGraw-Hill, 1990.

Hogg, R. V. and Craig, A. T. *Introduction to Mathematical Statistics*, 4th ed. New York: Macmillan, 1978.

Juran, J. M. and Gryna, F. M., Jr. *Quality Planning and Analysis*, 2d ed. New York: McGraw-Hill, 1980.

Kiefer, N. M. "Specification diagnostics based on Laguerre alternatives for econometric models of duration." *Journal of Econometrics*, 1985, *28*, pp. 135–154.

Lamberson, L. R. "Reliability tutorial." *American Society for Quality Control Quality Congress Transaction*, Baltimore, 1985, pp. 88–99.

Lindgren, B. W. *Statistical Theory*, 3d ed. New York: Macmillan, 1976. Chapters 2 and 3.

Martz, H. F. and Waller, R. A. *Bayesian Reliability Analysis*. New York: Wiley, 1982, pp. 1, 256.

Montgomery, D. C. *Introduction to Statistical Quality Control*, 2d ed. New York: Wiley, 1991.

Mood, A. M., Graybill, F. A., and Boes, D. C. *Introduction to the Theory of Statistics*, 3d ed. New York: McGraw-Hill, 1974. Chapter 3.

Murphy, A. H. and Winkler, R. L. "Probability forecasting in meteorology." *Journal of the American Statistical Association*, 1984, *79*, pp. 489–500.

Neter, J., Wasserman, W., and Whitmore, G. A. *Applied Statistics*, 4th ed. Boston: Allyn & Bacon, 1993.

Schick, G. J. "The search for a software reliability model." *Decision Sciences*, Oct. 1974, 5, p. 529.

Sims, S. H. and Templeton, J. G. C. "Steady state results for the *M/M* (*a*, *b*)/*c* batch-service system." *European Journal of Operational Research*, 1985, 21, pp. 260–267.

Winkler, R. L. and Hays, W. *Statistics: Probability, Inference, and Decision*, 2d ed. New York: Holt, Rinehart and Winston, 1975. Chapter 3.

Wright, J. L., Owen, F., and Pena, D. *Status of Mn/DOT's Weigh-in-Motion Program*. St. Paul: Minnesota Department of Transportation, Jan. 1983.

CHAPTER SIX

Sampling Distributions

Contents

Case Studies

Where We've Been

We have learned in earlier chapters that the objective of most statistical investigations is inference—that is, estimating some numerical descriptive measure (a parameter) of a population based on sample data. To estimate a population parameter, we use sample data to compute numerical descriptive measures (sample statistics) such as the sample mean or variance.

Where We're Going

Because sample measurements are observed values of random variables, the value of a sample statistic will vary in a random manner from sample to sample. In other words, since sample statistics are computed from random variables, they themselves are random variables, and they have probability distributions that are either discrete or continuous. The probability distribution of a sample statistic is called a *sampling distribution* because it characterizes the distribution of values of the statistic over a very large number of samples. Sampling distributions are the topic of this chapter. We will discuss why many sampling distributions tend to be approximately normal, and you will see how sampling distributions can be used to evaluate the accuracy of parameter estimates. Then in subsequent chapters we will show how to use sampling distributions to make inferences in a variety of business and industrial applications.

In Chapters 4 and 5 we assumed that we knew the probability distribution of a random variable, and based on this knowledge, we were able to compute the mean, variance, and probability that the random variable assumed specific values. However, in most practical applications, this information will not be available. To illustrate, in Example 4.8, we calculated the probability that the binomial random variable x—the number of 20 polled employees who favor unionization—assumed specific values. To do this, it was necessary to assume some value for p, the proportion of the employees in the population who favor unionization. Thus, for the purpose of illustration we assumed $p = .6$. Typically, however, the exact value of p would be unknown. In fact, the probable purpose of taking the poll was to estimate p. Similarly, when we modeled the in-city gas mileage of a certain type of automobile in Example 5.7, we used the normal probability distribution with an *assumed* mean and standard deviation of 27 and 3 miles per gallon, respectively. In reality, the true mean and standard deviation are unknown quantities that would have to be estimated.

Numerical quantities that describe probability distributions are called **parameters**. Thus, p, the probability of a success in a binomial experiment, and μ and σ, the mean and standard deviation of a normal distribution, are examples of parameters. Since probability distributions are used to characterize populations, it follows that parameters are also numerical descriptive measures of populations.

Definition 6.1

A **parameter** is a numerical descriptive measure of a population. It is calculated from the observations in the population.

Since it is almost always too costly and/or time-consuming to conduct a census of a population and compute its parameters, we use the information contained in a sample to make inferences about the parameters of a population. In order to make such inferences, we must compute **sample statistics** that will aid in making these inferences.

Definition 6.2

A **sample statistic** is a numerical descriptive measure of a sample. It is calculated from the observations in the sample.

Some examples of useful sample statistics we have already discussed are the sample mean, $\bar{x}$; sample median; sample variance, s^2; and the sample standard deviation, s. Before we can use these and other sample statistics to make inferences about population parameters, we have to be able to evaluate their properties. How can we decide which sample statistic contains the most information about a population parameter? One purpose of this chapter is to answer this question.

6.1 Introduction to Sampling Distributions

If we want to estimate a parameter of a population—say, the population mean, μ—there are a number of sample statistics that we could use for the estimate. Two possibilities are the sample mean, $\bar{x}$, and the sample median, m. Which of these do you think will provide a better estimate of μ?

Before answering this question, consider the following example: Toss a fair die and let x equal the number of dots showing on the up face. Suppose the die is tossed three times, producing the sample measurements 2, 2, 6. The sample mean is $\bar{x} = 3.33$ and the sample median is $m = 2$. Since the mean of x is $\mu = 3.5$, you can see that for this sample of three measurements, the sample mean $\bar{x}$ provides an estimate that falls closer to μ than does the sample median [see Figure 6.1(a)].

FIGURE 6.1 ► Comparing the sample mean ($\bar{x}$) and sample median (m) as estimators of the population mean (μ)

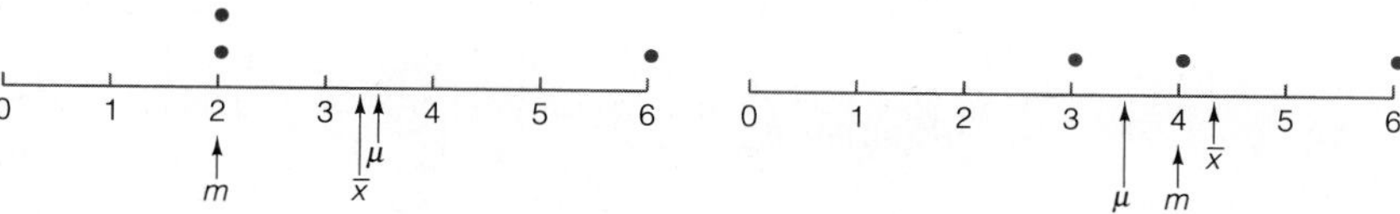

Now suppose we toss the die three more times and obtain the sample measurements, 3, 4, 6. The mean and median of this sample are $\bar{x} = 4.33$ and $m = 4$, respectively. This time m is closer to μ [see Figure 6.1(b)].

This simple example illustrates an important point: Neither the sample mean nor the sample median will *always* fall closer to the population mean. Consequently, we cannot compare these two sample statistics or, in general, any two sample statistics on the basis of their performance for a single sample. Instead, we need to recognize that sample statistics are themselves random variables because different samples can lead to different values for the sample statistics. As random variables, sample statistics must be judged and compared on the basis of their probability distributions—i.e., the collection of values and associated probabilities of each statistic that would be obtained if the sampling experiment were repeated a *very large number of times*. We will illustrate this concept with an example.

Suppose it is known that in a certain part of Canada the daily high temperature recorded for all past months of January has a mean of $\mu = 10°F$ and a standard deviation of $\sigma = 5°F$. Consider an experiment consisting of randomly selecting 25 daily high temperatures from the records of past months of January and calculating the sample mean, $\bar{x}$. If this experiment were repeated a very large number of times, the value of $\bar{x}$ would vary from sample to sample. For example, the first sample of 25 temperature measurements might have a mean of $\bar{x} = 9.8$; the second sample, a mean of $\bar{x} = 11.4$; the third sample, a mean of $\bar{x} = 10.5$; etc. If the sampling experiment were repeated a very large number of times and the resulting values of $\bar{x}$ were displayed in a histogram, the histogram would be the approximate probability distribution of $\bar{x}$. If $\bar{x}$ is a good estimator of μ, we would expect the values of $\bar{x}$ to cluster around μ as shown in Figure 6.2. This probability distribution is called a **sampling distribution** because it describes the potential outcomes of $\bar{x}$ in repeated sampling.

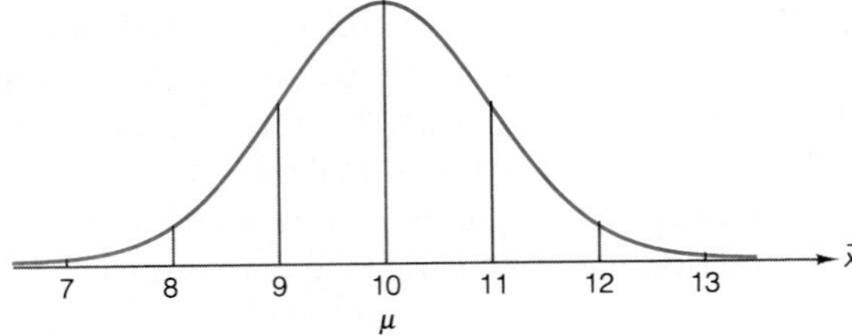

FIGURE 6.2 ► Sampling distribution for $\bar{x}$ based on a sample of $n = 25$ measurements: $\mu = 10$, $\sigma = 5$

In actual practice, the sampling distribution of a statistic is obtained mathematically or (approximately) by simulating the sampling on a computer using the procedure just described.

> **Definition 6.3**
>
> The **sampling distribution** for a sample statistic (calculated from a sample of n measurements) is the probability distribution of the statistic.

If $\bar{x}$ has been calculated from a sample of $n = 25$ measurements selected from a population with mean $\mu = 10$ and standard deviation $\sigma = 5$, the sampling distribution (Figure 6.2) provides information about the behavior of $\bar{x}$ in repeated sampling. For example, the probability that you will draw a sample of 25 measurements and obtain a value of $\bar{x}$ in the interval $9 \leq \bar{x} \leq 10$ will be the area under the sampling distribution over that interval.

Since the properties of a statistic are typified by its sampling distribution, it follows that, to compare two statistics, you compare their sampling distributions. For example, if you have two statistics, A and B, for estimating the same parameter (for purposes of illustration, suppose the parameter is the population variance σ^2) and if their sampling distributions are as shown in Figure 6.3, you would choose statistic A in preference to statistic B. You would make this choice because the sampling distribution for statistic

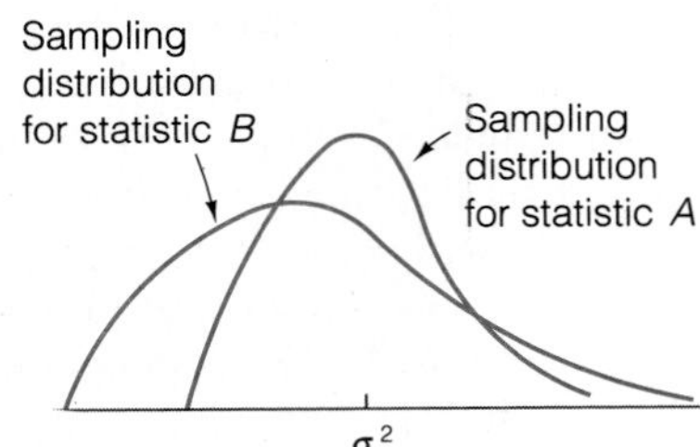

FIGURE 6.3 ► Two sampling distributions for estimating the population variance, σ^2

A centers over σ^2 and has less spread (variation) than the sampling distribution for statistic B. When you draw a single sample in a practical sampling situation, the probability is higher that statistic A will fall closer to σ^2.

Remember that in practice we will not know the numerical value of the unknown parameter, σ^2, so we will not know whether statistic A or statistic B is closer to σ^2 for particular samples. We have to rely on our theoretical knowledge of the sampling distributions to choose the better sample statistic and then use it sample after sample.

EXAMPLE 6.1

Consider a large population consisting only of the measurements 0, 3, and 12 and described by the probability distribution shown in the table. A random sample of $n = 3$ measurements is selected from the population.

x	0	3	12
$p(x)$	1/3	1/3	1/3

a. Find the sampling distribution of the sample mean $\bar{x}$.

b. Find the sampling distribution of the sample median m.

Solution

Every possible sample of $n = 3$ measurements is listed in Table 6.1, along with the sample mean and median. Because any one sample is as likely to be selected as any other (random sampling), the probability of observing any particular sample is 1/27. This probability is also listed in Table 6.1.

TABLE 6.1 Possible Samples of $n = 3$ Measurements

Possible Samples	$\bar{x}$	m	Probability	Possible Samples	$\bar{x}$	m	Probability
0, 0, 0	0	0	1/27	3, 3, 12	6	3	1/27
0, 0, 3	1	0	1/27	3, 12, 0	5	3	1/27
0, 0, 12	4	0	1/27	3, 12, 3	6	3	1/27
0, 3, 0	1	0	1/27	3, 12, 12	9	12	1/27
0, 3, 3	2	3	1/27	12, 0, 0	4	0	1/27
0, 3, 12	5	3	1/27	12, 0, 3	5	3	1/27
0, 12, 0	4	0	1/27	12, 0, 12	8	12	1/27
0, 12, 3	5	3	1/27	12, 3, 0	5	3	1/27
0, 12, 12	8	12	1/27	12, 3, 3	6	3	1/27
3, 0, 0	1	0	1/27	12, 3, 12	9	12	1/27
3, 0, 3	2	3	1/27	12, 12, 0	8	12	1/27
3, 0, 12	5	3	1/27	12, 12, 3	9	12	1/27
3, 3, 0	2	3	1/27	12, 12, 12	12	12	1/27
3, 3, 3	3	3	1/27				

a. From Table 6.1 you can see that $\bar{x}$ can assume the values 0, 1, 2, 3, 4, 5, 6, 8, 9, and 12. Because $\bar{x} = 0$ occurs in only one sample, $P(\bar{x} = 0) = 1/27$. Similarly, $\bar{x} =$

1 occurs in three samples: (0, 0, 3), (0, 3, 0), and (3, 0, 0). Therefore, $P(\bar{x} = 1) = 3/27 = 1/9$. Calculating the probabilities of the remaining values of $\bar{x}$ and arranging them in a table, we obtain the following probability distribution:

$\bar{x}$	0	1	2	3	4	5	6	8	9	12
$p(\bar{x})$	1/27	3/27	3/27	1/27	3/27	6/27	3/27	3/27	3/27	1/27

This is the sampling distribution for $\bar{x}$, because it specifies the probability associated with each possible value of $\bar{x}$.

b. In Table 6.1 you can see that the median m can assume one of the three values 0, 3, or 12. The value $m = 0$ occurs in seven different samples. Therefore, $P(m = 0) = 7/27$. Similarly, $m = 3$ occurs in thirteen samples and $m = 12$ occurs in seven samples. Therefore, the probability distribution (i.e., the sampling distribution) for the median m is as follows:

m	0	3	12
$p(m)$	7/27	13/27	7/27

Example 6.1 demonstrates the procedure for finding the exact sampling distribution of a statistic when the number of different samples that could be selected from the population is relatively small. In the real world, populations often consist of a large number of different values, making samples difficult (or impossible) to enumerate. When such a situation occurs, we may choose to obtain the approximate sampling distribution for a statistic by simulating the sampling over and over again and recording the proportion of times different values of the statistic occur. Example 6.2 illustrates this procedure.

EXAMPLE 6.2

Suppose we perform the following experiment: Take a sample of 11 measurements from the uniform distribution shown in Figure 6.4. Calculate the two sample statistics

$$\bar{x} = \text{Sample mean} = \frac{\sum_{i=1}^{11} x_i}{11}$$

m = Median = Sixth sample measurement when the 11 measurements are arranged in ascending order

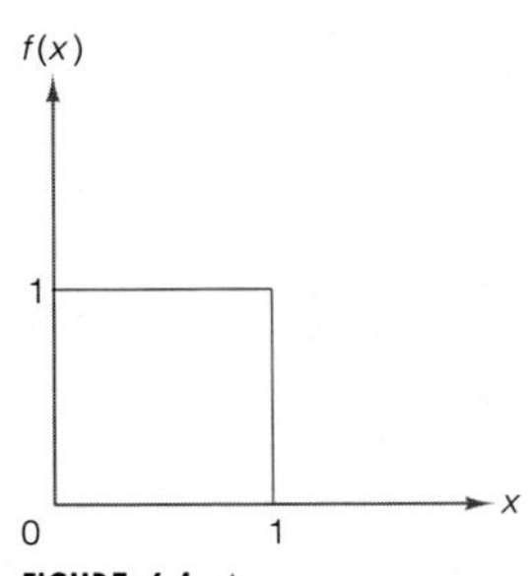

FIGURE 6.4 ▲
Uniform distribution from 0 to 1

In this particular example we *know* that the population mean is $\mu = .5$. The objective will be to find out which sample statistic contains more information about μ. We use a computer to generate 1,000 samples, each with $n = 11$ observations. Then, we compute $\bar{x}$ and m for each sample. Our goal is to find the resulting approximate sampling distributions for $\bar{x}$ and m.

Solution

The first 10 of the 1,000 samples generated are presented in Table 6.2. For each of the 1,000 samples we compute the sample mean, $\bar{x}$, and the sample median, m. For example, the first computer-generated sample from the uniform distribution (arranged in ascending order) contained the following measurements: .125, .138, .139, .217, .419, .506, .516, .757, .771, .786, .919. The sample mean, $\bar{x}$, and median, m, computed for this sample are

$$\bar{x} = \frac{.125 + .138 + \cdots + .919}{11} = .481$$

$$m = \text{Sixth ordered measurement} = .506$$

TABLE 6.2 First 10 Samples of $n = 11$ Measurements from a Uniform Distribution

Sample	Measurements										
1	.217	.786	.757	.125	.139	.919	.506	.771	.138	.516	.419
2	.303	.703	.812	.650	.848	.392	.988	.469	.632	.012	.065
3	.383	.547	.383	.584	.098	.676	.091	.535	.256	.163	.390
4	.218	.376	.248	.606	.610	.055	.095	.311	.086	.165	.665
5	.144	.069	.485	.739	.491	.054	.953	.179	.865	.429	.648
6	.426	.563	.186	.896	.628	.075	.283	.549	.295	.522	.674
7	.643	.828	.465	.672	.074	.300	.319	.254	.708	.384	.534
8	.616	.049	.324	.700	.803	.399	.557	.975	.569	.023	.072
9	.093	.835	.534	.212	.201	.041	.889	.728	.466	.142	.574
10	.957	.253	.983	.904	.696	.766	.880	.485	.035	.881	.732

The relative frequency histograms for $\bar{x}$ and m for the 1,000 samples of size $n = 11$ are shown in Figure 6.5.

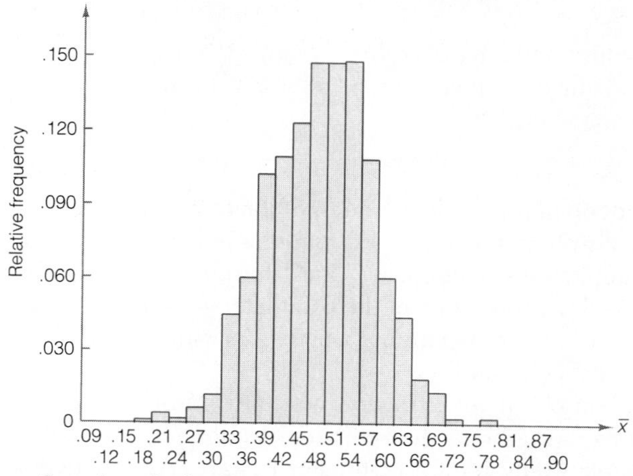

a. Sampling distribution for $\bar{x}$ (based on 1,000 samples of $n = 11$ measurements)

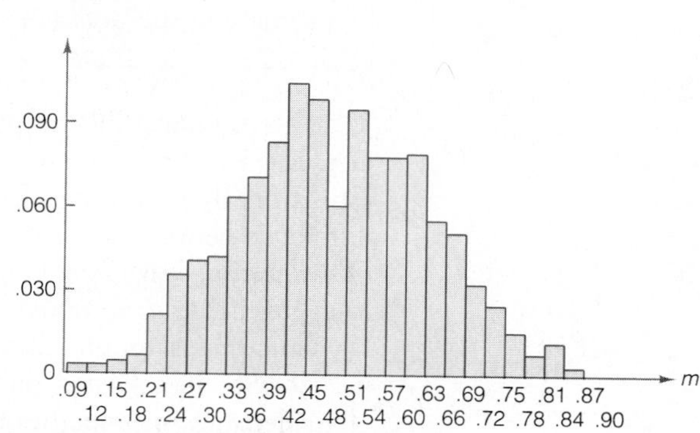

b. Sampling distribution for m (based on 1,000 samples of $n = 11$ measurements)

FIGURE 6.5 ▲ **Approximate sampling distributions for $\bar{x}$ and m, Example 6.2**

You can see that the values of $\bar{x}$ tend to cluster around μ to a greater extent than do the values of m. Thus, on the basis of the observed sampling distributions, we conclude that $\bar{x}$ contains more information about μ than m does—at least for samples of $n = 11$ measurements from the uniform distribution.

We will not always have to simulate repeated sampling on a computer to find sampling distributions. Many sampling distributions can be derived mathematically, but the theory necessary to do this is beyond the scope of this text. Consequently, when we need to know the properties of a statistic, we will present its sampling distribution and simply describe its properties. Several of the important properties of sampling distributions are discussed in the next section.

6.2 Properties of Sampling Distributions: Unbiasedness and Minimum Variance

A sample statistic that is used to make inferences about a population parameter is called a **point estimator**. A point estimator is a rule or formula that tells us how to use the sample data to calculate a single number that can be used as an **estimate** of the value of some population parameter. For example, the sample mean, $\bar{x}$, is a point estimator of the population mean, μ. Similarly, the sample variance, s^2, is a point estimator of the population variance, σ^2.

Definition 6.4

A **point estimator** of a population parameter is a rule or formula that tells us how to use the sample data to calculate a single number that can be used as an **estimate** of the population parameter.

Often, many different point estimators can be found to estimate the same parameter. Each will have a sampling distribution that provides information about the point estimator. By examining the sampling distribution, we can determine how large the difference between an estimate and the true value of the parameter—called the **error of estimation**—is likely to be. We can also determine whether an estimator is likely to overestimate or to underestimate a parameter.

Since the sampling distribution of a point estimator (or of any statistic) describes its behavior in repeated sampling, we look to the sampling distribution to identify characteristics or properties that we would want an estimator to have. As a first consideration, we would like the sampling distribution to center over the value of the parameter we want to estimate. One way to express centrality is in terms of the mean of the sampling distribution. Consequently, we say that a statistic is **unbiased** if its sampling distribution has a mean equal to the parameter it is intended to estimate.

When this occurs, the sampling distribution of the statistic will be centered over the parameter as shown in Figure 6.6(a). If the mean of a sampling distribution is not equal to the parameter it is intended to estimate, the statistic is said to be **biased**. The amount of bias is the difference between the mean of the sampling distribution and the value of the parameter you wish to estimate. The sampling distribution for a biased statistic is shown in Figure 6.6(b).

FIGURE 6.6 ►
Sampling distributions for unbiased and biased estimators

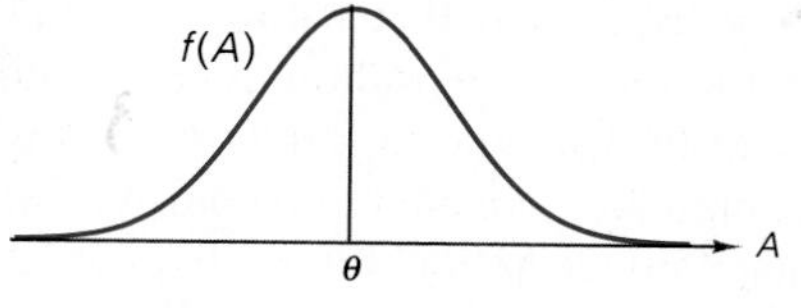

a. Unbiased sample statistic for the parameter θ

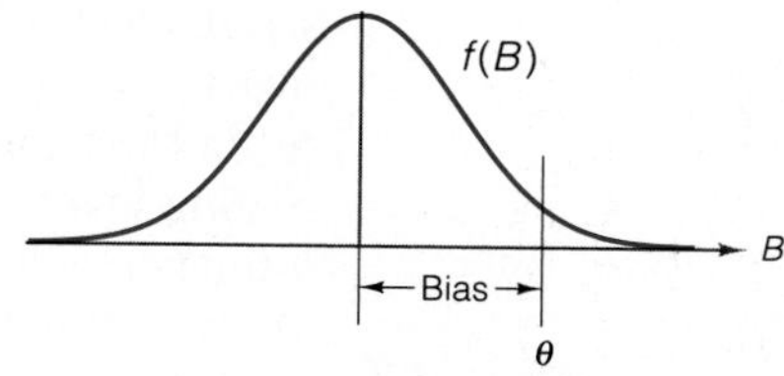

b. Biased sample statistic for the parameter θ

> **Definition 6.5**
>
> If a sample statistic has a sampling distribution with a mean equal to the population parameter the statistic is intended to estimate, the statistic is said to be an **unbiased** estimator of the parameter.
>
> If the mean of the sampling distribution is not equal to the parameter, the statistic is said to be a **biased** estimator of the parameter.

The standard deviation of a sampling distribution measures another important property of a statistic—the spread of the estimates generated by repeated sampling. Suppose two statistics, A and B, are both unbiased estimators of the population parameter called θ (theta). Note that θ could by any parameter, such as μ, σ^2, or σ. Since the means of the two sampling distributions are the same, we turn to their standard deviations to decide which will provide estimates that fall closer to the unknown population parameter we are estimating. Naturally, we will choose the sample statistic that has the smaller standard deviation. Figure 6.7 depicts sampling distribu-

FIGURE 6.7 ►
Sampling distributions for two unbiased estimators

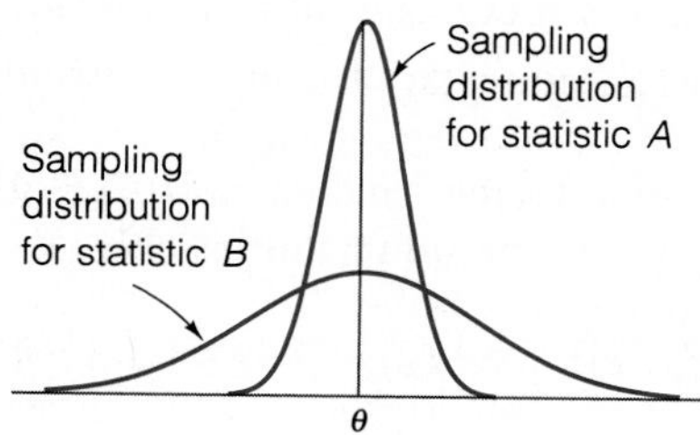

tions for A and B. Note that the standard deviation of the distribution of A is smaller than the standard deviation for B, indicating that over a large number of samples, the values of A cluster more closely around the unknown population parameter than do the values of B. Stated differently, the probability is high that A is closer to the parameter value than B is. Therefore, we would choose statistic A instead of statistic B as an estimator of θ.

In summary, to make an inference about a population parameter, use the sample statistic with a sampling distribution that is unbiased and has a small standard deviation (usually smaller than the standard deviation of other unbiased sample statistics). How to find this sample statistic will not concern us, because the "best" statistic for estimating a particular parameter is a matter of record. We will simply present an unbiased estimator with its standard deviation for each population parameter we consider. [*Note*: The standard deviation of the sampling distribution for a statistic is often called the **standard error** of the statistic.]

EXAMPLE 6.3

In Example 6.1, we found the sampling distributions of the sample mean $\bar{x}$ and the sample median m for random samples of $n = 3$ measurements from a population defined by the following probability distribution:

x	0	3	12
$p(x)$	$1/3$	$1/3$	$1/3$

The sampling distributions of $\bar{x}$ and m were found to be the following:

$\bar{x}$	0	1	2	3	4	5	6	8	9	12
$p(\bar{x})$	$1/27$	$3/27$	$3/27$	$1/27$	$3/27$	$6/27$	$3/27$	$3/27$	$3/27$	$1/27$

m	0	3	12
$p(m)$	$7/27$	$13/27$	$7/27$

a. Show that $\bar{x}$ is an unbiased estimator of μ in this situation.

b. Show that m is a biased estimator of μ in this situation.

Solution

a. The expected value of a discrete random variable x (see Section 4.3) is defined to be $E(x) = \sum xp(x)$, where the summation is over all values of x. Then

$$E(x) = \mu = \sum xp(x) = (0)(1/3) + (3)(1/3) + (12)(1/3) = 5$$

The expected value of the discrete random variable $\bar{x}$ is

$$E(\bar{x}) = \sum (\bar{x})p(\bar{x})$$

summed over all values of $\bar{x}$. Or

$$E(\bar{x}) = (0)(1/27) + (1)(3/27) + 2(3/27) + \cdots + (12)(1/27) = 5$$

Since $E(\bar{x}) = \mu$, we see that $\bar{x}$ is an unbiased estimator of μ.

b. The expected value of the sample median m is

$$E(m) = \sum mp(m) = (0)(7/27) + (3)(13/27) + (12)(7/27) = 4.56$$

Since the expected value of m is not equal to μ ($\mu = 5$), the sample median m is a biased estimator of μ.

EXAMPLE 6.4

Refer to Example 6.3 and find the standard deviations of the sampling distributions of $\bar{x}$ and m. Which statistic would appear to be a better estimator for μ?

Solution

The variance of the sampling distribution of $\bar{x}$ (we will denote it by the symbol $\sigma_{\bar{x}}^2$) is found to be

$$\sigma_{\bar{x}}^2 = E\{[\bar{x} - E(\bar{x})]^2\} = \sum (\bar{x} - \mu)^2 p(\bar{x})$$

where, from Example 6.3,

$$E(\bar{x}) = \mu = 5$$

Then

$$\sigma_{\bar{x}}^2 = (0 - 5)^2(1/27) + (1 - 5)^2(3/27) + (2 - 5)^2(3/27) + \cdots + (12 - 5)^2(1/27)$$
$$= 8.6667$$

and

$$\sigma_{\bar{x}} = \sqrt{8.6667} = 2.94$$

Similarly, the variance of the sampling distribution of m (we will denote it by σ_m^2) is

$$\sigma_m^2 = E\{[m - E(m)]^2\}$$

where, from Example 6.3, the expected value of m is $E(m) = 4.56$. Then

$$\sigma_m^2 = E\{[m - E(m)]^2\} = \sum [m - E(m)]^2 p(m)$$
$$= (0 - 4.56)^2(7/27) + (3 - 4.56)^2(13/27) + (12 - 4.56)^2(7/27) = 20.9136$$

and

$$\sigma_m = \sqrt{20.9136} = 4.57$$

Which statistic appears to be the better estimator for the population mean μ: the sample mean $\bar{x}$ or the median m? To answer this question, we compare the sampling distributions of the two statistics. The sampling distribution of the sample median m is biased (i.e., it is located to the left of the mean μ) and its standard deviation $\sigma_m = 4.57$ is much larger than the standard deviation of the sampling distribution of $\bar{x}$, $\sigma_{\bar{x}} = 2.94$. Consequently, the sample mean $\bar{x}$ would be a better estimator of the population mean μ, for the population in question, than would the sample median m.

6.3 The Sampling Distribution of the Sample Mean

Estimating the mean useful life of automobiles, the mean monthly sales for all automobile dealers in a large city, and the mean breaking strength of a new plastic are practical problems with something in common. In each, we are interested in making an inference about the mean, μ, of some population. Because many practical business problems involve estimating μ, it is particularly important to have a sample statistic that is a good estimator of μ. As we mentioned in Chapter 2, the sample mean, $\bar{x}$, is generally a good choice as an estimator of μ. The mean and standard deviation of the sampling distribution of this useful statistic are related to the mean, μ, and standard deviation, σ, of the sampled population as described in the box.

The Mean and Standard Deviation of The Sampling Distribution of $\bar{x}$

Regardless of the shape of the population relative frequency distribution,

1. The mean of the sampling distribution of $\bar{x}$ will equal μ, the mean of the sampled population; i.e., $\mu_{\bar{x}} = \mu$.
2. The standard deviation of the sampling distribution of $\bar{x}$ will equal σ, the standard deviation of the sampled population, divided by the square root of the sample size, n; i.e.,

$$\sigma_{\bar{x}} = \frac{\sigma^*}{\sqrt{n}}$$

The standard deviation $\sigma_{\bar{x}}$ is often referred to as the **standard error of the mean**.

*If the sample size, n, is large relative to the number, N, of elements in the population, (e.g., 5% or more), $\sigma/\sqrt{n}$ must be multiplied by a finite population correction factor, $\sqrt{(N - n)/(N - 1)}$. For most sampling situations, this correction factor will be close to 1 and can be ignored.

Thus, $\bar{x}$ is an unbiased estimator of the population mean μ [i.e., $E(\bar{x}) = \mu$], and the standard deviation of $\bar{x}$ is inversely related to the (square root of the) sample size (i.e., $\sigma_{\bar{x}} = \sigma/\sqrt{n}$).

For example, suppose the sampled population has the uniform probability distribution shown in Figure 6.8(a). The mean and standard deviation of this probability distribution are $\mu = .5$ and $\sigma = .29$ (refer to Section 5.2 for the formulas for μ and σ). Now suppose a sample of 11 measurements is selected from this population. The sampling distribution of the sample mean for samples of size 11 will also have a mean of .5, with a standard deviation

$$\sigma_{\bar{x}} = \frac{\sigma}{\sqrt{n}} = \frac{.29}{\sqrt{11}} = .09$$

(That is, the standard error of $\bar{x}$ is $\sigma/\sqrt{n} = .09$.)

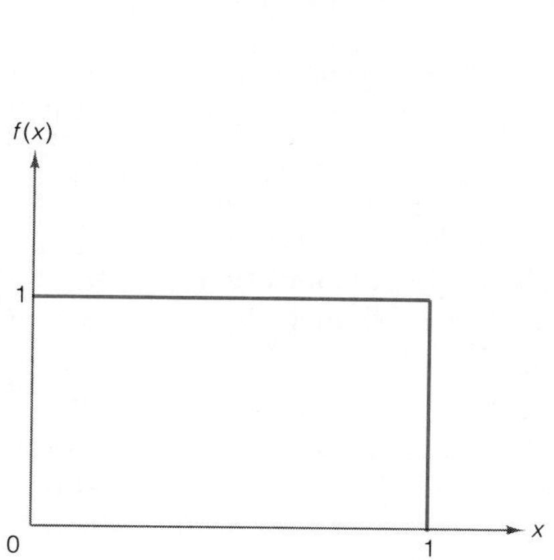

a. Relative frequency distribution of the sampled population

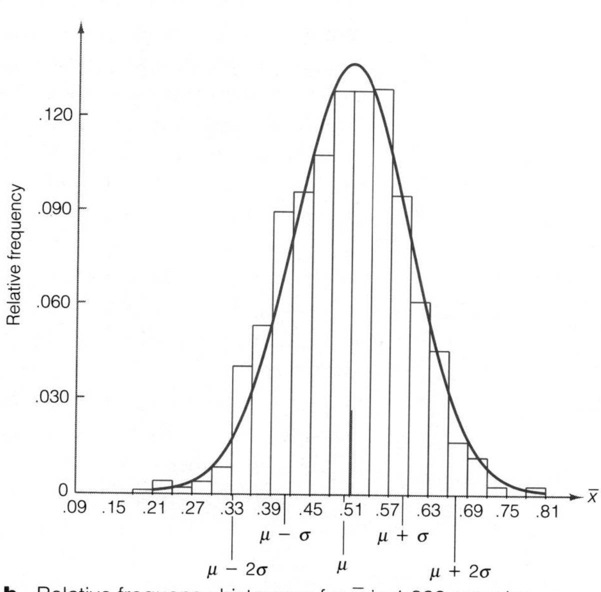

b. Relative frequency histogram for $\bar{x}$ in 1,000 samples of $n = 11$ measurements from a uniform distribution

FIGURE 6.8 ▲

What can be said about the shape of the sampling distribution of $\bar{x}$? Two important theorems provide this information.

Theorem 6.1

If a random sample of n observations is selected from a population with a normal distribution, the sampling distribution of $\bar{x}$ will be a normal distribution.

Theorem 6.2: The Central Limit Theorem

If a random sample of n observations is selected from a population (any population), then, when n is sufficiently large, the sampling distribution of $\bar{x}$ will be approximately a normal distribution. The larger the sample size, n, the better will be the normal approximation to the sampling distribution of $\bar{x}$.*

Thus, for sufficiently large samples, the sampling distribution of $\bar{x}$ will be approximately normal. How large must the sample size, n, be so that $\bar{x}$ has a normal sampling distribution? The answer depends on the shape of the relative frequency distribution of the sampled population, as shown in Figure 6.9.

Generally speaking, the greater the skewness of the distribution of the sampled population, the larger the sample size must be to obtain an adequate normal approximation to the sampling distribution of $\bar{x}$. But for some populations, particularly those with symmetric distributions, n may be fairly small and the sampling distribution of $\bar{x}$ will be approximately normal.

To demonstrate how small n can be and still achieve approximate normality for the sampling distribution of $\bar{x}$, recall Example 6.2, in which we generated 1,000 samples of $n = 11$ measurements from a uniform distribution. The relative frequency histogram for the 1,000 sample means is shown in Figure 6.8(b), and the normal probability distribution with a mean of .5 and a standard deviation of .09 is superimposed. You can see that this normal probability distribution approximates the computer-generated sampling distribution very well, even though the sample size is only $n = 11$.

The implications of the Central Limit Theorem are apparent by comparing the sampling distribution of $\bar{x}$ in Figure 6.8(b) with the distribution of the sampled population in Figure 6.8(a). In this situation, the population relative frequency distribution is not skewed and, as a result, a very small sample size proved large enough to apply the Central Limit Theorem. The sampling distribution of $\bar{x}$ is approximately normal, even though the distribution of the sampled population is decidedly nonnormal. You will also note that the mean of the sampling distribution is equal to the mean of the distribution of the sampled population, but that the variability of the sampling distribution is substantially less than the variability of the sampled population.

In most real-life applications, the shape of the population distribution will *not* be known. In such cases, we typically require $n \geq 30$ in order to invoke the Central Limit Theorem.

---

*Also, because of the Central Limit Theorem, the sum of a random sample of n observations, $\sum_{i=1}^{n} x_i$, will have a sampling distribution that will be approximately normal for large samples. This distribution will have a mean equal to $n\mu$ and a variance equal to $n\sigma^2$.

FIGURE 6.9 ► Sampling distribution of $\bar{x}$ for different populations and different sample sizes

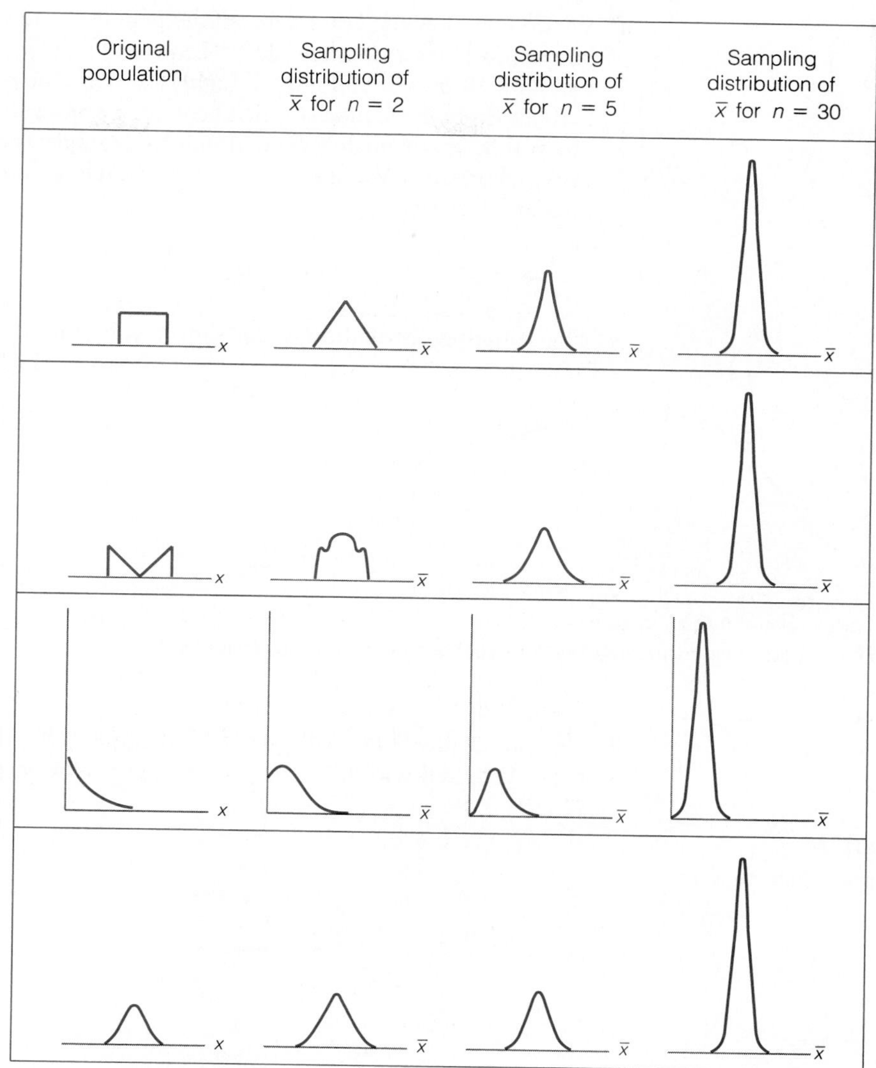

EXAMPLE 6.5

Suppose we have selected a random sample of $n = 25$ observations from a population with mean equal to 80 and standard deviation equal to 5. It is known that the population is not extremely skewed.

a. Sketch the relative frequency distributions for the population and for the sampling distribution of the sample mean, $\bar{x}$.

b. Find the probability that $\bar{x}$ will be larger than 82.

Solution

a. We do not know the exact shape of the population relative frequency distribution, but we do know that it should be centered about $\mu = 80$, its spread should be measured by $\sigma = 5$, and it is not highly skewed. One possibility is shown in Figure 6.10(a). From the Central Limit Theorem, we know that the sampling distribution of $\bar{x}$ will be approximately normal since the sampled population distribution is not extremely skewed. We also know that the sampling distribution will have mean and standard deviation

$$\mu_{\bar{x}} = \mu = 80 \qquad \text{and} \qquad \sigma_{\bar{x}} = \frac{\sigma}{\sqrt{n}} = \frac{5}{\sqrt{25}} = 1$$

The sampling distribution of $\bar{x}$ is shown in Figure 6.10(b).

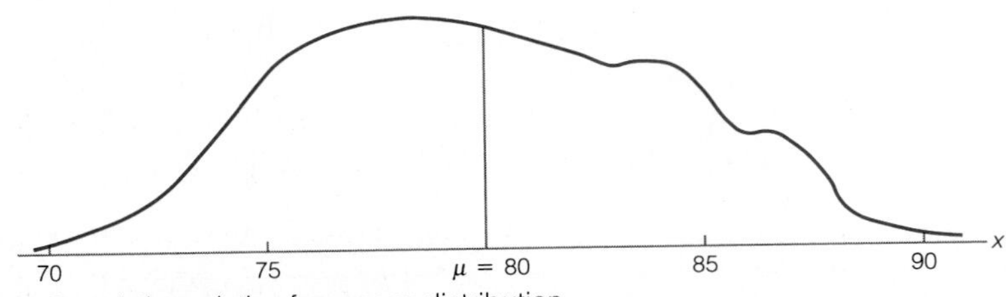

a. Population relative frequency distribution

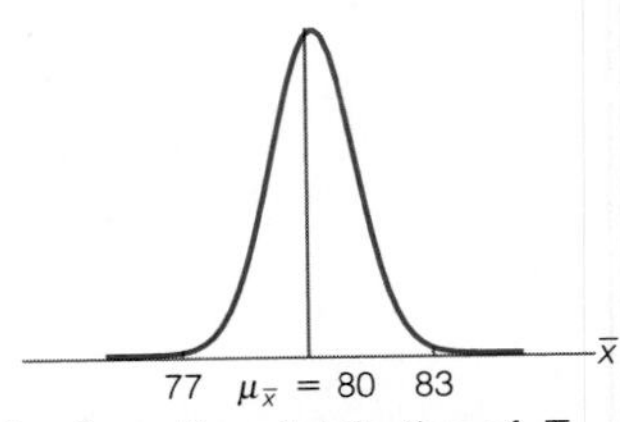

b. Sampling distribution of $\bar{x}$

FIGURE 6.10 ▲ A population relative frequency distribution and the sampling distribution for $\bar{x}$

b. The probability that $\bar{x}$ will exceed 82 is equal to the darker shaded area in Figure 6.11. To find this area, we need to find the z value corresponding to $\bar{x} = 82$. Recall

FIGURE 6.11 ▶
The sampling distribution of $\bar{x}$

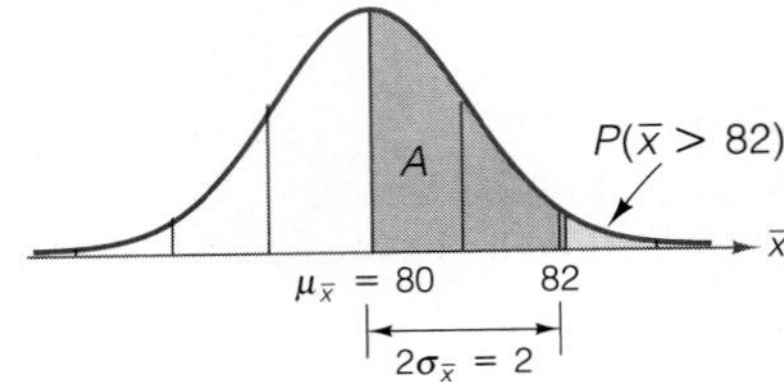

that the standard normal random variable z is the difference between any normally distributed random variable and its mean, expressed in units of its standard deviation. Since $\bar{x}$ is a normally distributed random variable with mean $\mu_{\bar{x}} = \mu$ and standard deviation $\sigma_{\bar{x}} = \sigma/\sqrt{n}$, it follows that the standard normal z value corresponding to the sample mean, $\bar{x}$, is

$$z = \frac{\text{(Normal random variable)} - \text{(Mean)}}{\text{Standard deviation}} = \frac{\bar{x} - \mu_{\bar{x}}}{\sigma_{\bar{x}}}$$

Therefore, for $\bar{x} = 82$, we have

$$z = \frac{\bar{x} - \mu_{\bar{x}}}{\sigma_{\bar{x}}} = \frac{82 - 80}{1} = 2$$

The area A in Figure 6.11 corresponding to $z = 2$ is given in the table of areas under the normal curve (see Table IV of Appendix B) as .4772. Therefore, the tail area corresponding to the probability that $\bar{x}$ exceeds 82 is

$$P(\bar{x} > 82) = P(z > 2) = .5 - .4772 = .0228$$

EXAMPLE 6.6

A manufacturer of automobile batteries claims that the distribution of the lifetimes of its best battery has a mean of 54 months and a standard deviation of 6 months. Suppose a consumer group decides to check the claim by purchasing a sample of 50 of these batteries and subjecting them to tests that determine their lifetimes.

a. Assuming the manufacturer's claim is true, describe the sampling distribution of the mean lifetime of a sample of 50 batteries.

b. Assuming the manufacturer's claim is true, what is the probability that the consumer group's sample has a mean lifetime of 52 months or less?

Solution

a. Even though we have no information about the shape of the probability distribution of the lifetimes of the batteries, we can use the Central Limit Theorem to deduce that the sampling distribution for a sample mean lifetime of 50 batteries is approximately normally distributed. Furthermore, the mean of this sampling distribution is the same as the mean of the sampled population, which is $\mu = 54$ months, according to the manufacturer's claim. Finally, the standard deviation of the sampling distribution is given by

$$\sigma_{\bar{x}} = \frac{\sigma}{\sqrt{n}} = \frac{6}{\sqrt{50}} = .85 \text{ month}$$

Thus, if we assume the claim is true, the sampling distribution of the mean lifetime of the 50 batteries is approximately normal with mean 54 months and standard deviation .85 month. The sampling distribution is shown in Figure 6.12.

FIGURE 6.12 ▶ Sampling distribution of $\bar{x}$ in Example 6.6

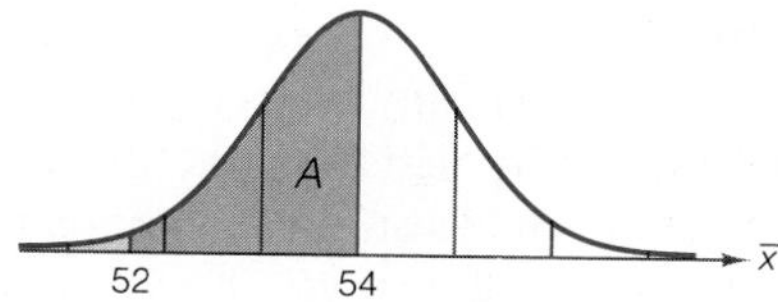

b. If the manufacturer's claim is true, the probability that the consumer group observes a mean battery lifetime of 52 months or less for their sample of 50 batteries, $P(\bar{x} \leq 52)$, is equivalent to the darker shaded area in Figure 6.12. Since the sampling distribution is approximately normal, we can find this area by computing the standard normal z value:

$$z = \frac{\bar{x} - \mu_{\bar{x}}}{\sigma_{\bar{x}}} = \frac{\bar{x} - \mu}{\sigma_{\bar{x}}} = \frac{52 - 54}{.85} = -2.35$$

where $\mu_{\bar{x}}$, the mean of the sampling distribution of $\bar{x}$, is equal to μ, the mean of the sampled population.

The area A shown in Figure 6.12 between $\bar{x} = 52$ and 54 is found in Table IV of Appendix B to be .4906, so the area to the left of 52 is

$$P(\bar{x} \leq 52) = .5 - A = .5 - .4906 = .0094$$

Thus, the probability that the consumer group will observe a sample mean of 52 or less is only .0094 if the manufacturer's claim is true. If the 50 tested batteries do result in a mean of 52 months or less, the consumer group will have strong evidence that the manufacturer's claim is untrue, because such an event is very unlikely to occur if the claim is true. (This is still another application of the rare event approach to statistical inference.)

In addition to providing a very useful approximation for the sampling distribution of a sample mean, the Central Limit Theorem offers an explanation for the fact that many relative frequency distributions of data are mound-shaped. Many of the macroscopic measurements we take in business research are really means or sums of many microscopic phenomena. For example, a company's sales for one year is the total of the many individual sales the company made during the year. Thus, the year's sales for a sample of similar companies may have a mound-shaped relative frequency distribution. Similarly, the length of time a construction company takes to complete a house might be viewed as the total of the time each of the large number of distinct jobs necessary to build the house takes to complete. The monthly profit of a firm can be viewed as the sum of the profits of all the transactions of the firm for that month. If we adopt viewpoints like these, the Central Limit Theorem offers some explanation for the frequent occurrence of mound-shaped distributions in nature.

CASE STUDY 6.1 / Evaluating the Condition of Rental Cars

In the summer of 1986, National Car Rental Systems, Inc., commissioned USAC Properties, Inc. [the performance testing/endorsement arm of the United States Automobile Club (USAC)] to conduct a survey of the general condition of the cars rented to the public by Hertz, Avis, National, and Budget Rent-a-Car.* National was interested in comparing the conditions of the cars it rented with those of the other leading car-rental companies.

Teams of USAC officials would evaluate each company's cars on appearance and cleanliness, accessory performance, mechanical functions, and vehicle safety using a demerit point system designed specifically for this survey. Each car would start with a perfect score of 0 points and would incur demerit points for each discrepancy noted by the inspectors. The number of demerits associated with a discrepancy would be based on the seriousness of the discrepancy.

*Information by personal communication with Rajiv Tandon, Corporate Vice President and General Manager of the Car Rental Division, National Car Rental Systems, Inc., Minneapolis, Minn.

If all cars in each company's fleet could be inspected and graded, one measure of the overall condition of a company's cars would be the mean of all scores received by the company, i.e., the company's *fleet mean score*. Such a census, however, besides being virtually impossible to conduct logistically, would be prohibitively expensive. It was therefore decided that the fleet mean score would have to be estimated. Accordingly, 10 major airports were randomly selected, and 10 cars from each company were randomly rented for inspection from each airport by USAC officials; i.e., a sample of size $n = 100$ cars from each company's fleet was drawn and inspected. In the analysis of USAC's inspection results, each company's mean score was used to estimate the company's unknown fleet mean score. (The use of a sample mean to estimate a population mean will be discussed in detail in Chapter 7). As we have seen in this chapter, $\bar{x}$ is a random variable with a sampling distribution that has a mean equal to the mean of the population from which the sample was drawn. Thus, in the context of this case study, the mean of the sampling distribution of $\bar{x}$ is the unknown fleet mean score. Since the sample size used by USAC was 100, the statisticians who evaluated USAC's inspection results were able to invoke the Central Limit Theorem and assume the sampling distribution of $\bar{x}$ to be approximately normally distributed. This assumption enabled comparisons of fleet mean scores for Hertz, Avis, National, and Budget Rent-a-Car to be made using the conventional large-sample testing procedures that are presented in Chapters 8 and 9.

CASE STUDY 6.2 / Reducing Investment Risk Through Diversification

In Case Study 4.2, it was noted that the variance of the monthly rate of return of a security is used by many investors as a measure of the risk or uncertainty involved in investing in the security. In this case study, we demonstrate that an investor can reduce investment risk by investing in more than one security—that is, by *diversifying* investments.

A number of studies by financial analysts have shown that the total risk (total variation) of a stock, as measured by the variance of the stock's rates of return over time, is comprised of two components: **systematic risk** and **unsystematic risk**. Systematic risk (systematic variability) is the portion of total risk caused by factors that simultaneously influence the prices of all stocks. Examples of such factors are changes in federal economic policies and changes in the national political climate. These factors explain why the prices of all stocks tend to move together over time (i.e., generally upward or generally downward). Unsystematic risk (unsystematic variability) is the portion of the total risk of a particular stock due to factors that influence the firm in question but generally do not influence other firms. Examples of such factors are labor strikes, management errors, and lawsuits (Francis, 1986). Although the proportions of systematic and unsystematic risk vary from firm to firm, it has been determined that, for many of the stocks listed on the New York Stock Exchange, systematic risk comprises about 25% and unsystematic risk comprises about 75% of the stock's total risk (Blume, 1971). As is demonstrated in this case study, an investor can use diversification to reduce the unsystematic portion of the total investment risk.

Suppose an investor is considering investing a total of \$5,000 in one or more of five different stocks. We denote the monthly rates of return of these stocks by r_1, r_2, r_3, r_4, and r_5. For simplicity, we assume these monthly returns are independent and identically distributed random variables with mean $\mu = 10\%$ and standard deviation $\sigma = 4\%$. Suppose the investor has narrowed the choice to two options: (1) invest \$5,000 in stock 1 or (2) invest \$1,000 in each of the five stocks.

Under the first option, the investor's monthly rate of return is r_1. Under the second option, since equal amounts of money were invested in each stock, the investor's monthly rate of return is $\bar{r} = \sum_{i=1}^{5} r_i/5$. If the first option is chosen, the investor's expected monthly rate of return, $E(r_i)$, is $\mu = 10\%$ and the risk, as measured by the variance of the stock's rate of return, is

$\sigma^2 = 16$. If the second option is chosen, the investor's expected monthly rate of return, $E(\bar{r})$, can be shown to be the same as in the first option. However, since the numerator of $\bar{r}$ is the sum of $n = 5$ independent and identically distributed random variables, each with mean μ and variance σ^2, the variance of $\bar{r}$ is $\sigma_{\bar{r}}^2 = \sigma^2/n$. Accordingly, the risk faced by the investor is $16/5 = 3.2$ and is lower than under the first option. Thus, the uncertainty faced by the investor is lower if the investor diversifies, rather than putting "all the eggs in one basket." For a more detailed discussion of risk reduction through diversification, see Sharpe and Alexander (1990) and Elton and Gruber (1987), or Alexander and Francis (1986).

6.4 The Relationship Between Sample Size and a Sampling Distribution

Suppose you draw two random samples from a population—one sample containing $n = 5$ observations and the second containing $n = 10$—and you want to compute $\bar{x}$ for each sample and use these statistics to estimate the population mean, μ. Intuitively, it would seem that the $\bar{x}$ based on the sample of 10 measurements would contain more information about μ than the $\bar{x}$ based on five measurements. But how is this larger sample size reflected in the sampling distribution of a statistic?

For the statistics you will encounter in this text, the variance of a statistic's sampling distribution will be inversely proportional to the sample size.* Or, since the standard deviation of $\bar{x}$ is equal to $\sigma/\sqrt{n}$, you can say that the standard deviation of the sampling distribution is proportional to $1/\sqrt{n}$. So, to reduce the standard deviation of the sampling distribution of a statistic to ½ its original value you will need 4 times as many observations in your sample ($1/\sqrt{n} = 1/\sqrt{4} = ½$). Or to reduce the standard deviation to ⅓ its original value, you will need 9 times as many observations.

The sampling distributions for the sample mean, $\bar{x}$, based on random samples from a normally distributed population, are shown in Figure 6.13 for $n = 1$, 4, and

FIGURE 6.13 ▶ Three sampling distributions of $\bar{x}$

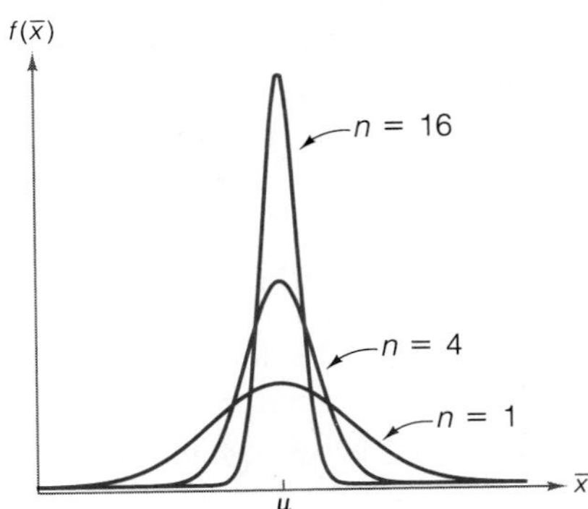

*Note that this is not true of all statistics, but it is true for most.

16 observations. The curve for $n = 1$ represents the probability distribution of the population. Those for $n = 4$ and $n = 16$ are sampling distributions of $\bar{x}$. Note how the distributions contract (variation decreases) as n increases from 1 to 4 to 16. The standard deviation of $\bar{x}$ based on $n = 16$ measurements is half the corresponding standard deviation of the distribution based on $n = 4$ measurements.

EXAMPLE 6.7

Consider the **Bernoulli** random variable x that can assume the values 0 and 1 with probabilities p and $q = (1 - p)$, respectively. The distribution is summarized in Table 6.3, where we have associated the term *Success* with the outcome $x = 1$, and *Failure* with the outcome $x = 0$. The Bernoulli random variable is just a binomial random variable with $n = 1$ trial.

TABLE 6.3 Bernoulli Distribution

Outcome	x	$p(x)$
Failure	0	q
Success	1	p

Suppose a random sample of n measurements is drawn from this Bernoulli distribution, and the sample mean $\bar{x}$ is calculated. Note that

$$\bar{x} = \frac{\sum_{i=1}^{n} x_i}{n}$$

and that $\sum_{i=1}^{n} x_i$ is the total number of successes (the number of 1's in the sample) in the random sample of n measurements (or *trials*). This sum is therefore a binomial random variable, with n trials and probability of success p. For example, the Bernoulli random variable might be the status of a single computer microchip (nondefective or defective), x the number of n such chips that are good, and $\bar{x}$ the fraction of nondefective (good) chips in a set of n.

a. Find the mean and standard deviation of the sampling distribution of $\bar{x}$.
b. Simulate the distribution of $\bar{x}$ using $p = .8$, and $n = 1$, 10, 25, and 100 by generating 1,000 samples for each sample size and creating a histogram of the 1,000 sample means.

Solution

a. The mean and variance of the Bernoulli random variable are

$$\mu = E(x) = 0(q) + 1(p) = p$$
$$\sigma^2 = E[(x - \mu)^2] = (0 - p)^2(q) + (1 - p)^2(p)$$
$$= p^2q + q^2p = pq(p + q) = pq$$

Note that these are the mean and standard deviation of a binomial random variable with $n = 1$, which is another description of a Bernoulli random variable.

We know that the mean $\bar{x}$ of a random sample is unbiased, so that

$$E(\bar{x}) = \mu = p$$

and the standard error is

$$\sigma_{\bar{x}} = \frac{\sigma}{\sqrt{n}} = \frac{\sqrt{pq}}{\sqrt{n}} = \sqrt{\frac{pq}{n}}$$

Because $\bar{x}$ provides an unbiased estimate of the probability of success p and has a standard error that decreases as the sample size increases, we use it to estimate p in subsequent chapters, where we refer to it as the sample fraction of successes, $\hat{p}$ (read "p hat"). We refer to its standard error as $\sigma_{\hat{p}}$.

b. The mean and standard error of $\bar{x}$ when $p = .8$ are

$$\mu_{\bar{x}} = p = .8$$

$$\sigma_{\bar{x}} = \sqrt{\frac{pq}{n}} = \sqrt{\frac{(.8)(.2)}{n}} = \frac{.4}{\sqrt{n}}$$

TABLE 6.4 Mean and Standard Error of $\bar{x}$ for Bernoulli Random Variable with $p = .8$

n	$\mu_{\bar{x}}$	$\sigma_{\bar{x}}$
1	.8	$\frac{.4}{\sqrt{1}} = .400$
10	.8	$\frac{.4}{\sqrt{10}} = .1265$
25	.8	$\frac{.4}{\sqrt{25}} = .0800$
100	.8	$\frac{.4}{\sqrt{100}} = .0400$

You can see in Table 6.4 how the standard error of $\bar{x}$ decreases as n increases.

The simulation of 1,000 sample means with each of the sample sizes given in Table 6.4 resulted in the sampling distributions shown in Figure 6.14. A relative frequency histogram is used to display each sampling distribution. Note that each sampling distribution more closely resembles a normal distribution than the previous one. As the Central Limit Theorem promises, the distribution of $\bar{x}$ becomes approximately normal for large n, and the approximation improves as n increases. Note too that the values of $\bar{x}$ cluster more closely around their mean (.8) as n is increased. We will make use of these properties to estimate p for binomial random variables in Chapter 7.

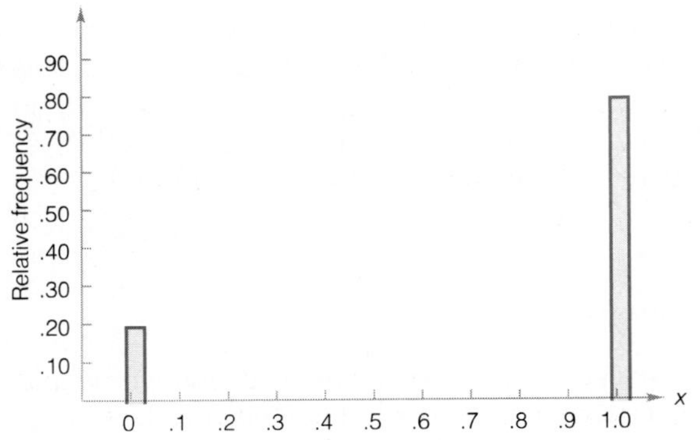

a. $n = 1$

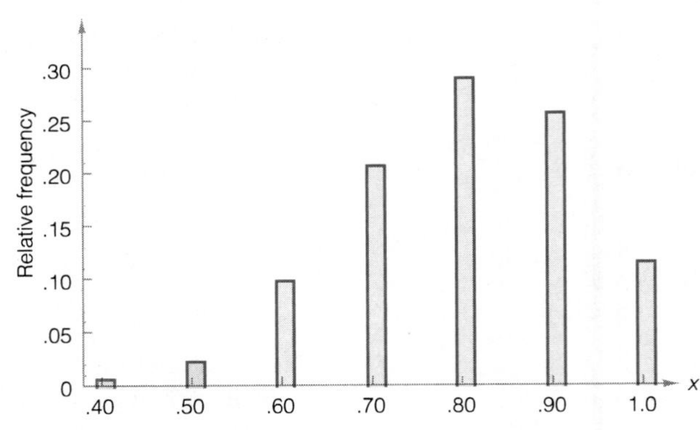

b. $n = 10$

FIGURE 6.14 ▲ Sampling distributions for Bernoulli sample means

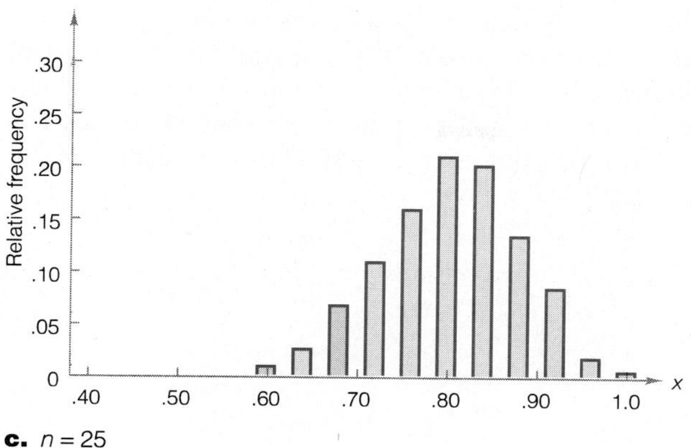

c. $n = 25$

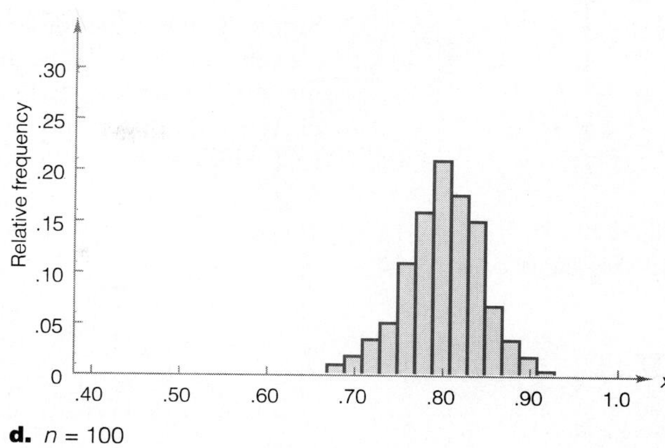

d. $n = 100$

EXAMPLE 6.8

Refer to Example 6.7. Suppose $n = 400$ microchips are to be sampled from a very large batch, of which a proportion p are good. The proportion of good chips in the sample will be determined by assigning a 1 to each good chip, a 0 to each defective chip, and calculating the sample mean of the 400 Bernoulli observations. Assuming that the approximate value of p is .8, what is the probability that $\bar{x}$ will fall within .03 of the exact value of p?

Solution

We first note (see Example 6.7) that the mean and standard deviation of $\bar{x}$ are

$$E(\bar{x}) = p \qquad \text{and} \qquad \sigma_{\bar{x}} = \sqrt{\frac{pq}{n}} = \sqrt{\frac{pq}{400}}$$

Using the approximate value of p to obtain an approximation for the standard deviation of the sampling distribution of $\bar{x}$, we find

$$\sigma_{\bar{x}} \approx \sqrt{\frac{(.8)(.2)}{400}} = .02$$

Next, the Central Limit Theorem implies that the distribution of $\bar{x}$ based on a sample of size 400 is approximately normal. The properties are summarized in Figure 6.15 on page 292. Note that the mean of the sampling distribution is p, so that the standard normal z value is given by the formula

$$z = \frac{\bar{x} - \mu_{\bar{x}}}{\sigma_{\bar{x}}} = \frac{\bar{x} - p}{.02}$$

Even though the value of p is unknown, we can find the probability that $\bar{x}$ is within .03 of p by first calculating the z value corresponding to the $\bar{x}$ value that is .03 greater than p:

$$z = \frac{(p + .03) - p}{.02} = \frac{.03}{.02} = 1.5$$

Similarly, the value .03 less than p is 1.5 standard deviations *below* the mean and has a z value of -1.5. Thus, the event that $\bar{x}$ falls within .03 of p is equivalent to the event that the standard normal random variable z is between -1.5 and 1.5. Using Table IV of Appendix B, we find the probability that $\bar{x}$ falls within 1.5 standard deviations of p is $2(.4332) = .8664$, the area shown shaded in Figure 6.15. The interpretation of this

FIGURE 6.15 ▶
Sampling distribution of $\bar{x}$

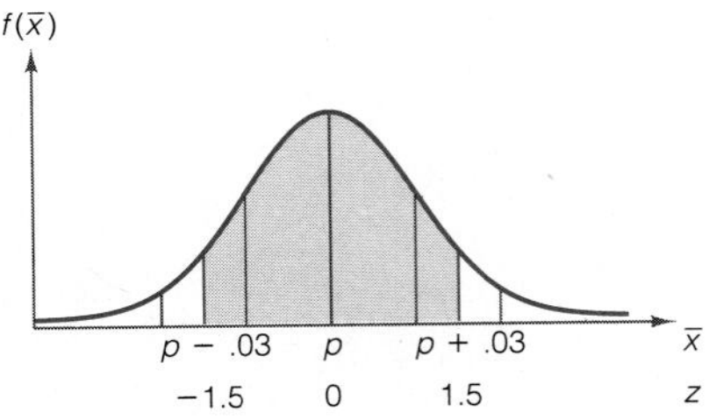

probability is that there is about an 87% chance that the proportion of good chips in the sample of 400 will fall within .03 of the exact proportion of good chips in the entire batch.

For most sampling distributions, the standard deviation of the distribution decreases as the sample size increases. We use this result in Chapters 7, 8, and 9 to help us determine the sample size needed to obtain a specified accuracy of estimation.

Summary

Many practical business problems require that an inference be made about some population **parameter** (for example, a mean, standard deviation, etc.). If we want to make this inference on the basis of sample information, we need to compute a **sample statistic** that contains information about the population parameter. The amount of such information is reflected in the sample statistic's **sampling distribution**—that is, the probability distribution of the sample statistic. The sampling distribution describes the behavior of the statistic in repeated sampling. We want a sample statistic that is an **unbiased** estimator of the population parameter and has a smaller variance than any other unbiased sample statistic.

When the population parameter of interest is the mean, μ, the sample mean, $\bar{x}$, provides an unbiased estimator with a standard deviation of $\sigma/\sqrt{n}$. In addition, the **Central Limit Theorem** assures us that the sampling distribution of the mean of a large sample will be approximately normally distributed, no matter what the shape of the relative frequency distribution of the sampled population.

The amount of information in a sample that is relevant to some population parameter is related to the sample size. Another way of saying this is that as n gets larger, the standard deviations of most sample statistics get smaller.

The sampling distributions for all the many statistics that can be computed from sample data could be discussed in detail, but this would delay discussion of the practical objective of this course—the role of statistical inference in business decision-making. Consequently, we will comment further on the sampling distributions of particular statistics when we use them as estimators in the following chapters.

Exercises 6.1–6.32

Note: Starred (*) exercises require the use of a computer.

Learning the Mechanics

6.1 Consider a sample statistic, A. As with all sample statistics, A is computed by utilizing a specified function (formula) of the sample measurements. (For example, if A were the sample mean, the specified formula would be to sum the measurements and divide by the number of measurements.)

a. Describe what we mean by the phrase "the sampling distribution of the sample statistic A."

b. Suppose A is to be used to estimate a population parameter α. Explain what is meant by the assertion: A is an unbiased estimator of α.

c. Consider another sample statistic, B. Assume that B is also an unbiased estimator of the population parameter α. How can we use the sampling distributions of A and B to decide which is the better estimator of α?

d. If the sample sizes on which A and B are based are large, can we apply the Central Limit Theorem and assert that the sampling distributions of A and B are approximately normal? Why or why not?

6.2 The standard deviation (or, as it is frequently called, the standard error) of the sampling distribution for the sample mean, $\bar{x}$, is equal to the standard deviation of the population from which the sample was selected, divided by the square root of the sample size. That is,

$$\sigma_{\bar{x}} = \frac{\sigma}{\sqrt{n}}$$

a. As the sample size is increased, what happens to the standard error of $\bar{x}$? Why is this property important?

b. Suppose that a sample statistic has a standard error that is not a function of the sample size. In other words, the standard error remains constant as n changes. What would this imply about the statistic as an estimator of a population parameter?

c. Suppose another unbiased estimator (call it A) of the population mean is a sample statistic with a standard error equal to

$$\sigma_{A} = \frac{\sigma}{\sqrt[3]{n}}$$

Which of the sample statistics, $\bar{x}$ or A, is preferable as an estimator of the population mean? Why?

d. Suppose that the population standard deviation σ is equal to 10 and that the sample size is 64. Calculate the standard errors of $\bar{x}$ and A. Assuming that the sampling distribution A is approximately normal, interpret the standard errors. Why is the assumption of (approximate) normality unnecessary for the sampling distribution of $\bar{x}$?

6.3 In each of the following cases, find the mean and standard deviation of the sampling distribution of the sample mean, $\bar{x}$, for a random sample of size n drawn from a population with mean μ and standard deviation σ:

a. $n = 16, \quad \mu = 8, \quad \sigma = 3$
b. $n = 15, \quad \mu = 200, \quad \sigma = 15$
c. $n = 100, \quad \mu = 50, \quad \sigma = 4$
d. $n = 110, \quad \mu = -10, \quad \sigma = 17$

6.4 A random sample of $n = 36$ observations is drawn from a normal population with mean equal to 10 and standard deviation equal to 12.

a. Give the mean and standard deviation of the (repeated) sampling distribution of $\bar{x}$.
b. Describe the shape of the sampling distribution of $\bar{x}$. Does your answer depend on the sample size?
c. Calculate the z-score corresponding to a value of $\bar{x} = 13.4$.
d. Calculate the z-score corresponding to $\bar{x} = 8.3$.

6.5 Refer to Exercise 6.4. Find the probability that:

a. $\bar{x}$ is larger than 11
b. $\bar{x}$ is less than 11
c. $\bar{x}$ is larger than 13.1
d. $\bar{x}$ falls between 9 and 11
e. $\bar{x}$ is less than 6.7

6.6 A random sample of 64 observations is to be drawn from a large population with mean 500 and standard deviation 80. Find the probability that:

a. $\bar{x} \leq 500$
b. $\bar{x} < 500$
c. $\bar{x} < 510$
d. $\bar{x} > 480$
e. $\bar{x} < 491.9$
f. $487 \leq \bar{x} \leq 525.3$
g. $488 \leq \bar{x} \leq 495.2$
h. $501.7 \leq \bar{x} \leq 511.7$

6.7 A random sample of 40 observations is to be drawn from a large population of measurements. It is known that 30% of the measurements in the population are 1's, 20% of the measurements are 2's, 20% are 3's, and 30% are 4's.

a. Give the mean and standard deviation of the sampling distribution of $\bar{x}$, the sample mean of the 40 observations.
b. Describe the sampling distribution of $\bar{x}$. Does your answer depend on the sample size?

***6.8** Use a statistical software package to generate 100 random samples of size $n = 2$ from a population characterized by a uniform probability distribution (Section 5.2) with $c = 0$ and $d = 10$. Compute $\bar{x}$ for each sample and plot a frequency distribution for the 100 values of $\bar{x}$. Repeat this process for $n = 5$, 10, 30, and 50. Explain how your plots illustrate the Central Limit Theorem.

***6.9** Use a statistical software package to generate 100 random samples of size $n = 2$ from a population characterized by a normal probability distribution with mean 100 and standard deviation 10. Compute $\bar{x}$ for each sample and plot a frequency distribution for the 100 values of $\bar{x}$. Repeat this process for $n = 5$, 10, 30, and 50. Explain how your plots illustrate the Central Limit Theorem.

6.10 Suppose a sample of $n = 50$ items is drawn from a population of manufactured products and the weight, x, of each item is recorded. Prior experience has shown that the weight has a probability distribution with $\mu = 6$ ounces and $\sigma = 2.5$ ounces. Then $\bar{x}$, the sample mean, will be approximately normally distributed (because of the Central Limit Theorem).

a. Calculate $\mu_{\bar{x}}$ and $\sigma_{\bar{x}}$.
b. What is the probability that the manufacturer's sample has a mean weight of between 5.75 and 6.25 ounces?
c. What is the probability that the manufacturer's sample has a mean weight of less than 5.5 ounces?
d. How would the sampling distribution of $\bar{x}$ change if the sample size, n, were increased from 50 to 100?

6.11 A population consists of four numbers, 1, 2, 2, and 3, marked on poker chips.

a. How many different samples of $n = 2$ chips could be selected (without replacement) from the population? List the possible samples.

b. Give the probability of selecting any one of these samples (assume that the sampling is random).

c. Calculate $\bar{x}$ for each of the samples in part **b**.

d. Calculate the probability associated with each of the possible values of $\bar{x}$.

e. Graph the population probability distribution and the sampling distribution of $\bar{x}$ (obtained in part **d**).

6.12 Suppose x equals the number of heads observed when a single coin is tossed (i.e., $x = 0$ or $x = 1$). The population corresponding to x is the set of 0's and 1's generated when the coin is tossed repeatedly a large number of times. Suppose we select $n = 3$ observations from this population (i.e., we toss the coin three times and observe the three values of x).

a. List the four different samples (combinations of 0's and 1's) that could be obtained.

b. Calculate the value of $\bar{x}$ for each of the samples.

c. List the values that $\bar{x}$ can assume, and find the probabilities of observing these values.

d. Construct a graph of the sampling distribution of $\bar{x}$.

6.13 A random sample of size n is to be drawn from a large population with mean 100 and standard deviation 10, and the sample mean, $\bar{x}$, is to be calculated. To see the effect of different sample sizes on the standard deviation of the sampling distribution of $\bar{x}$, plot $\sigma/\sqrt{n}$ against n for $n = 1, 5, 10, 20, 30, 40$, and 50.

6.14 Refer to Exercise 6.13. If you increase the sample size from $n = 5$ to $n = 25$, does the information in the sample mean, $\bar{x}$, pertinent to μ increase by the same amount as it does for an increase in sample size from $n = 30$ to $n = 50$? How is this answer shown in the graph you constructed for Exercise 6.13?

6.15 A random sample of size $n = 30$ is to be drawn from a large population with $\mu = 500$ and $\sigma = 200$.

a. What is the standard deviation of the sampling distribution of $\bar{x}$?

b. In order to reduce the standard deviation of $\bar{x}$ to 50% of the value in part **a**, how much larger would n need to be?

c. In order to reduce $\sigma_{\bar{x}}$ to 75% of the value in part **a**, how much larger would n need to be?

Applying the Concepts

6.16 A soft-drink bottler requires bottles with an internal pressure strength of at least 150 pounds per square inch (psi). A prospective bottle vendor claims that its production process yields bottles with a mean internal strength of 157 psi and a standard deviation of 3 psi. As part of its vendor surveillance, the bottler strikes an agreement with the vendor that permits the bottler to sample from the vendor's production process to verify the vendor's claim. The bottler randomly selected 40 bottles from the last 10,000 produced, measured the internal pressure strength of each, and found the mean strength for the sample to be 1.3 psi below the process mean cited by the vendor.

a. Assuming the vendor's claim to be true, what is the probability of obtaining a sample mean 1.3 psi or more below the process mean? What does your answer suggest about the validity of the vendor's claim?

b. If the process standard deviation were 3 psi (as claimed by the vendor), but the mean were 156 psi, would the observed sample result be more or less likely than in part **a**? What if the mean were 158 psi?

c. If the process mean were 157 psi as claimed, but the process standard deviation were 2 psi, would the observed sample result be more or less likely than in part **a**? What if the standard deviation were 6 psi?

6.17 Steven Hartley (1983) conducted an experiment to evaluate the forecasting skills of 140 retail buyers of two large midwestern retail organizations. In one part of the experiment, 61 of the buyers were given historical sales data for the previous 30 months and asked to forecast sales 6 months from now. For each buyer, Hartley calculated the difference between the actual number of units sold 6 months later and the buyer's forecast. This difference is sometimes called **forecast error** and is denoted here as x. In order to characterize the accuracy of this group of 61 buyers, Hartley calculated $\bar{x}$, the mean forecast error for the sample.

a. Assume the sample of 61 buyers was randomly selected from a large population of buyers whose forecast errors have a distribution with mean 10 and standard deviation 16. Describe the sampling distribution of $\bar{x}$.

b. Hartley found $\bar{x} = 13.49$. Given the information in part **a** about the population of buyers, is this a likely result? Explain.

c. Suppose $\bar{x} = 0$ and $s^2 = 1$. What do these statistics tell you about the accuracy of the forecasts of the 61 buyers?

6.18 In Case Study 6.2, an individual was considering investing \$1,000 in each of $n = 5$ different stocks. The monthly rate of return on each stock had mean $\mu = 10\%$ and standard deviation $\sigma = 4\%$. The investor's monthly rate of return for the portfolio of five stocks was $\bar{r} = \sum r_i/5$. It was shown that the variance of the investor's monthly rate of return was $\sigma_{\bar{r}}^2 = \sigma^2/n = 3.2$ and that this number is a measure of the risk faced by the investor.

a. If instead the individual were to invest \$1,000 in only three of the five stocks, would the risk faced by the investor increase or decrease? Explain.

b. Suppose \$1,000 was invested in each of 10 stocks with rate-of-return characteristics identical to those described above. Measure the risk faced by the investor and compare it to the risk associated with investing in just five of the stocks.

6.19 It was not until the 1950s that manufacturing companies began to measure and account for the costs of their quality functions. In general, quality costs are those associated with producing, identifying, avoiding, or repairing products that do not meet requirements. They are often divided into four categories: prevention costs, appraisal costs, internal failure costs, and external failure costs (Montgomery, 1991). This exercise is concerned with an inspection and evaluation activity that would generate appraisal costs.

A particular manufacturing process requires steel rods that are at least 3 meters in length. The rods are purchased in lots of 50,000. To determine whether the lot meets the required quality standards, 100 rods are randomly sampled from each incoming lot and the mean length of rods in the sample is calculated. The quality manager has decided to accept lots whose sample mean is 3.005 meters or more. Assume that the standard deviation of the rod lengths in a lot is .03 meter.

a. If in fact each lot has a mean length of 3 meters, what percentage of the lots received by the manufacturer will be returned to the vendor (i.e., the supplier)?

b. If in fact all of the rods in all of the lots received by the manufacturer are between 2.999 and 3.004 meters in length, what percentage of the lots will be returned to the vendor?

6.20 A local bank reported to the federal government that its 5,246 savings accounts have a mean balance of \$1,000 and a standard deviation of \$240. Government auditors have asked to randomly sample 64 of the bank's accounts in order to assess the reliability of the mean balance reported by the bank. The auditors say they will certify the bank's report only if the sample mean balance is within \$60 of the reported mean balance. What is the probability that the auditors will *not* certify the bank's report, even if the mean balance really is \$1,000? (Assume the standard deviation reported by the bank is accurate.)

6.21 A manufacturer produces safety jackets for competitive fencers. These jackets are rated by the minimum force, in newtons, that will allow a weapon to pierce the jacket. When this process is operating correctly, it produces jackets that have ratings with an average of 840 newtons and a standard deviation of 15 newtons. FIE, the international governing body for fencing, requires jackets to be rated at a minimum of 800 newtons. To check whether the process is operating correctly, a manager takes a sample of 50 jackets from the process, rates them, and calculates $\bar{x}$, the mean rating for jackets in the sample. She assumes that the standard deviation of the process is fixed, but is worried that the mean rating of the process may have changed.

a. What is the sampling distribution of $\bar{x}$ if the process is still operating correctly?

b. Suppose the manager's sample has a mean rating of 830 newtons. What is the probability of getting an $\bar{x}$ of 830 newtons or lower if the process is operating correctly?

c. Given the manager's assumption that the standard deviation of the process is fixed, what does your answer to part **b** suggest about the current state of the process (i.e., does it appear that the mean jacket rating is still 840 newtons)?

d. Now suppose that the mean of the process has not changed, but the standard deviation of the process has increased from 15 newtons to 45 newtons. What is the sampling distribution of $\bar{x}$ in this case? What is the probability of getting an $\bar{x}$ of 830 newtons or lower when $\bar{x}$ has this distribution?

6.22 A machine for filling cereal boxes can be set to dispense a mean weight of between 12 and 20 ounces. The standard deviation of the weight of boxes filled with the machine is 1 ounce and is independent of the mean. The product is shipped in cases of 36 boxes. Assume each case represents a random sample from the population of boxes filled by this machine at a fixed setting of the mean weight dispensed.

a. At what level should the mean be set so that 95% of all cases produced will have a mean weight of at least 16 ounces?

b. Repeat part **a**, this time assuming that the standard deviation of the weight of boxes filled by the machine is 2 ounces.

c. Repeat part **b**, but for cases of 64 boxes.

d. Do any of these answers depend on the distribution of weights for individual boxes filled by the machine? Why or why not?

6.23 To determine whether a metal lathe that produces machine bearings is properly adjusted, a sample of 36 bearings is collected and the diameter of each is measured. Assume the standard deviation of the diameter of the machine bearings is stable and equal to .001 inch. What is the probability that the mean diameter $\bar{x}$ of the sample will lie within .0001 inch of the process mean?

6.24 Refer to Exercise 6.23. Suppose the mean diameter of the bearings produced by the machine is supposed to be .5 inch. The company decides to use the sample mean (from Exercise 6.23) to decide whether the process is in control—i.e., whether it is producing bearings with a mean diameter of .5 inch. The machine will be considered out of control if the mean of the sample of $n = 36$ diameters is less than .4994 inch or larger than .5006 inch. If the true mean diameter of the bearings produced by the machine is .501 inch, what is the probability that the test will fail to imply that the process is out of control?

6.25 In 1981, Northern States Power (NSP), a private utility in Minnesota, mailed many of its customers an offer for an in-home energy audit designed to encourage households to conserve energy, but the response rate fell far short of expectations. To improve the response rate, Richard Weijo (1983) evaluated alternative means of making NSP customers aware of the audit program and of motivating them to make inexpensive energy-saving changes to their homes. As part of the study, two independent samples were selected from among the 9,900 customers scheduled to receive energy-audit offers by April 1982. The 280 households in the first sample received an inexpensive water-flow controller with their energy-audit offer, while the 348 households in the second sample received only the energy-audit offer. All of the households in both samples were surveyed by telephone and assigned a score from 0 to 5 based on how much they remembered about the energy-audit offers. To compare the effectiveness of the two types of mailings with respect to information recall, Weijo calculated $\bar{x}$ and $\bar{y}$ (the sample means of the scores for households receiving and not receiving the water-flow controller, respectively).

a. The scores received by the 280 households that were mailed the water-flow controller can be viewed as a sample from the population of scores that would exist if all 9,900 households had been sent the water-flow controller and then been interviewed. Suppose this hypothetical population of scores has standard deviation 1.2. Describe the sampling distribution of $\bar{x}$.

b. Suppose all 9,900 households had received only the energy-audit offer and were later scored with respect to their recall of information about the mailing. Assume that the standard deviation of this population of scores is 1.0, and describe the sampling distribution of $\bar{y}$.

c. How large would the sample of households receiving the water-flow controller have to be in order to have $\sigma_{\bar{x}} = \sigma_{\bar{y}}$?

6.26 The primary responsibility of government tax assessors is to estimate the market values of all properties in their respective jurisdictions. These estimates form the basis upon which property tax bills are determined. To estimate the mean market value for single-family homes in a particular county in 1994, a county tax assessor uses the mean sale price for all single-family homes that sold during 1994. Such a sample is typically treated by tax assessors as if it were a random sample from the population of properties in question (*Standard on Assessment-Ratio Studies*, 1980). Suppose 400 properties sold during 1994 and assume that the standard deviation of the market value for the population of properties is about $50,000.

a. What is the population in question?

b. What is the probability that the assessor's sample mean will overestimate the actual mean market value of the population?

c. What is the probability that the assessor's sample mean will fall within $4,000 of the actual mean market value for the population?

6.27 In any production process, some variation in the quality of the product is unavoidable. Variation in product can be divided into two categories: **variation due to special causes** and **variation due to common causes**. The former includes variation due to a specific worker or group of workers, a specific machine, or a specific local condition. The latter includes variation due to faults of the overall production system, such as poor working conditions for workers, use of raw materials of inferior quality, poor design of the product, etc. The discovery and removal of special causes is generally the responsibility of a person directly involved with the production operation in question. In contrast, common causes are the responsibility of management.

A production process in which all special causes of variation have been eliminated is said to be **stable** or **in statistical control**. The variation that remains is simply random variation. If the magnitude of the random variation is unacceptable to management, it can be reduced through the elimination of common causes (Deming, 1982, 1986).

It is common practice to monitor the variation in the quality characteristic of a product over time by plotting the quality characteristic on a **control chart**. For example, the amount of alkali in soap might be monitored each hour by randomly selecting from the production process and measuring the quantity of alkali in $n = 5$ test specimens of soap. The mean, $\bar{x}$, of the sample alkaline measurements would be plotted against time, as shown in the accompanying figure. If the process is in statistical control, $\bar{x}$ should assume a distribution with a mean equal to the process mean, μ, with standard deviation equal to the process standard deviation divided by the square root of the sample size, $\sigma_{\bar{x}} = \sigma/\sqrt{n}$. The control chart shown here includes a horizontal line to locate the process mean and two lines, called **control limits**, located $3\sigma_{\bar{x}}$ above and below μ. If $\bar{x}$ falls within the control limits, the process is deemed to be in control. If $\bar{x}$ is outside the limits, there is strong evidence that special causes of variation are present, and the process is deemed to be out of control. We discuss control charts in detail in Chapter 13.

When the production process is in statistical control, the percentage of alkali in a test specimen of soap follows approximately a normal distribution with $\mu = 2\%$ and $\sigma = 1\%$.

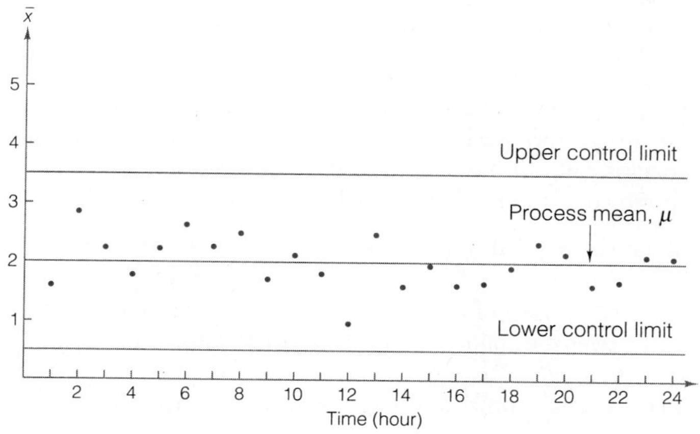

a. If $n = 4$, how far from μ should the upper and lower control limits be located?
b. If the process is in control, what is the probability that $\bar{x}$ will fall outside the control limits?
c. If the process mean shifts to $\mu = 3\%$ before the sample is drawn, what is the probability that the sample will lead to the (correct) conclusion that the process is out of control?

6.28 Refer to Exercise 6.27. The soap company has decided to set tighter control limits than the $3\sigma_{\bar{x}}$ limits described in Exercise 6.27. In particular, when the process is in control, the company is willing to risk a .10 probability that $\bar{x}$ falls outside the control limits.

a. The company still wants the control limits to be located at equal distances above and below the process mean and is still planning to use $n = 4$ measurements in each hourly sample. Where should the control limits be located?
b. Suppose the control limits developed in part **a** are implemented, but unknown to the company, μ is currently 3% (not 2%). What is the probability that $\bar{x}$ will fall outside the control limits if $n = 4$? If $n = 9$?

6.29 Refer to Exercise 6.28. To improve the sensitivity of control charts, **warning limits** are sometimes included on the chart along with the control limits. These limits are typically set at $\mu \pm 1.96\sigma_{\bar{x}}$. If two successive data points fall outside the warning limits, the process is deemed to be out of control (Wetherhill, 1977).

a. If the soap process is in control (i.e., it follows a normal distribution with $\mu = 2\%$ and $\sigma = 1\%$), what is the probability that the next value of $\bar{x}$ will fall outside the warning limits?
b. If the soap process is in control, how many of the next 40 values of $\bar{x}$ plotted on the control chart would be expected to fall above the upper warning limit?
c. If the soap process is in control, what is the probability that the next two values of $\bar{x}$ will fall below the lower warning limit?

6.30 A manufacturer of aluminum foil claims that its 75-foot roll has a mean length of 75.11 feet per roll and a standard deviation of .15 foot. To check this claim, a consumer group plans to purchase 36 of the company's 75-foot rolls, measure the length of each, and compute the sample mean length. Treat the sample as if it were a random sample from the population of rolls currently stocked by grocery stores.

a. Assuming the manufacturer's claim is true, describe the sampling distribution of the sample mean.

b. Assuming the manufacturer's claim is true, what is the probability that the sample mean will be less than 75 feet?

c. Suppose the sample mean actually equals 74.95 feet. Can this evidence be used to refute the manufacturer's claim? Explain.

6.31 A building contractor has decided to purchase a load of factory-reject aluminum siding if the average number of flaws per piece of siding in a sample of size 35 from the factory's reject pile is 2.1 or less. Suppose the number of flaws per piece of siding in the factory's reject pile has a Poisson probability distribution with a mean of 2.5. Find the probability that the contractor will not purchase a load of siding. [*Hint:* If x is a Poisson random variable with mean λ, then the variance of the random variable x is also equal to λ.]

6.32 The distribution of the number of loaves of bread sold per day by a large bakery over the past 5 years has a mean of 250 and a standard deviation of 45 loaves.

a. Describe the sampling distribution of the total number of loaves of bread sold in 30 randomly selected days. [*Hint:* See the footnote on page 282 that gives the application of the Central Limit Theorem to the sum of the measurements in a sample.]

b. What is the approximate probability that the total number of loaves sold in 30 randomly selected days is between 7,000 and 8,000?

c. What is the approximate probability that the total number of loaves sold in 30 randomly selected days is greater than 8,100 loaves?

Using the Computer

Calculate the mean and standard deviation for the 1,000 zip codes' median household incomes (Appendix C). We will treat these quantities as the population mean μ and standard deviation σ for this exercise.

a. Draw 100 random samples of $n = 20$ observations from the 1,000 zip codes' median household incomes. Select the samples with replacement, i.e., replace each measurement before selecting the next.* Calculate the 100 sample means. Generate a stem-and-leaf display or a relative frequency histogram for the 100 means. Then count the number of the 100 sample means that fall in the intervals $\mu \pm \sigma/\sqrt{n}$, $\mu \pm 2\sigma/\sqrt{n}$, and $\mu \pm 3\sigma/\sqrt{n}$. How do the graphical description and the percentage of means falling in the intervals agree with a normal distribution having mean μ and standard deviation $\sigma/\sqrt{n}$?

b. Repeat part **a** using a sample size of $n = 50$. Is the sampling distribution of the sample means closer to normal for the larger sample size? Why?

*In this and some future exercises we specify "sampling with replacement" to simulate the sampling from very large populations. This avoids the use of finite population correction factors (Chapter 20) for samples from (relatively) small populations.

References

Alexander, G. J. and Francis, J. C. *Portfolio Analysis*. Englewood Cliffs, N.J.: Prentice-Hall, Inc., 1986. Chapters 4 and 5.

Blume, M. "On the assessment of risk." *Journal of Finance*, Mar. 1971, *26*, pp. 1–10.

Deming, W. E. *Out of the Crisis*. Cambridge,Mass.: MIT Center for Advanced Engineering Study, 1986.

Deming, W. E. *Quality, Productivity, and Competitive Position*. Cambridge, Mass.: MIT Center for Advanced Engineering Study, 1982. Chapter 7.

Elton, E. J. and Gruber, M. J. *Modern Portfolio Theory and Investment Analysis*, 3rd ed. New York: Wiley, 1987.

Francis, J. C. *Investments: Analysis and Management*, 4th ed. New York: McGraw-Hill, 1986.

Hartley, S. W. *Judgmental Sales Forecasting: An Experimental Investigation of Task Structure and Environmental Complexity*. Unpublished Ph.D. dissertation, University of Minnesota, 1983.

Hogg, R. V. and Craig, A. T. *Introduction to Mathematical Statistics*, 4th ed. New York: Macmillan, 1978. Chapter 4.

Lindgren, B. W. *Statistical Theory*, 3rd ed. New York: Macmillan, 1976. Chapter 2.

Montgomery, D. C. *Introduction to Statistical Quality Control*, 2d ed. New York: Wiley, 1991.

Neter, J., Wasserman, W., and Whitmore, G. A. *Applied Statistics*, 2d ed. Boston: Allyn & Bacon, 1983. Chapters 8 and 9.

Sharpe, W. F. and Alexander, G. J. *Investments*, 4th ed. Englewood Cliffs, N.J.: Prentice-Hall, 1990. Chapter 7.

Standard on Assessment-Ratio Studies. Chicago: International Association of Assessing Officers, 1980.

Weijo, R. O. *Evaluating Information, Incentive, and Door-to-door Interventions for the MECS Energy Audit Program: A Theoretical Application of the Petty and Cacippo Elaboration Likelihood Model*. Unpublished Ph.D. dissertation, University of Minnesota, 1983.

Wetherill, G. B. *Sampling Inspection and Quality Control*, 2d ed. New York: Chapman and Hall, 1977. Chapter 3.

Winkler, R. L. and Hays, W. *Statistics: Probability, Inference, and Decision*, 2d ed. New York: Holt, Rinehart and Winston, 1975. Chapter 5.

CHAPTER SEVEN

Inferences Based on a Single Sample: Estimation

Contents

Case Study

Where We've Been

In the preceding chapters we learned that populations are characterized by numerical descriptive measures (called *parameters*), and that decisions about their values are based on sample statistics computed from sample data. Since statistics vary in a random manner from sample to sample, inferences based on them will be subject to uncertainty. This property is reflected in the sampling (probability) distribution of a statistic.

Where We're Going

This chapter puts all the preceding material into practice; that is, we estimate population means and proportions based on a single sample selected from the population of interest. Most important, we use the sampling distribution of a sample statistic to assess the reliability of the estimate.

7

The estimation of the mean gas mileage for a new car model, the estimation of the expected life of a computer monitor, and the estimation of the mean yearly sales for companies in the steel industry are problems with a common element: In each case, we are interested in estimating the mean of a population of measurements. This important problem constitutes the primary topic of this chapter.

You will see that different techniques are used for estimating a mean, depending on whether a sample contains a large or small number of measurements. Regardless, our objectives remain the same. We want to use the sample information to estimate the mean and to assess the reliability of the estimate.

In Sections 7.1 and 7.2, we consider a method of estimating a population mean using a large sample and develop a formula to determine just how large the sample must be to achieve a specified degree of reliability. In Section 7.3, we show how a mean can be estimated when only a small sample is available. Finally, estimation of binomial probabilities (e.g., population proportions) and the determination of sample sizes necessary to make reliable estimates are covered in Sections 7.4 and 7.5, respectively.

7.1 Large-Sample Estimation of a Population Mean

Suppose a large credit corporation wants to estimate the average amount of money owed by its delinquent debtors—i.e., debtors who are more than 2 months behind in payment. To accomplish this objective, the company plans to sample 100 of its delinquent accounts and to use the sample mean, $\bar{x}$, of the amounts overdue to estimate μ, the mean for *all* delinquent accounts. The sample mean $\bar{x}$ represents a **point estimator** of the population mean μ (Definition 6.4). How can we assess the accuracy of this point estimator?

Recall that for sufficiently large samples the sampling distribution of the sample mean is approximately normal, as shown in Figure 7.1. Now, suppose you plan to take a sample of $n = 100$ measurements and calculate the following interval:

$$\bar{x} \pm 2\sigma_{\bar{x}} = \bar{x} \pm \frac{2\sigma}{\sqrt{n}}$$

FIGURE 7.1 ► Sampling distribution of $\bar{x}$

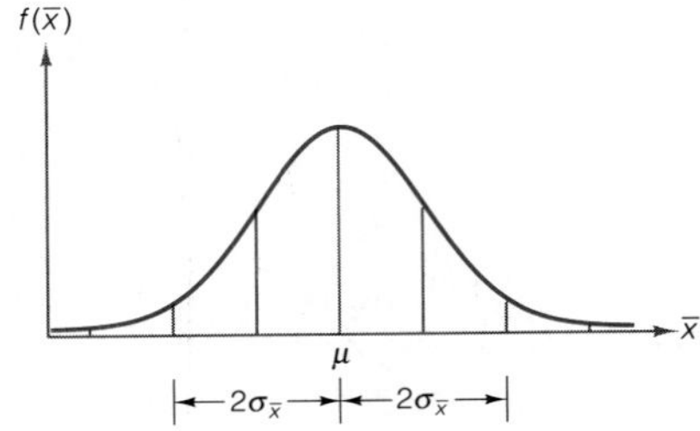

That is, you will form an interval 4 standard deviations wide: from 2 standard deviations below the sample mean to 2 standard deviations above the sample mean. What is the chance, *before we have drawn the sample*, that this interval will enclose μ, the population mean?

To answer this question, refer to Figure 7.1. If the 100 measurements yield a value of $\bar{x}$ that falls between the two lines on either side of μ—i.e., within 2 standard deviations of μ—then the interval $\bar{x} \pm 2\sigma_{\bar{x}}$ will contain μ. If $\bar{x}$ falls outside either of these boundaries, then the interval $\bar{x} \pm 2\sigma_{\bar{x}}$ will not contain μ. Since the area under the normal curve (the sampling distribution of $\bar{x}$) between these boundaries is approximately .95 (more precisely, from Table IV of Appendix B, the area is .9544), we know that the interval $\bar{x} \pm 2\sigma_{\bar{x}}$ will contain μ with a probability approximately equal to .95.

To illustrate, suppose that the sum and the sum of squared deviations for the sample of debts of 100 delinquent accounts are

$$\sum_{i=1}^{n} x_i = \$23{,}300 \qquad \sum_{i=1}^{n} (x_i - \bar{x})^2 = 801{,}900$$

We first calculate the sample mean and standard deviation:

$$\bar{x} = \frac{\sum_{i=1}^{100} x_i}{n} = \frac{23{,}300}{100} = \$233 \qquad s = \sqrt{\frac{\sum_{i=1}^{100} (x_i - \bar{x})^2}{n-1}} = \sqrt{\frac{801{,}900}{99}} = \$90$$

We next substitute for $\bar{x}$ in

$$\bar{x} \pm 2\sigma_{\bar{x}} = 233 \pm 2\left(\frac{\sigma}{\sqrt{100}}\right)$$

But now we face a problem. You can see that without knowing the standard deviation, σ, of the original population—i.e., the standard deviation of the amounts of *all* delinquent accounts—we cannot calculate this interval. However, since we have a large sample ($n = 100$ measurements), we can approximate the interval by using the sample standard deviation, s, to approximate σ. Thus,

$$\bar{x} \pm 2\left(\frac{\sigma}{\sqrt{100}}\right) \approx \bar{x} \pm 2\left(\frac{s}{\sqrt{100}}\right) = 233 \pm 2\left(\frac{90}{10}\right) = 233 \pm 18$$

That is, we estimate the mean amount of delinquency for all accounts to fall within the interval (\$215, \$251).

Can we be sure that μ, the true mean amount due, is within the interval (\$215, \$251)? We cannot be certain, but we can be reasonably confident that it is. This confidence is derived from the knowledge that if we were to repeatedly draw samples of 100 accounts from this group of delinquent accounts and form the interval $\bar{x} \pm 2\sigma_{\bar{x}}$ each time, approximately 95% of the intervals would contain μ. We have no way of knowing (without looking at *all* the delinquent accounts) whether our sample interval is one of the 95% that contain μ or one of the 5% that do not, so we simply state that we are 95% confident our interval ($215, $251) contains μ. Thus, we have given an interval estimate of the mean delinquency per account and a measure of the reliability of the estimate.

The formula that tells us how to calculate an interval estimate based on sample data is called an **interval estimator**. The probability .95, that measures the confidence that we can place in the interval estimate is called a **confidence coefficient**. The percentage, 95%, is called the **confidence level** for the interval estimate.

Definition 7.1

An **interval estimator** is a formula that tells us how to use sample data to calculate an interval that estimates a population parameter.

Definition 7.2

The **confidence coefficient** is the probability that an interval estimator encloses the population parameter—that is, the relative frequency with which the interval estimator encloses the population parameter when the estimator is used repeatedly a very large number of times. The **confidence level** is the confidence coefficient expressed as a percentage.

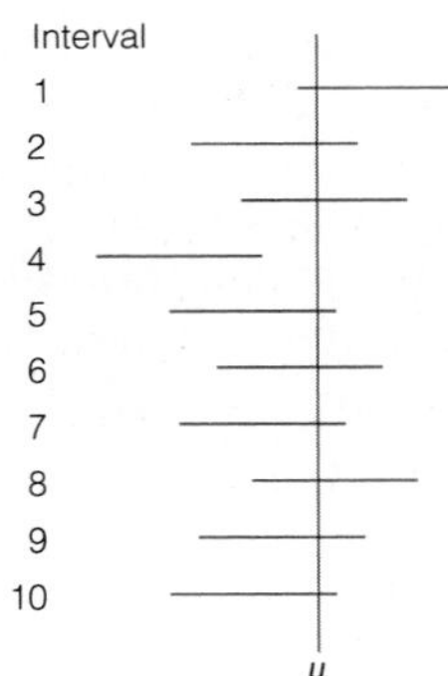

FIGURE 7.2 ▲ Interval estimators for μ: 10 samples

The foregoing describes how an interval can be used to estimate a population mean. When we use an interval estimator, we can usually calculate the probability that the interval estimation *process* will result in an interval that contains the true value of the population mean. That is, the probability that the interval contains the parameter in repeated usage is usually known. Figure 7.2 shows what happens when a number of samples are drawn from a population and a confidence interval for μ is calculated from each. The location of μ is indicated by the vertical line in the figure. Ten confidence intervals, each based on one of 10 samples, are shown as horizontal line segments. Note that the confidence intervals move from sample to sample—sometimes containing μ and other times missing μ. If our confidence level is 95%, then in the long run, 95% of our sample confidence intervals will contain μ.

Suppose you wish to choose a **confidence coefficient** other than .95. Notice that in Figure 7.1 the confidence coefficient .95 is equal to the total area under the sampling distribution, less .05 of the area; this .05 is divided equally between the two tails. Using this idea, we can construct a confidence interval with any desired confidence coefficient by increasing or decreasing the area (call it α) assigned to the tails of the sampling distribution (see Figure 7.3). For example, if we place $\alpha/2$ in each tail and if $z_{\alpha/2}$ is the z value such that the area $\alpha/2$ will lie to its right, then the confidence interval with confidence coefficient $(1 - \alpha)$ is

$$\bar{x} \pm z_{\alpha/2}\sigma_{\bar{x}}$$

FIGURE 7.3 ▶ Locating $z_{\alpha/2}$ on the standard normal curve

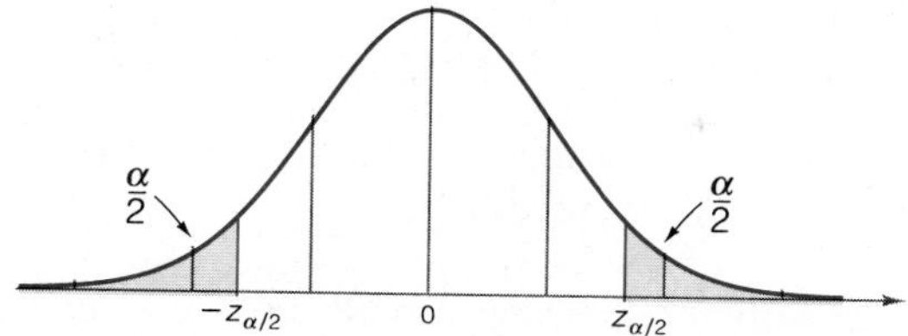

Large-Sample 100(1 − α)% Confidence Interval for μ

$$\bar{x} \pm z_{\alpha/2}\sigma_{\bar{x}}$$

where $z_{\alpha/2}$ is the z value with an area $\alpha/2$ to its right (see Figure 7.3) and $\sigma_{\bar{x}} = \sigma/\sqrt{n}$. The parameter σ is the standard deviation of the sampled population and n is the sample size. When σ is unknown (as is almost always the case) and n is large (say, $n \geq 30$), the value of σ can be approximated by the sample standard deviation, s.

To illustrate, for a confidence coefficient of .90, $(1 - \alpha) = .90$, $\alpha = .10$, $\alpha/2 = .05$, and $z_{.05}$ is the z value that locates .05 in one tail of the sampling distribution. Recall that Table IV of Appendix B (and inside front cover) gives the areas between the mean and a specified z value. Since the total area to the right of the mean is .50, $z_{.05}$ will be the z value corresponding to the tabulated area to the right of the mean equal to .4500. This z value is $z_{.05} = 1.645$. Confidence coefficients used in practice (in reports and published articles) usually range from .90 to .99. The most commonly used confidence coefficients with corresponding values of α and $z_{\alpha/2}$ are shown in Table 7.1.

TABLE 7.1 Commonly Used Values of $z_{\alpha/2}$

Confidence Level $100(1 - \alpha)$	α	$\alpha/2$	$z_{\alpha/2}$
90%	.10	.05	1.645
95%	.05	.025	1.96
99%	.01	.005	2.575

EXAMPLE 7.1

Unoccupied seats on flights cause the airlines to lose revenue. Suppose a large airline wants to estimate its average number of unoccupied seats per flight over the past year. To accomplish this, the records of 225 flights are randomly selected and the number of unoccupied seats is noted for each of the sampled flights. The sample mean and standard deviation are

$$\bar{x} = 11.6 \text{ seats} \qquad s = 4.1 \text{ seats}$$

Estimate μ, the mean number of unoccupied seats per flight during the past year, using a 90% confidence interval.

Solution

The general form of the large-sample 90% confidence interval for a population mean is

$$\bar{x} \pm z_{\alpha/2}\sigma_{\bar{x}} = \bar{x} \pm z_{.05}\sigma_{\bar{x}} = \bar{x} \pm 1.645\left(\frac{\sigma}{\sqrt{n}}\right)$$

For the 225 records sampled, we have

$$11.6 \pm 1.645\left(\frac{\sigma}{\sqrt{225}}\right)$$

Since we do not know the value of σ (the standard deviation of the number of unoccupied seats per flight for all flights of the year), we use our best approximation, the sample standard deviation s. Then the 90% confidence interval is, approximately,

$$11.6 \pm 1.645\left(\frac{4.1}{\sqrt{225}}\right) = 11.6 \pm .45$$

or from 11.15 to 12.05. That is, the airline can be 90% confident that the mean number of unoccupied seats per flight was between 11.15 and 12.05 during the sampled year.

Remember that the 90% confidence comes from the knowledge that the interval estimation process will result in an interval that contains μ 90% of the time in repeated sampling. We do not know whether this particular interval (11.15, 12.05) is one of the 90% that contain μ, or one of the 10% that do not.

The interpretation of confidence intervals for a population mean is summarized in the box.

Interpretation of a Confidence Interval for a Population Mean

When we form a $100(1 - \alpha)\%$ confidence interval for μ, we usually express our confidence in the interval with a statement such as, "We can be $100(1 - \alpha)\%$ confident that μ lies between the lower and upper bounds of the confidence interval," where for a particular application we substitute the appropriate numerical values for the confidence, and the lower and upper bounds. **The statement reflects our confidence in the estimation process rather than in the particular interval that is calculated from sample data.** We know that repeated application of the same procedure will result in different lower and upper bounds on the interval. Furthermore, we know that $100(1 - \alpha)\%$ of the resulting intervals will contain μ. There is (usually) no way to determine whether a particular interval is one of those that contain μ, or one that does not. However, unlike point estimators, confidence intervals have some measure of reliability, the confidence coefficient, associated with them, and for that reason are generally preferred to point estimators.

CASE STUDY 7.1 / Dancing to the Customer's Tune: The Need to Assess Customer Preferences

We sometimes lose sight of the fact that many firms were customer-oriented long before the quality movement swept through the U.S. in the 1980s and early 1990s. The following quotations were extracted from the Dec. 13, 1976, issue of *Business Week*:

"We're dancing to the tune of the customer as never before," says J. Janvier Wetzel, vice-president for sales promotion at Los Angeles–based Broadway Department Stores. "With population growth down to a trickle compared with its previous level, we're no longer spoiled with instant success every time we open a new store. Traditional department stores are locked in the biggest competitive battle in their history."

The nation's retailers are becoming uncomfortably aware that today's operating environment is vastly different from that of the 1960s. Population growth is slowing, a growing singles market is emerging, family formations are coming at later ages, and more women are embarking on careers. Of the 71 million households in the U.S. today, the dominant consumer buying segment is families headed by persons over 45. But by 1980 this group will have lost its majority status to the 25 to 40 year-old group. Merchants must now reposition their stores to attract these new customers.

To do so retailers are using market research to ferret out new purchasing attitudes and lifestyles and then translating this into customer buying segments. Data on demographics, psychographics (measurement of attitudes), and lifestyles are being fed into retailers' computers so they can make marketing decisions based on actual spending patterns and estimate their inventory needs with less risk.

To stock its various departments with the type and style of goods that appeal to its potential group of customers, a downtown department store should be interested in estimating the average age of downtown shoppers, not shoppers in general. Suppose a downtown department store questions 49 downtown shoppers concerning their age (the offer of a small gift certificate may help convince shoppers to respond to such questions). The sample mean and standard deviation are found to be 40.1 and 8.6, respectively. The store could then estimate μ, the mean age of all downtown shoppers, with a 95% confidence interval as follows:

$$\begin{aligned}\bar{x} \pm 1.96\left(\frac{s}{\sqrt{n}}\right) &= 40.1 \pm 1.96\left(\frac{8.6}{\sqrt{49}}\right)\\ &= 40.1 \pm 2.4\end{aligned}$$

Thus, the department store should gear its sales to the segment of consumers with average age between 37.7 and 42.5.

Exercises 7.1–7.15

Learning the Mechanics

7.1 Find $z_{\alpha/2}$ for each of the following:
a. $\alpha = .10$ b. $\alpha = .01$ c. $\alpha = .05$ d. $\alpha = .20$

7.2 What is the confidence level of each of the following confidence intervals for μ?
a. $\bar{x} \pm 1.96\sigma/\sqrt{n}$ b. $\bar{x} \pm 1.645\sigma/\sqrt{n}$ c. $\bar{x} \pm 2.575\sigma/\sqrt{n}$
d. $\bar{x} \pm 1.282\sigma/\sqrt{n}$ e. $\bar{x} \pm .99\sigma/\sqrt{n}$

7.3 A random sample of 50 observations from a population produced the following summary statistics:

$$\sum x = 850 \qquad \sum (x_i - \bar{x})^2 = 4,720$$

a. Find a 95% confidence interval for μ. b. Interpret the confidence interval you found in part **a**.

7.4 A random sample of 70 observations from a normally distributed population produced a sample mean of 26.2 and a standard deviation of 4.1.
a. Find a 95% confidence interval for μ. b. Find a 99% confidence interval for μ.
c. What happens to the width of a confidence interval as the sample size is held fixed and the value of the confidence coefficient is increased?
d. Would your confidence intervals of parts **a** and **b** be valid if the distribution of the original population were not normal? Explain.

7.5 Refer to Exercise 7.4. Suppose the sample contained only 40 observations.
a. Recompute the 95% confidence interval for μ.
b. What is the effect on the width of a confidence interval of reducing the sample size as the confidence coefficient remains fixed? Of increasing the sample size as the confidence coefficient remains fixed?

7.6 Explain what is meant by the statement "We are 95% confident that our interval estimate contains μ."

7.7 Explain the difference between an interval estimator and a point estimator for μ.

7.8 Describe the relationship between the size of the confidence coefficient and the width of the confidence interval.

7.9 Describe the relationship between the sample size and the width of the confidence interval.

Applying the Concepts

7.10 To assess the magnitude of recent rent increases in metropolitan Minneapolis, an apartment referral service randomly sampled and interviewed 32 apartment building owners. They obtained the following data concerning the change in rent for two-bedroom apartments between May 1989 and May 1990:

1.13%	2.03%	1.79%	1.00%	−.45%	2.03%	1.41%	.34%
2.61	2.42	2.06	−1.07	1.25	2.27	.61	.68
−1.22	1.00	5.20	.62	.00	.05	1.82	1.39
7.33	3.32	.68	3.59	.90	1.42	1.80	.00

a. Carefully describe the population from which the sample was drawn.
b. Use a 95% confidence interval to estimate the mean percentage change in rent for two-bedroom apartments.
c. In constructing the confidence interval of part **b**, was it necessary to assume that the population was normally distributed? Explain.

7.11 A major department store chain is interested in estimating the average amount its credit card customers spent on their first visit to the chain's new store in the mall. Fifty credit card accounts were randomly sampled and analyzed with the following results: $\bar{x} = \$62.56$ and $s^2 = 400$.
a. Identify the population the department store chain is interested in learning about.
b. Which population parameter does the chain wish to estimate?
c. Construct a 90% confidence interval for the parameter identified in part **b**.

7.12 At the end of 1989, 1990, and 1991, the average prices of a share of stock on the New York Stock Exchange (NYSE) were \$36.51, \$31.08, and \$37.27, respectively (*Statistical Abstract of the United States: 1992*, p. 512). To investigate the average share price at the end of 1992, a random sample of 30 NYSE stocks was drawn. Their closing prices on December 31, 1992, are listed (by their NYSE abbreviations) in the table.

Stock	*Price*	*Stock*	*Price*	*Stock*	*Price*
RowanCos	7⅞	LaZ Boy	25⅞	Supervalu	31⅛
Huntway	2	Tektronix	20⅛	Tambrands	64⅛
ChemBank	38⅝	Giantlnd	5⅜	Crane	23⅝
Advest	5⅞	CalMat	22½	Ameritech	71¼
Caterpillar	53⅝	Omnicom	41¼	Goodyear	68⅜
Sears	45½	HewlettPk	69⅞	WellsF	76⅜
UnElec	37⅜	StrtGlob	12⅛	Boeing	40⅛
Travelers	27¼	VanDorn	20⅜	BankAmer	46½
Rockwell	29	MinnMngMfg	100⅝	Zurnlnd	39⅜
HomeShop	8¼	HondaMotor	20¾	Equimark	8⅛

Source: *Wall Street Journal*, January 4, 1993.

a. Using a 90% confidence interval, estimate the average price of a share of stock at the end of 1992.
b. Use the latest edition of the *Statistical Abstract of the United States* to find the actual average price of a stock on the NYSE at the end of 1992. Is this figure in agreement with your confidence interval? Explain. If not, provide a possible explanation for the disagreement.

7.13 An auditor was hired to verify the accuracy of a company's new billing system. The auditor randomly sampled 35 invoices produced since the system was installed. Each invoice was compared against the relevant internal records to determine by how much the invoice was in error. The amount of the error, x, was defined as $(A - I)$, where A is the actual amount owed the company and I is the amount indicated on the invoice. The auditor found that $\bar{x} = \$1$ and $s = \$124$.
a. Identify the population the auditor studied.
b. Describe the variable that the auditor measured.
c. Construct a 98% confidence interval for the mean error per invoice.
d. Interpret the confidence interval.
e. Comment on the accuracy of the billing system.

7.14 A process has been developed that can transform ordinary iron into a kind of super-iron called *metallic glass* ("One Answer to Imports," 1981). Metallic glass is 3 to 4 times as strong as the toughest steel alloys, but it becomes brittle at very high temperatures. To estimate the mean temperature, μ, at which a particular type of metallic glass becomes brittle, 36 pieces of the metallic glass were randomly sampled from a recent production run. Each piece was independently subjected to higher and higher temperatures until it became brittle. The temperature at which brittleness was first noticed was recorded for each piece in the sample. The following results were obtained: $\bar{x} = 480°\text{F}$, $s = 11°\text{F}$. Use a 90% confidence interval to estimate μ. Interpret your confidence interval.

7.15 Nasser Arshadi and Edward Lawrence (1984) investigated the profiles (i.e., career patterns, social backgrounds, and so forth) of the top executives in the U.S. banking industry. They sampled 96 executives and found that 80% studied business or economics and that 45% had a graduate degree. With respect to the number of years of service, x, at the same bank, the group had a mean of 23.43 years and a standard deviation of 10.82 years.

a. Construct a 90% confidence interval for $E(x) = \mu$.
b. Interpret your interval in the context of the problem.
c. What assumption(s) was it necessary to make in order to construct the confidence interval of part **a**?
d. Is your interval estimate for $E(x)$ also an interval estimate for $E(\bar{x})$? Explain. [*Hint:* See Chapter 6.]

7.2 Determining the Sample Size Necessary to Estimate a Population Mean

Typically, an analyst must plan the sampling study that produces the data used to estimate a population mean. Perhaps the most important design decision faced by the analyst is to determine the size of the sample. We will show in this section that the appropriate sample size for estimating a population mean depends on the desired reliability.

To see this, consider the example from Section 7.1, in which we estimated the mean overdue amount for all delinquent accounts in a large credit corporation. A sample of 100 delinquent accounts produced an estimate $\bar{x}$ that was within \$18 of the true mean amount due, μ, for all delinquent accounts at the 95% confidence level. That is, the 95% confidence interval for μ was \$36 wide when 100 accounts were sampled. This is illustrated in Figure 7.4(a).

Now suppose that we want to estimate μ to within \$5 with 95% confidence. That is, we want to narrow the width of the confidence interval from \$36 to \$10, as shown in Figure 7.4(b). How much will the sample size have to be increased to accomplish this? If we want the estimator $\bar{x}$ to be within \$5 of μ, we must have

$$1.96\sigma_{\bar{x}} = 5 \quad \text{or, equivalently,} \quad 1.96\left(\frac{\sigma}{\sqrt{n}}\right) = 5$$

The necessary sample size is obtained by solving this equation for n. To do this we need an approximation for σ. We have an approximation from the initial sample of 100 accounts—namely, the sample standard deviation, $s = 90$. Thus,

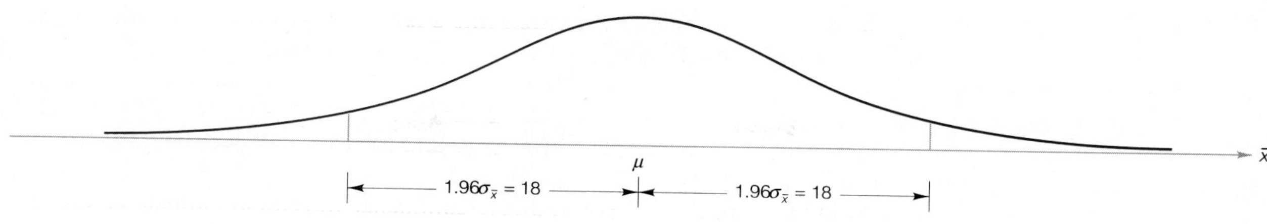

a. $n = 100$

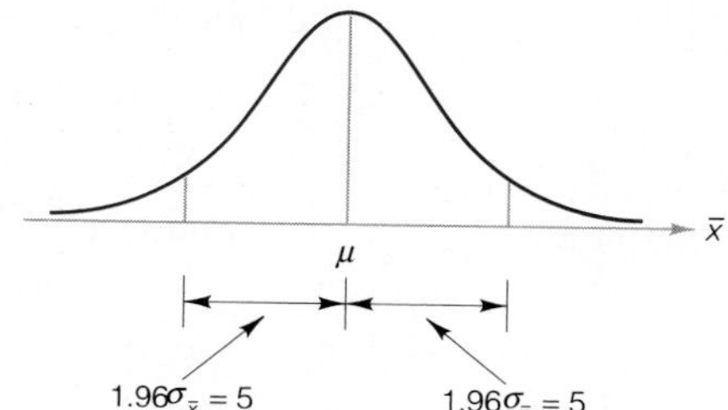

b. $n = 1{,}300$

FIGURE 7.4 ▲
Relationship between sample size and width of confidence interval for delinquent creditors example

$$1.96\left(\frac{\sigma}{\sqrt{n}}\right) \approx 1.96\left(\frac{s}{\sqrt{n}}\right) = 1.96\left(\frac{90}{\sqrt{n}}\right) = 5$$

$$\sqrt{n} = \frac{1.96(90)}{5} = 35.28$$

$$n = (35.28)^2 = 1{,}244.68$$

Approximately 1,245 accounts will have to be sampled to estimate the mean overdue amount μ to within \$5 with (approximately) 95% confidence. The confidence interval resulting from a sample of this size will be approximately \$10 wide [see Figure 7.4(b)].

In general, we can express the reliability associated with a confidence interval for the population mean μ in one of two equivalent ways: We can specify the bound, B, within which we want to estimate μ with $100(1 - \alpha)\%$ confidence. This bound B then is equal to the half-width of the confidence interval, as shown in Figure 7.5. Equivalently, we can specify the total width, W, of the $100(1 - \alpha)\%$ confidence interval for μ, also shown in Figure 7.5. Note that $W = 2B$.

FIGURE 7.5 ►
Specifying the bound *B* or total width *W* for a confidence interval

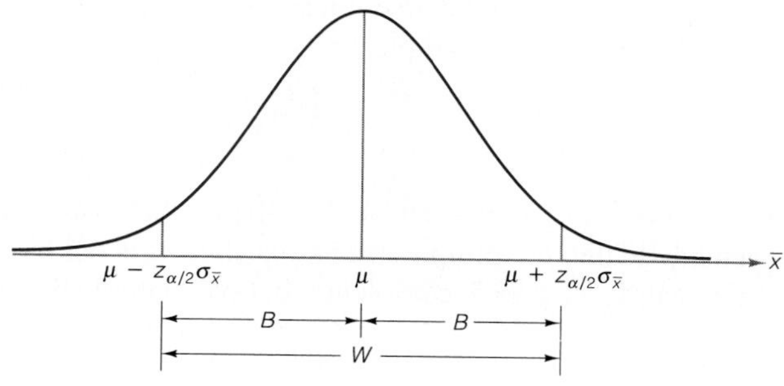

The procedure for finding the sample size necessary to estimate μ to within a given bound B or with a total interval width W is given in the box.

Sample Size Determination for $100(1 - \alpha)\%$ Confidence Intervals for μ

To estimate μ to within a bound B or, equivalently, with a confidence interval of total width W, with $100(1 - \alpha)\%$ confidence, the required sample size is found by solving one of the following equations for n:

$$z_{\alpha/2}\left(\frac{\sigma}{\sqrt{n}}\right) = B \qquad z_{\alpha/2}\left(\frac{\sigma}{\sqrt{n}}\right) = \frac{W}{2}$$

The solution can be written in terms of either B or W:

$$n = \frac{(z_{\alpha/2})^2\sigma^2}{B^2} \qquad n = \frac{4(z_{\alpha/2})^2\sigma^2}{W^2}$$

The value of σ is almost always unknown. It can be estimated by the standard deviation, s, from a prior sample. Alternatively, we may approximate the range R of observations in the population, and (conservatively) estimate $\sigma \approx R/4$. In any case, you should round the value of n you obtain *upward* to ensure that the sample size will be sufficient to achieve the specified reliability.

EXAMPLE 7.2

Suppose the manufacturer of official NFL footballs uses a machine to inflate the new balls to a pressure of 13.5 pounds. When the machine is properly calibrated, the mean inflation pressure is 13.5, but uncontrollable factors cause the pressures of individual footballs to vary randomly from about 13.3 to 13.7 pounds. For quality control purposes, the manufacturer wishes to estimate the true mean inflation pressure for a particular shipment of balls with a 99% confidence interval that is only .05 pound wide. What sample size should be used?

Solution

For a 99% confidence interval we have $z_{\alpha/2} = z_{.005} = 2.575$, and to estimate σ we note that the range of observations is $R = 13.7 - 13.3 = .4$ and use $\sigma \approx R/4 = .1$. Thus, in order to have a confidence interval of width $W = .05$, we use the formula derived in the box to find the sample size n:

$$n = \frac{4(z_{\alpha/2})^2\sigma^2}{W^2} \approx \frac{4(2.575)^2(.1)^2}{(.05)^2} = 106.09$$

We round this up to $n = 107$ and, realizing that σ was approximated by $R/4$, we might even advise that the sample size be specified as $n = 110$ to be more certain of attaining the objective of a 99% confidence interval with width $W = .05$ pound or less.

Sometimes the formulas will lead to a solution that indicates a small sample size is sufficient to achieve the confidence interval goal. As we will see in Section 7.3, the procedures and assumptions for making inferences from small samples differ from those for large samples. Therefore, if the formulas yield a small sample size ($n < 30$), one simple strategy is to select a sample size $n \geq 30$.

The cost of sampling must also be considered when the sample size is being determined. Although more complex formulas can be derived to take sampling costs into account, these are beyond the scope of this text. For our purposes, it is sufficient to realize that a sampling budget may prove to be a restriction on the sample size, and therefore on the reliability of the confidence interval.

Exercises 7.16–7.25

Learning the Mechanics

7.16 If you wish to estimate a population mean correct to within a bound $B = .3$ with probability .95 and you know from prior sampling that σ^2 is approximately equal to 7.2, how many observations would have to be included in your sample?

7.17 If you wish to estimate a population mean with a 95% confidence interval of width $W = .3$ and you know from prior sampling that σ^2 is approximately to 7.2, how many observations would have to be included in your sample? Compare your answer to that for Exercise 7.16. Explain the difference.

7.18 Suppose you wish to estimate the mean of a normal population using a 99% confidence interval and you know from prior information that $\sigma^2 \approx 1$.

a. To see the effect of the sample size on the width of the confidence interval, calculate the width W of the confidence interval for $n = 9, 36, 81, 400$.

b. Plot the width as a function of sample size n on graph paper. Connect the points by a smooth curve and note how the width decreases as n increases.

7.19 Suppose you wish to estimate a population mean correct to within a bound $B = .20$ with probability equal to .90. You do not know σ^2, but you know that the observations will range in value between 30 and 34.

a. Find the approximate sample size that will produce the desired accuracy of the estimate. You wish to be conservative to ensure that the sample size will be ample to achieve the desired accuracy of the estimate. [*Hint:* Using your knowledge of data variation from Section 2.6, assume that the range of the observations will equal 4σ.]

b. Calculate the approximate sample size making the less conservative assumption that the range of the observations is equal to 6σ.

7.20 It costs you \$10 to draw a sample of size $n = 1$ and measure the attribute of interest. You have a budget of \$1,500.

a. Do you have sufficient funds to estimate the population mean for the attribute of interest with a 95% confidence interval 5 units in width? Assume $\sigma = 14$.

b. If a 90% confidence level were used, would your answer to part **a** change? Explain.

Applying the Concepts

7.21 It costs more to produce defective items—since they must be scrapped or reworked—than it does to produce nondefective items. This simple fact suggests that manufacturers should ensure the quality of their products by perfecting their production processes rather than through inspection of finished products (Deming, 1986). To better understand a particular metal-stamping process, a manufacturer wishes to estimate the mean length of items produced by the process during the past 24 hours.

a. How many parts should be sampled in order to estimate the population mean to within .1 mm with 80% confidence? Previous studies of this machine have indicated that the standard deviation of lengths produced by the stamping operation is about 2 mm.

b. Time permits the use of a sample size no larger than 100. If an 80% confidence interval for μ is constructed using $n = 100$, will it be wider or narrower than the interval that would have been obtained using the sample size determined in part **a**? Explain.

c. If management requires that μ be estimated to within .1 mm and that a sample of size no more than 100 be used, what is (approximately) the maximum confidence level that could be attained for a confidence interval that meets management's specifications?

7.22 A large food-products company receives about 100,000 phone calls a year from consumers on its toll-free number. A computer monitors and records how many rings it takes for an operator to answer, how much time each caller spends "on hold," and other data. However, the reliability of the monitoring system has been called into question by the operators and their labor union. As a check on the computer system, approximately how many calls should be manually monitored during the next year to estimate the true mean time that callers spend on hold to within 3 seconds with 95% confidence? Answer this question for the following values of the standard deviation of waiting times (in seconds): 10, 20, and 30.

7.23 Suppose a department store wants to estimate μ, the average age of the customers of its contemporary apparel department, correct to within 2 years with probability equal to .95. Approximately how large a sample would be required? [*Note:* Management does not know the standard deviation σ but estimates that the ages of its customers range from 15 to 45. Use a conservative approximation for σ to calculate n.]

7.24 Mortgage companies and banks use professional appraisers to determine the value of properties (e.g., single-family homes) that their customers offer as security when applying for a loan. In a year's time, a real estate appraiser may appraise 500 properties. Norwest Mortgage requires its appraisers to keep track of their turnaround time for each property—the time it takes an appraiser from receipt of the appraisal request to the submission of the completed appraisal to a loan officer. How large a sample of properties would be needed to estimate the mean turnaround time for a particular appraiser during the previous 12 months with an 80% confidence interval of width no more than 2 days? The turnaround times range from 2 days to 18 days.

7.25 The United States Golf Association (USGA) tests all new brands of golf balls to assure that they meet USGA specifications. One test is intended to measure the average distance travelled when the ball is hit by a machine called "Iron Byron," a name inspired by the swing of the famous golfer Byron Nelson. Suppose the USGA wants to estimate the mean distance for a new brand with a 90% confidence interval of width 2 yards. Assume that past tests have indicated that the standard deviation of the distances "Iron Byron" hits golf balls is approximately 10 yards. How many golf balls should be hit by "Iron Byron" to achieve the desired accuracy in estimating the mean?

7.3 Small-Sample Estimation of a Population Mean

One of the items of interest to an investor in a company is a forecast of the company's annual earnings per share for the next year. Recall that earnings per share is computed by dividing the total annual earnings of the company by the total number of shares of stock outstanding. One way of trying to project the earnings per share is to ask the opinion of several experts, thus obtaining a sample of projections for the particular company. Then, this sample of opinions can be used to make an inference about the mean projected earnings per share, μ, of all stock analysts. However, time and cost restrictions would probably limit the sample of opinions to a small number, and thus the large-sample estimation technique of Section 7.1 may not be applicable.

Many inferences in business must be made on the basis of very limited information—i.e., **small samples**. When an inference is to be made about a population mean, μ, small samples have two immediate problematic effects:

PROBLEM 1

The shape of the sampling distribution of the sample mean $\bar{x}$ now depends on the shape of the population that is sampled. Because the Central Limit Theorem applies only to large samples, we can no longer assume that the sampling distribution of $\bar{x}$ is approximately normal.

PROBLEM 2

Although it is still true that $\sigma_{\bar{x}} = \sigma/\sqrt{n}$, the sample standard deviation s may provide a poor approximation of the population standard deviation σ when the sample size is small.

Solution to Problem 1

According to Theorem 6.1, the sampling distribution of $\bar{x}$ will be normal (approximately normal) even for small samples *if the population being sampled is normal (approximately normal).*

Solution to Problem 2

Instead of using the standard normal statistic

$$z = \frac{\bar{x} - \mu}{\sigma_{\bar{x}}} = \frac{\bar{x} - \mu}{\sigma/\sqrt{n}}$$

which requires knowledge of, or a good approximation to, σ, we define and use the statistic

$$t = \frac{\bar{x} - \mu}{s/\sqrt{n}}$$

in which the sample standard deviation, s, replaces the population standard deviation, σ.

The distribution of the **t statistic** in repeated sampling was discovered by W. S. Gosset, a scientist in the Guinness brewery in Ireland, who published his discovery in 1908 under the pen name of Student. The main result of Gosset's work is that if we are sampling from a normal distribution, the t statistic will have a sampling distribution very much like that of the z statistic: mound-shaped, symmetric, and with mean 0. The primary difference between the sampling distributions of t and z is that the t statistic is more variable than the z, which follows intuitively when you realize that t contains two random quantities ($\bar{x}$ and s), while z contains only one ($\bar{x}$).

The actual increase in variability in the sampling distribution of t depends on the sample size, n. In particular, the smaller the value of n, the more variable will be the sampling distribution of t. A convenient way of expressing this dependence is to say that the t statistic has $(n - 1)$ **degrees of freedom (df)**. Recall that the quantity $(n - 1)$ is the divisor that appears in the formula for s^2. This number plays a key role in the sampling distribution of s^2 and appears in discussions of other statistics in later chapters.

In Figure 7.6, we show both the sampling distribution of z and the sampling distribution of a t statistic with 4 df. You can see that the increased variability of the t statistic means that the t value, t_α, that locates an area α in the upper tail of the t distribution will be larger than the corresponding value z_α. Values of t that will be used in forming small-sample confidence intervals for μ are given in Table VI of Appendix B (and inside back cover). A partial reproduction of this table is shown in Table 7.2. Note that t_α values are listed for degrees of freedom from 1 to 29, where α refers to the tail area to the right of t_α. For example, if we want the t value with an area of .025 to its right and 4 df, we look in the table under the column $t_{.025}$ for the entry in the row corresponding to 4 df. This entry is $t_{.025} = 2.776$, as shown in Figure 7.7. The corresponding standard normal z-score is $z_{.025} = 1.96$.

FIGURE 7.6 ▶ **Standard normal (z) distribution and t distribution with 4 df**

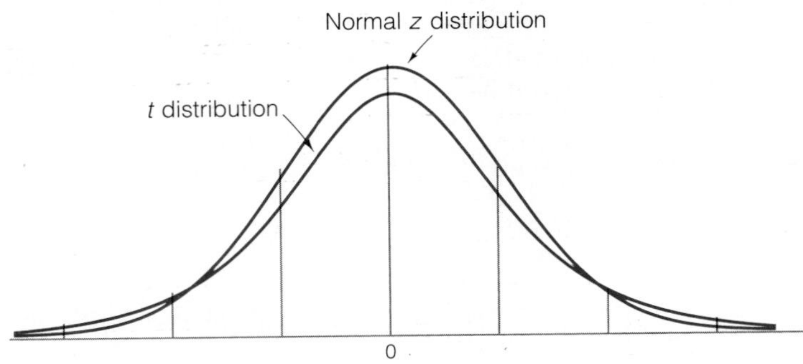

Note that the last row of Table VI, where df = infinity, contains the standard normal z values. This follows from the fact that as the sample size n grows very large, s becomes closer to σ, and thus t becomes closer in distribution to z. In fact, when df = 29, there is little difference between corresponding tabulated values of z and t. Thus, we choose the arbitrary cutoff of $n = 30$ (df = 29) to distinguish between the large- and small-sample inferential techniques.

TABLE 7.2 Reproduction of Part of Table VI of Appendix B: Critical Values of the t Statistic

Degrees of Freedom	$t_{.100}$	$t_{.050}$	$t_{.025}$	$t_{.010}$	$t_{.005}$	$t_{.001}$	$t_{.0005}$
1	3.078	6.314	12.706	31.821	63.657	318.31	636.62
2	1.886	2.920	4.303	6.965	9.925	22.326	31.598
3	1.638	2.353	3.182	4.541	5.841	10.213	12.924
4	1.533	2.132	2.776	3.747	4.604	7.173	8.610
5	1.476	2.015	2.571	3.365	4.032	5.893	6.869
6	1.440	1.943	2.447	3.143	3.707	5.208	5.959
7	1.415	1.895	2.365	2.998	3.499	4.785	5.408
8	1.397	1.860	2.306	2.896	3.355	4.501	5.041
9	1.383	1.833	2.262	2.821	3.250	4.297	4.781
10	1.372	1.812	2.228	2.764	3.169	4.144	4.587
11	1.363	1.796	2.201	2.718	3.106	4.025	4.437
12	1.356	1.782	2.179	2.681	3.055	3.930	4.318
13	1.350	1.771	2.160	2.650	3.012	3.852	4.221
14	1.345	1.761	2.145	2.624	2.977	3.787	4.140
15	1.341	1.753	2.131	2.602	2.947	3.733	4.073
⋮	⋮	⋮	⋮	⋮	⋮	⋮	⋮
∞	1.282	1.645	1.960	2.326	2.576	3.090	3.291

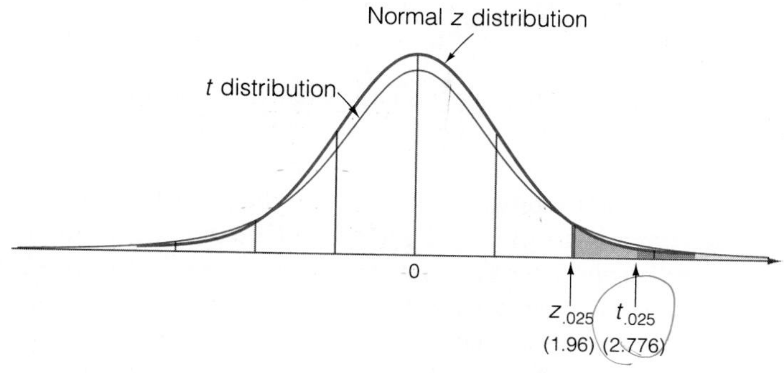

FIGURE 7.7 ► The $t_{.025}$ value in a t distribution with 4 df and the corresponding $z_{.025}$ value

Returning to the projected earnings per share example, suppose we can get a sample of five expert opinions about next year's earnings per share for a stock. We calculate the mean and standard deviation of these five projections to be

$$\bar{x} = \$2.63 \qquad s = \$.72$$

How can we use this information to construct a 95% confidence interval for μ, the mean projected earnings per share for all stock analysts in the population?

First, we know that we are dealing with a sample too small to assume that the sample mean $\bar{x}$ is approximately normally distributed by the Central Limit Theorem.

That is, we cannot "automatically" conclude that $\bar{x}$ is normally distributed from the Central Limit Theorem when the sample size is small. Instead, we must assume the measured variable, in this case the earnings per share projection, is normally distributed in order for the distribution of $\bar{x}$ to be normal.

Second, unless we are fortunate enough to know the population standard deviation σ, which in this case represents the standard deviation of *all* the analysts' projections of the stock's earnings per share, we cannot use the standard normal z statistic to form our confidence interval for μ. Instead, we must use the t distribution, with $(n - 1)$ degrees of freedom.

In this case, $n - 1 = 4$ df, and the t value is found in the table (Table 7.2 or inside back cover) to be

$$t_{.025} = 2.766 \quad \text{with 4 df}$$

Recall that the large-sample confidence interval would have been of the form

$$\bar{x} \pm z_{\alpha/2}\sigma_{\bar{x}} = \bar{x} \pm z_{\alpha/2}\frac{\sigma}{\sqrt{n}} = \bar{x} \pm z_{.025}\frac{\sigma}{\sqrt{n}}$$

where 95% is the desired confidence level. To form the interval for a small sample *from a normal distribution, we simply substitute t for z and s for σ in the preceding formula:*

$$\bar{x} \pm t_{\alpha/2}\frac{s}{\sqrt{n}}$$

Substituting the numerical values, we get

$$2.63 \pm (2.776)\frac{(.72)}{\sqrt{5}} = 2.63 \pm .89$$

or \$1.77 to \$3.52 per share. That is, we can be 95% confident that the mean of *all* analysts' earnings per share projections for this stock is between \$1.77 and \$3.52. As with our large-sample interval estimates, our confidence is in the process, not in this particular interval. We know that the procedure we used will produce an interval that contains the true mean, μ, 95% of the time we utilize it, *assuming that the population of earnings-per-share projections from which our sample was selected is normally distributed*. The latter assumption is necessary for the small-sample interval to be valid.

What price did we pay for having to utilize a small sample to make the inference? First, we had to assume the underlying population is normally distributed; if the assumption is invalid, our interval might also be invalid.* Second, we had to form the interval using a t value of 2.776 rather than a z value of 1.96, resulting in a wider interval to achieve the same 95% level of confidence. If the interval from \$1.77 to \$3.52 is too wide to be of use, then we know how to remedy the situation: increase the number of analysts sampled in order to decrease the interval width (on average).

---

*By *invalid* we mean that the probability that the procedure will yield an interval that contains μ is not equal to $(1 - \alpha)$. Generally, if the underlying population is approximately normal, then the confidence coefficient will approximate the probability that the interval contains μ.

The procedure for forming a small-sample confidence interval is summarized in the accompanying box.

Small-Sample Confidence Interval for μ*

$$\bar{x} \pm t_{\alpha/2}\left(\frac{s}{\sqrt{n}}\right)$$

where $t_{\alpha/2}$ is based on $(n - 1)$ degrees of freedom.

Assumption: The relative frequency distribution of the sampled population is approximately normal.

EXAMPLE 7.3

Some quality control experiments require **destructive sampling** in order to measure some particular characteristic of the product. For example, suppose a manufacturer of printers for personal computers wishes to estimate the mean number of characters printed before the printhead fails. The cost of destructive sampling often dictates small samples. Suppose the printer manufacturer tests $n = 15$ printheads and calculates the following statistics:

$$\bar{x} = 1.23 \text{ million characters} \qquad s = .27 \text{ million characters}$$

Form a 99% confidence interval for the mean number of characters printed before the head fails.

Solution

If we assume that the number of characters printed before printhead failure is normally distributed, we can use the t statistic to form the confidence interval. We use a confidence coefficient of .99 and degrees of freedom $n - 1 = 14$ to find in Table VI of Appendix B:

$$t_{\alpha/2} = t_{.005} = 2.977$$

Thus, the small sample forces us to assume normality and extend the interval almost 3 standard deviations (of $\bar{x}$) on each side of the sample mean in order to form the 99% confidence interval. For these data, the interval is

$$\bar{x} \pm t_{.005}\left(\frac{s}{\sqrt{n}}\right) = 1.23 \pm 2.977\left(\frac{.27}{\sqrt{15}}\right) = 1.23 \pm .21 \quad \text{or} \quad (1.02, 1.44)$$

*The procedure given in the box assumes that the population standard deviation σ is unknown, which is almost always the case. If σ is known, we can form the small-sample confidence interval just as we would a large-sample confidence interval using a standard normal z value instead of t. However, we must still assume that the underlying population is approximately normal.

Thus, the manufacturer can be 99% confident that the printhead has a mean life between 1.02 and 1.44 million characters. If the manufacturer were to advertise that the mean life of its printheads is (at least) 1 million characters, the interval would support such a claim. Our confidence is derived from the knowledge that 99% of the intervals formed in repeated applications of this procedure will contain μ.

We have emphasized throughout this section that the assumption of a normally distributed population is necessary for making small-sample inferences about μ when using the t statistic. While many business phenomena do have approximately normal distributions, it is also true that many have distributions that are not normal or even mound-shaped. Empirical evidence acquired over the years has shown that the t distribution is rather insensitive to moderate departures from normality. That is, the use of the t statistic when sampling from mound-shaped populations generally produces credible results; however, for cases in which the distribution is distinctly nonnormal, *nonparametric methods* should be used. Nonparametric statistics are the subject of Chapter 17.

Exercises 7.26–7.37

Learning the Mechanics

7.26 Explain the differences in the sampling distributions of $\bar{x}$ for large and small samples under the following assumptions:

a. The variable of interest, x, is normally distributed.
b. Nothing is known about the distribution of the variable x.

7.27 Suppose you have selected a random sample of $n = 5$ measurements from a normal distribution. Compare the standard normal z values with the corresponding t values if you were forming the following confidence intervals:

a. 80% confidence interval **b.** 90% confidence interval **c.** 95% confidence interval
d. 98% confidence interval **e.** 99% confidence interval
f. Use the table values you obtained in parts **a–e** to sketch the z and t distributions. What are the similarities and differences?

7.28 Let t_0 be a particular value of t. Use Table VI of Appendix B to find t_0 values such that the following statements are true:

a. $P(t \geq t_0) = .025$ where df = 11 **b.** $P(t \geq t_0) = .01$ where df = 9
c. $P(t \leq t_0) = .005$ where df = 6 **d.** $P(t \leq t_0) = .05$ where df = 18

7.29 Let t_0 be a particular value of t. Use Table VI of Appendix B to find t_0 values such that the following statements are true:

a. $P(-t_0 < t < t_0) = .95$ where df = 10 **b.** $P(t \leq -t_0 \text{ or } t \geq t_0) = .05$ where df = 10
c. $P(t \leq t_0) = .05$ where df = 10 **d.** $P(t \leq -t_0 \text{ or } t \geq t_0) = .05$ where df = 20
e. $P(t \leq -t_0 \text{ or } t \geq t_0) = .01$ where df = 5

7.30 The following random sample was selected from a normal distribution: 5, 6, 2, 8, 4

a. Construct a 90% confidence interval for the population mean μ.

b. Construct a 95% confidence interval for the population mean μ.

c. Construct a 99% confidence interval for the population mean μ.

d. Assume that the sample mean $\bar{x}$ and sample standard deviation s remain exactly the same as those you just calculated, but that they are based on a sample of $n = 20$ observations rather than $n = 6$ observations. Repeat parts **a–c**. What is the effect of increasing the sample size on the width of the confidence intervals?

7.31 The following sample of 24 measurements was selected from a population that is approximately normally distributed:

91	80	99	110	95	106	78	121
106	100	97	82	100	83	115	104
114	118	96	101	79	130	94	101

a. Construct an 80% confidence interval for the population mean.

b. Construct a 95% confidence interval for the population mean and compare the width of this interval with that of part **a**.

c. Carefully interpret each of the confidence intervals, and explain why the 80% confidence interval is narrower.

Applying the Concepts

7.32 Health insurers and the federal government are both putting pressure on hospitals to shorten the average length of stay (LOS) of their patients. A random sample of 20 hospitals in one state had a mean LOS in 1990 of 3.8 days, and a standard deviation of 1.2 days.

a. Use a 90% confidence interval to estimate the population mean LOS for the state's hospitals in 1990.

b. Interpret the interval in terms of this application.

c. What is meant by the phrase "90% confidence interval"?

7.33 Named for the section of the 1978 Internal Revenue Code that authorized it, a 401(k) plan permits employees to shift part of their before-tax salaries into investments such as mutual funds. Employers typically match 50% of the employee's contribution up to about 6% of salary (Paré, 1992). Concerned with what it believed was a low employee-participation rate in its 401(k) plan, a particular Fortune 500 company sampled 11 other Fortune 500 companies and asked for their 401(k) participation rates. The following answers were obtained (in percentages):

80	76	81	77	82	80	85	60	80	79	82

a. Use a 95% confidence interval to estimate the mean participation rate for Fortune 500 companies.

b. Interpret the interval in the context of this problem.

c. What assumption is necessary to ensure the validity of this confidence interval?

d. If the value 60% in the above data set were instead 80%, how would the center and width of confidence interval you constructed in part **a** be affected?

7.34 A **mortgage** is a type of loan that is secured by a designated piece of property. If the borrower defaults on the loan, the lender can sell the property to recover the outstanding debt. The home mortgage is the most important type of personal loan in the United States. In a home mortgage, the borrower pledges the home in question as security for the loan (Sharpe and Alexander, 1990). A federal bank examiner is interested in estimating the mean outstanding principal balance of all home mortgages foreclosed by the bank due to default by the borrower during the last 3 years. A random sample of 12 foreclosed mortgages yielded the following data (in dollars):

95,982	81,422	39,888	46,836	66,899	69,110
59,200	62,331	105,812	55,545	56,635	72,123

a. Describe the population from which the bank examiner collected the sample data. What characteristic must this population possess to enable us to construct a confidence interval for the mean outstanding principal balance using the method described in this section?
b. Construct a 90% confidence interval for the mean of interest.
c. Carefully interpret your confidence interval in the context of the problem.

7.35 A study indicated that the cost of hiring an employee (excluding salary) ranges from about \$1,500 for a secretary to more than \$40,000 for a manager (Dessler, 1986). In order to estimate its mean cost of hiring an entry-level secretary, a large corporation randomly selected eight of the entry-level secretaries it hired during the last 2 years and determined the costs (in dollars) involved in hiring each. The following data were obtained:

2,100	1,650	1,315	2,035	2,245	1,980	1,700	2,190

Assume that the population from which these data were sampled is approximately normally distributed.
a. Describe the population from which the corporation collected the sample data.
b. Use a 90% confidence interval to estimate the mean of interest to the corporation.
c. How wide is the confidence interval you constructed in part **b**? Would a 95% confidence interval be wider or narrower? Explain.

7.36 Private colleges and universities rely on money contributed by individuals and corporations for their operating expenses. Much of this money is put into a fund called an **endowment**, and the college spends only the interest earned by the fund. A recent survey of eight private colleges in the United States revealed the following endowments (in millions of dollars): 60.2, 47.0, 235.1, 490.0, 122.6, 177.5, 95.4, and 220.0. Estimate the mean endowment for private colleges, using a 95% confidence interval. List any assumptions you make.

7.37 One of the concerns of U.S. industry as we enter the mid-1990s is the increasing cost of health insurance for its workers. A random sample of 23 small companies (companies with less than \$10 million in annual revenues) that offer paid health insurance as a benefit was selected. The mean health insurance cost per worker per month was \$135, and the standard deviation was \$32.
a. Use a 95% confidence interval to estimate the mean cost per worker per month for all small companies.
b. What assumption is necessary to ensure the validity of the confidence interval?
c. What is meant by the phrase "95% confidence interval"?

7.4 Large-Sample Estimation of a Binomial Probability

Many market studies are conducted by companies with the objective of determining the fraction of buyers of a particular product who prefer the company's brand. For example, a tobacco company may conduct a market study by sampling and interviewing 1,000 smokers to determine their brand preference. The objective of the survey is to estimate the proportion of all smokers who smoke the company's brand. The number, x, of the 1,000 sampled who smoke the company's brand is a binomial random variable. (See Section 4.4 for a description of the binomial experiment.) The probability, p, that a smoker prefers the company's brand is the parameter to be estimated. That is, we want to estimate the proportion p of all smokers who prefer the company's brand.

How would you estimate the probability, p, of success in a binomial experiment? One logical answer is to use the proportion of successes in the sample. That is, we can estimate p by calculating

$$\hat{p} = \frac{\text{Number of successes in the sample}}{\text{Number of trials}} = \frac{x}{n}$$

Thus, if 313 of the 1,000 smokers were found to smoke the company's brand, we would estimate the proportion p of all smokers who prefer the brand to be

$$\hat{p} = \frac{x}{n} = \frac{313}{1{,}000} = .313$$

To determine the reliability of the estimator $\hat{p}$, we need to know its sampling distribution. That is, if we were to draw samples of 1,000 smokers over and over again, each time calculating a new estimate $\hat{p}$, what would be the frequency distribution of all the $\hat{p}$ values? The answer lies in viewing $\hat{p}$ as the average or mean number of successes per trial over the n trials. Thus, if each success in the sample is assigned a value equal to 1 and each failure is assigned a 0, then the sum of all n sample observations is x, the total number of successes; and $\hat{p} = x/n$ is the average or mean number of successes per trial in the n trials. The Central Limit Theorem tells us that for any population the relative frequency distribution of the sample mean is approximately normal for sufficiently large samples. We demonstrated the properties of $\hat{p}$ in Example 6.7. The sampling distribution of $\hat{p}$ has the characteristics indicated in Figure 7.8 and listed in the next box.

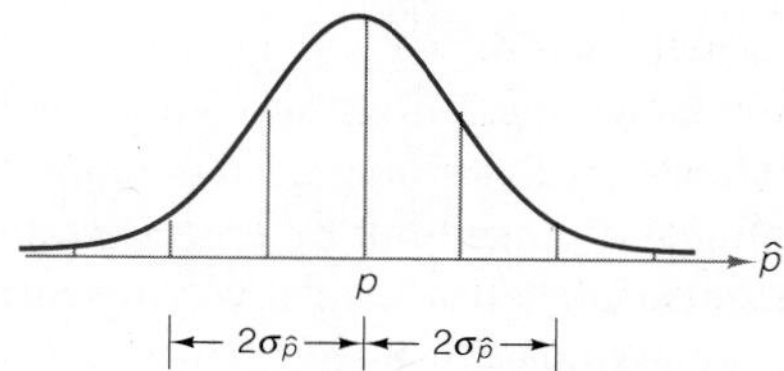

FIGURE 7.8 ▶
Sampling distribution of $\hat{p}$

Sampling Distribution of $\hat{p}$

1. The mean of the sampling distribution of $\hat{p}$ is p; i.e., $\hat{p}$ is an unbiased estimator of p.

Continued

2. The standard deviation of the sampling distribution of $\hat{p}$ is $\sqrt{pq/n}$; i.e., $\sigma_{\hat{p}} = \sqrt{pq/n}$, where $q = 1 - p$.
3. For large samples, the sampling distribution of $\hat{p}$ is approximately normal. A sample size will be considered large if the interval $\hat{p} \pm 3\sigma_{\hat{p}}$ does not include 0 or 1. [*Note:* This requirement is almost equivalent to that given in Section 5.5 for approximating a binomial distribution with a normal one. The main difference is that we assumed p to be known in Section 5.5; now we are trying to make inferences about an unknown p, so we use $\hat{p}$ to estimate p in checking the adequacy of the normal approximation.]

The fact that $\hat{p}$ is a mean number of successes per trial allows us to form confidence intervals in a manner that is completely analogous to that used for large-sample inferences about μ:

Large-Sample Confidence Interval for p

$$\hat{p} \pm z_{\alpha/2}\sigma_{\hat{p}} = \hat{p} \pm z_{\alpha/2}\sqrt{\frac{pq}{n}} \approx \hat{p} \pm z_{\alpha/2}\sqrt{\frac{\hat{p}\hat{q}}{n}}$$

where $\hat{p} = \frac{x}{n}$ and $\hat{q} = 1 - \hat{p}$

[*Note:* When n is large, we can use $\hat{p}$ to approximate the value of p in the formula for $\sigma_{\hat{p}}$.]

Thus, if 313 of 1,000 smokers smoke the company's brand, a 95% confidence interval for the proportion of all smokers who prefer the company's brand is

$$\hat{p} \pm z_{\alpha/2}\sigma_{\hat{p}} = .313 \pm 1.96\sqrt{\frac{pq}{1{,}000}}$$

where $q = 1 - p$. Just as we needed an approximator for σ in calculating a large-sample confidence interval for μ, we now need an approximation for p. As Table 7.3 shows, the approximation for p need not be especially accurate, because the value of pq needed for the confidence interval is relatively insensitive to changes in p. Therefore, we can use $\hat{p}$ to approximate p. Keeping in mind that $\hat{q} = 1 - \hat{p}$, we substitute these values into the formula for the confidence interval:

TABLE 7.3 Values of pq for Several Different p Values

p	pq
.5	.25
.6 or .4	.24
.7 or .3	.21
.8 or .2	.16
.9 or .1	.09

$$\hat{p} \pm 1.96\sqrt{\frac{pq}{1{,}000}} \approx \hat{p} \pm 1.96\sqrt{\frac{\hat{p}\hat{q}}{1{,}000}}$$

$$= .313 \pm 1.96\sqrt{\frac{(.313)(.687)}{1{,}000}} = .313 \pm .029$$

$$= (.284, .342)$$

The company can be 95% confident that the interval from 28.4% to 34.2% contains the true percentage of all smokers who prefer its brand. That is, in repeated construction of confidence intervals, approximately 95% of all samples would produce confidence intervals that enclose p. Note that confidence intervals for p are interpreted in the same manner as confidence intervals for μ.

EXAMPLE 7.4

Many public polling agencies conduct surveys to determine the current consumer sentiment concerning the state of the economy. For example, the Bureau of Economic and Business Research (BEBR) at the University of Florida conducts quarterly surveys to gauge consumer sentiment in the Sunshine State. Suppose that BEBR randomly samples 484 consumers and finds that 257 are optimistic about the state of the economy. Use a 90% confidence interval to estimate the proportion of all consumers in Florida who are optimistic about the state of the economy. Based on the confidence interval, can BEBR infer that the majority of Florida consumers are optimistic about the economy?

Solution

The number, x, of the 484 sampled consumers who are optimistic about the Florida economy is a binomial random variable if we can assume that the sample was randomly selected from the population of Florida consumers and that the poll was conducted identically for each sampled consumer.

The point estimate of the proportion of Florida consumers who are optimistic about the economy is

$$\hat{p} = \frac{x}{n} = \frac{257}{484} = .531$$

We first check to be sure that the sample size is sufficiently large that the normal distribution provides a reasonable approximation for the sampling distribution of $\hat{p}$. We check the 3-standard-deviation interval around $\hat{p}$:

$$\hat{p} \pm 3\sigma_{\hat{p}} \approx \hat{p} \pm 3\sqrt{\frac{\hat{p}\hat{q}}{n}}$$

$$= .531 \pm 3\sqrt{\frac{(.531)(.469)}{484}} = .531 \pm .068 = (.463, .599)$$

Since this interval is wholly contained in the interval (0, 1), we may conclude that the normal approximation is reasonable.

We now proceed to form the 90% confidence interval for p, the true proportion of Florida consumers who are optimistic about the state of the economy:

$$\hat{p} \pm z_{\alpha/2}\sigma_{\hat{p}} = \hat{p} \pm z_{\alpha/2}\sqrt{\frac{pq}{n}} \approx \hat{p} \pm z_{\alpha/2}\sqrt{\frac{\hat{p}\hat{q}}{n}}$$

$$= .531 \pm 1.645\sqrt{\frac{(.531)(.469)}{484}} = .531 \pm .037 = (.494, .568)$$

Thus, we can be 90% confident that the proportion of all Florida consumers who are confident about the economy is between .494 and .568. As always, our confidence stems from the fact that 90% of all similarly formed intervals will contain the true proportion p and not from any knowledge about whether this particular interval does.

Can we conclude that the majority of Florida consumers are optimistic about the economy based on this interval? If we wished to use this interval to infer that a majority is optimistic, the interval would have to support the inference that p exceeds .5—that is, that more than 50% of the Florida consumers are optimistic about the economy. Note that the interval contains some values below .5 (as low as .494) as well as some above .5 (as high as .568). Therefore, we cannot conclude that the true value of p exceeds .5 based on this 90% confidence interval.

In Example 7.4, we used a confidence interval to make an inference about whether the true value of p exceeds .5. That is, we used the sample to *test* whether p exceeds .5. When we want to use sample information to test the value of a population parameter, we usually use a *test of hypothesis*, the subject of Chapter 8.

We conclude Chapter 7 by showing how to determine the sample size necessary to make inferences about a binomial probability with a specified reliability.

Exercises 7.38–7.48

Learning the Mechanics

7.38 Describe the sampling distribution of $\hat{p}$ based on the large samples of size n. That is, give the mean, the standard deviation, and the (approximate) shape of the distribution of $\hat{p}$ when large samples of size n are (repeatedly) selected from a binomial distribution with probability of success p.

7.39 Explain the meaning of the phrase "$\hat{p}$ is an unbiased estimator of p."

7.40 A random sample of size $n = 225$ yielded $\hat{p} = .46$.

a. Is the sample size large enough to use the methods of this section to construct a confidence interval for p? Explain.

b. Construct a 95% confidence interval for p.

c. Interpret the 95% confidence interval.

d. Explain what is meant by the phrase "95% confidence interval."

7.41 A random sample of size $n = 121$ yielded $\hat{p} = .88$.

a. Is the sample size large enough to use the methods of this section to construct a confidence interval for p? Explain.

b. Construct a 90% confidence interval for p.

c. What assumption is necessary to assure the validity of this confidence interval?

7.42 For the binomial sample information summarized in each part, indicate whether the sample size is large enough to use the methods of this chapter to construct a confidence interval for p.

a. $n = 400$, $\hat{p} = .10$ **b.** $n = 50$, $\hat{p} = .10$ **c.** $n = 20$, $\hat{p} = .5$ **d.** $n = 20$, $\hat{p} = .3$

7.43 A random sample of 50 consumers taste-tested a new snack food. Their responses were coded (0: do not like; 1: like; 2: indifferent) and listed:

1 0 0 1 2 0 1 1 0 0 0 1 0 2 0 2 2 0 0 1 1 0 0 0 0
1 0 2 0 0 0 1 0 0 1 0 0 1 0 1 0 2 0 0 1 1 0 0 0 1

a. Use an 80% confidence interval to estimate the proportion of consumers who like the snack food.
b. Provide a statistical interpretation for the confidence interval constructed in part **a**.

Applying the Concepts

7.44 A random sample of 122 Illinois law firms was selected to determine their degree of computer usage (Wentling, 1988). This survey showed that 76 of the firms used microcomputers (PCs).
a. Use a 95% confidence interval to estimate the proportion of all Illinois law firms who used microcomputers at the time of the survey.
b. Interpret the interval in the terms of this application.
c. What is meant by the phrase "95% confidence interval"?
d. Do you think this interval provides an estimate for the proportion of all U.S. law firms who were using microcomputers at the time of the survey? Why or why not?

7.45 For decades, U.S. companies have tied a portion of the compensation of many upper-management employees to the performance of the firm. In the last few years, however, a growing number of companies have begun offering similar performance incentives to all of their employees. What instigated this change in pay structures? Of 46 companies (surveyed by the American Compensation Association) that had modified their traditional pay structures, 26 reported making the modification in response to profound market changes in the last decade: global competition, consumer demand for high-quality goods and services, etc. They needed new ways to improve performance and cut costs (Alters, 1992). For the population of U.S. firms that have switched to nontraditional pay structures, estimate the proportion that have done so in response to market forces (rather than to growth, downsizing, or some other reason). Use a 90% confidence interval and specify whatever assumptions are necessary to assure the validity of the estimate.

7.46 Following a year-long investigation begun in 1985, a congressional subcommittee concluded that inflated home appraisals are responsible, in part, for many defaulted home mortgages. One insurer sampled 300 defaulted home mortgages and found that 120 of them involved defective appraisals (Harney, 1986).
a. Assume that the 300 defaulted mortgages were randomly sampled from all defaulted mortgages during 1985. Use a 90% confidence interval to estimate the proportion of all defaulted home mortgages in 1985 that involved a defective home appraisal.
b. How wide is the confidence interval you constructed in part **a**? Would an 80% confidence interval be wider or narrower than your interval of part **a**? Explain.

7.47 In February and March 1985, the Gallup Organization conducted telephone interviews with a random sample of 258 owners and managers of small U.S. businesses (i.e., firms with 20 or more employees but with less than $50 million in sales). Forty percent of those interviewed were under 45 years of age and 30% have worked at four or more companies (Graham, 1985).
a. Describe the population of interest to the Gallup Organization.
b. Is the sample size large enough to construct a confidence interval for the proportion of owners and managers who are under 45 years of age? Explain.
c. Construct a 95% confidence interval for the proportion of interest in part **b**.

d. If a 95% confidence interval were used to estimate the proportion of owners and managers who have worked at four or more companies, would the interval be wider, narrower, or the same width as the interval estimate of part **c**? Explain.

7.48 Shoplifting is an escalating problem for retailers. Recently, one New York City store randomly selected 500 shoppers and observed them while they were in the store. Two in 25 were seen stealing. How accurate is this estimate? To help you answer this question, construct a 95% confidence interval for p, the proportion of all the store's customers who are shoplifters.

7.5 Determining the Sample Size Necessary to Estimate a Binomial Probability

We showed in Section 7.2 that samples can be designed to estimate a population mean μ with a specified degree of reliability. An analogous situation exists for estimating a binomial probability p. For example, in Section 7.4 a tobacco company used a sample of 1,000 smokers to calculate a 95% confidence interval for the proportion of smokers who preferred its brand, obtaining the interval .284 to .342. Note that the total width of the interval is about .06 (.342 − .284 ≈ .06). Suppose the company wants to estimate its market share more precisely, say with a 95% confidence interval having a width of .03.

The company wants a confidence interval width W of .03. This corresponds to a half-width, or bound B on the estimate of p, of $B = W/2 = .015$. The sample size n to generate such an interval is found by solving the following equation for n:

$$z_{\alpha/2}\sigma_{\hat{p}} = B \qquad \text{or} \qquad z_{\alpha/2}\sqrt{\frac{pq}{n}} = .015 \quad \text{(see Figure 7.9)}$$

FIGURE 7.9 ► Specifying the total width W (or bound B) of a confidence interval for a binomial probability p

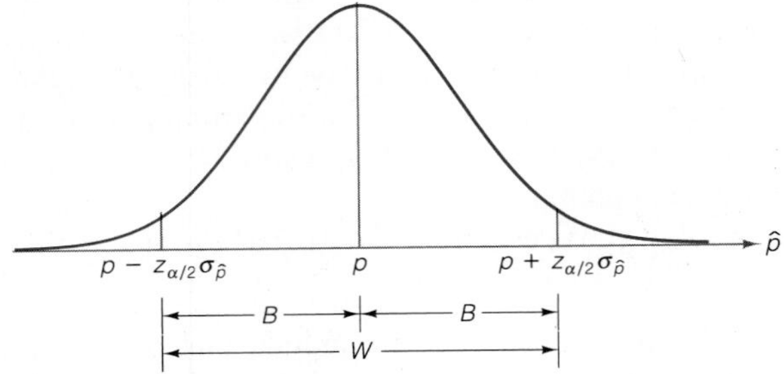

Since a 95% confidence interval is desired, the appropriate z value is $z_{\alpha/2} = z_{.025} = 1.96$. We must approximate the value of the product pq before we can solve the equation for n. As shown in Table 7.3, the closer the values of p and q to .5, the larger the product pq. Thus, to find a conservatively large sample size that will generate a confidence interval with the specified reliability, we generally choose an approximation of p close to .5. In the case of the tobacco company, however, we have an initial sample estimate of $\hat{p} = .313$. A conservatively large estimate of pq can therefore be obtained by using, say, $p = .35$. We now substitute into the equation and solve for n:

$$1.96\sqrt{\frac{(.35)(.65)}{n}} = .015$$

$$n = \frac{(1.96)^2(.35)(.65)}{(.015)^2} = 3{,}884.28 \approx 3{,}885$$

The company must sample about 3,885 smokers to estimate the percentage who prefer its brand with a 95% confidence interval of width .03.

The procedure for finding the sample size necessary to estimate a binomial probability p to within a given bound B or with a total interval width W is given in the box.

Sample Size Determination for $100(1 - \alpha)\%$ Confidence Interval for p

In order to estimate a binomial probability p to within a bound B or, equivalently, with a confidence interval of total width W with $100(1 - \alpha)\%$ confidence, the required sample size is found by solving one of the following equations for n:

$$z_{\alpha/2}\sqrt{\frac{pq}{n}} = B \qquad z_{\alpha/2}\sqrt{\frac{pq}{n}} = \frac{W}{2}$$

The solution can be written in terms of either B or W:

$$n = \frac{(z_{\alpha/2})^2(pq)}{B^2} \qquad n = \frac{4(z_{\alpha/2})^2(pq)}{W^2}$$

The value of the product pq is usually unknown. It can be estimated by using the sample fraction of successes, $\hat{p}$, from a prior sample. Alternatively, remember from Table 7.3 that the value of pq is at its maximum when p equals .5, so that you can obtain conservatively large values of n by approximating p by .5, or values close to .5. In any case, you should round the value of n you obtain *upward* to ensure that the sample size will be sufficient to achieve the specified reliability.

EXAMPLE 7.5

Since the deregulation of the telephonic communications industry, many firms have begun manufacturing telephones. Suppose one large manufacturer that entered the market quickly has an initial problem with excessive customer complaints and consequent returns of the phones for repair or replacement. The manufacturer wants to estimate the magnitude of the problem in order to estimate potential warranty costs. How many telephones should be sampled from the warehouse and checked in order to estimate the fraction defective p to within .01 with 90% confidence?

Solution

In order to estimate p to within a bound of .01, we set the half-width of the confidence interval equal to $B = .01$, as shown in Figure 7.10.

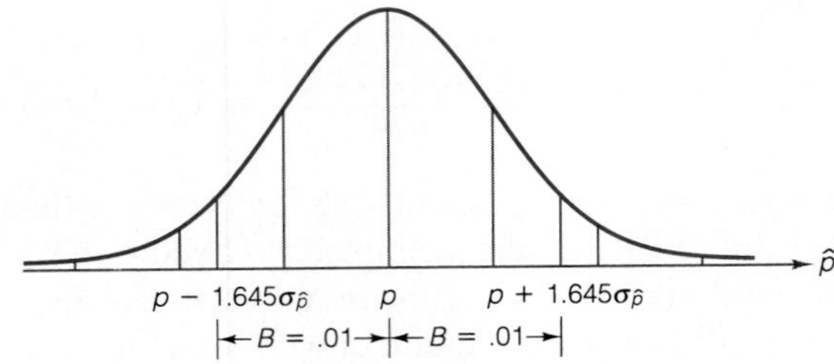

FIGURE 7.10 ► Specified reliability for estimate of fraction defective in Example 7.5

The equation for the sample size n requires an estimate of the product pq. We could most conservatively estimate $pq = .25$ (i.e., use $p = .5$), but this may be overly conservative when estimating a fraction defective (as consumers, we hope so!). A value of .1, corresponding to 10% defective, will probably be conservatively large for this application. The solution is, therefore,

$$n = \frac{(z_{\alpha/2})^2(pq)}{B^2} \approx \frac{(1.645)^2(.1)(.9)}{(.01)^2} = 2{,}435.4 \approx 2{,}436$$

Thus, the manufacturer should sample 2,436 telephones in order to estimate the fraction defective p to within .01 with 90% confidence. Remember that this answer depends on our approximation for pq, where we used .09. If the fraction defective is closer to . 05 than .10, we can use a sample of 1,286 telephones (check this) to estimate p to within .01 with 90% confidence.

The cost of sampling will play an important role in the final determination of the sample size to be selected to estimate a binomial probability p. Although more complex formulas can be derived to balance the reliability and cost considerations, we will solve for the necessary sample size and note that the sampling budget may be a limiting factor. Consult the references for a more complete treatment of this problem.

Exercises 7.49–7.56

Learning the Mechanics

7.49 In each case, find the approximate sample size necessary to estimate a binomial proportion p correct to within .03 with probability equal to .90.

a. Assume you know p is near .8.

b. Assume you have no knowledge of the value of p, but you want to be certain that your sample is large enough to achieve the specified accuracy for the estimate.

7.50 In each case, find the approximate sample size required to construct a 95% confidence interval for p that has width .08.

a. Assume p is near .2. b. Assume you have no prior knowledge about p.

7.51 The following is a 90% confidence interval for p: (.26, .54). How large was the sample used to construct this interval?

Applying the Concepts

7.52 According to estimates made by the General Accounting Office, the Internal Revenue Service (IRS) answered 18.3 million telephone inquiries during the 1986 tax season and 17% of the IRS offices provided answers that were wrong. These estimates were based on data collected from sample calls to numerous IRS offices (Roper, 1986). How many IRS offices should be randomly selected and contacted in order to estimate the proportion of IRS offices that fail to correctly answer questions about gift taxes with a 90% confidence interval of width .06?

7.53 While corporate executives are probably not as highly stressed as air traffic controllers or inner-city police, research has indicated that they are among the more highly pressured work groups. In order to estimate p, the proportion of managers who perceive themselves to be frequently under stress, Hall and Savery (1986) sampled 532 managers in Western Australian corporations. One hundred ninety of these managers fell into the "high-stress" group. Assume that random sampling was used in this study. Was the sample size large enough to estimate p to within .03 with 95% confidence? Explain.

7.54 The quality director of a firm that manufactures videotapes has recommended that the percentage of defective tapes produced each hour be monitored using 90% confidence intervals. Based on previous testing and consumer complaints, it is believed that about 3% of the tapes produced each hour are defective. How many tapes should be sampled and tested each hour for the sample proportion to be within .01 of the true hourly percentage defective with 90% confidence?

7.55 Suppose you are a retailer and you want to estimate the proportion of your customers who are shoplifters. You decide to select a sample of shoppers and check closely to determine whether they steal any merchandise while in the store. Suppose experience suggests that the percentage of customers who are shoplifters is near 5%. How many customers should you include in your sample if you want to estimate the proportion of shoplifters in your store correct to within .02 with 90% confidence?

7.56 Some quality control testing involves destructive sampling, i.e., the test to determine whether the item is defective destroys the product. This type of sampling is generally expensive, and high costs often prohibit large sample sizes. For example, suppose the National Highway Safety Administration (NHSA) wishes to determine the proportion of new tires that will fail when subjected to hard braking at a speed of 60 miles per hour. NHSA can obtain the tires for \$25 (wholesale price) each. Suppose the budget for the experiment is \$10,000, and NHSA wishes to estimate the percentage that will fail to within .02 with 95% confidence. If the entire \$10,000 can be spent on tires (i.e., ignoring other costs), and assuming that the true fraction that will fail is approximately .05, can NHSA attain its goal while staying within the budget? Explain.

Summary

From the beginning we have stressed that the theme of this text is the use of sample information to make inferences about the parameters of a population. For the first seven chapters, we developed the tools necessary to perform inference-making procedures. In Chapter 7 we discussed the inferential procedure called **estimation**.

Estimation involves both **point estimates** and **interval estimates**. Point estimates are single points calculated from a sample to estimate a population parameter. Interval estimates utilize **confidence coefficients** to express the reliability of the estimate. The confidence coefficient is the probability that the estimation procedure will generate an interval that contains the parameter being estimated. Thus, confidence is expressed in terms of the long-run performance of the estimation procedure.

We used the standard normal z distribution to develop large-sample interval estimates for the population mean μ and the binomial probability p. The t distribution was used to form confidence intervals for the mean of normal distributions when n is small.

Supplementary Exercises 7.57–7.71

7.57 In each of the following instances, determine whether you would use a z statistic, a t statistic, or neither to form a 95% confidence interval; then look up the appropriate z or t value in the tables.

a. Random sample of size $n = 23$ from a normal distribution with unknown mean μ and standard deviation σ

b. Random sample of size $n = 135$ from a normal distribution with unknown mean μ and standard deviation σ

c. Random sample of size $n = 10$ from a normal distribution with unknown mean μ and known standard deviation $\sigma = 5$

d. Random sample of size $n = 73$ from a distribution about which nothing is known

e. Random sample of size $n = 12$ from a distribution about which nothing is known

7.58 Let t_0 represent a particular value of t from Table VI of Appendix B. Find the table values such that the following statements are true:

a. $P(t \leq t_0) = .05$ where df $= 20$

b. $P(t \geq t_0) = .005$ where df $= 9$

c. $P(t \leq -t_0$ or $t \geq t_0) = .10$ where df $= 8$

d. $P(t \leq -t_0$ or $t \geq t_0) = .01$ where df $= 17$

7.59 A large New York City bank is interested in estimating (1) the proportion of weeks in which it processes more than 100,000 checks and (2) the average number of checks it processes per week. The bank maintains records of x, the number of checks processed each week. Suppose the bank records the number of checks, x, processed per week for 50 weeks randomly sampled from among the past 6 years. Define each of the following *in the context of the problem*:

a. $\bar{x}$ **b.** $\hat{p}$ **c.** σ_x **d.** μ_x **e.** n **f.** $\sigma_{\bar{x}}$ **g.** p **h.** s_x

7.60 A company is interested in estimating μ, the mean number of days of sick leave taken by all its employees. The firm's statistician selects at random 100 personnel files and notes the number of sick days taken by each employee. The following sample statistics are computed:

$$\bar{x} = 12.2 \text{ days} \qquad s = 10 \text{ days}$$

a. Estimate μ using a 90% confidence interval.

b. How many personnel files would the statistician have to select in order to estimate μ to within 2 days with 99% confidence?

7.61 A firm's president, vice presidents, department managers, and others use financial data generated by the firm's accounting system to help them make decisions regarding such things as pricing, budgeting, and plant expansion. To provide reasonable certainty that the system provides reliable data, internal auditors periodically perform various checks of the system (Taylor and Glezen, 1979). Suppose an internal auditor is interested in determining the proportion of sales invoices in a population of 5,000 sales invoices for which the "total sales" figure is in error. She plans to estimate the true proportion of invoices in error based on a random sample of size 100.

a. Assume that the population of invoices is numbered from 1 to 5,000 and that every invoice ending with a 0 is in error (i.e., 10% are in error). Use the random number table (Table I in Appendix B) to draw a random sample of 100 invoices from the population of 5,000 invoices. For example, random number 456 stands for invoice number 456. List the invoice numbers in your sample and indicate which of your sampled invoices are in error (i.e., those ending in a 0).

b. Use the results of your sample of part **a** to construct a 90% confidence interval for the true proportion of invoices in error.

c. Recall that the true population proportion of invoices in error is equal to .1. Compare the true proportion with the estimate of the true proportion you developed in part **b**. Does your confidence interval include the true proportion?

7.62 When companies employ control charts to monitor the quality of their products, a series of small samples is typically used to determine if the process is "in control" during the period of time in which each sample is selected. (We cover quality control charts in Chapter 13.) Suppose a concrete-block manufacturer samples nine blocks per hour and tests the breaking strength of each. During one hour's test the mean and standard deviation are 985.6 pounds per square inch (psi) and 22.9 psi, respectively.

a. Construct a 99% confidence interval for the mean breaking strength of blocks produced during the hour in which the sample was selected.

b. The process is to be considered "out of control" if the mean strength differs from 1,000 psi. What would you conclude based on the confidence interval constructed in part **a**?

c. Repeat parts **a** and **b** using a 90% confidence interval.

d. The manufacturer wants to be reasonably certain that the process is really out of control before shutting down the process and trying to determine the problem. Which interval, the 99% or 90% confidence interval, is more appropriate for making the decision? Explain.

e. Which assumptions are necessary to assure the validity of the confidence intervals?

7.63 The Internal Revenue Service is conducting an audit of the 10,000 outlets of a large fast-food chain. It is interested in determining the average error in reported income last year for all outlets in the chain. The size of the chain precludes a census (an audit of all 10,000 outlets), so 100 outlets are randomly selected and audited. Let x = error in reported income = (actual income − reported income) for a given firm. The audits yielded the following statistics:

$$\bar{x} = \$12,522 \qquad s = \$4,000$$

a. Construct a 95% confidence interval for the mean error in reported income per outlet.
b. What does the confidence interval from part **a** reflect regarding the chain's income-reporting behavior last year?

7.64 A market researcher wants to select one sample to estimate both μ, the average age of people living within 5 miles of a proposed shopping mall site, and p, the proportion of people within that 5-mile radius who are between 20 and 40 years of age. He wants to estimate μ with a 95% confidence interval that is no more than 6 years wide and p with a 90% confidence interval of width no greater than .1. It is known from previous studies of this population that the standard deviation of the ages in the population is 10 years, and it is believed that p is near .4. How large a sample does the researcher need to draw to construct confidence intervals for both μ and p that satisfy the above specifications?

7.65 Recycling has received increasing emphasis among environmental concerns as a means of dealing with the significant growth in the trash and garbage of our "throw-away" society. To estimate the degrees of awareness about recycling in one major city, a random sample of 346 households was selected, and 212 were found to use available recycling facilities. Use a 95% confidence interval to estimate the proportion of households in the city who are using available recycling facilities.

7.66 In 1988, tire sales in the United States reached a historic high ("Bounce for Synthetic Rubber: Tires Are Back in the Fast Lane," 1989). Tires made of synthetic rubber are accounting for much of this growth. Suppose one manufacturer is testing a new synthetic rubber design. Twenty tires of the new design are produced and subjected to wear tests. The results indicate that the mean wear for the test tires is 42,250 miles, and the standard deviation is 4,355 miles.
a. What is the point estimate of the true mean wear for the new design?
b. Construct a 90% confidence interval estimate for the mean wear associated with the new design.
c. Which method of estimation is better, point estimation or interval estimation? Why?

7.67 Refer to Exercise 7.66. Suppose that 200 tires rather than 20 were tested, and assume that the sample mean and standard deviation for the 200 tires remain the same: $\bar{x}$ = 42,250 miles, s = 4,355 miles. Repeat parts **a–c** of Exercise 7.66 and comment on the similarities and differences in your answers.

7.68 In 1987, a case of salmonella (bacterial) poisoning was traced to a particular brand of ice cream bar, and the manufacturer removed its bars from the market. Despite this response, many consumers refused to purchase any brand of ice cream bars for some period of time after the event (McClave, personal communication). One manufacturer conducted a survey of consumers 6 months after the poisoning. A sample of 244 ice cream bar consumers was contacted, and 23 consumers indicated that they would not purchase ice cream bars because of the potential for food poisoning.
a. What is the point estimate of the true fraction of the entire market who refuse to purchase bars 6 months after the poisoning?
b. Is the sample size large enough to use the normal approximation for the sampling distribution of the estimator of the binomial probability? Justify your response.

c. Construct a 95% confidence interval for the true proportion of the market who still refuse to purchase ice cream bars 6 months after the event. What assumptions are necessary to assure the validity of the interval?
d. Interpret both the point estimate and confidence interval in terms of this application.

7.69 Refer to Exercise 7.68. Suppose it is now 1 year after the poisoning was traced to ice cream bars. The manufacturer wishes to estimate the proportion who will still not purchase bars using a 95% confidence interval of width .04. How many consumers should be sampled?

7.70 Medicaid health assistance programs are administered by the individual states, even though part of the funding is federal. The federal government requires that the states perform regular audits in order to assure that payments are accurate. One Florida hospital was audited by the Florida Department of Health and Rehabilitative Services (HRS) in 1989, and a random sample of 25 Medicaid claims was selected. The sample mean of the claims was $34.76, and the standard deviation was $11.34.
a. Use a 99% confidence interval to estimate the mean of all claims submitted by this hospital.
b. What assumptions are necessary to assure the validity of this confidence interval?

7.71 Refer to Exercise 7.70.
a. How many claims must be sampled if the HRS wants to estimate the mean size of the hospital's claims to within $1.00 using a 99% confidence interval?
b. If a sample of this size were to be selected and a 99% confidence interval constructed, what assumptions would be necessary to assure the validity of the interval?

On Your Own

Choose a population pertinent to your major area of interest that has an unknown mean (or, if the population is binomial, that has an unknown proportion of success). For example, a marketing major may be interested in the proportion of consumers who prefer a particular product. An advertising major might want to estimate the proportion of the television viewing audience who regularly watch a particular program. An economics major may want to estimate the mean monthly expenditure of college students on food.

Define the parameter you want to estimate and conduct a **pilot study** to obtain an initial estimate of the parameter of interest and, more important, an estimate of the variability associated with the estimator. A pilot study employs a small sample (perhaps 20 to 30 observations) to gain some information about the population of interest. The purpose is to help plan more elaborate future studies. Based on the results of your pilot study, determine the sample size necessary to estimate the parameter to within a reasonable bound (of your choice) with a 95% confidence interval.

Using the Computer

Refer to **Using the Computer** in Chapter 6. Recall the values of the "population" mean μ and standard deviation σ for the 1,000 zip code income measurements. Suppose our objective is to sample from this population and to estimate the mean μ using a 95% confidence interval.

a. Determine the sample size n_1 necessary to estimate μ to within $2,000 with 95% confidence. Then generate 100 95% confidence intervals by repeatedly drawing samples of size n_1 (with replacement) from the 1,000 measurements, and using the sample statistics to form a confidence interval. Treat σ as unknown when forming the confidence intervals. What percentage of the confidence intervals contain μ?

b. Determine the sample size n_2 necessary to estimate μ to within \$500 with 95% confidence. Then generate 100 95% confidence intervals by repeatedly drawing samples of size n_2 (with replacement) from the 1,000 measurements, and using the sample statistics to form a confidence interval. Treat σ as unknown when forming the confidence intervals. What percentage of the confidence intervals contain μ?

c. Repeat parts **a** and **b** using an 80% confidence interval. Compare the results.

References

Alters, D. "Teamwork means a share of the gain." *Star Tribune* (Minneapolis), June 29, 1992.

Arshadi, N. and Lawrence, E. C. "A characteristic appraisal of top bank executives." *Journal of Retail Banking*, Winter 1983–1984, Vol. 5, No. 4, pp. 19–25.

"Bounce for synthetic rubber: Tires are back in the fast lane." *Chemical Week*, Apr. 19, 1989, pp. 40–44.

Deming, W. E. *Out of the Crisis*. Cambridge, Mass.: M.I.T. Center for Advanced Study of Engineering, 1986.

Dessler, G. "Personnel tests gain in popularity." *St. Paul Pioneer Press and Dispatch*, Mar. 17, 1986.

Graham, E. "The entrepreneurial mystique." *Wall Street Journal*, May 20, 1985.

Hall, K. and Savery, L. K. "Tight rein, more stress." *Harvard Business Review*, Jan.–Feb. 1986, pp. 160–164.

Harney, K. "The nation's housing." *Minneapolis Star and*

"One answer to imports: Wonder-iron." *Fortune*, Feb. 9, 1981, p. 71.

Paré, T. P. "Is your 401(k) plan good enough?" *Fortune*, Dec. 28, 1992, pp. 78–83.

Roper, J. E. "Survey: 17% of IRS phone answers wrong." *Minneapolis Star and Tribune*, July 26, 1986.

Sharpe, W. F. and Alexander, G. J. *Investments*, 4th ed. Englewood Cliffs, N.J.: Prentice-Hall, Inc., 1990.

Taylor, D. H. and Glezen, G. W. *Auditing, Integrated Concepts and Procedures*. New York: Wiley, 1979, p. 3.

U.S. Bureau of the Census. *Statistical Abstract of the United States: 1992*. Washington, D.C.: U.S. Government Printing Office, 1992.

Wentling, R. M. "Master the software explosion." *Legal Assistant Today*, Mar./Apr. 1988, Vol. 5, No. 4, pp. 44–51.

CHAPTER EIGHT

Inferences Based on a Single Sample: Tests of Hypotheses

Contents

Case Studies

Where We've Been

We showed how to use sample information to estimate population parameters in Chapter 7. The sampling distribution of a statistic was used to assess the reliability of an estimate, which was expressed in terms of a confidence interval.

Where We're Going

We now show how to utilize sample information to test whether a population parameter is less than, equal to, or greater than a specified value. This type of inference is called a *test of hypothesis*. We show how to conduct a test of hypothesis about a population mean μ and a binomial probability p. But just as with estimation, we stress the measurement of the reliability of the inference. An inference without a measure of reliability is little more than a guess.

Suppose you want to determine whether the mean waiting time in the drive-through line of a fast-food restaurant exceeds 5 minutes, whether the mean breaking strength of sewer pipe exceeds 2,400 pounds per foot, or whether the majority of consumers are optimistic about the economy. In each case you are interested in making an inference about how the value of a parameter relates to a specific numerical value: Is it less than, equal to, or greater than the specified number? This type of inference is called a **test of hypothesis**, and testing hypotheses is the subject of this chapter.

We introduce the elements of a test of hypothesis in Section 8.1. We then show how to conduct a large-sample test of hypothesis about a population mean in Sections 8.2 and 8.3. In Section 8.4 we utilize small samples to conduct tests about means. Large-sample tests about binomial probabilities (e.g., population proportions) are the subject of Section 8.5, and some advanced methods for determining the reliability of a test are covered in the optional Section 8.6.

8.1 The Elements of a Test of Hypothesis

Suppose building specifications in a certain city require that the average breaking strength of residential sewer pipe be more than 2,400 pounds per foot of length (that is, per lineal foot). Each manufacturer who wants to sell pipe in this city must demonstrate that its product meets the specification. Note that we are again interested in making an inference about the mean, μ, of a population. However, in this example, we are less interested in estimating the value of μ than we are in **testing a hypothesis** about its value. That is, we want to decide whether the mean breaking strength of the pipe exceeds 2,400 pounds per lineal foot.

The method used to reach a decision is based on the rare event concept explained in earlier chapters. We define two hypotheses: (1) The **null hypothesis** is that which represents the status quo to the party performing the sampling experiment—the hypothesis that will not be rejected unless the data provide convincing evidence that it is false. (2) The **alternative, or research, hypothesis** is that which will be accepted only if the data provide convincing evidence of its truth. From the point of view of the city conducting the tests, the null hypothesis is that the manufacturer's pipe does *not* meet specifications unless the tests provide convincing evidence otherwise. The null and alternative hypotheses are therefore:

Null hypothesis (H_0): $\mu \leq 2{,}400$ (i.e., the manufacturer's pipe does not meet specifications)

Alternative (research) hypothesis (H_a): $\mu > 2{,}400$ (i.e., the manufacturer's pipe meets specifications)

How can the city decide when enough evidence exists to conclude that the manufacturer's pipe meets specifications? Since the hypotheses concern the value of the population mean μ, it is reasonable to use the sample mean $\bar{x}$ to make the inference, just as we did when forming confidence intervals for μ in Sections 7.1 and 7.2. The city will conclude that the pipe meets specifications only when the sample mean $\bar{x}$ convincingly indicates that the population mean exceeds 2,400 pounds per lineal foot.

"Convincing" evidence in favor of the alternative hypothesis will exist when the value of $\bar{x}$ exceeds 2,400 by an amount that cannot be readily attributed to sampling variability. To decide, we compute a **test statistic**, which will be the z value that measures the distance between the value of $\bar{x}$ and the value of μ specified in the null hypothesis. When the null hypothesis contains more than one value of μ, as in this case (H_0: $\mu \leq 2{,}400$), we use the value of μ closest to the values specified in the alternative hypothesis. The idea is that if the hypothesis that μ *equals* 2,400 can be rejected in favor of $\mu > 2{,}400$, then μ *less than or equal to* 2,400 can certainly be rejected. Thus, the test statistic is

$$z = \frac{\bar{x} - 2{,}400}{\sigma_{\bar{x}}} = \frac{\bar{x} - 2{,}400}{\sigma/\sqrt{n}}$$

Note that a value of $z = 1$ means that $\bar{x}$ is 1 standard deviation above $\mu = 2{,}400$; a value of $z = 1.5$ means that $\bar{x}$ is 1.5 standard deviations above $\mu = 2{,}400$, etc. How large must z be before the city can be convinced that the null hypothesis can be rejected in favor of the alternative, and conclude that the pipe meets specifications?

If you examine Figure 8.1, you will note that the chance of observing a value of $\bar{x}$ more than 1.645 standard deviations above 2,400 is only .05, *if in fact the true mean μ is 2,400*. Thus, if the sample mean is more than 1.645 standard deviations above 2,400, either H_0 is true and a relatively rare event has occurred (probability .05 or less) or H_a is true and the population mean exceeds 2,400. Since we would most likely reject the notion that a rare event has occurred, we would reject the null hypothesis ($\mu \leq 2{,}400$) and conclude that the alternative hypothesis ($\mu > 2{,}400$) is true. What is the probability that this procedure will lead us to a wrong decision?

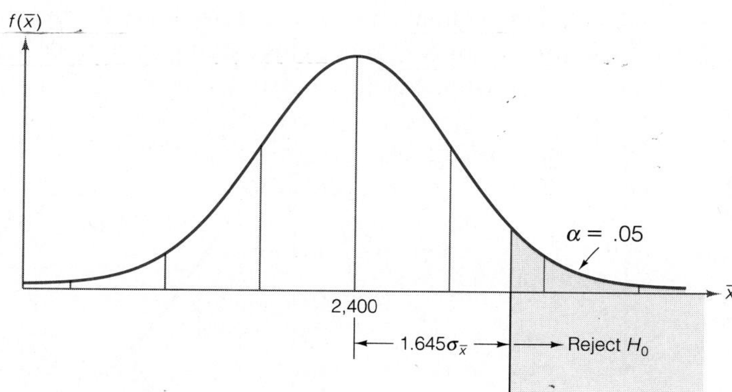

FIGURE 8.1 ► The sampling distribution of $\bar{x}$, assuming $\mu = 2{,}400$

Deciding that the null hypothesis is false when in fact it is true is called a **Type I error**. As indicated in Figure 8.1, the probability of making a Type I error—that is, deciding in favor of the alternative hypothesis when in fact the null hypothesis is true—is only $\alpha = .05$. That is,

$$\alpha = P(\text{Type I error})$$
$$= P(\text{Rejecting the null hypothesis when in fact the null hypothesis is true})$$

In our example,

$$\alpha = P(z > 1.645 \text{ when in fact } \mu = 2{,}400) = .05$$

Therefore, the test can be summarized as follows:

Null and alternative hypotheses: H_0: $\mu = 2{,}400$ $\quad$ H_a: $\mu > 2{,}400$

Test statistic: $z = \dfrac{\bar{x} - 2{,}400}{\sigma_{\bar{x}}}$

Rejection region: $z > 1.645$ for $\alpha = .05$

To illustrate the use of the test, suppose we tested 50 sections of sewer pipe and found the mean and standard deviation for these 50 measurements of breaking strength to be

$$\bar{x} = 2{,}460 \text{ pounds per lineal foot} \qquad s = 200 \text{ pounds per lineal foot}$$

As in the case of estimation, we can use s to approximate σ when s is calculated from a large set of sample measurements.

The test statistic is

$$z = \frac{\bar{x} - 2{,}400}{\sigma_{\bar{x}}} = \frac{\bar{x} - 2{,}400}{\sigma/\sqrt{n}} \approx \frac{\bar{x} - 2{,}400}{s/\sqrt{n}}$$

Substituting $\bar{x} = 2{,}460$, $n = 50$, and $s = 200$, we have

$$z \approx \frac{2{,}460 - 2{,}400}{200/\sqrt{50}} = \frac{60}{28.28} = 2.12$$

Therefore, the sample mean lies $2.12\sigma_{\bar{x}}$ above the hypothesized value of μ, 2,400, as shown in Figure 8.2. Since this value of z exceeds 1.645, it falls in the rejection

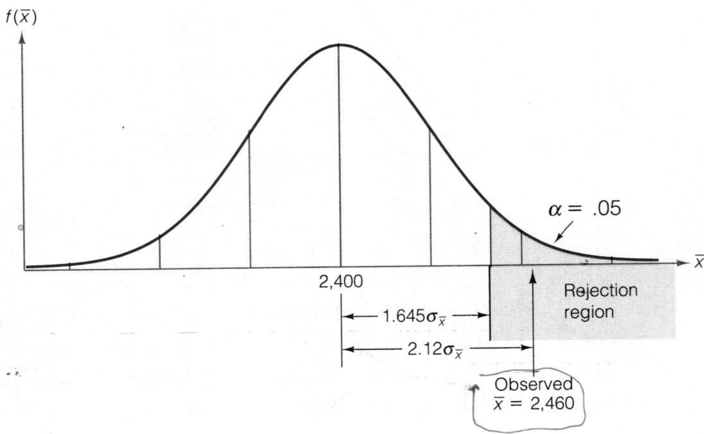

FIGURE 8.2 ► Location of test statistic when $\bar{x} = 2{,}460$

region. That is, we reject the null hypothesis that $\mu = 2{,}400$ and conclude that $\mu > 2{,}400$. Thus, it appears that the company's pipe has a mean strength that exceeds 2,400 pounds per lineal foot.

How much faith can be placed in this conclusion? What is the probability that our statistical test would lead us to reject the null hypothesis (and conclude that the company's pipe met the city's specifications) when in fact the null hypothesis was true? The answer is "$\alpha = .05$." That is, we selected the level of risk, α, of making a Type I error when we constructed the test. Thus, the chance is only 1 in 20 that our test would lead us to conclude the manufacturer's pipe satisfied the city's specifications when in fact the pipe does *not* meet specifications.

Now, suppose the sample mean breaking strength for the 50 sections of sewer pipe turned out to be $\bar{x} = 2{,}430$ pounds per lineal foot. Assuming the sample standard deviation is still $s = 200$, the test statistic is

$$z = \frac{2{,}430 - 2{,}400}{200/\sqrt{50}} = \frac{30}{28.28} = 1.06$$

Therefore, the sample mean $\bar{x} = 2{,}430$ is only 1.06 standard deviations above the null hypothesized value of $\mu = 2{,}400$. As shown in Figure 8.3, this value does not fall in the rejection region ($z > 1.645$). Therefore, we cannot reject H_0 using $\alpha = .05$. Even though the sample mean exceeds the city's specification of 2,400 by 30 pounds per lineal foot, it does not exceed the specification by enough to provide *convincing* evidence that the *population mean* exceeds 2,400.

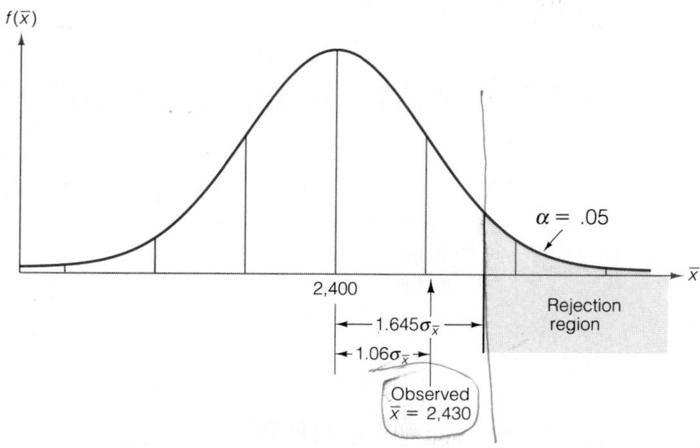

FIGURE 8.3 ► Location of test statistic when $\bar{x} = 2{,}430$

Should we accept the null hypothesis H_0: $\mu \le 2{,}400$, and conclude that the manufacturer's pipe does not meet specifications? To do so would be to risk a **Type II error**—that of concluding that the null hypothesis is true (the pipe does not meet specifications) when in fact it is false (the pipe does meet specifications). We will denote the probability of committing a Type II error by β, and we will show in optional Section 8.6 that β is often difficult to determine precisely. Rather than make a decision (accept H_0) for which the probability of error (β) is unknown, we avoid the potential

Type II error by avoiding the conclusion that the null hypothesis is true. Instead, we will simply state that *the sample evidence is insufficient to reject* H_0 *at* $\alpha = .05$. Since the null hypothesis is the "status-quo" hypothesis, the effect of not rejecting H_0 is to maintain the status quo. In our pipe-testing example, the effect of having insufficient evidence to reject the null hypothesis that the pipe does not meet specifications is probably to prohibit the utilization of the manufacturer's pipe unless and until there is sufficient evidence that the pipe does meet specifications. That is, until the data indicate convincingly that the null hypothesis is false, we usually maintain the status quo implied by its truth.

Table 8.1 summarizes the four possible outcomes of a test of hypothesis. The "true state of nature" columns in Table 8.1 refer to the fact that either the null hypothesis H_0 is true, or the alternative hypothesis H_a is true. Note that the true state of nature is unknown to the researcher conducting the test. The "decision" rows in Table 8.1 refer to the action of the researcher, assuming that he or she will either conclude that H_0 is true or that H_a is true based on the results of the sampling experiment. Note that a Type I error can be made *only* when the alternative hypothesis is accepted (equivalently, when the null hypothesis is rejected) and a Type II error can be made *only* when the null hypothesis is accepted. Our policy will be to make a decision only when we know the probability of making the error that corresponds to that decision. Since α is usually specified by the analyst, we will generally be able to reject H_0 (accept H_a) when the sample evidence supports that decision. However, since β is usually not specified, we will generally avoid the decision to accept H_0, preferring instead to state that the sample evidence is insufficient to reject H_0 when the test statistic is not in the rejection region.

TABLE 8.1 Decisions and Consequences for a Test of Hypothesis

		True State of Nature	
		H_0 True	H_a True
Decision	H_0 True	Correct decision	Type II error (probability β)
	H_a True	Type I error (probability α)	Correct decision

Elements of a Test of Hypothesis

1. *Null hypothesis* (H_0): A theory about the values of one or more population parameters. The theory generally represents the status quo, which we do not reject unless we have strong contradictory evidence.
2. *Alternative* (*research*) *hypothesis* (H_a): A theory that contradicts the null hypothesis. The theory generally represents that which we will accept only when sufficient supporting evidence exists.

3. *Test statistic*: A sample statistic used to decide whether to reject the null hypothesis.
4. *Rejection region*: The numerical values of the test statistic for which the null hypothesis will be rejected. The rejected region is chosen so that the probability is α that it will contain the test statistic when the null hypothesis is true, thereby leading to a Type I error. The value of α is usually chosen to be small (e.g., .01, .05, or .10), and is referred to as the **level of significance** of the test.
5. *Assumptions*: Any assumptions made about the population(s) being sampled should be clearly stated.
6. *Sample and calculate test statistic*: The sample is drawn and the numerical value of the test statistic is determined.
7. *Conclusion*:
   a. If the numerical value of the test statistic falls in the rejection region, we reject the null hypothesis and conclude that the alternative hypothesis is true. We know that the hypothesis-testing process will lead to this conclusion incorrectly (Type I error) only $100\alpha\%$ of the time when H_0 is true.
   b. If the test statistic does not fall in the rejection region, we do not reject H_0. Thus, we reserve judgment about which hypothesis is true. We do not conclude that the null hypothesis is true, because we do not (in general) know the probability β that our test procedure will lead to an incorrect acceptance of H_0 (Type II error).*

The elements of a test of hypothesis are summarized in the preceding box. Note that the first four elements are all specified *before* the sample is drawn. In no case will the results of the sample be used to determine the hypotheses: the data are collected to test the (predetermined) hypotheses, not to formulate them.

CASE STUDY 8.1 / Statistics Is Murder!

The jury trial of an accused murderer is analogous to the statistical hypothesis-testing process. Each of the elements of a test of hypothesis applies to the jury system of deciding the guilt or innocence of the accused:

1. H_0: The null hypothesis in a jury trial is that the accused is not guilty. The status-quo hypothesis in the American system of justice is innocence, which is assumed to be true until proven otherwise.

*In many practical business applications of hypothesis testing, nonrejection leads management to behave as if the null hypothesis were accepted. Accordingly, the distinction between acceptance and nonrejection is frequently blurred in practice. We discuss the issues connected with the acceptance of the null hypothesis and the calculation of β in more detail in (optional) Section 8.6.

2. H_a: The alternative hypothesis is guilt, which is accepted only when sufficient evidence exists to establish its truth.
3. *Test statistic*: The test statistic in a trial is the final vote of the jury, i.e., the number of the jury members who vote "guilty."
4. *Rejection region*: In a murder trial the jury vote must be unanimous in favor of guilt before the null hypothesis of innocence is rejected in favor of the alternative hypothesis of guilt. Thus, for a 12-member jury trial, the rejection region is $x = 12$, where x is the number of "guilty" votes.
5. *Assumption*: The primary assumption made in jury trials concerns the method of selecting the jury. The jury is assumed to represent a random sample of citizens who have no prejudice concerning the case.
6. *Sample and calculate the test statistic*: Sample data are generated through the trial jury's deliberations. The final vote of the jury is analogous to the calculation of the test statistic.
7. *Conclusion*:
   a. If the vote of the jury is unanimous in favor of guilt, the null hypothesis of innocence is rejected and the court concludes that the accused murderer is guilty. Although the court does not, in general, know the probability α that the conclusion is in error, the system relies on the belief that the value is made very small by requiring a unanimous vote before guilt is concluded.
   b. Any vote other than a unanimous one for guilt results in the court reserving judgment about the hypotheses, either by declaring the accused "not guilty," or by declaring a mistrial and repeating the "test" with a new jury. (The latter is analogous to collecting more data and repeating a statistical test of hypothesis.) The court never accepts the null hypothesis by declaring the accused "innocent," perhaps recognizing both that innocence is the "status-quo" hypothesis and does not need to be proved, and that the probability β of incorrectly concluding innocence may be large.

As in the case of tests of statistical hypotheses, we may never know whether the verdict in a murder trial is correct. Instead, we rely on the knowledge that the trial procedure will lead to incorrect conclusions (especially, guilt when the accused is in fact innocent) in only a very small percentage of trials.

Exercises 8.1–8.10

Learning the Mechanics

8.1 Which hypothesis, the null or the alternative, is the status-quo hypothesis? Which is the research hypothesis?

8.2 Which elements of a test of hypothesis are used to decide whether to reject the null hypothesis?

8.3 What is the level of significance of a test of a hypothesis?

8.4 What is the difference between Type I and Type II errors in hypothesis testing? How do α and β relate to Type I and Type II errors?

8.5 List the four possible combinations of decisions and true states of nature for a test of hypothesis.

8.6 Why do we (generally) reject the null hypothesis when the test statistic falls in the rejection region, but do not accept the null hypothesis when the test statistic does not fall in the rejection region?

8.7 If you test a hypothesis and reject the null hypothesis in favor of the alternative hypothesis, does your test prove that the alternative hypothesis is correct? Explain.

Applying the Concepts

8.8 In 1895 an Italian criminologist, Cesare Lombroso, proposed that blood pressure be used to test for truthfulness. In the 1930s, William Marston added the measurements of respiration and perspiration to the process and called his machine the **polygraph**—or lie detector. Today, its use in screening job applicants is on the rise in both industry and government (Dujack, 1986). But how well does it work? Physicians Michael Phillips, Allan Brett, and John Beary subjected the polygraph to the same careful testing given to medical diagnostic tests. They found that if 1,000 people were subjected to the polygraph and 500 told the truth and 500 lied, the polygraph would indicate that approximately 185 of the truth-tellers were liars and that approximately 120 of the liars were truth-tellers ("Lie Detectors Can Make a Liar of You," *Discover*, June 1986, p. 7).

a. In the application of a polygraph test, an individual is presumed to be a truth-teller (H_0) until "proven" a liar (H_a). In this context, what is a Type I error? A Type II error?

b. According to Phillips, Brett, and Beary, what is the probability (approximately) that a polygraph test will result in a Type I error? A Type II error?

8.9 Pharmaceutical companies spend millions of dollars annually on research and development of new drugs to benefit society. After a new drug is formulated, the pharmaceutical company must subject it to lengthy and involved testing before receiving the necessary permission from the Food and Drug Administration (FDA) to market the drug. The pharmaceutical company must provide substantial evidence that a new drug is safe prior to receiving FDA approval, so that the FDA can confidently certify the safety of the drug.

a. If the new drug testing were to be placed in a test-of-hypothesis framework, would the null hypothesis be that the drug is safe or unsafe? The alternative hypothesis?

b. Given the choice of null and alternative hypotheses in part **a**, describe Type I and Type II errors in terms of this application. Define α and β in terms of this application.

c. If the FDA wants to be very confident that the drug is safe before permitting it to be marketed, is it more important that α or β be small? Explain.

8.10 Sometimes monetary costs can be assigned to the commission of Type I and Type II errors. A comparison of these costs helps to decide how large or small to set α—the probability over which the tester usually has control. Suppose you are the owner of a small company that produces a single product and that has averaged sales of \$1,000,000 per year for the past 3 years. One of your employees has conducted a survey and recommends you switch to a different product immediately. If you change your product and he is correct, sales will be \$1,010,000 per year. If you change and he is wrong, your company's yearly sales will drop to \$400,000. Assume your sales will remain at \$1,000,000 per year if you stay with the current product. You are testing with the following hypotheses: H_0: The current product is better; H_a: The new product is better.

a. Define Type I and Type II errors in terms of this example.

b. What is the cost over the next three years of a Type I error? Of a Type II error?

c. Based on your answer to part **b**, at which of the following levels would you set α: .01, .05, .10? Explain.

8.2 Large-Sample Test of Hypothesis About a Population Mean

In Section 8.1 we learned that the null and alternative hypotheses form the basis for a test of hypothesis inference. The null and alternative hypotheses may take one of several forms. In the sewer pipe example, we tested the null hypothesis that the population mean strength of the pipe is less than or equal to 2,400 pounds per lineal foot, against the alternative hypothesis that the mean strength exceeds 2,400. That is, we tested

H_0: $\mu \le 2{,}400$
H_a: $\mu > 2{,}400$

This is a **one-tailed** (or **one-sided**) statistical test because the alternative hypothesis specifies that the population parameter (the population mean μ, in this example) is strictly greater than a specified value (2,400, in this example). If the null hypothesis had been H_0: $\mu \ge 2{,}400$ and the alternative hypothesis had been H_a: $\mu < 2{,}400$, the test would still be one-sided, because the parameter is still specified to be on "one side" of the null hypothesis value.

Some statistical investigations seek to show that the population parameter is *either larger or smaller* than some specified value. Such an alternative hypothesis is called a **two-tailed** (or **two-sided**) hypothesis.

While alternative hypotheses are always specified as strict inequalities, such as $\mu < 2{,}400$, $\mu > 2{,}400$, or $\mu \ne 2{,}400$, null hypotheses are usually specified as equalities, such as $\mu = 2{,}400$. Even when the null hypothesis is an inequality, such as $\mu \le 2{,}400$, we specify H_0: $\mu = 2{,}400$, reasoning that if sufficient evidence exists to show that H_a: $\mu > 2{,}400$ is true when tested against H_0: $\mu = 2{,}400$, then surely sufficient evidence exists to reject $\mu < 2{,}400$ as well. Therefore, the null hypothesis is specified as the value of μ closest to a one-sided alternative hypothesis, and as the only value *not* specified in a two-tailed alternative hypothesis. The steps for selecting the null and alternative hypotheses are summarized in the box.

Steps for Selecting the Null and Alternative Hypotheses

1. Select the **alternative hypothesis** as that which the sampling experiment is intended to establish. The alternative hypothesis will assume one of three forms:

Form	*Example*
One-tailed, upper-tailed	H_a: $\mu > 2{,}400$
One-tailed, lower-tailed	H_a: $\mu < 2{,}400$
Two-tailed	H_a: $\mu \ne 2{,}400$

2. Select the **null hypothesis** as the status quo, that which will be presumed true unless the sample evidence conclusively establishes the alternative hypothesis. The null hypothesis will be specified as the parameter value closest to the alternative in one-tailed tests, and as the complementary (or only unspecified) value to the alternative in two-tailed tests.

   Example: H_0: $\mu = 2{,}400$

The rejection region for a two-tailed test differs from that for a one-tailed test. When we are trying to detect departure from the null hypothesis value in *either* direction, we must establish a rejection region in both tails of the sampling distribution of the test statistic. Figure 8.4(a) and (b) show the one-tailed rejection regions for lower- and upper-tailed tests, respectively. The two-tailed rejection region is illustrated in Figure 8.4(c). Note that a rejection region is established in each tail of the sampling distribution for a two-tailed test.

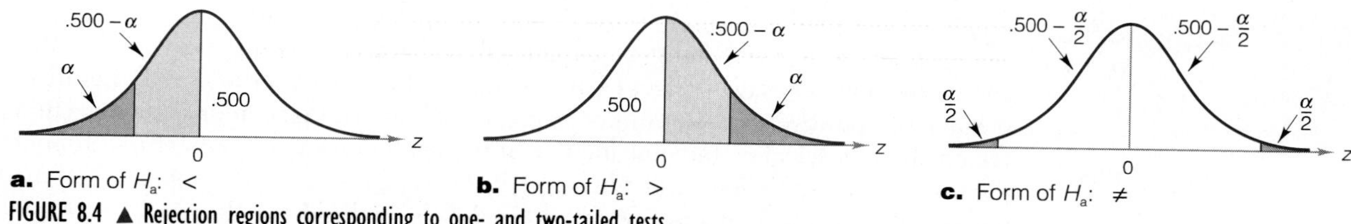

FIGURE 8.4 ▲ Rejection regions corresponding to one- and two-tailed tests

The rejection regions corresponding to typical values selected for α are shown in Table 8.2 for one- and two-tailed tests. Note that the smaller α you select, the more evidence (the larger z) is required before you can reject H_0.

TABLE 8.2 Rejection Regions for Common Values of α

	Alternative Hypothesis		
	Lower-Tailed	Upper-Tailed	Two-tailed
$\alpha = .10$	$z < -1.28$	$z > 1.28$	$z < -1.645$ or $z > 1.645$
$\alpha = .05$	$z < -1.645$	$z > 1.645$	$z < -1.96$ or $z > 1.96$
$\alpha = .01$	$z < -2.33$	$z > 2.33$	$z < -2.576$ or $z > 2.576$

EXAMPLE 8.1

A manufacturer of cereal wants to test the performance of one of its filling machines. The machine is designed to discharge a mean amount of $\mu = 12$ ounces per box, and the manufacturer wants to detect any departure from this setting. This quality control study calls for sampling 100 boxes to determine whether the machine is performing to specifications. Set up a test of hypothesis for this study, using $\alpha = .01$.

Solution

Since the manufacturer wishes to detect a departure from the setting of $\mu = 12$ in either direction, $\mu < 12$ or $\mu > 12$, we conduct a two-tailed statistical test. Following the procedure for selecting the null and alternative hypotheses, we specify as the alternative hypothesis that the mean differs from 12 ounces, since detecting the machine's departure from specifications is the purpose of the quality control study. The null hypothesis is the presumption that the fill machine is operating properly unless the sample data indicate otherwise.

Thus,

H_0: $\mu = 12$
H_a: $\mu \neq 12$ (i.e., $\mu < 12$ or $\mu > 12$)

The test statistic measures the number of standard deviations between the observed value of $\bar{x}$ and the null hypothesized value $\mu = 12$:

Test statistic: $z = \dfrac{\bar{x} - 12}{\sigma_{\bar{x}}}$

The rejection region must be designated to detect a departure from $\mu = 12$ in *either* direction, so we will reject H_0 for values of z that are either too small (negative) or too large (positive). To determine the precise values of z that comprise the rejection region, we first select α, the probability that the test will lead to incorrect rejection of the null hypothesis. Then we divide α equally between the lower and upper tails of the distribution of z, as shown in Figure 8.5. In this example, $\alpha = .01$, so $\alpha/2 = .005$ is placed in each tail. These tail areas correspond to $z = -2.576$ and $z =. 2.576$, respectively (from Table 8.2):

Rejection region: $z < -2.576$ or $z > 2.576$ (Figure 8.5)

Assumptions: Since the sample size is large enough ($n > 30$), the Central Limit Theorem will apply, and no assumptions need to be made about the population of fill measurements. The sampling distribution of the sample mean fill of 100 boxes will be approximately normal regardless of the distribution of the individual boxes' fills.

FIGURE 8.5 ▶
Two-tailed rejection region: $\alpha = .01$

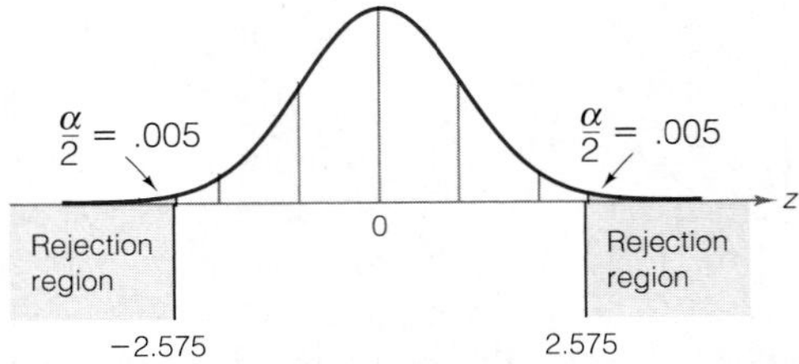

Note that the test in Example 8.1 is set up *before* the sample is drawn; the data are not used to develop the test. Evidently, the manufacturer does not want to disrupt the filling process to adjust the machine unless the sample data provide very convincing evidence that it is not meeting specifications, because the value of α has been set quite low at .01. If the sample evidence results in the rejection of H_0, the manufacturer can be 99% confident that the machine needs adjustment.

Once the test is set up, we are ready to perform the sampling experiment and conduct the test. The test is performed in Example 8.2.

EXAMPLE 8.2

Refer to the quality control test set up in Example 8.1. Suppose the sample yields the following results:

$$n = 100 \text{ observations} \qquad \bar{x} = 11.85 \text{ ounces} \qquad s = .5 \text{ ounce}$$

Use these data to conduct the test of hypothesis.

Solution

Since the test is completely specified in Example 8.1, we simply substitute the sample statistics into the test statistic:

$$z = \frac{\bar{x} - 12}{\sigma_{\bar{x}}} = \frac{\bar{x} - 12}{\sigma/\sqrt{n}} = \frac{11.85 - 12}{\sigma/\sqrt{100}}$$

$$\approx \frac{11.85 - 12}{s/10} = \frac{-.15}{.5/10} = -3.0$$

The implication is that the sample mean, 11.85, is (approximately) 3 standard deviations below the null hypothesized value of 12.0 in the sampling distribution of $\bar{x}$. You can see in Figure 8.5 that this value of z is in the lower-tail rejection region, which consists of all values of $z < -2.575$. These sample data provide sufficient evidence to reject H_0 and conclude, at the $\alpha = .01$ level of significance, that the mean fill differs from the specification of $\mu = 12$ ounces. It appears that the machine is, on average, underfilling the boxes.

Two final points about the test of hypothesis in Example 8.2 apply to all statistical tests:

1. Since z is less than -2.576, it is tempting to state our conclusion at a significance level lower than $\alpha = .01$. We resist the temptation because the level of α is determined *before* the sampling experiment is performed. If we decide that we are willing to tolerate a 1% Type I error rate, the result of the sampling experiment should have no effect on that decision. *In general, the same data should not be used both to determine the significance level and to conduct the test.*
2. When we state our conclusion at the .01 level of significance, we are referring to the failure rate of the *procedure*, not the result of this particular test. We know that the test procedure will lead to the rejection of the null hypothesis only 1% of the time when in fact $\mu = 12$. *Therefore, when the test statistic falls in the rejection region, we infer that the alternative $\mu \neq 12$ is true and express our confidence in the procedure by quoting the α level of significance—that is, the probability that the test leads to a rejection of the null hypothesis when it is true.*

The set-up of a large-sample test of hypothesis about a population mean is summarized in the box. Both the one- and two-tailed tests are shown.

Large-Sample Test of Hypothesis About μ

One-Tailed Test	*Two-Tailed Test*
H_0: $\mu = \mu_0$*	H_0: $\mu = \mu_0$*
H_a: $\mu < \mu_0$ (or H_a: $\mu > \mu_0$)	H_a: $\mu \neq \mu_0$
Test statistic: $z = \dfrac{\bar{x} - \mu_0}{\sigma_{\bar{x}}}$	*Test statistic*: $z = \dfrac{\bar{x} - \mu_0}{\sigma_{\bar{x}}}$
Rejection region: $z < -z_\alpha$ (or $z > z_\alpha$ when H_a: $\mu > \mu_0$)	*Rejection region*: $z < -z_{\alpha/2}$ or $z > z_{\alpha/2}$
where z_α is chosen so that $P(z > z_\alpha) = \alpha$	where $z_{\alpha/2}$ is chosen so that $P(z > z_{\alpha/2}) = \alpha/2$

Assumptions: No assumptions need to be made about the probability distribution of the population, because the Central Limit Theorem assures us that, for large samples, the test statistic will be approximately normally distributed regardless of the shape of the underlying probability distribution of the population.

**Note:* μ_0 is the symbol for the numerical value assigned to μ under the null hypothesis.

Once the test has been set up, the sample is drawn and the test statistic is calculated. The next box contains possible conclusions for a test of hypothesis, depending on the result of the sampling experiment.

Possible Conclusions for a Test of Hypothesis

1. If the calculated test statistic falls in the rejection region, reject H_0 and conclude that the alternative hypothesis H_a is true. State that you are rejecting H_0 at the α level of significance. Remember that the significance level refers to the Type I error rate for the testing process, not the particular result of a single test.
2. If the test statistic does not fall in the rejection region, conclude that the sample data do not provide sufficient evidence to reject H_0 at the α level of significance. [Generally, we will not "accept" the null hypothesis unless the probability β of a Type II error has been calculated (see optional Section 8.6).]

CASE STUDY 8.2 / Statistical Quality Control, Part I

In Exercise 6.27 we described a graphical device, known as a control chart, that can be used in business operations to monitor the variation over time in the quality of products and services being produced. The control chart was developed by Walter A. Shewhart of Bell Telephone Laboratories in 1924 and has become a basic tool of quality control engineers and operations managers the world over. Japan's emergence as an industrial superpower is due in part to their early adoption and refinement of quality control techniques, such as the control chart (Duncan, 1986). In this case study and Case Study 8.3, we expand the discussion of control charts and demonstrate that they are simply vehicles for conducting hypothesis tests. We discuss control charts in detail in Chapter 13.

Suppose it is desired to monitor the pitch diameter of the threads on a particular aircraft fitting. When the process is in control, the pitch diameters follow a normal distribution with mean μ_0 and standard deviation σ_0. Recall from Exercise 6.27 that such monitoring can be accomplished by (1) sampling n items from the production process at regular time intervals, (2) measuring the pitch diameter of each item sampled, and (3) plotting the mean diameter of each sample, $\bar{x}$, on a control chart like that in Figure 8.6. Such a control chart is called an **$\bar{x}$-chart**. If a value of $\bar{x}$ falls above the upper control limit or below the lower control limit, there is strong evidence that the process is out of control—i.e., that the quality of the product being produced does not meet established standards. Otherwise, the process is deemed to be in control.

In the language of hypothesis testing, this decision process can be described as follows:

1. There are two hypotheses of interest:

$$H_0: \text{ Process is in control, } \mu = \mu_0$$
$$H_a: \text{ Process is out of control, } \mu \neq \mu_0$$

2. The test statistic used to investigate these hypotheses is $\bar{x}$.
3. The upper and lower control limits define the rejection region for the test.
4. Since the control limits are located at $\mu_0 \pm 3\sigma_{\bar{x}}$, the probability of committing a Type I error is $\alpha = .0026$. (Why?)

Thus, each time an analyst plots a sample mean on an $\bar{x}$-chart and observes where it falls in relation to the control limits, the analyst is conducting a two-tailed hypothesis test.

FIGURE 8.6 ►
$\bar{x}$-chart

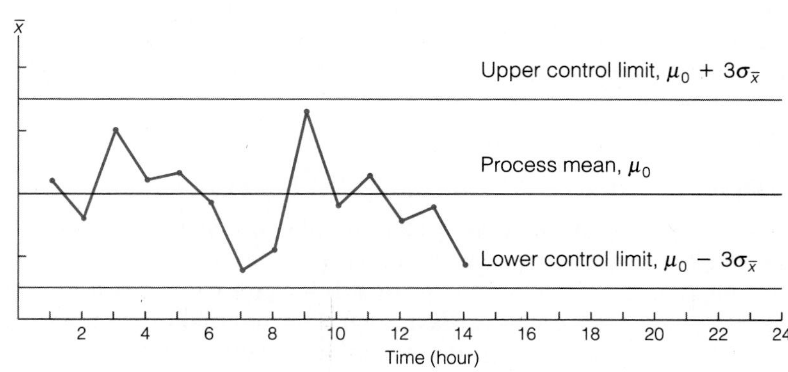

Exercises 8.11–8.19

Learning the Mechanics

8.11 For each of the following rejection regions, sketch the sampling distribution of z and indicate the location of the rejection region:

a. $z > 1.96$ b. $z > 1.645$ c. $z > 2.575$
d. $z < -1.29$ e. $z < -1.645$ or $z > 1.645$ f. $z < -2.575$ or $z > 2.575$

g. For each of the rejection regions specified in parts **a–f**, what is the probability that a Type I error will be made?

8.12 A random sample of 64 observations produced the following sums:

$$\sum_{i=1}^{64} x_i = 59.65 \qquad \sum_{i=1}^{64} (x_i - \bar{x})^2 = 8.63$$

a. Test the null hypothesis that $\mu = 1.00$ against the alternative hypothesis that $\mu < 1.00$ using $\alpha = .10$. Interpret the results of the test.

b. Test the null hypothesis that $\mu = 1.00$ against the alternative hypothesis that $\mu \neq 1.00$ using $\alpha = .10$. Interpret the results of the test.

8.13 Suppose you are interested in conducting the following statistical test:

$$H_0: \mu = 255 \qquad H_a: \mu > 255$$

and you have decided to use the following decision rule: "Reject H_0 if the sample mean of a random sample of 81 items is more than 270." Assume the standard deviation of the population is 63.

a. Express the decision rule in terms of z.

b. Find α, the probability of making a Type I error, for this decision rule.

8.14 A random sample of 70 observations produced the following sums:

$$\sum x_i = 47.46 \qquad \sum x_i^2 = 40.81$$

a. Test the null hypothesis that $\mu = .60$ against the alternative hypothesis that $\mu > .60$. Test using $\alpha = .05$. Interpret the results.

b. Test the null hypothesis that $\mu = .60$ against the alternative hypothesis that $\mu \neq .60$. Test using $\alpha = .05$. Interpret the results.

Applying the Concepts

8.15 A telephone company's records indicate that private customers pay an average of \$23.14 per month for long-distance telephone calls; the standard deviation of the amounts paid for long-distance calls is \$10.48.

a. If a random sample of 50 bills is taken, what is the probability that the sample mean is greater than \$25?

b. If a random sample of 100 bills is taken, what is the probability that the sample mean is greater than \$25?

c. The telephone company suspects that the mean amount paid per month per customer for long-distance service has increased. A random sample of 100 customers' bills during a given month produced a sample mean of \$27.21 expended for long-distance calls. Do the data indicate that the mean level of the amounts billed per month for long-distance telephone calls has increased from \$23.14? Test using $\alpha = .05$.

8.16 In quality control applications of hypothesis testing, the null and alternative hypotheses are frequently specified as

H_0: The production process is performing satisfactorily.

H_a: The process is performing in an unsatisfactory manner.

Accordingly, α is sometimes referred to as the **producer's risk**, while β is called the **consumer's risk** (Montgomery, 1991). An injection molder produces plastic golf tees. The process is designed to produce tees with a mean weight of .250 ounce. To investigate whether the injection molder is operating satisfactorily, 40 tees were randomly sampled from the last hour's production. Their weights (in ounces) are listed below:

.247	.251	.254	.253	.253	.248	.253	.255	.256	.252
.253	.252	.253	.256	.254	.256	.252	.251	.253	.251
.253	.253	.248	.251	.253	.256	.254	.250	.254	.255
.249	.250	.254	.251	.251	.255	.251	.253	.252	.253

a. Do the data provide sufficient evidence to conclude that the process is not operating satisfactorily? Test using $\alpha = .05$.

b. In the context of this problem, explain why it makes sense to call α the producer's risk and β the consumer's risk.

8.17 The Environmental Protection Agency (EPA) estimated that the 1994 Polaris automobile obtains a mean of 35 miles per gallon (mpg) on the highway. However, the company that manufactures the car claims that the EPA has underestimated the Polaris' mileage. To support its assertion, the company selects 36 1994 Polaris cars and records the mileage obtained for each car over a driving course similar to the one used by the EPA. The following data resulted:

$\bar{x} = 37.3$ mpg $\qquad s = 6.4$ mpg

a. If the auto manufacturer wishes to show that the mean mpg for 1994 Polaris autos is greater than 35 mpg, what should the alternative hypothesis be? The null hypothesis?

b. Do the data provide sufficient evidence to support the auto manufacturer's claim? Test using $\alpha = .05$. List any assumptions you make in conducting the test.

8.18 The introduction of printed circuit boards (PCBs) in the 1950s revolutionized the electronics industry. However, solder-joint defects on PCBs have always been a problem. A single PCB may contain thousands of solder joints. Until the 1980s the only way of checking the quality of solder joints was by visual inspection. Because of the low reliability of visual inspection, some manufacturers required each of their joints to be inspected four times by four different people. Now both X-ray and laser technologies are available for use in inspection (Streeter, 1986). A particular manufacturer of laser-based inspection equipment claims that its product can inspect on average at least 10 solder joints per second when the joints are spaced .1 inch apart. The equipment was tested by a potential buyer on 48 different PCBs. In each case, the equipment was operated for exactly 1 second. The numbers of solder joints inspected on each run are as follows:

10	9	10	10	11	9	12	8	8	9	6	10	7	10	11	9
9	13	9	10	11	10	12	8	9	9	9	7	12	6	9	10
10	8	7	9	11	12	10	0	10	11	12	9	7	9	9	10

a. The potential buyer wants to know whether the sample data refute the manufacturer's claim. Specify the null and alternative hypotheses that the buyer should test.
b. In the context of this exercise, what is a Type I error? A Type II error?
c. Conduct the hypothesis test described in part **a** and interpret the results in the context of this exercise. Use $\alpha = .05$.

8.19 A company has devised a new ink-jet cartridge for its plain-paper fax machine that it believes has a longer lifetime (on average) than the one currently being produced. To investigate its length of life, 225 of the new cartridges were tested by counting the number of high-quality printed pages each was able to produce. The sample mean and standard deviation were determined to be 1,511.4 pages and 35.7 pages, respectively. The historical average lifetime for cartridges produced by the current process is 1,502.5 pages; the historical standard deviation is 97.3 pages.
a. What are the appropriate null and alternative hypotheses to test whether the mean lifetime of the new cartridges exceeds that of the old cartridges?
b. Use $\alpha = .005$ to conduct the test in part **a**. Do the new cartridges have an average lifetime that is statistically significantly longer than the cartridges currently in production?
c. Does the difference in average lifetimes appear to be of practical significance from the perspective of the consumer? Explain.
d. Should the apparent decrease in the standard deviation in lifetimes associated with the new cartridges be viewed as an improvement over the old cartridges? Explain.

8.3 Observed Significance Levels: *p*-Values

According to the statistical test procedure described in Section 8.2, the value of α and, correspondingly, the rejection region are selected prior to conducting the test, and the conclusion is stated in terms of rejecting or not rejecting the null hypothesis. A second method of presenting the results of a statistical test reports the extent to which the test statistic disagrees with the null hypothesis and leaves to the reader the task of deciding whether to reject the null hypothesis. This measure of disagreement is called the **observed significance level** (or ***p*-value**) for the test.

Definition 8.1

The **observed significance level**, or ***p*-value**, for a specific statistical test is the probability, assuming H_0 is true, of observing a value of the test statistic that is at least as contradictory to the null hypothesis (and as supportive of the alternative hypothesis) as the actual one computed from the sample data.

For example, the value of the test statistic computed for the sample of $n = 50$ sections of sewer pipe was $z = 2.12$. Since the test was one-tailed—i.e., H_a: $\mu >$ 2,400—values of the test statistic even more contradictory to H_0 than the one observed would be values larger than $z = 2.12$. Therefore, the observed significance level (p-value) for this test is

$$p\text{-value} = P(z \geq 2.12)$$

or, equivalently, the area under the standard normal curve to the right of $z = 2.12$ (see Figure 8.7).

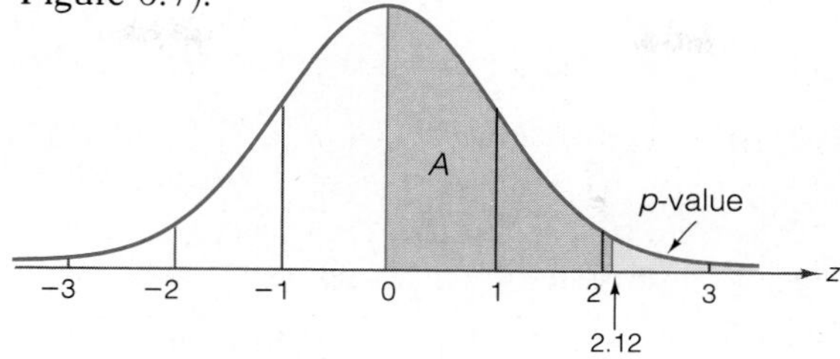

FIGURE 8.7 ► **Finding the *p*-value for an upper-tail test when $z = 2.12$**

The area A in Figure 8.7 is given in Table IV of Appendix B as .4830. Therefore, the upper-tail area corresponding to $z = 2.12$ is

$$p\text{-value} = .5 - .4830 = .0170$$

Consequently, we say these test results are highly significant; that is, they disagree rather strongly with the null hypothesis, H_0: $\mu = 2{,}400$, and favor H_a: $\mu > 2{,}400$. The probability of observing a z value at least as large as 2.12 is only .0170, if in fact H_0 is true.

If you are inclined to select $\alpha = .05$ for this test, then you would reject the null hypothesis because the *p*-value for the test, .0170, is less than .05. In contrast, if you choose $\alpha = .01$, you would not reject the null hypothesis because the *p*-value for the test is larger than .01. Thus, the use of the observed significance level is identical to the test procedure described in the preceding sections except that the choice of α is left to the reader.

The steps for calculating the *p*-value corresponding to a test statistic for a population mean are given in the box.

Steps for Calculating the *p*-Value for a Test of Hypothesis

1. Determine the value of the test statistic z from the sample data.
2. a. If the test is one-tailed, the *p*-value is equal to the tail area beyond z in the same direction as the alternative hypothesis. Thus, if the alternative hypothesis is of the form "$>$," the *p*-value is the area to the right of, or above, the observed z value. Conversely, if the alternative is of the form "$<$," the *p*-value is the area to the left of, or below, the observed z value.
   b. If the test is two-tailed, the *p*-value is equal to twice the tail area beyond the observed z value in the direction of the sign of z. That is, if z is positive, the *p*-value is twice the area to the right of, or above, the observed z value. Conversely, if z is negative, the *p*-value is twice the area to the left of, or below, the observed z value.

EXAMPLE 8.3

Find the observed significance level for the test of the mean filling weight in Examples 8.1 and 8.2.

Solution

Example 8.1 presented a two-tailed test of the hypothesis

$$H_0: \quad \mu = 12 \text{ ounces}$$

against the alternative hypothesis

$$H_a: \quad \mu \neq 12 \text{ ounces}$$

The observed value of the test statistic in Example 8.2 was $z = -3.0$, and any value of z less than -3.0 or greater than $+3.0$ (because this is a two-tailed test) would be even more contradictory to H_0. Therefore, the observed significance level for the test is

$$p\text{-value} = P(z < -3.0 \quad \text{or} \quad z > +3.0)$$

Thus, we calculate the area below the observed z value, $z = -3.0$, and double it. From Table IV of Appendix B, we find that $P(z < -3.0) = .5 - .4987 = .0013$. Therefore, the p-value for this two-tailed test is

$$\begin{aligned} 2P(z < -3.0) &= 2(.0013) \\ &= .0026 \end{aligned}$$

We can interpret this p-value as a strong indication that the machine is not filling the boxes according to specifications since we would observe a test statistic this extreme or more extreme only 26 in 10,000 times if the machine were meeting specifications ($\mu = 12$). The extent to which the mean differs from 12 could be better determined by calculating a confidence interval for μ.

When publishing the results of a statistical test of hypothesis in journals, case studies, reports, etc., many researchers make use of p-values. Instead of selecting α and then conducting a test as outlined in this chapter, the researcher will compute and report the value of the approximate test statistic and its associated p-value. It is left to the reader of the report to judge the significance of the result—i.e., the reader must determine whether to reject the null hypothesis in favor of the alternative hypothesis, based on the reported p-value. This p-value is often referred to as the **observed significance level** of the test. Usually, the null hypothesis will be rejected if the observed significance level is *less* than the fixed significance level, α, chosen by the reader. The inherent advantages of reporting test results in this manner are twofold: (1) readers are able to draw their own conclusions about the reported hypothesis test by choosing α themselves and comparing it to the reported p-value and (2) a measure of the degree of significance of the test result (i.e., the p-value) is provided.

How to Decide Whether to Reject H_0 Using Reported p-Values

1. Choose the maximum value of α you are willing to tolerate.
2. If the observed significance level (p-value) of the test is less than the chosen value of α, then reject the null hypothesis. Otherwise, do not reject the null hypothesis.

Exercises 8.20–8.33

Learning the Mechanics

8.20 If a hypothesis test were conducted using $\alpha = .05$, for which of the following p-values would the null hypothesis be rejected?
a. .06 b. .10 c. .01 d. .001 e. .251 f. .042

8.21 For each α and observed significance level (p-value) pair, indicate whether the null hypothesis would be rejected.
a. $\alpha = .05$, p-value $= .10$ b. $\alpha = .10$, p-value $= .05$ c. $\alpha = .01$, p-value $= .001$
d. $\alpha = .025$, p-value $= .05$ e. $\alpha = .10$, p-value $= .45$

8.22 Explain the difference between statistical significance and practical significance.

8.23 An analyst tested the null hypothesis $\mu \geq 34$ against the alternative hypothesis that $\mu < 34$. The analyst reported a p-value of .083. What is the smallest value of α for which the null hypothesis would be rejected?

8.24 In a test of H_0: $\mu \leq 178$ against H_a: $\mu > 178$, the sample data yielded the test statistic $z = 1.83$. Find the p-value for the test.

8.25 In a test of H_0: $\mu = 178$ against H_a: $\mu \neq 178$, the sample data yielded the test statistic $z = 1.77$. Find the p-value for the test.

8.26 In a test of H_0: $\mu \geq 178$ against H_a: $\mu < 178$, the sample data yielded the test statistic $z = -2.13$. Find the observed significance level of the test.

8.27 In a test of the hypothesis H_0: $\mu = 35$ versus H_a: $\mu > 35$, a sample of $n = 100$ observations possessed mean $\bar{x} = 35.7$ and standard deviation $s = 3.1$. Find and interpret the p-value for this test.

8.28 In a test of the hypothesis H_0: $\mu = 70$ versus H_a: $\mu \neq 70$, a sample of $n = 50$ observations possessed mean $\bar{x} = 69.4$ and standard deviation $s = 2.9$. Find and interpret the p-value for this test.

Applying the Concepts

8.29 The manufacturer of an over-the-counter pain reliever claims that its product brings pain relief to headache sufferers in less than 3.5 minutes, on average. To be able to make this claim in its television advertisements, the manufacturer was required by a particular television network to present statistical evidence in support of the claim. The manufacturer reported that for a random sample of 50 headache sufferers, the mean time to relief was 3.3 minutes and the standard deviation was 66 seconds.

a. Do these data support the manufacturer's claim? Test using $\alpha = .05$.
b. Report the p-value of the test.
c. In general, do large p-values or small p-values support the manufacturer's claim? Explain.

8.30 Refer to Exercise 8.16, in which the performance of an injection molder that produces plastic golf tees was investigated. Find the p-value for the test and interpret its value.

8.31 The *Chronicle of Higher Education Almanac* (Aug. 1992) reported that for the 1990–1991 academic year, 4-year private colleges charged students an average of \$9,083 for tuition and fees, whereas at 4-year public colleges the average was \$1,888. Suppose that for 1992–1993 a random sample of 30 private colleges yielded the following data on tuition and fees: $\bar{x} = \$9,667$ and $s = \$1,721$. Assume that \$9,083 is the population mean for 1990–1991.

a. Specify the null and alternative hypotheses you would use to investigate whether the mean amount for tuition and fees in 1992–1993 was significantly larger (in the statistical sense) than it was in 1990–1991.
b. The data are submitted to a statistical software package, and the following results are obtained:

```
Z = 1.86          P-VALUE = .0314
```

Check these calculations, and interpret the result.
c. Explain the difference between statistical significance and practical significance in the context of this exercise.

8.32 Florida's housing market remains strong due to the steady stream of new residents fleeing harsh northern winters. The state association of realtors claims that during the last 6 months the mean sale price of a new home in Florida was \$101,335. A realtor who believes this figure is too low obtains a random sample of 40 sale prices from a list of all homes sold in Florida during the last 6 months. The sample mean and standard deviation were: $\bar{x} = \$104,754$, $s = \$10,254$.

a. The realtor wishes to conduct an hypothesis test to substantiate her belief. Identify the null and alternative hypotheses of interest to the realtor.
b. Find the observed significance level for this test and interpret its value.

8.33 Recall Exercise 8.19, in which we tested the null hypothesis H_0: $\mu \leq 1,502.5$ versus the alternative H_a: $\mu >$ 1,502.5, where μ is the mean lifetime of new cartridges for fax machines. The sample data were analyzed using a statistical software package, with the following results:

```
Z = 1.37          P-VALUE = .0853
```

a. Interpret the results.
b. If the test were two-tailed, how would the p-value change? Interpret the two-tailed p-value.

8.4 Small-Sample Test of Hypothesis About a Population Mean

A manufacturer of computer disk drives monitors the retail prices of its drives in order to gauge the market. For one type of drive the list price is \$750, and the manufacturer wishes to know whether the current mean retail price differs from the list price. Seventeen retail establishments are sampled, and their current prices for the drive are determined. The mean and standard deviation for the 17 retail prices are calculated:

$$\bar{x} = \$732 \qquad s = \$38$$

Does this sample provide sufficient evidence to conclude that the mean retail price differs from the list price of \$750?

This inference can be placed in a test of hypothesis framework. We establish the null hypothesized value as the list price and then utilize a two-tailed alternative to test whether the true mean retail price differs from the list price:

$$H_0\text{:} \quad \mu = \$750$$
$$H_a\text{:} \quad \mu \neq \$750$$

Recall from Section 7.3 that when we are faced with making inferences about a population mean using the information in a small sample, two problems emerge:

1. The normality of the sampling distribution for $\bar{x}$ does not follow from the Central Limit Theorem when the sample size is small. We must assume that the distribution of measurements from which the sample was selected is approximately normally distributed in order to assure the approximate normality of the sampling distribution of $\bar{x}$.
2. If the population standard deviation σ is unknown, as is usually the case, then we cannot assume that s will provide a good approximation for σ when the sample size is small. Instead, we must use the t distribution rather than the standard normal z distribution to make inferences about the population mean μ.

Therefore, for the test statistic of a small-sample test of a population mean, we use the t statistic:

$$\textit{Test statistic:} \quad t = \frac{\bar{x} - \mu_0}{s/\sqrt{n}} = \frac{\bar{x} - 750}{s/\sqrt{n}}$$

where μ_0 is the null hypothesized value of the population mean, μ. In our example, $\mu_0 = \$750$.

To find the rejection region we must specify the value of α, the probability that the test will lead to rejection of the null hypothesis when it is true, and then consult the t table (Table VI of Appendix B or inside back cover). Using $\alpha = .05$, the two-tailed rejection region is:

Rejection region: $t_{\alpha/2} = t_{.025} = 2.120$ with $n - 1 = 16$ degrees of freedom.
Reject H_0 if $t < -2.120$ or $t > 2.120$.

The rejection region is shown in Figure 8.8.

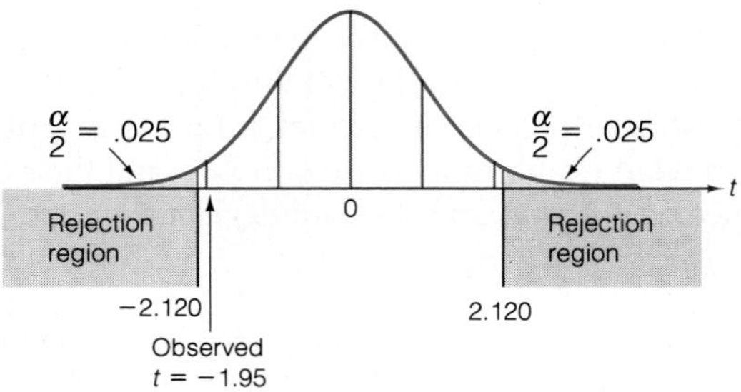

FIGURE 8.8 ► Two-tailed rejection region for small-sample *t*-test

We are now prepared to calculate the test statistic and reach a conclusion:

$$t = \frac{\bar{x} - \mu_0}{s/\sqrt{n}} = \frac{732 - 750}{38/\sqrt{17}} = \frac{-18}{9.22} = -1.95$$

Since the calculated value of t does not fall in the rejection region (Figure 8.8), we cannot reject H_0 at the $\alpha = .05$ level of significance. Thus, based on the sample evidence, the manufacturer should not conclude that the mean retail price differs from the \$750 list price based on the sample evidence.

We summarize the technique for conducting a small-sample test of hypothesis about a population mean in the box.

Small-Sample Test of Hypothesis About μ

One-Tailed Test	*Two-Tailed Test*
H_0: $\mu = \mu_0$	H_0: $\mu = \mu_0$
H_a: $\mu < \mu_0$ (or H_a: $\mu > \mu_0$)	H_a: $\mu \neq \mu_0$
Test statistic: $t = \frac{\bar{x} - \mu_0}{s/\sqrt{n}}$	*Test statistic:* $t = \frac{\bar{x} - \mu_0}{s/\sqrt{n}}$
Rejection region: $t < -t_\alpha$ (or $t > t_\alpha$ when H_a: $\mu > \mu_0$)	*Rejection region:* $t < -t_{\alpha/2}$ or $t > t_{\alpha/2}$

where t_α and $t_{\alpha/2}$ are based on $(n - 1)$ degrees of freedom.

Assumption: A random sample is selected from a population with a relative frequency distribution that is approximately normal.

Remember, the basic assumption necessary for the use of the t statistic is that the sampled population has a relative frequency distribution that is approximately normal.

What can be done if you know that the population relative frequency distribution is decidedly nonnormal, say highly skewed?

> **What Can Be Done If the Population Relative Frequency Distribution Departs Greatly From Normal?**
>
> *Answer:* Use the nonparametric statistical methods of Chapter 17.

EXAMPLE 8.4

A major car manufacturer wants to test a new engine to determine whether it meets new air pollution standards. The mean emission, μ, of all engines of this type must be less than 20 parts per million of carbon. Ten engines are manufactured for testing purposes, and the mean and standard deviation of the emissions for this sample of engines are determined to be

$$\bar{x} = 17.1 \text{ parts per million} \qquad s = 3.0 \text{ parts per million}$$

Do the data supply sufficient evidence to allow the manufacturer to conclude that this type of engine meets the pollution standard? Assume that the manufacturer is willing to risk a Type I error with probability equal to $\alpha = .01$.

Solution

The manufacturer wants to establish the alternative hypothesis that the mean emission level, μ, for all engines of this type is less than 20 ppm. The elements of this small-sample one-tailed test are

H_0: $\mu = 20$

H_a: $\mu < 20$

Test statistic: $t = \dfrac{\bar{x} - 20}{s/\sqrt{n}}$

Assumption: The relative frequency distribution of the population of emission levels for all engines of this type is approximately normal.

Rejection region: For $\alpha = .01$ and df $= n - 1 = 9$, the one-tailed rejection region (see Figure 8.9) is $t < -t_{.01} = -2.821$.

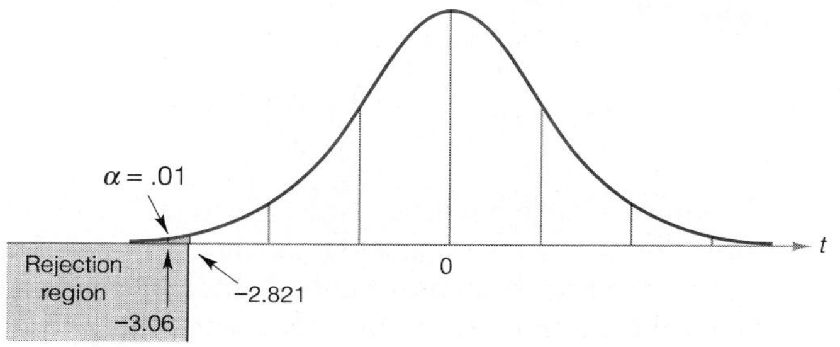

FIGURE 8.9 ► Rejection region for Example 8.4

We now calculate the test statistic:

$$t = \frac{\bar{x} - 20}{s/\sqrt{n}} = \frac{17.1 - 20}{3.0/\sqrt{10}} = -3.06$$

Since the calculated t falls in the rejection region (see Figure 8.9), the manufacturer concludes that $\mu < 20$ parts per million and the new engine type meets the pollution standard. Are you satisfied with the reliability associated with this inference? The probability is only $\alpha = .01$ that the test would support the alternative hypothesis if in fact it was false.

EXAMPLE 8.5

Find the observed significance level for the test in Example 8.4.

Solution

The test performed in Example 8.4 was a one-tailed test, in which H_0: $\mu = 20$ would be rejected in favor of H_a: $\mu < 20$ for values of t in the lower tail of the t distribution. Since the value of t computed from the sample data was $t = -3.06$, the observed significance level (or p-value) for the test is equal to the probability that t would assume a value less than or equal to -3.06, if in fact H_0 were true. This is equal to the area in the lower tail of the t distribution (shaded in Figure 8.10). To find this area—i.e., the p-value for the test—we consult the t table in Table VI of Appendix B. Unlike the table of areas under the normal curve, Table VI gives only the t values corresponding to the areas .100, .050, .025, .010, .005, .001, and .0005. Therefore, we can only approximate the p-value for the test. Since the observed t value was based on 9 degrees of freedom, we use the df = 9 row in Table VI and move across the row until we reach the t value that is closest to the observed $t = -3.06$. [*Note*: We ignore the minus sign.] The t values corresponding to p-values of .010 and .005 are 2.821 and 3.250, respectively. Since the observed t value falls between $t_{.010}$ and $t_{.005}$, the p-value for the test lies between .010 and .005.

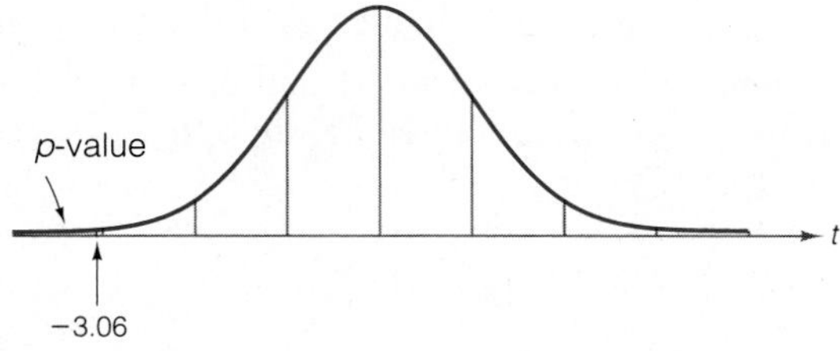

FIGURE 8.10 ► **The observed significance level for the test in Example 8.4**

Small-sample inferences typically require more assumptions and provide less information about the population parameter than do large-sample inferences. Nevertheless, the t test is a viable method of testing a hypothesis about a population mean of a normal distribution when only a small number of observations is available.

Exercises 8.34–8.46

Learning the Mechanics

8.34 In what ways are the distributions of the z-test statistic and t-test statistic alike? How do they differ?

8.35 Under what circumstances should you use the t distribution in testing a hypothesis about a population mean?

8.36 For each of the following rejection regions, sketch the sampling distribution of t and indicate the location of the rejection region on your sketch:
a. $t > 1.440$ where df = 6 b. $t < -1.782$ where df = 12
c. $t < -2.060$ or $t > 2.060$ where df = 25

8.37 For each of the rejection regions defined in Exercise 8.36, what is the probability that a Type I error will be made?

8.38 The following sample of five measurements was randomly selected from a normally distributed population: 11.78, 11.14, 9.81, 9.25, 8.75, 10.34.
a. Test the null hypothesis that the mean of the population is 11 against the alternative hypothesis, $\mu < 11$. Use $\alpha = .05$.
b. Test the null hypothesis that the mean of the population is 11 against the alternative hypothesis, $\mu \neq 11$. Use $\alpha = .05$.

8.39 Find the observed significance level for each test in Exercise 8.38.

8.40 Refer to Exercise 8.38.
a. Find a 95% confidence interval for μ.
b. Give the value of t that would be used to form a 90% confidence interval for μ.
c. Describe a practical situation that would motivate you to form a confidence interval for μ rather than testing a hypothesis about μ.

8.41 A statistical software package is used to conduct a t-test for the null hypothesis H_0: $\mu = 1{,}000$ versus the alternative hypothesis H_a: $\mu > 1{,}000$ based on a sample of 17 observations. The software's output is:

```
T = 1.894          P-VALUE = .0382
```

a. Most software does not indicate what assumptions are necessary for the validity of a statistical procedure. What assumptions are necessary for the validity of this procedure?
b. Interpret the results of the test.
c. Suppose the alternative hypothesis had been the two-tailed H_a: $\mu \neq 1{,}000$. If the t statistic were unchanged, then what would the p-value be for this test? Interpret the p-value for the two-tailed test.

Applying the Concepts

8.42 In any bottling process, a manufacturer will lose money if the bottles contain either more or less than is claimed on the label. Suppose a quality manager for a catsup company is interested in testing whether the mean number of ounces of catsup per family-size bottle differs from the labeled amount of 20 ounces. The manager samples nine bottles, measures the weight of their contents, and finds that $\bar{x} = 19.7$ ounces and $s = .3$ ounce.

a. Does the sample evidence indicate that the catsup dispensing machine needs adjustment? Test at $\alpha = .05$.
b. What is the p-value for the hypothesis test you conducted in part **a**?
c. What assumptions are necessary so that the procedure used in part **a** is valid?
d. Find a 90% confidence interval for the mean number of ounces of catsup being dispensed.

8.43 A company purchases large quantities of naphtha in 50-gallon drums. Because the purchases are ongoing, small shortages in the drums can represent a sizable loss to the company. The weights of the drums vary slightly from drum to drum, so the weight of the naphtha is measured after removing it from the drums. Suppose the company samples the contents of 20 drums, measures the naphtha in each, and calculates $\bar{x} = 49.70$ gallons and $s = .32$ gallon. Do the sample statistics provide sufficient evidence to indicate that the mean fill per 50-gallon drum is less than 50 gallons? Use $\alpha = .10$. List your assumptions.

8.44 To instill customer loyalty, airlines, hotels, rental car companies, and credit card companies (among others) have initiated **frequency marketing programs** that reward their regular customers. In the United States alone, 30 million people are members of the frequent flier programs of the airline industry (Rice, 1993). A large fast-food restaurant chain wished to explore the profitability of such a program. They randomly selected 12 of their 1,200 restaurants nationwide and instituted a frequency program that rewarded customers with a $5.00 gift certificate after every 10 meals purchased at full price. They ran the trial program for 3 months. The restaurants not in the sample had an average increase in profits of $1,047.34 over the previous 3 months, whereas the restaurants in the sample had the following changes in profit:

$2,232.90	$ 545.47	$3,440.70	$1,809.10	$6,552.70	$4,798.70
$2,965.00	$2,610.70	$3,381.30	$1,591.40	$2,376.20	−$2,191.00

Note that the last number is negative, representing a decrease in profits.

a. Specify the appropriate null and alternative hypotheses for determining whether the mean profit change for restaurants with frequency programs is significantly greater (in a statistical sense) than $1,047.34.
b. Conduct the test of part **b** using $\alpha = .05$. Does it appear that the frequency program would be profitable for the company if adopted nationwide?

8.45 In a nationwide survey of 1,000 men and women conducted by Caldwell Davis Partners (an advertising agency), it was found that two-thirds of the people surveyed perceived themselves as younger than their actual chronological age. This may explain the recent failures of a line of food advertised for senior citizens and a shampoo directed at "hair over 40." However, men and women under 30 generally perceived themselves as older than their actual age (Nemy, 1982). A researcher randomly sampled 10 college students under the age of 30 and asked them how old they were and how old they perceived themselves to be. The results are shown in the table:

Chronological Age:	20	19	25	22	26	19	18	20	20	21
Perceived Age:	22	21	30	25	22	19	20	18	21	21

a. Do the sample data support the survey's findings with respect to the perceptions of men and women under age 30? Test using $\alpha = .10$.
b. What assumption must hold for the procedure you used in part **a** to be valid?

8.46 The Occupational Safety and Health Act of 1970 (OSHA) allows issuance of engineering standards to assure safe workplaces for all Americans. In 1975, the standards for exposure to arsenic in smelters, herbicide production facilities, and other places where arsenic is used were reviewed, and the previous maximum allowable level of .5 milligram per cubic meter of air was reduced to .004. Suppose smelters at two plants are being investigated to determine whether they are meeting OSHA standards. Two analyses of the air are made at each plant, and the results (in milligrams per cubic meter of air) are shown in the table.

Plant 1		Plant 2	
Observation	Arsenic Level	Observation	Arsenic Level
1	.01	1	.05
2	.005	2	.09

a. What are the appropriate null and alternative hypotheses if we wish to test whether the plants meet the current OSHA standard?

b. These data are analyzed by statistical software, with the following results:

```
PLANT 1        T = 1.40        P-VALUE = .200
PLANT 2        T = 3.30        P-VALUE = .094
```

Check the calculations of the t statistics and p-values.

c. Interpret the results of the two tests.

8.5 Large-Sample Test of Hypothesis About a Binomial Probability

Inferences about proportions (or percentages) are often made in the context of the probability, p, of "success" for a binomial distribution. We showed how to use large samples from binomial distributions to form confidence intervals for p in Section 7.5. We now consider tests of hypotheses about a binomial probability p.

For example, consider the problem of *insider trading* in the stock market. Insider trading is the buying and selling of stock by an individual privy to inside information in a company, usually a high-level executive in the firm. The Securities and Exchange Commission (SEC) imposes strict guidelines about insider trading so that all investors can have equal access to information that may affect the stock's price. An investor wishing to test the effectiveness of the SEC guidelines monitors the market for a period of a year and records the number of times a stock price increases the day following a significant purchase of stock by an insider. For a total of 576 such transactions, the stock increased the following day 327 times. Does this sample provide evidence that the stock price may be affected by insider trading?

We first view this as a binomial experiment, with the 576 transactions as the trials, and success representing an increase in the stock's price the following day. Let p

represent the probability that the stock price will increase following a large insider purchase. If the insider purchase has no effect on the stock price (that is, if the information available to the insider is identical to that available to the general market), then the investor expects the probability of a stock increase to be the same as that of a decrease, or $p = .5$. On the other hand, if insider trading affects the stock price (indicating that the market has not fully accounted for the information known to the insiders), then the investor expects the stock either to decrease or to increase more than half the time following significant insider transactions; that is, $p \neq .5$.

We can now place the problem in the context of a test of hypothesis:

$$H_0\text{: } p = .5$$
$$H_a\text{: } p \neq .5$$

Recall that the sample proportion, $\hat{p}$, is really just the sample mean of the outcomes of the individual binomial trials, and as such is approximately normally distributed (for large samples) according to the Central Limit Theorem. Thus, for large samples we can use the standard normal z as the test statistic:

$$\textit{Test statistic: } z = \frac{(\text{Sample proportion} - \text{Null hypothesized proportion})}{\text{Standard deviation of sample proportion}} = \frac{\hat{p} - p_0}{\sigma_{\hat{p}}}$$

where we use the symbol p_0 to represent the null hypothesized value of p.

Rejection region: We use the standard normal distribution to find the appropriate rejection region for the specified value of α. Using $\alpha = .05$, the two-tailed rejection region is

$$z < -z_{\alpha/2} = -z_{.025} = -1.96 \quad \text{or} \quad z > z_{\alpha/2} = z_{.025} = 1.96$$

See Figure 8.11.

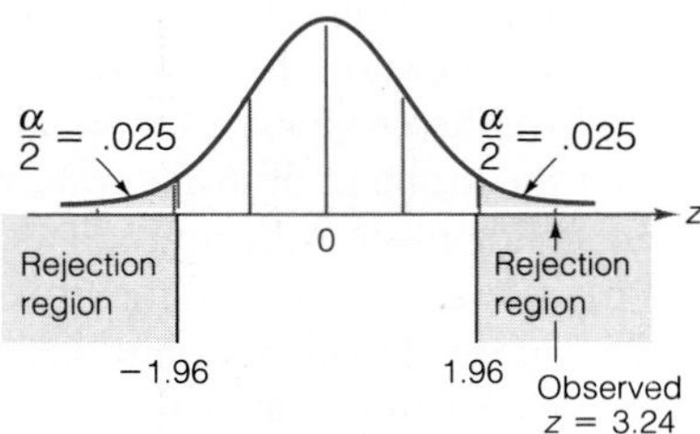

FIGURE 8.11 ► **Rejection region for insider trading example**

We are now prepared to calculate the value of the test statistic. Before doing so, we want to be sure that the sample size is large enough to assure that the normal approximation for the sampling distribution of $\hat{p}$ is reasonable. To check this, we calculate a 3-standard-deviation interval around the null hypothesized value, p_0, which is assumed to be the true value of p until our test procedure might prove otherwise. Recall that $\sigma_{\hat{p}} = \sqrt{(pq)/n}$ and that we will need an estimate of the product (pq) in order to calculate a numerical value of the test statistic z. Since the null hypothesized value is generally the "accepted until proven otherwise" value, we use the value of p_0q_0 (where $q_0 = 1 - p_0$) to estimate pq in the calculation of z. Thus,

$$\sigma_{\hat{p}} = \sqrt{\frac{pq}{n}} \approx \sqrt{\frac{p_0q_0}{n}} = \sqrt{\frac{(.5)(.5)}{576}} = .021$$

and the 3-standard-deviation interval around p_0 is

$$p_0 \pm 3\sigma_{\hat{p}} \approx .5 \pm 3(.021) = (.437, .563)$$

As long as this interval does not contain 0 or 1 (i.e., is completely contained in the interval 0 to 1), as is the case here, then the normal distribution will provide a reasonable approximation for the sampling distribution of $\hat{p}$.

Returning to the hypothesis test at hand, the proportion of the sampled transactions that resulted in a stock increase is

$$\hat{p} = \frac{327}{576} = .568$$

Finally, we calculate the number of standard deviations (the z value) between the sampled and hypothesized value of the binomial probability:

$$z = \frac{\hat{p} - p_0}{\sigma_{\hat{p}}} \approx \frac{\hat{p} - p_0}{\sqrt{(p_0q_0)/n}} = \frac{.568 - .5}{.021} = \frac{.068}{.021} = 3.24$$

The implication is that the observed sample proportion is (approximately) 3.24 standard deviations above the null hypothesized probability, .5 (Figure 8.11). Therefore, we reject the null hypothesis, concluding at the .05 level of significance that the true probability of an increase or decrease in a stock's price differs from .5 the day following significant insider purchase of the stock. It appears that an insider purchase significantly *increases* the probability that the stock price will increase the following day.

The test of hypothesis about a binomial probability p is summarized in the box. Note that the procedure is entirely analogous to that used for conducting large-sample tests about a population mean.

Large-Sample Test of Hypothesis About p

One-Tailed Test

H_0: $p = p_0$

H_a: $p < p_0$
(or H_a: $p > p_0$)

where p_0 = the hypothesized value of p

Test statistic: $z = \dfrac{\hat{p} - p_0}{\sigma_{\hat{p}}}$

Two-Tailed Test

H_0: $p = p_0$

H_a: $p \neq p_0$

Test statistic: $z = \dfrac{\hat{p} - p_0}{\sigma_{\hat{p}}}$

Continued

where $\sigma_{\hat{p}} = \sqrt{[p_0(q_0)]/n}$, and, as usual, we assume H_0 is true until convinced otherwise.

Rejection region: $z < -z_\alpha$ (or $z > z_\alpha$ when H_a: $p > p_0$)

Rejection region: $z < -z_{\alpha/2}$ or $z > z_{\alpha/2}$

Assumption: The experiment is binomial, and the sample size is large enough that the interval $p_0 \pm 3\sigma_{\hat{p}}$ does not include 0 or 1.

EXAMPLE 8.6

The reputations (and hence, sales) of many businesses can be severely damaged by shipments of manufactured items that contain a large percentage of defectives. For example, a manufacturer of alkaline batteries may want to be reasonably certain that less than 5% of the batteries are defective. Suppose 300 batteries are randomly selected from a very large shipment, each is tested, and 10 defective batteries are found. Does this provide sufficient evidence for the manufacturer to conclude that the fraction defective in the entire shipment is less than .05? Use $\alpha = .01$.

Solution

Before conducting the test of hypothesis, we check to determine whether the sample size is large enough to use the normal approximation for the sampling distribution of $\hat{p}$. To do this, the following interval is constructed:

$$p_0 \pm 3\sigma_{\hat{p}} = p_0 \pm 3\sqrt{\frac{p_0q_0}{n}} = .05 \pm 3\sqrt{\frac{(.05)(.95)}{300}}$$
$$= .05 \pm .04 \quad \text{or} \quad (.01, .09)$$

Since the interval lies within the interval (0, 1), the normal approximation will be adequate.

The objective of our analysis is to determine whether there is sufficient evidence to indicate that p is less than .05. Consequently, we will test the null hypothesis that $p = .05$ against the alternative hypothesis that $p < .05$. The elements of the test are

H_0: $p = .05$

H_a: $p < .05$

Test statistic: $z = \dfrac{\hat{p} - p_0}{\sigma_{\hat{p}}}$

Rejection region: $z < -z_{.01} = -2.33$ (see Figure 8.12)

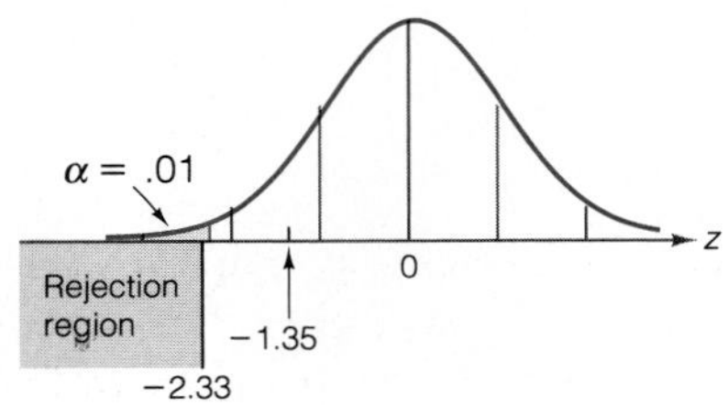

FIGURE 8.12 ► Rejection region for Example 8.6

We now calculate the test statistic:

$$z = \frac{\hat{p} - .05}{\sigma_{\hat{p}}} \approx \frac{(10/300) - .05}{\sqrt{p_0 q_0/n}} = \frac{.033 - .05}{\sqrt{p_0 q_0/300}}$$

Substituting the values for p_0 and q_0 into the z statistic, we obtain

$$z \approx \frac{-.017}{\sqrt{(.05)(.95)/300}} = \frac{-.017}{.0126} = -1.35$$

As shown in Figure 8.12, the calculated z value does not fall in the rejection region. Therefore, there is insufficient evidence at the .01 level of significance to indicate that the shipment contains fewer than 5% defective batteries.

EXAMPLE 8.7

In Example 8.6 we found that we did not have sufficient evidence, at the $\alpha = .01$ level of significance, to indicate that the fraction defective, p, of alkaline batteries was less than $p = .05$. How strong was the weight of evidence favoring the alternative hypothesis (H_a: $p < .05$)? Find the observed significance level for the test.

Solution

The computed value of the test statistic was $z = -1.35$. Therefore, for this one-tailed test,

$$\text{Observed significance level} = P(z \leq -1.35)$$

This lower-tail area is shown in Figure 8.13. The area A between $z = 0$ and $z = 1.35$ is given in Table IV of Appendix B as .4115. Therefore, the observed significance level is $.5 - .4115 = .0885$. Note that this probability is quite small. That is, the probability of observing a z value as small as or smaller than -1.35 is only .0885. Therefore, we would reject H_0 if we choose $\alpha = .10$ (since the observed significance level is less than .10), but we would not reject H_0 (the conclusion of Example 8.6) if we choose $\alpha = .05$ or $\alpha = .01$.

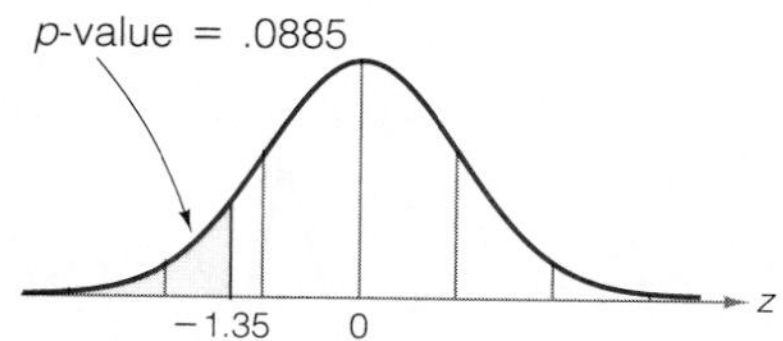

FIGURE 8.13 ► The observed significance level for Example 8.7

Small-sample test procedures are also available for p. These are omitted from our discussion because most surveys conducted in business use samples that are large enough to employ the large-sample tests presented in this section.

CASE STUDY 8.3 / Statistical Quality Control, Part 2

In complicated assembly operations (such as railway-car assembly), many quality variables could be measured (e.g., strength of welds, degree of corrosion, and number of paint flaws), and in principle, each could be monitored over time using control charts, as described in Case Study 8.2. In some situations, however, an alternative, simpler procedure may be more appropriate. For example, n finished products could be randomly sampled at regular time intervals, inspected for defects, and simply classified as being defective or nondefective products. Then $\hat{p}$, the proportion of defectives in each sample, could be determined and plotted on a control chart called a ***p*-chart**, as illustrated in Figure 8.14. In this way, the proportion of defective products produced and, therefore, product quality and the current capability of the production process could be monitored over time (Wetherill, 1977; Montgomery, 1991).

In order to construct the control chart shown in Figure 8.14, it is necessary to know (i.e., have a good estimate for) p_0, the proportion of defectives produced when the process is operating properly (i.e., in control). Then, assuming n is large enough to use the normal distribution to approximate the sampling distribution of $\hat{p}$, the control limits are located $3\sigma_{\hat{p}}$ above and below p_0. If a value of $\hat{p}$ falls above the upper limit, it is a signal that the process is turning out more defectives than usual and may be out of control. But what is the significance of the lower control limit? Why should the manufacturer be concerned if fewer defectives than usual are being produced? Two important reasons follow (Caplen, 1970):

1. It may be an indication that the inspector is not performing his or her job carefully and may be missing defectives that normally would be identified. As a result, defective products may be sold to customers.
2. If the inspector is performing adequately, it may be an indication that the production process is temporarily better. If so, the low $\hat{p}$ value signals management to begin a search for the causes for the improvement. If found, it may be possible to improve the production process permanently.

As in Case Study 8.2, this control chart procedure for monitoring the proportion of defective products is nothing more than a two-tailed hypothesis test with the following elements:

FIGURE 8.14 ►
***p*-chart for proportion defective**

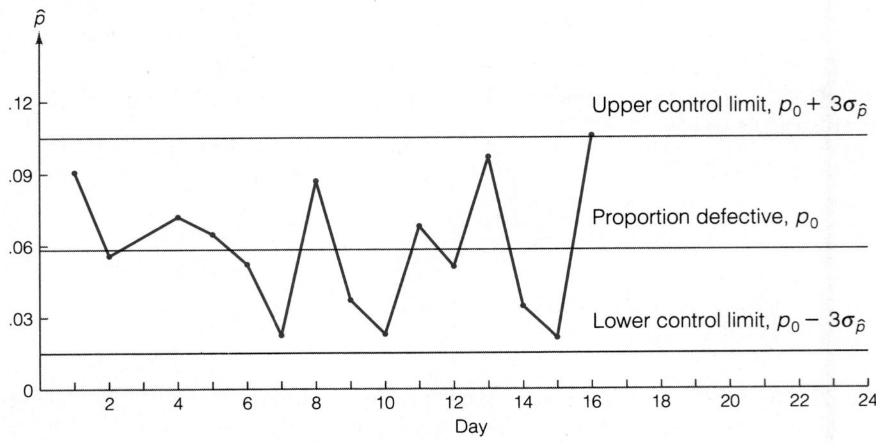

1. H_0: Process is in control, $p = p_0$
   H_a: Process is out of control, $p \neq p_0$
2. The test statistic is $\hat{p}$.
3. The rejection region is defined by the upper and lower control limits.
4. Since the control limits are located at $p_0 \pm 3\sigma_{\hat{p}}$, the probability of committing a Type I error is approximately $\alpha = .0026$. (Why?)

We discuss p-charts in more detail in Chapter 13.

Exercises 8.47–8.57

Learning the Mechanics

8.47 For the binomial sample sizes and null hypothesized values of p in each part, determine whether the sample size is large enough to use the normal approximation methodology presented in this section to conduct a test of the null hypothesis H_0: $p = p_0$.

a. $n = 900$, $p_0 = .975$ b. $n = 125$, $p_0 = .01$ c. $n = 40$, $p_0 = .75$
d. $n = 15$, $p_0 = .75$ e. $n = 12$, $p_0 = .62$

8.48 Suppose a random sample of 100 observations from a binomial population gave a value of $\hat{p} = .63$ and you wish to test the null hypothesis that the population parameter p is equal to .70 against the alternative hypothesis that $p < .70$.

a. Noting that $\hat{p} = .63$, what does your intuition tell you? Does the value of $\hat{p}$ appear to contradict the null hypothesis?
b. Use the large-sample z test to test H_0: $p = .70$ against the alternative hypothesis H_a: $p < .70$ at the $\alpha = .05$ significance level. How do the test results compare with your intuitive decision from part **a**?
c. Find and interpret the p-value for the hypothesis test you conducted in part **b**.

8.49 Suppose the sample in Exercise 8.48 has produced $\hat{p} = .83$ and we wish to test H_0: $p = .9$ against the alternative hypothesis H_a: $p < .9$.

a. Calculate the value of the z statistic for this test.
b. Notice that $\hat{p} - p_0 = .83 - .9 = -.07$ is the same as for Exercise 8.48. Considering this, why is the absolute value of z for this exercise larger than that calculated in Exercise 8.48?
c. Complete the test using $\alpha = .05$ and interpret the results.
d. Find the observed significance level for your hypothesis test, and interpret its value.

8.50 A statistics student used a computer program to test the null hypothesis H_0: $p = .5$ against the one-tailed alternative H_a: $p > .5$. A sample of 500 observations are input into the computer, which returns the following result:

```
Z = 0.44          P-VALUE = 0.3300
```

a. The student concludes, based on the p-value, that there is a 33% chance that the alternative hypothesis is true. Do you agree? If not, correct the interpretation.

b. How would the p-value change if the alternative hypothesis were two-tailed, H_a: $p \neq .5$? Interpret this p-value.

Applying the Concepts

8.51 A random sample of 50 consumers taste-tested a new snack food. Their responses were coded (0: do not like; 1: like; 2: indifferent) and listed as follows:

1 0 0 1 2 0 1 1 0 0 0 1 0 2 0 2 2 0 0 1
1 0 0 0 0 1 0 2 0 0 0 1 0 0 1 0 0 1 0 1
0 2 0 0 1 1 0 0 0 1

a. Test H_0: $p = .5$ against H_a: $p > .5$, where p is the proportion of customers who do not like the snack food. Use $\alpha = .10$.
b. Report the observed significance level of your test.

8.52 According to a spokesperson for General Mills, the company's cents-off coupon offers are designed to get people to buy its products and its refund offers (money returned with proof of repeated purchases) are designed to encourage people to continue buying its products. In a national survey conducted by the Nielsen Clearing House in 1975, 65% of the respondents indicated that they used cents-off coupons when grocery shopping. In a 1980 survey, the Nielsen organization found that 76% of those surveyed used cents-off coupons (*Minneapolis Star*, Nov. 29, 1981). Suppose the 1980 survey consisted of a random sample of 100 shoppers, of whom 76 indicated that they used cents-off coupons.
a. Does the 1980 sample provide sufficient evidence that the percentage of shoppers using cents-off coupons exceeds 65%? Test using $\alpha = .05$.
b. Is the sample size large enough to use the inferential procedures presented in this section? Explain.
c. Find the observed significance level for the test you conducted in part **a**, and interpret its value.

8.53 In gambling, a *system* is a strategy for playing a game that is thought to improve one's chances of winning. A small gambling school teaches customers how to play casino blackjack. Their published advertisements claim that they ". . . teach a system that guarantees winning results on average." A customer who lost thousands of dollars while using the system decided to sue the school for false advertising. To test the school's claim, the customer's lawyer commissioned a computer simulation study. The computer "played" 1,000,000 independent games of blackjack with the "player" following the school's system and the "dealer" obeying the standard house rules. The "player" won 497,584 of the simulated games.
a. Letting p represent the long-run proportion of games won by players using the system, test H_0: $p \geq .50$ versus H_a: $p < .50$ using $\alpha = .01$.
b. Do the results of this test contradict the advertised claim? Explain.
c. Suppose the simulation had been run only 10,000 times. Would the results of the test be the same if the "player" won the same proportion of the simulated games (i.e., 4,976)?
d. List any assumptions that you made in answering parts **a** and **c**.

8.54 Refer to Exercise 8.53. Find the observed significance levels for the tests you conducted in parts **a** and **c** and interpret their values.

8.55 Marketing research has been defined by the American Marketing Association as the "systematic gathering, recording, and analyzing of data about problems relating to the marketing of goods and services" (*Report of*

Definitions Committee of the American Marketing Association, 1961). Companies may have their own marketing research departments, or they may use the services of a marketing research firm. The marketing research department of a large West Coast manufacturer of facial tissue paper was charged with the responsibility of determining customer preferences regarding the softness of its newly developed product (brand A) relative to the industry leader (brand B). A random sample of 250 customers was selected. Each customer was asked to rank the softness of brands A and B: 146 ranked brand A softer and 104 ranked brand B softer.

a. Do the data indicate that brand A is perceived as softer than brand B? Test using $\alpha = .05$.

b. Find the p-value for the test and interpret its value.

8.56 A producer of frozen orange juice claims that 52% of all orange-juice drinkers prefer its product. To test the validity of this claim, a competitor samples 200 orange-juice drinkers and finds that only 95 prefer the producer's brand.

a. What are the appropriate null and alternative hypotheses to test the producer's claim?

b. When these data are analyzed by a statistical software program, the following results are obtained:

```
Z = -1.27          P-VALUE = .1020
```

Check the computer calculations.

c. Interpret the results.

d. How would the p-value change if the test were two-tailed rather than one-tailed?

8.57 A major videocassette rental chain is considering opening a new store in an area that currently does not have any such stores. The chain will open a store if there is evidence that at least 5,000 of the 20,000 households in the area are equipped with videocassette recorders (VCRs). It conducts a telephone poll of 300 randomly selected households in the area and finds that 96 have VCRs. Is this enough evidence to conclude that more than 5,000 of the homes in the area are equipped with VCRs? Test at $\alpha = .01$.

8.6 Calculating Type II Error Probabilities: More About β (Optional)

In our introduction to hypothesis testing in Section 8.1, we showed that the probability of committing a Type I error, α, can be controlled by the selection of the rejection region for the test. Thus, when the test statistic falls in the rejection region and we make the decision to reject the null hypothesis, we do so knowing the error rate for incorrect rejections of H_0. The situation corresponding to accepting the null hypothesis, and thereby risking a Type II error, is not generally as controllable. For that reason, we adopted a policy of nonrejection of H_0 when the test statistic does not fall in the rejection region, rather than risking an error of unknown magnitude.

To see how β, the probability of a Type II error, can be calculated for a test of hypothesis, recall the example in Section 8.2 in which a city tests a manufacturer's pipe to see whether it meets the requirement that the mean strength exceeds 2,400 pounds per lineal foot. The setup for the test is as follows:

H_0: $\mu = 2{,}400$

H_a: $\mu > 2{,}400$

Test statistic: $z = \dfrac{\bar{x} - 2{,}400}{\sigma/\sqrt{n}}$

Rejection region: $z > 1.645$ for $\alpha = .05$

Figure 8.15(a) shows the rejection region for the **null distribution**—that is, the distribution of the test statistic assuming the null hypothesis is true. The area in the rejection region is .05, and this area represents α, the probability that the test statistic leads to rejection of H_0 when, in fact, H_0 is true.

The Type II error probability β is calculated assuming that the null hypothesis is false, because it is defined as the *probability of accepting H_0 when it is false*. Since H_0 is false for any value of μ exceeding 2,400, one value of β exists for each possible value of μ greater than 2,400 (an infinite number of possibilities). Figure 8.15(b), (c), and (d) show three of the possibilities, corresponding to alternative hypothesis values of μ equal to 2,425, 2,450, and 2,475, respectively. Note that β is the area in the *nonrejection* (or *acceptance*) *region* in each of these distributions, and that β decreases as the true value of μ moves farther from the null-hypothesized value of $\mu = 2{,}400$. This is sensible, because the probability of incorrectly accepting the null hypothesis should decrease as the distance between the null and alternative values of μ increases.

In order to calculate the value of β for a specific value of μ in H_a, we proceed as follows:

1. First, calculate the value of $\bar{x}$ that corresponds to the border between the acceptance and rejection regions. For the sewer pipe example, that is the value of $\bar{x}$ that lies 1.645 standard deviations above $\mu = 2{,}400$ in the sampling distribution of $\bar{x}$. Denoting this value by $\bar{x}_0$, corresponding to the largest value of $\bar{x}$ that supports the null hypothesis, we find (recalling that $s = 200$ and $n = 50$):

$$\begin{aligned}\bar{x}_0 &= \mu_0 + 1.645\sigma_{\bar{x}} = 2{,}400 + 1.645\left(\frac{\sigma}{\sqrt{n}}\right)\\ &\approx 2{,}400 + 1.645\left(\frac{s}{\sqrt{n}}\right) = 2{,}400 + 1.645\left(\frac{200}{\sqrt{50}}\right)\\ &= 2{,}400 + 1.645(28.28) = 2{,}446.5\end{aligned}$$

2. Next, for a particular alternative distribution corresponding to a value of μ, denoted by μ_a, we calculate the z value corresponding to $\bar{x}_0$, the border between the rejection and acceptance regions. We then use this z value and Table IV of Appendix B to determine the area in the *acceptance* region under the alternative distribution. This area is the value of β corresponding to the particular alternative μ_a. For example, for the alternative $\mu_a = 2{,}425$, we calculate

$$\begin{aligned}z &= \frac{\bar{x}_0 - 2{,}425}{\sigma_{\bar{x}}} = \frac{\bar{x}_0 - 2{,}425}{\sigma/\sqrt{n}}\\ &\approx \frac{\bar{x}_0 - 2{,}425}{s/\sqrt{n}} = \frac{2{,}446.5 - 2{,}425}{28.28} = .76\end{aligned}$$

FIGURE 8.15 ▶
Values of α and β for various values of μ

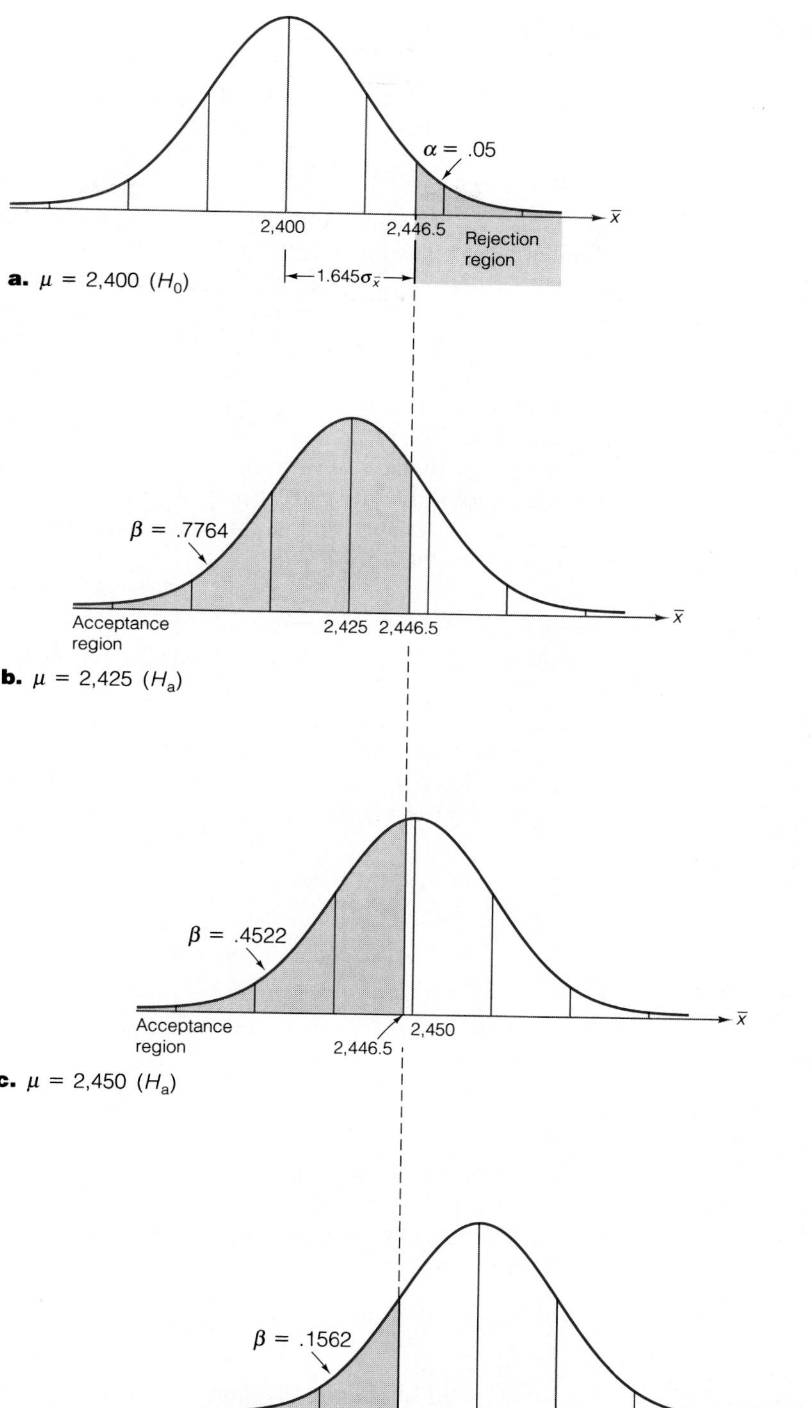

Note in Figure 8.15(b) that the area in the acceptance region is the area to the left of $z = .76$. This area is

$$\beta = .5 + .2764 = .7764$$

Thus, the probability that the test procedure will lead to an incorrect acceptance of the null hypothesis $\mu = 2{,}400$ when in fact $\mu = 2{,}425$ is about .78. As the average strength of the pipe increases to 2,450, the value of β decreases to .4522 [Figure 8.15(c)]. If the mean strength is further increased to 2,475, the value of β is further decreased to .1562 [Figure 8.15(d)]. Thus, even if the true mean strength of the pipe exceeds the minimum specification by 75 pounds per lineal foot, the test procedure will lead to an incorrect acceptance of the null hypothesis (rejection of the pipe) approximately 16% of the time. The upshot is that the pipe must be manufactured so that the mean strength well exceeds the minimum requirement if the manufacturer wants the probability of its acceptance by the city to be large (i.e., β to be small).

The steps for calculating β for a large-sample test about a population mean are summarized in the box.

Steps for Calculating β For a Large-Sample Test About μ

1. Calculate the value(s) of $\bar{x}$ corresponding to the border(s) of the rejection region. There will be one border value for a one-tailed test and two for a two-tailed test. The formula is one of the following, corresponding to a test with level of significance α:

   Upper-tailed test: $\bar{x}_0 = \mu_0 + z_\alpha \sigma_{\bar{x}} \approx \mu_0 + z_\alpha\left(\frac{s}{\sqrt{n}}\right)$

   Lower-tailed test: $\bar{x}_0 = \mu_0 - z_\alpha \sigma_{\bar{x}} \approx \mu_0 - z_\alpha\left(\frac{s}{\sqrt{n}}\right)$

   Two-tailed test: $\bar{x}_{0,\text{L}} = \mu_0 - z_{\alpha/2} \sigma_{\bar{x}} \approx \mu_0 - z_{\alpha/2}\left(\frac{s}{\sqrt{n}}\right)$

   $\bar{x}_{0,\text{U}} = \mu_0 + z_{\alpha/2} \sigma_{\bar{x}} \approx \mu_0 + z_{\alpha/2}\left(\frac{s}{\sqrt{n}}\right)$

2. Specify the value of μ_a in the alternative hypothesis for which the value of β is to be calculated. Then convert the border value(s) of $\bar{x}_0$ to z value(s) using the alternative distribution with mean μ_a. The general formula for the z value is

   $$z = \frac{\bar{x}_0 - \mu_a}{\sigma_{\bar{x}}}$$

   Sketch the alternative distribution (centered at μ_a), and shade the area above the acceptance (nonrejection) region. Use the z statistic(s) and Table IV of Appendix B to find the shaded area, which is β.

Following the calculation of β for a particular value of μ_a, you should interpret the value in the context of the hypothesis-testing application. It is often useful to interpret the value of $1 - \beta$, which is known as the **power of the test**, corresponding to a particular alternative, μ_a. Since β is the probability of accepting the null hypothesis when the alternative hypothesis is true with $\mu = \mu_a$, $1 - \beta$ is the probability of the complementary event, or the probability of rejecting the null hypothesis when the alternative $\mu = \mu_a$ is true. That is, the power $1 - \beta$ measures the likelihood that the test procedure will lead to the *correct* decision (reject H_0) for a particular value of the mean in the alternative hypothesis.

Definition 8.2

The **power** of a test is the probability that the test will correctly lead to the rejection of the null hypothesis for a particular value of μ in the alternative hypothesis. The power is equal to $1 - \beta$ for the particular alternative considered.

For example, in the sewer pipe example we found that $\beta = .7764$ when $\mu = 2{,}425$. This is the probability that the test leads to the (incorrect) acceptance of the null hypothesis when $\mu = 2{,}425$. Or, equivalently, the power of the test is $1 - .7764 = .2236$, which means that the test will lead to the (correct) rejection of the null hypothesis only 22% of the time when the pipe exceeds specifications by 25 pounds per lineal foot. When the manufacturer's pipe has a mean strength of 2,475 (that is, 75 pounds per lineal foot in excess of specifications), the power of the test increases to $1 - .1562 = .8438$. That is, the test will lead to the acceptance of the manufacturer's pipe 84% of the time if $\mu = 2{,}475$.

EXAMPLE 8.8

Recall the quality control study in Examples 8.1 and 8.2, in which we tested to determine whether a cereal box filling machine was deviating from the specified mean fill of $\mu = 12$ ounces. The test setup is repeated here:

H_0: $\mu = 12$

H_a: $\mu \neq 12$ (i.e., $\mu < 12$ or $\mu > 12$)

Test statistic: $z = \dfrac{\bar{x} - 12}{\sigma_{\bar{x}}}$

Rejection region: $z < -1.96$ or $z > 1.96$ for $\alpha = .05$
$z < -2.575$ or $z > 2.575$ for $\alpha = .01$

Note that two rejection regions have been specified corresponding to values of $\alpha = .05$ and $\alpha = .01$, respectively. Assume that $n = 100$ and $s = .5$.

a. Suppose the machine is underfilling the boxes by an average of .1 ounce, i.e., $\mu = 11.9$. Calculate the values of β corresponding to the two rejection regions. Discuss the relationship between the values of α and β.

b. Calculate the power of the test for each of the rejection regions when $\mu = 11.9$.

Solution

a. We first consider the rejection region corresponding to $\alpha = .05$. The first step is to calculate the border values of $\bar{x}$ corresponding to the two-tailed rejection region, $z < -1.96$ or $z > 1.96$:

$$\bar{x}_{0,\mathrm{L}} = \mu_0 - 1.96\sigma_{\bar{x}} \approx \mu_0 - 1.96\left(\frac{s}{\sqrt{n}}\right) = 12.0 - 1.96\left(\frac{.5}{10}\right) = 11.902$$

$$\bar{x}_{0,\mathrm{U}} = \mu_0 + 1.96\sigma_{\bar{x}} \approx \mu_0 + 1.96\left(\frac{s}{\sqrt{n}}\right) = 12.0 + 1.96\left(\frac{.5}{10}\right) = 12.098$$

These border values are shown in Figure 8.16(a).

FIGURE 8.16 ▶ Calculation for β for Example 8.8

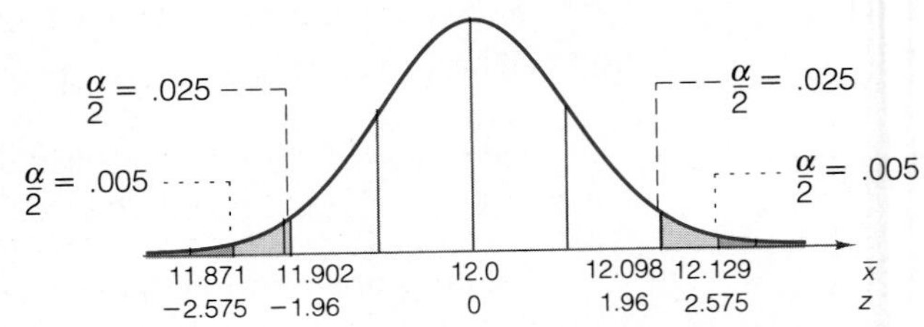

a. $\mu = 12$ (H_0)
Two rejection regions
$\alpha = .05$ and $\alpha = .01$

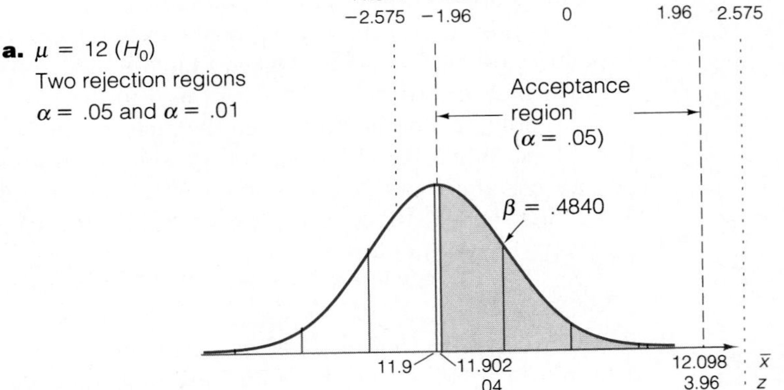

b. $\mu = 11.9$ (H_a)
β for $\alpha = .05$ rejection region

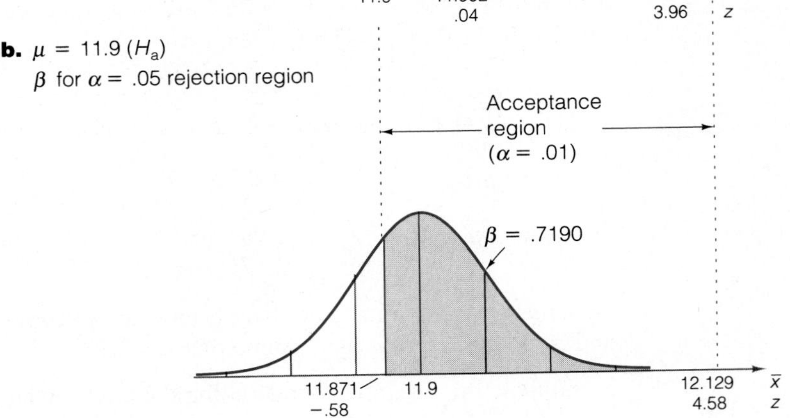

c. $\mu = 11.9$ (H_a)
β for $\alpha = .01$ rejection region

Next, we convert these values to z values in the alternative distribution with $\mu_a = 11.9$:

$$z_L = \frac{\bar{x}_{0,L} - \mu_a}{\sigma_{\bar{x}}} \approx \frac{11.902 - 11.9}{.05} = .04$$

$$z_U = \frac{\bar{x}_{0,U} - \mu_a}{\sigma_{\bar{x}}} \approx \frac{12.098 - 11.9}{.05} = 3.96$$

These z values are shown in Figure 8.16(b), and you can see that the acceptance (or nonrejection) region is the area between them. Using Table IV of Appendix B, we find that the area between $z = 0$ and $z = .04$ is .0160, and the area between $z = 0$ and $z = 3.96$ is (approximately) .5 (since $z = 3.96$ is off the scale of Table IV). Then the area between $z = .04$ and $z = 3.96$ is approximately

$$\beta = .5 - .0160 = .4840$$

Thus, the test with $\alpha = .05$ will lead to a Type II error about 48% of the time when the machine is underfilling by .1 ounce.

For the rejection region corresponding to $\alpha = .01$, $z < -2.575$ or $z > 2.575$, we find

$$\bar{x}_{0,L} = 12.0 - 2.575\left(\frac{.5}{10}\right) = 11.871$$

$$\bar{x}_{0,U} = 12.0 + 2.575\left(\frac{.5}{10}\right) = 12.129$$

These border values of the rejection region are shown in Figure 8.16(c).

Converting these to z values in the alternative distribution with $\mu_a = 11.9$, we find $z_L = -.58$ and $z_U = 4.58$. The area between these values is approximately

$$\beta = .2190 + .5 = .7190$$

Thus, the chance that the test procedure with $\alpha = .01$ will lead to an incorrect acceptance of H_0 is about 72%.

Note that the value of β increases from .4840 to .7190 when we decrease the value of α from .05 to .01. This is a general property of the relationship between α and β: *as α is decreased (increased), β is increased (decreased).*

b. The power is defined to be the probability of (correctly) rejecting the null hypothesis when the alternative is true. When $\mu = 11.9$ and $\alpha = .05$, we find

$$\text{Power} = 1 - \beta = 1 - .4840 = .5160$$

When $\mu = 11.9$ and $\alpha = .01$,

$$\text{Power} = 1 - \beta = 1 - .7190 = .2810$$

You can see that the power of the test is decreased as the level of α is decreased. This means that as the probability of incorrectly rejecting the null hypothesis is decreased, the probability of correctly accepting the null hypothesis for a given alternative is also decreased. Thus, the value of α must be selected carefully, with the realization that a test is made less powerful to detect departures from the null hypothesis when the value of α is decreased.

We have shown that the probability of committing a Type II error, β, is inversely related to α (Example 8.8), and that the value of β decreases as the value of μ moves farther from the null hypothesis value (sewer pipe example). The sample size n also affects β. Remember that the standard deviation of the sampling distribution of $\bar{x}$ is inversely proportional to the (square root of) the sample size: $\sigma_{\bar{x}} = \sigma/\sqrt{n}$. Thus, as illustrated in Figure 8.17, the variability of both the null and alternative sampling distributions is decreased as n is increased. If the value of α is specified and remains fixed, the value of β decreases as n increases, as illustrated in Figure 8.17(b). Conversely, the power of the test for a given alternative hypothesis is increased as the sample size is increased.

FIGURE 8.17 ►
Relationship between α, β, and n

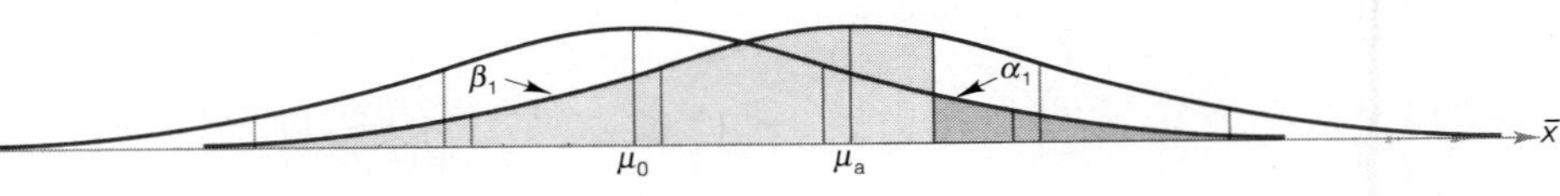

a. Small n

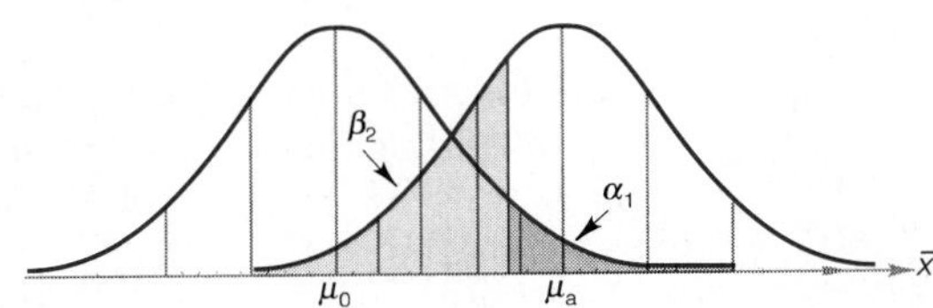

b. Large n, fixed α ($\beta_2 < \beta_1$)

The properties of β and power are summarized in the box.

Properties of β and Power

1. The value of β decreases and the power increases as the distance between the null and alternative values of μ increases (see Figure 8.15).
2. The value of β increases and the power decreases as the value of α is decreased (see Figure 8.16).
3. The value of β decreases and the power increases as the sample size is increased, assuming α remains fixed (see Figure 8.17).

Exercises 8.58–8.69

Learning the Mechanics

8.58 What is the relationship between β, the probability of committing a Type II error, and the power of a test?

8.59 List three factors that increase the power of a test.

8.60 Suppose you want to test H_0: $\mu = 500$ against H_a: $\mu > 500$ using $\alpha = .05$. The population in question is normally distributed with standard deviation 100. A random sample of size $n = 25$ will be used.

- **a.** Sketch the sampling distribution of $\bar{x}$ assuming that H_0 is true.
- **b.** Find the value of $\bar{x}_0$, that value of $\bar{x}$ above which the null hypothesis will be rejected. Indicate the rejection region on your graph of part **a**. Shade the area above the rejection region and label it α.
- **c.** On your graph of part **a**, sketch the sampling distribution of $\bar{x}$ if $\mu = 550$. Shade the area under this distribution that corresponds to the probability that $\bar{x}$ falls in the nonrejection region when $\mu = 550$. Label this area β.
- **d.** Find β.
- **e.** Compute the power of this test for detecting the alternative H_a: $\mu = 550$.

8.61 Refer to Exercise 8.60.

- **a.** If $\mu = 575$ instead of 550, what is the probability that the hypothesis test will incorrectly fail to reject H_0? That is, what is β?
- **b.** If $\mu = 575$, what is the probability that the test will correctly reject the null hypothesis? That is, what is the power of the test?
- **c.** Compare β and the power of the test when $\mu = 575$ to the values you obtained in Exercise 8.60 for $\mu = 550$. Explain the differences.

8.62 It is desired to test H_0: $\mu = 75$ against H_a: $\mu < 75$ using $\alpha = .10$. The population in question is uniformly distributed with standard deviation 15. A random sample of size 49 will be drawn from the population.

- **a.** Describe the (approximate) sampling distribution of $\bar{x}$ under the assumption that H_0 is true.
- **b.** Describe the (approximate) sampling distribution of $\bar{x}$ under the assumption that the population mean is 70.
- **c.** If μ were really equal to 70, what is the probability that the hypothesis test would lead the investigator to commit a type II error?
- **d.** What is the power of this test for detecting the alternative H_a: $\mu = 70$?

8.63 Refer to Exercise 8.62. If the true value of the population mean is $\mu = 73$, what is the power of the test? How does it compare with the power when $\mu = 70$?

8.64 Refer to Exercises 8.62 and 8.63.

- **a.** Find β for each of the following values of the population mean: 74, 72, 70, 68, 66.
- **b.** Plot each value of β you obtained in part **a** against its associated population mean. Show β on the vertical axis and μ on the horizontal axis. Draw a curve through the five points on your graph.
- **c.** Use your graph of part **b** to find the approximate probability that the hypothesis test will lead to a type II error when $\mu = 73$. Compare your answer to the result you obtained in Exercise 8.63.
- **d.** Convert each of the β values you calculated in part **a** to the power of the test at the specified value of μ. Plot the power on the vertical axis against μ on the horizontal axis. Compare the β graph of part **b** to the **power curve** of this part.

e. Examine the graphs of parts **b** and **d**. Explain what they reveal about the relationships among the distance between the true mean μ and the null hypothesized mean μ_0, the value of β, and the power.

8.65 Suppose you want to conduct the two-tailed test of H_0: $\mu = 30$ against H_a: $\mu \neq 30$ using $\alpha = .05$. A random sample of size 121 will be drawn from the population in question. Assume the population has a standard deviation equal to 1.2.

a. Describe the sampling distribution of $\bar{x}$ under the assumption that H_0 is true.
b. Describe the sampling distribution of $\bar{x}$ under the assumption that $\mu = 29.8$.
c. If μ were really equal to 29.8, find the value of β associated with the test.
d. Find the value of β for the alternative H_a: $\mu = 30.4$.

Applying the Concepts

8.66 If a manufacturer (the vendee) buys all items of a particular type from a particular vendor, the manufacturer is practicing *sole sourcing*. Sole sourcing is a purchasing policy that is generally recognized as an important component of a firm's quality system. One of the major benefits of sole sourcing for the vendee is the improved communication that results from the closer vendee/vendor relationship (Treleven, 1986). As part of a sole sourcing arrangement, a vendor agreed to periodically supply its vendee with sample data from its production process. The vendee uses the data to investigate whether the mean length of rods produced by the vendor's production process is truly 5.0 millimeters or more, as claimed by the vendor and desired by the vendee.

a. If the production process has a standard deviation of .01 millimeter, the vendor supplies $n = 100$ items to the vendee, and the vendee uses $\alpha = .05$ in testing H_0: $\mu = 5.0$ millimeters against H_a: $\mu < 5.0$ millimeters, what is the probability that the vendee's test will fail to reject the null hypothesis when in fact $\mu = 4.9975$ millimeters? What is the name given to this type of error?
b. Refer to part **a**. What is the probability that the vendee's test will reject the null hypothesis when in fact $\mu = 5.0$? What is the name given to this type of error?
c. What is the power of the test to detect a departure of .0025 millimeter below the specified mean rod length of 5.0 millimeters?

8.67 Refer to Exercise 8.18 in which the performance of a particular type of laser-based inspection equipment was investigated. Assume that the standard deviation of the number of solder joints inspected on each run is 1.2. If $\alpha = .05$ is used in conducting the hypothesis test of interest using a sample of 48 circuit boards, and if the true mean number of solder joints that can be inspected is really equal to 9.5, what is the probability that the test will result in a Type II error?

8.68 Refer to Exercise 8.17, in which the alternative hypothesis that the mean miles per gallon achieved by 1994 Polaris automobiles exceeds 35 is tested against the null hypothesis that the mean is 35 (or less). A sample of 36 automobiles were tested; assume that the resulting standard deviation of $s = 6$ is a good estimate of the true standard deviation.

a. Calculate the power of the test for the mean values of 35.5, 36.0, 36.5, 37.0, and 37.5.
b. Plot the power of the test on the vertical axis against the mean on the horizontal axis. Draw a curve through the points.
c. Use the power curve of part **b** to estimate the power for the mean value $\mu = 36.75$. Calculate the power for this value of μ, and compare it to your approximation.
d. Use the power curve to approximate the power of the test when $\mu = 40$. If the true value of the mean mpg for this model is really 40, what (approximately) are the chances that the test will fail to reject the null hypothesis that the mean is 35?

8.69 Refer to Exercise 8.68. Show what happens to the power curve when the sample size is increased from $n = 36$ to $n = 100$. Assume that the standard deviation is $\sigma = 6$.

Summary

We have extended the concept of making inferences from using samples to estimate parameters of a distribution to performing tests of hypotheses about these parameters. The essential elements of a test of hypothesis are the **null** and **alternative hypotheses**, the **test statistic**, the **rejection region**, the **calculation of the test statistic**, and the **conclusion**.

The null hypothesis is the *status-quo* hypothesis, the hypothesis not rejected until contradicted by sufficient sample evidence. The alternative is the *research hypothesis*, the hypothesis that will not be accepted until convincingly established by sample information. The test statistic is a value calculated from the sample information and utilized to decide whether to reject the null hypothesis. The rejection region is the collection of values of the test statistic that will cause us to reject the null hypothesis.

We design the test to have a specified **Type I error probability**, α, so that we know the probability of rejecting H_0 when H_0 is true. When the test statistic falls in the rejection region, we infer that the null hypothesis is false (that is, the alternative is true) at the α level of significance. If the test statistic does not fall in the rejection region, then we fail to reject the null hypothesis. We do not accept the null hypothesis in this case unless we know the **Type II error probability**, β.

We have covered large-sample tests involving the population mean μ and the binomial probability p, both using the standard normal z as a test statistic. We have also discussed the small-sample test involving the mean of a normal distribution using the t statistic.

Supplementary Exercises 8.70–8.91

Note: Starred (*) exercises refer to the optional section.

Learning the Mechanics

8.70 Which of the elements of a test of hypothesis can and should be specified *prior* to analyzing the data that are to be utilized to conduct the test?

8.71 Complete the following statement: The smaller the p-value associated with a test of hypothesis, the stronger the support for the ________ hypothesis. Explain your answer.

8.72 Specify the differences between a large-sample and small-sample test of hypothesis about a population mean μ. Focus on the assumptions and test statistics.

8.73 Medical tests have been developed to detect most serious diseases. A medical test is designed to minimize the probability that it will produce a "false positive" or a "false negative." A false positive refers to a positive test result when, in fact, the individual does not have the disease, whereas a false negative is a negative test result for an individual who does have the disease.

a. If we treat a medical test for a disease as a statistical test of hypothesis, what are the null and alternative hypotheses for the medical test?
b. What are the Type I and Type II errors for the test? Relate each to false positives and false negatives.
c. Considering which of the errors has more serious consequences, is it more important to minimize α or β? Explain.

8.74 If the rejection of the null hypothesis of a particular test would cause your firm to go out of business, would you want α to be small or large? Explain.

8.75 In conducting a hypothesis test, if you select a very small value for α, will β tend to be large or small? Explain.

Applying the Concepts

8.76 The EPA sets a limit of 5 parts per million (ppm) on PCB (a dangerous substance) in water. A major manufacturing firm producing PCB for electrical insulation discharges small amounts from the plant. The company management, attempting to control the amount of PCB in its discharge, has given instructions to halt production if the mean amount of PCB in the effluent exceeds 3 ppm. A random sample of 50 water specimens produced the following statistics: $\bar{x} = 3.1$ ppm, $s = .5$ ppm.

a. Do these statistics provide sufficient evidence to halt the production process? Use $\alpha = .01$.
b. If you were the plant manager, would you want to use a large or a small value for α for the test in part **a**? Explain.
c. Find the p-value for the test and interpret its value.

***8.77** Refer to Exercise 8.76.

a. In the context of the problem, define a Type II error.
b. Calculate β for the test described in part **a** of Exercise 8.76 assuming that the true mean is $\mu = 3.1$ ppm.
c. What is the power of the test to detect the effluent's departure from the standard of 3.0 ppm when the mean is 3.1 ppm?
d. Repeat parts **b** and **c** assuming that the true mean is 3.2 ppm. What happens to the power of the test as the plant's mean PCB departs further from the standard?

***8.78** Refer to Exercises 8.76 and 8.77.

a. Suppose an α value of .05 is used to conduct the test. Does this change favor the manufacturer? Explain.
b. Determine the value of β and the power for the test when $\alpha = .05$ and $\mu = 3.1$.
c. What happens to the power of the test when α is increased?

8.79 According to the Internal Revenue Service (IRS), in 1980, 4,414 taxpayers reported earning more than \$1 million in adjusted gross income. By 1990 that number had grown to 63,642. Interestingly, as the number of millionaires rose dramatically, their charitable contributions nose-dived from an average of \$207,089 in 1980 to \$83,929 in 1989 (Farhi, 1992). Will this trend continue into the mid 1990s? The IRS sampled six 1993 tax returns of millionaires and found the following charitable contributions (in thousands of dollars):

91.1 103.2 62.9 150.4 209.5 31.7

a. Do the sample data indicate that annual charitable contributions by millionaires have increased since 1989? Conduct the test using $\alpha = .05$.

b. What assumptions are necessary to assure the validity of the test conducted in part **a**? Comment on the validity of the assumptions.

8.80 During the 1980s, testing products for quality during production became a way of life for most successful businesses. Although first applications of quality testing were primarily in the manufacturing sector, companies in the service sector have increasingly turned to quality improvement techniques to compete effectively. For example, suppose a drive-in bank wants to assure that the mean waiting time for its customers will be less than 2 minutes. A random sample of 20 drive-in customers is selected daily, and the waiting times are recorded and analyzed. On one busy day the sample results were: $\bar{x} = 112$ seconds, $s = 28$ seconds. Can the bank safely conclude that the mean waiting time was less than 2 minutes on this day? Use $\alpha = .05$.

8.81 Refer to Exercise 8.80. What is the approximate observed significance level associated with this test?

8.82 In the past, a chemical company produced 880 pounds of a certain type of plastic per day. Now, using a newly developed and less expensive process, the mean daily yield of plastic for the first 50 days of production was 871 pounds, and the standard deviation was 21 pounds.

a. Do the data provide sufficient evidence to indicate that the mean daily yield for the new process is less than for the old procedure? (Test using $\alpha = .01$.)

b. What assumptions must you make in order to use the statistical test you employed?

8.83 Refer to Exercise 8.82. Find and interpret the p-value for the test conducted.

8.84 Refer to Exercise 8.82. Calculate the probability β that the test fails to reject the null hypothesis that the new process has a mean daily yield of $\mu = 880$ when in fact the true mean is $\mu = 875$. What is the power of the test to determine that $\mu < 880$ when $\mu = 875$?

8.85 A discount store chain claims that its steel-belted radial tires are more resistant to wear than those of a major tire company. The following experiment was performed to test this claim. On each of 40 cars, one discount tire and one rubber company tire were mounted on the real axle. After each car was driven 8,000 miles, the tires were inspected for wear. Suppose the tires of the discount chain show less wear on 32 of the cars. What would you conclude about the discount chain's claim? Why?

8.86 The *beta coefficients* of stocks are a measure of their volatility (or risk) relative to the market as a whole. Stocks with beta coefficients greater than 1 generally bear greater risk (more volatile) than the market, whereas stocks with beta coefficients less than 1 are less risky (less volatile) than the overall market (Sharpe and Alexander, 1990). A random sample of 15 high-technology stocks was selected at the end of 1990, and the mean and standard deviation of the beta coefficients were calculated: $\bar{x} = 1.23$, $s = .37$.

a. Set up the appropriate null and alternative hypotheses to test that the average high-technology stock is riskier than the market as a whole.

b. Establish the appropriate test statistic and rejection region for the test. Use $\alpha = .10$.

c. What assumptions are necessary to assure the validity of the test?

d. Calculate the test statistic and state your conclusion.

e. What is the approximate p-value associated with this test? Interpret it.

8.87 Refer to Exercise 8.86. When the data were analyzed with a statistical software package, the output was:

```
T = 2.408          P-VALUE = 0.0304
```

The p-value corresponds to a two-tailed test of the null hypothesis $\mu = 1$.

a. Interpret the p-value of the computer output.
b. If the alternative hypothesis of interest is $\mu > 1$, what is the appropriate p-value of the test?

8.88 One study of gambling newsletters that purport to improve a bettor's odds of winning bets on NFL football games indicates that the newsletters' betting schemes were not profitable (Sauer et al., 1988). Suppose a random sample of 50 games is selected to test one gambling newsletter. Following the newsletter's recommendations, 30 of the 50 games produced winning wagers. Test to see whether the newsletter can be said to increase the odds of winning significantly over what one could expect by selecting winners at random. Use $\alpha = .05$.

8.89 Refer to Exercise 8.88. Calculate and interpret the p-value for the test.

8.90 Refer to Exercise 8.88.
a. Describe a Type II error in terms of this application.
b. Calculate the probability β of a Type II error for this test assuming that the newsletter really does increase the probability of winning a wager on an NFL game to $p = .55$
c. Suppose the number of games sampled is increased from 50 to 100. How does this affect the probability of the Type II error in part **b**?

8.91 A random sample of 200 residents in one large city is selected, and each is asked whether a new increase in the property tax would be favored if the income were used for public education. A total of 113 indicate support for the tax.
a. What are the appropriate null and alternative hypotheses to test whether a majority of residents in the city favor the tax?
b. The data are analyzed using a statistical software package, with the following output:

```
Z = 1.84          P-VALUE = .0329
```

Interpret the results.
c. What assumptions, if any, are necessary to assure the validity of this test?

On Your Own

The **efficient market** theory postulates that the best predictor of a stock's price at some point in the future is the current price of the stock (with some adjustments for inflation and transaction costs, which we shall assume to be negligible for the purpose of this exercise). To test this theory, select a random sample of 25 stocks on the New York Stock Exchange and record the closing prices on the last day of each of two recent consecutive months. Calculate the increase or decrease in the stock price over the 1 month period.

a. Define μ as the mean change in price of all stocks over a 1-month period. Set up the appropriate null and alternative hypothesis in terms of μ.
b. Use the sample of the 25 stock price differences to conduct the test of hypothesis established in part **a**. Use $\alpha = .05$.

c. Tabulate the number of rejections and nonrejections of the null hypothesis obtained by all the students in your class. If the null hypothesis were true, how many rejections of the null hypothesis would you expect among those in your class? How does this expectation compare with the actual number of nonrejections? What does the result of this exercise indicate about the efficient market theory?

Using the Computer

Refer to **Using the Computer** in Chapters 6 and 7. Let μ_0 be the mean of the population of 1,000 zip codes you selected (that is, μ_0 is the "true" value of the population mean). In each of the following scenarios you will be testing the null hypothesis H_0: $\mu = \mu_0$ against the two-tailed alternative hypothesis H_a: $\mu \neq \mu_0$. Thus, in this exercise you know that the null hypothesis is true.

a. Select 100 samples of size 50 (with replacement) from the 1,000 zip codes, and conduct the test for each sample using $\alpha = .05$. In how many of the tests did the sample lead to an incorrect rejection of the null hypothesis? How does this compare with what you expected to occur? Repeat the exercise using $\alpha = .10$.
b. Select 100 samples of size 20 (with replacement) from the 1,000 zip codes, and conduct the test for each sample using $\alpha = .05$. In how many of the tests did the sample lead to an incorrect rejection of the null hypothesis? How does this compare with what you expected to occur? Repeat the exercise using $\alpha = .10$.

References

Caplen, R. *A Practical Approach to Quality Control*. London: Business Books, 1970. Chapter 15.

Dujack, S. R. "Science leaves no doubt: The polygraph lies." *Minneapolis Star and Tribune*, July 31, 1986.

Duncan, A. J. *Quality Control and Industrial Statistics*, 5th ed. Homewood, Ill.: Richard D. Irwin, 1986. Chapter 1.

Farhi, P. "Multiplying millionaires." *Star Tribune* (Minneapolis), July 14, 1992.

Montgomery, D. C. *Introduction to Statistical Quality Control*, 2d ed. New York: Wiley, 1991.

Nemy, E. "Survey says two-thirds of Americans see themselves as younger than they are." *Minneapolis Tribune*, Dec. 19, 1982.

Report of Definitions Committee of the American Marketing Association. Chicago: American Marketing Association, 1961.

Rice, F. "Be a smarter frequent flier." *Fortune*, Feb. 22, 1993, pp. 108–112.

Sauer, R. D., Brajer, V., Ferris, S. P., and Marr, M. W. "Hold your bets: Another look at the efficiency of the gambling market for National Football League games." *Journal of Political Economy*, Feb. 1988, Vol. 96, No. 1, pp. 260–273.

Sharpe, W. F. and Alexander, G. J. *Investments*, 4th ed. Englewood Cliffs, N.J.: Prentice-Hall, 1990.

Streeter, J. P. "Solder joint inspection using a laser inspector." *Quality Congress Transactions*. Milwaukee: American Society for Quality Control, 1986. pp. 507–515.

Treleven, M. "Sole sourcing from the vendor side."*Quality Congress Transactions*. Milwaukee: American Society for Quality Control, 1986. pp. 584–590.

Wetherill, G. B. *Sampling, Inspection, and Quality Control*, 2d ed. London: Chapman and Hall, 1977. Chapter 3.

Willis, R. E. and Chervany, N. L. *Statistical Analysis and Modeling for Management Decision-making*. Belmont, Calif.: Wadsworth, 1974. Chapters 8 and 11.

CHAPTER NINE

Inferences Based on Two Samples: Estimation and Tests of Hypotheses

Contents

Case Studies

Where We've Been

The two methods for making statistical inferences, estimation and tests of hypotheses, were presented in Chapters 7 and 8. Confidence intervals and tests of hypotheses based on single samples were used to make inferences about sampled populations. We gave confidence intervals and tests of hypotheses concerning a population mean, μ, and a binomial proportion, p, and learned how to select the sample size necessary to obtain a specified amount of information concerning a parameter.

Where We're Going

Now that we have learned to make inferences about a single population, it is natural that we would want to compare two populations. We may want to compare the mean costs per pound in the manufacture of two drugs or the mean shelf lives of two industrial products. We may also wish to compare two population proportions, say the proportions of consumers who prefer a product before and after an advertising campaign. How to decide whether differences exist in population means or proportions and how to estimate these differences is the subject of this chapter.

9

9.1 Large-Sample Inferences About the Difference Between Two Population Means: Independent Sampling

Suppose a chain of department stores is considering two suburbs of a large city as alternatives for locating a new store. The final decision about which location to choose will be based on a comparison of the mean incomes of families living in the two suburbs.* The store is to be located in the suburb that has the higher mean income per household.

Let μ_1 represent the mean income of families in suburb 1 and μ_2 represent the mean income of families in suburb 2. Then our objective is to make an inference about $(\mu_1 - \mu_2)$, the difference between the mean incomes for the two suburbs.

Suppose independent samples of 100 households are randomly selected from each suburb, and the mean incomes, $\bar{x}_1$ and $\bar{x}_2$, are calculated for the two samples. An intuitively appealing estimator for $(\mu_1 - \mu_2)$ is the difference between the sample means, $(\bar{x}_1 - \bar{x}_2)$. The performance of this estimator in repeated sampling is summarized by the properties of its sampling distribution (see Figure 9.1).

FIGURE 9.1 ► Sampling distribution of $(\bar{x}_1 - \bar{x}_2)$

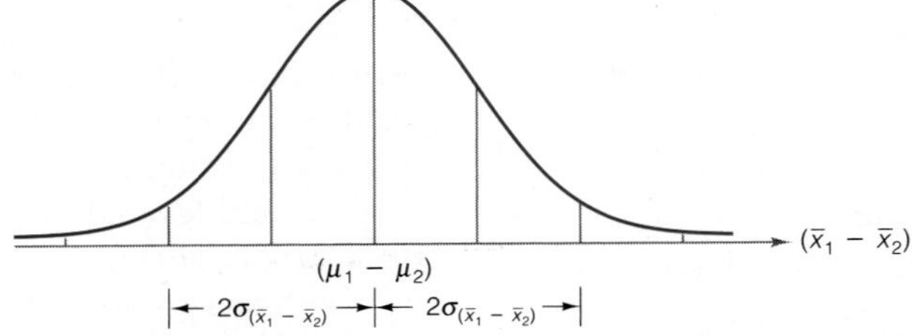

Properties of the Sampling Distribution $(\bar{x}_1 - \bar{x}_2)$

1. The mean of the sampling distribution of $(\bar{x}_1 - \bar{x}_2)$ is $(\mu_1 - \mu_2)$.
2. If the two samples are independent, the standard deviation of the sampling distribution is

$$\sigma_{(\bar{x}_1 - \bar{x}_2)} = \sqrt{\frac{\sigma_1^2}{n_1} + \frac{\sigma_2^2}{n_2}}$$

   where σ_1^2 and σ_2^2 are the variances of the two populations being sampled, and n_1 and n_2 are the respective sample sizes. We also refer to $\sigma_{(\bar{x}_1 - \bar{x}_2)}$ as the **standard error** of the statistic $(\bar{x}_1 - \bar{x}_2)$.
3. The sampling distribution of $(\bar{x}_1 - \bar{x}_2)$ is approximately normal for *large* samples by the Central Limit Theorem.

*Assume that the incomes within a suburb are moderately homogeneous, hence the distributions are not heavily skewed, so the mean would be a satisfactory measure of central tendency for the data.

Since the shape of the sampling distribution is approximately normal for large samples, we can use the z statistic to make inferences about $(\mu_1 - \mu_2)$, just as we did for a single mean. The procedures for forming confidence intervals and testing hypotheses are summarized in the boxes. Note the similarity of these procedures to their counterparts for a single mean (Sections 7.1 and 8.2).

Large-Sample Confidence Interval for $(\mu_1 - \mu_2)$

$$(\bar{x}_1 - \bar{x}_2) \pm z_{\alpha/2}\sigma_{(\bar{x}_1-\bar{x}_2)} = (\bar{x}_1 - \bar{x}_2) \pm z_{\alpha/2}\sqrt{\frac{\sigma_1^2}{n_1} + \frac{\sigma_2^2}{n_2}}$$

Assumptions: The two samples are randomly selected in an independent manner from the two populations. The sample sizes, n_1 and n_2, are large enough so that $\bar{x}_1$ and $\bar{x}_2$ each have approximately normal sampling distributions and so that s_1^2 and s_2^2 provide good approximations to σ_1^2 and σ_2^2. This will generally be true if $n_1 \geq 30$ and $n_2 \geq 30$.

Large-Sample Test of Hypothesis for $(\mu_1 - \mu_2)$

One-Tailed Test	*Two-Tailed Test*
H_0: $(\mu_1 - \mu_2) = D_0$	H_0: $(\mu_1 - \mu_2) = D_0$
H_a: $(\mu_1 - \mu_2) < D_0$	H_a: $(\mu_1 - \mu_2) \neq D_0$
[or H_a: $(\mu_1 - \mu_2) > D_0$]	

where D_0 = Hypothesized difference between the means (this is often 0)

Test statistic: $z = \dfrac{(\bar{x}_1 - \bar{x}_2) - D_0}{\sigma_{(\bar{x}_1-\bar{x}_2)}}$	*Test statistic*: $z = \dfrac{(\bar{x}_1 - \bar{x}_2) - D_0}{\sigma_{(\bar{x}_1-\bar{x}_2)}}$

where $\sigma_{(\bar{x}_1-\bar{x}_2)} = \sqrt{\dfrac{\sigma_1^2}{n_1} + \dfrac{\sigma_2^2}{n_2}}$

Rejection region: $z < -z_\alpha$	*Rejection region*: $z < -z_{\alpha/2}$
[or $z > z_\alpha$ when H_a: $(\mu_1 - \mu_2) > D_0$]	or $z > z_{\alpha/2}$

Assumptions: Same as for the large-sample confidence interval.

Suburb 1	*Suburb 2*
$\bar{x}_1$ = \$38,750	$\bar{x}_2$ = \$35,150
s_1 = \$3,200	s_2 = \$2,700
n_1 = 100	n_2 = 100

For example, suppose the means and standard deviations of the incomes of sampled households from the two suburbs are as shown. Then to form a 95% confidence interval for the difference $(\mu_1 - \mu_2)$ between the true mean suburban incomes, we calculate

$$(\bar{x}_1 - \bar{x}_2) \pm 1.96\sqrt{\frac{\sigma_1^2}{n_1} + \frac{\sigma_2^2}{n_2}} = (38{,}750 - 35{,}150) \pm 1.96\sqrt{\frac{\sigma_1^2}{100} + \frac{\sigma_2^2}{100}}$$

To complete the calculations for this confidence interval, we must estimate σ_1^2 and σ_2^2. Since the samples are both relatively large, the sample variances s_1^2 and s_2^2 will provide reasonable approximations. Thus, our interval is approximately

$$3{,}600 \pm 1.96\sqrt{\frac{(3{,}200)^2}{100} + \frac{(2{,}700)^2}{100}} = 3{,}600 \pm 821 = (2{,}779,\ 4{,}421)$$

When this estimation procedure is used, confidence intervals of this type will enclose the difference in population means, $(\mu_1 - \mu_2)$, 95% of the time. Therefore, we are reasonably confident that the mean income of households in suburb 1 is between \$2,779 and \$4,421 higher than the mean income of households in suburb 2. Based on this information, the department store chain should build the new store in suburb 1.

EXAMPLE 9.1

In the late 1980s and early 1990s the United States and Japan engaged in intense negotiations regarding restrictions on trade between the two countries. One of the claims made repeatedly by U.S. officials was that many Japanese manufacturers price their goods higher in Japan than in the United States, in effect subsidizing low prices in the United States by extremely high prices in Japan. The basis of the U.S. argument was that Japan was able to do that only by keeping competitive U.S. goods from reaching the Japanese marketplace.

An economist decided to test the hypothesis that higher retail prices were being charged for Japanese automobiles in Japan than in the United States. She obtained random samples of 50 retail sales in the United States and 30 retail sales in Japan over the same time period and for the same model of automobile, converted the Japanese sales prices from yen to dollars using current conversion rates, and obtained the summary shown in the table. Do these data provide sufficient evidence for the economist to conclude that the mean sales price for this model is higher in Japan than in the United States?

U.S. Sales	*Japanese Sales*
$n_1 = 50$	$n_2 = 30$
$\bar{x}_1 = \$11{,}545$	$\bar{x}_2 = \$12{,}243$
$s_1 = \$1{,}989$	$s_2 = \$1{,}843$

Solution

We can best answer this question by performing a test of hypothesis. Defining μ_1 as the mean sales price in the United States and μ_2 as the mean sales price in Japan during this period and for this model of automobile, we want to test whether the data support the alternative (research) hypothesis that $\mu_2 > \mu_1$ [i.e., that $(\mu_1 - \mu_2) < 0$]. Thus, we test the null hypothesis, $(\mu_1 - \mu_2) = 0$. The evidence necessary to reject this hypothesis in favor of the alternative is a sufficiently large negative value of the difference between the sample means, $(\bar{x}_1 - \bar{x}_2)$.

The elements of the test are as follows:

H_0: $(\mu_1 - \mu_2) = 0$ (i.e., $\mu_1 = \mu_2$; note that $D_0 = 0$ for this hypothesis test)

H_a: $(\mu_1 - \mu_2) < 0$ (i.e., $\mu_1 < \mu_2$)

Test statistic: $z = \dfrac{(\bar{x}_1 - \bar{x}_2) - D_0}{\sigma_{(\bar{x}_1 - \bar{x}_2)}} = \dfrac{(\bar{x}_1 - \bar{x}_2) - 0}{\sigma_{(\bar{x}_1 - \bar{x}_2)}}$

Rejection region: $z < -z_\alpha = -1.645$ (see Figure 9.2)

FIGURE 9.2 ►
Rejection region for Example 9.1

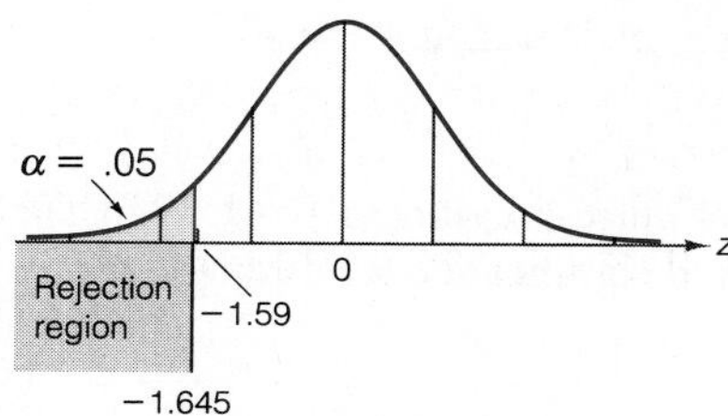

Assuming the samples from the United States and Japan are independent, we now calculate

$$z = \frac{(\bar{x}_1 - \bar{x}_2) - 0}{\sigma_{(\bar{x}_1 - \bar{x}_2)}} = \frac{(11{,}545 - 12{,}243)}{\sqrt{\dfrac{\sigma_1^2}{n_1} + \dfrac{\sigma_2^2}{n_2}}}$$

$$\approx \frac{-698}{\sqrt{\dfrac{s_1^2}{n_1} + \dfrac{s_2^2}{n_2}}} = \frac{-698}{\sqrt{\dfrac{(1{,}989)^2}{50} + \dfrac{(1{,}843)^2}{30}}} = \frac{-698}{438.57} = -1.59$$

As you can see in Figure 9.2, the calculated z value does not fall in the rejection region. The samples do not provide sufficient evidence, at $\alpha = .05$, for the economist to conclude that the mean retail price in Japan exceeds that in the United States.

EXAMPLE 9.2

Find the observed significance level for the test from Example 9.1.

Solution

The alternative hypothesis in Example 9.1, H_a: $(\mu_1 - \mu_2) < 0$, required a lower one-tailed test using

$$z = \frac{\bar{x}_1 - \bar{x}_2}{\sigma_{(\bar{x}_1 - \bar{x}_2)}}$$

as a test statistic. Since the value of z calculated from the sample data was -1.59, the observed significance level (p-value) for the test is the probability of observing a value of z at least as contradictory to the null hypothesis as $z = -1.59$; i.e.,

$$p\text{-value} = P(z \leq -1.59)$$

This probability is computed assuming H_0 is true and is equal to the shaded area shown in Figure 9.3.

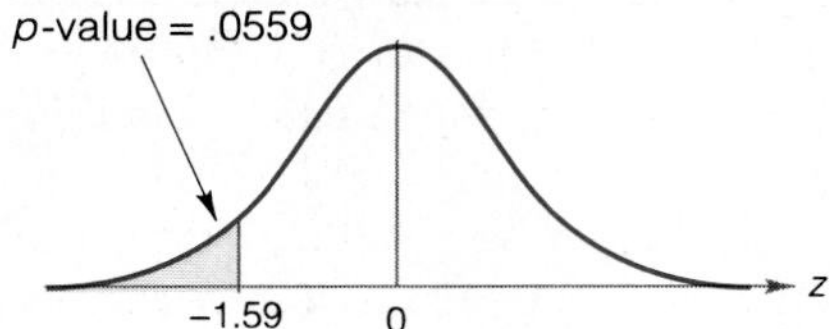

FIGURE 9.3 ▶ **The observed significance level for Example 9.1**

The tabulated area corresponding to $z = 1.59$ in Table IV of Appendix B is .4441. Therefore, the observed significance level for the test is

$$\begin{aligned} p\text{-value} &= .5 - .4441 \\ &= .0559 \end{aligned}$$

You will recall that in Example 9.1, we chose $\alpha = .05$ as the probability of a Type I error and consequently did not reject H_0; that is, we did not find sufficient evidence to indicate that $(\mu_1 - \mu_2) < 0$. If the results of the test had been presented in terms of an observed significance level and left for us to interpret, we might not have reached such an inflexible conclusion. Observing a value of z as small as $z = -1.59$ is an improbable event (rare event) if, in fact, $\mu_1 = \mu_2$. Since the probability is fairly small (.0559)—in fact, quite close to .05—we would conclude that there is some evidence to suggest that $\mu_1 < \mu_2$. Naturally, we would be more certain of this conclusion if the observed significance level were smaller, say, .05, .01, or, better yet, .001. However, the practical question to be answered is not whether the test results are statistically significant but whether the difference between μ_1 and μ_2 is large enough to be significant from a practical business perspective. To shed light on this question, we will wish to estimate the difference, $(\mu_1 - \mu_2)$.

EXAMPLE 9.3

Find a 95% confidence interval for the difference in mean retail prices in the United States and Japan of Example 9.1 and discuss the implications of the confidence interval.

Solution

The 95% confidence interval for $(\mu_1 - \mu_2)$ is

$$(\bar{x}_1 - \bar{x}_2) \pm z_{\alpha/2}\sqrt{\frac{\sigma_1^2}{n_1} + \frac{\sigma_2^2}{n_2}}$$

Once again, we will substitute s_1^2 and s_2^2 for σ_1^2 and σ_2^2 because these quantities will provide good approximations to σ_1^2 and σ_2^2 for samples as large as $n_1 = 50$ and $n_2 = 30$. Then, the 95% confidence interval for $(\mu_1 - \mu_2)$ is

$$(11{,}545 - 12{,}243) \pm 1.96\sqrt{\frac{(1.989)^2}{50} + \frac{(1{,}843)^2}{30}} = -698 \pm 859.60$$

Thus, we estimate the difference in mean retail prices to fall in the interval −$1,557.60 to $161.60. In other words, we estimate that μ_2, the mean retail price in Japan, could be larger than μ_1, the mean retail price in the United States, by as much as $1,557.60, or it could be less than μ_1 by as much as $161.60.

What is the practical interpretation of all this? By all appearances, the Japanese retail price for this model car was indeed higher than the U.S. retail price during the same time period. However, the data collected were insufficient to provide statistical support for this conclusion at the specified level of significance.

What can be done further to refine and sharpen the analysis? The first and most obvious improvement would be to increase the sample sizes: Sample sizes of 50 and 30 yield a rather wide confidence interval (total width of $1,719.20) for estimating the difference between the means. Increasing the sample sizes will decrease the standard error $\sigma_{(\bar{x}_1 - \bar{x}_2)}$; hence, the width of the confidence interval will also decrease.

A second, less obvious, refinement to the analysis would be somehow to decrease the rather large standard deviations associated with the samples, s_1 and s_2. Both exceed $1,800, indicating that the retail prices contained within each sample vary widely. Perhaps the options available on this model result in significant price discrepancies, or perhaps the dealerships in urban settings have significantly different pricing strategies from those in rural settings. *Regression analysis* is a useful statistical methodology for comparing population means while controlling for other factors affecting the variability of the measurements, such as options and geographic location in this example. Regression analysis is the subject of Chapters 10–12.

Exercises 9.1–9.15

Learning the Mechanics

9.1 The purpose of this exercise is to compare the variation of $\bar{x}_1$ and $\bar{x}_2$ with the variation of $(\bar{x}_1 - \bar{x}_2)$.

a. Suppose the first sample is selected from a population with mean $\mu_1 = 150$ and variance $\sigma_1^2 = 900$. Within what range should the sample mean vary about 95% of the time in repeated samples of 100 measurements from this distribution? That is, construct an interval extending 2 standard deviations of $\bar{x}_1$ on each side of μ_1.

b. Suppose the second sample is selected independently of the first from a second population with mean $\mu_2 = 150$ and variance $\sigma_2^2 = 1{,}600$. Within what range should the sample mean vary about 95% of the time in repeated samples of 100 measurements from this distribution? That is, construct an interval extending 2 standard deviations of $\bar{x}_2$ on each side of μ_2.

c. Now consider the difference between the two sample means, $(\bar{x}_1 - \bar{x}_2)$. What are the mean and standard deviation of the sampling distribution of $(\bar{x}_1 - \bar{x}_2)$?
d. Within what range should the difference in sample means vary about 95% of the time in repeated independent samples of 100 measurements each from the two populations? That is, construct an interval extending 2 standard deviations of $(\bar{x}_1 - \bar{x}_2)$ on each side of $(\mu_1 - \mu_2)$.
e. What, in general, can be said about the variability of the difference between independent sample means relative to the variability of the individual sample means?

9.2 Independent random samples of 64 observations each are chosen from two normal populations with $\mu_1 = 12$, $\mu_2 = 10$, $\sigma_1 = 4$, and $\sigma_2 = 3$. Let $\bar{x}_1$ and $\bar{x}_2$ denote the two sample means.
a. Give the mean and standard deviation of the sampling distribution of $\bar{x}_1$.
b. Give the mean and standard deviation of the sampling distribution of $\bar{x}_2$.
c. Find the mean and standard deviation of the sampling distribution of $(\bar{x}_1 - \bar{x}_2)$.
d. Will the statistic $(\bar{x}_1 - \bar{x}_2)$ be normally distributed? Explain.

9.3 Refer to Exercise 9.2.
a. Give the z-score corresponding to $(\bar{x}_1 - \bar{x}_2) = 2.25$.
b. Find the probability that $(\bar{x}_1 - \bar{x}_2)$ is larger than 2.25.
c. Find the probability that $(\bar{x}_1 - \bar{x}_2)$ is less than 2.25.
d. Find the probability that $(\bar{x}_1 - \bar{x}_2)$ is less than $-.8$ or larger than .8.

9.4 Two independent random samples have been selected, 100 observations from population 1 and 100 from population 2. Sample means $\bar{x}_1 = 15.5$ and $\bar{x}_2 = 26.6$ were obtained. From previous experience with these populations, it is known that the variances are $\sigma_1^2 = 9$ and $\sigma_2^2 = 16$.
a. Find $\sigma_{(\bar{x}_1 - \bar{x}_2)}$.
b. Sketch the approximate sampling distribution for $(\bar{x}_1 - \bar{x}_2)$ assuming $(\mu_1 - \mu_2) = 10$.
c. Locate the observed value of $(\bar{x}_1 - \bar{x}_2)$ on the graph you drew in part **b**. Does it appear that this value contradicts the null hypothesis H_0: $(\mu_1 - \mu_2) = 10$?
d. Use the z table on the inside front cover to determine the rejection region for the test of the hypothesis H_0: $(\mu_1 - \mu_2) = 10$ against H_a: $(\mu_1 - \mu_2) \neq 10$. Use $\alpha = .05$.
e. Conduct the hypothesis test of part **d** and interpret your results.

9.5 Refer to Exercise 9.4. Construct a 95% confidence interval for $(\mu_1 - \mu_2)$. Interpret the interval. Which inference provides more information about the value of $(\mu_1 - \mu_2)$—the test of hypothesis in Exercise 9.4 or the confidence interval in this exercise?

9.6 Two independent random samples have been selected, 100 from population 1 and 150 from population 2. Sample means $\bar{x}_1 = 1{,}025$ and $\bar{x}_2 = 1{,}039$ were obtained. The sample standard deviations are $s_1 = 10$ and $s_2 = 12$.
a. Describe the sampling distribution of $(\bar{x}_1 - \bar{x}_2)$. Assume that $(\mu_1 - \mu_2) = -5$.
b. Test H_0: $(\mu_1 - \mu_2) = -5$ against H_a: $(\mu_1 - \mu_2) < -5$, using $\alpha = .01$. Interpret the results of your test.
c. Report and interpret the p-value of your test.

9.7 Independent random samples are selected from two populations and are used to test the hypothesis H_0: $(\mu_1 - \mu_2) = 0$ against the alternative H_a: $(\mu_1 - \mu_2) \neq 0$. A total of 233 observations from population 1 and 312 from population 2 are analyzed with a statistical software package, with the results shown.

```
X̄1 = 473            X̄2 = 485
S1 = 84             S2 = 93
Z = -1.576          P-VALUE = .1150
```

a. Interpret the results of the computer analysis.
b. If the alternative hypothesis had been H_a: $(\mu_1 - \mu_2) < 0$, how would the p-value change? Interpret the p-value for this one-tailed test.

Applying the Concepts

9.8 Wansley, Roenfeldt, and Cooley (1983) compared the profiles of a sample of 44 firms that merged during 1975–1976 with those of a sample of 44 firms that did not merge. The table displays information obtained on the firms' price–earnings ratios.

	Merged Firms	*Nonmerged Firms*
Sample mean	7.295	14.666
Sample standard deviation	7.374	16.089

a. The analysis indicated that merged firms generally have smaller price–earnings ratios. Do you agree? Test using $\alpha = .05$.
b. Report and interpret the p-value of the test you conducted in part **a**.
c. What assumption(s) was it necessary to make in order to perform the test in part **a**?
d. Do you think that the distributions of the price–earnings ratios for the populations from which these samples were drawn are normally distributed? Why or why not? [*Hint*: Note the relative values of the sample means and standard deviations.]

9.9 A paper company conducted an experiment to compare the mean time to unload shipments of logs for two different unloading procedures. Two samples of 50 trucks each were unloaded using a new method and the company's current method. The objective of the experiment is to determine whether the new method will reduce the mean unloading time. The sample means and standard deviations are shown in the table.

New Method	*Current Method*
$n_1 = 50$	$n_2 = 50$
$\bar{x}_1 = 25.4$ minutes	$\bar{x}_2 = 27.3$ minutes
$s_1 = 3.1$ minutes	$s_2 = 3.7$ minutes

a. Do the data provide sufficient evidence to indicate that the mean unloading time for the new method is less than the mean unloading time for the method currently in use? Test using $\alpha = .05$.
b. Give the observed significance level for the test.

9.10 Refer to Exercise 9.9. Find a 90% confidence interval for the difference in mean unloading times between the two methods.

9.11 For over a decade, Tennant Co., a Minnesota manufacturer of industrial floor-cleaning machines, has been using quality circles to help improve its product. The term *quality circles* describes a process in which groups comprising both white-collar and blue-collar employees attempt to solve quality, productivity, or work environment problems.

When Tennant began its quality circles program in 1979, a sample of finished machines was found to have an average of 4.2 defects per machine. Because of such defects, the company had to pay its employees $16.6 million for the 39,600 hours of labor required to rework the defective machines. In 1982, a sample of machines revealed a substantial improvement in quality. The mean number of defectives was reduced to 1.3 defects per machine, and rework was down to 3,500 hours of labor (Marcotty, 1983a, 1983b). Assume that each sample of machines was randomly selected and was of size 100. Further, assume that the sample standard deviations for 1979 and 1983 were 2.0 and 1.1, respectively.

a. Although the decline in the average number of defects per machine between 1979 and 1982 appears to be significant from a managerial perspective, is it statistically significant? To answer the question, conduct the appropriate hypothesis test and report and interpret the observed significance level of the test.

b. In the context of the problem, describe the Type I and Type II errors associated with your hypothesis test of part **a**.

9.12 As part of a study of participative management, George H. Hines (1974) sampled workers from two types of New Zealand sociocultural backgrounds: those who believe in the existence of a class system and those who believe that they live and work in a classless society. Each worker in the sampling was selected from a work environment with a high degree of participatory management. Do workers who consider themselves to be social equals with their management superiors possess other levels of job satisfaction than those workers who see themselves as socially different from management? Each worker in the independent random samples answered this question by rating his or her job satisfaction on a scale of 1 (poor) to 7 (excellent). Based on the results shown in the table, what can you say about differences in job satisfaction for the two different sociocultural types of workers? Use $\alpha = .10$.

	Believe Class System Exists	*Believe No Class System Exists*
Sample size	175	277
Mean	5.42	5.19
Standard deviation	1.24	1.17

9.13 An experiment has been conducted to compare the productivity of two machines. Machine 1 produced an average of 51.4 items per hour and a standard deviation of $s_1 = 2.1$ for 35 randomly selected hours during the past 2 weeks. Machine 2 produced an average of 49.5 items per hour and a standard deviation of $s_2 = 1.8$ for 45 randomly selected hours during the past 2 weeks.

a. Describe the populations being compared.

b. Do the samples provide sufficient evidence at $\alpha = .10$ to conclude that machine 1 produces more items per hour, on the average, than machine 2?

c. Report the p-value for the test you conducted in part **b**.

9.14 Refer to Exercise 9.13. Construct a 95% confidence interval for $(\mu_1 - \mu_2)$. Would a 99% confidence interval be narrower or wider than the one you constructed? Why?

9.15 The mean or median value of a community's single-family homes is sometimes used as an indicator of local wealth or "buying power." Such information may be useful to retail merchants looking for communities in which to locate their businesses or deciding where to concentrate their advertising. A retailer believes that the mean price of a single-family home in Edina, Minnesota, is substantially more than in St. Louis Park, Minnesota. If this is confirmed by a statistical hypothesis test the retailer will increase her advertising budget for the Edina area. The table contains the asking prices of a random sample of 30 homes up for sale in Edina in late February 1993 and a random sample of 30 homes also on the market then in St. Louis Park.

Edina			*St. Louis Park*		
$215,000	179,900	134,600	$134,900	89,900	180,000
205,000	225,000	179,750	84,900	153,500	94,900
214,900	145,000	165,900	90,000	34,900	87,900
178,500	329,500	164,900	110,000	109,000	89,900
349,900	142,900	119,000	78,700	103,900	69,500
145,900	140,900	159,900	74,900	104,900	66,900
82,000	189,500	179,000	69,900	87,900	54,900
402,500	203,900	160,000	149,900	120,000	69,900
217,500	672,900	116,700	116,900	86,500	73,900
138,000	205,900	129,900	84,900	89,900	65,800

Source: Minneapolis Area Association of Realtors, 1993.

a. Describe the populations from which the sample data were selected.
b. What are the parameters of interest to the retailer?
c. For Edina, $\bar{x}_1 = \$203,142$ and $s_1 = \$112,285$. For St. Louis Park, $\bar{x}_2 = \$94,300$ and $s_2 = \$30,687$. Do the data indicate that the mean price of a home in Edina is more than $60,000 greater than the mean price in St. Louis Park? Test using $\alpha = .05$.
d. Use a 95% confidence interval to estimate the true difference in the mean prices for Edina and St. Louis Park.

9.2 Small-Sample Inferences About the Difference Between Two Population Means: Independent Sampling

Suppose a television network wanted to determine whether major sports events or first-run movies attract more viewers in the prime-time hours. It selected 28 prime-time evenings; of these, 13 had programs devoted to major sports events and the remaining 15 had first-run movies. The number of viewers (estimated by a television viewer rating firm) was recorded for each program. If μ_1 is the mean number of sports viewers per evening of sports programming and μ_2 is the mean number of movie viewers per evening of movie programming, we want to detect a difference between μ_1 and μ_2—if such a difference exists. Therefore, we want to test the null hypothesis

$$H_0\text{: } (\mu_1 - \mu_2) = 0$$

against the alternative hypothesis

$$H_a\text{: } (\mu_1 - \mu_2) \neq 0 \qquad \text{(i.e., either } \mu_1 > \mu_2 \text{ or } \mu_2 > \mu_1\text{)}$$

Since the sample sizes are small, s_1^2 and s_2^2 will be unreliable estimates of σ_1^2 and σ_2^2 and the z test statistic will be inappropriate for the test. But, as in the case of a single mean (Section 8.4), we can construct a Student's t statistic. This statistic (formula to be given subsequently) has the familiar t distribution described in Chapter 7. **To use the t statistic, both sampled populations must be approximately normally distributed with equal population variances, and the random samples must be selected independently of each other.** The normality and equal variances assumptions would imply relative frequency distributions for the populations that would appear as shown in Figure 9.4. We will assume that the distributions of the two populations of numbers of television viewers will approximately satisfy these assumptions.

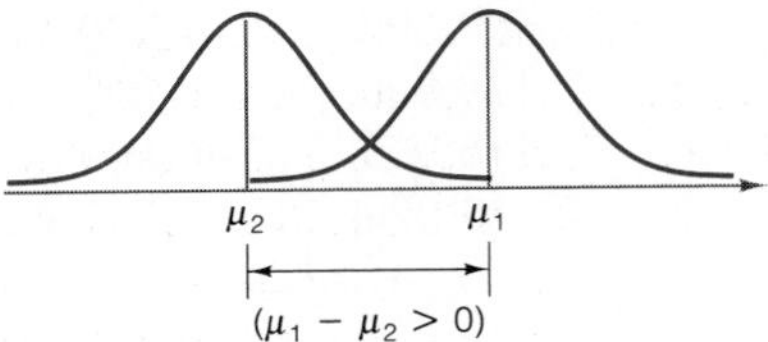

FIGURE 9.4 ▶ Assumptions for the two-sample t: (1) normal populations (2) equal variances

Since we assume the two populations have equal variances ($\sigma_1^2 = \sigma_2^2 = \sigma^2$), it is reasonable to combine the sum of squares of deviations from the two samples to construct a pooled sample estimator of σ^2 for use in the t statistic. Thus, if s_1^2 and s_2^2 are the two sample variances (each estimating the variance σ^2 common to both populations), the pooled estimator of σ^2, denoted as s_p^2, is

$$s_p^2 = \frac{\overbrace{\sum_{i=1}^{n_1} (x_i - \bar{x}_1)^2}^{\text{From sample 1}} + \overbrace{\sum_{i=1}^{n_2} (x_i - \bar{x}_2)^2}^{\text{From sample 2}}}{n_1 + n_2 - 2}$$

or

$$s_p^2 = \frac{(n_1 - 1)s_1^2 + (n_2 - 1)s_2^2}{(n_1 - 1) + (n_2 - 1)} = \frac{(n_1 - 1)s_1^2 + (n_2 - 1)s_2^2}{n_1 + n_2 - 2}$$

where x_1 represents a measurement from sample 1 and x_2 represents a measurement from sample 2. Recall that the term **degrees of freedom** was defined in Section 7.3 as 1 less than the sample size. Thus, in this case we have $(n_1 - 1)$ degrees of freedom for sample 1 and $(n_2 - 1)$ degrees of freedom for sample 2. Since we are pooling the information on σ^2 obtained from both samples, the degrees of freedom associated with the pooled variance s_p^2 is equal to the sum of the degrees of freedom for the two samples, namely, the denominator of s_p^2—i.e., $(n_1 - 1) + (n_2 - 1) = n_1 + n_2 - 2$.

Note that the second formula given for s_p^2 shows that the pooled variance is simply a *weighted average* of the two sample variances, s_1^2 and s_2^2. The weight given each variance is proportional to its degrees of freedom. If the two variances have the same number of degrees of freedom (i.e., if the sample sizes are equal), then the pooled variance is a simple average of the two sample variances. The result is an average or pooled variance that is a better estimate of σ_2 than either s_1^2 or s_2^2 alone.

To obtain the small-sample test statistic for testing H_0: $(\mu_1 - \mu_2) = D_0$, substitute the pooled estimate of σ^2 into the formula for the two-sample z statistic (Section 9.1) to obtain

$$t = \frac{(\bar{x}_1 - \bar{x}_2) - D_0}{\sqrt{s_p^2\left(\frac{1}{n_1} + \frac{1}{n_2}\right)}}$$

It can be shown that this statistic, like the t statistic of Chapter 7, has a t distribution, but with $(n_1 + n_2 - 2)$ degrees of freedom.

We will use the television viewer example to outline the final steps for this t test. The hypothesized difference in mean number of viewers is $D_0 = 0$. The rejection region will be two-tailed and will be based on a t distribution with $(n_1 + n_2 - 2)$ or $(13 + 15 - 2) = 26$ df. Letting $\alpha = .05$, the rejection region for the test would be

$$t < -t_{\alpha/2} \quad \text{or} \quad t > t_{\alpha/2}$$

The value for $t_{.025}$ with df = 26 given in Table VI of Appendix B is 2.056. Thus, the rejection region for the television example is

$$t < -2.056 \quad \text{or} \quad t > 2.056$$

This rejection region is shown in Figure 9.5.

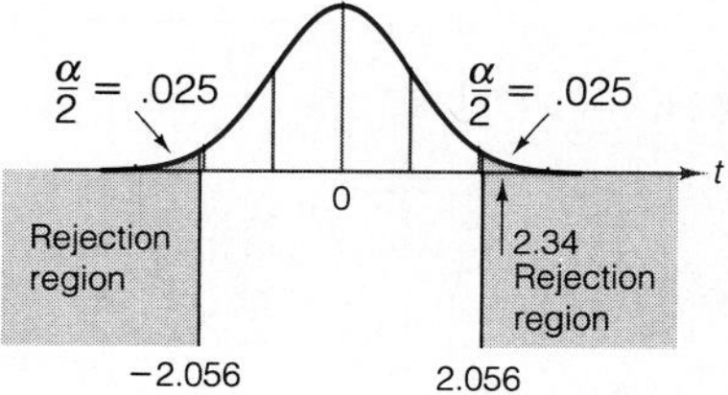

FIGURE 9.5 ▶ Rejection region for two-tailed t test: $\alpha = .05$, df = 26

Now, suppose the television network's samples produce the results shown in the table:

Sports	*Movie*
$n_1 = 13$	$n_2 = 15$
$\bar{x}_1 = 6.8$ million	$\bar{x}_2 = 5.3$ million
$s_1 = 1.8$ million	$s_2 = 1.6$ million

We must assume that these independent samples were selected from normal distributions (i.e., the number of viewers per evening is normally distributed for each type of program) with equal variances. [*Note*: Although the sample estimates of variance are not equal, the assumption that the population variances are equal may still be valid. We will present a method for checking this assumption statistically in Section 9.3.]

We now calculate

$$s_p^2 = \frac{(n_1 - 1)s_1^2 + (n_2 - 1)s_2^2}{n_1 + n_2 - 2}$$

$$= \frac{(13 - 1)(1.8)^2 + (15 - 1)(1.6)^2}{13 + 15 - 2} = \frac{74.72}{26} = 2.87$$

Then,

$$t = \frac{(\bar{x}_1 - \bar{x}_2) - D_0}{\sqrt{s_p^2\left(\frac{1}{n_1} + \frac{1}{n_2}\right)}} = \frac{(6.8 - 5.3) - 0}{\sqrt{2.87\left(\frac{1}{13} + \frac{1}{15}\right)}} = \frac{1.5}{.64} = 2.34$$

Since the observed value of t, $t = 2.34$, falls in the rejection region (see Figure 9.5), the samples provide sufficient evidence to indicate that the mean numbers of viewers differ for major sports events and first-run movies shown in prime time. Or, we say that the test results are statistically significant at the $\alpha = .05$ level of significance. Because the rejection was in the positive or upper tail of the t distribution, the indication is that the mean number of viewers for sports events exceeds that for movies.

The t statistic can also be used to construct a confidence interval for the difference between population means. Both the confidence interval and the test of hypothesis procedures are summarized in the boxes.

Small-Sample Confidence Interval for $(\mu_1 - \mu_2)$

$$(\bar{x}_1 - \bar{x}_2) \pm t_{\alpha/2}\sqrt{s_p^2\left(\frac{1}{n_1} + \frac{1}{n_2}\right)}$$

where $s_p^2 = \dfrac{(n_1 - 1)s_1^2 + (n_2 - 1)s_2^2}{n_1 + n_2 - 2}$

and $t_{\alpha/2}$ is based on $(n_1 + n_2 - 2)$ df.

Assumptions:
1. Both sampled populations have relative frequency distributions that are approximately normal.
2. The population variances are equal.
3. The samples are randomly and independently selected from the populations.

Small-Sample Test of Hypothesis for $(\mu_1 - \mu_2)$ (Independent Samples)

One-Tailed Test	*Two-Tailed Test*
H_0: $(\mu_1 - \mu_2) = D_0$ H_a: $(\mu_1 - \mu_2) < D_0$ [or H_a: $(\mu_1 - \mu_2) > D_0$]	H_0: $(\mu_1 - \mu_2) = D_0$ H_a: $(\mu_1 - \mu_2) \neq D_0$
Test statistic: $t = \dfrac{(\bar{x}_1 - \bar{x}_2) - D_0}{\sqrt{s_p^2\left(\dfrac{1}{n_1} + \dfrac{1}{n_2}\right)}}$	*Test statistic:* $t = \dfrac{(\bar{x}_1 - \bar{x}_2) - D_0}{\sqrt{s_p^2\left(\dfrac{1}{n_1} + \dfrac{1}{n_2}\right)}}$
Rejection region: $t < -t_\alpha$ [or $t > t_\alpha$ when H_a: $(\mu_1 - \mu_2) > D_0$] where $t_{\alpha/2}$ is based on $(n_1 + n_2 - 2)$ df.	*Rejection region:* $t < -t_{\alpha/2}$ or $t > t_{\alpha/2}$ where t_α is based on $(n_1 + n_2 - 2)$ df.

Assumptions: Same as for the small-sample confidence interval for $(\mu_1 - \mu_2)$.

EXAMPLE 9.4

Suppose you want to estimate the difference in annual operating costs for automobiles with automatic transmissions and those with standard transmissions. Using the records of a local automobile club, you randomly select 8 owners of cars with automatic transmissions and 12 owners of cars with standard transmissions who have purchased their cars within the last 2 years. Each of the 20 owners keeps accurate records of the amount spent on operating his or her car (including gasoline, oil, repairs, etc.) for a 12-month period. All costs are recorded on a per-thousand-mile basis to adjust for differences in mileage driven during the 12-month period. The results are summarized in the table. Estimate the true difference $(\mu_1 - \mu_2)$ in the mean operating costs per thousand miles between cars with automatic and cars with standard transmissions. Use a 90% confidence interval.

Automatic	*Standard*
$n_1 = 8$	$n_2 = 12$
$\bar{x}_1 = 156.96$	$\bar{x}_2 = 152.73$
$s_1 = \$4.85$	$s_2 = \$6.35$

Solution

The objective of this experiment is to obtain a 90% confidence interval for $(\mu_1 - \mu_2)$. To use the small-sample confidence interval for $(\mu_1 - \mu_2)$, the following assumptions must be satisfied:

1. The operating cost per thousand miles is normally distributed for cars with both automatic and standard transmissions. Since these costs are averages (because we

observe them on a per-thousand-mile basis), the Central Limit Theorem lends credence to this assumption.

2. The variance in cost is the same for the two types of cars. Under these circumstances, we might expect the variation in costs from automobile to automobile to be about the same for both types of transmissions.
3. The samples are randomly and independently selected from the two populations. We have randomly chosen 20 different owners for the two samples in such a way that the cost measurement for one owner is not dependent on the cost measurement for any other owner. Therefore, this assumption would be valid.

The first step in performing the test is to calculate the pooled estimate of variance:

$$s_p^2 = \frac{(n_1 - 1)s_1^2 + (n_2 - 1)s_2^2}{n_1 + n_2 - 2}$$

$$= \frac{(8 - 1)(4.85)^2 + (12 - 1)(6.35)^2}{8 + 12 - 2} = 33.7892$$

where s_p^2 possesses $(n_1 + n_2 - 2) = (8 + 12 - 2) = 18$ df. Then, the 90% confidence interval for $(\mu_1 - \mu_2)$, the difference in mean operating costs for the two types of automobiles, is

$$(\bar{x}_1 - \bar{x}_2) \pm t_{\alpha/2}\sqrt{s_p^2\left(\frac{1}{n_1} + \frac{1}{n_2}\right)} = (156.96 - 152.73) \pm t_{.05}\sqrt{33.7892\left(\frac{1}{8} + \frac{1}{12}\right)}$$

$$= 4.23 \pm 1.734(2.653) = 4.23 \pm 4.60$$

$$= (-.37, 8.83)$$

This means that, with 90% confidence, we estimate the difference in mean operating costs per thousand miles between cars with automatic transmissions and those with standard transmissions to fall in the interval from −\$.37 to \$8.83. In other words, we estimate the mean operating costs for cars with automatic transmissions to be anywhere from \$.37 less than to \$8.83 more than the operating costs per thousand miles for cars with standard transmissions. Although the sample means seem to suggest that cars with automatic transmissions cost more to operate, there is insufficient evidence to indicate that $(\mu_1 - \mu_2)$ differs from 0 because the interval includes 0 as a possible value for $(\mu_1 - \mu_2)$. To show a difference in mean operating costs (if it exists), it would be necessary to increase the sample sizes and thereby narrow the width of the confidence interval for $(\mu_1 - \mu_2)$.

The two-sample t statistic is a powerful tool for comparing population means when the assumptions are satisfied. It has also been shown to retain its usefulness when the sampled populations are only approximately normally distributed. Furthermore, when the sample sizes are equal, the assumption of equal population variances can be relaxed. That is, when $n_1 = n_2$, σ_1^2 and σ_2^2 can be quite different and the test statistic will still have (approximately) a Student's t distribution. When the assumptions are not satisfied, other statistical tests are available. For example, the nonparametric statistical tests discussed in Chapter 17 may be appropriate.

What Can Be Done If the Assumptions Are Not Satisfied?

Answer: If you are concerned that your assumptions are not satisfied, use the Wilcoxon rank sum test for independent samples to test for a shift in population distributions. (See Chapter 17.)

Exercises 9.16–9.30

Learning the Mechanics

9.16 To use the t statistic to test for differences in the means of two populations, what assumptions must be made about the two populations? About the two samples?

9.17 Two populations are described in each of the following cases. In which cases would it be appropriate to apply the small-sample t test to investigate the difference between the population means?

a. Population 1: Normal distribution with variance σ_1^2
 Population 2: Skewed to the right with variance $\sigma_2^2 = \sigma_1^2$

b. Population 1: Normal distribution with variance σ_1^2
 Population 2: Normal distribution with variance $\sigma_2^2 \neq \sigma_1^2$

c. Population 1: Skewed to the left with variance σ_1^2
 Population 2: Skewed to the left with variance $\sigma_2^2 = \sigma_1^2$

d. Population 1: Normal distribution with variance σ_1^2
 Population 2: Normal distribution with variance $\sigma_2^2 = \sigma_1^2$

e. Population 1: Uniform distribution with variance σ_1^2
 Population 2: Uniform distribution with variance $\sigma_2^2 = \sigma_1^2$

9.18 In the t tests of this section, σ_1^2 and σ_2^2 are assumed to be equal. Thus, we say $\sigma_1^2 = \sigma_2^2 = \sigma^2$. Why is a pooled estimator of σ^2 used instead of either s_1^2 or s_2^2?

9.19 Assume that $\sigma_1^2 = \sigma_2^2 = \sigma^2$. Calculate the pooled estimator of σ^2 for each of the following cases:

a. $s_1^2 = 120$, $s_2^2 = 100$, $n_1 = n_2 = 25$ b. $s_1^2 = 12$, $s_2^2 = 20$, $n_1 = 20$, $n_2 = 10$

c. $s_1^2 = .15$, $s_2^2 = .20$, $n_1 = 6$, $n_2 = 10$ d. $s_1^2 = 3{,}000$, $s_2^2 = 2{,}500$, $n_1 = 16$, $n_2 = 17$

e. Note that the pooled estimate is a weighted average of the sample variances. Which of the variances does the pooled estimate fall nearer in each of the above cases?

9.20 Independent random samples from normal populations produced the results shown in the table.

Sample 1	2.1	3.4	1.4	3.0	2.9	3.3
Sample 2	3.4	3.0	4.2	3.8	3.5	

a. Calculate the pooled estimate of σ^2.
b. Do the data provide sufficient evidence to indicate that $\mu_2 > \mu_1$? Test using $\alpha = .10$.
c. Find the approximate observed significance level for the test, and interpret its value.

9.21 Refer to Exercise 9.20.
a. Find a 90% confidence interval for $(\mu_1 - \mu_2)$. Interpret the confidence interval.
b. Which of the two inferential procedures, the test of hypothesis in Exercise 9.20 or the confidence interval in this exercise, provides more information about $(\mu_1 - \mu_2)$? Justify your answer.

9.22 Independent random samples from two normal populations produced the results shown in the table.

Sample 1	*Sample 2*
$n_1 = 12$	$n_2 = 16$
$\bar{x}_1 = 35$	$\bar{x}_2 = 43$
$s_1 = 4.2$	$s_2 = 3.7$

a. Calculate the pooled estimate of σ^2.
b. Do these data provide sufficient evidence to indicate that $\mu_1 \neq \mu_2$? Test using $\alpha = .05$.
c. Find the approximate p-value for the test, and interpret its value.

9.23 Refer to Exercise 9.22. Find a 95% confidence interval for $(\mu_1 - \mu_2)$. Interpret the confidence interval.

9.24 Independent random samples from approximately normal populations produced the results shown in the table.

Sample 1								*Sample 2*							
52	33	42	44	41	50	44	51	52	43	47	56	62	53	61	50
45	38	37	40	44	50	43		56	52	53	60	50	48	60	55

a. Do the data provide sufficient evidence to conclude that $(\mu_2 - \mu_1) > 10$? Test using $\alpha = .01$.
b. Construct a 98% confidence interval for $(\mu_2 - \mu_1)$. Interpret your result.

Applying the Concepts

9.25 While cable television companies in Minnesota are prohibited from holding exclusive rights to an area, the laws do not demand that a company face competition (Gross, 1993). Many subscribers feel that these *de facto* monopolies exploit consumers by charging excessive monthly cable fees. Suppose a congressional subcommittee considering regulation of the cable industry investigates whether cable rates are higher in areas with no competition than in areas with competition. They randomly sample basic rates for six cable companies that have no competition and for six companies that face competition (but not from each other). The observed rates are shown in the table.

No competition	\$18.44	\$26.88	\$22.87	\$25.78	\$23.34	\$27.52
Competition	\$18.95	\$23.74	\$17.25	\$20.14	\$18.98	\$20.14

a. What are the appropriate null and alternative hypotheses to test the research hypothesis of the subcommittee?

b. Conduct the test of part **a** using $\alpha = .05$. Report and interpret the approximate significance level of the test.

c. What assumptions are necessary to assure the validity of the test? Why does it matter that none of the companies in the sample compete against each other?

9.26 An industrial plant wants to determine which of two types of fuel—gas or electric—will produce more useful energy at a lower cost. One measure of economical energy production, the *plant investment per delivered quad*, is calculated by taking the amount of money (in dollars) invested in the particular utility by the plant and dividing by the delivered amount of energy (in quadrillion British thermal units). The smaller this ratio, the less an industrial plant pays for its delivered energy.

Random samples of 11 plants using electrical utilities and 16 plants using gas utilities were taken, and the plant investment per quad was calculated for each. The data produced the results shown in the table.

	Electric	*Gas*
Sample size	11	16
Mean investment/quad (billions)	\$22.5	\$17.5
Variance	17.5	15

a. Do these data provide sufficient evidence at $\alpha = .05$ to indicate a difference in the average investment per quad between the plants using gas and those using electrical utilities?

b. Find a 90% confidence interval for $(\mu_1 - \mu_2)$. Give a practical interpretation of this interval.

9.27 Marketing strategists would like to be able to predict consumer response to new products and their accompanying promotional schemes. To this end, studies that examine the differences between buyers and nonbuyers of a product are of interest. One such study conducted by Shuchman and Riesz (1975) was aimed at characterizing the purchasers and nonpurchasers of Crest toothpaste. Purchasers were defined as households that converted to Crest following its endorsement by the Council on Dental Therapeutics of the American Dental Association on August 1, 1960, and remained "loyal" to Crest until at least April 1963. Nonpurchasers were defined as households that did not convert to Crest during the same time period. Using demographic data collected from a sample of 499 purchasers and 499 nonpurchasers, Shuchman and Riesz demonstrated that both the mean household size (number of persons) and mean household income were significantly larger for purchasers than for nonpurchasers. A similar study utilized random samples of size 20 and yielded the data shown in the table on the age of the householder primarily responsible for buying toothpaste.

Purchasers										*Nonpurchasers*									
34	35	23	44	52	46	28	48	28	34	28	22	44	33	55	63	45	31	60	54
33	52	41	32	34	49	50	45	29	59	53	58	52	52	66	35	25	48	59	61

a. Do the data present sufficient evidence to conclude there is a difference in the mean age of purchasers and nonpurchasers? Use $\alpha = .10$.

b. What assumptions are necessary in order to answer part **a**?

c. Give the observed significance level for the test, and interpret its value.

d. Find a 90% confidence interval for the difference between the mean ages of purchasers and nonpurchasers.

9.28 One way corporations raise money for expansion is to issue *bonds*, loan agreements to repay the purchaser a specified amount with a fixed rate of interest paid periodically over the life of the bond. The sale of bonds is usually handled by an underwriting firm. In a study described in the *Harvard Business Review* (July–Aug. 1979), D. Logue and R. Rogalski asked, "Does It Pay to Shop for Your Bond Underwriter?" The reason for the question is that the price of a bond may rise or fall after its issuance. Therefore, whether a corporation receives the market price for a bond depends on the skill of its underwriter. The mean change in the prices of 27 bonds handled over a 12-month period by one underwriter and in the prices of 23 bonds handled by another are shown.

	Underwriter 1	*Underwriter 2*
Sample size	27	23
Sample mean	−.0491	−.0307
Sample variance	.009800	.002465

a. Do the data provide sufficient evidence to indicate a difference in the mean change in bond prices handled by the two underwriters? Test using $\alpha = .05$.
b. Find a 95% confidence interval for the mean difference for the two underwriters, and interpret it.

9.29 With the emergence of Japan as an industrial superpower, U.S. businesses are looking closely at Japanese management styles and philosophies. Some of the credit for the high quality of Japanese products is attributed to the Japanese system of permanent employment for their workers. In the United States, high job turnover rates are common in many industries and are associated with high product defect rates. High turnover rates mean U.S. plants have more inexperienced workers who are unfamiliar with the company's product lines than Japan has. In a study of the air conditioner industry in Japan and the United States, David Garvin (1983) reported that the difference in the average annual turnover rate of workers between U.S. plants and Japanese plants was 3.1%. In another study, five Japanese and five U.S. plants that manufacture air conditioners were randomly sampled; their turnover rates are listed in the table.

U.S. Plants	*Japanese Plants*
7.11%	3.52%
6.06	2.02
8.00	4.91
6.87	3.22
4.77	1.92

a. Do the data provide sufficient evidence to indicate that the mean annual percentage turnover for U.S. plants exceeds the corresponding mean percentage for Japanese plants? Test using $\alpha = .05$.
b. Report and interpret the observed significance level of the test you conducted in part **a**.
c. List any assumptions you made in conducting the hypothesis test of part **a**. Comment on their validity for this application.

9.30 *Sales quotas* are volume objectives assigned to specific sales units, such as regions, districts, or salespersons' territories. Sometimes to achieve manufacturing efficiency or long-term goals, sales managers set quotas for specific products at challenging levels. The underlying idea is that setting challenging quotas and attaching significant rewards will direct salespersons' efforts along desired paths (Winer, 1973). The Universal Products Company (real company, fictitious name) manufactures and markets electronic and electromechanical industrial equipment. It has a sales force of over 1,000, organized in ten districts and 135 branch offices.

Salespersons have sales quotas on two specific products, Dataprinters and Micromagnetics, as well as an overall sales volume quota. Many salespersons have complained that having to make the quota on Dataprinters takes so much time that it keeps them from generating a higher overall sales volume. To determine how reducing the Dataprinter quota would affect total sales volume, Winer compared the sales volumes of a sample of branch offices whose salespersons all worked under the standard quota with the sales volumes of a sample of branch offices whose salespersons all were given a lower Dataprinter quota. Data were collected for a 7-month period and are reported in the table in terms of total sales per worker-month (in thousands of dollars).

Branch	*Lower Quota*	*Standard Quota*
1		17.7
2	15.6	
3		15.1
4	14.0	
5		12.3
6		12.0
7	11.2	
8	11.0	
9		10.5
10	10.3	
11		10.0
12	9.4	

a. What are the appropriate null and alternative hypotheses for testing whether salespersons on the two types of quotas differ in their mean sales per worker-month? Define any symbols you use.

b. The data in the table are submitted to a statistical software program, with results as shown. Interpret this output. What do you conclude about the test set up in part a?

```
MEAN DIFFERENCE = -1.02
T-VALUE = -0.66     DF = 10     TWO-TAILED P-VALUE = .525
```

c. The same software package yielded the output shown. Interpret the interval. Does its width imply that little or much information about the difference in mean sales is contained in these data? How could the amount of information be increased?

```
90% CONFIDENCE INTERVAL: (-3.814, 1.781)
```

9.3 Comparing Two Population Variances: Independent Random Samples

Suppose you want to use the two-sample t statistic to compare the mean production levels of two paper mills. However, you are concerned that the assumption of equal variances of the production levels for the two plants may be unrealistic. It would be helpful to have a statistical procedure to check the validity of this assumption.

The common statistical procedure for comparing population variances σ_1^2 and σ_2^2 makes an inference about the ratio, σ_1^2/σ_2^2, based on the ratio of the sample variances, s_1^2/s_2^2. Thus, we will attempt to support the alternative hypothesis that the ratio σ_1^2/σ_2^2 differs from 1 (i.e., the variances are unequal) by testing the null hypothesis that the ratio equals 1 (i.e., the variances are equal).

$$H_0\text{: } \frac{\sigma_1^2}{\sigma_2^2} = 1 \qquad (\sigma_1^2 = \sigma_2^2)$$

$$H_a\text{: } \frac{\sigma_1^2}{\sigma_2^2} \neq 1 \qquad (\sigma_1^2 \neq \sigma_2^2)$$

We will use the test statistic $F = s_1^2/s_2^2$.

To establish a rejection region for the test statistic, we need to know how s_1^2/s_2^2 is distributed in repeated samples. That is, we need to know the sampling distribution of s_1^2/s_2^2. As you will see, the sampling distribution of s_1^2/s_2^2 is based on two of the assumptions already required for the t test, namely:

1. The two sampled populations are normally distributed.
2. The samples are randomly and independently selected from their respective populations.

When these assumptions are satisfied and when the null hypothesis is true (i.e., $\sigma_1^2 = \sigma_2^2$), the sampling distribution of $F = s_1^2/s_2^2$ is the ***F* distribution** with $\nu_1 = (n_1 - 1)$ numerator degrees of freedom and $\nu_2 = (n_2 - 1)$ denominator degrees of freedom. The shape of the F distribution will depend on the degrees of freedom associated with s_1^2 and s_2^2—i.e., $(n_1 - 1)$ and $(n_2 - 1)$. An F distribution with $\nu_1 = 7$ and $\nu_2 = 9$ df is shown in Figure 9.6. As you can see, the distribution is skewed to the right.

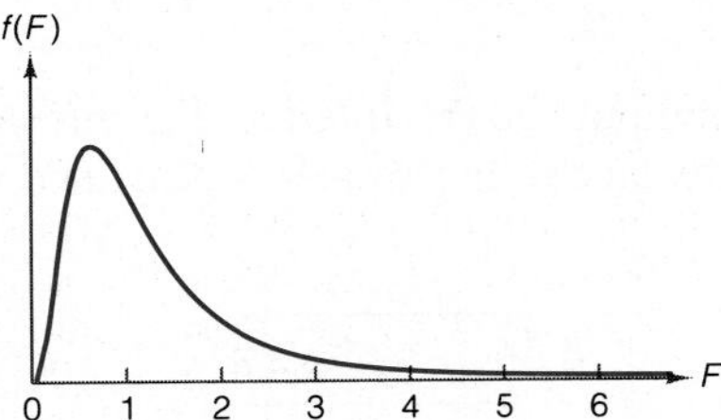

FIGURE 9.6 ▶
An *F* distribution with 7 and 9 df

We need to be able to find F values corresponding to the tail areas of this distribution in order to establish the rejection region for our test of hypothesis because, when the population variances are unequal, we expect the ratio F of the sample variances to be either very large or very small. The upper-tail F values can be found in Tables VII, VIII, IX, and X of Appendix B. Table VIII is partially reproduced in Table 9.1. It gives F values that correspond to $\alpha = .05$ upper-tail areas for different degrees of freedom. The columns of Tables VII, VIII, IX, and X correspond to various degrees of freedom for the numerator sample variance, s_1^2, whereas the rows correspond

TABLE 9.1 Reproduction of Part of Table VIII of Appendix B: Percentage Points of the F Distribution, $\alpha = .05$

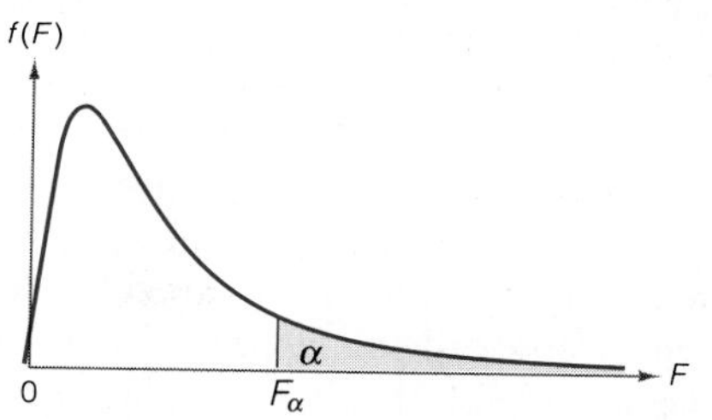

ν_2 \ ν_1 (Numerator Degrees of Freedom)	1	2	3	4	5	6	7	8	9
Denominator Degrees of Freedom 1	161.4	199.5	215.7	224.6	230.2	234.0	236.8	238.9	240.5
2	18.51	19.00	19.16	19.25	19.30	19.33	19.35	19.37	19.38
3	10.13	9.55	9.28	9.12	9.01	8.94	8.89	8.85	8.81
4	7.71	6.94	6.59	6.39	6.26	6.16	6.09	6.04	6.00
5	6.61	5.79	5.41	5.19	5.05	4.95	4.88	4.82	4.77
6	5.99	5.14	4.76	4.53	4.39	4.28	4.21	4.15	4.10
7	5.59	4.74	4.35	4.12	3.97	3.87	3.79	3.73	3.68
8	5.32	4.46	4.07	3.84	3.69	3.58	3.50	3.44	3.39
9	5.12	4.26	3.86	3.63	3.48	3.37	3.29	3.23	3.18
10	4.96	4.10	3.71	3.48	3.33	3.22	3.14	3.07	3.02
11	4.84	3.98	3.59	3.36	3.20	3.09	3.01	2.95	2.90
12	4.75	3.89	3.49	3.25	3.11	3.00	2.91	2.85	2.80
13	4.67	3.81	3.41	3.18	3.03	2.92	2.83	2.77	2.71
14	4.60	3.74	3.34	3.11	2.96	2.85	2.76	2.70	2.65

to the degrees of freedom for the denominator sample variance, s_2^2. Thus, if the numerator degrees of freedom are 7 and the denominator degrees of freedom are 9, we look in the seventh column and ninth row of Table VIII to find $F_{.05} = 3.29$. As shown in Figure 9.7, $\alpha = .05$ is the tail area to the right of 3.29 in the F distribution with 7 and 9 df. That is, if $\sigma_1^2 = \sigma_2^2$, the probability that the F statistic will exceed 3.29 is $\alpha = .05$.

FIGURE 9.7 ▶ An F distribution for 7 and 9 df: $\alpha = .05$

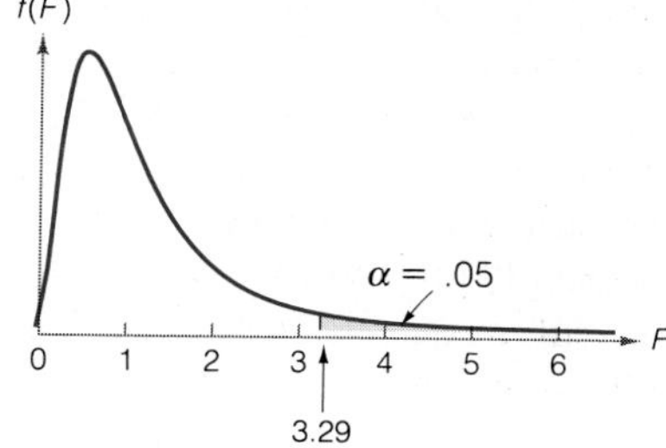

Suppose we want to compare the variability in production for two paper mills and we have obtained the results shown in the table.

Sample 1	*Sample 2*
$n_1 = 13$ days	$n_2 = 18$ days
$\bar{x}_1 = 26.3$ production units	$\bar{x}_2 = 19.7$ production units
$s_1 = 8.2$ production units	$s_2 = 4.7$ production units

To form the rejection region for a two-tailed F test we want to make certain that the upper tail is used because only the upper-tail values of F are shown in Tables VII, VIII, IX, and X. To accomplish this, *we will always place the larger sample variance in the numerator of the F test statistic.* This has the effect of doubling the tabulated value for α, since we double the probability that the F ratio will fall in the upper tail by always placing the larger sample variance in the numerator. That is, we establish a one-tailed rejection region by putting the larger variance in the numerator rather than establishing rejection regions in both tails.

Thus, for our production example, we have a numerator s_1^2 with $\nu_1 = n_1 - 1 = 12$ and a denominator of s_2^2 with $\nu_2 = n_2 - 1 = 17$. Therefore, the test statistic will be

$$F = \frac{\text{Larger sample variance}}{\text{Smaller sample variance}} = \frac{s_1^2}{s_2^2}$$

and we will reject $H_0\colon \sigma_1^2 = \sigma_2^2$ for $\alpha = .10$ when the calculated value of F exceeds the tabulated value:

$$F_{.05} = 2.38 \qquad \text{(see Figure 9.8)}$$

FIGURE 9.8 ▶
Rejection region for production example: *F* distribution

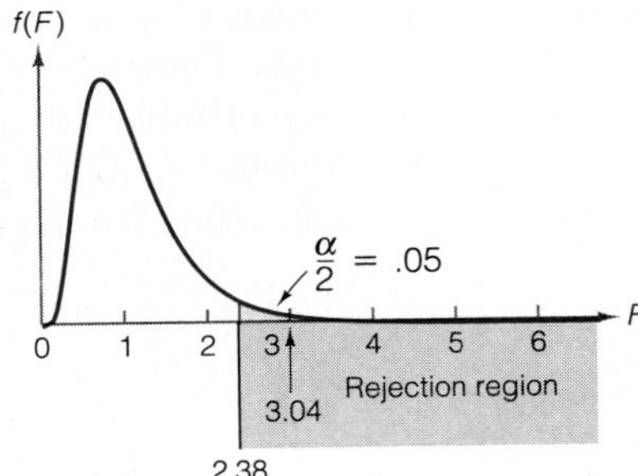

Now, what do the data tell us? We calculate

$$F = \frac{s_1^2}{s_2^2} = \frac{(8.2)^2}{(4.7)^2} = 3.04$$

and compare it to the rejection region shown in Figure 9.8. You can see that the F value 3.04 falls in the rejection region, therefore the data provide sufficient evidence to indicate that the population variances differ. Consequently, we would be reluctant to use the two-sample t statistic to compare the population means because the assumption of equal population variances is questionable.

What would you have concluded if the value of F calculated from the samples had not fallen in the rejection region? Would you have concluded that the null hypothesis of equal variances was true? No, because then you risk the possibility of a Type II error (accepting H_0 when H_a is true) without knowing the value of β, the probability of accepting H_0: $\sigma_1^2 = \sigma_2^2$ if in fact it is false. Since we will not consider the calculation of β for specific alternatives in this text, when the F statistic does not fall in the rejection region we simply conclude that insufficient sample evidence exists to refute the null hypothesis that $\sigma_1^2 = \sigma_2^2$.

The F test for equal population variances is summarized in the box.

F Test for Equal Population Variances

One-Tailed Test	*Two-Tailed Test*
H_0: $\sigma_1^2 = \sigma_2^2$ H_a: $\sigma_1^2 < \sigma_2^2$ (or H_a: $\sigma_1^2 > \sigma_2^2$)	H_0: $\sigma_1^2 = \sigma_2^2$ H_a: $\sigma_1^2 \neq \sigma_2^2$
Test statistic: $F = \frac{s_2^2}{s_1^2}$ when H_a: $\sigma_1^2 < \sigma_2^2$ $\left(\text{or } F = \frac{s_1^2}{s_2^2} \text{ when } H_a\text{: } \sigma_1^2 > \sigma_2^2\right)$	*Test statistic:* $F = \frac{\text{Larger sample variance}}{\text{Smaller sample variance}}$ $= \frac{s_1^2}{s_2^2}$ when $s_1^2 > s_2^2$ $\left(\text{or } F = \frac{s_2^2}{s_1^2} \text{ when } s_2^2 > s_1^2\right)$
Rejection region: $F > F_\alpha$ where F_α is based on $\nu_1 = n_2 - 1$ and $\nu_2 = n_1 - 1$ df	*Rejection region:* $F > F_{\alpha/2}$ when $s_1^2 > s_2^2$ where $F_{\alpha/2}$ is based on $\nu_1 = n_1 - 1$ and $\nu_2 = n_2 - 1$ df
(or $F > F_\alpha$ when H_a: $\sigma_1^2 > \sigma_2^2$, where F_α is based on $\nu_1 = n_1 - 1$ and $\nu_2 = n_2 - 1$ df)	(or $F > F_{\alpha/2}$ when $s_2^2 > s_1^2$, where $F_{\alpha/2}$ is based on $\nu_1 = n_2 - 1$ and $\nu_2 = n_1 - 1$ df)

Assumptions:
1. Both sampled populations are normally distributed.
2. The samples are random and independent.

EXAMPLE 9.5

Refer to Example 9.4, in which we used the two-sample t statistic to compare the mean operating costs of cars with automatic and standard transmissions. Use the F test to check the assumption that the population variances are equal. Use $\alpha = .10$.

Solution

The elements of the test are as follows:

$$H_0\text{: } \frac{\sigma_1^2}{\sigma_2^2} = 1 \qquad H_a\text{: } \frac{\sigma_1^2}{\sigma_2^2} \neq 1$$

$$\textit{Test statistic:} \quad F = \frac{\text{Larger sample variance}}{\text{Smaller sample variance}} = \frac{s_2^2}{s_1^2}$$

To find the rejection region, we proceed as follows: The number of degrees of freedom in the numerator is $\nu_1 = n_2 - 1 = 11$; the number of degrees of freedom in the denominator is $\nu_2 = n_1 - 1 = 7$. Thus, from Table VIII of Appendix B we find the rejection region:

$$F > F_{\alpha/2} = F_{.05} \approx 3.60$$

(Since no table value is given for $\nu_1 = 11$, we average the entries at 10 and 12 to obtain $F_{.05} \approx 3.60$.) We now calculate

$$F = \frac{s_2^2}{s_1^2} = \frac{(6.35)^2}{(4.85)^2} = 1.71$$

This F value is not in the rejection region. Therefore, there is insufficient evidence at the $\alpha = .10$ level to refute the assumption of equal population variances.

The following example shows that the F statistic is sometimes used to compare population variances in their own right rather than just to check the validity of an assumption.

EXAMPLE 9.6

Suppose an investor wants to compare the risks associated with two different stocks, where the risk of a given stock is measured by the variation in daily price changes. Suppose we obtain a random sample of 25 daily price changes for stock 1 and 25 for stock 2. The sample results are summarized in the table. Compare the risks associated with the two stocks by testing the null hypothesis that the variances of the price changes for the stocks are equal. Use $\alpha = .10$.

Stock 1	*Stock 2*
$n_1 = 25$	$n_2 = 25$
$\bar{x}_1 = .250$	$\bar{x}_2 = .125$
$s_1 = .76$	$s_2 = .46$

Solution

Since we wish to detect a difference in population variances, we will want to detect either $\sigma_1^2 > \sigma_2^2$ or $\sigma_2^2 > \sigma_1^2$. Therefore, we choose as the alternative (research) hypothesis, H_a: $\sigma_1^2 \neq \sigma_2^2$, and will conduct this two-tailed test:

$$H_0: \frac{\sigma_1^2}{\sigma_2^2} = 1 \qquad H_a: \frac{\sigma_1^2}{\sigma_2^2} \neq 1$$

Test statistic: $F = \dfrac{\text{Larger sample variance}}{\text{Smaller sample variance}} = \dfrac{s_1^2}{s_2^2}$

Assumptions:
1. The changes in daily prices for each stock have relative frequency distributions that are approximately normal.
2. The samples are randomly and independently selected from a set of daily stock reports.

Rejection region: $F > F_{\alpha/2} = F_{.05} = 1.98$
where $F_{.05}$ possesses $\nu_1 = n_1 - 1 = 24$ and $\nu_2 = n_2 - 1 = 24$ df.

We calculate

$$F = \frac{s_1^2}{s_2^2} = \frac{(.76)^2}{(.46)^2} = 2.73$$

The calculated F exceeds the rejection value of 1.98. Therefore, we conclude that the variances of daily price changes differ for the two stocks. It appears that the risk, as measured by the variance of daily price changes, is greater for stock 1 than for stock 2. How much reliability can we place in this inference? Only one time in 10 (since $\alpha = .10$), on the average, would this statistical test lead us to conclude erroneously that σ_1^2 and σ_2^2 were different when in fact they were equal.

EXAMPLE 9.7

Find the approximate observed significance level for the F test in Example 9.6.

Solution

Since the observed value of the F statistic in Example 9.6 was 2.73, the observed significance level for the test would equal the probability of observing a value of F at least as contradictory to H_0: $\sigma_1^2 = \sigma_2^2$ as $F = 2.73$, if in fact H_0 were true. Since we give the F tables in Appendix B only for values of α equal to .10, .05, .025, and .01, we can only approximate the observed significance level. Checking Tables VII, VIII, IX, and X, we find $F_{.05} = 1.98$, $F_{.025} = 2.27$, and $F_{.01} = 2.66$. Since the observed value of $F = 2.73$ slightly exceeds $F_{.01}$, the observed significance level for the test will be slightly less than

$$\text{Approximate } p\text{-value} = 2(.01) = .02$$

Note that we doubled the tail probability shown in Table X because this was a two-tailed test.

We have presented the F test as a test of hypothesis of the equality of variances—i.e., $\sigma_1^2 = \sigma_2^2$. Although this is the most common application of the test, it can also be used to test a hypothesis that the ratio of the population variances is equal to some specified value, H_0: $\sigma_1^2/\sigma_2^2 = k$. The test would be conducted exactly the same way as a test of hypothesis concerning the equality of variances except that we would use the test statistic

$$F = \frac{s_1^2}{s_2^2}\left(\frac{1}{k}\right)$$

What Do You Do If the Assumption of Normal Population Distributions Is Not Satisfied?

Answer: The F test is much more sensitive to departures from normality than the t test for comparing population means discussed in Section 9.2. If you have doubts about the normality of the population frequency distributions, you can use a nonparametric method for comparing the two population variances. A method can be found in the references listed at the end of this chapter.

Exercises 9.31–9.41

Learning the Mechanics

9.31 Under what conditions is the sampling distribution of s_1^2/s_2^2 an F distribution?

9.32 Use Tables VII, VIII, IX, and X of Appendix B to find each of the following F values:
a. $F_{.05}$ where $\nu_1 = 9$ and $\nu_2 = 6$
b. $F_{.01}$ where $\nu_1 = 18$ and $\nu_2 = 14$
c. $F_{.025}$ where $\nu_1 = 11$ and $\nu_2 = 4$
d. $F_{.10}$ where $\nu_1 = 20$ and $\nu_2 = 5$

9.33 Given ν_1 and ν_2, find the following probabilities:
a. $\nu_1 = 2$, $\nu_2 = 30$, $P(F \geq 5.39)$
b. $\nu_1 = 24$, $\nu_2 = 10$, $P(F < 2.74)$
c. $\nu_1 = 7$, $\nu_2 = 1$, $P(F \leq 236.8)$
d. $\nu_1 = 40$, $\nu_2 = 40$, $P(F > 2.11)$

9.34 For each of the following cases, identify the rejection region that should be used to test H_0: $\sigma_1^2 = \sigma_2^2$ against H_a: $\sigma_1^2 > \sigma_2^2$. Use $\nu_1 = 30$ and $\nu_2 = 20$.
a. $\alpha = .10$ b. $\alpha = .05$ c. $\alpha = .025$ d. $\alpha = .01$

9.35 For each of the following cases, identify the rejection region that should be used to test H_0: $\sigma_1^2 = \sigma_2^2$ against H_a: $\sigma_1^2 \neq \sigma_2^2$. Use $\nu_1 = 10$ and $\nu_2 = 12$.
a. $\alpha = .20$ b. $\alpha = .10$ c. $\alpha = .05$ d. $\alpha = .02$

9.36 Independent random samples were selected from each of two normally distributed populations, $n_1 = 12$ from population 1 and $n_2 = 27$ from population 2. The means and variances for the two samples are shown in the table.

Sample 1	Sample 2
$n_1 = 12$	$n_2 = 27$
$\bar{x}_1 = 31.7$	$\bar{x}_2 = 37.4$
$s_1^2 = 3.87$	$s_2^2 = 8.75$

a. Do the data provide sufficient evidence to indicate a difference between the population variances? Test using $\alpha = .10$.
b. Find and interpret the approximate p-value for the test.

9.37 Independent random samples were selected from each of two normally distributed populations, $n_1 = 6$ from population 1 and $n_2 = 5$ from population 2. The data are shown in the table.

Sample 1	3.1	4.4	1.2	1.7	.7	3.4
Sample 2	2.3	1.4	3.7	8.9	5.5	

a. Do these data provide sufficient evidence to indicate a difference between the population variances? Use $\alpha = .05$.
b. Find and interpret the approximate observed significance level for the test.

Applying the Concepts

9.38 Tests of product quality can be completely automated or can be conducted using human inspectors or human inspectors aided by mechanical devices. While human inspection is frequently the most economical alternative, it can lead to serious inspection error problems (Benson and Ohta, 1986). Numerous studies have demonstrated that inspectors rarely are able to detect as many as 85% of the defective items that they inspect and that performance varies across inspectors (Sinclair, 1978). To evaluate the performance of inspectors, a quality manager had a sample of 12 novice inspectors evaluate 200 finished products. The same 200 items were evaluated by 12 experienced inspectors. The quality of each item—whether defective or nondefective—was known to the manager. The table lists the number of inspection errors (classifying a defective item as nondefective, or vice versa) made by each inspector.

Novice Inspectors	30	35	26	40	36	20	45	31	33	29	21	48
Experienced Inspectors	31	15	25	19	28	17	19	18	24	10	20	21

a. Prior to conducting this experiment the manager believed the variance in inspection errors was lower for experienced inspectors than for novice inspectors. Do the sample data support her belief? Test using $\alpha = .05$.
b. What is the approximate p-value of the test you conducted in part **a**?

9.39 Suppose your firm has been experimenting with two different physical arrangements of its assembly line. It has been determined that both arrangements yield approximately the same average number of finished units per day. To obtain greater process control you suggest that the arrangement with the smaller variance in the number of finished units produced per day be permanently adopted. Two independent random samples yield the results shown in the table on page 420. Do the samples provide sufficient evidence at $\alpha = .10$ to conclude that the variances in the number of units produced per day differ for the two arrangements? If so, which arrangement would you choose? If not, what would you suggest the firm do?

Assembly Line 1	*Assembly Line 2*
$n_1 = 21$ days	$n_2 = 21$ days
$s_1^2 = 1{,}432$	$s_2^2 = 3{,}761$

9.40 The quality control department of a paper company measures the brightness (a measure of reflectance) of finished paper periodically throughout the day. Two instruments that are available to measure the paper specimens are subject to error, but they can be adjusted so the mean readings for a control paper specimen are the same for both instruments. Suppose you are concerned about the precision of the two instruments—namely, that instrument 2 is less precise than instrument 1. To check this theory, five measurements of a single paper sample are made on both instruments. The data are shown in the table. Do the data provide sufficient evidence to indicate that instrument 2 is less precise than instrument 1? Test using $\alpha = .05$.

Instrument 1	29	28	30	28	30
Instrument 2	26	34	30	32	28

9.41 The computerization of the telephone operator's workplace has significantly shortened the time necessary to serve a customer. Thus, while the number of telephone calls made in this country increased from 67 billion in 1950 to 310 billion in 1980, the number of telephone operators decreased from 244,000 to 128,000. However, the computer not only speeds the operator's work, it monitors the operator's work as well—somewhat like a supervisor. This monitoring has led some telephone company offices to issue a guideline of 30 seconds as the maximum amount of time an operator should spend completing operator-assisted calls (Serrin, 1983). To study the effect of this guideline on the mean time to complete a call, a researcher planned to estimate the difference in the mean time between offices that do and offices that do not issue the guideline to its operators. The completion times (in seconds) shown in the table were sampled from the computer-collected data that results from monitoring the operators.

Guideline	31.2	33.5	27.2	23.5	28.8	26.3	27.3	25.9	20.3	26.8
No Guideline	30.0	22.6	35.9	41.3	28.9	21.4	39.8	45.3	37.1	25.6

a. The researcher knows that to use the two-sample t statistic in constructing a confidence interval, the assumption of equal population variances should first be examined. Test the equality of the population variances using $\alpha = .05$. What does your test indicate about the appropriateness of using the two-sample t in this situation?

b. List any assumptions you made in conducting the hypothesis test of part **a**.

c. In the context of this problem, describe the Type I and Type II errors associated with your hypothesis test of part **a**.

d. What is the approximate p-value of the test you conducted in part **a**?

9.4 Inferences About the Difference Between Two Population Means: Paired Difference Experiments

Suppose you want to compare the mean daily sales of two restaurants located in the same city. If you were to record the restaurants' total sales for each of 12 days (2 work weeks), the results might appear as shown in Table 9.2.

TABLE 9.2 Daily Sales for Two Restaurants

Day	Restaurant 1	Restaurant 2
1 (Monday)	\$ 759	\$ 678
2 (Tuesday)	981	933
3 (Wednesday)	1,005	918
4 (Thursday)	1,449	1,302
5 (Friday)	1,905	1,782
6 (Saturday)	2,073	1,971
7 (Monday)	693	639
8 (Tuesday)	873	825
9 (Wednesday)	1,074	999
10 (Thursday)	1,338	1,281
11 (Friday)	1,932	1,827
12 (Saturday)	2,106	2,049
	$\bar{x}_1 = \$1,349.00$	$\bar{x}_2 = \$1,267.00$
	$s_1 = \$530.07$	$s_2 = \$516.03$

Test the null hypothesis that the mean daily sales, μ_1 and μ_2, for the two restaurants are equal against the alternative hypothesis that they differ; i.e.,

$$H_0: \quad (\mu_1 - \mu_2) = 0 \qquad H_a: \quad (\mu_1 - \mu_2) \neq 0$$

Using the two-sample t statistic (Section 9.2) we would calculate

$$s_p^2 = \frac{(n_1 - 1)s_1^2 + (n_2 - 1)s_2^2}{n_1 + n_2 - 2}$$

$$= \frac{(12 - 1)(530.07)^2 + (12 - 1)(516.03)^2}{12 + 2 - 2} = 273,630.6$$

and

$$t = \frac{(\bar{x}_1 - \bar{x}_2) - 0}{\sqrt{s_p^2\left(\frac{1}{n_1} + \frac{1}{n_2}\right)}} = \frac{(1,349.00 - 1,267.00)}{\sqrt{273,630.6\left(\frac{1}{12} + \frac{1}{12}\right)}} = \frac{82.0}{213.54} = .38$$

This small t value will not lead to rejection of H_0 when compared to the t distribution with $n_1 + n_2 - 2 = 22$ df, even if α were chosen as large as .20 ($t_{\alpha/2} = t_{.10} = 1.321$). Thus, we might conclude that insufficient evidence exists to infer that there is a difference in mean daily sales for the two restaurants.

However, if you examine the data in Table 9.2 more closely, you will find this conclusion difficult to accept. The sales of restaurant 1 exceed those of restaurant 2 *for every one of the 12 days*. This, in itself, is strong evidence to indicate that μ_1 differs from μ_2, and we will subsequently confirm this fact. Why, then, was the t test unable to detect this difference?

The cause of this apparent inconsistency is that the two-sample t is inappropriate, because the assumption of independent samples is invalid. If you examine the pairs of daily sales, you will note that the sales of the two restaurants tend to rise and fall together over the days of the week. This pattern suggests a very strong daily dependence between the two samples and a violation of the assumption of independence required for the two-sample t test of Section 9.2. In this particular situation, note the *large variation within samples* (reflected by the large value of s_p^2) in comparison to the *small difference between the sample means*. Because s_p^2 was so large, the t test was unable to detect a possible difference between μ_1 and μ_2.

Now, consider a valid method to analyze the data of Table 9.2. We add to this table a column of differences between the daily sales of the restaurants, and thus form Table 9.3. We can regard these daily differences in sales as a sample of all daily differences, past and present. Then we can use this sample to make inferences about the mean, μ_D, of the population of differences, *which is equal to the difference* $(\mu_1 - \mu_2)$: i.e., the mean of the population (sample) of differences equals the difference between the population (sample) means. Thus, our test becomes

$$H_0\text{:}\quad \mu_D = 0 \qquad (\text{i.e., } \mu_1 - \mu_2 = 0)$$
$$H_a\text{:}\quad \mu_D \neq 0 \qquad (\text{i.e., } \mu_1 - \mu_2 \neq 0)$$

TABLE 9.3 Daily Sales and Differences for Two Restaurants

Day	Restaurant 1	Restaurant 2	(Restaurant 1 − Restaurant 2)
1 (Monday)	\$ 759	\$ 678	\$ 81
2 (Tuesday)	981	933	48
3 (Wednesday)	1,005	918	87
4 (Thursday)	1,449	1,302	147
5 (Friday)	1,905	1,782	123
6 (Saturday)	2,073	1,971	102
7 (Monday)	693	639	54
8 (Tuesday)	873	825	48
9 (Wednesday)	1,074	999	75
10 (Thursday)	1,338	1,281	57
11 (Friday)	1,932	1,827	105
12 (Saturday)	2,106	2,049	57
			$\bar{x}_D = \$82.00$
			$s_D = \$32.00$

The test statistic is a one-sample t (Section 8.4), since we are now analyzing a single sample of differences:

$$\textit{Test statistic:}\quad t = \frac{\bar{x}_D - 0}{s_D/\sqrt{n_D}}$$

where

$\bar{x}_D$ = Sample mean of differences
s_D = Sample standard deviation of differences
n_D = Number of differences

Assumptions: 1. The population of differences in daily sales is approximately normally distributed.
2. The sample differences are randomly selected from a population of differences.

To find the rejection region, we first choose $\alpha = .05$. Then we will reject H_0 if

$$t < -t_{.025} \quad \text{or} \quad t > t_{.025}$$

where $t_{.025}$ is based on $(n_D - 1)$ degrees of freedom.

Referring to Table VI of Appendix B, we find the t value corresponding to $\alpha/2 = .025$ and $n_D - 1 = 12 - 1 = 11$ df to be $t_{.025} = 2.201$. Thus, the null hypothesis will be rejected if $t < -2.201$ or $t > 2.201$. Note that the number of degrees of freedom has decreased from $n_1 + n_2 - 2 = 22$ to 11 by using the *paired difference experiment* rather than the two independent random samples design.

Now calculate

$$t = \frac{\bar{x}_D - 0}{s_D/\sqrt{n_D}} = \frac{82.00}{32.00/\sqrt{12}} = 8.88$$

Because this computed value of t falls in the rejection region, we conclude that the difference in mean daily sales for the two restaurants differs from 0. The fact that $\bar{x}_1 - \bar{x}_2 = \bar{x}_D = \$82.00$ strongly suggests that the mean daily sales for restaurant 1 exceed the mean daily sales for restaurant 2.

This kind of experiment, in which observations are paired and the differences analyzed, is called a **paired difference experiment**. In many cases a paired difference experiment can provide more information about the difference between population means than an independent samples experiment. The differencing removes the variability due to the dimension on which the observations are paired. For instance, in the restaurant example, the day-to-day variability in daily sales is removed by analyzing the differences between the restaurants' daily sales. The removal of the variability due to this extra dimension is called **blocking**, and the paired difference experiment is a simple example of a **randomized block experiment**. In our example, the days represent the blocks.

Some other examples for which the paired difference experiment might be appropriate are the following:

1. To compare the performance of two automobile salespeople, we might test a hypothesis about the difference $(\mu_1 - \mu_2)$ in their respective mean monthly sales. If we randomly choose n_1 months of salesperson 1's sales and independently choose n_2 months of salesperson 2's sales, the month-to-month variability caused by the seasonal nature of new car sales might inflate s_p^2 and prevent the two-sample t statistic from detecting a difference between μ_1 and μ_2, if such a difference actually

exists. However, by taking the difference in monthly sales for the two salespeople for each of n months, we eliminate the month-to-month variability (seasonal variation) in sales, and the probability of detecting a difference between μ_1 and μ_2, if a difference exists, is increased.

2. Suppose you want to estimate the difference $(\mu_1 - \mu_2)$ in mean price between two major brands of premium gasoline. If you were to choose two independent random samples of stations for each brand, the variability in price due to geographic location may be large. To eliminate this source of variability, you could choose pairs of stations, one station for each brand, in close geographic proximity and use the sample of differences between the prices of the brands to make an inference about $(\mu_1 - \mu_2)$.
3. Suppose a college placement center wants to estimate the difference $(\mu_1 - \mu_2)$ in mean starting salaries for men and women graduates who seek jobs through the center. If it independently samples men and women, the starting salaries may vary due to their different college majors and differences in grade-point averages. To eliminate these sources of variability, the placement center could match male and female job-seekers according to their majors and grade-point averages. Then the difference between the starting salaries of each pair in the sample could be used to make an inference about $(\mu_1 - \mu_2)$.

The hypothesis-testing and confidence interval procedures based on a paired difference experiment are summarized in the next two boxes.

Paired Difference Test of Hypothesis

One-Tailed Test

H_0: $(\mu_1 - \mu_2) = D_0$,
i.e., $\mu_D = D_0$

H_a: $(\mu_1 - \mu_2) < D_0$,
i.e., $\mu_D < D_0$
[or H_a: $(\mu_1 - \mu_2) > D_0$,
i.e., $\mu_D > D_0$]

Test statistic: $t = \dfrac{\bar{x}_D - D_0}{s_D/\sqrt{n_D}}$

Rejection region: $t < -t_\alpha$
[or $t > t_\alpha$ when
H_a: $(\mu_1 - \mu_2) > D_0$]

where t_α has $(n_D - 1)$ df.

Two-Tailed Test

H_0: $(\mu_1 - \mu_2) = D_0$,
i.e., $\mu_D = D_0$

H_a: $(\mu_1 - \mu_2) \neq D_0$,
i.e., $\mu_D \neq D_0$

Test statistic: $t = \dfrac{\bar{x}_D - D_0}{s_D/\sqrt{n_D}}$

Rejection region: $t < -t_{\alpha/2}$
or $t > t_{\alpha/2}$

where $t_{\alpha/2}$ has $(n_D - 1)$ df.

Assumptions:
1. The relative frequency distribution of the population of differences is normal.
2. The differences are randomly selected from the population of differences.

Paired Difference Confidence Interval

$$\bar{x}_D \pm t_{\alpha/2} \frac{s_D}{\sqrt{n_D}}$$

where $t_{\alpha/2}$ has $(n_D - 1)$ df.

Assumption: Same as for the paired difference test (previous box).

EXAMPLE 9.8

A paired difference experiment is conducted to compare the starting salaries of male and female college graduates who find jobs. Pairs are formed by choosing a male and a female with the same major and similar grade-point averages. Suppose a random sample of 10 pairs is formed in this manner, and the starting annual salary of each person is recorded. The results are shown in Table 9.4. Test to see whether there is evidence that the mean starting salary, μ_1, for males exceeds the mean starting salary, μ_2, for females. Use $\alpha = .05$.

TABLE 9.4 Starting Salaries

Pair	Male	Female	Difference (Male − Female)
1	$24,300	$23,800	$ 500
2	26,500	26,600	−100
3	25,400	24,800	600
4	23,500	23,500	0
5	28,500	27,600	900
6	22,800	23,000	−200
7	24,500	24,200	300
8	26,200	25,100	1,100
9	23,400	23,200	200
10	24,200	23,500	700

Solution

Since we are interested in determining whether the data indicate that μ_1 exceeds μ_2—i.e., whether the mean starting salary for men exceeds the mean starting salary for women—we will choose a one-sided alternative (research) hypothesis. Then the elements of the paired difference test are

H_0: $\mu_D = 0$ $\quad (\mu_1 - \mu_2 = 0)$

H_a: $\mu_D > 0$ $\quad (\mu_1 - \mu_2 > 0)$

Test statistic: $t = \dfrac{\bar{x}_D - 0}{s_D/\sqrt{n_D}}$

Assumption: The relative frequency distribution for the population of differences is normal.

Since the test is upper-tailed, we will reject H_0 if $t > t_\alpha = t_{.05} = 1.833$, where t_α is based on $n_D - 1 = 9$ df. The rejection region is shown in Figure 9.9.

FIGURE 9.9 ▶
Rejection region for Example 9.8

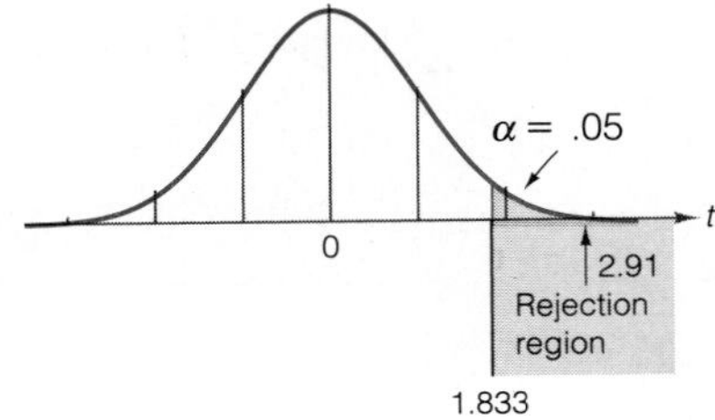

Using x_{Di} to represent the ith difference measurement, we now calculate

$$\sum_{i=1}^{10} x_{Di} = 500 + (-100) + \cdots + 700 = 4{,}000$$

and

$$\sum_{i=1}^{10} x_{Di}^2 = 3{,}300{,}000$$

Then,

$$\bar{x}_D = \frac{\sum_{i=1}^{10} x_{Di}}{10} = \frac{4{,}000}{10} = 400$$

$$s_D^2 = \frac{\sum_{i=1}^{n_D} (x_{Di} - \bar{x}_D)^2}{n_D - 1} = \frac{\sum_{i=1}^{10} x_{Di}^2 - \left(\sum_{i=1}^{10} x_{Di}\right)^2 \Big/ 10}{9}$$

$$= \frac{3{,}300{,}000 - (4{,}000)^2/10}{9} = 188{,}888.89$$

$$s_D = \sqrt{s_D^2} = 434.61$$

Substituting these values into the formula for the test statistic, we find that

$$t = \frac{\bar{x}_D - 0}{s_D/\sqrt{n_D}} = \frac{400}{434.61/\sqrt{10}} = \frac{400}{137.44} = 2.91$$

As you can see in Figure 9.9, the calculated t falls in the rejection region. Thus, we conclude at $\alpha = .05$ that the mean starting salary for males exceeds the mean starting salary for females.

One measure of the amount of information about $(\mu_1 - \mu_2)$ gained by using a paired difference experiment rather than an independent samples experiment in Example 9.8 is the relative widths of the confidence intervals obtained by the two methods. A 95% confidence interval for $(\mu_1 - \mu_2)$ using the paired difference experiment is

$$\bar{x}_D \pm t_{\alpha/2}\frac{s_D}{\sqrt{n_D}}$$

where t_α has $n_D - 1 = 9$ df. Substituting into this formula, we obtain

$$400 \pm t_{.025}\frac{434.61}{\sqrt{10}} = 400 \pm 2.262\,\frac{434.61}{\sqrt{10}} = 400 \pm 310.88$$
$$\approx 400 \pm 311 = (\$89,\ \$711)$$

If we analyzed the same data as though they were from an independent samples experiment,* we would first calculate the following quantities:

Males	*Females*
$\bar{x}_1 = \$24{,}930$	$\bar{x}_2 = \$24{,}530$
$s_1^2 = 3{,}009{,}000$	$s_2^2 = 2{,}331{,}222.22$

Then

$$s_p^2 = \frac{(n_1 - 1)s_1^2 + (n_2 - 1)s_2^2}{n_1 + n_2 - 2} = \frac{9(3{,}009{,}000) + 9(2{,}331{,}222.22)}{18}$$
$$= 2{,}670{,}111.11$$

where s_p^2 is based on $(n_1 + n_2 - 2) = (10 + 10 - 2) = 18$ df.

The 95% confidence interval is

$$(\bar{x}_1 - \bar{x}_2) \pm t_{\alpha/2}\sqrt{s_p^2\left(\frac{1}{n_1} + \frac{1}{n_2}\right)} = 400 \pm t_{.025}\sqrt{2{,}670{,}111.11\left(\frac{1}{10} + \frac{1}{10}\right)}$$
$$= 400 \pm 2.101\sqrt{2{,}670{,}111.11\left(\frac{1}{10} + \frac{1}{10}\right)}$$
$$= 400 \pm 1{,}535.35$$
$$\approx 400 \pm 1{,}535 = (-\$1{,}135,\ \$1{,}935)$$

The confidence interval for the independent sampling experiment is about five times wider than the corresponding paired difference confidence interval. Blocking out the variability due to differences in majors and grade-point averages significantly increases the information about the difference in mean male and female starting salaries

*This is done only to indicate how much more information is obtained by a paired design versus an unpaired design. Actually, for an experiment designed using pairing, an unpaired analysis is invalid because the assumption of independent samples is not satisfied.

by providing a much more accurate (smaller confidence interval for the same confidence coefficient) estimate of $(\mu_1 - \mu_2)$.

You may wonder whether a paired difference experiment is always superior to an independent samples experiment. The answer is: most of the time, but not always. We sacrifice half the degrees of freedom in the t statistic when a paired difference design is used instead of an independent samples design. This is a *loss* of information, and unless this loss is more than compensated for by the reduction in variability obtained by blocking (pairing), the paired difference experiment will result in a net loss of information about $(\mu_1 - \mu_2)$. Thus, we should be convinced that the pairing will significantly reduce variability before performing the paired difference experiment. Most of the time this will happen.

One final note: The pairing of the observations is determined *before* the experiment is performed (that is, by the *design* of the experiment). **A paired difference experiment is never obtained by pairing the sample observations after the measurements have been acquired**. Such is the stuff of which statistical lies are made!

What Do You Do When the Assumption of a Normal Distribution for the Population of Differences Is Not Satisfied?

Answer: Use the Wilcoxon signed rank test for the paired difference design. (See Chapter 17.)

CASE STUDY 9.1 / Comparing Salaries for Equivalent Work

Procedures to maintain comparable salaries between federal white-collar workers and those in the private sector are mandated by federal law. William M. Smith (1976) discussed one statistical mechanism used in complying with this law. An annual survey, the *National Survey for Professional, Administrative, Technical, and Clerical Pay*, is conducted by the Bureau of Labor Statistics. Salary information is collected for approximately 85 work-level categories ranging from clerical to administrative positions in 20 occupations.

The design of the survey resembles a paired difference design because the occupation and experience of employees in the private and government sectors are matched as closely as possible before a comparison is made. Test statistics like the paired difference t can be used to compare the mean salaries for the two sectors. Then, if the data indicate that the mean salaries differ for certain levels, the need for an adjustment is indicated.

Smith points out that presidential or legislative intervention often prevents the adjustments indicated by the data from being enacted. Regardless, the results of the survey are valuable as a salary guide for the private sector and to those performing general economic analyses.

Exercises 9.42–9.53

Learning the Mechanics

9.42 A paired difference experiment yielded n_D pairs of observations. In each case, what is the rejection region for testing H_0: $\mu_D = 2$ against H_a: $\mu_D > 2$?

a. $n_D = 12$, $\alpha = .05$ **b.** $n_D = 24$, $\alpha = .10$
c. $n_D = 4$, $\alpha = .025$ **d.** $n_D = 8$, $\alpha = .01$

9.43 A paired difference experiment yielded the data shown in the table.

Person	*Before*, x_1	*After*, x_2
1	82	92
2	60	72
3	55	57
4	97	104
5	79	89

a. Compute $\bar{x}_D$ and s_D. **b.** Demonstrate that $\bar{x}_D = \bar{x}_1 - \bar{x}_2$.
c. Is there sufficient evidence to conclude that $\mu_1 \neq \mu_2$? Test using $\alpha = .05$.
d. Find and interpret the approximate p-value for the paired difference test.
e. What assumptions are necessary so the paired difference test will be valid?

9.44 Refer to Exercise 9.43. Construct a 95% confidence interval for μ_D.

9.45 A paired difference experiment produced the following data:

$$n_D = 18 \qquad \bar{x}_1 = 92 \qquad \bar{x}_2 = 95.5 \qquad \bar{x}_D = -3.5 \qquad s_D^2 = 21$$

a. Determine the values of t for which the null hypothesis $(\mu_1 - \mu_2) = 0$ would be rejected in favor of the alternative hypothesis $(\mu_1 - \mu_2) < 0$. Use $\alpha = .10$.
b. Conduct the paired difference test described in part **a**. Draw the appropriate conclusions.
c. What assumptions are necessary so that the paired difference test will be valid?
d. Construct a 90% confidence interval for the mean difference μ_D.
e. Which of the two inferential procedures, the confidence interval of part **d** or the test of hypothesis of part **b**, provides more information about the difference between the population means?

9.46 Frequently, a paired difference experiment provides more information about the difference between two population means than an independent random samples experiment. Explain. When may it not?

9.47 A paired difference experiment yielded the data shown in the table.

Pair	1	2	3	4	5	6	7
x	24.4	86.3	43.2	69.5	93.0	29.9	67.3
y	12.3	74.3	34.4	61.6	84.3	23.4	59.9

a. Test H_0: $\mu_D = 10$ against H_a: $\mu_D \neq 10$, where $\mu_D = (\mu_x - \mu_y)$. Use $\alpha = .05$.
b. Report and interpret the p-value for the test you conducted in part **a**.

Applying the Concepts

9.48 Do people generally receive the asking price when they sell their home? The accompanying table lists the asking price and sale price for a random sample of 10 1993 single-family home sales in Eden Prairie, Minnesota.

House	*Asked*	*Received*	*House*	*Asked*	*Received*
1	$312,900	$305,600	6	122,855	121,000
2	142,900	142,900	7	164,900	159,900
3	136,700	135,000	8	74,900	74,900
4	134,900	125,700	9	129,900	132,000
5	196,900	200,000	10	150,000	146,500

Source: Minneapolis Area Association of Realtors, 1993.

a. Explain why the samples are not independent.
b. What is the dimension on which the samples are paired or matched?
c. Defining D as (Price asked − Price received), $\sum x_{D_i} = \$23{,}355$ and $\sum x_{D_i}^2 = 195{,}531{,}025$. Conduct a test of hypothesis that will help answer the question of interest for people in Eden Prairie who sold their homes in 1993. Use $\alpha = .05$.

9.49 *Time* (Dec. 7, 1981) reported that many corporations now encourage their employees to "get things off their chests" by communicating directly to management on hot-line phones, through the mail, or in face-to-face meetings. For example, American Express Corp. instituted a mail program that guaranteed its employees anonymity (if desired) and an answer from the responsible person (including the company chairperson) within 10 days. A similar program at IBM generated an average of 13,000 letters a year from IBM's 195,000 U.S. employees. Such programs reflect the constant search by American business for ways to improve worker productivity.

Before instituting an employee complaint/suggestion program in its manufacturing plant, a company randomly sampled eight workers and measured their productivity in terms of the number of items produced per day. A year after the start of the complaint program the productivity of seven of these workers was reevaluated. (The eighth had been promoted in the interim.) The productivity data are shown in the table.

Employee ID Number	*August 1992*	*August 1993*
1011	10	9
0033	9	11
0998	12	14
0006	8	9
1802	10	9
0246	11	14
0777	14	–
1112	11	13

a. Do the data provide evidence that the complaint/suggestion program has helped to increase worker productivity? Test using $\alpha = .10$, and clearly state any assumptions you make in conducting the test.
b. Discuss how the 1-year gap between productivity evaluations could weaken the results of the study.

9.50 Among the better known and most frequently enforced antitrust laws are those against price fixing (Sherman Act, Section One) and monopoly or attempt to monopolize (Sherman Act, Section Two). Other antitrust violations include price discrimination, retail price maintenance, and tying the sale of a product to the purchase of another product. Alan R. Beckenstein, H. Landis Gabel, and Karlene Roberts (1983) surveyed 188 Fortune 500 industrial companies and found that, on average (without adjusting for inflation), these companies "spent 3½ times the amount on antitrust investigations, legal fees, fines, damages, court costs, and out-of-court settlements in the 1970s as they did in the 1960s." Further, on average, these companies faced 3.22 more antitrust litigations in the 1970s than in the 1960s. Beckenstein, Gabel, and Roberts observed that companies must now consider the possible legal consequences of their actions more than they used to. They concluded that with today's legal risks managers at all levels should be helping to design and implement strategies to comply with the antitrust laws. Another survey of 10 Fortune 500 companies yielded the data in the table about the number of litigations faced by the firms in the 1960s and 1970s.

Firm	*1960s*	*1970s*	*Firm*	*1960s*	*1970s*
1	10	10	6	7	6
2	8	12	7	6	11
3	9	8	8	9	12
4	7	16	9	8	11
5	8	14	10	7	12

a. Do the data provide sufficient evidence to reject the researchers' claim? Conduct a test of hypothesis and use the p-value of your test to help you answer this question.
b. Use a 90% confidence interval to estimate the difference in the mean number of antitrust litigations per firm in the 1970s and the 1960s. Compare your results to the reported survey finding.

9.51 A manufacturer of automobile shock absorbers was interested in comparing the durability of its shocks with that of the shocks produced by its biggest competitor. To make the comparison, one of the manufacturer's and one of the competitor's shocks were randomly selected and installed on the rear wheels of each of six cars. After the cars had been driven 20,000 miles, the strength of each test shock was measured, coded, and recorded. Results of the examination are shown in the table.

Car Number	*Manufacturer's Shock*	*Competitor's Shock*
1	8.8	8.4
2	10.5	10.1
3	12.5	12.0
4	9.7	9.3
5	9.6	9.0
6	13.2	13.0

a. Do the data present sufficient evidence to conclude that there is a difference in the mean strength of the two types of shocks after 20,000 miles of use? Use $\alpha = .05$.
b. Find the approximate observed significance level for the test, and interpret its value.
c. What assumptions are necessary to apply a paired difference analysis to the data?
d. Construct a 95% confidence interval for $(\mu_1 - \mu_2)$. Interpret the confidence interval.

9.52 Suppose the data in Exercise 9.51 are based on independent random samples.
a. Do the data provide sufficient evidence to indicate a difference between the mean strengths for the two types of shocks? Use $\alpha = .05$.
b. Construct a 95% confidence interval for $(\mu_1 - \mu_2)$. Interpret your result.

c. Compare the confidence intervals you obtained in Exercise 9.51 and in part **b** of this exercise. Which is wider? To what do you attribute the difference in width? Assuming in each case that the appropriate assumptions are satisfied, which interval provides you with more information about $(\mu_1 - \mu_2)$? Explain.
d. Are the results of an unpaired analysis valid if the data come from a paired experiment?

9.53 A *pupillometer* is a device used to observe changes in pupil dilations as the eye is exposed to different visual stimuli. Since there is a direct correlation between the amount an individual's pupil dilates and his or her interest in the stimuli, marketing organizations sometimes use pupillometers to help them evaluate potential consumer interest in new products, alternative package designs, and other factors. The Design and Market Research Laboratories of the Container Corporation of America used a pupillometer to evaluate consumer reaction to different silverware patterns for a client (McGuire, 1973). Suppose 15 consumers were chosen at random, and each was shown two silverware patterns. Their pupillometer readings are shown in the table (in millimeters).

Consumer	*Pattern 1*	*Pattern 2*	*Consumer*	*Pattern 1*	*Pattern 2*
1	1.00	.80	9	.98	.91
2	.97	.66	10	1.46	1.10
3	1.45	1.22	11	1.85	1.60
4	1.21	1.00	12	.33	.21
5	.77	.81	13	1.77	1.50
6	1.32	1.11	14	.85	.65
7	1.81	1.30	15	.15	.05
8	.91	.32			

a. What are the appropriate null and alternative hypotheses to test whether the mean amount of pupil dilation differs for the two patterns? Define any symbols you use.
b. The data were analyzed using a statistical software package, with the results shown. Interpret these results. Do you think that the p-value is exactly equal to zero? Explain.

```
MEAN DIFFERENCE (1 - 2) = .239
T = 5.76      DF = 14     TWO-TAILED P-VALUE = .000
```

c. Another part of the computer output was the confidence interval. Interpret the interval.

```
95% CONFIDENCE INTERVAL: (.150, .328)
```

d. Is the paired difference design used for this study preferable to an independent samples design? For independent samples we could select 30 consumers, divide them into two groups of 15, and show each group a different pattern. Explain your preference.

9.5 Inferences About the Difference Between Two Population Proportions: Independent Binomial Experiments

Suppose a manufacturer of camper vans wants to compare the potential market for its products in the northeastern United States to the market in the southeastern United States. Such a comparison would help the manufacturer decide where to concentrate sales efforts. Using telephone directories, the company randomly chooses 1,000 households in the northeast (NE) and 1,000 households in the southeast (SE) and determines whether each household plans to buy a camper within the next 5 years. The objective is to use this sample information to make an inference about the difference $(p_1 - p_2)$ between the proportion p_1 of *all* households in the NE and the proportion p_2 of *all* households in the SE that plan to purchase a camper within 5 years.

The two samples represent independent binomial experiments (see Section 4.4 for the characteristics of binomial experiments), with the binomial random variables x_1 and x_2 being the numbers of the 1,000 sampled households in each area that indicate they will purchase a camper within 5 years. The results of the sampling can be summarized as shown in the table. We can now calculate the *sample* proportions $\hat{p}_1$ $\hat{p}_2$ of the households in the NE and SE, respectively, that are prospective buyers:

NE	SE
$n_1 = 1{,}000$	$n_2 = 1{,}000$
$x_1 = 42$	$x_2 = 24$

$$\hat{p}_1 = \frac{x_1}{n_1} = \frac{42}{1{,}000} = .042 \qquad \hat{p}_2 = \frac{x_2}{n_2} = \frac{24}{1{,}000} = .024$$

The difference between the sample proportions, $(\hat{p}_1 - \hat{p}_2)$, is an intuitively appealing point estimator of the difference between the population parameters, $(p_1 - p_2)$. For our example, the estimate is

$$(\hat{p}_1 - \hat{p}_2) = .042 - .024 = .018$$

To judge the reliability of the estimator $(\hat{p}_1 - \hat{p}_2)$, we must know something about its behavior in repeated sampling from the two populations. That is, we need to know the sampling distribution of $(\hat{p}_1 - \hat{p}_2)$. Properties of the sampling distribution are given in the box on page 434. Remember that $\hat{p}_1$ and $\hat{p}_2$ can be viewed as means of the number of successes in the respective samples so that the Central Limit Theorem will apply when the sample sizes are large.

Since the distribution of $(\hat{p}_1 - \hat{p}_2)$ in repeated sampling is approximately normal, we can use the z statistic to derive confidence intervals for $(p_1 - p_2)$ or to test a hypothesis about $(p_1 - p_2)$. For the camper example, a 95% confidence interval for the difference $(p_1 - p_2)$ is

$$(\hat{p}_1 - \hat{p}_2) \pm 1.96\sigma_{(\hat{p}_1 - \hat{p}_2)} = (\hat{p}_1 - \hat{p}_2) \pm 1.96\sqrt{\frac{p_1q_1}{n_1} + \frac{p_2q_2}{n_2}}$$

The quantities p_1q_1 and p_2q_2 must be estimated in order to complete the calculation of the standard deviation, $\sigma_{(\hat{p}_1 - \hat{p}_2)}$, and hence of the confidence interval. In Section 7.4, we showed that the value of pq is relatively insensitive to the value chosen

Properties of the Sampling Distribution of $(\hat{p}_1 - \hat{p}_2)$

1. If the sample sizes n_1 and n_2 are large (see Section 7.4 for a guideline), the sampling distribution of $(\hat{p}_1 - \hat{p}_2)$ is approximately normal.
2. The mean of the sampling distribution if $(\hat{p}_1 - \hat{p}_2)$ is $(p_1 - p_2)$; i.e.,

$$E(\hat{p}_1 - \hat{p}_2) = p_1 - p_2$$

Thus, $(\hat{p}_1 - \hat{p}_2)$ is an unbiased estimator of $(p_1 - p_2)$.

3. The standard deviation of the sampling distribution of $(\hat{p}_1 - \hat{p}_2)$ is

$$\sigma_{(\hat{p}_1 - \hat{p}_2)} = \sqrt{\frac{p_1 q_1}{n_1} + \frac{p_2 q_2}{n_2}}$$

to approximate p. Therefore, $\hat{p}_1\hat{q}_1$ and $\hat{p}_2\hat{q}_2$ will provide satisfactory estimates of $p_1 q_1$ and $p_2 q_2$, respectively. Then

$$(\hat{p}_1 - \hat{p}_2) \pm 1.96\sqrt{\frac{p_1 q_1}{n_1} + \frac{p_2 q_2}{n_2}} \approx (\hat{p}_1 - \hat{p}_2) \pm 1.96\sqrt{\frac{\hat{p}_1\hat{q}_1}{n_1} + \frac{\hat{p}_2\hat{q}_2}{n_2}}$$

$$= (.042 - .024) \pm 1.96\sqrt{\frac{(.042)(.958)}{1{,}000} + \frac{(.024)(.976)}{1{,}000}}$$

$$= .018 \pm .016 = (.002, .034)$$

Thus, we are 95% confident that the interval (.002, .034) contains $(p_1 - p_2)$. It appears that there are between .2% and 3.4% more households in the NE than in the SE that plan to purchase campers in the next 5 years.

Large-Sample $100(1 - \alpha)\%$ Confidence Interval for $(p_1 - p_2)$

$$(\hat{p}_1 - \hat{p}_2) \pm z_{\alpha/2}\,\sigma_{(\hat{p}_1 - \hat{p}_2)} = (\hat{p}_1 - \hat{p}_2) \pm z_{\alpha/2}\sqrt{\frac{p_1 q_1}{n_1} + \frac{p_2 q_2}{n_2}}$$

$$\approx (\hat{p}_1 - \hat{p}_2) \pm z_{\alpha/2}\sqrt{\frac{\hat{p}_1\hat{q}_1}{n_1} + \frac{\hat{p}_2\hat{q}_2}{n_2}}$$

Assumptions: The two samples are independent random samples from binomial distributions. Both samples should be large enough that the normal distribution provides an adequate approximation to the sampling distribution of $\hat{p}_1$ and $\hat{p}_2$ (see Section 7.4).

The z statistic,

$$z = \frac{(\hat{p}_1 - \hat{p}_2) - (p_1 - p_2)}{\sigma_{(\hat{p}_1 - \hat{p}_2)}}$$

is used to test the null hypothesis that $(p_1 - p_2)$ equals some specified difference—say, D_0. For the special case where $D_0 = 0$—i.e., where we want to test the null hypothesis that $(p_1 - p_2) = 0$ (or equivalently, that $p_1 = p_2$)—the best estimate of $p_1 = p_2 = p$ is obtained by dividing the total number of successes $(x_1 + x_2)$ for the two samples by the total number of observations $(n_1 + n_2)$; i.e.,

$$\hat{p} = \frac{x_1 + x_2}{n_1 + n_2} \quad \text{or} \quad \hat{p} = \frac{n_1\hat{p}_1 + n_2\hat{p}_2}{n_1 + n_2}$$

The second equation shows that $\hat{p}$ is a weighted average of $\hat{p}_1$ and $\hat{p}_2$, with the larger sample receiving more weight. If the sample sizes are equal, then $\hat{p}$ is a simple average of the two sample proportions of successes.

We now substitute the weighted average $\hat{p}$ for both p_1 and p_2 in the formula for the standard deviation of $(\hat{p}_1 - \hat{p}_2)$:

$$\sigma_{(\hat{p}_1 - \hat{p}_2)} = \sqrt{\frac{p_1 q_1}{n_1} + \frac{p_2 q_2}{n_2}} \approx \sqrt{\frac{\hat{p}\hat{q}}{n_1} + \frac{\hat{p}\hat{q}}{n_2}} = \sqrt{\hat{p}\hat{q}\left(\frac{1}{n_1} + \frac{1}{n_2}\right)}$$

The test is summarized in the box.

Large-Sample Test of Hypothesis About $(p_1 - p_2)$

One-Tailed Test

H_0: $(p_1 - p_2) = 0$*

H_a: $(p_1 - p_2) < 0$

[or H_a: $(p_1 - p_2) > 0$]

Test statistic: $z = \dfrac{(\hat{p}_1 - \hat{p}_2)}{\sigma_{(\hat{p}_1 - \hat{p}_2)}}$

Rejection region: $z < -z_\alpha$

[or $z > z_\alpha$ when H_a: $(p_1 - p_2) > 0$]

Two-Tailed Test

H_0: $(p_1 - p_2) = 0$

H_a: $(p_1 - p_2) \neq 0$

Test statistic: $z = \dfrac{(\hat{p}_1 - \hat{p}_2)}{\sigma_{(\hat{p}_1 - \hat{p}_2)}}$

Rejection region: $z < -z_{\alpha/2}$ or $z > z_{\alpha/2}$

Note: $\sigma_{(\hat{p}_1 - \hat{p}_2)} = \sqrt{\dfrac{p_1 q_1}{n_1} + \dfrac{p_2 q_2}{n_2}} \approx \sqrt{\hat{p}\hat{q}\left(\dfrac{1}{n_1} + \dfrac{1}{n_2}\right)}$

where $\hat{p} = \dfrac{x_1 + x_2}{n_1 + n_2} = \dfrac{n_1\hat{p}_1 + n_2\hat{p}_2}{n_1 + n_2}$

Assumption: Same as for large-sample confidence interval for $(p_1 - p_2)$. (See previous box.)

*The test can be adapted to test for a difference $D_0 \neq 0$. Because most applications call for a comparison of p_1 and p_2, implying $D_0 = 0$, we will confine our attention to this case.

EXAMPLE 9.9

A consumer agency wants to determine whether there is a difference between the proportions of the two leading automobile models that need major repairs (more than \$500) within 2 years of their purchase. A sample of 400 2-year owners of model 1 are contacted, and a sample of 500 2-year owners of model 2 are contacted. The numbers x_1 and x_2 of owners who report that their cars needed major repairs within the first 2 years are 53 and 78, respectively. Test the null hypothesis that no difference exists between the proportions in populations 1 and 2 needing major repairs against the alternative that a difference does exist. Use $\alpha = .10$.

Solution

If we define p_1 and p_2 as the true proportions of model 1 and model 2 owners, respectively, whose cars need major repairs within 2 years, the elements of the test are

$$H_0: \ (p_1 - p_2) = 0 \qquad H_a: \ (p_1 - p_2) \neq 0$$

Test statistic: $z = \dfrac{(\hat{p}_1 - \hat{p}_2)}{\sigma_{(\hat{p}_1 - \hat{p}_2)}}$

Rejection region: $(\alpha = .10)$: $z > z_{\alpha/2} = z_{.05} = 1.645$ or $z < -z_{\alpha/2} = -z_{.05} = -1.645$ (see Figure 9.10)

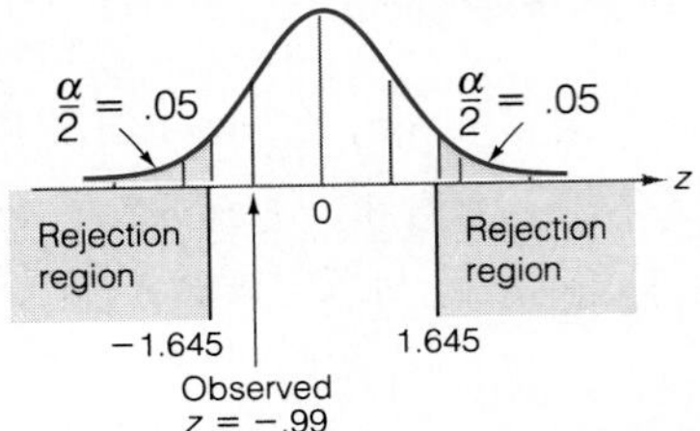

FIGURE 9.10 ► Rejection region for Example 9.9

We now calculate

$$z = \frac{(\hat{p}_1 - \hat{p}_2)}{\sigma_{(\hat{p}_1 - \hat{p}_2)}} = \frac{(\hat{p}_1 - \hat{p}_2)}{\sqrt{\dfrac{p_1 q_1}{n_1} + \dfrac{p_2 q_2}{n_2}}} \approx \frac{(\hat{p}_1 - \hat{p}_2)}{\sqrt{\hat{p}\hat{q}\left(\dfrac{1}{n_1} + \dfrac{1}{n_2}\right)}}$$

where

$$\hat{p}_1 = \frac{x_1}{n_1} = \frac{53}{400} = .1325$$

$$\hat{p}_2 = \frac{x_2}{n_2} = \frac{78}{500} = .1560$$

and

$$\hat{p} = \frac{n_1\hat{p}_1 + n_2\hat{p}_2}{n_1 + n_2} = \frac{400(.1325) + 500(.1560)}{400 + 500} = .1456$$

Thus, $\hat{p}$ is a weighted average of $\hat{p}_1$ and $\hat{p}_2$, with more weight given the larger sample of model 2 owners. We substitute to obtain the (approximate) z test statistic:

$$z \approx \frac{.1325 - .1560}{\sqrt{(.1456)(.8544)\left(\frac{1}{400} + \frac{1}{500}\right)}} = \frac{-.0235}{.0237} = -.99$$

The samples provide insufficient evidence at $\alpha = .10$ to detect a difference between the proportions of the two models that need repairs within 2 years. Even though 2.35% more sampled owners of model 2 found major repairs, this difference is only .99 standard deviation ($z = -.99$) from the hypothesized zero difference between the true proportions.

EXAMPLE 9.10

Find the observed significance level for the test in Example 9.9.

Solution

The observed value of z for this two-tailed test was $z = -.99$. Therefore, the observed significance level is

$$p\text{-value} = P(z < -.99 \text{ or } z > .99)$$

This probability is equal to the shaded area shown in Figure 9.11. The area corresponding to $z = .99$ is given in Table IV of Appendix B as .3389. Therefore, the

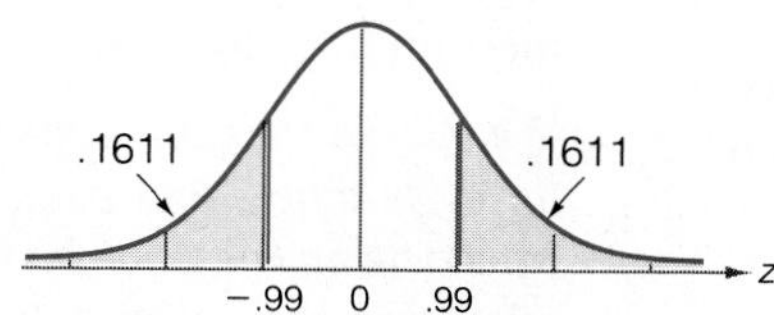

FIGURE 9.11 ► The observed significance level for the test of Example 9.9

observed significance level for the test, the sum of the two shaded tail areas under the standard normal curve, is

$$p\text{-value} = 2(.5 - .3389) = .3222$$

The probability of observing a z as large as .99 or less than $-.99$ if in fact $p_1 = p_2$ is .3222. This large p-value indicates that there is little or no evidence of a difference between p_1 and p_2.

CASE STUDY 9.2 / Hotel Room Interviewing—Anxiety and Suspicion

Writing in the *Sloan Management Review,* Lois Kaufman and John Wolf (1982) describe the typical approach used by sales managers for attracting and interviewing prospective sales representatives:

To maintain product distribution and enhance customer service, corporate sales personnel are typically assigned to a district sales office that is tied to a regional office by telephone or telex. The managers who are responsible for staffing sales territories usually operate with considerable autonomy. When hiring sales representatives, they usually advertise for these positions in local newspapers, and traditionally invite applicants to interviews which are held in hotel rooms.

The purpose of the article was to examine the effects of the hotel interview site on prospective sales representatives, and on women in particular.

As part of their study, Kaufman and Wolf asked a sample of 74* female college students from Rutgers University, Montclair State College, and Union College whether they would agree to a job interview in a room at a local hotel; 62% said they would. Another sample of 74 college women was asked whether they would agree to a job interview in a room of a local office building, and 98% said yes. The authors used a one-tailed hypothesis test to examine the following hypotheses:

$$H_0: \; p_1 - p_2 = 0 \qquad H_a: \; p_1 - p_2 > 0$$

where

p_1 = Proportion of female college students who, if offered a job interview in a room of a local office building, would say they would attend the interview

p_2 = Proportion of female college students who, if offered a job interview in a room of a local hotel, would say they would attend the interview

They obtained an observed significance level of less than .05 for their z statistic and concluded that the proportion of women who would agree to an interview in an office building is significantly greater than the proportion willing to interview in a hotel room.

A similar but less extreme result was obtained when college men were asked the same questions. For the women $\hat{p}_1 - \hat{p}_2 = .36$, whereas for the men $\hat{p}_1 - \hat{p}_2 = .09$. Based on these results and other information supplied by the men and women who participated in their study, Kaufman and Wolf explained the study's findings as follows:

Both men and women find hotel rooms more stressful than offices, but women are even more anxious than men about the prospect of interviewing in hotels. Most respondents said they would take precautions if they interviewed in hotels. Hotel room hiring was described with such words as "shady," "suspicious," "secretive," and "fishy."

Kaufman and Wolf summarized the implications of their findings by saying:

We believe that companies should reexamine their hiring practices and recognize that hotel room interviewing may not be the best approach to recruiting and hiring talented employees. Companies should consider using college campuses, local placement agencies, civic centers, libraries, or government buildings as possible interview sites. Although companies may not intend to discourage applicants when they interview in hotels, our survey showed that many women as well as some men will not attend an interview held in a hotel.

---

*Sample sizes were obtained via personal communication from Professor Kaufman.

Exercises 9.54–9.71

Learning the Mechanics

9.54 What are the characteristics of a binomial experiment?

9.55 The quantities $\hat{p}_1$ and $\hat{p}_2$ have been defined as x_1/n_1 and x_2/n_2, respectively. What assumptions do we make about x_1 and x_2? Describe the sampling distributions of $\hat{p}_1$ and $\hat{p}_2$, assuming n_1 and n_2 are large.

9.56 In each case, determine whether the sample sizes are large enough to conclude that the sampling distribution of $(\hat{p}_1 - \hat{p}_2)$ is approximately normal.

a. $n_1 = 12,\quad n_2 = 14,\quad \hat{p}_1 = .42,\quad \hat{p}_2 = .57$
b. $n_1 = 12,\quad n_2 = 14,\quad \hat{p}_1 = .92,\quad \hat{p}_2 = .86$
c. $n_1 = 30,\quad n_2 = 30,\quad \hat{p}_1 = .70,\quad \hat{p}_2 = .73$
d. $n_1 = 100,\quad n_2 = 250,\quad \hat{p}_1 = .93,\quad \hat{p}_2 = .97$
e. $n_1 = 125,\quad n_2 = 200,\quad \hat{p}_1 = .08,\quad \hat{p}_2 = .12$

9.57 In each case, find the values of z for which H_0: $(p_1 - p_2) = 0$ would be rejected in favor of H_a: $(p_1 - p_2) < 0$.

a. $\alpha = .01$ b. $\alpha = .025$ c. $\alpha = .05$ d. $\alpha = .10$

9.58 Random samples of $n_1 = 200$ and $n_2 = 220$ from two binomial populations produced $x_1 = 47$ and $x_2 = 72$ successes, respectively.

a. Compute the sample proportions, $\hat{p}_1$ and $\hat{p}_2$.
b. Find the values of z for which the null hypothesis H_0: $(p_1 - p_2) = 0$ would be rejected in favor of the alternative hypothesis H_a: $(p_1 - p_2) \neq 0$. Use $\alpha = .10$.
c. Test the hypotheses described in part **b**. Interpret your results.
d. Find and interpret the p-value for the test.
e. What assumptions must be satisfied so the test will be valid?

9.59 Refer to Exercise 9.58. Construct a 98% confidence interval for $(p_1 - p_2)$. Interpret your confidence interval.

9.60 A random sample of size $n_1 = 1{,}000$ from population 1 and a random sample of size $n_2 = 1{,}000$ from population 2 yielded $x_1 = 290$ and $x_2 = 343$ successes, respectively.

a. Given H_0: $(p_1 - p_2) = 0$ and H_a: $(p_1 - p_2) < 0$, find the values of z for which the null hypothesis would be rejected in favor of the alternative hypothesis. Use $\alpha = .025$.
b. Conduct the test described above. Interpret the results.
c. Find the observed significance level of the hypothesis test, and interpret its value.
d. What assumptions must be satisfied so the test will be valid?

9.61 Refer to Exercise 9.60. Construct and interpret an 80% confidence interval for $(p_1 - p_2)$.

9.62 Random samples of size $n_1 = 55$ and $n_2 = 65$ were drawn from populations 1 and 2, respectively. The samples yielded $\hat{p}_1 = .7$ and $\hat{p}_2 = .6$. Test H_0: $(p_1 - p_2) = 0$ against H_a: $(p_1 - p_2) > 0$ using $\alpha = .05$.

Applying the Concepts

9.63 Despite company policies allowing unpaid family leave for new fathers, many men fear that exercising this option would be held against them by their superiors. (*Minneapolis Star Tribune*, February 14, 1993). In a random sample of 100 male workers planning to become fathers, 35 agreed with the statement, "If I knew there would be no repercussions, I would choose to participate in the family leave program after the birth of a son or daughter." However, of 96 men who became fathers in the previous 16 months, only nine participated in the program.

a. Specify the appropriate null and alternative hypotheses to test whether the sample data provide sufficient evidence to reject the hypothesis that the proportion of new fathers participating in the program is the same as the proportion that would like to participate. Define any symbols you use.
b. Are the sample sizes large enough to conclude that the sampling distribution of $(\hat{p}_1 - \hat{p}_2)$ is approximately normal?
c. Conduct the hypothesis test using $\alpha = .05$. Report the observed significance level of the test.
d. What assumptions must be satisfied for the test to be valid?

9.64 Women are filling managerial positions in increasing numbers, although whether they progress fast enough it still debated. Does marriage hinder a woman's career progression more than a man's? A recent article (Stroh, Brett, and Reilly, 1992) investigated this and other questions relating to managerial careers of men and women in today's work force. In a random sample of 795 male managers and 223 female managers from 20 Fortune 500 corporations, 86% of the male managers and 45% of the female managers were married.

a. Use a 95% confidence interval to estimate the difference between the proportions of men and women managers who are married.
b. Interpret the interval. State your conclusion in terms of the population from which the sample was drawn.

9.65 What makes entrepreneurs different from chief executive officers (CEOs) of Fortune 500 companies? The *Wall Street Journal* hired the Gallup Organization to investigate this question. For the study, entrepreneurs were defined as chief executive officers of companies listed by *Inc.* magazine as among the 500 fastest-growing smaller companies in the United States. The Gallup Organization sampled 207 CEOs of Fortune 500 companies and 153 entrepreneurs. They obtained the results shown in the table.

	Fortune 500 CEOs	*Entrepreneurs*
Age: Under 45 years old	19	96
Education: Completed 4 years of college	195	116
Employment record: Have been fired or dismissed from a job	19	47

Source: Graham, E. "The entrepreneurial mystique." *Wall Street Journal*, May 20, 1985.

a. In each of the three areas—age, education, and employment record—are the sample sizes large enough to use the inferential methods of this section to investigate the differences between Fortune 500 CEOs and entrepreneurs? Justify your answer.
b. Do the data indicate that Fortune 500 CEOs and entrepreneurs differ in terms of education? Test using $\alpha = .05$.
c. What assumption(s) must be satisfied in order for your test of part **b** to be valid?

9.66 Refer to Exercise 9.65. Use a 95% confidence interval to estimate the difference in the proportions of Fortune 500 CEOs and entrepreneurs who are under 45 years of age.

9.67 Refer to Exercise 9.65

a. Test to determine whether the data indicate that the fractions of CEOs and entrepreneurs who have been fired or dismissed from a job differ at the $\alpha = .01$ level of significance.
b. Construct a 99% confidence interval for the difference between the fractions of CEOs and entrepreneurs who have been fired or dismissed from a job.

c. Which inferential procedure provides more information about the difference between employment records, the test of hypothesis of part **a** or the confidence interval of part **b**? Explain.

9.68 To aid in the development of new products and to guide their marketing programs, food producers attempt to identify the taste preferences of various segments of the population. A study by Professor Susan Schiffman of the Duke University Center for the Study of Aging and Human Development showed that a person's ability to identify food by smell and taste decreases with increasing age. As a result, she recommends adding simulated odors to the food of older people to improve its flavor. Part of Professor Schiffman's experiment involved asking a random sample of older persons and a random sample of college students to smell, taste, and identify a variety of foods that had been blended to prevent identification by "feel." Subjects were blindfolded during the experiment (Meier, 1980). Suppose that blended apple was correctly identified by 81 of 100 students and by 51 of 100 older people.

a. Would these data support the conclusion that the ability to identify food decreases with age? Test using $\alpha = .05$.

b. Report the p-value of the hypothesis test, and interpret its value.

c. What assumption must be satisfied so the hypothesis test of part **a** will be valid? Why was this assumption necessary?

9.69 Moving companies (home movers, etc.) are required by the government to publish a Carrier Performance Report each year. One of the descriptive statistics they must include is the annual percentage of shipments on which a \$50 or greater claim for loss or damage was filed. Suppose company A and company B each decide to estimate this figure by sampling their records, and they obtain the data shown in the table.

	Company A	*Company B*
Total shipments delivered	9,542	6,631
Number of shipments with a claim ≥ \$50	1,653	501

a. Estimate the true proportion of shipments on which a claim of \$50 or more was made against company A. Use an estimate that reflects its reliability.

b. Repeat part **a** for company B.

c. Use a 95% confidence interval to estimate the true difference in the proportions of shipments that result in claims being made against company A and company B.

9.70 Refer to Exercise 9.69. Test the null hypothesis that no difference exists between the true percentage of shipments resulting in claims made against company A and company B against the alternative hypothesis that a difference does exist. Use $\alpha = .05$. Do your test results indicate that one carrier is superior to the other? If so, which one? Explain how you arrived at your conclusion.

9.71 If α were set at .01 in Exercise 9.70, would you be more or less likely to reject the null hypothesis if in fact it is true? Explain.

9.6 Determining the Sample Size

You can find the appropriate sample size to estimate the difference between two population parameters with a specified degree of reliability by using the method described in Sections 7.2 and 7.5. That is, to estimate the difference between two parameters correct to within B units with probability $(1 - \alpha)$, set $z_{\alpha/2}$ standard deviations of the sampling distribution of the estimator equal to B. Then solve for the sample size. To do this, you have to specify a particular ratio between n_1 and n_2. Most often, you will want to have equal sample sizes—i.e., $n_1 = n_2 = n$. We will illustrate the procedure with two examples.

EXAMPLE 9.11

The sales manager for a chain of supermarkets wants to determine whether store location, management, and other factors produce a difference in the mean meat purchase per customer (zero purchases to be excluded) at two different stores. The manager would like the estimate of the difference in mean meat purchase per customer to be correct to within \$2.00 with probability equal to .95. If the two sample sizes are to be equal, find $n_1 = n_2 = n$, the number of customer meat sales to be randomly selected from each store.

Solution

To solve the problem, you have to know something about the variation in the dollar amount of meat sales per customer. Suppose you know that the sales have a range of approximately \$30 at each store. Then you could approximate $\sigma_1 = \sigma_2 = \sigma$ by letting the range equal 4σ, and

$$4\sigma \approx \$30 \qquad \text{so} \qquad \sigma \approx \$7.50$$

The next step is to solve the equation

$$z_{\alpha/2}\sqrt{\frac{\sigma_1^2}{n_1} + \frac{\sigma_2^2}{n_2}} = B$$

for n, where $n = n_1 = n_2$. Since we want the estimate to lie within $B = \$2.00$ of $(\mu_1 - \mu_2)$ with probability equal to .95, $z_{\alpha/2} = z_{.025} = 1.96$. Then, letting $\sigma_1 = \sigma_2 = 7.5$ and solving for n, we have

$$1.96\sqrt{\frac{(7.5)^2}{n} + \frac{(7.5)^2}{n}} = 2.00$$

$$1.96\sqrt{\frac{2(7.5)^2}{n}} = 2.00$$

$$n = 108.05 \approx 109 \quad \text{(rounding upward)}$$

Consequently, you will have to randomly sample 109 meat sales per store to estimate the difference in mean meat sales per customer correct to within \$2.00 with probability approximately equal to .95.

EXAMPLE 9.12

A production supervisor suspects a difference exists between the proportions of defective items produced by two different machines. Experience has shown that the proportion defective for the two machines is in the neighborhood of .03. If the supervisor wants to estimate the difference in the proportions using a 95% confidence interval of width .01, how many items must be randomly sampled from the production of each machine? (Assume that the supervisor wants $n_1 = n_2 = n$.)

Solution

For the specified level of reliability, $z_{\alpha/2} = z_{.025} = 1.96$. Then, letting $p_1 = p_2 = .03$ and $n_1 = n_2 = n$, we find the required sample size per machine by solving the following equation for n:

$$z_{\alpha/2}\sqrt{\frac{p_1q_1}{n_1} + \frac{p_2q_2}{n_2}} = \frac{W}{2}$$

where W is the desired width of the confidence interval, i.e., $W = .01$. Substituting and solving for n, we obtain

$$1.96\sqrt{\frac{(.03)(.97)}{n} + \frac{(.03)(.97)}{n}} = .005$$

$$1.96\sqrt{\frac{2(.03)(.97)}{n}} = .005$$

$$n = 8{,}943.2$$

You can see that this may be a tedious sampling procedure. If the supervisor insists on estimating $(p_1 - p_2)$ correct to within .005 with probability equal to .95, approximately 9,000 items will have to be inspected for each machine.

You can see from the calculations in Example 9.12 that $\sigma_{(\hat{p}_1 - \hat{p}_2)}$ (and hence the solution, $n_1 = n_2 = n$) depends on the actual (but unknown) values of p_1 and p_2.

In fact, the solution for $n_1 = n_2 = n$ is largest when $p_1 = p_2 = .5$. Therefore, if we have no prior information on the approximate values of p_1 and p_2, we use $p_1 = p_2 = .5$ in the formula for $\sigma_{(\hat{p}_1 - \hat{p}_2)}$. If p_1 and p_2 really are close to .5, then the values of n_1 and n_2 that you have calculated will be appropriate. If p_1 and p_2 differ substantially from .5, then your solutions for n_1 and n_2 will be larger than needed. Consequently, using $p_1 = p_2 = .5$ when solving for n_1 and n_2 is a conservative procedure because the sample sizes n_1 and n_2 will be at least as large as (and probably larger than) needed.

The procedures for determining sample sizes necessary for estimating $(\mu_1 - \mu_2)$ or $(p_1 - p_2)$ for the case $n_1 = n_2$ are given in the box.

Determination of Sample Size for Two-Sample Procedures

1. To estimate $(\mu_1 - \mu_2)$ to within a given bound B with probability $(1 - \alpha)$ or, equivalently, with a $100(1 - \alpha)\%$ confidence interval of width $W = 2B$, use the following formula to solve for equal sample sizes that will achieve the desired reliability:

$$n_1 = n_2 = \frac{(z_{\alpha/2})^2(\sigma_1^2 + \sigma_2^2)}{B^2} = \frac{4(z_{\alpha/2})^2(\sigma_1^2 + \sigma_2^2)}{W^2}$$

Continued

You will need to substitute estimates for the values of σ_1^2 and σ_2^2 before solving for the sample size. These estimates might be sample variances s_1^2 and s_2^2 from prior sampling (e.g., a pilot sample) or from an educated (and conservatively large) guess based on the range, i.e., $s \approx R/4$.

2. To estimate $(p_1 - p_2)$ to within a given bound B with probability $(1 - \alpha)$ or, equivalently, with a $100(1 - \alpha)\%$ confidence interval of width $W = 2B$, use the following formula to solve for equal sample sizes that will achieve the desired reliability:

$$n_1 = n_2 = \frac{(z_{\alpha/2})^2(p_1q_1 + p_2q_2)}{B^2} = \frac{4(z_{\alpha/2})^2(p_1q_1 + p_2q_2)}{W^2}$$

You will need to substitute estimates for the values of p_1 and p_2 before solving for the sample size. These estimates might be based on prior samples, obtained from educated guesses, or (most conservatively) specified as $p_1 = p_2 = .5$.

Exercises 9.72–9.81

Learning the Mechanics

9.72 Suppose you want to estimate the difference between two population means correct to within 1.8 with probability .95. If prior information suggests that the population variances are approximately equal to $\sigma_1^2 = \sigma_2^2 = 14$ and you want to select independent random samples of equal size from the populations, how large should the sample sizes, n_1 and n_2, be?

9.73 A pollster wants to estimate the difference between the proportions of men and women who favor a particular national political candidate using a 90% confidence interval of width .04. Suppose the pollster has no prior information about the proportions. If equal numbers of men and women are to be polled, how large should the samples be?

9.74 You want to estimate the difference between two population proportions correct to within .04 with confidence coefficient equal to .90. You think both p_1 and p_2 are near .4, and you want to select samples of equal size from the two populations. Find the required sample sizes, n_1 and n_2.

9.75 Enough money has been budgeted to collect independent random samples of size $n_1 = n_2 = 100$ from populations 1 and 2 in order to estimate $(\mu_1 - \mu_2)$. Prior information indicates that $\sigma_1 = \sigma_2 = 10$. Have sufficient funds been allocated to construct a 90% confidence interval for $(\mu_1 - \mu_2)$ of width 4 or less? Justify your answer.

Applying the Concepts

9.76 Is housework hazardous to your health? A recent study (Rogot, Sorlie, and Johnson, 1992) compares the life expectancies for 25-year-old white women who are in the labor force to those who are housewives. How large a sample must be taken from each group to be 95% confident that the estimate of difference in life expectancies for the two groups is within 1 year of the true difference in life expectancies? Assume that equal sample sizes will be selected from the two groups, and that the standard deviation for both groups is approximately 15 years.

9.77 Nationally televised home shopping was introduced in 1985. Overnight it became the hottest craze in television programming. By December 1986 there were 34 home shopping cable services (Covert, 1986), and now nearly every cable service carries at least one such channel. Who uses these home shopping services? Are the shoppers primarily men or women? Suppose you want to estimate the difference in the proportions of men and women who say they have used or expect to use televised home shopping using an 80% confidence interval of width .06 or less.

a. Approximately how many people should be included in your samples?

b. Suppose you want to obtain individual estimates for the two proportions of interest. Will the sample size found in part **a** be large enough to provide estimates of each proportion correct to within .02 with probability equal to .90? Justify your response.

9.78 In Exercise 9.41 you should have rejected the hypothesis that $\sigma_1^2 = \sigma_2^2$ and concluded that it is inappropriate to use the two-sample t statistic in making inferences about $(\mu_1 - \mu_2)$. In such cases, inferences about $(\mu_1 - \mu_2)$ can still be made if data are plentiful enough so that large-sample procedures such as those in Section 9.1 can be used. Accordingly, how many additional completion times should be sampled in order that the difference in the mean completion times for operators working with and without the 30-second guideline can be estimated to within 2 seconds with probability .80? Assume equal sample sizes are desired.

9.79 Refer to Exercise 9.68 and the study of whether a person's ability to identify food by smell and taste decreases with increasing age. How large should the samples be to estimate the difference in the proportion of students and the proportion of older people who are able to identify blended apple to within .05 with probability .90?

9.80 Rat damage is a costly nuisance in the production of sugarcane. One aspect of the problem that has been investigated by the U.S. Department of Agriculture concerns the optimal place to locate rat poison. To be most effective, should the poison be in the middle of the field or on the outer perimeter? One way to answer this question is to determine where more damage occurs. If damage is measured by the proportion of cane stalks that have been damaged by rats, how many stalks from each section of the field should be sampled in order to estimate the true difference between the proportions of stalks damaged in the two sections to within .02 with probability .95?

9.81 Even though Japan is an economic superpower, Japanese workers are in many ways worse off than their U.S. and European counterparts. For example, in 1991 the estimated average housing space per person (in square feet) was 665.2 in the United States, 400.4 in Germany, and only 269 in Japan (Berg, 1993). Next year a team of economists and sociologists from the United Nations plans to reestimate the difference in the mean housing space per person for U.S. and Japanese workers. Assume that equal sample sizes will be used for each country and that the standard deviation is 35 square feet for Japan and 80 for the United States. How many people should be sampled in each country to estimate the difference to within 10 square feet with 95% confidence?

Summary

We have presented various techniques for using the information in two samples to make inferences about the difference between population parameters. As you would expect, we can make reliable inferences with fewer assumptions about the sampled populations when the sample sizes are large. When we cannot take large samples from the populations, the **two-sample *t* statistic** permits us to use the limited sample information to make inferences about the **difference between means** when the assumptions of normality and equal population variances are at least approximately true. The **paired difference experiment** offers the possibility of increasing the information about $(\mu_1 - \mu_2)$ by pairing similar observational units to control variability. In designing a paired difference experiment, we expect that the reduction in variability will more than compensate for the loss in degrees of freedom.

Two other inferential procedures for making comparisons between population parameters were presented in this chapter. The ***F* test** was used to compare two population variances, σ_1^2 and σ_2^2. This test is useful in checking the assumption of equal population variances, an assumption that is essential to the independent samples *t* test (and confidence interval) for a comparison of two population means. The *F* test can also be used to compare the variances of two populations when these variances assume practical importance as a measure of risk, error, etc.

This chapter concluded with a comparison of two binomial parameters, p_1 and p_2. Practical examples of such comparisons are numerous; they frequently appear in the analysis of business surveys. A company might want to compare the proportion of consumers who prefer a new product A to a new (or old) product B. Or, the comparison might occur in a production setting when a manufacturer wants to compare the fractions of defectives that emerge from two production lines.

Supplementary Exercises 9.82–9.103

Note: In each problem, state the assumptions necessary for the procedure to be valid.

9.82 The threat of earthquakes is a part of life for homeowners in California. Scientists have been warning about the "big one" for decades. A recent article considered some factors involved when California homeowners purchase earthquake insurance, including the proximity to a major earthquake fault. Surveys were mailed to residents in four California counties. The data collected are shown in the table.

	Contra Costa	*Santa Clara*	*Los Angeles*	*San Bernardino*
Sample Size	521	556	337	372
Number with Earthquake Insurance	117	222	133	109

Source: Palm, R., and Hodgson, M. "Earthquake insurance: Mandated disclosure and homeowner response in California." *Annals of the Association of American Geographers*, June 1992, Vol. 82, No. 2, pp. 207–221.

a. Los Angeles County is the closest of the four to a major earthquake fault. Calculate 95% confidence intervals for the difference in the proportion of earthquake-insured residents in Los Angeles County and each of the other counties.

b. Do these results support the contention that closer proximities to major earthquake faults result in higher proportions of earthquake-insured residents?

9.83 Refer to Exercise 9.82. How large must the samples from Los Angeles and San Bernardino counties be to estimate the difference between earthquake-insured proportions to within .03 with 95% confidence?

9.84 Was the average amount spent by firms in the electronics industry on company-sponsored research and development (R&D) higher in 1989 than in 1988? The table below lists R&D expenditures (in millions of dollars) for a sample of firms in the electronics industry.

Firm	*1988*	*1989*
Adams-Russell	6.3	5.1
Harris	116.9	104.0
Aydin	8.5	6.6
Andrew	14.1	17.0
Compudyn	2.3	1.8
Raytheon	271.0	274.7
Varian Associates	80.2	83.1
General Instr.	37.5	46.9

Source: *Business Week*, Special Issue on Innovation in America, 1989, 1990.

a. A securities analyst who follows the electronics industry believes R&D expenditures have increased. Do the data support the analyst's belief? Test using $\alpha = .10$.

b. What are the Type I and Type II errors associated with your hypothesis test of part **a**?

c. What assumptions must hold in order for your test of part **a** to be valid?

9.85 Refer to Exercise 9.84. Use a 95% confidence interval to estimate the mean difference between 1989 and 1988 R&D expenditures. Interpret the interval.

9.86 When new instruments are developed to perform chemical analyses of products (food, medicine, etc.), they are usually evaluated with respect to two criteria: accuracy and precision. *Accuracy* refers to the ability of the instrument to identify correctly the nature and amounts of a product's components. *Precision* refers to the consistency with which the instrument will identify the components of the same material in repeated analyses. Thus, a large variability in the identification of the components of a single batch of a product indicates a lack of precision. Suppose a pharmaceutical firm is considering two brands of an instrument designed to identify the components of certain drugs. As part of a comparison of precision, 10 test-tube samples of a well-mixed batch of a drug are selected and then five are analyzed by instrument A and five by instrument B. The data shown in the table are the percentages of the primary component of the drug given by the instruments. Do these data provide evidence of a difference in the precision of the two machines? Use $\alpha = .10$.

A	B
43	46
48	49
37	43
52	41
45	48

9.87 A procedure developed by Tele-Research, Inc., evaluates the effectiveness of newly developed television commercials prior to their release. A study using this procedure (Jenssen, 1966) randomly selected 392 shoppers as they entered a large Los Angeles supermarket and asked their preferences for several product brands. One was brand XYZ, whose new television commercial was the object of the study. Ostensibly in exchange for their time, the shoppers were given a packet of ten cents-off coupons for various products sold in the supermarket, including XYZ. The coupons could be used only in that store and only on that day. A second sample of 387 shoppers was given the same interview, but was also asked to watch four television commercials in a trailer parked outside the supermarket. One commercial was the newly developed ad for XYZ. Following the viewing, the shoppers were asked for their reactions to the commercials, then were given the same packet of coupons. Of the 392 shoppers not exposed to the television commercials, 57 redeemed the coupon for XYZ. Of the 387 shoppers who saw XYZ's commercial, 84 redeemed the XYZ coupon.

a. Do the sample data provide sufficient evidence to conclude that the new XYZ commercial motivates shoppers to purchase the XYZ brand? Use $\alpha = .05$.
b. Find and interpret the observed significance level for the test.

9.88 List the assumptions necessary for each of the following inferential techniques:

a. Large-sample inferences about the difference $(\mu_1 - \mu_2)$ between population means using a two-sample z statistic
b. Small-sample inferences about $(\mu_1 - \mu_2)$ using an independent samples design and a two-sample t statistic
c. Small-sample inferences about $(\mu_1 - \mu_2)$ using a paired difference design and a single-sample t statistic to analyze the differences
d. Large-sample inferences about the difference $(p_1 - p_2)$ between binomial proportions using a two-sample z statistic

9.89 Advertising companies often try to characterize the average user of a client's product so ads can be targeted at particular segments of the buying community. A new movie is about to be released, and the advertising company wants to determine whether to aim the ad campaign at people under or over 25 years of age. It plans to arrange an advance showing of the movie to an audience from each group, then obtain an opinion about the movie from each individual. How many individuals should be included in each sample if the advertising company wants to estimate the difference in the proportions of viewers in each age group who will like the movie to within .05 with 90% confidence? Assume the sample size for each group will be the same and about half of each group will like the movie.

9.90 How does gender affect what type of advertising proves most effective? Because numerous studies have shown that males tend to be more competitive with others than with themselves, an advertising researcher created two versions of a soft drink ad (Prokash, 1990):

Ad 1: Four men compete in racquetball.

Ad 2: One man competes against himself in racquetball.

The author hypothesized that men would find the first ad more effective. To test this hypothesis, 43 males were shown both ads and asked to measure their attitude toward the advertisement (*Aad*), their attitude toward the brand of soft drink (*Ab*), and their intention to purchase the soft drink (*Intention*). Each variable was measured on a 7-point scale, with higher scores indicating a more favorable attitude. The results are in the table.

	Aad	*Ab*	*Intention*
Ad 1 sample mean	4.465	3.311	4.366
Ad 2 sample mean	4.150	2.902	3.813
Level of significance	$p = .091$	$p = .032$	$p = .050$

a. What are the appropriate null and alternative hypotheses to test the author's research hypothesis? Define any symbols you use.
b. Do you think this is an independent samples experiment or a paired difference experiment? Explain.
c. Interpret the p-value for each test.
d. What assumptions are necessary for the validity of the tests?

9.91 An important interaction occurs when a consumer dissatisfied with a purchase returns to the retailer to obtain satisfaction. The action taken by the retailer, however, may not conform to the consumer's expectations, and the resulting frustration and ill will benefit neither party. Ronald Dornoff and Clint Tankersley (1975) conducted a study to test the hypothesis that differences exist between retailers' and consumers' perceptions regarding actions taken by retailers in market transactions. A random sample of 300 consumers selected from the Cincinnati Metropolitan Area Telephone Directory was asked via mail questionnaire to react to scenarios like the following:

> A customer calls the retailer to report that her refrigerator purchased 2 weeks ago is not cooling properly and that all the food has spoiled.
>
> Action that should be taken by the retailer: The customer should be reimbursed for the value of the spoiled food.

One hundred usable questionnaires were returned. The same questionnaire was presented in person to 100 managers and assistant managers of a random sample of 40 retail establishments drawn from the yellow pages of the Cincinnati Telephone Directory. For the preceding scenario, 89 consumers agreed with the action prescribed for the retailer, 3 disagreed, and 8 had no opinion. Thirty-seven retailers (managers or assistant managers) agreed with the prescribed action, 54 disagreed, and 9 had no opinion.

a. Use a 95% confidence interval to estimate the difference in the proportions of consumers and retailers who agree with the action prescribed. Draw appropriate conclusions regarding the hypothesis of interest to Dornoff and Tankersley.
b. What assumption(s), if any, must be made in constructing the confidence interval?
c. Discuss the implications of the composition of the sample of retailers for the validity of the conclusions you made in part **a**. How would you improve the sampling procedure?

9.92 In 1986, the American Society for Quality Control commissioned the Gallup Organization to explore the attitudes, beliefs, and experiences of upper-level executives in U.S. businesses with respect to the quality and quality practices related to their companies' products and services. One of the questions was: "Poor quality—as measured by repair, rework and scrap costs, lost sales, and so on—is said to cost American business billions of dollars annually. How much does poor quality cost your company, as a percent of gross sales?" The table describes the responses (rounded to the nearest percent) of 387 service company executives and 311 industrial company executives.

	Service Companies	*Industrial Companies*
Less than 5%	45%	47%
5%–10%	24	23
11%–19%	5	12
20%–29%	3	6
30%–49%	2	2
50% or more	1	0
Don't know	20	10

Source: "Gallup survey: Top executives talk quality." *Quality Progress*, Dec. 1986, pp. 49–54.

a. Use a 90% confidence interval to estimate the proportion of all executives in U.S. service companies who believe poor quality costs their firm 10% or less of gross sales.

b. Use a 90% confidence interval to estimate the difference between the proportion of all executives in service companies and the proportion of all executives in industrial companies who believe poor quality costs their companies 10% or less of gross sales.

c. What assumptions must hold in order for your confidence intervals of parts **a** and **b** to be valid?

9.93 Refer to Exercise 9.92. Do the data provide sufficient evidence to indicate that a difference exists between the proportion of service company executives and the proportion of industrial company executives who do not know how much poor quality costs their companies? Test using $\alpha = .10$.

9.94 Management training programs are often instituted to teach supervisory skills and thereby increase productivity. Suppose a company psychologist administers a set of examinations to each of ten supervisors before such a training program begins and then administers similar examinations at the end of the program. The examinations are designed to measure supervisory skills, with higher scores indicating increased skill. The results of the tests are shown in the table.

Supervisor	*Pre-Test*	*Post-Test*	*Supervisor*	*Pre-Test*	*Post-Test*
1	63	78	6	72	85
2	93	92	7	91	99
3	84	91	8	84	82
4	72	80	9	71	81
5	65	69	10	80	87

a. Do the data provide evidence that the training program is effective in increasing supervisory skills, as measured by the examination scores? Use $\alpha = .10$.

b. Find and interpret the approximate p-value for the test.

9.95 A recent study (Jackson, Jackson, and Newmiller, 1992) investigated how much Americans worry about product tampering. Random samples of male and female consumers were asked to rate their concern about product tampering on a scale of 1 (little or no concern) to 9 (very concerned).

a. What are the appropriate null and alternative hypotheses to determine whether a difference exists in the mean level of concern about product tampering between men and women? Define any symbols you use.
b. Some of the statistics reported are shown here. Interpret these results.

```
MEAN SCORES: MEN = 3.209        WOMEN = 3.923
          Z = -2.69     TWO-TAILED P-VALUE = .0072
```

c. What assumptions are necessary to assure the validity of this test?

9.96 You have been offered similar jobs in two different locales. To help decide which job to accept, you want to compare the cost of living in both cities. One of your primary concerns is the cost of housing, so you obtain a newspaper from each locale and study the housing prices in the classified ads. One convenient way to get a general idea of prices is to compute the prices per square foot. This is done by dividing the price of the house by its heated area (in square feet). Random samples of 63 ads in locale 1 and 78 in locale 2 produce the results shown in the table. Is there evidence that the mean housing price per square foot differs in the two locales? Use $\alpha = .01$.

Locale 1	*Locale 2*
$\bar{x}_1 = \$50.40$ per square foot	$\bar{x}_2 = \$53.70$ per square foot
$s_1 = \$4.50$ per square foot	$s_2 = \$5.30$ per square foot

9.97 Many computer programs are available to conduct two-sample tests of hypotheses to compare the means of two populations, for both independent and paired samples. Most of these report both the test statistic and the observed significance level for the test, but some report only the observed significance level of the test. Suppose you use one of these programs to test the null hypothesis H_0: $(\mu_1 - \mu_2) = 0$ versus the alternative H_a: $(\mu_1 - \mu_2) \neq 0$ with independent samples of size 12 and 10, respectively. Using $\alpha = .05$, what conclusions would you reach for the following observed significance levels reported by the program?
a. P-VALUE = .0429 b. P-VALUE = .1984 c. P-VALUE = .0001
d. P-VALUE = .0344 e. P-VALUE = .0545 f. P-VALUE = .9633
g. Always make sure that the program is performing the calculations correctly, especially with an unfamiliar program. Even programs that perform the calculations correctly usually do not remind you of the assumptions necessary for the validity of the procedure. What assumptions are necessary for this test?

9.98 The relationships among physical fitness, stress, and worker productivity have been the subject of much research during the past decade. One sports psychology researcher focused on the relationship between level of fitness and stress among employees of companies offering health and fitness programs. Random samples of employees were selected from each of three fitness-level categories, and each person was evaluated for signs of stress. The resulting data are shown.

Fitness Level	*Sample Size*	*Proportion with Signs of Stress*
Poor	242	.155
Average	212	.133
Good	95	.108

Source: Tucker, L. A. "Physical fitness and psychological distress." *International Journal of Sports Psychology*, July–Sept. 1990, Vol. 21, pp. 185–201.

a. What are the appropriate null and alternative hypotheses to test whether a greater proportion of employees in the poor fitness category show signs of stress than those in the average fitness category? Define any symbols you use.
b. Conduct the test constructed in part **a** using $\alpha = .10$. Interpret the result.
c. How would your null and alternative hypotheses change if you wanted to compare the proportions showing signs of stress in the poor and good fitness categories?
d. To conduct the test in part **c**, the data were analyzed by a statistical software package with the results shown. Based on these results, would you agree that fitness level has no bearing on whether an individual shows signs of stress? Explain.

```
Z = 1.11          P-VALUE = .1335
```

9.99 Refer to Exercise 9.98. Even though the sample proportions differed by nearly .05 (that is, almost 5% more of those in poor condition exhibited signs of stress than those in good condition), the p-value was not small enough to reject the null hypothesis that the corresponding population proportions are equal.
a. How large would the samples have to be to estimate the difference in the proportions showing signs of stress to within .04 with 95% confidence? Assume equal sample sizes for the two groups, and remember that the first sample selected resulted in proportions of .155 and .108 for the poor and good fitness levels, respectively.
b. Suppose samples of the size you calculated in part **a** were selected from each group, and the sample proportions again turned out to be .155 and .108. Test the null hypothesis H_0: $(p_1 - p_2) = 0$ against the alternative H_a: $(p_1 - p_2) > 0$, where p_1 is the proportion of all employees in the poor fitness category who show signs of stress and p_2 is the proportion of all employees having good fitness who show signs of stress. Calculate and interpret the observed significance level.

9.100 An economist wants to investigate the difference in unemployment rates between an urban industrial community and a university community in the same state. She interviews 525 potential members of the work force in the industrial community and 375 in the university community. Of these, 47 and 22, respectively, are unemployed. Use a 95% confidence interval to estimate the difference in unemployment rates in the two communities.

9.101 Smoke detectors are highly recommended safety devices for early fire detection in homes and businesses. It is vital that the devices be nondefective. Suppose 100 brand A smoke detectors are tested and 12 fail to emit a warning signal. Subjected to the same test, 15 out of 90 brand B detectors fail to operate. Form a 90% confidence interval to estimate the difference in the fractions of defective smoke detectors produced by the two companies. Interpret this confidence interval.

9.102 Does the time of day during which one works affect job satisfaction? A recent study examined differences in job satisfaction between day-shift and night-shift nurses. Satisfaction with the hours of work, free time away from work, and breaks during work was measured. The table shows the mean scores for each measure of job satisfaction (higher scores indicate increased satisfaction), along with the observed significance level comparing the means for the day-shift and night-shift samples:

	Mean Satisfaction		
	Day Shift	*Night Shift*	*p-Value*
Hours of work	3.91	3.56	.813
Free time	2.55	1.72	.047
Breaks	2.53	3.75	.0073

Source: Barton, J., and Folkard, S. "The response of day and night nurses to their work schedules." *Occupational Psychology*, Sept. 1991, Vol. 64, pp. 207–218.

a. Specify the null and alternative hypotheses if we wish to test whether a difference in job satisfaction exists between day-shift and night-shift nurses on each of the three measures. Define any symbols you use.
b. Interpret the p-value for each test (each p-value in the table is two-tailed).
c. Assume that each test is based on a small sample of nurses from each group. What assumptions are necessary in order for the tests to be valid?

9.103 Japanese laborers worked 2,120 hours on average in 1991, while their American counterparts worked only 1,940 hours. The Japanese government, hoping to avoid a trade war and recognizing the lower quality of life for Japanese workers, has set a goal of 1,800 hours per year on average for its workers (Berg, 1993). A large Japanese company wants to test whether the gap in average hours worked in 1993 differed from 1991. A sample of 50 American workers had a sample mean of 1,933 hours worked and a sample standard deviation of 38 hours. A sample of 60 Japanese workers had a mean of 2,097 hours worked and a standard deviation of 43 hours.

a. What are the null and alternative hypotheses of interest?
b. Test the null hypothesis and report the p-value. Is there evidence that the difference in average hours between U.S. and Japanese workers has changed?
c. What assumptions must be made for the methodology you employed in part **b** to be valid?
d. Construct a 95% confidence interval to estimate the true difference in average hours between American and Japanese workers in 1993. Does this confidence interval cover the hypothesized value of D_0 in part **a**?

On Your Own

Many stock market indexes, such as the Dow Jones Industrial Average, act both as indicators of stock market trends and as economic indicators. One way of comparing economic conditions at the end of two consecutive years is to estimate the difference in the mean closing prices of all stocks on the New York Stock Exchange. We outline two methods of sampling to estimate the difference in mean closing prices on the last day of market operations for two consecutive years, 1992 and 1993.

Method I: Two Independent Samples

Step 1 Obtain lists of the closing prices of all stocks on the New York Stock Exchange for the last operating days of 1992 and 1993. (Any library will have these available.)

Step 2 Using a table of random numbers, randomly choose 15 stocks from the 1992 list and record the closing price of each.

Step 3 Again refer to a table of random numbers and choose a second (independent) sample of 15 closing prices from the 1993 list.

Step 4 Using the two samples of closing prices, form a 95% confidence interval for the true difference in mean closing prices for the two years.

Method 2: Paired Samples

Step 1 Same as for method 1.

Step 2 Same as for method 1.

Step 3 Obtain the 1993 closing prices for the *same stocks as those used in 1992.*

Step 4 Using this set of paired observations, form a 95% confidence interval for the true mean difference in closing prices for the two years.

Before actually collecting any data, state which method you think will provide more information (and why). Then, to compare the two methods, first perform the entire experiment outlined in method 1. After you have completed this, obtain the 1993 closing prices for the *same* stocks as the 1992 stocks analyzed and complete step 4 of method 2.

Which method provided a narrower confidence interval and thus more information on this performance of the experiment? Does this agree with your preliminary answer?

Using the Computer

Select two census regions from Appendix C, and consider the sports purchasing index. A marketing firm wants to target one of the two regions for a sports magazine marketing campaign.

a. Treat the sports index measurements for the zip codes in the regions as a random sample of sports index measurements from all the zip codes for the regions. Test the null hypothesis that the populations' mean sports purchasing indexes are equal using $\alpha = .01$, and place a 99% confidence interval on the true difference between the mean purchasing index for the two regions.

b. Repeat part **a** using the following pairs of α and confidence levels for the tests and confidence intervals: (.05, 95%), (.10, 90%), and (.20, 80%). Describe what happens to the tests and confidence intervals as α is increased and the confidence level is decreased. Which do you think is more informative, the tests or the confidence intervals? Explain.

References

Beckenstein, A. R., Gabel, H. L., and Roberts, K. "An executive's guide to antitrust compliance." *Harvard Business Review*, Sept.–Oct. 1983, pp. 94–102.

Benson, P. G., and Ohta, H. "Classifying sensory inspectors with heterogeneous inspection-error probabilities." *Journal of Quality Technology*, Apr. 1986, Vol. 18, No. 2, pp. 79–90.

Berg, S. "Land of rising fun?" *Minneapolis Star Tribune*, Jan. 31, 1993.

Covert, C. "Television viewers snapping up Home Shopping Network." *Minneapolis Star and Tribune*, Dec. 16, 1986.

Dornoff, R. J., and Tankersley, C. B. "Perceptual differences in market transactions: A source of customer frustration," *Journal of Consumer Affairs*, Summer 1975, Vol. 9, pp. 97–103.

Garvin, D. A. "Quality on the line." *Harvard Business Review*, Sept.–Oct. 1983, pp. 65–75.

Gibbons, J. D. *Nonparametric Statistical Inference*, 2d ed. New York: McGraw-Hill, 1985.

Gross, S. "Cable TV—An industry in flux." *Minneapolis Star Tribune*, Jan. 10, 1993.

Hines, G. H. "Sociocultural influences on employee expectancy and participative management." *Academy of Management Journal*, 1974, Vol. 17, No. 2.

Hollander, M., and Wolfe, D. A. *Nonparametric Statistical Methods*. New York: Wiley, 1973.

Jackson, G. B., Jackson, R. W., and Newmiller, C. E., Jr. "Consumer demographics and reaction to product tampering." *Journal of Psychology and Marketing*, Jan. 1992, Vol. 9, No. 1, pp. 45–57.

Jenssen, W. J. "Sales effects of TV, radio, and print advertising." *Journal of Advertising Research*, June 1966, Vol. 6, pp. 2–7.

Kaufman, L., and Wolf, J. "Hotel room interviewing—anxiety and suspicion." *Sloan Management Review*, Spring 1982, Vol. 23, No. 3.

"Many fathers are reluctant to take time for family leave." *Minneapolis Star Tribune*, Feb. 14, 1993.

Marcotty, J. "Quantity of quality circles proves they're no fad." *Minneapolis Tribune*, Nov. 20, 1983a.

Marcotty, J. "Tennant tightens up loose screws." *Minneapolis Tribune*, Nov. 20, 1983b.

McGuire, E. P. *Evaluating New Product Proposals*. New York: National Industrial Conference Board, 1973. Pp. 54–55.

Meier, P. "Taste: It's in the buds—and they don't improve with age." *Minneapolis Tribune*, June 8, 1980.

Neter, J., Wasserman, W., and Whitmore, G. A. *Applied Statistics*, 2d ed. Boston: Allyn & Bacon, 1982. Chapter 13.

Rogot, E., Sorlie, P. D., and Johnson, N. T. "Life expectancy by employment status, income, and education in the national longitudinal mortality study." *Public Health Reports*, July–Aug. 1992, Vol. 107, No. 4, p. 457.

Prokash, V. "Sex roles and advertising preferences." *Advertising Research*, May/June 1990, pp. 43–50.

Serrin, W. "Technology takes toll on operators." *Minneapolis Tribune*, Nov. 27, 1983.

Shuchman, A., and Riesz, P. C. "Correlates of persuasibility: The Crest case." *Journal of Marketing Research*, Feb. 1975, Vol. 12, pp. 7–11.

Sinclair, M. A. "A collection of modelling approaches for visual inspection in industry." *International Journal of Production Research*, 1978, Vol. 16, No. 4, pp. 275–292.

Smith, W. M. "Federal pay procedures and the comparability survey." *Monthly Labor Review*, Aug. 1976, pp. 27–31.

Stroh, L. K., Brett, J. M., and Reilly, A. H. "All the right stuff: A comparison of female and male managers' career progression." *Applied Psychology*, 1992, Vol. 77, No. 3, pp. 251–260.

Wansley, J. W., Roenfeldt, R. L., and Cooley, P. L. "Abnormal returns from merger profiles." *Journal of Financial and Quantitative Analysis*, June 1983, Vol. 18, No. 2, pp. 149–162.

Winer, L. "The effect of product sales quotas on sales force productivity." *Journal of Marketing Research*, May 1973, Vol. 10, p. 180.

Winkler, R. L., and Hays, W. L. *Statistics: Probability, Inference, and Decision*, 2d ed. New York: Holt, Rinehart and Winston, 1975. Chapter 6.

CHAPTER TEN

Simple Linear Regression

Contents

Case Studies

Where We've Been

The answers to many questions that arise in business require knowledge about the mean of a population or about the difference between two population means. Estimating and testing hypotheses about means, or the difference between two means, were the subjects of Chapters 7, 8, and 9.

Where We're Going

Suppose you want to predict the assessed value of a house in a particular community. Using the methods of Chapter 7, you could select a random sample of houses from the community and use the mean of their assessed values to predict the assessed value of the house of interest to you. But using this procedure would ignore the information contained in easily observed variables that are related to assessed house value—namely, the square feet of floor space, number of bathrooms, age of the house, etc. In this chapter we consider the problem of relating the mean value of a single dependent variable y (for example, assessed house value) to a single independent variable x (say, square feet of floor space) using a linear relationship. The more complex problem of relating y to many independent variables is the topic of Chapter 11.

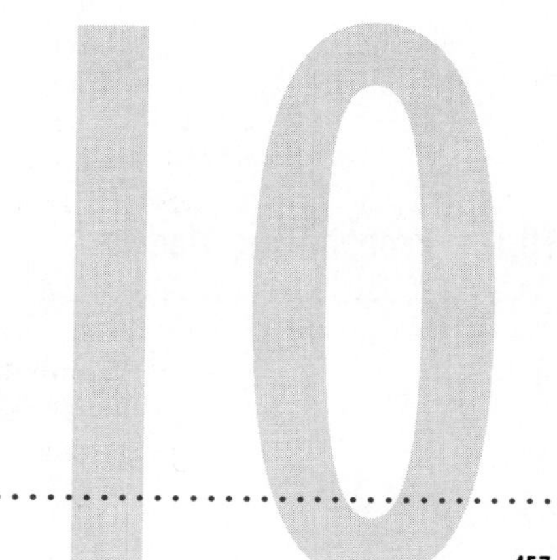

In the previous three chapters we described methods for making inferences about population means. The mean of a population was treated as a **constant**, and we showed how to use sample data to estimate or to test hypotheses about this constant mean. In many business applications, the mean of a population is not viewed as a constant but instead as a variable. For example, the mean sales price of residences in a large city during 1990 can be treated as a constant and might be equal to \$150,000. But we might also treat the mean sales price as a variable that depends on the square feet of living space in the residence. For example, the relationship might be

Mean sales price = \$30,000 + \$60 · (Square feet)

The formula implies that the mean sales price of 1,000-square-foot homes is \$90,000, the mean sales price of 2,000-square-foot homes is \$150,000, and the mean sales price of 3,000-square-foot homes is \$210,000.

What do we gain by treating the mean as a variable rather than a constant? In many practical applications we are dealing with highly variable data, data for which the standard deviation is so large that a constant mean is almost "lost" in a sea of variability. For example, if the mean residential sales price is \$150,000 but the standard deviation is \$75,000, then the actual sales prices will vary considerably, and the mean price is not a very meaningful or useful characterization of the price distribution. On the other hand, if the mean sales price is treated as a variable that depends on the square feet of living space, the standard deviation of sales prices for any given size of home might be only \$10,000. In this case, the mean price provides a much better characterization of sales prices when it is treated as a variable rather than a constant.

In this chapter we discuss situations in which the mean of the population is treated as a variable, dependent on the value of another variable. The preceding example of residential sales price depending on the square feet of living space is one illustration. Other examples are the mean sales revenue of a firm depending on the advertising expenditure, the mean price of a specific stock depending on the Dow Jones Industrial Stock Average, and the mean monthly production of automobiles depending on the total number of sales in the previous month.

In this chapter we discuss the simplest of all models relating a population mean to another variable, using the **straight-line model**. We show how to use sample data to estimate the straight-line relationship between the mean value of one variable, y, as it relates to a second variable, x. The methodology of estimating and using a straight-line relationship we call **simple linear regression analysis**.

10.1 Probabilistic Models

An important consideration in merchandising a product is the amount of money spent on advertising. Suppose you want to model the monthly sales revenue of an appliance store as a function of the monthly advertising expenditure. The first question to be answered is this: Do you think an exact relationship exists between these two variables? That is, can the exact value of sales revenue be predicted if the advertising expenditure is specified? We think you will agree this is not possible for several reasons. Sales depend on many variables other than advertising expenditure—for example, time of year, state of general economy, inventory, and price structure. However, even if many variables are included in the model (the topic of Chapter 11), it is still unlikely that we can predict the monthly sales *exactly*. There will almost certainly be some variation in sales due strictly to **random phenomena** that cannot be anticipated or explained. We will refer to all unexplained variation in sales—caused by important but unincluded variables or by unexplainable random phenomena—as **random error**.

If we construct a model that hypothesizes an exact relationship between variables, it is called a **deterministic model**. For example, if we believe that monthly sales revenue y will be exactly 10 times the monthly advertising expenditure x, we write

$$y = 10x$$

to represent a **deterministic** relationship between the variables y and x.

On the other hand, if we believe that the model should be constructed to allow for random error, then we hypothesize a **probabilistic model**. This includes both a deterministic component and a random error component. For example, if we hypothesize that the sales y is related to advertising x by

$$y = 10x + \text{Random error}$$

we are hypothesizing a **probabilistic** relationship between y and x. Note that the deterministic component of this probabilistic model is $10x$.

General Form of a Probabilistic Model

$$y = \text{Deterministic component} + \text{Random error}$$

where y is the variable of interest.

As you will see, the random error plays an important role in testing hypotheses and finding confidence intervals for the deterministic portion of the model and enables us to estimate the magnitude of the error of prediction when the model is used to predict some value of y to be observed in the future.

We begin with the simplest of probabilistic models—**a first-order linear model**, which graphs as a straight line. The elements of the straight-line model are summarized in the next box.

A First-Order (Straight-Line) Model

$$y = \beta_0 + \beta_1 x + \epsilon$$

where

y = **Dependent** or **response variable** (variable to be modeled)
x = **Independent*** or **predictor variable** (variable used as a predictor of y)
ϵ (epsilon) = Random error component
β_0 (beta zero) = y-intercept of the line—i.e., point at which the line intercepts or cuts through the y-axis (see Figure 10.1)
β_1 (beta one) = Slope of the line—i.e., amount of increase (or decrease) in the deterministic component of y for every 1-unit increase in x (see Figure 10.1)

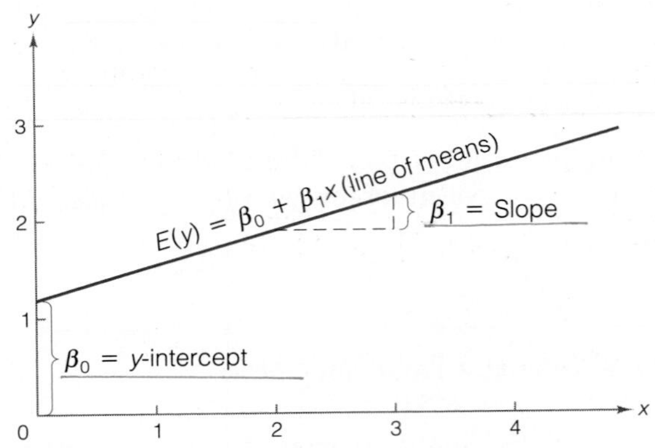

FIGURE 10.1 ► The straight-line model

In the probabilistic model the deterministic component is referred to as the **line of means** because the mean of y, $E(y)$, is equal to the straight-line component of the model. That is,

$$E(y) = \beta_0 + \beta_1 x$$

Note that the Greek symbols β_0 and β_1 represent the y-intercept and slope of the model. They are population parameters that would be known only if we had access to the entire population of (x, y) measurements. Together with a specific value of the

*The word *independent* should not be interpreted in a probabilistic sense. The phrase *independent variable* is used in regression analysis to refer to a predictor variable for the response y.

independent variable x, they determine the mean value of y, which is just a specific point on the line of means (Figure 10.1).

The values of β_0 and β_1 are unknown in almost all practical applications of regression analysis. The process of developing a model, estimating the unknown parameters, and using the model can be viewed as the five-step procedure shown in the box.

Step 1 Hypothesize the deterministic component of the model that relates the mean, $E(y)$, to the independent variable x (Section 10.1).

Step 2 Use the sample data to estimate unknown parameters in the model (Section 10.2).

Step 3 Specify the probability distribution of the random error term, and estimate the standard deviation of this distribution (Sections 10.3 and 10.4).

Step 4 Statistically evaluate the usefulness of the model (Sections 10.5, 10.6, and 10.7).

Step 5 When satisfied that the model is useful, use it for prediction, estimation, and other purposes (Section 10.8).

In this chapter only the straight-line model is discussed; more complex models are addressed in Chapters 11 and 12.

Exercises 10.1 – 10.8

Learning the Mechanics

10.1 In each case, graph the line that passes through the given points.
a. (2, 0) and (5, 5) **b.** (0, 0) and (4, 3) **c.** (0, −2) and (7, 6) **d.** (−1, −3) and (4, 4)

10.2 The equation for a straight line (deterministic) is

$$y = \beta_0 + \beta_1 x$$

If the line passes through the point (−2, 1), then $x = -2$, $y = 1$ must satisfy the equation; i.e.,

$$1 = \beta_0 + \beta_1(-2)$$

Similarly, if the line passes through the point (6, 6), then $x = 6$, $y = 6$ must satisfy the equation; i.e.,

$$6 = \beta_0 + \beta_1(6)$$

Use these two equations to solve for β_0 and β_1, and find the equation of the line that passes through the points (−2, 1) and (6, 6).

10.3 Refer to Exercise 10.2. Find the equations of the lines that pass through the points listed in Exercise 10.1.

10.4 Plot the following lines:
a. $y = 4 + x$ b. $y = -4 + x$ c. $y = 4 + 2x$
d. $y = -2x$ e. $y = x$ f. $y = .50 + .75x$

10.5 Give the slope and y-intercept for each of the lines defined in Exercise 10.4.

10.6 Why do we generally prefer a probabilistic model to a deterministic model? Give examples for which the two types of models might be appropriate.

10.7 What is the line of means?

10.8 If a straight-line probabilistic relationship relates the mean $E(y)$ to an independent variable x, does it imply that every value of the variable y will always fall exactly on the line of means? Why or why not?

10.2 Fitting the Model: The Method of Least Squares

After the straight-line model has been hypothesized as being appropriate for relating the mean $E(y)$ to the independent variable x, the next step is to collect data and to estimate the (unknown) population parameters—the y-intercept β_0 and the slope β_1.

To begin with a simple example, suppose an appliance store conducts a 5-month experiment to determine the effect of advertising on sales revenue. The results are shown in Table 10.1. (The number of measurements is small, and the measurements themselves are unrealistically simple to avoid arithmetic confusion in this initial example.) The relationship between sales revenue y and advertising expenditure x is hypothesized to follow a first-order linear model, that is,

$$y = \beta_0 + \beta_1 x + \epsilon$$

The question is this: How can we best use the information in the sample of five observations in Table 10.1 to estimate the unknown y-intercept β_0 and slope β_1?

TABLE 10.1 Advertising–Sales Data

Month	Advertising Expenditure x (\$100s)	Sales Revenue y (\$1,000s)
1	1	1
2	2	1
3	3	2
4	4	2
5	5	4

To gain information on the approximate values of these parameters, it is helpful to construct a **scattergram** for the sample data. Such a plot, also called a **scatter plot** or **scatter diagram**, locates each of the five data points on a graph, as in Figure 10.2. Note that the scattergram suggests a general tendency for y to increase as x increases. If you place a ruler on the scattergram, you will see that a line may be drawn through three of the five points, as shown in Figure 10.3. To obtain the equation of this visually fitted line, note that the line intersects the y-axis at $y = -1$, so the y-intercept is -1. Also, y increases exactly 1 unit for every 1-unit increase in x, indicating that the slope is $+1$. Therefore, the equation is

$$\tilde{y} = -1 + 1(x) = -1 + x$$

where $\tilde{y}$ is used to denote the value of y predicted from the visually fitted model.

FIGURE 10.2 ▶
Scattergram for data in Table 10.1

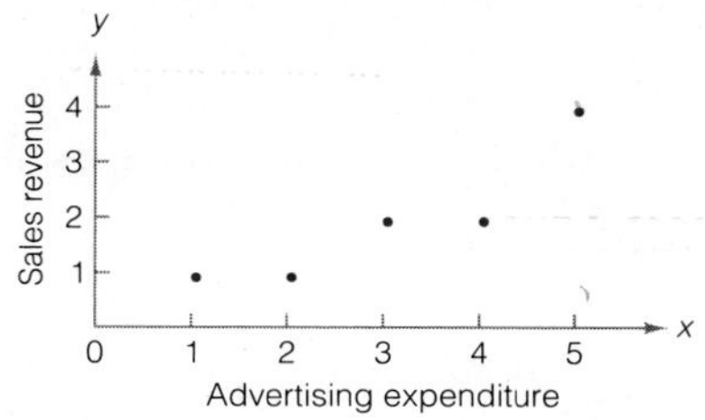

FIGURE 10.3 ▶
Visual straight-line fit to the data

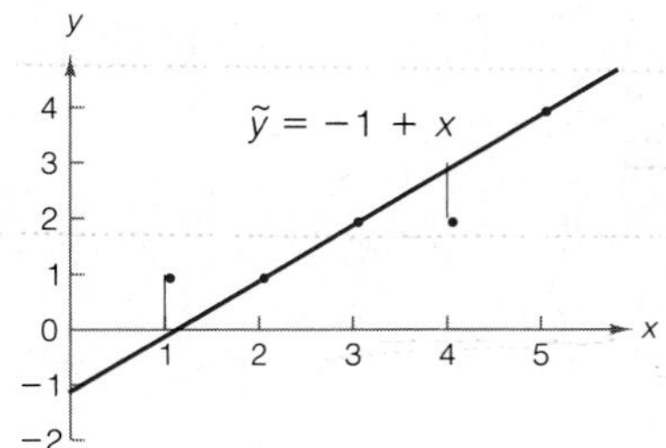

One way to decide quantitatively how well a straight line fits a set of data is to note the extent to which the data points deviate from the line. For example, to evaluate the visual model in Figure 10.3 we calculate the **deviations**—i.e., the differences between the observed and the predicted values of y. These deviations, or **errors**, are the vertical distances between observed and predicted values (see Figure 10.3). The observed and predicted values of y, their differences, and their squared differences are shown in Table 10.2 on page 464. Note that the **sum of errors** equals 0 and the **sum of squares of the errors (SSE)** is equal to 2.

You can see by shifting the ruler around the graph that it is possible to find many lines for which the sum of the errors is equal to 0, but it can be shown that there is one (and only one) line for which the *SSE is a minimum*. This line is called by various names, including the **least squares line**, the **regression line**, the **least squares prediction equation**, or the **fitted line**. We will typically refer to it as the least squares line.

TABLE 10.2 Comparing Observed and Predicted Values for the Visual Model

x	y	$\tilde{y} = -1 + x$	$(y - \tilde{y})$	$(y - \tilde{y})^2$
1	1	0	$(1 - 0) = 1$	1
2	1	1	$(1 - 1) = 0$	0
3	2	2	$(2 - 2) = 0$	0
4	2	3	$(2 - 3) = -1$	1
5	4	4	$(4 - 4) = 0$	0
			Sum of errors = 0	Sum of squared errors (SSE) = 2

To find the least squares line for a set of data, assume that we have a sample of n data points consisting of values of x and y, say $(x_1, y_1), (x_2, y_2), \ldots, (x_n, y_n)$. For example, the $n = 5$ data points shown in Table 10.2 are (1, 1), (2, 1), (3, 2), (4, 2), and (5, 4). The fitted line, which we calculate using the five data points, is represented as

$$\hat{y} = \hat{\beta}_0 + \hat{\beta}_1 x$$

The "hats" can be read as "estimator of." Thus, $\hat{y}$ is an estimator of the mean value of y, $E(y)$, and $\hat{\beta}_0$ and $\hat{\beta}_1$ are estimators of β_0 and β_1, respectively.

For a given data point, say the point (x_i, y_i), the observed value of y is y_i and the predicted value of y would be obtained by substituting x_i into the prediction equation:

$$\hat{y}_i = \hat{\beta}_0 + \hat{\beta}_1 x_i$$

And the deviation of the ith value of y from its predicted value is

$$y_i - \hat{y}_i = y_i - (\hat{\beta}_0 + \hat{\beta}_1 x_i)$$

Then the sum of squares of the deviations of the y values about their predicted values for all the n data points is

$$\text{SSE} = \sum_{i=1}^{n} [y_i - (\hat{\beta}_0 + \hat{\beta}_1 x_i)]^2$$

The quantities $\hat{\beta}_0$ and $\hat{\beta}_1$ that make the SSE a minimum are called the **least squares estimates** of the population parameters β_0 and β_1, and the prediction equation $\hat{y} = \hat{\beta}_0 + \hat{\beta}_1 x$, as noted above, is called the *least squares line.*

Definition 10.1

The **least squares line** is one that has a smaller sum of squared errors than any other straight-line model.

The values of $\hat{\beta}_0$ and $\hat{\beta}_1$ that minimize the SSE (proof omitted) are given by the formulas in the accompanying box.*

Formulas for the Least Squares Estimates

Slope: $\hat{\beta}_1 = \frac{SS_{xy}}{SS_{xx}}$ y-intercept: $\hat{\beta}_0 = \bar{y} - \hat{\beta}_1\bar{x}$

where

$$SS_{xy} = \sum_{i=1}^{n} x_i y_i - \frac{\left(\sum_{i=1}^{n} x_i\right)\left(\sum_{i=1}^{n} y_i\right)}{n}$$

$$SS_{xx} = \sum_{i=1}^{n} x_i^2 - \frac{\left(\sum_{i=1}^{n} x_i\right)^2}{n}$$

n = Sample size

Preliminary computations for finding the least squares line for the advertising–sales example are contained in Table 10.3.

TABLE 10.3 Preliminary Computations for the Advertising–Sales Example

	x_i	y_i	x_i^2	$x_i y_i$
	1	1	1	1
	2	1	4	2
	3	2	9	6
	4	2	16	8
	5	4	25	20
Totals	$\Sigma x_i = 15$	$\Sigma y_i = 10$	$\Sigma x_i^2 = 55$	$\Sigma x_i y_i = 37$

*Students who are familiar with calculus should note that the values of $\hat{\beta}_0$ and $\hat{\beta}_1$ that minimize SSE = $\sum_{i=1}^{n} (y_i - \hat{y}_i)^2$ are obtained by setting the two partial derivatives $\partial SSE/\partial\beta_0$ and $\partial SSE/\partial\beta_1$ equal to 0. The *sample* solutions to these equations are denoted by $\hat{\beta}_0$ and $\hat{\beta}_1$, where the ^ (hat) denotes that these are sample estimates of the true population intercept β_0 and slope β_1. The solutions to these two equations yield the formulas shown in the box.

We can now calculate*

$$SS_{xy} = \sum x_i y_i - \frac{\left(\sum x_i\right)\left(\sum y_i\right)}{5} = 37 - \frac{(15)(10)}{5} = 37 - 30 = 7$$

$$SS_{xx} = \sum x_i^2 - \frac{\left(\sum x_i\right)^2}{5} = 55 - \frac{(15)^2}{5} = 55 - 45 = 10$$

Then, the slope of the least squares line is

$$\hat{\beta}_1 = \frac{SS_{xy}}{SS_{xx}} = \frac{7}{10} = .7$$

and the y-intercept is

$$\hat{\beta}_0 = \bar{y} - \hat{\beta}_1\bar{x} = \frac{\sum y_i}{5} - \hat{\beta}_1\frac{\left(\sum x_i\right)}{5}$$

$$= \frac{10}{5} - (.7)\frac{15}{5} = 2 - (.7)(3) = 2 - 2.1 = -.1$$

The least squares line is thus

$$\hat{y} = \hat{\beta}_0 + \hat{\beta}_1 x = -.1 + .7x$$

The graph of this line is shown in Figure 10.4.

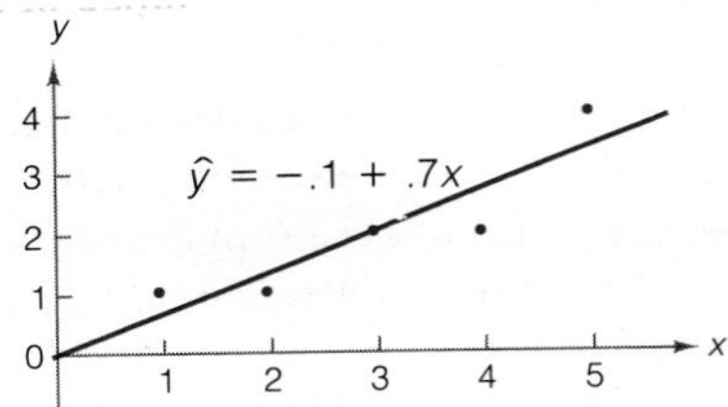

FIGURE 10.4 ▶ The line $\hat{y} = -.1 + .7x$ fit to the data

The observed and predicted values of y, the deviations of the y values about their predicted values, and the squares of these deviations are shown in Table 10.4. Note that the sum of squares of the deviations (SSE) is 1.10, and (as we would expect) this is less than the SSE = 2.0 obtained in Table 10.2 for the visually fitted line.

It is important that you be able to interpret the intercept and slope in terms of the data being utilized to fit the model. In the sales–advertising example, the estimated y-intercept, $\hat{\beta}_0$, is $-.1$. This value would seem to imply that the estimated mean sales are equal to $-.1$, or $-\$100$, when the advertising expenditure, x, is equal to \$0. Since negative sales are not possible, this seems to make the model nonsensical. However, *the model parameters should be interpreted only within the sampled range of the*

*Since summations are used extensively from this point on, we omit the limits on Σ when the summation includes all the measurements in the sample; i.e., when the symbol is $\sum_{i=1}^{n}$, we write Σ.

TABLE 10.4 Comparing Observed and Predicted Values for the Least Squares Model

x	y	$\hat{y} = -.1 + .7x$	$(y - \hat{y})$	$(y - \hat{y})^2$
1	1	.6	$(1 - .6) = .4$	.16
2	1	1.3	$(1 - 1.3) = -.3$	.09
3	2	2.0	$(2 - 2.0) = 0$	.00
4	2	2.7	$(2 - 2.7) = -.7$	.49
5	4	3.4	$(4 - 3.4) = .6$	.36
			Sum of errors = 0	SSE = 1.10

independent variable—in this case, for advertising expenditures between 1 ($100) and 5 ($500). Thus, the *y*-intercept—which is defined at $x = 0$ ($0 advertising expenditure)—is not within the range of the sampled values of x and is not subject to meaningful interpretation.

The slope, $\hat{\beta}_1$, of the least squares line was calculated to be .7. The implication is that for every unit increase of x, the mean value of y is estimated to increase by .7 unit. In terms of this example, for every $100 increase in advertising, the mean sales are estimated to increase by $700 *over the sampled range of advertising from* $100 *to* $500. Thus, the model does not imply that increasing the advertising expenditure from $500 to $1,000 will result in an increase in mean sales of $3,500 because the range of x in the sample does not extend to $1,000 ($x = 10$). Be careful to interpret the estimated parameters only within the sampled range of x.

Even when the interpretations of the estimated parameters are meaningful, it should be remembered that they are only estimates based on the sample. As such, their values will typically change in repeated sampling. How much confidence do we have that the estimated slope $\hat{\beta}_1$ accurately estimates the true slope β_1? This requires statistical inference, in the form of confidence intervals and tests of hypotheses, which we address in Section 10.5.

To summarize, we define the best-fitting straight line to be the one that minimizes the sum of squared errors around the line, and we call it the least squares line. We should interpret the least squares line only within the sampled range of the independent variable. In subsequent sections we show how to make statistical inferences about the model.

Exercises 10.9–10.18

Learning the Mechanics

10.9 The table on page 468 is similar to Tables 10.3 and 10.4. Columns 3 and 4 are for the preliminary computations to find the least squares line for the given pairs of x and y values. After the least squares line has been obtained, columns 5, 6, and 7 are used to compare the observed and predicted values of y and to calculate the SSE.

x_i	y_i	x_i^2	x_iy_i	$\hat{y} =$	$(y - \hat{y})$	$(y - \hat{y})^2$
7	2					
4	4					
6	2					
2	5					
1	7					
1	6					
3	5					
Totals $\Sigma x_i =$	$\Sigma y_i =$	$\Sigma x_i^2 =$	$\Sigma x_iy_i =$		$\Sigma(y - \hat{y}) =$	$SSE = \Sigma(y - \hat{y})^2 =$

a. Complete columns 3 and 4 of the table and calculate the totals for columns 1–4.
b. Find SS_{xy}.
c. Find SS_{xx}.
d. Find $\hat{\beta}_1$.
e. Find $\bar{x}$ and $\bar{y}$.
f. Find $\hat{\beta}_0$.
g. Find the least squares line and write it at the top of column 5.
h. Complete columns 5, 6 and 7 of the table.

10.10 Refer to Exercise 10.9.
a. Plot the least squares line on a scattergram of the data.
b. Plot this line on the same graph: $\hat{y} = 14 - 2.5x$.
c. Show that SSE is larger for the line of part **b** than it is for the least squares line.

10.11 Construct a scattergram for the data in the table.

x	.5	1	1.5
y	2	1	3

a. Plot these two lines on your scattergram:

$$y = 3 - x \quad \text{and} \quad y = 1 + x$$

b. Which of these lines would you choose to characterize the relationship between x and y? Explain.
c. Show that the sum of errors for both of these lines equals 0.
d. Which of these lines has the smaller SSE?
e. Find the least squares line for the data, and compare it to the two lines described in part **a**.

10.12 Consider the following pairs of measurements:

x	8	5	4	6	2	5	3
y	1	3	6	3	7	2	5

a. Construct a scattergram for the data.
b. What does the scattergram suggest about the relationship between x and y?
c. Given that $SS_{xx} = 23.4286$, $SS_{xy} = -23.2857$, $\bar{y} = 3.8571$, and $\bar{x} = 4.7143$, calculate the least squares estimates of β_0 and β_1.

d. Plot the least squares line on your scattergram. Does the line appear to fit the data well? Explain.
e. Interpret the y-intercept and slope of the least squares line. Over what range of x are these interpretations meaningful?

10.13 Suppose that $n = 100$ recent residential home sales in a city are used to fit a least squares straight-line model relating the sales price, y, to the square feet of living space, x. Homes in the sample range from 1,500 square feet to 4,000 square feet of living space, and the resulting least squares equation is

$$\hat{y} = -30{,}000 + 70x$$

a. What is the underlying hypothesized probabilistic model for this application? What does it imply about the relationship between the mean sales price and living space?
b. Identify the least squares estimates of the y-intercept and slope of the model.
c. Interpret the least squares estimate of the y-intercept. Is it meaningful for this application? Explain.
d. Interpret the least squares estimate of the slope of the model. Over what range of x is the interpretation meaningful?
e. Use the least squares model to estimate the mean sales price of a 3,000-square-foot home. Is the estimate meaningful? Explain.
f. Use the least squares model to estimate the mean sales price of a 5,000-square-foot home. Is the estimate meaningful? Explain. [*Note:* We show how to measure the statistical reliability of these least squares estimates in subsequent sections.]

Applying the Concepts

10.14 Individuals who report perceived wrongdoing of a corporation or public agency are known as *whistle blowers*. Two researchers developed an index to measure the extent of retaliation against a whistle blower. The index was based on the number of forms of reprisal actually experienced, the number of forms of reprisal threatened, and the number of people within the organization (e.g., coworkers or immediate supervisor) who retaliated against them. The table lists the retaliation index (higher numbers indicate more extensive retaliation) and salary for a sample of 15 whistle blowers from federal agencies.

Retaliation Index	*Salary*	*Retaliation Index*	*Salary*
301	\$62,000	535	\$15,800
550	36,500	455	44,000
755	17,600	615	46,600
327	20,000	700	12,100
500	30,100	650	62,000
377	35,000	630	21,000
290	47,500	360	11,900
452	54,000		

Source: Data adapted from Near, J. P., and Miceli, M. P. "Retaliation against whistle blowers: Predictors and effects." *Journal of Applied Psychology*, 1986, Vol. 71, No. 1, pp. 137–145.

a. Construct a scattergram for the data. Does it appear that the extent of retaliation increases, decreases, or stays the same with an increase in salary? Explain.
b. Use the method of least squares to fit a straight line to the data.

c. Graph the least squares line on your scattergram. Does the least squares line support your answer to the question in part **a**? Explain.
d. Interpret the y-intercept, $\hat{\beta}_0$, of the least squares line in terms of this application. Is the interpretation meaningful?
e. Interpret the slope, $\hat{\beta}_1$, of the least squares line in terms of this application. Over what range of x is this interpretation meaningful?

10.15 Due primarily to the price controls of the Organization of Petroleum Exporting Countries (OPEC), a cartel of crude oil suppliers, the price of crude oil rose dramatically from the mid-1970s to the mid-1980s. As a result, motorists saw an upward spiral in gasoline prices. The data in the table are typical prices for a gallon of regular leaded gasoline and a barrel of crude oil (refiner acquisition cost) for the indicated years, and $\Sigma y = 1{,}521$, $\Sigma x = 330.28$, $\Sigma y^2 = 153{,}735$, $\Sigma x^2 = 7{,}824.1822$, and $\Sigma xy = 34{,}259.58$.

Year	Gasoline y (¢/gal.)	Crude Oil x (\$/bbl.)	Year	Gasoline y (ct/gal.)	Crude Oil x (\$/bbl.)
1975	57	10.38	1983	116	28.99
1976	59	10.89	1984	113	28.63
1977	62	11.96	1985	112	26.75
1978	63	12.46	1986	86	14.55
1979	86	17.72	1987	90	17.90
1980	119	28.07	1988	90	14.67
1981	131	35.24	1989	100	17.97
1982	122	31.87	1990	115	22.23

Source: U.S. Bureau of the Census. *Statistical Abstract of the United States: 1982–1992.*

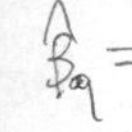

a. Use the data to calculate the least squares line that describes the relationship between the price of a gallon of gasoline and the price of a barrel of crude oil.
b. Plot your least squares line on a scattergram of the data. Does your least squares line appear to be an appropriate characterization of the relationship between y and x? Explain.
c. If the price of crude oil fell to \$15 per barrel, to what level (approximately) would the price of regular gasoline fall? Justify your response.

10.16 A car dealer is interested in modeling the relationship between the number of cars sold by the firm each week and the average number of salespeople who work on the showroom floor per day during the week. The dealer believes the relationship between the two variables can best be described by a straight line. The sample data shown in the table were supplied by the car dealer.

Week of	*Cars Sold, y*	*Average Salespeople on Duty, x*
January 30	20	6
June 29	18	6
March 2	10	4
October 26	6	2
February 7	11	3

a. Construct a scattergram for the data.
b. Assuming the relationship between the variables is best described by a straight line, use the method of least squares to estimate the y-intercept and the slope of the line.

c. Plot the least squares line on your scattergram.
d. Interpret the least squares estimates of the y-intercept and slope of the line of means. Are the interpretations meaningful?
e. According to your least squares line, approximately how many cars should the dealer expect to sell in a week if an average of five salespeople is on the showroom floor each day? [*Note:* A measure of the reliability of these predictions is discussed in Section 10.8.]

10.17 Baseball wisdom says if you can't hit, you can't win. But is the number of games won by a major league baseball team in a season related to the team's batting average? The accompanying table shows the number of games won and the batting averages for the 14 teams in the American League for the 1991 season.

Team	*Games Won, y*	*Team Batting Average, x*	*Team*	*Games Won, y*	*Team Batting Average, x*
Cleveland	57	.254	Baltimore	67	.254
New York	71	.256	California	81	.255
Boston	84	.269	Milwaukee	83	.271
Toronto	91	.257	Seattle	83	.255
Texas	85	.270	Kansas City	82	.264
Detroit	84	.247	Oakland	84	.248
Minnesota	95	.280	Chicago	87	.262

Source: *Official Major League Baseball 1992 Stat Book*. Major League Baseball Properties, Inc., and the editors of *The Baseball Encyclopedia*, New York.

a. If you were to model the relationship between the mean (or expected) number of games won by a major league team and the team's batting average x, using a straight line, would you expect the slope of the line to be positive or negative? Explain.
b. Construct a scattergram for the data. Does the pattern revealed by the scattergram agree with your answer to part **a**?
c. Given that $\Sigma y = 1{,}134$, $\Sigma x = 3.642$, $\Sigma y^2 = 93{,}110$, $\Sigma x^2 = .948622$, and $\Sigma xy = 295.54$, fit a simple linear regression model to the data.
d. Graph the least squares line on your scattergram. Does your least squares line seem to fit the points on your scattergram?
e. Why might the mean (or expected) number of games won not appear to be strongly related to a team's batting average?

10.18 The 1990 winner of Britain's Best Factory Award, NCR Ltd., manufactures automated teller machines. NCR management attributes its success to Total Quality Management (TQM) strategies, including reducing inventory levels through just-in-time (JIT) inventory ordering policies, instructing operators to stop the assembly line when quality problems are detected, and using only high-quality suppliers. This last strategy helped NCR cut the number of suppliers from 430 to 180 between 1980 and 1989. It saw a corresponding increase in the percentage of products that passed final inspection, from 40% to 98%.

	1980	*1982*	*1985*	*1987*	*1989*
Number of Suppliers, x	430	395	360	270	180
Products Passing Inspection, y (%)	40	60	80	88	98

Source: Lee-Mortimer, A. "Best of the best." *Total Quality Management*, Dec. 1990, p. 317.

a. Given that $SS_{xx} = 41,180.0$, $SS_{xy} = -8,582.0$, $\bar{y} = 73.2$, and $\bar{x} = 327.0$, find the least squares line relating y to x.
b. Plot the data and graph the least squares line as a check on your calculations.
c. Interpret the least squares estimates $\hat{\beta}_0$ and $\hat{\beta}_1$ in the context of this problem.

10.3 Model Assumptions

In Section 10.2, we assumed that the probabilistic model relating the firm's sales revenue y to advertising dollars x is

$$y = \beta_0 + \beta_1 x + \epsilon$$

We also recall that the least squares estimate of the deterministic component of the model $\beta_0 + \beta_1 x$ is

$$\hat{y} = \hat{\beta}_0 + \hat{\beta}_1 x = -.1 + .7x$$

Now we turn our attention to the random component ϵ of the probabilistic model and its relation to the errors in estimating β_0 and β_1. We use a probability distribution to characterize the behavior of ϵ. We will see how the probability distribution of ϵ determines how well the regression model describes the relationship between the dependent variable y and the independent variable x.

We make four basic assumptions about the general form of the probability distribution of ϵ:

Assumption 1 The mean of the probability distribution of ϵ is 0. That is, the average of the values of ϵ over an infinitely long series of experiments is 0 for each setting of the independent variable x. This assumption implies that the mean value of y, $E(y)$, for a given value of x is $E(y) = \beta_0 + \beta_1 x$.

Assumption 2 The variance of the probability distribution of ϵ is constant for all values of the independent variable, x. For our straight-line model, this assumption means that the variance of ϵ is equal to a constant, say σ^2, for all values of x.

Assumption 3 The probability distribution of ϵ is normal.

Assumption 4 The values of ϵ associated with any two observed values of y are independent. That is, the value of ϵ associated with one value of y has no effect on the values of ϵ associated with other y values.

The implications of the first three assumptions can be seen in Figure 10.5, which shows distributions of errors for three particular values of x—namely, x_1, x_2, and x_3. Note that the relative frequency distributions of the errors are normal, with a mean of 0 and a constant variance σ^2 (all the distributions shown have the same amount of spread or variability). The straight line shown in Figure 10.5 is the line of means. It indicates the mean value $E(y)$ for a given value of x and is given by the equation

$$E(y) = \beta_0 + \beta_1 x$$

FIGURE 10.5 ▶
The probability distribution of ϵ

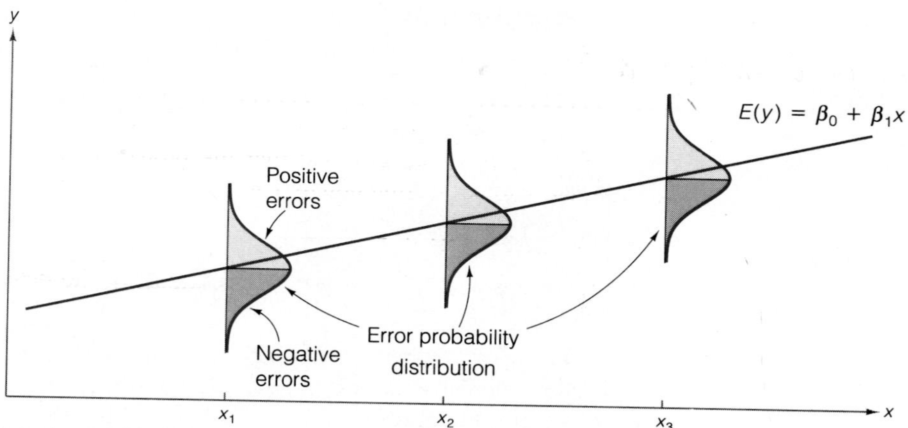

These assumptions make it possible for us to develop measures of reliability for the least squares estimators and to develop hypothesis tests for examining the usefulness of the least squares line. Various techniques exist for checking the validity of these assumptions, and there are remedies to be applied when they appear to be invalid. We discuss some of these techniques and remedies in Chapters 11 and 12.

In actual practice, the assumptions need not hold exactly in order for least squares estimators to be useful. The assumptions will be satisfied adequately for many applications encountered in business.

10.4 An Estimator of σ^2

It seems reasonable to assume that the greater the variability of the random error ϵ (which is measured by its variance σ^2), the greater will be the errors in the estimation of the model parameters β_0 and β_1 and in the error of prediction when $\hat{y}$ is used to predict y for some value of x. Consequently, you should not be surprised, as we proceed through this chapter, to find that σ^2 appears in the formulas for all confidence intervals and test statistics that we use.

In most practical situations, σ^2 will be unknown, and we must use our data to estimate its value. The best estimate of σ^2, denoted by s^2, is obtained by dividing the sum of squares of deviations,

$$\text{SSE} = \sum (y_i - \hat{y}_i)^2$$

by the number of degrees of freedom (df) associated with this quantity. We use 2 df to estimate the y-intercept and slope in the straight-line model, leaving $(n - 2)$ df for the error variance estimation (see the formulas in the box).

Estimation of σ^2

$$s^2 = \frac{\text{SSE}}{\text{Degrees of freedom for error}} = \frac{\text{SSE}}{n-2}$$

where

$$\text{SSE} = \sum (y_i - \hat{y}_i)^2 = \text{SS}_{yy} - \hat{\beta}_1 \text{SS}_{xy}$$

$$\text{SS}_{yy} = \sum (y_i - \bar{y})^2 = \sum y_i^2 - \frac{\left(\sum y_i\right)^2}{n}$$

Warning: When performing these calculations, you may be tempted to round the calculated values of SS_{yy}, $\hat{\beta}_1$, and SS_{xy}. Be certain to carry at least six significant figures for each of these quantities to avoid substantial errors in the calculation of SSE.

In the advertising–sales example, we previously calculated SSE = 1.10 for the least squares line $\hat{y} = -.1 + .7x$. Recalling that there were $n = 5$ data points, we have $n - 2 = 5 - 2 = 3$ df for estimating σ^2. Thus,

$$s^2 = \frac{\text{SSE}}{n-2} = \frac{1.10}{3} = .367$$

is the estimated variance, and

$$s = \sqrt{.367} = .61$$

is the estimated standard deviation of ϵ.

You may be able to obtain an intuitive feeling for s by recalling the interpretation given to a standard deviation in Chapter 2 and remembering that the least squares line estimates the mean value of y for a given value of x. Since s measures the spread of the distribution of y values about the least squares line, we should not be surprised to find that most of the observations lie within $2s$ or $2(.61) = 1.22$ of the least squares line. For this simple example (only five data points), all five data points fall within $2s$ of the least squares line. In Section 10.8, we use s to evaluate the error of prediction when the least squares line is used to predict a value of y to be observed for a given value of x.

Exercises 10.19–10.26

Learning the Mechanics

10.19 Suppose you fit a least squares line to 26 data points and calculate SSE = 8.34. Find s^2, the estimator of σ^2, the variance of the random error term ϵ.

10.20 Calculate SSE and s^2 for each of the following cases:

a. $n = 15$, $SS_{yy} = 7{,}492.91$, $SS_{xy} = 261.43$, $\hat{\beta}_1 = 26.97$

b. $n = 87$, $\Sigma y^2 = 10{,}235.76$, $\Sigma y = 672.60$, $SS_{xy} = 340.85$, $\hat{\beta}_1 = .04$

c. $n = 9$, $\Sigma(y_i - \bar{y})^2 = 38{,}377.81$, $SS_{xy} = -12{,}824.93$, $SS_{xx} = 8{,}221.11$

10.21 Refer to Exercises 10.9 and 10.12. Calculate s^2 and s for the least squares lines obtained in those exercises.

10.22 Visually compare the given scattergrams. If a least squares line were determined for each data set, which do you think would have the smallest variance, s^2? Explain.

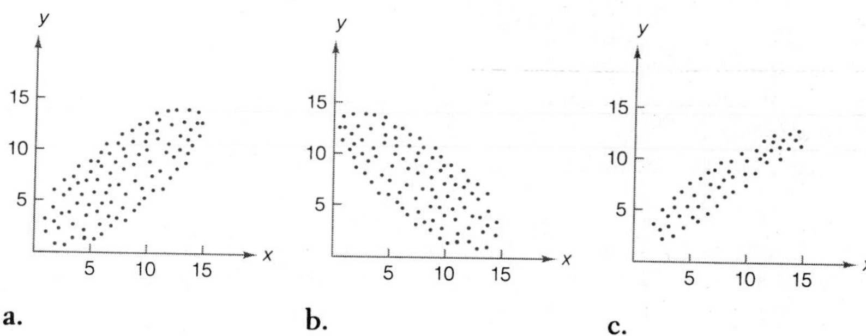

Applying the Concepts

10.23 Prior to the 1970s the developing countries played a small role in world trade because their own economic policies hindered integration with the world economy. However, many of these countries have since changed their policies and vastly improved their importance to the global economy (Krueger, 1992). The table provides data for investigating the relationship between developing countries and industrial countries in annual import levels (given in billions of U.S. dollars).

	1950	*1960*	*1970*	*1980*	*1989*
Industrial Countries' Imports, x	39.8	85.4	226.9	1,370.2	2,237.9
Developing Countries' Imports, y	21.1	40.1	75.6	556.4	819.4

a. Given that $SS_{xx} = 3{,}809{,}368.452$, $SS_{xy} = 1{,}419{,}492.796$, $SS_{yy} = 531{,}174.148$, $\bar{x} = 792.04$, and $\bar{y} = 302.52$, fit a least squares line to the data. Plot the data points and graph the least squares line as a check on your calculations.

b. According to your least squares line, approximately what would you expect annual imports for developing countries to be if annual imports for industrial countries were $1,600 billion?
c. Calculate SSE and s^2.
d. Interpret the standard deviation s in the context of this problem.

10.24 To improve the quality of the output of any production process, we must first understand the capabilities of the process (Deming, 1986). In a particular manufacturing process, the useful life of a cutting tool is related to the speed at which the tool is operated. If we understand this relationship we can predict when the tool should be replaced and how many spare tools should be available. The data in the table were derived from life tests for the two brands of cutting tools currently in use.

Cutting Speed	Useful Life (hrs.)	
(meters/min.)	Brand A	Brand B
30	4.5	6.0
30	3.5	6.5
30	5.2	5.0
40	5.2	6.0
40	4.0	4.5
40	2.5	5.0
50	4.4	4.5
50	2.8	4.0
50	1.0	3.7
60	4.0	3.8
60	2.0	3.0
60	1.1	2.4
70	1.1	1.5
70	.5	2.0
70	3.0	1.0

a. Construct a scattergram for each brand of cutting tool.
b. For each brand, use the method of least squares to model the relationship between useful life and cutting speed.
c. Find SSE, s^2, and s for each least squares line.
d. For a cutting speed of 70 meters per minute, find $\hat{y} \pm 2s$ for each least squares line.
e. For which brand would you feel more confident in using the least squares line to predict useful life for a given cutting speed? Explain.

10.25 Although the cable television industry could provide viewers with 100 or more channels, a study by the A. C. Nielsen Co. suggests that viewers may not want so many. The Nielsen survey indicates that as television channels increase, the *percentage* of channels viewed for 10 minutes a week or more declines (Landro and Mayer, 1982). In a similar study, 20 households were sampled, and the number of channels available to each household was recorded. In addition, each household was asked to monitor its television viewing for 1 week and report the number of channels watched for 10 minutes or more. The results appear in the accompanying table.

Household	Number of Channels Available	Number of Channels Watched ≥ 10 Minutes	Household	Number of Channels Available	Number of Channels Watched ≥ 10 Minutes
1	12	6	11	25	10
2	29	10	12	8	6
3	4	3	13	5	4
4	20	8	14	10	4
5	40	12	15	16	9
6	5	3	16	4	4
7	6	5	17	5	1
8	4	4	18	45	13
9	14	8	19	35	5
10	20	6	20	50	10

a. Do these data tend to support the Nielsen findings? Find the appropriate least squares line and use it to justify your answer.
b. Plot your least squares line on a scattergram of the data.
c. Calculate SSE, s^2, and s. For the given number of channels available to a particular household, within what approximate bounds would you expect this least squares line to be able to predict the percentage of channels watched for 10 minutes or more?

10.26 A much larger proportion of U.S. teenagers work while attending high school than was the case a decade ago, and this proportion exceeds that in Japan, Germany, and Sweden. The change was fueled by the growth of the service sector after World War II, the rise of the fast-food industry in the 1960s and 1970s, and the larger number of teenage girls entering the work force. Because heavy workloads often result in poor classroom performance and lower grades, many states are tightening the child labor laws. A 1991 study of high school students in California and Wisconsin showed that those who worked only a few hours per week had the highest grade-point averages. The table shows GPAs and work hours per week for a sample of five students.

Grade-Point Average, y	2.93	3.00	2.86	3.04	2.66
Hours Worked per Week, x	12	0	17	5	21

Source: Adapted from Waldman, S., and Springen, K. "Too old, too fast." *Newsweek*, Nov. 16, 1992, p. 80.

a. Given that $SS_{xx} = 294.00000$, $SS_{xy} = -4.55000$, $SS_{yy} = .08968$, $\bar{x} = 11.00000$, and $\bar{y} = 2.89800$, fit a least squares line to the data.
b. Plot the data and graph the least squares line.
c. Predict the GPA of a high school student who works 10 hours per week. Repeat this for one who works 16 hours per week.
d. Calculate SSE, s^2, and s.
e. Within what approximate distance do you expect your predictions in part **c** to fall from the true GPA earned by the high school student? [*Note:* A more precise measure of reliability for these predictions is discussed in Section 10.8.]

10.5 Assessing the Usefulness of the Model: Making Inferences About the Slope β_1

Refer again to the data of Table 10.1 and suppose that the appliance store's sales revenue is *completely unrelated* to the advertising expenditure. What could be said about the values of β_0 and β_1 in the hypothesized probabilistic model

$$y = \beta_0 + \beta_1 x + \epsilon$$

if x contributes no information for the prediction of y? The implication is that the mean of y—i.e., the deterministic part of the model $E(y) = \beta_0 + \beta_1 x$—does not change as x changes. Regardless of the value of x, you always predict the same value of y. In the straight-line model, this means that the true slope, β_1, is equal to 0. Therefore, to test the null hypothesis that the linear model contributes no information for the prediction of y against the alternative hypothesis that the linear model is useful for predicting y, we test

$$H_0\colon\ \beta_1 = 0 \qquad H_a\colon\ \beta_1 \neq 0$$

If the data support the alternative hypothesis, we conclude that x does contribute information for the prediction of y using the straight-line model [although the true relationship between $E(y)$ and x could be more complex than a straight line]. Thus, in effect, this is a test of the usefulness of the hypothesized model.

The appropriate test statistic is found by considering the sampling distribution of $\hat{\beta}_1$, the least squares estimator of the slope β_1.

Sampling Distribution of $\hat{\beta}_1$

If the four assumptions about ϵ (see Section 10.3) are satisfied, then the sampling distribution of $\hat{\beta}_1$, the least squares estimator of slope, will be normal with mean β_1 (the true slope) and standard deviation

$$\sigma_{\hat{\beta}_1} = \frac{\sigma}{\sqrt{SS_{xx}}} \quad \text{(see Figure 10.6)}$$

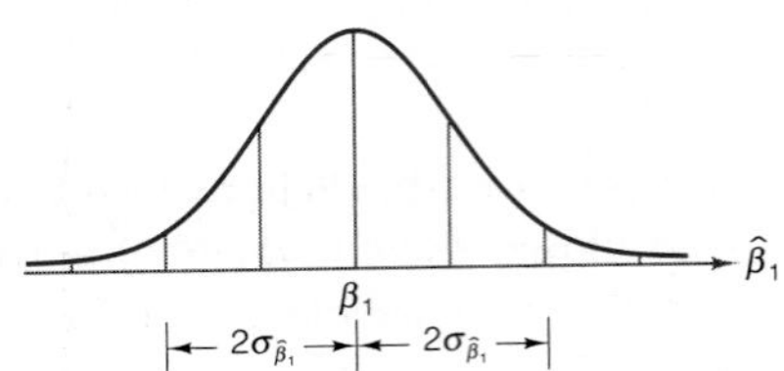

FIGURE 10.6 ► Sampling distribution of $\hat{\beta}_1$

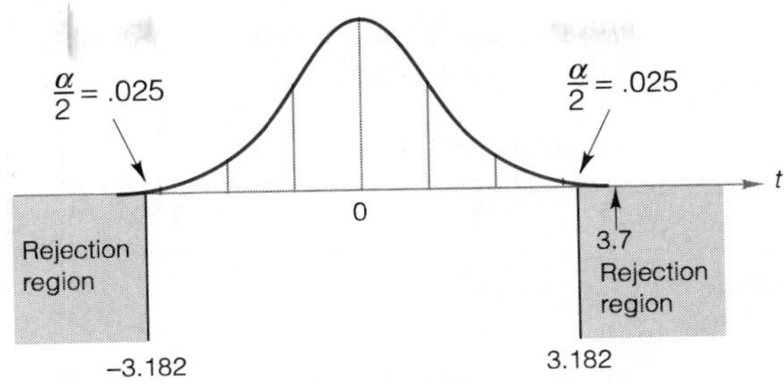

FIGURE 10.7 ▶ Rejection region and calculated t value for testing whether the slope $\beta_1 = 0$

What conclusion can be drawn if the calculated t value does not fall in the rejection region? We know from previous discussions of the philosophy of hypothesis testing that such a t value does *not* lead us to accept the null hypothesis. That is, we do not conclude that $\beta_1 = 0$. Additional data might indicate that β_1 differs from 0, or a more complex relationship may exist between y and x, requiring the fitting of a model other than the straight-line model. We discuss several such models in Chapter 11.

Another way to make inferences about the slope β_1 is to estimate it using a confidence interval. This interval is formed as shown in the box.

A 100(1 − α)% Confidence Interval for the Slope β_1

$$\hat{\beta}_1 \pm t_{\alpha/2}\, s_{\hat{\beta}_1} \qquad \text{where} \quad s_{\hat{\beta}_1} = \frac{s}{\sqrt{\text{SS}_{xx}}}$$

and $t_{\alpha/2}$ is based on $(n - 2)$ df.

For the advertising–sales example, a 95% confidence interval for the slope β_1 is

$$\hat{\beta}_1 \pm t_{.025}\, s_{\hat{\beta}_1} = .7 \pm 3.182\left(\frac{s}{\sqrt{\text{SS}_{xx}}}\right) = .7 \pm 3.182\left(\frac{.61}{\sqrt{10}}\right) = .7 \pm .61$$

Thus, we estimate with 95% confidence that the interval from .09 to 1.31 includes the slope parameter β_1. In terms of this example, the implication is that we can be 95% confident that the *true* mean increase in sales per additional \$100 advertising expenditure is between \$90 and \$1,310. This inference is meaningful only over the sampled range of x—that is, from \$100 to \$500 advertising expenditure.

Since all the values in this interval are positive, it appears that β_1 is positive and that the mean of y, $E(y)$, increases as x increases. However, the rather large width of the confidence interval reflects the small number of data points (and, consequently, a lack of information) in the experiment. Particularly bothersome is the fact that the lower end of the confidence interval implies that we are not even recovering our additional expenditure, since a \$100 increase in advertising may produce as little as a \$90 increase in mean sales. If we wish to tighten this interval, we need to increase the sample size.

Since σ will usually be unknown, the appropriate test statistic is generally a Student's t statistic, formed as follows:

$$t = \frac{\hat{\beta}_1 - \text{Hypothesized value of } \beta_1}{s_{\hat{\beta}_1}} = \frac{\hat{\beta}_1 - 0}{s/\sqrt{SS_{xx}}}$$

$$\text{where} \quad s_{\hat{\beta}_1} = \frac{s}{\sqrt{SS_{xx}}}$$

Note that we have substituted the estimator s for σ, and then formed $s_{\hat{\beta}_1}$, the estimator of the standard deviation of $\hat{\beta}_1$, by dividing s by $\sqrt{SS_{xx}}$. The number of degrees of freedom associated with this t statistic is the same as the number of degrees of freedom associated with s. Recall that this is $(n - 2)$ when the hypothesized model is a straight line (see Section 10.4).

The setup of our test of the usefulness of the straight-line model is summarized in the box.

A Test of Model Usefulness

One-Tailed Test	*Two-Tailed Test*
H_0: $\beta_1 = 0$	H_0: $\beta_1 = 0$
H_a: $\beta_1 < 0$ (or H_a: $\beta_1 > 0$)	H_a: $\beta_1 \neq 0$
Test statistic: $t = \frac{\hat{\beta}_1}{s_{\hat{\beta}_1}} = \frac{\hat{\beta}_1}{s/\sqrt{SS_{xx}}}$	*Test statistic:* $t = \frac{\hat{\beta}_1}{s_{\hat{\beta}_1}} = \frac{\hat{\beta}_1}{s/\sqrt{SS_{xx}}}$
Rejection region: $t < -t_\alpha$ (or $t > t_\alpha$)	*Rejection region:* $t < -t_{\alpha/2}$ or $t > t_{\alpha/2}$
where t_α is based on $(n - 2)$ df.	where $t_{\alpha/2}$ is based on $(n - 2)$ df.

Assumptions: The four assumptions about ϵ listed in Section 10.3.

For the advertising–sales example, we will choose $\alpha = .05$ and, since $n = 5$, df $= (n - 2) = 5 - 2 = 3$. Then the rejection region for the two-tailed test is

$$t < -t_{.025} = -3.182 \quad \text{or} \quad t > t_{.025} = 3.182$$

We previously calculated $\hat{\beta}_1 = .7$, $s = .61$, and $SS_{xx} = 10$. Thus,

$$t = \frac{\hat{\beta}_1}{s/\sqrt{SS_{xx}}} = \frac{.7}{.61/\sqrt{10}} = \frac{.7}{.19} = 3.7$$

Since this calculated t value falls in the upper-tail rejection region (see Figure 10.7 on page 480), we reject the null hypothesis and conclude that the slope β_1 is not 0. The sample evidence indicates that x contributes information for the prediction of y when a linear model is used to characterize the relationship between sales revenue and advertising.

Exercises 10.27–10.38

Learning the Mechanics

10.27 Construct both a 95% and a 90% confidence interval for β_1 for each of the following cases:

a. $\hat{\beta}_1 = 31, \quad s = 3, \quad SS_{xx} = 35, \quad n = 10$

b. $\hat{\beta}_1 = 64, \quad SSE = 1{,}960, \quad SS_{xx} = 30, \quad n = 14$

c. $\hat{\beta}_1 = -8.4, \quad SSE = 146, \quad SS_{xx} = 64, \quad n = 20$

10.28 Consider the following pairs of observations:

x	1	4	3	2	5	6	0
y	1	3	3	1	4	7	2

a. Construct a scattergram for the data.

b. Use the method of least squares to fit a straight line to the seven data points in the table.

c. Plot the least squares line on your scattergram of part **a**.

d. Specify the null and alternative hypotheses you would use to test whether the data provide sufficient evidence to indicate that x contributes information for the (linear) prediction of y.

e. What is the test statistic that should be used in conducting the hypothesis test of part **d**? Specify the degrees of freedom associated with the test statistic.

f. Conduct the hypothesis test of part **d** using $\alpha = .05$.

10.29 Refer to Exercise 10.28. Construct an 80% and a 98% confidence interval for β_1.

10.30 Do the accompanying data provide sufficient evidence to conclude that a straight line is useful for characterizing the relationship between x and y?

y	4	2	4	3	2	4
x	1	6	5	3	2	4

10.31 Suppose that $n = 100$ recent residential home sales in a city are used to fit a least squares straight-line model relating the sales price y to the square feet of living space x. The 100 homes range from 1,500 to 4,000 square feet of living space, and the resulting least squares equation is $\hat{y} = -30{,}000 + 70x$.

a. When the null hypothesis that the true slope is zero is tested, the resulting test statistic is $t = 6.572$. Give the approximate p-value of this test, and interpret the result in the context of this application.

b. The 95% confidence interval for the slope is calculated to be 49.1 to 90.9. Interpret this interval in the context of this application. What can be done to obtain a narrower confidence interval?

Applying the Concepts

10.32 From an observational study of five chief executives, Mintzberg (1973) identified 10 roles found in all managerial jobs: figurehead, leader, liaison, monitor, disseminator, spokesperson, entrepreneur, disturbance handler, resource allocator, and negotiator. An observational study of 19 managers from a medium-sized manufacturing plant (Luthans, Rosenkrantz, and Hennessey, 1985) extended Mintzberg's work by investigating which activities *successful* managers actually perform. A success index considered the manager's length of time in the organization and his or her rank within the firm; the higher the index, the more successful the manager. The table presents similar data that can be used to determine whether managerial success can be partly explained by the extent of network-building interactions with people outside the manager's work unit. Such interactions include phone and face-to-face meetings with customers and suppliers, attending external meetings, and doing public relations work.

Manager	Success Index, y	Interactions With Outsiders, x	Manager	Success Index, y	Interactions With Outsiders, x
1	40	12	11	70	20
2	73	71	12	47	81
3	95	70	13	80	40
4	60	81	14	51	33
5	81	43	15	32	45
6	27	50	16	50	10
7	53	42	17	52	65
8	66	18	18	30	20
9	25	35	19	42	21
10	63	82			

a. Construct a scattergram for the data.
b. Given $SS_{yy} = 7{,}006.6316$, $SS_{xx} = 10{,}824.5263$, $SS_{xy} = 2{,}561.2632$, $\bar{y} = 54.5789$, and $\bar{x} = 44.1579$, use the method of least squares to find a prediction equation for managerial success.
c. Find SSE, s^2, and s for your prediction equation. Interpret the standard deviation s in the context of this problem.
d. Plot the least squares line on your scattergram of part **a**. Does the number of interactions with outsiders seem to help predict managerial success? Explain.
e. Conduct a formal statistical hypothesis test to answer the question posed in part **d**. Use $\alpha = .05$.
f. Construct a 95% confidence interval for β_1. Interpret the interval in the context of the problem.

10.33 The expenses involved in a manufacturing operation may be categorized as being for *raw material, direct labor,* and *overhead*. Direct labor refers to the persons employed to transform the raw materials into the finished product. Overhead refers to all expenses other than those for raw materials and direct labor that are involved with running the factory (e.g., supervisory labor, maintenance of equipment, and office supplies) (Gray and Johnston, 1977). A manufacturer of 10-speed racing bicycles is interested in estimating the relationship between its monthly factory overhead and the total number of bicycles produced per month. The estimate will be used to help develop the manufacturing budget for next year. The data in the table have been collected for the previous 12 months.

Month	Production Level (1,000s of units)	Overhead ($1,000s)	Month	Production Level (1,000 of units)	Overhead ($1,000s)
1	16.9	41.4	7	16.3	37.5
2	15.6	35.0	8	15.5	37.0
3	17.4	38.3	9	23.4	47.9
4	11.6	29.5	10	28.4	55.6
5	17.7	39.6	11	27.1	53.1
6	17.6	37.4	12	19.2	40.6

a. Find the least squares prediction equation relating monthly overhead y to monthly production level x.
b. Does the straight-line model contribute information for predicting overhead costs? Test at $\alpha = .05$.
c. Which of the four assumptions we make about the random error ϵ may be inappropriate in this problem? Explain.

10.34 During June, July, and early August of 1981, 10 bids were made by DuPont, Seagram, and Mobil to take over Conoco. Finally, on August 5, DuPont announced success. The total value of the offer accepted by Conoco was \$7.54 billion, making it the largest takeover in the history of American business at that time. A study of the Conoco takeover used regression analysis to examine whether movements in the rate of return of the contending companies' common stock could be explained by movements in the return rate of the stock market as a whole. The model was $y = \beta_0 + \beta_1 x + \epsilon$, where y is the daily rate of return of a stock, x is the daily rate of return of the stock market as a whole (based on Standard & Poor's 500 Composite Index), and ϵ is believed to satisfy the assumptions of Section 10.3. (This model is known in the finance literature as the *market model*. Note that the parameter β_1 reflects the sensitivity of the stock's return rate to movements in the stock market as a whole.) Daily data from early 1979 through 1980 ($n = 504$), yielded the least squares lines shown in the table for the four firms in question. The t statistics and p-values associated with the values of β_1 are also shown.

Firm	*Estimated Market Model*	*t Statistics*	*p-Values*
Conoco	$\hat{y} = .0010 + 1.40x$	$t = 21.93$	.000
DuPont	$\hat{y} = -.0005 + 1.21x$	$t = 18.76$	.000
Mobil	$\hat{y} = .0010 + 1.62x$	$t = 16.21$	.000
Seagram	$\hat{y} = .0013 + .76x$	$t = 6.05$	.000

Source: Ruback, R. S. *Sloan Management Review*, "The Conoco takeover and stockholder returns." Winter 1982, Vol. 23, pp. 13–33.

a. For each of the models, test H_0: $\beta_1 = 0$ versus H_a: $\beta_1 \neq 0$. Use $\alpha = .01$. Draw the appropriate conclusion regarding the usefulness of the market model in each case.
b. If the rate of return of Standard & Poor's 500 Composite Index increased by .10, how much change would occur in the mean rate of return for Conoco's common stock? For Seagram's?
c. Which of the two stocks, Conoco or Seagram, appears to be more responsive to changes in the market as a whole? Explain.

10.35 Refer to Exercise 10.14, in which the extent of retaliation against whistle blowers was investigated. Since salary is a reasonably good indicator of a person's power within an organization, the data of Exercise 10.14 can be used to investigate whether the extent of retaliation is related to the power of the whistle blower in the organization. The researchers were unable to reject the hypothesis that the extent of retaliation is unrelated to power. Do you agree? Test using $\alpha = .05$.

10.36 Buyers are often influenced by bulk advertising of a particular product. For example, if a product that sells for 25¢ is advertised at 2/50¢, 3/75¢, or 4/\$1, some people think they are getting a bargain. To test this theory, a store manager advertised an item for equal periods of time at five different bulk rates and achieved the sales volumes listed in the table on page 484. Do the data provide sufficient evidence to indicate that mean sales increase as the bulk number increases? Test at $\alpha = .05$.

Advertised Number in Bulk Sale, x	Volume Sold, y
1	27
2	36
3	34
4	63
5	52

10.37 Do the data in Exercise 10.25 support the theory that the expected percentage of channels watched for 10 or more minutes decreases as the number of channels available increases? Test using $\alpha = .10$. Comment on the assumptions necessary for the validity of the test.

10.38 The precision of $\hat{\beta}_1$ as an estimator of β_1 is generally measured by its standard deviation $\sigma_{\hat{\beta}_1}$. In general, the larger the value of $\sigma_{\hat{\beta}_1}$, the wider (less precise) are confidence intervals for β_1; the smaller the value of $\sigma_{\hat{\beta}_1}$, the narrower (more precise) are confidence intervals for β_1.

a. Examine the formula for $\sigma_{\hat{\beta}_1}$ and explain how the observed values of the independent variable influence the size of $\sigma_{\hat{\beta}_1}$.

b. Sometimes it is possible to obtain data for a regression study by setting the independent variable x at different levels and observing the resulting values of the dependent variable y. For example, suppose a supermarket chain is studying the relationship between the sales of a product and how many square feet of display space it is given. Data for such a study will be generated by utilizing display areas of different sizes in different stores and observing the resulting sales. If you were designing such a study, how would your answer to part **a** influence the choice of display area sizes?

10.6 Correlation: Another Measure of the Usefulness of the Model

The claim is often made that the crime rate and the unemployment rate are "highly correlated." Another popular belief is that the Gross Domestic Product (GDP) and the rate of inflation are "correlated." Some people even believe that the Dow Jones Industrial Average and the lengths of fashionable skirts are "correlated." Thus, the term **correlation** implies a relationship between two variables.

The **Pearson product moment correlation coefficient** r, defined in the box, provides a quantitative measure of the strength of the linear relationship between x and y, just as does the least squares slope $\hat{\beta}_1$. However, unlike the slope, the correlation coefficient r is *scaleless.* **The value of r is always between -1 and $+1$, no matter what the units of x and y are.**

Note that r is computed from the same quantities used in fitting the least squares line. Since both r and $\hat{\beta}_1$ provide information about the utility of the model, it is not surprising that there is a similarity in their computational formulas. In particular, note that SS_{xy} appears in the numerators of both expressions and, since both denominators are always positive, r and $\hat{\beta}_1$ will always be of the same sign (either both positive or both negative). A value of r near or equal to 0 implies little or no linear relationship between the values of y and x that were observed in the sample. In contrast, the closer r is to

Definition 10.2

The **Pearson product moment coefficient of correlation *r*** is a measure of the strength of the linear relationship between two variables x and y. It is computed (for a sample of n measurements on x and y) as follows:

$$r = \frac{SS_{xy}}{\sqrt{SS_{xx}SS_{yy}}}$$

1 or -1, the stronger the linear relationship between y and x. And if $r = 1$ or $r = -1$, all the points fall exactly on the least squares line. Positive values of r imply that y increases as x increases; negative values imply that y decreases as x increases. Each of these situations is portrayed in Figure 10.8.

FIGURE 10.8 ▶
Values of *r* and their implications

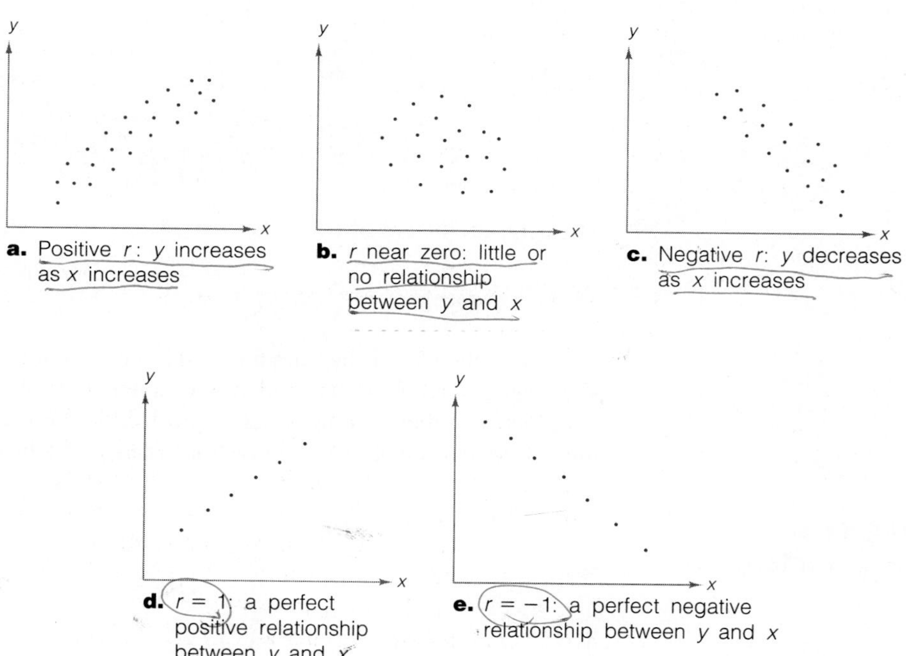

y
x
a. Positive r: y increases as x increases
b. r near zero: little or no relationship between y and x
c. Negative r: y decreases as x increases
d. r = 1: a perfect positive relationship between y and x
e. r = −1: a perfect negative relationship between y and x

EXAMPLE 10.1

A firm wants to know the correlation between the size of its sales force and its yearly sales revenue. The records for the past 10 years are examined, and the results listed in Table 10.5 (page 486) are obtained. Calculate the coefficient of correlation r for the data.

TABLE 10.5 Sales Force–Revenue Data, Example 10.1

Year	Salespeople x	Sales y ($100,000s)	Year	Salespeople x	Sales y ($100,000s)
1981	15	1.35	1986	29	2.93
1982	18	1.63	1987	30	3.41
1983	24	2.33	1988	32	3.26
1984	22	2.41	1989	35	3.63
1985	25	2.63	1990	38	4.15

Solution

We need to calculate SS_{xy}, SS_{xx}, and SS_{yy}:

$$SS_{xy} = \sum x_i y_i - \frac{\left(\sum x_i\right)\left(\sum y_i\right)}{10} = 800.62 - \frac{(268)(27.73)}{10} = 57.456$$

$$SS_{xx} = \sum x_i^2 - \frac{\left(\sum x_i\right)^2}{10} = 7{,}668 - \frac{(268)^2}{10} = 485.6$$

$$SS_{yy} = \sum y_i^2 - \frac{\left(\sum y_i\right)^2}{10} = 83.8733 - \frac{(27.73)^2}{10} = 6.97801$$

Then the coefficient of correlation is

$$r = \frac{SS_{xy}}{\sqrt{SS_{xx}SS_{yy}}} = \frac{57.456}{\sqrt{(485.6)(6.97801)}} = \frac{57.456}{58.211} = .99$$

Thus, the size of the sales force and sales revenue are very highly correlated—at least over the past 10 years. The implication is that a strong positive linear relationship exists between these variables (see Figure 10.9). We must be careful, however, not to jump to unwarranted conclusions. For instance, the firm may be tempted to conclude

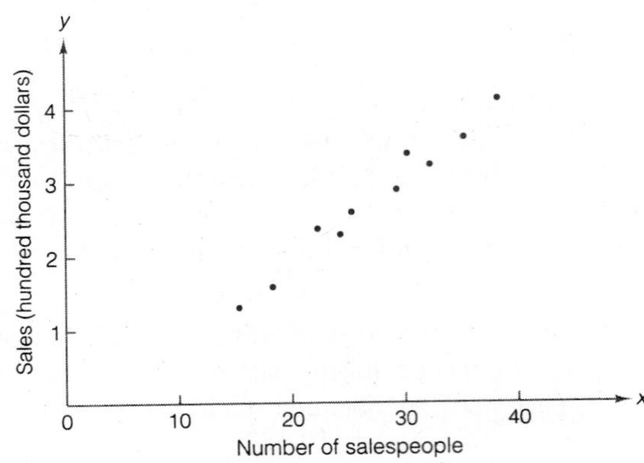

FIGURE 10.9 ►
Scattergram for Example 10.1

that the best thing it can do to increase sales is to hire lots of new salespeople. The implication of such a conclusion is that there is a *causal* relationship between the two variables. However, **high correlation does not imply causality**. The fact is that many things have probably contributed both to the increase in the size of the sales force and to the increase in sales revenue. The firm's expertise has undoubtedly grown, the rate of inflation has increased (so that 1990 dollars are not worth as much as 1981 dollars), and perhaps the scope of products and services sold by the firm has widened. We must be careful not to infer a causal relationship on the basis of high sample correlation. The only safe conclusion when a high correlation is observed in the sample data is that a linear trend may exist between x and y.

The correlation coefficient r measures the correlation between x values and y values in the sample, and a similar coefficient of correlation exists for the population from which the data points were selected. The **population correlation coefficient** is denoted by the symbol ρ (rho). As you might expect, ρ is estimated by the corresponding sample statistic, r. Or, rather than estimating ρ, we might want to test the hypothesis H_0: $\rho = 0$ against H_a: $\rho \neq 0$—i.e., test the hypothesis that x contributes no information for the prediction of y using the straight-line model against the alternative that the two variables are at least linearly related. We performed the identical test in Section 10.5 when we tested H_0: $\beta_1 = 0$ against H_a: $\beta_1 \neq 0$. That is, the null hypothesis H_0: $\rho = 0$ is equivalent to the hypothesis H_0: $\beta_1 = 0$.* When we tested the null hypothesis H_0: $\beta_1 = 0$ in connection with the advertising–sales example, the data led to a rejection of the hypothesis for $\alpha = .05$. This implies that the null hypothesis of a zero linear correlation between the two variables (advertising and sales) can also be rejected at $\alpha = .05$. The only real difference between the least squares slope $\hat{\beta}_1$ and the coefficient of correlation r is the measurement scale. Therefore, the information they provide about the usefulness of the least squares model is to some extent redundant. For this reason, we will use the slope to make inferences about the existence of a positive or negative linear relationship between the two variables.

10.7 The Coefficient of Determination

Another way to measure the usefulness of the model is to measure the contribution of x in predicting y. To accomplish this, we calculate how much the errors of prediction of y were reduced by using the information provided by x. To illustrate, consider the sample shown in the scattergram of Figure 10.10(a) on page 488. If we assume that x contributes no information for the prediction of y, the best prediction for a value of y

*The correlation test statistic that is equivalent to $t = \hat{\beta}_1/s_{\hat{\beta}_1}$ is

$$t = \frac{r}{\sqrt{(1 - r^2)/(n - 2)}}$$

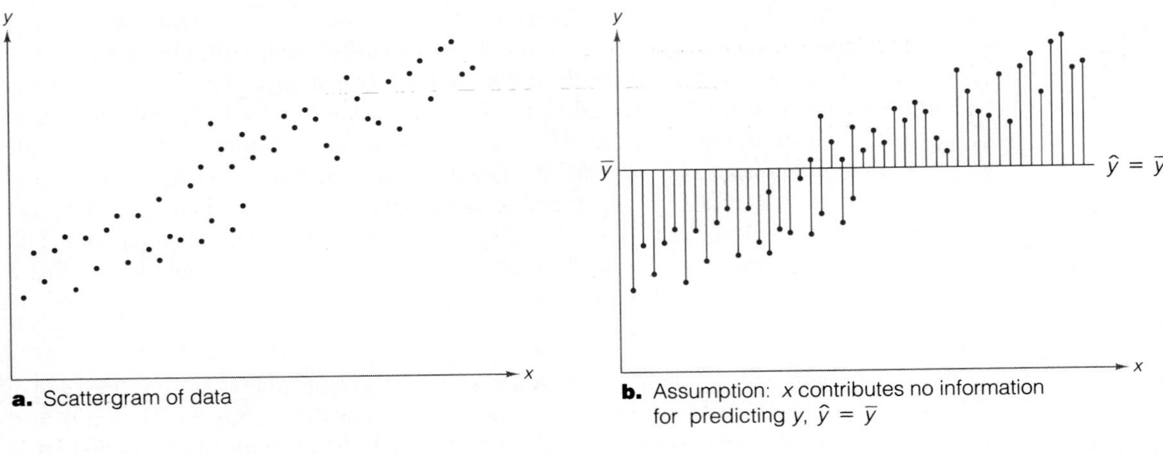
y
x
a. Scattergram of data
y
$\bar{y}$
$\hat{y} = \bar{y}$
x
b. Assumption: x contributes no information for predicting y, $\hat{y} = \bar{y}$

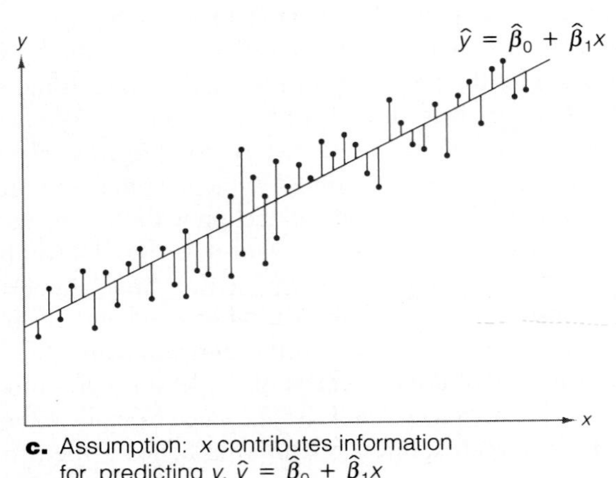
y
$\hat{y} = \hat{\beta}_0 + \hat{\beta}_1 x$
x
c. Assumption: x contributes information for predicting y, $\hat{y} = \hat{\beta}_0 + \hat{\beta}_1 x$

FIGURE 10.10 ▲
A comparison of the sum of squares of deviations for two models

is the sample mean $\bar{y}$, which is shown as the horizontal line in Figure 10.10(b). The vertical line segments in Figure 10.10(b) are the deviations of the points about the mean $\bar{y}$. Note that the sum of squares of deviations for the prediction equation $\hat{y} = \bar{y}$ is

$$SS_{yy} = \sum (y_i - \bar{y})^2$$

Now suppose you fit a least squares line to the same set of data and locate the deviations of the points about the line as shown in Figure 10.10(c). Compare the deviations about the prediction lines in parts (b) and (c) of Figure 10.10. You can see that

1. If x contributes little or no information for the prediction of y, the sums of squares of deviations for the two lines,

$$SS_{yy} = \sum (y_i - \bar{y})^2 \quad \text{and} \quad SSE = \sum (y_i - \hat{y}_i)^2$$

will be nearly equal.

2. If x does contribute information for the prediction of y, the SSE will be smaller than SS_{yy}. In fact, if all the points fall on the least squares line, then SSE = 0.

Then, the reduction in the sum of squares of deviations that can be attributed to x, expressed as a proportion of SS_{yy}, is

$$\frac{SS_{yy} - SSE}{SS_{yy}}$$

Note that SS_{yy} is the "total sample variation" of the observations around the sample mean $\bar{y}$, and that SSE is the remaining "unexplained sample variability" after fitting the line $\hat{y}$. Thus, the difference (SS_{yy} − SSE) is the "explained sample variability" attributable to the linear relationship with x. Then a verbal description of the proportion is

$$\frac{SS_{yy} - SSE}{SS_{yy}} = \frac{\text{Explained sample variability}}{\text{Total sample variability}} = \frac{\text{Proportion of total sample variability}}{\text{explained by the linear relationship}}$$

It can be shown that this proportion is equal to the square of the simple linear coefficient of correlation r (the Pearson product moment coefficient of correlation).

Definition 10.3

The **coefficient of determination** is the square of the coefficient of correlation. It represents the proportion of the total sample variability around $\bar{y}$ that is explained by the linear relationship between y and x. We compute it as

$$r^2 = \frac{SS_{yy} - SSE}{SS_{yy}} = 1 - \frac{SSE}{SS_{yy}}$$

Note that r^2 is always between 0 and 1, because r is between −1 and +1. Thus, an r^2 of .60 means that the sum of squares of deviations of the y values about their predicted values has been reduced 60% by the use of the least squares equation $\hat{y}$, instead of $\bar{y}$, to predict y.

EXAMPLE 10.2

Calculate the coefficient of determination for the advertising–sales example. The data are repeated in Table 10.6 (page 490).

TABLE 10.6

Advertising Expenditure x (\$100s)	Sales Revenue y (\$1,000s)
1	1
2	1
3	2
4	2
5	4

Solution

We first calculate

$$SS_{yy} = \sum y_i^2 - \frac{\left(\sum y_i\right)^2}{5} = 26 - \frac{(10)^2}{5} = 26 - 20 = 6$$

From previous calculations, we know $SSE = \sum (y_i - \hat{y}_i)^2 = 1.10$. Then, the coefficient of determination is given by

$$r^2 = \frac{SS_{yy} - SSE}{SS_{yy}} = \frac{6.0 - 1.1}{6.0} = \frac{4.9}{6.0} = .82$$

So we know that in using the advertising expenditure x to predict y with the least squares line

$$\hat{y} = -.1 + .7x$$

the total sum of squares of deviations of the five sample y values about their predicted values has been reduced 82% by the use of $\hat{y}$, instead of $\bar{y}$, to predict y.

CASE STUDY 10.1 / Estimating the Cost of a Construction Project

As evidenced by the cost overruns of public building projects, the initial estimate of the ultimate cost of a structure is often rather poor. These estimates usually rely on a precise definition of the proposed building in terms of working drawings and specifications. Thus, cost estimators generally use deterministic models that do not take random error into account, so no measure of reliability is possible for their deterministic estimates. Crandall and Cedercreutz (1976) favor a probabilistic model to make cost estimates. They use regression models to relate cost to independent variables like volume, amount of glass, and floor area, explaining that "one of the principal merits of the least squares regression model, for the purpose of preliminary cost estimating, is the method of dealing with anticipated error." Specifically, when random error is anticipated, "statistical methods, such as regression analysis, attack the problem head on."

Crandall and Cedercreutz initially focused on the cost of mechanical work (heating, ventilating, and plumbing), since this part of the total cost is generally difficult to predict. Conventional cost estimates rely heavily on the amount of ductwork and piping used in construction, but this information is not precisely known until too late to be of use to the cost estimator. One of several models discussed was a simple linear model relating mechanical cost to floor area. Based on the data associated with 26 factory and warehouse buildings, the least squares prediction equation given in Figure 10.11 was found. It was concluded that floor area and mechanical cost are linearly related, since the t statistic (for testing H_0: $\beta_1 = 0$) equalled 3.61, which is significant with an α as small as .002. Thus, floor area should help predict the mechanical cost of a factory or warehouse. In addition, the regression model enables the reliability of the predicted cost to be assessed.

The value of the coefficient of determination r^2 was .35. This tells us that only 35% of the variation among mechanical costs is accounted for by the differences in floor areas. Since there is only one independent variable in the model, this relatively small value of r^2 should not be too surprising. If other variables related to mechanical cost were included in the model, they would probably account for a significant portion of the remaining 65% of the variation in mechanical cost not explained by floor area. In the next two chapters, we describe how to build regression models using multiple independent variables.

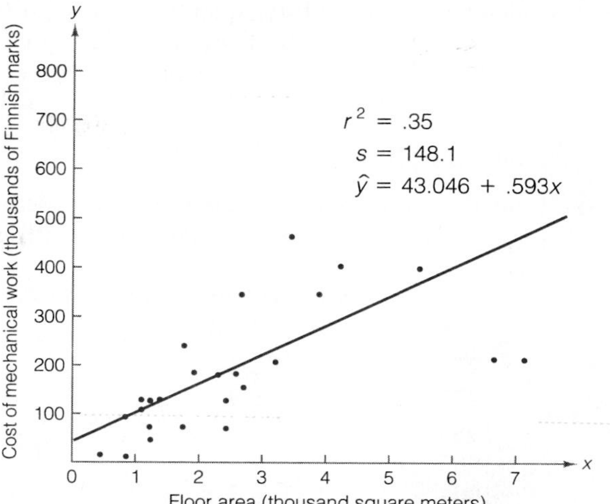

FIGURE 10.11 ▲ Simple linear model relating cost to floor area

Exercises 10.39–10.51

Learning the Mechanics

10.39 Explain what each of the following sample correlation coefficients tells you about the relationship between the x and y values in the sample:

a. $r = 1$ b. $r = -1$ c. $r = 0$ d. $r = .90$ e. $r = .10$ f. $r = -.88$

10.40 Describe the slope of the least squares line if:

a. $r = .7$ b. $r = -.7$ c. $r = 0$ d. $r^2 = .64$

10.41 Construct a scattergram for each data set. Then calculate r and r^2 for each data set. Interpret their values.

a.

x	−2	−1	0	1	2
y	−2	1	2	5	6

b.

x	−2	−1	0	1	2
y	6	5	3	2	0

c.

x	1	2	2	3	3	3	4
y	2	1	3	1	2	3	2

d.

x	0	1	3	5	6
y	0	1	2	1	0

10.42 Calculate r^2 for the least squares line in each of the following exercises. Interpret their values.
a. Exercise 10.9 **b.** Exercise 10.12 **c.** Exercise 10.28

Applying the Concepts

10.43 Find the correlation coefficient and the coefficient of determination for the data listed in the table and interpret your results. Can a causal relationship be inferred? Explain.

Year	*Hunting Licenses Sold, x (millions)*	*Divorces and Annulments, y (millions)*
1970	22.2	.71
1975	25.9	1.04
1980	27.0	1.19
1983	28.9	1.16
1984	28.5	1.17
1985	27.7	1.19
1986	27.9	1.18
1987	28.8	1.17
1988	30.0	1.17

Source: U.S. Bureau of the Census. *Statistical Abstract of the United States: 1992*, pp. 90, 238.

10.44 Find the correlation coefficient and the coefficient of determination for the sample data on Gross Domestic Product (GDP) and new housing starts listed in the table and interpret your results. For these data, $SS_{xx} = 27{,}144{,}855.71$, $SS_{yy} = 1{,}132{,}806.4$, $SS_{xy} = -91{,}861.52$, $\bar{x} = 3{,}688.91$, and $\bar{y} = 1{,}431.2$.

Year	*GDP, x ($ billions)*	*Housing Starts, y (1,000s)*	Year	*GDP, x ($ billions)*	*Housing Starts, y (1,000s)*
1970	1,010.7	1,434	1985	4,038.7	1,742
1975	1,585.9	1,160	1986	4,268.6	1,805
1979	2,488.6	1,745	1987	4,539.9	1,620
1980	2,708.0	1,292	1988	4,900.4	1,488
1981	3,030.6	1,084	1989	5,244.0	1,376
1982	3,149.6	1,062	1990	5,513.8	1,193
1983	3,405.0	1,703	1991	5,672.6	1,014
1984	3,777.2	1,750			

Source: U.S. Bureau of the Census. *Statistical Abstract of the United States: 1992*, pp. 428, 710.

10.45 Data on monthly sales, y, price per unit during the month, x_1, and amount spent on advertising, x_2, for a product are shown for a 5-month period in the table. Based on this sample, which variable—price or advertising expenditure—appears to provide more information about sales? Explain.

Month	*Monthly Sales* y *(1,000s)*	*Unit Price* x_1	*Advertising Costs* x_2 *(100s)*
June	$40	$ .85	$6.0
July	50	.76	5.0
August	55	.75	8.0
September	30	1.00	7.5
October	45	.80	5.5

10.46 A *negotiable certificate of deposit* is a marketable receipt for funds deposited in a bank for a specified period of time at a specified rate of interest (Cook, 1977). The accompanying table lists the end-of-quarter interest rate for 3-month certificates of deposit from January 1980 through December 1991 with the concurrent end-of-quarter values of Standard & Poor's 500 Stock Composite Average (an indicator of stock market activity). For these data, $\Sigma x = 452.96$, $\Sigma y = 10{,}937.82$, $\Sigma x^2 = 4{,}780.03$, $\Sigma y^2 = 2{,}889{,}220.05$, and $\Sigma xy = 93{,}654.35$. Find the coefficient of determination and the correlation coefficient for the data, and interpret your results.

Year	Quarter	Interest Rate, x	S&P 500, y	Year	Quarter	Interest Rate, x	S&P 500, y
1980	I	17.57	104.69	1986	I	7.24	238.90
	II	8.49	114.24		II	6.73	250.84
	III	11.29	125.46		III	5.71	231.32
	IV	18.65	135.76		IV	6.04	242.17
1981	I	14.43	136.00	1987	I	6.17	291.70
	II	16.90	131.21		II	6.94	304.00
	III	16.84	116.18		III	7.37	321.83
	IV	12.49	122.55		IV	7.66	247.08
1982	I	14.21	111.96	1988	I	6.63	258.89
	II	14.46	109.61		II	7.51	273.50
	III	10.66	120.42		III	8.23	271.91
	IV	8.66	135.28		IV	9.25	277.72
1983	I	8.69	152.96	1989	I	10.09	294.87
	II	9.20	168.11		II	9.20	317.98
	III	9.39	166.07		III	8.78	349.15
	IV	9.69	164.93		IV	8.32	353.40
1984	I	10.08	159.18	1990	I	8.27	339.94
	II	11.34	153.18		II	8.33	358.02
	III	11.29	166.10		III	8.08	306.05
	IV	8.60	167.24		IV	7.96	330.22
1985	I	9.02	180.66	1991	I	6.71	375.22
	II	7.44	191.85		II	6.01	371.16
	III	7.93	182.08		III	5.70	387.86
	IV	7.80	211.28		IV	4.91	417.09

Source: *Standard & Poor's Statistical Service, Current Statistics*. Standard & Poor's Corporation, 1992.

10.47 The Minnesota Department of Transportation installed a computerized weigh-in-motion scale in the concrete surface of eastbound Interstate 494 in Bloomington, Minnesota. The system monitors traffic continuously and can distinguish among 13 types of vehicles (car, five-axle semi, five-axle twin trailer, etc.). The purpose of the system is to provide traffic counts and weights for use in the planning and design of roadways.

A *calibration study* was undertaken to determine whether the scale's readings correspond with the static weights of the vehicles being monitored. After some preliminary comparisons using a two-axle six-tire truck carrying different loads, calibration adjustments were made in the software of the weigh-in-motion system and the scales were reevaluated. The table shows the trial data in thousands of pounds.

Trial	*Static Weight, x*	*Initial Motion Reading, y_1*	*Adjusted Reading, y_2*
1	27.9	26.0	27.8
2	29.1	29.9	29.1
3	38.0	39.5	37.8
4	27.0	25.1	27.1
5	30.3	31.6	30.6
6	34.5	36.2	34.3
7	27.8	25.1	26.9
8	29.6	31.0	29.6
9	33.1	35.6	33.0
10	35.5	40.2	35.0

Source: Adapted from data in Wright, J. L., Owen, F., and Pena, D. "Status of MN/DOT's weigh-in-motion program." St. Paul: Minnesota Department of Transportation, Jan. 1983.

a. Construct two scattergrams, one of y_1 versus x and the other of y_2 versus x.
b. Use the scattergram of part **a** to evaluate the performance of the weigh-in-motion scale both before and after the calibration adjustment.
c. Calculate the correlation coefficient for both sets of data, and interpret their values. Explain how these correlation coefficients can be used to evaluate the weigh-in-motion scale.
d. Suppose the sample correlation coefficient for y_2 and x were 1. Could this happen if the static weights and the weigh-in-motion readings disagreed? Explain.

10.48 A problem of economic and social concern in the United States is the importation and sale of illicit drugs. The data in the table were collected by the Florida attorney general's office in an attempt to relate the incidence of drug seizures and drug arrests to the characteristics of Florida counties. Given are the number, y, of drug arrests per county in 1982, the density, x_1, of the county (population per square mile), and the number, x_2, of law enforcement employees. To simplify calculations, only 10 counties are described.

	County									
	1	*2*	*3*	*4*	*5*	*6*	*7*	*8*	*9*	*10*
Population Density, x_1	169	68	278	842	18	42	112	529	276	613
Law Enforcement Employees, x_2	498	35	772	5,788	18	57	300	1,762	416	520
Arrests in 1982, y	370	44	716	7,416	25	50	189	1,097	256	432

a. Fit a least squares line to relate the number, y, of drug arrests per county in 1982 to the county population density, x_1.
b. We might expect the mean number of arrests to increase as the population density increases. Do the data support this theory? Test using $\alpha = .05$.
c. Calculate the coefficient of determination for this regression analysis and interpret its value.

10.49 Repeat parts **a**, **b**, and **c** of Exercise 10.48 using the number, x_2, of county law enforcement employees as the independent variable.

d. Which least squares line has the lower SSE?
e. Which independent variable explains more of the variation in y? Explain.

10.50 Refer to Exercise 10.48.
a. Calculate the correlation coefficient r between the county population density x_1 and the number of law enforcement employees x_2.
b. Does the correlation between x_1 and x_2 differ significantly from 0? Test using $\alpha = .05$.

10.51 A firm's *demand curve* describes the quantity of its product that can be sold at various prices, other things being equal (McConnell and Brue, 1990). Over the period of a year, a tire company varied the price of a radial tire to estimate its demand curve. When the price was set very low or very high, few tires sold. The latter result the firm understood; the former it attributed to consumer misperception that low price implies poor quality. The data in the table describe the tire's sales over the experimental period.

Price x (\$)	Sales y (100s)
20	13
35	57
45	85
60	43
70	17

a. Calculate a least squares line to approximate the firm's demand curve.
b. Construct a scattergram and plot your least squares line as a check on your calculations.
c. Test H_0: $\beta_1 = 0$ using a two-tailed test and $\alpha = .05$. Draw the appropriate conclusion in the context of the problem.
d. Does the nonrejection of H_0 in part **c** imply that no relationship exists between tire price and sales volume? Explain.
e. Calculate the coefficient of determination for the least squares line of part **a** and interpret its value in the context of the problem.

10.8 Using the Model for Estimation and Prediction

If we are satisfied that a useful model has been found to describe the relationship between sales revenue and advertising, we are ready to accomplish the original objectives for building the model: using it to estimate or to predict sales on the basis of advertising dollars spent.

The most common uses of a probabilistic model can be divided into two categories. The first is the use of the model for estimating the mean value of y, $E(y)$ for a specific value of x.

For our example, we may want to estimate the mean sales revenue for *all* months during which \$400 ($x = 4$) is expended on advertising.

The second use of the model entails predicting a particular y value for a given value of x.

That is, if we decide to expend \$400 next month, we want to predict the firm's sales revenue for that month.

In the case of estimating a mean value of y, we are attempting to estimate the mean result of a very large number of experiments at the given x value. In the second case, we are trying to predict the outcome of a single experiment at the given x value.

In which of these model uses do you expect to have more success; i.e., which value—the mean or individual value of y—can we estimate (or predict) with more accuracy?

Before answering this question, we first consider the problem of choosing an estimator (or predictor) of the mean (or individual) y value. We will use the least squares model

$$\hat{y} = \hat{\beta}_0 + \hat{\beta}_1 x$$

both to estimate the mean value of y and to predict a particular value of y for a given value of x. For our example, we found

$$\hat{y} = -.1 + .7x$$

so that the estimated mean value of sales revenue for all months when $x = 4$ (advertising = \$400) is

$$\hat{y} = -.1 + .7(4) = 2.7$$

or \$2,700 (the units of y are thousands of dollars). The same value is used to predict the y value when $x = 4$. That is, both the estimated mean value and the predicted value of y equal $\hat{y} = 2.7$ when $x = 4$, as shown in Figure 10.12.

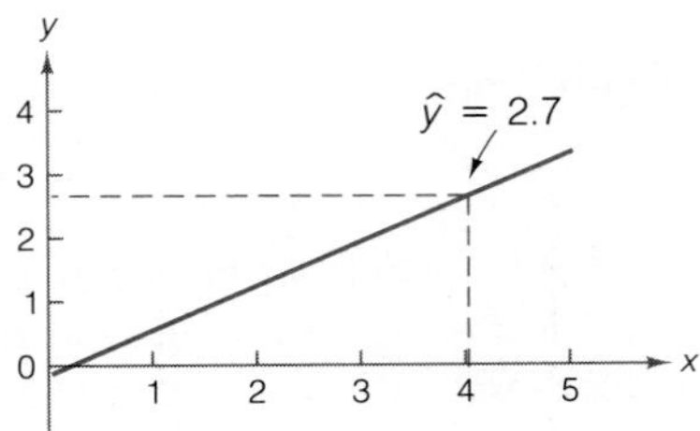

FIGURE 10.12 ▶ **Estimated mean value and predicted individual value of sales revenue y for $x = 4$**

The difference in these two model uses lies in the relative accuracy of the estimate and the prediction. These accuracies are best measured by using the sampling errors of the least squares line when it is used as an estimator and as a predictor, respectively. These errors are reflected in the standard deviations given in the box.

Sampling Errors for the Estimator of the Mean of y and the Predictor of an Individual y

1. The standard deviation of the sampling distribution of the estimator $\hat{y}$ of the mean value of y at a particular value of x, say x_p, is

$$\sigma_{\hat{y}} = \sigma\sqrt{\frac{1}{n} + \frac{(x_p - \bar{x})^2}{SS_{xx}}}$$

where σ is the standard deviation of the random error ϵ.

2. The standard deviation of the prediction error for the predictor $\hat{y}$ of an individual y value for $x = x_p$ is

$$\sigma_{(y-\hat{y})} = \sigma\sqrt{1 + \frac{1}{n} + \frac{(x_p - \bar{x})^2}{SS_{xx}}}$$

where σ is the standard deviation of the random error ϵ.

The true value of σ is rarely known. Thus, we estimate σ by s and calculate the estimation and prediction intervals as shown in the following boxes.

A $100(1 - \alpha)\%$ Confidence Interval for the Mean Value of y for $x = x_p$

$$\hat{y} \pm t_{\alpha/2}(\text{Estimated standard deviation of } \hat{y})$$

or

$$\hat{y} \pm t_{\alpha/2}\, s\sqrt{\frac{1}{n} + \frac{(x_p - \bar{x})^2}{SS_{xx}}}$$

where $t_{\alpha/2}$ is based on $(n - 2)$ df.

A $100(1 - \alpha)\%$ Prediction Interval for an Individual y for $x = x_p$

$$\hat{y} \pm t_{\alpha/2}[\text{Estimated standard deviation of } (y - \hat{y})]$$

or

$$\hat{y} \pm t_{\alpha/2}\, s\sqrt{1 + \frac{1}{n} + \frac{(x_p - \bar{x})^2}{SS_{xx}}}$$

where $t_{\alpha/2}$ is based on $(n - 2)$ df.

EXAMPLE 10.3

Find a 95% confidence interval for mean monthly sales when the appliance store spends \$400 on advertising.

Solution

For a \$400 advertising expenditure, $x_p = 4$ and, since $n = 5$, df $= n - 2 = 3$. Then the confidence interval for the mean value of y is

$$\hat{y} \pm t_{\alpha/2}\, s\sqrt{\frac{1}{n} + \frac{(x_p - \bar{x})^2}{SS_{xx}}} \quad \text{or} \quad \hat{y} \pm t_{.025}\, s\sqrt{\frac{1}{5} + \frac{(4 - \bar{x})^2}{SS_{xx}}}$$

Recall that $\hat{y} = 2.7$, $s = .61$, $\bar{x} = 3$, and $SS_{xx} = 10$. From Table VI in Appendix B, $t_{.025} = 3.182$. Thus, we have

$$2.7 \pm (3.182)(.61)\sqrt{\frac{1}{5} + \frac{(4 - 3)^2}{10}} = 2.7 \pm (3.182)(.61)(.55)$$
$$= 2.7 \pm 1.1 \quad \text{or} \quad (1.6, 3.8)$$

We are 95% confident that the interval from \$1,600 to \$3,800 encloses the mean sales revenue when the store expends \$400 a month on advertising. Note that we used a small amount of data for purposes of illustration in fitting the least squares line and that the width of the interval could be decreased by using a larger number of data points.

EXAMPLE 10.4

Predict the monthly sales for next month if a \$400 expenditure is to be made on advertising. Use a 95% prediction interval.

Solution

To predict the sales for a particular month for which $x_p = 4$, we calculate the 95% prediction interval as

$$\hat{y} \pm t_{\alpha/2}\, s\sqrt{1 + \frac{1}{n} + \frac{(x_p - \bar{x})^2}{SS_{xx}}} = 2.7 \pm (3.182)(.61)\sqrt{1 + \frac{1}{5} + \frac{(4 - 3)^2}{10}}$$
$$= 2.7 \pm (3.182)(.61)(1.14)$$
$$= 2.7 \pm 2.2 \quad \text{or} \quad (.5, 4.9)$$

Therefore, we predict with 95% confidence that the sales next month will fall in the interval from \$500 to \$4,900.

As with the confidence interval for the mean value of y (Example 10.3), the prediction interval for y (Example 10.4) is quite large. This is because we have chosen a small number of data points to fit the least squares line. The width of the prediction interval could be reduced by using more data points.

A comparison of the confidence interval for the mean value of y and the prediction interval for some future value of y for a \$400 advertising expenditure ($x = 4$) is illustrated in Figure 10.13. It is important to note that the prediction interval for an individual value of y will always be wider than the confidence interval for a mean value of y. You can see this by examining the formulas for the two intervals, and you can see it in Figure 10.13.

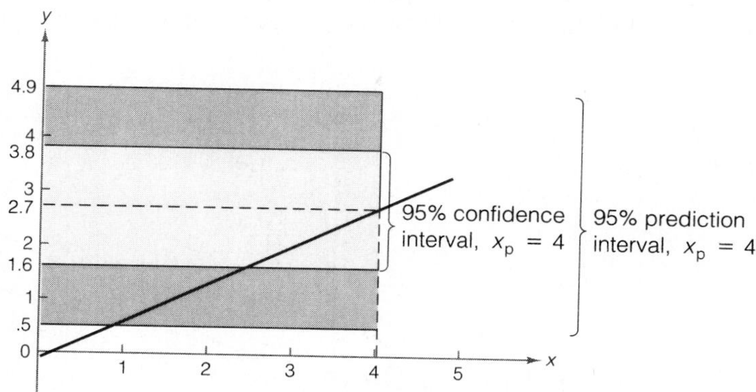

FIGURE 10.13 ► A 95% confidence interval for mean sales and a prediction interval for sales when $x = 4$

The error in estimating the mean value of y, $E(y)$, for a given value of x, say x_p, is the distance between the least squares line and the true line of means, $E(y) = \beta_0 + \beta_1 x$. This error, $[\hat{y} - E(y)]$, is shown in Figure 10.14. In contrast, the error $(y_p - \hat{y})$ in predicting some future value of y is the sum of two errors—the error of estimating the mean of y, $E(y)$, shown in Figure 10.14, plus the random error that is a component of the value of y to be predicted (see Figure 10.15, page 500). Consequently, the error of predicting a particular value of y will be larger than the error of estimating the mean value of y for a particular value of x.

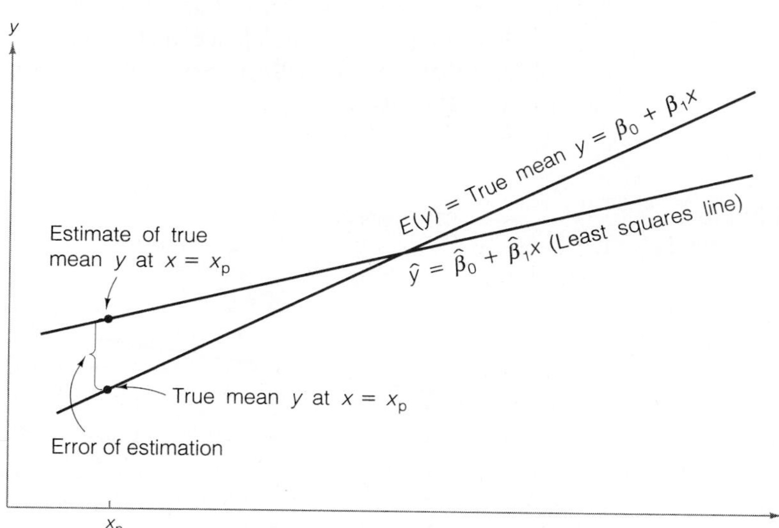

FIGURE 10.14 ► Error of estimating the mean value of y for a given value of x

Note from their formulas that both the error of estimation and the error of prediction take their smallest values when $x_p = \bar{x}$. The farther x_p lies from x, the larger will be the errors of estimation and prediction. You can see why this is true by noting the deviations for different values of x_p between the line of means $E(y) = \beta_0 + \beta_1 x$ and the predicted line of means $\hat{y} = \hat{\beta}_0 + \hat{\beta}_1 x$ shown in Figure 10.15. The deviation is larger at the extremities of the interval where the largest and smallest values of x in the data set occur.

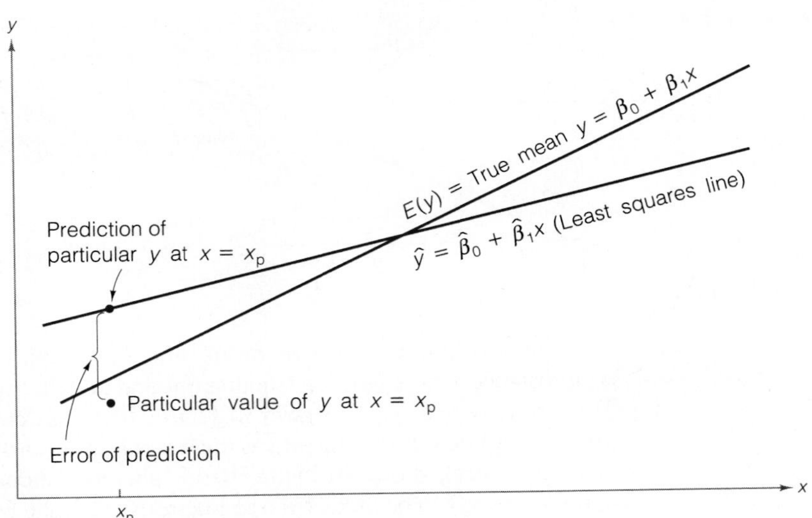

FIGURE 10.15 ▶ Error of predicting a future value of y for a given value of x

Both the confidence intervals for the mean values and the prediction intervals for individual values are depicted over the entire range of the regression line in Figure 10.16. You can see that the confidence interval is always narrower than the prediction interval, and that they are both narrowest at the mean $\bar{x}$, increasing steadily as the distance $|x - \bar{x}|$ increases.

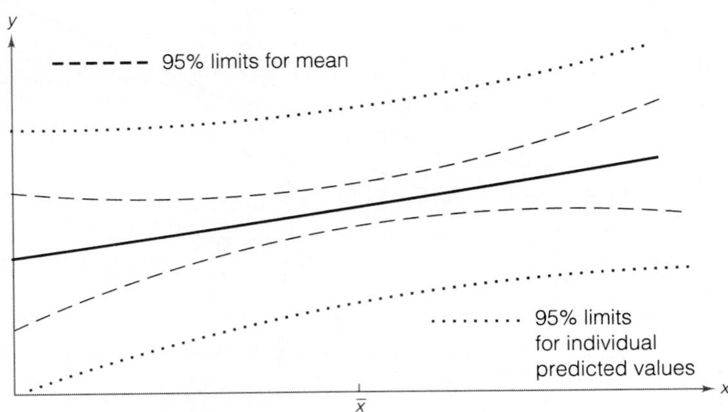

FIGURE 10.16 ▶ Confidence intervals for mean values and prediction intervals for individual values

The confidence interval width grows smaller as n is increased; thus, in theory, you can obtain as precise an estimate of the mean value of y as desired (at any given x) by selecting a large enough sample. The prediction interval for a particular value of y also grows smaller as n increases, but there is a lower limit on its width. If you examine the formula for the prediction interval (page 497), you will see that the interval can get no smaller than $\hat{y} \pm z_{\alpha/2}\sigma$.* Thus, the only way to obtain more accurate predictions for individual values of y is to reduce the standard deviation of the regression model, σ. This can be accomplished only by improving the model, either by using a curvilinear (rather than linear) relationship with x, or by adding new independent variables to the model, or both. Methods of improving the model are discussed in Chapters 11 and 12.

Exercises 10.52–10.61

Learning the Mechanics

10.52 Consider the following pairs of measurements:

x	−2	0	2	4	6	8	10
y	0	3	2	3	8	10	11

a. Construct a scattergram for these data.
b. Find the least squares line and plot it on your scattergram.
c. Find s^2.
d. Find a 90% confidence interval for the mean value of y when $x_p = 3$. Plot the upper and lower bounds of the confidence interval on your scattergram.
e. Find a 90% prediction interval for y when $x_p = 3$. Plot the upper and lower bounds of the prediction interval on your scattergram.
f. Compare the widths of the intervals you constructed in parts **d** and **e**. Which is wider and why?

10.53 Consider the following pairs of measurements:

x	4	6	0	5	2	3	2	6	2	1
y	3	5	−1	4	3	2	0	4	1	1

For these data, $SS_{xx} = 38.900$, $SS_{yy} = 33.600$, $SS_{xy} = 32.8$, and $\hat{y} = -.414 + .843x$.
a. Construct a scattergram for the data.
b. Plot the least squares line on your scattergram.
c. Use a 95% confidence interval to estimate the mean value of y when $x_p = 6$. Plot the upper and lower bounds of the interval on your scattergram.
d. Repeat part **c** for $x_p = 3.2$ and $x_p = 0$.
e. Compare the widths of the three confidence intervals you constructed in parts **c** and **d** and explain why they differ.

*The result follows from the facts that, for large n, $t_{\alpha/2} \approx z_{\alpha/2}$, $s \approx \sigma$, and the last two terms under the radical in the standard error of the predictor are approximately 0.

10.54 Refer to Exercise 10.53.

a. Using no information about x, estimate and calculate a 95% confidence interval for the mean value of y. [*Hint:* Use the one-sample t methodology of Section 7.3.]

b. Plot the estimated mean value and the confidence interval as horizontal lines on your scattergram of Exercise 10.53.

c. Compare the confidence intervals you calculated in parts **c** and **d** of Exercise 10.53 with the one you calculated in part **a** here. Does x seem to contribute information about the mean value of y?

d. Check the answer you gave in part **c** with a statistical test of the null hypothesis H_0: $\beta_1 = 0$ versus H_a: $\beta_1 \neq 0$. Use $\alpha = .05$.

10.55 In fitting a least squares line to $n = 10$ data points, the following quantities were computed: $SS_{xx} = 32$, $\bar{x} = 3$, $SS_{yy} = 26$, $\bar{y} = 4$, and $SS_{xy} = 28$.

a. Find the least squares line.
b. Graph the least squares line.
c. Calculate SSE.
d. Calculate s^2.
e. Find a 95% confidence interval for the mean value of y when $x_p = 2.5$.
f. Find a 95% prediction interval for y when $x_p = 4$.

Applying the Concepts

10.56 Many variables influence the sales of existing single-family homes. One of these is the interest rate charged for mortgage loans. Shown in the table are the total number of existing single-family homes sold annually, y, and the average annual conventional mortgage interest rate, x, for 1982–1991.

Year	*Homes Sold* y *(1,000s)*	*Interest Rate* x *(%)*	*Year*	*Homes Sold* y *(1,000s)*	*Interest Rate* x *(%)*
1982	1,990	14.8	1987	3,526	8.9
1983	2,719	12.3	1988	3,594	9.0
1984	2,868	12.0	1989	3,346	9.8
1985	3,214	11.2	1990	3,211	9.8
1986	3,565	9.8	1991	3,220	9.2

Source: U.S. Bureau of the Census. *Statistical Abstract of the United States: 1992*, pp. 502, 712.

a. Plot the data points on graph paper.

b. Find the least squares line relating y to x. As a check on your calculations, plot the line on your graph from part **a** to see if the line appears to model the relationship between y and x.

c. Do the data provide sufficient evidence to indicate that mortgage interest rates contribute information for the prediction of annual sales of existing single-family homes? Use $\alpha = .05$.

d. Calculate r and r^2, and interpret their values.

e. Find a 90% confidence interval for the mean annual number of existing single-family homes sold if the average annual mortgage interest rate is 10.0%.

f. Find a 90% prediction interval for the annual number of existing single-family homes sold if the average annual mortgage interest rate is 10.0%.

g. Explain why the widths of the intervals found in parts **e** and **f** differ.

10.57 The reasons given by workers for quitting their jobs generally fall into two categories: (1) seeking or taking a different job; or (2) withdrawing from the labor force. Economic theory suggests that wages and quit rates are related. Evidence regarding this relationship is important for understanding how labor markets operate (Koch and Ragan, 1986). The table lists quit rates (quits per 100 employees) and the average hourly wage in a sample of 15 manufacturing industries during 1986. For these data $\Sigma x_i = 129.050$, $\Sigma x_i^2 = 1{,}178.960$, $\Sigma y_i = 28.200$, $\Sigma y_i^2 = 64.340$, and $\Sigma x_i y_i = 218.806$.

Industry	*Quit Rate*	*Average Wage*	*Industry*	*Quit Rate*	*Average Wage*
1	1.4	\$ 8.20	9	2.0	\$ 7.99
2	.7	10.35	10	3.8	5.54
3	2.6	6.18	11	2.3	7.50
4	3.4	5.37	12	1.9	6.43
5	1.7	9.94	13	1.4	8.83
6	1.7	9.11	14	1.8	10.93
7	1.0	10.59	15	2.0	8.80
8	.5	13.29			

a. Construct a scattergram for these data.
b. Use the method of least squares to model the quit rate as a function of average hourly wage.
c. Do the data present sufficient evidence to conclude that average hourly wage rate contributes useful information for the prediction of quit rates? What does your model suggest about the relationship between quit rates and wages?
d. Find a 95% prediction interval for the quit rate in an industry with an average hourly wage of \$9.00.
e. Estimate the mean quit rate for industries with an average hourly wage of \$6.50. Use a 95% confidence interval.

10.58 Refer to Exercise 10.15. Find a 95% confidence interval for the mean price of a gallon of gasoline when crude oil costs \$20 per barrel.

10.59 Managers are an important part of any organization's resource base. Accordingly, the organization should be just as concerned about forecasting its future managerial needs as it is with forecasting its needs for, say, the natural resources used in its production processes. According to William F. Glueck (1977), a common forecasting procedure is to model the relationship between sales and the number of managers needed, since "the demand for managers is the result of the increases and decreases in the demand for products and services that an enterprise offers its customers and clients." To develop this relationship, the data shown in the table are collected from a firm's records. For these data $\Sigma x_i = 540$, $\Sigma x_i^2 = 19{,}098$, $\Sigma y_i = 423$, $\Sigma y_i^2 = 10{,}617$, and $\Sigma x_i y_i = 14{,}069$.

Month	*Units Sold, x*	*Managers, y*	*Month*	*Units Sold, x*	*Managers, y*
3/88	5	10	9/90	30	22
6/88	4	11	12/90	31	25
9/88	8	10	3/91	36	30
12/88	7	10	6/91	38	30
3/89	9	9	9/91	40	31
6/89	15	10	12/91	41	31
9/89	20	11	3/92	51	32
12/89	21	17	6/92	40	30
3/90	25	19	9/92	48	32
6/90	24	21	12/92	47	32

a. Use simple linear regression to model the relationship between the number of managers and the number of units sold.
b. Plot your least squares line on a scattergram of the data. Does it appear that the relationship between y and x is linear? If not, does it appear that your least squares model will provide a useful approximation to the relationship? Explain.
c. Test the usefulness of your model. Use $\alpha = .05$. State your conclusion in the context of the problem.
d. The company projects that next May it will sell 39 units. Use your least squares model to construct (1) a 90% confidence interval for the mean number of managers needed when the sales level is 39 and (2) a 90% prediction interval for the number of managers needed next May.
e. Compare the widths and interpret the two intervals you constructed in part **d**.

10.60 Refer to Exercise 10.24.
a. Use a 90% confidence interval to estimate the mean useful life of brand A cutting tools when the cutting speed is 45 meters per minute. Repeat for brand B. Compare the width of the two intervals, and comment on the reasons for any difference.
b. Use a 90% prediction interval to predict the useful life of a brand A cutting tool when the cutting speed is 45 meters per minute. Repeat for brand B. Compare the widths of the two intervals to each other, and to the two intervals you calculated in part **a**. Explain any differences.
c. Note that the estimation and prediction you performed in parts **a** and **b** were for a value of x that was not included in the original sample; that is, the value $x = 45$ was not part of the sample. However, the value is within the range of x values in the sample, so that the regression model spans the x value for which the estimation and prediction were made. In such situations, estimation and prediction represent *interpolations*.

Suppose you were asked to predict the useful life for a cutting speed of $x = 100$ meters per minute. Since the given value of x is outside the range of the sample x values, the prediction is an example of *extrapolation*. Predict the useful life of a brand A cutting tool that is operated at 100 meters per minute, and construct a 95% prediction interval for the actual useful life of the tool. What additional assumption do you have to make in order to assure the validity of an extrapolation?

10.61 Explain the implications of using a regression model for estimating the mean value of y for a given value of $x = x_p$ as x_p gets farther from the mean $\bar{x}$. What happens to the width of the confidence interval?

10.9 Simple Linear Regression: An Example

In the previous sections we have presented the basic elements necessary to fit and use a straight-line regression model. In this final section we assemble these elements by applying them in an example.

Suppose a fire insurance company wants to relate the amount of fire damage in major residential fires to the distance between the residence and the nearest fire station. The study is to be conducted in a large suburb of a major city; a sample of 15 recent fires in this suburb is selected. The amount of damage y and the distance x between the fire and the nearest fire station are recorded for each fire. The results are given in Table 10.7.

TABLE 10.7 Fire Damage Data

Distance from Fire Station	Fire Damage	Distance from Fire Station	Fire Damage
x (miles)	y ($1,000s)	x (miles)	y ($1,000s)
3.4	26.2	2.6	19.6
1.8	17.8	4.3	31.3
4.6	31.3	2.1	24.0
2.3	23.1	1.1	17.3
3.1	27.5	6.1	43.2
5.5	36.0	4.8	36.4
.7	14.1	3.8	26.1
3.0	22.3		

Step 1 First, we hypothesize a model to relate fire damage y to the distance x from the nearest fire station. We will hypothesize a straight-line probabilistic model:

$$y = \beta_0 + \beta_1 x + \epsilon$$

Step 2 Next, we use the data to estimate the unknown parameters in the deterministic component of the hypothesized model. We make some preliminary calculations:

$$\text{SS}_{xx} = \sum x_i^2 - \frac{\left(\sum x_i\right)^2}{15} = 196.16 - \frac{(49.2)^2}{15}$$
$$= 196.160 - 161.376 = 34.784$$

$$\text{SS}_{yy} = \sum y_i^2 - \frac{\left(\sum y_i\right)^2}{15} = 11{,}376.48 - \frac{(396.2)^2}{15}$$
$$= 11{,}376.480 - 10{,}464.96267 = 911.517334$$

$$\text{SS}_{xy} = \sum x_i y_i - \frac{\left(\sum x_i\right)\left(\sum y_i\right)}{15} = 1{,}470.65 - \frac{(49.2)(396.2)}{15}$$
$$= 1{,}470.650 - 1{,}299.536 = 171.114$$

Then the least squares estimate of the slope β_1 and intercept β_0 are

$$\hat{\beta}_1 = \frac{\text{SS}_{xy}}{\text{SS}_{xx}} = \frac{171.114}{34.784} = 4.919331$$

$$\hat{\beta}_0 = \bar{y} - \hat{\beta}_1 \bar{x} = \frac{396.2}{15} - 4.919331\left(\frac{49.2}{15}\right)$$
$$= 26.413333 - (4.919331)(3.28) = 26.413333 - 16.135406$$
$$= 10.277927$$

And the least squares equation is

$$\hat{y} = 10.278 + 4.919x$$

This prediction equation is graphed in Figure 10.17, along with a plot of the data points.

FIGURE 10.17 ► Least squares model for the fire damage data

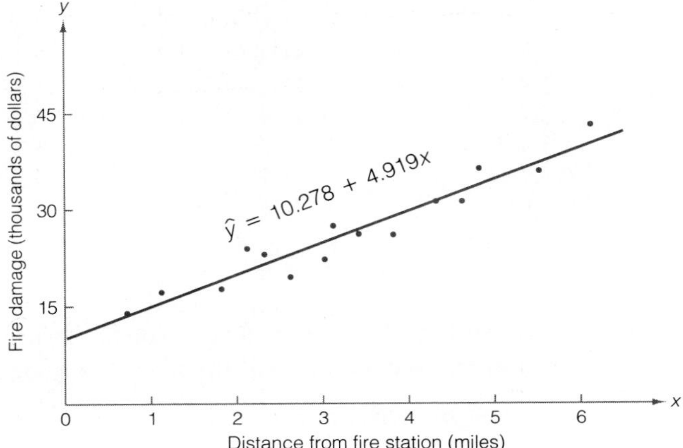

The least squares estimate of the slope β_1 is 4.919, which implies that the estimated mean damage increases by \$4,919 per additional mile from the fire station. This interpretation is valid over the range of x, or from .7 to 6.1 miles from the station. The estimated y-intercept, $\hat{\beta}_0 = 10.278$, has the interpretation that a fire 0 miles from the fire station has an estimated mean damage of \$10,278. Although this would seem to apply to the fire station itself, remember that the y-intercept can be interpreted meaningfully only if $x = 0$ is within the sampled range of the independent variable.

Step 3 Now, we specify the probability distribution of the random error component, ϵ. The assumptions about the distribution will be identical to those listed in Section 10.3. Although we know that these assumptions are not completely satisfied (they rarely are for any practical problem), we are willing to assume they are approximately satisfied for this example. We have to estimate the variance σ^2 of ϵ, so we calculate

$$\text{SSE} = \sum (y_i - \hat{y}_i)^2 = \text{SS}_{yy} - \hat{\beta}_1\text{SS}_{xy}$$

where the last expression represents a shortcut formula for SSE. Thus,

$$\begin{aligned}\text{SSE} &= 911.517334 - (4.919331)(171.114)\\ &= 911.517334 - 841.766405 = 69.750929^*\end{aligned}$$

*For problems where rounding is necessary, at least six significant figures should be carried for these quantities. Otherwise, the calculated value of SSE may be substantially in error.

To estimate σ^2, we divide SSE by the degrees of freedom available for error, $n - 2$. Thus,

$$s^2 = \frac{\text{SSE}}{n - 2} = \frac{69.750929}{15 - 2} = 5.3655$$

$$s = \sqrt{5.3655} = 2.32$$

Step 4 We can now check the usefulness of the hypothesized model—that is, whether x really contributes information for the prediction of y using the straight-line model. First test the null hypothesis that the slope β_1 is zero—i.e., that there is no linear relationship between fire damage and the distance from the nearest fire station—against the alternative that the slope β_1 differs from zero. We test

$$H_0\text{: } \beta_1 = 0 \qquad H_a\text{: } \beta_1 \neq 0$$

Test statistic: $t = \dfrac{\hat{\beta}_1 - 0}{s_{\hat{\beta}_1}} = \dfrac{\hat{\beta}_1}{s/\sqrt{\text{SS}_{xx}}}$

Assumptions: Those made about ϵ in Section 10.3.

For $\alpha = .05$, we will reject H_0 if $t > t_{\alpha/2}$ or $t < -t_{\alpha/2}$, where for $n = 15$, $\text{df} = n - 2 = 15 - 2 = 13$ and $t_{.025} = 2.160$. We then calculate the t statistic:

$$t = \frac{\hat{\beta}_1}{s_{\hat{\beta}_1}} = \frac{\hat{\beta}_1}{s/\sqrt{\text{SS}_{xx}}} = \frac{4.919}{2.32/\sqrt{34.784}} = \frac{4.919}{.393} = 12.5$$

This large t value leaves little doubt that the distance between the fire and the fire station contributes information for the prediction of fire damage. Furthermore, it appears (as we might suspect) that fire damage increases with distance.

We gain additional information about the relationship by forming a confidence interval for the slope β_1. A 95% confidence interval is

$$\hat{\beta}_1 \pm t_{.025}\, s_{\hat{\beta}_1} = 4.919 \pm (2.160)(.393)$$
$$= 4.919 \pm .849 = (4.070,\ 5.768)$$

We estimate that the interval from \$4,070 to \$5,768 encloses the mean increase (β_1) in fire damage per additional mile of distance from the fire station.

Another measure of the usefulness of the model is the coefficient of correlation r:

$$r = \frac{\text{SS}_{xy}}{\sqrt{\text{SS}_{xx}\text{SS}_{yy}}} = \frac{171.114}{\sqrt{(34.784)(911.517)}} = \frac{171.114}{178.062} = .96$$

The high correlation provides further support for our conclusion that β_1 differs from 0; it appears that fire damage and distance from the fire station are highly correlated.

The coefficient of determination is $r^2 = (.96)^2 = .92$, which implies that 92% of the sum of squares of deviations in the sample of y values about $\bar{y}$ is explained by the distance x between the fire and the fire station. All signs point to a strong linear relationship between y and x.

Step 5 We are now prepared to use the least squares model. Suppose the insurance company wants to predict the fire damage if a major residential fire were to occur 3.5 miles from the nearest fire station; i.e., $x_p = 3.5$. The predicted value is

$$\hat{y} = \hat{\beta}_0 + \hat{\beta}_1 x_p = 10.278 + (4.919)(3.5) = 10.278 + 17.216 = 27.5$$

(we round to the nearest tenth to be consistent with the units of the original data in Table 10.7). If we want a 95% prediction interval, we calculate

$$\hat{y} \pm t_{.025}\, s\sqrt{1 + \frac{1}{n} + \frac{(x_p - \bar{x})^2}{SS_{xx}}} = 27.5 \pm (2.16)(2.32)\sqrt{1 + \frac{1}{15} + \frac{(3.5 - 3.28)^2}{34.784}}$$
$$= 27.5 \pm (2.16)(2.32)\sqrt{1.0681}$$
$$= 27.5 \pm 5.2 \quad \text{or} \quad (22.3, 32.7)$$

The model yields a 95% prediction interval for fire damage in a major residential fire 3.5 miles from the nearest station of \$22,300 to \$32,700.

One Caution Before Closing We would not use this prediction model to make predictions for homes less than .7 mile or more than 6.1 miles from the nearest fire station. A look at the data in Table 10.7 reveals that all the x values fall between .7 and 6.1. It is dangerous to use the model to make predictions outside the region in which the sample data fall. A straight line might not provide a good model for the relationship between the mean value of y and x for values of x beyond the range of the sample data.

10.10 Using the Computer for Simple Linear Regression

All the examples of simple linear regression that we have presented thus far have required rather tedious calculations involving SS_{yy}, SS_{xx}, SS_{xy}, and so forth. Even with the use of a calculator, the process is laborious and susceptible to error. Fortunately, the use of computers can significantly reduce the labor involved in regression calculations. In this section we introduce the regression output from one statistical software package, the SAS System. Though this is just one of the many statistical packages available that provide simple linear regression output, most produce essentially the same quantities, differing only in format and labeling. The regression output of other packages is presented in Chapter 11.

The SAS output for the fire damage example of Section 10.9 is presented in Figure 10.18. We have shaded the parts of the printout corresponding to most of the key simple linear regression quantities introduced in this chapter.

First, the estimates of the y-intercept and the slope are found about halfway down the printout on the left-hand side, under the column labeled **Parameter Estimate** and

FIGURE 10.18 ►
SAS printout for fire damage regression analysis

FIRE DAMAGE EXAMPLE
STRAIGHT-LINE MODEL WITH PREDICTION INTERVALS

Model: MODEL1
Dep Variable: Y

Analysis of Variance

Source	DF	Sum of Squares	Mean Square	F Value	Prob>F
Model	1	841.76636	841.76636	156.886	0.0001
Error	13	69.75098	5.36546		
C Total	14	911.51733			

Root MSE	2.31635	R-Square	0.9235
Dep Mean	26.41333	Adj R-Sq	0.9176
C.V.	8.76961		

Parameter Estimates

Variable	DF	Parameter Estimate	Standard Error	T for H0: Parameter=0	Prob > |T|
INTERCEP	1	10.277929	1.42027781	7.237	0.0001
X	1	4.919331	0.39274775	12.525	0.0001

Obs	X	Y	Predict Value	Residual	Lower95% Predict	Upper95% Predict
1	3.4	26.2000	27.0037	-0.8037	21.8344	32.1729
2	1.8	17.8000	19.1327	-1.3327	13.8141	24.4514
3	4.6	31.3000	32.9068	-1.6068	27.6186	38.1951
4	2.3	23.1000	21.5924	1.5076	16.3577	26.8271
5	3.1	27.5000	25.5279	1.9721	20.3573	30.6984
6	5.5	36.0000	37.3342	-1.3342	31.8334	42.8351
7	0.7	14.1000	13.7215	0.3785	8.1087	19.3342
8	3	22.3000	25.0359	-2.7359	19.8622	30.2097
9	2.6	19.6000	23.0682	-3.4682	17.8678	28.2686
10	4.3	31.3000	31.4311	-0.1311	26.1908	36.6713
11	2.1	24.0000	20.6085	3.3915	15.3442	25.8729
12	1.1	17.3000	15.6892	1.6108	10.1999	21.1785
13	6.1	43.2000	40.2858	2.9142	34.5906	45.9811
14	4.8	36.4000	33.8907	2.5093	28.5640	39.2175
15	3.8	26.1000	28.9714	-2.8714	23.7843	34.1585
16 *	3.5	.	27.4956	.	22.3239	32.6672

Sum of Residuals	-3.73035E-14
Sum of Squared Residuals	69.7510
Predicted Resid SS (Press)	93.2117

in the rows labeled **INTERCEP** and **X**, respectively. The values are $\hat{\beta}_0 = 10.277929$ and $\hat{\beta}_1 = 4.919331$. When rounded to three decimal places, these quantities agree with our calculations in Section 10.9.

Next, we find the measures of variability: SSE, s^2, and s. They are shaded in the upper portion of the printout. SSE is found under the column heading **Sum of Squares** and in the row labeled **Error**: SSE = 69.75098. The estimate of the error variance σ^2 is under the column heading **Mean Square** and in the row labeled **Error**: $s^2 = 5.36546$. The estimate of the standard deviation σ is directly to the right of the heading **Root MSE**: $s = 2.31635$. Again, all values (after rounding) agree with the corresponding quantities we calculated in Section 10.9.

The coefficient of determination is shown (shaded) to the right of the heading **R-Square** in the upper portion of the printout: $r^2 = .9235$. Again, to two decimal places, this agrees with our calculation in Section 10.9. The coefficient of correlation r is not given on the printout.

The t statistic for testing H_0: $\beta_1 = 0$ versus H_a: $\beta_1 \neq 0$ is given (shaded) in the center of the page under the column heading **T for H0: Parameter = 0** in the row corresponding to **X**. The value $t = 12.525$ agrees with our computed value when we used the formula

$$t = \frac{\hat{\beta}_1}{s_{\hat{\beta}_1}} = \frac{\hat{\beta}_1}{s/\sqrt{SS_{xx}}}$$

To determine which hypothesis this test statistic supports, we can establish a rejection region using the t table (Table VI of Appendix B), just as we did in Section 10.9. However, the printout makes this unnecessary, because the observed significance level, or p-value, is shown (shaded) immediately to the right of the t statistic, under the column heading **Prob > |T|**. Remember that if the observed significance level is less than the α value you select, then the test statistic supports the alternative hypothesis at that level. For example, if we select $\alpha = .05$ in this example, the observed significance level of .0001 given on the printout indicates that we should reject H_0. We can conclude that there is sufficient evidence at $\alpha = .05$ to infer that a linear relationship between fire damage and distance from the station is useful for predicting damage.

If you wish to conduct a one-tailed test, the observed significance level is half that given on the printout (assuming the sign of the test statistic agrees with the alternative hypothesis). Thus, if we were testing H_0: $\beta_1 = 0$ versus H_a: $\beta_1 > 0$ in this example, the t statistic is positive, agreeing with H_a, so that the observed significance level would be $\frac{1}{2}(.0001) = .00005$.

Predicted y values and the corresponding prediction intervals are given in the lower portion of the SAS printout. To find the 95% prediction interval for the fire damage y when the distance from the fire station is $x = 3.5$ miles, first locate the value 3.5 in the column labeled **X** (the last value in the column). The prediction is given in the center column labeled **Predict Value** in the row corresponding to 3.5:

$$\hat{y} = 27.4956$$

The lower and upper confidence bounds are given in the columns headed **Lower** and **Upper 95% Predict**, respectively:

Lower = 22.3239   Upper = 32.6672

Again, after rounding all values agree with our calculations in Section 10.9.

Although much more information is given on the SAS printout, we have discussed only those aspects that have been presented in the chapter. In the next two chapters we expand our discussion to include other important components of the SAS printout, as well as other statistical packages. The point here is that the computer can alleviate much of the burden of calculation involved in a regression analysis, and enable us to spend more time on the interpretation of the model. The time spent in learning how to read computer regression output is a good investment.

Exercises 10.62–10.63

Learning the Mechanics

10.62 Refer to Exercise 10.14, in which the extent of retaliation is related to the salary of 15 whistle blowers using a least squares model. A SAS printout for this regression model is given.

```
                    RETALIATION INDEX VS. SALARY

Model: MODEL1
Dep Variable: R_INDEX

                        Analysis of Variance

                      Sum of          Mean
  Source      DF     Squares        Square       F Value       Prob>F

  Model        1   20853.93900   20853.93900       0.925       0.3538
  Error       13  293208.46100   22554.49700
  C Total     14  314062.40000

       Root MSE      150.18155     R-Square       0.0664
       Dep Mean      499.80000     Adj R-Sq      -0.0054
       C.V.           30.04833

                        Parameter Estimates

                   Parameter       Standard     T for H0:
 Variable   DF      Estimate          Error   Parameter=0     Prob > |T|

 INTERCEP    1    575.028672    87.31829499         6.585         0.0001
 SALARY      1     -0.002186     0.00227386        -0.962         0.3538
```

Continued

Obs	SALARY	R_INDEX	Predict Value	Residual	Lower95% Mean	Upper95% Mean
1	62000	301.0	439.5	-138.5	280.1	598.8
2	36500	550.0	495.2	54.7770	410.8	579.6
3	17600	755.0	536.5	218.5	418.9	654.2
4	20000	327.0	531.3	-204.3	421.6	641.0
5	30100	500.0	509.2	-9.2163	422.8	595.6
6	35000	377.0	498.5	-121.5	414.7	582.3
7	47500	290.0	471.2	-181.2	365.6	576.8
8	54000	452.0	457.0	-4.9600	329.4	584.6
9	15800	535.0	540.5	-5.4827	416.5	664.5
10	44000	455.0	478.8	-23.8246	382.7	574.9
11	46600	615.0	473.1	141.9	370.2	576.1
12	12100	700.0	548.6	151.4	410.6	686.5
13	62000	650.0	439.5	210.5	280.1	598.8
14	21000	630.0	529.1	100.9	422.6	635.7
15	11900	360.0	549.0	-189.0	410.3	687.7

a. Find the least squares estimates of the intercept β_0 and the slope β_1.
b. Identify the values of SSE, s^2, and s. Based on the standard deviation, how accurately do you expect to be able to predict the retaliation index using the salary of the whistle blower?
c. Find and interpret the coefficient of determination r^2.
d. Find the t statistic and its observed significance level for testing the usefulness of the model. Interpret the values.
e. Find and interpret the 95% confidence interval for the mean retaliation index of whistle blowers with salaries of \$35,000.

10.63 Refer to Exercise 10.32, in which an index of managerial success was related to the number of interactions with outsiders. The SAS simple linear regression printout for these data is shown.

Source	DF	Sum of Squares	Mean Square	F Value	Prob>F
Model	1	606.03751	606.03751	1.610	0.2216
Error	17	6400.59407	376.50553		
C Total	18	7006.63158			

Root MSE	19.40375	R-Square	0.0865
Dep Mean	54.57895	Adj R-Sq	0.0328
C.V.	35.55171		

Parameter Estimates

Variable	DF	Parameter Estimate	Standard Error	T for H0: Parameter=0	Prob > |T|
INTERCEP	1	44.130454	9.36159293	4.714	0.0002
INTERACT	1	0.236617	0.18650103	1.269	0.2216

a. Find the least squares equation, and compare it to the one you calculated in Exercise 10.32.
b. Find the standard deviation s, and interpret the value in terms of this exercise.
c. Find and interpret r^2.

d. Is there sufficient evidence to indicate that the model is useful for predicting y? What is the observed significance level of the test?

Summary

We have introduced an extremely useful tool in this chapter—**the method of least squares** for fitting a prediction equation to a set of data. The application of this methodology, along with the associated inferential procedures, is referred to as **regression analysis**. In five steps we showed how to use sample data to build a model relating a dependent variable y to a single independent variable x:

1. Hypothesize a **probabilistic model**. In this chapter, we confined our attention to the **first-order (straight-line) model**, $y = \beta_0 + \beta_1 x + \epsilon$.
2. Use the method of least squares to estimate the unknown parameters in the **deterministic component**, $\beta_0 + \beta_1 x$. The least squares estimates yield a model $\hat{y} = \hat{\beta}_0 + \hat{\beta}_1 x$ with a **sum of squared errors** (SSE) that is smaller than the SSE for any other straight-line model.
3. Specify the probability distribution of the **random error component**, ϵ.
4. Assess the usefulness of the hypothesized model. Included here are making inferences about the **slope**, β_1; calculating the **coefficient of correlation**, r; and calculating the **coefficient of determination**, r^2.
5. Finally, if we are satisfied with the model, we can use it to **estimate the mean y value**, $E(y)$, for a given x value and/or to **predict an individual y value** for a specific value of x.

Supplementary Exercises 10.64–10.75

10.64 In fitting a least squares line to $n = 15$ data points, the following quantities were computed: $SS_{xx} = 55$, $SS_{yy} = 198$, $SS_{xy} = -88$, $\bar{x} = 17$, and $\bar{y} = 35$.

a. Find the least squares line. **b.** Graph the least squares line.
c. Calculate SSE. **d.** Calculate s^2.
e. Find a 90% confidence interval for β_1. Interpret this interval.
f. Find a 90% confidence interval for the mean value of y when $x_p = 15$.
g. Find a 90% prediction interval for y when $x_p = 15$.

10.65 Consider the following data:

y	5	1	3
x	5	1	3

a. Construct a scattergram for the data.
b. It is possible to find many lines for which $\Sigma(y - \hat{y}) = 0$. For this reason, the criterion $\Sigma(y - \hat{y}) = 0$ is not used for identifying the "best" fitting straight line. Find two lines for which $\Sigma(y - \hat{y}) = 0$.
c. Find the least squares line.
d. Compare the value of SSE for the least squares line to that of the two lines you found in part **b**. What principle of the method of least squares is demonstrated by this comparison?

10.66 Emotional exhaustion, or *burnout*, is a significant problem for people with careers in the field of human services. It seriously affects productivity and feelings of job satisfaction. Leiter and Meechan (1986) used regression analysis to investigate the relationship between burnout and aspects of the human services professional's job and job-related behavior. To measure emotional exhaustion, they used the Maslach Burnout Inventory, a questionnaire. One of the independent variables considered was the proportion of social contacts with individuals who belong to a person's work group. They called this variable *concentration*. The table lists the values of the emotional exhaustion index (higher values indicate greater exhaustion) and concentration for a sample of 25 human services professionals who work in a large public hospital. For this data set, $SS_{yy} = 1{,}800{,}417.44$, $SS_{xx} = 14{,}026.16$, $SS_{xy} = 124{,}348.520$, $\bar{y} = 578.32$, and $\bar{x} = 68.560$.

Exhaustion Index, y	*Concentration, x*	*Exhaustion Index, y*	*Concentration, x*
100	20%	493	86%
525	60	892	83
300	38	527	79
980	88	600	75
310	79	855	81
900	87	709	75
410	68	791	77
296	12	718	77
120	35	684	77
501	70	141	17
920	80	400	85
810	92	970	96
506	77		

a. Construct a scattergram for the data. Do the variables x and y appear to be related?
b. Calculate the correlation coefficient for the data and interpret its value.
c. Use the method of least squares to estimate the straight-line model relating emotional exhaustion and concentration, and plot it on your scattergram.
d. Calculate the coefficient of determination for your least squares line and interpret it.
e. Test the usefulness of the straight-line relationship with concentration for predicting burnout. Use $\alpha = .05$.
f. Is there evidence that the correlation between emotional exhaustion and concentration differs from 0? Does your conclusion mean that concentration causes emotional exhaustion? Explain.
g. Use a 95% prediction interval to forecast the level of emotional exhaustion for a human services professional for whom 80% of all social contacts are within her work group. Interpret the prediction in the context of the problem.

h. Use a 95% confidence interval to estimate the mean exhaustion level for all professionals who have 80% of their social contacts within their work groups. Interpret the interval, and compare its width with that of the prediction interval in part **g**.

10.67 *Work standards* specify time, cost, and efficiency norms for the performance of work tasks. They are typically used to monitor job performance. In the distribution center of McCormick and Co., Inc., data were collected to develop work standards for the time to assemble or fill customer orders. The table contains data for a random sample of 9 orders.

Time (*mins.*)	*Order Size* (*cases*)
27	36
15	34
71	255
35	103
8	4
60	555
3	6
10	60
10	96

Source: Boyle, D., Ray, B. A., and Kahan, G. "Work standards—the quality way." *Production and Inventory Management Journal*, Second Quarter, 1991, p. 67.

a. Construct a scattergram for these data and interpret it.
b. Fit a least squares line to these data using time as the dependent variable.
c. In general, we would expect the mean time to fill an order to increase with the size of the order. Do the data support this theory? Test using $\alpha = .05$.
d. Find a 95% confidence interval for the mean time to fill an order consisting of 150 cases.

10.68 Refer to Exercise 10.32, in which managerial success y was modeled as a function of the number of contacts x a manager makes with people outside his or her work unit during a specific period of time. Recall that $SS_{yy} = 7{,}006.6316$, $SS_{xx} = 10{,}824.5263$, $SS_{xy} = 2{,}561.2632$, $\bar{y} = 54.5789$, and $\bar{x} = 44.1579$.

a. A manager observed for 2 weeks made 55 contacts with people outside her work unit. Predict the value of the manager's success index. Use a 90% prediction interval.
b. A second manager observed for 2 weeks made 110 contacts outside his work unit. Give two reasons why caution should be exercised in using the least squares model developed from the given data set to construct a prediction interval for this manager's success index.
c. In the context of this problem, determine the value of x whose associated prediction interval for y is the narrowest.

10.69 The preferred mode of long-distance travel has changed dramatically since 1950. The table shows the number of passengers (in millions) carried by scheduled airlines and railroads in the United States over the period from 1950 to 1990. For these data $\Sigma y = 3{,}917$, $\Sigma x = 5{,}011$, $\Sigma y^2 = 1{,}278{,}301$, $\Sigma x^2 = 1{,}628{,}915$, and $\Sigma xy = 1{,}165{,}664$.

Year	Air Passengers, y	Rail Passengers, x	Year	Air Passengers, y	Rail Passengers, x
1950	19	488	1977	240	276
1955	42	433	1978	275	262
1960	62	327	1979	317	274
1965	103	306	1980	297	280
1970	169	289	1985	382	275
1974	208	275	1988	455	325
1975	205	270	1989	454	330
1976	223	272	1990	466	329

Source: U.S. Bureau of the Census. *Statistical Abstract of the United States: 1982, 1992.*

a. Find the correlation coefficient and the coefficient of determination for the data in the table and interpret their values.
b. Do the data provide sufficient evidence to indicate that x and y are correlated? Test using $\alpha = .05$.

10.70 Sometimes it is known from theoretical considerations that the straight-line relationship between two variables, x and y, passes through the origin of the xy-plane. Consider the relationship between the total weight of a shipment of 50-pound bags of flour, y, and the number of bags in the shipment, x. Since a shipment containing $x = 0$ bags (i.e., no shipment at all) has a total weight of $y = 0$, a straight-line model of the relationship between x and y should pass through the point $x = 0$, $y = 0$. In such a case you could assume $\beta_0 = 0$ and characterize the relationship between x and y with the model

$$y = \beta_1 x + \epsilon$$

The least squares estimate of β_1 for this model is

$$\hat{\beta}_1 = \frac{\sum x_i y_i}{\sum x_i^2}$$

From the records of past flour shipments, 15 shipments were randomly chosen and the data shown in the table were recorded.

Weight of Shipment	50-Pound Bags in Shipment	Weight of Shipment	50-Pound Bags in Shipment
5,050	100	7,162	150
10,249	205	24,000	500
20,000	450	4,900	100
7,420	150	14,501	300
24,685	500	28,000	600
10,206	200	17,002	400
7,325	150	16,100	400
4,958	100		

a. Find the least squares line for the given data under the assumption that $\beta_0 = 0$. Plot the least squares line on a scattergram of the data.

b. Find the least squares line for the given data using the model

$$y = \beta_0 + \beta_1 x + \epsilon$$

(i.e., do not restrict β_0 to equal 0). Plot this line on the same scatterplot you constructed in part **a**.

c. Refer to part **b**. Why might $\hat{\beta}_0$ be different from 0 even though the true value of β_0 is known to be 0?

d. The estimated standard error of $\hat{\beta}_0$ is equal to

$$s\sqrt{\frac{1}{n} + \frac{\bar{x}^2}{SS_{xx}}}$$

Use the t statistic

$$t = \frac{\hat{\beta}_0 - 0}{s\sqrt{\frac{1}{n} + \frac{\bar{x}^2}{SS_{xx}}}}$$

to test the null hypothesis $H_0\colon \beta_0 = 0$ against the alternative $H_a\colon \beta_0 \neq 0$. Use $\alpha = .10$. Should you include β_0 in your model?

10.71 As a result of the increase in suburban shopping centers, many downtown stores are suffering financially. A downtown department store thinks that increased advertising might help lure more shoppers into the area. To study the effect of advertising on sales, records were obtained for several midyear months during which the store varied advertising expenditures. Those records are shown in the table.

Advertising Expense x *(\$1,000s)*	*Sales* y *(\$1,000s)*
.9	30
1.1	34
.8	32
1.2	37
.7	31

a. Estimate the coefficient of correlation between sales and advertising expenditures.

b. Do the data provide sufficient evidence to indicate a nonzero correlation between sales y and advertising expense x?

c. Can you use the results of parts **a** and **b** to conclude that additional advertising expense will *cause* sales to increase? Why or why not?

10.72 *Comparable worth* is a compensation plan designed to eliminate pay inequities among jobs of similar worth. A number of state and municipal governments have adopted such plans, and some unions have sought comparable worth clauses for their contracts. To develop such a plan, a sample of benchmark jobs is evaluated and assigned points (x) based on factors such as responsibility, skill, effort, and working conditions. A market survey is conducted to determine the market rates or salaries (y) of the benchmark jobs. A regression analysis is then used to characterize the relationship between salary and job evaluation points (Scholl and Cooper, 1991). The table gives job evaluation points and salaries for a set of 21 benchmark jobs.

Points, x	Salary, y	Job	Points, x	Salary, y	Job
970	$15,704	Electrician	1865	$17,341	Registered nurse
500	13,984	Semi-skilled laborer	1065	15,194	Licensed practical nurse
370	14,196	Motor equipment operator	880	13,614	Principal clerk typist
220	13,380	Janitor	340	12,594	Clerk typist
250	13,153	Laborer	540	13,126	Senior clerk stenographer
1350	18,472	Senior engineering technician	490	12,958	Senior clerk typist
470	14,193	Senior janitor	940	13,894	Principal clerk stenographer
2040	20,642	Revenue agent	600	13,380	Institutional attendant
370	13,614	Engineering aide	805	15,559	Eligibility technician
1200	16,869	Electrician supervisor	220	13,844	Cook's helper
820	15,184	Senior maintenance technician			

a. Construct a scattergram for these data. What does it suggest about the relationship between salary and job evaluation points?

b. A straight-line model was fit to these data using the SAS system, and the printout is given here. Identify and interpret the least squares equation.

```
Dependent Variable: Y

                         Analysis of Variance

                           Sum of          Mean
Source           DF       Squares        Square        F Value        Prob>F

Model             1  66801750.334  66801750.334         74.670        0.0001
Error            19  16997968.904  894629.94232
C Total          20  83799719.238

    Root MSE       945.84879      R-square        0.7972
    Dep Mean     14804.52381      Adj R-sq        0.7865
    C.V.             6.38892

                         Parameter Estimates

                     Parameter       Standard     T for H0:
Variable   DF         Estimate          Error   Parameter=0      Prob > |T|

INTERCEP    1            12024   382.31829064        31.449          0.0001
X           1         3.581616     0.41448305         8.641          0.0001

                  Dep Var    Predict    Std Err   Lower95%   Upper95%
Obs        X            Y      Value    Predict    Predict    Predict   Residual

  1      970      15704.0    15497.8    221.447    13464.6    17531.0      206.2
  2      500      13984.0    13814.5    236.070    11774.1    15854.9      169.5
  3      370      14196.0    13348.9    266.420    11292.1    15405.6      847.1
  4      220      13380.0    12811.6    309.502    10728.6    14894.6      568.4
  5      250      13153.0    12919.1    300.351    10842.0    14996.2      233.9
  6     1350      18472.0    16858.8    314.833    14772.4    18945.3     1613.2
  7      470      14193.0    13707.0    242.349    11663.4    15750.6      486.0
  8     2040      20642.0    19330.2    562.933    17026.4    21633.9     1311.8
  9      370      13614.0    13348.9    266.420    11292.1    15405.6      265.1
 10     1200      16869.0    16321.6    270.968    14262.3    18380.9      547.4
 11      820      15184.0    14960.6    207.190    12934.0    16987.2      223.4
 12     1865      17341.0    18703.4    496.163    16467.8    20938.9    -1362.4
 13     1065      15194.0    15838.1    238.553    13796.4    17879.8     -644.1
 14      880      13614.0    15175.5    210.818    13147.2    17203.7    -1561.5
 15      340      12594.0    13241.4    274.451    11180.1    15302.7     -647.4
 16      540      13126.0    13957.7    228.483    11921.1    15994.4     -831.7
 17      490      12958.0    13778.6    238.109    11737.2    15820.1     -820.6
 18      940      13894.0    15390.4    217.251    13359.1    17421.6    -1496.4
 19      600      13380.0    14172.6    218.972    12140.6    16204.7     -792.6
 20      805      15559.0    14906.9    206.741    12880.4    16933.3      652.1
 21      220      13844.0    12811.6    309.502    10728.6    14894.6     1032.4
 22      800            .    14888.9    206.632    12862.6    16915.3          .
```

c. Interpret the value of r^2 for this least squares equation.

d. Is there sufficient evidence to conclude that a straight-line model provides useful information about the relationship in question? Interpret the p-value for this test.

e. A job outside the set of benchmark jobs is evaluated and receives a score of 800 points. Under the comparable worth plan, what is a reasonable range within which a fair salary for this job should be found?

10.73 In the late 1970s and early 1980s, the prices of single-family homes in the United States rose faster than the rate of inflation. As a result, many investors directed their funds to the housing market as a hedge against inflation. One way an investor can assess the value of a specific house is to compare it to recent sale prices for similar houses. Another popular approach, according to Cho and Reichert (1980), uses a regression analysis to model the relationship between price and the variables that influence price. Independent variables that could be utilized are total living area, number of rooms, number of baths, age of property, and so on. Of these factors, Cho and Reichart indicate that total living area provides the most information for determining the worth of a house. The table below lists the final selling price and total living area for a sample of 24 homes in the same geographic area that were sold during the last 3 months of 1990. The regression printout using a straight-line model to relate price to area for these data is given on page 520.

Area (sq. ft.)	*Price*
2,100	$150,000
1,455	114,900
1,630	106,500
2,600	195,000
1,210	75,500
1,857	126,600
2,000	135,400
2,400	178,650
2,256	145,100
1,290	62,600
2,332	168,200
1,725	138,100
3,000	205,000
1,400	79,400
2,750	200,000
2,900	215,100
2,500	180,800
1,535	120,900
1,333	70,000
2,455	165,200
3,010	185,000
2,180	160,000
1,870	119,900
1,582	99,900

a. Construct a scattergram for the data.

b. Find the least squares line and plot it on your scattergram.

Source	DF	Sum of Squares	Mean Square	F Value	Prob>F
Model	1	42825909688	42825909688	247.857	0.0001
Error	22	3801265207.6	172784782.16		
C Total	23	46627174896			

Root MSE	13144.76254	R-Square	0.9185
Dep Mean	141572.91667	Adj R-Sq	0.9148
C.V.	9.28480		

Parameter Estimates

Variable	DF	Parameter Estimate	Standard Error	T for H0: Parameter=0	Prob > |T|
INTERCEP	1	-15124	10308.485043	-1.467	0.1565
AREA	1	76.174547	4.83848421	15.743	0.0001

Obs	AREA	PRICE	Predict Value	Residual	Lower95% Mean	Upper95% Mean
1	2100	150000	144842	5157.9	139261	150423
2	1455	114900	95709	19190.5	87495.9	103923
3	1630	106500	109040	-2540.0	102017	116064
4	2600	195000	182929	12070.7	175142	190717
5	1210	75500.0	77046.7	-1546.7	66887.4	87206.1
6	1857	126600	126332	268.3	120416	132247
7	2000	135400	137225	-1824.6	131631	142819
8	2400	178650	167694	10955.6	161152	174237
9	2256	145100	156725	-11625.3	150814	162637
10	1290	62600.0	83140.7	-20540.7	73642.8	92638.6
11	2332	168200	162515	5685.4	156304	168725
12	1725	138100	116277	21823.4	109791	122763
13	3000	205000	213399	-8399.2	202423	224376
14	1400	79400.0	91519.9	-12119.9	82892.2	100148
15	2750	200000	194356	5644.5	185450	203261
16	2900	215100	205782	9318.3	195657	215906
17	2500	180800	175312	5488.1	168190	182433
18	1535	120900	101803	19096.5	94160.9	109446
19	1333	70000.0	86416.2	-16416.2	77264.5	95568
20	2455	165200	171884	-6684.0	165035	178733
21	3010	185000	214161	-29160.9	203098	225224
22	2180	160000	150936	9064.0	145236	156636
23	1870	119900	127322	-7421.9	121449	133195
24	1582	99900	105384	-5483.7	98056	112711
25	2200	.	152460	.	146713	158206

c. Find r^2 and interpret its value in the context of the problem.

d. Do the data provide evidence that living area contributes information for predicting the price of a home? Use $\alpha = .05$.

e. Find a 95% confidence interval for β_1. Does your confidence interval support the conclusion you reached in part **d**? Explain. (Note that the standard error of $\hat{\beta}_1$ is given on the printout in the column headed **Standard Error** and in the row corresponding to **AREA**.)

f. Find the observed significance level for the test in part **d**, and interpret its value.

g. Estimate the mean selling price for homes with a total living area of 2,200 square feet. Use a 95% confidence interval.

10.74 The table lists the 1991 sales y (in millions of dollars) and number of employees x (in thousands) for a random sample of 20 Fortune 500 companies. For this data set, $SS_{yy} = 1,756,002,651.75$, $SS_{xx} = 28,385.22$, $SS_{xy} = 3,881,560.4$, $\bar{y} = 7,740.75$, and $\bar{x} = 39$. Perform a regression analysis that follows the five steps presented in this chapter; be sure to state all assumptions you make. Give a prediction interval for a Fortune 500 company with 50,000 employees.

Company	*Sales, y*	*Employees, x*	*Company*	*Sales, y*	*Employees, x*
Procter & Gamble	27,406	94.0	Texaco	37,551	40.2
H. B. Fuller	855	5.6	Westinghouse Electric	12,794	113.7
Johnson & Johnson	12,447	82.7	Tribune	2,035	12.9
Mattel	1,650	12.5	Allied-Signal	11,882	98.3
Intel	4,779	25.1	Polaroid	2,096	12.0
New York Times	1,703	10.1	Leggett & Platt	1,082	10.4
GAF	926	4.1	Gillette	4,706	31.2
Coltec Industries	1,373	11.4	Georgia-Pacific	11,524	57.0
General Mills	7,153	108.1	Compaq Computer	3,271	10.0
ITT Rayonier	979	3.0	Merck	8,603	37.7

Source: "The Fortune 500." *Fortune*, April 20, 1992, pp. 259–284.

10.75 The *prime interest rate* is the rate banks charge their best commercial customers for short-term loans. The table contains mean annual prime interest rates and the corresponding number of business failures.

	1970	*1973*	*1976*	*1979*	*1982*	*1985*	*1988*	*1991*
Interest, x (%)	7.91	8.02	6.84	12.67	14.86	9.94	9.32	8.47
Failures, y (1,000s)	10.7	9.3	9.6	7.6	24.9	57.2	57.1	87.2

Sources: *Moody's Bank and Finance Manual* 1992, Vol. 1, and *Business Failure Record*. Dun and Bradstreet Corp.

a. Find the least squares regression line relating number of business failures y, to prime interest rates x.
b. What do the values r, r^2, and s suggest about the usefulness of the fitted model for predicting business failures?
c. Conduct the appropriate hypothesis test to determine whether the prime rate contributes information for the prediction of business failures.

On Your Own

The Gross Domestic Product (GDP) is one of the nation's best-known economic indicators. Many economists have developed models to forecast future values of the GDP. Surely many variables should be included if an accurate prediction is to be made. For the moment, however, consider the simple case of choosing one important variable to include in a simple straight-line model for GDP.

First, list three independent variables (x_1, x_2, and x_3) that you think might be (individually) strongly related to the GDP. Next, obtain 10 yearly values (preferably for the last 10 years) of the three independent variables and the GDP.*

a. Use the least squares formulas given in this chapter to fit three straight-line models—one for each independent variable—for predicting the GDP.
b. Interpret the sign of the estimated slope coefficient, $\hat{\beta}_1$, in each case, and test the utility of each model by testing H_0: $\beta_1 = 0$ against H_a: $\beta_1 \neq 0$.
c. Calculate the coefficient of determination r^2 for each model. Which of the independent variables predicts the GDP best over the 10 sample years when a straight-line model is used? Is this variable necessarily best in general (i.e., for all years)? Explain.

Using the Computer

Suppose we want to model the relationship between the median household income y and the percentage of college graduates x in a zip code using a straight-line model.

a. Draw a random sample of 100 zip codes from the 1,000 described in Appendix C, and extract y and x for each zip code sampled. Use a software package that includes a regression program to obtain the least squares fit for the model of interest.
1. Graph the fitted model.
2. Interpret the estimated intercept and slope.
3. Interpret the estimated standard deviation of the error term.
4. Evaluate the usefulness of the model.
5. Estimate the mean median income for all zip codes having 15% college graduates using a 95% confidence interval.
6. Predict the median income for a particular zip code having 15% college graduates using a 95% prediction interval.

b. Repeat part **a** using the entire set of 1,000 zip codes. Compare the results to those you obtained in part **a**.

References

Cho, C. C., and Reichert, A. "An application of multiple regression analysis for appraising single-family housing values." *Business Economics*, Jan. 1980, Vol. 15, pp. 47–52.

Cook, T. Q. (ed.) *Instruments of the Money Market*, 4th ed. Richmond, Va.: Federal Reserve Bank of Richmond, 1977.

Crandall, J. S., and Cedercreutz, M. "Preliminary cost estimates for mechanical work." *Building Systems Design*, Oct.–Nov. 1976, Vol. 73, pp. 35–51.

Deming, W. E. *Out of the Crisis*. Cambridge, Mass.: MIT Center For Advanced Engineering Study, 1986.

*The assumption that the random errors are independent is debatable for time series data. For the purposes of illustration, we assume they are approximately independent. The problem of dependent errors is discussed in Chapter 15.

Draper, N., and Smith, H. *Applied Regression Analysis*, 2d ed. New York: Wiley, 1981.

Glueck, W. F. *Management*. Hinsdale, Ill.: Dryden Press, 1977. p. 274.

Gray, J., and Johnston, K. S. *Accounting and Management Action*, 2d ed. New York: McGraw-Hill, 1977. Pp. 267–268.

Koch, P. D., and Ragan, J. F., Jr. "Investigating the causal relationship between quits and wages: An exercise in comparative dynamics." *Economic Inquiry*, Jan. 1986, Vol. 24, pp. 61–83.

Krueger, A. O. "Global trade prospects for the developing countries." *World Economy*, July 1992, Vol. 15, No. 4, p. 457.

Landro, L., and Mayer, J. "Cable-TV viewing study dims prospect of large increase in number of channels." *Wall Street Journal*, Nov. 16, 1982.

Leiter, M. P., and Meechan, K. A. "Role structure and burnout in the field of human services." *Journal of Applied Behavioral Science*, 1986, Vol. 22, No. 1, pp. 47–52.

Luthans, F., Rosenkrantz, S. A., and Hennessey, H. W. "What do successful managers really do? An observational study of managerial activities." *Journal of Applied Behavioral Science*, Aug. 1985, Vol. 21, No. 3, pp. 255–270.

McConnell, C. R., and Brue, S. L. *Economics*. New York: McGraw-Hill, 1990.

Miller, R. B., and Wichern, D. W. *Intermediate Business Statistics: Analysis of Variance, Regression, and Time Series*. New York: Holt, Rinehart and Winston, 1977. Chapter 5.

Mintzberg, H. *The Nature of Managerial Work*. New York: Harper and Row, 1973.

Neter, J., Wasserman, W., and Kutner, M. H. *Applied Linear Regression Models*, 2d ed. Homewood, Ill.: Richard D. Irwin, 1989.

Scholl, R. W., and Cooper, E. "The use of job evaluation to eliminate gender based pay differentials." *Public Personnel Management*, Spring 1991, Vol. 20, No. 1, p. 1.

Weisberg, S. *Applied Linear Regression*, 2d ed. New York: Wiley, 1985.

Younger, M. S. *First Course in Linear Regression*. Boston: Duxbury, 1985.

CHAPTER ELEVEN

Multiple Regression

Contents

Case Studies

Where We've Been

In Chapter 10 we demonstrated how to model the relationship between a dependent variable y and an independent variable x using a straight line. We fit the straight line to the data points, and used r and r^2 to measure the strength of the relationship between y and x. We used the resulting prediction equation to estimate the mean value of y or to predict some future value of y for a given value of x.

Where We're Going

This chapter converts the basic concept of Chapter 10 into a powerful and useful estimation and prediction device by modeling the mean value of y as a function of two or more independent variables. This will enable you to model a response y (for instance, the assessed value of a house) as a function of quantitative variables (such as floor space and age of the house) or as a function of qualitative variables (such as type of construction and location). As in the case of simple linear regression, multiple regression analysis includes fitting the model to a data set, testing the usefulness of the model, and using it for estimation and prediction. We also use the model to predict some particular value of y to be observed in the future.

11.1 A Multiple Regression Analysis: The Model and the Procedure

Most practical applications of regression use models that are more complex than the first-order (straight-line) model. For example, a realistic probabilistic model for monthly sales revenue would include more than just advertising expenditures in order to provide a good predictive model for sales. Factors such as season, inventory on hand, size of sales force, and price are a few of the many variables that might influence sales. Thus, we would want to incorporate these and other potentially important independent variables into the model if we need to make accurate predictions.

Probabilistic models that include terms involving x^2, x^3 (or higher-order terms), or more than one independent variable are called **multiple regression models**. The general form of these models is

$$y = \beta_0 + \beta_1 x_1 + \beta_2 x_2 + \cdots + \beta_k x_k + \epsilon$$

The dependent variable y is now written as a function of k independent variables, $x_1, x_2, \ldots, x_k$. The random error term is added to make the model probabilistic rather than deterministic. The value of the coefficient β_i determines the contribution of the independent variable x_i, given that the other x variables are held constant. β_0 is the y-intercept. The coefficients $\beta_0, \beta_1, \ldots, \beta_k$ will usually be unknown, because they represent population parameters.

At first glance it might appear that the regression model shown above would not allow for anything other than straight-line relationships between y and the independent variables, but this is not true. Actually, $x_1, x_2, \ldots, x_k$ can be functions of variables

Step 1 Hypothesize the deterministic component of the model. This component relates the mean, $E(y)$, to the independent variables, $x_1, x_2, \ldots, x_k$. This involves the choice of the independent variables to be included in the model (Chapter 12).

Step 2 Use the sample data to estimate the unknown model parameters $\beta_0, \beta_1, \beta_2, \ldots, \beta_k$ (Section 11.2).

Step 3 Specify the probability distribution of the random error term, ϵ, and estimate the standard deviation of this distribution, σ (Sections 11.3 and 11.9).

Step 4 Statistically evaluate the usefulness of the model (Sections 11.4 and 11.5).

Step 5 When satisfied that the model is useful, use it for prediction, estimation, and other purposes (Section 11.6).

as long as the functions do not contain unknown parameters. For example, the dollar sales, y, of new housing in a region could be a function of the independent variables

$$x_1 = \text{Mortgage interest rate}$$
$$x_2 = (\text{Mortgage interest rate})^2 = x_1^2$$
$$x_3 = \text{Unemployment rate in the region}$$

and so on. You could even insert a cyclical term (if it would be useful) of the form $x_4 = \sin t$, where t is a time variable. The multiple regression model is quite versatile and can be made to model many different types of response variables.

As shown in the box on the previous page, we use the same steps to develop the multiple regression model as we used for the simple regression model. Although we introduce several different types of models in this chapter, we defer formal discussion of model building (step 1) until Chapter 12.

CASE STUDY 11.1 / Predicting Corporate Executive Compensation

Towers, Perrin, Forster & Crosby (TPF&C), an international management consulting firm, has developed a unique and interesting application of multiple regression analysis. Many firms are interested in evaluating their management salary structure, and TPF&C uses multiple regression models to accomplish this salary evaluation. The Compensation Management Service, as TPF&C calls it, measures both the internal and external consistency of a company's pay policies to determine whether they reflect management's intent.

The dependent variable y used to represent executive compensation is annual salary. The independent variables used to explain salary structure include the executive's age, education, rank, and bonus eligibility; number of employees under the executive's direct supervision; as well as variables that describe the company for which the executive works, such as annual sales, profit, and total assets.

The initial step in developing models for executive compensation is to obtain a sample of executives from various client firms, which TPF&C calls the Compensation Data Bank. The data for these executives are used to estimate the model coefficients (the β parameters), and these estimates are then substituted into the linear model to form a prediction equation. To predict a particular executive's compensation, TPF&C substitutes into the prediction equation the values of the independent variables that pertain to the executive (age, rank, etc.). This application of multiple regression analysis is developed more fully in Section 11.7.

11.2 Fitting the Model: The Method of Least Squares

The method of fitting multiple regression models is identical to that of fitting the first-order (straight-line) model—namely, the method of least squares. That is, we choose the estimated model

$$\hat{y} = \hat{\beta}_0 + \hat{\beta}_1 x_1 + \cdots + \hat{\beta}_k x_k$$

that minimizes

$$\text{SSE} = \sum (y_i - \hat{y}_i)^2$$

As in the case of the straight-line model, the sample estimates $\hat{\beta}_0, \hat{\beta}_1, \ldots, \hat{\beta}_k$ will be obtained as solutions to a set of simultaneous linear equations.*

The primary difference between fitting the simple and multiple regression models is computational difficulty. The $(k + 1)$ simultaneous linear equations that must be solved to find the $(k + 1)$ estimated coefficients $\hat{\beta}_0, \hat{\beta}_1, \ldots, \hat{\beta}_k$ are often difficult (sometimes impossible) to solve with a calculator. Consequently, we resort to the use of computers. Many computer packages have been developed to fit a multiple regression model by the method of least squares. We will present output from several of the more popular computer packages, commencing, in our first example, with the computer output from the SAS System. Since the SAS regression output is similar to that of most other package regression programs, you should have little trouble interpreting regression output from other packages as you encounter them in future examples and exercises, at your computer center, or in using a microcomputer.

To illustrate, suppose we theorize that monthly electrical usage y in all-electric homes is related to the size x of the home by the model $y = \beta_0 + \beta_1 x + \beta_2 x^2 + \epsilon$. To estimate the unknown parameters β_0, β_1, and β_2, values of y and x were collected for each of 10 homes during one month. The data are shown in Table 11.1.

TABLE 11.1

Size of Home x (square feet)	Monthly Usage y (kilowatt-hours)	Size of Home x (square feet)	Monthly Usage y (kilowatt-hours)
1,290	1,182	1,840	1,711
1,350	1,172	1,980	1,804
1,470	1,264	2,230	1,840
1,600	1,493	2,400	1,956
1,710	1,571	2,930	1,954

Notice that we include a term involving x^2 in the model above because we expect curvature in the graph of the response model relating y to x. The term involving x^2 is called a **second-order**, or **quadratic**, term. Figure 11.1 illustrates that the electrical usage appears to increase in a curvilinear manner with the size of the home. This provides some support for the inclusion of the second-order term x^2 in the model.

Part of the output from the SAS multiple regression routine for the data in Table 11.1 is reproduced in Figure 11.2. The least squares estimates of the β parameters

*Students who are familiar with calculus should note that $\hat{\beta}_0, \hat{\beta}_1, \ldots, \hat{\beta}_k$ are the solutions to the set of equations $\partial\text{SSE}/\partial\hat{\beta}_0 = 0$, $\partial\text{SSE}/\partial\hat{\beta}_1 = 0, \ldots, \partial\text{SSE}/\partial\hat{\beta}_k = 0$. The solution, given in matrix notation, is presented in many of the texts listed in the references at the end of the chapter.

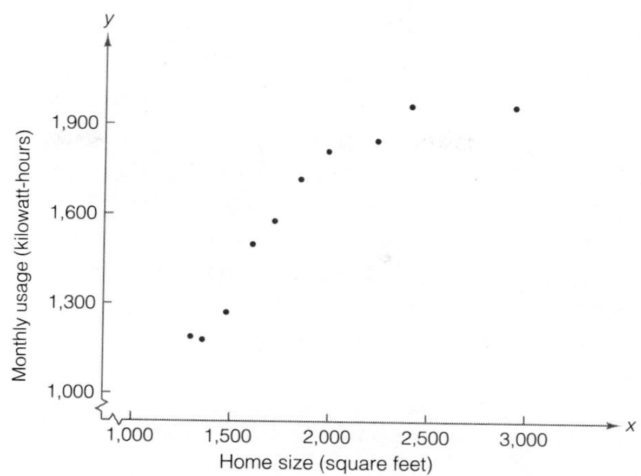

FIGURE 11.1 ► Scattergram of the home size–electrical usage data

SOURCE	DF	SUM OF SQUARES	MEAN SQUARE	F VALUE	PR > F
MODEL	2	831069.54637065	415534.77318533	189.71	0.0001
ERROR	7	15332.55362935	2190.36480419		ROOT MSE
CORRECTED TOTAL	9	846402.10000000		R-SQUARE	46.8013333
				0.981885	

PARAMETER	ESTIMATE	T FOR H0: PARAMETER = 0	PR > |T|	STD ERROR OF ESTIMATE
INTERCEPT	-1216.14388700	-5.01	0.0016	242.80636850
X	2.39893018	9.76	0.0001	0.24583560
X*X	-0.00045004	-7.62	0.0001	0.00005908

FIGURE 11.2 ▲ SAS computer printout for the home size–electrical usage data

appear in the column labeled **ESTIMATE**. You can see that $\hat{\beta}_0 = -1{,}216.1$, $\hat{\beta}_1 = 2.3989$, and $\hat{\beta}_2 = -.00045$. Therefore, the equation that minimizes the SSE for the data is

$$\hat{y} = -1{,}216.1 + 2.3989x - .00045x^2$$

The minimum value of SSE, 15,332.6, also appears in the printout. [*Note:* Much detail on the printout has not yet been discussed. We will continue throughout this chapter to shade the aspects of the printout that are under discussion.]

Note that the graph of the multiple regression model (Figure 11.3 on page 530, a response curve) provides a good fit to the data of Table 11.1. Furthermore, the small value of $\hat{\beta}_2$ does *not* imply that the curvature is insignificant, since the numerical value of $\hat{\beta}_2$ is dependent on the scale of the measurements. We test the contribution of the second-order coefficient $\hat{\beta}_2$ in Section 11.4.

The interpretation of the estimated coefficients in a quadratic model must be undertaken cautiously. First, the estimate of the y-intercept, $\hat{\beta}_0$, can be meaningfully interpreted only if the range of the independent variable includes zero—i.e., if $x = 0$

FIGURE 11.3 ▶
Least squares model for the home size–electrical usage data

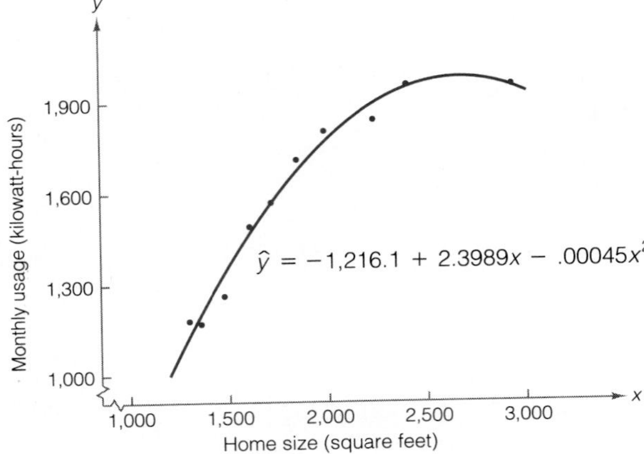

is included in the sampled range of x. In the electrical usage example, $\hat{\beta}_0 = -1{,}216.1$, which would seem to imply that the estimated electrical usage is negative when $x = 0$, but this zero point represents a home with 0 square feet. The zero point is, of course, not in the range of the sample (the least value of x is 1,290 square feet), and thus the interpretation of $\hat{\beta}_0$ is not meaningful.

The estimated coefficient of x is $\hat{\beta}_1$, but it no longer represents a slope in the presence of the quadratic term x^2.* The estimated coefficient of the linear term x does not, in general, have a meaningful practical interpretation in the quadratic model.

The sign of the coefficient, $\hat{\beta}_2$, of the quadratic term, x^2, is the indicator of whether the curve is concave downward ("mound"-shaped) or concave upward ("bowl"-shaped). A negative $\hat{\beta}_2$ implies downward concavity, as in the electrical usage example (Figure 11.3), and a positive $\hat{\beta}_2$ implies upward concavity. Rather than interpreting the numerical value of $\hat{\beta}_2$ itself, we utilize a graphical representation of the model, as in Figure 11.3, to describe the model.

Note that Figure 11.3 implies that the estimated electrical usage is leveling off as the home sizes increase beyond 2,500 square feet. In fact, the convexity of the model would lead to decreasing usage estimates if we were to display the model out to 4,000 square feet and beyond (see Figure 11.4). However, model interpretations are not meaningful outside the range of the independent variable, which has a maximum value of 2,930 square feet in this example. Thus, although the model appears to support the hypothesis that the *rate of increase* per square foot *decreases* as the home sizes approach the high end of the sampled values, the conclusion that usage will actually begin to decrease for very large homes is a *misuse* of the model, since no homes of 3,000 square feet or more were included in the sample.

---

*For students with knowledge of calculus, note that the slope of the quadratic model is the first derivative $\partial y/\partial x = \beta_1 + 2\beta_2 x$. Thus, the slope varies as a function of x, rather than remaining constant as in the straight-line model.

FIGURE 11.4 ►
Potential misuse of model

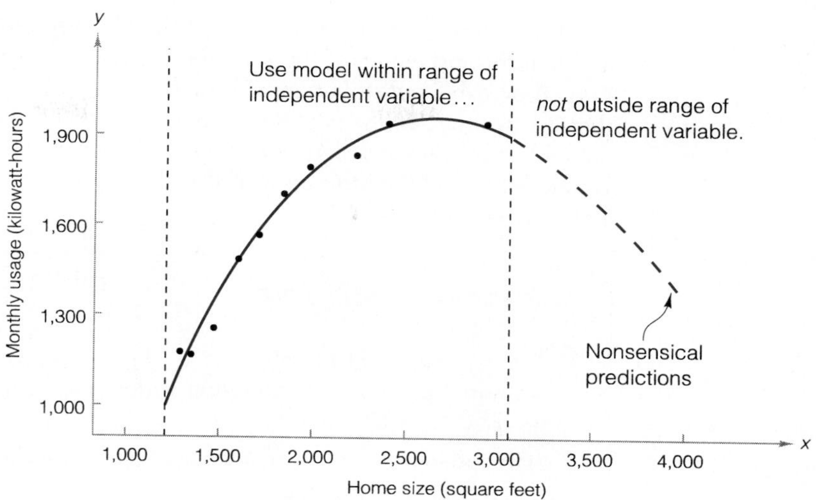

All the interpretations of the electrical usage model were based on the sample estimates of unknown model parameters. As we have throughout, we want to measure the reliability of these estimations. Additionally, we want to use the model to estimate usage for homes not included in the sample and to measure the reliability of these estimates. All this requires an assessment of the error in the model—i.e., of the probability distribution of the random error component, ϵ. This is the subject of the next section.

11.3 Model Assumptions

We noted in Section 11.1 that the multiple regression model is of the form

$$y = \beta_0 + \beta_1 x_1 + \beta_2 x_2 + \cdots + \beta_k x_k + \epsilon$$

where y is the response variable that you wish to predict; $\beta_0, \beta_1, \ldots, \beta_k$ are parameters with unknown values; $x_1, x_2, \ldots, x_k$ are information-contributing variables that are measured without error; and ϵ is a random error component. Since $\beta_0, \beta_1, \ldots, \beta_k$ and $x_1, x_2, \ldots, x_k$ are nonrandom, the quantity

$$\beta_0 + \beta_1 x_1 + \beta_2 x_2 + \cdots + \beta_k x_k$$

represents the deterministic portion of the model. Therefore, y is composed of two components—one fixed and one random—and, consequently, y is a random variable.

$$y = \overbrace{\beta_0 + \beta_1 x_1 + \cdots + \beta_k x_k}^{\text{Deterministic portion of model}} + \overbrace{\epsilon}^{\text{Random error}}$$

We assume (as in Chapter 10) that the random error can be positive or negative and that for any setting of the x values, $x_1, x_2, \ldots, x_k$, the random error ϵ has a normal probability distribution with mean equal to 0 and variance equal to σ^2. Further, we assume that the random errors associated with any (and every) pair of y values are probabilistically independent. That is, the error ϵ associated with any one y value is independent of the error associated with any other y value. These assumptions are summarized in the box.

Assumptions for Random Error ϵ

1. For any given set of values of $x_1, x_2, \ldots, x_k$, the random error ϵ has a normal probability distribution with mean equal to 0 and variance equal to σ^2.
2. The random errors are independent (in a probabilistic sense).

Note that σ^2 represents the variance of the random error ϵ. As such, σ^2 is an important measure of the usefulness of the model for the estimation of the mean of y and the prediction of actual values of y. If $\sigma^2 = 0$, all the random errors equal 0 and the predicted values, $\hat{y}$, are identical to $E(y)$; i.e., $E(y)$ is estimated without error. In contrast, a large value of σ^2 implies large (absolute) values of ϵ and larger deviations between the predicted values, $\hat{y}$, and the mean value, $E(y)$. Consequently, the larger the value of σ^2, the greater the error in estimating the model parameters $\beta_0, \beta_1, \ldots, \beta_k$ and the error in predicting a value of y for a specific set of values of $x_1, x_2, \ldots, x_k$. Thus, σ^2 plays a major role in making inferences about $\beta_0, \beta_1, \ldots, \beta_k$, in estimating $E(y)$, and in predicting y for specific values of $x_1, x_2, \ldots, x_k$.

Since the variance σ^2 of the random error ϵ is rarely known, we must use the results of the regression analysis to estimate its value. You will recall that σ^2 is the variance of the probability distribution of the random error ϵ for a given set of values for $x_1, x_2, \ldots, x_k$. Hence, it is the mean value of the squares of the deviations of the y values (for given values of $x_1, x_2, \ldots, x_k$) about the mean value $E(y)$.* Since the predicted value $\hat{y}$ estimates $E(y)$ for each of the data points, it seems natural to use

$$\text{SSE} = \sum (y_i - \hat{y}_i)^2$$

to construct an estimator of σ^2.

For example, in the second-order (quadratic) model describing electrical usage as a function of home size, we found that SSE = 15,332.6. We now want to use this quantity to estimate the variance of ϵ. Recall that the estimator for the straight-line model was $s^2 = \text{SSE}/(n - 2)$, and note that the denominator is (n − Number of estimated β parameters), which is $(n - 2)$ in the first-order (straight-line) model. Since

*Since $y = E(y) + \epsilon$, ϵ is equal to the deviation $y - E(y)$. Also, by definition, the variance of a random variable is the expected value of the square of the deviation of the random variable from its mean. According to our model, $E(\epsilon) = 0$. Therefore, $\sigma^2 = E[\epsilon - E(\epsilon)]^2 = E[\epsilon - 0]^2 = E(\epsilon^2) = E[y - E(y)]^2$.

we must estimate one more parameter, β_2, for the second-order model, the estimator of σ^2 is

$$s^2 = \frac{\text{SSE}}{n - 3}$$

That is, the denominator becomes $(n - 3)$ because there are now three β parameters in the model. The numerical estimate for this example is

$$s^2 = \frac{\text{SSE}}{10 - 3} = \frac{15{,}332.6}{7} = 2{,}190.36$$

In many computer printouts and textbooks, s^2 is called the **mean square for error (MSE)**. This estimate of σ^2 is shown in the column titled **MEAN SQUARE** in the SAS printout in Figure 11.2.

The units of the estimated variance are squared units of the dependent variable, y. In the electrical usage example, the units of s^2 are (kilowatt-hours)^2. This makes meaningful interpretation of s^2 difficult, so we use the standard deviation s to provide a more meaningful measure of variability. In this example,

$$s = \sqrt{(2{,}190.36)} = 46.8$$

which is given on the computer printout in Figure 11.2. under **ROOT MSE**. One useful interpretation of the estimated standard deviation s is that the interval $\pm 2s$ will provide a rough approximation of the accuracy with which the model will predict future values of y for given values of x. Thus, in the electrical usage example, we expect the model to provide predictions of electrical usage to within about $\pm 2s = \pm 93.6$ kilowatt-hours.*

For the general multiple regression model

$$y = \beta_0 + \beta_1 x_1 + \beta_2 x_2 + \cdots + \beta_k x_k + \epsilon$$

we must estimate the $(k + 1)$ parameters $\beta_0, \beta_1, \beta_2, \ldots, \beta_k$. Thus, the estimator of σ^2 is SSE divided by the quantity (n − Number of estimated β parameters).

We will use the estimator of σ^2 both to check the utility of the model (Sections 11.4 and 11.5) and to provide a measure of reliability of predictions and estimates when the model is used for those purposes (Section 11.6). Thus, you can see that the estimation of σ^2 plays an important part in the development and use of a regression model.

Estimator of σ^2 for Multiple Regression Model with k Independent Variables

$$s^2 = \text{MSE} = \frac{\text{SSE}}{n - \text{Number of estimated } \beta \text{ parameters}} = \frac{\text{SSE}}{n - (k + 1)}$$

*The $\pm 2s$ approximation improves as the sample size is increased. We provide more precise methodology for the construction of prediction intervals in Section 11.6.

11.4 Estimating and Testing Hypotheses About the β Parameters

Sometimes the individual β parameters in a model have particular practical significance, and we want to estimate their values or test hypotheses about them. For example, if electrical usage y is related to home size x by the straight-line relationship

$$y = \beta_0 + \beta_1 x_1 + \epsilon$$

then β_1 has a very practical interpretation. As you saw in Chapter 10, β_1 is the increase in mean kilowatt-hours of electrical usage, y, for a 1-square-foot increase in home size, x.

As proposed in the preceding sections, suppose that the electrical usage y is related to home size x by the quadratic model

$$y = \beta_0 + \beta_1 x + \beta_2 x^2 + \epsilon$$

Then the mean value of y for a given value of x is

$$E(y) = \beta_0 + \beta_1 x + \beta_2 x^2$$

What is the practical interpretation of β_2? As noted earlier, the parameter β_2 measures the curvature of the response curve shown in Figure 11.5. If $\beta_2 > 0$, the slope of the curve will increase as x increases, as shown in Figure 11.5(a). If $\beta_2 < 0$, the slope of the curve will decrease as x increases, as shown in Figure 11.5(b).

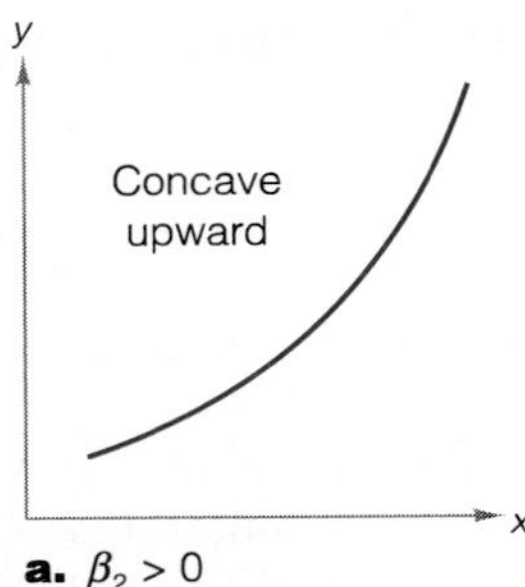

a. $\beta_2 > 0$

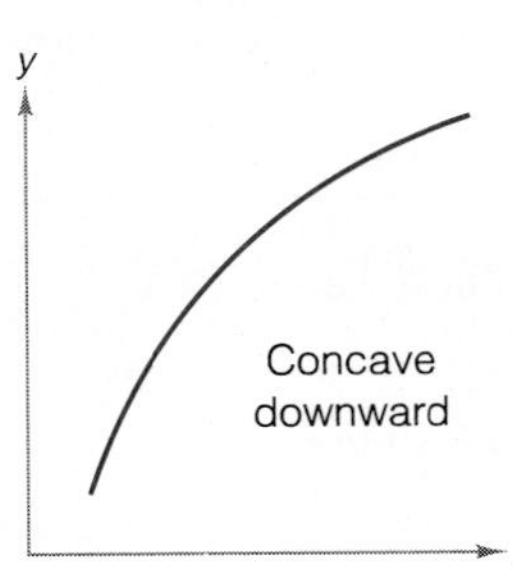

b. $\beta_2 < 0$

FIGURE 11.5 ▲ The interpretation of β_2 for a second-order model

Intuitively, we would expect the electrical usage y to rise almost proportionally to home size x. Then, eventually, as the size of the home increases, the increase in electrical usage for a 1-unit increase in home size might begin to decrease. Thus, a forecaster of electrical usage would want to determine whether this type of curvature actually is present in the response curve, or, equivalently, the forecaster would want to test the null hypothesis

$$H_0: \beta_2 = 0 \quad \text{(no curvature in the response curve)}$$

against the alternative hypothesis

$$H_a: \beta_2 < 0 \quad \text{(downward concavity in the response curve)}$$

A test of this hypothesis can be performed using a Student's t test.

The t test utilizes a test statistic analogous to that used to make inferences about the slope of the straight-line model (Section 10.5). The t statistic is formed by dividing the sample estimate, $\hat{\beta}_2$, of the parameter, β_2, by the estimated standard deviation of the sampling distribution of $\hat{\beta}_2$:

$$\textit{Test statistic:} \quad t = \frac{\hat{\beta}_2}{s_{\hat{\beta}_2}}$$

We use the symbol $s_{\hat{\beta}_2}$ to represent the estimated standard deviation of $\hat{\beta}_2$. The formula for computing $s_{\hat{\beta}_2}$ is very complex, but its computation is performed automatically as part of most standard multiple regression computer analyses. Thus, most computer packages list the estimated standard deviation $s_{\hat{\beta}_i}$ for each estimated model coefficient $\hat{\beta}_i$. In addition, they usually give the calculated t values for testing H_0: $\hat{\beta}_i = 0$ for each coefficient in the model.

The rejection region for the test is found in exactly the same way as the rejection regions for the t tests in Chapters 7, 8, and 9. That is, we consult Table VI in Appendix B to obtain an upper-tail value of t. This is a value, t_α, such that $P(t > t_\alpha) = \alpha$. We can then use this value to construct rejection regions for either one- or two-tailed tests. To illustrate, in the electrical usage example the error degrees of freedom are $(n - 3) = 7$, the denominator of the estimate of σ^2. Then the rejection region for a one-tailed test with $\alpha = .05$ is

Rejection region: $t < -t_\alpha$: $\alpha = .05$, df $= 7$

$t < -1.895$ (see Figure 11.6)

FIGURE 11.6 ►
Rejection region for test of β_2

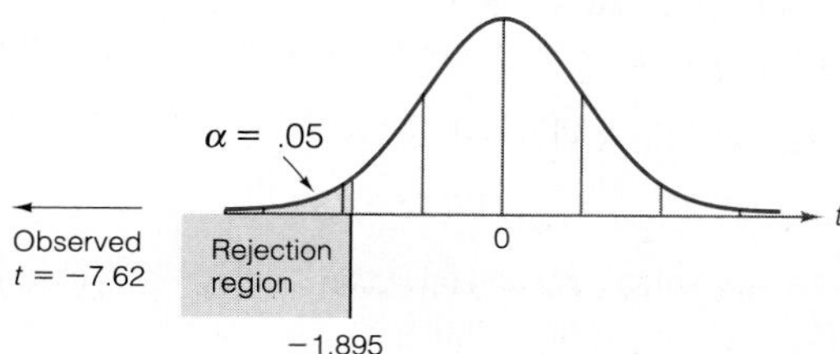

In Figure 11.7 we again show a portion of the SAS printout for the electrical usage example. The following quantities are shaded:

1. The estimated coefficients, $\hat{\beta}_0$, $\hat{\beta}_1$, and $\hat{\beta}_2$
2. The SSE
3. The MSE (estimate of σ^2, the variance of ϵ)

SOURCE	DF	SUM OF SQUARES	MEAN SQUARE	F VALUE	PR > F
MODEL	2	831069.54637065	415534.77318533	189.71	0.0001
ERROR	7	15332.55362935	2190.36480419		ROOT MSE
CORRECTED TOTAL	9	846402.10000000		R-SQUARE	46.8013333
				0.981885	

PARAMETER	ESTIMATE	T FOR H0: PARAMETER = 0	PR > |T|	STD ERROR OF ESTIMATE
INTERCEPT	-1216.14388700	-5.01	0.0016	242.80636850
X	2.39893018	9.76	0.0001	0.24583560
X*X	-0.00045004	-7.62	0.0001	0.00005908

FIGURE 11.7 ▲
SAS output for the home size–electrical usage data

The estimated standard deviations for the estimated model coefficients appear under the column labeled **STD ERROR OF ESTIMATE**. The t statistics for testing the null hypotheses that the coefficients $\beta_0, \beta_1, \ldots, \beta_k$ individually equal 0 appear under the column headed **T FOR H0: PARAMETER = 0**. The t value corresponding to the test of the null hypothesis H_0: $\beta_2 = 0$ is the last one in the column, i.e., $t = -7.62$. Since this value falls in the rejection region (i.e., it is less than -1.895), we conclude that the second-order term $\beta_2 x^2$ makes an important contribution to the prediction model of electrical usage.

The SAS printout shown in Figure 11.7 also lists the two-tailed observed significance levels (or p-values) for each t value. These values appear under the column headed **PR > |T|**. The observed significance level .0001 corresponds to the quadratic term, and this implies that we would reject H_0: $\beta_2 = 0$ in favor of H_a: $\beta_2 \neq 0$ at any α level larger than .0001. Since our alternative hypothesis was one-sided, H_a: $\beta_2 < 0$, the observed significance level is half that given in the printout; i.e., ½(.0001) = .00005. Thus, there is very strong evidence that the mean electrical usage increases more slowly per square foot for large houses than for small houses.

We can also form a 95% confidence interval for the parameter β_2 as follows:

$$\hat{\beta}_2 \pm t_{\alpha/2}\, s_{\hat{\beta}_2} = -.000450 \pm (2.365)(.0000591)$$

Test of an Individual Parameter Coefficient in the Multiple Regression Model

$$y = \beta_0 + \beta_1 x_1 + \beta_2 x_2 + \cdots + \beta_k x_k + \epsilon$$

One-Tailed Test

H_0: $\beta_i = 0$*
H_a: $\beta_i > 0$
(or $\beta_i < 0$)

Test statistic: $t = \dfrac{\hat{\beta}_i}{s_{\hat{\beta}_i}}$

Rejection region: $t > t_\alpha$
(or $t < -t_\alpha$)

Two-Tailed Test

H_0: $\beta_i = 0$*
H_a: $\beta_i \neq 0$

Test statistic: $t = \dfrac{\hat{\beta}_i}{s_{\hat{\beta}_i}}$

Rejection region: $t > t_{\alpha/2}$
or $t < -t_{\alpha/2}$

where

n = Number of observations
k = Number of independent variables in the model

and $t_{\alpha/2}$ is based on $[n - (k + 1)]$ df.

Assumptions: See page 532 for the assumptions about the probability distribution of the random error component ϵ.

*To test the null hypothesis that a parameter, β_i, equals some value other than 0, say H_0: $\beta_i = \beta_{i0}$, use the test statistic $t = (\hat{\beta}_i - \beta_{i0})/s_{\hat{\beta}_i}$. All other aspects of the test are as described in the box.

or (−.000590, −.000310). Note that the t value 2.365 corresponds to $\alpha/2 = .025$ and $(n - 3) = 7$ df. This interval constitutes a 95% confidence interval for β_2, the rate of change in curvature in mean electrical usage as home size is increased. Note that all values in the interval are negative, providing strong support for the conclusion of our test, although the test was one-tailed, whereas the confidence interval is two-tailed.

Note that the computer printout in Figure 11.7 also provides the t test statistic and corresponding two-tailed p-values for the tests of H_0: $\beta_0 = 0$ and H_0: $\beta_1 = 0$. Since the interpretation of these parameters is not practically meaningful for this model, the tests are not of interest.

Testing a hypothesis about a single β parameter that appears in any multiple regression model is accomplished in exactly the same manner as described for the second-order electrical usage model. The form of the t test is shown in the box.

EXAMPLE 11.1

A collector of antique grandfather clocks believes that the price received for the clocks at an antique auction increases with the age of the clocks and with the number of bidders. Thus, the model hypothesized is

$$y = \beta_0 + \beta_1 x_1 + \beta_2 x_2 + \epsilon$$

where

y = Auction price $\quad x_1$ = Age of clock (years) $\quad x_2$ = Number of bidders

A sample of 32 auction prices of grandfather clocks, along with their ages and the number of bidders, is given in Table 11.2.

TABLE 11.2 Auction Price Data

Age	Bidders	Price	Age	Bidders	Price
x_1	x_2	y	x_1	x_2	y
127	13	$1,235	170	14	$2,131
115	12	1,080	182	8	1,550
127	7	845	162	11	1,884
150	9	1,522	184	10	2,041
156	6	1,047	143	6	854
182	11	1,979	159	9	1,483
156	12	1,822	108	14	1,055
132	10	1,253	175	8	1,545
137	9	1,297	108	6	729
113	9	946	179	9	1,792
137	15	1,713	111	15	1,175
117	11	1,024	187	8	1,593
137	8	1,147	111	7	785
153	6	1,092	115	7	744
117	13	1,152	194	5	1,356
126	10	1,336	168	7	1,262

The model

$$y = \beta_0 + \beta_1 x_1 + \beta_2 x_2 + \epsilon$$

is fit to the data, and a portion of the SAS printout is shown in Figure 11.8. Test the hypothesis that the mean auction price increases as the number of bidders increases (and age is held constant), i.e., $\beta_2 > 0$. Use $\alpha = .05$.

SOURCE	DF	SUM OF SQUARES	MEAN SQUARE	F VALUE	PR > F
MODEL	2	4277159.70340504	2138579.85170252	120.65	0.0001
ERROR	29	514034.51534496	17725.32811534		ROOT MSE
CORRECTED TOTAL	31	4791194.21875000	R-SQUARE		133.13650181
			0.892713		

PARAMETER	ESTIMATE	T FOR H0: PARAMETER = 0	PR > |T|	STD ERROR OF ESTIMATE
INTERCEPT	-1336.72205214	-7.71	0.0001	173.35612607
X1	12.73619884	14.11	0.0001	0.90238049
X2	85.81513260	9.86	0.0001	8.70575681

FIGURE 11.8 ▲ SAS printout for Example 11.1

Solution

The hypothesis of interest concerns the parameter β_2. Specifically,

$$H_0\colon\ \beta_2 = 0 \qquad H_a\colon\ \beta_2 > 0$$

Test statistic: $t = \dfrac{\hat{\beta}_2}{s_{\hat{\beta}_2}}$

Rejection region: For $\alpha = .05$, $t > t_{.05}$
where df $= n - (k + 1) = 32 - 3 = 29$ or $t > 1.699$ (see Figure 11.9)

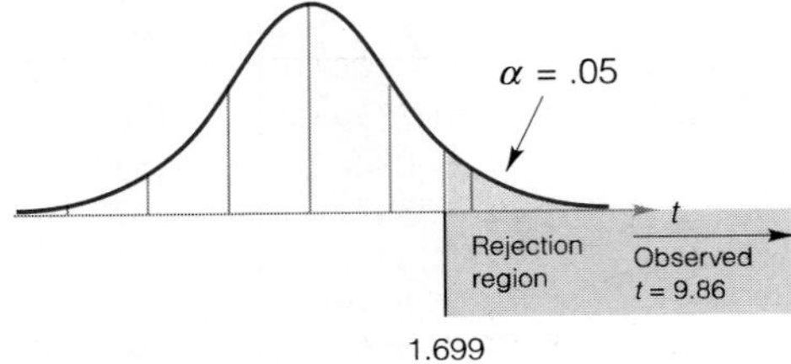

FIGURE 11.9 ► Rejection region for H_0: $\beta_2 = 0$

The calculated t value, $t = 9.86$, is indicated in Figure 11.8. This value exceeds 1.699 and therefore falls in the rejection region. Thus, the collector can conclude that the mean auction price of the clocks increases as the number of bidders increases, when age is held constant.

Note that the values $\hat{\beta}_1 = 12.74$ and $\hat{\beta}_2 = 85.82$ (shaded in Figure 11.8) are both slopes and are easily interpreted. We estimate that the mean auction price increases \$12.74 per year of age of the clock, for a fixed number of bidders, and that the mean price increases by \$85.82 per additional bidder for a fixed age clock.

Be careful not to try to interpret the estimated intercept $\hat{\beta}_0 = -1{,}336.72$ in the same way we interpreted $\hat{\beta}_1$ and $\hat{\beta}_2$ in Example 11.1. You might think that this implies a negative price for clocks 0 years of age with 0 bidders. However, these zeros are meaningless numbers in this example, since the ages range from 108 to 194 and the number of bidders ranges from 5 to 15. Keep in mind that we are modeling y within the range of values observed for the predictor variables and that interpretations of the models for values of the independent variables outside their sampled ranges can be very misleading.

Exercises 11.1–11.10

Note: Starred (*) exercises require the use of a computer.

Learning the Mechanics

11.1 The SAS System was used to fit the model $y = \beta_0 + \beta_1 x_1 + \beta_2 x_2 + \epsilon$ to $n = 20$ data points and the accompanying printout was obtained.

```
Dep Variable: Y

                        Analysis of Variance

                          Sum of          Mean
Source         DF        Squares        Square      F Value        Prob>F

Model           2   128329.27624   64164.63812        7.223        0.0054
Error          17   151015.72376    8883.27787
C Total        19   279345.00000

    Root MSE         94.25114      R-Square        0.4594
    Dep Mean        360.50000      Adj R-Sq        0.3958
    C.V.             26.14456

                        Parameter Estimates

                    Parameter       Standard    T for H0:
Variable   DF        Estimate          Error   Parameter=0    Prob > |T|

INTERCEP    1      506.346067    45.16942487        11.210        0.0001
X1          1     -941.900226   275.08555975        -3.424        0.0032
X2          1     -429.060418   379.82566485        -1.130        0.2743
```

a. What are the sample estimates of β_0, β_1, and β_2? Interpret each.
b. What is the least squares prediction equation?
c. Find SSE, MSE, and s. Interpret the standard deviation in the context of the problem.
d. Test H_0: $\beta_1 = 0$ against H_a: $\beta_1 \neq 0$. Use $\alpha = .05$.
e. Use a 95% confidence interval to estimate β_2. Interpret the p-value corresponding to the estimate $\hat{\beta}_2$. Does the confidence interval support your interpretation?

11.2 Suppose you fit the multiple regression model

$$y = \beta_0 + \beta_1 x_1 + \beta_2 x_2 + \beta_3 x_3 + \epsilon$$

to $n = 30$ data points and obtained the following result:

$$\hat{y} = 3.4 - 4.6x_1 + 2.7x_2 + .93x_3$$

The estimated standard errors of $\hat{\beta}_2$ and $\hat{\beta}_3$ are 1.86 and .29, respectively.

a. Test the null hypothesis H_0: $\beta_2 = 0$ against the alternative hypothesis H_a: $\beta_2 \neq 0$. Use $\alpha = .05$.
b. Test the null hypothesis H_0: $\beta_3 = 0$ against the alternative hypothesis H_a: $\beta_3 \neq 0$. Use $\alpha = .05$.
c. The null hypothesis H_0: $\beta_2 = 0$ is not rejected. In contrast, the null hypothesis H_0: $\beta_3 = 0$ is rejected. Explain how this can happen even though $\hat{\beta}_2 > \hat{\beta}_3$.

11.3 Suppose you fit the second-order model,

$$y = \beta_0 + \beta_1 x + \beta_2 x^2 + \epsilon$$

to $n = 25$ data points. Your estimate of β_2 is $\hat{\beta}_2 = .47$, and the estimated standard error of the estimate is $s_{\hat{\beta}_2} = .15$.

a. Test the null hypothesis that the mean value of y is related to x by the (*first-order*) linear model

$$E(y) = \beta_0 + \beta_1 x$$

(H_0: $\beta_2 = 0$) against the alternative hypothesis that the true relationship is given by the quadratic model (a *second-order* linear model)

$$E(y) = \beta_0 + \beta_1 x + \beta_2 x^2$$

(H_a: $\beta_2 \neq 0$). Use $\alpha = .05$.

b. Suppose you wanted to determine only whether the quadratic curve opens upward, i.e., as x increases, the slope of the curve increases. Give the test statistic and the rejection region for the test for $\alpha = .05$. Do the data support the theory that the slope of the curve increases as x increases? Explain.

11.4 How is the number of degrees of freedom available for estimating σ^2, the variance of ϵ, related to the number of variables in a regression model?

Applying the Concepts

11.5 Economists have two major types of data available to them: *time series data* and *cross-sectional data*. For example, an economist estimating a consumption function, say, household food consumption, as a function of household income and household size, might measure the variables of interest for one sample of households at one point in time. In this case, the economist is using *cross-sectional data*. If instead, the economist is interested in how total consumption in the United States is related to national income, the economist probably would track these variables over time, using *time series data* (Wonnacott and Wonnacott, 1979). The data in the table were collected for a random sample of 25 households in Washington, D.C., during 1990, and therefore are cross-sectional data.

Household	1990 Food Consumption ($1,000s)	1990 Income ($1,000s)	Household Size (Dec. 1990)	Household	1990 Food Consumption ($1,000s)	1990 Income ($1,000s)	Household Size (Dec. 1990)
1	3.2	31.1	4	14	3.1	85.2	2
2	2.4	20.5	2	15	4.5	35.6	9
3	3.8	42.3	4	16	3.5	68.5	3
4	1.9	18.9	1	17	4.0	10.5	5
5	2.5	26.5	2	18	3.5	21.6	4
6	3.0	29.8	4	19	1.8	29.9	1
7	2.6	24.3	3	20	2.9	28.6	3
8	3.2	38.1	4	21	2.6	20.2	2
9	3.9	52.0	5	22	3.6	38.7	5
10	1.7	16.0	1	23	2.8	11.2	3
11	2.9	41.9	3	24	4.5	14.3	7
12	1.7	9.9	1	25	3.5	16.9	5
13	4.5	33.1	7				

a. It has been hypothesized that household food consumption y is related to household income x_1 and to household size x_2 as follows:

$$y = \beta_0 + \beta_1 x_1 + \beta_2 x_2 + \epsilon$$

The SAS computer printout for fitting the model to the data is shown. Give the least squares prediction equation.

```
DEPENDENT VARIABLE: FOOD

SOURCE                   DF  SUM OF SQUARES   MEAN SQUARE F VALUE

MODEL                     2     15.46228509    7.73114255  100.80
ERROR                    22      1.68731491    0.07669613
CORRECTED TOTAL          24     17.14960000                PR > F

                                                           0.0001

R-SQUARE                 C.V.       ROOT MSE     FOOD MEAN

0.901612               8.9221     0.27694067    3.10400000

                                  T FOR HO:     PR > |T|    STD ERROR OF
PARAMETER        ESTIMATE       PARAMETER=0                  ESTIMATE

INTERCEPT      1.43260377              9.76       0.0001     0.14673751
INCOME         0.00999062              3.15       0.0046     0.00316806
SIZE           0.37928986             13.68       0.0001     0.02772472
```

b. Do the data provide sufficient evidence to conclude that the mean food consumption increases with household income? Test using $\alpha = .01$.

c. As a check on your conclusion of part **b**, construct a scattergram of household food consumption versus household income. Does the plot support your conclusion in part **b**? Explain.

d. In Chapter 10, we used the method of least squares to fit a straight line to a set of data points that were plotted in two dimensions. In this exercise, we are fitting a plane to a set of points plotted in three dimensions. We are attempting to determine the plane, $\hat{y} = \hat{\beta}_0 + \hat{\beta}_1 x_1 + \hat{\beta}_2 x_2$, that, according to the principle of least squares, best fits the data points. Sketch the least squares plane you developed in part **a**. Be sure to label all three axes of your graph.

11.6 Henry and Haynes (1978) report that in the mid-1970s Data Resources, Inc. (DRI), a firm that supplies economic information analyses and advice to government, industry, and financial institutions, used multiple regression to develop a model that characterized a bank's demand for mortgage loans. Working with quarterly time series data on the variables described here, DRI obtained the following least squares model:

$$\hat{y} = \underset{(3.5)}{37{,}350.40} + \underset{(5.5)}{.61x_1} - \underset{(-2.8)}{155.74x_2} + \underset{(4.5)}{19{,}934.7x_3} - \underset{(-3.9)}{5{,}354.9x_4} + \underset{(2.7)}{5{,}317.61x_5}$$

where

y = Mortgage loan demand (in dollars)

x_1 = Seasonally adjusted mortgage loans outstanding during previous period

x_2 = Deposits at mutual savings banks and savings and loan associations

x_3 = Average number of housing starts per month

x_4 = State's rate of unemployment

x_5 = Conventional mortgage loan interest rate

The numbers in parentheses are the t statistics associated with the estimates of the model coefficients above them. Assume $n = 28$. The preceding model, along with one for the demand for commercial loans and another for the demand for installment loans, became the heart of the bank's loan-demand forecasting system.

a. Find the estimated standard deviation of $\hat{\beta}_2$.

b. Prior to fitting the model, DRI hypothesized that β_2 should be negative because x_2 represents an alternative source of mortgage money available to the consumer. Do the data support this hypothesis? Test using $\alpha = .01$. State your conclusion in the context of the problem.

c. Report the approximate p-value of your test.

d. Provide an economic explanation for why it is reasonable to expect β_4 to be negative.

11.7 Running a manufacturing operation efficiently requires knowledge of the time it takes employees to manufacture the product, otherwise the cost of making the product cannot be determined. Furthermore, management would not be able to establish an effective incentive plan for its employees because it would not know how to set work standards (Chase and Aquilano, 1992). Estimates of production time are frequently obtained using time studies. The data in the accompanying table came from a recent time study of a sample of 15 employees performing a particular task on an automobile assembly line.

a. The SAS computer printout for fitting the model $y = \beta_0 + \beta_1 x + \beta_2 x^2 + \epsilon$ is shown on the facing page. Find the least squares prediction equation.

b. Plot the fitted equation on a scattergram of the data. Is there sufficient evidence to support the inclusion of the quadratic term in the model? Explain.

c. Test the null hypothesis $\beta_2 = 0$ against the alternative that $\beta_2 \neq 0$. Use $\alpha = .01$. Does the quadratic term make an important contribution to the model?

Completion Time, y (mins.)	*Experience, x (mos.)*	*Completion Time, y (mins.)*	*Experience, x (mos.)*
10	24	17	3
20	1	18	1
15	10	16	7
11	15	16	9
11	17	17	7
19	3	18	5
11	20	10	20
13	9		

```
DEPENDENT VARIABLE: Y

SOURCE                     DF    SUM OF SQUARES    MEAN SQUARE    F VALUE

MODEL                       2      156.11947722    78.05973861      65.59
ERROR                      12       14.28052278     1.19004356     PR > F
CORRECTED TOTAL            14      170.40000000                    0.0001

R-SQUARE                  C.V.         ROOT MSE         Y MEAN

0.916194                7.3709       1.09089118    14.80000000

                                T FOR H0:         PR > |T|    STD ERROR OF
PARAMETER         ESTIMATE     PARAMETER=0                     ESTIMATE

INTERCEPT      20.09110757          27.72          0.0001      0.72470507
X              -0.67052219          -4.33          0.0010      0.15470634
X*X             0.00953474           1.51          0.1576      0.00632580
```

d. Your conclusion in part **c** should have been to drop the quadratic term from the model. Do so and fit the *reduced model*, $y = \beta_0 + \beta_1 x + \epsilon$, to the data.

e. Define β_1 in the context of this exercise. Find a 90% confidence interval for β_1 in the reduced model of part **d**.

11.8 A researcher wished to investigate the effects of several factors on production line supervisors' attitudes toward handicapped workers. A study was conducted involving 40 randomly selected supervisors. The response y, a supervisor's attitude toward handicapped workers, was measured with a standardized attitude scale. Independent variables used in the study were

$$x_1 = \begin{cases} 1 & \text{if the supervisor is female} \\ 0 & \text{if the supervisor is male} \end{cases} \qquad \text{(This is called a dummy variable.)}$$

x_2 = Number of years of experience in a supervisory job

The researcher fit the model

$$y = \beta_0 + \beta_1 x_1 + \beta_2 x_2 + \beta_3 x_2^2 + \epsilon$$

to the data with the following results:

$$\hat{y} = 50 + 5x_1 + 5x_2 - .1x_2^2 \qquad s_{\hat{\beta}_3} = .03$$

a. Is there sufficient evidence to indicate that the quadratic term in years of experience, x_2^2, is useful for predicting attitude score? Use $\alpha = .05$.

b. Sketch the predicted attitude score $\hat{y}$ as a function of the number of years of experience x_2 for male supervisors ($x_1 = 0$). Next, substitute x_1 into the least squares equation and thereby obtain a plot of the prediction equation for female supervisors. [*Note:* For both males and females, plotting $\hat{y}$ for $x_2 = 0, 2, 4, 6, 8$, and 10 will produce a good picture of the prediction equations. The vertical distance between the males' and females' prediction curves is the same for all values of x_2.]

11.9 An employer found that factory workers who are with the company longer tend to invest more in the company's investment program per year than workers newer to the company. The following model is believed to be adequate for modeling the relationship of annual amount invested y to years working for the company x:

$$y = \beta_0 + \beta_1 x + \beta_2 x^2 + \epsilon$$

The employer checks the records for a sample of 50 factory employees for a previous year and fits this model to get $\hat{\beta}_2 = .0015$ and $s_{\hat{\beta}_2} = .000712$. Test to determine whether the employer can conclude that $\beta_2 > 0$. Use $\alpha = .05$.

***11.10** The owner of an apartment building in Minneapolis believed that her 1990 property tax bill was too high due to an overassessment of the property's value by the city tax assessor. The owner hired an independent real estate appraiser to investigate the appropriateness of the city's assessment. The appraiser used regression analysis to explore the relationship between the sale prices of apartment buildings sold in Minneapolis during 1990 and various characteristics of the properties. Twenty-five buildings were randomly sampled from all apartment buildings sold during 1990. The table lists the data collected by the appraiser. The real estate appraiser hypothesized that the sale price (i.e., market value) of an apartment building is related to the other variables in the table according to the model $y = \beta_0 + \beta_1 x_1 + \beta_2 x_2 + \beta_3 x_3 + \beta_4 x_4 + \beta_5 x_5 + \epsilon$.

a. Fit the real estate appraiser's model to the data in the table. Report the least squares prediction equation.

b. Find the standard deviation of the regression model and interpret its value in the context of this problem.

c. Do the data provide sufficient evidence to conclude that value increases with the number of units in an apartment building? Report the observed significance level, and reach a conclusion using $\alpha = .05$.

d. Interpret the value of $\hat{\beta}_1$ in terms of these data. Remember that your interpretation must recognize the presence of the other variables in the model.

e. Construct a scattergram of sale price versus age. What does your scattergram suggest about the relationship between these variables?

f. Test H_0: $\beta_2 = 0$ against H_a: $\beta_2 < 0$ using $\alpha = .01$. Interpret the result in the context of the problem. Does the result agree with your observation in part **e**? Why is it reasonable to conduct a one-tailed rather than a two-tailed test of this null hypothesis?

g. What is the observed significance level of the hypothesis test of part **f**?

Code	Sale Price	Apartment Units	Age of Structure	Lot Size	Parking Spaces	Gross Building Area
	y (\$)	x_1	x_2 (yrs.)	x_3 (sq. ft.)	x_4	x_5 (sq. ft.)
0229	90,300	4	82	4,635	0	4,266
0094	384,000	20	13	17,798	0	14,391
0043	157,500	5	66	5,913	0	6,615
0079	676,200	26	64	7,750	6	34,144
0134	165,000	5	55	5,150	0	6,120
0179	300,000	10	65	12,506	0	14,552
0087	108,750	4	82	7,160	0	3,040
0120	276,538	11	23	5,120	0	7,881
0246	420,000	20	18	11,745	20	12,600
0025	950,000	62	71	21,000	3	39,448
0015	560,000	26	74	11,221	0	30,000
0131	268,000	13	56	7,818	13	8,088
0172	290,000	9	76	4,900	0	11,315
0095	173,200	6	21	5,424	6	4,461
0121	323,650	11	24	11,834	8	9,000
0077	162,500	5	19	5,246	5	3,828
0060	353,500	20	62	11,223	2	13,680
0174	134,400	4	70	5,834	0	4,680
0084	187,000	8	19	9,075	0	7,392
0031	155,700	4	57	5,280	0	6,030
0019	93,600	4	82	6,864	0	3,840
0074	110,000	4	50	4,510	0	3,092
0057	573,200	14	10	11,192	0	23,704
0104	79,300	4	82	7,425	0	3,876
0024	272,000	5	82	7,500	0	9,542

11.5 Checking the Usefulness of a Model: R^2 and the Analysis of Variance F Test

Conducting t tests on each β parameter in a model is not a good way to determine whether a model is contributing information for the prediction of y. If we were to conduct a series of t tests to determine whether the independent variables are contributing to the predictive relationship, it is very likely that we would make one or more errors in deciding which terms to retain in the model and which to exclude. For example, suppose that all the β parameters (except β_0) are in fact equal to 0. Although the probability of concluding that any *single* β parameter differs from 0 is only α, the probability of rejecting *at least one* of a set of null hypotheses when each is true is much higher. You can see why this is true by considering the following analogy. The probability of observing a head on a single toss of a coin is .5, but the probability of observing *at least one* head in five tosses of a coin is .97. Thus, in multiple regression models for which a large number of independent variables are being considered, conducting a series of t tests may include a large number of insignificant variables and exclude some useful ones. If we want to test the usefulness of a multiple regression model, we will need a *global* test (one that encompasses all the β parameters). We would also like to find some statistical quantity that measures how well the model fits the data.

We commence with the easier problem—finding a measure of how well a linear model fits a set of data. For this we use the multiple regression equivalent of r^2, the coefficient of determination for the straight-line model (Chapter 10). Thus, we define the **sample multiple coefficient of determination, R^2**, as

$$R^2 = 1 - \frac{\sum (y_i - \hat{y}_i)^2}{\sum (y_i - \bar{y})^2} = 1 - \frac{\text{SSE}}{\text{SS}_{yy}} = \frac{\text{SS}_{yy} - \text{SSE}}{\text{SS}_{yy}} = \frac{\text{Explained variability}}{\text{Total variability}}$$

where $\hat{y}_i$ is the predicted value of y_i for the model. Just as for the simple linear model, R^2 is a sample statistic that represents the fraction of the sample variation of the y values (measured by SS_{yy}) that is attributable to the regression model. Thus, $R^2 = 0$ implies a complete lack of fit of the model to the data, and $R^2 = 1$ implies a perfect fit, with the model passing through every sample data point. In general, the larger the value of R^2, the better the model fits the data.

To illustrate, the value $R^2 = .982$ for the electrical usage example is indicated in Figure 11.10. This very high value of R^2 implies that 98.2% of the sample variation in electrical usage is attributable to, or explained by, the independent variable (home size) x. Thus, R^2 is a sample statistic that tells how well the model fits the data, and thereby represents a measure of the usefulness of the model.

The fact that R^2 is a sample statistic implies that it can be used to make inferences about the usefulness of the model for predicting y values for specific settings of the independent variables. In particular, for the electrical usage data, the test

```
SOURCE               DF    SUM OF SQUARES      MEAN SQUARE    F VALUE      PR > F

MODEL                 2   831069.54637065  415534.77318533     189.71      0.0001
ERROR                 7    15332.55362935    2190.36480419               ROOT MSE
CORRECTED TOTAL       9   846402.10000000                    R-SQUARE   46.8013333

                                                             0.981885

                                         T FOR H0:      PR > |T|    STD ERROR OF
PARAMETER              ESTIMATE      PARAMETER = 0                   ESTIMATE

INTERCEPT         -1216.14388700             -5.01        0.0016     242.80636850
X                     2.39893018              9.76        0.0001       0.24583560
X*X                  -0.00045004             -7.62        0.0001       0.00005908
```

FIGURE 11.10 ▲
SAS printout for electrical usage example

H_0: $\beta_1 = \beta_2 = 0$

H_a: At least one of the parameters β_1 and β_2 is nonzero

would formally test the global usefulness of the model.

The test statistic used to test this hypothesis is an F statistic. While several equivalent versions of the F statistic can be used—here we show two of them—we will usually rely on the computer to calculate the F statistic:

$$\textit{Test statistic:}\quad F = \frac{(\mathrm{SS}_{yy} - \mathrm{SSE})/k}{\mathrm{SSE}/[n - (k + 1)]} = \frac{R^2/k}{(1 - R^2)/[n - (k + 1)]}$$

Both these formulas indicate that the F statistic is the ratio of the *explained* variability divided by the model degrees of freedom to the *unexplained* variability divided by the error degrees of freedom. Thus, the larger the proportion of the total variability accounted for by the model, the larger the F statistic.

To determine when the ratio becomes large enough that we can confidently reject the null hypothesis and conclude that the model is more useful than no model at all for predicting y, we compare the calculated F statistic to a tabled F value with k df in the numerator and $[n - (k + 1)]$ df in the denominator. Tables of the F distribution for various values of α are given in Tables VII, VIII, IX, and X of Appendix B.

$$\textit{Rejection region:}\quad F > F_\alpha \qquad \text{where } \nu_1 = k \text{ df}, \ \nu_2 = n - (k + 1) \text{ df}$$

For the electrical usage example, $n = 10$, $k = 2$, $n - (k + 1) = 7$, and $\alpha = .05$. Consequently, we reject H_0: $\beta_1 = \beta_2 = 0$ if

$$F > F_{.05} \quad \text{where } \nu_1 = 2, \ \nu_2 = 7$$

or

$$F > 4.74 \quad \text{(see Figure 11.11)}$$

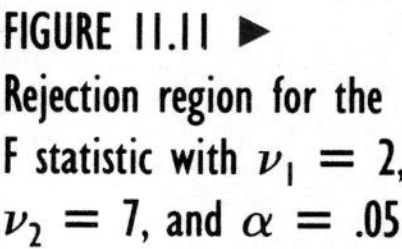

FIGURE 11.11 ▶ Rejection region for the F statistic with $\nu_1 = 2$, $\nu_2 = 7$, and $\alpha = .05$

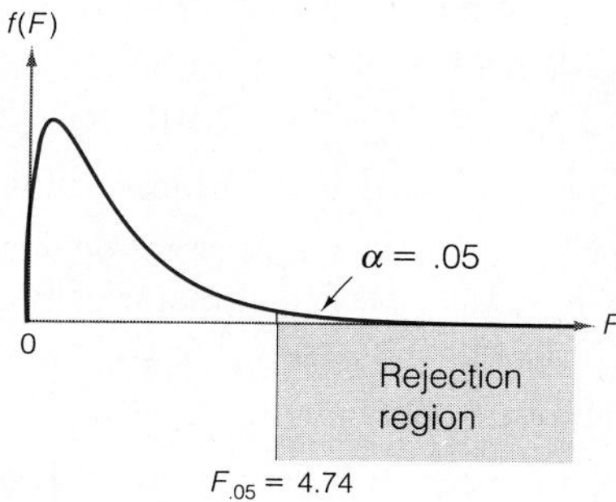

From the computer printout (Figure 11.10), we find that the computed F is 189.71. Since this value greatly exceeds the tabulated value of 4.74, we conclude that at least one of the model coefficients β_1 and β_2 is nonzero. Therefore, this global F test indicates that the second-order model $y = \beta_0 + \beta_1 x + \beta_2 x^2 + \epsilon$ is useful for predicting electrical usage.

The F statistic is also given as a part of most regression printouts, usually in a portion of the printout called "Analysis of Variance." This is an appropriate descriptive term, since the F statistic relates the explained and unexplained portions of the total variance of y. For example, the elements of the SAS computer printout in Figure 11.10 that lead to the calculation of the F value are

$$\text{F VALUE} = \frac{\text{SUM OF SQUARES(MODEL)/DF(MODEL)}}{\text{SUM OF SQUARES(ERROR)/DF(ERROR)}} = \frac{\text{MEAN SQUARE(MODEL)}}{\text{MEAN SQUARE(ERROR)}}$$

From Figure 11.10 we see that **F VALUE** = 189.71. Note, too, that the observed significance level for the F statistic is given under the heading of **PR > F** as .0001, which means that we would reject the null hypothesis H_0: $\beta_1 = \beta_2 = 0$ at any α value greater than .0001.

The analysis of variance F test for testing the usefulness of the model is summarized in the box.

Testing Global Usefulness of the Model: The Analysis of Variance *F* Test

H_0: $\beta_1 = \beta_2 = \cdots = \beta_k = 0$
(All model terms are unimportant for predicting y)

H_a: At least one $\beta_i \neq 0$
(At least one model term is useful for predicting y)

Test statistic:

$$F = \frac{(\text{SS}_{yy} - \text{SSE})/k}{\text{SSE}/[n - (k + 1)]} = \frac{\text{R}^2/k}{(1 - \text{R}^2)/[n - (k + 1)]} = \frac{\text{MEAN SQUARE(MODEL)}}{\text{MEAN SQUARE(ERROR)}}$$

where n is the sample size and k is the number of terms in the model.

Rejection region: $F > F_\alpha$, with k numerator degrees of freedom and $[n - (k + 1)]$ denominator degrees of freedom.

Assumptions: The standard regression assumptions about the random error component (Section 11.3).

Caution: A rejection of the null hypothesis leads to the conclusion (at the α level of significance) that the model is useful. However, "useful" does not necessarily mean "best." Another model may prove even more useful in terms of providing more reliable estimates and predictions. **This global F test is usually regarded as a test that the model *must* pass to merit further consideration.**

EXAMPLE 11.2

Refer to Example 11.1, in which an antique collector modeled the auction price y of grandfather clocks as a function of the age of the clock, x_1, and the number of bidders, x_2. The hypothesized model was

$$y = \beta_0 + \beta_1 x_1 + \beta_2 x_2 + \epsilon$$

A sample of 32 observations was obtained, with the results summarized in the SAS printout repeated in Figure 11.12. Discuss the coefficient of determination R^2 for this example and then conduct the global F test of model usefulness using $\alpha = .05$.

SOURCE	DF	SUM OF SQUARES	MEAN SQUARE	F VALUE	PR > F
MODEL	2	4277159.70340504	2138579.85170252	120.65	0.0001
ERROR	29	514034.51534496	17725.32811534		ROOT MSE
CORRECTED TOTAL	31	4791194.21875000	R-SQUARE		133.13650181
			0.892713		

PARAMETER	ESTIMATE	T FOR H0: PARAMETER = 0	PR > |T|	STD ERROR OF ESTIMATE
INTERCEPT	-1336.72205214	-7.71	0.0001	173.35612607
X1	12.73619884	14.11	0.0001	0.90238049
X2	85.81513260	9.86	0.0001	8.70575681

FIGURE 11.12 ▲
SAS printout for Example 11.2

Solution

The R^2 value is .89 (see Figure 11.12). This implies that 89% of the variation of the sample y values (the auction prices) about their mean can be explained by the least squares model. We now test

H_0: $\beta_1 = \beta_2 = 0$ [*Note:* $k = 2$]

H_a: At least one of the two model coefficients is nonzero

Test statistic: $F = \dfrac{R^2/k}{(1 - R^2)/[n - (k + 1)]}$

Rejection region: $F > F_\alpha$ where $\nu_1 = k$ df, $\nu_2 = n - (k + 1)$ df

For this example, $n = 32$, $k = 2$, and $n - (k + 1) = 32 - 3 = 29$. Then, for $\alpha = .05$, we will reject H_0: $\beta_1 = \beta_2 = 0$ if $F > F_{.05}$—i.e., if $F > 3.33$ (obtained from Table VIII in Appendix B). The computed value of the F test statistic is 120.65 (see Figure 11.12). Since this value of F falls in the rejection region ($F = 120.65$ greatly exceeds $F_{.05} = 3.33$), the data provide strong evidence that at least one of the model coefficients is nonzero. The model appears to be useful for predicting auction prices.

Can we be sure that the best prediction model has been found if the global F test indicates that a model is useful? Unfortunately, we cannot. The addition of other independent variables may further improve the usefulness of the model, as Example 11.3 indicates

EXAMPLE 11.3

Refer to Examples 11.1 and 11.2. Suppose the collector, having observed many auctions, believes that the *rate of increase* of the auction price with age will be driven upward by a large number of bidders. Thus, instead of a relationship like that shown in Figure 11.13(a), where the rate of increase in price with age is the same for any number of bidders, the collector believes the relationship is like that shown in Figure 11.13(b). Note that as the number of bidders increases from 5 to 15, the slope of the price versus age line increases. When the slope of the relationship between y and one independent variable (x_1) depends on the value of a second independent variable (x_2), as is the case here, we say that x_1 and x_2 **interact**. (Further discussion of *interaction* is given in Chapter 12.) A model that accounts for this type of interaction is written

$$y = \beta_0 + \beta_1x_1 + \beta_2x_2 + \beta_3x_1x_2 + \epsilon$$

FIGURE 11.13 ► **Examples of no interaction and interaction models**

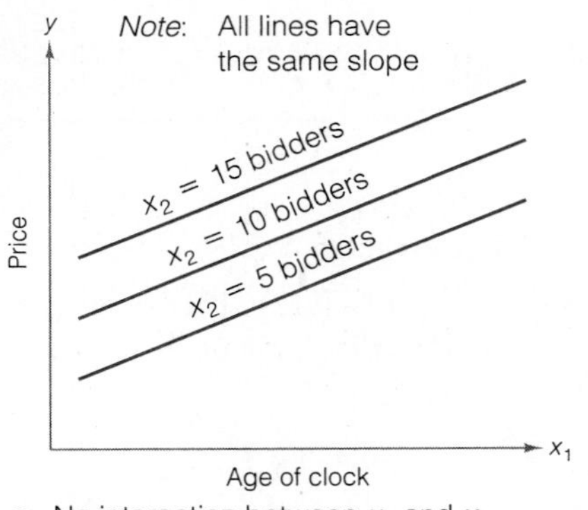

a. No interaction between x_1 and x_2

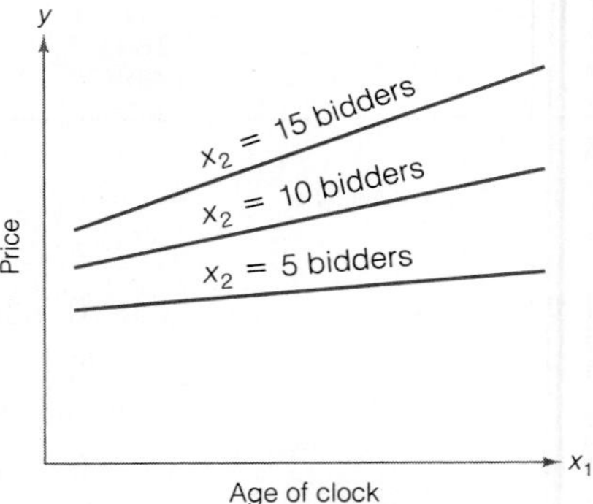

b. Interaction between x_1 and x_2

Note that the increase in the mean price, $E(y)$, for each 1-year increase in age, x_1, is no longer given by the constant, β_1, but is now $\beta_1 + \beta_3x_2$. That is, the amount $E(y)$ increases for each 1-unit increase in x_1 is *dependent on the number of bidders*, x_2. Thus, the two variables x_1 and x_2 interact to affect y.

The 32 data points listed in Table 11.2 were used to fit the first-order model with interaction. A portion of the SAS printout is shown in Figure 11.14.

Test the hypothesis that the price–age slope increases as the number of bidders increases—i.e., that age x_1 and number of bidders x_2 interact positively.

Solution

The model is

$$y = \beta_0 + \beta_1x_1 + \beta_2x_2 + \beta_3x_1x_2 + \epsilon$$

and the hypothesis of interest to the collector concerns the parameter β_3. Specifically,

SOURCE	DF	SUM OF SQUARES	MEAN SQUARE	F VALUE	PR > F
MODEL	3	4572547.98717668	1524182.66239223	195.19	0.0001
ERROR	28	218646.23157332	7808.79398476		ROOT MSE
CORRECTED TOTAL	31	4791194.21875000		R-SQUARE	88.36738077
				0.954365	

PARAMETER	ESTIMATE	T FOR H0: PARAMETER = 0	PR > |T|	STD ERROR OF ESTIMATE
INTERCEPT	322.75435309	1.10	0.2806	293.32514660
X1	0.87328775	0.43	0.6688	2.01965115
X2	-93.40991991	-3.14	0.0039	29.70767946
X1*X2	1.29789828	6.15	0.0001	0.21102602

FIGURE 11.14 ▲
Portion of the SAS printout for the model with interaction

H_0: $\beta_3 = 0$ $\quad H_a$: $\beta_3 > 0$

Test statistic: $t = \dfrac{\hat{\beta}_3}{s_{\hat{\beta}_3}}$

Rejection region: For $\alpha = .05$, $t > t_{.05}$ where df $= n - (k + 1)$

In this example, $n = 32$, $k = 3$, df $= n - (k + 1) = 32 - 4 = 28$, and thus, $t_{.05} = 1.701$.

The t value corresponding to $\hat{\beta}_3$ is indicated in Figure 11.14. The value, $t = 6.15$, exceeds 1.701, and therefore falls in the rejection region. Thus, the collector can conclude that the rate of change in the mean price of the clocks with age increases as the number of bidders increases—i.e., x_1 and x_2 interact positively. It appears that the interaction term should be included in the model.

One Note of Caution Although the coefficient of x_2 is negative ($\hat{\beta}_2 = -93.41$), this does *not* imply that the auction price decreases as the number of bidders increases. Since interaction is present, the rate of change (slope) of mean auction price with the number of bidders *depends on* x_1, the age of the clock. Thus, for example, the estimated rate of change of y with x_2 for a 150-year-old clock is

$$\text{Estimated } x_2 \text{ slope} = \hat{\beta}_2 + \hat{\beta}_3 x_1 = -93.41 + 1.30(150) = 101.60$$

In other words, we estimate that the auction price of a 150-year-old clock will *increase* by \$101.60 for every additional bidder. Although this rate of increase will vary as x_1 is changed, it will remain positive for the range of values of x_1 included in the sample. Extreme care is needed in interpreting the signs and sizes of coefficients in a multiple regression model.

To summarize the discussion in this section, the value of R^2 is an indicator of how well the prediction equation fits the data. More importantly, it can be used in the F statistic to determine whether the data provide sufficient evidence to indicate that the model contributes information for the prediction of y. Intuitive evaluations of the contribution of the model based on the computed value of R^2 must be examined with care. The value of R^2 will increase as more and more variables are added to the model. Consequently, you could force R^2 to take a value very close to 1 even though the model contributes no information for the prediction of y. In fact, R^2 will equal 1 when the number of terms in the model equals the number of data points. Therefore, you should not rely solely on the value of R^2 to tell you whether the model is useful for predicting y. Use the F test.

Exercises 11.11–11.21

Note: *Starred (*) exercises require the use of a computer.*

Learning the Mechanics

11.11 The model $y = \beta_0 + \beta_1 x + \beta_2 x^2 + \epsilon$ was fit to $n = 19$ data points with the results shown in the accompanying printout.

Dep Variable: Y

Analysis of Variance

Source	DF	Sum of Squares	Mean Square	F Value	Prob>F
Model	2	24.22335	12.11167	65.478	0.0001
Error	16	2.95955	0.18497		
C Total	18	27.18289			

Root MSE	0.43008	R-Square	0.8911
Dep Mean	3.56053	Adj R-Sq	0.8775
C.V.	12.07921		

Parameter Estimates

Variable	DF	Parameter Estimate	Standard Error	T for H0: Parameter=0	Prob > |T|
INTERCEP	1	0.734606	0.29313351	2.506	0.0234
X	1	0.765179	0.08754136	8.741	0.0001
XSQ	1	-0.030810	0.00452890	-6.803	0.0001

a. Find R^2 and interpret its value.
b. Test the null hypothesis that $\beta_1 = \beta_2 = 0$ against the alternative hypothesis that at least one of β_1 and β_2 are nonzero. Calculate the test statistic using the two formulas given in this section, and compare your results to each other and to that given on the printout. Use $\alpha = .05$ and interpret the result of your test.
c. Find the observed significance level for this test on the printout, and interpret it.
d. Test H_0: $\beta_2 = 0$ against H_a: $\beta_2 \neq 0$. Use $\alpha = .05$ and interpret the result of your test. Report and interpret the observed significance level of the test.

11.12 Suppose you fit the model

$$y = \beta_0 + \beta_1 x_1 + \beta_2 x_2 + \beta_3 x_1 x_2 + \beta_4 x_1^2 + \beta_5 x_2^2 + \epsilon$$

to $n = 30$ data points, and you obtain

$$\text{SSE} = .46 \qquad R^2 = .87$$

a. Do the values of SSE and R^2 suggest that the model provides a good fit to the data? Explain.
b. Is the model of any use in predicting y? Using $\alpha = .05$, test the null hypothesis that $E(y) = \beta_0$, i.e.,

$$H_0\colon \ \beta_1 = \beta_2 = \cdots = \beta_5 = 0$$

against the alternative hypothesis

$$H_a\colon \ \text{At least one of the parameters } \beta_1, \beta_2, \ldots, \beta_5 \text{ is nonzero}$$

11.13 Suppose you fit the model

$$y = \beta_0 + \beta_1 x_1 + \beta_2 x_2 + \epsilon$$

to $n = 20$ data points and obtain

$$\sum (y_i - \hat{y}_i)^2 = 12.37 \qquad \sum (y_i - \bar{y})^2 = 23.75$$

a. Construct an analysis of variance table for this regression analysis, using the printout in Exercise 11.11 as a model. Be sure to include the sources of variability, the degrees of freedom, the sums of squares, the mean squares, and the F statistic. Calculate R^2 for the regression analysis.
b. Test the null hypothesis that $\beta_1 = \beta_2 = 0$ against the alternative hypothesis that at least one of the parameters differs from 0. Calculate the test statistic in two different ways and compare the results. Use $\alpha = .05$ to reach a conclusion about whether the model contributes information for the prediction of y.

11.14 If the analysis of variance F test leads to the conclusion that at least one of the model parameters is nonzero, can you conclude that the model is the best predictor for the dependent variable y? Can you conclude that all of the terms in the model are important for predicting y? What is the appropriate conclusion?

Applying the Concepts

11.15 Much research (and, for that matter, much litigation) has been conducted on the disparity between the salary levels of men and women. Some recent research examined salaries for a sample of 191 Illinois managers using a regression analysis with the following independent variables:

$$x_1 = \text{Gender of manager} = \begin{cases} 1 & \text{if male} \\ 0 & \text{if not} \end{cases} \qquad x_2 = \text{Race of manager} = \begin{cases} 1 & \text{if white} \\ 0 & \text{if not} \end{cases}$$

$$x_3 = \text{Education level (in years)} \qquad x_4 = \text{Tenure with firm (in years)}$$

$$x_5 = \text{Number of hours worked per week}$$

The regression results are shown in the table on page 554 for the sample of $n = 191$; $R^2 = .240$.

a. Write the hypothesized model that was used, and interpret each of its β parameters.
b. Write the least squares equation that estimates the model in part **a**, and interpret each of the β estimates.
c. Interpret the value of R^2. Is the model useful for predicting annual salary? Test using $\alpha = .05$.

Variable	$\hat{\beta}$	*p-Value*
x_1	12.774	<.05
x_2	.713	>.10
x_3	1.519	<.05
x_4	.320	<.05
x_5	.205	<.05
Constant	15.491	–

Source: Reskin, B. F., and Ross, C. E. "Jobs, authority, and earnings among managers: The continuing significance of sex," *Work and Occupations*, Nov. 1992, Vol. 19, No. 4.

d. Test to determine whether the gender variable indicates that male managers are paid more than female managers, even after adjusting for and holding constant the other four factors in the model. Use $\alpha = .05$. [*Note:* The p-values given in the table are two-tailed.]

e. Why would one want to adjust for these other factors prior to conducting a test for salary discrimination?

f. Discuss how the interaction between gender (x_1) and tenure with the firm (x_4) might affect the results of the analysis. [*Note:* We will discuss the use and implications of this type of interaction term more fully in Chapter 12.]

11.16 Hoping for a larger share of the fine food market, researchers for a meat-processing firm that prepares meats for exclusive restaurants are working to improve the quality of its hickory-smoked hams. One of their studies concerns the effect of time spent in the smokehouse on the flavor of the ham. Hams kept in the smokehouse for varying amounts of time were taste tested by a panel of 10 food experts. The following model was thought to be appropriate by the researchers:

$$y = \beta_0 + \beta_1 t + \beta_2 t^2 + \epsilon$$

where

y = Mean of the taste scores for the 10 experts

t = Time in the smokehouse (hours)

Using a sample of 20 hams, the following least squares model was obtained:

$$\hat{y} = 20.3 + 5.2t - .0025t^2$$

where $s_{\hat{\beta}_2} = .0011$. The coefficient of determination is $R^2 = .79$.

a. Is there evidence to indicate that the overall model is useful? Test at $\alpha = .05$.

b. Is there evidence to indicate that the quadratic term is important in this model? Test at $\alpha = .05$.

11.17 Describing how multiple regression can be used by accountants in cost analysis, G. J. Benston (1966, p. 658) points out that multiple regression models can shed light on "the factors that cause costs to be incurred and the magnitudes of their effects." The independent variables of such a regression model are the factors believed to be related to cost, the dependent variable. Assuming no interactions between the independent variables, the estimates of the coefficients of the regression model provide measures of the magnitude of the factors' effects on cost. In some instances, however, Benston finds it desirable to use physical units instead of cost as the dependent variable in a cost analysis. This would be the case if most of the cost associated with the activity of interest is a function of some physical unit, such as hours of labor. The advantage of this approach is that the regression model will provide estimates of the number of labor hours required under different circumstances and these hours can then be costed at the current labor rate.

The sample data shown in the table have been collected from a firm's accounting and production records to provide cost information about the firm's shipping department. The variables for which data were collected were suggested by Benston.

Week	Labor	Pounds Shipped	Percentage of Units Shipped by Truck	Average Shipment Weight
	y (hrs.)	x_1 (1,000s)	x_2	x_3 (lbs.)
1	100	5.1	90	20
2	85	3.8	99	22
3	108	5.3	58	19
4	116	7.5	16	15
5	92	4.5	54	20
6	63	3.3	42	26
7	79	5.3	12	25
8	101	5.9	32	21
9	88	4.0	56	24
10	71	4.2	64	29
11	122	6.8	78	10
12	85	3.9	90	30
13	50	3.8	74	28
14	114	7.5	89	14
15	104	4.5	90	21
16	111	6.0	40	20
17	110	8.1	55	16
18	100	2.9	64	19
19	82	4.0	35	23
20	85	4.8	58	25

The SAS computer printout for fitting the model $y = \beta_0 + \beta_1 x_1 + \beta_2 x_2 + \beta_3 x_3 + \epsilon$ to the data is shown on page 556.

a. Find the least squares prediction equation.

b. Use an F test to investigate the usefulness of the model specified in part **a**. Use $\alpha = .01$, and state your conclusion in the context of the problem.

c. Test H_0: $\beta_2 = 0$ versus H_a: $\beta_2 \neq 0$ using $\alpha = .05$. What do the results of your test suggest about the magnitude of the effects of x_2 on labor costs?

d. Find R^2, and interpret its value in the context of the problem.

e. If shipping department employees are paid \$7.50 per hour, how much less, on average, will it cost the company per week if the average number of pounds per shipment increases from a level of 20 to 21? Assume that x_1 and x_2 remain unchanged. Your answer is an estimate of what is known in economics as the *expected marginal cost* associated with a 1-pound increase in x_3.

f. With what approximate precision can this model be used to predict the hours of labor? [*Note:* The precision of multiple regression predictions is discussed in Section 11.6.]

g. Can regression analysis alone indicate what factors *cause* costs to increase? Explain.

```
DEPENDENT VARIABLE: LABOR

SOURCE                         DF    SUM OF SQUARES      MEAN SQUARE    F VALUE

MODEL                           3     5158.31382780    1719.43794260      17.87
ERROR                          16     1539.88617220      96.24288576    PR > F
CORRECTED TOTAL                19     6698.20000000                      0.0001

R-SQUARE                     C.V.         ROOT MSE       LABOR MEAN

0.770104                  10.5148       9.81034585      93.30000000

                                   T FOR HO:      PR > |T|    STD ERROR OF
PARAMETER         ESTIMATE      PARAMETER=0                    ESTIMATE

INTERCEPT     131.92425208             5.13        0.0001     25.69321439
WEIGHT          2.72608977             1.20        0.2483      2.27500488
TRUCK           0.04721841             0.51        0.6199      0.09334856
AVGSHIP        -2.58744391            -4.03        0.0010      0.64281819
```

11.18 Because the coefficient of determination R^2 always increases when a new independent variable is added to the model, it may be tempting to include many variables in a model to force R^2 to be near 1. However, doing so reduces the degrees of freedom available for estimating σ^2, which adversely affects our ability to make reliable inferences. As an example, suppose you want to use 18 economic indicators to predict next year's GNP. You fit the model

$$y = \beta_0 + \beta_1 x_1 + \beta_2 x_2 + \cdots + \beta_{17} x_{17} + \beta_{18} x_{18} + \epsilon$$

where y = GNP and $x_1, x_2, \ldots, x_{18}$ are indicators. Only 20 years of data ($n = 20$) are used to fit the model, and you obtain $R^2 = .95$. Test to determine whether this impressive-looking R^2 is large enough to infer that the model is useful—i.e., that at least one term in the model is important for predicting GNP. Use $\alpha = .05$.

11.19 Refer to Exercise 11.8. The dependent variable being modeled was production line supervisors' scores on a test designed to measure attitudes toward handicapped workers. The independent variables were the sex of the supervisor ($x_1 = 1$ if female, 0 if male) and the supervisor's years of experience (x_2). Suppose the model proposed adds an interaction between x_1 and x_2; i.e.,

$$y = \beta_0 + \beta_1 x_1 + \beta_2 x_2 + \beta_3 x_2^2 + \beta_4 x_1 x_2 + \epsilon$$

This model is fit to the same data (40 observations) as in Exercise 11.8 with the results

$$\hat{y} = 50 + 5x_1 + 6x_2 - .2x_2^2 - x_1 x_2 \qquad s_{\hat{\beta}_4} = .02 \qquad R^2 = .87$$

a. Interpret the value of R^2.

b. Is there sufficient evidence to indicate that this model is useful for predicting attitude score? Test H_0: $\beta_1 = \beta_2 = \beta_3 = \beta_4 = 0$ using $\alpha = .05$.

c. Is there evidence that the interaction between sex and years of experience is useful in the prediction model?

d. Sketch the predicted attitude score $\hat{y}$ as a function of the number of years of experience x_2 for males ($x_1 = 0$). Next, substitute $x_1 = 1$ into the least squares equation to obtain a plot of the prediction equation for females. Compare these sketches with those obtained when the model without interaction was fit in

Exercise 11.8. [*Note:* For both sexes, plotting $\hat{y}$ for $x_2 = 0, 2, 4, 6, 8$, and 10 will produce a good picture of the prediction equations. The interaction term allows the vertical distance between the males' and females' prediction curves to change as x_2 changes.]

11.20 A researcher used regression analysis to investigate the determinants of survival size of nonprofit hospitals. For a given sample of hospitals, survival size y is defined as the largest size hospital (in terms of number of beds) exhibiting growth in market share over a specific time interval. Suppose 10 states are randomly selected and the survival size for all nonprofit hospitals in each state is determined for the time periods 1988–1989 and 1991–1992, yielding two observations per state. The 20 survival sizes are listed in the table along with the following data for each state, for the second year in each time interval:

x_1 = Percentage of beds that are in for-profit hospitals
x_2 = Ratio of the number of persons enrolled in health maintenance organizations to the number of persons covered by hospital insurance
x_3 = State populations (in thousands)
x_4 = Percent of state that is urban

It is hypothesized that the model characterizing the relationship between survival size y and the four x variables is

$$y = \beta_0 + \beta_1 x_1 + \beta_2 x_2 + \beta_3 x_3 + \beta_4 x_4 + \epsilon$$

State	*Time*	y	x_1	x_2	x_3	x_4
1	1	370	.13	.09	5,800	89
1	2	390	.15	.09	5,955	87
2	1	455	.08	.11	17,648	87
2	2	450	.10	.16	17,895	85
3	1	500	.03	.04	7,332	79
3	2	480	.07	.05	7,610	78
4	1	550	.06	.005	11,731	80
4	2	600	.10	.005	11,790	81
5	1	205	.30	.12	2,932	44
5	2	230	.25	.13	3,100	45
6	1	425	.04	.01	4,148	36
6	2	445	.07	.02	4,205	38
7	1	245	.20	.01	1,574	25
7	2	200	.30	.01	1,560	28
8	1	250	.07	.08	2,471	38
8	2	275	.08	.10	2,511	38
9	1	300	.09	.12	4,060	52
9	2	290	.12	.20	4,175	54
10	1	280	.10	.02	2,902	37
10	2	270	.11	.05	2,925	38

Source: Adapted from Bays, C. W. "The determinants of hospital size: A survival analysis." *Applied Economics*, 1986, Vol. 18, pp. 359–377.

The model was fit to the data in the table using the SAS System with the results shown.

```
Dep Variable: Y

                          Analysis of Variance

                        Sum of          Mean
Source        DF       Squares        Square       F Value        Prob>F

Model          4  246537.05939   61634.26485        28.180        0.0001
Error         15   32807.94061    2187.19604
C Total       19  279345.00000

    Root MSE          46.76747      R-Square        0.8826
    Dep Mean         360.50000      Adj R-Sq        0.8512
    C.V.              12.97295

                          Parameter Estimates

                   Parameter        Standard      T for H0:
Variable   DF       Estimate           Error    Parameter=0     Prob > |T|

INTERCEP    1     295.327091     40.17888737          7.350         0.0001
X1          1    -480.837576    150.39050364         -3.197         0.0060
X2          1    -829.464955    196.47303539         -4.222         0.0007
X3          1       0.007934      0.00355335          2.233         0.0412
X4          1       2.360769      0.76150774          3.100         0.0073
```

a. Report the least squares prediction equation.
b. Find the regression standard deviation s and interpret its value in the context of the problem.
c. Use an F test to investigate the usefulness of the hypothesized model. Report the observed significance level, and use $\alpha = .025$ to reach your conclusion.
d. Prior to collecting the data it was hypothesized that increases in the number of for-profit hospital beds would decrease the survival size of nonprofit hospitals. Do the data support this hypothesis? Testing using $\alpha = .05$.

***11.21** A company that services copy machines is interested in developing a regression model that will assist in personnel planning. It needs a model that describes the relationship between the time spent on a preventive maintenance service call to a customer, y, and two independent variables: the number of copy machines to be serviced, x_1, and the service person's experience in preventive maintenance, x_2. Company records were sampled and the data in the table were obtained.

Hours of Maintenance	*Number of Copy Machines*	*Months of Experience*	*Hours of Maintenance*	*Number of Copy Machines*	*Months of Experience*
1.0	1	12	1.8	1	1
3.1	3	8	11.5	10	10
17.0	10	5	9.3	5	2
14.0	8	2	6.0	4	6
6.0	5	10	12.2	10	8

a. Fit the model $y = \beta_0 + \beta_1 x_1 + \beta_2 x_2 + \epsilon$ to the data.
b. Investigate whether the model is useful. Test using $\alpha = .10$.
c. Find R^2 for the fitted model. Interpret your result.
d. Fit the model $y = \beta_0 + \beta_1 x_1 + \beta_2 x_2 + \beta_3 x_1 x_2 + \epsilon$ to the data.
e. Find R^2 for the model of part **d**.
f. Explain why you should not rely solely on a comparison of the two R^2 values for drawing conclusions about which model is more useful for predicting y.
g. Do the data provide sufficient evidence to indicate that the interaction term, $x_1 x_2$, contributes information for the prediction of y? [*Hint:* Test H_0: $\beta_3 = 0$.]
h. Can you be certain that the model you selected in part **f** is the best model to use in predicting maintenance time? Explain.

11.6 Using the Model for Estimation and Prediction

In Section 10.8 we discussed the use of the least squares line for estimating the mean value of y, $E(y)$, for some particular value of x, say $x = x_p$. We also showed how to use the same fitted model to predict, when $x = x_p$, some value of y to be observed in the future. Recall that the least squares line yielded the same value for both the estimate of $E(y)$ and the prediction of some future value of y. That is, both are the result of substituting x_p into the prediction equation $\hat{y} = \beta_0 + \beta_1 x$ and calculating $\hat{y}$. There the equivalence ends. The confidence interval for the mean $E(y)$ is narrower than the prediction interval for y because of the additional uncertainty attributable to the random error ϵ when predicting some future value of y.

These same concepts carry over to the multiple regression model. For example, suppose we want to estimate the mean electrical usage for a given home size, say $x_p = 1{,}500$ square feet. Assuming the quadratic model represents the true relationship between electrical usage and home size, we want to estimate

$$E(y) = \beta_0 + \beta_1 x_p + \beta_2 x_p^2 = \beta_0 + \beta_1(1{,}500) + \beta_2(1{,}500)^2$$

Substituting into the least squares prediction equation yields the following estimate of $E(y)$:

$$\begin{aligned}\hat{y} &= \hat{\beta}_0 + \hat{\beta}_1(1{,}500) + \hat{\beta}_2(1{,}500)^2 \\ &= -1{,}216.144 + 2.3989(1{,}500) - .00045004(1{,}500)^2 = 1{,}369.7\end{aligned}$$

To form a confidence interval for the mean, we need to know the standard deviation of the sampling distribution for the estimator $\hat{y}$. For multiple regression models, the form of this standard deviation is rather complex. However, some regression packages allow us to obtain the confidence intervals for mean values of y at any given setting of the independent variables. A portion of the SAS output for the electrical usage example is shown in Figure 11.15 (page 560). The mean value and corresponding 95% confidence interval for $x_p = 1{,}500$ are shown in the columns labeled **PREDICTED**

VALUE, **LOWER 95% CL FOR MEAN**, and **UPPER 95% CL FOR MEAN**. Note that $\hat{y} = 1{,}369.7$, which agrees with our earlier calculation. The 95% confidence interval for the true mean of y is shown to be 1,325.0 to 1,414.3 (see Figure 11.16).

FIGURE 11.15 ▶ SAS printout for estimated mean value and corresponding confidence interval for $x_p = 1{,}500$

X	PREDICTED VALUE	LOWER 95% CL FOR MEAN	UPPER 95% CL FOR MEAN
1500	1369.66088739	1324.98831001	1414.33346477

FIGURE 11.16 ▶ Confidence interval for mean electrical usage

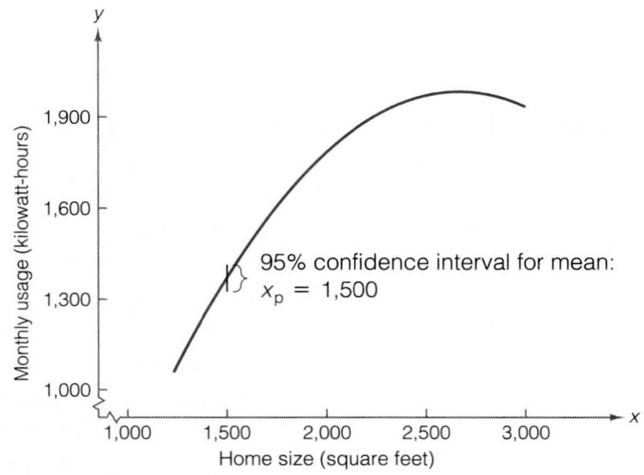

If we were interested in predicting the electrical usage for a particular 1,500-square-foot home, $\hat{y} = 1{,}369.7$ would be used as the predicted value. However, the prediction interval for a particular value of y will be wider than the confidence interval for the mean value. This is reflected by the printout shown in Figure 11.17, which gives the predicted value of y and corresponding 95% prediction interval for $x_p = 1{,}500$. This prediction interval, which extends from 1,250.3 to 1,489.0, is shown in Figure 11.18.

FIGURE 11.17 ▶ SAS printout for predicted value and corresponding prediction interval for $x_p = 1{,}500$

X	PREDICTED VALUE	LOWER 95% CL INDIVIDUAL	UPPER 95% CL INDIVIDUAL
1500	1369.66088739	1250.31627944	1489.00549533

Unfortunately, not all computer packages have the capability to produce confidence intervals for means and prediction intervals for particular y values. This is a rather serious oversight, since the estimation of mean values and the prediction of particular values represent the culmination of our model building efforts: using the model to make inferences about the dependent variable y.

FIGURE 11.18 ► **Prediction interval for electrical usage**

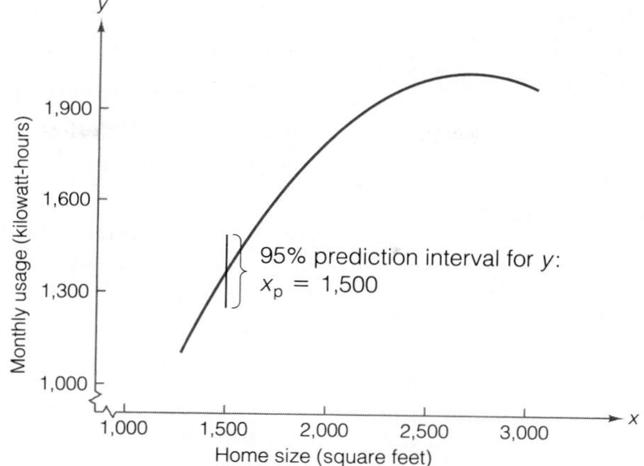

CASE STUDY 11.2 / Predicting the Sales of Crest Toothpaste

Knowing the determinants of demand for its product would help a company focus its marketing efforts on the appropriate market segments (e.g., middle-income families with school-age children). In addition, knowing the relationship between those determinants and product sales would assist the company in predicting sales and, therefore, in its planning throughout the organization. Using a model developed by Roger D. Carlson (1981) as her starting point, Carolyn I. Allmon (1982) employed multiple regression analysis to investigate the determinants of demand for Crest toothpaste and to predict Crest's sales.

Allmon modeled this year's sales of Crest as a function of current advertising expenditures, personal disposable income, and the ratio of Crest's advertising budget to that for Colgate toothpaste, Crest's closest competitor. Specifically, Allmon hypothesized the model

$$y = \beta_0 + \beta_1 x_1 + \beta_2 x_2 + \beta_3 x_3 + \epsilon$$

where

y = Sales of Crest toothpaste, computed by multiplying Crest's market share by the total market (thousands of dollars)

x_1 = Advertising budget for Crest (thousands of dollars)

x_2 = U.S. personal disposable income (billions of dollars)

x_3 = Ratio of Crest's advertising budget to Colgate's advertising budget (rounded to nearest hundredth)

Allmon hypothesized that the effect of each independent variable on sales is positive—i.e., that β_1, β_2, and β_3 have positive signs. Her belief that β_1 is positive follows from the fact that Procter and Gamble, the firm that manufactures Crest, has advertised Crest heavily for many years. It would be unlikely to do this if advertising did not help sales. That β_2 should be positive follows from a study of toothbrushing frequency which showed that as family income increases, family members brush their teeth more frequently. Allmon argued that since Colgate and Crest are competitors, sales of Colgate should affect sales of Crest and vice versa. Therefore, the larger x_3 is, the higher y should be. Accordingly, β_3 should be positive.

The model was estimated using the method of least squares on data collected for the years 1967–1979. The results were

$$\begin{array}{llll} \hat{y} = 30{,}626 & + 3.8932x_1 & + 86.519x_2 & - 29{,}607x_3 \\ (19{,}808) & (2.0812) & (18.693) & (23{,}822) \\ 1.5461 & 1.8706 & 4.6283 & -1.2429 \end{array}$$

$R^2 = .9575 \qquad F = 67.595$

where the numbers in parentheses are the standard errors of the coefficient estimates and the numbers below them are the associated t statistics.

Since $F = 67.595 > F_{.01} = 6.99$ (with $\nu_1 = 3$, $\nu_2 = 9$), Allmon concluded that at least one of the model coefficients is nonzero and that the model is useful for predicting sales. Next, she conducted one-sided, upper-tailed hypothesis tests to determine whether the data supported her expectations about the positive influence of each independent variable on sales. Comparing $t_{.05} = 1.833$ (df = 9) to the t statistics given, we can see that the data support her hypotheses with respect to β_1 and β_2, but that the null hypothesis H_0: $\beta_3 = 0$ cannot be rejected. Thus, she concluded,

> *. . . my model indicates that income is a significant determinant of sales as is advertising expenditure in the current year. The competition of Colgate dental cream via its advertising expenditure appears not to affect Crest sales significantly. This information could be used by Procter and Gamble in the formulation of their advertising policy, for example, in determining the segment of the market to which most advertising should be directed.*

Allmon's primary purpose for developing the model was prediction. As part of her test of the predictive power of the model, Allmon substituted the values for x_1, x_2, and x_3 in 1980 (which were known) into the fitted model and calculated $\hat{y}$. Her model predicted sales of \$251,060,340 and actual sales were \$245,000,000. Allmon's model did substantially better than Carlson's model, the model she had set out to refine, which predicted sales of \$275,447,450.

In concluding her report, Allmon pointed out several limitations of the model. Among them were the following:

1. It was developed from annual data. Accordingly, the results should be viewed cautiously, since the cumulative effect of advertising on sales may last less than a year.
2. The sample is quite small. Even one more data point may substantially change the coefficient estimates and overall model fit.
3. None of the data were obtained from primary sources (such as Procter and Gamble). All data except disposable income were collected from the publication *Advertising Age*, a secondary data source.

11.7 Multiple Regression: An Example

Let us return to the executive compensation example introduced in Case Study 11.1. Recall that the management consultant firm of Towers, Perrin, Forster & Crosby (TPF&C) uses a multiple regression model to project executive salaries. Suppose the list of independent variables given in Table 11.3 is to be used to build a model for the salaries of corporate executives.

Step 1 The first step is to hypothesize a model relating executive salary to the independent variables listed in Table 11.3. TPF&C have found that executive compensation models that use the logarithm of salary as the dependent variable are better predictive models than those using the salary as the dependent variable. This is probably because salaries tend to be incremented in *percentages* rather than dollar values. When a dependent variable undergoes percent-

TABLE 11.3 List of Independent Variables for Executive Compensation Example

Independent Variable	Description
x_1	Years of experience
x_2	Years of education
x_3	1 if male; 0 if female
x_4	Number of employees supervised
x_5	Corporate assets ($ millions)
x_6	x_1^2
x_7	x_3x_4

age changes as the independent variables are varied, the logarithm of the dependent variable will be more suitable as a dependent variable. The model we propose is

$$y = \beta_0 + \beta_1x_1 + \beta_2x_2 + \beta_3x_3 + \beta_4x_4 + \beta_5x_5 + \beta_6x_6 + \beta_7x_7 + \epsilon$$

where y = log(Executive salary), $x_6 = x_1^2$ (second-order term in years of experience), and $x_7 = x_3x_4$ (cross product or interaction term between sex and number of employees supervised). The variable x_3 is a **dummy variable**; it is used to describe an independent variable that is not measured on a numerical scale, but instead is **qualitative (categorical)** in nature. Sex is such a variable, since its values, male and female, are categories rather than numbers. Thus, we assign the value $x_3 = 1$ if the executive is male, $x_3 = 0$ if the executive is female. (For more detail on the use and interpretation of dummy variables, see Chapter 12.) The interaction term, x_3x_4, allows for the possibility that the relationship between the number of employees supervised, x_4, and corporate salary is dependent on sex, x_3. For example, as the number of supervised employees increases, with all other factors being equal, a woman's salary might rise more rapidly than a man's. (The concept of interaction is also explained in more detail in Chapter 12.)

Step 2 Now, we estimate the model coefficients $\beta_0, \beta_1, \ldots, \beta_7$. Suppose that a sample of 100 executives is selected, and the variables y and $x_1, x_2, \ldots, x_7$ are recorded (or, in the case of x_6 and x_7, calculated). The sample is then used as input for a computer regression routine; the SAS output is shown in Figure 11.19 (page 564). The least squares model is

$$\hat{y} = 9.88 + .045x_1 + .033x_2 + .119x_3 + .00033x_4 + .0020x_5 - .00072x_6 + .00031x_7$$

Step 3 The next step is to specify the probability distribution of ϵ, the random error component. We assume that ϵ is normally distributed, with a mean of 0 and a constant variance σ^2. Furthermore, we assume that the errors are independent. In Section 11.9 we describe procedures that can be used to investigate the

SOURCE	DF	SUM OF SQUARES	MEAN SQUARE	F VALUE	PR > F
MODEL	7	27.06425564	3.86632223	1819.30	0.0001
ERROR	92	0.19551523	0.00212517		ROOT MSE
CORRECTED TOTAL	99	27.25977087		R-SQUARE	0.0460995
				0.992828	

PARAMETER	ESTIMATE	T FOR H0: PARAMETER = 0	PR > |T|	STD ERROR OF ESTIMATE
INTERCEPT	9.87878688	192.49	0.0001	0.05132104
X1 (EXPERIENCE)	0.04460301	26.83	0.0001	0.00166257
X2 (EDUCATION)	0.03326230	12.31	0.0001	0.00270306
X3 (SEX)	0.11892473	6.89	0.0001	0.01724977
X4 (EMPLOYEES SUPERVISED)	0.00033216	19.97	0.0001	0.00001664
X5 (ASSETS)	0.00201021	73.25	0.0001	0.00002744
X6 (= X1*X1)	-0.00071702	-15.11	0.0001	0.00004746
X7 (= X3*X4)	0.00031244	16.16	0.0001	0.00001933

FIGURE 11.19 ▲
SAS printout for executive compensation example

appropriateness of these assumptions. The estimate of the variance, σ^2, is given in the SAS printout as

$$s^2 = \text{MSE} = \frac{\text{SSE}}{n - (k + 1)} = \frac{\text{SSE}}{100 - (7 + 1)} = .0021$$

Step 4 We now want to see how well the model predicts salaries. First, note that $R^2 = .993$. This implies that 99.3% of the variation in y (logarithm of salaries) for these 100 sampled executives is accounted for by the model. The significance of this can be tested using the F statistic:

H_0: $\beta_1 = \beta_2 = \cdots = \beta_7 = 0$

H_a: At least one of the model coefficients is nonzero

Test statistic: $F = \dfrac{R^2/k}{(1 - R^2)/[n - (k + 1)]} = 1{,}819.30$

We could establish a rejection region using the F tables, but the printout makes this unnecessary. The observed significance level for the test statistic $F = 1{,}819.30$ is given on the printout as .0001. We therefore conclude that the model does contribute information for predicting executive salaries. There is sufficient evidence that at least one of the β parameters in the model differs from 0.

We may be particularly interested in whether the data provide evidence that the mean salary of executives increases as the asset value of the company increases, when all other variables (experience, education, etc.) are held constant. In other words, we may want to know whether the data provide sufficient evidence to show that $\beta_5 > 0$. We use the following test:

H_0: $\beta_5 = 0$ $\quad H_a$: $\beta_5 > 0$

$$\textit{Test statistic:} \quad t = \frac{\hat{\beta}_5}{s_{\hat{\beta}_5}} = 73.25$$

The t statistic, 73.25, is given on the printout (Figure 11.19), along with its significance level of .0001. We could compare the t statistic with the appropriate tabled value for a specified α, but the extremely small significance level (and large t value) make this unnecessary. There is sufficient evidence to conclude that the mean salary of executives increases as the company assets increase, when all other variables are held constant.

Step 5 The culmination of the modeling effort is to use the model for estimation and prediction. Suppose a firm is trying to determine fair compensation for an executive with the characteristics shown in Table 11.4. The least squares model can be used to obtain a predicted value for the logarithm of salary. That is,

$$\hat{y} = \hat{\beta}_0 + \hat{\beta}_1(12) + \hat{\beta}_2(16) + \hat{\beta}_3(0) + \hat{\beta}_4(400) + \hat{\beta}_5(160.1) + \hat{\beta}_6(144) + \hat{\beta}_7(0)$$

TABLE 11.4 Values of Independent Variables for a Particular Executive

x_1 = 12 years of experience	x_5 = \$160.1 million (the firm's asset value)
x_2 = 16 years of education	$x_6 = x_1^2 = 144$
x_3 = 0 (female)	$x_7 = x_3x_4 = 0$
x_4 = 400 employees supervised	

This predicted value is given in Figure 11.20, a partial reproduction of the SAS regression printout for this problem: $\hat{y} = 11.298$. The 95% prediction interval is also given: from 11.203 to 11.392. To predict the salary of an executive with these characteristics we take the antilogarithm of these values. That is, the predicted salary is $e^{11.298}$ = \$80,700 (rounded to the nearest hundred) and the 95% prediction interval is from $e^{11.203}$ to $e^{11.392}$ (or from \$73,400 to \$88,600). Thus, an executive with the characteristics given in Table 11.4 should be paid between \$73,400 and \$88,600 to be consistent with the sample data.

X1	X2	X3	X4	X5	X6	X7	PREDICTED VALUE	LOWER 95% CL INDIVIDUAL	UPPER 95% CL INDIVIDUAL
12	16	0	400	160.1	144	0	11.29766682	11.20298295	11.39235070

FIGURE 11.20 ▲
SAS printout for executive compensation problem

11.8 Statistical Computer Programs

There are a number of different statistical program packages; some of the most popular are BMDP, Minitab, SAS, and SPSS (see the references at the end of the chapter). You may have access to one or more of these packages at your computer center.

The multiple regression programs for these packages may differ in what they can do, how they do it, and the appearance of their computer printouts, but all of them print the basic outputs needed for regression analysis. For example, some will compute confidence intervals for $E(y)$ and prediction intervals for y. Others will not. Some test the null hypotheses that the individual β parameters equal 0 using Student's t tests, while others use F tests.* But all give the least squares estimates, the values of SSE, s^2, etc.

To illustrate, the Minitab, SAS, and SPSS regression analysis computer printouts for Example 11.3 are shown in Figure 11.21. For that example, we fit the model

$$y = \beta_0 + \beta_1 x_1 + \beta_2 x_2 + \beta_3 x_1 x_2 + \epsilon$$

to $n = 32$ data points. The variables in the model were

$$y = \text{Auction price}$$
$$x_1 = \text{Age of clock (years)}$$
$$x_2 = \text{Number of bidders}$$

Notice that the Minitab printout in Figure 11.21(a) gives the prediction equation at the top. The independent variables, shown in the prediction equation and listed at the left side of the printout, are **AGE**, **BIDDERS**, and **AGE-BID**. Each program treats the product **AGE-BID** = **AGE** × **BIDDERS** as a third independent variable, which must be computed before the fitting begins. For this reason, the prediction equation will always appear on the printout as first-order even though some of the independent variables shown in the prediction equation may actually be the squares or cross products of other independent variables. The SPSS printout in Figure 11.21(c) (page 568) lists the independent variables in a different order: **AGE_BID**, **AGE**, and **BIDDERS**. The order is determined statistically rather than by the user, and the user must therefore pay attention to the name of each variable when interpreting the SPSS printout. Note that SPSS also prints the intercept last rather than first, using the label of **(Constant)**. Minitab prints the intercept first, but with no label.

The estimates of the regression coefficients appear opposite the identifying variable in the Minitab column titled **COEFFICIENT**, in the SAS column **Parameter Estimate**, and in the SPSS column **B**. Compare the estimates given in these three columns. Note that the Minitab printout gives the estimates with less accuracy (fewer decimal places) than the SAS and SPSS printouts. (Ignore the column titled **Beta** in the SPSS printout. These are standardized estimates and will not be discussed in this text.)

---

*A two-tailed Student's t test based on ν df is equivalent to an F test where the F statistic has 1 df in the numerator and ν df in the denominator. See Section 11.4.

FIGURE 11.21 ▶ Computer printouts for Example 11.3

(a) Minitab regression

```
THE REGRESSION EQUATION IS
PRICE = 323 + 0.87 AGE - 93.4 BIDDERS + 1.30 AGE-BID

                                  ST. DEV.      T-RATIO =
              COEFFICIENT         OF COEF.      COEF/S.D.
                 322.8             293.3          1.10
AGE              0.873             2.020          0.43
BIDDERS         -93.41             29.71         -3.14
AGE-BID          1.2979            0.2110         6.15

S = 88.37

R-SQUARED = 95.4 PERCENT
R-SQUARED = 94.9 PERCENT, ADJUSTED FOR D.F.

ANALYSIS OF VARIANCE

 DUE TO       DF          SS          MS=SS/DF
REGRESSION     3       4572548         1524183
RESIDUAL      28        218646            7809
TOTAL         31       4791194
```

(b) SAS regression

```
Dep Variable: PRICE

                         Analysis of Variance

                         Sum of           Mean
Source        DF        Squares          Square        F Value        Prob>F

Model          3  4572547.9872  1524182.6624        195.188        0.0001
Error         28  218646.23157    7808.79398
C Total       31  4791194.2187

      Root MSE        88.36738        R-Square        0.9544
      Dep Mean      1327.15625        Adj R-Sq        0.9495
      C.V.             6.65840

                         Parameter Estimates

                   Parameter         Standard      T for H0:
Variable   DF      Estimate            Error      Parameter=0      Prob > |T|

INTERCEP    1     322.754353    293.32514660           1.100          0.2806
AGE         1       0.873288      2.01965115           0.432          0.6688
BIDDERS     1     -93.409920     29.70767946          -3.144          0.0039
AGE_BID     1       1.297898      0.21102602           6.150          0.0001
```

The estimated standard errors of the estimates are given in the Minitab column titled **ST. DEV. OF COEF.**, in the SAS column **Standard Error**, and in the SPSS column **SE B**.

The values of the test statistics for testing H_0: $\beta_i = 0$, where $i = 1, 2, 3$, are shown in the Minitab column titled **T-RATIO = COEF/S.D.**, in the SAS column **T for H0:**

FIGURE 11.21 ▶
(c) SPSS regression

```
Equation Number 1    Dependent Variable..    PRICE

Beginning Block Number  1.  Method:  Enter
   AGE       BIDDERS  AGE_BID


Variable(s) Entered on Step Number
   1..     AGE_BID
   2..     AGE
   3..     BIDDERS


Multiple R              .97692
R Square                .95436
Adjusted R Square       .94948
Standard Error        88.36738

Analysis of Variance
                        DF       Sum of Squares        Mean Square
Regression               3        4572547.98718     1524182.66239
Residual                28         218646.23157        7808.79398

F =       195.18797          Signif F =  .0000

------------------ Variables in the Equation ------------------

Variable               B           SE B         Beta           T  Sig T

AGE_BID          1.29790         .21103      1.37032       6.150  .0000
AGE               .87329        2.01965       .06085        .432  .6688
BIDDERS        -93.40992       29.70768      -.67471      -3.144  .0039
(Constant)     322.75435      293.32515                    1.100  .2806
```

Parameter = 0, and in the SPSS column **T**. Note that the computed t values shown are identical (except for the number of decimal places), but Minitab does not give the observed significance level of the test. Consequently, to draw conclusions from the Minitab printout, you must compare the computed values of t with the critical values given in a t table (Table VI in Appendix B). In contrast, the SAS printout gives the observed significance level for each t test in the column titled **Prob > |T|**, and the SPSS printout in the column **Sig T**. Note that these observed significance levels have been computed assuming that the tests are two-tailed. The observed significance levels for one-tailed tests would equal half of these values.

The Minitab printout gives the value SSE = 218,646 under the **ANALYSIS OF VARIANCE** column headed **SS** and in the row identified as **RESIDUAL**. The value of s^2 = 7,809 is shown in the same row under the column headed **MS = SS/DF**, and the degrees of freedom, **DF**, appears in the same row as 28. The corresponding values are shown at the top of the SAS printout in the row labeled **Error** and in the columns designated as **Sum of Squares**, **Mean Square**, and **DF**, respectively. These quantities appear with similar headings in the SPSS printout, but **Error** is called **Residual** in SPSS. The standard deviation s for the regression equation is given on each printout, but with different labels. Minitab shows **S =** 88.37 in the center of the printout; SAS labels it **Root MSE** in the left center of the printout. SPSS calls the standard deviation the **Standard Error** and prints it in the center of the printout. All the estimates are identical, except for the number of decimal places.

The value of R^2, as defined in Section 11.5, is given in the Minitab printout as **95.4 PERCENT** (we defined this quantity as a ratio where $0 \leq R^2 \leq 1$). It is given in the SAS printout as 0.9544, and shown in the left column of the SPSS printout as 0.95436. (Ignore the quantities shown in the Minitab printout as R^2 **ADJUSTED FOR D.F.**, in the SAS printout as **Adj R Sq**, and in the SPSS printout as **Adjusted R Square**. These quantities are adjusted for the degrees of freedom associated with the total SS and SSE and are not used or discussed in this text.)

The F statistic for testing the usefulness of the model (Section 11.5)—i.e., testing the null hypothesis that all model parameters (except β_0) equal zero—is shown under the title **F VALUE** as 195.188 at the top of the SAS printout. In addition, the SAS printout gives the observed significance level of this F test under **Prob > F** as 0.0001. This F value, 195.18797, is also printed at the center of the SPSS printout with the observed significance level given as **Signif F**. The F statistic for testing the usefulness of the model is not given in the Minitab printout. If you are using Minitab and wish to obtain the value of this statistic, you must compute it using the analysis of variance formula given in Section 11.5:

$$F = \frac{\text{MEAN SQUARE(MODEL)}}{\text{MEAN SQUARE(ERROR)}}$$

These quantities are given in the Minitab printout under the column marked **MS = SS/DF**. The rows are labeled **REGRESSION** and **RESIDUAL**, which are synonymous with **MODEL** and **ERROR**, respectively. Thus,

$$F = \frac{1{,}524{,}183}{7{,}809} = 195.18$$

a value that agrees with the values given in the SAS and SPSS printouts. The logic behind this test and other tests of hypotheses concerning sets of the β parameters is discussed further in Section 12.4.

Although we will henceforth use the three popular statistical packages discussed in this section (SAS, Minitab, and SPSS), there are many others available. The rapid expansion of the microcomputer market has been accompanied by significant growth in the availability of statistical software. It is important to evaluate any new package carefully to assure that the results it produces are complete and correct. Fortunately, there is usually much similarity in the computer printouts produced by the packages, so you will find it relatively easy to learn how to read a new regression printout once you become familiar with those used in this text. We will intermix the different packages in the examples and exercises to help familiarize you with different formats.

11.9 Residual Analysis: Checking the Regression Assumptions

When we apply regression analysis to a set of data, we never know for certain whether the assumptions of Section 11.3 are satisfied. How far can we deviate from the assumptions and still expect regression analysis to yield results that will have the reliability stated in this chapter? How can we detect departures (if they exist) from the assumptions of Section 11.3 and what can we do about them? We provide some partial answers to these questions in this section and will direct you to further discussion in succeeding chapters.

Remember from Section 11.3 that

$$y = \beta_0 + \beta_1 x_1 + \beta_2 x_2 + \cdots + \beta_k x_k + \epsilon$$

where ϵ is a random error. The first assumption we made was that the mean value of ϵ for *any* given set of values of $x_1, x_2, \ldots, x_k$ is $E(\epsilon) = 0$. One consequence of this assumption is that the mean $E(y)$ for a specific set of values of $x_1, x_2, \ldots, x_k$ is

$$E(y) = \beta_0 + \beta_1 x_1 + \beta_2 x_2 + \cdots + \beta_k x_k$$

That is,

$$y = \underbrace{E(y)}_{\substack{\text{Mean value of } y \\ \text{for specific values} \\ \text{of } x_1,\, x_2,\, \ldots,\, x_k}} + \underbrace{\epsilon}_{\substack{\text{Random} \\ \text{error}}}$$

The second consequence of the assumption is that the least squares estimators of the model parameters, $\beta_0, \beta_1, \beta_2, \ldots, \beta_k$, will be unbiased regardless of the remaining assumptions that we attribute to the random errors and their probability distributions.

The properties of the sampling distributions of the parameter estimators $\hat{\beta}_0, \hat{\beta}_1, \ldots, \hat{\beta}_k$ will depend on the remaining assumptions that we specify concerning the probability distributions of the random errors. Recall that we assumed that for any given set of values of $x_1, x_2, \ldots, x_k$, ϵ has a normal probability distribution with mean equal to 0 and variance equal to σ^2. Also, we assumed that the random errors are probabilistically independent.

It is unlikely that these assumptions are ever satisfied exactly in a practical application of regression analysis. Fortunately, experience has shown that least squares regression analysis produces reliable statistical tests, confidence intervals, and prediction intervals as long as the departures from the assumptions are not too great. In this section and in Chapter 12 we present some methods for determining whether the data indicate significant departures from the assumptions.

Because the assumptions all concern the random error component ϵ of the model, the first step is to estimate the random error. Since the actual random error associated with a particular value of y is the difference between the actual y value and its unknown mean, we estimate the error by the difference between the actual y value and the *estimated* mean. This estimated error is called a **regression residual**, or simply the

residual, and is denoted by $\hat{\epsilon}$. The actual error ϵ and residual $\hat{\epsilon}$ are shown in Figure 11.22.

$$\begin{aligned}\epsilon &= \text{Actual random error}\\ &= (\text{Actual } y \text{ value}) - (\text{Mean of } y)\\ &= y - E(y) = y - (\beta_0 + \beta_1 x_1 + \beta_2 x_2 + \cdots + \beta_k x_k)\\ \hat{\epsilon} &= \text{Estimated random error (residual)}\\ &= (\text{Actual } y \text{ value}) - (\text{Estimated mean of } y)\\ &= y - \hat{y} = y - (\hat{\beta}_0 + \hat{\beta}_1 x_1 + \hat{\beta}_2 x_2 + \cdots + \hat{\beta}_k x_k)\end{aligned}$$

Since the true mean of y (i.e., the true regression model) is not known, the actual random error cannot be calculated. However, because the residual is based on the estimated mean (the least squares regression model), it can be calculated and used to estimate the random error and to check the regression assumptions. Such checks are generally referred to as **residual analyses**. Some useful properties of residuals are given in the box.

FIGURE 11.22 ►
Actual random error and regression residual

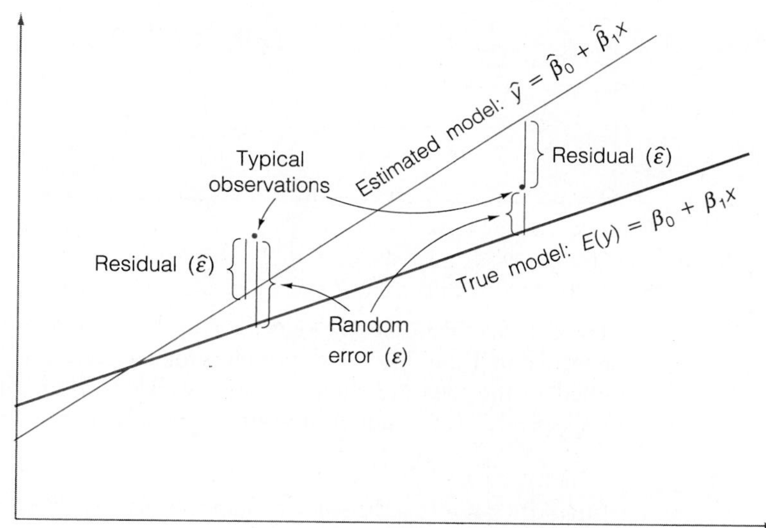

The following examples show how the analysis of regression residuals can be used to verify the assumptions associated with the model, and to improve the model when the assumptions do not appear to be satisfied. Although the residuals can be calculated and plotted by hand, we rely on the computer for these tasks in the examples and exercises. Most statistical computer packages now include residual analyses as a standard component of their regression modeling programs.

Properties of Regression Residuals

1. A residual is equal to the difference between the observed y value and its estimated (regression) mean:

$$\text{Residual} = y - \hat{y}$$

2. The mean of the residuals is equal to 0. This property follows from the fact that the sum of the differences between the observed y values and the least squares regression model is equal to 0.

$$\sum (\text{Residuals}) = \sum (y - \hat{y}) = 0$$

3. The standard deviation of the residuals is equal to the standard deviation of the fitted regression model, s. This property follows from the fact that the sum of the squared residuals is equal to SSE, which when divided by the error df is equal to the variance of the fitted regression model, s^2. The square root of the variance is both the standard deviation of the residuals and the standard deviation of the regression model.

$$\sum (\text{Residuals})^2 = \sum (y - \hat{y})^2 = \text{SSE}$$

$$s = \sqrt{\frac{\sum (\text{Residuals})^2}{n - (k + 1)}} = \sqrt{\frac{\text{SSE}}{n - (k + 1)}}$$

EXAMPLE 11.4

The data for the home size–electrical usage example used throughout this chapter are repeated in Table 11.5. SAS printouts for a straight-line model and a quadratic model fitted to the data are shown in Figures 11.23(a) and 11.23(b) on pages 523–524, respectively. The residuals from these models are shaded in the printouts. The resid-

TABLE 11.5 Home Size–Electrical Usage Data

Size of Home x (square feet)	Monthly Usage y (kilowatt-hours)	Size of Home x (square feet)	Monthly Usage y (kilowatt-hours)
1,290	1,182	1,840	1,711
1,350	1,172	1,980	1,804
1,470	1,264	2,230	1,840
1,600	1,493	2,400	1,956
1,710	1,571	2,930	1,954

FIGURE 11.23 ▶
SAS printouts for electrical usage example

(a) Straight-line model

```
Dep Variable: Y

                         Analysis of Variance

                         Sum of          Mean
   Source       DF      Squares        Square        F Value       Prob>F

   Model         1 703957.18342 703957.18342          39.536       0.0002
   Error         8 142444.91658  17805.61457
   C Total       9 846402.10000

      Root MSE       133.43768        R-Square        0.8317
      Dep Mean      1594.70000        Adj R-Sq        0.8107
      C.V.             8.36757

                         Parameter Estimates

                      Parameter         Standard      T for H0:
Variable     DF        Estimate            Error    Parameter=0      Prob > |T|

INTERCEP      1      578.927752     166.96805715          3.467          0.0085
X             1        0.540304       0.08592981          6.288          0.0002

                                           Predict
                     Obs           Y         Value      Residual

                       1      1182.0        1275.9      -93.9204
                       2      1172.0        1308.3        -136.3
                       3      1264.0        1373.2        -109.2
                       4      1493.0        1443.4       49.5852
                       5      1571.0        1502.8       68.1517
                       6      1711.0        1573.1         137.9
                       7      1804.0        1648.7         155.3
                       8      1840.0        1783.8       56.1935
                       9      1956.0        1875.7       80.3417
                      10      1954.0        2162.0        -208.0

Sum of Residuals                           0
Sum of Squared Residuals         142444.9166
```

uals are then plotted on the vertical axis against the variable x, size of home, on the horizontal axis in Figures 11.24(a) and 11.24(b), respectively (page 575).

a. Verify that each residual is equal to the difference between the observed y value and the estimated mean value, y.

b. Analyze the residual plots.

Solution

a. For the straight-line model the residual is calculated for the first y value as follows:

$$\begin{aligned}\text{Residual} &= (\text{Observed } y \text{ value}) - (\text{Estimated mean})\\ &= y - \hat{y} = 1{,}182 - 1{,}275.9 = -93.9\end{aligned}$$

FIGURE 11.23 ▶
(b) Quadratic model

```
Dep Variable: Y

                        Analysis of Variance

                      Sum of         Mean
Source        DF     Squares        Square          F Value       Prob>F

Model          2 831069.54637 415534.77319          189.710       0.0001
Error          7  15332.55363   2190.36480
C Total        9 846402.10000

     Root MSE       46.80133     R-Square           0.9819
     Dep Mean     1594.70000     Adj R-Sq           0.9767
     C.V.            2.93480

                       Parameter Estimates

                   Parameter          Standard       T for H0:
Variable    DF      Estimate             Error     Parameter=0      Prob > |T|

INTERCEP     1   -1216.143887     242.80636850          -5.009          0.0016
X            1       2.398930       0.24583560           9.758          0.0001
XSQ          1      -0.000450       0.00005908          -7.618          0.0001

                                     Predict
                Obs          Y         Value     Residual

                  1     1182.0        1129.6      52.4359
                  2     1172.0        1202.2     -30.2136
                  3     1264.0        1337.8     -73.7916
                  4     1493.0        1470.0      22.9586
                  5     1571.0        1570.1       0.9359
                  6     1711.0        1674.2      36.7685
                  7     1804.0        1769.4      34.5998
                  8     1840.0        1895.5     -55.4654
                  9     1956.0        1949.1       6.9431
                 10     1954.0        1949.2       4.8287

Sum of Residuals                -2.27374E-12
Sum of Squared Residuals          15332.5536
```

where the estimated mean is the first number (shaded) in the column labeled **Predict Value** on the SAS printout in Figure 11.23(a). Similarly, the residual for the first y value using the quadratic model is

$$\text{Residual} = 1{,}182 - 1{,}129.6 = 52.4$$

Both residuals agree (after rounding) with the first values given in the column labeled **Residual** in Figures 11.23(a) and 11.23(b), respectively. Although the residuals both correspond to the same observed y value, 1,182, they differ because the estimated mean value changes depending on whether the straight-line model or quadratic model is used. Similar calculations produced the remaining residuals.

b. The plot of the residuals for the straight-line model, Figure 11.24(a), reveals a nonrandom pattern. The residuals exhibit a mound shape, with the residuals for the

FIGURE 11.24 ▶
Residual plots for electrical usage example

(a) Straight-line model

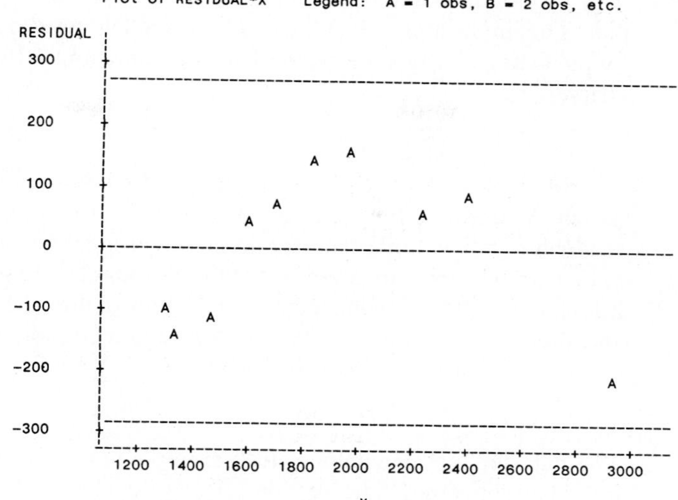

(b) Quadratic model

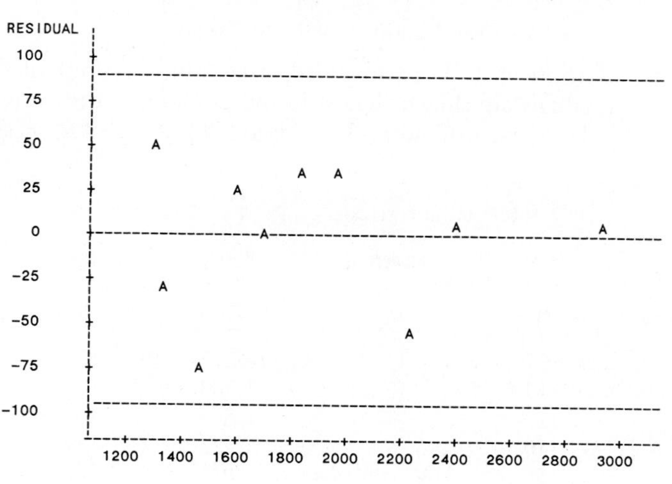

small values of x below the horizontal 0 (mean of the residuals) line, the residuals corresponding to the middle values of x above the 0 line, and the residual for the largest value of x again below the 0 line. The indication is that the mean value of the random error ϵ *within* each of these ranges of x (small, medium, large) may not be equal to 0. Such a pattern usually indicates that curvature needs to be added to the model.

When the second-order term is added to the model, the nonrandom pattern disappears. In Figure 11.24(b), the residuals appear to be randomly distributed around the 0 line, as expected. Note, too, that the ± 2 standard deviation lines are at about

± 95 on the quadratic residual plot, compared to (about) ± 275 on the straight-line plot. The implication is that the quadratic model provides a considerably better model for predicting electrical usage, verifying our conclusions from previous analyses in this chapter.

Residual analyses are also useful for detecting one or more observations that deviate significantly from the regression model. We expect approximately 95% of the residuals to fall within 2 standard deviations of the 0 line and all or almost all of them to lie within 3 standard deviations of their mean of 0. Residuals that are extremely far from the 0 line and disconnected from the bulk of the other residuals are called **outliers** and should receive special attention from the regression analyst. You will recall that outliers were first introduced in Chapter 2.

EXAMPLE 11.5

The data for the grandfather clock example are repeated in Table 11.6, with one important difference: The auction price of the clock at the top of the second column has been changed from \$2,131 to \$1,131 (shaded in Table 11.6). The interaction model

$$E(y) = \beta_0 + \beta_1 x_1 + \beta_2 x_2 + \beta_3 x_1 x_2$$

is again fit to these (modified) data, with the printout shown in Figure 11.25. The residuals are shown shaded in the printout, and then plotted against the number of bidders, x_2, in Figure 11.26 (page 578). Analyze the residual plot.

TABLE 11.6 Auction Price Data

Age x_1	Bidders x_2	Price y	Age x_1	Bidders x_2	Price y
127	13	\$1,235	170	14	\$1,131
115	12	1,080	182	8	1,550
127	7	845	162	11	1,884
150	9	1,522	184	10	2,041
156	6	1,047	143	6	854
182	11	1,979	159	9	1,483
156	12	1,822	108	14	1,055
132	10	1,253	175	8	1,545
137	9	1,297	108	6	729
113	9	946	179	9	1,792
137	15	1,713	111	15	1,175
117	11	1,024	187	8	1,593
137	8	1,147	111	7	785
153	6	1,092	115	7	744
117	13	1,152	194	5	1,356
126	10	1,336	168	7	1,262

FIGURE 11.25 ►
Printout for grandfather clock example with altered data

```
THE REGRESSION EQUATION IS
 PRICE = - 511 + 8.16 AGE + 19.7 BIDDERS + 0.320 AGE-BID

                              ST. DEV.     T-RATIO =
             COEFFICIENT      OF COEF.     COEF/S.D.
                 -510.5        664.8         -0.77
AGE               8.160        4.577          1.78
BIDDERS           19.74        67.33          0.29
AGE-BID          0.3197       0.4783          0.67

S = 200.3

R-SQUARED = 73.0 PERCENT
R-SQUARED = 70.1 PERCENT, ADJUSTED FOR D.F.

ANALYSIS OF VARIANCE

 DUE TO       DF          SS       MS=SS/DF
REGRESSION     3     3029141        1009714
RESIDUAL      28     1123116          40111
TOTAL         31     4152256
```

ROW	AGE	Y PRICE	PRED. Y VALUE	ST.DEV. PRED. Y	RESIDUAL
1	127	1235.0	1310.3	59.2	-75.3
2	115	1080.0	1106.0	62.0	-26.0
3	127	845.0	948.2	61.0	-103.2
4	150	1522.0	1322.8	37.1	199.2
5	156	1047.0	1180.2	60.2	-133.2
6	182	1979.0	1831.8	82.8	147.2
7	156	1822.0	1597.9	61.8	224.1
8	132	1253.0	1186.1	39.7	66.9
9	137	1297.0	1179.3	39.0	117.7
10	113	946.0	914.4	58.5	31.6
11	137	1713.0	1560.6	78.3	152.4
12	117	1024.0	1072.8	53.0	-48.8
13	137	1147.0	1115.8	44.2	31.2
14	153	1092.0	1149.9	58.9	-57.9
15	117	1152.0	1187.1	69.6	-35.1
16	126	1336.0	1117.9	43.4	218.1
17	170	1131.0	1914.0	116.5	-783.0
18	182	1550.0	1598.1	62.7	-48.1
19	162	1884.0	1598.3	56.9	285.7
20	184	2041.0	1776.6	70.6	264.4
21	143	854.0	1049.2	58.8	-195.2
22	159	1483.0	1422.1	40.5	60.9
23	108	1055.0	1130.6	97.8	-75.6
24	175	1545.0	1523.0	55.3	22.0
25	108	729.0	696.4	99.5	32.6
26	179	1792.0	1642.9	57.5	149.1
27	111	1175.0	1223.7	107.1	-48.7
28	187	1593.0	1651.7	68.5	-58.7
29	111	785.0	781.9	80.7	3.1
30	115	744.0	823.5	75.4	-79.5
31	194	1356.0	1481.4	133.3	-125.4
32	168	1262.0	1374.6	57.6	-112.6

FIGURE 11.26 ▶ Residual plot vs. number of bidders

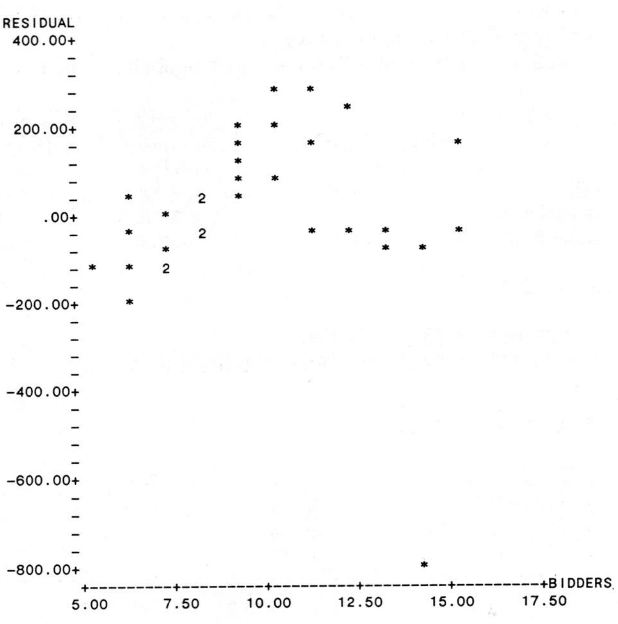

Solution

The residual plot dramatically reveals the one altered measurement. Note that one of the two residuals at $x_2 = 14$ bidders falls more than 3 standard deviations below 0. Note that no other residual lies more than 2 standard deviations from 0.

What do we do with outliers once we identify them? First, we try to determine the cause. Were the data entered into the computer incorrectly? Was the observation recorded incorrectly when the data were collected? If so, correct the observation and rerun the program. Or perhaps the observation is not representative of the conditions you are trying to model. For example, the low price may be attributable to extreme damage or to inferior quality compared to the other clocks. In these cases we probably would exclude the observation from the analysis. Often you may be unable to determine the cause of the outlier. Even so, you may want to rerun the regression analysis excluding the outlier, to assess the effect of that observation on the results of the analysis.

Figure 11.27 shows the printout when the outlier observation is excluded from the grandfather clock analysis, and Figure 11.28 shows the new plot of the residuals versus the number of bidders. Now only one residual lies beyond 2 standard deviations from 0, and none lies beyond 3 standard deviations. Also, the model statistics indicate a much better model without the outlier. Most notably, the standard deviation s has decreased from 200.3 to 85.28, indicating a model that will provide more precise estimates and predictions (narrower confidence and prediction intervals) for clocks like those in the reduced sample. Remember, though, that if you decide to remove the outlier from the analysis, and in fact it belongs to the same population as the rest of the sample, the resulting model may provide misleading estimates and predictions.

FIGURE 11.27 ► Minitab printout with outlier excluded

```
THE REGRESSION EQUATION IS
PRICE-2 = 476 - 0.46 AGE-2 - 114 BIDDERS2 + 1.48 AGE-BID2

                                         ST. DEV.      T-RATIO =
                  COEFFICIENT            OF COEF.      COEF/S.D.
                       475.8              296.2           1.61
AGE-2                 -0.465              2.093          -0.22
BIDDERS2             -114.19              31.03          -3.68
AGE-BID2              1.4775             0.2280           6.48

S = 85.28

R-SQUARED = 95.2 PERCENT
R-SQUARED = 94.7 PERCENT, ADJUSTED FOR D.F.

ANALYSIS OF VARIANCE

 DUE TO        DF          SS         MS=SS/DF
REGRESSION      3     3927844          1309282
RESIDUAL       27      196341             7272
TOTAL          30     4124186
```

FIGURE 11.28 ► Minitab residual plot

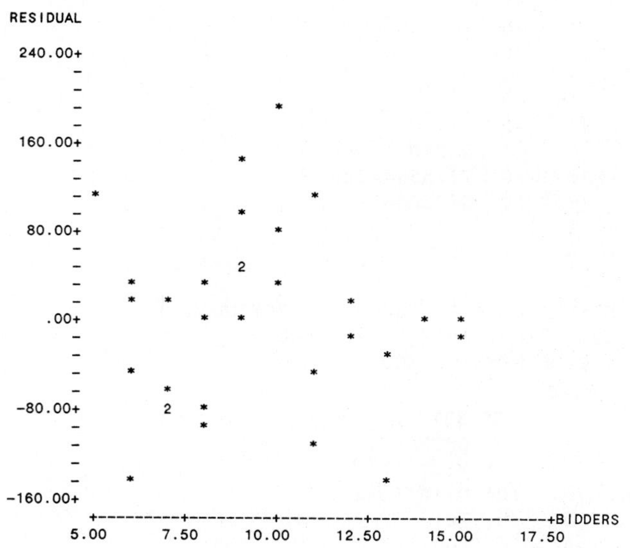

Outlier analysis is another example of testing the assumption that the expected (mean) value of the random error component is 0, since this assumption is in doubt for the error terms corresponding to the outliers. The last example in this section checks the assumption of the normality of the random error component.

EXAMPLE 11.6

Refer to Example 11.5. Use a stem-and-leaf display (Section 2.2) to plot the frequency distribution of the residuals in the grandfather clock example, both before and after the outlier residual is removed. Analyze the plots and determine whether the assumption of normality of the error distribution is reasonable.

Solution

The stem-and-leaf displays for the two sets of residuals are constructed using Minitab as shown in Figure 11.29. Note that the outlier appears to skew the frequency distribution in Figure 11.29(a), whereas the stem-and-leaf display in Figure 11.29(b) appears to be more mound-shaped. Although the displays do not provide formal statistical tests of normality, they do provide a descriptive display. Relative frequency histograms can also be used to check the normality assumption. In this example the normality assumption appears to be more plausible after the outlier is removed. Consult the references for methods to conduct statistical tests of normality using the residuals.

FIGURE 11.29* ► Stem-and-leaf displays for grandfather clock example

(a) Outlier included

```
STEM-AND-LEAF DISPLAY OF RESIDUAL
LEAF DIGIT UNIT =    10.0000
1 2 REPRESENTS 120.

        STEM  LEAF

   1     -7   8
   1     -6
   1     -5
   1     -4
   1     -3
   1     -2
   6     -1   93210
  16     -0   7775544432
  16      0   0233366
   9      1   14459
   4      2   1268
```

(b) Outlier excluded

```
STEM-AND-LEAF DISPLAY OF RESIDUAL
LEAF DIGIT UNIT =    10.0000
1 2 REPRESENTS 120.

   3    -1*  331
   9    -0.  987765
  (7)   -0*  4321000
  15    +0*  011223344
   6    +0.  79
   4     1*  004
   1     1.  9
```

*Recall that the left column of the Minitab printout shows the number of measurements at least as extreme as the stem. In Figure 11.29(a) the 6 corresponding to the STEM = −1 means that six measurements are less than or equal to −100. If a number in the leftmost column is enclosed in parentheses, it is the number of measurements in that row, and that row contains the median.

A summary of the residual analyses presented in this section, to check the assumption that the random error ϵ is normally distributed with mean 0, is presented in the box. Using the residuals to check the remaining assumptions of constant variance and independence will be presented in Chapters 12 and 15, respectively. Residual analysis is a useful tool for the regression analyst, not only to check the assumptions, but also to provide information about how the model can be improved.

Steps in a Residual Analysis

1. Calculate and plot the residuals against each of the independent variables, preferably with the assistance of a computer.
2. Analyze each plot, looking for curvature—either a mound or bowl shape. Both shapes are distinguished by groups of residuals at the low and high values of the x variable on one side of the 0 line, and the residuals for the medium x values on the opposite side of the 0 line. This shape signals the need for a curvature term in the model. Try a second-order term for the variable against which the residuals are plotted.
3. Examine the residual plots for outliers. Draw lines on the residual plots at 3 standard deviations below and above the 0 line. Residuals outside the 3 standard deviation lines are potential outliers. Determine whether each outlier can be explained as an error in data collection or transcription, or belongs to a population different from that of the remainder of the sample, or simply represents an unusual observation. If the observation is an error, fix it or remove it. Even if cause cannot be determined, you may want to rerun the regression analysis without the observation to determine its effect on the analysis.
4. Plot a frequency distribution of the residuals, using a stem-and-leaf display or a histogram. Check to see if obvious departures from normality exist. Extreme skewness of the frequency distribution may indicate the need for a transformation of the dependent variable, a topic we discuss in Chapter 12.

Exercises 11.22–11.31

Note: Starred (*) exercises require the use of a computer.

Learning the Mechanics

11.22 When a multiple regression model is used for estimating the mean of the dependent variable and for predicting a particular value of y, which will be narrower, the confidence interval for the mean or the prediction interval for the particular y value? Explain.

11.23 Refer to Exercise 11.1, in which the model

$$y = \beta_0 + \beta_1 x_1 + \beta_2 x_2 + \epsilon$$

was fit to $n = 20$ data points. The Minitab regression printout for these data is shown.

```
THE REGRESSION EQUATION IS
Y = 506 - 942 X1 - 429 X2

                                   ST. DEV.      T-RATIO =
              COEFFICIENT          OF COEF.      COEF/S.D.
                  506.35            45.17          11.21
X1               -941.9            275.1           -3.42
X2               -429.1            379.8           -1.13

S = 94.25

R-SQUARED = 45.9 PERCENT
R-SQUARED = 39.6 PERCENT, ADJUSTED FOR D.F.

ANALYSIS OF VARIANCE

 DUE TO       DF              SS        MS=SS/DF
REGRESSION     2          128329          64165
RESIDUAL      17          151016           8883
TOTAL         19          279345
```

a. Find the least squares prediction equation.
b. Find and interpret the standard deviation of the regression model.
c. Calculate the F statistic. Does the model contribute information for predicting y? Test at $\alpha = .05$.
d. Place a 90% confidence interval on the coefficient β_2.
e. Find and interpret the coefficient of determination R^2.

11.24 Repeat parts a–e of Exercise 11.23 using the SPSS printout for the same data set given here.

```
Multiple R              .67779
R Square                .45939
Adjusted R Square       .39579
Standard Error        94.25114

Analysis of Variance
                         DF      Sum of Squares        Mean Square
Regression                2        128329.27624        64164.63812
Residual                 17        151015.72376         8883.27787

F =          7.22308          Signif F =   .0054

------------------ Variables in the Equation ------------------

Variable                B            SE B         Beta          T   Sig T

X1             -941.90023       275.08556      -.61728     -3.424   .0032
X2             -429.06042       379.82566      -.20365     -1.130   .2743
(Constant)      506.34607        45.16942                  11.210   .0000
```

11.25 Refer to Exercise 11.11, in which a quadratic model was fit to $n = 19$ data points. Two residual plots are shown here—one corresponding to the fitting of a straight-line model to the data, and the other to the quadratic model fit in Exercise 11.11. Analyze the two plots. Is the need for a quadratic term evident from the residual plot for the straight-line model? Does your conclusion agree with your test of the quadratic term in Exercise 11.11?

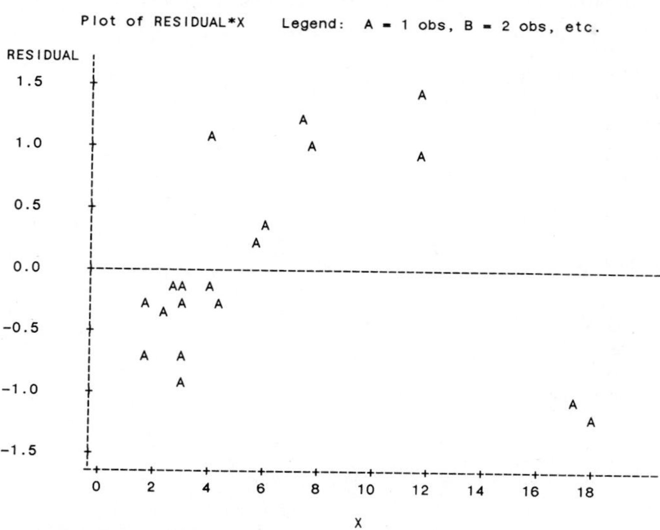

▲ Residual plot for straight-line model

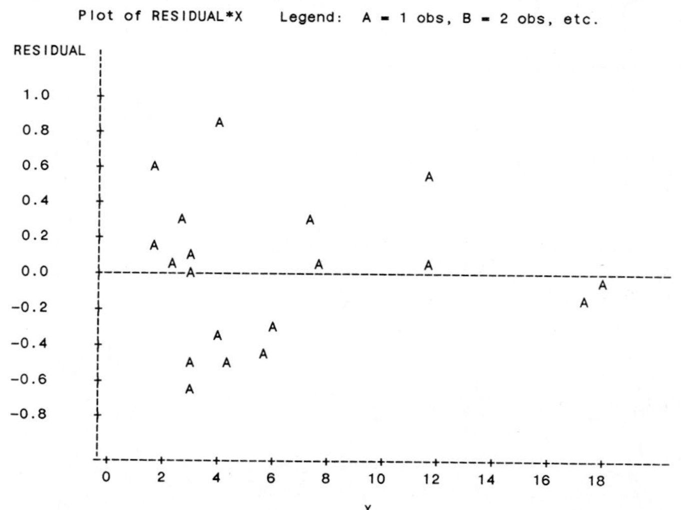

▲ Residual plot for quadratic model

Applying the Concepts

11.26 Refer to Exercise 11.5, in which the 1990 food consumption expenditure y of a sample of 25 households in Washington, D.C., was related to the household income x_1 and size of household x_2 by the model

$$y = \beta_0 + \beta_1 x_1 + \beta_2 x_2 + \epsilon$$

Plots of the residuals from this model are shown—one versus x_1 and one versus x_2. Analyze the plots. Is there visual evidence of a need for a quadratic term in either x_1 or x_2? Explain.

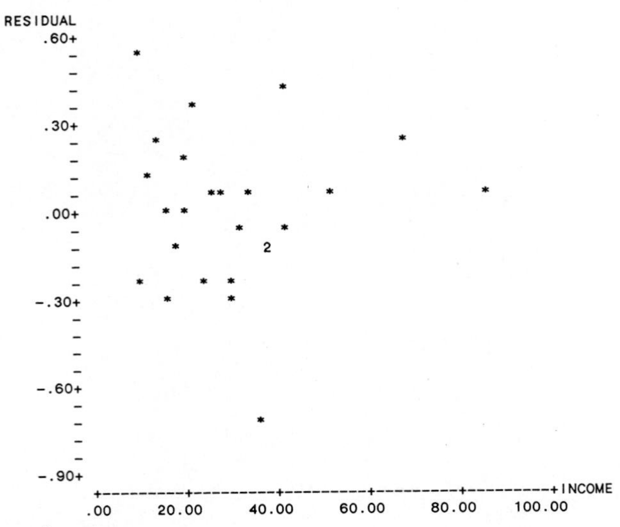

▲ $\hat{\epsilon}$ versus x_1

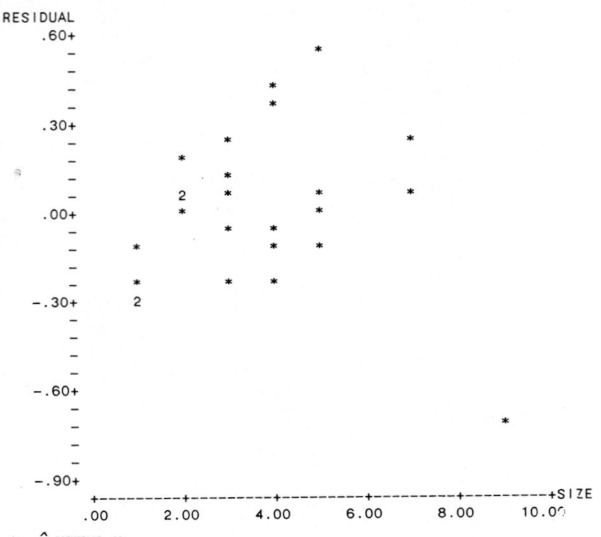

▲ $\hat{\epsilon}$ versus x_2

11.27 Refer to Exercise 11.26. Suppose a 26th household is added to the sample, with the following characteristics:

Food consumption: 6.5 *Income:* 62.3 *Household size:* 5

Two Minitab printouts are shown—one for the same model fit to the first 25 observations, the second fit to all 26 observations.

```
THE REGRESSION EQUATION IS
 FOOD = 1.43 + 0.00999 INCOME + 0.379 SIZE

                                     ST. DEV.      T-RATIO =
             COEFFICIENT             OF COEF.      COEF/S.D.
                 1.4326               0.1467           9.76
 INCOME         0.009991             0.003168           3.15
 SIZE            0.37929              0.02772          13.68

  S = 0.2769

  R-SQUARED = 90.2 PERCENT
  R-SQUARED = 89.3 PERCENT, ADJUSTED FOR D.F.

  ANALYSIS OF VARIANCE

   DUE TO       DF            SS        MS=SS/DF
  REGRESSION     2       15.4623          7.7311
  RESIDUAL      22        1.6873          0.0767
  TOTAL         24       17.1496
```

▲ Model fit to 25 observations

```
THE REGRESSION EQUATION IS
FOOD = 1.16 + 0.0187 INCOME + 0.406 SIZE

                                     ST. DEV.      T-RATIO =
COLUMN       COEFFICIENT             OF COEF.      COEF/S.D.
                 1.1554               0.2882           4.01
INCOME          0.018735             0.006034          3.11
SIZE             0.40577              0.05551          7.31

S = 0.5581

R-SQUARED = 74.6 PERCENT
R-SQUARED = 72.4 PERCENT, ADJUSTED FOR D.F.

ANALYSIS OF VARIANCE

 DUE TO       DF            SS        MS=SS/DF
REGRESSION     2        21.076          10.538
RESIDUAL      23         7.163           0.311
TOTAL         25        28.239
```

▲ Model fit to 26 observations

a. Record the least squares estimates of the model parameters for each model, and note differences in the estimates. Interpret each estimate.
b. Find and interpret the standard deviation for each model.
c. Conduct the analysis of variance F test for each model using $\alpha = .05$.
d. Place a 95% confidence interval on the mean rate of change in food consumption per additional person in the household for each model, assuming household income is constant.
e. According to the results of parts **a–d**, how much influence does the additional observation have on the model?

11.28 Refer to Exercises 11.26 and 11.27. Residual plots against household income and size of household for the models corresponding to 26 observations are shown.

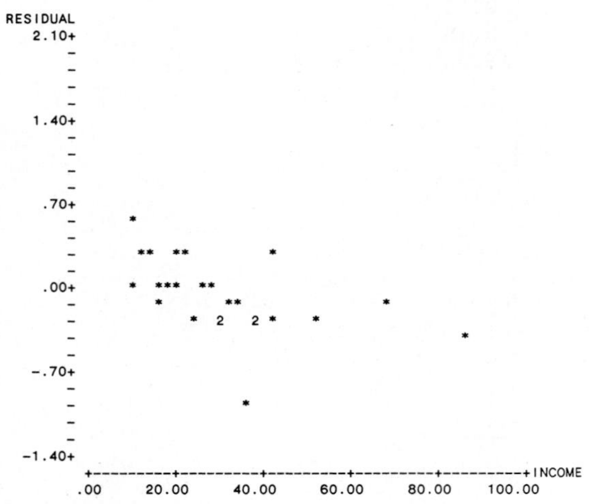

▲ $\hat{\epsilon}$ versus x_1

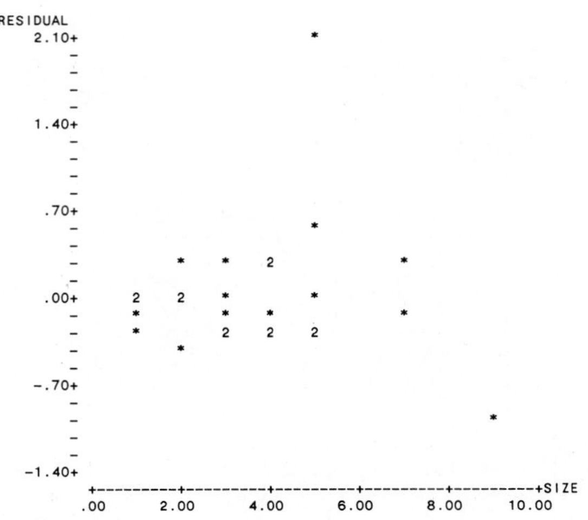

▲ $\hat{\epsilon}$ versus x_2

a. Analyze the plots. Are there any outliers? If so, identify them.
b. What are possible explanations for any outliers you identified in part a?

11.29 Refer to Exercises 11.26–11.28. The accompanying stem-and-leaf displays represent the frequency distributions of the residuals for the two data sets, one with $n = 25$ and one with $n = 26$. Analyze the displays, especially with regard to the normality assumption.

```
STEM-AND-LEAF DISPLAY OF RESIDUAL
LEAF DIGIT UNIT =     .1000
1 2 REPRESENTS 1.2


   1   -0S   7
   1   -0F
   6   -0T   32222
  11   -0*   11100
 (8)   +0*   00000001
   6   +0T   2223
   2   +0F   45
```

▲ Display for 25 residuals

```
STEM-AND-LEAF DISPLAY OF RESIDUAL
LEAF DIGIT UNIT =     .1000
1 2 REPRESENTS 1.2


    1    -0. 9
 (16)    -0* 4333222211110000
    9    +0* 0022223
    2    +0. 6
    1     1*
    1     1.
    1     2* 1
```

▲ Display for 26 residuals

11.30 Refer to Exercise 11.7, in which a quadratic model was used to relate the time to complete a task, y, to the months of experience, x, for a sample of 15 employees on an automobile assembly line. The SPSS printouts for both straight-line and quadratic models are shown here and on page 588.

```
Multiple R              .94886
R Square                .90033
Adjusted R Square       .89266
Standard Error         1.14301


Analysis of Variance
                      DF        Sum of Squares        Mean Square
Regression             1             153.41583          153.41583
Residual              13              16.98417            1.30647


F =      117.42733          Signif F =   .0000


------------------ Variables in the Equation ------------------

Variable                 B           SE B         Beta          T   Sig T

X                  -.44494         .04106     -.94886    -10.836   .0000
(Constant)        19.27908         .50788                 37.960   .0000
```

▲ SPSS printout for straight-line model

```
Multiple R              .95718
R Square                .91619
Adjusted R Square       .90223
Standard Error         1.09089

Analysis of Variance
                    DF        Sum of Squares        Mean Square
Regression           2             156.11948           78.05974
Residual            12              14.28052            1.19004

F =       65.59402        Signif F =   .0000

------------------ Variables in the Equation ------------------

Variable                  B          SE B         Beta          T   Sig T

X                   -.67052        .15471     -1.42992     -4.334   .0010
XSQ           9.534744E-03 6.32580E-03       .49728      1.507   .1576
(Constant)         20.09111        .72471                  27.723   .0000
```

▲ SPSS printout for quadratic model

a. Find and interpret the standard deviation of each regression model in the context of this application.
b. Find and interpret R^2 for each model.
c. Conduct the analysis of variance F test for each model. Use $\alpha = .05$ and interpret the results.
d. Test the quadratic term in the second-order model. Use $\alpha = .05$. Does the quadratic term for experience appear to provide information for the prediction of time to complete the task?

11.31 Refer to Exercise 11.30. Residual plots for the straight-line and quadratic models are shown. Analyze the plots. Does your residual analysis support the statistical test you conducted in part **d** of Exercise 11.30?

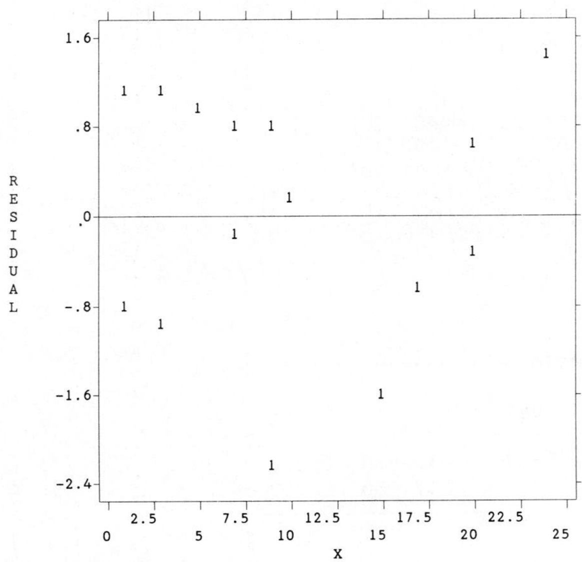

▲ Straight-line model

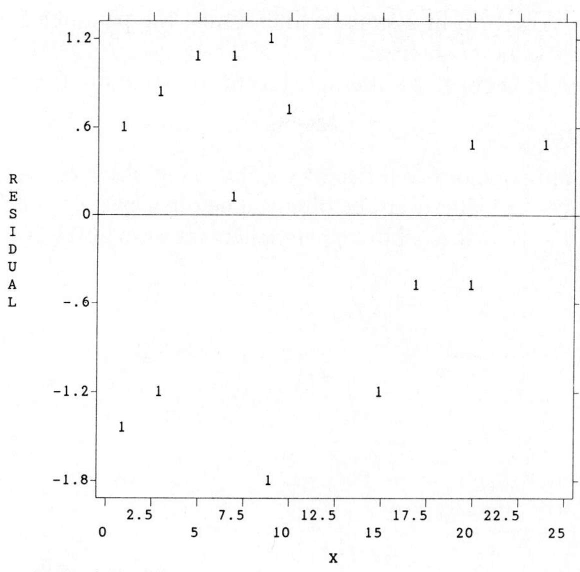

▲ Quadratic model

11.10 Some Pitfalls: Estimability, Multicollinearity, and Extrapolation

There are several problems you should be aware of when constructing a prediction model for some response, y. A few of the most important are discussed in this section.

Problem 1: Parameter Estimability

Suppose you want to fit a model relating a firm's monthly profit y to the advertising expenditure x. You propose the first-order model $E(y) = \beta_0 + \beta_1 x$. Now, suppose you have 3 months of data, and the firm spent \$1,000 on advertising during each month. The data are shown in Figure 11.30. Can you see the problem? The parameters of the line cannot be estimated when all the data are concentrated at a single x value. Recall

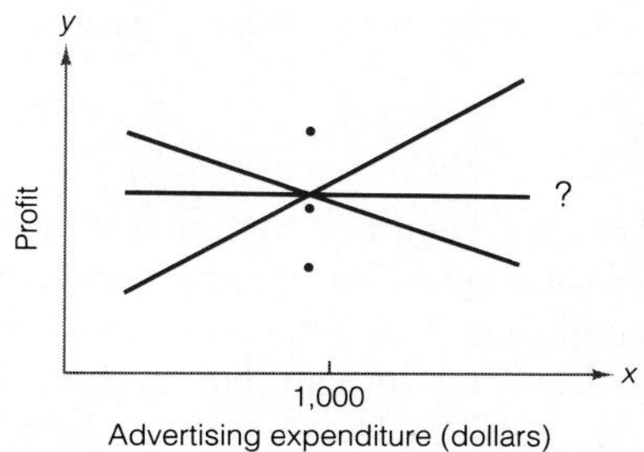

FIGURE 11.30 ▶ Profit and advertising expenditure data: 3 months

that it takes two points (x values) to fit a straight line. Thus, the parameters are not estimable when only one x value is observed.

A similar problem would occur if we attempted to fit the second-order model

$$E(y) = \beta_0 + \beta_1 x + \beta_2 x^2$$

to a set of data for which only one *or two* different x values were observed (see Figure 11.31). At least three different x values must be observed before a second-order model can be fit to a set of data (that is, before all three parameters are estimable). In general,

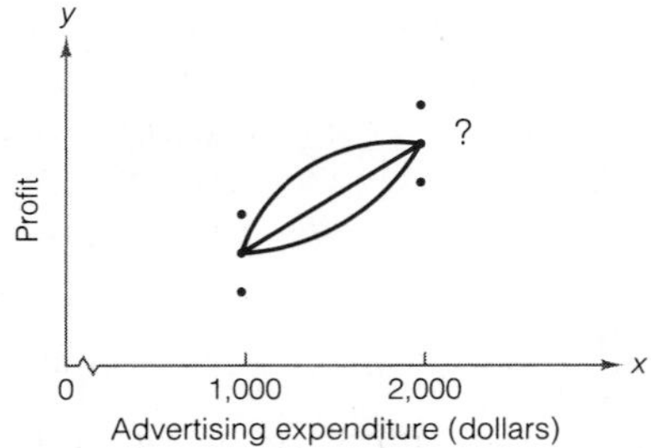

FIGURE 11.31 ▶ Only two x values observed; second-order model is not estimable

the number of levels of x must be at least one more than the order of the polynomial in x that you want to fit.

Since most business variables are not controlled by the researcher, the independent variables will usually be observed at a sufficient number of levels to permit estimation of the model parameters. However, when the computer program you use suddenly refuses to fit a model, the problem is probably *inestimable parameters.*

Problem 2: Multicollinearity

Often, two or more of the independent variables used in the model for $E(y)$ contribute redundant information. That is, the independent variables are correlated with each other. For example, suppose we want to construct a model to estimate the gasoline mileage of trucks as a function of their load x_1 and their horsepower x_2. In general, we expect heavy loads to require greater horsepower and to result in lower mileage ratings. Thus, although both x_1 and x_2 contribute information for the prediction of mileage rating, some of the information is overlapping because x_1 and x_2 are correlated.

If the model

$$E(y) = \beta_0 + \beta_1 x_1 + \beta_2 x_2$$

were fit to a set of data, we might find that the t values for both $\hat{\beta}_1$ and $\hat{\beta}_2$ (the least squares estimates) were nonsignificant. However, the F test for H_0: $\beta_1 = \beta_2 = 0$ would probably be highly significant. The tests may seem to be contradictory, but really they are not. The t tests indicate that the contribution of one variable, say x_1 = Load, is not significant after the effect of x_2 = Horsepower has been discounted (because x_2 is also in the model). The significant F test, on the other hand, tells us that at least one of the two variables is making a contribution to the prediction of y (i.e., β_1, β_2, or both differ from 0). In fact, both are probably contributing, but the contribution of one overlaps with that of the other.

When highly correlated independent variables are present in a regression model, the results may be confusing. The researcher may want to include only one of the variables in the final model. One way of deciding which variable to include is by using **stepwise regression**, a topic discussed in Chapter 12. Generally, only one (or a small number) of a set of *multicollinear* independent variables will be included in the regression model by a stepwise regression procedure. This procedure tests the parameter associated with each variable in the presence of all the variables already in the model. For example, if at one step, the variable Truck load is included as a significant variable in the prediction of the mileage rating, then the variable Horsepower will probably never be added in a future step. Thus, if a set of independent variables is thought to be multicollinear, some screening by stepwise regression may be helpful.

Problem 3: Prediction Outside the Experimental Region

By the late 1960s, many economists had developed highly technical models to relate the state of the economy to various economic indexes and other independent variables. Many of these models were multiple regression models, where, for example, the dependent variable y might be next year's growth in GNP and the independent variables might include this year's rate of inflation, this year's Consumer Price Index, and other factors. In other words, the model might be constructed to predict next year's economy using this year's knowledge.

Unfortunately, these models almost unanimously failed to predict the recession in the early 1970s. What went wrong? One problem was that the regression models were used to predict y for values of the independent variables that were *outside the region in which the model was developed*. For example, the inflation rate in the late 1960s, when the models were developed, ranged from 6% to 8%. When the double-digit inflation of the early 1970s became a reality, some researchers used the same old models to try to predict future growth in GNP. As you can see in Figure 11.32, the model can be very accurate for predicting y when x is in the range of experimentation, but the use of the model outside that range is a dangerous practice.

FIGURE 11.32 ▶
Using a regression model outside the experimental region

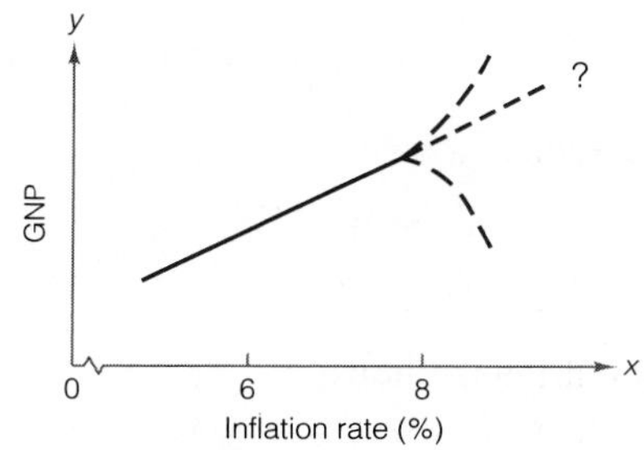

Problem 4: Correlated Errors

Another problem associated with using a regression model to predict an economic variable y based on independent variables $x_1, x_2, \ldots, x_k$ arises from the fact that the data are frequently *time series*. That is, the values of both the dependent and independent variables are observed sequentially over a period of time. Typically, such observations tend to be *correlated* over time, which in turn often causes the prediction errors of the regression model to be correlated. Thus, the assumption of independent errors is violated, and the model tests and prediction intervals are no longer valid. One solution to this problem is to construct a time series model; this is the subject of Chapter 15.

Summary

We have discussed some of the methodology of **multiple regression analysis**, a technique for modeling a dependent variable y as a function of several independent variables $x_1, x_2, \ldots, x_k$. The steps we follow in constructing and using multiple regression models are much the same as those for the simple straight-line models:

1. The form of the probabilistic model is hypothesized.
2. The model coefficients are estimated using the method of least squares.
3. The probability distribution of ϵ is verified and σ^2 is estimated.
4. The utility of the model is checked.
5. If the model is deemed useful, it may be used to make estimates and to predict values of y to be observed in the future.

We have covered steps 2–5 in this chapter, assuming that the model was specified. The most important topic, model building—step 1—is discussed in Chapter 12. Additional material on these topics can be found in the references at the end of the chapter.

Supplementary Exercises 11.32–11.56

Note: Starred (*) exercises require the use of a computer.

11.32 Suppose you used Minitab to fit the model

$$y = \beta_0 + \beta_1 x_1 + \beta_2 x_2 + \epsilon$$

to $n = 15$ data points and you obtained the printout shown.

```
THE REGRESSION EQUATION IS
Y =     90.1 -   1.84 X1 +   .285 X2

                                   ST. DEV.    T-RATIO =
                   COEFFICIENT    OF COEF.     COEF/S.D.

        --              90.1         23.1          3.90
X1     C2            -1.836         .367         -5.01
X2     C3              .285         .231          1.24

THE ST. DEV. OF Y ABOUT REGRESSION LINE IS
S =       10.7
WITH (   15- 3) =   12 DEGREES OF FREEDOM

R-SQUARED = 91.6 PERCENT
R-SQUARED = 90.2 PERCENT, ADJUSTED FOR D.F.

ANALYSIS OF VARIANCE

 DUE TO          DF        SS     MS=SS/DF

REGRESSION        2    14801.       7400.
RESIDUAL         12     1364.        114.
TOTAL            14    16165.
```

a. What is the least squares prediction equation?

b. Find R^2 and interpret its value.

c. Is there sufficient evidence to indicate that the model is useful for predicting y? Conduct an F test using $\alpha = .05$.

d. Test the null hypothesis H_0: $\beta_1 = 0$ against the alternative hypothesis H_a: $\beta_1 \neq 0$. Test using $\alpha = .05$. Draw the appropriate conclusions.

e. Find the standard deviation of the regression model, and interpret it.

11.33 Since the Great Depression of the 1930s, the link between the suicide rate and the state of the economy has been much researched. Recent work on this link applied regression analysis to data collected from a 45-year-period on the following variables:

y = Suicide rate

x_1 = Unemployment rate

x_2 = Percentage of females in the labor force

x_3 = Divorce rate

x_4 = Log of gross national product (GNP)

x_5 = Annual percent change in GNP

One of the models explored by the authors (Yang, Stack, and Lester, 1992) was a multiple regression model relating y to linear terms in x_1 through x_5. The following least squares model resulted (the observed significance levels of the β estimates are beneath the estimates) and $R^2 = .45$:

$$\hat{y} = .002 + .0204x_1 + (-.0231)x_2 + .0765x_3 + .2760x_4 + .0018x_5$$

(.002) (.02) (>.10) (>.10) (>.10)

a. Interpret the value of R^2. Is there sufficient evidence to indicate that the model is useful for predicting the suicide rate? Use $\alpha = .05$.
b. Interpret each of the coefficients in the model, and each of the corresponding significance levels.
c. Is there sufficient evidence to indicate that unemployment rate is a useful predictor of the suicide rate? Use $\alpha = .05$.
d. Discuss each of the following terms with respect to potential problems with the above model: curvature (second-order terms), interaction, and multicollinearity.

11.34 Several states now require all high school seniors to pass an achievement test before they can graduate. On the test, the seniors must demonstrate their familiarity with basic verbal and mathematical skills. Suppose the educational testing company that creates and administers these exams wants to model the score, y, on one of its exams as a function of the student's IQ, x_1, and socioeconomic status (SES). The SES is a *categorical (qualitative) variable* with three levels: low, medium, and high. As we will demonstrate in Chapter 12, two *dummy (indicator) variables* are needed to describe a qualitative independent variable with three levels. Thus, we define

$$x_1 = \begin{cases} 1 & \text{if SES is medium} \\ 0 & \text{if SES is low or high} \end{cases} \qquad x_3 = \begin{cases} 1 & \text{if SES is high} \\ 0 & \text{if SES is low or medium} \end{cases}$$

Data were collected for a random sample of 60 seniors who took the test, and the model

$$E(y) = \beta_0 + \beta_1 x_1 + \beta_2 x_2 + \beta_3 x_3$$

was fit to the data, with the results shown in the SAS printout.

SOURCE	DF	SUM OF SQUARES	MEAN SQUARE	F VALUE	PR > F
MODEL	3	12268.56439492	4089.52146497	188.33	0.0001
ERROR	56	1216.01893841	21.71462390		ROOT MSE
CORRECTED TOTAL	59	13484.58333333		R-SQUARE	
				0.909822	4.65989527

PARAMETER	ESTIMATE	T FOR HO: PARAMETER=0	PR > |T|	STD ERROR OF ESTIMATE
INTERCEPT	-13.06166081	-3.21	0.0022	4.07101383
X1	0.74193946	17.56	0.0001	0.04224805
X2	18.60320572	12.49	0.0001	1.48895324
X3	13.40965415	8.97	0.0001	1.49417069

a. Identify the least squares equation.
b. Interpret the value of R^2 and test to determine whether the data provide sufficient evidence to indicate that this model is useful for predicting achievement test scores.
c. Sketch the relationship between predicted achievement test score and IQ for the three levels of SES. [*Note*: Three graphs of $\hat{y}$ versus x_1 must be drawn: the first for the low SES model ($x_2 = x_3 = 0$), the second for the medium SES model ($x_2 = 1$, $x_3 = 0$), and the third for the high SES model ($x_2 = 0$, $x_3 = 1$). The increase in predicted achievement test score per unit increase in IQ is the same for all three levels of SES; i.e., all three lines are parallel.]

11.35 Refer to Exercise 11.34. We now use the same data to fit the model

$$E(y) = \beta_0 + \beta_1 x_1 + \beta_2 x_2 + \beta_3 x_3 + \beta_4 x_1 x_2 + \beta_5 x_1 x_3$$

Thus, we now add the interaction between IQ and SES to the model. The SAS printout for this model is shown.

SOURCE	DF	SUM OF SQUARES	MEAN SQUARE	F VALUE	PR > F
MODEL	5	12515.10021009	2503.02004202	139.42	0.0001
ERROR	54	969.48312324	17.95339117		
CORRECTED TOTAL	59	13484.58333333		R-SQUARE	ROOT MSE
				0.928104	4.23714422

PARAMETER	ESTIMATE	T FOR H0: PARAMETER = 0	PR > |T|	STD ERROR OF ESTIMATE
INTERCEPT	0.60129643	0.11	0.9096	5.26818519
X1	0.59526252	10.70	0.0001	0.05563379
X2	-3.72536406	-0.37	0.7115	10.01967496
X3	-16.23196444	-1.90	0.0631	8.55429931
X1*X2	0.23492147	2.29	0.0260	0.10263908
X1*X3	0.30807756	3.53	0.0009	0.08739554

a. Identify the least squares prediction equation.
b. Interpret the value of R^2 and test to determine whether the data provide sufficient evidence to indicate that this model is useful for predicting achievement test scores.
c. Sketch the relationship between predicted achievement test score and IQ for the three levels of SES. [*Note*: The interaction terms in the model allow nonparallelism among the three SES models; i.e., the mean increase in achievement test score per unit increase in IQ differs for the three levels of SES. To determine whether the interaction between IQ and SES is contributing to the prediction of achievement test score, we must test the null hypothesis H_0: $\beta_4 = \beta_5 = 0$ against the alternative that at least one of the coefficients of the interaction terms is nonzero. The method for testing portions of a regression model involving more than one β parameter (but less than all of them) is discussed in Chapter 12.]

***11.36** Refer to Exercise 11.10, in which regression analysis was used to explore the valuation of apartment buildings in Minneapolis.
a. Verify that number of apartment units, x_1, is highly correlated with gross building area, x_5.
b. Eliminate x_1 from the appraiser's model and refit the model to the data. Compare the standard deviations of the two models.
c. Use an F test to investigate the usefulness of this model. Use $\alpha = .05$.
d. Use the least squares equation of part **b** to estimate the value of an apartment building that is 50 years old with gross area 20,000 square feet, lot size 15,000 square feet, and 10 parking spaces. Use a computer program to compute a 95% confidence interval for the model's estimate.

11.37 Suppose you have developed a regression model to explain the relationship between y and x_1, x_2, and x_3. The ranges of the variables you used to develop your model are $10 \le y \le 100$, $5 \le x_1 \le 55$, $.5 \le x_2 \le 1$, and $1{,}000 \le x_3 \le 2{,}000$. Explain why you would have more confidence using your prediction equation to predict y when $x_1 = 30$, $x_2 = .6$, and $x_3 = 1{,}300$ than when $x_1 = 60$, $x_2 = .4$, and $x_3 = 900$.

11.38 Plastics made under different environmental conditions differ in strength. A scientist would like to know which combination of temperature and pressure yields a plastic with a high breaking strength. A small preliminary experiment was conducted at two pressure levels and two temperature levels. The model proposed was

$$E(y) = \beta_0 + \beta_1 x_1 + \beta_2 x_2 + \beta_3 x_1 x_2$$

where

$$y = \text{Breaking strength (pounds)}$$
$$x_1 = \text{Temperature (°F)} \qquad x_2 = \text{Pressure (pounds per square inch)}$$

A sample of $n = 16$ observations yielded

$$\hat{y} = 226.8 + 4.9x_1 + 1.2x_2 - .7x_1x_2$$

with

$$s_{\hat{\beta}_1} = 1.11 \qquad s_{\hat{\beta}_2} = .27 \qquad s_{\hat{\beta}_3} = .34$$

Do the data indicate there is an interaction between temperature and pressure? Test using $\alpha = .05$.

11.39 A large government agency wants to predict how many people it will hire in the next year to fill the 30 positions currently open. Historically, the agency has been unable to fill all its job openings. It was decided to model the number of positions filled in a year, y, as a function of the number of positions open, x_1, and the recruiting budget for the year in dollars, x_2 (e.g., for advertising and travel expenses). A random sample of 10 years of recruiting records was drawn from the agency's 30 years of records. The model

$$E(y) = \beta_0 + \beta_1 x_1 + \beta_2 x_2$$

was fit to the data using the Minitab regression computer program package. The results shown in the printout were obtained.

```
THE REGRESSION EQUATION IS
Y =    .0562 +   .273 X1  + .0006 X2

                                     ST. DEV.     T-RATIO =
                     COEFFICIENT     OF COEF.     COEF/S.D.

        --                  .056         .902           .06
X1      C2                 .2733        .0971          2.81
X2      C3               .000560      .000129          4.34

THE ST. DEV. OF Y ABOUT REGRESSION LINE IS
S =          1.33
WITH (   10- 3) =    7 DEGREES OF FREEDOM

R-SQUARED = 97.9 PERCENT
R-SQUARED = 97.3 PERCENT, ADJUSTED FOR D.F.

ANALYSIS OF VARIANCE

 DUE TO          DF        SS     MS=SS/DF

REGRESSION        2   583.18        291.59
RESIDUAL          7    12.42          1.77
TOTAL             9   595.60
```

a. Identify the least squares prediction equation.
b. Is there sufficient evidence to indicate that the model contributes information for predicting the number y of positions that will be filled? Conduct an F test using $\alpha = .05$.
c. Test the null hypothesis H_0: $\beta_2 = 0$ against the alternative hypothesis H_a: $\beta_2 \neq 0$ using $\alpha = .05$. Interpret the results of your test in the context of the problem.
d. Use the least squares prediction equation to predict how many of the 30 positions the agency will fill next year if the recruiting budget is \$10,000.
e. Which (if any) of the assumptions we make about ϵ in a regression analysis are likely to be violated in this problem? Explain.

11.40 A metropolitan bus company wants to know whether changes in numbers of bus riders are related to changes in gasoline prices. By using information in the company files and gasoline price information from fuel distributors, the company plans to fit the model

$$y = \beta_0 + \beta_1 x_1 + \beta_2 x_2 + \beta_3 x_1 x_2 + \epsilon$$

where

x_1 = Average wholesale price for regular gasoline in a given month

$$x_2 = \begin{cases} 1 & \text{if the bus travels a city route only} \\ 0 & \text{if the bus travels a suburb–city route} \end{cases}$$

y = Total number of riders in a bus over the month

a. For the preceding model, how would you test to determine whether the relationship between the mean number of riders and gasoline price is different for the two types of bus routes?
b. Suppose 12 months of data are kept, and the least squares model is

$$\hat{y} = 500 + 50x_1 + 5x_2 - 10x_1x_2$$

Graph the predicted relationship between number of riders and gasoline price for city buses and for suburb–city buses. Compare the slopes.

c. If $s_{\hat{\beta}_3} = 3.0$, do the data indicate that gasoline price affects the number of riders differently for city and suburb–city buses? Use $\alpha = .05$.

11.41 During the winter months a sample of 100 homes is taken to obtain information concerning the relationship between kilowatt usage y and total window and glass area x (measured as a percentage of the total wall area). A second-order model, $y = \beta_0 + \beta_1 x + \beta_2 x^2 + \epsilon$, was used to model this relationship. The multiple coefficient of determination for the data was .24. Test whether the data indicate that the model contributes information for the prediction of y. Use $\alpha = .05$. [*Hint:* Use the methods of Section 11.5.]

11.42 Most companies institute rigorous safety programs to ensure employee safety. Suppose accident reports over the last year at a company are sampled, and the number of hours the employee had worked before the accident occurred, x, and the amount of time the employee lost from work, y, are recorded. A quadratic model is proposed to investigate a fatigue hypothesis that more serious accidents occur near the end of workdays than occur near the beginning. Thus, the proposed model is

$$E(y) = \beta_0 + \beta_1 x + \beta_2 x^2$$

A total of 60 accident reports are examined and part of the computer printout is shown on page 598.

SOURCE	DF	SUM OF SQUARES	MEAN SQUARE	F VALUE
MODEL	2	112.110	56.055	1.28
ERROR	57	2496.201	43.793	R-SQUARE
TOTAL	59	2608.311		.0430

a. Do the data support the fatigue hypothesis? Use $\alpha = .05$ to test whether the proposed model is useful in predicting the lost work time, y.
b. Does the result of the test in part **a** necessarily mean that no fatigue factor exists? Explain.

11.43 Refer to Exercise 11.42. Suppose the company persists in using the quadratic model despite its apparent lack of usefulness. The fitted model is

$$\hat{y} = 12.3 + .25x - .0033x^2$$

where $\hat{y}$ is the predicted time lost (days) and x is the number of hours worked prior to an accident.
a. Use the model to predict the number of days missed by an employee who has an accident after 6 hours of work.
b. Suppose the 95% prediction interval for the predicted value in part **a** is determined to be (1.35, 26.01). Interpret this interval. Does this interval support your conclusion about this model in Exercise 11.42?

11.44 To increase the motivation and productivity of workers, an electronics manufacturer decides to experiment with a new pay incentive structure at one of two plants. The experimental plan will be tried at plant A for 6 months, whereas workers at plant B will remain on the original pay plan. To evaluate the effectiveness of the new plan, the average assembly time for part of an electronic system was measured for employees at both plants at the beginning and end of the 6-month period. Suppose the model proposed was:

$$y = \beta_0 + \beta_1 x_1 + \beta_2 x_2 + \epsilon$$

where

y = Assembly time (hours) at end of 6-month period
x_1 = Assembly time (hours) at beginning of 6-month period

$$x_2 = \begin{cases} 1 & \text{if plant A} \\ 0 & \text{if plant B} \end{cases} \quad \text{(dummy variable)}$$

A sample of $n = 42$ observations yielded

$$\hat{y} = .11 + .98x_1 - .53x_2$$

where

$$s_{\hat{\beta}_1} = .231 \qquad s_{\hat{\beta}_2} = .48$$

Test to see whether, after allowing for the effect of initial assembly time, plant A had a lower mean assembly time than plant B. Use $\alpha = .01$. [*Note:* When the {0, 1} coding is used to define a dummy variable, the coefficient of the variable represents the difference between the mean response at the two levels represented by the variable. Thus, the coefficient β_2 is the difference in mean assembly time between plant A and plant B at the end of the 6-month period, and $\hat{\beta}_2$ is the sample estimator of that difference.]

11.45 A company that relies on door-to-door sales wants to determine the relationship, if any, between the proportion of customers who buy its product, y, and two independent variables: price, x_1, and years of experience of the salesperson, x_2. Twenty salespeople employed by the company are randomly assigned to sell the products, five to each of four prices, ranging from \$1.98 to \$5.98. Each salesperson makes a sales presentation to 30 prospects, and the percentage of sales is recorded. The 20 observations are used to fit the model

$$y = \beta_0 + \beta_1 x_1 + \beta_2 x_2 + \epsilon$$

The least squares model is

$$\hat{y} = -.30 - .010x_1 + .10x_2$$

with $s_{\hat{\beta}_1} = .0030$, $s_{\hat{\beta}_2} = .025$, and $R^2 = .86$.

a. Interpret the values of $\hat{\beta}_1$ and $\hat{\beta}_2$.

b. Is there sufficient evidence to conclude that the overall model is useful for predicting y? Use $\alpha = .05$.

c. Do the data support the hypothesis that as the price of the product is increased the mean proportion of buyers will decrease?

d. Is there evidence that as sales experience increases the mean proportion of buyers increases?

11.46 Refer to Exercise 11.45. Suppose it is claimed that the least squares model cannot be correct, since $\hat{\beta}_0 = -.30$, and a negative proportion of buyers is clearly impossible. How do you refute this argument?

11.47 To determine whether extra personnel are needed for the day, the owners of a water adventure park would like to find a model that would allow them to predict the day's attendance each morning based on the day of the week and weather conditions. The model is of the form

$$E(y) = \beta_0 + \beta_1 x_1 + \beta_2 x_2 + \beta_3 x_3$$

where

$$y = \text{Daily attendance}$$

$$x_1 = \begin{cases} 1 & \text{if weekend} \\ 0 & \text{otherwise} \end{cases} \quad \text{(dummy variable)}$$

$$x_2 = \begin{cases} 1 & \text{if sunny} \\ 0 & \text{if overcast} \end{cases} \quad \text{(dummy variable)}$$

$$x_3 = \text{Predicted daily high temperature (°F)}$$

After collecting 30 days of data, the owners obtained the least squares model

$$\hat{y} = -105 + 25x_1 + 100x_2 + 10x_3$$

with $s_{\hat{\beta}_1} = 10$, $s_{\hat{\beta}_2} = 30$, and $s_{\hat{\beta}_3} = 4$. Also, $R^2 = .65$.

a. Interpret the estimated model coefficients.

b. Is there sufficient evidence to conclude that this model is useful for the prediction of daily attendance? Use $\alpha = .05$.

c. Is there enough evidence to conclude that mean attendance rises on weekends? Use $\alpha = .10$.

d. Use the model to predict the attendance on a sunny weekday with a predicted high of 95° F.

e. Suppose the 90% prediction interval for part **d** is (645, 1,245). Interpret this interval.

11.48 Refer to Exercise 11.47. The owners of the water adventure park are advised that the prediction model could probably be improved if interaction terms were added. In particular, it is thought that the *rate* that mean attendance increases as predicted high temperature increases will be greater on weekends than on weekdays. The model therefore proposed is

$$E(y) = \beta_0 + \beta_1 x_1 + \beta_2 x_2 + \beta_3 x_3 + \beta_4 x_1 x_3$$

The same 30 days of data are used to obtain the least squares model

$$\hat{y} = 250 - 700x_1 + 100x_2 + 5x_3 + 15x_1 x_3$$

with $s_{\hat{\beta}_4} = 3.0$ and $R^2 = .96$.

a. Graph the predicted day's attendance y against the day's predicted high temperature x_3 for a sunny weekday and for a sunny weekend day. Plot both on the same graph for x_3 between 70° and 100°F. Note that the slope for the weekend day is greater. Interpret this.

b. Do the data indicate that the interaction term is a useful addition to the model? Use $\alpha = .05$.

c. Use this model to predict the attendance for a sunny weekday with a predicted high temperature of 95°F.

d. Suppose the 90% prediction interval for part **c** is (800, 850). Compare this with the prediction interval for the model without interaction in Exercise 11.47, part **e**. Do the relative widths of the prediction intervals support or refute your conclusion about the usefulness of the interaction term (part **b**)?

11.49 Refer to Exercise 11.48. The owners, noting that $\hat{\beta}_1 = -700$, conclude the model is ridiculous because it seems to imply that the mean attendance will be 700 less on weekends than on weekdays. Refute their argument.

11.50 Many students must work part-time to help finance their college education. A survey of 100 students was completed at a university to determine whether the number of hours worked per week, x, was affecting their grade-point averages, y. A quadratic model was proposed:

$$y = \beta_0 + \beta_1 x + \beta_2 x^2 + \epsilon$$

The 100 observations yielded the least squares model

$$\hat{y} = 2.8 - .005x - .0002x^2$$

with $R^2 = .12$ and $s = .5$.

a. Do these statistics indicate that the model is useful in explaining grade-point averages? Use $\alpha = .05$.

b. Interpret the values of R^2 and s in terms of this application of regression analysis. Do you think the quadratic relationship between y and x is strong? Approximately how precisely would you expect to be able to predict a particular student's grade-point average if you knew the number of hours he or she worked per week?

c. If you changed the model to a straight-line model relating grade-point average to number of hours worked, would the value of R^2 increase, thereby providing a better prediction model for grade-point average? Explain.

***11.51** A manufacturer of electronic components for the computer industry believes that the yearly sales for one of its products can best be modeled as a function of the mean price of the product over the year, the mean price of its competitors' products over the year, and the mean number of salespeople per month over the year. Since much of the increase in sales revenue and prices over time may be due simply to inflation, it is advisable to eliminate the effects of inflation on the data before developing a model for annual sales. One way to do this is to express all sales revenue and price data in terms of constant dollars (i.e., inflation-adjusted dollars). We can convert each year's sales and price data to constant (1967) dollars by dividing each year's value by the Producer Price Index and multiplying the result by 100. The Producer Price Index is designed to measure average changes in prices over time of all commodities, at all stages of processing, produced or imported for sale in primary markets in the United States. Data on the four variables of interest to the manufacturer along with the Producer Price Index are presented in the table. [*Note:* Indexes such as the Producer Price Index and the transformation of current dollars to constant dollars are topics discussed in more detail in Chapter 14.]

Year	*Sales Revenue* y_1 *(1,000s)*	*Mean Number of Sales People* x_1	*Mean Price of Product* x_2	*Mean Price of Competitors' Products* x_3	*Producer Price Index*[a]
1967	$ 50	5	$30	$25	100.0
1968	120	7	30	26	102.5
1969	140	11	33	28	106.5
1970	135	16	34	30	110.4
1971	163	16	33	31	114.0
1972	233	16	36	34	119.1
1973	241	21	40	37	134.7
1974	255	27	45	42	160.1
1975	286	26	50	48	174.9
1976	330	30	53	54	183.0
1977	389	33	58	58	194.2
1978	425	36	60	61	209.3
1979	445	38	71	72	235.6
1980	472	37	80	81	268.8
1981	501	37	90	93	293.4
1982	510	38	92	92	299.3
1983	490	36	92	90	303.1
1984	505	37	94	94	310.3

[a]Source: U.S. Bureau of the Census. *Statistical Abstract of the United States: 1986*, p. 471.

a. Express the sales revenue data in terms of constant 1967 dollars. Call this new variable y_2. Do the same for the price data for the product of interest and the competitors' price data. Call these new variables x_4 and x_5, respectively.

b. Use the method of least squares to fit the model

$$y_2 = \beta_0 + \beta_1 x_1 + \beta_2 x_4 + \beta_3 x_5 + \epsilon$$

to the data. Describe what your least squares equation suggests about the relationship between y_2 and each of the three independent variables.

c. Find R^2 and interpret its value in the context of this problem.

d. In using an F test to investigate the usefulness of the model in part **b**, what are the null and alternative hypotheses?

e. Conduct the hypothesis test of part **d**. Use $\alpha = .05$ and interpret your results in the context of the problem. What is the observed significance level of your F test?

f. Test H_0: $\beta_2 = 0$ versus H_a: $\beta_2 < 0$ using $\alpha = .05$. Interpret the results of your test in the context of the problem.
g. Use a 95% confidence interval to estimate β_1. Interpret the results in the context of the problem.
h. Predict sales revenue (in 1967 dollars) for the product for a year in which the mean number of salespeople is 35, the mean price of the product is \$90, the mean price of competitors' products is \$92, and the Producer Price Index is 315.0.
i. Explain why it would not be advisable to use the least squares equation of part **b** to predict y_2 when $x_1 > 38$, $x_4 >$ \$31, and $x_5 >$ \$32.

11.52 Many colleges and universities develop regression models for predicting the grade-point average (GPA) of incoming freshmen to help make admission decisions. Although most models use many independent variables to predict GPA, we will illustrate by choosing two variables:

x_1 = Verbal score on college entrance examination (percentile)
x_2 = Mathematics score on college entrance examination (percentile)

The data in the table are obtained for a random sample of 40 freshmen at one college.

Verbal x_1	*Math* x_2	GPA y	*Verbal* x_1	*Math* x_2	GPA y	*Verbal* x_1	*Math* x_2	GPA y
81	87	3.49	83	76	3.75	98	67	2.73
68	99	2.89	64	66	2.70	97	80	3.27
57	86	2.73	83	72	3.15	77	90	3.47
100	49	1.54	93	54	2.28	49	54	1.30
54	83	2.56	74	59	2.92	39	81	1.22
82	86	3.43	51	75	2.48	87	69	3.23
75	74	3.59	79	75	3.45	70	95	3.82
58	98	2.86	81	62	2.76	57	89	2.93
55	54	1.46	50	69	1.90	74	67	2.83
49	81	2.11	72	70	3.01	87	93	3.84
64	76	2.69	54	52	1.48	90	65	3.01
66	59	2.16	65	79	2.98	81	76	3.33
80	61	2.60	56	78	2.58	84	69	3.06
100	85	3.30						

The SPSS printout corresponding to the model

$$y = \beta_0 + \beta_1 x_1 + \beta_2 x_2 + \epsilon$$

is shown on the facing page.
a. Interpret the least squares estimates $\hat{\beta}_1$ and $\hat{\beta}_2$ in the context of this application.
b. Interpret the standard deviation and the coefficient of determination of the regression model in the context of this application.
c. Is this model useful for predicting GPA? Conduct a statistical test to justify your answer.
d. Sketch the relationship between predicted GPA, $\hat{y}$, and verbal score, x_1, for the following mathematics scores: x_2 = 60, 75, and 90.

Multiple R	.82527		
R Square	.68106		
Adjusted R Square	.66382		
Standard Error	.40228		

Analysis of Variance

	DF	Sum of Squares	Mean Square
Regression	2	12.78595	6.39297
Residual	37	5.98755	.16183

F = 39.50530 Signif F = .0000

------------------ Variables in the Equation ------------------

Variable	B	SE B	Beta	T	Sig T
X1	.02573	4.02357E-03	.59719	6.395	.0000
X2	.03361	4.92751E-03	.63702	6.822	.0000
(Constant)	-1.57054	.49375		-3.181	.0030

11.53 Refer to Exercise 11.52. The residuals from the first-order model are plotted versus x_1 below and versus x_2 on page 604. Analyze the two plots, and determine whether visual evidence exists that curvature (a quadratic term) for either x_1 or x_2 should be added to the model.

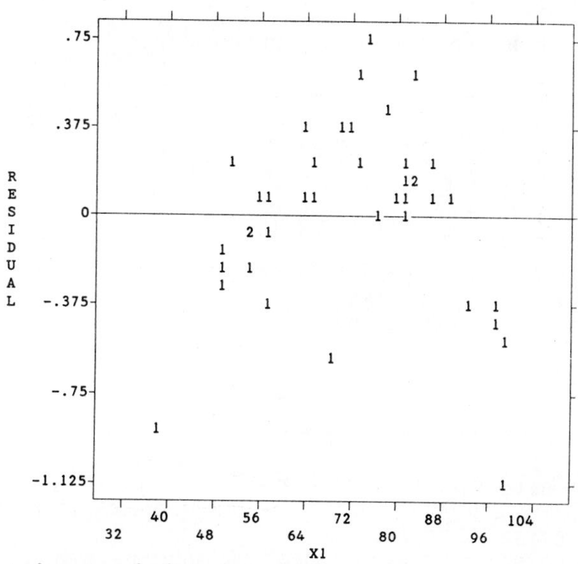

▲ $\hat{\epsilon}$ versus x_1 for first-order model

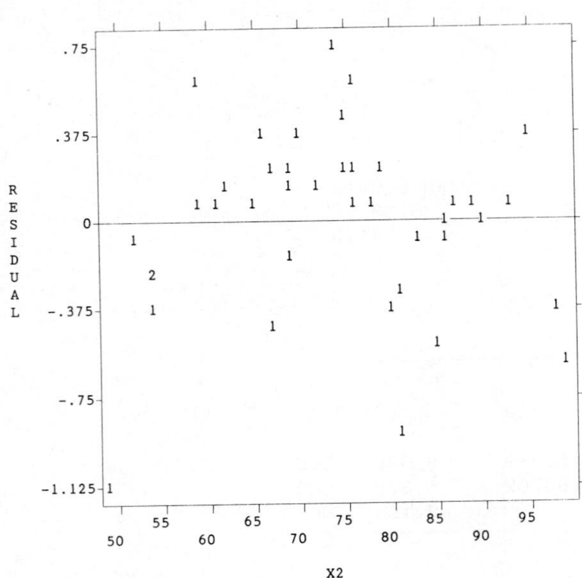

▲ $\hat{\epsilon}$ versus x_2 for first-order model

11.54 Refer to Exercises 11.52 and 11.53. The complete second-order model

$$y = \beta_0 + \beta_1 x_1 + \beta_2 x_2 + \beta_3 x_1^2 + \beta_4 x_2^2 + \beta_5 x_1 x_2 + \epsilon$$

is fit to the data given in Exercise 11.52. The resulting SPSS printout is reproduced here.

```
Multiple R                .96777
R Square                  .93657
Adjusted R Square         .92724
Standard Error            .18714

Analysis of Variance
                     DF        Sum of Squares          Mean Square
Regression            5              17.58274              3.51655
Residual             34               1.19076               .03502

F =       100.40901           Signif F =    .0000

------------------ Variables in the Equation ------------------

Variable                  B          SE B          Beta          T   Sig T

X1                   .16681        .02124       3.87132      7.852   .0000
X2                   .13760        .02673       2.60754      5.147   .0000
X1SQ           -1.10825E-03   1.17288E-04      -3.71359     -9.449   .0000
X2SQ           -8.43267E-04   1.59423E-04      -2.37284     -5.290   .0000
X1X2           2.410891E-04   1.43974E-04        .49600      1.675   .1032
(Constant)         -9.91676       1.35441                   -7.322   .0000
```

a. Compare the standard deviations of the first- and second-order regression models. With what relative precision will these two models predict GPA?
b. Test whether this model is useful for predicting GPA. Use $\alpha = .05$.
c. Test whether the interaction term, $\beta_5 x_1 x_2$, is important for the prediction of GPA. Use $\alpha = .10$. Note that this term permits the distance between three mathematics score curves for GPA versus verbal score to change as the verbal score changes.

11.55 Refer to Exercises 11.52–11.54. The residuals of the second-order model are plotted against both x_1 and x_2. Compare the residual plots to those of the first-order model in Exercise 11.53. Given the analyses you have performed in Exercises 11.52–11.55, which of the two models do you think is preferable as a predictor of GPA: the first- or second-order model? [*Note:* In Chapter 12 we show how to conduct a statistical test to compare two models.]

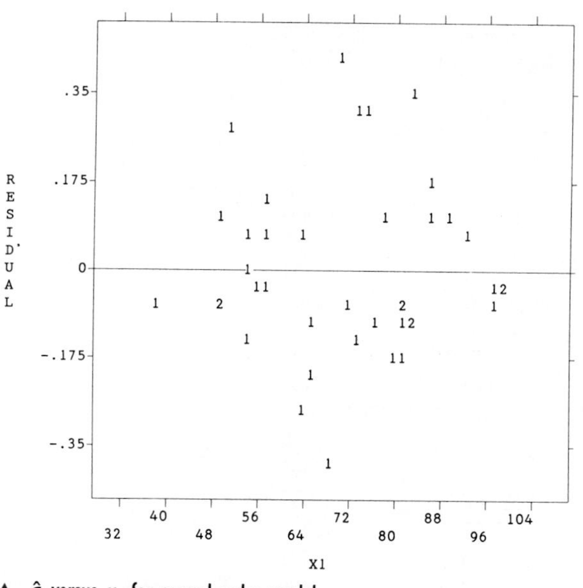

▲ $\hat{\epsilon}$ versus x_1 for second-order model

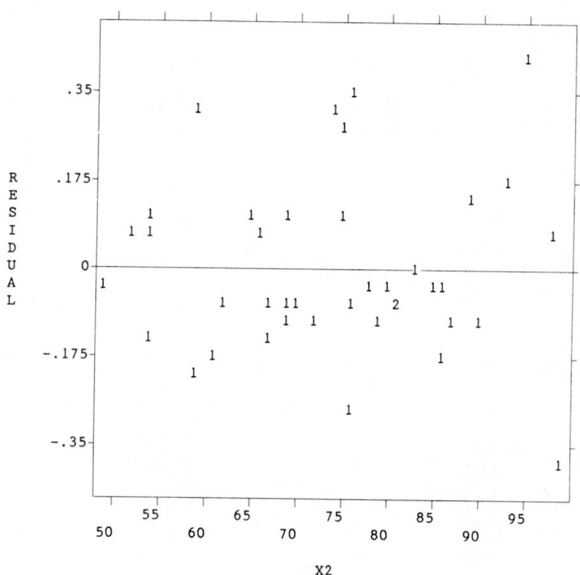

▲ $\hat{\epsilon}$ versus x_2 for second-order model

11.56 An economist is interested in estimating the demand function for passenger car motor fuel in the United States. While demand is clearly a function of many variables, the economist initially wants to model demand as a function of consumer income and the price of gasoline and plans to estimate the model using yearly time series data over the period 1965–1983. The measure of income will be the average gross weekly earnings per year for production or nonsupervisory workers on private nonagricultural payrolls. These data are available from the Bureau of Labor Statistics (BLS). However, since both earnings and the cost of living increased over this period, the BLS earnings data do not reflect actual purchasing power. Accordingly, the earnings data—originally expressed in *current dollars* (i.e., the number of dollars actually earned)—are converted to 1967 dollars by dividing each figure by the Consumer Price Index (CPI) for that year and multiplying the result by 100. (This procedure is described in detail in Chapter 14.) The resulting earnings data—called *real earnings*—are the number of dollars that would have to have been earned in 1967 to equal the purchasing power of current year weekly earnings. These data appear in the table (page 606).

Data on the actual price of gasoline and the relative price of gasoline (the CPI for gasoline divided by the CPI for all items) also appear in the table. The *relative price* data reflect the price of gasoline relative to the prices of other consumer goods. The economist believes the relative price may have a substantial influence on demand. For example, even though actual gasoline prices increase, consumers may actually increase their demand for gasoline if the relative price decreases.

Finally, data are collected on motor fuel consumed per year and the population of the United States. The population data are included so that the effects of the growing U.S. population on demand for motor fuel can be removed if desired. This is done by dividing the total motor fuel consumed in a year by the population that year. The result is called *per capita* motor fuel consumption. By using per capita consumption as the measure of demand, the economist can distinguish the effects on demand of factors such as price and income from the effects of population growth (Blair and Kenny, 1982).

Year	Car Fuel Consumed	U.S. Population	Average Gross Real Weekly Earnings	Price of Regular Gasoline	Relative Price of Gasoline
	(billion gals.)	(millions)	(1967 $)	($/gal.)	
1965	50.3	194.3	101.01	.32	1.004
1966	53.31	196.6	101.67	.33	.998
1967	55.11	198.7	101.84	.34	1.000
1968	58.52	200.7	103.39	.34	.973
1969	62.45	202.7	104.38	.36	.954
1970	65.8	205.1	103.04	.36	.908
1971	69.51	207.1	104.95	.36	.876
1972	73.5	209.9	109.26	.37	.859
1973	78.0	211.9	109.23	.40	.887
1974	74.2	213.9	104.78	.53	1.083
1975	76.5	216.0	101.45	.57	1.060
1976	78.8	218.0	102.90	.59	1.043
1977	80.7	220.2	104.13	.62	1.037
1978	83.8	222.6	104.25	.63	1.005
1979	80.2	225.1	101.15	.86	1.222
1980	73.7	227.7	95.26	1.19	1.496
1981	71.7	230.0	93.69	1.31	1.508
1982	72.8	232.3	92.45	1.22	1.346
1983	73.4	234.5	94.07	1.16	1.261

Source: U.S. Bureau of the Census. *Statistical Abstract of the United States*, various years; average gross real weekly earnings from U.S. Department of Labor, Bureau of Labor Statistics, *Employment and Earnings*, Dec. 1986, p. 79.

a. Explain what is being measured by the yearly per capita motor fuel consumption variable.
b. Initially, the economist hypothesizes that per capita motor fuel consumption is a linear function of the relative price of gasoline. Use the method of least squares to estimate this model.
c. Investigate the usefulness of the preliminary model. Use $\alpha = .05$. Also, find R^2.
d. Next, the economist would like to expand the model described in part **b** to include a second independent variable, average gross real weekly earnings. Use the method of least squares to fit this expanded model.
e. Investigate the usefulness of the expanded model of part **d**. Use $\alpha = .05$. Also, find R^2.

f. From an economic (or intuitive) perspective, do the signs of the estimated coefficients of the expanded model seem to be appropriate? Explain. [*Note:* The model will be modified in Exercise 12.57 to resolve this problem.]
g. Using the expanded model, estimate the mean per capita demand for motor fuel when average gross weekly earnings are $110.36 and the relative price of a gallon of gasoline is 1.521. What reservations (if any) do you have about the goodness of your estimate?
h. Calculate the residuals of the expanded model, and plot them against both independent variables. Check for evidence of curvature and for outliers.

On Your Own

This is a continuation of the **On Your Own** presented in Chapter 10, in which you selected three independent variables as predictors of the Gross Domestic Product and obtained 10 years of data for each. Now fit the following multiple regression model using an available computer package:

$$y = \beta_0 + \beta_1 x_1 + \beta_2 x_2 + \beta_3 x_3 + \epsilon$$

where

y = Gross Domestic Product
x_1 = First variable you chose
x_2 = Second variable you chose
x_3 = Third variable you chose

a. Compare the coefficients $\hat{\beta}_1$, $\hat{\beta}_2$, and $\hat{\beta}_3$ to their corresponding slope coefficients in the Chapter 10 **On Your Own**, where you fit three separate straight-line models. How do you account for the differences?
b. Calculate the coefficient of determination R^2, and conduct the F test of the null hypothesis H_0: $\beta_1 = \beta_2 = \beta_3 = 0$. What is your conclusion?

If the independent variables you chose are themselves highly correlated, you may encounter some results that are difficult to explain. For example, the coefficients $\hat{\beta}_1$, $\hat{\beta}_2$, and $\hat{\beta}_3$ may assume signs that are the opposite of what you expected. Or you may get a highly significant F value in part **b**, but the individual t statistics for the x_1, x_2, and x_3 may all be nonsignificant. This phenomenon—a high correlation between the independent variables in a regression model—is *multicollinearity*. This topic was discussed in Section 11.10.

Using the Computer

Suppose we wish to model the relationship of the median income y to the percent of college graduates x_1 and the percent of women in the work force x_2 in a zip code.

a. Draw a random sample of 100 zip codes from the 1,000 described in Appendix C, and extract y, x_1, and x_2 for each zip code sampled. Use a software package that includes a regression program to obtain the least squares fit of the following models:

$$E(y) = \beta_0 + \beta_1 x_1 + \beta_2 x_2$$
$$E(y) = \beta_0 + \beta_1 x_1 + \beta_2 x_2 + \beta_3 x_1 x_2$$

1. Interpret the estimated β parameters of the models.
2. Interpret the estimated standard deviations of the error terms.
3. Evaluate the usefulness of the models. Is there evidence of interaction between percent of college graduates and percent of women in the work force? Which is the preferred model? Why?
4. Use the preferred model to estimate the mean median income for all zip codes having 20% college graduates and 40% women in the work force using a 95% confidence interval.
5. Use the same model to predict the median income for a particular zip code having 20% college graduates and 40% women in the work force using a 95% prediction interval.
6. Plot the residuals against x_1 and x_2 for each model. Check for evidence that curvature is present for either x_1 or x_2.
7. Plot a histogram or stem-and-leaf display of the residuals for each model. Comment on the assumption of normality for the random error component.
8. Use the residual plots to determine whether outliers exist.

b. Repeat part **a** using the entire set of 1,000 zip codes. Compare the results to those you obtained in part **a**.

References

Allmon, C. I. "Advertising and sales relationships for toothpaste: Another look." *Business Economics*, Sept. 1982, Vol. 17, pp. 55–61.

Benston, G. J. "Multiple regression analysis of cost behavior." *Accounting Review*, Oct. 1966, Vol. 41, pp. 657–672.

Blair, R. D., and Kenny, L. W. *Microeconomics for Managerial Decision Making*. New York: McGraw-Hill, 1982. Chapter 3.

Carlson, R. D. "Advertising and sales relationships for toothpaste." *Business Economics*, Sept. 1981, Vol. 16, pp. 36–39.

Chase, R. B., and Aquilano, N. J. *Production and Operations Management*, 6th ed. Homewood, Ill.: Richard D. Irwin, 1992.

Chatterjee S., and Price, B. *Regression Analysis by Example*, 2nd ed. New York: Wiley, 1991.

Dixon, W. J., Brown, M. B., Engelman, L., Frane, J. W., Hill, M. A., Jennrich, R. I., and Toporek, J. D. *BMDP Statistical Software*. Berkeley: University of California Press, 1983.

Draper, N. R., and Smith, H. *Applied Regression Analysis*, 2nd ed. New York: Wiley, 1981.

Henry, W. R., and Haynes, W. W. *Managerial Economics: Analysis and Cases*, 4th ed. Dallas: Business Publications, 1978. Chapter 5.

Miller, R. B., and Wichern, D. W. *Intermediate Business Statistics: Analysis of Variance, Regression, and Time Series*. New York: Holt, Rinehart and Winston, 1977. Chapters 6–8.

Minitab Reference Manual, Release 8. Minitab, Inc., State College, Penn., 1991.

Neter, J., Wasserman, W., and Kutner, M. *Applied Linear Regression Models*, 2nd ed. Homewood, Ill.: Richard D. Irwin, 1989.

Nie, N., Hull, C. H., Jenkins, J. G., Steinbrenner, K., and Bent, D. H. *Statistical Package for the Social Sciences*, 2nd ed. New York: McGraw-Hill, 1975.

SAS User's Guide: Statistics, Version 5 ed. SAS Institute, Inc., Cary, N.C., 1985.

Weisberg, S. *Applied Linear Regression*, 2nd ed. New York: Wiley, 1985.

Winkler, R. L., and Hays, W. L. *Statistics: Probability, Inference, and Decision*, 2nd ed. New York: Holt, Rinehart and Winston, 1975. Chapter 10.

Wonnacott, R. J., and Wonnacott, T. H. *Econometrics*, 2nd ed. New York: Wiley, 1979. Chapter 6.

Yang, B., Stack, S., and Lester, D. "Suicide and unemployment: Predicting the smoothed trend and yearly fluctuations." *Socio-Economics*, Spring 1992, Vol. 21, No. 1.

Younger, M. S. *First Course in Linear Regression*, 2nd ed. Boston: Duxbury, 1985.

CHAPTER TWELVE

Introduction to Model Building

Contents

Case Studies

Where We've Been

One of the most important topics in applied statistics, regression analysis, was presented in Chapters 10 and 11. Simple linear regression, using a straight line to model the relationship between a dependent variable y and a single independent variable x, was the topic of Chapter 10. Multiple regression, relating a dependent variable to any number of independent variables, was the topic of Chapter 11. In both chapters, we learned how to fit regression models to a set of data and how to use the model to estimate the mean value of y or to predict a future value of y for a given value of x.

Where We're Going

In Chapters 10 and 11, an important problem was circumvented—the selection or construction of a model that is appropriate for the given data. No matter how much you know about regression analysis or how well you can fit a model to a set of data and interpret the results, the information will be of little value if you choose an inappropriate model to relate the mean value of y to the independent variables. The process of choosing a reasonable model and using the data to modify and improve it is called *model building*. We introduce you to this topic in Chapter 12.

12

As we indicated in Chapters 10 and 11, the first step in constructing a regression model is to hypothesize the form of the deterministic portion of the probabilistic model. Model construction, or **model building**, is the key to the success (or failure) of regression analysis. If the regression model does not reflect, at least approximately, the true nature of the relationship between the mean response $E(y)$ and the independent variables $x_1, x_2, \ldots, x_k$, the modeling effort will usually be unrewarded.

By *model building*, we mean developing a model that will provide a good fit to a set of data and that will give good estimates of the mean value of y and good predictions of future values of y for given values of the independent variables.

In the following sections, we present several useful models for relating a response y to one or more predictor variables. In addition, we provide tests and procedures for determining which model is appropriate.

12.1 The Two Types of Independent Variables: Quantitative and Qualitative

In Chapter 2, we defined two types of variables that may arise in business applications: quantitative and qualitative. In regression analysis, the dependent variable will always be quantitative, but the independent variables may be either quantitative or qualitative. As you will see, the way an independent variable enters the model depends on its type.

Recall from Chapter 2 that quantitative variables are measured on interval or ratio scales, whereas qualitative variables are measured on nominal or ordinal scales. Thus, the Gross Domestic Product, prime interest rate, number of defects in a product, and kilowatt-hours of electricity used per day are all examples of quantitative variables. On the other hand, gender, race, style of packaging, and job title are all examples of qualitative variables.

Quantitative variables usually assume numerical values, whereas qualitative variables usually assume nonnumerical values. The possible values of independent variables are also referred to as **levels**. For example, the style of packaging might have three possible levels: A, B, and C. Even if we designate these values as 1, 2, and 3, the numbers still represent categories, and the variable is still qualitative. Sometimes a variable we normally think of as quantitative, like income, is reported as *low*, *medium*, or *high*, which makes it qualitative for that application. Thus, the distinction between quantitative and qualitative variables is often a function of the type of information available.

EXAMPLE 12.1

In Chapter 11 we considered the problem of predicting executive salaries as a function of several independent variables. Consider the following four independent variables that may affect executive salaries:

a. Number of years of experience
b. Gender of the employee

c. Firm's net asset value
d. Rank of the employee

For each of these independent variables, give its type and describe the levels, or values, you would expect each to assume.

Solution

a. The independent variable for the number of years of experience is quantitative because its values are numerical. We would expect to observe levels ranging from 0 to 40 (approximately) years.
b. The independent variable for gender is qualitative because its values can be described only by the nonnumerical labels "female" and "male."
c. The independent variable for the firm's net asset value is quantitative, with a large number of possible levels corresponding to the various firms' net asset values.
d. Suppose the independent variable for the rank of the employee has three possible levels: supervisor, assistant vice president, and vice president. Since this is a nominal scale, rank is a qualitative independent variable.

Quantitative independent variables are treated differently from qualitative variables in regression modeling. In the next section, we will begin our discussion of how quantitative variables are used in the modeling effort.

12.2 Models with a Single Quantitative Independent Variable

The most common linear models relating y to a single quantitative independent variable x are those derived from a polynomial expression of the type shown in the box. Specific models, obtained by assigning particular values to p, are given in the following paragraphs.

Formula for a pth-Order Polynomial with One Quantitative Independent Variable

$$E(y) = \beta_0 + \beta_1 x + \beta_2 x^2 + \beta_3 x^3 + \cdots + \beta_p x^p$$

where p is an integer and $\beta_0, \beta_1, \ldots, \beta_p$ are unknown parameters that must be estimated.

I. First-Order Model

$$E(y) = \beta_0 + \beta_1 x$$

Comments on model parameters

β_0: y-intercept β_1: Slope of the line

General comments The first-order model is used when you expect the rate of change in y per unit change in x to remain fairly stable over the range of values of x for which you wish to predit y (see Figure 12.1). Many relationships between $E(y)$ and x are curvilinear, but the curvature over the range of values of x for which you wish to predict y may be very slight. When this occurs, a first-order (straight-line) model should provide a good fit to your data.

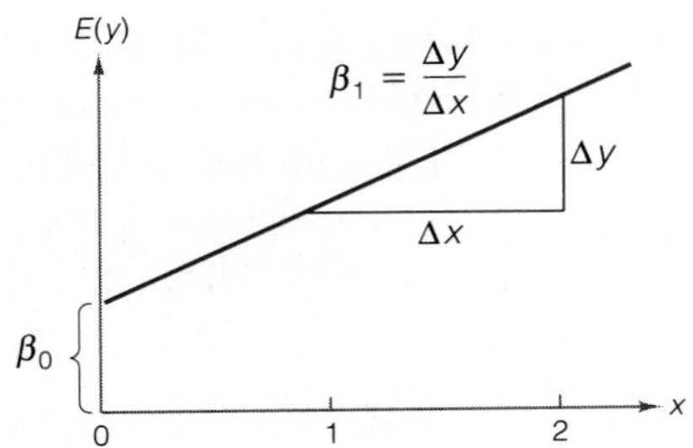

FIGURE 12.1 ► Graph of a first-order model

2. Second-Order Model

$$E(y) = \beta_0 + \beta_1 x + \beta_2 x^2$$

Comments on model parameters

β_0: y-intercept

β_1: Changing the value of β_1 shifts the parabola to the right or left; increasing the value of β_1 causes the parabola to shift to the left

β_2: Rate of curvature

General comments A second-order model traces a parabola, one that opens either downward ($\beta_2 < 0$) or upward ($\beta_2 > 0$), as shown in Figure 12.2. Since most relationships possess some curvature, a second-order model is often a good choice to relate y to x.

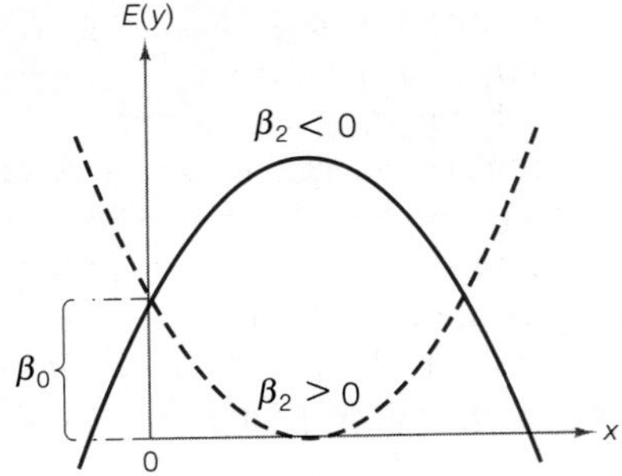

FIGURE 12.2 ► Graphs of two second-order models

3. Third-Order Model

$$E(y) = \beta_0 + \beta_1 x + \beta_2 x^2 + \beta_3 x^3$$

Comments on model parameters

β_0: y-intercept

β_3: The magnitude of β_3 controls the rate of reversal of the curvature

General comments Reversals in curvature are not common, but such relationships can be modeled by third- and higher-order polynomials. As can be seen in Figure 12.2, a second-order model contains no reversals in curvature. The slope continues to either increase or decrease as x increases and produces either a trough or a peak. A third-order model (see Figure 12.3) contains one reversal in curvature and produces one peak and one trough. In general, a graph of a pth-order polynomial will contain a total of $(p - 1)$ peaks and troughs.

Most functional relationships in nature seem to be smooth (except for random error), that is, they are not subject to rapid and irregular reversals in direction. Consequently, the second-order polynomial model is perhaps the most useful of those described above. To develop a better understanding of how this model is used, consider Example 12.2.

FIGURE 12.3 ▶ **Graphs of two third-order models**

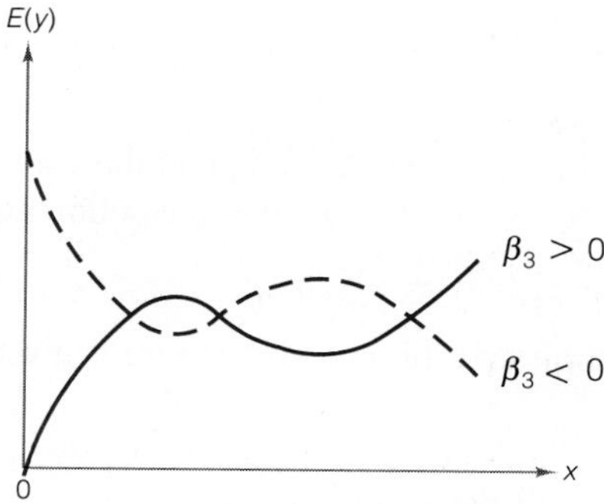

EXAMPLE 12.2

Power companies must be able to predict the peak power load at their various stations in order to operate effectively. The *peak power load* is the maximum amount of power that must be generated each day to meet demand.

Suppose a power company located in the southern part of the United States decides to model daily peak power load y as a function of the daily high temperature x, and the model is to be constructed for the summer months when demand is greatest. Although we would expect the peak power load to increase as the high temperature increases, the *rate* of increase in $E(y)$ might also increase as x increases. That is, a 1-unit increase in high temperature from 100° to 101°F might result in a larger increase in power demand than would a 1-unit increase from 80° to 81°F. Therefore, we postulate the second-order model

$$E(y) = \beta_0 + \beta_1 x + \beta_2 x^2$$

and we expect β_2 to be positive.

A random sample of 25 summer days is selected, and the data are shown in Table 12.1. Fit a second-order model using these data, and test the hypothesis that the power load increases at an increasing *rate* with temperature—i.e., that $\beta_2 > 0$.

TABLE 12.1 Power Load Data

Temperature (°F)	Peak Load (megawatts)	Temperature (°F)	Peak Load (megawatts)	Temperature (°F)	Peak Load (megawatts)
94	136.0	106	178.2	76	100.9
96	131.7	67	101.6	68	96.3
95	140.7	71	92.5	92	135.1
108	189.3	100	151.9	100	143.6
67	96.5	79	106.2	85	111.4
88	116.4	97	153.2	89	116.5
89	118.5	98	150.1	74	103.9
84	113.4	87	114.7	86	105.1
90	132.0				

Solution

The SAS printout shown in Figure 12.4 gives the least squares fit of the second-order model using the data in Table 12.1. The prediction equation is

$$\hat{y} = 385.048 - 8.293x + .05982x^2$$

A plot of this equation and the observed values is given in Figure 12.5.

```
SOURCE            DF   SUM OF SQUARES     MEAN SQUARE     F VALUE      PR > F

MODEL              2   15011.77199776   7505.88599888      259.69      0.0001
ERROR             22     635.87840224     28.90356374    R-SQUARE    ROOT MSE
CORRECTED TOTAL   24   15647.65040000
                                                         0.959363  5.37620347

                                     T FOR H0:          PR > |T|   STD ERROR OF
PARAMETER         ESTIMATE          PARAMETER = 0                   ESTIMATE

INTERCEPT      385.04809323              6.98            0.0001    55.17243578
TEMP            -8.29252680             -6.38            0.0001     1.29904502
TEMP*TEMP        0.05982337              7.93            0.0001     0.00754855
```

FIGURE 12.4 ▲
Portion of the SAS printout for the second-order model of Example 12.2

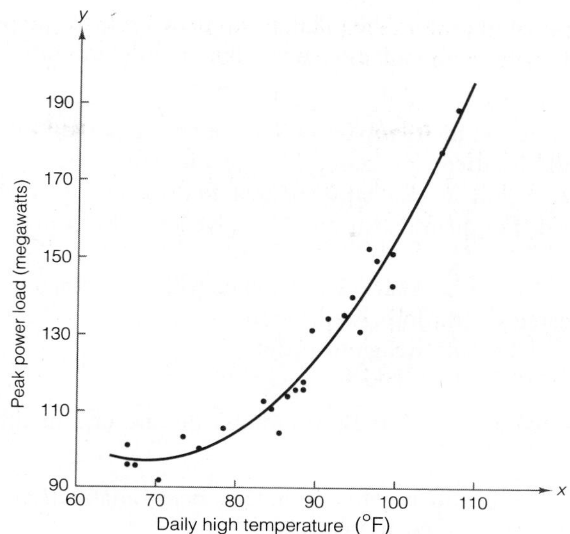

FIGURE 12.5 ▶
Plot of the observations and the second-order least squares fit

We now test whether the sample value $\hat{\beta}_2 = .05982$ is large enough to conclude *in general* that the power load increases at an increasing rate with temperature:

$$H_0\text{: } \beta_2 = 0 \qquad H_a\text{: } \beta_2 > 0$$

$$\textit{Test statistic: } \quad t = \frac{\hat{\beta}_2}{s_{\hat{\beta}_2}}$$

For $\alpha = .05$, $n = 25$, and $k = 2$, we reject H_0 if $t > t_{.05}$ where $t_{.05} = 1.717$ (from Table VI of Appendix B) is based on $n - (k + 1) = 22$ degrees of freedom. From Figure 12.4 the calculated value of t is 7.93. Since this value exceeds $t_{.05} = 1.717$, we reject H_0 at $\alpha = .05$ and conclude that the mean power load increases at an increasing rate with temperature.

Exercises 12.1 – 12.16

Note: Starred (*) exercises require the use of a computer.

Learning the Mechanics

12.1 The process of quality planning and control requires a continuous interaction between the customer and the manufacturer. To design, monitor, and control the quality of a product, the manufacturer needs to know which attributes of the product most influence the customer's decisions to buy and use the product (Schroeder, 1993). The following quality attributes were identified for a product:

a. Weight **b.** Color **c.** Age **d.** Hardness **e.** Package design

Regression analysis can be used to explore the relationship between these variables and sales of the product. Classify each of the variables as quantitative or qualitative and describe the levels that the variables may assume.

12.2 Companies keep personnel files on their employees that contain important information on each individual's background. The data could be used, for example, to predict employee performance ratings. Identify the independent variables listed below as qualitative or quantitative. For qualitative variables, suggest several levels that might be observed. For quantitative variables, give a range of values (levels) for which the variable might be observed.

a. Age
b. Years of experience with the company
c. Highest educational degree
d. Job classification
e. Marital status
f. Religious preference
g. Salary
h. Gender

12.3 Which of the asumptions about ϵ (Section 10.3) prohibit the use of a qualitative variable as a dependent variable?

12.4 The graphs depict pth-order polynomials with one independent variable. For each graph, identify the order of the polynomial. Find the value of β_0 and β_1 for each.

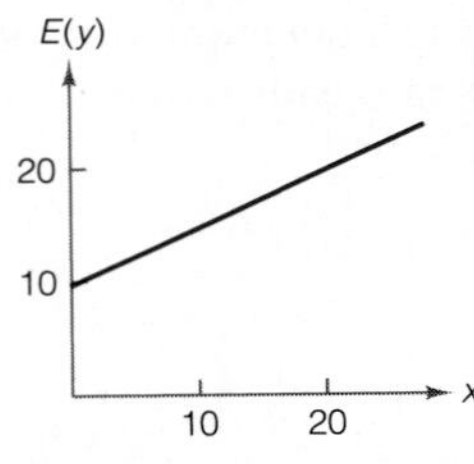

a.

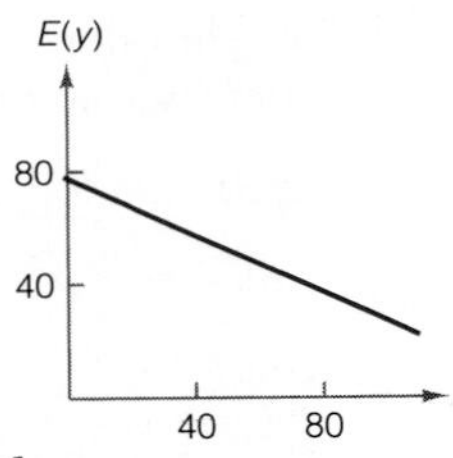

b.

12.5 The graphs depict pth-order polynomials for one independent variable. For each graph, identify the order of the polynomial, the value of β_0, and the sign of β_2.

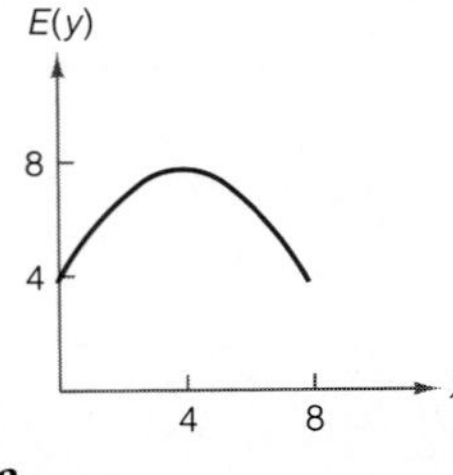

a.

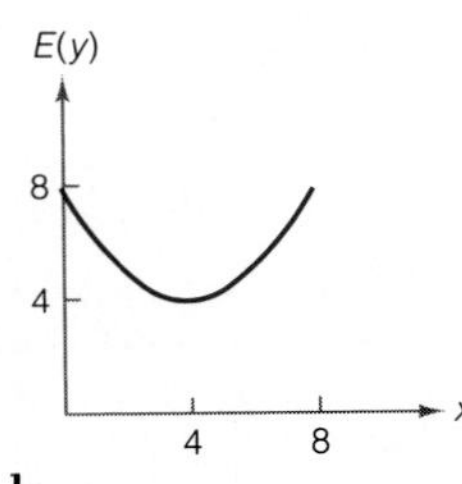

b.

12.6 Graph the following polynomials and identify the order of each on your graph.

a. $E(y) = 2 + 3x$
b. $E(y) = 2 + 3x^2$
c. $E(y) = 1 + 2x + 2x^2 + x^3$
d. $E(y) = 2x + 2x^2 + x^3$
e. $E(y) = 2 - 3x^2$
f. $E(y) = -2 + 3x$

12.7 Suppose $E(y)$ can best be modeled by a second-order polynomial in x, where x is a quantitative variable. Write the probabilistic model for y.

12.8 Minitab was used to fit the model

$$y = \beta_0 + \beta_1 x + \beta_2 x^2 + \epsilon$$

to $n = 17$ data points. The accompanying printout was obtained, in which **C21** is y, **C22** is x, and **C23** is x^2.

```
THE REGRESSION EQUATION IS
C21 = 63.1 + 1.17 C22 - 0.0374 C23

                                 ST. DEV.       T-RATIO =
             COEFFICIENT         OF COEF.       COEF/S.D.
                63.11              11.92           5.30
C22             1.1683             0.8276          1.41
C23            -0.03742            0.01234        -3.03

S = 9.289

R-SQUARED = 86.0 PERCENT
R-SQUARED = 84.0 PERCENT, ADJUSTED FOR D.F.

ANALYSIS OF VARIANCE

 DUE TO      DF          SS         MS=SS/DF
REGRESSION    2        7434.4        3717.2
RESIDUAL     14        1208.1          86.3
TOTAL        16        8642.5
```

a. Find and graph the least squares prediction equation.
b. Specify the null and alternative hypotheses you would use to determine whether the mean of y decreases at an increasing rate as x increases.
c. Conduct the hypothesis test of part **c** using $\alpha = .05$. Interpret your result.

Applying the Concepts

12.9 A company that sells and services copy machines conducted a study to relate the number of service calls per month required for a particular brand of table-top copier to the age of the copier (in months). All copiers used in the study were utilized by their owners to produce between 10,000 and 12,000 copies per month. The company suspects that new and old copiers require more service calls than those of middle age.

a. Based on this information, propose an appropriate model relating the mean number of service calls per month to the copier's age. Define all variables in your proposed model.
b. Indicate whether you think β_0 and β_2 assume positive or negative values, and explain the reasons for your decisions.

12.10 Supose you want to model the appraised value of a house, y, as a function of the number, x, of square feet of living space it contains. Regardless of how the appraisal is to be used, we would expect y to increase as x increases. In some instances, particularly if the appraisal is for tax purposes, the rate of increase in y decreases as x increases.

a. Write a suitable linear model to relate y to x.
b. Specify the signs of the coefficients in your model so they agree with the given verbal explanation of the relationship between y and x. Sketch the relationship on graph paper.

12.11 The strength of a certain plastic is thought to be related to the amount of pressure used to produce the plastic. Researchers believe that, as pressure is increased, the strength of the plastic increases until at some point increases in pressure become detrimental to strength. Write a model to relate the strength y of the plastic to pressure x that reflects the above beliefs. Sketch the model.

12.12 An economist has proposed the following model to describe the relationship between the number of items produced per day (output) and the number of hours of labor expended per day (input) in a particular production process:

$$y = \beta_0 + \beta_1 x + \beta_2 x^2 + \epsilon$$

where

y = Number of items produced per day
x = Number of hours of labor per day

A portion of the Minitab computer printout that results from fitting this model to a sample of 25 weeks of production data is reproduced here. Do the data provide sufficient evidence to indicate that the *rate* of increase in mean output per unit increase of input decreases as the input increases? Test using $\alpha = .05$.

```
THE REGRESSION EQUATION IS
Y =   -6.17   +   2.04 X1    -  .0323 X2

                                ST. DEV.    T-RATIO =
                   COEFFICIENT  OF COEF.    COEF/S.D.

     --             -6.173        1.666       -3.71
X1   C2              2.036         .185       11.02
X2   C3             -.03231       .00489      -6.60

THE ST. DEV. OF Y ABOUT REGRESSION LINE IS
S =        1.243
WITH (  25- 3) =   22 DEGREES OF FREEDOM

R-SQUARED = 95.5 PERCENT
R-SQUARED = 95.1 PERCENT, ADJUSTED FOR D.F.

ANALYSIS OF VARIANCE

 DUE TO        DF         SS      MS=SS/DF

REGRESSION      2    718.168   359.084
RESIDUAL       22     33.992     1.545
TOTAL          24    752.160
```

12.13 A company is considering adopting an assembly line schedule of 4 10-hour days per week instead of 5 8-hour days. Management believes that the effect of fatigue due to longer afternoons of work might increase assembly times to an unsatisfactory level, however. An experiment with the 4-day week is planned in which time studies will be conducted on some of the workers during the afternoons. It is believed that an adequate model of the relationship between assembly time, y, and time since lunch, x, should allow for the average assembly time to decrease for a while after lunch (as workers get back in the groove) before it starts to increase as the workers tire.

a. Propose a model to relate $E(y)$ and x that would reflect management's belief. Define all terms in your model.

b. Sketch the shape of the function described by your hypothesized model.

*12.14 Many studies confirm that manufacturing productivity increases as a result of learning. As more units are produced, the work force becomes more proficient at the task and each successive unit takes less time to complete, up to a point. Generally, productivity steadily rises following the initiation of a manufacturing process, then levels off and stabilizes. However, some evidence suggests this may not be the case in less developed countries. The table presents data collected on the sacking production of the Victory Jute Mill, Chittagong, Bangladesh, a jute spinning and weaving plant.

Age of Mill, (yrs.)	*Productivity (kgs./hr.)*	*Age of Mill, (yrs.)*	*Productivity, (kgs./hr.)*
2	4.08	12	4.57
3	4.18	13	4.56
4	4.42	14	4.59
5	4.48	15	4.51
6	4.55	16	4.46
7	4.58	22	4.32
8	4.63	23	4.28
9	4.65	24	4.24
10	4.67	25	4.17
11	4.66	26	4.13

Source: Kibria, M. G., and Tisdell, C. A. "International comparisons of learning curves and productivity." *Management International Review*, 1985, Vol. 25, No. 4, pp. 66–72.

a. Construct a scattergram for these data. What does your scattergram reveal about the productivity of the mill as it aged?

b. Based on the pattern revealed in your scattergram, propose a model to relate productivity y to the age of the mill x.

c. Fit your model of part **b** to the data. Report the least squares prediction equation and plot it on the scattergram of part **a**.

d. Do the data provide sufficient evidence to conclude that the model proposed in part **b** provides information for the prediction of productivity? Test using $\alpha = .05$.

*12.15 The Federal Reserve System (FRS) was established in 1913 to provide central banking facilities for the United States. It is often referred to as the "banker's bank." One of its major responsibilities is to control the flow of the nation's money supply in order to facilitate orderly economic growth. One way this is done is through buying and selling government securities. The sale of securities to the public draws money from the commercial banking system; the purchase of securities by the FRS from the public moves money into the commercial banking system. This ebb and flow of the money supply affects the level of interest rates (the prices paid for borrowed money) in the economy (*Federal Reserve System*, 1963). Data on the money supply and the prime interest rate are given in the table on page 620. The money supply, M1, is the total sum of all currency, demand deposits, traveler's checks, and other checkable deposits.

a. Fit a straight line to the data using the prime interest rate as the dependent variable y. Compute R^2.

b. Based on the results of part **a**, describe in words the apparent relationship between interest rates and M1 growth.

Month	*Prime Rate*	*M1 ($ billions)*	*Month*	*Prime Rate*	*M1 ($ billions)*
1/90	10.5	795.4	1/91	10	826.7
2/90	10	801.1	2/91	9.5	836.4
3/90	10	804.7	3/91	9	843.0
4/90	10	807.7	4/91	9	842.1
5/90	10	807.5	5/91	9	851.6
6/90	10	811.5	6/91	8.5	858.4
7/90	10	810.7	7/91	8.5	859.5
8/90	10	816.5	8/91	8.5	866.1
9/90	10	821.8	9/91	8.5	870.0
10/90	10	821.2	10/91	8	879.1
11/90	10	823.3	11/91	8	890.3
12/90	10	825.4	12/91	7.5	896.7

Source: *Economic Report of the President*, Feb. 1992, pp. 373, 379.

c. Plot the least squares line on a scattergram of the data. Does it apear that a second-order model might better explain the variation in interest rates?
d. Fit a second-order model to the data and compute R^2.
e. Plot the prediction equation you developed in part **d** on a scattergram of the data.
f. Do the data provide sufficient evidence to conclude that $\beta_2 \neq 0$? Test using $\alpha = .05$. Draw the appropriate conclusions regarding the usefulness of the second-order model relative to the first-order model of part **a** for explaining the variation in interest rates.

12.16 Automobile accidents result in a tragic loss of life as well as a serious dollar loss to the nation's economy. The table shows the number of highway deaths (to the nearest hundred), y, and the number of licensed vehicles (in hundreds of thousands), x, for the years 1956–1991. (The years are coded 1–36 for convenience.) During the years 1974–1987 (years 19–32 in the table), a nationwide 55-mile-per-hour speed limit was in effect.

Year	*Highway Deaths, y*	*Licensed Vehicles, x_1*	Year	*Highway Deaths, y*	*Licensed Vehicles, x_1*	Year	*Highway Deaths, y*	*Licensed Vehicles, x_1*
1	39.6	65.2	13	54.9	103.1	25	53.1	161.6
2	38.7	67.6	14	55.8	107.4	26	51.4	164.1
3	37.0	68.8	15	54.6	111.2	27	45.8	165.2
4	37.9	72.1	16	54.4	116.3	28	44.5	169.4
5	38.1	74.5	17	56.3	122.3	29	46.2	172.0
6	38.1	76.4	18	55.5	129.8	30	45.9	177.1
7	40.8	79.7	19	46.4	134.9	31	47.9	181.4
8	43.6	83.5	20	45.9	137.9	32	48.3	183.9
9	47.7	87.3	21	47.0	143.5	33	49.1	189.0
10	49.2	91.8	22	49.5	148.8	34	47.6	191.7
11	53.0	95.9	23	52.4	153.6	35	46.8	192.9
12	52.9	98.9	24	53.5	159.6	36	43.5	194.9

Source: *Accident Facts*. National Safety Council, 1992, pp. 72, 73.

a. Write a second-order model relating the number, y, of highway deaths for a year to the number, x_1, of licensed vehicles.

b. The SAS computer printout for fitting the model to the data is shown here. Is there sufficient evidence to indicate that the model provides information for the prediction of the number of annual highway deaths? Test using $\alpha = .05$.

```
Model: MODEL1
Dependent Variable: Y

                         Analysis of Variance

                        Sum of          Mean
Source          DF     Squares        Square      F Value       Prob>F

Model            2   698.43643     349.21822       25.247       0.0001
Error           33   456.45996      13.83212
C Total         35  1154.89639

    Root MSE        3.71916      R-square        0.6048
    Dep Mean       47.58056      Adj R-sq        0.5808
    C.V.            7.81655

                         Parameter Estimates

                   Parameter        Standard      T for H0:
Variable   DF       Estimate           Error     Parameter=0      Prob > |T|

INTERCEP    1      -1.280177      6.99594102          -0.183          0.8559
X1          1       0.790965      0.11699897           6.760          0.0001
X1SQ        1      -0.002884      0.00044821          -6.435          0.0001
```

c. Give the observed significance level for the test of part **b**, and interpret it.

d. Does the second-order term contribute information for the prediction of y? Test using $\alpha = .05$.

e. Give the observed significance level for the test of part **d**, and interpret it.

12.3 Models with Two Quantitative Independent Variables

I. First-Order Model

$$E(y) = \beta_0 + \beta_1 x_1 + \beta_2 x_2$$

Comments on model parameters

β_0: y-intercept, the value of $E(y)$ when $x_1 = x_2 = 0$

β_1: Change in $E(y)$ for a 1-unit increase in x_1, when x_2 is held fixed

β_2: Change in $E(y)$ for a 1-unit increase in x_2, when x_1 is held fixed

General comments The graph in Figure 12.6 (page 622) traces a **response surface** (in contrast to the **response curve** used to relate $E(y)$ to a *single* quantitative variable). In particular, a first-order model relating $E(y)$ to two independent quantitative variables, x_1 and x_2, graphs as a plane in three-dimensional space. The plane traces the value of $E(y)$ for every combination of values (x_1, x_2) that correspond to points in the x_1, x_2

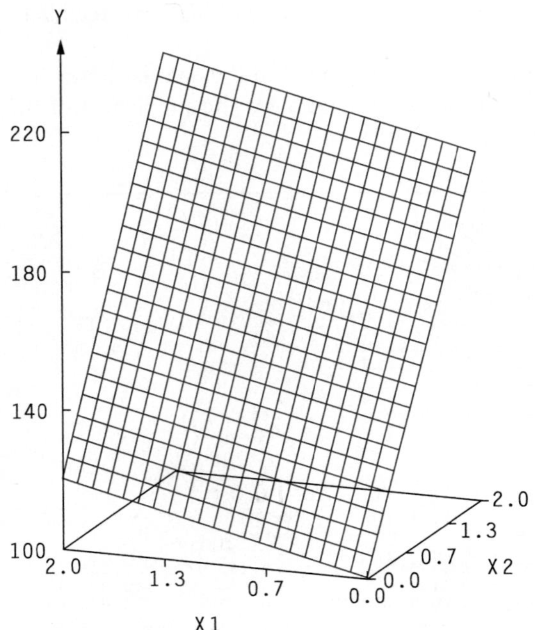

FIGURE 12.6 ▶ Computer-generated graph of a first-order model

plane. Many response surfaces in the real world are well behaved (smooth), and they have curvature. Consequently, a first-order model is appropriate only if the response surface is fairly flat over the x_1, x_2 region that is of interest to you.

The assumption that a first-order model will adequately characterize the relationship between $E(y)$ and the variables x_1 and x_2 is equivalent to assuming that x_1 and x_2 do not interact; that is, you assume that the effect on $E(y)$ of a change in x_1 (for a fixed value of x_2) is the same regardless of the value of x_2 (and vice versa). Thus, no interaction implies that the effect of changes in one variable (say x_1) on $E(y)$ is *independent* of the value of the second variable (say x_2). For example, if we assign values to x_2 in a first-order model, the graph of $E(y)$ as a function of x_1 would produce parallel lines as shown in Figure 12.7. These lines, called **contour lines**, show the contours of the surface when it is sliced by three planes, each of which is parallel to the $E(y)$, x_1 plane, at distances $x_2 = 1$, 2, and 3 from the origin.

Definition 12.1

Two variables x_1 and x_2 are said to **interact** if the change in $E(y)$ for a 1-unit change in x_1 (when x_2 is held fixed) is dependent on the value of x_2.

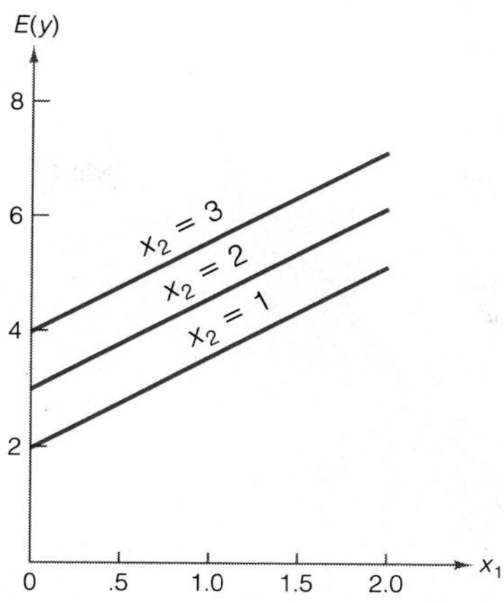

FIGURE 12.7 ► A graph indicating no interaction between x_1 and x_2

2. An Interaction Model (Second-Order)

$$E(y) = \beta_0 + \beta_1 x_1 + \beta_2 x_2 + \beta_3 x_1 x_2$$

Comments on model parameters

β_0: y-intercept, the value of $E(y)$ when $x_1 = x_2 = 0$

β_1 *and* β_2: Changing β_1 and β_2 causes the surface to shift along the x_1- and x_2-axes

β_3: Controls the rate of twist in the ruled surface (see Figure 12.8, page 624)

General comments This model is said to be second-order because the order of the highest-order term ($x_1 x_2$) in x_1 and x_2 is 2; i.e., the sum of the exponents of x_1 and x_2 equals 2. This interaction model traces a ruled surface in a three-dimensional space (Figure 12.8). You could produce such a surface by placing a pencil perpendicular to a line and moving it along the line, while rotating it around the line. The resulting surface would appear as a twisted plane. A graph of $E(y)$ as a function of x_1 for given values of x_2 (say $x_2 = 1$, 2, and 3) produces nonparallel contour lines (see Figure 12.9 on page 624), thus indicating that the change in $E(y)$ for a given change in x_1 is dependent on the value of x_2 and, therefore, that x_1 and x_2 interact. Interaction is an extremely important concept because it is easy to get in the habit of fitting first-order models and individually examining the relationships between $E(y)$ and each of a set of independent variables, $x_1, x_2, \ldots, x_k$. Such a procedure is meaningless when interaction exists (which is, at least to some extent, almost always the case), and it can lead to gross errors in interpretation. For example, suppose the relationship between $E(y)$ and x_1 and x_2 is as shown in Figure 12.9 and that you have observed y for each of the $n = 9$ combinations of values of x_1 and x_2 ($x_1 = 1, 2, 3$ and $x_2 = 1, 2, 3$). If you fit a first-order model in x_1 and x_2 to the data, the fitted plane would be (except for random error) approximately parallel to the x_1, x_2 plane, thus suggesting that x_1 and x_2 contribute very

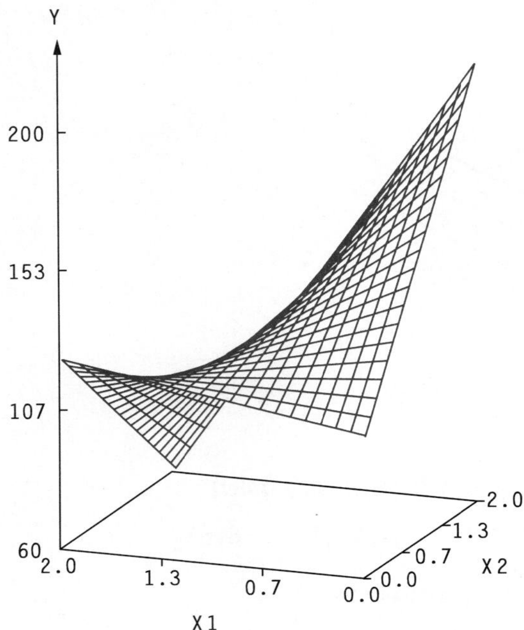

FIGURE 12.8 ► Computer-generated graph for an interaction model (second-order)

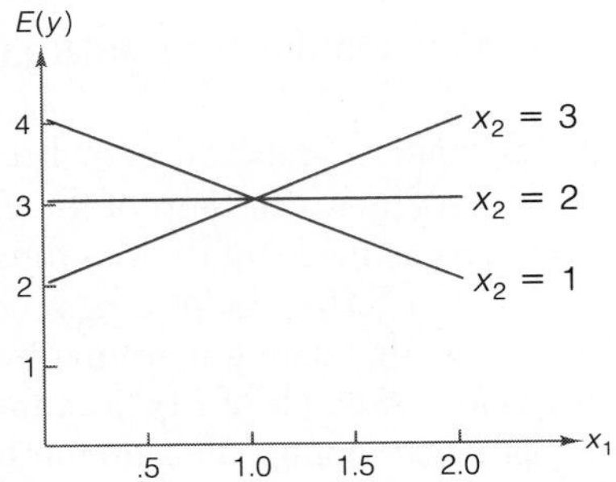

FIGURE 12.9 ► A graph indicating interaction between x_1 and x_2

little information about $E(y)$. That this is not the case is clearly indicated by the figure. Fitting a first-order model to the data would not allow for the twist in the true surface and would therefore give a false impression of the relationship between $E(y)$ and x_1 and x_2. The procedure for detecting interaction between two independent variables can be seen by examining the model. The interaction model differs from the noninteraction first-order model only in the inclusion of the $\beta_3 x_1 x_2$ term:

$$\textit{Interaction model:} \quad E(y) = \beta_0 + \beta_1 x_1 + \beta_2 x_2 + \beta_3 x_1 x_2$$

$$\textit{First-order model:} \quad E(y) = \beta_0 + \beta_1 x_1 + \beta_2 x_2$$

Therefore, to test for the presence of interaction, we test the hypotheses

H_0: $\beta_3 = 0$ (no interaction)

H_a: $\beta_3 \neq 0$ (interaction)

using the familiar Student's t test of Section 11.4.

3. A Complete Second-Order Model

$$E(y) = \beta_0 + \beta_1 x_1 + \beta_2 x_2 + \beta_3 x_1 x_2 + \beta_4 x_1^2 + \beta_5 x_2^2$$

Comments on model parameters

β_0: y-intercept, the value of $E(y)$ when $x_1 = x_2 = 0$

β_1 *and* β_2: Changing β_1 and β_2 causes the surface to shift along the x_1- and x_2-axes

β_3: The value of β_3 controls the rotation of the surface

β_4 *and* β_5: Signs and values of these parameters control the type of surface and the rates of curvature

The following three types of surfaces may be produced by a second-order model:

β_4 *and* β_5 *positive:* A paraboloid that opens upward; Figure 12.10 (a)

β_4 *and* β_5 *negative:* A paraboloid that opens downward; Figure 12.10 (b)

β_4 *and* β_5 *differ in sign:* A saddle-shaped surface; Figure 12.10(c)

FIGURE 12.10 ▶ Graphs of three second-order surfaces

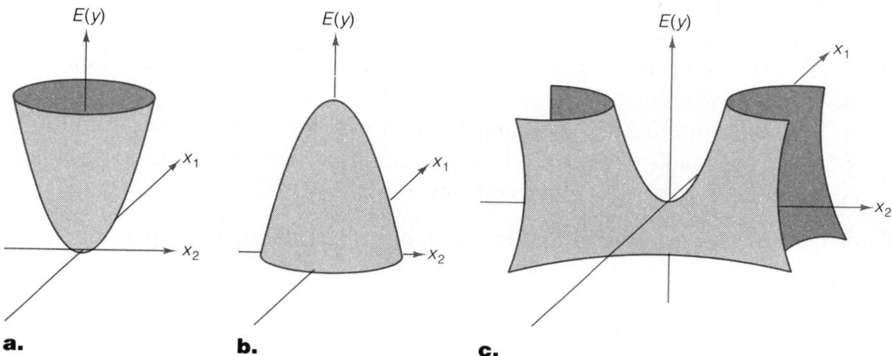

General comments A complete second-order model is the three-dimensional equivalent of a second-order model with a single quantitative variable. Instead of tracing parabolas, it traces paraboloids and saddle surfaces. Since you fit only a portion of the complete surface to the data, a complete second-order model provides a large variety of gently curving surfaces. It is a good choice for a model if you expect curvature in the response surface relating $E(y)$ to x_1 and x_2.

Exercises 12.17–12.26

Note: Starred (*) exercises require the use of a computer.

Learning the Mechanics

12.17 Consider an application in which you are trying to relate a response y to two independent variables, x_1 and x_2.

a. Write a first-order model relating $E(y)$ to x_1 and x_2.

b. Modify the model you constructed in part **a** to include an interaction term.

c. Modify the model you constructed in part **b** to make it a complete second-order model.

12.18 Suppose the true relationship between $E(y)$ and the quantitative independent variables x_1 and x_2 is described by the first-order model

$$E(y) = 4 - x_1 + 2x_2$$

a. Describe the corresponding response surface.

b. Plot the contour lines of the response surface for $x_1 = 2, 3, 4$, where $0 \le x_2 \le 5$.

c. Plot the contour lines of the response surface for $x_2 = 2, 3, 4$, where $0 \le x_1 \le 5$.

d. Use the contour lines you plotted in parts **b** and **c** to explain how changes in the settings of x_1 and x_2 affect $E(y)$.

e. Use your graph from part **b** to determine how much $E(y)$ changes when x_1 is changed from 4 to 2 and, simultaneously, x_2 is changed from 1 to 2.

12.19 Suppose the true relationship between $E(y)$ and the quantitative independent variables x_1 and x_2 is

$$E(y) = 4 - x_1 + 2x_2 + x_1x_2$$

a. Identify the order of the model.

b. Describe the corresponding response surface.

c. Plot the contour lines of the response surface for $x_1 = 0, 1, 2$, where $0 \le x_2 \le 5$.

d. Explain why the contour lines you plotted in part **c** are not parallel.

e. Use the contour lines you plotted in part **c** to explain how changes in the settings of x_1 and x_2 affect $E(y)$.

f. Use your graph from part **c** to determine how much $E(y)$ changes when x_1 is changed from 2 to 0 and, simultaneously, x_2 is changed from 4 to 5.

12.20 What does it mean to say that two variables affect the mean response $E(y)$ independently of each other?

12.21 Minitab was used to fit the model

$$y = \beta_0 + \beta_1x_1 + \beta_2x_2 + \beta_3x_1x_2 + \epsilon$$

to $n = 15$ data points. The results are shown in the printout, where **C1** is y, **C2** is x_1, **C3** is x_2, and **C4** is x_1x_2.

a. What is the prediction equation for the response surface?

b. Describe the geometric form of the response surface of part **a**.

c. Plot the prediction equation for the case when $x_2 = 1$. Do this twice more on the same graph for the cases when $x_2 = 3$ and $x_2 = 5$.

d. Explain what it means to say that x_1 and x_2 interact. Explain why your graph of part **c** suggests that x_1 and x_2 interact.

```
THE REGRESSION EQUATION IS
C1 = - 2.55 + 3.82 C2 + 2.63 C3 - 1.29 C4

                                   ST. DEV.      T-RATIO =
              COEFFICIENT          OF COEF.      COEF/S.D.
                 -2.550             1.142          -2.23
C2                3.8150            0.5286          7.22
C3                2.6300            0.3443          7.64
C4               -1.2850            0.1594         -8.06

S = 0.7127

R-SQUARED = 85.6 PERCENT
R-SQUARED = 81.6 PERCENT, ADJUSTED FOR D.F.

ANALYSIS OF VARIANCE

 DUE TO        DF          SS          MS=SS/DF
REGRESSION      3        33.149         11.050
RESIDUAL       11         5.587          0.508
TOTAL          14        38.736
```

e. Specify the null and alternative hypotheses you would use to test whether x_1 and x_2 interact.
f. Conduct the hypothesis test of part **e** using $\alpha = .01$.

Applying the Concepts

12.22 The Department of Energy wants to develop a regression model to help forecast annual gasoline consumption in the United States, y. It decides to model $E(y)$ as a function of two independent variables:

x_1 = Number of cars in use during year

x_2 = Number of trucks in use during year

a. Identify the independent variables as quantitative or qualitative.
b. Write the first-order model for $E(y)$.
c. Write the complete second-order model for $E(y)$.
d. With respect to the model in part **c**, specify the null and alternative hypotheses you would use to test whether the second-order terms contribute information for the prediction of annual gasoline consumption.

12.23 Some corporations, instead of owning a fleet of cars, use a rental agency. Sometimes it is more economical to rent new cars for a year than to buy new cars each year. A major rental agency wants to develop a model that will allow it to estimate the average annual cost to the prospective customer of renting cars, y, as a function of two independent variables:

x_1 = Number of cars rented

x_2 = Average number of miles driven per car during the year (in thousands)

a. Identify the independent variables as quantitative or qualitative.
b. Write the first-order model for $E(y)$.

c. Write a model for $E(y)$ that contains all first-order and interaction terms. Sketch typical response curves showing the mean cost $E(y)$ versus the average mileage driven, x_2, for different values of x_1. (Assume that x_1 and x_2 interact.)

d. Write the complete second-order model for $E(y)$.

12.24 Refer to Exercise 12.23. Suppose the model from part **c** is fit, with the result

$$\hat{y} = 1 + .05x_1 + x_2 + .05x_1x_2$$

(The units of $\hat{y}$ are thousands of dollars.) Graph the estimated cost $\hat{y}$ as a function of the average number of miles driven, x_2, over the range $x_2 = 10$ to $x_2 = 50$ (10,000–50,000 miles) for $x_1 = 1$, 5, and 10. Do these functions agree (approximately) with the graphs you drew for Exercise 12.23, part **c**?

12.25 An economist is interested in modeling the relationship between quarterly sales of central air conditioning systems (in thousands) for single-family homes in the United States and two quantitative independent variables, housing starts in the previous quarter (in thousands) and the Gross Domestic Product (in billions of 1972 dollars). Data were collected, and a model was fit. The displayed portion of the resulting Minitab printout describes the least squares prediction equation. In the prediction equation, $x_3 = x_1^2$ and $x_4 = x_2^2$.

```
THE REGRESSION EQUATION IS
Y =      149. + .472 X1 - .0993 X2
    - .0005 X3 + .0000 X4

                                 ST. DEV.    T-RATIO =
                  COEFFICIENT    OF COEF.    COEF/S.D.
        --              148.5       224.5          .66
X1    C2                 .472        .126         3.74
X2    C3                -.099        .162         -.61
X3    C12           -.000535      .000359        -1.49
X4    C13           .0000153     .0000294          .52

THE ST. DEV. OF Y ABOUT REGRESSION LINE IS
S =        6.884
WITH (   16-5) = 11 DEGREES OF FREEDOM

R-SQUARED = 92.0 PERCENT
R-SQUARED = 89.1 PERCENT, ADJUSTED FOR D.F.

ANALYSIS OF VARIANCE

 DUE TO         DF          SS    MS=SS/DF

REGRESSION       4     6002.08     1500.52
RESIDUAL        11      521.36       47.40
TOTAL           15     6523.44
```

a. Write the prediction equation for the response surface.

b. Describe the geometric form of the response surface of part **a**.

c. Do the data provide sufficient evidence to conclude that the model hypothesized by the economist is useful for predicting quarterly sales of central air conditioning systems? Test using $\alpha = .01$.

d. Does it appear that the variation in air conditioner sales could be adequately explained by a less complex regression model? Explain. [*Note:* In the next section we discuss a formal procedure for making this inference.]

***12.26** A supermarket chain is interested in exploring the relationship between the sales of its store-brand vegetables, y, the amount spent on promotion of the vegetables in local newspapers, x_1, and the amount of shelf space allocated to the brand; x_2. One of the chain's supermarkets was randomly selected and over a 20-week period x_1 and x_2 were varied as reported in the table.

Week	*Sales*	*Advertising*	*Space (sq. ft.)*	*Week*	*Sales*	*Advertising*	*Space (sq. ft.)*
1	$2,010	$201	75	11	$5,005	$ 996	75
2	1,850	205	50	12	2,500	625	50
3	2,400	355	75	13	3,005	860	50
4	1,575	208	30	14	3,480	1,012	50
5	3,550	590	75	15	5,500	1,135	75
6	2,015	397	50	16	1,995	635	30
7	3,908	820	75	17	2,390	837	30
8	1,870	400	30	18	4,390	1,200	50
9	4,877	997	75	19	2,785	990	30
10	2,190	515	30	20	2,989	1,205	30

a. Fit this model to the data:

$$y = \beta_0 + \beta_1 x_1 + \beta_2 x_2 + \beta_3 x_1 x_2 + \epsilon$$

b. Conduct an F test to investigate the overall usefulness of this model. Use $\alpha = .05$.
c. Test for the presence of interaction between advertising costs and shelf space. Use $\alpha = .05$.
d. Explain what it means to say that advertising expenditures and shelf space interact.
e. Explain how you could be misled by using a first-order model instead of an interaction model to explain how advertising expenditure and shelf space influence sales.

12.4 Model Building: Testing Portions of a Model

The presentation of models with one and with two quantitative independent variables raises a very general question. Do certain terms in the model contribute more information than others for the prediction of y?

To illustrate, suppose you have collected data on a response y and two quantitative independent variables x_1 and x_2, and you are considering the use of either a first-order or a second-order model to relate $E(y)$ to x_1 and x_2. Will the second-order model provide better predictions of y than the first-order model? To answer this question, examine the two models (page 630) and note that the second-order model contains all the terms contained in the first-order model plus three additional terms, those involving β_3, β_4, and β_5.

First-order model: $E(y) = \beta_0 + \beta_1 x_1 + \beta_2 x_2$

$$\textit{Second-order model:}\quad E(y) = \beta_0 + \beta_1 x_1 + \beta_2 x_2 + \overbrace{\beta_3 x_1 x_2 + \beta_4 x_1^2 + \beta_5 x_2^2}^{\text{Second-order terms}}$$

Therefore, asking whether the second-order model contributes more information for the prediction of y than the first-order model is equivalent to asking whether at least one of the parameters, β_3, β_4, and β_5, differs from 0—i.e., whether the terms involving β_3, β_4, and β_5 should be retained in the model. Therefore, to test whether the second-order terms should be included in the model, we test the null hypothesis

$$H_0\text{:}\quad \beta_3 = \beta_4 = \beta_5 = 0$$

(i.e., the second-order terms do not contribute information for the prediction of y) against the alternative hypothesis

H_a: At least one of the parameters, β_3, β_4, or β_5, differs from 0

(i.e., at least one of the second-order terms contributes information for the prediction of y).

The procedure for conducting this test is intuitive. First, we use the method of least squares to fit the first-order model—the **reduced** model—and calculate the corresponding sum of squares for error, SSE_r (the sum of squares of the deviations between observed and predicted values). Next, we find the second-order model—the **complete** model—and calculate its sum of squares for error, SSE_c. Then, we compare SSE_r to SSE_c. If the second-order terms contribute to the model, then SSE_c should be much smaller than SSE_r, and the difference $(SSE_r - SSE_c)$ will be large. The larger the difference, the greater the weight of evidence that the second-order (complete) model provides better predictions of y than does the first-order (reduced) model.

The sum of squares for error will always decrease when new terms are added to the model. The question is whether this decrease is large enough to conclude that it is due to more than just an increase in the number of model terms and to chance. To test the null hypothesis that the parameters of the second-order terms β_3, β_4, and β_5 simultaneously equal 0, we use an F statistic calculated as

$$F = \frac{\text{Drop in SSE/Number of } \beta \text{ parameters being tested}}{s^2 \text{ for the complete second-order model}} = \frac{(SSE_r - SSE_c)/3}{SSE_c/[n - (5 + 1)]}$$

When the assumptions listed in Section 11.3 about the error term ϵ are satisfied and the β parameters for the second-order terms are all 0 (i.e., when H_0 is true), this F statistic has an F distribution with $\nu_1 = 3$ and $\nu_2 = n - 6$ degrees of freedom. Note that ν_1 is the number of β parameters being tested and ν_2 is the number of degrees of freedom associated with s^2 in the complete second-order model.

If the second-order terms *do* contribute to the model (i.e., if H_a is true), we expect the F statistic to be large. Thus, we use a one-tailed test and reject H_0 when F exceeds some critical value, F_α, as shown in Figure 12.11. A summary of the steps used in testing the null hypothesis that a set of model parameters are all equal to 0 is shown in the next box.

FIGURE 12.11 ►
Rejection region for the F test of H_0: $\beta_3 = \beta_4 = \beta_5 = 0$

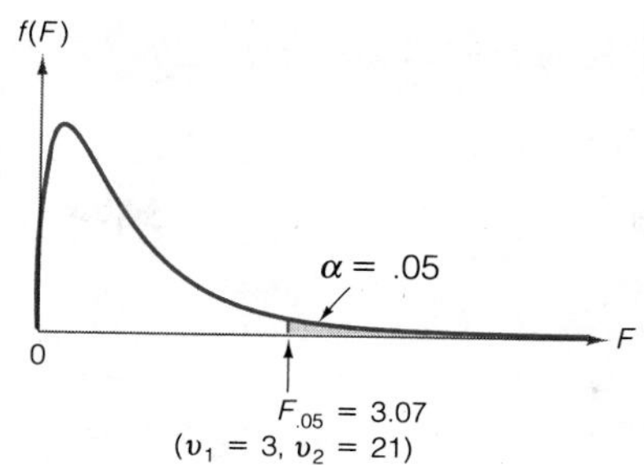

F Test for Testing the Null Hypothesis: Set of β Parameters Equals 0

Reduced model: $E(y) = \beta_0 + \beta_1 x_1 + \cdots + \beta_g x_g$

Complete model: $E(y) = \beta_0 + \beta_1 x_1 + \cdots + \beta_g x_g + \beta_{g+1} x_{g+1} + \cdots + \beta_k x_k$

H_0: $\beta_{g+1} = \beta_{g+2} = \cdots = \beta_k = 0$

H_a: At least one of the β parameters under test is nonzero

Test statistic: $F = \dfrac{(\text{SSE}_r - \text{SSE}_c)/(k - g)}{\text{SSE}_c/[n - (k + 1)]}$

where SSE_r = Sum of squared errors for the reduced model
SSE_c = Sum of squared errors for the complete model
$k - g$ = Number of β parameters specified in H_0
$k + 1$ = Number of β parameters in the complete model
n = Sample size

Rejection region: $F > F_\alpha$

where $\nu_1 = k - g$ = Degrees of freedom for the numerator
$\nu_2 = n - (k + 1)$ = Degrees of freedom for the denominator

EXAMPLE 12.3

Many companies manufacture products (e.g., steel, paint, gasoline) that are at least partially chemically produced. In many instances, the quality of the finished product is a function of the temperature and pressure at which the chemical reactions take place. Suppose you want to model the quality y of a product as a function of the temperature x_1 and the pressure x_2 at which it is produced. Four inspectors independently assign a quality score between 0 and 100 to each product, and then the quality y is calculated by averaging the four scores. An experiment is conducted by varying temperature between 80°–100°F and pressure between 50–60 pounds per square inch (psi). The resulting data are given in Table 12.2 on page 632.

TABLE 12.2 Temperature, Pressure, and Quality of the Finished Product

x_1 (°F)	x_2 (psi)	y	x_1 (°F)	x_2 (psi)	y	x_1 (°F)	x_2 (psi)	y
80	50	50.8	90	50	63.4	100	50	46.6
80	50	50.7	90	50	61.6	100	50	49.1
80	50	49.4	90	50	63.4	100	50	46.4
80	55	93.7	90	55	93.8	100	55	69.8
80	55	90.9	90	55	92.1	100	55	72.5
80	55	90.9	90	55	97.4	100	55	73.2
80	60	74.5	90	60	70.9	100	60	38.7
80	60	73.0	90	60	68.8	100	60	42.5
80	60	71.2	90	60	71.3	100	60	41.4

a. Fit a second-order model to the data.

b. Sketch the response surface.

c. Do the data provide sufficient evidence to indicate that the second-order terms contribute information for the prediction of y?

Solution

a. The complete second-order model is

$$E(y) = \beta_0 + \beta_1x_1 + \beta_2x_2 + \beta_3x_1x_2 + \beta_4x_1^2 + \beta_5x_2^2$$

The data in Table 12.2 were used to fit this model, and a portion of the SAS output is shown in Figure 12.12. The least squares prediction equation is

$$\hat{y} = -5{,}127.90 + 31.10x_1 + 139.75x_2 - .146x_1x_2 - .133x_1^2 - 1.14x_2^2$$

```
SOURCE              DF   SUM OF SQUARES     MEAN SQUARE    F VALUE       PR > F

MODEL                5   8402.26453714   1680.45290743     596.32       0.0001
ERROR               21     59.17842582      2.81802028   R-SQUARE     ROOT MSE
CORRECTED TOTAL     26   8461.44296296                              1.67869601
                                                         0.993006

                                      T FOR H0:       PR > |T|   STD ERROR OF
PARAMETER              ESTIMATE    PARAMETER = 0                   ESTIMATE

INTERCEPT        -5127.89907417          -46.49     0.0001    110.29601483
X1                  31.09638889           23.13     0.0001      1.34441322
X2                 139.74722222           44.50     0.0001      3.14005411
X1*X2               -0.14550000          -15.01     0.0001      0.00969196
X1*X1               -0.13338889          -19.46     0.0001      0.00685325
X2*X2               -1.14422222          -41.74     0.0001      0.02741299
```

FIGURE 12.12 ▲ Portion of the SAS printout for Example 12.3

b. A three-dimensional graph of this prediction model is shown in Figure 12.13. Note that the mean quality seems to be greatest for temperatures of about 85°–90°F and

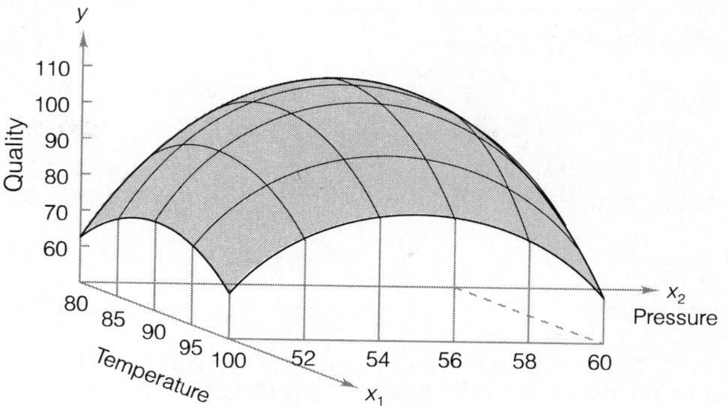

FIGURE 12.13 ▶ Plot of second-order least squares model for Example 12.3

for pressures of about 55–57 pounds per square inch.* Further experimentation in these ranges might lead to a more precise determination of the optimal temperature–pressure combination.

c. To determine whether the data provide sufficient evidence to indicate that the second-order terms contribute information for the prediction of y, we test

$$H_0\colon \quad \beta_3 = \beta_4 = \beta_5 = 0$$

against the alternative hypothesis

H_a: At least one of the parameters, β_3, β_4, or β_5, differs from 0

The first step in conducting the test is to drop the second-order terms out of the complete (second-order) model and fit the reduced model

$$E(y) = \beta_0 + \beta_1 x_1 + \beta_2 x_2$$

to the data. The SAS computer printout for this procedure is shown in Figure 12.14, on page 634. You can see that the sums of squares for error, given in Figures 12.12 and 12.14 for the complete and reduced models, respectively, are

$$\text{SSE}_c = 59.17842582 \qquad \text{SSE}_r = 6{,}671.50851852$$

and that the estimated variance for the complete model is

$$s^2 = 2.81802028$$

Recall that $n = 27$, $k = 5$, and $g = 2$. Therefore, the calculated value of the F statistic, based on $\nu_1 = k - g = 3$ and $\nu_2 = n - (k + 1) = 21$ df is

$$F = \frac{(\text{SSE}_r - \text{SSE}_c)/(k - g)}{\text{SSE}_c/[n - (k + 1)]} = \frac{(\text{SSE}_r - \text{SSE}_c/(k - g)}{s^2}$$

*Students with knowledge of calculus should note that we can solve for the exact temperature and pressure that maximize quality in the least squares model by solving $\partial\hat{y}/\partial x_1 = 0$ and $\partial\hat{y}/\partial x_2 = 0$ for x_1 and x_2. These estimated optimal values are $x_1 = 86.25$°F and $x_2 = 55.58$ psi. Remember, however, that these represent only sample estimates of the coordinates for the optimal value.

FIGURE 12.14 ► SAS computer printout for the reduced (first-order) model in Example 12.3

```
DEPENDENT VARIABLE: Y

SOURCE                  DF    SUM OF SQUARES     MEAN SQUARE     F VALUE

MODEL                    2    1789.93444444    894.96722222        3.22
ERROR                   24    6671.50851852    277.97952160      PR > F
CORRECTED TOTAL         26    8461.44296296                      0.0577

R-SQUARE               C.V.         ROOT MSE          Y MEAN

0.211540            24.8984      16.67271788     66.96296296

                                T FOR HO:      PR > |T|    STD ERROR OF
PARAMETER        ESTIMATE    PARAMETER=0

INTERCEPT    106.08518519           1.90       0.0700     55.94500427
X1            -0.91611111          -2.33       0.0285      0.39297973
X2             0.78777778           1.00       0.3262      0.78595946
```

where $\nu_1 = k - g$ is equal to the number of parameters involved in H_0. Therefore,

$$F = \frac{(6{,}671.50851852 - 59.17842582)/3}{2.81802028} = 782.1$$

The final step in the test is to compare this computed value of F with the tabulated value based on $\nu_1 = 3$ and $\nu_2 = 21$ degrees of freedom. If we choose $\alpha = .05$, $F_{.05} = 3.07$. Since the computed value of F falls in the rejection region (see Figure 12.11)—i.e., it exceeds $F_{.05} = 3.07$—we reject H_0 and conclude that at least one of the second-order terms contributes information for the prediction of y. In other words, the data support the contention that the curvature we see in the response surface is not due simply to random variation in the data. The second-order model does appear to provide better predictions of y than a first-order model.

Example 12.3 demonstrates the motivation for testing a hypothesis that each one of a set of β parameters equals 0, and it also demonstrates the procedure. Other applications of this test appear in the following sections.

Exercises 12.27–12.36

Note: Starred (*) exercises require the use of a computer.

Learning the Mechanics

12.27 Suppose you fit the regression model

$$y = \beta_0 + \beta_1 x_1 + \beta_2 x_2 + \beta_3 x_1 x_2 + \beta_4 x_1^2 + \beta_5 x_2^2 + \epsilon$$

to $n = 30$ data points and you wish to test H_0: $\beta_3 = \beta_4 = \beta_5 = 0$.

a. What would the alternative hypothesis be?
b. Explain in detail how you would find the quantities necessary to compute the F statistic for your hypothesis test.
c. How many numerator and denominator degrees of freedom are associated with your F statistic?

12.28 Refer to Exercise 12.27. Suppose you fit the complete and reduced models for the hypothesis test described and obtain $SSE_c = 215.2$ and $SSE_r = 246.1$. Conduct the hypothesis test and interpret the results of your test. Use $\alpha = .05$.

12.29 Explain why the F test used to compare complete and reduced models is a one-tailed, upper-tailed test.

12.30 Minitab was used to fit the complete model

$$y = \beta_0 + \beta_1 x_1 + \beta_2 x_2 + \beta_3 x_3 + \beta_4 x_4 + \epsilon$$

to $n = 20$ data points. The resulting printout is shown, where **C1** is the dependent variable y, and **C2–C5** are the independent variables x_1–x_4 respectively. The independent variables x_3 and x_4 were then dropped from the above model and Minitab was used to fit the resulting reduced model. The reduced model printout is shown on page 636.

```
THE REGRESSION EQUATION IS
C1 = 14.6 - 0.611 C2 + 0.439 C3 - 0.080 C4 - 0.064 C5

                           ST. DEV.      T-RATIO =
         COEFFICIENT       OF COEF.      COEF/S.D.
             14.575          4.887          2.98
C2          -0.6113         0.1775         -3.44
C3           0.4388         0.2199          2.00
C4          -0.0796         0.1083         -0.74
C5          -0.0636         0.1247         -0.51

S = 3.190

R-SQUARED = 84.5 PERCENT
R-SQUARED = 80.3 PERCENT, ADJUSTED FOR D.F.

ANALYSIS OF VARIANCE

 DUE TO       DF        SS        MS=SS/DF
REGRESSION     4      831.09       207.77
RESIDUAL      15      152.66        10.18
TOTAL         19      983.75
```

▲ **Complete model**

a. Report the least squares prediction equations for the complete and reduced models.
b. Find SSE_c and SSE_r. Interpret each of these quantities.
c. How many β parameters are in the complete model? The reduced model?
d. Specify the null and alternative hypotheses you would use to investigate whether the complete model contributes more information for the prediction of y than the reduced model.
e. Conduct the hypothesis test of part **d**. Use $\alpha = .05$.
f. What is the aproximate p-value of the test of part **e**?

```
THE REGRESSION EQUATION IS
C1 = 14.0 - 0.642 C2 + 0.396 C3

                               ST. DEV.     T-RATIO =
          COEFFICIENT          OF COEF.     COEF/S.D.
               13.968            4.626         3.02
C2            -0.6422           0.1675        -3.84
C3             0.3959           0.2061         1.92

S = 3.072

R-SQUARED = 83.7 PERCENT
R-SQUARED = 81.8 PERCENT, ADJUSTED FOR D.F.

ANALYSIS OF VARIANCE

 DUE TO        DF          SS           MS=SS/DF
REGRESSION      2        823.31           411.66
RESIDUAL       17        160.44             9.44
TOTAL          19        983.75
```

▲ Reduced model, Exercise 12.30

Applying the Concepts

12.31 A large research and development company rates the performance of its technical staff once a year. Each person is rated on a scale of 0 to 100 by his or her immediate supervisor, and this merit rating is used to help determine the size of the person's pay raise for the coming year. The company's personnel department is interested in developing a regression model to help forecast the merit rating that an applicant for a technical position will receive after being with the company 3 years. The company proposes to use the following model to forecast the merit ratings of applicants who have just completed their graduate studies and have no prior related job experience:

$$E(y) = \beta_0 + \beta_1 x_1 + \beta_2 x_2 + \beta_3 x_1 x_2 + \beta_4 x_1^2 + \beta_5 x_2^2$$

where

y = Applicant's merit rating after 3 years

x_1 = Applicant's grade-point average (GPA) in graduate school

x_2 = Applicant's verbal score on the Graduate Record Examination (GRE; percentile)

A random sample of $n = 40$ employees who have been with the company more than 3 years was selected. Each employee's merit rating after 3 years, graduate school GPA, and the percentile in which the verbal GRE score fell were recorded. The preceding model was fit to these data and a portion of the resulting computer printout is shown.

```
SOURCE  DF  SUM OF SQUARES  MEAN SQUARE
MODEL    5        4911.56        982.31
ERROR   34        1830.44         53.84
TOTAL   39        6742.00      R-SQUARE
                                  0.729
```

The reduced model, $E(y) = \beta_0 + \beta_1 x_1 + \beta_2 x_2$, was fit to the same data and the resulting computer printout is partially reproduced here.

SOURCE	DF	SUM OF SQUARES	MEAN SQUARE
MODEL	2	3544.84	1772.42
ERROR	37	3197.16	86.41
TOTAL	39	6742.00	R-SQUARE
			0.526

a. Identify the null and alternative hypotheses for a test to determine whether the complete model contributes information for the prediction of y.
b. Identify the null and alternative hypotheses for a test to determine whether a second-order model contributes more information than a first-order model for the prediction of y.
c. Conduct the hypothesis test you described in part **a**. Test using $\alpha = .05$. Draw the appropriate conclusions in the context of the problem.
d. Conduct the hypothesis test you described in part **b**. Test using $\alpha = .05$. Draw the appropriate conclusions in the context of the problem.

***12.32** A firm wants to forecast its yearly sales in each of its sales regions. The firm decides to base its forecasts on regional population size and its yearly regional advertising expenditures. The population data in the accompanying table were obtained from the Bureau of the Census, and the advertising and sales data are from the firm's internal records.

Region	*Sales (1,000s of units)*	*Population (1,000s)*	*Advertising Costs ($1,000s)*
1	65	200	8
2	80	210	10
3	85	205	9
4	100	300	8.5
5	108	320	12
6	114	290	10
7	40	90	6
8	45	85	8
9	150	450	9
10	42	87	9
11	220	480	13
12	200	500	15

a. Fit a complete second-order model to the data.
b. Is the complete second-order model useful for forecasting sales? Test using $\alpha = .05$.
c. The firm is planning to market its product in a new sales region next year. The region has a population of 400,000, and the firm plans to spend $12,000 on advertising. Use the fitted model you obtained in part **a** to forecast next year's sales in this new region. [*Note:* We would want to express this estimate as a prediction interval, but its computation is beyond the scope of this text. The procedure is described in the references at the end of the chapter. You may also find that it can be obtained using your statistical computer package.]

12.33 Refer to Exercise 12.32, in which a firm would like to develop a regression model to forecast its yearly sales in each of its sales regions. The model under consideration is a complete second-order model:

$$E(y) = \beta_0 + \beta_1 x_1 + \beta_2 x_2 + \beta_3 x_1 x_2 + \beta_4 x_1^2 + \beta_5 x_2^2$$

where

y = Yearly regional sales $\quad x_1$ = Population of sales region
x_2 = Yearly regional advertising expenditures

The following is a portion of the computer printout that results from fitting this model to the $n = 12$ data points given in Exercise 12.32.

SOURCE	DF	SUM OF SQUARES	MEAN SQUARE
MODEL	5	38638.97	7727.79
ERROR	6	159.94	26.66
TOTAL	11	38798.91	R-SQUARE
			0.996

The reduced first-order model, $E(y) = \beta_0 + \beta_1 x_1 + \beta_2 x_2$, was fit to the same data and the resulting computer printout is partially reproduced here.

SOURCE	DF	SUM OF SQUARES	MEAN SQUARE
MODEL	2	36704.5	18352.2
ERROR	9	2094.4	232.7
TOTAL	11	38798.9	R-SQUARE
			0.946

Is there sufficient evidence to conclude that a second-order model contributes more information for the prediction of y than a first-order model? Test using $\alpha = .05$.

12.34 Refer to Exercise 12.22 in which the Department of Energy wants to develop a regression model to help forecast annual gasoline consumption in the United States. The complete and reduced models for the test that you described in part **d** of Exercise 12.22 were fit to $n = 25$ data points. The resulting values of SSE_r and SSE_c were 1,065.9 and 400.6, respectively.

a. Conduct the test to determine whether the data present sufficient evidence to indicate interaction between x_1 and x_2. Test using $\alpha = .05$.

b. Find the approximate observed significance level for the test in part **a**.

12.35 In Exercise 11.17 we found that a first-order model was useful for explaining the variation in the number of hours worked per week in the shipping department of a particular firm. The data used to fit the model are repeated in the accompanying table.

a. Write a complete second-order model for the data.

b. The SAS computer printout for fitting the second-order model of part **a** is shown on page 639. Find the prediction equation.

Week	Labor y (hrs.)	Pounds Shipped x_1 (1,000s)	Percentage of Units Shipped by Truck x_2	Shipment Weight x_3 (lbs.)
1	100	5.1	90	20
2	85	3.8	99	22
3	108	5.3	58	19
4	116	7.5	16	15
5	92	4.5	54	20
6	63	3.3	42	26
7	79	5.3	12	25
8	101	5.9	32	21
9	88	4.0	56	24
10	71	4.2	64	29
11	122	6.8	78	10
12	85	3.9	90	30
13	50	3.8	74	28
14	114	7.5	89	14
15	104	4.5	90	21
16	111	6.0	40	20
17	110	8.1	55	16
18	100	2.9	64	19
19	82	4.0	35	23
20	85	4.8	58	25

```
DEPENDENT VARIABLE: LABOR

SOURCE                        DF   SUM OF SQUARES    MEAN SQUARE    F VALUE

MODEL                          9   6043.40529897    671.48947766     10.25
ERROR                         10    654.79470103     65.47947010    PR > F
CORRECTED TOTAL               19   6698.20000000                    0.0006

R-SQUARE                   C.V.        ROOT MSE     LABOR MEAN

0.902243                 8.6730      8.09193859    93.30000000

                                    T FOR HO:      PR > |T|   STD ERROR OF
PARAMETER             ESTIMATE    PARAMETER=0                   ESTIMATE

INTERCEPT          655.80722615          2.90       0.0159    226.28033719
WEIGHT             -57.32665806         -1.73       0.1138     33.08376144
TRUCK               -3.38961628         -1.77       0.1079      1.91983172
AVGSHIP            -28.27072935         -2.66       0.0238     10.61901566
WEIGHT*TRUCK         0.22403392          1.52       0.1583      0.14691511
WEIGHT*AVGSHIP       2.20142243          2.44       0.0348      0.90189596
TRUCK*AVGSHIP        0.08858647          2.23       0.0496      0.03966598
WEIGHT*WEIGHT        0.45327631          0.34       0.7414      1.33580032
TRUCK*TRUCK          0.00371216          0.92       0.3800      0.00404162
AVGSHIP*AVGSHIP      0.20833049          1.64       0.1315      0.12683455
```

c. Do the data provide sufficient evidence to indicate that the model is useful for predicting y? Test using $\alpha = .05$.
d. Is there sufficient evidence to indicate that the second-order terms are useful for predicting y? Test using $\alpha = .05$.

***12.36** In Exercise 11.10, a real estate appraiser used regression analysis to explore the relationship between the sale prices of apartments and various characteristics of the apartments. The data are repeated in the accompanying table.

Code	Sale Price y (\$)	Apartment Units x_1	Age of Structure x_2 (yrs.)	Lot Size x_3 (sq. ft.)	Parking Spaces x_4	Gross Building Area x_5 (sq. ft.)
0229	90,300	4	82	4,635	0	4,266
0094	384,000	20	13	17,798	0	14,391
0043	157,500	5	66	5,913	0	6,615
0079	676,200	26	64	7,750	6	34,144
0134	165,000	5	55	5,150	0	6,120
0179	300,000	10	65	12,506	0	14,552
0087	108,750	4	82	7,160	0	3,040
0120	276,538	11	23	5,120	0	7,881
0246	420,000	20	18	11,745	20	12,600
0025	950,000	62	71	21,000	3	39,448
0015	560,000	26	74	11,221	0	30,000
0131	268,000	13	56	7,818	13	8,088
0172	290,000	9	76	4,900	0	11,315
0095	173,200	6	21	5,424	6	4,461
0121	323,650	11	24	11,834	8	9,000
0077	162,500	5	19	5,246	5	3,828
0060	353,500	20	62	11,223	2	13,680
0174	134,400	4	70	5,834	0	4,680
0084	187,000	8	19	9,075	0	7,392
0031	155,700	4	57	5,280	0	6,030
0019	93,600	4	82	6,864	0	3,840
0074	110,000	4	50	4,510	0	3,092
0057	573,200	14	10	11,192	0	23,704
0104	79,300	4	82	7,425	0	3,876
0024	272,000	5	82	7,500	0	9,542

Source: Robinson Appraisal Co., Inc., Mankato, Minnesota.

a. Fit a first-order model to the data. (You may already have done this in Exercise 11.10.)
b. Do the data provide sufficient evidence to conclude that the model of part **a** is useful for predicting sale price? Test using $\alpha = .05$.
c. Drop x_3 and x_4 from the model of part **a** and refit the model to the data.
d. Do the data provide sufficient evidence to conclude that the model of part **c** is useful for predicting sale price? Test using $\alpha = .05$.

12.5 Models with One Qualitative Independent Variable

Suppose we want to write a model for the mean profit $E(y)$ per sales dollar of a construction company as a function of the sales engineer who estimates and bids on a job. (For simplicity, we will ignore other independent variables that might affect the response.) There are three sales engineers, Adams, Brown, and Clark, so Sales engineer is a single qualitative variable set at three levels, corresponding to Adams, Brown, and Clark. Note that with a qualitative independent variable, we cannot attach a quantitative meaning to a given level. All we can do is describe it.

To simplify our notation, let μ_A be the mean profit per sales dollar for Adams, and let μ_B and μ_C be the corresponding mean profits for Brown and Clark. Our objective is to write a single prediction equation that will give the mean value of y for the three sales engineers. This can be done as follows:

$$E(y) = \beta_0 + \beta_1 x_1 + \beta_2 x_2$$

where

$$x_1 = \begin{cases} 1 & \text{if Brown is the sales engineer} \\ 0 & \text{if Brown is not the sales engineer} \end{cases}$$

$$x_2 = \begin{cases} 1 & \text{if Clark is the sales engineer} \\ 0 & \text{if Clark is not the sales engineer} \end{cases}$$

The variables x_1 and x_2 are not meaningful independent variables as for the case of the models with quantitative independent variables. Instead, they are **dummy** (or **indicator**) **variables** that make the model function. To see how they work, let $x_1 = 0$ and $x_2 = 0$. This condition will apply when we are seeking the mean response for Adams (neither Brown nor Clark is the sales engineer; hence, it must be Adams). Then the mean value of y when Adams is the sales engineer is

$$\mu_A = E(y) = \beta_0 + \beta_1(0) + \beta_2(0) = \beta_0$$

This tells us that the mean profit per sales dollar for Adams is β_0. In other words, $\beta_0 = \mu_A$.

Now suppose we want to represent the mean response $E(y)$ when Brown is the sales engineer. Checking the dummy variable definitions, we see that we should let $x_1 = 1$ and $x_2 = 0$:

$$\mu_B = E(y) = \beta_0 + \beta_1 x_1 + \beta_2 x_2 = \beta_0 + \beta_1(1) + \beta_2(0) = \beta_0 + \beta_1$$

or, since $\beta_0 = \mu_A$,

$$\mu_B = \mu_A + \beta_1$$

Then it follows that the interpretation of β_1 is

$$\beta_1 = \mu_B - \mu_A$$

which is the difference in the mean profit per sales dollar between Brown and Adams.

Finally, if we want the mean value of y when Clark is the sales engineer, we let $x_1 = 0$ and $x_2 = 1$:

$$\mu_C = E(y) = \beta_0 + \beta_1(0) + \beta_2(1) = \beta_0 + \beta_2$$

or, since $\beta_0 = \mu_A$,

$$\mu_C = \mu_A + \beta_2$$

Then it follows that the interpretation of β_2 is

$$\beta_2 = \mu_C - \mu_A$$

Note that we are able to describe *three levels* of the qualitative variable with only *two dummy variables*. This is because the mean of the base level (Adams, in this case) is accounted for by the intercept β_0.

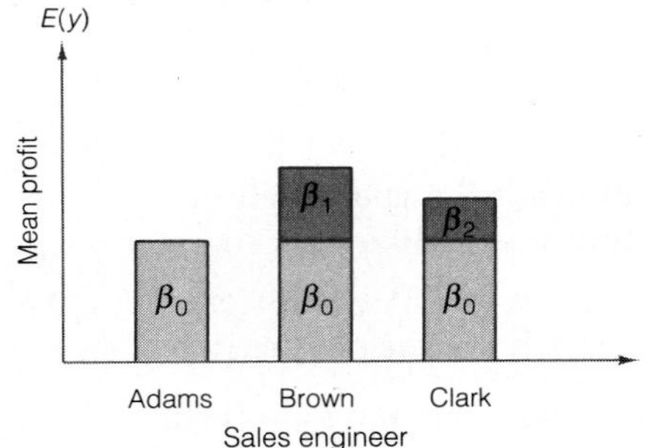

FIGURE 12.15 ▶ Bar chart comparing $E(y)$ for the three sales engineers

Since Sales engineer is a qualitative variable, we will use a bar graph to show the value of mean profit $E(y)$ for the three levels of Sales engineer (see Figure 12.15). In particular, note that the height of the bar, $E(y)$, for each level of Sales engineer is equal to the sum of the model parameters shown in the preceding equations. You can see that the height of the bar corresponding to Adams is β_0; i.e., $E(y) = \beta_0$. Similarly, the heights of the bars corresponding to Brown and Clark are $E(y) = \beta_0 + \beta_1$ and $E(y) = \beta_0 + \beta_2$, respectively.*

Now, carefully examine the model with a single qualitative independent variable at three levels, because we use exactly the same pattern for any number of levels. Also, the interpretation of the parameters is always the same.

One level is selected as the base level (we used Adams as level A). Then for the 1–0 system of coding† for the dummy variables,

---

*Either β_1 or β_2, or both, could be negative. If, e.g., β_1 were negative, the height of Brown's bar would be less (instead of more) than the height of Adams' bar by the amount β_1. Figure 12.15 is constructed assuming β_1 and β_2 are positive quantities.

†We do not have to use a 1–0 system of coding for the dummy variables. Any two-value system will work, but the interpretation given to the model parameters will depend on the code. Using the 1–0 system makes the model parameters easy to interpret.

$$\mu_A = \beta_0$$

The coding for all dummy variables is as follows: To represent the mean value of y for a particular level, let that dummy variable equal 1; otherwise, the dummy variable is set equal to 0. Using this system of coding, we have

$$\mu_B = \beta_0 + \beta_1$$
$$\mu_C = \beta_0 + \beta_2$$
$$\vdots$$

Because $\mu_A = \beta_0$, any other model parameter will represent the difference in means between that level and the base level:

$$\beta_1 = \mu_B - \mu_A$$
$$\beta_2 = \mu_C - \mu_A$$
$$\vdots$$

Procedure for Writing a Model with One Qualitative Independent Variable with k Levels

Always use one less dummy variable than the number of levels of the qualitative variable. Thus, for a qualitative variable with k levels, we use $k - 1$ dummy variables:

$$y = \beta_0 + \beta_1 x_1 + \beta_2 x_2 + \cdots + \beta_{k-1} x_{k-1} + \epsilon$$

where x_i is the dummy variable for level $(i + 1)$ and

$$x = \begin{cases} 1 & \text{if } y \text{ is observed at level } (i+1) \\ 0 & \text{otherwise} \end{cases}$$

Then, for this system of coding

$$\begin{aligned} \mu_A &= \beta_0 & \text{and} \quad \beta_1 &= \mu_B - \mu_A \\ \mu_B &= \beta_0 + \beta_1 & \beta_2 &= \mu_C - \mu_A \\ \mu_C &= \beta_0 + \beta_2 & \beta_3 &= \mu_D - \mu_A \\ \mu_D &= \beta_0 + \beta_3 & &\vdots \\ &\vdots & & \end{aligned}$$

Level 1, or the A level, is the *base level* for this coding system.

EXAMPLE 12.4

Suppose a large chain of department stores wants to compare the mean dollar amounts owed by its delinquent credit card customers in three annual income groups: under \$12,000, \$12,000–\$25,000, and over \$25,000. A sample of ten customers with delinquent accounts is selected from each group and the amount owed by each is recorded, as shown in Table 12.3. Do the data provide sufficient evidence to indicate that the mean dollar amounts owed by customers differ for the three income groups?

TABLE 12.3 Income Class: Dollars Owed

	Category 1	Category 2	Category 3
	Under $12,000	$12,000–$25,000	Over $25,000
	$148	$513	$335
	76	264	643
	393	433	216
	520	94	536
	236	535	128
	134	327	723
	55	214	258
	166	135	380
	415	280	594
	153	304	465
Totals	$2,296	$3,099	$4,278

Solution Note that Income is ordinarily a quantitative variable, but in this example only the group (low, medium, high) in which the customer's income falls is known. We therefore treat Income as a qualitative variable (measured on an ordinal scale). For a three-level qualitative variable, we need two dummy variables in the regression model. The model relating $E(y)$ to this single qualitative variable, Income level, is

$$E(y) = \beta_0 + \beta_1 x_1 + \beta_2 x_2$$

where

$$x_1 = \begin{cases} 1 & \text{if income level 2} \\ 0 & \text{if not} \end{cases} \qquad x_2 = \begin{cases} 1 & \text{if income level 3} \\ 0 & \text{if not} \end{cases}$$

and

$$\beta_1 = \mu_2 - \mu_1 \qquad \beta_2 = \mu_3 - \mu_1$$

where μ_1, μ_2, and μ_3 are the mean responses for income categories 1, 2, and 3, respectively. Testing the null hypothesis that the means for the three income levels are equal, i.e., $\mu_1 = \mu_2 = \mu_3$, is equivalent to testing

$$H_0: \beta_1 = \beta_2 = 0$$

because if $\beta_1 = \mu_2 - \mu_1 = 0$ and $\beta_2 = \mu_3 - \mu_1 = 0$, then μ_1, μ_2, and μ_3 must be equal. The alternative hypothesis is

H_a: At least one of the parameters, β_1 or β_2, differs from 0 which implies that at least two of the three means (μ_1, μ_2, and μ_3) differ.

There are two ways to conduct this test. We can fit the complete model shown previously and the reduced model (deleting the terms involving β_1 and β_2),

$$E(y) = \beta_0$$

then conduct the F test described in the preceding section (we leave this as an exercise for you). Or, we can use the F test of the complete model (Section 11.5), which tests the null hypothesis that all parameters in the model, except β_0, equal 0.

Either way you conduct the test, you will obtain the same computed value of F, the value shown on the SAS printout for a test of the complete model. The SAS printout for fitting the complete model,

$$E(y) = \beta_0 + \beta_1 x_1 + \beta_2 x_2$$

is shown in Figure 12.16 and the value of the F statistic for testing the complete model, $F = 3.48$, is shaded. If we choose $\alpha = .05$, we will reject H_0: $\beta_1 = \beta_2 = 0$ because the computed value $F = 3.48$ has an observed significance level of .0452 (also shaded). We therefore reject H_0 and conclude that at least one of the parameters, β_1 or β_2, differs from 0. Or, equivalently, we conclude that the data provide sufficient evidence to indicate that the mean indebtedness does vary from one income group to another.

```
DEPENDENT VARIABLE: Y

SOURCE                   DF    SUM OF SQUARES       MEAN SQUARE    F VALUE

MODEL                     2   198772.46666667    99386.23333333       3.48
ERROR                    27   770670.90000000    28543.36666667    PR > F
CORRECTED TOTAL          29   969443.36666667                       0.0452

R-SQUARE                 C.V.        ROOT MSE            Y MEAN

0.205038              52.3978    168.94782232      322.43333333

                                T FOR HO:      PR > |T|    STD ERROR OF
PARAMETER        ESTIMATE     PARAMETER=0                   ESTIMATE

INTERCEPT    229.60000000           4.30       0.0002      53.42599243
X1            80.30000000           1.06       0.2973      75.55576307
X2           198.20000000           2.62       0.0141      75.55576307
```

FIGURE 12.16 ▲ SAS computer printout for Example 12.4

We need to make two additional comments about Example 12.4. First, regression analysis is not the only way to analyze these data. Another procedure for calculating the value of the F statistic, known as an *analysis of variance*, is described in Chapter 16. Second, if you choose to analyze the data by fitting complete and reduced models (Section 12.4), you will find that the least squares estimate of β_0 in the reduced model,

$$E(y) = \beta_0$$

is $\bar{y}$, the mean of all $n = 30$ observations, and the sum of squares for error for the reduced model is

$$\text{SSE}_r = \sum (y_i - \hat{y}_i)^2 = \sum (y_i - \bar{y})^2 = 969{,}443.367$$

This value is shown in the SAS printout in Figure 12.16 as **SUM OF SQUARES** corresponding to **CORRECTED TOTAL**. We leave the remaining steps, calculating the drop in SSE and the resulting F statistic, to you. You will find that the value you obtain will be the same as the value of F shown in the SAS printout in Figure 12.16.

Exercises 12.37–12.48

Note: Starred (*) exercises require the use of a computer.

Learning the Mechanics

12.37 Write a regression model relating the mean value of y to a qualitative independent variable that can assume two levels. Interpret all the terms in the model.

12.38 Write a regression model relating $E(y)$ to a qualitative independent variable that can assume three levels. Interpret all the terms in the model.

12.39 The following model was used to relate $E(y)$ to a single qualitative variable with four levels:

$$E(y) = \beta_0 + \beta_1 x_1 + \beta_2 x_2 + \beta_3 x_3$$

where

$$x_1 = \begin{cases} 1 & \text{if level 2} \\ 0 & \text{if not} \end{cases} \qquad x_2 = \begin{cases} 1 & \text{if level 3} \\ 0 & \text{if not} \end{cases} \qquad x_3 = \begin{cases} 1 & \text{if level 4} \\ 0 & \text{if not} \end{cases}$$

This model was fit to $n = 30$ data points with the following result:

$$\hat{y} = 10.2 - 4x_1 + 12x_2 + 2x_3$$

Find estimates for $E(y)$ when the qualitative independent variable is set at each of the following levels.

a. Level 1 **b.** Level 2 **c.** Level 3 **d.** Level 4

e. Specify the null and alternative hypotheses you would use to test whether $E(y)$ is the same for all four levels of the independent variable.

12.40 Minitab was used to fit this model to $n = 15$ data points:

$$y = \beta_0 + \beta_1 x_1 + \beta_2 x_2 + \epsilon$$

where

$$x_1 = \begin{cases} 1 & \text{if level 2} \\ 0 & \text{if not} \end{cases} \qquad x_2 = \begin{cases} 1 & \text{if level 3} \\ 0 & \text{if not} \end{cases}$$

The resulting printout is shown. (Note that in the printout **C1** represents y, and **C2** and **C3** represent x_1 and x_2, respectively.)

```
THE REGRESSION EQUATION IS
C1 = 80.0 + 16.8 C2 + 40.4 C3

                                      ST. DEV.        T-RATIO =
              COEFFICIENT             OF COEF.        COEF/S.D.
                  80.000                4.082            19.60
C2                16.800                5.774             2.91
C3                40.400                5.774             7.00

S = 9.129

R-SQUARED = 80.5 PERCENT
R-SQUARED = 77.2 PERCENT, ADJUSTED FOR D.F.

ANALYSIS OF VARIANCE

 DUE TO          DF             SS              MS=SS/DF
REGRESSION        2          4118.9              2059.5
RESIDUAL         12          1000.0                83.3
TOTAL            14          5118.9
```

a. Report the least squares prediction equation.

b. Interpret the values of $\hat{\beta}_1$ and $\hat{\beta}_2$.

c. Interpret the following hypotheses in terms of μ_1, μ_2, and μ_3:

$$H_0\colon\ \beta_1 = \beta_2 = 0 \qquad H_a\colon \text{ At least one of the parameters } \beta_1 \text{ and } \beta_2 \text{ differs from } 0$$

d. Conduct the hypothesis test of part **c**.

Applying the Concepts

12.41 In 1990–1991, 4-year private colleges charged an average of \$9,083 for tuition and fees for the year; 4-year public colleges charged \$1,888 (*Chronicle of Higher Education Almanac*, Aug. 1992). In order to estimate the difference in the mean amounts charged for the 1993–1994 academic year, random samples of 40 private colleges and 40 public colleges were contacted and questioned about their tuition structures.

a. Which of the procedures described in Chapter 9 could be used to estimate the difference in mean charges between private and public colleges?

b. Propose a regression model involving the qualitative indepedent variable Type of college that could be used to investigate the difference between the means. Be sure to specify the coding scheme for the dummy variable in the model.

c. Explain how the regression model you developed in part **b** could be used to estimate the difference between the population means.

12.42 An independent testing laboratory was hired to compare the length of life (in months) of four brands of color television picture tubes, A, B, C, and D. Life data were obtained on 10 randomly selected picture tubes of each brand. Propose a regression model involving the qualitative independent variable Brand of tube to estimate the mean longevity of a tube. Interpret each term in your model.

12.43 The director of marketing of a company that sells business machines is interested in modeling mean monthly sales (in thousands of dollars) per salesperson, $E(y)$, as a function of the type of sales incentive plan that is in effect: commission only, straight salary, or salary plus commission on each sale. The director proposes the model

$$E(y) = \beta_0 + \beta_1 x_1 + \beta_2 x_2$$

where

$$x_1 = \begin{cases} 1 & \text{if straight salary} \\ 0 & \text{if otherwise} \end{cases}$$

$$x_2 = \begin{cases} 1 & \text{if salary plus commission} \\ 0 & \text{if otherwise} \end{cases}$$

A portion of the computer printout that results from using Minitab to fit this model to the sales data collected from a sample of 15 salespersons (five from each incentive plan) is shown here.

```
THE REGRESSION EQUATION IS
Y =     20.0 -   8.60 X1 + 3.80 X2

                                   ST. DEV.    T-RATIO =
                     COEFFICIENT   OF COEF.    COEF/S.D.

       --              20.000        2.898        6.90
X1    C21              -8.60         4.10        -2.10
X2    C22               3.80         4.10          .93

THE ST. DEV. OF Y ABOUT REGRESSION LINE IS
S =         6.481
WITH ( 15- 3) = 12 DEGREES OF FREEDOM

R-SQUARED = 44.5 PERCENT
R-SQUARED = 35.2 PERCENT, ADJUSTED FOR D.F.

ANALYSIS OF VARIANCE

 DUE TO        DF       SS     MS=SS/DF

REGRESSION      2   403.60     201.80
RESIDUAL       12   504.00      42.00
TOTAL          14   907.60
```

a. Do the data provide sufficient evidence to conclude that there is a difference in mean monthly sales among the three incentive plans? Test using $\alpha = .05$.

b. Use the least squares prediction equation to estimate the mean sales for salespersons working on a straight salary basis.

c. Use the least squares prediction equation to estimate the mean sales for salespersons working on a commission only basis. [*Note:* We would prefer to use confidence intervals for the estimates in parts **b** and **c**, but their calculation is beyond the scope of this text. This procedure can be found in the references at the end of the chapter.]

12.44 Refer to Exercise 12.43. Find a 90% confidence interval for the difference between the mean monthly sales for salespersons on salary plus commission versus those on commission only.

12.45 The manager of a supermarket wants to model the total weekly sales of beer, y, as a function of brand. This model will enable the manager to plan the store's inventory. The market carries three brands, B_1, B_2, and B_3.

a. What type of independent variable is Brand of beer?

b. Write the model relating mean weekly beer sales $E(y)$ as a function of Brand of beer. Use brand B_1 as the base level in coding the dummy variables. Be sure to explain any dummy variables you use.

c. Interpret the β parameters of your model in part **b**.

d. In terms of the model parameters, what are the mean weekly sales for brand B_3?

12.46 Refer to Exercise 12.45. Suppose the manager uses brand B_1 as the base level and obtains the prediction equation

$$\hat{y} = 450 + 60x_1 - 30x_2$$

where

$$x_1 = \begin{cases} 1 & \text{if brand } B_2 \\ 0 & \text{otherwise} \end{cases}$$

$$x_2 = \begin{cases} 1 & \text{if brand } B_3 \\ 0 & \text{otherwise} \end{cases}$$

a. What is the difference between the estimated mean weekly sales for brands B_2 and B_1?

b. What is the estimated mean weekly sales for brand B_2? [*Note*: We would generally form confidence intervals for the true means in order to assess the reliability of these estimates. Our objective in these exercises is to develop the ability to use the models to obtain the estimates.]

12.47 Five varieties of peas are being tested by a large agribusiness cooperative in Ohio to determine which is best suited for production. A field was divided into 20 plots, with each variety of peas, A–E, planted in four plots. The yields (in bushels) produced from each plot are shown in the table.

A	B	C	D	E
26.2	29.2	29.1	21.3	20.1
24.3	28.1	30.8	22.4	19.3
21.8	27.3	33.9	24.3	19.9
28.1	31.2	32.8	21.8	22.1

SAS was used to fit the model

$$y = \beta_0 + \beta_1 x_1 + \beta_2 x_2 + \beta_3 x_3 + \beta_4 x_4 + \epsilon$$

to these data, using the coding $x_1 = 1$ for pea A, $x_2 = 1$ for pea B, $x_3 = 1$ for pea C, and $x_4 = 1$ for pea D. The resulting printout is shown on page 650.

a. Write the model relating mean yield to pea variety, and interpret all the parameters in the model.

b. Report the least squares model from the SAS printout.

c. What null and alternative hypotheses are tested by the global F test for this model? Interpret the hypotheses both in terms of the β parameters and the mean yields for the five varieties of peas.

d. Test the hypotheses of part **c** using $\alpha = .05$.

e. Place a 95% confidence interval on the difference between the mean yields of varieties D and E.

```
Dep Variable: Y

                              Analysis of Variance

                           Sum of           Mean
Source          DF        Squares         Square        F Value          Prob>F

Model            4      342.04000       85.51000         23.966          0.0001
Error           15       53.52000        3.56800
C Total         19      395.56000

        Root MSE          1.88892       R-Square         0.8647
        Dep Mean         25.70000       Adj R-Sq         0.8286
        C.V.              7.34986

                              Parameter Estimates

                        Parameter        Standard      T for H0:
Variable     DF          Estimate           Error    Parameter=0     Prob > |T|

INTERCEP      1         20.350000      0.94445752         21.547         0.0001
X1            1          4.750000      1.33566463          3.556         0.0029
X2            1          8.600000      1.33566463          6.439         0.0001
X3            1         11.300000      1.33566463          8.460         0.0001
X4            1          2.100000      1.33566463          1.572         0.1367
```

***12.48** A firm's *debt-to-equity ratio* is a measure of the extent to which management is using borrowed funds. It is a measure of considerable importance to individuals and organizations that are potential lenders to the firm. A high debt-to-equity ratio signals that in case of default, the lender is unlikely to recover outstanding loans to the firm, since insufficient equity exists to cover all the firm's obligations. In addition, the higher the debt, the more funds that are required to service the debt. As a result, if business falls off, the firm may lack sufficient operating funds to meet debt service payments (Spiro, 1982). The debt-to-equity ratios for firms in four industries are given in the table for the fiscal year that ended during 1986.

Field	*Firm*	*Debt-to-equity*	*Field*	*Firm*	*Debt-to-equity*
Insurance	Chubb	.24	Publishing	Deluxe Check	.03
	Kemper	.09		New York Times	.32
	St. Paul Cos.	.12		Times Mirror	.51
	Lincoln National	.09		Dow Jones	.15
	USF&G	.00		Gannett	.73
	Aetna Life & Cas.	.07			
Electric utilities	Pacific G & E	.84	Banking	U.S. Bancorp	.54
	Houston Ind.	1.00		Sun Trust Banks	.29
	Florida Progress	1.03		Mellon Bank	.82
	Penn Power & Light	1.11		Michigan National	.13
	North States Power	.83		Southeast Banking	.41
	Ohio Edison	1.16			
	Orange & Rockland	.70			

Source: "39th Annual Report on American Industry," *Forbes*, Jan. 12, 1987.

a. Propose a regression model involving the qualitative independent variable Industry that could be used to investigate whether the mean debt-to-equity ratio varies among the four industries. Be sure to specify the coding scheme for the dummy variables in your model.
b. Test the null hypothesis that the mean debt-to-equity ratios are equal in the four industries. Use $\alpha = .05$.
c. Do the data provide sufficient evidence to conclude that the mean debt-to-equity ratios of the electric utilities industry and the insurance industry differ? Test using $\alpha = .10$.

12.6 Comparing the Slopes of Two or More Lines

Suppose you want to relate the mean monthly sales $E(y)$ of a company to monthly advertising expenditure x, for three advertising media—say, newspaper, radio, and television—and you want to use first-order (straight-line) models to model the responses for all three media. Graphs of these three relationships might appear as shown in Figure 12.17.

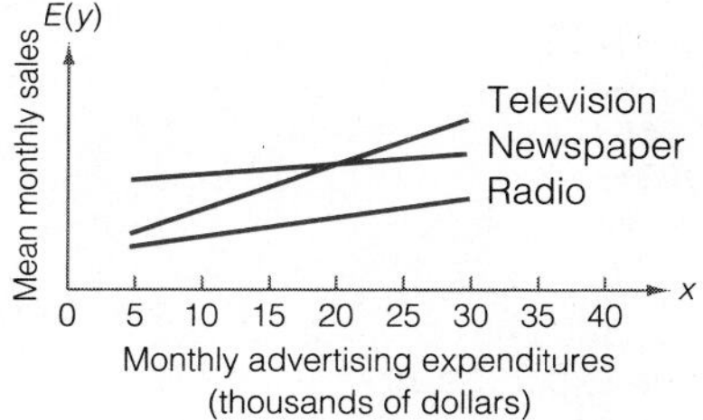

FIGURE 12.17 ▶ Graphs of the relationship between mean sales $E(y)$ and advertising expenditure x

Since the lines in Figure 12.17 are hypothetical, a number of practical business questions arise. Is one advertising medium as effective as any other; that is, do the three mean sales lines differ for the three media? Do the increases in mean sales per dollar increase in advertising differ for the three media; that is, do the slopes of the three lines differ? Note that each of the two business questions has been rephrased as a question about the parameters that define the three lines of Figure 12.17. To answer them, we must write a single linear statistical model that will characterize the three lines of Figure 12.17. Then the business questions can be answered by testing hypotheses about the model parameters.

In the preceding example, the response (monthly sales) is a function of *two* independent variables, one quantitative (advertising expenditure x) and one qualitative (type of medium). We examine the different models that can be constructed relating $E(y)$ to these two independent variables.

1. The straight-line relationship between mean sales $E(y)$ and advertising expenditure is the same for all three media. That is, a single line will describe the relationship between $E(y)$ and advertising expenditure x_1 for all the media; see Figure 12.18 (page 652).

$$E(y) = \beta_0 + \beta_1 x_1 \qquad \text{where} \quad x_1 = \text{Advertising expenditure}$$

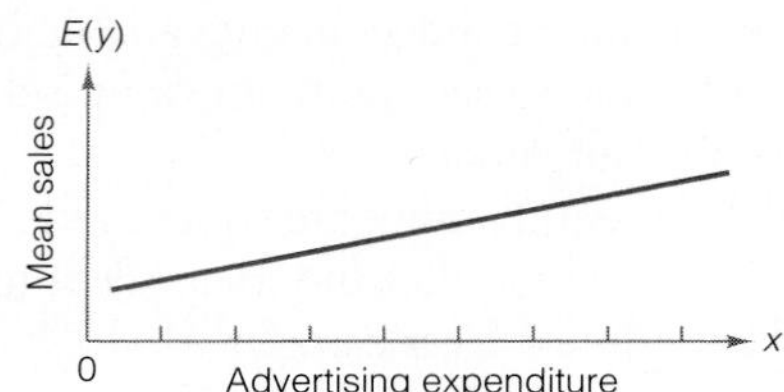

FIGURE 12.18 ▶ The relationship between $E(y)$ and x_1 is the same for all media.

2. The straight lines relating mean sales $E(y)$ to advertising expenditure x_1 differ from one medium to another, but the increase in mean sales per unit increase in dollar advertising expenditure, x_1, is the same for all media. That is, the lines are parallel but have different y-intercepts; see Figure 12.19.

$$E(y) = \beta_0 + \beta_1 x_1 + \beta_2 x_2 + \beta_3 x_3$$

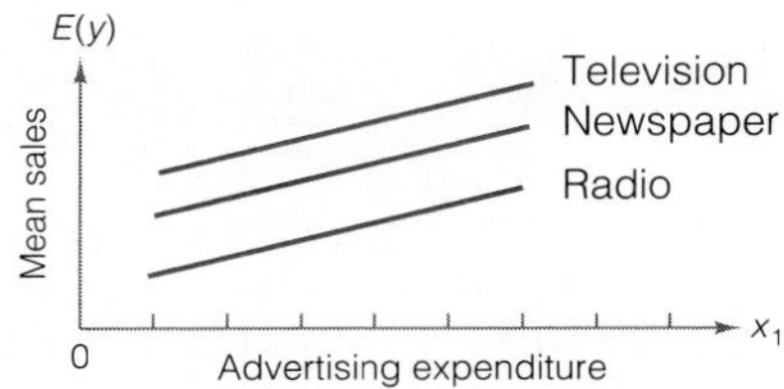

FIGURE 12.19 ▶ Parallel response lines for the three media

where

$$x_1 = \text{Advertising expenditure}$$

$$x_2 = \begin{cases} 1 & \text{if radio} \\ 0 & \text{if not} \end{cases} \qquad x_3 = \begin{cases} 1 & \text{if television} \\ 0 & \text{if not} \end{cases}$$

Notice that this model is essentially a combination of a first-order model with a single quantitative variable and the model with a single qualitative variable:

First-order model with single quantitative variable: $\quad E(y) = \beta_0 + \boxed{\beta_1 x_1}$

Model with single qualitative variable at three levels: $\quad E(y) = \beta_0 + \boxed{\beta_2 x_2 + \beta_3 x_3}$

where x_1, x_2, and x_3 are defined as before. The model described implies no interaction between the two independent variables, advertising expenditure x_1 and the qualitative variable Type of advertising medium. The change in $E(y)$ for a 1-unit increase in x_1 is identical (i.e., the slopes of the lines are equal) for all three advertising media. The terms corresponding to each of the independent variables are called **main effect terms** because they imply no interaction.

3. The straight lines relating mean sales $E(y)$ to advertising expenditure x_1 differ for the three advertising media. That is, the intercepts and slopes differ for the three lines; see Figure 12.20. As you will see, this interaction model is obtained by adding

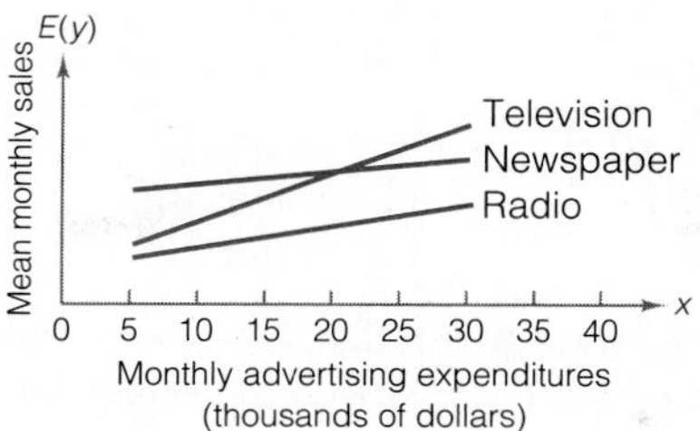

FIGURE 12.20 ▶ **Different response lines for the three media**

interaction terms (those involving the cross product terms, one each from each of the two independent variables):

$$E(y) = \beta_0 + \overbrace{\beta_1 x_1}^{\substack{\text{Main effect,}\\ \text{advertising}\\ \text{expenditure}}} + \overbrace{\beta_2 x_2 + \beta_3 x_3}^{\substack{\text{Main effect,}\\ \text{type of medium}}} + \overbrace{\beta_4 x_1 x_2 + \beta_5 x_1 x_3}^{\text{Interaction}}$$

Note that each of the preceding models is obtained by adding terms to model 1, the single first-order model used to model the responses for all three media. Model 2 is obtained by adding the main effect terms for the qualitative variable, Type of medium; and model 3 is obtained by adding the interaction terms to model 2.

Will a single line (Figure 12.18) characterize the responses for all three media, or do the three response lines differ as shown in Figure 12.20? A test of the null hypothesis that a single first-order model adequately describes the relationship between $E(y)$ and advertising expenditure x_1 for all three media is a test of the null hypothesis that the parameters of model 3, β_2, β_3, β_4, and β_5, equal 0; i.e.,

$$H_0\text{:} \quad \beta_2 = \beta_3 = \beta_4 = \beta_5 = 0$$

This hypothesis can be tested by fitting the complete model (model 3) and the reduced model (model 1) and conducting an F test, as described in Section 12.4.

Suppose we assume that the response lines for the three media will differ but wonder whether the data present sufficient evidence to indicate differences in the slopes of the lines. To test the null hypothesis that model 2 adequately describes the relationship between $E(y)$ and advertising expenditure x_1, we wish to test

$$H_0\text{:} \quad \beta_4 = \beta_5 = 0$$

that is, that the two independent variables, advertising expenditure x_1 and the qualitative variable, Type of medium, do not interact. This test can be conducted by fitting the complete model (model 3) and the reduced model (model 2), calculating the drop in the sum of squares for error, and conducting an F test.

EXAMPLE 12.5

Substitute the appropriate values of the dummy variables in model 3 to obtain the equations of the three response lines in Figure 12.20.

Solution

The complete model that characterizes the three lines in Figure 12.20 is

$$E(y) = \beta_0 + \beta_1 x_1 + \beta_2 x_2 + \beta_3 x_3 + \beta_4 x_1 x_2 + \beta_5 x_1 x_3$$

where

$$x_1 = \text{Advertising expenditure}$$

$$x_2 = \begin{cases} 1 & \text{if radio} \\ 0 & \text{if not} \end{cases} \qquad x_3 = \begin{cases} 1 & \text{if television} \\ 0 & \text{if not} \end{cases}$$

Examining the coding, you can see that $x_2 = x_3 = 0$ when the advertising medium is newspaper. Substituting these values into the expression for $E(y)$, we obtain the newspaper medium line.

Newspaper line:
$$E(y) = \beta_0 + \beta_1 x_1 + \beta_2(0) + \beta_3(0) + \beta_4 x_1(0) + \beta_5 x_1(0) = \beta_0 + \beta_1 x_1$$

Similarly, we substitute the appropriate values of x_2 and x_3 into the expression for $E(y)$ to obtain the radio medium line and the television medium line.

Radio line:
$$E(y) = \beta_0 + \beta_1 x_1 + \beta_2(1) + \beta_3(0) + \beta_4 x_1(1) + \beta_5 x_1(0) = \overbrace{\beta_0 + \beta_2}^{y\text{-intercept}} + \overbrace{(\beta_1 + \beta_4)}^{\text{Slope}} x_1$$

Television line:
$$E(y) = \beta_0 + \beta_1 x_1 + \beta_2(0) + \beta_3(1) + \beta_4 x_1(0) + \beta_5 x_1(1) = \overbrace{\beta_0 + \beta_3}^{y\text{-intercept}} + \overbrace{(\beta_1 + \beta_5)}^{\text{Slope}} x_1$$

If you were to fit model 3, obtain estimates of $\beta_0, \beta_1, \beta_2, \ldots, \beta_5$, and substitute them into the equations for the three media lines shown in Eample 12.5, you would obtain exactly the same prediction equations as you would obtain if you fit three separate straight lines, one to each of the three sets of media data. You may ask why we would not fit the three lines separately. Why fit a model (model 3) that combines all three lines into the same equation? The answer is that you need to use this procedure if you plan to use statistical tests to compare the three media lines. We need to be able to express a practical question about the lines in terms of a hypothesis that a set of parameters in the model equals 0. You could not do this if you were to perform three separate regression analyses and fit a line to each set of media data.

EXAMPLE 12.6

An industrial psychologist conducted an experiment to investigate the relationship between worker productivity and a measure of salary incentive for two manufacturing plants: A has union representation and B has nonunion representation. The productivity y per worker was measured by recording the number of machined castings that a worker could produce in a 4-week, 40-hour-per-week period. The incentive was the amount x_1 of bonus (in cents per casting) paid for all castings produced in excess of 1,000 per worker for the 4-week period. Nine workers were selected from each plant and three from each group of nine were assigned to receive a 20¢ bonus per casting, three a 30¢ bonus, and three a 40¢ bonus. The productivity data for the 18 workers, three for each plant type and incentive combination, are shown in the accompanying table.

	20¢/Casting Bonus			*30¢/Casting Bonus*			*40¢/Casting Bonus*		
Union plant	1,435	1,512	1,491	1,583	1,529	1,610	1,601	1,574	1,636
Nonunion plant	1,575	1,512	1,488	1,635	1,589	1,661	1,645	1,616	1,689

a. Plot the data points, and graph the prediction equations for the two productivity lines. Assume that the relationship between mean productivity and incentive is first-order.

b. Do the data provide sufficient evidence to indicate a difference in worker response to incentives between the two plants?

Solution

If we assume that a first-order model* is adequate to detect a change in mean productivity $E(y)$ as a function of incentive x_1, then the model that produces two productivity lines, one for each plant, is

$$E(y) = \beta_0 + \beta_1 x_1 + \beta_2 x_2 + \beta_3 x_1 x_2$$

where

$$x_1 = \text{Incentive} \qquad x_2 = \begin{cases} 1 & \text{if nonunion plant} \\ 0 & \text{if union plant} \end{cases}$$

a. The SAS printout for the regression analysis is shown in Figure 12.21. The prediction equation obtained by reading the parameter estimates from the printout is

```
DEPENDENT VARIABLE: Y

SOURCE                  DF   SUM OF SQUARES      MEAN SQUARE   F VALUE

MODEL                    3   57332.38888889   19110.79629630     11.46
ERROR                   14   23349.22222223    1667.80158730    PR > F
CORRECTED TOTAL         17   80681.61111112                     0.0005

R-SQUARE              C.V.         ROOT MSE           Y MEAN

0.710600            2.5901      40.83872656    1576.72222222

                                T FOR HO:     PR > |T|   STD ERROR OF
PARAMETER         ESTIMATE    PARAMETER=0                 ESTIMATE

INTERCEPT     1365.8333333          26.35     0.0001     51.83641257
X1               6.21666667          3.73     0.0022      1.66723403
X2              47.77777778          0.65     0.5251     73.30775769
X1*X2            0.03333333          0.01     0.9889      2.35782498
```

FIGURE 12.21 ▲
SAS computer printout for the complete model of Example 12.6

*Although the model contains a term involving x_1x_2, it is first-order (i.e., graphs as a straight line) in the quantitative variable x_1. The variable x_2 is a dummy that introduces or deletes terms in the model. The order of a term is determined only by the quantitative variables that appear in the term.

$$\hat{y} = 1{,}365.833 + 6.217x_1 + 47.778x_2 + .033x_1x_2$$

The prediction equation for the union plant can be obtained by substituting $x_2 = 0$ into the general prediction equation. Then

$$\begin{aligned}\hat{y} = \hat{\beta}_0 + \hat{\beta}_1x_1 + \hat{\beta}_2(0) + \hat{\beta}_3x_1(0) &= \hat{\beta}_0 + \hat{\beta}_1x_1 \\ &= 1{,}365.833 + 6.217x_1\end{aligned}$$

Similarly, the prediction equation for the nonunion plant is obtained by substituting $x_2 = 1$ into the general prediction equation. Then

$$\begin{aligned}\hat{y} &= \hat{\beta}_0 + \hat{\beta}_1x_1 + \hat{\beta}_2x_2 + \hat{\beta}_3x_1x_2 \\ &= \hat{\beta}_0 + \hat{\beta}_1x_1 + \hat{\beta}_2(1) + \hat{\beta}_3x_1(1) \\ &= \overbrace{(\hat{\beta}_0 + \hat{\beta}_2)}^{y\text{-intercept}} + \overbrace{(\hat{\beta}_1 + \hat{\beta}_3)}^{\text{Slope}}x_1 \\ &= (1{,}365.833 + 47.778) + (6.217 + .033)x_1 = 1{,}413.611 + 6.250x_1\end{aligned}$$

The graphs of these prediction equations are shown in Figure 12.22.

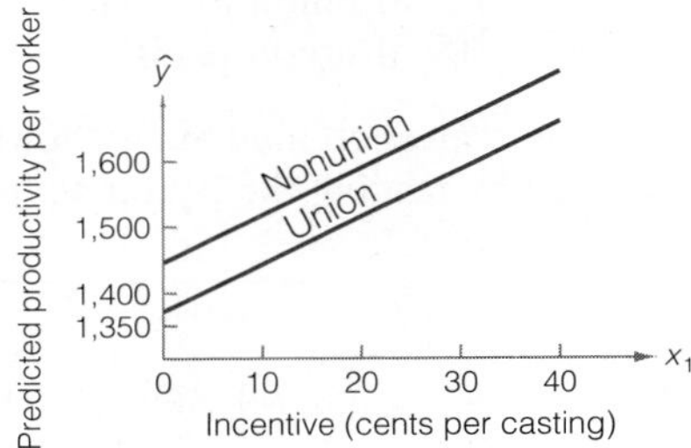

FIGURE 12.22 ► Graphs of the prediction equations for the two productivity lines of Example 12.6

b. To determine whether the data provide sufficient evidence to indicate a difference in worker response to incentives for the two plants, we test the null hypothesis that a *single* line characterizes the relationship between productivity per worker and the amount of incentive x_1 against the alternative hypothesis that we need two separate lines to characterize the relationship, one for each plant. If there is no difference in mean response $E(y)$ to x_1 between the two plants, then we do not need type of plant in the model; i.e., we do not need the terms involving x_2. Therefore, we wish to test

H_0: $\beta_2 = \beta_3 = 0$
H_a: At least one of the two parameters, β_2 or β_3, differs from 0

The SAS computer printout for fitting the reduced model

$$E(y) = \beta_0 + \beta_1x_1$$

to the data is shown in Figure 12.23. Reading SSE_c and SSE_r from Figures 12.21 and 12.23, respectively, we obtain

```
DEPENDENT VARIABLE: Y

SOURCE                 DF   SUM OF SQUARES      MEAN SQUARE   F VALUE

MODEL                   1   46625.33333333   46625.33333333     21.91
ERROR                  16   34056.27777778    2128.51736111   PR > F
CORRECTED TOTAL        17   80681.61111112                      0.0003

R-SQUARE            C.V.         ROOT MSE            Y MEAN

0.577893          2.9261      46.13585765     1576.72222222

                                T FOR HO:    PR > |T|   STD ERROR OF
PARAMETER         ESTIMATE   PARAMETER=0                  ESTIMATE

INTERCEPT   1389.72222222          33.56     0.0001    41.40819949
X1             6.23333333           4.68     0.0003     1.33182749
```

FIGURE 12.23 ▲
SAS computer printout for the reduced model of Example 12.6

Complete model: $\text{SSE}_c = 23{,}349.22$

Reduced model: $\text{SSE}_r = 34{,}056.28$

Drop in SSE: $\text{SSE}_r - \text{SSE}_c = 10{,}707.06$

The value of s^2 for the complete model, obtained from Figure 12.21, is $s^2 = 1{,}667.80$. Substituting these values, along with $k = 3$ and $g = 1$, into the formula for the F statistic yields

$$F = \frac{(\text{SSE}_r - \text{SSE}_c)/(k - g)}{s^2} = \frac{10{,}707.06/2}{1{,}667.80} = 3.21$$

The numerator degrees of freedom (the number of parameters involved in H_0) are 2 and the denominator degrees of freedom (the number associated with s^2 in the complete model) are 14. If we choose $\alpha = .05$, the tabulated value of $F_{.05}$, given in Table VIII of Appendix B, is 3.74. Since the computed value $F = 3.21$ is less than the tabulated value $F_{.05} = 3.74$, there is insufficient evidence (at $\alpha = .05$) to indicate a difference in worker response to incentives between the two plants. Therefore, there is no evidence to indicate that two different lines, one for each plant, are needed to describe the relationship between the mean productivity per worker $E(y)$ and the amount of incentive x_1.

EXAMPLE 12.7

Refer to Example 12.6 and explain how you would determine whether the data provide sufficient evidence to indicate that the incentive x_1 affects mean productivity.

Solution

If incentive did *not* affect mean productivity, we would not need terms involving x_1 in the model. Therefore, we would test the hypotheses

H_0: $\beta_1 = \beta_3 = 0$
H_a: At least one of the parameters, β_1 or β_3, differs from 0

We would fit the reduced model

$$E(y) = \beta_0 + \beta_2 x_2$$

to the data and find SSE_r. The values of SSE_c and s^2 for the complete model would be the same as those used in Example 12.6. Finally, we would calculate the value of the F statistic and compare it with a tabulated value of F based on $\nu_1 = 2$ and $\nu_2 = 14$ degrees of freedom. If the test leads to rejection of H_0, we have evidence to indicate that the increase in mean productivity that appears to be present in the graphs in Figure 12.22 is not due to random variation in the data.

Exercises 12.49–12.57

Note: Starred (*) exercises require the use of a computer.

Learning the Mechanics

12.49 Suppose you are interested in a business application with a response y, one quantitative independent variable x_1, and one qualitative variable at three levels.

a. Write a first-order model that relates the mean response $E(y)$ to the quantitative independent variable.
b. Add the main effect terms for the qualitative independent variable to the model of part **a**. Specify the coding scheme you use.
c. Add terms to the model of part **b** to allow for interaction between the quantitative and qualitative independent variables.
d. Under what circumstances will the response lines of the model in part **c** be parallel?
e. Under what circumstances will the model in part **c** have only one response line?

12.50 SAS was used to fit this model to $n = 15$ data points:

$$y = \beta_0 + \beta_1 x_1 + \beta_2 x_2 + \beta_3 x_3 + \epsilon$$

where x_1 is a quantitative variable and x_2 and x_3 are dummy variables describing a qualitative variable at three levels using the coding scheme

$$x_2 = \begin{cases} 1 & \text{if level 2} \\ 0 & \text{otherwise} \end{cases} \qquad x_3 = \begin{cases} 1 & \text{if level 3} \\ 0 & \text{otherwise} \end{cases}$$

The resulting printout is shown on the facing page.

a. What is the response line (equation) for $E(y)$ when $x_2 = x_3 = 0$? When $x_2 = 1$ and $x_3 = 0$? When $x_2 = 0$ and $x_3 = 1$?
b. What is the general least squares prediction equation for the preceding model?
c. What is the least squares prediction equation associated with level 1? Level 2? Level 3? Plot them on the same graph.

```
Dep Variable: Y

                         Analysis of Variance

                     Sum of          Mean
Source      DF      Squares        Square       F Value       Prob>F

Model        3   4747.70480    1582.56827        46.894       0.0001
Error       11    371.22853      33.74805
C Total     14   5118.93333

    Root MSE        5.80931      R-Square        0.9275
    Dep Mean       99.06667      Adj R-Sq        0.9077
    C.V.            5.86404

                        Parameter Estimates

                  Parameter        Standard     T for H0:
Variable  DF       Estimate           Error   Parameter=0    Prob > |T|

INTERCEP   1      44.802703      8.55817652         5.235        0.0003
X1         1       2.172673      0.50335248         4.316        0.0012
X2         1       9.412913      4.05315974         2.322        0.0404
X3         1      15.631532      6.81368981         2.294        0.0425
```

d. Specify the null and alternative hypotheses that should be used to investigate whether a difference exists between the response lines of the qualitative variable levels 1, 2, and 3.

e. Use the model with only the quantitative variable x_1 (printout shown) to conduct the hypothesis test of part **d**. Use $\alpha = .05$.

```
Dep Variable: Y

                         Analysis of Variance

                     Sum of          Mean
Source      DF      Squares        Square       F Value       Prob>F

Model        1   4522.90741    4522.90741        98.650       0.0001
Error       13    596.02592      45.84815
C Total     14   5118.93333

    Root MSE        6.77113      R-Square        0.8836
    Dep Mean       99.06667      Adj R-Sq        0.8746
    C.V.            6.83492

                        Parameter Estimates

                  Parameter        Standard     T for H0:
Variable  DF       Estimate           Error   Parameter=0    Prob > |T|

INTERCEP   1      33.904568      6.78960378         4.994        0.0002
X1         1       3.083380      0.31044102         9.932        0.0001
```

Applying the Concepts

12.51 The Florida Citrus Commission is interested in evaluating the performance of two orange juice extractors, brand A and brand B. It is believed that the size of the fruit used in the test may influence the juice yield (amount of juice per pound of oranges) obtained by the extractors. The commission wants to develop a regression model relating the mean juice yield $E(y)$ to the type of orange juice extractor (brand A or brand B) and the size of orange (diameter), x_1.

a. Identify the independent variables as qualitative or quantitative.
b. Write a model that describes the relationship between $E(y)$ and size of orange as two parallel lines, one for each brand of extractor.
c. Modify the model of part **b** to permit the slopes of the two lines to differ.
d. Sketch typical response lines for the model of part **b**. Do the same for the model of part **c**. Carefully label your graphs.
e. Specify the null and alternative hypotheses you would use to determine whether the model in part **c** provides more information for predicting yield than does the model in part **b**.
f. Explain how you would obtain the quantities necessary to compute the F statistic that would be used in testing the hypotheses you described in part **e**.

12.52 An economist plans to model the mean monthly demand $E(y)$ (in thousands of units) for a product as a function of the product's price (in dollars) and the season of the year. The model proposed is

$$E(y) = \beta_0 + \beta_1 x_1 + \beta_2 x_2 + \beta_3 x_3 + \beta_4 x_4$$

```
THE REGRESSION EQUATION IS
Y =      12.1 -   1.11 X1 +    3.94 X2
    +    7.17 X3 +   3.72 X4

                              ST. DEV.    T-RATIO =
              COEFFICIENT    OF COEF.    COEF/S.D.

     --           12.067       1.473        8.19
X1   C2           -1.113        .187       -5.93
X2   C3            3.942        .858        4.59
X3   C4            7.166        .815        8.80
X4   C5            3.724        .819        4.55

THE ST. DEV. OF Y ABOUT REGRESSION LINE IS
S =         1.135
WITH ( 16 - 5) =    11 DEGREES OF FREEDOM

R-SQUARED = 93.1 PERCENT
R-SQUARED = 90.6 PERCENT, ADJUSTED FOR D.F.

ANALYSIS OF VARIANCE

 DUE TO         DF        SS     MS=SS/DF

REGRESSION       4    191.590     47.897
RESIDUAL        11     14.160      1.287
TOTAL           15    205.750
```

▲ Complete model

```
THE REGRESSION EQUATION IS
Y =    17.4 -    1.35 X1

                                  ST. DEV.    T-RATIO =
                    COEFFICIENT   OF COEF.    COEF/S.D.

       --              17.394       2.863          6.08
X1     C2              -1.348        .403         -3.34

THE ST. DEV. OF Y ABOUT REGRESSION LINE IS
S =       2.859
WITH ( 16- 2) =     14 DEGREES OF FREEDOM

R-SQUARED = 44.4 PERCENT
R-SQUARED = 40.4 PERCENT, ADJUSTED FOR D.F.

ANALYSIS OF VARIANCE

 DUE TO        DF         SS    MS=SS/DF

REGRESSION      1     91.345     91.345
RESIDUAL       14    114.405      8.172
TOTAL          15    205.750
```

▲ **Reduced model**

where

$$x_1 = \text{Price} \qquad x_2 = \begin{cases} 1 & \text{if spring} \\ 0 & \text{otherwise} \end{cases}$$

$$x_3 = \begin{cases} 1 & \text{if summer} \\ 0 & \text{otherwise} \end{cases} \qquad x_4 = \begin{cases} 1 & \text{if fall} \\ 0 & \text{otherwise} \end{cases}$$

A portion of the Minitab computer printout that results from fitting this model to a sample of 16 months of sales data selected from the last 2 years is shown on page 660. The reduced model, $E(y) = \beta_0 + \beta_1 x_1$, was also fit to the same data, and the resulting computer printout is partially reproduced above.

a. Is there sufficient evidence to conclude that mean monthly demand depends on the season of the year? Test using $\alpha = .05$.

b. Find an estimate for $E(y)$ when the price is \$8.00 and it is summer. Interpret your result. [*Note*: We prefer to use a confidence interval for the estimate but its calculation is beyond the scope of this text. This procedure can be found in the references at the end of the chapter.]

***12.53** In Exercises 11.10 and 12.36 a real estate appraiser used regression analysis to explore the relationship between the sale prices and various characteristics of apartment buildings. Some of the data are reproduced in the table on page 662, along with data on the physical condition of each building (E: excellent; G: good; F: fair).

a. Write a model that describes the relationship between sale price and number of apartment units as three parallel lines, one for each level of physical condition. Be sure to specify the dummy variable coding scheme you use.

Code	Sale Price	Apartment Units	Building Condition
0229	$ 90,300	4	F
0094	384,000	20	G
0043	157,500	5	G
0079	676,200	26	E
0134	165,000	5	G
0179	300,000	10	G
0087	108,750	4	G
0120	276,538	11	G
0246	420,000	20	G
0025	950,000	62	G
0015	560,000	26	G
0131	268,000	13	F
0172	290,000	9	E
0095	$173,200	6	G
0121	323,650	11	G
0077	162,500	5	G
0060	353,500	20	F
0174	134,400	4	E
0084	187,000	8	G
0031	155,700	4	E
0019	93,600	4	F
0074	110,000	4	G
0057	573,200	14	E
0104	79,300	4	F
0024	272,000	5	E

Source: Robinson Appraisal Co., Inc., Mankato, Minnesota.

b. Plot y against x_1 (number of apartment units) for all buildings in excellent condition. On the same graph, plot y against x_1 for all buildings in good condition. Do this again for all buildings in fair condition. Does it appear that the model you specified in part **a** is appropriate? Explain.

c. Fit the model from part **a** to the data. Report the least squares prediction equation for each of the three building condition levels.

d. Plot the three prediction equations of part **c** on a scattergram of the data.

e. Do the data provide sufficient evidence to conclude that the relationship between the mean sale price and number of units differs depending on the physical condition of the apartments? Test using $\alpha = .05$.

12.54 A company is studying three safety programs, A, B, and C, in an attempt to reduce the number of work-hours lost due to accidents. Each program is to be tried at three of the company's nine factories, and the plan is to monitor the lost work-hours, y, for one year beginning 6 months after the new safety program is instituted.

a. Write a main effects model relating $E(y)$ to the lost work-hours, x_1, the year before the plan is instituted and to the type of program that is instituted.

b. In terms of the model parameters from part **a**, what hypothesis would you test to determine whether the mean work-hours lost differ for the three safety programs?

12.55 Refer to Exercise 12.54. After the three safety programs have been in effect for 18 months, the complete main effects model is fit to the $n = 9$ data points. With safety program A as the base level, the following results are obtained:

$$\hat{y} = -2.1 + .88x_1 - 150x_2 + 35x_3 \qquad \text{SSE}_c = 1{,}527.27$$

Then the reduced model $E(y) = \beta_0 + \beta_1 x_1$ is fit, with the result

$$\hat{y} = 15.3 + .84x_1 \qquad \text{SSE}_r = 3{,}113.14$$

Test whether the mean work-hours lost differ for the three programs. Use $\alpha = .05$.

12.56 An insurance company is experimenting with three training programs, A, B, and C, for its salespeople. The following main effects model is proposed:

$$E(y) = \beta_0 + \beta_1 x_1 + \beta_2 x_2 + \beta_3 x_3$$

where

y = Monthly sales (\$1,000) x_1 = Months of experience $x_2 = \begin{cases} 1 & \text{if program B} \\ 0 & \text{otherwise} \end{cases}$ $x_3 = \begin{cases} 1 & \text{if program C} \\ 0 & \text{otherwise} \end{cases}$

Training program A is the base level.

a. What hypothesis would you test to determine whether the mean monthly sales differ for salespeople trained by the three programs?

b. After experimenting with 50 salespeople over a 5-year period, the complete model is fit, with the result

$$\hat{y} = 10 + .5x_1 + 1.2x_2 - .4x_3 \qquad \text{SSE} = 140.5$$

Then the reduced model $E(y) = \beta_0 + \beta_1 x_1$ is fit to the same data, with the result

$$\hat{y} = 11.4 + .4x_1 \qquad \text{SSE} = 183.2$$

Test the hypothesis you formulated in part **a**. Use $\alpha = .05$.

***12.57** In Exercise 11.56, an economist modeled the per capita demand for passenger car motor fuel in the United States as a function of two quantitative independent variables, personal income and the relative price of a gallon of gasoline. In this exercise we explore the relationship between per capita demand for motor fuel, y, and only one of those variables, the relative price of a gallon of gasoline, x. The data are repeated in the table.

Year	*Car Fuel (billion gals.)*	*U.S. Population (millions)*	*Relative Price of Gasoline*
1965	50.3	194.3	1.004
1966	53.31	196.6	.998
1967	55.11	198.7	1.000
1968	58.52	200.7	.973
1969	62.45	202.7	.954
1970	65.8	205.1	.908
1971	69.51	207.1	.876
1972	73.5	209.9	.859
1973	78.0	211.9	.887
1974	74.2	213.9	1.083
1975	76.5	216.0	1.060
1976	78.8	218.0	1.043
1977	80.7	220.2	1.037
1978	83.8	222.6	1.005
1979	80.2	225.1	1.222
1980	73.7	227.7	1.496
1981	71.7	230.0	1.508
1982	72.8	232.3	1.346
1983	73.4	234.5	1.261

Source: U.S. Bureau of the Census. *Statistical Abstract of the United States*, various years.

a. Construct a scattergram of the data.
b. In 1973, the Organization of Petroleum Exporting Countries (OPEC) began manipulating the supply of oil—and therefore gasoline—which subsequently rocketed prices to record heights. This accounts for the unusual pattern you should have observed in the scattergram. Propose a first-order regression model that allows for differences in the relationship between y and x before and after OPEC began manipulating oil prices. [*Note*: Allow for differences in the slopes of the two regression lines.]
c. Fit your proposed model to the data.
d. Plot the two prediction equations associated with the least squares model obtained in part **c** on the scattergram of the data constructed in part **a**. Interpret these equations in the context of the problem.
e. Do the data provide sufficient evidence to conclude that the relationship between demand for motor fuel and the relative price of gasoline changed after 1973? Test using $\alpha = .05$.
f. Is there sufficient evidence to conclude that the slopes of the demand functions prior to 1973 and after 1973 differ? Test using $\alpha = .05$.

12.7 Comparing Two or More Response Curves

Suppose we think that the relationship between mean monthly sales, $E(y)$, and advertising expenditure, x_1 (Section 12.6), is second-order. The scenario for writing the models for this situation is as follows.

1. The mean sales curves are identical for all three advertising media. That is, a single second-order curve will suffice to describe the relationship between $E(y)$ and x_1 for all the media; see Figure 12.24.

$$E(y) = \beta_0 + \beta_1 x_1 + \beta_2 x_1^2 \qquad x_1 = \text{Advertising expenditure}$$

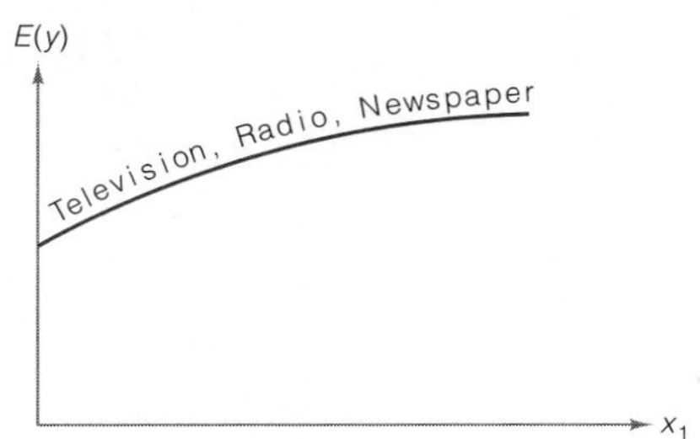

FIGURE 12.24 ► The relationship between $E(y)$ and x_1 is the same for all media

2. The response curves have the same shapes but different y-intercepts; see Figure 12.25.

$$E(y) = \beta_0 + \beta_1 x_1 + \beta_2 x_1^2 + \beta_3 x_2 + \beta_4 x_3$$

$$x_1 = \text{Advertising expenditure}$$

$$x_2 = \begin{cases} 1 & \text{if radio} \\ 0 & \text{if not} \end{cases} \qquad x_3 = \begin{cases} 1 & \text{if television} \\ 0 & \text{if not} \end{cases}$$

FIGURE 12.25 ▶
The response curves have the same shapes but different y-intercepts

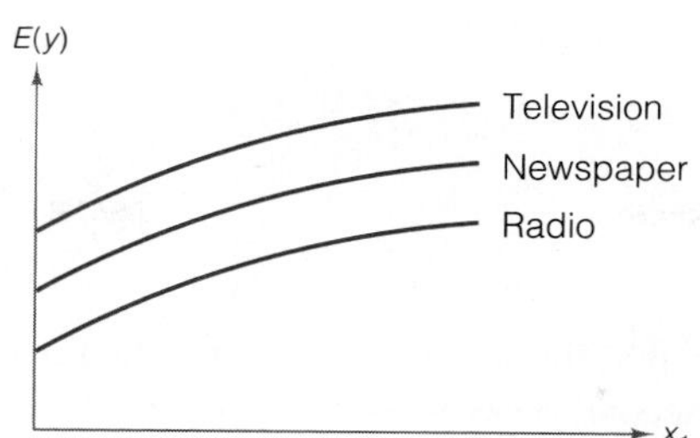

3. The response curves for the three advertising media are different—i.e., Advertising expenditure and Type of medium interact; see Figure 12.26.

$$E(y) = \beta_0 + \beta_1 x_1 + \beta_2 x_1^2 + \beta_3 x_2 + \beta_4 x_3 + \beta_5 x_1 x_2 + \beta_6 x_1 x_3 + \beta_7 x_1^2 x_2 + \beta_8 x_1^2 x_3$$

FIGURE 12.26 ▶
The response curves differ for the three media

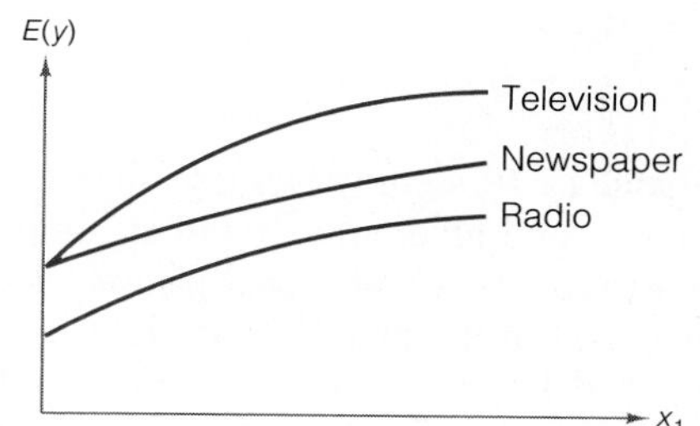

EXAMPLE 12.8

Give the equation of the second-order model for the radio advertising medium.

Solution

Model 3 characterizes the relationship between $E(y)$ and x_1 for the radio advertising medium (see the coding) when $x_2 = 1$ and $x_3 = 0$. Substituting these values into model 3, we obtain

$$\begin{aligned} E(y) &= \beta_0 + \beta_1 x_1 + \beta_2 x_1^2 + \beta_3 x_2 + \beta_4 x_3 + \beta_5 x_1 x_2 + \beta_6 x_1 x_3 + \beta_7 x_1^2 x_2 + \beta_8 x_1^2 x_3 \\ &= \beta_0 + \beta_1 x_1 + \beta_2 x_1^2 + \beta_3(1) + \beta_4(0) + \beta_5 x_1(1) + \beta_6 x_1(0) + \beta_7 x_1^2(1) + \beta_8 x_1^2(0) \\ &= (\beta_0 + \beta_3) + (\beta_1 + \beta_5)x_1 + (\beta_2 + \beta_7)x_1^2 \end{aligned}$$

EXAMPLE 12.9

What null hypothesis about the parameters of model 3 would you test if you wished to determine whether the second-order curves for the three media differ?

Solution

If the curves were identical, we would not need the independent variable Type of medium in the model; that is, we would delete all terms involving x_2 and x_3. This would produce model 1,

$$E(y) = \beta_0 + \beta_1 x_1 + \beta_2 x_1^2$$

and the null hypothesis would be

$$H_0: \ \beta_3 = \beta_4 = \beta_5 = \beta_6 = \beta_7 = \beta_8 = 0$$

EXAMPLE 12.10

Suppose we assume that the response curves for the three media differ but we want to know whether the second-order terms contribute information for the prediction of y. Or, equivalently, will a second-order model give better predictions than a first-order model?

Solution

The only difference between model 3 and a first-order model are those terms involving x_1^2. Therefore, the null hypothesis, "the second-order terms contribute no information for the prediction of y," is equivalent to

$$H_0: \ \beta_2 = \beta_7 = \beta_8 = 0$$

Examples 12.9 and 12.10 identify two tests that answer practical questions concerning a collection of second-order models. Other comparisons among the curves can be made by testing appropriate sets of model parameters (see the exercises).

The models described in the preceding sections provide only an introduction to statistical modeling. Models can be constructed to relate $E(y)$ to any number of quantitative and qualitative independent variables. You can compare response curves and surfaces for different levels of a qualitative variable or for different combinations of levels of two or more qualitative independent variables. A general explanation of how to write linear statistical models can be found in Chapter 6 of Mendenhall and Sincich (1993).

CASE STUDY 12.1 / Forecasting Peak-Hour Traffic Volume*

In designing future metropolitan roadways or redesigning existing roads, highway engineers rely heavily on traffic volume forecasts. Should the road have two, three, or four lanes in each direction? How should the on- and off-ramps be configured? Should the on-ramps be metered? Should one or more lanes be reversible to accommodate rush hour traffic? Should one or more lanes be restricted to carpools? The answers to such questions depend primarily on the traffic volumes the road must be able to handle during the peak travel times of the day—morning and evening rush hours.

Traffic forecasters at the Minnesota Department of Transportation use regression analysis to estimate weekday peak-hour traffic volumes on existing and pro-

*Personal communications from John Sem, Director; Allan E. Pint, State Traffic Forecast Engineer; and James Page, Sr., Transportation Planner, Traffic and Commodities Studies Section, Minnesota Department of Transportation, St. Paul, Minnesota.

FIGURE 12.27 ▶
Scattergram: Peak-hour traffic volume vs. total traffic volume for the day

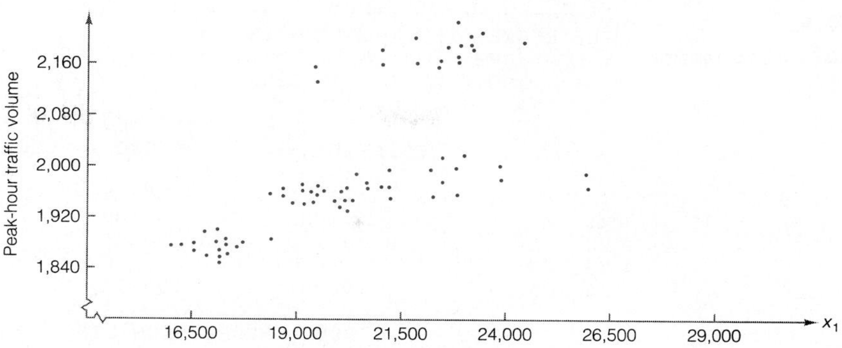

posed roadways. In particular, they model the traffic volume for the hour with the highest volume (this is the *peak hour*, typically 7–8 A.M.), y, as a function of the road's total traffic volume for the day, x_1. The model is developed using traffic data from existing roadways. Upon obtaining a forecast for the total weekday traffic volume for the road in question, the forecasters use the regression model to estimate (forecast) the mean peak-hour volume for the road.

For a recent project involving the redesign of a section of Interstate 494 in Bloomington, Minnesota, the forecasters collected data on peak-hour traffic volumes and weekday traffic volumes (using electronic sensors that count vehicles) at eight locations in the Minneapolis area believed to be similar to the one being redesigned. One of the characteristics that was common to the Interstate 494 site and the eight other locations was that the peak-hour volume at each was nearing its theoretical upper bound. A sample of $n = 72$ measurements was obtained and a scattergram of the data was constructed; see Figure 12.27.

The traffic forecasters were surprised to see the isolated group of observations at the top of the scattergram. They investigated and found that all these data points were collected on Interstate 35W at the intersection of 46th Street, on the south side of Minneapolis. It turned out that, while all locations in the sample were three-lane highways (including this location), this location was unique because the highway widens to four lanes just a short distance north of the electronic sensor. Accordingly, this location was *not* similar to the section of Interstate 494 being redesigned, as was originally thought. It was decided that the 18 measurements collected at this location should be kept in the sample (in order to maintain a large sample size for the estimation of σ^2), but that a separate peak-hour-volume response curve should be estimated for this location through the use of a dummy variable in the regression model.

Knowing that peak-hour traffic volumes have a theoretical upper bound and that the locations sampled were nearing their bounds, the forecasters hypothesized that a second-order model should be used to explain the variation in peak-hour volume. Since this hypothesis was also supported by the scattergram, they fit the following model to the $n = 72$ data points:

$$E(y) = \beta_0 + \beta_1 x_1 + \beta_2 x_1^2 + \beta_3 x_2 + \beta_4 x_1 x_2$$

where

$$x_2 = \begin{cases} 1 & \text{if Interstate 35W and 46th Street} \\ 0 & \text{otherwise} \end{cases}$$

The results are shown in the Minitab printout in Figure 12.28 (page 668). In order to investigate formally whether the relationship between peak-hour volume and total traffic volume for the day at the Interstate 35W location is different from that at the other locations, they eliminated the terms $\beta_3 x_2$ and $\beta_4 x_1 x_2$ from the model and fit the resulting reduced model to the same data, with the results shown in Figure 12.29 (page 668). From the information contained on the two printouts, the forecasters calculated

FIGURE 12.28 ► Peak-hour traffic volume: Complete model

```
THE REGRESSION EQUATION IS
Y =      779. + .104 X1 -   .0000 X2
  +      98.0 X3 +   .0029 X4

                                    ST. DEV.    T-RATIO=
                    COEFFICIENT     OF COEF.   COEF/S.D.

   --                    779.0         141.8        5.49
X1   C2                  .1039         .0136        7.63
X2   C3           -.000002220     .000000324       -6.84
X3   C4                   98.0          75.6        1.30
X4   C5                 .00291        .00332         .88

THE ST. DEV. OF Y ABOUT REGRESSION LINE IS
S =        15.47
WITH (   72- 5) =   67 DEGREES OF FREEDOM

R-SQUARED = 97.2 PERCENT
R-SQUARED = 97.0 PERCENT, ADJUSTED FOR D.F.

ANALYSIS OF VARIANCE

 DUE TO       DF          SS    MS=SS/DF

REGRESSION     4    555747.6    138936.9
RESIDUAL      67     16031.3       239.3
TOTAL         71    571778.9
```

FIGURE 12.29 ► Peak-hour traffic volume: Reduced model

```
THE REGRESSION EQUATION IS
Y =      197. +   .149 X1 -  .0000 X2
                                 ST. DEV.   T-RATIO =
                  COEFFICIENT    OF COEF.   COEF/S.D.

   --                197.4         578.9         .34
X1   C2              .1492         .0555        2.69
X2   C3         -.00000295     .00000132       -2.24

THE ST. DEV. OF Y ABOUT REGRESSION LINE IS
S =       65.45
WITH (   72- 3) = 69 DEGREES OF FREEDOM

R-SQUARED = 48.3 PERCENT
R-SQUARED = 46.8 PERCENT, ADJUSTED FOR D.F.

ANALYSIS OF VARIANCE

 DUE TO       DF        SS    MS=SS/DF

REGRESSION     2    276177     138088
RESIDUAL      69    295602       4284
TOTAL         71    571779
```

$$F = \frac{(295{,}602 - 16{,}031.3)/2}{16{,}031.3/67} = 584.2$$

and compared it to $F_{.01} \approx 4.96$, where $\nu_1 = 2$ and $\nu_2 = 67$. They concluded that the two response curves were different. In addition, the forecasters noted that β_1 was significantly greater than 0 and that β_2 was significantly less than 0—as would be expected, since peak-hour volume has an upper bound. Accordingly, for estimating and predicting peak-hour volumes at the Interstate 494 site, the forecasters set $x_2 = 0$ to obtain the prediction equation

$$\hat{y} = 779 + .1039x_1 - .00000222x_1^2$$

Exercises 12.58–12.66

Note: *Starred (*) exercises require the use of a computer.*

Learning the Mechanics

12.58 Consider an application for which you want to relate the response variable y to one quantitative variable and one qualitative variable at three levels.

a. Write a complete second-order model that relates $E(y)$ to the quantitative variable.

b. Add the main effect terms for the qualitative variable (at three levels) to the model of part **a**.

c. Add terms to the model of part **b** to allow for interaction between the quantitative and qualitative independent variables.

12.59 Refer to Exercise 12.58, part **c**.

a. Under what circumstances will the response curves of the model have the same shape but different y-intercepts?

b. Under what circumstances will the response curves of the model be parallel lines?

c. Under what circumstances will the response curves of the model be identical?

12.60 Write a model for $E(y)$ in which there are two independent variables, one quantitative and one qualitative with four levels. Construct the model so the associated response curves are second-order and the independent variables do not interact.

12.61 Minitab was used to fit the following model to $n = 25$ data points:

$$y = \beta_0 + \beta_1x_1 + \beta_2x_1^2 + \beta_3x_2 + \beta_4x_3 + \beta_5x_1x_2 + \beta_6x_1x_3 + \beta_7x_1^2x_2 + \beta_8x_1^2x_3 + \epsilon$$

where x_1 is a quantitative variable and

$$x_2 = \begin{cases} 1 & \text{if level 2} \\ 0 & \text{otherwise} \end{cases} \qquad x_3 = \begin{cases} 1 & \text{if level 3} \\ 0 & \text{otherwise} \end{cases}$$

The resulting printout is shown on page 670, where **C1** represents y, and **C2–C9** represent the eight independent variable terms in the model, in the same order as specified above.

Minitab was also used to fit the reduced model

$$y = \beta_0 + \beta_1x_1 + \beta_2x_1^2 + \epsilon$$

to the same data set. The printout is shown on page 670 also.

```
THE REGRESSION EQUATION IS
C1 = 48.8 - 3.36 C2 + 0.0749 C3 - 2.36 C4 - 7.60 C5 + 3.71 C6
         + 2.66 C7 - 0.0183 C8 - 0.0372 C9

                                    ST. DEV.      T-RATIO =
            COEFFICIENT             OF COEF.      COEF/S.D.
                 48.780               4.186          11.65
C2              -3.3618              0.4708          -7.14
C3              0.07492             0.01088           6.89
C4               -2.364               6.922          -0.34
C5               -7.595               6.102          -1.24
C6               3.7144              0.8295           4.48
C7               2.6590              0.6469           4.11
C8             -0.01826             0.02175          -0.84
C9             -0.03724             0.01453          -2.56

S = 3.190

R-SQUARED = 98.9 PERCENT
R-SQUARED = 98.4 PERCENT, ADJUSTED FOR D.F.
CONTINUE?

ANALYSIS OF VARIANCE

 DUE TO        DF            SS          MS=SS/DF
REGRESSION      8       15256.9            1907.1
RESIDUAL       16         162.9              10.2
TOTAL          24       15419.8
```

▲ **Complete model**

```
THE REGRESSION EQUATION IS
C1 = 33.9 + 0.86 C2 - 0.0110 C3

                                    ST. DEV.      T-RATIO =
            COEFFICIENT             OF COEF.      COEF/S.D.
                  33.93               20.74           1.64
C2                0.862               2.218           0.39
C3             -0.01102             0.05103          -0.22

S = 26.10

R-SQUARED =   2.8 PERCENT
R-SQUARED =    .0 PERCENT, ADJUSTED FOR D.F.

ANALYSIS OF VARIANCE

 DUE TO        DF            SS          MS=SS/DF
REGRESSION      2         433.7             216.9
RESIDUAL       22       14986.0             681.2
TOTAL          24       15419.8
```

▲ **Reduced model**

a. What is the general least squares prediction equation for the complete model?

b. What is the equation of the response curve for $E(y)$ when $x_2 = 0$ and $x_3 = 0$? When $x_2 = 1$ and $x_3 = 0$? When $x_2 = 0$ and $x_3 = 1$?

c. On the same graph, plot the least squares prediction equations associated with level 1, with level 2, and with level 3.
d. Specify the null and alternative hypotheses that should be used to investigate whether the second-order response curves for the three levels differ.
e. Conduct the hypothesis test of part **d**. Use $\alpha = .05$.

Applying the Concepts

*12.62 An operations manager is interested in modeling $E(y)$, the expected length of time per month (in hours) that a machine will be shut down for repairs as a function of the type of machine (001 or 002) and the age of the machine (in years). Data were obtained on $n = 20$ machine breakdowns and the manager has proposed the model

$$E(y) = \beta_0 + \beta_1 x_1 + \beta_2 x_1^2 + \beta_3 x_2$$

where

x_1 = Age of machine $\quad x_2 = \begin{cases} 1 & \text{if machine type 001} \\ 0 & \text{if machine type 002} \end{cases}$

Down Time (hrs./mon.)	*Machine Type*	*Machine Age (yrs.)*
10	001	1.0
20	001	2.0
30	001	2.7
40	001	4.1
9	001	1.2
25	001	2.5
19	001	1.9
41	001	5.0
22	001	2.1
12	001	1.1
10	002	2.0
20	002	4.0
30	002	5.0
44	002	8.0
9	002	2.4
25	002	5.1
20	002	3.5
42	002	7.0
20	002	4.0
13	002	2.1

a. Interpret each of the β parameter estimates.
b. Use the data to test the null hypothesis that $\beta_1 = \beta_2 = 0$. Test using $\alpha = .10$. Interpret the result.
c. Test the null hypothesis H_0: $\beta_1 = 0$. Use $\alpha = .05$ and interpret the result.

12.63 An equal rights group has charged that women are being discriminated against in terms of the salary structure in a state university system. It is thought that a complete second-order model will be adequate to describe the relationship between salary and years of experience for both groups. A sample is to be taken from the records for faculty members (all of equal rank) within the system and the following model is to be fit:

$$E(y) = \beta_0 + \beta_1 x_1 + \beta_2 x_1^2 + \beta_3 x_2 + \beta_4 x_1 x_2 + \beta_5 x_1^2 x_2$$

where

$$y = \text{Annual salary (in thousands of dollars)}$$

$$x_1 = \text{Experience (years)} \qquad x_2 = \begin{cases} 1 & \text{if female} \\ 0 & \text{if male} \end{cases}$$

[*Note*: In practice, we would include other variables in the model. We include only two here to simplify the exercise.]

a. What hypothesis would you test to determine whether the *rate* of increase of mean salary with experience is different for males and females?

b. What hypothesis would you test to determine whether there are differences in mean salaries that are attributable to sex?

12.64 Refer to Exercise 12.16, where we presented data on the number of highway deaths and the number of licensed vehicles on the road for the years 1956–1991. We mentioned that the number of deaths y may also have been affected by the existence of the national 55-mile-per-hour speed limit during the years 1974–1987 (i.e., years 19–32). Define the dummy variable

$$x_2 = \begin{cases} 1 & \text{if 55-mile-per-hour speed limit was in effect} \\ 0 & \text{if not} \end{cases}$$

a. Introduce the variable x_2 into the second-order model of Exercise 12.16 to account for the presence or absence of the 55-mile-per-hour speed limit in a given year. Include terms involving the interaction between x_2 and x_1.

b. Refer to your model for part **a**. Sketch on a single piece of graph paper your visualization of the two response curves, the second-order curves relating y to x_1 before and after the imposition of the 55-mile-per-hour speed limit.

c. Suppose that x_1 and x_2 do not interact. How would that affect the graphs of the two response curves of part **b**?

d. Refer to part **c**. Suppose that x_1 and x_2 do interact. How would this affect the graphs of the two response curves?

12.65 In Exercise 12.16, we fit a second-order model to data relating the number y of U.S. highway deaths per year to the number x_1 of licensed vehicles on the road. In Exercise 12.64 we added a qualitative variable x_2 to account for the presence or absence of the 55-mile-per-hour national speed limit. The accompanying SAS computer printout gives the results of fitting the model

$$E(y) = \beta_0 + \beta_1 x_1 + \beta_2 x_1^2 + \beta_3 x_2 + \beta_4 x_1 x_2 + \beta_5 x_1^2 x_2$$

to the data. Use the printout and the printout for Exercise 12.16 to determine whether the data provide sufficient evidence to indicate that the qualitative variable (speed limit) contributes information for the prediction of the annual number of highway deaths. Test using $\alpha = .05$. Discuss the practical implications of your test results.

```
Model: MODEL1
Dependent Variable: Y

                         Analysis of Variance

                          Sum of          Mean
Source          DF       Squares        Square       F Value       Prob>F

Model            5     977.80898     195.56180        33.130       0.0001
Error           30     177.08741       5.90291
C Total         35    1154.89639

    Root MSE       2.42959     R-square       0.8467
    Dep Mean      47.58056     Adj R-sq       0.8211
    C.V.           5.10627

                         Parameter Estimates

                    Parameter        Standard     T for H0:
Variable  DF         Estimate           Error    Parameter=0    Prob > |T|

INTERCEP   1       -21.890972      5.56165987         -3.936        0.0005
X1         1         1.146053      0.09420319         12.166        0.0001
X1SQ       1        -0.004108      0.00035254        -11.652        0.0001
X2         1       -93.369172     75.75677533         -1.232        0.2273
X1X2       1         0.935585      0.95870236          0.976        0.3369
X1SQX2     1        -0.002450      0.00301237         -0.813        0.4224
```

***12.66** *Productivity* is typically evaluated by dividing a measure of system output by a measure of the inputs to the system. Some examples of productivity measures are Sales/Salesperson, (Yards of carpet laid)/(Number of carpet layers), Shipments/[(Direct labor) + (Indirect labor) + (Materials)]. Notice that productivity can be improved by producing greater output with the same inputs or by producing the same output with fewer inputs. In manufacturing operations, productivity ratios generally vary with the volume of output produced (Schroeder, 1993). The production data in the table have been collected for a random sample of months at the three regional plants of a manufacturing firm. Each plant manufactures the same product.

North Plant		*South Plant*		*West Plant*	
Productivity Ratio	*Output*	*Productivity Ratio*	*Output*	*Productivity Ratio*	*Output*
1.30	1,000	1.43	1,015	1.61	501
.90	400	1.50	925	.74	140
1.21	650	.91	150	1.19	303
.75	200	.99	222	1.88	930
1.32	850	1.33	545	1.72	776
1.29	600	1.15	402	1.39	400
1.18	756	1.51	709	1.86	810
1.10	500	1.01	176	.99	220
1.26	925	1.24	392	.79	160
.93	300	1.49	699	1.59	626
.81	258	1.37	800	1.82	640
1.12	590	1.39	660	.91	190

a. Construct a scattergram for the data. Plot the data from the north plant using dots, from the south plant using small circles, and from the west plant using small triangles.

b. Visually fit each plant's response curve to the scattergram.

c. Based on the results of part **b**, propose a second-order regression model that could be used to estimate the relationship between productivity and volume for the three plants.

d. Fit the model you proposed in part **c** to the data.

e. Is there sufficient evidence to conclude that the productivity response curves for the three plants differ? Test using $\alpha = .05$.

f. Do the data provide sufficient evidence to conclude that the second-order model contributes more information for the prediction of productivity than a first-order model? Test using $\alpha = .05$.

g. Next month, 890 units are scheduled to be produced at the west plant. Use the model you developed in part **d** to predict next month's productivity ratio at the west plant.

12.8 Model Building: Stepwise Regression

The problem of predicting executive salaries was discussed in Chapter 11. Perhaps the biggest problem in building a model to describe executive salaries is choosing the important independent variables to be included in the model. The list of potentially important independent variables is extremely long, and we need some objective method of screening out those that are not important.

The problem of deciding which of a large set of independent variables to include in a model is common. Trying to determine which variables influence the profit of a firm, affect product quality, or are related to the state of the economy are only a few examples.

A systematic approach to building a model with a large number of independent variables is difficult because the interpretation of multivariable interactions and higher-order polynomials is tedious. We therefore turn to a screening procedure known as **stepwise regression**.

The most commonly used stepwise regression procedure, available in most popular computer packages, works as follows: The user first identifies the response, y, and the set of potentially important independent variables, $x_1, x_2, \ldots, x_k$, where k will generally be large. (Note that this set of variables could represent both first- and higher-order terms, as well as any interaction terms that might be important information contributors.) The response and independent variables are then entered into the computer, and the stepwise procedure begins.

Step 1 The computer fits all possible one-variable models of the form

$$E(y) = \beta_0 + \beta_1 x_i$$

to the data. For each model, the test of the hypotheses

$$H_0\colon\ \beta_1 = 0 \qquad H_a\colon\ \beta_1 \neq 0$$

is conducted using the t (or the equivalent F) test for a single β parameter. The independent variable that produces the largest (absolute) t value is declared the best one-variable predictor of y. Call this independent variable x_1.

Step 2 The stepwise program now begins to search through the remaining $(k - 1)$ independent variables for the best two-variable model of the form

$$E(y) = \beta_0 + \beta_1 x_1 + \beta_2 x_i$$

This is done by fitting all two-variable models containing x_1 and each of the other $(k - 1)$ options for the second variable x_i. The t values for the test H_0: $\beta_2 = 0$ are computed for each of the $(k - 1)$ models (corresponding to the remaining independent variables x_i, $i = 2, 3, \ldots, k$), and the variable having the largest t is retained. Call this variable x_2.

At this point, some computer packages diverge in methodology. The better packages now go back and check the t value of $\hat{\beta}_1$ *after* $\hat{\beta}_2 x_2$ *has been added to the model*. If the t value has become nonsignificant at some specified α level (say $\alpha = .10$), the variable x_1 is removed and a search is made for the independent variable with a β parameter that will yield the most significant t value in the presence of $\hat{\beta}_2 x_2$. Other packages do not recheck the significance of $\hat{\beta}_1$, but proceed directly to step 3.

The best-fitting plane may yield a different value for $\hat{\beta}_1$ than that obtained in step 1, because $\hat{\beta}_1$ and $\hat{\beta}_2$ may be correlated. Thus, both the value of $\hat{\beta}_1$ and, therefore, its significance will usually change from step 1 to step 2. For this reason, the computer packages that recheck the t values at each step are preferred.

Step 3 The stepwise procedure now checks for a third independent variable to include in the model with x_1 and x_2. That is, we seek the best model of the form

$$E(y) = \beta_0 + \beta_1 x_1 + \beta_2 x_2 + \beta_3 x_i$$

To do this, we fit all the $(k - 2)$ models using x_1, x_2, and each of the $(k - 2)$ remaining variables, x_i, as a possible x_3. The criterion is again to include the independent variable with the largest t value. Call this best third variable x_3.

The better programs now recheck the t values corresponding to the x_1 and x_2 coefficients, replacing the variables where values have become nonsignificant. This procedure is continued until no additional independent variables can be found that yield significant t values (at the specified α level) in the presence of the variables already in the model.

The result of the stepwise procedure is a model containing only those terms with t values that are significant at the specified α level. Thus, in most practical situations, only several of the large number of independent variables will remain. However, it is very important *not* to jump to the conclusion that all the independent variables important for predicting y have been identified or that the unimportant independent variables have been eliminated. Remember, the stepwise procedure is using only *sample estimates* of the true model coefficients (β's) to select the important variables. An extremely large number of single β parameter t tests have been conducted, and the probability is very high that one or more errors have been made in including or excluding variables. That is, we have very probably included some unimportant independent variables in the model (Type I errors) and eliminated some important ones (Type II errors).

There is a second reason why we might not have arrived at a good model. When we choose the variables to be included in the stepwise regression, we may often omit high-order terms (to keep the number of variables manageable). Consequently, we may have initially omitted several important terms from the model. Thus, we should recognize stepwise regression for what it is—an objective screening procedure.

Now, we consider interactions and quadratic terms (for quantitative variables) among variables screened by the stepwise procedure. It would be best to develop this response surface model with a second set of data independent of that used for the screening, so the results of the stepwise procedure can be partially verified with new data. However, this is not always possible because in many business modeling situations only a small amount of data is available.

Remember, do not be deceived by the impressive looking t values that result from the stepwise procedure—it has retained only the independent variables with the largest t values. Also, if you have used a main effects model for your stepwise procedure, remember that it may be greatly improved by the addition of interaction and quadratic terms.

EXAMPLE 12.11

In Section 11.7, we fit a multiple regression model for executive salaries as a function of experience, education, sex, and other factors. A preliminary step in the construction of this model was the determination of the most important independent variables. Ten independent variables were considered, as shown in Table 12.4. It would be very difficult to construct a second-order model with 10 independent variables. Therefore, use the sample of 100 executives from Section 11.7 to decide which of the 10 variables should be included in the construction of the final model for executive salaries.

TABLE 12.4 Independent Variables in the Executive Salary Example

Independent Variable	Description	Type
x_1	Experience (years)	Quantitative
x_2	Education (years)	Quantitative
x_3	Sex (1 if male, 0 if female)	Qualitative
x_4	Number of employees supervised	Quantitative
x_5	Corporate assets (millions of dollars)	Quantitative
x_6	Board member (1 if yes, 0 if no)	Qualitative
x_7	Age (years)	Quantitative
x_8	Company profits (past 12 months, millions of dollars)	Quantitative
x_9	Has international responsibility (1 if yes, 0 if no)	Qualitative
x_{10}	Company's total sales (past 12 months, millions of dollars)	Quantitative

Solution

We use stepwise regression with the main effects of the 10 independent variables to identify the most important variables. The dependent variable y is the natural logarithm of the executive salaries. The SAS stepwise regression printout is shown in Figure 12.30 (pages 677–679). Note that the first variable included in the model is x_4,

```
STEP 1
    VARIABLE X4 ENTERED                  R-SQUARE = 0.42071677

                              DF          SUM OF SQUARES  MEAN SQUARE        F   PROB>F

      REGRESSION               1             11.46854285  11.46864285    71.17   0.0001
      ERROR                   98             15.79112802   0.16113396
      TOTAL                   99             27.25977087

                                B VALUE        STD ERROR                     F   PROB>F

      INTERCEPT              10.20077500
      X4 (EMPLOYEES SUPERVISED) 0.00057284      0.00006790                 71.17   0.0001

STEP 2
    VARIABLE X5 ENTERED                  R-SQUARE = 0.78299675

                              DF          SUM OF SQUARES  MEAN SQUARE        F   PROB>F

      REGRESSION               2             21.34431198  10.67215599   175.00   0.0001
      ERROR                   97              5.91545889   0.06098411
      TOTAL                   99             27.25977087

                                B VALUE        STD ERROR                     F   PROB>F

      INTERCEPT               9.87702903
      X4 (EMPLOYEES SUPERVISED) 0.00058353      0.00004178                195.06   0.0001
      X5 (ASSETS)             0.00183730      0.00014438                161.94   0.0001

STEP 3
    VARIABLE X4 ENTERED                  R-SQUARE = 0.89667614

                              DF          SUM OF SQUARES  MEAN SQUARE        F   PROB>F

      REGRESSION               3             24.44318616   8.14772872   277.71   0.0001
      ERROR                   96              2.81658471   0.02933942
      TOTAL                   99             27.25977087
```

FIGURE 12.30 ▲
SAS stepwise regression computer printout for Example 12.11

```
                                B VALUE        STD ERROR                     F   PROB>F

INTERCEPT                      9.66449288
X1 (EXPERIENCE)                0.01870784     0.00182032                105.62   0.0001
X4 (EMPLOYEES SUPERVISED)      0.00055251     0.00002914                359.59   0.0001
X5 (ASSETS)                    0.00191195     0.00010041                362.60   0.0001

STEP 4
   VARIABLE X3 ENTERED         R-SQUARE = 0.94815717

                              DF           SUM OF SQUARES   MEAN SQUARE      F   PROB>F

REGRESSION                     4              25.84654710    6.46163678  434.37   0.0001
ERROR                         95               1.41322377    0.01487604
TOTAL                         99              27.25977087

                                B VALUE        STD ERROR                     F   PROB>F

INTERCEPT                      9.40077349
X1 (EXPERIENCE)                0.02074868     0.00131310                249.68   0.0001
X3 (SEX)                       0.30011726     0.03089939                 94.34   0.0001
X4 (EMPLOYEES SUPERVISED)      0.00055288     0.00002075                710.15   0.0001
X5 (ASSETS)                    0.00190876     0.00007150                712.74   0.0001

STEP 5
   VARIABLE X2 ENTERED         R-SQUARE = 0.96039323

                              DF           SUM OF SQUARES   MEAN SQUARE      F   PROB>F

REGRESSION                     5              26.18009940    5.23601988  455.87   0.0001
ERROR                         94               1.07967147    0.01148587
TOTAL                         99              27.25977087

                                B VALUE        STD ERROR                     F   PROB>F

INTERCEPT                      8.85387930
X1 (EXPERIENCE)                0.02141724     0.00116047                340.61   0.0001
X2 (EDUCATION)                 0.03315807     0.00615303                 29.04   0.0001
X3 (SEX)                       0.31927842     0.02738298                135.95   0.0001
X4 (EMPLOYEES SUPERVISED)      0.00056061     0.00001829                939.84   0.0001
X5 (ASSETS)                    0.00193684     0.00006304                943.98   0.0001
```

FIGURE 12.30 ▲
(continued)

```
STEP 6
    VARIABLE X6 ENTERED                         R-SQUARE = 0.96100666

                                      DF           SUM OF SQUARES  MEAN SQUARE         F   PROB>F

            REGRESSION                 6              26.19682148   4.36613691    382.00   0.0001
            ERROR                     93               1.06294939   0.01142956
            TOTAL                     99              27.25977087

                                         B VALUE       STD ERROR                       F   PROB>F

            INTERCEPT                 8.87509152
            X1 (EXPERIENCE)           0.02133460      0.00115963                  338.48   0.0001
            X2 (EDUCATION)            0.03272195      0.00614851                   28.32   0.0001
            X3 (SEX)                  0.31093801      0.02817264                  121.81   0.0001
            X4 (EMPLOYEES SUPERVISED) 0.00055820      0.00001835                  925.32   0.0001
            X5 (ASSETS)               0.00193764      0.00006289                  949.31   0.0001
            X6 (BOARD)                0.03866226      0.03196369                    1.46   0.2295

STEP 7
    VARIABLE X6 REMOVED                         R-SQUARE = 0.96039323

                                      DF           SUM OF SQUARES  MEAN SQUARE         F   PROB>F

            REGRESSION                 5              26.18009940   5.23601988    455.87   0.0001
            ERROR                     94               1.07967147   0.01148587
            TOTAL                     99              27.25977087

                                         B VALUE       STD ERROR                       F   PROB>F

            INTERCEPT                 8.85387930
             X1 (EXPERIENCE)          0.02141724      0.00116047                  340.61   0.0001
             X2 (EDUCATION)           0.03315807      0.00615303                   29.04   0.0001
             X3 (SEX)                 0.31927842      0.02738298                  135.95   0.0001
             X4 (EMPLOYEES SUPERVISED) 0.00056061     0.00001829                  939.84   0.0001
             X5 (ASSETS)              0.00193684      0.00006304                  943.98   0.0001
```

FIGURE 12.30 ▲ (continued)

Number of employees supervised. At the second step x_5, Corporate assets, enters the model. At the sixth step x_6, a dummy variable for the qualitative variable Board member or not, is brought into the model. However, because the significance (.2295) of the F statistic (SAS uses the $F = t^2$ statistic in the stepwise procedure rather than the t statistic) for x_6 is greater than the preassigned $\alpha = .10$, x_6 is removed from the model. Thus, at step 7 the procedure indicates that the five-variable model including x_1, x_2, x_3, x_4, and x_5 is best. That is, none of the other independent variables can meet the $\alpha = .10$ criterion for admission to the model.

Thus, in our final modeling effort (Section 11.7) we concentrated on these five independent variables, and determined that several second-order terms were important in the prediction of executive salaries.

CASE STUDY 12.2 / A Statistical Model for Land Appraisal

New factors and a lack of knowledge about the importance of factors that affect value continue to complicate the job of the rural appraiser. In order to provide knowledge on the subject, this article reports on and evaluates a study in which multiple linear regression equations were used to evaluate and quantify factors affecting value. . . . It is believed that the findings obtained with these equations, and the relationships they indicate, will be of value to the appraiser.

The authors of this statement, James O. Wise and H. Jackson Dover (1974), use stepwise regression to identify a number of important factors (variables) that can be used to predict rural property values. They obtained their results by analyzing a sample of 105 cases from seven counties in Georgia. Part of their findings are duplicated in Table 12.5. The variable names are listed in the order in which the stepwise regression procedure identified their importance, and the t values found at each step are given for each variable. Note that both qualitative and quantitative variables were included. Since each qualitative variable is at two levels, only one main effect term could be included in the model for each factor.

TABLE 12.5 Stepwise Regression Analysis of Price per Acre

Variable Name	t Value
Residential land (yes–no)	10.466
Seedlings and saplings (number)	6.692
Percent ponds (percent)	4.141
Distance to state park (miles)	3.985
Branches or springs (yes–no)	3.855
Site index (ratio)	3.160
Size (acres)	1.142
Farmland (yes–no)	2.288

Since there were 105 cases in the study, a large number of degrees of freedom are associated with each t statistic (first 103, then 102, etc.). Thus, we should compare the value of the test statistic to a corresponding z value (1.645 for $\alpha = .10$ and the two-sided alternative hypothesis H_a: $\beta_i \neq 0$) when we judge the importance of each variable. Although Wise and Dover imply that the variable Size is important, we might not include it, since the t value is only 1.142.

Summary

Although this chapter provides only an introduction to the very important topic of **model building**, it enables you to construct many interesting and useful models for business phenomena. You can build on this foundation and, with experience, develop competence in this fascinating area of statistics. Successful model building requires a delicate blend of knowledge of the phenomenon being modeled, geometry, and formal statistical testing.

The first step in model building is to identify the response variable y and a set of independent variables. Each independent variable is then classified as either **quantitative** or **qualitative**, and **dummy variables** are defined to represent the qualitative independent variables. If the total number of independent variables is large, you may want to use **stepwise regression** to screen out those that do not seem important for the prediction of y.

When the number of independent variables is manageable, the model builder is ready to begin a systematic effort. At least **second-order models**, those containing **two-way interactions** and **quadratic terms** in the quantitative variables, should be considered. Remember that a model with no interaction terms implies that each of the independent variables affects the response independently of the other independent variables. Quadratic terms add curvature to the contour lines when $E(y)$ is plotted as a function of the independent variable. The F test for testing a set of β parameters aids in deciding the final form of the prediction model.

Many problems can arise in building a regression model, and the process is often tedious and frustrating. However, the end result of a careful and determined modeling effort is very rewarding—you will have a better understanding of the independent variables that influence the dependent variable y, and you will have a predictive model for y.

Supplementary Exercises 12.67–12.82

Note: Starred (*) exercises require the use of a computer.

12.67 Investors are interested in knowing the relationship between the behavior of a mutual fund and the behavior of the stock market as a whole. Researchers in finance have hypothesized that the model that appropriately characterizes this relationship is

$$E(y) = \beta_0 + \beta_1 x$$

where

y = Monthly rate of return of a mutual fund

x = Monthly rate of return of the stock market as a whole as measured by the monthly rate of return to a market index such as Standard & Poor's 500 Composite Index

The value of β_1 in the model is referred to as the mutual fund's *beta coefficient*. Assuming the preceding model is true, investors can predict how the returns of an individual mutual fund will react to changes in the behavior of the market. For example, if $\beta_1 > 1$, the implication is that the return to the mutual fund will be greatly influenced by the behavior of the market and will move in the same direction as the change in the market return. If $0 \leq \beta_1 < 1$, the return to the mutual fund will be less sensitive to changes in market behavior but will also move in the same direction as the change in the market return.

In studying mutual funds, Alexander and Stover (1980) included a dummy variable in the above model to determine whether the beta coefficient for an individual mutual fund depends on whether the market is moving generally upward (a *bull market*) or generally downward (a *bear market*).

a. Modify the regression model (as Alexander and Stover did) to reflect the possibility that $E(y)$ may depend on whether the market is bullish or bearish. Include an interaction term in your model and carefully define the dummy variable coding scheme you use.

b. Using the model you developed in part **a**, describe the differences that may exist between the response curves of $E(y)$ under bull and bear markets.

c. Specify the hypothesis you would test to determine whether a mutual fund's beta coefficient is different during bull and bear markets.

d. Specify the hypothesis you would test to determine whether $E(y)$ should be characterized as $E(y) = \beta_0 + \beta_1 x$ or as in the modified model you developed in part **a**.

12.68 The audience for a product's advertising can be divided into four segments according to the degree of exposure received as a result of the advertising. These segments are groups of consumers who receive very high (VH), high (H), medium (M), or low (L) exposure to the advertising. A company is interested in exploring whether its advertising effort affects its product's market share. Accordingly, the company identifies 24 sample groups of consumers who have been exposed to its advertising, six groups at each exposure level. Then, the company determines its product's market share within each group.

a. Write a regression model that expresses the company's market share as a function of advertising exposure level. Define all terms in your model, and list any assumptions you make about them.

b. Did you include interaction terms in your model? Why or why not?

c. How many degrees of freedom are associated with the F test you would use to test the overall usefulness of the model you constructed in part **a**?

12.69 One of the distinguishing features of banks as compared with other financial service institutions is their well-developed system for permitting convenient, personal access to the services they offer. One indication of the demand for these services is the number of trips individuals make to their bank each year. Murphy and Stock (1983) employed multiple regression analysis to investigate the determinants of household trips to the bank in the state of Oklahoma. In the summer of 1979, personal interviews were conducted with a random sample of 597 residents of Oklahoma. The interviewees were asked questions about their yearly banking activities, including number of trips to the bank per year and number of miles to the bank. At the same time, data were collected on other variables, such as number of cars in the household and kind of job (if any) held by the interviewee. The method of least squares was used to develop the following model:

$$\begin{aligned}\hat{y} = 18.40 &+ \underset{(3.696)}{2.02x_1} - \underset{(-3.788)}{.254x_2} + \underset{(1.637)}{1.65x_3} + \underset{(1.138)}{1.124x_4} - \underset{(-.63)}{.06x_5} + \underset{(2.15)}{.00017x_6} - \underset{(-3.052)}{8.679x_7} - \underset{(-.958)}{3.21x_8} - \underset{(-.407)}{1.29x_9} - \underset{(-2.705)}{11.59x_{10}} + \underset{(3.239)}{.092x_{11}}\end{aligned}$$

$R^2 = .1836 \qquad F = 11.96$

where

y = Number of trips to the bank per year

x_1 = Number of people in household x_2 = Miles to bank

x_3 = Number of cars in household x_4 = Years of education

x_5 = Miles to work x_6 = Total family income

$x_7 = \begin{cases} 1 & \text{if not employed} \\ 0 & \text{otherwise} \end{cases}$ $x_8 = \begin{cases} 1 & \text{if white-collar job} \\ 0 & \text{otherwise} \end{cases}$

$x_9 = \begin{cases} 1 & \text{if blue-collar job} \\ 0 & \text{otherwise} \end{cases}$ $x_{10} = \begin{cases} 1 & \text{if farm-related job} \\ 0 & \text{otherwise} \end{cases}$

x_{11} = Number of shopping trips per year for purposes other than banking

The numbers in parentheses are the t statistics associated with the coefficient estimates above them.

a. Identify which independent variables in this model are qualitative and which are quantitative.

b. Murphy and Stock use four dummy variables to describe the kind of work done. How many different levels can the variable Kind of work assume? List them.

c. Interpret the value of R^2 in the context of the problem.

d. Test the usefulness of this model for explaining the variation in y. Use $\alpha = .01$.

e. Specify the null and alternative hypotheses you would use to test whether the trips-to-the-bank response surface is the same regardless of the kind of work done by the interviewee.

12.70 A fast-food restaurant chain is interested in modeling the mean weekly sales of a restaurant, $E(y)$, as a function of the weekly traffic flow on the street where the restaurant is located and the city in which the restaurant is located. The table contains data that were collected on 24 restaurants in four cities. The model that has been proposed is

$$E(y) = \beta_0 + \beta_1 x_1 + \beta_2 x_2 + \beta_3 x_3 + \beta_4 x_4$$

where

x_1 = Traffic flow

$x_2 = \begin{cases} 1 & \text{if city 1} \\ 0 & \text{otherwise} \end{cases}$ $x_3 = \begin{cases} 1 & \text{if city 2} \\ 0 & \text{otherwise} \end{cases}$ $x_4 = \begin{cases} 1 & \text{if city 3} \\ 0 & \text{otherwise} \end{cases}$

City	*Traffic* x_1 *(1,000s)*	*Weekly Sales* y *($1,000s)*	*City*	*Traffic* x_1 *(1,000s)*	*Weekly Sales* y *($1,000s)*
1	59.3	6.3	3	75.8	8.2
1	60.3	6.6	3	48.3	5.0
1	82.1	7.6	3	41.4	3.9
1	32.3	3.0	3	52.5	5.4
1	98.0	9.5	3	41.0	4.1
1	54.1	5.9	3	29.6	3.1
1	54.4	6.1	3	49.5	5.4
1	51.3	5.0	4	73.1	8.4
1	36.7	3.6	4	81.3	9.5
2	23.6	2.8	4	72.4	8.7
2	57.6	6.7	4	88.4	10.6
2	44.6	5.2	4	23.2	3.3

A portion of the computer printout that results from fitting the model to the data is shown.

```
Dep Variable: Y

                          Analysis of Variance

                        Sum of          Mean
Source        DF       Squares        Square        F Value          Prob>F

Model          4     116.65552      29.16388        222.173          0.0001
Error         19       2.49407       0.13127
C Total       23     119.14958

     Root MSE        0.36231      R-Square        0.9791
     Dep Mean        5.99583      Adj R-Sq        0.9747
     C.V.            6.04265

                          Parameter Estimates

                      Parameter        Standard     T for H0:
Variable    DF         Estimate           Error    Parameter=0      Prob > |T|

INTERCEP     1         1.083388      0.32100795          3.375          0.0032
X1           1         0.103673      0.00409449         25.320          0.0001
X2           1        -1.215762      0.20538681         -5.919          0.0001
X3           1        -0.530757      0.28481946         -1.863          0.0779
X4           1        -1.076525      0.22650014         -4.753          0.0001
```

The reduced model, $E(y) = \beta_0 + \beta_1 x_1$, was also fit to the same data and the resulting computer printout is partially reproduced here.

```
SOURCE              DF  SUM OF SQUARES  MEAN SQUARE
MODEL                1        111.3423     111.3423
ERROR               22          7.8073       0.3549
CORRECTED TOTAL     23        119.1496     R-SQUARE
                                              0.934
```

a. Test the null hypothesis that $\beta_1 = \beta_2 = \beta_3 = \beta_4 = 0$ using $\alpha = .05$. Interpret the results of your test.
b. Is mean weekly sales $E(y)$ dependent on the city where a restaurant is located? Test using $\alpha = .05$. Interpret the results of your test.
c. Describe the nature of the response lines that the complete model, $E(y) = \beta_0 + \beta_1 x_1 + \beta_2 x_2 + \beta_3 x_3 + \beta_4 x_4$, would generate. Does the model imply interaction between city and traffic flow?
d. Use the prediction equation based on the complete model to graph the response lines that relate predicted weekly sales $\hat{y}$ to traffic flow x_1 (for each of the four cities). Do the graphed response lines suggest an interaction between city and traffic flow?
e. Write a model that includes interaction between city and car traffic flow.
*f. Fit the model of part **e** to the data.
g. Do the data provide sufficient evidence to indicate that the slopes of the lines differ? Test using $\alpha = .05$.

12.71 One factor that must be considered in developing a shipping system that is beneficial to both the customer and the seller is time of delivery. A manufacturer of farm equipment can ship its products by either rail or truck. Quadratic models are thought to be adequate in relating time of delivery to distance traveled for both modes of transportation. Consequently, it has been suggested that the model to be fit is

$$E(y) = \beta_0 + \beta_1 x_1 + \beta_2 x_1^2 + \beta_3 x_2 + \beta_4 x_1 x_2 + \beta_5 x_1^2 x_2$$

where

y = Shipping time

x_1 = Distance to be shipped $\qquad x_2 = \begin{cases} 1 & \text{if rail} \\ 0 & \text{if truck} \end{cases}$

a. What hypothesis would you test to determine whether the data indicate that the quadratic distance terms are useful in the model—i.e., whether curvature is present in the relationship between mean delivery time and distance?

b. What hypothesis would you test to determine whether there is a difference in mean delivery time by rail and by truck?

12.72 Refer to Exercise 12.71. Suppose the model is fit to a total of 50 observations on delivery time. The sum of squared errors is SSE = 226.12. Then, the reduced model

$$E(y) = \beta_0 + \beta_1 x_1 + \beta_2 x_1^2$$

is fit to the same data, and SSE = 259.34. Test whether the data indicate that the mean delivery time differs from rail and truck deliveries. Use $\alpha = .05$.

***12.73** Refer to Exercise 12.68, where a company was examining the relationship between its market share and its advertising effort. The data in the table were obtained by the company.

Market Share Within Group	*Exposure Level*	*Market Share Within Group*	*Exposure Level*	*Market Share Within Group*	*Exposure Level*
10.1	L	11.2	M	11.9	H
10.3	L	10.9	M	12.9	H
10.0	L	10.8	M	10.7	VH
10.3	L	11.0	M	10.8	VH
10.2	L	12.2	H	11.0	VH
10.5	L	12.1	H	10.5	VH
10.6	M	11.8	H	10.8	VH
11.0	M	12.6	H	10.6	VH

a. Fit the model you constructed in part **a** of Exercise 12.68 to the data.

b. Is there evidence to suggest that the firm's expected market share differs for different levels of advertising exposure? Test using $\alpha = .05$.

12.74 To make a product more appealing to the consumer, an automobile manufacturer is experimenting with a new type of paint that is supposed to help the car maintain its new-car look. The durability of this paint depends on the length of time the car body is in the oven after its has been painted. In the initial experiment, three groups of ten car bodies each were baked for three different lengths of time—12, 24, and 36 hours—at the standard temperature setting. Then, the paint finish of each of the 30 cars was analyzed to determine a durability rating y.

a. Write a second-order model relating the mean durability $E(y)$ to the length of baking.

b. Could a third-order model be fit to the data? Explain.

12.75 Refer to Exercise 12.74. Suppose the Research and Development Department develops three new types of paint to be tested. Thus, 90 cars are to be tested—30 for each type of paint—in the manner described in Exercise 12.74. Write a model that describes $E(y)$ as a function of the type of paint and bake time. Assume that the independent variables interact.

***12.76** In Exercise 12.67, we discussed the relationship between the behavior of an individual mutual fund and the behavior of the stock market as a whole. The table lists the monthly rates of return for the Dreyfus Fund (a mutual fund) and the monthly rates of return for Standard & Poor's 500 Composite Index (S&P) for the period January 1966 to December 1971. The bear market periods were from January 1966 through September 1966 and from December 1968 through May 1970. The bull market periods were from October 1966 through November 1968 and from June 1970 through December 1971 (Alexander and Stover, 1980).

Month	*Dreyfus*	S&P	*Month*	*Dreyfus*	S&P	*Month*	*Dreyfus*	S&P	*Month*	*Dreyfus*	S&P
1/66	.008	.006	7/67	.073	.047	1/69	−.001	−.007	7/70	.051	.075
2/66	.067	−.013	8/67	−.019	−.007	2/69	−.070	−.043	8/70	.051	.051
3/66	−.008	−.021	9/67	.010	.034	3/69	.015	.036	9/70	.047	.035
4/66	.021	.022	10/67	−.029	−.028	4/69	.014	.023	10/70	−.026	−.010
5/66	−.074	−.049	11/67	.011	.007	5/69	−.003	.003	11/70	.040	.054
6/66	.024	−.015	12/67	.025	.028	6/69	−.066	−.054	12/70	.046	.058
7/66	−.011	−.012	1/68	−.063	−.043	7/69	−.059	−.059	1/71	.044	.042
8/66	.099	−.073	2/68	−.042	−.026	8/69	.057	.045	2/71	.025	.014
9/66	−.011	−.005	3/68	.022	.011	9/69	−.001	−.024	3/71	.031	.038
10/66	.015	.049	4/68	.100	.083	10/69	.050	.046	4/71	.035	.038
11/66	.079	.010	5/68	.021	.016	11/69	−.027	−.030	5/71	−.034	−.037
12/66	.008	.000	6/68	.001	.011	12/69	−.027	−.018	6/71	−.008	.002
1/67	.086	.080	7/68	−.038	−.017	1/70	−.083	−.074	7/71	−.026	−.040
2/67	.010	.007	8/68	.021	.016	2/70	−.044	.059	8/71	.045	.041
3/67	.041	.041	9/68	.056	.040	3/70	−.007	.003	9/71	−.028	−.006
4/67	.048	.044	10/68	.010	.009	4/70	−.110	−.089	10/71	−.039	−.040
5/67	−.039	−.048	11/68	.062	.053	5/70	−.060	−.055	11/71	.019	.003
6/67	.027	.019	12/68	−.025	−.040	6/70	−.042	−.048	12/71	.075	.090

Sources: Standard & Poor's Composite Index returns from Ibbotson, R. G., and Singuefield, R. A. *Stocks, Bonds, Bills, and Inflation: The Past (1926–1976) and the Future (1977–2000)*. Financial Analysts Research Foundation, 1977. Dreyfus returns from *The Wall Street Journal*.

a. Fit the model you developed in part **a** of Exercise 12.67 to the data shown in the table.

b. Using the fitted model, estimate the Dreyfus Fund's beta coefficient for bull markets. Estimate the corresponding parameter for bear markets. Describe the relative responsiveness of the mutual fund to the market during bullish and bearish periods.

c. Test the hypothesis you specified in part **c** of Exercise 12.67. Draw the appropriate conclusions. Test using $\alpha = .05$.

d. Test the hypothesis you specified in part **d** of Exercise 12.67. Draw the appropriate conclusions. Test using $\alpha = .05$.

12.77 To model the relationship between y, a dependent variable, and x, an independent variable, a researcher has taken one measurement on y at each of five different x values. Drawing on her mathematical expertise, the researcher realizes that she can fit the fourth-order polynomial model

$$E(y) = \beta_0 + \beta_1 x + \beta_2 x^2 + \beta_3 x^3 + \beta_4 x^4$$

and it will pass exactly through all five points, yielding SSE = 0. The researcher, delighted with the "excellent" fit of the model, eagerly sets out to use it to make inferences. What problems will she encounter in attempting to make inferences?

12.78 Due to an increase in gasoline prices, many service stations are offering self-service gasoline at reduced prices. Suppose an oil company wants to model the mean monthly gasoline sales $E(y)$ of its affiliated stations as a function of the type of service offered: self-service, full service, or both.

a. How many dummy variables will be needed to describe the qualitative independent variable Type of service?

b. Write the main effects model relating $E(y)$ to the type of service. Describe the coding of the dummy variables.

***12.79** In Case Study 11.2, we described the first-order regression model developed by Carolyn I. Allmon for predicting the sales of Crest toothpaste. The data she used to estimate the model are presented in the table. Using these data and the procedures you learned in this chapter, attempt to develop a second-order model that, according to the appropriate F test, explains more of the variation in sales than Allmon's model. Describe each step of your model building process.

Year	*Crest Sales ($1,000s)*	*Advertising Budget ($1,000s)*	*Advertising Ratio*	*Income ($ billions)*	Year	*Crest Sales ($1,000s)*	*Advertising Budget ($1,000s)*	*Advertising Ratio*	*Income ($ billions)*
1966	86,250	–	–	–	1974	126,000	18,250	1.27	998.3
1967	105,000	16,300	1.25	547.9	1975	162,000	17,300	1.07	1,096.1
1968	105,000	15,800	1.34	593.4	1976	191,625	23,000	1.17	1,194.4
1969	121,600	16,000	1.22	638.9	1977	189,000	19,300	1.07	1,311.5
1970	113,750	14,200	1.00	695.3	1978	210,000	23,056	1.54	1,462.9
1971	113,750	15,000	1.15	751.8	1979	224,250	26,000	1.59	1,641.7
1972	128,925	14,000	1.13	810.3	1980	245,000	28,000	1.56	1,821.7
1973	142,500	15,400	1.05	914.5					

Source: Allmon, C. I. "Advertising and sales relationships for toothpaste: Another look." *Business Economics*, Sept. 1982, Vol. 17, p. 58.

12.80 Many service companies must accurately estimate their costs before a job is begun in order to be able to make a profit. A heating and plumbing contractor, for example, may base cost estimates for new homes on the total area of the house and whether central air conditioning is to be installed.

a. Write a main effects model relating the mean cost of material and labor, $E(y)$, to the area and central air conditioning variables.

b. Write a complete second-order model for the mean cost as a function of the same two variables.

c. What hypothesis would you test to determine whether the second-order terms are useful for predicting mean cost?

d. Explain how you would compute the F statistic needed to test the hypothesis of part **c**.

12.81 Refer to Exercise 12.80. The contractor samples 25 recent jobs and fits both the complete second-order model (part **b**) and the reduced main effects model (part **a**), so that a test can be conducted to determine whether the additional comlexity of the second-order model is necessary. The resulting SSE and R^2 values are

Main effects: SSE = 8.548 and $R^2 = .950$

Second-order: SSE = 6.133 and $R^2 = .964$

a. Is there sufficient evidence to conclude that the second-order terms are important for predicting the mean cost? Use $\alpha = .05$.

b. Suppose the contractor decides to use the main effects model to predict costs. Use the global F test (Section 11.5) to determine whether the main effects model is useful for predicting costs.

***12.82** The table lists the $n = 72$ observations used by the Minnesota Department of Transportation to develop the peak-hour traffic volume model described in Case Study 12.1. Observations 55–72 are from Interstate 35W at 46th Street.

Observation Number	*Peak-Hour Volume*	*24-Hour Volume*	*Observation Number*	*Peak-Hour Volume*	*24-Hour Volume*	*Observation Number*	*Peak-Hour Volume*	*24-Hour Volume*
1	1,990.94	20,070	25	1,923.87	18,184	49	1,978.72	24,249
2	1,989.63	21,234	26	1,922.79	16,926	50	1,975.29	23,321
3	1,986.96	20,633	27	1,917.64	19,062	51	1,973.55	22,842
4	1,986.96	20,676	28	1,916.17	18,043	52	1,973.91	20,626
5	1,983.78	19,818	29	1,916.17	18,043	53	1,972.92	26,166
6	1,983.13	19,931	30	1,916.13	16,691	54	1,966.65	21,755
7	1,982.47	19,266	31	1,912.49	17,339	55	2,120.00	20,250
8	1,981.53	19,658	32	1,912.49	17,339	56	2,140.00	20,251
9	1,979.83	19,203	33	1,909.98	17,867	57	2,160.00	21,852
10	1,979.83	19,958	34	1,907.04	17,773	58	2,186.52	23,511
11	1,978.40	19,152	35	1,907.46	17,678	59	2,180.29	22,431
12	1,978.90	21,651	36	1,905.14	18,024	60	2,174.03	23,734
13	1,977.38	20,198	37	1,902.37	17,405	61	2,174.03	23,734
14	1,972.87	20,508	38	2,017.76	23,517	62	2,167.97	23,387
15	1,964.45	19,783	39	2,009.38	23,017	63	2,160.02	24,885
16	1,962.85	20,815	40	2,007.10	22,808	64	2,160.54	23,332
17	1,964.26	20,105	41	2,007.28	23,152	65	2,159.72	23,838
18	1,961.85	20,500	42	2,004.17	24,352	66	2,155.61	23,662
19	1,961.26	19,593	43	1,997.58	20,939	67	2,147.93	22,948
20	1,958.97	20,818	44	1,994.53	21,822	68	2,147.93	22,948
21	1,943.78	17,480	45	1,984.70	22,918	69	2,147.85	23,551
22	1,927.83	17,768	46	1,984.01	21,129	70	2,144.23	21,637
23	1,928.36	17,659	47	1,983.17	21,674	71	2,142.41	23,543
24	1,925.65	18,357	48	1,982.02	26,148	72	2,137.39	22,594

a. Use the data to replicate the model building process described in Case Study 12.1

b. Compare your results of part **a** with those presented in Figures 12.27–12.29.

c. Calculate and plot the residuals of the model in a stem-and-leaf display or a box plot. Evaluate the assumption of normality of the random error component, and determine whether any apparent outliers exist.

On Your Own

We continue our **On Your Own** theme from Chapters 10 and 11. Remember that you selected three independent variables related to the annual GDP. Now, increase your list of three variables to include approximately 10 that you feel would be useful in predicting the GDP. Obtain data for as many years as possible for the new list of variables and the GDP. Use a stepwise regression program to choose the important variables among those you have listed. To test your intuition, list the variables in the order you think they will be selected before you conduct the analysis. How does your list compare with the stepwise regression results?

After the group of 10 variables has been narrowed to a smaller group of variables by the stepwise analysis, try to improve the model by including interactions and quadratic terms. Be sure to consider the meaning of each interaction or quadratic term before adding it to the model—a quick sketch can be very helpful. See if you can systematically construct a useful model for predicting the GDP. You might want to hold out the last several years of data to test the predictive ability of your model after it is constructed. (As noted in Section 12.8, using the same data to construct *and* to evaluate predictive ability can lead to invalid statistical tests and a false sense of security.)

Using the Computer

Consider the relationship between the mean median household income y of a zip code, the percent of college graduates x_1 in the zip code, and the census region of the country to which the zip code belongs. Randomly select 50 zip codes from each region.

a. Fit each model listed to your sample:

1. Complete second-order model
2. Complete second-order model *without* the qualitative variable census region
3. Main effects plus interaction (complete first-order) model
4. Main effects model including both percent college graduates and census region
5. Main effects model including only percent college graduates

b. Test the complete second-order model (1) against each of the reduced models, (2) and (3). Be sure to write the hypotheses you are testing in terms of this exercise. Which of the models is preferred as a result of your testing? If model (1) is preferred, proceed to part **c**. If model (2) is preferred, test it against model (5). If model (3) is preferred, test it against model (4). Select the preferred model among the five as a result of the testing.

c. Interpret the preferred model. Discuss the value of R^2 and the standard deviation of the model. Plot the predicted value of mean median income against the percent of college graduates, using a different plotting symbol for each census region (assuming the variable census region is in your preferred model). Interpret your plot.

d. Use a prediction interval to predict the mean median income for a zip code in the South region with 20% college graduates.

e. What assumptions are necessary to ensure the validity of the inferences in parts **b–d**? Plot the residuals of the preferred model against the predicted value of mean income. Interpret the pattern. Does the plot indicate a need to transform the dependent variable?

f. Repeat parts **a–e** using a multiplicative model—that is, using the logarithmic transformation of mean median income as the dependent variable. Be careful to recognize the transformation when you interpret the model and perform the prediction. Which model is to be preferred, the additive or multiplicative? Explain.

g. Repeat this exercise using all the measurements in the data base, rather than your sample. Compare the results to those based on your sample.

References

Alexander, G. J., and Stover, R. D. "Consistency of mutual fund performance during varying market conditions." *Journal of Economics and Business*, Spring 1980, Vol. 32, pp. 219–226.

Chatterjee, S., and Price, B. *Regression Analysis by Example*, 2nd ed. New York: Wiley, 1991.

Churchill, G. A., Jr., Ford, N. M., and Walker, O. C., Jr. *Sales Force Management*, 2nd ed. Homewood, Ill.: Richard D. Irwin, 1985.

Draper, N., and Smith, H. *Applied Regression Analysis*, 2nd ed. New York: Wiley, 1981.

Federal Reserve System: Purpose and Functions. Washington, D.C.: Board of Governors of the Federal Reserve System, 1963.

Graybill, F. A. *Theory and Application of the Linear Model*. North Scituate, Mass.: Duxbury, 1976.

Mendenhall, W. *Introduction to Linear Models and the Design and Analysis of Experiments*. Belmont, Calif.: Wadsworth, 1968.

Mendenhall, W., and Sincich, T. *A Second Course in Business Statistics: Regression Analysis*, 4th ed. San Francisco: Dellen, 1993.

Miller, R. B., and Wichern, D. W. *Intermediate Business Statistics: Analysis of Variance, Regression, and Time Series*. New York: Holt, Rinehart and Winston, 1977. Chapters 6–8.

Murphy, N. B., and Stock, D. R. "Determinants of the use of banking facilities: Trips to the bank in Oklahoma." *Review of Regional Economics and Business*, Oct. 1983, Vol. 8, pp. 33–35.

Myers, R. H. *Classical Modern Regression Analysis with Applications*. Boston: PWS-Kent, 1990.

Neter, J., Wasserman, W., and Kutner, M. *Applied Linear Regression Models*, 2nd ed. Homewood, Ill.: Richard D. Irwin, 1989.

Schroeder, R. G. *Operations Management: Decision Making in the Operations Function*, 4th ed. New York: McGraw-Hill, 1993.

Spiro, H. T. *Finance for the Non-Financial Manager*, 2nd ed. New York: Wiley, 1982. Chapter 16.

Weisberg, S. *Applied Linear Regression*, 2nd ed. New York: Wiley, 1985.

Winkler, R. L., and Hays, W. L. *Statistics: Probability, Inference and Decision*, 2nd ed. New York: Holt, Rinehart and Winston, 1975. Chapter 10.

Wise, J. O., and Dover, H. J. "An evaluation of a statistical method of appraising rural property." *Appraisal Journal*, Jan. 1974, Vol. 42, pp. 103–113.

Wittink, D. R. *The Application of Regression Analysis*. Boston: Allyn and Bacon, 1988.

Younger, M. S. *A First Course in Linear Regression*, 2nd ed. Boston, Mass.: Duxbury, 1985.

CHAPTER THIRTEEN

Methods for Quality Improvement

Contents

Case Study

Where We've Been

In Chapters 7–9, we described methods for making inferences about populations based on sample data. In Chapters 10–12 we focused on modeling relationships between variables using regression analysis.

Where We're Going

In this chapter and the two that follow, we turn our attention to processes. Recall from Chapter 1 that a process is a series of actions or operations that transform inputs to outputs. This chapter describes methods for improving processes and the quality of the output they produce. The next two chapters are concerned with modeling and forecasting the output of processes.

Over the last two decades U.S. firms have been seriously challenged by products of superior quality from overseas, particularly from Japan. Japan currently produces 26% of the cars sold in the United States, and some authorities predict this will climb to 40% within a decade. In 1989, for the first time, the top-selling car in the United States was made in Japan: the Honda Accord. Although it's an American invention, virtually all VCRs are produced in Japan. Only one U.S. firm still manufactures televisions; the rest are made in Japan.

To meet this competitive challenge, more and more U.S. firms—both manufacturing and service firms—have begun quality-improvement initiatives of their own. Many of these firms now stress the management of quality in all phases and aspects of their business, from the design of their products to production, distribution, sales, and service.

Broadly speaking, quality-improvement programs are concerned with (1) finding out what it is that the customer wants, (2) translating those wants into a product or service design, and (3) producing a product or service that meets or exceeds the specifications of the design. In this chapter we focus primarily on the third of these three areas and its major problem—product and service variation.

Variation is inherent in the output of all production and service processes. No two parts produced by a given machine are the same; no two transactions performed by a given bank teller are the same. Why is this a problem? With variation in output comes variation in the quality of the product or service. If this variation is unacceptable to customers, sales are lost, profits suffer, and the firm may not survive.

The existence of this ever-present variation has made statistical methods and statistical training vitally important to industry. In this chapter we present some of the tools and methods currently employed by firms worldwide to monitor and reduce product and service variation.

Before introducing you to these quality-improvement methods, some background is needed to help you understand, appreciate, and apply the methods appropriately. We begin with a brief history of the quality movement in the United States (Section 13.1), then we discuss what the word *quality* means as it is used in industry (Section 13.2). Next, we look at the concepts of systems, processes, and process variation (Sections 13.3 and 13.4). We conclude the background material with some management principles that guide firms in their quality-improvement efforts (Section 13.5).

13.1 History of the Quality Movement in the United States

In the 1700s and 1800s products were designed and built by master craftspeople or by apprentices who were trained and closely supervised by master artisans. These artisans took great pride in their work. They knew what their customers wanted and, since goods were produced in small volumes, were able to put substantial effort and care into each item they produced. Product quality was not really an issue; it was a natural result of the work ethic.

Things changed, however, with the advent of mass production in the late 1800s and early 1900s. Production processes became more complex. The assembly line was invented. No longer was a single, well-trained individual responsible for all aspects of production. As a result, the quality of the output of manufacturing processes became more variable and less dependable. At the same time, however, this new industrial age demanded interchangeable parts of absolutely uniform quality. To solve this problem, manufacturers instituted formal inspection programs for their finished goods. The job of quality inspector was created.

Initially, such programs involved the inspection of 100% of the goods produced. But this is a costly and inefficient way to sort good products from bad. In 1908 Western Electric, the manufacturing arm of the company now known as the American Telephone and Telegraph Company (AT&T), began using primitive sampling techniques for this purpose. In the 1920s and 1930s at Bell Telephone Laboratories, AT&T's newly formed research and development unit, significant advances were made in sampling techniques for quality inspection. Harold Dodge and Harry Romig did much of the pioneering work in this area, which has come to be known as **acceptance sampling**.

In the mid-1920s, Walter Shewhart, another researcher from Bell Laboratories, made perhaps the most significant breakthrough of this century for the improvement of manufactured products. He recognized that variation in manufactured products is inevitable and that it can be understood, monitored, and controlled by using statistical procedures. He developed a simple graphical technique—called a **control chart**—for determining whether product variation is within acceptable limits. This method provides guidance for when to adjust or change a production process and when to leave it alone. It can be used at the end of the production process, or most significantly, at different points *within* the process (i.e., farther "upstream" in the production process). The Bell System's use of control charts and inspection sampling plans yielded significant improvements in telephone equipment and services. In addition, inspection costs fell, the proportion of defective units produced fell, and having to rework fewer defective units made the workers more productive.

Despite the successes enjoyed by AT&T, neither acceptance sampling techniques nor control charts were widely adopted outside the Bell System. It was not until World War II spurred the country's need for mass quantities of war munitions that, at the insistence of the U.S. War Department, acceptance sampling came into wider use. However, only in the late 1970s and early 1980s, as U.S. manufacturers felt threatened by the superior quality of Japanese goods, did control charts finally gain widespread use in the United States.

In the last few years quality practitioners have begun to expand the scope of their efforts upstream from production into the product design stage. Emphasis has been placed on "designing quality into the product." This is done by designing a product that the customer wants and that the firm's production processes are capable of producing in a condition that conforms to customer specifications. With this recent interest in product design, another long-established set of statistical tools has come to the forefront in the quality movement—the tools of experimental design. We discuss the design of experiments in Chapter 16.

But the quality movement encompasses more than statistical tools. Since World War II much progress has been made in the development of quality-management philosophies and methods for implementing them. Joseph M. Juran, a management consultant who spent many years working in quality programs at Western Electric, argues that quality is management's responsibility and that it should be treated as a discipline or functional area like finance or marketing.

Armand Feigenbaum, once a quality expert at General Electric and now a management consultant, advocates a systematic approach to quality improvement that involves all areas and departments of a company, not just manufacturing. His approach is known as **total quality control** (TQC) or **total quality management** (TQM).

W. Edwards Deming, a statistician who had studied with Walter Shewhart of Bell Laboratories, developed a philosophy of quality management that complements and supports the statistical methods of quality improvement. His philosophy, which is summarized in his well-known "14 Points," is based on the principle that a firm's quality-improvement efforts should be continual and never-ending. He argues that as quality improves, costs decrease because fewer products require rework, fewer mistakes are made, fewer delays are incurred, etc. Thus, worker productivity improves. The firm's higher-quality, lower-cost products capture a larger market share, which ensures the continued survival of the firm. We present Deming's 14 points in Case Study 13.1. Like the statistical methods for quality improvement previously mentioned, Deming's, Juran's, and Feigenbaum's approaches to quality management were also generally ignored by U.S. managers until they were threatened by the high-quality products from Japan in the 1970s and 1980s.

But how is it that the Japanese became quality leaders? What inspired their concern for quality? In part, it was the statistical and managerial expertise exported to Japan from the United States following World War II. At the end of the war Japan faced the difficult task of rebuilding its economy. To this end, a group of engineers, each of whom had worked for Western Electric or Bell Laboratories, was assigned by the Allied command to assist the Japanese in improving the quality of their communication systems. These engineers taught the Japanese the statistical quality-control methods that had been developed in the United States. Then, in 1950, the Japanese Union of Scientists and Engineers invited Deming to present a series of lectures on statistical quality-improvement methods to hundreds of Japanese researchers, plant managers, and engineers. During his stay in Japan he also met with many of the top executives of Japan's largest companies. At the time, Japan was notorious for the inferior quality of its products. Deming told the executives that by listening to what consumers wanted and by applying statistical methods in the production of those goods, they could export high-quality products that would find markets all over the world.

The Japanese also benefited from the advice of Juran, who is credited with helping the Japanese to implement management practices that support and complement the statistical methodologies taught to them by Deming. They also read and adopted Feigenbaum's total quality control concept.

Why didn't the United States adopt a similar focus on quality after the war? Following World War II, while Japan and Europe were rebuilding, the United States had very little competition in the world market. Consequently, the emphasis was on production—getting the goods out the door. Many firms could not keep up with the demand for their products. Quality control was little more than end-of-the-line inspection. Profits were so large that it seemed pointless to worry about the costs of scrap and rework. Of course, this all changed in the 1970s.

Interestingly, in 1951 the Japanese established the **Deming Prize** to be given annually to companies with significant accomplishments in the area of quality. In 1989, for the first time, the Deming Prize was given to a U.S. company—Florida Power and Light Company.

13.2 Quality: What Is It?

Before describing various tools and methods that can be used to monitor and improve the quality of products and services, we need to consider what is meant by the term **quality**. Quality can be defined from several different perspectives. To the engineers and scientists who design products, quality typically refers to the amount of some ingredient or attribute possessed by the product. For example, high-quality ice cream contains a large amount of butterfat. High-quality rugs have a large number of knots per square inch. A high-quality shirt or blouse has 22 to 26 stitches per inch.

To managers, engineers, and workers involved in the production of a product (or the delivery of a service), quality usually means conformance to requirements, or the degree to which the product or service conforms to its design specifications. For example, in order to fit properly, the cap of a particular molded plastic bottle must be between 1.0000 inch and 1.0015 inches in diameter. Caps that do not conform to this requirement are considered to be of inferior quality. For an example in a service operation, consider the service provided to customers in a fast-food restaurant. A particular restaurant has been designed to serve customers within 2 minutes of the time their order is placed. If it takes more than 2 minutes, the service does not conform to specifications and is considered to be of inferior quality. Using this production-based interpretation of quality, well-made products are high quality; poorly made products are low quality. Thus, a well-made Rolls Royce and a well-made Chevrolet Nova are both high-quality cars.

Although quality can be defined from either the perspective of the designers or the producers of a product, in the final analysis both definitions should be derived from the needs and preferences of the *user* of the product or service. A firm that produces goods that no one wants to purchase cannot stay in business. Deming likes to say, "The consumer is the most important part of the production line. Quality should be aimed at the needs of the consumer, present and future." We define quality accordingly.

Definition 13.1

The **quality** of a good or service is indicated by the extent to which it satisfies the needs and preferences of its users.

Consumers' needs and wants shape their perceptions of quality. Thus, to produce a high-quality product, it is necessary to study the needs and wants of consumers. This is typically one of the major functions of a firm's marketing department. Once the necessary consumer research has been conducted, it is necessary to translate consumers' desires into a product design. This design must then be translated into a production plan and production specifications that, if properly implemented, will turn out a product with characteristics that will satisfy users' needs and wants. In short, consumer perceptions of quality play a role in all phases and aspects of a firm's operations.

But what product characteristics are consumers looking for? What is it that influences users' perceptions of quality? This is the kind of knowledge that firms need in order to develop and deliver high-quality goods and services. David A. Garvin (1988) summarizes the basic elements of quality in what he calls the **eight dimensions of quality**.

The Eight Dimensions of Quality

1. **Performance**: The primary operating characteristics of the product. For an automobile, these would include acceleration, handling, smoothness of ride, gas mileage, etc.
2. **Features**: The "bells and whistles" that supplement the product's basic functions. Examples include CD players and digital clocks on cars and the frequent-flyer mileage and free drinks offered by airlines.
3. **Reliability**: Reflects the probability that the product will not operate properly within a given period of time.
4. **Conformance**: The extent or degree to which a product meets pre-established standards. This is reflected in, for example, a pharmaceutical manufacturer's concern that the plastic bottles it orders for its drugs have caps that are between 1.0000 and 1.0015 inches in diameter, as specified in their order.
5. **Durability**: The life of the product. If repair is possible, durability relates to the length of time a product can be used before replacement is judged to be preferable to continued repair.
6. **Serviceability**: The ease of repair, speed of repair, and competence and courtesy of the repair staff.
7. **Aesthetics**: How a product looks, feels, sounds, smells, or tastes.
8. **Other perceptions that influence judgments of quality**: Such factors as a firm's reputation and the images of the firm and its products that are created through advertising.

In order to design and produce products of high quality, it is necessary to translate the characteristics described in the box into product attributes that can be built into the product by the manufacturer. That is, user preferences must be interpreted in terms of product variables over which the manufacturer has control. For example, in considering the performance characteristics of a particular brand of wooden pencil, users may indicate a preference for being able to use the pencil for longer periods between sharpenings. The manufacturer may translate this performance characteristic into one or more measurable physical characteristics such as wood hardness, lead hardness, and lead composition. As we will see later in the chapter, besides being used to design high-quality products, such variables are used in the process of monitoring and improving quality during production.

13.3 Processes

Much of this textbook focuses on methods for using sample data drawn from a population to learn about that population. In Chapters 13–15, however, our attention is not on populations, but on processes—such as manufacturing processes—and the output that they generate. In this chapter our concern is with monitoring and improving processes. In the next two chapters we consider modeling and forecasting the output of processes. But first, what is a *process*?

Definition 13.2

A **process** is a series of actions or operations that transforms inputs to outputs. A process produces output over time.

Processes can be organizational or personal in nature. Organizational processes are those associated with organizations such as businesses and governments. Perhaps the best example is a manufacturing process, which consists of a series of operations, performed by people and machines, whereby inputs such as raw materials and parts are converted into finished products (the outputs). Examples include automobile assembly lines, oil refineries, and steel mills. Personal processes are those associated with your private life. The series of steps you go through each morning to get ready for school or work can be thought of as a process. Through turning off the alarm clock, showering, dressing, eating, and opening the garage door, you transform yourself from a sleeping person to one who is ready to interact with the outside world. Figure 13.1 (page 698) presents a general description of a process and its inputs.

It is useful to think of processes as *adding value* to the inputs of the process. Manufacturing processes, for example, are designed so that the value of the outputs to potential customers exceeds the value of the inputs—otherwise the firm would have no demand for its products and would not survive.

Understanding the difference between a process and a project is important. A *process* is executed over and over, thereby producing a series of products over time. A *project* is typically a one-time activity that produces one product. When a family builds a new house, they have undertaken a project. However, the contractor who builds the

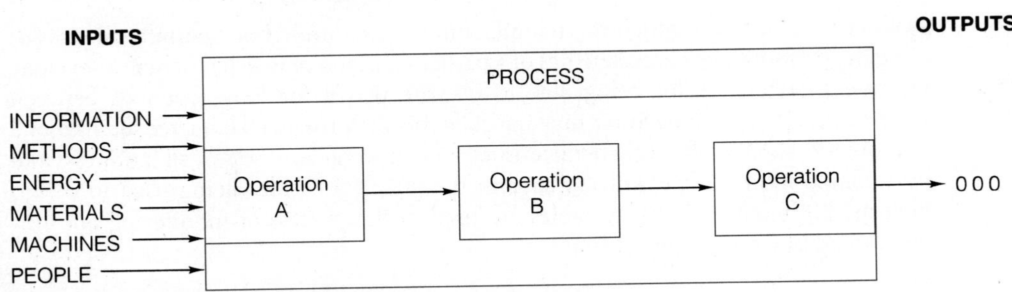

FIGURE 13.1 ▲ Graphical depiction of a process and its inputs

house is concerned with the process of construction, since she produces many houses over time.

The most effective tool for documenting and understanding processes is the *flowchart*. While several different sets of symbols have been proposed for use in flowcharts, we prefer the set defined in the next box. An example of a flowchart is presented in Figure 13.2.

Flowchart Symbols

[Rectangle] An operation or activity

[Diamond] A decision, evaluation, or computation (used for branching)

[Parallelogram] Input or output

[Document shape] Used when the input or output is a document

[Arrow] Indicates the direction of flow

[Rounded rectangle] Used to indicate the beginning and the end of the process

[Circle] Connector: used to indicate the location of the continued flowchart

FIGURE 13.2 ►
Flowchart for turning on a television; adapted from Brassard (1989, p. 265)

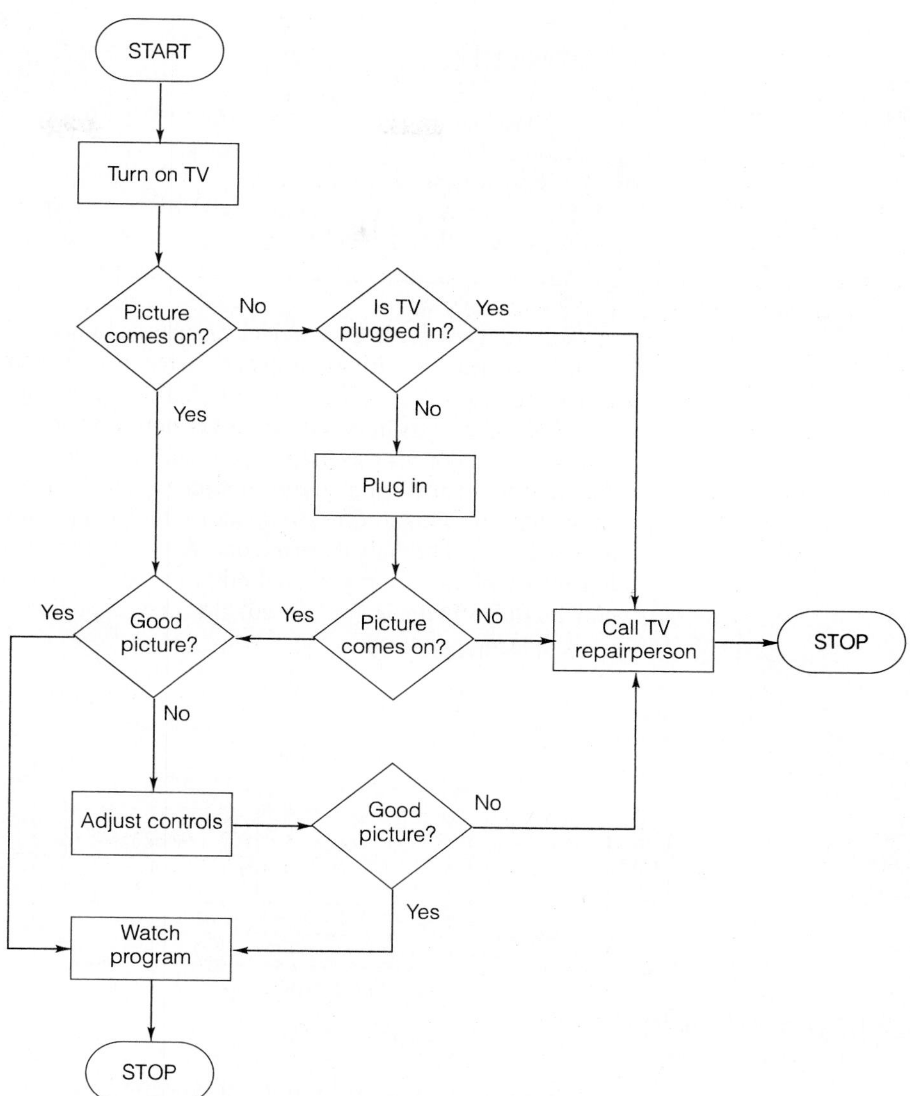

13.4 Systems and Systems Thinking

To understand what causes variation in process output and how processes and their output can be improved, we must understand the role that processes play in *systems*. Thus, we continue our study of processes with an introduction to some basic systems concepts. We begin by defining a system.

Definition 13.3

A **system** is a collection or arrangement of interacting processes that has an ongoing purpose or mission. A system receives inputs from its environment, transforms those inputs to outputs, and delivers them to its environment. In order to survive, a system uses feedback (i.e., information) from its environment to understand and adapt to changes in its environment.

Figure 13.3 presents a model of a basic system. As an example of a system, consider a manufacturing company. It has a collection of interacting processes—marketing research, engineering, purchasing, receiving, production, sales, distribution, billing, etc. Its mission is to make money for its owners, to provide high-quality working conditions for its employees, and to stay in business. The firm receives raw materials and parts (inputs) from outside vendors which, through its production processes, it transforms to finished goods (outputs). The finished goods are distributed to its customers. Through its marketing research, the firm "listens" (receives feedback from) its customers and potential customers in order to change or adapt its processes and products to meet (or exceed) the needs, preferences, and expectations of the marketplace.

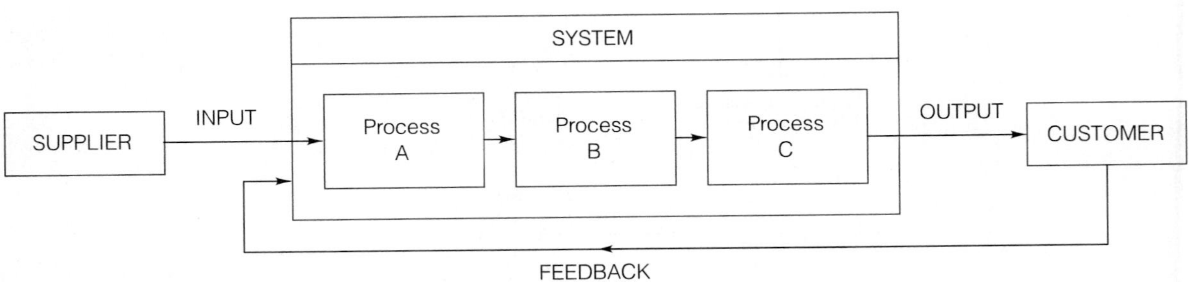

FIGURE 13.3 ▲ Model of a basic system

Since systems are collections of processes, the various types of system inputs are the same as those listed in Figure 13.1 for processes. System outputs are products or services. These outputs may be physical objects made, assembled, repaired, or moved by the system; or they may be symbolical, such as information, ideas, or knowledge. For example, a brokerage house supplies customers with information about stocks and bonds and the markets where they are traded.

One of a system's most important types of information input is environmental feedback—in particular, **customer feedback**. It is through feedback that a system learns about its environment and how the environment reacts to its outputs. A system uses feedback to change or modify its goals, processes, and output to better achieve its mission.

Let's take a closer look at the processes that make up the system described in Figure 13.3. The processes perform the operations that transform the inputs. Each process does its share of the transformation; each adds it share of value to the inputs. Process A receives inputs from the environment, transforms them, and passes the result or product on to Process B. Process B transforms them and passes the result on to Process C, and so on. For example, a firm's warehouse might receive pallets of raw materials from an outside supplier, break down the pallets into manageable portions, and pass these portions on to the production department. The production department manufactures finished products and passes them on to the shipping department. Thus, a system has not only external suppliers and customers (outside the firm), but internal customers and suppliers as well. Process A is Process B's supplier; Process B is Process A's customer, and so on.

Each of the system's processes may itself be composed of a collection of processes. For example, the warehouse may have a receiving process, a storage process, an unpacking process, and a distribution process. Thus, each process of the original system may itself be a system. We call such systems *subsystems*. Two subsystems are shown in Figure 13.4. In general, systems are hierarchical in nature—that is, any system may be contained in a wider system and may itself contain subsystems.

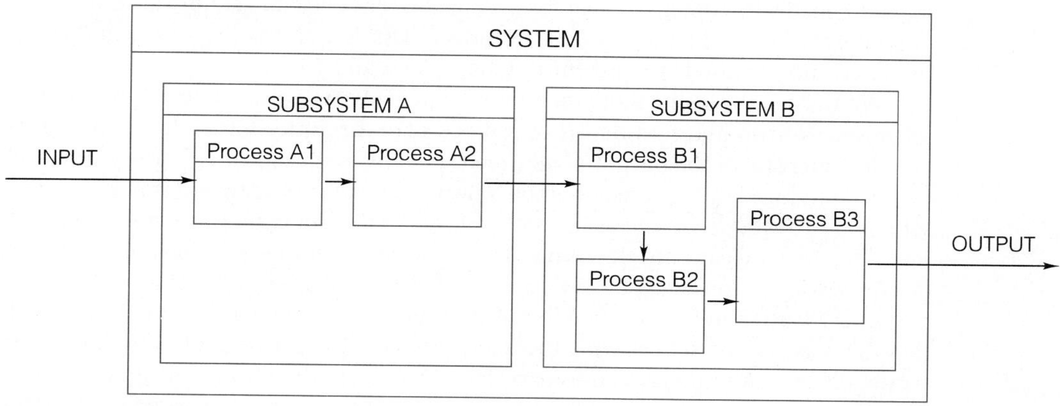

FIGURE 13.4 ▲ Subsystems

In describing or analyzing a system we may begin with a basic model in which the system's processes are concealed, as in Figure 13.5 on page 702. Such a system is referred to as a *black box*. Similarly, an individual process whose operations are not specified is a black box.

> **Definition 13.4**
>
> A process whose steps or operations (i.e., subprocesses or subsystems) are unknown or unspecified is called a **black box**.

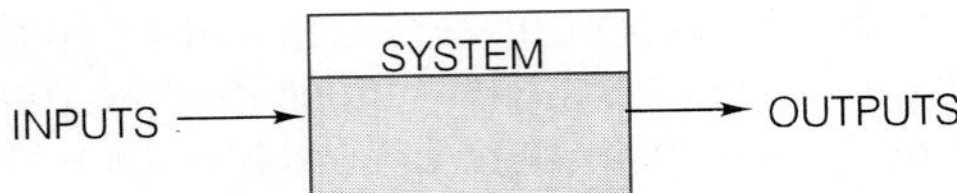

FIGURE 13.5 ►
A black box system

Sometimes, we "open" a black box and describe the subsystems within, as in Figures 13.3 and 13.4. In other cases, we may be satisfied with or have to settle for a simple black box description of the system. Such would be the case if we were unwilling or unable to describe adequately the contents of the box.

As an example of the latter case, consider the system that generates the closing prices of a particular common stock that is traded on the New York Stock Exchange. Prices are the output of the system. The inputs to the system include such items as the price of the stock in previous time periods, information about the national economy as well as the world economy, political news items, news of scientific discoveries, and business news. The processes of this system include those of institutional investors, individual investors, retail brokers, specialists on the stock exchange floor, and so on. The point is that the system is very complex. Depending on the purpose of our analysis, it may not be worth attempting to model the processes of the system and their relationships. For example, if we were interested in forecasting the stock's price, we would probably be content to treat the system as a black box and simply to analyze the output of the box—the stream of closing prices. This is precisely the approach used by the forecasting methods we present in Chapters 14 and 15.

We want to make one more important point about systems and the output of their processes. **No two items produced by a process are the same. Variability is an inherent characteristic of the output of all processes.** This is illustrated in Figure 13.6. No two cars produced by the same assembly line are the same: No two windshields are the same; no two wheels are the same; no two tires are the same; no two lug nuts are the same; no two hubcaps are the same. The same thing can be said for processes that deliver services. Consider the services offered at the teller windows of a bank to two customers waiting in two lines. Will they wait in line the same amount of time? Will they be serviced by tellers with the same degree of expertise and with the same personalities? Assuming the customers' transactions are the same, will they take the same amount of time to execute? The answer to all these questions is no.

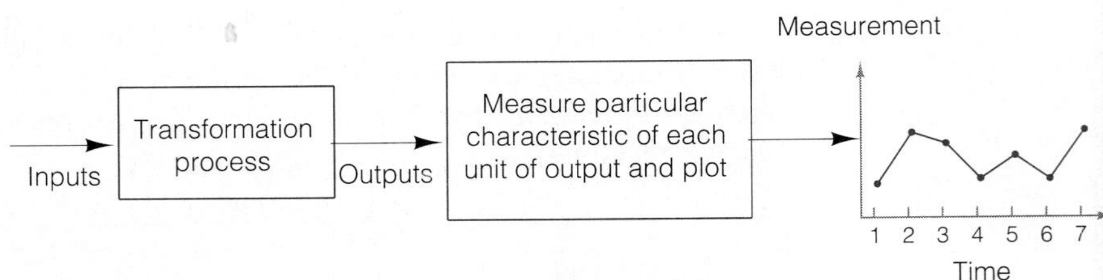

FIGURE 13.6 ►
Output variation

That the final output of a system is variable can be understood by considering the implications of Figure 13.4. The characteristics of the final product are influenced by all the subsystems, each subsystem perhaps having many subsystems of its own. The output of each subsystem is in turn influenced by many factors. For example, the quality of the raw materials entering Subsystem A may not be consistent. As a result the output from Subsystem A, which is the input to Subsystem B, is inconsistent. It follows that the output of Subsystem B and of every other subsystem "downstream" from Subsystem A will vary. Other factors that vary and cause output to vary in quality include worker expertise, worker motivation, training programs, instructions from supervisors, methods used in production, the age and quality of machinery, the layout of physical facilities, and the weather. In general, variation in output is caused by the six factors listed in the box.

The Six Major Sources of Process Variation

1. People
2. Machines
3. Materials
4. Methods
5. Measurement
6. Environment

With all these factors influencing each subsystem, it is not surprising that measures of final output display the kind of variation shown in Figure 13.6. Moreover, because *measurement is itself a process* it is subject to the same sources of variation. Thus, part of the variation revealed in the plot of Figure 13.6 may be due to measurement error.

Awareness of this ever-present process variation has made training in statistical thinking and statistical methods highly valued by industry. By **statistical thinking** we mean the knack of recognizing variation, and exploiting it in problem solving and decision-making.

Before concluding our discussion of systems, we want to recognize two important properties of systems:

1. The performance of a system depends on the performance of each of its processes.
2. The way any one process affects system performance depends on the performance of at least one other process.

In simpler terms, *a system is an indivisible whole.* You can take a system apart and optimize the individual performance of each process in isolation, but the total performance of the system still will not be optimal. For example, suppose the performances of the marketing and production functions of a firm were each independently optimized. It might turn out that marketing sells more products than the production department can make within a reasonable period of time. The result: unhappy customers and a loss of market share for the firm—a suboptimal performance.

The problem is that focusing attention on isolated processes tends to ignore the relationships between processes. These relationships are as important as the processes themselves; the relationships are what make a collection of processes a system.

Why are such systems principles important for us to understand in a statistics course? Because as we try to improve a process using statistical methods—the focus of the remainder of this chapter—we risk losing sight of the system in which the process is embedded. *Always keep in mind that the ultimate goal of process improvement is system improvement*. Any action taken on a process must therefore be considered in light of its effect on the system as a whole.

This section has served to introduce basic systems concepts. The application of these concepts to the understanding and management of organizations and their processes is called **systems thinking**. The major benefits of systems thinking are listed in the accompanying box.

Major Benefits of Systems Thinking

1. Systems thinking focuses attention on the processes through which things get done rather than on final outcomes. The effort of correcting problems after they occur can thus be redirected toward preventing them in advance.
2. It fosters recognition of the fact that variation is an inherent characteristic of the output of all processes and all organizations.
3. It facilitates the understanding of complex organizations.
4. It encourages consideration of the relationships between processes and between organizational levels in designing, evaluating, and managing organizations.
5. It encourages the investigation of all processes and levels of the system hierarchy in searching for problems such as the causes of unnacceptable variation in system output.
6. The generic, flexible nature of the system's framework makes it possible to construct models of systems that are broader than an individual organization (i.e., that contain the individual organization) or narrower than an individual organization (i.e., that are contained in the individual organization).

In line with point 6, quality management experts often advise modeling an organization, its customers, and its suppliers as parts of the same system—allies with complementary goals. In fact, this has been a key point in W. Edwards Deming's (1986) philosophy of quality management. It was systems thinking—along with statistical methods for quality improvement—that Deming taught to the Japanese in 1950. He used the systems diagram shown in Figure 13.7 to emphasize how much the quality of a firm's products and services relies on thinking of the customer and suppliers as part of the firm's extended system.

In the next section we focus on an important aspect of the management of systems—*process management*.

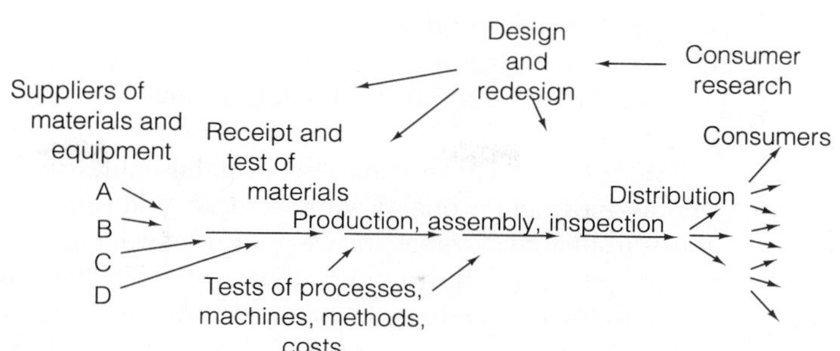

FIGURE 13.7 ►
Deming's figure, "Production Viewed as a System"

13.5 Process Management

As we noted at the beginning of this chapter, early attempts to improve the quality of products focused primarily on the inspection of outgoing products. Inspectors sorted finished goods into those that were acceptable for shipment to customers, those that needed to be reworked (i.e., fixed) before shipment, and those that should be scrapped. In effect, manufacturers tried to inspect quality into their products.

The production process itself was frequently treated as a black box. Work in process was generally not monitored. Data on the operation of internal subsystems of production processes were not collected. The data generated through inspection of process output were usually the only quality-related data that were collected and analyzed by manufacturers. As a result, it was these data—and customer feedback—that management had to rely on to determine when and why their production processes were not operating properly. This approach to the management of quality, the **detection–reaction approach**, is illustrated in Figure 13.8. It has fallen into disfavor primarily because the data received by management are collected too far downstream

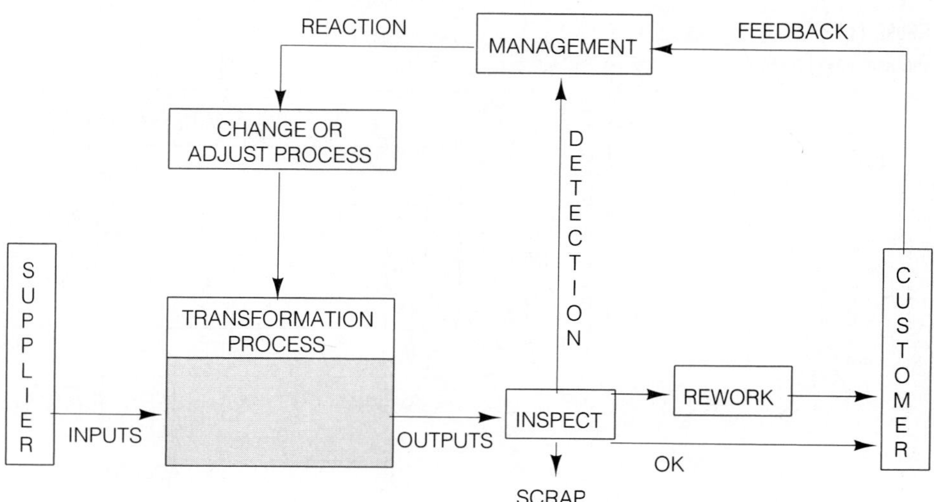

FIGURE 13.8 ►
The old way: Detection–reaction

from the cause of the quality problems. As a result, quality problems are not discovered until well after they arise. Further, even when the data reveal the existence of a problem, they are typically insufficient for management to diagnose what is wrong with the process.

With the advent of management philosophies such as Deming's that advocate a systems approach to quality improvement, and spurred on by the recent competitive threats from abroad, more and more firms now focus their quality-improvement efforts farther upstream in the production process. They have shifted from inspection of final output to the monitoring and control of subsystems within the production process and even as far upstream as the product-design process. In this chapter we are concerned primarily with the first of these two new areas of focus. Our focus, however, is not limited to the monitoring of manufacturing processes. The methods we present can be applied to *any* process within the firm, including the process of receiving and recording customer orders, the process of paying bills, the process of designing new products, the processes involved with typing and distributing internal memos, and even the process of servicing customer complaints.

This change in quality-management activities means that instead of treating their transformation processes as black boxes that generate a stream of data (as in Figure 13.6), firms have opened the boxes and are monitoring, controlling, and improving the subsystems within. In this approach to quality management, quality problems are identified and corrected as they arise, instead of through the inspection of finished products. **The goal is the prevention of the production of inferior-quality products. Properly applied, this approach makes the inspection of process output unnecessary.**

The **preventative style** of quality management is illustrated in Figure 13.9. The graphs at the bottom of the figure are control charts. We describe control charts in detail later in the chapter and show you how to use them to monitor processes.

This modern approach to producing high-quality goods and services involves the active management of processes, which we will refer to as **process management**.

FIGURE 13.9 ►
The new way: Prevention

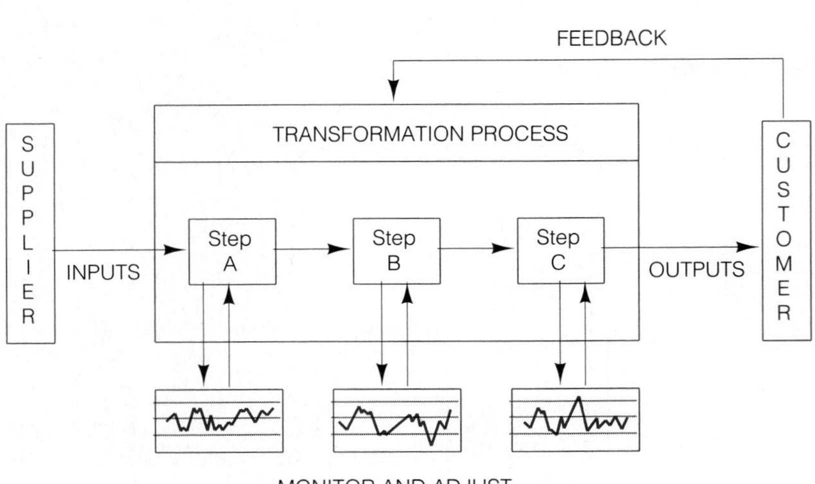

Definition 13.5

Process management is the monitoring, controlling, and improving of components and/or subsystems of a process for the purpose of improving the quality of the output of the process.

A full discussion of modern process management is beyond the scope of this book. However, we briefly summarize some of its fundamentals in the accompanying box. They are written for application to an existing process.

Fundamentals of Process Management

1. Establish the **objective** or **purpose** of the process.
2. Identify the suppliers and customers of the process to reveal the **boundaries** of the process. The suppliers and customers may be internal to the organization or external.
3. Establish **ownership** of the process: the person closest to the process who has the power to change the process. This is the person who is responsible for the process and whose participation and support are vital to the success of quality-improvement initiatives.
4. Identify the **subsystems** of the process.
5. Use a **flowchart** to specify in detail the flow of the operations within the process and its subsystems: the series of steps that transforms inputs to outputs.
6. Simplify the process and eliminate factors that distract workers.
7. Having defined the process, identify **listening points** within each subsystem—opportunities to monitor the performance of various aspects of the process.
8. Implement measurement procedures at the listening points and **continuously monitor** the process.
9. Establish ways to **listen continuously to the customers and potential customers** of the process. Without market research and customer feedback, products and services cannot keep pace with changing needs and preferences.
10. Establish ways to **listen continuously to suppliers' processes**. For suppliers external to the firm, establish strong relationships with those that deliver high-quality goods or services.
11. Continuously **analyze and interpret** the data received in items 8, 9, and 10.
12. Based on the analysis in item 11, act as necessary to change or refine the process so that it yields higher-quality outputs. **Quality improvement should be a never-ending goal that is continuously pursued.**

The hallmarks of process management are (1) a strong customer orientation, (2) purposeful data collection from both inside and outside the process, and (3) continuous improvement of quality. Process management can and should be practiced in every area of the firm. These fundamentals are just as applicable and useful in accounting, finance, and marketing or service operations such as banks and restaurants as they are in production areas.

CASE STUDY 13.1 / Deming's 14 Points

One of W. Edwards Deming's major contributions to the quality movement that is spreading across the major industrialized nations of the world was his recognition that statistical (and other) process improvement methods cannot succeed without the proper organizational climate and culture. Accordingly, he proposed 14 guidelines that, if followed, transform the organizational climate to one in which process-management efforts can flourish. These 14 points are, in essence, Deming's philosophy of management. He argues convincingly that all 14 should be implemented, not just certain subsets. We list all 14 points here, adding clarifying statements where needed. For a fuller discussion of these points, see Deming (1986), Gitlow et al. (1989), Walton (1986), and Joiner and Goudard (1990).

1. **Create constancy of purpose toward improvement of product and service, with the aim to become competitive and to stay in business, and to provide jobs.** The organization must have a clear goal or purpose. Everyone in the organization must be encouraged to work toward that goal day in and day out, year after year.
2. **Adopt the new philosophy.** Reject detection–rejection management in favor of a customer-oriented, preventative style of management in which never-ending quality improvement is the driving force.
3. **Cease dependence on inspection to achieve quality.** It is because of poorly designed products and excessive process variation that inspection is needed. If quality is designed into products and process management is used in their production, mass inspection of finished products will not be necessary.
4. **End the practice of awarding business on the basis of price tag.** Do not simply buy from the lowest bidder. Consider the quality of the supplier's products along with the supplier's price. Establish long-term relationships with suppliers based on loyalty and trust. Move toward using a single supplier for each item needed.
5. **Improve constantly and forever the system of production and service, to improve quality and productivity, and thus constantly decrease costs.**
6. **Institute training.** Workers are often trained by other workers who were never properly trained themselves. The result is excessive process variation and inferior products and services. This is not the workers' fault; no one has told them how to do their jobs well.
7. **Institute leadership.** Supervisors should help the workers to do a better job. Their job is to lead, not to order workers around or to punish them.
8. **Drive out fear, so that everyone may work effectively for the company.** Many workers are afraid to ask questions or to bring problems to the attention of management. Such a climate is not conducive to producing high-quality good and services. People work best when they feel secure.
9. **Break down barriers between departments.** Everyone in the organization must work together as a team. Different areas within the firm should have complementary, not conflicting, goals. People across the organization must realize that they are all part of the same system. Pooling their resources

to solve problems is better than competing against each other.

10. **Eliminate slogans, exhortations, and arbitrary numerical goals and targets for the work force which urge the workers to achieve new levels of productivity and quality.** Simply asking the workers to improve their work is not enough; they must be shown *how* to improve it. Management must realize that significant improvements can be achieved only if management takes responsibility for quality and makes the necessary changes in the design of the system in which the workers operate.
11. **Eliminate numerical quotas.** Quotas are purely quantitative (e.g., number of pieces to produce per day); they do not take quality into consideration. When faced with quotas, people attempt to meet them at any cost, regardless of the damage to the organization.
12. **Remove barriers that rob employees of their pride of workmanship.** People must be treated as human beings, not commodities. Working conditions must be improved, including the elimination of poor supervision, poor product design, defective materials, and defective machines. These things stand in the way of workers' performing up to their capabilities and producing work they are proud of.
13. **Institute a vigorous program of education and self-improvement.** Continuous improvement requires continuous learning. Everyone in the organization must be trained in the modern methods of quality improvement, including statistical concepts and interdepartmental teamwork. Top management should be the first to be trained.
14. **Take action to accomplish the transformation.** Hire people with the knowledge to implement the 14 points. Build a critical mass of people committed to transforming the organization. Put together a top management team to lead the way. Develop a plan and an organizational structure that will facilitate the transformation.

13.6 Statistical Control

For the rest of this chapter we turn our attention to control charts—graphical devices used for monitoring process variation, for identifying when to take action to improve the process, and for assisting in diagnosing the causes of process variation. Control charts are the tool of choice for continuously monitoring processes (point 8 of the fundamentals of process management, page 707). Before we go into the details of control chart construction and use, however, it is important that you have a fuller understanding of process variation. To this end, we discuss patterns of variation in this section.

As was discussed in Chapter 2, the proper graphical method for describing the variation of process output is a **time series plot**, sometimes called a **run chart**. Recall that in a time series plot the measurements of interest are plotted against time or are plotted in the order in which the measurements were made, as in Figure 13.10, page 710. Whenever you face the task of analyzing data that were generated over time, your first reaction should be to plot them. The human eye is one of our most sensitive statistical instruments. Take advantage of that sensitivity by plotting the data and allowing your eyes to seek out patterns in the data.

Let's begin thinking about process variation by examining the plot in Figure 13.10 more closely. The measurements, taken from a paint-manufacturing process, are the weights of 50 1-gallon cans of paint that were consecutively filled by the same filling

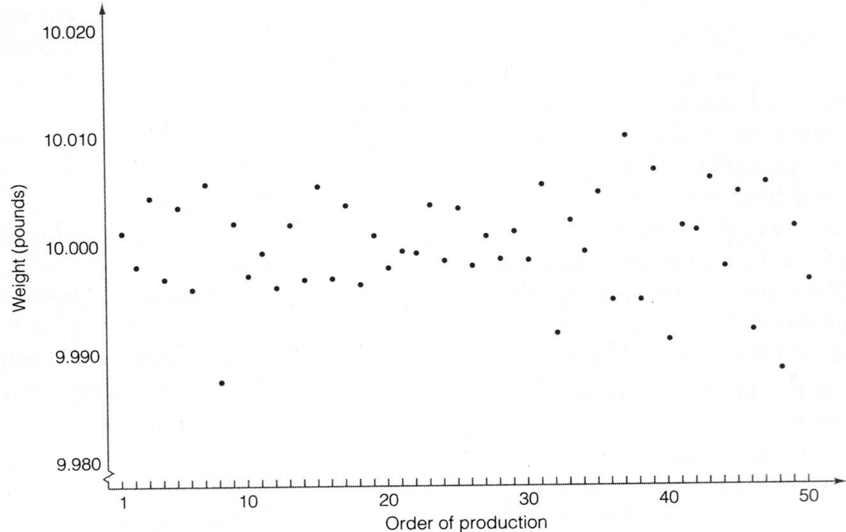

FIGURE 13.10 ▶ Time series plot of fill weights for 50 consecutively produced gallon cans of paint

head (nozzle). The weights were plotted in the order of production. Do you detect any systematic, persistent patterns in the sequence of weights? For example, do the weights tend to drift steadily upward or downward over time? Do they oscillate—high, then low, then high, then low, etc.?

To assist your visual examination of this or any other time series plot, Roberts (1991) recommends enhancing the basic plot in two ways. First, compute (or simply estimate) the mean of the set of 50 weights and draw a horizontal line on the graph at the level of the mean. This **centerline** gives you a point of reference in searching for patterns in the data. Second, using straight lines, connect each of the plotted weights in the order in which they were produced. This helps display the sequence of the measurements. Both enhancements are shown in Figure 13.11.

Now do you see a pattern in the data? Successive points alternate up and down, high then low, in an **oscillating sequence**. In this case, the points alternate above and below the centerline. This pattern was caused by a valve in the paint-filling machine that tended to stick in a partially closed position every other time it operated.

Other patterns of process variation are shown in Figure 13.12. We discuss several of them later.

In trying to describe process variation and diagnose its causes, it helps to think of the sequence of measurements of the output variable (e.g., weight, length, number of defects) as having been generated in the following way:

1. At any point in time, the output variable of interest can be described by a particular probability distribution (or relative frequency distribution). This distribution describes the possible values that the variable can assume and their likelihood of occurrence. Three such distributions are shown in Figure 13.13 (page 712).

FIGURE 13.11 ▶
An enhanced version of the paint fill time series

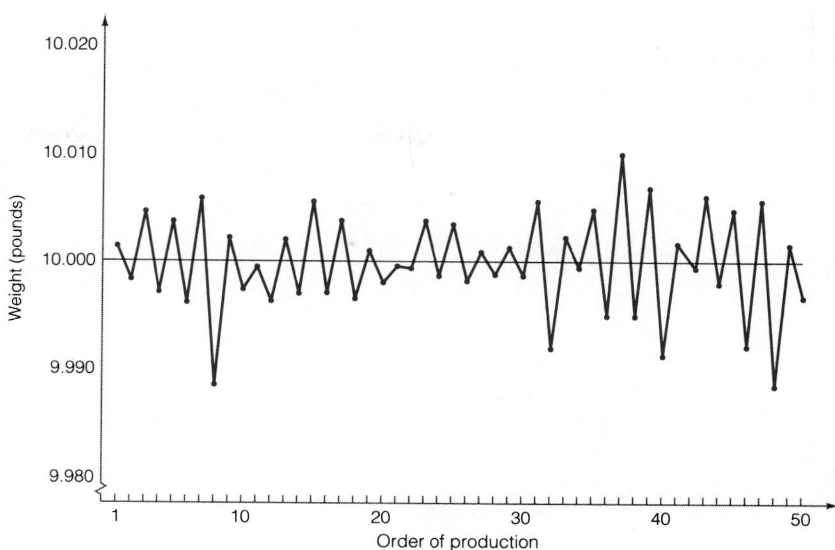

FIGURE 13.12 ▶
Patterns of process variation: Some examples

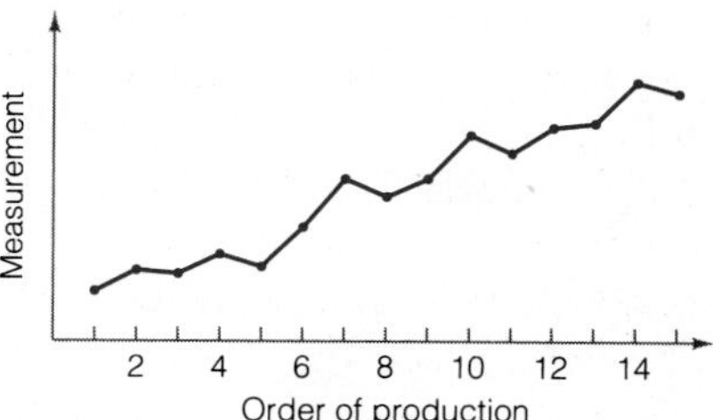

a. Uptrend

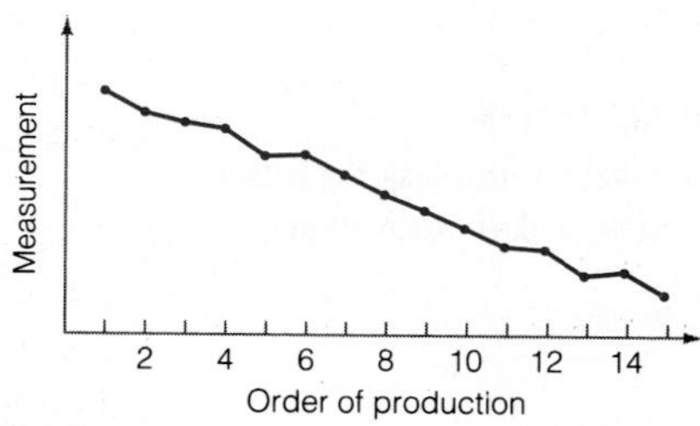

b. Downtrend

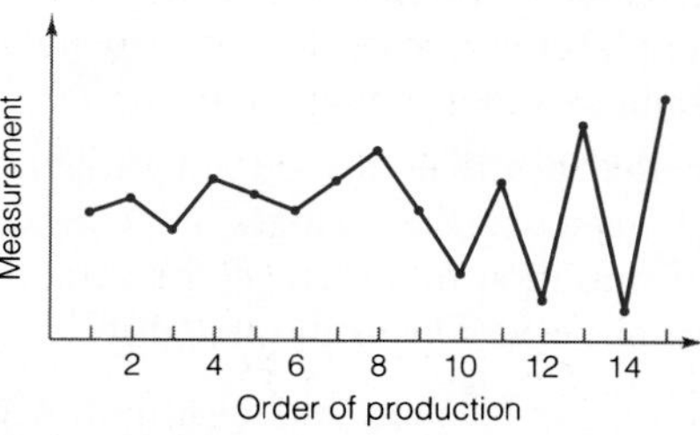

c. Increasing variance

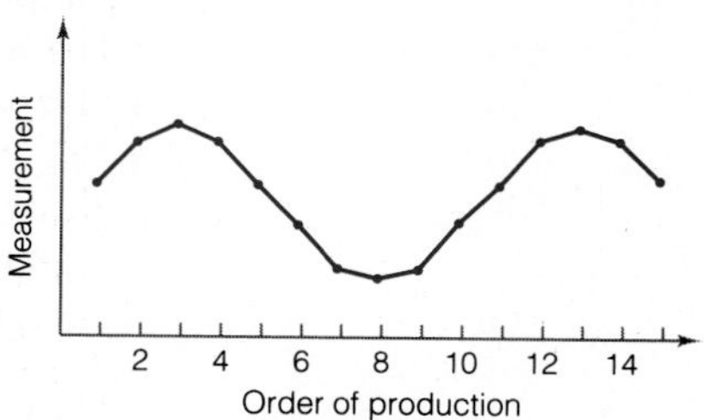

d. Cyclical

(*continued*)

FIGURE 13.12 ▶
Continued

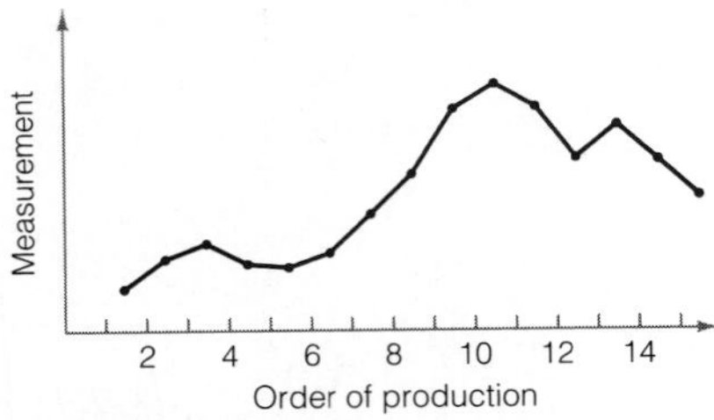

e. Meandering

f. Shock/Freak/Outlier

g. Level shift

FIGURE 13.13 ▶
Distributions describing one output variable at three points in time

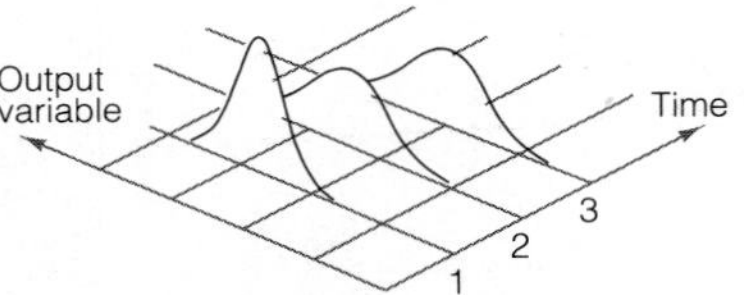

2. The particular value of the output variable that is realized at a given time can be thought of as being generated or produced according to the distribution described in point 1. (Alternatively, the realized value can be thought of as being generated by a random sample of size $n = 1$ from a population of values whose relative frequency distribution is that of point 1.)
3. The distribution that describes the output variable may change over time. For simplicity, we characterize the changes as being of three types: the mean (i.e., location) of the distribution may change; the variance (i.e., shape) of the distribution may change; or both. This is illustrated in Figure 13.14.

In general, when the output variable's distribution changes over time, we refer to this as a change in the *process*. Thus, if the mean shifts to a higher level, we say that the process mean has shifted. Accordingly, we sometimes refer to the distribution of the output variable as simply the **distribution of the process**, or the **output distribution of the process**.

FIGURE 13.14 ►
Types of changes in output variables

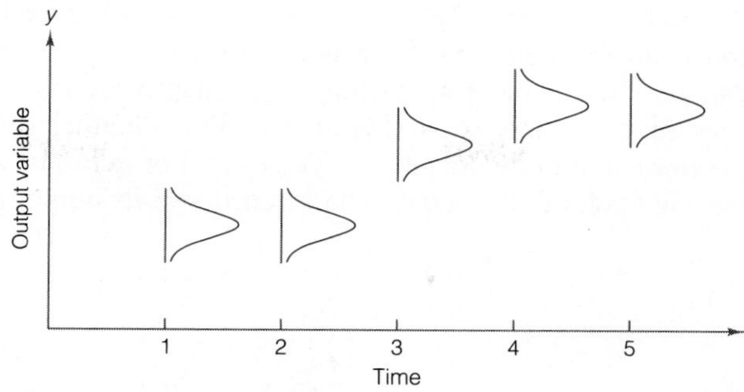

a. Change in mean (i.e., location)

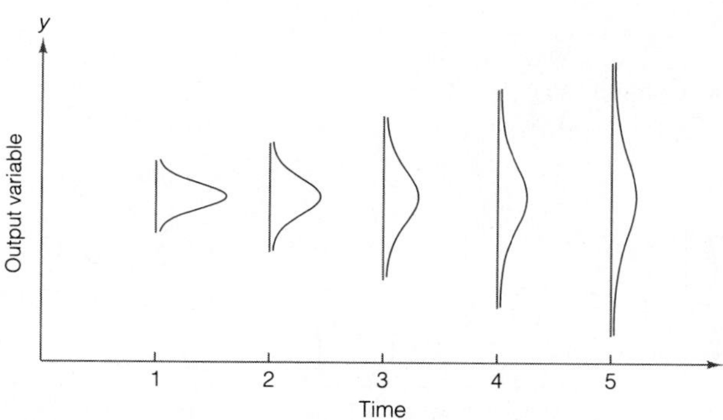

b. Change in variance (i.e., shape)

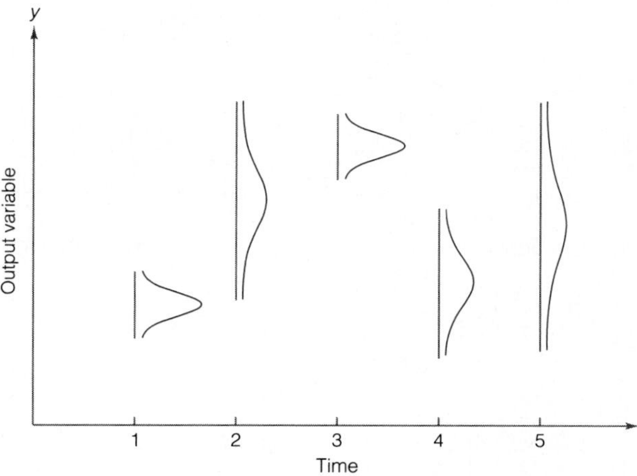

c. Change in mean and variance

Let's reconsider the patterns of variation in Figure 13.12 and model them using this conceptualization. This is done in Figure 13.15. The uptrend of Figure 13.12(a) can be characterized as resulting from a process whose mean is gradually shifting upward over time, as in Figure 13.15(a). Gradual shifts like this are a common phenomenon in manufacturing processes. For example, as a machine wears out (e.g., cutting blades dull), certain characteristics of its output gradually change.

FIGURE 13.15 ►
Patterns of process variation described by changing distributions

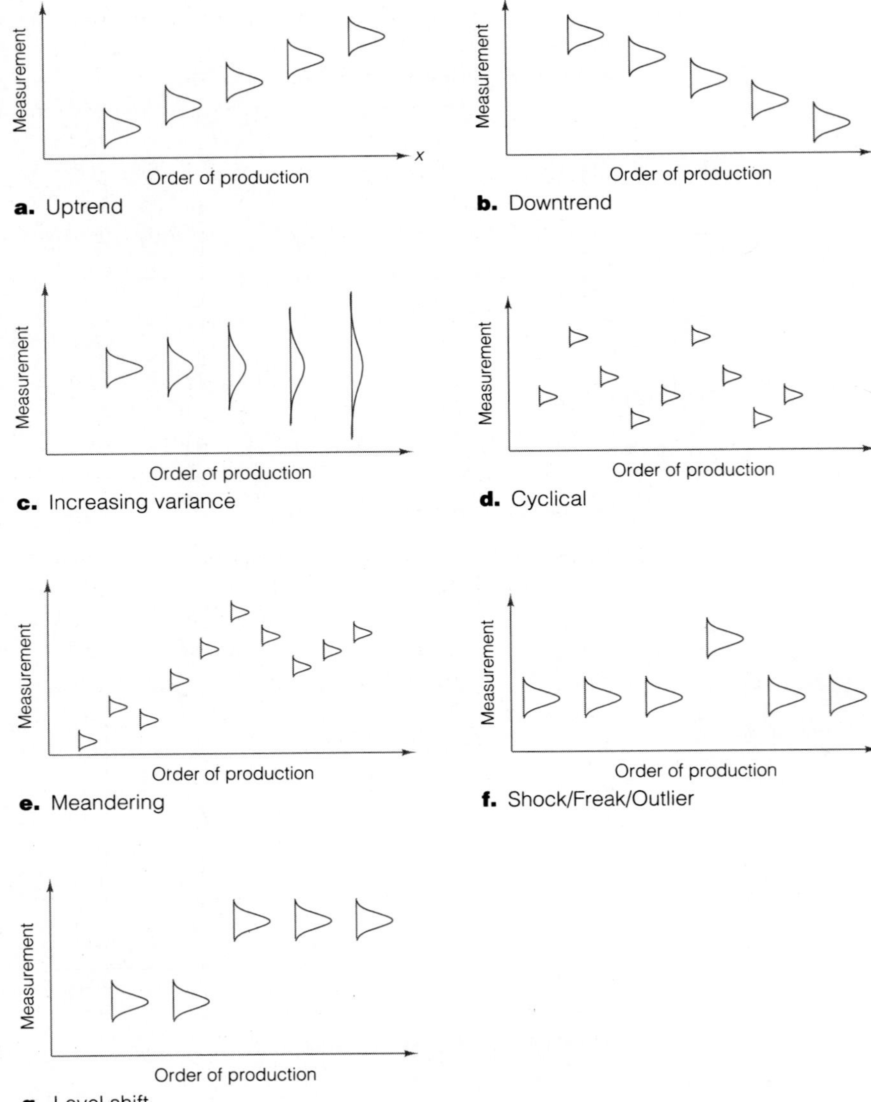

The pattern of increasing dispersion in Figure 13.12(c) can be thought of as resulting from a process whose mean remains constant but whose variance increases over time, as shown in Figure 13.15(c). This type of deterioration in a process may be the result of worker fatigue. At the beginning of a shift, workers—whether they be typists, machine operators, waiters, or managers—are fresh and pay close attention to every item that they process. But as the day wears on, concentration may wane and the workers may become more and more careless or more easily distracted. As a result, some items receive more attention than other items, causing the variance of the workers' output to increase.

The sudden shift in the level of the measurements in Figure 13.12(g) can be thought of as resulting from a process whose mean suddenly increases but whose variance remains constant, as shown in Figure 13.15(g). This type of pattern may be caused by such things as a change in the quality of raw materials used in the process or bringing a new machine or new operator into the process.

One thing that all these examples have in common is that the distribution of the output variable *changes over time*. In such cases, we say the process lacks **stability**. We formalize the notion of stability in the following definition.

Definition 13.6

A process whose output distribution does *not* change over time is said to be in a state of **statistical control**, or simply **in control**. If it does change, it is said to be **out of statistical control**, or simply **out of control**.

Figure 13.16 illustrates a sequence of output distributions for both an in-control and an out-of-control process. To see what the pattern of measurements looks like on a time series plot for a process that is in statistical control, consider Figure 13.17 on page 716. These data are from the same paint-filling process we described earlier, but the sequence of measurements was made *after* the faulty valve was replaced. Notice that there are no discernible persistent, systematic patterns in the sequence of measurements such as those in Figures 13.11 and 13.12(a)–(g). Nor, are there level shifts or transitory shocks as in Figures 13.12(e)–(f). This "patternless" behavior is called **ran-**

FIGURE 13.16 ►
Comparison of in-control and out-of-control processes

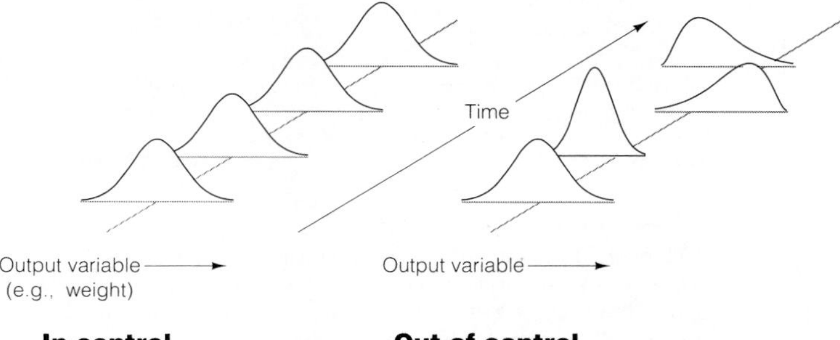

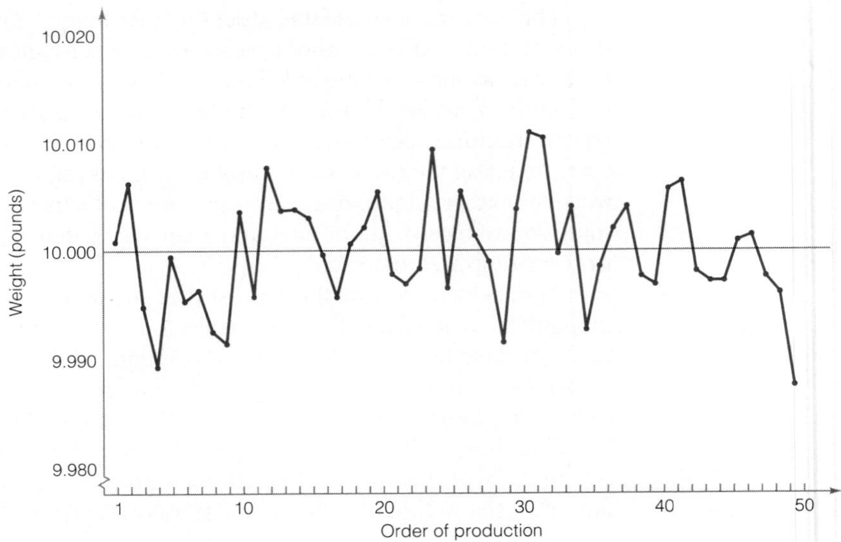

FIGURE 13.17 ► Time series plot of 50 consecutive paint can fills collected after replacing faulty valve

dom behavior. The output of processes that are in statistical control exhibits random behavior. Thus, even the output of stable processes exhibits variation.

If a process is in control and remains in control, its future will be like its past. Accordingly, the process is predictable, in the sense that its output will stay within certain limits. This cannot be said about an out-of-control process. As illustrated in Figure 13.18, with most out-of-control processes you have no idea what the future pattern of output from the process may look like.* You simply do not know what to

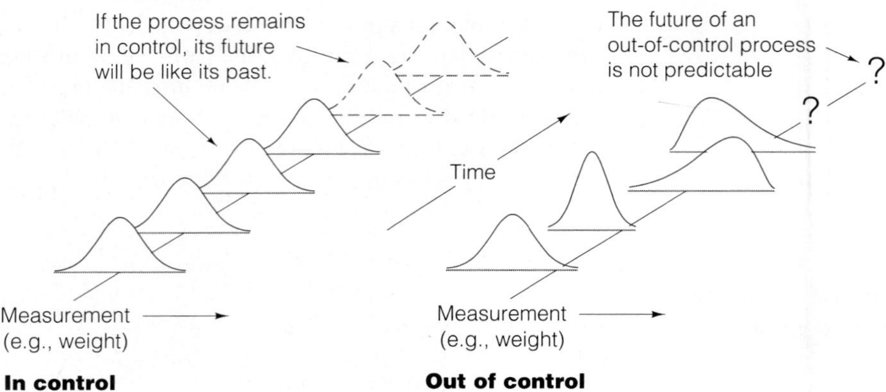

FIGURE 13.18 ► In-control processes are predictable; out-of-control processes are not.

*The output variables of in-control processes may follow approximately normal distributions, as in Figure 13.16 and 13.18, or they may not. But any in-control process will follow the *same* distribution over time. Do not misinterpret the use of normal distributions in many figures in this chapter as indicating that all in-control processes follow normal distributions.

expect from the process. Consequently, a business that operates out-of-control processes runs the risk of (1) providing inferior quality products and services to its internal customers (people within the organization who use the outputs of the processes) and (2) selling inferior products and services to its external customers. In short, it risks losing its customers and threatens its own survival.

One of the fundamental goals of process management is to identify out-of-control processes, to take actions to bring them into statistical control, and to keep them in a state of statistical control. The series of activities used to attain this goal is referred to as **statistical process control**.

Definition 13.7

The process of monitoring and eliminating variation in order to *keep* a process in a state of statistical control or to *bring* a process into statistical control is called **statistical process control (SPC)**.

Everything discussed in this section and the remaining sections of this chapter is concerned with statistical process control. We now continue our discussion of statistical control.

The variation that is exhibited by processes that are in control is said to be due to *common causes of variation*.

Definition 13.8

Common causes of variation are the methods, materials, machines, personnel, and environment that make up a process and the inputs required by the process. Common causes are thus attributable to the design of the process. Common causes affect all output of the process and may affect everyone who participates in the process.

The total variation that is exhibited by an in-control process is due to many different common causes, most of which affect process output in very minor ways. In general, however, each common cause has the potential to affect every unit of output produced by the process. Examples of common causes include the lighting in a factory or office, the grade of raw materials required, and the extent of worker training. Each of these factors can influence the variability of the process output. Poor lighting can cause workers to overlook flaws and defects that they might otherwise catch. Inconsistencies in raw materials can cause inconsistencies in the quality of the finished product. The extent of the training provided to workers can affect their level of expertise and, as a result, the quality of the products and services for which they are responsible.

Since common causes are, in effect, designed into a process, the level of variation that results from common causes is viewed as being representative of the capability of the process. If that level is too great (i.e., if the quality of the output varies too much), the process must be redesigned (or modified) to eliminate one or more common causes of variation. Since process redesign is the responsibility of management, *the elimination of common causes of variation is typically the responsibility of management*, not of the workers.

Processes that are out of control exhibit variation that is the result of both common causes and **special causes of variation**.

Definition 13.9

Special causes of variation (sometimes called **assignable causes**) are events or actions that are not part of the process design. Typically, they are transient, fleeting events that affect only local areas or operations within the process (e.g., a single worker, machine, or batch of materials) for a brief period of time. Occasionally, however, such events may have a persistent or recurrent effect on the process.

Examples of special causes of variation include a worker accidentally setting the controls of a machine improperly, a worker becoming ill on the job and continuing to work, a particular machine slipping out of adjustment, and a negligent supplier shipping a batch of inferior raw materials to the process.

In the latter case, the pattern of output variation may look like Figure 13.12(f). If instead of shipping just one bad batch the supplier continued to send inferior materials, the pattern of variation might look like Figure 13.12(g). The output of a machine that is gradually slipping out of adjustment might yield a pattern like Figure 13.12(a), (b), or (c). All these patterns owe part of their variation to common causes and part to the noted special causes. In general, we treat any pattern of variation other than a random pattern as due to both common and special causes.*

Since the effects of special causes are frequently localized within a process, *special causes can often be diagnosed and eliminated by workers or their immediate supervisor.* Occasionally, they must be dealt with by management, as in the case of a negligent or deceitful supplier.

It is important to recognize that **most processes are not naturally in a state of statistical control**. As Deming (1986, p. 322) observed:

> *Stability [i.e., statistical control] is seldom a natural state. It is an achievement, the result of eliminating special causes one by one . . . leaving only the random variation of a stable process.*

---

*For certain processes (e.g., those affected by seasonal factors), a persistent systematic pattern—such as the cyclical pattern of Figure 13.12(d)—is an inherent characteristic. In these special cases, some analysts treat the cause of the systematic variation as a common cause. This type of analysis is beyond the scope of this text. We refer the interested reader to Alwan and Roberts (1988).

Process improvement first requires the identification, diagnosis, and removal of special causes of variation. Removing all special causes puts the process in a state of statistical control. Further improvement of the process then requires the identification, diagnosis, and removal of common causes of variation. The effects on the process of the removal of special and common causes of variation are illustrated in Figure 13.19.

FIGURE 13.19 ►
The effects of eliminating causes of variation.

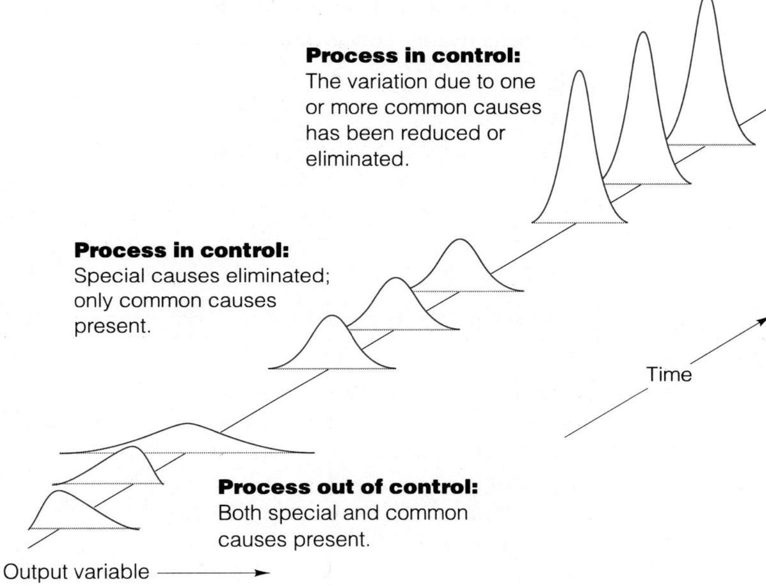
Process in control:
The variation due to one or more common causes has been reduced or eliminated.
Process in control:
Special causes eliminated; only common causes present.
Time
Process out of control:
Both special and common causes present.
Output variable

In the remainder of this chapter, we introduce you to some of the methods of statistical process control. In particular, we address how control charts help us determine whether a given process is in control.

13.7 The Logic of Control Charts

We use control charts to help us differentiate between process variation due to common causes and special causes. That is, we use them to determine whether a process is under statistical control (only common causes present) or not (both common and special causes present). Being able to differentiate means knowing when to take action to find and remove special causes and when to leave the process alone. If you take actions to remove special causes that do not exist—this is called tampering with the process—you may actually end up increasing the variation of the process and, thereby, hurting the quality of the output.

In general, control charts are useful for evaluating the past performance of a process and for monitoring its current performance. We can use them to determine whether a process was in control during, say, the past two weeks or to determine whether the process is remaining under control from hour to hour or minute to minute. In the latter case, our goal is the swiftest detection and removal of any special causes of variation that might arise. Keep in mind that **the primary goal of quality improvement activities is variance reduction**.

In this chapter we show you how to construct and use control charts for both quantitative and qualitative quality variables. Important quantitative variables include such things as weight, width, and time. An important qualitative variable is product status: defective or nondefective.

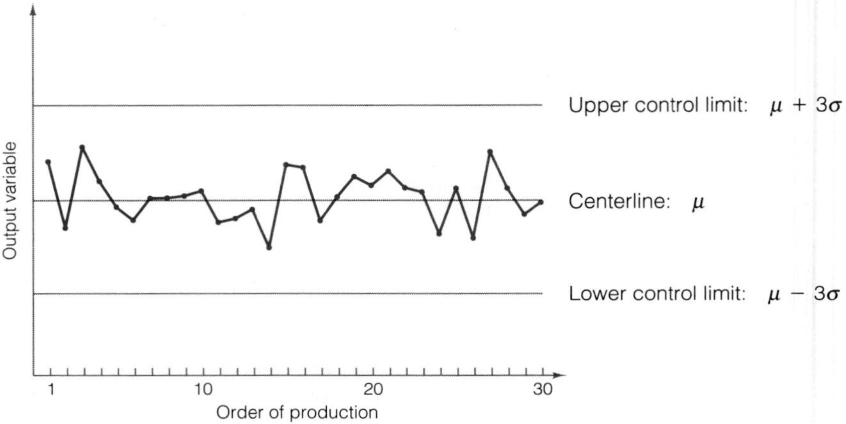

FIGURE 13.20 ► A control chart

An example of a control chart is shown in Figure 13.20. It is simply a time series plot of the individual measurements of a quality variable (i.e., an output variable), to which a centerline and two other horizontal lines called **control limits** have been added. The centerline represents the mean of the process (i.e., the mean of the quality variable) *when the process is in a state of statistical control*. The **upper control limit** and the **lower control limit** are positioned so that *when the process is in control* the probability of an individual value of the output variable falling outside the control limits is very small. Most practitioners position the control limits a distance of three standard deviations from the centerline (i.e., from the process mean) and refer to them as **3-sigma limits**. If the process is in control and following a normal distribution, the probability of an individual measurement falling outside the control limits is .0027 (less than 3 chances in 1,000). This is shown in Figure 13.21.

As long as the individual values stay between the control limits, the process is considered to be under control, meaning that no special causes of variation are influencing the output of the process. If one or more values fall outside the control limits, either a **rare event** has occurred or the process is out of control. Following the rare-event approach to inference described earlier in the text, such a result is interpreted as evidence that the process is out of control and that actions should be taken to eliminate the special causes of variation that exist.

FIGURE 13.21 ▶
The probability of observing a measurement beyond the control limits when the process is in control

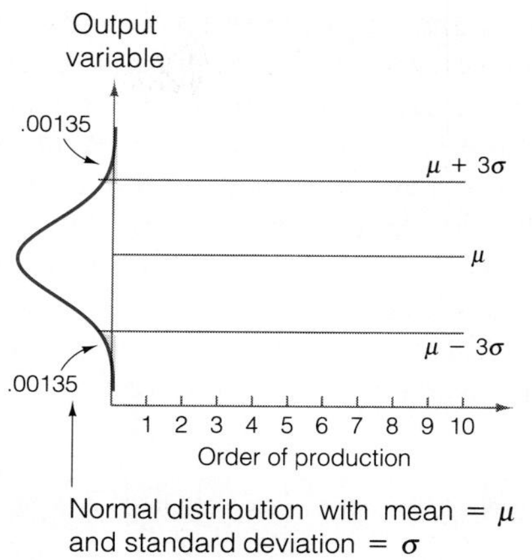

Other evidence to indicate that the process is out of control may be present on the control chart. For example, if we observe any of the patterns of variation shown in Figure 13.12, we can conclude the process is out of control *even if all the points fall between the control limits*. In general, any persistent, systematic variation pattern (i.e., any nonrandom pattern) is interpreted as evidence that the process is out of control. We discuss this in detail in the next section.

In Chapter 8 we described how to make inferences about populations using hypothesis-testing techniques. What we do in this section should seem quite similar. Although our focus now is on making inferences about a *process* rather than a *population*, we are again testing hypotheses. In this case, we test

H_0: Process is under control $\quad$ H_a: Process is out of control

Each time we plot a new point and see whether it falls inside or outside of the control limits, we are running a two-sided hypothesis test. The control limits function as the critical values for the test.

What we learned in Chapter 9 about the types of errors that we might make in running a hypothesis test holds true in using control charts as well. Any time we reject the hypothesis that the process is under control and conclude that the process is out of control, we run the risk of making a Type I error (rejecting the null hypothesis when the null is true). Anytime we conclude (or behave as if we conclude) that the process is in control, we run the risk of a Type II error (accepting the null hypothesis when the alternative is true). There is nothing magical or mystical about control charts. Just as in any hypothesis test, the conclusion suggested by a control chart may be wrong.

One of the main reasons that 3-sigma control limits are used (rather than 2-sigma or 1-sigma limits, for example) is the small Type I error probability associated with their use. The probability we noted previously of an individual measurement falling outside the control limits—.0027—is a Type I error probability. Since we interpret a sample point that falls beyond the limits as a signal that the process is out of control, the use of 3-sigma limits yields very few signals that are "false alarms."

To make these ideas more concrete, we will construct and interpret a control chart for the paint-filling process discussed in Section 13.6. Our intention is simply to help you better understand the logic of control charts. Structured, step-by-step descriptions of how to construct control charts will be given in later sections.

The sample measurements from the paint-filling process, presented in Table 13.1, were previously plotted in Figure 13.17. We use the mean and standard deviation of the sample, $\bar{x} = 9.9997$ and $s = .0053$, to estimate the mean and the standard deviation of the process. Although these are estimates, in using and interpreting control charts we treat them *as if* they were the actual mean μ and standard deviation σ of the process. This is standard practice in control charting.

TABLE 13.1 Fill Weights of 50 Consecutively Produced Cans of Paint

1. 10.0008	11. 9.9957	21. 9.9977	31. 10.0107	41. 10.0054
2. 10.0062	12. 10.0076	22. 9.9968	32. 10.0102	42. 10.0061
3. 9.9948	13. 10.0036	23. 9.9982	33. 9.9995	43. 9.9978
4. 9.9893	14. 10.0037	24. 10.0092	34. 10.0038	44. 9.9969
5. 9.9994	15. 10.0029	25. 9.9964	35. 9.9925	45. 9.9969
6. 9.9953	16. 9.9995	26. 10.0053	36. 9.9983	46. 10.0006
7. 9.9963	17. 9.9956	27. 10.0012	37. 10.0018	47. 10.0011
8. 9.9925	18. 10.0005	28. 9.9988	38. 10.0038	48. 9.9973
9. 9.9914	19. 10.0020	29. 9.9914	39. 9.9974	49. 9.9958
10. 10.0035	20. 10.0053	30. 10.0036	40. 9.9966	50. 9.9873

The centerline of the control chart, representing the process mean, is drawn so that it intersects the vertical axis at 9.9997, as shown in Figure 13.22. The upper control limit is drawn at a distance of $3s = 3(.0053) = .0159$ above the centerline, and the lower control limit is $3s = .0159$ below the centerline. Then the 50 sample weights are plotted on the chart in the order that they were generated by the paint-filling process.

As can be seen in Figure 13.22, all the weight measurements fall within the control limits. Further, there do not appear to be any systematic nonrandom patterns in the data such as displayed in Figures 13.11 and 13.12. Accordingly, we are unable to conclude that the process is out of control. That is, we are unable to reject the null hypothesis that the process is in control. However, instead of using this formal hypothesis-testing language in interpreting control chart results, we prefer simply to say that the data suggest or indicate that the process is in control. We do this, however, with the full understanding that the probability of a Type II error is generally unknown in control chart applications and that we might be wrong in our conclusion. What we are really saying when we conclude that the process is in control is that *the data indicate that it is better to behave as if the process were under control than to tamper with the process.*

We have portrayed the control chart hypothesis test as testing "in control" versus "out of control." Another way to look at it is this: When we compare the weight of an

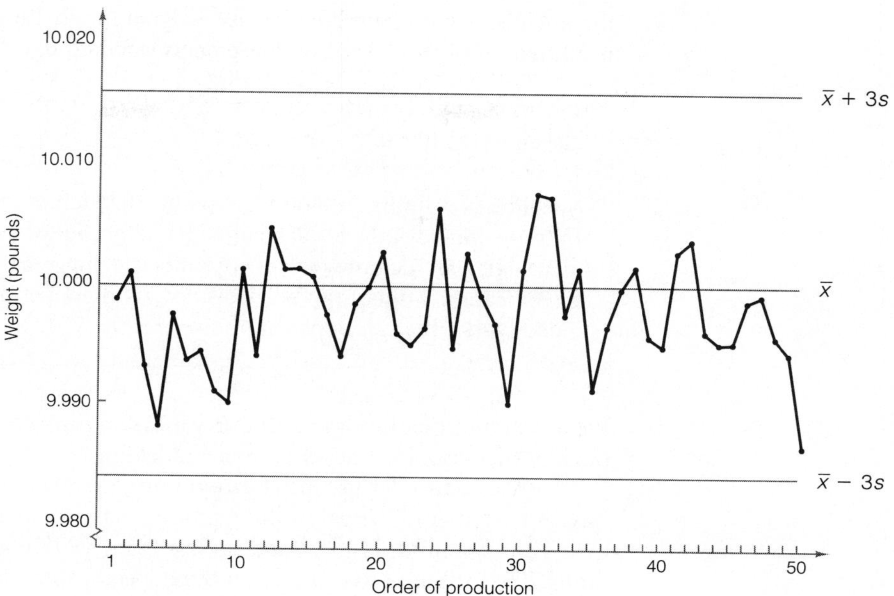

FIGURE 13.22 ▶
Control chart of fill weights for 50 consecutive paint can fills

individual can of paint to the control limits, we are conducting the following two-tailed hypothesis test:

$$H_0: \quad \mu = 9.9997 \qquad H_a: \quad \mu \neq 9.9997$$

where 9.9997 is the centerline of the control chart. The control limits delineate the two rejection regions for this test. Accordingly, with each weight measurement that we plot and compare to the control limits, we are testing whether the process mean (the mean fill weight) has changed. Thus, what the control chart is monitoring is the mean of the process. **The control chart leads us to accept or reject statistical control on the basis of whether the mean of the process has changed or not**. This type of process instability is illustrated in the top graph of Figure 13.14. In the paint-filling process example, the process mean apparently has remained constant over the period in which the sample weights were collected.

Other types of control charts—one of which we will describe in Section 13.10—help us determine whether the *variance* of the process has changed, as in the middle and bottom graphs of Figure 13.14.

The control chart we have just described is called an **individuals chart**, or an ***x*-chart**. The term *individuals* refers to the fact that the chart uses individual measurements to monitor the process—that is, measurements taken from individual units of process output. This is in contrast to plotting sample means on the control chart, for example, as we do in the next section.

Students sometimes confuse control limits with product **specification limits**. We have already explained control limits, which are a function of the natural variability of the process. Assuming we always use 3-sigma limits, the position of the control limits is a function of the size of σ, the process standard deviation.

Definition 13.10

Specification limits are boundary points that define the acceptable values for an output variable (i.e., for a quality characteristic) of a particular product or service. They are determined by customers, management, and product designers. Specification limits may be two-sided, with upper and lower limits, or one-sided, with either an upper or a lower limit.

Process output that falls inside the specification limits is said to **conform to specifications**. Otherwise it is said to be **nonconforming**.

Unlike control limits, specification limits are not dependent on the process in any way. A customer of the paint-filling process may specify that all cans contain no more than 10.005 pounds of paint and no less than 9.995 pounds. These are specification limits. The customer has reasons for these specifications but may have no idea whether the supplier's process can meet them. Both the customer's specification limits and the control limits of the supplier's paint-filling process are shown in Figure 13.23. Do you think the customer will be satisfied with the quality of the product received? We don't. Although some cans are within the specification limits, most are not, as indicated by the shaded region on the figure.

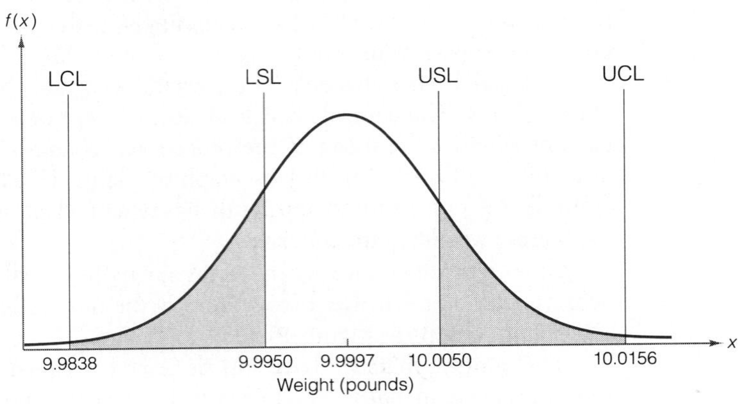

FIGURE 13.23 ▶ Comparison of control limits and specification limits

13.8 A Control Chart for Monitoring the Mean of a Process: The $\bar{x}$-Chart

In the last section we introduced you to the logic of control charts by focusing on a chart that reflected the variation in individual measurements of process output. We used the chart to determine whether the process mean had shifted. The control chart we present in this section—the $\bar{x}$-chart—is also used to detect changes in the process mean, but it does so by monitoring the variation in the mean of samples that have been drawn from the process. That is, instead of plotting individual measurements on the control chart, in this case we plot sample means. Because of the additional information reflected in sample means (because each sample mean is calculated from n individual measurements), the $\bar{x}$-chart is more sensitive than the individuals chart for detecting changes in the process mean.

In practice, the $\bar{x}$-chart is rarely used alone. It is typically used in conjunction with a chart that monitors the variation of the process, usually a chart called an R-chart. The $\bar{x}$- and R-charts are the most widely used control charts in industry. Used in concert, these charts make it possible to determine whether a process has gone out of control because the variation has changed or because the mean has changed. We present the R-chart in the next section, at the end of which we discuss their simultaneous use. For now, we focus only on the $\bar{x}$-chart. **Consequently, we assume throughout this section that the process variation is stable.***

Figure 13.24 provides an example of an $\bar{x}$-chart. As with the individuals chart, the centerline represents the mean of the process and the upper and lower control limits are positioned a distance of three standard deviations from the mean. However, since the chart is tracking sample means rather than individual measurements, the relevant standard deviation is the standard deviation of $\bar{x}$, not σ, the standard deviation of the output variable.

FIGURE 13.24 ▶
$\bar{x}$-Chart

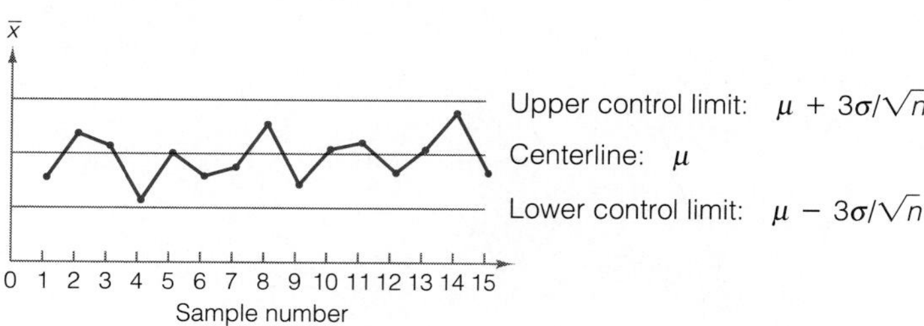

*_To the instructor:_ Technically, the R-chart should be constructed and interpreted before the $\bar{x}$-chart. However, in our experience, students more quickly grasp control-chart concepts if they are familiar with the underlying theory. We begin with $\bar{x}$-charts because their underlying theory was presented in Chapters 6–8.

If the process were in statistical control, the sequence of $\bar{x}$'s plotted on the chart would exhibit random behavior between the control limits. Only if a rare event occurred or if the process went out of control would a sample mean fall beyond the control limits.

To better understand the justification for having control limits that involve $\sigma_{\bar{x}}$, consider the following. The $\bar{x}$-chart is concerned with the variation in $\bar{x}$ which, as we saw in Chapter 6, is described by $\bar{x}$'s sampling distribution. But what is $\bar{x}$'s sampling distribution? If the process is in control and its output variable x is characterized at each point in time by a normal distribution with mean μ and standard deviation σ, the distribution of $\bar{x}$ (i.e., $\bar{x}$'s sampling distribution) also follows a normal distribution with mean μ at each point in time. But, as we saw in Chapter 6, its standard deviation is $\sigma_{\bar{x}} = \sigma/\sqrt{n}$. The control limits of the $\bar{x}$-chart are determined from and interpreted with respect to the sampling distribution of $\bar{x}$, not the distribution of x. These points are illustrated in Figure 13.25.*

In order to construct an $\bar{x}$-chart, you should have at least 20 samples of n items each, where $n \geq 2$. This will provide sufficient data to obtain reasonably good esti-

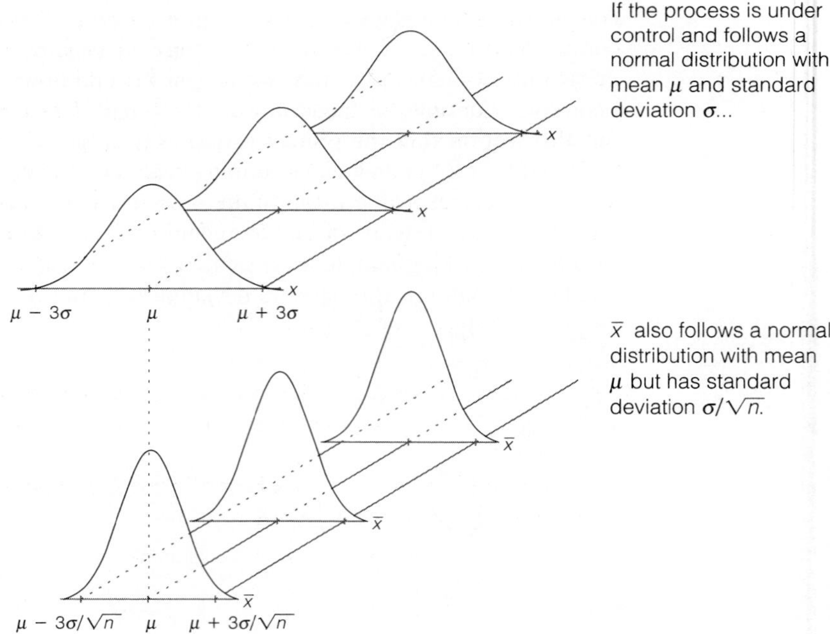
If the process is under control and follows a normal distribution with mean μ and standard deviation σ...
x
x
x
μ − 3σ
μ
μ + 3σ
x̄ also follows a normal distribution with mean μ but has standard deviation σ/√n.
x̄
x̄
x̄
μ − 3σ/√n
μ
μ + 3σ/√n

FIGURE 13.25 ► The sampling distribution of $\bar{x}$

*The sampling distribution of $\bar{x}$ can also be approximated using the central limit theorem (Chapter 6). That is, when the process is under control and $\bar{x}$ is to be computed from a large sample from the process ($n \geq 30$), the sampling distribution will be approximately normally distributed with mean μ and standard deviation $\sigma/\sqrt{n}$. Even for samples as small as 4 or 5, the sampling distribution of $\bar{x}$ will be approximately normal as long as the distribution of x is reasonably symmetric and roughly bell-shaped.

mates of the mean and variance of the process. The centerline, which represents the mean of the process, is determined as follows:

$$\textit{Centerline:} \quad \bar{\bar{x}} = \frac{\bar{x}_1 + \bar{x}_2 + \cdots + \bar{x}_k}{k}$$

where k is the number of samples of size n from which the chart is to be constructed and $\bar{x}_i$ is the sample mean of the ith sample. Thus $\bar{\bar{x}}$ is an estimator of μ.

The control limits are positioned as follows:

$$\textit{Upper control limit:} \quad \bar{\bar{x}} + \frac{3\sigma}{\sqrt{n}}$$

$$\textit{Lower control limit:} \quad \bar{\bar{x}} - \frac{3\sigma}{\sqrt{n}}$$

Since σ, the process standard deviation, is virtually always unknown, it must be estimated. This can be done in several ways. One approach involves calculating the standard deviations for each of the k samples and averaging them. Another involves using the sample standard deviation s from a large sample that was generated while the process was believed to be in control. We employ a third approach, however—the one favored by industry. It has been shown to be as effective as the other approaches for sample sizes of $n = 10$ or less, the sizes most often used in industry.

This approach utilizes the ranges of the k samples to estimate the process standard deviation, σ. Recall from Chapter 2 that the range, R, of a sample is the difference between the maximum and minimum measurements in the sample. It can be shown that dividing the mean of the k ranges, $\bar{R}$, by the constant d_2, obtains an unbiased estimator for σ. [For details, see Ryan (1989).] The estimator, denoted by $\hat{\sigma}$, is calculated as follows:

$$\hat{\sigma} = \frac{\bar{R}}{d_2} = \frac{R_1 + R_2 + \cdots + R_k}{k}\left(\frac{1}{d_2}\right)$$

where R_i is the range of the ith sample and d_2 is a constant that depends on the sample size. Values of d_2 for samples of size $n = 2$ to $n = 25$ can be found in Appendix B, Table XVII.

Substituting $\hat{\sigma}$ for σ in the formulas for the upper control limit (UCL) and the lower control limit (LCL), we get

$$\text{UCL:} \quad \bar{\bar{x}} + \frac{3\left(\dfrac{\bar{R}}{d_2}\right)}{\sqrt{n}} \qquad \text{LCL:} \quad \bar{\bar{x}} - \frac{3\left(\dfrac{\bar{R}}{d_2}\right)}{\sqrt{n}}$$

Notice that $(\bar{R}/d_2)/\sqrt{n}$ is an estimator of $\sigma_{\bar{x}}$. The calculation of these limits can be simplified by creating the constant

$$A_2 = \frac{3}{d_2\sqrt{n}}$$

Then the control limits can be expressed as

$$\text{UCL: } \bar{\bar{x}} + A_2\bar{R} \qquad \text{LCL: } \bar{\bar{x}} - A_2\bar{R}$$

where the values for A_2 for samples of size $n = 2$ to $n = 25$ can be found in Appendix B, Table XVII.

The degree of sensitivity of the $\bar{x}$-chart to changes in the process mean depends on two decisions that must be made in constructing the chart.

The Two Most Important Decisions in Constructing an $\bar{x}$-Chart

1. The sample size, n, must be determined.
2. The frequency with which samples are to be drawn from the process must be determined (e.g., once an hour, once each shift, or once a day).

In order to quickly detect process change, we try to choose samples in such a way that the change in the process mean occurs *between* samples, not *within* samples (i.e., not during the period when a sample is being drawn). In this way, every measurement in the sample before the change will be unaffected by the change and every measurement in the sample following the change will be affected. The result is that the $\bar{x}$ computed from the latter sample should be substantially different from that of the former sample—a signal that something has happened to the process mean.

Definition 13.11

Samples whose size and frequency have been designed to make it likely that process changes will occur between, rather than within, the samples are referred to as **rational subgroups**.

Rational Subgrouping Strategy

The samples (rational subgroups) should be chosen in a manner that:

1. Gives the maximum chance for the *measurements* in each sample to be similar (i.e., to be affected by the same sources of variation).
2. Gives the maximum chance for the *samples* to differ (i.e., be affected by at least one different source of variation).

The following example illustrates the concept of **rational subgrouping**. An operations manager suspects that the quality of the output in a manufacturing process may differ from shift to shift because of the preponderance of newly hired workers on the night shift. The manager wants to be able to detect such differences quickly, using an $\bar{x}$-chart. Following the rational subgrouping strategy, the control chart should be constructed with samples that are drawn *within* each shift. None of the samples should span shifts. That is, no sample should contain, say, the last three items produced by shift 1 and the first two items produced by shift 2. In that way, the measurements in each sample would be similar, but the $\bar{x}$'s would reflect differences between shifts.

The secret to designing an effective $\bar{x}$-chart is to anticipate the *types of special causes of variation* that might affect the process mean. Then purposeful rational subgrouping can be employed to construct a chart that is sensitive to the anticipated cause or causes of variation.

The preceding discussion and example focused primarily on the timing or frequency of samples. Concerning the size of the samples, practitioners typically work with samples of size $n = 4$ to $n = 10$ consecutively produced items. Using small samples of consecutively produced items helps to ensure that the measurements in each sample will be similar (i.e., affected by the same causes of variation).

Constructing an $\bar{x}$-Chart: A Summary

1. Using a rational subgrouping strategy, collect at least 20 samples (subgroups), each of size $n \geq 2$.
2. Calculate the mean and range for each sample.
3. Calculate the mean of the sample means, $\bar{\bar{x}}$, and the mean of the sample ranges, $\bar{R}$:

$$\bar{\bar{x}} = \frac{\bar{x}_1 + \bar{x}_2 + \cdots + \bar{x}_k}{k} \qquad \bar{R} = \frac{R_1 + R_2 + \cdots + R_k}{k}$$

where k = The number of samples (i.e., subgroups)
$\bar{x}_i$ = The sample mean for the ith sample
R_i = The range of the ith sample

4. Plot the centerline and control limits:

Centerline: $\bar{\bar{x}}$
Upper control limit: $\bar{\bar{x}} + A_2\bar{R}$ *Lower control limit:* $\bar{\bar{x}} - A_2\bar{R}$

where A_2 is a constant that depends on n. Its values are given in Appendix B, Table XVII, for samples of size $n = 2$ to $n = 25$.

5. Plot the k sample means on the control chart in the order that the samples were produced by the process.

When interpreting a control chart, it is convenient to think of the chart as consisting of six zones, as shown in Figure 13.26 (page 730). Each zone is one standard

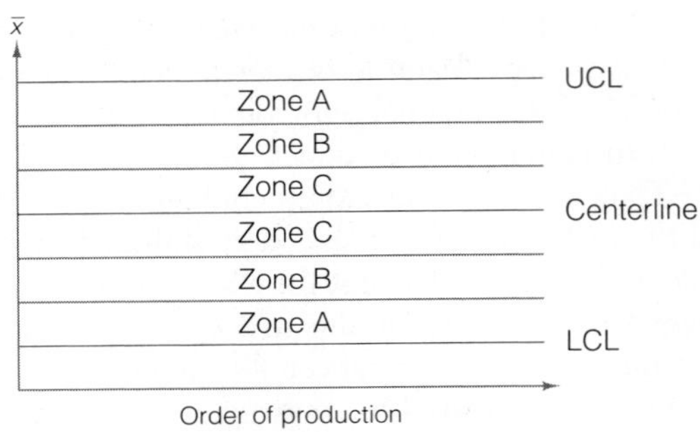

FIGURE 13.26 ► The zones of a control chart

deviation wide. The two zones within one standard deviation of the centerline are called **C zones**; the regions between one and two standard deviations from the centerline are called **B zones**; and the regions between two and three standard deviations from the centerline are called **A zones**. The box describes how to construct the *zone boundaries* for an $\bar{x}$-chart.

Practitioners use six simple rules that are based on these zones to help determine when a process is out of control. The six rules are summarized in Figure 13.27 on page 732. They are referred to as **pattern-analysis rules**.

Rule 1 is the familiar point-beyond-the-control-limit rule that we have mentioned several times. The other rules all help to determine when the process is out of control *even though all the plotted points fall within the control limits*. That is, the other rules help to identify nonrandom patterns of variation that have not yet broken through the control limits (or may never break through).

All the patterns shown in Figure 13.27 are **rare events** under the assumption that the process is under control. To see this, let's assume that the process is under control and follows a normal distribution. We can then easily work out the probability that an individual point will fall in any given zone. (We dealt with this type of problem in Chapters 4 and 5.) Just focusing on one side of the centerline, you can show that the probability of a point falling beyond Zone A is .00135, in Zone A is .02135, in Zone B is .1360, and in Zone C is .3413. Of course, the same probabilities apply to both sides of the centerline.

From these probabilities we can determine the likelihood of various patterns of points. For example, let's evaluate rule 1. The probability of observing a point outside the control limits (i.e., above the upper control limit or below the lower control limit) is $.00135 + .00135 = .0027$. This is clearly a rare event.

As another example, rule 5 indicates that the observation of two out of three points in a row in Zone A or beyond is a rare event. Is it? The probability of being in Zone A or beyond is $.00135 + .02135 = .0227$. We can use the binomial distribution (Chapter 5) to find the probability of observing 2 out of 3 points in or beyond Zone A. The binomial probability $P(x = 2)$ when $n = 3$ and $p = .0227$ is .0015. Again, this is clearly a rare event.

In general, when the process is in control and normally distributed, the probability of any one of these rules *incorrectly* signaling the presence of special causes of variation

Constructing Zone Boundaries for an $\bar{x}$-Chart

The zone boundaries can be constructed in either of the following ways:

1. Using the 3-sigma control limits:

$$\textit{Upper A–B boundary:} \quad \bar{\bar{x}} + \frac{2}{3}(A_2\bar{R})$$

$$\textit{Lower A–B boundary:} \quad \bar{\bar{x}} - \frac{2}{3}(A_2\bar{R})$$

$$\textit{Upper B–C boundary:} \quad \bar{\bar{x}} + \frac{1}{3}(A_2\bar{R})$$

$$\textit{Lower B–C boundary:} \quad \bar{\bar{x}} - \frac{1}{3}(A_2\bar{R})$$

2. Using the estimated standard deviation of $\bar{x}$, $(\bar{R}/d_2)/\sqrt{n}$:

$$\textit{Upper A–B boundary:} \quad \bar{\bar{x}} + 2\left[\frac{\left(\frac{\bar{R}}{d_2}\right)}{\sqrt{n}}\right]$$

$$\textit{Lower A–B boundary:} \quad \bar{\bar{x}} - 2\left[\frac{\left(\frac{\bar{R}}{d_2}\right)}{\sqrt{n}}\right]$$

$$\textit{Upper B–C boundary:} \quad \bar{\bar{x}} + \left[\frac{\left(\frac{\bar{R}}{d_2}\right)}{\sqrt{n}}\right]$$

$$\textit{Lower B–C boundary:} \quad \bar{\bar{x}} - \left[\frac{\left(\frac{\bar{R}}{d_2}\right)}{\sqrt{n}}\right]$$

is less than .005, or 5 chances in 1,000. If all of the first four rules are applied, the overall probability of a false signal is about .01. If all six of the rules are applied, the overall probability of a false signal rises to .02, or 2 chances in 100. These three probabilities can be thought of as Type I error probabilities. Each indicates the probability of incorrectly rejecting the null hypothesis that the process is in a state of statistical control.

Explanation of the possible causes of these nonrandom patterns is beyond the scope of this text. We refer the interested reader to AT&T's *Statistical Quality Control Handbook* (1956).

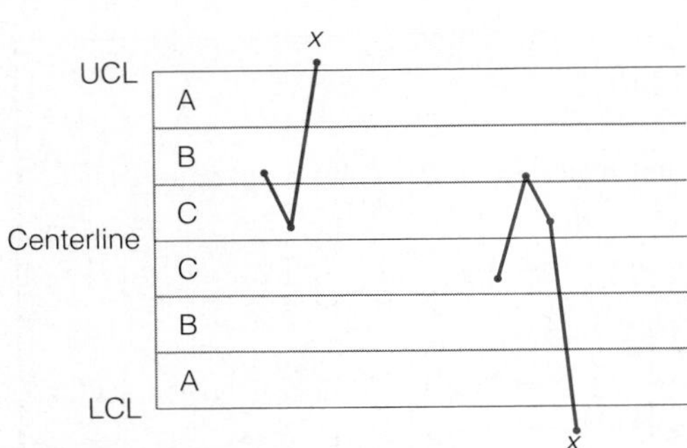

Rule1: One point beyond Zone A

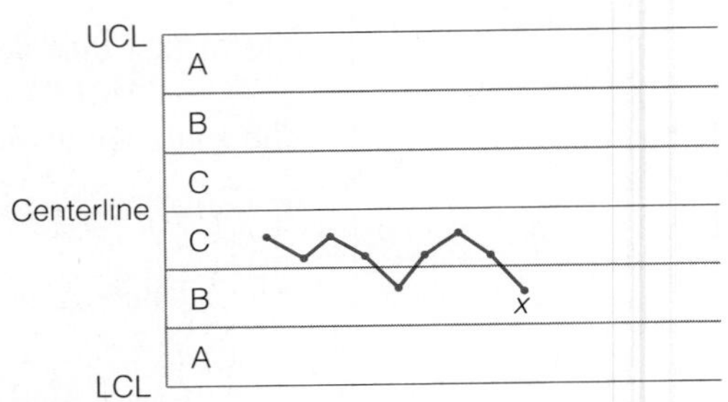

Rule 2: Nine points in a row in Zone C or beyond

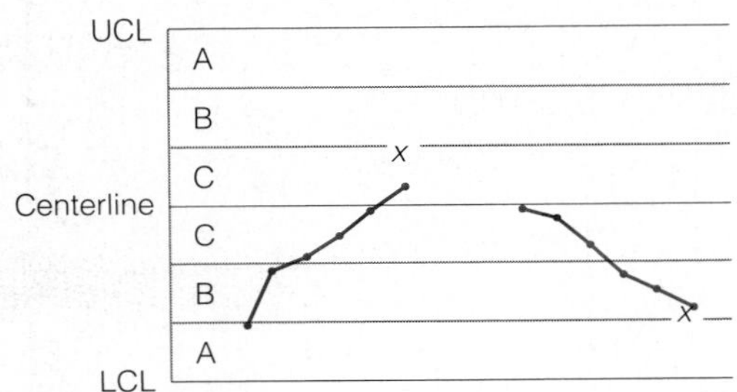

Rule 3: Six points in a row steadily increasing or decreasing

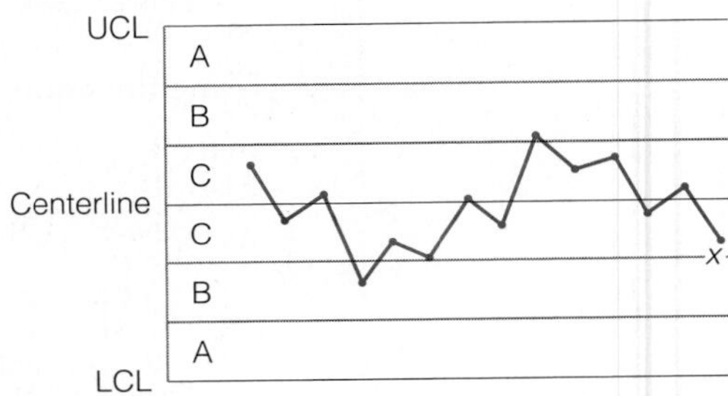

Rule 4: Fourteen points in a row alternating up and down

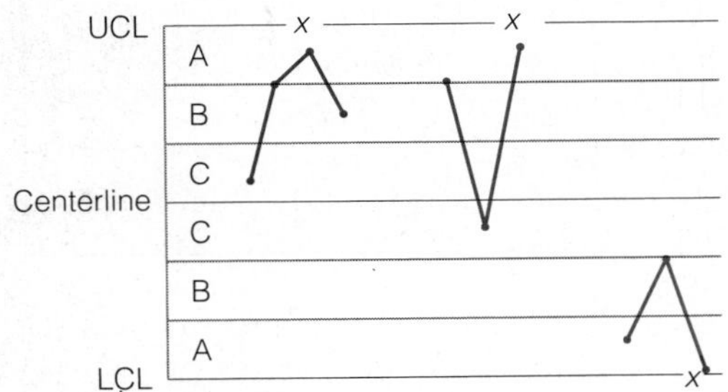

Rule 5: Two out of three points in a row in Zone A or beyond

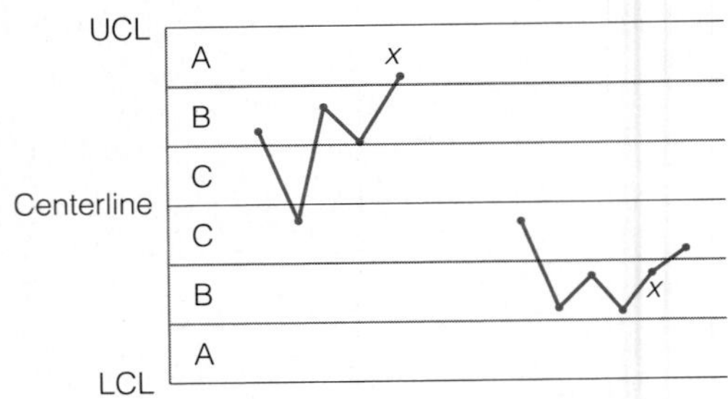

Rule 6: Four out of five points in a row in Zone B or beyond

Rules 1, 2, 5, and 6 should be applied separately to the upper and lower halves of the control chart. Rules 3 and 4 should be applied to the whole chart.

FIGURE 13.27 ▲
Pattern-analysis rules for detecting the presence of special causes of variation

We use these rules again in the next section when we interpret the R-chart.

Interpreting an $\bar{x}$-Chart

1. The **process is out of control** if one or more sample means fall beyond the control limits or if any of the other five patterns of variation of Figure 13.27 are observed. Such signals are an indication that one or more special causes of variation are affecting the process mean. We must identify and eliminate them to bring the process into control.
2. The **process is treated as being in control** if none of the previously noted out-of-control signals are observed. Processes that are in control should not be tampered with. However, if the level of variation is unacceptably high, common causes of variation should be identified and eliminated.

Assumption: The variation of the process is stable. (If it were not, the control limits of the $\bar{x}$-chart would be meaningless, since they are a function of the process variation. The R-chart, presented in the next section, is used to investigate this assumption.)

In theory, the centerline and control limits should be developed using samples that were collected during a period in which the process was in control. Otherwise, they will not be representative of the variation of the process (or, in the present case, the variation of $\bar{x}$) when the process is in control. However, we will not know whether the process is in control until after we have contructed a control chart. Consequently, when a control chart is first constructed, the centerline and the control limits are treated as **trial values**. If the chart indicates that the process was in control during the period when the sample data were collected, then the centerline and control limits become "official" (i.e., no longer treated as trial values). It is then appropriate to extend the control limits and the centerline to the right and to use the chart to monitor future process output.

However, if in applying the pattern-analysis rules of Figure 13.27 it is determined that the process was out of control while the sample data were being collected, the trial values (i.e., the trial chart) should, in general, not be used to monitor the process. The points on the control chart that indicate that the process is out of control should be investigated to see if any special causes of variation can be identified. A graphical method that can be used to facilitate this investigation—a *cause-and-effect diagram*—is described in Section 13.11. If special causes of variation are found, (1) they should be eliminated, (2) any points on the chart determined to have been influenced by the special causes—whether inside or outside the control limits—should be discarded, and (3) *new* trial centerline and control limits should be calculated from the remaining data. However, the new trial limits may still indicate that the process is out of control. If so, repeat these three steps until all points fall within the control limits.

If special causes cannot be found and eliminated, the severity of the out-of-control indications should be evaluated and a judgment made as to whether (1) the out-of-control points should be discarded anyway and new trial limits constructed, (2) the original trial limits are good enough to be made official, or (3) new sample data should be collected to construct new trial limits.

EXAMPLE 13.1

Let's return to the paint-filling process described in Sections 13.6 and 13.7. Suppose instead of sampling 50 consecutive gallons of paint from the filling process to develop a control chart, it was decided to sample five consecutive cans once each hour for the next 25 hours. The sample data are presented in Table 13.2. This sampling strategy (rational subgrouping) was selected because several times a month the filling head in question becomes clogged. When that happens, the head dispenses less and less paint over the course of the day. However, the pattern of decrease is so irregular that minute-to-minute or even half-hour-to-half-hour changes are difficult to detect.

TABLE 13.2 Twenty-five Samples of Size 5 from the Paint-Filling Process

Sample	Measurements					Mean	Range
1	10.0042	9.9981	10.0010	9.9964	10.0001	9.99995	.0078
2	9.9950	9.9986	9.9948	10.0030	9.9938	9.99704	.0092
3	10.0028	9.9998	10.0086	9.9949	9.9980	10.00082	.0137
4	9.9952	9.9923	10.0034	9.9965	10.0026	9.99800	.0111
5	9.9997	9.9883	9.9975	10.0078	9.9891	9.99649	.0195
6	9.9987	10.0027	10.0001	10.0027	10.0029	10.00141	.0042
7	10.0004	10.0023	10.0024	9.9992	10.0135	10.00358	.0143
8	10.0013	9.9938	10.0017	10.0089	10.0001	10.00116	.0151
9	10.0103	10.0009	9.9969	10.0103	9.9986	10.00339	.0134
10	9.9980	9.9954	9.9941	9.9958	9.9963	9.99594	.0039
11	10.0013	10.0033	9.9943	9.9949	9.9999	9.99874	.0090
12	9.9986	9.9990	10.0009	9.9947	10.0008	9.99882	.0062
13	10.0089	10.0056	9.9976	9.9997	9.9922	10.00080	.0167
14	9.9971	10.0015	9.9962	10.0038	10.0022	10.00016	.0076
15	9.9949	10.0011	10.0043	9.9988	9.9919	9.99822	.0124
16	9.9951	9.9957	10.0094	10.0040	9.9974	10.00033	.0137
17	10.0015	10.0026	10.0032	9.9971	10.0019	10.00127	.0061
18	9.9983	10.0019	9.9978	9.9997	10.0029	10.00130	.0051
19	9.9977	9.9963	9.9981	9.9968	10.0009	9.99798	.0127
20	10.0078	10.0004	9.9966	10.0051	10.0007	10.00212	.0112
21	9.9963	9.9990	10.0037	9.9936	9.9962	9.99764	.0101
22	9.9999	10.0022	10.0057	10.0026	10.0032	10.00272	.0058
23	9.9998	10.0002	9.9978	9.9966	10.0060	10.00009	.0094
24	10.0031	10.0078	9.9988	10.0032	9.9944	10.00146	.0134
25	9.9993	9.9978	9.9964	10.0032	10.0041	10.00015	.0077

a. Explain the logic behind the rational subgrouping strategy that was used.

b. Construct an $\bar{x}$-chart for the process using the data in Table 13.2.

c. What does the chart suggest about the stability of the filling process (whether the process is in or out of statistical control)?

d Should the control limits be used to monitor future process output?

Solution

a. The samples are far enough apart in time to detect hour-to-hour shifts or changes in the mean amount of paint dispensed, but the individual measurements that make up each sample are close enough together in time to ensure that the process has changed little, if at all, during the time the individual measurements were made. Overall, the rational subgrouping employed affords the opportunity for process changes to occur between samples and therefore show up on the control chart as differences between the sample means.

b. Twenty-five samples ($k = 25$ subgroups), each containing $n = 5$ cans of paint, were collected from the process. The first step after collecting the data is to calculate the 25 sample means and sample ranges needed to construct the $\bar{x}$-chart. The mean and range of the first sample are

$$\bar{x} = \frac{10.0042 + 9.9981 + 10.0010 + 9.9964 + 10.0001}{5} = 9.99995$$

$$R = 10.0042 - 9.9964 = .0078$$

All 25 means and ranges are displayed in Table 13.2

Next, we calculate the mean of the sample means and the mean of the sample ranges:

$$\bar{\bar{x}} = \frac{9.99995 + 9.99704 + \cdots + 10.00015}{25} = 9.9999$$

$$\bar{R} = \frac{.0078 + .0092 + \cdots + .0077}{25} = .01028$$

The centerline of the chart is positioned at $\bar{\bar{x}} = 9.9999$. To determine the control limits, we need the constant A_2, which can be found in Table XVII of Appendix B. For $n = 5$, $A_2 = .577$. Then

$$\text{UCL:} \quad \bar{\bar{x}} + A_2\bar{R} = 9.9999 + .577(.01028) = 10.0058$$

$$\text{LCL:} \quad \bar{\bar{x}} - A_2\bar{R} = 9.9999 - .577(.01028) = 9.9940$$

After positioning the control limits on the chart, we plot the 25 sample means in the order of sampling and connect the points with straight lines. The resulting trial $\bar{x}$-chart is shown in Figure 13.28, page 736.

c. To check the stability of the process, we use the six pattern-analysis rules for detecting special causes of variation, which were presented in Figure 13.27. To apply most of these rules requires identifying the A, B, and C zones of the control

FIGURE 13.28 ▶
$\bar{x}$-chart for the paint-filling process

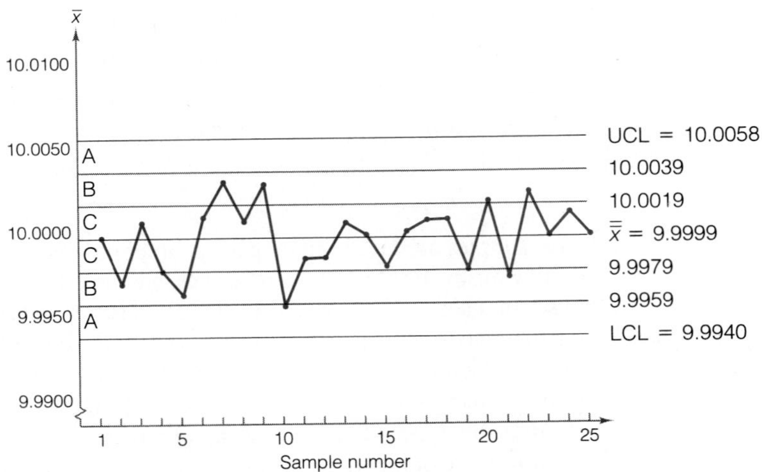

chart. These are indicated in Figure 13.28. We describe how they were constructed below.

The boundary between the A and B zones is two standard deviations from the centerline, and the boundary between the B and C zones is one standard deviation from the centerline. Thus, using $A_2\bar{R}$ and the 3-sigma limits previously calculated, we locate the A, B, and C zones above the centerline:

$$\begin{aligned}\text{A–B boundary} &= \bar{\bar{x}} + \tfrac{2}{3}(A_2\bar{R}) \\ &= 9.9999 + \tfrac{2}{3}(.577)(.01028) = 10.0039\end{aligned}$$

$$\begin{aligned}\text{B–C boundary} &= \bar{\bar{x}} + \tfrac{1}{3}(A_2\bar{R}) \\ &= 9.9999 + \tfrac{1}{3}(.577)(.01028) = 10.0019\end{aligned}$$

Similarly, the zones below the centerline are located:

$$\text{A–B boundary} = \bar{\bar{x}} + \tfrac{2}{3}(A_2\bar{R}) = 9.9959$$
$$\text{B–C boundary} = \bar{\bar{x}} + \tfrac{1}{3}(A_2\bar{R}) = 9.9979$$

A careful comparison of the six pattern-analysis rules with the sequence of sample means yields no out-of-control signals. All points are inside the control limits and there appear to be no nonrandom patterns within the control limits. That is, we can find no evidence of a shift in the process mean. Accordingly, we conclude that the process is in control.

d. Since the process was found to be in control during the period in which the samples were drawn, the trial control limits constructed in part **b** can be considered official. They should be extended to the right and used to monitor future process output.

EXAMPLE 13.2

Ten new samples of size $n = 5$ were drawn from the paint-filling process of the previous example. The sample data, including sample means and ranges, are shown in Table 13.3. Investigate whether the process remained in control during the period in which the new sample data were collected.

Solution

We begin by simply extending the control limits, centerline, and zone boundaries of the control chart in Figure 13.28 to the right. Next, beginning with sample number 26, we plot the 10 new sample means on the control chart and connect them with straight lines. This extended version of the control chart is shown in Figure 13.29.

Now that the control chart has been prepared, we apply the six pattern-analysis rules for detecting special causes of variation (Figure 13.27) to the new sequence of

TABLE 13.3 Ten Additional Samples of Size 5 from the Paint-Filling Process

Sample	Measurements					Mean	Range
1	10.0019	9.9981	9.9952	9.9976	9.9999	9.99841	.0067
2	10.0041	9.9982	10.0028	10.0040	9.9971	10.00125	.0070
3	9.9999	9.9974	10.0078	9.9971	9.9923	9.99890	.0155
4	9.9982	10.0002	9.9916	10.0040	9.9916	9.99713	.0124
5	9.9933	9.9963	9.9955	9.9993	9.9905	9.99498	.0088
6	9.9915	9.9984	10.0053	9.9888	9.9876	9.99433	.0177
7	9.9912	9.9970	9.9961	9.9879	9.9970	9.99382	.0091
8	9.9942	9.9960	9.9975	10.0019	9.9912	9.99614	.0107
9	9.9949	9.9967	9.9936	9.9941	10.0071	9.99726	.0135
10	9.9943	9.9969	9.9937	9.9912	10.0053	9.99626	.0141

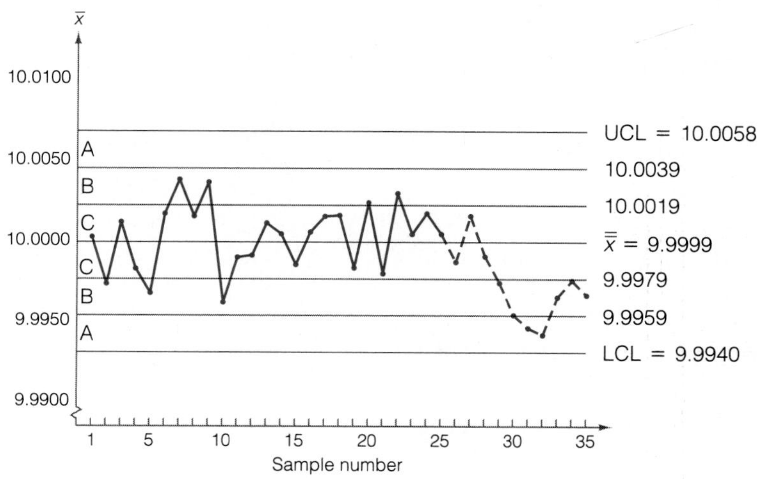

FIGURE 13.29 ▶ **Extended $\bar{x}$-chart for paint-filling process**

sample means. No points fall outside the control limits, but we notice six points in a row that steadily decrease (samples 27–32). Rule 3 says that if we observe six points in a row steadily increasing or decreasing, that is an indication of the presence of special causes of variation.

Notice that if you apply the rules from left to right along the sequence of sample means, the decreasing pattern also triggers signals from Rules 5 (samples 29–31) and 6 (samples 28–32).

These signals lead us to conclude that the process has gone out of control. Apparently, the filling head began to clog about the time that either sample 26 or 27 was drawn from the process. As a result, the mean of the process (the mean fill weight dispensed by the process) began to decline.

Exercises 13.1–13.15

Learning the Mechanics

13.1 What is a control chart? Describe its use.

13.2 Explain why rational subgrouping should be used in constructing control charts.

13.3 When a control chart is first constructed, why are the centerline and control limits treated as trial values?

13.4 Which process parameter is an $\bar{x}$-chart used to monitor?

13.5 Even if all the points on an $\bar{x}$-chart fall between the control limits, the process may be out of control. Explain.

13.6 What must be true about the variation of a process before an $\bar{x}$-chart is used to monitor the mean of the process? Why?

13.7 Use the six pattern-analysis rules described in Figure 13.27 to determine whether the process being monitored with the accompanying $\bar{x}$-chart is out of statistical control.

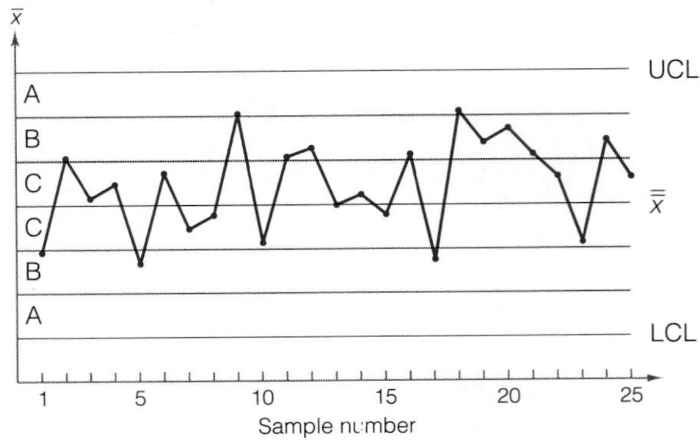

13.8 Is the process for which the accompanying $\bar{x}$-chart was constructed affected by only special causes of variation, only common causes of variation, or both? Explain.

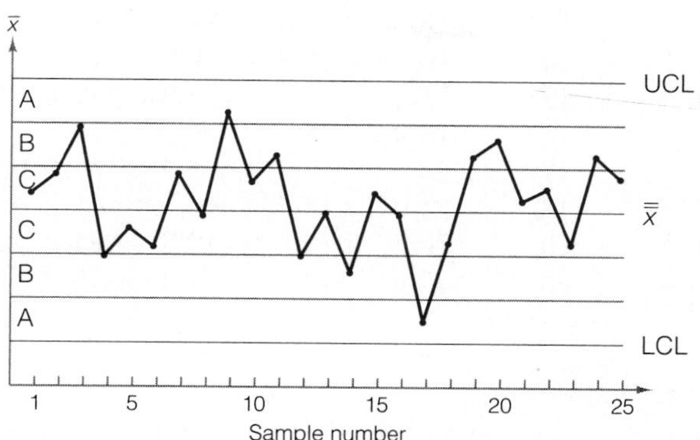

13.9 Use Table XVII in Appendix B to find the value of A_2 for each of the following sample sizes.
a. $n = 3$ b. $n = 10$ c. $n = 22$

13.10 Twenty-five samples of size $n = 5$ were collected to construct an $\bar{x}$-chart. The accompanying sample means and ranges were calculated for these data.

Sample	$\bar{x}$	R	*Sample*	$\bar{x}$	R	*Sample*	$\bar{x}$	R
1	80.2	7.2	10	85.3	7.1	18	75.9	9.9
2	79.1	9.0	11	77.7	9.8	19	78.1	6.0
3	83.2	4.7	12	82.3	10.7	20	81.4	7.4
4	81.0	5.6	13	79.5	9.2	21	81.7	10.4
5	77.6	10.1	14	83.1	10.2	22	80.9	9.1
6	81.7	8.6	15	79.6	7.8	23	78.4	7.3
7	80.4	4.4	16	80.0	6.1	24	79.6	8.0
8	77.5	6.2	17	83.2	8.4	25	81.6	7.6
9	79.8	7.9						

a. Calculate the mean of the sample means, $\bar{\bar{x}}$, and the mean of the sample ranges, $\bar{R}$.
b. Calculate and plot the centerline and the upper and lower control limits for the $\bar{x}$-chart.
c. Calculate and plot the A, B, and C zone boundaries of the $\bar{x}$-chart.
d. Plot the 25 sample means on the $\bar{x}$-chart and use the six pattern-analysis rules to determine whether the process is under statistical control.

13.11 The data on page 740 were collected for the purpose of constructing an $\bar{x}$-chart.
a. Calculate $\bar{x}$ and R for each sample.
b. Calculate $\bar{\bar{x}}$ and $\bar{R}$.
c. Calculate and plot the centerline and the upper and lower control limits for the $\bar{x}$-chart.
d. Calculate and plot the A, B, and C zone boundaries of the $\bar{x}$-chart.
e. Plot the 20 sample means on the $\bar{x}$-chart. Is the process in control? Justify your answer.

Sample	*Measurements*				*Sample*	*Measurements*			
1	19.4	19.7	20.6	21.2	11	22.7	21.2	21.5	19.5
2	18.7	18.4	21.2	20.7	12	20.1	20.6	21.0	20.2
3	20.2	18.8	22.6	20.1	13	19.7	18.6	21.2	19.1
4	19.6	21.2	18.7	19.4	14	18.6	21.7	17.7	18.3
5	20.4	20.9	22.3	18.6	15	18.2	20.4	19.8	19.2
6	17.3	22.3	20.3	19.7	16	18.9	20.7	23.2	20.0
7	21.8	17.6	22.8	23.1	17	20.5	19.7	21.4	17.8
8	20.9	17.4	19.5	20.7	18	21.0	18.7	19.9	21.2
9	18.1	18.3	20.6	20.4	19	20.5	19.6	19.8	21.8
10	22.6	21.4	18.5	19.7	20	20.6	16.9	22.4	19.7

Applying the Concepts

13.12 The central processing unit (CPU) of a microcomputer is a computer chip containing millions of transistors. Connecting the transistors are slender circuit paths only .5 to .85 micron wide. To understand how narrow these paths are, consider that a micron is a millionth of a meter, and a human hair is 70 microns wide (Wood, 1992). A manufacturer of CPU chips knows that if the circuit paths are not .5–.85 micron wide, a variety of problems will arise in the chips' performance. The manufacturer sampled four CPU chips six times a day (every 90 minutes from 8:00 A.M. until 4:30 P.M.) for five consecutive days and measured the circuit path widths. These data and Minitab were used to construct the $\bar{x}$-chart shown.

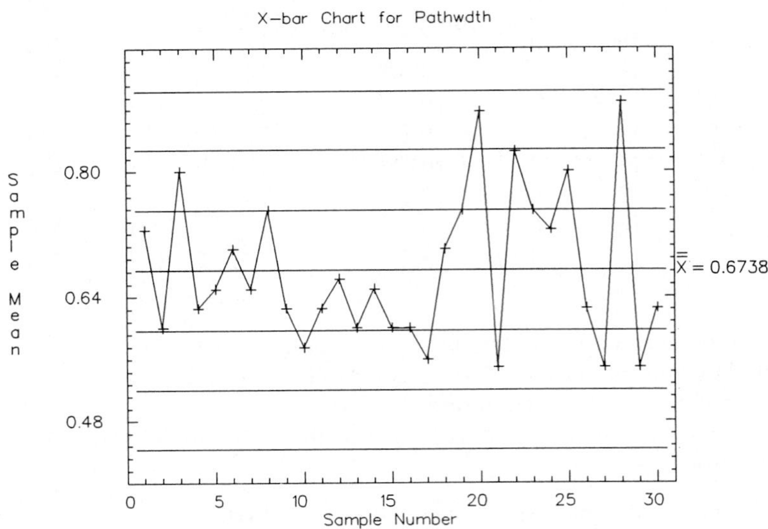

a. Assuming that $\bar{R} = .3162$, calculate the chart's upper and lower control limits, the upper and lower A–B boundaries, and the upper and lower B–C boundaries.

b. What does the chart suggest about the stability of the process used to put circuit paths on the CPU chip? Justify your answer.

c. Should the control limits be used to monitor future process output? Explain.

13.13 A machine at K-Company fills boxes with bran flakes cereal. The target weight for the filled boxes is 24 ounces. The company would like to use an $\bar{x}$-chart to monitor the performance of the machine. To develop the control chart, the company decides to sample and weigh five consecutive boxes of cereal five times each day (at 8:00 and 11:00 A.M. and 2:00, 5:00, and 8:00 P.M.) for four consecutive days. The data are presented in the table.

Day	*Weight of Cereal Boxes (ozs.)*					*Day*	*Weight of Cereal Boxes (ozs.)*				
1	24.02	23.91	24.12	24.06	24.13	11	24.10	23.90	24.11	23.98	23.95
2	23.89	23.98	24.01	24.00	23.91	12	24.01	24.07	23.93	24.09	23.98
3	24.11	24.02	23.99	23.79	24.04	13	24.14	24.07	24.08	23.98	24.02
4	24.06	23.98	23.95	24.01	24.11	14	23.91	24.04	23.89	24.01	23.95
5	23.81	23.90	23.99	24.07	23.96	15	24.03	24.04	24.01	23.98	24.10
6	23.87	24.12	24.07	24.01	23.99	16	23.94	24.07	24.12	24.00	24.02
7	23.88	24.00	24.05	23.97	23.97	17	23.88	23.94	23.91	24.06	24.07
8	24.01	24.03	23.99	23.91	23.98	18	24.11	23.99	23.90	24.01	23.98
9	24.06	24.02	23.80	23.79	24.07	19	24.05	24.04	23.97	24.08	23.95
10	23.96	23.99	24.03	23.99	24.01	20	24.02	23.96	23.95	23.89	24.04

a. Construct an $\bar{x}$-chart from the given data.
b. What does the chart suggest about the stability of the filling process (whether the process is in or out of statistical control)? Justify your answer.
c. Should the control limits be used to monitor future process output? Explain.
d. Two shifts of workers run the filling operation. Each day the second shift takes over at 3:00 P.M. Will the rational subgrouping strategy used by K-Company facilitate or hinder the identification of process variation caused by differences in the two shifts? Explain.

13.14 A precision parts manufacturer produces bolts for use in military aircraft. Ideally, the bolts should be 37 centimeters in length. The company sampled four consecutively produced bolts each hour on the hour for 25 consecutive hours and measured them using a computerized precision instrument. The data are presented here.

Hour	*Bolt Lengths (cm)*				*Hour*	*Bolt Lengths (cm)*			
1	37.03	37.08	36.90	36.88	14	37.08	37.07	37.10	37.04
2	36.96	37.04	36.85	36.98	15	37.03	37.04	36.89	37.01
3	37.16	37.11	36.99	37.01	16	36.95	36.98	36.90	36.99
4	37.20	37.06	37.02	36.98	17	36.97	36.94	37.14	37.10
5	36.81	36.97	36.91	37.10	18	37.11	37.04	36.98	36.91
6	37.13	36.96	37.01	36.89	19	36.88	36.99	37.01	36.94
7	37.07	36.94	36.99	37.00	20	36.90	37.15	37.09	37.00
8	37.01	36.91	36.98	37.12	21	37.01	36.96	37.05	36.96
9	37.17	37.03	36.90	37.01	22	37.09	36.95	36.93	37.12
10	36.91	36.99	36.87	37.11	23	37.00	37.02	36.95	37.04
11	36.88	37.10	37.07	37.03	24	36.99	37.07	36.90	37.02
12	37.06	36.98	36.90	36.99	25	37.10	37.03	37.01	36.90
13	36.91	37.22	37.12	37.03					

a. What process is the manufacturer interested in monitoring?
b. Construct an $\bar{x}$-chart from the data.

c. Does the chart suggest that special causes of variation are present? Justify your answer.
d. Provide an example of a special cause of variation that could potentially affect this process. Do the same for a common cause of variation.
e. Should the control limits be used to monitor future process output? Explain.

13.15 A walk-in freezer thermostat at a restaurant is set at 5°F. Because of the perishability of the food in the freezer, the restaurant's manager has decided to begin monitoring the temperature inside the freezer. In order to establish the centerline and control limits of an $\bar{x}$-chart, the manager used a precision thermometer to take sample temperature readings at five randomly chosen times per day for 20 days. The data are presented here.

Day	*Temperature Readings (°F)*					*Day*	*Temperature Readings (°F)*				
1	5.22	5.29	5.11	4.95	4.78	11	5.10	5.17	4.65	5.20	4.93
2	4.40	4.41	4.63	6.03	4.83	12	4.86	4.97	5.35	4.81	4.72
3	5.11	5.43	4.90	4.55	5.23	13	5.03	3.91	5.16	4.80	4.98
4	5.65	4.24	5.09	4.82	5.50	14	5.16	4.80	5.61	5.06	4.91
5	4.68	5.92	4.71	4.67	4.75	15	4.92	4.33	5.25	4.93	4.81
6	5.01	5.26	6.10	5.20	5.25	16	5.21	5.17	4.79	4.80	5.30
7	5.20	4.99	5.15	5.96	5.35	17	3.82	5.19	5.24	4.60	4.78
8	4.30	4.91	5.03	4.97	4.80	18	4.33	4.88	5.21	5.18	5.07
9	5.45	5.62	6.11	5.13	4.90	19	5.16	5.83	5.29	4.75	4.77
10	5.06	5.13	4.95	5.59	5.80	20	5.26	4.61	5.34	5.40	4.71

a. What process is the manager interested in monitoring?
b. What parameter of that process can be monitored using an $\bar{x}$-chart?
c. Construct an $\bar{x}$-chart from the data.
d. What does the chart suggest about the stability of the temperature in the freezer unit? Explain.
e. Should the control limits be used to monitor future temperatures? Explain.
f. Critique the rational subgrouping strategy employed by the restaurant manager.

13.9 A Control Chart for Monitoring the Variation of a Process: The *R*-Chart

Recall from Section 13.6 that a process may be out of statistical control because its mean or variance or both are changing over time (see Figure 13.14). The $\bar{x}$-chart of the previous section is used to detect changes in the process mean. The control chart we present in this section—the R-chart—is used to detect changes in process variation.

The primary difference between the $\bar{x}$-chart and the R-chart is that instead of plotting *sample means* and monitoring their variation, we plot and monitor the variation of *sample ranges*. Changes in the behavior of the sample range signal changes in the variation of the process.

We could also monitor process variation by plotting *sample standard deviations*. That is, we could calculate s for each sample (i.e., each subgroup) and plot them on a control chart known as an **s-chart**. In this chapter, however, we focus on just the R-chart because (1) when using samples of size 9 or less, the s-chart and the R-chart reflect about the same information, and (2) the *R*-chart is used much more widely by practitioners than is the *s*-chart (primarily because the sample range is easier to calculate and interpret than the sample standard deviation). For more information about *s*-charts, see the references at the end of the chapter.

The underlying logic and basic form of the *R*-chart are similar to the $\bar{x}$-chart. In monitoring $\bar{x}$, we use the standard deviation of $\bar{x}$ to develop 3-sigma control limits. Now, since we want to be able to determine when *R* takes on unusually large or small values, we use the standard deviation of *R*, σ_R, to construct 3-sigma control limits. The centerline of the $\bar{x}$-chart represents the process mean μ or, equivalently, the mean of the sampling distribution of $\bar{x}$, $\mu_{\bar{x}}$. Similarly, the centerline of the *R*-chart represents μ_R, the mean of the sampling distribution of R. These points are illustrated in the *R*-chart of Figure 13.30.

FIGURE 13.30 ▶
***R*-chart**

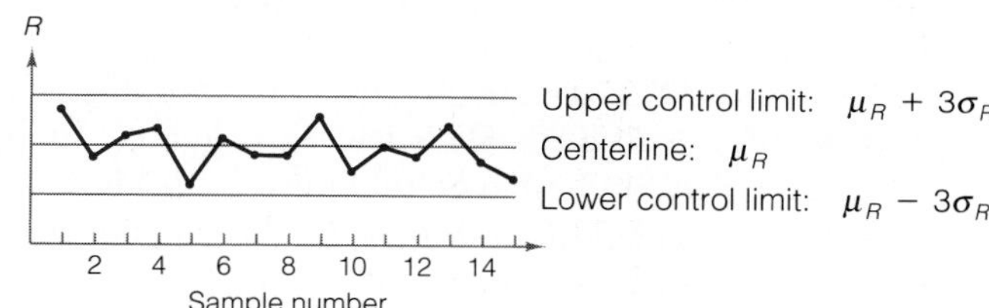

As with the $\bar{x}$-chart, you should have at least 20 samples of *n* items each ($n \geq 2$) to construct an *R*-chart. This will provide sufficient data to obtain reasonably good estimates of μ_R and σ_R. Rational subgrouping is again used for determining sample size and frequency of sampling.

The centerline of the *R*-chart is positioned as follows:

$$\textit{Centerline:}\quad \bar{R} = \frac{R_1 + R_2 + \cdots + R_k}{k}$$

where *k* is the number of samples of size *n* and R_i is the range of the *i*th sample. $\bar{R}$ is an estimate of μ_R.

In order to construct the control limits, we need an estimator of σ_R. The estimator recommended by Montgomery (1991) and Ryan (1989) is

$$\hat{\sigma}_R = d_3\left(\frac{\bar{R}}{d_2}\right)$$

where d_2 and d_3 are constants whose values depend on the sample size, *n*. Values for d_2 and d_3 for samples of size $n = 2$ to $n = 25$ are given in Table XVII of Appendix B.

The control limits are positioned as follows:

$$\textit{Upper control limit:}\quad \bar{R} + 3\hat{\sigma}_R = \bar{R} + 3d_3\left(\frac{\bar{R}}{d_2}\right)$$

$$\textit{Lower control limit:}\quad \bar{R} - 3\hat{\sigma}_R = \bar{R} - 3d_3\left(\frac{\bar{R}}{d_2}\right)$$

Notice that $\overline{R}$ appears twice in each control limit. Accordingly, we can simplify the calculation of these limits by factoring out $\overline{R}$:

$$\text{UCL: } \overline{R}\left(1 + \frac{3d_3}{d_2}\right) = \overline{R}D_4 \qquad \text{LCL: } \overline{R}\left(1 - \frac{3d_3}{d_2}\right) = \overline{R}D_3$$

where

$$D_4 = \left(1 + \frac{3d_3}{d_2}\right) \qquad D_3 = \left(1 - \frac{3d_3}{d_2}\right)$$

The values for D_3 and D_4 have been tabulated for samples of size $n = 2$ to $n = 25$ and can be found in Appendix B, Table XVII.

For samples of size $n = 2$ through $n = 6$, D_3 is negative, and the lower control limit falls below zero. Since the sample range cannot take on negative values, such a control limit is meaningless. Thus, when $n \le 6$ the R-chart contains only one control limit, the upper control limit.

Although D_3 is actually negative for $n \le 6$, the values reported in Table XVII in Appendix B are all zeros. This has been done to discourage the inappropriate construction of negative lower control limits. If the lower control limit is calculated using $D_3 = 0$, you obtain $D_3\overline{R} = 0$. This should be interpreted as indicating that the R-chart has no lower 3-sigma control limit.

Constructing an *R*-Chart: A Summary

1. Using a rational subgrouping strategy, collect at least 20 samples (i.e., subgroups), each of size $n \ge 2$.
2. Calculate the range of each sample.
3. Calculate the mean of the sample ranges, $\overline{R}$:

$$\overline{R} = \frac{R_1 + R_2 + \cdots + R_k}{k}$$

where k = The number of samples (i.e., subgroups)
R_i = The range of the ith sample

4. Plot the centerline and control limits:

Centerline: $\overline{R}$ *Upper control limit:* $\overline{R}D_4$ *Lower control limit:* $\overline{R}D_3$

where D_3 and D_4 are constants that depend on n. Their values can be found in Appendix B, Table XVII. When $n \le 6$, $D_3 = 0$, indicating that the control chart does not have a lower control limit.

5. Plot the k sample ranges on the control chart in the order that the samples were produced by the process.

We interpret the completed R-chart in basically the same way as we did the $\bar{x}$-chart. We look for indications that the process is out of control. Those indications include points that fall outside the control limits as well as any nonrandom patterns of variation that appear between the control limits. To help spot nonrandom behavior, we include the A, B, and C zones (described in the previous section) on the *R*-chart. The next box describes how to construct the zone boundaries for the *R*-chart. It requires only rules 1 through 4 of Figure 13.27, because rules 5 and 6 are based on the assumption that the statistic plotted on the control chart follows a normal (or nearly normal) distribution, whereas *R*'s distribution is skewed to the right.*

Constructing Zone Boundaries for an *R*-Chart

The simplest method of construction uses the estimator of the standard deviation of R, which is $\hat{\sigma}_R = d_3(\overline{R}/d_2)$:

Upper A–B *boundary*: $\overline{R} + 2d_3\left(\frac{\overline{R}}{d_2}\right)$

Lower A–B *boundary*: $\overline{R} - 2d_3\left(\frac{\overline{R}}{d_2}\right)$

Upper B–C *boundary*: $\overline{R} + d_3\left(\frac{\overline{R}}{d_2}\right)$

Lower B–C *boundary*: $\overline{R} - d_3\left(\frac{\overline{R}}{d_2}\right)$

Note: Whenever $n \leq 6$ the R-chart has no lower 3-sigma control limit. However, the lower A–B and B–C boundaries can still be plotted if they are nonnegative.

Interpreting an *R*-Chart

1. The **process is out of control** if one or more sample ranges fall beyond the control limits (rule 1) or if any of the three patterns of variation described by rules 2, 3, and 4 (Figure 13.27) are observed. Such signals indicate that one or more special causes of variation are influencing the *variation* of the process. These causes should be identified and eliminated to bring the process into control.
2. The **process is treated as being in control** if none of the noted out-of-control signals are observed. Processes that are in control should not be tampered with. However, if the level of variation is unacceptably high, common causes of variation should be identified and eliminated.

*Some authors (e.g., Kane, 1989) apply all six pattern-analysis rules as long as $n \geq 4$.

As with the $\bar{x}$-chart, the centerline and control limits should be developed using samples that were collected during a period in which the process was in control. Accordingly, when an R-chart is first constructed, the centerline and the control limits are treated as **trial values** (see Section 13.8) and are modified, if necessary, before being extended to the right and used to monitor future process output.

EXAMPLE 13.3

Refer to Example 13.1.

a. Construct an R-chart for the paint-filling process.
b. What does the chart indicate about the stability of the filling process during the time when the data were collected?
c. Is it appropriate to use the control limits constructed in part **a** to monitor future process output?

Solution

a. The first step after collecting the data is to calculate the range of each sample. For the first sample the range is

$$R = 10.0042 - 9.9964 = .0078$$

All 25 sample ranges appear in Table 13.2 (page 734).

Next, calculate the mean of the ranges:

$$\bar{R} = \frac{.0078 + .0092 + \cdots + .0077}{25} = .01028$$

The centerline of the chart is positioned at $\bar{R} = .01028$. To determine the control limits, we need the constants D_3 and D_4, which can be found in Table XVII of Appendix B. For $n = 5$, $D_3 = 0$ and $D_4 = 2.115$. Since $D_3 = 0$, the lower 3-sigma control limit is negative and is not included on the chart. The upper control limit is calculated as follows:

$$\text{UCL:} \quad \bar{R}D_4 = (.01028)(2.115) = .0217$$

After positioning the upper control limit on the chart, we plot the 25 sample ranges in the order of sampling and connect the points with straight lines. The resulting trial R-chart is shown in Figure 13.31.

b. To facilitate our examination of the R-chart, we plot the four zone boundaries. Recall that in general the A–B boundaries are positioned two standard deviations from the centerline and the B–C boundaries are one standard deviation from the centerline. In the case of the R-chart, we use the estimated standard deviation of R, $\hat{\sigma}_R = d_3(\bar{R}/d_2)$, and calculate the boundaries:

$$\textit{Upper A–B boundary:} \quad \bar{R} + 2d_3\left(\frac{\bar{R}}{d_2}\right) = .01792$$

$$\textit{Lower A–B boundary:} \quad \bar{R} - 2d_3\left(\frac{\bar{R}}{d_2}\right) = .00264$$

FIGURE 13.31 ▶
***R*-chart for the paint-filling process**

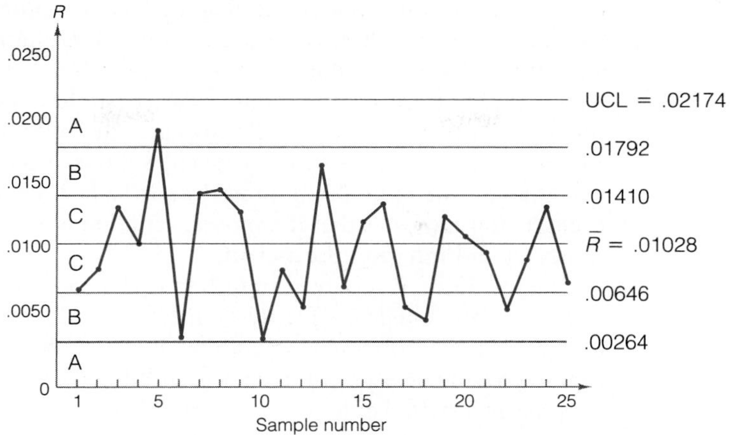

$$\textit{Upper B–C boundary:}\quad \overline{R} + d_3\left(\frac{\overline{R}}{d_2}\right) = .01410$$

$$\textit{Lower B–C boundary:}\quad \overline{R} - d_3\left(\frac{\overline{R}}{d_2}\right) = .00646$$

where (from Table XVII of Appendix B) for $n = 5$, $d_2 = 2.326$ and $d_3 = .864$. Notice in Figure 13.31 that the lower A zone is slightly narrower than the upper A zone. This occurs because the lower 3-sigma control limit (the usual lower boundary of the lower A zone) is negative.

All the plotted R values fall below the upper control limit. This is one indication that the process is under control (i.e., is stable). However, we must also look for patterns of points that would be unlikely to occur if the process were in control. To assist us with this process, we use pattern-analysis rules 1–4 (Figure 13.27). None of the rules signal the presence of special causes of variation. Accordingly, we conclude that it is reasonable to treat the process—in particular, the variation of the process—as being under control during the period in question. Apparently, no significant special causes of variation are influencing the variation of the process.

c. Yes. Since the variation of the process appears to be in control during the period when the sample data were collected, the control limits appropriately characterize the variation in R that would be expected when the process is in a state of statistical control.

In practice, the $\bar{x}$-chart and the R-chart are not used in isolation, as our presentation so far might suggest. Rather, they are used together to monitor the mean (i.e., the location) of the process and the variation of the process simultaneously. In fact, many practitioners plot them on the same piece of paper.

One important reason for dealing with them as a unit is that the control limits of the $\bar{x}$-chart are a function of R. That is, the control limits depend on the variation of the process. (Recall that the control limits are $\bar{x} \pm A_2\bar{R}$.) Thus, if the process variation is out of control, the control limits of the $\bar{x}$-chart have little meaning. This is because when the process variation is changing (as in the bottom two graphs of Figure 13.14), any single estimate of the variation (such as $\bar{R}$ or s) is not representative of the process. Accordingly, **the appropriate procedure is to first construct and then interpret the *R*-chart. If it indicates that the process variation is in control, then it makes sense to construct and interpret the $\bar{x}$-chart.**

Figure 13.32 is reprinted from Kaoru Ishikawa's classic text on quality-improvement methods, *Guide to Quality Control* (1986). It illustrates how particular changes in a process over time may be reflected in $\bar{x}$- and R-charts. At the top of the figure, running across the page, is a series of probability distributions A, B, and C that describe the process (i.e., the output variable) at different points in time. In practice, we never have this information. For this example, however, Ishikawa worked with a known process (i.e., with its given probabilistic characterization) to illustrate how sample data from a known process might behave.

FIGURE 13.32 ▶
Combined $\bar{x}$- and *R*-chart (Ishikawa, 1986)

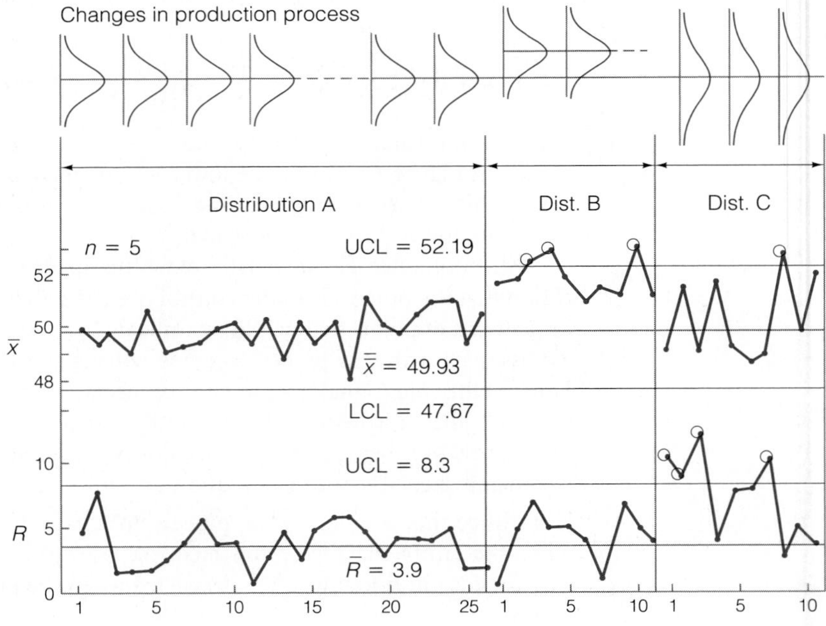

The control limits for both charts were constructed from $k = 25$ samples of size $n = 5$. These data were generated by Distribution A. The 25 sample means and ranges were plotted on the $\bar{x}$- and R-charts, respectively. Since the distribution did not change over this period of time, it follows from the definition of statistical control that the process was under control. If you did not know this—as would be the case in practice—what would you conclude from looking at the control charts? (Remember, always interpret the R-chart before the $\bar{x}$-chart.) Both charts indicate that the process is under control. Accordingly, the control limits are made official and can be used to monitor future output, as is done next.

Toward the middle of the figure, the process changes. The mean shifts to a higher level. Now the output variable is described by Distribution B. The process is out of control. Ten new samples of size 5 are sampled from the process. Since the variation of the process has not changed, the R-chart should indicate that the variation remains stable. This is, in fact, the case. All points fall below the upper control limit. As we would hope, it is the $\bar{x}$-chart that reacts to the change in the mean of the process.

Then the process changes again (Distribution C). This time the mean shifts back to its original position, but the variation of the process increases. The process is still out of control but this time for a different reason. Checking the R-chart first, we see that it has reacted as we would hope. It has detected the increase in the variation. Given this R-chart finding, the control limits of the $\bar{x}$-chart become inappropriate (as described before) and we would not use them. Notice, however, how the sample means react to the increased variation in the process. This increased variation in $\bar{x}$ is consistent with what we know about the variance of $\bar{x}$. It is directly proportional to the variance of the process, $\sigma_{\bar{x}}^2 = \sigma^2/n$.

Keep in mind that what Ishikawa did in this example is exactly the opposite of what we do in practice. In practice we use sample data and control charts to make inferences about changes in unknown process distributions. Here, for the purpose of helping you to understand and interpret control charts, known process distributions were changed to see what would happen to the control charts.

Exercises 13.16–13.25

Learning the Mechanics

13.16 What characteristic of a process is an R-chart designed to monitor?

13.17 In practice, $\bar{x}$- and R-charts are used together to monitor a process. However, the R-chart should be interpreted before the $\bar{x}$-chart. Why?

13.18 Use Table XVII in Appendix B to find the values of D_3 and D_4 for each of the following sample sizes.
a. $n = 4$ **b.** $n = 12$ **c.** $n = 24$

13.19 Construct and interpret an R-chart for the data in Exercise 13.10.

a. Calculate and plot the upper control limit and, if appropriate, the lower control limit.

b. Calculate and plot the A, B, and C zone boundaries on the R-chart.

c. Plot the sample ranges on the R-chart and use pattern-analysis rules 1–4 of Figure 13.27 to determine whether the process is under statistical control.

13.20 Construct and interpret an R-chart for the data in Exercise 13.11.

a. Calculate and plot the upper control limit and, if appropriate, the lower control limit.
b. Calculate and plot the A, B, and C zone boundaries on the R-chart.
c. Plot the sample ranges on the R-chart and determine whether the process is in control.

13.21 Construct and interpret an R-chart and an $\bar{x}$-chart from the following sample data. Remember to interpret the R-chart *before* the $\bar{x}$-chart.

Sample	*Measurements*								
	1	2	3	4	5	6	7	$\bar{x}$	R
1	20.1	19.0	20.9	22.2	18.9	18.1	21.3	20.07	4.1
2	19.0	17.9	21.2	20.4	20.0	22.3	21.5	20.33	4.4
3	22.6	21.4	21.4	22.1	19.2	20.6	18.7	20.86	3.9
4	18.1	20.8	17.8	19.6	19.8	21.7	20.0	19.69	3.9
5	22.6	19.1	21.4	21.8	18.4	18.0	19.5	20.11	4.6
6	19.1	19.0	22.3	21.5	17.8	19.2	19.4	19.76	4.5
7	17.1	19.4	18.6	20.9	21.8	21.0	19.8	19.80	4.7
8	20.2	22.4	22.0	19.6	19.6	20.0	18.5	20.33	3.9
9	21.9	24.1	23.1	22.8	25.6	24.2	25.2	23.84	3.7
10	25.1	24.3	26.0	23.1	25.8	27.0	26.5	25.40	3.9
11	25.8	29.2	28.5	29.1	27.8	29.0	28.0	28.20	3.4
12	28.2	27.5	29.3	30.7	27.6	28.0	27.0	28.33	3.7
13	28.2	28.6	28.1	26.0	30.0	28.5	28.3	28.24	4.0
14	22.1	21.4	23.3	20.5	19.8	20.5	19.0	20.94	4.3
15	18.5	19.2	18.0	20.1	22.0	20.2	19.5	19.64	4.0
16	21.4	20.3	22.0	19.2	18.0	17.9	19.5	19.76	4.1
17	18.4	16.5	18.1	19.2	17.5	20.9	19.6	18.60	4.4
18	20.1	19.8	22.3	22.5	21.8	22.7	23.0	21.74	3.2
19	20.0	17.5	21.0	18.2	19.5	17.2	18.1	18.79	3.8
20	22.3	18.2	21.5	19.0	19.4	20.5	20.0	20.13	4.1

Applying the Concepts

13.22 Refer to Exercise 13.12, where the desired circuit path widths were .5 to .85 micron. The manufacturer sampled four CPU chips six times a day (every 90 minutes from 8:00 A.M. until 4:30 P.M.) for five consecutive days. The path widths were measured and used to construct the R-chart shown.

a. Calculate the chart's upper and lower control limits.
b. What does the R-chart suggest about the presence of special causes of variation during the time when the data were collected?
c. Should the control limit(s) be used to monitor future process output? Explain.
d. How many different R values are plotted on the control chart? Notice how most of the R values fall along three horizontal lines. What could cause such a pattern?

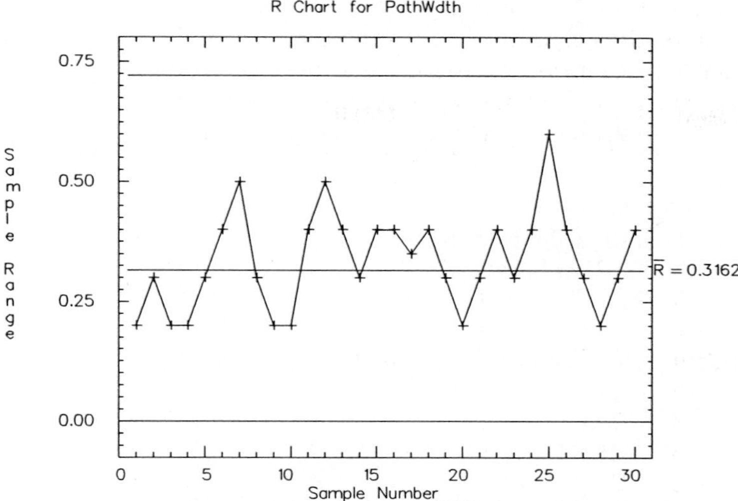

13.23 A soft-drink bottling company is interested in monitoring the amount of cola injected into 16-ounce bottles by a particular filling head. The process is entirely automated and operates 24 hours a day. At 6 A.M. and 6 P.M. each day, a new dispenser of carbon dioxide capable of producing 20,000 gallons of cola is hooked up to the filling machine. In order to monitor the process using control charts, the company decided to sample five consecutive bottles of cola each hour beginning at 6:15 A.M. (i.e., 6:15 A.M., 7:15 A.M., 8:15 A.M., etc.). The data for the first day are given here.

Sample	*Measurements*					*Sample*	*Measurements*				
1	16.01	16.03	15.98	16.00	16.01	13	15.96	16.00	16.01	16.00	15.98
2	16.03	16.02	15.97	15.99	15.99	14	15.98	16.01	16.02	15.99	15.99
3	15.98	16.00	16.03	16.04	15.99	15	15.99	16.03	16.00	15.98	16.01
4	16.00	16.03	16.02	15.98	15.98	16	16.02	16.02	16.01	15.97	16.00
5	15.97	15.99	16.03	16.01	16.04	17	16.01	16.05	15.99	15.99	16.03
6	16.01	16.03	16.04	15.97	15.99	18	15.98	16.03	16.04	15.98	16.01
7	16.04	16.05	15.97	15.96	16.00	19	15.97	15.96	15.99	15.99	16.01
8	16.02	16.05	16.03	15.97	15.98	20	16.03	16.01	16.04	15.96	15.99
9	15.97	15.99	16.02	16.03	15.95	21	15.99	16.03	15.97	16.05	16.03
10	16.00	16.01	15.95	16.04	16.06	22	15.98	15.95	16.07	16.01	16.04
11	15.95	16.04	16.07	15.93	16.03	23	15.99	16.06	15.95	16.03	16.07
12	15.98	16.07	15.94	16.08	16.02	24	16.00	16.01	16.08	15.94	15.93

a. Will the rational subgrouping strategy that was used enable the company to detect variation in fill caused by differences in the carbon dioxide dispensers? Explain.

b. Construct an *R*-chart from the data.

c. What does the *R*-chart indicate about the stability of the filling process during the time when the data were collected? Justify your answer.

d. Should the control limit(s) be used to monitor future process output? Explain.

e. Given your answer to part **c**, should an $\bar{x}$-chart be constructed from the given data? Explain.

13.24 Refer to the data in Exercise 13.13.
a. Construct an R-chart for the process.
b. What does the R-chart suggest about the presence of special causes of variation during the time when the data were collected?
c. Should the control limit(s) be used to monitor future process output? Explain.

13.25 Refer to the data in Exercise 13.14.
a. Construct an R-chart for the production process.
b. What does the R-chart indicate about the stability of the process?
c. Should the control limits be used to monitor future output? Explain.

13.10 A Control Chart for Monitoring the Proportion of Defectives Generated by a Process: The p-Chart

Among the dozens of different control charts that have been proposed by researchers and practitioners, the $\bar{x}$- and R-charts are by far the most popular for use in monitoring **quantitative** output variables such as time, length, and weight. Among the charts developed for use with **qualitative** output variables, the chart we introduce in this section is the most popular. Called the ***p*-chart**, it is used when the output variable is categorical (i.e., measured on a nominal scale). With the p-chart, the proportion, p, of units produced by the process that belong to a particular category (e.g., defective or nondefective; successful or unsuccessful; early, on-time, or late) can be monitored.

The p-chart is typically used to monitor the proportion of defective units produced by a process (i.e., the proportion of units that do not conform to specification). This proportion is used to characterize a process in the same sense that the mean and variance are used to characterize a process when the output variable is quantitative. Examples of process proportions that are monitored in industry include the proportion of billing errors made by credit-card companies; the proportion of nonfunctional semiconductor chips produced; and the proportion of checks that a bank's magnetic ink character-recognition system is unable to read.

As is the case for the mean and variance, the process proportion can change over time. For example, it can drift upward or downward or jump to a new level. In such cases, the process is out of control. **As long as the process proportion remains constant, the process is in a state of statistical control.**

As with the other control charts presented in this chapter, the p-chart has a centerline and control limits that are determined from sample data. After k samples of size n are drawn from the process, each unit is classified (e.g., defective or nondefective), the proportion of defective units in each sample—$\hat{p}$—is calculated, the centerline and control limits are determined using this information, and the sample proportions are plotted on the p-chart. It is the variation in the $\hat{p}$'s over time that we monitor and interpret. Changes in the behavior of the $\hat{p}$'s signal changes in the process proportion, p.

The p-chart is based on the assumption that the number of defectives observed in each sample is a binomial random variable. What we have called the process proportion is really the binomial probability, p. (We discussed binomial random variables in Chapter 4.) When the process is in a state of statistical control, p remains constant over time. Variation in $\hat{p}$—as displayed on a p-chart—is used to judge whether p is stable.

To determine the centerline and control limits for the p-chart we need to know $\hat{p}$'s sampling distribution. We described the sampling distribution of $\hat{p}$ in Section 7.4. Recall that

$$\hat{p} = \frac{\text{Number of defective items in the sample}}{\text{Number of items in the sample}} = \frac{x}{n}$$

$$\mu_{\hat{p}} = p$$

$$\sigma_{\hat{p}} = \sqrt{\frac{p(1-p)}{n}}$$

and that for large samples $\hat{p}$ is approximately normally distributed. Thus, if p were known, the centerline would be p and the 3-sigma control limits would be $p \pm 3\sqrt{p(1-p)/n}$. However, since p is unknown, it must be estimated from the sample data. The appropriate estimator is $\bar{p}$, the overall proportion of defective units in the nk units sampled:

$$\bar{p} = \frac{\text{Total number of defective units in all } k \text{ samples}}{\text{Total number of units sampled}}$$

To calculate the control limits of the p-chart, substitute $\bar{p}$ for p in the preceding expression for the control limits, as illustrated in Figure 13.33.

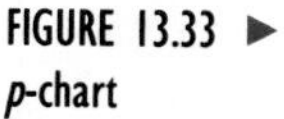

FIGURE 13.33 ▶
p-chart

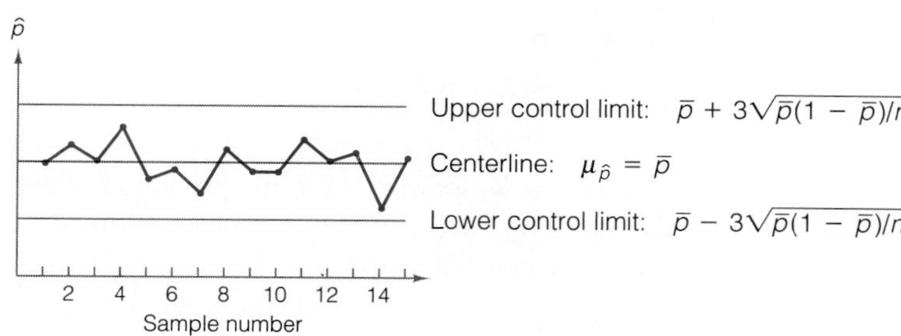

In constructing a p-chart it is advisable to use a much larger sample size than is typically used for $\bar{x}$- and R-charts. Most processes that are monitored in industry have relatively small process proportions, often less than .05 (i.e., less than 5% of output is nonconforming). In those cases, if a small sample size is used, say $n = 5$, samples drawn from the process would likely not contain any nonconforming output. As a result, most, if not all, $\hat{p}$'s would equal zero.

We present a rule of thumb that can be used to determine a sample size large enough to avoid this problem. This rule will also help protect against ending up with a negative lower control limit, a situation that frequently occurs when both p and n are small. See Montgomery (1991) or Duncan (1986) for further details.

Sample-Size Determination

Choose n such that $n > \dfrac{9(1 - p_0)}{p_0}$

where

n = Sample size
p_0 = An estimate (perhaps judgmental) of the process proportion p

For example, if p is thought to be about .05, the rule indicates that samples of at least size 171 should be used in constructing the p-chart:

$$n > \frac{9(1 - .05)}{.05} = 171$$

In the next three boxes we summarize how to construct a p-chart and its zone boundaries and how to interpret a p-chart.

Constructing a p-Chart: A Summary

1. Using a rational subgrouping strategy, collect at least 20 samples, each of size

$$n > \frac{9(1 - p_0)}{p_0}$$

where p_0 is an estimate of p, the proportion defective (i.e., nonconforming) produced by the process. p_0 can be determined from sample data (i.e., $\hat{p}$) or may be based on expert opinion.

2. For each sample, calculate $\hat{p}$, the proportion of defective units in the sample:

$$\hat{p} = \frac{\text{Number of defective items in the sample}}{\text{Number of items in the sample}}$$

3. Plot the centerline and control limits:

$$\textit{Centerline:}\quad \bar{p} = \frac{\text{Total number of defective units in all } k \text{ samples}}{\text{Total number of units in all } k \text{ samples}}$$

$$\textit{Upper control limit:}\quad \bar{p} + 3\sqrt{\frac{\bar{p}(1 - \bar{p})}{n}}$$

Lower control limit: $\bar{p} - 3\sqrt{\dfrac{\bar{p}(1-\bar{p})}{n}}$

where k is the number of samples of size n and $\bar{p}$ is the overall proportion of defective units in the nk units sampled. $\bar{p}$ is an estimate of the unknown process proportion p.

4. Plot the k sample proportions on the control chart in the order that the samples were produced by the process.

Constructing Zone Boundaries for a *p*-Chart

Upper A–B *boundary:* $\bar{p} + 2\sqrt{\dfrac{\bar{p}(1-\bar{p})}{n}}$

Lower A–B *boundary:* $\bar{p} - 2\sqrt{\dfrac{\bar{p}(1-\bar{p})}{n}}$

Upper B–C *boundary:* $\bar{p} + \sqrt{\dfrac{\bar{p}(1-\bar{p})}{n}}$

Lower B–C *boundary:* $\bar{p} - \sqrt{\dfrac{\bar{p}(1-\bar{p})}{n}}$

Note: When the lower control limit is negative, it should not be plotted on the control chart. However, the lower zone boundaries can still be plotted if they are nonnegative.

Interpreting a *p*-Chart

1. The **process is out of control** if one or more sample proportions fall beyond the control limits (rule 1) or if any of the three patterns of variation described by rules 2, 3, and 4 (Figure 13.27) are observed. Such signals indicate that one or more special causes of variation are influencing the process proportion, p. These causes should be identified and eliminated in order to bring the process into control.
2. The **process is treated as being in control** if none of the above noted out-of-control signals are observed. Processes that are in control should not be tampered with. However, if the level of variation is unacceptably high, common causes of variation should be identified and eliminated.

As with the $\bar{x}$- and R-charts, the centerline and control limits should be developed using samples that were collected during a period in which the process was in control. Accordingly, when a p-chart is first constructed, the centerline and the control limits should be treated as *trial values* (see Section 13.8) and, if necessary, modified before being extended to the right on the control chart and used to monitor future process output.

EXAMPLE 13.4

A manufacturer of auto parts is interested in implementing statistical process control in several areas within its warehouse operation. The manufacturer wants to begin with the order-assembly process. Too frequently orders received by customers contain the wrong items or too few items.

For each order received, parts are picked from storage bins in the warehouse, labeled, and placed on a conveyor belt system. Since the bins are spread over a 3-acre area, items that are part of the same order may be placed on different spurs of the conveyor belt system. Near the end of the belt system all spurs converge and a worker sorts the items according to the order they belong to. That information is contained on the labels that were placed on the items by the pickers.

The workers have identified three errors that cause shipments to be improperly assembled: (1) pickers pick from the wrong bin, (2) pickers mislabel items, and (3) the sorter makes an error.

The firm's quality manager has implemented a sampling program in which 90 assembled orders are sampled each day and checked for accuracy. An assembled order is considered nonconforming (defective) if it differs in any way from the order placed by the customer. To date, 25 samples have been evaluated. The resulting data are shown in Table 13.4.

a. Construct a p-chart for the order-assembly operation.
b. What does the chart indicate about the stability of the process?
c. Is it appropriate to use the control limits and centerline constructed in part **a** to monitor future process output?

Solution

a. The first step in constructing the p-chart after collecting the sample data is to calculate the sample proportion for each sample. For the first sample,

$$\hat{p} = \frac{\text{Number of defective items in the sample}}{\text{Number of items in the sample}} = \frac{12}{90} = .13333$$

All the sample proportions are displayed in Table 13.4. Next, calculate the proportion of defective items in the total number of items sampled:

$$\bar{p} = \frac{\text{Total number of defective items}}{\text{Total number of items sampled}} = \frac{292}{2250} = .12978$$

TABLE 13.4 Twenty-Five Samples of Size 90 from the Warehouse Order-Assembly Process

Sample	Size	Defective Orders	Sample Proportion
1	90	12	.13333
2	90	6	.06666
3	90	11	.12222
4	90	8	.08888
5	90	13	.14444
6	90	14	.15555
7	90	12	.13333
8	90	6	.06666
9	90	10	.11111
10	90	13	.14444
11	90	12	.13333
12	90	24	.26666
13	90	23	.25555
14	90	22	.24444
15	90	8	.08888
16	90	3	.03333
17	90	11	.12222
18	90	14	.15555
19	90	5	.05555
20	90	12	.13333
21	90	18	.20000
22	90	12	.13333
23	90	13	.14444
24	90	4	.04444
25	90	6	.06666
Totals	2,250	292	

The centerline is positioned at $\bar{p}$, and $\bar{p}$ is used to calculate the control limits:

$$\bar{p} \pm 3\sqrt{\frac{\bar{p}(1-\bar{p})}{n}} = .12978 \pm 3\sqrt{\frac{.12978(1-.12978)}{90}}$$

$$= .12978 \pm .10627$$

UCL: .23605

LCL: .02351

After plotting the centerline and the control limits, plot the 25 sample proportions in the order of sampling and connect the points with straight lines. The completed control chart is shown in Figure 13.34 on page 758.

FIGURE 13.34 ▶
p-chart for order-assembly process

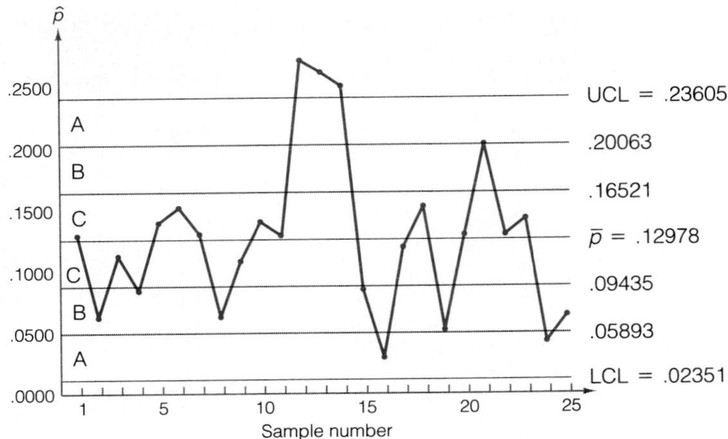

b. To assist our examination of the control chart, we add the one- and two-standard-deviation zone boundaries. The boundaries are located by substituting $\bar{p} = .12978$ into the following formulas:

$$\textit{Upper A–B boundary:}\quad \bar{p} + 2\sqrt{\frac{\bar{p}(1-\bar{p})}{n}} = .20063$$

$$\textit{Upper B–C boundary:}\quad \bar{p} + \sqrt{\frac{\bar{p}(1-\bar{p})}{n}} = .16521$$

$$\textit{Lower A–B boundary:}\quad \bar{p} - 2\sqrt{\frac{\bar{p}(1-\bar{p})}{n}} = .05893$$

$$\textit{Lower B–C boundary:}\quad \bar{p} - \sqrt{\frac{\bar{p}(1-\bar{p})}{n}} = .09435$$

Because three of the sample proportions fall above the upper control limit (rule 1), there is strong evidence that the process is out of control. None of the nonrandom patterns of rules 2, 3, and 4 (Figure 13.27) are evident. The process proportion appears to have increased dramatically somewhere around sample 12.

c. Because the process was apparently out of control during the period in which sample data were collected to build the control chart, it is not appropriate to continue using the chart. The control limits and centerline are not representative of the process when it is in control. The chart must be revised before it is used to monitor future output.

In this case, the three out-of-control points were investigated and it was discovered that they occurred on days when a temporary sorter was working in place of the regular sorter. Actions were taken to ensure that in the future better-trained temporary sorters would be available.

Since the special cause of the observed variation was identified and eliminated, all sample data from the three days the temporary sorter was working were dropped from the data set and the centerline and control limits were recalculated:

$$\textit{Centerline:} \quad \bar{p} = \frac{223}{1980} = .11263$$

$$\textit{Control limits:} \quad \bar{p} \pm 3\sqrt{\frac{\bar{p}(1-\bar{p})}{n}} = .11263 \pm 3\sqrt{\frac{.11263(.88737)}{90}}$$

$$= .11263 \pm .09997$$

UCL: .21259 LCL: .01266

The revised zones are calculated by substituting $\bar{p} = .11263$ in the following formulas:

$$\textit{Upper A–B boundary:} \quad \bar{p} + 2\sqrt{\frac{\bar{p}(1-\bar{p})}{n}} = .17927$$

$$\textit{Upper B–C boundary:} \quad \bar{p} + \sqrt{\frac{\bar{p}(1-\bar{p})}{n}} = .14595$$

$$\textit{Lower A–B boundary:} \quad \bar{p} - 2\sqrt{\frac{\bar{p}(1-\bar{p})}{n}} = .04598$$

$$\textit{Lower B–C boundary:} \quad \bar{p} - \sqrt{\frac{\bar{p}(1-\bar{p})}{n}} = .07931$$

The revised control chart appears in Figure 13.35. Notice that now all sample proportions fall within the control limits. These limits can now be treated as official, extended to the right on the chart, and used to monitor future orders.

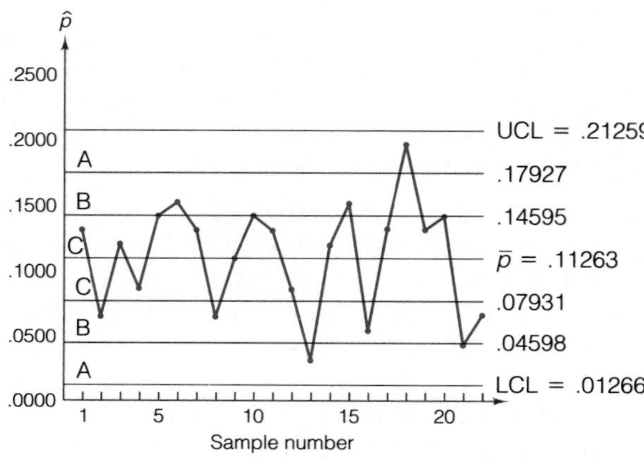

FIGURE 13.35 ▶ **Revised *p*-chart for order-assembly process**

Exercises 13.26–13.34

Learning the Mechanics

13.26 What characteristic of a process is a p-chart designed to monitor?

13.27 The proportion of defective items generated by a manufacturing process is believed to be 8%. In constructing a p-chart for the process, determine how large the sample size should be to avoid ending up with a negative lower control limit.

13.28 To construct a p-chart for a manufacturing process, 25 samples of size 200 were drawn from the process. The number of defectives in each sample is listed here.

Sample	*Sample Size*	*Defectives*	*Sample*	*Sample Size*	*Defectives*
1	200	16	14	200	12
2	200	14	15	200	14
3	200	9	16	200	11
4	200	11	17	200	8
5	200	15	18	200	7
6	200	8	19	200	12
7	200	12	20	200	15
8	200	16	21	200	9
9	200	17	22	200	16
10	200	13	23	200	13
11	200	15	24	200	11
12	200	10	25	200	10
13	200	9			

a. Calculate the proportion defective in each sample.
b. Calculate and plot $\bar{p}$ and the upper and lower control limits for the p-chart.
c. Calculate and plot the A, B, and C zone boundaries on the p-chart.
d. Plot the sample proportions on the p-chart and connect them with straight lines.
e. Use the pattern-analysis rules 1–4 for detecting the presence of special causes of variation (Figure 13.27) to determine whether the process is out of control.

13.29 To construct a p-chart, 20 samples of size 150 were drawn from a process. The proportion of defective items found in each of the samples is listed in the accompanying table.

Sample	*Proportion Defective*	*Sample*	*Proportion Defective*	*Sample*	*Proportion Defective*
1	.03	8	.05	15	.07
2	.05	9	.07	16	.06
3	.10	10	.06	17	.07
4	.02	11	.07	18	.02
5	.08	12	.04	19	.05
6	.09	13	.06	20	.03
7	.08	14	.05		

a. Calculate and plot the centerline and the upper and lower control limits for the p-chart.
b. Calculate and plot the A, B, and C zone boundaries on the p-chart.
c. Plot the sample proportions on the p-chart.
d. Is the process under control? Explain.
e. Should the control limits and centerline of part **a** be used to monitor future process output? Explain.

13.30 In each of the following cases, use the sample-size formula on page 754 to determine a sample size large enough to avoid constructing a p-chart with a negative lower control limit.

a. $p_0 = .01$ b. $p_0 = .05$ c. $p_0 = .10$ d. $p_0 = .20$

Applying the Concepts

13.31 A manufacturer produces disks for personal computers. From past experience the production manager believes that 1% of the disks are defective. The company collected a sample of the first 1,000 disks manufactured after 4:00 P.M. every other day for a month. The disks were analyzed for defects, then these data and Minitab were used to contruct a p-chart.

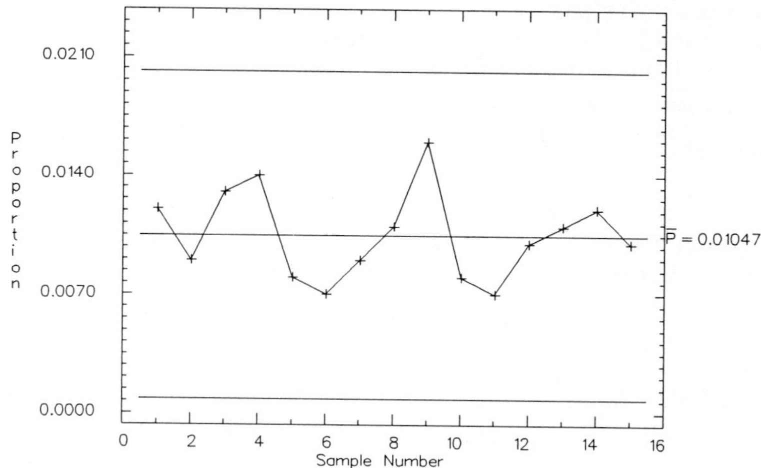

a. From a statistical perspective, is a sample size of 1,000 adequate for constructing the p-chart? Explain.
b. Calculate the chart's upper and lower control limits.
c. What does the p-chart suggest about the presence of special causes during the time when the data were collected?
d. Critique the rational subgrouping strategy used by the disk manufacturer.

13.32 Goodstone Tire & Rubber Company is interested in monitoring the proportion of defective tires generated by the production process at its Akron, Ohio, production plant. The company's chief engineer believes that the proportion is about 7%. Because the tires are destroyed during the testing process, the company would like to keep the number of tires tested to a minimum. However, the engineer would also like to use a p-chart with a positive lower control limit. A positive lower control limit makes it possible to determine when the process has generated an unusually small proportion of defectives. Such an occurrence is good news and would signal the engineer to look for causes of the superior performance. That information can be used to improve the production process. Using the sample-size formula on page 754, the chief engineer recommended that the company randomly sample and test 120 tires from each day's production. To date, 20 samples have been taken. The data are presented here.

Sample	*Sample Size*	*Defectives*	*Sample*	*Sample Size*	*Defectives*
1	120	11	11	120	10
2	120	5	12	120	12
3	120	4	13	120	8
4	120	8	14	120	6
5	120	10	15	120	10
6	120	13	16	120	5
7	120	9	17	120	10
8	120	8	18	120	10
9	120	10	19	120	3
10	120	11	20	120	8

a. Use the sample-size formula to show how the chief engineer arrived at the recommended sample size of 120.
b. Construct a p-chart for the tire-production process.
c. What does the chart indicate about the stability of the process? Explain.
d. Is it appropriate to use the control limits to monitor future process output? Explain.
e. Is the p-chart you constructed in part **b** capable of signaling hour-to-hour changes in p? Explain.

13.33 With the recent increase in competition among credit-card companies, the banks issuing VistaCard are interested in improving customer service as a means of retaining current customers and attracting new ones. The banks targeted a reduction in customer billing errors as a top priority. All customer accounts are processed at a central billing center in Eureka, California. The director of the billing center believes that billing errors may run as high as 4%. A program was implemented at the center to check a random sample of 250 customer bills each day for accuracy. To date, 24 samples have been evaluated. The data are presented in the table.

Sample	*Sample Size*	*Bills with Errors*	*Sample*	*Sample Size*	*Bills with Errors*
1	250	11	13	250	12
2	250	9	14	250	4
3	250	8	15	250	9
4	250	14	16	250	10
5	250	16	17	250	13
6	250	12	18	250	19
7	250	7	19	250	18
8	250	17	20	250	10
9	250	14	21	250	16
10	250	3	22	250	19
11	250	11	23	250	8
12	250	6	24	250	7

a. Construct a p-chart for the customer billing operation.
b. What does the chart indicate about the stability of the process? Explain.
c. Should the control limits be used to monitor the billing process in the future? Explain.

13.34 A manufacturing company makes hemostats for hospital emergency rooms. The company is interested in implementing statistical process control procedures in its production operation. The production manager believes that the proportion of defective hemostats generated by the process is about 3%. The company collected one sample of 300 consecutively manufactured hemostats each day for 20 days. Each day sampling began at 10:00 A.M. The data are given here.

Sample	*Sample Size*	*Defectives*	*Sample*	*Sample Size*	*Defectives*
1	300	8	11	300	12
2	300	6	12	300	11
3	300	11	13	300	14
4	300	15	14	300	8
5	300	12	15	300	7
6	300	11	16	300	3
7	300	9	17	300	9
8	300	6	18	300	11
9	300	5	19	300	10
10	300	4	20	300	6

a. Use the sample-size formula on page 754 to show that a sample size of 300 is large enough to prevent the lower control limit of the p-chart from being negative.
b. Construct a p-chart for the manufacturing process.
c. Does the chart indicate the presence of special causes of variation? Explain.
d. Provide an example of a special cause of variation that could potentially affect this process. Do the same for a common cause of variation.
e. Should the control limits and centerline be used to monitor future process output? If not, what should be done to develop a centerline and control limits that can be used to monitor future output?

13.11 Diagnosing Causes of Variation

Statistical process control (SPC) consists of three major activities or phases: (1) monitoring process variation, (2) diagnosing causes of variation, and (3) eliminating those causes. A more detailed description of SPC is shown in Figure 13.36, page 764, which depicts SPC as a quality improvement cycle. In the monitoring phase, statistical signals from the process are evaluated in order to uncover opportunities to improve the process. This is the phase we have dealt with in Sections 13.7–13.10. We turn our attention now to the diagnosis phase.

The diagnosis phase is the critical link in the SPC improvement cycle. The monitoring phase simply identifies *whether* problems exist; the diagnosis phase identifies *what* the problems are. If the monitoring phase detected the presence of special causes of variation (i.e., an out-of-control signal was observed on a control chart), the diagnosis phase is concerned with tracking down the underlying cause or causes. If no special causes were detected in the monitoring phase (i.e., the process is under statistical control) and further improvement in the process is desired, the diagnosis phase concentrates on uncovering common causes of variation.

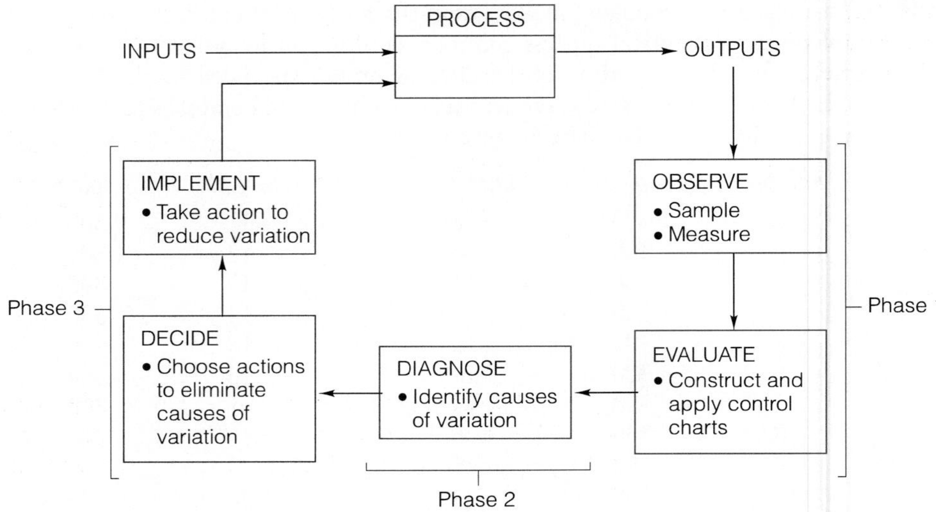

FIGURE 13.36 ▶ SPC viewed as a quality improvement cycle

It is important to recognize that the achievement of process improvement requires more than the application of statistical tools such as control charts. This is particularly evident in the diagnosis phase. The diagnosis of causes of variation requires expert knowledge about the process in question. Just as you would go to a physician to diagnose a pain in your back, you would turn to people who work in the process or to engineers or analysts with process expertise to help you diagnose the causes of process variation.

Several methods have been developed for assisting process experts with process diagnosis, including *flowcharting* (Section 13.3) and the simple but powerful graphical tool called *Pareto analysis* (Chapter 2). Another graphical method, the **cause-and-effect diagram**, is described in this section. A fourth methodology, *experimental design*, is the topic of Chapter 16.

The cause-and-effect diagram was developed by Kaoru Ishikawa of the University of Tokyo in 1943. As a result, it is also known as an *Ishikawa diagram*. The cause-and-effect diagram facilitates the construction of causal chains that explain the occurrence of events, problems, or conditions. It is often constructed through brainstorming sessions involving a small group of process experts. It has been employed for decades by Japanese firms, but was not widely applied in the United States until the mid-1980s.

The basic framework of the cause-and-effect diagram is shown in Figure 13.37. In the right-hand box in the figure, we record the effect whose cause(s) we want to diagnose. For instance, Figures 13.38 and 13.39 (page 766), record two examples of effects to be diagnosed: why the wrong meals are being served to hospital patients, and the reasons for high variation in the fill weights of 20-pound bags of dry dog food.

Examining Figure 13.37, we see the branches of the cause-and-effect diagram, which represent the major factors that influence the process and that could be responsible for the effect. These are often taken to be the six universal sources of process

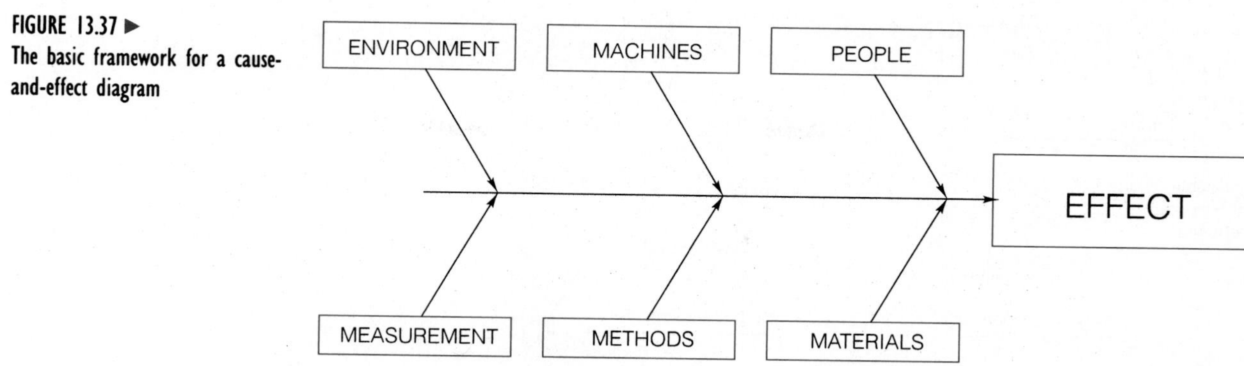

FIGURE 13.37 ▶ The basic framework for a cause-and-effect diagram

variation that we described in Section 13.4: people, machines, materials, methods, measurement, and environment. Notice that in the examples of Figures 13.38 and 13.39 these categories were tailored to fit the process in question. The set of categories

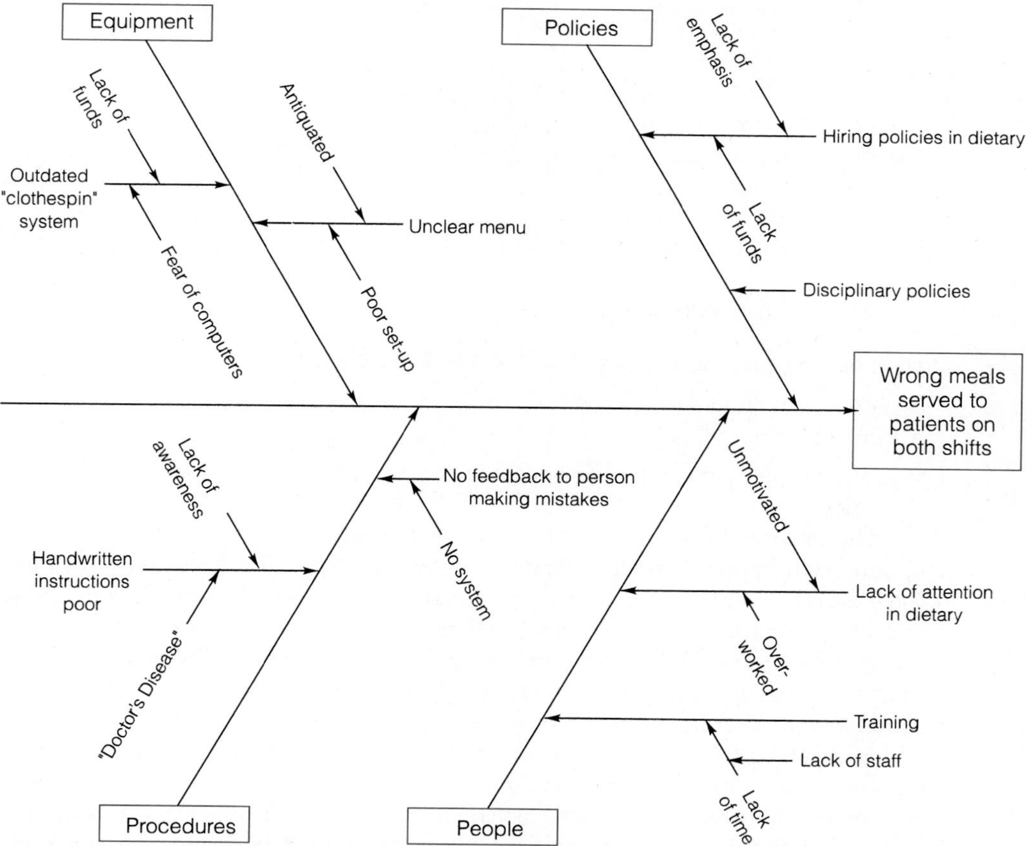

FIGURE 13.38 ▲ Cause-and-effect diagram for the meal service at a hospital [Source: *The Memory Jogger*, The Growth Opportunity Alliance of Greater Lawrence, Lawrence, Mass., 1985. p. 24.]

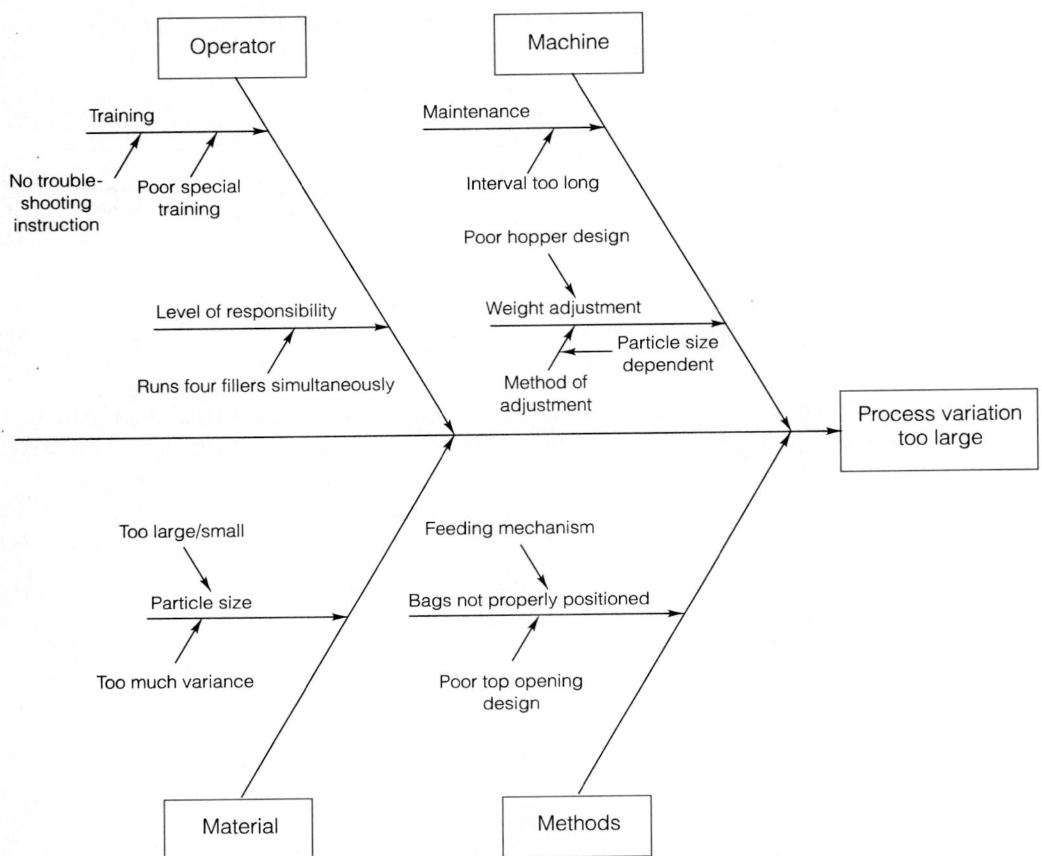

FIGURE 13.39 ▲ Cause-and-effect diagram for the filling process for 20-lb. bags of dog food; from DeVor, Chang, and Southerland (1992, p. 191)

must be broad enough to include virtually all possible factors that could influence the process. It is less important how many categories are used or how you label the categories.

The cause-and-effect diagram is constructed using effect-to-cause reasoning. That is, you begin by specifying the effect of interest and then move backward to identify *potential* causes of the effect. After a potential cause has been identified, you treat it as an effect and try to find its cause, and so forth. The result is a **causal chain.** A completed cause-and-effect diagram typically contains many causal chains. These chains help us to track down causes whose eradication will reduce, improve, or eliminate the effect in question.

After setting up the basic framework for the cause-and-effect diagram and recording the effect of interest in the box on the right, you construct the causal chains, proceeding backward from general potential causes to increasingly specific causes. Begin by choosing one of the universal cause categories—say, people—and asking, "What factors related to people could cause the effect in question?" In the hospital example of Figure 13.38, two factors were identified: (1) insufficient training and (2) lack of attention to detail by workers in the dietary department. Each of these causes is written on a twig of the People branch. Next, each cause is treated as an effect and an attempt is made to identify its cause. That is, we look for subcauses. The insufficient training was blamed on a lack of available time to train the hospital staff. This in turn was blamed on a shortage of hospital personnel. Thus, the Training twig has a Lack of time twig attached to it, which in turn has a Lack of staff twig attached. Multiple causal chains like this should be constructed for each branch of the cause-and-effect diagram, as illustrated in Figures 13.38 and 13.39.

Once completed, the various causal chains of the cause-and-effect diagram must be evaluated (often subjectively) to identify one or more factors thought most likely to be causes of the effect in question. Then actions can be chosen and implemented (see Figure 13.36) to eliminate the causes and improve the process.

Besides facilitating process diagnosis, cause-and-effect diagrams serve to document the causal factors that may potentially affect a process and to communicate that information to others in the organization. The cause-and-effect diagram is a very flexible tool that can be applied in a variety of situations. It can be used as a formal part of the SPC improvement cycle, as suggested above, or simply as a means of investigating the causes of organizational problems, events, or conditions. It can also help select the appropriate process variables to monitor with control charts.

Summary

We learned that a **process** is a series of actions or operations that transform inputs to outputs. In this chapter we focused on methods for monitoring and improving processes. The methods discussed can be applied to any process, whether a personal process or an organizational process. The primary target of the methods is **variation** in process output. Variation is inherent in all processes and is caused by **people**, **machines**, **materials**, **methods**, **measurement**, and the **environment**. Consumers' perceptions of the quality of a particular product or service are strongly influenced by the amount of variation in that product or service. Variation may plague the **performance**, **reliability**, **conformance**, **durability**, and **serviceability** of a product. Other product dimensions that influence customers' perceptions of quality include the product's **features** and **aesthetics**.

We introduced you to **process management** and argued that a **preventative** style of process management is preferred to a **detection–reaction** approach. The hallmarks of good process management are **purposeful data collection** and **continuous improvement**.

We learned that unless a process is brought into *statistical control* its full capability will remain unknown. The set of methods and activities used to bring a process into control and keep it in control is called **statistical process control**. Processes that are in control reflect only **common causes of variation**; processes that are out of control reflect both common and **special causes of variation**.

Control charts are used to identify the presence of special causes of variation. Control charts help us decide when to take action to find and remove special causes and when to leave the process alone. The **$\bar{x}$-chart** monitors the mean of the process; the ***R*-chart** monitors the variation of the process; and the ***p*-chart** monitors the proportion defective produced by the process.

In constructing a control chart, we try to anticipate the special causes of variation that might affect the process and employ a **rational subgrouping strategy** that makes the chart sensitive to the anticipated variation. In interpreting a control chart, we consider the process out of control if one or more points fall beyond the **control limits** or if any of the other five sequences of points shown in the **pattern-analysis rules** of Figure 13.27 are observed between the control limits.

We learned that statistical process control can be viewed as a quality improvement cycle in which we **observe and evaluate** process data, **diagnose** causes of variation, **decide** what action to take to eliminate the cause of the variation, and **implement** the action to improve the process. Repetition of this cycle will ensure continuous improvement of the process. We saw that **cause-and-effect diagrams** are helpful in the diagnosis phase of the cycle.

Supplementary Exercises 13.35–13.61

13.35 Define *quality* and lists its important dimensions.

13.36 What is a system? Give an example of a system with which you are familiar, and describe its inputs, outputs, and transformation process.

13.37 What is a process? Give an examle of an organizational process and a personal process.

13.38 Select a personal process that you would like to better understand or to improve and construct a flowchart for it.

13.39 Describe the six major sources of process variation.

13.40 What is process management?

13.41 What is systems thinking and how does it benefit process management?

13.42 Compare and contrast the detection–reaction and preventative styles of process management.

13.43 Select an organizational process that you would like to improve.

a. Construct a process flowchart.
b. What are the inputs and outputs of the process?
c. Analyze the flowchart and suggest one or more ways that the process can be simplified.
d. Choose one operation in the process (i.e., one of the rectangles in the flowchart of part **a**) and identify the inputs and outputs of the operation. (Note that this operation is a *subprocess* of the process described in your flowchart.)

e. Define a variable associated with the subprocess of part **d** that would be useful to monitor in managing the subprocess.

f. Construct a cause-and-effect diagram for the variable you defined in part **e** to help explain why it takes on different values over time. Label the effect box "Variation in x," where x is your chosen variable.

13.44 Select a problem, event, or condition whose cause or causes you would like to diagnose. Construct a cause-and-effect diagram that would facilitate your diagnosis.

13.45 What are the two properties of a system that make it an "indivisible whole"?

13.46 Why is it important for managers who implement statistical process control to have a basic understanding of systems theory?

13.47 In estimating a population mean μ using a sample mean $\bar{x}$, why is it likely that $\bar{x} \neq \mu$? Construct a cause-and-effect diagram for the effect $\bar{x} \neq \mu$.

13.48 Construct a cause-and-effect diagram to help explain why customer waiting time at the drive-in window of a fast-food restaurant is a variable.

13.49 Processes that are in control are predictable; out-of-control processes are not. Explain.

13.50 Compare and contrast special and common causes of variation.

13.51 Explain the role of the control limits of a control chart.

13.52 Explain the difference between control limits and specification limits.

13.53 A process is under control and follows a normal distribution with mean 100 and standard deviation 10. In constructing a standard $\bar{x}$-chart for this process, the control limits are set three standard deviations from the mean—i.e., $100 + 3(10/\sqrt{n})$. The probability of observing an $\bar{x}$ outside the control limits is $.00135 + .00135 = .0027$. Suppose it is desired to construct a control chart that signals the presence of a potential special cause of variation for less extreme values of $\bar{x}$. How many standard deviations from the mean should the control limits be set such that the probability of the chart falsely indicating the presence of a special cause of variation is .10 rather than .0027?

13.54 Consider the following time series data.

Order of Production	*Weight (grams)*	*Order of Production*	*Weight (grams)*
1	6.0	9	6.5
2	5.0	10	9.0
3	7.0	11	3.0
4	5.5	12	11.0
5	7.0	13	3.0
6	6.0	14	12.0
7	8.0	15	2.0
8	5.0		

a. Construct a time series plot. Be sure to connect the points and add a centerline.

b. Which type of variation pattern in Figure 13.13 (page 712) best describes the pattern revealed by your plot?

13.55 The accompanying length measurements were made on 20 consecutively produced pencils.

Order of Production	*Length (inches)*	*Order of Production*	*Length (inches)*
1	7.47	11	7.57
2	7.48	12	7.56
3	7.51	13	7.55
4	7.49	14	7.58
5	7.50	15	7.56
6	7.51	16	7.59
7	7.48	17	7.57
8	7.49	18	7.55
9	7.48	19	7.56
10	7.50	20	7.58

a. Construct a time series plot. Be sure to connect the plotted points and add a centerline.
b. Which type of variation pattern in Figure 13.13 best describes the pattern shown in your plot?

13.56 Use the appropriate pattern-analysis rules (page 732) to determine whether the process being monitored by the accompanying control chart is under the influence of special causes of variation.

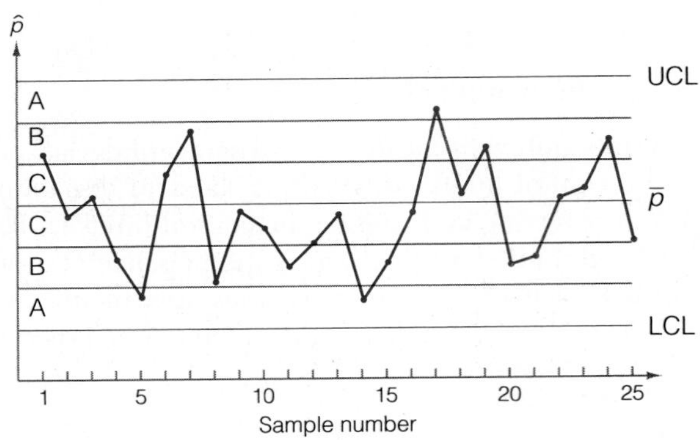

13.57 A company that manufactures plastic molded parts believes it is producing an unusually large number of defects. To investigate this suspicion, each shift drew seven random samples of 200 parts, visually inspected each part to determine whether it was defective, and tallied the primary type of defect present (Hart, 1992). These data are presented in the table.

a. From a statistical perspective, are the number of samples and the sample size of 200 adequate for constructing a p-chart for these data? Explain.
b. Construct a p-chart for this manufacturing process.
c. Should the control limits be used to monitor future process output? Explain.
d. Suggest a strategy for identifying the special causes of variation that may be present.

Sample	*Shift*	*# of Defects*	*Crack*	*Burn*	*Dirt*	*Blister*	*Trim*
			Type of Defect				
1	1	4	1	1	1	0	1
2	1	6	2	1	0	2	1
3	1	11	1	2	3	3	2
4	1	12	2	2	2	3	3
5	1	5	0	1	0	2	2
6	1	10	1	3	2	2	2
7	1	8	0	3	1	1	3
8	2	16	2	0	8	2	4
9	2	17	3	2	8	2	2
10	2	20	0	3	11	3	3
11	2	28	3	2	17	2	4
12	2	20	0	0	16	4	0
13	2	20	1	1	18	0	0
14	2	17	2	2	13	0	0
15	3	13	3	2	5	1	2
16	3	10	0	3	4	2	1
17	3	11	2	2	3	2	2
18	3	7	0	3	2	2	0
19	3	6	1	2	0	1	2
20	3	8	1	1	2	3	1
21	3	9	1	2	2	2	2

13.58 A hospital has used control charts continuously since 1978 to monitor the quality of its nursing care. A set of 363 scoring criteria, or standards, are applied at critical points in the patients' stay to determine whether the patients are receiving beneficial nursing care. Auditors regularly visit each hospital unit, sample two patients, and evaluate their care. The auditors review patients' records; interview the patients, the nurse, and the head nurse; and observe the nursing care given (Cheng, 1992). The data in the table were collected over a 3-month period for a newly opened unit of the hospital.

Sample	*Scores*	*Sample*	*Scores*	*Sample*	*Scores*
1	345,341	8	344,344	15	345,329
2	331,328	9	359,334	16	358,351
3	343,355	10	346,361	17	353,352
4	351,352	11	360,355	18	334,340
5	360,348	12	325,335	19	341,335
6	342,336	13	350,348	20	358,345
7	328,331	14	336,337		

a. Construct an R-chart for the nursing care process.

b. Construct an $\bar{x}$-chart for the nursing care process.

c. Should the control charts of parts **a** and **b** be used to monitor future process output? Explain.

d. The hospital would like all scores to exceed 335 (their specification limit). Over the 3-month periods, what proportion of the sampled patients received care that did not conform to the hospital's requirements?

13.59 AirExpress, an overnight mail service, is concerned about the operating efficiency of the package-sorting department at its Toledo, Ohio, terminal. The company would like to monitor the time it takes for packages to be put in outgoing delivery bins from the time they are received. The sorting department operates 6 hours per day, from 6 P.M. to midnight. The company randomly sampled four packages during each hour of operation during four consecutive days. The time for each package to move through the system, in minutes, is given.

Sample	*Transit Time (mins.)*				*Sample*	*Transit Time (mins.)*			
1	31.9	33.4	37.8	26.2	13	24.6	29.9	31.8	37.9
2	29.1	24.3	33.2	36.7	14	30.6	36.0	40.2	30.8
3	30.3	31.1	26.3	34.1	15	29.7	33.2	34.9	27.6
4	39.6	29.4	31.4	37.7	16	24.1	26.8	32.7	29.0
5	27.4	29.7	36.5	33.3	17	29.4	31.6	35.2	27.6
6	32.7	32.9	40.1	29.7	18	31.1	33.0	29.6	35.2
7	30.7	36.9	26.8	34.0	19	27.0	29.0	35.1	25.1
8	28.4	24.1	29.6	30.9	20	36.6	32.4	28.7	27.9
9	30.5	35.5	36.1	27.4	21	33.0	27.1	26.2	35.1
10	27.8	29.6	29.0	34.1	22	33.2	41.2	30.7	31.6
11	34.0	30.1	35.9	28.8	23	26.7	35.2	39.7	31.5
12	25.5	26.3	34.8	30.0	24	30.5	36.8	27.9	28.6

a. Construct an $\bar{x}$-chart from these data. In order for this chart to be meaningful, what assumption must be made about the variation of the process? Why?
b. What does the chart suggest about the stability of the package-sorting process? Explain.
c. Should the control limits be used to monitor future process output? Explain.

13.60 Officials at Mountain Airlines are interested in monitoring the length of time customers must wait in line to check in at their airport counter in Reno, Nevada. In order to develop a control chart, five customers were sampled each day for 20 days. The data, in minutes, are presented here.

Sample	*Waiting Time (mins.)*					*Sample*	*Waiting Time (mins.)*				
1	3.2	6.7	1.3	8.4	2.2	11	3.2	2.9	4.1	5.6	.8
2	5.0	4.1	7.9	8.1	.4	12	2.4	4.3	6.7	1.9	4.8
3	7.1	3.2	2.1	6.5	3.7	13	8.8	5.3	6.6	1.0	4.5
4	4.2	1.6	2.7	7.2	1.4	14	3.7	3.6	2.0	2.7	5.9
5	1.7	7.1	1.6	.9	1.8	15	1.0	1.9	6.5	3.3	4.7
6	4.7	5.5	1.6	3.9	4.0	16	7.0	4.0	4.9	4.4	4.7
7	6.2	2.0	1.2	.9	1.4	17	5.5	7.1	2.1	.9	2.8
8	1.4	2.7	3.8	4.6	3.8	18	1.8	5.6	2.2	1.7	2.1
9	1.1	4.3	9.1	3.1	2.7	19	2.6	3.7	4.8	1.4	5.8
10	5.3	4.1	9.8	2.9	2.7	20	3.6	.8	5.1	4.7	6.3

a. Construct an *R*-chart from these data.
b. What does the *R*-chart suggest about the stability of the process? Explain.
c. Explain why the *R*-chart should be interpreted prior to the $\bar{x}$-chart.
d. Construct an $\bar{x}$-chart from these data.
e. What does the $\bar{x}$-chart suggest about the stability of the process? Explain.
f. Should the control limits for the *R*-chart and $\bar{x}$-chart be used to monitor future process output? Explain.

13.61 A company called CRW runs credit checks for a large number of banks and insurance companies. Credit history information is typed into computer files by trained administrative assistants. The company is interested in monitoring the proportion of credit histories that contain one or more data-entry errors. Based on her experience with the data-entry operation, the director of the data-processing unit believes that the proportion of histories with data-entry errors is about 6%. CRW audited 150 randomly selected credit histories each day for 20 days. The sample data are presented here.

Sample	*Sample Size*	*Histories with Errors*	*Sample*	*Sample Size*	*Histories with Errors*
1	150	9	11	150	7
2	150	11	12	150	6
3	150	12	13	150	12
4	150	8	14	150	10
5	150	10	15	150	11
6	150	6	16	150	7
7	150	13	17	150	6
8	150	9	18	150	12
9	150	11	19	150	14
10	150	5	20	150	10

a. Use the sample size formula on page 754 to show that a sample size of 150 is large enough to prevent the lower control limit of the p-chart they plan to construct from being negative.
b. Construct a p-chart for the data-entry process.
c. What does the chart indicate about the presence of special causes of variation? Explain.
d. Provide an example of a special cause of variation that could potentially affect this process. Do the same for a common cause of variation.
e. Should the control limits be used to monitor future credit histories produced by the data-entry operation? Explain.

On Your Own

Choose a process that is of interest to you and that you know something about. It can be an organizational process, such as those used in the examples of this chapter, or a personal process. Examples of personal processes include cooking, changing diapers, studying for exams, making a cup of coffee, balancing a checkbook, and so on.

a. Describe the objective or purpose of the process.
b. Identify the suppliers and customers of the process and the input(s) and output(s) of the process.
c. Identify the owner of the process. That is, identify the person closest to the process with the power to change the design of the process.
d. Specify the flow of operations within the process using a flowchart.
e. Identify two listening points within the process. That is, identify opportunities for monitoring the performance of various aspects of the process. Define the variables to be monitored.
f. Construct a cause-and-effect diagram that identifies the potential causes of variation in one of the variables you defined in part **e**.
g. Indicate which of the three control charts discussed in this chapter, if any, are appropriate for monitoring the variables you identified above. Justify your answer.

References

Ackoff, R. "The Second Industrial Revolution," speech transcript, undated.

Alwan, L. C., and Roberts, H. V. "Time-series modeling for statistical process control." *Journal of Business and Economic Statistics*, 1988, Vol. 6, pp. 87–95.

Banks, J. *Principles of Quality Control*. New York: Wiley, 1989.

Brassard, M. *The Memory Jogger Plus+*. Methuen, Mass.: GOAL/QPC, 1989.

Checkland, P. *Systems Thinking, Systems Practice*. New York, Wiley, 1981.

Cheng, T. C. E. "A case study of hospital quality assurance." *International Journal of Quality and Reliability Management*, 1992, Vol. 9, No. 4, p. 21.

Deming, W. E. *Out of the Crisis*. Cambridge, Mass.: M.I.T. Center for Advanced Engineering Study, 1986.

DeVor, R. E., Chang, T., and Southerland, J. W. *Statistical Quality Design and Control*. New York: Macmillan, 1992.

Duncan, A. J. *Quality Control and Industrial Statistics*. Homewood, Ill.: Irwin, 1986.

Feigenbaum, A. V. *Total Quality Control*, 3rd ed. New York: McGraw-Hill, 1983.

Fuller, T. "Eliminating complexity from work: Improving productivity by enhancing quality." *National Productivity Review*, 1985.

Garvin, D. A. *Managing Quality*. New York: Free Press/Macmillan, 1988.

Gitlow, H., Gitlow, S., Oppenheim, A., and Oppenheim, R. *Tools and Methods for the Improvement of Quality*. Homewood, Ill.: Irwin, 1989.

Grant, E. L., and Leavenworth, R. S. *Statistical Quality Control*, 6th ed. New York: McGraw-Hill, 1988.

Hart, Marilyn K. "Quality tools for improvement." *Production and Inventory Management Journal*, First Quarter 1992, Vol. 33, No. 1, p. 59.

Ishikawa, K. *Guide to Quality Control*, 2nd ed. White Plains, N.Y.: Kraus International Publications, 1986.

Joiner, B. L., and Goudard, M. A. "Variation, management, and W. Edwards Deming." *Quality Process*, Dec. 1990, pp. 29–37.

Juran, J. M. *Juran of Planning for Quality*. New York: Free Press/Macmillan, 1988.

Juran, J. M., and Gryna, F. M., Jr. *Quality Planning and Analysis*, 2nd ed. New York: McGraw-Hill, 1980.

Kane, V. E. *Defect Prevention*. New York: Marcel Dekker, 1989.

Latzko, W. J. *Quality and Productivity for Bankers and Financial Managers*, New York: Marcel Dekker, 1986.

Melan, E. H. "Process management in service and administrative operations." *Quality Progress*, June 1985, pp. 52–59.

Moen, R. D., Nolan, T. W., and Provost, L. P. *Improving Quality Through Planned Experimentation*. New York: McGraw-Hill, 1991.

Montgomery, D. C. *Introduction to Statistical Quality Control*, 2nd ed. New York: Wiley, 1991.

Nelson, L. S. "The Shewhart control chart—Tests for special causes." *Journal of Quality Technology*, Oct. 1984, Vol. 16, No. 4, pp. 237–239.

Optner, S. L. *Systems Analysis for Business and Industrial Problem Solving*. Englewood Cliffs, N.J.: Prentice-Hall, 1965.

Roberts, H. V. *Data Analysis for Managers*, 2nd ed. Redwood City, Calif.: Scientific Press, 1991.

Rosander, A. C. *Applications of Quality Control in the Service Industries*. New York: Marcel Dekker, 1985.

Rummler, G. A., and Brache, A. P. *Improving Performance: How to Manage the White Space on the Organization Chart*. San Francisco: Jossey-Bass, 1991.

Ryan, T. P. *Statistical Methods for Quality Improvement*. New York: Wiley, 1989.

Statistical Quality Control Handbook. Indianapolis, Ind.: AT&T Technologies, Select Code 700-444 (Inquiries: 800-432-6600); originally published by Western Electric Company, 1956.

The Ernst and Young Quality Improvement Consulting Group. *Total Quality: An Executive's Guide for the 1990s*. Homewood, Ill.: Dow-Jones Irwin, 1990.

Wadsworth, H. M., Stephens, K. S., and Godfrey, A. B. *Modern Methods for Quality Control and Improvement*. New York: Wiley, 1986.

Walton, M. *The Deming Management Method*. New York: Dodd, Mead, & Company, 1986.

Wheeler, D. J., and Chambers, D. S. *Understanding Statistical Process Control*. Knoxville, Tenn.: Statistical Process Controls, Inc., 1986.

Wood, L. "MIPS, BIPS, and super-chips." *Compute*, 1992, Vol. 14, No. 9, pp. 58–66.

CHAPTER FOURTEEN

Time Series: Index Numbers and Descriptive Analyses

Contents

Case Study

Where We've Been

In the last chapter we began our study of processes. We learned that variation in the output of processes is inevitable. However, we saw that by looking inside processes rather than treating them as "black boxes," we can often identify the causes of process variation. We examined both managerial and statistical methods for continuously improving processes and the quality of their output. We focused on control charts, which are statistical tools for monitoring output variation and determining when action should be taken to improve a process.

Where We're Going

In this chapter we turn from discussing methods for improving processes to methods for better understanding the output of processes. The methods of this chapter treat the process as a black box and focus exclusively on the data stream generated by a process over time, i.e., on time series data. By studying the time series data generated by a process, we can learn much about both the past and future behavior of the process. We begin with index numbers, a type of time series data often used to characterize the performance of one of the most complex processes—our economy. In the next chapter we discuss methods for forecasting future values of process output.

14

In the previous chapter we were concerned with improving processes. We saw that to do so requires making changes within or redesigning the process. In general, improvement requires knowledge of the actions and operations that comprise the process. In this chapter and the next, our concern is not with the improvement of the internal workings of processes but with describing and predicting the output of processes.

The processes in question may be manufacturing and service processes such as those highlighted in the previous chapter, or they may be extremely complex processes like an entire business organization, an industry, or even the U.S. economy. These more complex processes are more typically referred to as **systems** (see Section 13.4).

The process outputs on which we focus are the streams of data generated by processes (and systems) over time. Recall from Chapters 2 and 13 that such data streams are called **time series** or **time series data**. For example, businesses generate time series data such as weekly sales, quarterly earnings, and yearly profits that can be used to describe and evaluate the performance of the business. The U.S. economy can be thought of as a system that generates streams of data that include the Gross Domestic Product, the Consumer Price Index, and the unemployment rate.

The methods of this chapter and the next focus exclusively on the time series data generated by a process. Properly analyzed, these data reveal much about the past and future behavior of the process. Time series data, like other types of data we have discussed in previous chapters, are subjected to two kinds of analyses: **descriptive** and **inferential**. Descriptive analyses, the topic of this chapter, use graphical and numerical techniques to provide a clear understanding of any patterns that are present in the time series. After graphing the data, you will often want to use it to make inferences about the future values of the time series, i.e., you will want to **forecast** future values. For example, once you understand the past and present trends of the Dow Jones Industrial Average, you would probably want to forecast its future trend before making decisions about buying and selling stocks. Since significant amounts of money may be riding on the accuracy of your forecasts, you would be interested in measures of their reliability. Forecasts and their measures of reliability are examples of **inferential techniques** in time series analysis. Inferential techniques are the topic of Chapter 15.

14.1 Index Numbers: An Introduction

The most common technique for characterizing a business or economic time series is to compute **index numbers**. Index numbers measure how a time series changes over time. Change is measured relative to a preselected time period, called the **base period**.

> **Definition 14.1**
>
> An **index number** is a number that measures the change in a variable over time relative to the value of the variable during a specific **base period**.

Two types of indexes dominate business and economic applications: **price** and **quantity indexes**. Price indexes measure changes in the price of a commodity or group of commodities over time. The Consumer Price Index (CPI) is a price index because it measures price changes of a group of commodities that are intended to reflect typical purchases of American consumers. On the other hand, an index constructed to measure the change in the total number of automobiles produced annually by American manufacturers would be an example of a quantity index.

Methods of calculating index numbers range from very simple to extremely complex, depending on the numbers and types of commodities represented by the index. The next two sections provide details on the calculation and interpretation of several important types of index numbers.

14.2 Simple Index Numbers

When an index number is based on the price or quantity of a single commodity, it is called a **simple index number**.

Definition 14.2

A **simple index number** is based on the relative changes (over time) in the price or quantity of a single commodity.

For example, consider the price of silver between 1975 and 1990, shown in Table 14.1. To construct a simple index to describe the relative changes in silver prices, we must first choose a **base period**. The choice is important because the price for all other

TABLE 14.1 Silver Prices, 1975–1990

Year	Price ($/oz.)	Year	Price ($/oz.)
1975	4.42	1983	11.44
1976	4.35	1984	8.14
1977	4.62	1985	6.14
1978	5.40	1986	5.47
1979	11.09	1987	7.01
1980	20.64	1988	6.53
1981	10.52	1989	5.50
1982	7.95	1990	4.82

Source: U.S. Bureau of the Census. *Statistical Abstract of the United States, 1992.*

periods will be compared with the price during the base period. We select 1975 as the base period, a time just preceding the period of rapid economic inflation associated with dramatic oil price increases.

To calculate the simple index number for a particular year, we divide that year's price by the price during the base year and multiply the result by 100. Thus, for the 1980 silver price index number, we calculate

$$\text{1980 index number} = \left(\frac{\text{1980 silver price}}{\text{1975 silver price}}\right)100 = \left(\frac{20.64}{4.42}\right)100$$
$$= 466.97$$

Similarly, the index number for 1990 is

$$\text{1990 index number} = \left(\frac{\text{1990 silver price}}{\text{1975 silver price}}\right)100 = \left(\frac{4.82}{4.42}\right)100$$
$$= 109.05$$

The index number for the base period is always 100. In our example, we have

$$\text{1975 index number} = \left(\frac{\text{1975 silver price}}{\text{1975 silver price}}\right)100 = 100$$

Thus, the silver price has risen by 366.97% (the difference between the 1980 and 1975 index numbers) between 1975 and 1980, and by only 9.05% between 1975 and 1990. The simple index numbers for silver between 1975 and 1990 are given in Table 14.2 and are portrayed graphically in Figure 14.1. The steps for calculating simple index numbers are summarized in the next box.

TABLE 14.2 Simple Index Numbers for Silver Prices (Base 1975)

Year	Index
1975	100.00
1976	98.42
1977	104.52
1978	122.17
1979	250.90
1980	466.97
1981	238.01
1982	179.86
1983	258.82
1984	184.16
1985	138.91
1986	123.76
1987	158.60
1988	147.74
1989	124.43
1990	109.05

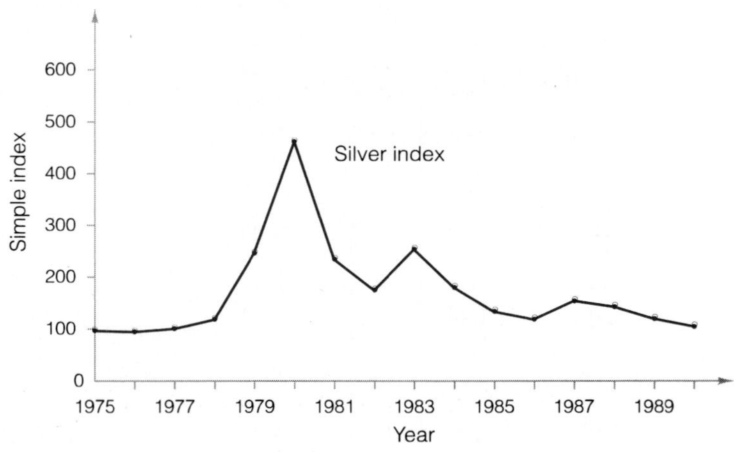

FIGURE 14.1 ▲
Graph of silver price index, 1975–1990

Steps for Calculating a Simple Index Number

1. Obtain the prices or quantities for the commodity over the time period of interest.
2. Select a base period.
3. Calculate the index number for each period according to the formula

$$\text{Index number at time } t = \left(\frac{\text{Time series value at time } t}{\text{Time series value at base period}}\right)100$$

$$\text{Symbolically,} \quad I_t = \left(\frac{Y_t}{Y_0}\right)100$$

where I_t is the index number at time t, Y_t is the time series value at time t, and Y_0 is the time series value at the base period.

EXAMPLE 14.1

Crude oil prices (in dollars per barrel) between 1975 and 1990 are shown in Table 14.3. Construct a simple index for foreign crude oil prices using 1975 as the base period and portray the index on the same graph as the silver price index (Table 14.2).

TABLE 14.3 Crude Oil Prices, 1975–1990

Year	Price ($/bbl.)	Year	Price ($/bbl.)
1975	7.67	1983	26.19
1976	8.19	1984	25.88
1977	8.57	1985	24.09
1978	9.00	1986	12.51
1979	12.64	1987	15.40
1980	21.59	1988	12.58
1981	31.77	1989	15.86
1982	28.52	1990	20.03

Source: U.S. Bureau of the Census. *Statistical Abstract of the United States*, 1992.

Solution

Represent the foreign crude oil price at time t by Y_t, the price during the base period (1975) by Y_0, and the simple index number at time t by I_t. Then

$$I_t = \left(\frac{Y_t}{Y_0}\right)100$$

TABLE 14.4 Simple Index Numbers for Crude Oil Prices (Base 1975)

Year	Index
1975	100.00
1976	106.78
1977	111.73
1978	117.34
1979	164.80
1980	281.49
1981	414.21
1982	371.84
1983	341.46
1984	337.42
1985	314.08
1986	163.10
1987	200.78
1988	164.02
1989	206.78
1990	261.15

For example, the index number for 1980 is

$$I_{1980} = \left(\frac{Y_{1980}}{Y_0}\right)100 = \left(\frac{21.59}{7.67}\right)100 = 281.49$$

The interpretation is that crude oil prices increased by 181.49% between 1975 and 1980. All the index numbers for foreign crude oil prices are similarly calculated and are given in Table 14.4. The graphs of the silver and crude oil price indexes are combined in Figure 14.2. Since the same base period was used for both simple indexes, the two graphs intersect at the 1975 base period, where both indexes have a value of 100.

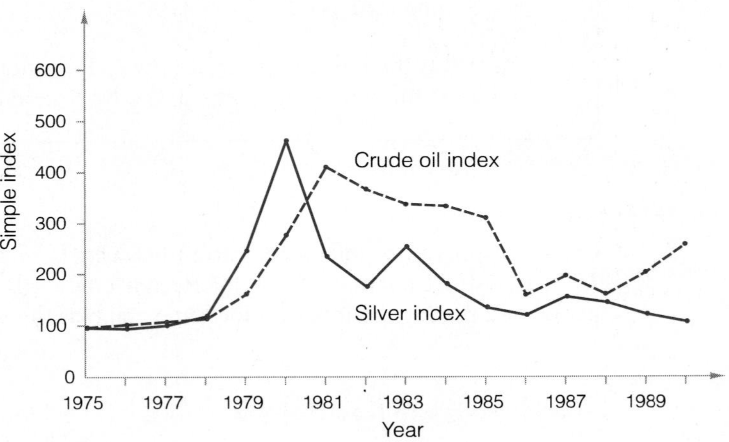

FIGURE 14.2 ▲
Simple indexes for silver and crude oil prices (base 1975)

Portraying two indexes on the same graph (as we did in Example 14.1) indicates another reason indexes are calculated: Two or more commodities' relative price (or quantity) changes can be compared, even though the units of measurement are different (dollars per ounce and dollars per barrel, in our examples). In other words, apples and oranges *can* be compared, so long as you use index numbers to represent them. Figure 14.2 reveals that, during the period 1975–1980, the price of crude oil escalated more rapidly than that of silver, relative to the 1975 price. Also, the very high silver prices in 1980 are obvious on the graph. Note that silver prices began to fall in 1981, but the crude oil prices did not begin their decline until 1983. Interestingly, in the late 1980s silver continued to fall, until silver prices were almost back to their 1975 level by 1990, while crude oil prices remained at a level more than twice that of their 1975 value. Of course, these index numbers provide only a *descriptive* comparison of the two time series. Any inferential implications of such a comparison will require a blend of inferential statistical analysis and economic theory.

Exercises 14.1 – 14.8

Learning the Mechanics

14.1 Explain in words how to construct a simple index.

14.2 The table lists the U.S. median annual family income during the period 1975–1990. It also contains several values for each of two simple indexes for median family income.

Year	*Income*	*Base 1975 Index*	*Base 1980 Index*
1975	$13,719		65.26
1976	14,958	109.03	
1977	16,009	116.69	
1978	17,640		
1979	19,587		93.17
1980	21,023	153.24	
1981	22,388		
1982	23,433		111.46
1983	24,580		
1984	26,433	192.67	
1985	27,735	202.16	
1986	29,458		140.12
1987	30,970		147.31
1988	32,191		153.12
1989	34,213	249.38	
1990	35,353	257.69	

Source: U.S. Bureau of the Census. *Statistical Abstract of the United States, 1992.*

a. Calculate the missing values of each simple index.
b. Interpret the 1990 value for each index.

14.3 The table describes U.S. beer production (in millions of barrels) for the period 1970–1989. Use 1977 as the base period to compute the simple index for this time series.

Year	*Beer*	*Year*	*Beer*	*Year*	*Beer*
1970	133.1	1977	170.5	1984	192.2
1971	137.4	1978	179.1	1985	194.3
1972	141.3	1979	184.2	1986	194.4
1973	148.6	1980	194.1	1987	195.9
1974	156.2	1981	193.7	1988	197.4
1975	160.6	1982	196.2	1989	197.8
1976	163.7	1983	195.4		

Source: U.S. Bureau of the Census. *Statistical Abstract of the United States, 1992.*

14.4 Refer to Exercise 14.3. Is this an example of a quantity index or a price index? Explain.

14.5 Refer to Exercise 14.3. Recompute the simple index using 1980 as the base period. Plot the two indexes on the same graph.

Applying the Concepts

14.6 The stock of Abbott Laboratories has had the yearly closing prices shown in the table.

Year	*Closing Price*	*Year*	*Closing Price*
1973	$49.875	1980	$56.500
1974	50.750	1981	27.000
1975	41.250	1982	38.750
1976	49.125	1983	45.250
1977	56.500	1984	41.750
1978	33.750	1985	68.375
1979	41.125	1986	45.625

Source: *Daily Stock Price Record*, NYSE, 1973–1986; *Security Owner's Stock Guide*, Standard & Poor's, Jan. 1987.

a. Using 1973 as the base period, calculate the simple index for this stock's yearly closing price between 1973 and 1986.
b. By what percentage did the stock price increase or decrease between 1973 and 1986? Between 1978 and 1986?

14.7 The table lists the price of natural gas (in cents per million BTUs) between 1982 and 1990.

Year	*Price*	*Year*	*Price*
1982	222.2	1987	150.2
1983	232.3	1988	152.4
1984	239.9	1989	152.7
1985	225.7	1990	155.4
1986	174.8		

Source: U.S. Bureau of the Census. *Statistical Abstract of the United States*, 1992.

a. Using 1982 as the base period, calculate and plot the simple index for the price of natural gas from 1982 through 1990.
b. Use the simple index to interpret the trend in the price of natural gas in the late 1980s.
c. Is the index you constructed in part **a** a price or quantity index? Explain.

14.8 Civilian employment is broadly classified by the federal government into two categories—agricultural and nonagricultural. Employment figures (in thousands of workers) for farm and nonfarm categories for the years 1975–1991 are given in the table.

Year	Farm	Nonfarm	Year	Farm	Nonfarm	Year	Farm	Nonfarm
1975	3,408	82,438	1981	3,368	97,030	1987	3,208	109,232
1976	3,331	85,421	1982	3,401	96,125	1988	3,169	111,800
1977	3,283	88,734	1983	3,383	97,450	1989	3,199	114,142
1978	3,387	92,661	1984	3,321	101,685	1990	3,186	114,728
1979	3,347	95,477	1985	3,179	103,971	1991	3,233	113,644
1980	3,364	95,938	1986	3,163	106,434			

Source: U.S. Bureau of the Census. *Statistical Abstract of the United States, 1992.*

a. Compute simple indexes for each of the two time series using 1975 as the base period.
b. Which segment has shown the greater percentage change in employment over the period shown?
c. Are these indexes price or quantity indexes? Explain.

14.3 Composite Index Numbers

A **composite index number** represents combinations of the prices or quantities of several commodities. For example, suppose you want to construct an index for the total number of sales of the three major automobile manufacturers in the United States. The first step is to collect data on the sales of each manufacturer during the period in which you are interested. The total sales of automobiles by each manufacturer between 1972 and 1985 are shown in Table 14.5. To summarize the information from all three time series in a single index, we add the sales of each manufacturer for each year. That is, we form a new time series consisting of the total number of automobiles sold by the three manufacturers.

TABLE 14.5 Sales of Automobiles by Three Manufacturers (Thousands)

Year	General Motors	Ford	Chrysler	Year	General Motors	Ford	Chrysler
1972	7,790.52	5,593.04	2,192.00	1979	8,993.00	5,810.30	1,796.00
1973	8,683.80	5,871.00	2,423.00	1980	7,101.00	4,328.45	1,225.00
1974	6,690.00	5,258.93	2,015.00	1981	6,762.00	4,313.18	1,283.00
1975	6,629.00	4,577.77	1,773.00	1982	6,244.00	4,254.90	1,182.00
1976	8,568.00	5,304.44	2,371.00	1983	7,769.00	4,934.23	1,493.96
1977	9,068.00	6,422.30	2,328.00	1984	8,256.35	5,584.65	2,034.35
1978	9,482.00	6,462.06	2,212.00	1985	9,305.00	5,550.50	2,157.37

Source: *Moody's Industrial Manual,* 1986.

We now construct a simple index for the *total* of the three series. Selecting 1977 as the base year, we divide each total by the 1977 total sales. The resulting **simple composite index** is shown in Table 14.6, page 784.

TABLE 14.6 Simple Composite Index for Total Automobiles Sold by Three Manufacturers

Year	Index	Year	Index
1972	87.41	1979	93.16
1973	95.28	1980	71.02
1974	78.37	1981	69.36
1975	72.85	1982	65.56
1976	91.16	1983	79.68
1977	100.00	1984	89.10
1978	101.90	1985	95.48

Definition 14.3

A **simple composite index** is a simple index for a time series consisting of the total price or total quantity of two or more commodities.

EXAMPLE 14.2

One of the primary uses of index numbers is to characterize changes in stock prices over time. Stock market indexes have been constructed for many different types of companies and industries, and several composite indexes have been developed to characterize all stocks. These indexes are reported on a daily basis in the news media (e.g., Standard and Poor's 500 Stocks Index and Dow Jones 65 Stocks Index).

Consider the monthly prices given in Table 14.7 for four high-technology company stocks listed on the New York Stock Exchange between 1984 and 1986. To see how this type of stock fared as the market began to rally in the mid-1980s, construct a simple composite index using January 1984 as the base period. Graph the index, and comment on its implications.

Solution

First, we calculate the total of the four stock prices each month. These totals are shown in Table 14.7. Then the simple composite index is calculated by dividing each monthly total by the January 1984 total. The index values are given in Table 14.8 (page 786), and a graph of the simple composite index is shown in Figure 14.3 on page 786.

The plot of the simple composite index for these high-technology stocks shows that their performance was rather flat through 1984 and most of 1985. The index begins a dramatic increase in November 1985, peaking at 23.12% over the January 1984 value

TABLE 14.7 Monthly Prices of Four High-Technology Company Stocks

Month	Bell Industries	Xerox	Harris	IBM	Total
Jan. 1984	$29.875	$44.000	$38.500	$114.125	$226.500
Feb.	24.750	41.125	29.375	110.250	205.500
Mar.	26.500	41.250	31.000	114.000	212.750
Apr.	25.625	40.500	32.000	113.750	211.875
May	24.000	37.250	28.000	107.750	197.000
June	23.000	38.375	25.125	105.750	192.250
July	21.375	33.750	25.125	110.750	191.000
Aug.	27.125	38.375	30.375	123.750	219.625
Sept.	24.375	37.625	26.750	124.250	213.000
Oct.	23.500	35.500	25.750	124.625	209.375
Nov.	22.625	37.250	28.875	121.750	210.500
Dec.	22.125	37.875	27.125	123.125	210.250
Jan. 1985	22.875	43.375	32.250	136.375	234.875
Feb.	24.625	45.375	30.500	134.000	234.500
Mar.	22.500	43.375	28.250	127.000	221.125
Apr.	21.750	45.500	25.500	126.500	219.250
May	22.250	50.000	27.250	128.625	228.125
June	20.875	52.625	28.375	123.750	225.625
July	23.500	53.875	28.625	131.375	237.375
Aug.	23.375	51.750	26.125	126.625	227.875
Sept.	22.250	50.375	23.500	123.875	220.000
Oct.	24.500	50.375	24.500	129.875	229.250
Nov.	26.875	60.125	26.125	139.750	252.875
Dec.	26.500	59.750	27.250	155.500	269.000
Jan. 1986	24.500	64.250	26.625	151.500	266.875
Feb.	26.625	70.625	30.750	150.875	278.875
Mar.	26.875	68.000	28.250	149.125	272.250
Apr.	25.625	60.000	32.250	156.250	274.125
May	26.125	61.250	33.000	152.375	272.750
June	22.000	56.125	33.250	146.500	257.875
July	19.375	53.250	28.875	132.500	234.000
Aug.	18.750	57.000	29.750	138.750	244.250
Sept.	17.625	51.500	28.625	134.500	232.250
Oct.	18.000	54.750	29.750	123.625	226.125
Nov.	20.250	60.500	31.500	127.125	239.375
Dec.	20.500	60.000	29.750	120.000	230.250

Source: *Daily Stock Price Record*, NYSE 1983–1986; *Standard & Poor's Security Owner's Stock Guide*, Aug. 1986–Jan. 1987.

in February 1986. However, while the rest of the market continued to enjoy new highs throughout 1986, these high-technology stocks receded to about the same level as January 1984 by the end of 1986.

TABLE 14.8 Simple Composite Index of Stock Prices

1984	Index	1985	Index	1986	Index
Jan.	100.00	Jan.	103.70	Jan.	117.83
Feb.	90.73	Feb.	103.53	Feb.	123.12
Mar.	93.93	Mar.	97.63	Mar.	120.20
Apr.	93.54	Apr.	96.80	Apr.	121.03
May	86.98	May	100.72	May	120.42
June	84.88	June	99.61	June	113.85
July	84.33	July	104.80	July	103.31
Aug.	96.96	Aug.	100.61	Aug.	107.84
Sept.	94.04	Sept.	97.13	Sept.	102.54
Oct.	92.44	Oct.	101.21	Oct.	99.83
Nov.	92.94	Nov.	111.64	Nov.	105.68
Dec.	92.83	Dec.	118.76	Dec.	101.66

FIGURE 14.3 ►
Graph of simple composite index of four stocks

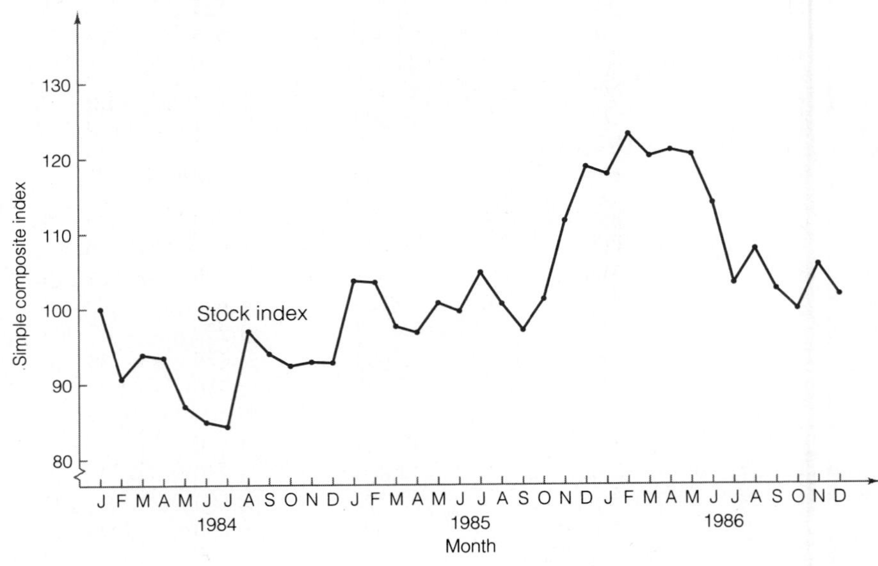

A simple composite price index has a major drawback: The quantity of the commodity that is purchased during each period is not taken into account. Only the price totals are used to calculate the index. We can remedy this situation by constructing a **weighted composite price index**.

Definition 14.4

A **weighted composite price index** weights the prices by quantities purchased prior to calculating totals for each time period. The weighted totals are then used to compute the index in the same way that the unweighted totals are used for simple composite indexes.

Since the quantities purchased change from time period to time period, the choice of which time period's quantities to use as the basis for the weighted composite index is an important one. A **Laspeyres index** uses the base period quantities as weights. The rationale is that the prices at each time period should be compared as if the same quantities were purchased each period as were purchased during the base period. This method measures price inflation (or deflation) by fixing the purchase quantities at their base period values. The method for calculating a Laspeyres index is given in the box.

Steps for Calculating a Laspeyres Index

1. Collect price information for each of the k price series to be used in the composite index. Denote these series by $P_{1t}, P_{2t}, \ldots, P_{kt}$.
2. Select a base period. Call this time period t_0.
3. Collect purchase quantity information for the base period. Denote the k quantities by $Q_{1t_0}, Q_{2t_0}, \ldots, Q_{kt_0}$.
4. Calculate the weighted totals for each time period according to the formula

$$\sum_{i=1}^{k} Q_{it_0} P_{it}$$

5. Calculate the Laspeyres index, I_t, at time t by taking the ratio of the weighted total at time t to the base period weighted total and multiplying by 100. That is,

$$I_t = \frac{\sum_{i=1}^{k} Q_{it_0} P_{it}}{\sum_{i=1}^{k} Q_{it_0} P_{it_0}} \times 100$$

EXAMPLE 14.3

The January prices for the four high-technology company stocks are given in Table 14.9. Suppose that, in January 1984, an investor purchased the quantities shown in the table. [*Note:* Only January prices and quantities are used to simplify the example. The same methods can be applied to calculate the index for other months.] Calculate the Laspeyres index for the investor's portfolio of high-technology stocks using January 1984 as the base period.

TABLE 14.9 January Prices of High-Technology Stocks with Quantities Purchased

	Bell Industries	Xerox	Harris	IBM
Shares purchased	500	100	100	1,000
January 1984 price	\$29.875	\$44.000	\$38.500	\$114.125
January 1985 price	\$22.875	\$43.375	\$32.250	\$136.375
January 1986 price	\$24.500	\$64.250	\$26.625	\$151.500

Solution

First, we calculate the weighted price totals for each time period, using the January 1984 quantities as weights. Thus,

$$\begin{aligned}\text{January 1984 weighted total} &= \sum_{i=1}^{4} Q_{i,1984}P_{i,1984}\\ &= 500(29.875) + 100(44.000) + 100(38.500) + 1{,}000(114.125)\\ &= 137{,}312.5\end{aligned}$$

$$\begin{aligned}\text{January 1985 weighted total} &= \sum_{i=1}^{4} Q_{i,1984}P_{i,1985}\\ &= 500(22.875) + 100(43.375) + 100(32.250) + 1{,}000(136.375)\\ &= 155{,}375.0\end{aligned}$$

$$\begin{aligned}\text{January 1986 weighted total} &= \sum_{i=1}^{4} Q_{i,1984}P_{i,1986}\\ &= 500(24.500) + 100(64.250) + 100(26.625) + 1{,}000(151.500)\\ &= 172{,}837.5\end{aligned}$$

Then the Laspeyres index is calculated by multiplying the ratio of each weighted total to the base period weighted total by 100. Thus,

$$I_{1984} = \frac{\sum_{i=1}^{4} Q_{i,1984}P_{i,1984}}{\sum_{i=1}^{4} Q_{i,1984}P_{i,1984}} \times 100 = 100$$

$$I_{1985} = \frac{\sum_{i=1}^{4} Q_{i,1984}P_{i,1985}}{\sum_{i=1}^{4} Q_{i,1984}P_{i,1984}} \times 100 = \frac{155{,}375.0}{137{,}312.5} = 113.2$$

$$I_{1986} = \frac{\sum_{i=1}^{4} Q_{i,1984}P_{i,1986}}{\sum_{i=1}^{4} Q_{i,1984}P_{i,1984}} \times 100 = \frac{172{,}837.5}{137{,}312.5} = 125.9$$

The implication is that these stocks were worth 13.2% more to the investor in January 1985 than in January 1984 and 25.9% more in January 1986.

The Laspeyres index is appropriate when the base period quantities are reasonable weights to apply to all time periods. This is the case in applications such as that described in Example 14.3, where the base period quantities represent actual quantities of stock purchased and held for some period of time. Laspeyres indexes are also appropriate when the base period quantities remain reasonable approximations of purchase quantities in subsequent periods. However, it can be misleading when the relative purchase quantities change significantly from those in the base period.

Probably the best-known Laspeyres index is the all-items Consumer Price Index (CPI). This monthly composite index is made up of hundreds of item prices, and the U.S. Bureau of Labor Statistics (BLS) sampled over 30,000 families' purchases in 1982–1984 to determine the base period quantities. Thus, beginning in 1987, the all-items CPI published each month reflects quantities purchased in 1982–1984 by a sample of families across the United States. However, as prices increase for some commodities more quickly than for others, consumers tend to substitute less expensive commodities where possible. For example, as automobile and gasoline prices rapidly inflated in the mid-1970s, consumers began to purchase smaller cars. The net effect of using the base period quantities for the CPI is to overestimate the effect of inflation on consumers, because the quantities are fixed at levels that will actually change in response to price changes.

There are several solutions to the problem of purchase quantities that change relative to those of the base period. One is to change the base period regularly, so that the quantities are regularly updated. A second solution is to compute the index at each time period by using the purchase quantities of that period, rather than those of the base period. A **Paasche index** is calculated by using price totals weighted by the purchase quantities of the period the index value represents. The steps for calculating a Paasche index are given in the box.

Steps for Calculating a Paasche Index

1. Collect price information for each of the k price series to be used in the composite index. Denote these series by $P_{1t}, P_{2t}, \ldots, P_{kt}$.
2. Select a base period. Call this time period t_0.
3. Collect purchase quantity information for every period. Denote the k quantities for period t by $Q_{1t}, Q_{2t}, \ldots, Q_{kt}$.
4. Calculate the Paasche index for time t by multiplying the ratio of the weighted total at time t to the weighted total at time t_0 (base period) by 100, where the weights used are the purchase quantities for time period t. Thus,

$$I_t = \frac{\sum_{i=1}^{k} Q_{it} P_{it}}{\sum_{i=1}^{k} Q_{it} P_{it_0}} \times 100$$

EXAMPLE 14.4

The January prices and volumes (actual quantities purchased) in thousands of shares for the four high-technology company stocks are shown for 1984, 1985, and 1986 in Table 14.10. Calculate and interpret the Paasche index, using January 1984 as the base period.

TABLE 14.10 January Prices and Volumes of High-Technology Company Stocks

	Bell Industries		Xerox		Harris		IBM	
	Price	Volume	Price	Volume	Price	Volume	Price	Volume
January 1984	\$29.875	229.7	\$44.000	4,843.2	\$38.500	1,377.5	\$114.125	27,215.2
January 1985	22.875	487.4	43.375	11,869.0	32.250	2,834.7	136.375	36,521.8
January 1986	24.500	167.3	64.250	7,772.9	26.625	1,651.6	151.500	31,936.9

Source: *Daily Stock Price Record*, NYSE, Jan. 1984–1986.

Solution

The key to calculating a Paasche index is to remember that the weights (purchase quantities) change for each time period. Thus,

$$I_{1984} = \frac{\sum_{i=1}^{4} Q_{i,1984} P_{i,1984}}{\sum_{i=1}^{4} Q_{i,1984} P_{i,1984}} \times 100 = 100$$

$$I_{1985} = \frac{\sum_{i=1}^{4} Q_{i,1985} P_{i,1985}}{\sum_{i=1}^{4} Q_{i,1985} P_{i,1984}} \times 100 = \frac{5{,}598{,}047}{4{,}813{,}983} \times 100 = 116.3$$

$$I_{1986} = \frac{\sum_{i=1}^{4} Q_{i,1986} P_{i,1986}}{\sum_{i=1}^{4} Q_{i,1986} P_{i,1984}} \times 100 = \frac{5{,}385{,}922}{4{,}055{,}391} \times 100 = 132.8$$

The implication is that 1985 prices represent a 16.3% increase over 1984 prices, assuming the purchase quantities were at January 1985 levels for *both* periods. Similarly, the 1986 index value of 132.8 implies a 32.8% increase when purchase quantities are at the January 1986 level.

The Paasche index is most appropriate when you want to compare current prices to base period prices at *current* purchase levels. However, there are several major problems associated with the Paasche index. First, it requires that purchase quantities be known for every time period. This rules out a Paasche index for applications such as the CPI because the time and monetary resource expenditures required to collect quantity information are considerable. (Recall that more than 30,000 families were sampled to estimate purchase quantities in 1982–1984.) A second problem is that, although each period is compared to the base period, it is difficult to compare the index at two other periods because the quantities used are different for each period. For example, for the four high-technology stocks in Example 14.4, we calculated index values of 116.3 in 1985 and 132.8 in 1986. Although this apparently represents an increase of 16.5% from 1985 to 1986, these two index values are determined using different quantities, and therefore, the change in the index is affected by changes in both prices *and* quantities. This fact makes it difficult to interpret the change in a Paasche index between periods when neither is the base period.

Although there are other types of indexes that use different weighting factors, the Laspeyres and Paasche indexes are the most popular composite indexes. Depending on the primary objective in constructing an index, one of them will probably be suitable for most purposes.

CASE STUDY 14.1 / The Consumer Price Index: CPI-U and CPI-W

The Consumer Price Index (CPI), first published by the U.S. Bureau of Labor Statistics (BLS) in 1919, is the country's principal measure of price changes. One major use of the CPI is as an indicator of inflation, through which the success or failure of government economic policies can be monitored. A second major use of the CPI is to escalate income payments. Millions of workers have escalator clauses in their collective bargaining contracts that call for increases in wage rates based on increases in the CPI. In addition, the incomes of Social Security beneficiaries and retired military and federal civil service employees are tied to the CPI. It has been estimated that a 1% increase in the CPI can trigger an increase of over $1 billion in income payments.

Since 1978, the BLS has published two national, all-items indexes: the new CPI-U and the traditional CPI-W. The CPI-U measures the price change of a constant market basket of goods and services that are representative of the purchases of all urban residents—approximately 80% of the U.S. population. The CPI-W measures the price change of a constant market basket of goods and services that are representative of the purchases of urban wage earners and clerical workers—approximately 50% of all urban residents. Until 1988 the base period for both indexes was 1967. Then the base period was shifted to 1982–1984 to coincide with the updating of the base-period quantities (see page 789). The CPI-U is the index typically reported by the press and broadcast media. The CPI-W is the index used in the escalator clauses of most labor contracts and government benefit programs. In addition to these two national indexes, the BLS publishes CPI-U and CPI-W indexes for each of 27 metropolitan areas. The national indexes and the metropolitan indexes are reported monthly in the BLS's *CPI Detailed Report*.

The market basket of goods priced by both the CPI-U and the CPI-W includes a homeownership component. Accounting for over 30% of the overall weight of the indexes, the homeownership component influences the indexes more than food, energy, or medical care. This component includes the costs associated with purchasing a home (the price of the home and mortgage interest), as well as the cost of property taxes, property insurance, and maintenance and repairs. During the 1970s and early 1980s, the use of these quantities to measure the cost of homeownership met with much criticism. Two arguments were made by critics (U.S. Department of Labor, 1980, p. 2):

1. Since the CPI is used to measure the change in purchasing power for the purpose of escalating income or determining the rate of inflation, it "should not include the impact of rising prices on the value of assets such as houses. Just as the CPI excludes changes in the value of stocks and bonds, . . . the change in the asset value of the house (appreciation or depreciation) and the cost of equity in holding that asset should be distinguished from the change in the cost of the shelter provided by the house. It is the cost of consuming the shelter provided by the house—not the investment aspects of homeownership—which should be reflected in an index used to keep real income constant."
2. The CPI overstates the rate of inflation because "it uses *current* house prices and *current* mortgage interest rates. . . . The CPI should not measure the costs of purchasing the base period houses in today's prices and today's mortgage interest rates, but rather the CPI should measure what people are actually paying for housing."

In response to these criticisms, the BLS developed and experimented with an entirely new approach to measuring the cost of housing. As a result, instead of explicitly including in the market basket the homeownership costs described above, the BLS now recommends that a *rental equivalency* component be included. This approach assumes that a household's cost of consuming the flow of services from the housing unit

FIGURE 14.4 ► Changes in the consumer price index for all urban consumers: Official (CPI-U) and experimental rental equivalence (CPI-U-X1) measures

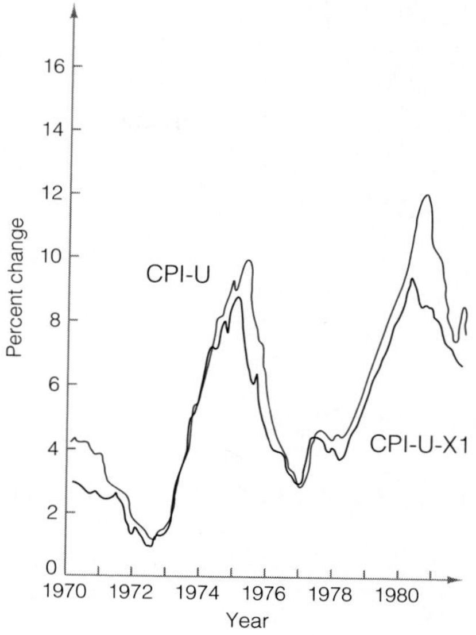

can be represented by the income that the household could receive by renting the home to someone else. This rental equivalency approach to measuring homeownership costs was implemented in an experimental version of the CPI-U called the CPI-U-X1.

Figure 14.4 shows the movement that the CPI-U and CPI-U-X1 would have displayed over the period 1970–1981. [*Note*: Percentage changes are calculated using 12 months of unadjusted data. See Gillingham and Lane (1982, p. 13).] Notice that the two indexes generally move together, but that the CPI-U-X1 tends to stay below the CPI-U, particularly during the periods of high mortgage interest rates in 1970, 1974–1975, and 1978–1981. If, as some critics charge, the CPI-U overstates inflation during periods of high mortgage interest rates, it appears that the experimental CPI-U-X1 would provide a better measure of inflation. In January 1983, the BLS changed the official CPI-U to include the rental equivalency approach to measuring homeownership costs and it made a similar change in the CPI-W in 1985.

Exercises 14.9–14.17

Learning the Mechanics

14.9 Explain in words how to calculate the following types of indexes:

a. Simple composite index **b.** Weighted composite index
c. Laspeyres index **d.** Paasche index

14.10 Using 1983 as the base, compute and plot the simple composite index for the volume of futures trading (in millions of contracts) for the commodities listed in the table on page 794.

Year	*Grain*	*Energy Products*	*Currencies*
1983	17.8	3.2	11.4
1984	15.9	4.9	16.7
1985	10.7	7.0	16.4
1986	10.3	11.5	19.1
1987	10.9	20.3	19.9
1988	15.9	26.3	21.2
1989	15.9	31.3	25.7
1990	17.0	35.2	27.2
1991	16.6	31.8	28.8

Source: U.S. Bureau of the Census. *Statistical Abstract of the United States, 1992.*

14.11 Explain in words the difference between Laspeyres and Paasche indexes.

Applying the Concepts

14.12 The gross domestic product (GDP) is the total national output of goods and services valued at market prices. As such, the GDP is a commonly used barometer of the U.S. economy. One component of the GDP is personal consumption expenditures, which is itself the sum of expenditures for durable goods, nondurable goods, and services. The GDP for these components (in billions of dollars) is shown in the table, in five-year increments from 1960 to 1980 and annually from 1980 to 1991.

Year	*Durables*	*Nondurables*	*Services*	*Year*	*Durables*	*Nondurables*	*Services*
1960	$ 43.5	$153.1	$ 135.9	1984	$317.9	$ 873.0	$1269.4
1965	63.5	191.9	189.2	1985	352.9	919.4	1395.1
1970	85.3	270.4	290.8	1986	389.6	952.2	1508.8
1975	134.3	416.0	474.5	1987	403.7	1011.1	1637.4
1980	212.5	682.9	852.7	1988	437.1	1073.8	1785.2
1981	228.5	744.2	953.5	1989	459.8	1146.9	1911.2
1982	236.5	772.3	1050.4	1990	465.9	1217.7	2059.0
1983	275.0	817.8	1164.7	1991	445.2	1251.9	2191.9

Source: U.S. Bureau of the Census. *Statistical Abstract of the United States, 1992.*

a. Using these three component values, construct a simple composite index for the personal consumption component of GDP. Use 1970 as the base year.

b. Suppose we want to update the index by using 1980 as the base year. Update the index using only the index values you calculated in part **a**, without referring to the original data.

14.13 Refer to Exercise 14.12, in which a personal consumption expenditure index was constructed. Graph the personal consumption expenditure index for the years 1960–1991 (remember that the data are in five-year increments through 1980 and annual thereafter), first using 1970 as the base year and then using 1980 as the base year. What effect does changing the base year have on the graph of this index?

14.14 Refer to Exercise 14.12. Suppose the output quantities in 1970, measured in billions of units purchased, are as follows:

Durable goods: 10.9

Nondurable goods: 14.02

Services: 42.6

Use the outputs to calculate the Laspeyres index from 1960 to 1991 (same increments as in Exercise 14.12) with 1970 as the base period.

14.15 Refer to Exercises 14.12 and 14.14. Plot the simple composite index and Laspeyres index on the same graph. Comment on the differences between the two indexes.

14.16 The level of price and production of metals in the United States is one measure of the strength of the industrial economy. The table lists the 1984 prices (in dollars per ton) and production (in tons) for three metals important to U.S. industry.

Month	*Copper*		*Pig Iron*		*Lead*	
	Price	*Production*	*Price*	*Production*	*Price*	*Production*
Jan.	$1,361.6	100.7	$213	4,311	$530.0	46.1
Feb.	1,399.0	95.1	213	4,497	520.0	47.0
Mar.	1,483.6	104.0	213	5,083	529.0	51.0
Apr.	1,531.6	95.6	213	5,077	540.0	23.0
May	1,431.2	103.3	213	5,166	531.0	26.5
June	1,383.8	106.9	213	4,565	580.0	13.5
July	1,326.8	95.9	213	4,329	642.8	27.4
Aug.	1,328.8	96.7	213	4,057	602.6	25.8
Sept.	1,307.8	95.7	213	3,473	513.6	20.5
Oct.	1,278.4	89.1	213	3,739	480.8	24.6
Nov.	1,354.2	100.5	213	3,817	528.4	21.5
Dec.	1,305.2	96.9	213	3,694	462.2	27.9

Source: *Standard & Poor's Basic Statistics: Metals*, 1985.

a. Compute simple composite price and quantity indexes for the 12-month period, using January as the base period.
b. Compute the Laspeyres price index for the 12-month period using January as the base period.
c. Plot the simple composite and Laspeyres indexes on the same graph. Comment on the differences.

14.17 Refer to Exercise 14.16.

a. Compute the Paasche price index for metals for the 12-month period using January as the base period.
b. Plot the Laspeyres and Paasche indexes on the same graph. Comment on the differences.
c. Compare the Laspeyres and Paasche index values for September and December. Which index is more appropriate for describing the change in this 4-month period? Explain.

14.4 Exponential Smoothing

As you have seen in the previous sections, index numbers are useful for describing trends and changes in time series. However, time series often have such irregular fluctuations that trends are difficult to describe. Index numbers can be misleading in such cases because the series is changing so rapidly. Methods for removing the rapid fluctuations in a time series so the general trend can be seen are called **smoothing** techniques.

Exponential smoothing is one type of weighted average that assigns positive weights to past and current values of the time series. A single weight, w, called the **exponential smoothing constant**, is selected so that w is between 0 and 1. Then the exponentially smoothed series, E_t, is calculated as follows:

$$E_1 = Y_1$$
$$E_2 = wY_2 + (1 - w)E_1$$
$$E_3 = wY_3 + (1 - w)E_2$$
$$\vdots$$
$$E_t = wY_t + (1 - w)E_{t-1}$$

Thus, the exponentially smoothed value at time t assigns the weight w to the current series value and the weight $(1 - w)$ to the previous smoothed value.

For example, consider the Dow Jones Industrial Average (DJIA) time series in Table 14.11. Suppose we want to calculate the exponentially smoothed series using a smoothing constant of $w = .3$. The calculations proceed as follows:

$$E_{1961} = Y_{1961} = 731.14$$
$$E_{1962} = .3Y_{1962} + (1 - .3)E_{1961} = .3(652.10) + .7(731.14) = 707.43$$
$$E_{1963} = .3Y_{1963} + (1 - .3)E_{1962} = .3(762.95) + .7(707.43) = 724.09$$
$$\vdots$$

All the exponentially smoothed values corresponding to $w = .3$ are given in Table 14.11. Note that no values are lost at either end of the smoothed series.

The DJIA and exponentially smoothed DJIA are graphed in Figure 14.5. Like many averages, the exponentially smoothed series changes less rapidly than the time series itself. The choice of w affects the smoothness of E_t. The smaller (closer to 0) is the value of w, the smoother is E_t. Since small values of w give more weight to the past values of the time series, the smothed series is not affected by rapid changes in the current values and, therefore, appears smoother than the original series. Conversely, choosing w near 1 yields an exponentially smoothed series that is much like the original series. That is, large values of w give more weight to the current value of the time series so the smoothed series looks like the original series. This concept is illustrated in Figure 14.6, page 798. The steps for calculating an exponentially smoothed series are given in the box on page 798.

TABLE 14.11 Dow Jones Industrial Average (1961–1991) with Exponential Smoothing ($w = .3$)

Year	DJIA	Exponentially Smoothed DJIA	Year	DJIA	Exponentially Smoothed DJIA
1961	731.14	731.14	1977	835.15	866.90
1962	652.10	707.43	1978	805.01	848.33
1963	762.95	724.09	1979	838.74	845.45
1964	874.13	769.10	1980	963.99	881.01
1965	969.26	829.15	1981	899.01	886.41
1966	785.69	816.11	1982	1,046.54	934.45
1967	905.11	842.81	1983	1,258.64	1,031.71
1968	943.75	873.09	1984	1,211.57	1,085.67
1969	800.36	851.27	1985	1,546.67	1,223.97
1970	838.92	847.57	1986	1,895.95	1,425.56
1971	884.76	858.73	1987	2,276.02	1,680.70
1972	950.71	886.32	1988	2,060.83	1,794.74
1973	923.88	897.59	1989	2,508.04	2,008.73
1974	759.37	856.12	1990	2,678.91	2,209.78
1975	802.49	840.03	1991	2,929.30	2,425.64
1976	974.92	880.50			

Source: *The Dow Jones Investor's Handbook*, 1991; *Wall Street Journal*, Jan. 2, 1987.

FIGURE 14.5 ▶ Exponentially smoothed values ($w = .3$) for the DJIA

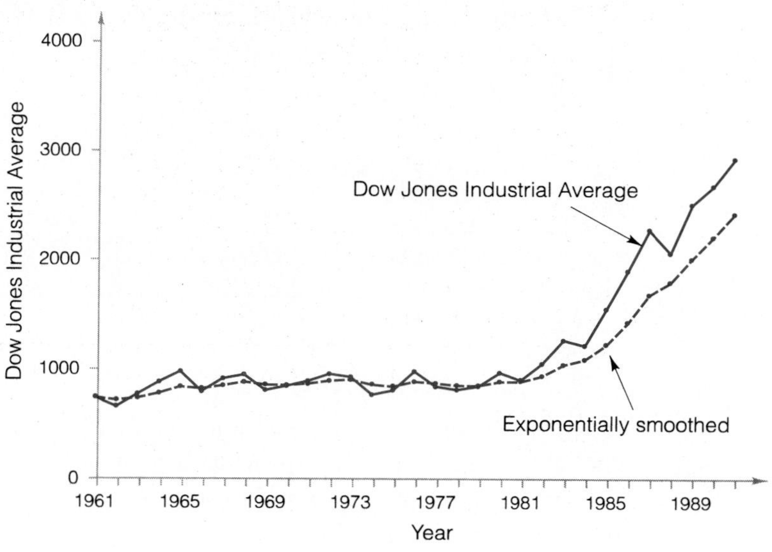

FIGURE 14.6 ▶
Exponentially smoothed values ($w = .3$ and $w = .7$) for the DJIA

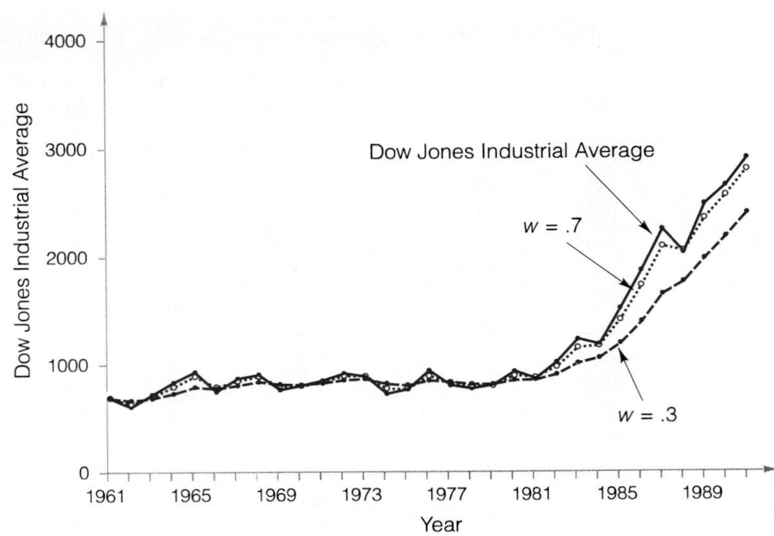

Steps for Calculating an Exponentially Smoothed Series

1. Select an exponential smoothing constant, w, between 0 and 1. Remember that small values of w give less weight to the current value of the series and yield a smoother series. Larger choices of w assign more weight to the current value of the series and yield a more variable series.
2. Calculate the exponentially smoothed series E_t from the original time series Y_t as follows:

$$
\begin{aligned}
E_1 &= Y_1 \\
E_2 &= wY_2 + (1 - w)E_1 \\
E_3 &= wY_3 + (1 - w)E_2 \\
&\vdots \\
E_t &= wY_t + (1 - w)E_{t-1}
\end{aligned}
$$

EXAMPLE 14.5

Consider the IBM common stock price from January 1990 to December 1992, shown in Table 14.12. Create the exponentially smoothed series using $w = .5$, and plot both series.

TABLE 14.12 IBM Stock Prices and Exponentially Smoothed Series ($w = .5$)

1990	IBM	Smoothed	1991	IBM	Smoothed	1992	IBM	Smoothed
Jan.	98.125	98.125	Jan.	107.500	108.594	Jan.	95.375	94.581
Feb.	103.750	100.938	Feb.	137.375	122.984	Feb.	89.750	92.165
Mar.	107.000	103.969	Mar.	126.625	124.805	Mar.	89.625	90.895
Apr.	107.125	105.547	Apr.	106.875	115.840	Apr.	88.500	89.698
May	115.625	110.586	May	102.750	109.295	May	91.625	90.661
June	120.375	115.480	June	100.375	104.835	June	94.125	92.393
July	121.000	118.240	July	99.250	102.042	July	99.250	95.822
Aug.	105.000	111.620	Aug.	98.375	100.209	Aug.	88.375	92.098
Sept.	104.375	107.998	Sept.	102.875	101.542	Sept.	85.500	88.799
Oct.	99.250	103.624	Oct.	104.250	102.896	Oct.	72.875	80.837
Nov.	112.625	108.124	Nov.	96.250	99.573	Nov.	64.875	72.856
Dec.	111.250	109.687	Dec.	88.000	93.786	Dec.	56.125	64.491

Source: U.S. Bureau of the Census. *Statistical Abstract of the United States, 1992.*

Solution

To create the exponentially smoothed series with $w = .5$, we calculate

$$E_1 = Y_1 = 98.125$$
$$E_2 = wY_2 + (1 - w)E_1 = .5(103.750) + .5(98.125) = 100.938$$
$$\vdots$$
$$E_{36} = wY_{36} + (1 - w)E_{35} = .5(56.125) + .5(72.856) = 64.491$$

The plot of the original and exponentially smoothed series is shown in Figure 14.7.

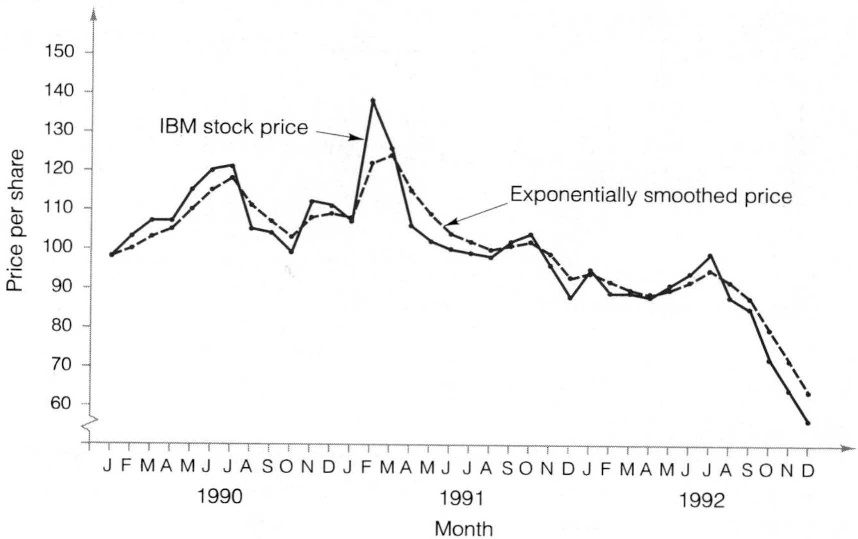

FIGURE 14.7 ▶ IBM stock price and exponentially smoothed price ($w = .5$)

The smoothed series provides a good picture of the general trend of the original series. Note, too, that the exponentially smoothed series will be less sensitive to any short-term deviations of the prices from the trend. The downward trend in price in 1992 is particularly well portrayed; this price decline resulted in billions of dollars in losses to the many institutional investors (pension plans, mutual funds, etc.) that hold significant amounts of IBM stock.

One of the primary uses of exponential smoothing is to forecast future values of a time series. Because only current and past values of the time series are used in exponential smoothing, it is easily adapted to forecasting. We demonstrate this application of exponentially smoothed series in Chapter 16.

Exercises 14.18–14.23

Learning the Mechanics

14.18 Describe the effect of selecting an exponential constant of $w = .2$. Of $w = .8$. Which will produce a smoother trend?

14.19 The table lists the number (in millions) of Chevrolet passenger cars sold by General Motors to automotive dealers in the United States and Canada from 1977 to 1985.

Year	*Sales*	*Exponentially Smoothed Sales* ($w = .5$)
1977	2.133	
1978	2.349	2.241
1979	2.233	2.237
1980	1.740	
1981	1.444	
1982	.986	
1983	1.289	
1984	1.455	
1985	4.882	

Source: *Moody's Industrial Manual*, 1986, p. 1401.

a. Calculate the missing values in the exponentially smoothed series using $w = .5$.
b. Graph the time series and the exponentially smoothed series on the same graph.

14.20 Refer to Exercise 14.3.
a. Calculate the exponentially smoothed series for U.S. beer production for the period 1970–1989 using $w = .2$.
b. Calculate the exponentially smoothed series using $w = .8$.
c. Plot the two exponentially smoothed series ($w = .2$ and $w = .8$) on the same graph.

Applying the Concepts

14.21 The price of gold is used by some financial analysts as a barometer of investors' expectations of inflation, with the price of gold tending to increase as concerns about inflation increase. The following table shows the average annual price of gold (in dollars per ounce) from 1975 through 1990.

Year	*Price*	*Year*	*Price*
1975	$161	1983	$424
1976	125	1984	361
1977	148	1985	318
1978	194	1986	368
1979	308	1987	448
1980	613	1988	438
1981	460	1989	383
1982	376	1990	387

Source: U.S. Bureau of the Census. *Statistical Abstract of the United States, 1992.*

a. Compute an exponentially smoothed series for the gold price time series for the period from 1975 to 1990 using a smoothing coefficient of $w = .8$.

b. Plot the original series and the exponentially smoothed series on the same graph.

14.42 Refer to Exercise 14.8. Using $w = .4$, compute an exponentially smoothed series for each of the two time series: agricultural and nonagricultural employment. Plot each smoothed series.

14.23 There has been phenomenal growth in the transportation sector of the economy since 1960. The personal consumption expenditure figures (in billions of dollars) are given in the table.

Year	*Expenditure on Transportation*	*Year*	*Expenditure on Transportation*	*Year*	*Expenditure on Transportation*
1960	$42.4	1969	$ 75.7	1977	$179.3
1961	44.8	1970	80.6	1978	198.1
1962	47.4	1971	92.3	1979	219.4
1963	49.5	1972	105.4	1980	236.6
1964	54.3	1973	114.6	1981	261.5
1965	58.4	1974	117.9	1982	267.3
1966	60.4	1975	129.4	1983	291.9
1967	63.3	1976	155.2	1984	319.5
1968	69.3				

Source: U.S. Bureau of the Census. *Statistical Abstract of the United States: 1986.*

a. Compute exponentially smoothed values of this personal consumption time series using the smoothing constants $w = .2$ and $w = .8$.

b. Plot the actual series and the two smoothed series on the same graph. Comment on the trend in personal consumption expenditure on transportation in the 1970s and early 1980s as compared to the 1960s.

Summary

Time series are observations made sequentially over time. **Index numbers** measure the changes in a time series or group of time series. **Simple index numbers** are based on a single series, while **composite index numbers** measure changes in several series simultaneously. Price indexes that are weighted by purchase quantities are **weighted composite indexes**. **Laspeyres indexes** use weights that are base-period purchase quantities, while **Paasche indexes** use the current period purchase quantities as weights.

Smoothing techniques are used to make it easier to discern trends in time series. **Exponential smoothing** combines past and current values of the series.

Supplementary Exercises 14.24–14.34

14.24 The U.S. steel industry was the object of much economic attention in the 1970s and 1980s due to the increasing market share of imported steel, the effects of several recessions, and other economic woes. Prices (in cents per pound) of three varieties of U.S. steel are given in the table for the period 1976–1989.

Year	*Cold Rolled*	*Hot Rolled*	*Galvanized*	*Year*	*Cold Rolled*	*Hot Rolled*	*Galvanized*
1976	14.51	12.20	16.07	1983	26.36	22.23	28.43
1977	16.44	13.79	18.10	1984	28.15	23.75	30.30
1978	18.43	15.53	20.47	1985	28.15	23.75	30.30
1979	20.25	17.05	22.32	1986	25.65	21.15	30.30
1980	21.91	18.46	23.88	1987	27.38	21.64	30.49
1981	23.90	20.15	26.88	1988	28.15	21.50	31.05
1982	24.65	20.80	26.75	1989	28.15	21.50	31.05

Source: *Standard & Poor's Statistics: Metals,* 1991.

a. Compute the exponentially smoothed series corresponding to each of the price series using the smoothing constant $w = .5$.

b. Plot the prices and their exponentially smoothed series on the same graph.

14.25 Refer to Exercise 14.24.

a. Calculate a simple composite index for the three steel price series using 1980 as the base period.

b. Is the index a price index or a quantity index?

c. What information would you need in order to calculate a Laspeyres index with a base period of 1980? A Paasche index with a base period of 1980?

14.26 Foreign exchange rates, the values of foreign currency in U.S. dollars, are important to investors and international travelers. The table lists the monthly foreign exchange rates of the British pound (in U.S. dollars per pound) for 1984 and 1985.

Month	*1984*	*1985*	*Month*	*1984*	*1985*
Jan.	1.41	1.13	July	1.32	1.38
Feb.	1.44	1.10	Aug.	1.31	1.39
Mar.	1.45	1.13	Sept.	1.26	1.36
Apr.	1.43	1.23	Oct.	1.23	1.42
May	1.39	1.25	Nov.	1.24	1.44
June	1.37	1.28	Dec.	1.18	1.44

Source: Standard & Poor's Statistical Service. *Current Statistics*, 1986.

a. Calculate a simple index for the foreign exchange rate series using January 1984 as the base period.
b. Plot the index, and use the plot to identify the best time for a U.S. traveler to visit Britain during this period.

14.27 Refer to Exercise 14.23. Using 1975 as the base period, compute a simple index for the personal consumption series.

14.28 A major portion of total consumer credit is extended in the categories of automobile loans, mobile home loans, and revolving credit. Amounts outstanding (in billions of dollars) for the period 1980–1991 are given in the table.

Year	*Automobile*	*Mobile Home*	*Revolving*
1980	112.0	18.7	55.1
1981	119.0	20.1	61.1
1982	125.9	22.6	66.5
1983	143.6	23.6	79.1
1984	173.6	25.9	100.3
1985	210.2	26.8	121.8
1986	247.4	27.1	135.9
1987	265.9	25.9	153.1
1988	284.2	25.3	174.1
1989	290.7	22.5	199.1
1990	284.6	21.0	220.1
1991	267.9	19.1	234.5

Source: U.S. Bureau of the Census. *Statistical Abstract of the United States*, 1992.

a. Calculate a simple composite index using 1980 as the base period.
b. Compute a simple composite index for the series using 1985 as the base period.
c. Are the indexes constructed in parts **a** and **b** price or quantity indexes?
d. Compute a simple index for revolving credit loans using 1980 as the base. Plot the simple index and the composite index from part **a** on the same graph.
e. In 1986 Congress passed tax legislation that made the interest from most loans nondeductible. Analysts expected a shift in credit to home equity loans, a kind of revolving credit, the interest on which would remain deductible in most cases. Interpret the graph from part **d** in terms of this expectation.

14.29 Refer to Exercise 14.28.

a. Using a smoothing constant of $w = .3$, calculate an exponentially smoothed series corresponding to the simple composite index.

b. Plot the simple composite index and the exponentially smoothed series on the same graph. Comment on the relative smoothness of the three series.

14.30 Refer to Exercise 14.28. Assume that in 1990 the number of outstanding loans of each type is as follows:

Automobile: 400,000 *Mobile home*: 100,000 *Revolving credit*: 1,000,000

a. Calculate a Laspeyres index for 1980–1991 using 1990 as the base and the quantities given above.

b. Which category of credit is given most weight in the calculation of the Laspeyres index?

14.31 Refer to Exercise 14.28. Suppose the numbers of outstanding loans in each category from 1985 to 1991 are as shown in the table.

Year	*Automobile*	*Mobile Home*	*Revolving*
1985	340,000	75,000	500,000
1986	350,000	80,000	600,000
1987	325,000	75,000	850,000
1988	350,000	90,000	900,000
1989	380,000	95,000	990,000
1990	400,000	100,000	1,000,000
1991	425,000	110,000	1,200,000

a. Calculate the Paasche index for 1985–1991 using 1990 as a base and the quantities given in the table.

b. Compare the simple composite index (from Exercise 14.28), the Laspeyres index (from Exercise 14.30), and the Paasche index. Explain why the 1985 values are different for each index, and interpret each.

14.32 Three of many indicators used for measuring the level of economic activity are the index of net business formation, the index of industrial production, and the index of new private housing units authorized by local building permits. End-of-year values of these indicators for the period 1970–1984 are given in the table.

Year	*Index of Net Business Formation*	*Index of Industrial Production*	*Index of New Private Housing Units*	*Year*	*Index of Net Business Formation*	*Index of Industrial Production*	*Index of New Private Housing Units*
1970	106.4	107.8	111.0	1978	128.2	144.1	156.3
1971	108.5	109.6	158.8	1979	128.3	152.5	135.1
1972	115.9	119.7	182.4	1980	122.4	147.0	100.0
1973	114.9	129.8	158.4	1981	118.6	151.0	83.9
1974	109.2	129.3	103.6	1982	113.2	138.6	82.2
1975	107.0	117.8	89.8	1983	114.8	147.6	131.8
1976	115.6	130.5	119.0	1984	117.1	163.3	135.4
1977	123.2	138.2	153.8				

Source: U.S. Bureau of the Census, *Statistical Abstract of the United States: 1986*, pp. 521, 725; *Annual U.S. Economic Data*, Federal Reserve Bank of St. Louis, 1985, p. 15.

a. Calculate a simple composite index for the three indicator series, using 1970 as the base period.
b. Compute the simple index for each of the three indicator series, using a base period of 1970 for each. Plot the three simple indexes and the simple composite index on the same graph.

14.33 Refer to Exercise 14.32.
a. Calculate an exponentially smoothed series corresponding to the business formation index using $w = .2$. Using $w = .8$.
b. Plot on a graph both the business formation index and the two exponentially smoothed series from part a. Which exponential smoothing constant yields a smoother series? Explain.

14.34 The number of dollars a person receives in a year is referred to as his or her **monetary** (or **money**) **income**. This figure can be adjusted to reflect the purchasing power of the dollars received relative to the purchasing power of dollars in some base period. The result is called a person's **real income**. Monetary income and real income can be compared to determine, for example, whether an increase in a person's monetary income truly reflects an increase in his or her purchasing power. The Consumer Price Index (CPI) can be used to adjust monetary income to obtain real income (in terms of 1967 dollars). To compute your real income for a specific year, simply divide your monetary income for the year by that year's CPI and multiply by 100. The table lists the CPI for each year during the period 1970–1991.

Year	*CPI*	*Year*	*CPI*	*Year*	*CPI*
1970	116.3	1978	195.4	1985	322.2
1971	121.3	1979	217.4	1986	328.1
1972	125.3	1980	246.8	1987	340.1
1973	133.1	1981	272.4	1988	354.2
1974	147.7	1982	289.1	1989	371.3
1975	161.2	1983	298.4	1990	391.3
1976	170.5	1984	311.1	1991	407.8
1977	181.5				

Source: U.S. Bureau of the Census. *Statistical Abstract of the United States, 1992.*

a. Suppose your monetary income increased from $20,000 in 1970 to $60,000 in 1991. What were your real incomes in 1970 and 1991? Were you able to buy more goods and services in 1970 or 1991? Explain.
b. What monetary income would have been required in 1991 to provide equivalent purchasing power to a 1970 monetary income of $20,000?

On Your Own

Select a time series of interest to you. It should have at least 24 consecutive values (years, quarters, months, etc.).

a. Calculate a simple index for the series using the first value as the base. Plot the index. How much does the series change from the first value to the last?

b. Calculate and plot the exponentially smoothed series with $w = .3$. Repeat the process using $w = .7$.

c. What do the smoothed values of the index indicate about the long-term trend of the series?

References

Box, G. E. P., and Jenkins, G. M. *Time Series Analysis: Forecasting and Control*, 2nd ed. San Francisco: Holden-Day, 1977.

Gillingham, R., and Lane, W. "Changing the treatment of shelter costs for homeowners in the CPI." *Monthly Labor Review*, June 1982, pp. 9–14.

Gross, C. W. and Patterson, R. J. *Business Forecasting*, 2nd ed. Boston: Houghton Mifflin, 1983.

Nelson, C. R. *Applied Time Series Analysis for Managerial Forecasting*. San Francisco: Holden-Day, 1983.

U.S. Department of Labor. *BLS Handbook of Methods. Vol. II. The Consumer Price Index*. Bureau of Labor Statistics, Bulletin 2134-2, Apr. 1984.

U.S. Department of Labor. *The Consumer Price Index: Concepts and Content over the Years*. Bureau of Labor Statistics, Report 517, May 1978.

U.S. Department of Labor. *CPI Issues*. Bureau of Labor Statistics, Report S93, Feb. 1980.

Willis, R. E. A *Guide to Forecasting for Planners*. Englewood Cliffs, N.J.: Prentice-Hall, 1987.

CHAPTER FIFTEEN

Time Series: Models and Forecasting

Contents

Case Study

Where We've Been

In Chapter 14 we discussed methods for describing time series. Index numbers were used to describe changes in a time series; exponential smoothing was introduced to describe trends.

Where We're Going

In Chapter 15 we use mathematical models (such as the regression models of Chapters 10–12) to describe time series. These models range in complexity from the relatively simple exponential smoothing model to time series models that account for correlation between values observed at different points in time. The primary objective of constructing these models is to use them for forecasting future values of the time series.

In Chapter 14 we showed how to use various *descriptive* techniques to obtain a picture of the behavior of a time series. Now we want to expand our coverage to include techniques that will let us make statistical inferences about the time series. These *inferential* techniques are generally focused on the problem of *forecasting* future values of the time series, and we will discuss several methods for predicting the future with something other than a crystal ball. Unlike fortune tellers, we will show how to provide measures of reliability for our forecasts. Nevertheless, because we are trying to predict a value outside the region of the sample data, forecasting remains a precarious type of statistical inference.

We find that separating the time series into basic components often assists in the modeling and forecasting process. We discuss these time series components in Section 15.1. We then present several popular forecasting techniques in Sections 15.2–15.5. Finally, in Section 15.6 we show how to determine whether the regression residuals are correlated.

15.1 Time Series Components

Before forecasts of future values of a time series can be made, some type of model that can be projected into the future must be used to describe the series. Time series models range in complexity from **descriptive models**, such as the exponential smoothing models discussed in Chapter 14, to **inferential models**, such as the combinations of regression and specialized time series models to be discussed in this chapter. Whether the model is simple or complex, the objective is the same: to produce accurate forecasts of future values of the time series.

Many different algebraic representations of time series models have been proposed. One of the most widely used is an **additive model*** of the form

$$Y_t = T_t + C_t + S_t + R_t$$

The **secular trend**, T_t, also known as the **long-term trend**, is a time series component that describes the long-term movements of Y_t. For example, if you want to characterize the secular trend of the production of automobiles since 1930, you would show T_t as an upward-moving time series over the period from 1930 to the present. This does not imply that the automobile production series has always moved upward from month to month and from year to year, but it does mean the long-term trend has been an increase over that period of time.

The **cyclical effect**, C_t, generally describes fluctuations of the time series about the secular trend that are attributable to business and economic conditions. For example, the closing Dow Jones Industrial Average (DJIA) on the last business day of the year for the years 1961–1991 is given in Table 15.1. You can see in Figure 15.1 that it has a

*Another useful form is the *multiplicative model*: $Y_t = T_tC_tS_tR_t$. This can be changed to an additive form by taking natural logarithms—i.e., $\ln Y_t = \ln T_t + \ln C_t + \ln S_t + \ln R_t$.

TABLE 15.1 Dow Jones Industrial Average for 1961–1991 with Exponential Smoothing ($w = .3$)

Year	DJIA	Exponentially Smoothed DJIA	Year	DJIA	Exponentially Smoothed DJIA	Year	DJIA	Exponentially Smoothed DJIA
1961	731.14	731.14	1972	950.71	886.32	1982	1,046.54	934.45
1962	652.10	707.43	1973	923.88	897.59	1983	1,258.64	1,031.71
1963	762.95	724.09	1974	759.37	856.12	1984	1,211.57	1,085.67
1964	874.13	769.10	1975	802.49	840.03	1985	1,546.67	1,223.97
1965	969.26	829.15	1976	974.92	880.50	1986	1,895.95	1,425.56
1966	785.69	816.11	1977	835.15	866.90	1987	2,276.02	1,680.70
1967	905.11	842.81	1978	805.01	848.33	1988	2,060.83	1,794.74
1968	943.75	873.09	1979	838.74	845.45	1989	2,508.04	2,008.73
1969	800.36	851.27	1980	963.99	881.01	1990	2,678.91	2,209.78
1970	838.92	847.57	1981	899.01	886.41	1991	2,929.30	2,425.64
1971	884.76	858.73						

Source: *The Dow Jones Investor's Handbook*, 1991; *Wall Street Journal*, Jan. 2, 1987.

FIGURE 15.1 ► Secular trend for the exponentially smoothed DJIA ($w = .3$)

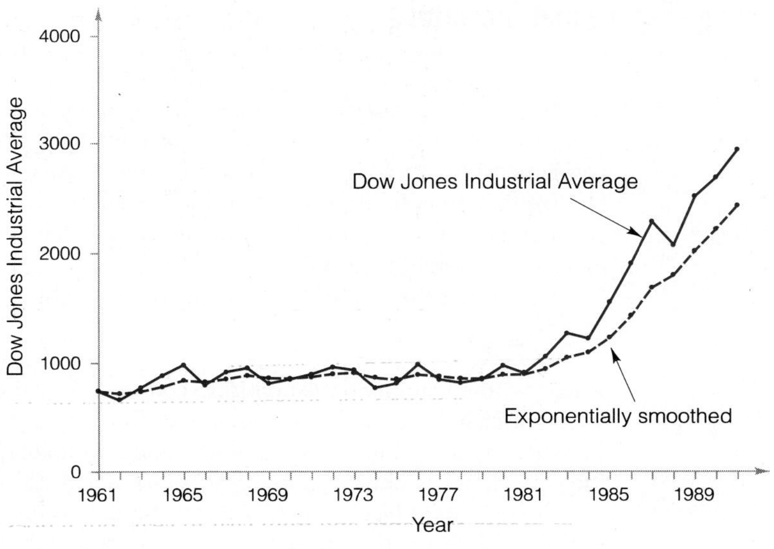

generally increasing secular trend. However, during periods of recession, the DJIA tends to lie below the secular trend, while in times of general economic expansion, it lies above the long-term trend line.

The **seasonal effect**, S_t, describes the fluctuations in the time series that recur during specific time periods. For example, quarterly power loads for a utility company tend to be highest in the summer months (quarter III), with another smaller peak in the winter months (quarter I). The spring and fall (quarters II and IV) seasonal effects are negative, meaning that the series tends to lie below the long-term trend line during those quarters.

The **residual effect**, R_t, is what remains of Y_t after the secular, cyclical, and seasonal components have been removed. Part of the residual effect may be attributable to unpredictable rare events (earthquake, presidential assassination, etc.) and part to the randomness of human actions. In any case, the presence of the residual component makes it impossible to forecast the future values of a time series without error. Thus, the presence of the residual effect emphasizes a point we first made in Chapter 10 in connection with regression models: No business phenomena should be described by deterministic models. All realistic business models, time series or otherwise, should include a residual component.

Each of the four components contributes to the determination of the value of Y_t at each time period. Although it will not always be possible to characterize each component separately, the component model provides a useful theoretical formulation that helps the time series analyst achieve a better understanding of the phenomena affecting the path followed by the time series.

15.2 Forecasting: Exponential Smoothing

In Chapter 14, we discussed exponential smoothing as a method for describing a time series that involved removing the irregular fluctuations. In terms of the time series components discussed in the previous section, exponential smoothing tends to deemphasize (or "smooth") most of the residual effects. This, coupled with the fact that exponential smoothing uses only past and current values of the series, makes it a useful tool for forecasting time series.

Recall that the formula for exponential smoothing is

$$E_t = wY_t + (1 - w)E_{t-1}$$

where w, the **exponential smoothing constant**, is a number between 0 and 1. The selection of w controls the smoothness of E_t. A choice near 0 places more emphasis (weight) on *past* values of the time series and, therefore, yields a smoother series. On the other hand, a choice near 1 gives more weight to *current* values of the series.

Suppose the objective is to forecast the next value of the time series, Y_{t+1}. The **exponentially smoothed forecast** for Y_{t+1} is simply the smoothed value at time t:

$$F_{t+1} = E_t$$

where F_{t+1} is the **forecast** of Y_{t+1}. To help interpret this forecast formula, substitute the smoothing formula for E_t:

$$\begin{aligned} F_{t+1} &= E_t = wY_t + (1 - w)E_{t-1} \\ &= wY_t + (1 - w)F_t \\ &= F_t + w(Y_t - F_t) \end{aligned}$$

Note that we have substituted F_t for E_{t-1}, since the forecast for time t is the smoothed value for time $(t - 1)$. The final equation provides insight into the exponential smoothing forecast: The forecast for time $(t + 1)$ is equal to the forecast for time t, F_t, plus a correction for the error in the forecast for time t, $(Y_t - F_t)$. This is why the exponentially smoothed forecast is called an **adaptive forecast**—the forecast for time $(t + 1)$ is explicitly adapted for the error in the forecast for time t.

Because exponential smoothing consists of averaging past and present values, the smoothed values will tend to lag behind the series when a long-term trend exists. In addition, the averaging tends to smooth any seasonal component. Therefore, exponentially smoothed forecasts are appropriate only when the trend and seasonal components are relatively insignificant. Since the exponential smoothing model assumes that the time series has little or no trend or seasonal component, the forecast F_{t+1} is used to forecast not only Y_{t+1} but also *all* future values of Y_t. That is, the forecast for two time periods ahead is

$$F_{t+2} = F_{t+1}$$

and for three time periods ahead is

$$F_{t+3} = F_{t+2} = F_{t+1}$$

The exponential smoothing forecasting technique is summarized in the box.

Calculation of Exponentially Smoothed Forecasts

1. Given the observed time series $Y_1, Y_2, \ldots, Y_t$, first calculate the exponentially smoothed values $E_1, E_2, \ldots, E_t$ using

$$\begin{aligned} E_1 &= Y_1 \\ E_2 &= wY_2 + (1 - w)E_1 \\ &\vdots \\ E_t &= wY_t + (1 - w)E_{t-1} \end{aligned}$$

2. Use the last smoothed value to forecast the next time series value:

$$F_{t+1} = E_t$$

3. Assuming that Y_t is relatively free of trend and seasonal components, use the same forecast for all future values of Y_t:

$$\begin{aligned} F_{t+2} &= F_{t+1} \\ F_{t+3} &= F_{t+1} \\ &\vdots \end{aligned}$$

Two important points must be made about exponentially smoothed forecasts:

1. The choice of w is crucial. If you decide that w will be small (near 0), you will obtain a smooth, slowly changing series of forecasts. On the other hand, the

selection of a large value of w (near 1) will yield more rapidly changing forecasts that depend mostly on the current values of the series. In general, several values of w should be tried to determine how sensitive the forecast series is to the choice of w. Forecasting experience will provide the best basis for the choice of w for a particular application.

2. The further into the future you forecast, the less certain you can be of accuracy. Since the exponentially smoothed forecast is constant for all future values, any changes in trend or seasonality are not taken into account. However, the uncertainty associated with future forecasts applies not only to exponentially smoothed forecasts, but also to all methods of forecasting. In general, time series forecasting should be confined to the short term.

EXAMPLE 15.1

The annual Dow Jones Industrial Averages from 1961 to 1988 are given in Table 15.2, along with the exponentially smoothed values using $w = .3$ and $w = .7$. Use the exponential smoothing technique to forecast the values from 1989 to 1991 using both $w = .3$ and $w = .7$.

TABLE 15.2 Dow Jones Industrial Average (1961–1988) with Exponentially Smoothed Values

Year	DJIA	Exponentially Smoothed		Year	DJIA	Exponentially Smoothed	
		($w = .3$)	($w = .7$)			($w = .3$)	($w = .7$)
1961	731.14	731.14	731.14	1975	802.49	840.03	804.44
1962	652.10	707.43	675.81	1976	974.92	880.50	923.78
1963	762.95	724.09	736.81	1977	835.15	866.90	861.74
1964	874.13	769.10	832.93	1978	805.01	848.33	822.03
1965	969.26	829.15	928.36	1979	838.74	845.45	833.73
1966	785.69	816.11	828.49	1980	963.99	881.01	924.91
1967	905.11	842.81	882.12	1981	899.01	886.41	906.78
1968	943.75	873.09	925.26	1982	1,046.54	934.45	1,004.61
1969	800.36	851.27	837.83	1983	1,258.64	1,031.71	1,182.43
1970	838.92	847.57	838.59	1984	1,211.57	1,085.67	1,202.83
1971	884.76	858.73	870.91	1985	1,546.67	1,223.97	1,443.52
1972	950.71	886.32	926.77	1986	1,895.95	1,425.56	1,760.22
1973	923.88	897.59	924.75	1987	2,276.02	1,680.70	2,020.88
1974	759.37	856.12	808.98	1988	2,060.83	1,794.74	2,048.85

Source: *The Dow Jones Investor's Handbook*, 1991.

Solution

First, we calculate the exponentially smoothed forecasts using $w = .3$. Following the steps outlined in the box, we find

$$F_{1989} = E_{1988} = 1{,}794.74$$
$$F_{1990} = F_{1989} = 1{,}794.74$$
$$F_{1991} = F_{1990} = F_{1989} = 1{,}794.74$$

The same steps are repeated using $w = .7$, and both sets of forecasts are shown in Table 15.3. Also shown are the actual Dow Jones Industrial Averages from 1989 to 1991. The **forecast error**, defined as the actual value minus the forecast value, is given for each exponentially smoothed forecast.

Notice that the one-step-ahead forecasts for 1989 have considerably smaller forecast errors than the two- and three-step-ahead forecasts for 1990 and 1991.

TABLE 15.3 Dow Jones Industrial Average (1989–1991): Actual vs. Forecast Values

Year	Actual	Forecast ($w = .3$)	Forecast Error	Forecast ($w = .7$)	Forecast Error
1989	2,508.04	1,794.74	713.30	2,048.85	459.19
1990	2,678.91	1,794.74	884.17	2,048.85	630.16
1991	2,929.30	1,794.74	1,134.56	2,048.85	880.45

Source: *The Dow Jones Investor's Handbook*, 1991; *Wall Street Journal*, Jan. 2, 1987.

Neither the $w = .3$ nor the $w = .7$ forecast projects the 1989–1991 upturn in the DJIA, because exponentially smoothed forecasts implicitly assume no trend exists in the time series. This example dramatically illustrates the risk associated with anything other than very short-term forecasting.

Many time series have long-term, or secular, trends. For such series the exponentially smoothed forecast is inappropriate for all but the very short term. In the next section we present an extension of the exponentially smoothed forecast—the *Holt-Winters forecast*—that allows for secular trend in the forecasts.

15.3 Forecasting Trends: The Holt-Winters Forecasting Model

The exponentially smoothed forecasts for the Dow Jones Industrial Average in Section 15.2 have large forecast errors, in part because they do not recognize the trend in the time series. In this section we present an extension of the exponential smoothing method of forecasting that explicitly recognizes the trend in a time series. The **Holt-Winters forecasting model** consists of both an exponentially smoothed component (E_t) and a trend component (T_t). The trend component is used in the calculation of the exponentially smoothed value. The following equations show that both E_t and T_t are weighted averages:

$$E_t = wY_t + (1 - w)(E_{t-1} + T_{t-1})$$
$$T_t = v(E_t - E_{t-1}) + (1 - v)T_{t-1}$$

Note that the equations require *two* smoothing constants, w and v, each of which is between 0 and 1. As before, w controls the smoothness of E_t; a choice near 0 places more emphasis on past values of the time series, while a value of w near 1 gives more weight to current values of the series, and deemphasizes the past.

The trend component of the series is estimated **adaptively**, using a weighted average of the most recent change in the level, represented by $(E_t - E_{t-1})$, and the trend estimate, represented by T_{t-1}, from the previous period. A choice of the weight v near 0 places more emphasis on the past estimates of trend, while a choice of v near 1 gives more weight to the current change in level.

The calculation of the Holt-Winters components, which proceeds much like the exponential smoothing calculations, is summarized in the box.

Steps for Calculating Components of the Holt-Winters Model

1. Select an exponential smoothing constant w between 0 and 1. Small values of w give less weight to the current values of the time series, and more weight to the past. Larger choices assign more weight to the current value of the series.
2. Select a trend smoothing constant v between 0 and 1. Small values of v give less weight to the current changes in the level of the series, and more weight to the past trend. Larger values assign more weight to the most recent trend of the series and less to past trends.
3. Calculate the two components, E_t and T_t, from the time series Y_t beginning at time $t = 2$ as follows:*

$$
\begin{aligned}
E_2 &= Y_2 \\
T_2 &= Y_2 - Y_1 \\
E_3 &= wY_3 + (1 - w)(E_2 + T_2) \\
T_3 &= v(E_3 - E_2) + (1 - v)T_2 \\
&\vdots \\
E_t &= wY_t + (1 - w)(E_{t-1} + T_{t-1}) \\
T_t &= v(E_t - E_{t-1}) + (1 - v)T_{t-1}
\end{aligned}
$$

[*Note:* E_1 and T_1 are not defined.]

EXAMPLE 15.2

The yearly sales data for a firm's first 35 years of operation are given in Table 15.4. Calculate the Holt-Winters exponential smoothing and trend components for this time series using $w = .7$ and $v = .5$. Show the data and the exponential smoothing component E_t on the same graph.

*The calculation begins at time $t = 2$ rather than at $t = 1$ because the first two observations are needed to obtain the first estimate of trend, T_2.

TABLE 15.4 A Firm's Yearly Sales Revenue (Thousands of Dollars)

t	Y_t	t	Y_t	t	Y_t
1	4.8	13	48.4	25	100.3
2	4.0	14	61.6	26	111.7
3	5.5	15	65.6	27	108.2
4	15.6	16	71.4	28	115.5
5	23.1	17	83.4	29	119.2
6	23.3	18	93.6	30	125.2
7	31.4	19	94.2	31	136.3
8	46.0	20	85.4	32	146.8
9	46.1	21	86.2	33	146.1
10	41.9	22	89.9	34	151.4
11	45.5	23	89.2	35	150.9
12	53.5	24	99.1		

Solution

Following the formulas for the Holt-Winters components given in the box, we calculate

$$E_2 = Y_2 = 4.0$$
$$T_2 = Y_2 - Y_1 = 4.0 - 4.8 = -.8$$
$$E_3 = .7Y_3 + (1 - .7)(E_2 + T_2) = .7(5.5) + .3(4.0 - .8) = 4.8$$
$$T_3 = .5(E_3 - E_2) + (1 - .5)T_2 = .5(4.8 - 4.0) + .5(-.8) = 0$$
$$\vdots$$

All the E_t and T_t values are given in Table 15.5, and a graph of Y_t and E_t is shown in Figure 15.2 on page 816. Note that the trend component T_t measures the general

TABLE 15.5 Holt-Winters Components for Sales Data

Month	Sales Y_t	E_t ($w = .7$)	T_t ($v = .5$)	Month	Sales Y_t	E_t ($w = .7$)	T_t ($v = .5$)	Month	Sales Y_t	E_t ($w = .7$)	T_t ($v = .5$)
1	4.8	–	–	13	48.4	50.5	1.1	25	100.3	100.1	3.8
2	4.0	4.0	−.8	14	61.6	58.6	4.6	26	111.7	109.4	6.5
3	5.5	4.8	.0	15	65.6	64.9	5.4	27	108.2	110.5	3.8
4	15.6	12.4	3.8	16	71.4	71.1	5.8	28	115.5	115.1	4.2
5	23.1	21.0	6.2	17	83.4	81.5	8.1	29	119.2	119.3	4.2
6	23.3	24.5	4.8	18	93.6	92.4	9.5	30	125.2	124.7	4.8
7	31.4	30.8	5.6	19	94.2	96.5	6.8	31	136.3	134.2	7.2
8	46.0	43.1	8.9	20	85.4	90.8	.5	32	146.8	145.2	9.1
9	46.1	47.9	6.9	21	86.2	87.7	−1.2	33	146.1	148.5	6.2
10	41.9	45.8	2.4	22	89.9	88.9	−.1	34	151.4	152.4	5.0
11	45.5	46.3	1.4	23	89.2	89.1	.1	35	150.9	152.9	2.7
12	53.5	51.8	3.5	24	99.1	96.1	3.6				

FIGURE 15.2 ▶ Sales data and Holt-Winters exponentially smoothed series

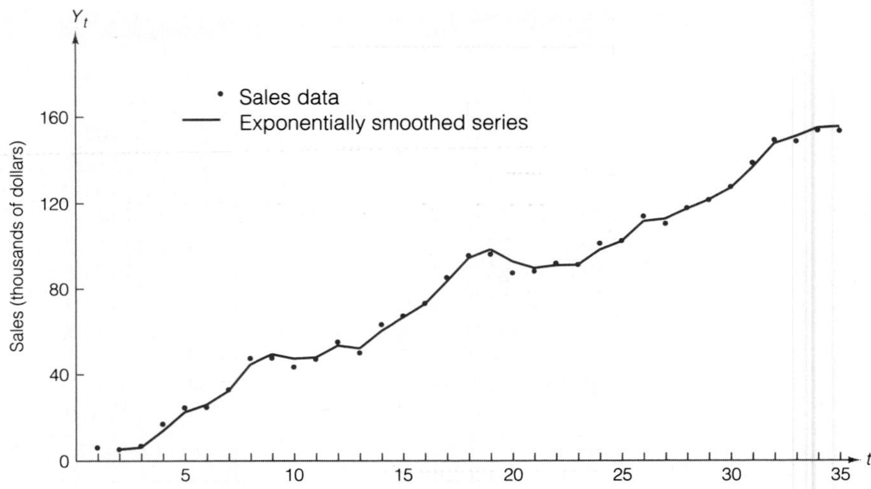

upward trend in Y_t. The choice of $v = .5$ gives equal weight to the most recent trend and to past trends in the sales of the firm. The result is that the exponential smoothing component E_t provides a smooth, upward-trending description of the firm's sales.

Our objective is to use the Holt-Winters exponentially smoothed series to forecast the future values of the time series. For the one-step-ahead forecast, this is accomplished by adding the most recent exponentially smoothed component to the most recent trend component. That is, the forecast at time $(t + 1)$, given observed values up to time t, is

$$F_{t+1} = E_t + T_t$$

The idea is that we are constructing the forecast by combining the most recent smoothed estimate, E_t, with the estimate of the expected increase (or decrease) attributable to trend, T_t.

The forecast for two steps ahead is similar, except that we add estimated trend for *two* periods:

$$F_{t+2} = E_t + 2T_t$$

Similarly, for the k-step-ahead forecast, we add the estimated increase (or decrease) in trend over k periods:

$$F_{t+k} = E_t + kT_t$$

The Holt-Winters forecasting methodology is summarized in the box.

Holt-Winters Forecasting

1. Calculate the exponentially smoothed and trend components, E_t and T_t, for each observed value of Y_t $(t \geq 2)$ using the formulas given in the previous box.
2. Calculate the one-step-ahead forecast using

$$F_{t+1} = E_t + T_t$$

3. Calculate the k-step-ahead forecast using

$$F_{t+k} = E_t + kT_t$$

EXAMPLE 15.3

Refer to Example 15.2 and Table 15.5, where we listed the firm's 35 yearly sales figures, along with the Holt-Winters components using $w = .7$ and $v = .5$. Use the Holt-Winters forecasting technique to forecast the firm's annual sales in years 36–40.

Solution

For year 36 we calculate

$$F_{36} = E_{35} + T_{35} = 152.9 + 2.7 = 155.6$$

The forecast 2 years ahead is

$$F_{37} = E_{35} + 2T_{35} = 152.9 + 2(2.7) = 158.3$$

For years 38–40 we find

$$F_{38} = 152.9 + 3(2.7) = 161.0$$
$$F_{39} = 152.9 + 4(2.7) = 163.7$$
$$F_{40} = 152.9 + 5(2.7) = 166.4$$

These forecasts are displayed in Figure 15.3. Note that the upward trend in the forecast is a result of the Holt-Winters estimated trend component.

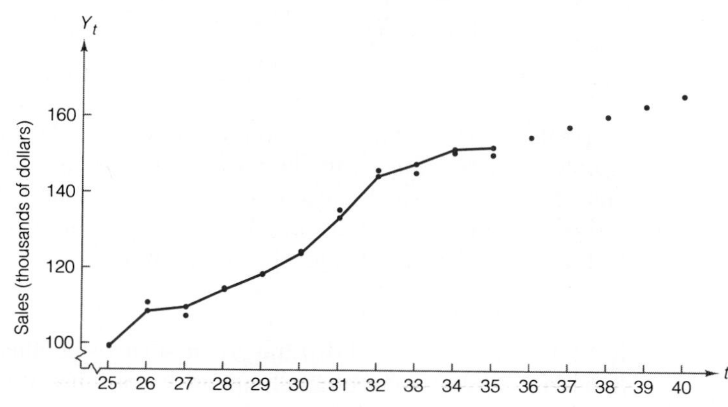

FIGURE 15.3 ▶ Holt-Winters sales forecasts, years 36–40

The selection of $w = .7$ and $v = .5$ as the smoothing and trend weights for the sales forecasts in Example 15.3 was based on the objectives of assigning more weight to recent series values in the exponentially smoothed component, and of assigning equal weights to the recent and past trend estimates. However, you may want to try several different combinations of weights when using the Holt-Winters forecasting model so that you can assess the sensitivity of the forecasts to the choice of weights. Experience with the particular time series and Holt-Winters forecasts will help in the selection of w and v in a practical application.

You can use the forecast errors to evaluate the accuracy of the forecast, and to aid in the selection of both the forecasting methodology to be utilized and the parameters of the forecast formula (e.g., the weights in the exponentially smoothed or Holt-Winters forecasts). Two popular measures of forecast accuracy are the **mean absolute deviation (MAD)** and the **root mean squared error (RMSE)** of the forecasts. Their formulas are given in the box.

Measures of Forecast Accuracy

1. The **mean absolute deviation (MAD)** is defined as the mean absolute difference between the forecast and actual values of the time series:

$$\text{MAD} = \frac{\sum_{t=1}^{N} |F_t - Y_t|}{N}$$

where N is the number of forecasts used for evaluation.

2. The **root mean squared error (RMSE)** is defined as the square root of the mean squared difference between the forecast and actual values of the time series:

$$\text{RMSE} = \sqrt{\frac{\sum_{t=1}^{N} (F_t - Y_t)^2}{N}}$$

Note that both measures require one or more actual values of the time series against which to compare the forecasts. Thus, we can either wait several time periods until the observed values are available, or we can hold out several of the values at the end of the time series, not using them to model the time series, but saving them for evaluating the forecasts obtained from the model.

EXAMPLE 15.4

In Section 15.2 we found that the exponentially smoothed forecasts of the Dow Jones Industrial Average were generally unsatisfactory, because they did not recognize the trend in the time series. We now try the Holt-Winters model to account for the trend. Table 15.6 shows the original series through 1988, with two Holt-Winters models: one with a trend weight $v = .2$ and the other with trend weight $v = .8$. Both have smoothing weight $w = .5$. Use the time period 1989–1991 and apply the MAD and RMSE criteria to evaluate the two forecasting models.

TABLE 15.6 Dow Jones Industrial Average (1961–1988) with Two Holt-Winters Forecasting Models

Year	DJIA	Holt-Winters Model 1		Holt-Winters Model 2	
		$E(w = .5)$	$T(v = .2)$	$E(w = .5)$	$T(v = .8)$
1961	731.14	–	–	–	–
1962	652.10	652.10	−79.04	652.10	−79.04
1963	762.95	668.01	−60.05	668.01	−3.08
1964	874.13	741.04	−33.43	769.53	80.60
1965	969.26	838.43	−7.27	909.69	128.25
1966	785.69	808.43	−11.82	911.82	27.35
1967	905.11	850.86	−.97	922.14	13.73
1968	943.75	896.82	8.42	939.81	16.88
1969	800.36	852.80	−2.07	878.52	−45.65
1970	838.92	844.83	−3.25	835.90	−43.23
1971	884.76	863.17	1.07	838.71	−6.39
1972	950.71	907.47	9.72	891.51	40.96
1973	923.88	920.53	10.38	928.18	37.52
1974	759.37	845.14	−6.77	862.54	−45.01
1975	802.49	820.43	−10.36	810.01	−51.02
1976	974.92	892.50	6.13	866.95	35.35
1977	835.15	866.89	−.22	868.73	8.49
1978	805.01	835.84	−6.39	841.11	−20.39
1979	838.74	834.10	−5.46	829.73	−13.18
1980	963.99	896.31	8.08	890.27	45.79
1981	899.01	901.70	7.54	917.54	30.97
1982	1,046.54	977.89	21.27	997.52	70.18
1983	1,258.64	1,128.90	47.22	1,163.17	146.56
1984	1,211.57	1,193.84	50.76	1,260.65	107.29
1985	1,546.67	1,395.64	80.97	1,457.31	178.78
1986	1,895.95	1,686.28	122.90	1,766.02	282.73
1987	2,276.02	2,042.60	169.59	2,162.38	373.64
1988	2,060.83	2,136.51	154.45	2,298.42	183.56

Solution

The first step is to calculate the Holt-Winters forecasts for 1989–1991 corresponding to the two models. The formula

$$F_{t+k} = E_t + kT_t$$

is used to obtain the forecasts given in Table 15.7, which also presents the actual DJIA values and the forecast errors.

TABLE 15.7 Forecasts and Forecast Errors for Two Holt-Winters Models: Dow Jones Industrial Average (1989–1991)

Year	DJIA	Model 1		Model 2	
	Actual	Forecast	Error	Forecast	Error
1989	2,508.04	2,290.96	−217.08	2,320.07	−187.97
1990	2,678.91	2,445.41	−233.50	2,503.63	−175.28
1991	2,929.30	2,599.86	−329.44	2,687.19	−242.11

To compare the two models, we first calculate the mean absolute deviations:

$$\text{MAD (Model 1)} = \frac{|-217.08| + |-233.50| + |-329.44|}{3} = 260.00$$

$$\text{MAD (Model 2)} = \frac{|-187.97| + |-175.28| + |-242.11|}{3} = 201.79$$

Next, we calculate the root mean squared errors:

$$\text{RMSE (Model 1)} = \sqrt{\frac{(-217.08)^2 + (-233.50)^2 + (-329.44)^2}{3}} = 264.68$$

$$\text{RMSE (Model 2)} = \sqrt{\frac{(-187.97)^2 + (-175.28)^2 + (-242.11)^2}{3}} = 203.86$$

Note that both criteria lead to the same conclusion: Model 2 provides more accurate predictions of the Dow Jones Industrial Average for the period 1989–1991 than does Model 1. Note that the only difference between the two Holt-Winters models is the value of the trend weight: $v = .2$ for Model 1 and $v = .8$ for Model 2. The higher value in Model 2 gives more weight to the short-term trend than to the long-term trend. For the 3 years in question the model sensitive to the short-term trend better forecasts the market's rally in the mid-1980s than does the model that is more sensitive to the long-term trend of the market.

Criteria such as MAD and RMSE for assessing forecast accuracy require special care in interpretation. The number of time periods included in the evaluation is critical to the decision about which model is preferred. The choice also depends on how many time periods ahead the analyst plans to forecast. Once these decisions are made, the criteria can be used to measure forecast accuracy. Nevertheless, only *inferential* models, such as regression models, that have explicit random error components can be used to estimate the reliability of a forecast *before* the actual value of the time series is observed. We discuss an inferential model in the next section.

Exercises 15.1–15.9

Learning the Mechanics

15.1 State two criteria for evaluating the accuracy of an exponentially smoothed forecast.

15.2 The table lists the number (in thousands) of new privately owned housing units started each year during the period 1975–1991. Some exponentially smoothed values ($w = .6$) are also shown.

Year	*Starts*	*Smoothed*	*Year*	*Starts*	*Smoothed*
1975	1,160		1984	1,750	
1976	1,538		1985	1,742	
1977	1,987	1,746.92	1986	1,805	1,763.59
1978	2,020	1,910.77	1987	1,620	1,677.44
1979	1,745		1988	1,488	
1980	1,292		1989	1,376	1,451.11
1981	1,084		1990	1,193	
1982	1,062	1,137.32	1991	1,014	
1983	1,703	1,476.73			

Source: U.S. Bureau of the Census. *Statistical Abstract of the United States, 1992.*

a. Calculate the missing values of the exponentially smoothed housing-starts series.
b. Plot the housing-starts series and the exponentially smoothed series on the same graph.
c. Use the exponentially smoothed data from 1975–1990 to forecast the number of housing starts in 1991. What is the error associated with this forecast?

15.3 The U.S. beer production (in millions of barrels) for the years 1970–1989 is given in the table.

Year	*Beer*	*Year*	*Beer*	*Year*	*Beer*
1970	133.1	1977	170.5	1984	192.2
1971	137.4	1978	179.1	1985	194.3
1972	141.3	1979	184.2	1986	194.4
1973	148.6	1980	194.1	1987	195.9
1974	156.2	1981	193.7	1988	197.4
1975	160.6	1982	196.2	1989	197.8
1976	163.7	1983	195.4		

Source: U.S. Bureau of the Census. *Statistical Abstract of the United States, 1992.*

a. Using the 1970–1986 values to forecast the 1987–1989 production using simple exponential smoothing with $w = .3$. With $w = .7$. Calculate the forecast errors associated with each.
b. Use the Holt-Winters model with $w = .7$ and $v = .3$ to forecast the 1987–1989 production. Repeat with $w = .3$ and $v = .7$. Calculate the forecast errors associated with each.
c. Compare the two simple exponential smoothing forecasts to the two Holt-Winters forecasts using the MAD and RMSE criteria. Does there appear to be a trend in the beer production time series? Which forecasting model is likely to be more appropriate?

15.4 Refer to part **a** of Exercise 15.3. Use the 1970–1989 values to forecast the 1990 production, using exponential smoothing with $w = .3$ and then with $w = .7$. Can the errors be measured or estimated for these 1990 forecasts? Explain.

Applying the Concepts

15.5 Standard & Poor's 500 Stock Composite Average (S&P 500) is a stock market index. Like the Dow Jones Industrial Average, it is an indicator of stock market activity. The table contains end-of-quarter values of the S&P 500 for the years 1978–1992.

Year	*Quarter*	*S&P 500*	*Year*	*Quarter*	*S&P 500*	*Year*	*Quarter*	*S&P 500*
1978	I	88.82	1983	I	151.9	1988	I	265.7
	II	97.66		II	166.4		II	270.7
	III	103.9		III	167.2		III	268.0
	IV	96.11		IV	164.4		IV	276.5
1979	I	100.1	1984	I	157.4	1989	I	292.7
	II	101.7		II	153.1		II	323.7
	III	108.6		III	166.1		III	347.3
	IV	107.8		IV	164.5		IV	348.6
1980	I	104.7	1985	I	179.4	1990	I	338.5
	II	114.6		II	188.9		II	360.4
	III	126.5		III	184.1		III	315.4
	IV	133.5		IV	207.3		IV	328.8
1981	I	133.2	1986	I	232.3	1991	I	372.3
	II	132.3		II	245.3		II	378.3
	III	118.3		III	238.3		III	387.2
	IV	123.8		IV	248.6		IV	388.5
1982	I	110.8	1987	I	292.5	1992	I	407.36
	II	109.7		II	301.4		II	408.27
	III	122.4		III	318.7		III	418.48
	IV	139.4		IV	241.0		IV	435.64

Source: *Standard & Poor's Statistical Service, Security Price Index Record*, 1992; *Standard & Poor's Current Statistics*, Jan. 1993.

a. Use $w = .7$ to smooth the series from 1978 through 1991. Then forecast the four quarterly values in 1992 using *only* the information through the fourth quarter of 1991.
b. Compute the forecast errors for the 1992 forecasts.

c. Repeat parts **a** and **b** using $w = .3$.
d. Use the Holt-Winters methodology with $w = .7$ and $v = .5$ to forecast the 1992 quarterly values. Repeat with $w = .3$ and $v = .5$.

15.6 Refer to Exercise 15.5. Use the MAD and RMSE criteria to compare the four models' forecasts of the four 1992 S&P 500 values.

15.7 Refer to Exercise 15.5.
a. Use the 1987–1992 values to forecast the quarterly 1993 values using simple exponential smoothing with $w = .3$ and $w = .7$.
b. Calculate the forecasts using the Holt-Winters model with $w = .3$ and $v = .5$. Repeat with $w = .7$ and $v = .5$.
c. Is there any way to know which of the four sets of forecasts for 1993 is best? On the basis of the criteria you applied to compare the forecasts for 1992, which model would you choose for 1993?

15.8 The fluctuation of gold and other precious metal prices in the mid-1980s and early 1990s was a reflection of the strength or weakness of the U.S. dollar and subsequent turmoil among European currencies. The accompanying table shows monthly gold prices from January 1986 to December 1992.

Month	*1986*	*1987*	*1988*	*1989*	*1990*	*1991*	*1992*
Jan.	345.8	409.8	478.0	404.0	411.5	384.9	355.7
Feb.	339.6	402.9	442.9	387.8	418.5	365.1	355.2
Mar.	363.4	409.9	444.7	390.1	394.4	347.2	346.0
Apr.	340.8	440.4	453.3	384.4	375.5	359.6	339.8
May	342.8	461.7	452.2	371.3	370.4	358.1	338.5
June	343.1	451.2	452.7	367.6	353.6	368.0	342.0
July	349.2	452.4	439.0	374.9	382.0	369.7	338.9
Aug.	375.8	462.6	432.8	364.9	396.5	341.9	344.2
Sept.	419.4	461.9	414.1	361.9	390.8	368.4	346.7
Oct.	423.9	466.8	408.1	366.9	382.0	360.1	345.6
Nov.	398.7	467.9	421.3	392.3	383.0	344.2	336.3
Dec.	392.6	487.7	419.8	409.2	379.3	380.3	335.5

Source: *Standard & Poor's Statistics: Metals*, 1991; *Standard & Poor's Current Statistics*, 1993.

a. Use exponential smoothing with $w = .5$ to calculate monthly smoothed values from January 1986 to December 1991. Then forecast the twelve 1992 monthly gold prices.
b. Calculate the forecast errors for 1992. What is the trend in errors as the time distance increases?
c. Calculate 12 one-step-ahead forecasts for 1992 by updating the exponentially smoothed values with each month's actual value, and then forecasting the next month's value.
d. Calculate each month's one-step-ahead forecast error, and compare them with the forecast errors in part b. What does the comparison indicate about the difference between short-term and long-term forecasting?
e. Repeat parts **a–d** using the Holt-Winters technique with $w = .5$ and $v = .5$.

15.9 Refer to Exercise 15.8. Two models were used to forecast the monthly 1992 gold prices: an exponential smoothing model with $w = .5$ and a Holt-Winters model with $w = .5$ and $v = .5$.
a. Use the MAD and RMSE criteria to evaluate the two models' accuracy for forecasting the 1992 values using only the 1986–1991 data.

b. Use the MAD and RMSE criteria to evaluate the two models' accuracy when making the 12 one-step-ahead forecasts for 1992, updating the models with each month's actual value before forecasting the next month's value (parts **c** and **d** of Exercise 15.8).

15.4 Forecasting Trends: Simple Linear Regression

Perhaps the simplest **inferential** forecasting model is one with which you are familiar: the simple linear regression model. A straight-line model is used to relate the time series, Y_t, to time, t, and the least squares line is used to forecast future values of Y_t.

Suppose a firm is interested in forecasting its sales revenues for each of the next 5 years. To make such forecasts and assess their reliability, a time series model must be constructed. Refer again to the yearly sales data for a firm's 35 years of operation, given in Table 15.4 (page 815). A plot of the data (Figure 15.4) reveals a linearly increasing trend, so the model

$$E(Y_t) = \beta_0 + \beta_1 t$$

FIGURE 15.4 ▶
Plot of sales data

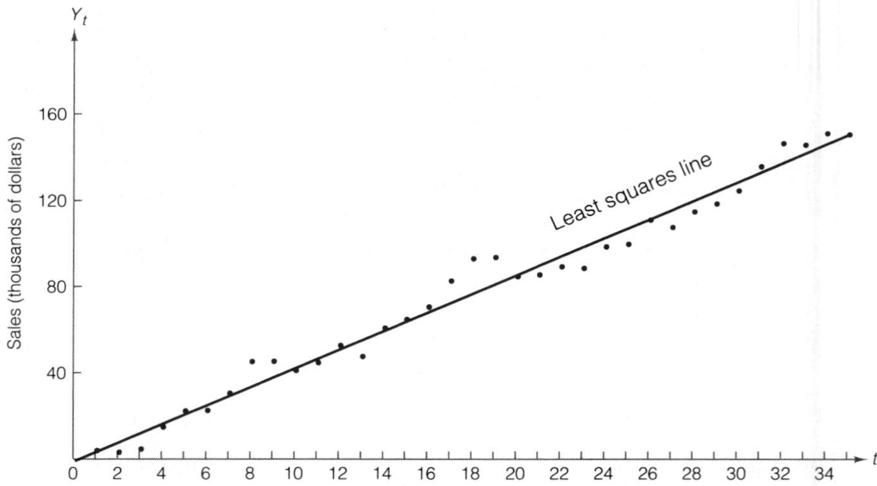

seems plausible for the secular trend. Fitting the model using the method of least squares (see Section 10.2), we obtain the least squares model

$$\hat{Y}_t = \hat{\beta}_0 + \hat{\beta}_1 t = .4015 + 4.2956t \quad \text{with} \quad \text{SSE} = 1{,}345.45$$

This least squares line is shown in Figure 15.4, and the SAS printout is given in Figure 15.5.

We can now forecast sales for years 36–40. The forecasts of sales and the corresponding 95% prediction intervals are shown in the printout. For example, for $t = 36$, we have

$$\hat{Y}_{36} = 155.0$$

SOURCE	DF	SUM OF SQUARES	MEAN SQUARE	F VALUE	PR > F
MODEL	1	65875.20816807	65875.20816807	1615.72	0.0001
ERROR	33	1345.45354622	40.77131958	R-SQUARE	ROOT MSE
CORRECTED TOTAL	34	67220.66171429		0.979985	6.38524233

PARAMETER	ESTIMATE	T FOR HO: PARAMETER = 0	PR > |T|	STD ERROR OF ESTIMATE
INTERCEPT	0.40151261	0.18	0.8567	2.20570829
T	4.29563025	40.20	0.0001	0.10686692

T	PREDICTED VALUE	LOWER 95% CL INDIVIDUAL	UPPER 95% CL INDIVIDUAL
36	155.04420168	141.30017574	168.78822762
37	159.33983193	145.53232286	173.14734101
38	163.63546218	149.76135290	177.50957147
39	167.93109244	153.98731054	181.87487434
40	172.22672269	158.21024159	186.24320379

FIGURE 15.5 ▲ SAS printout for least squares fit (straight line) to Y_t = Sales

with the 95% prediction interval (141.3, 168.8). Similarly, we can obtain the forecasts and prediction intervals for years 37–40. The observed sales, forecast sales, and prediction intervals are shown in Figure 15.6. Although it is not easily perceptible in the figure, the prediction intervals widen as we attempt to forecast further into the future (see the printout in Figure 15.5). This agrees with the intuitive notion that short-term forecasts should be more reliable than long-term forecasts.

There are two problems associated with forecasting time series using a least squares model.

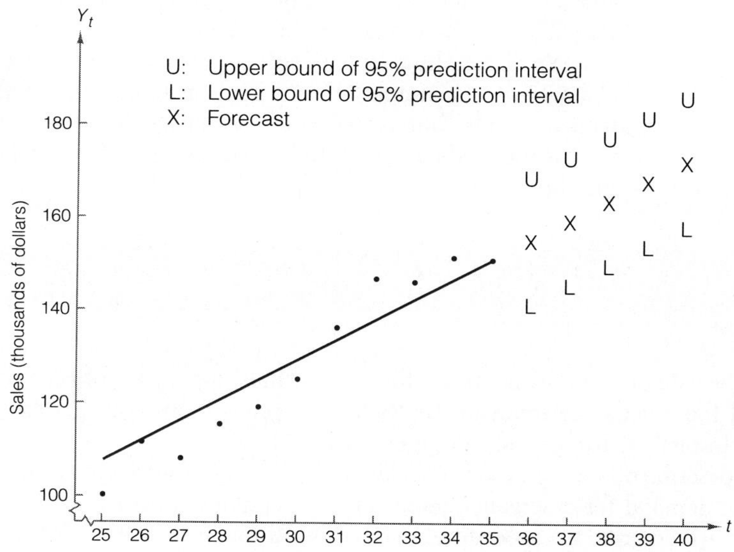

FIGURE 15.6 ► Observed (years 1–35) and forecast (years 36–40) sales using the straight-line model

Problem 1

We are using the least squares model to forecast values outside the region of observation of the independent variable, t. That is, we are forecasting for values of t between 36 and 40, but the observed sales are for t values between 1 and 35. As we noted in Chapters 10–12, it is extremely risky to use a least squares regression model for prediction outside the experimental region.

Problem 1 obviously cannot be avoided. Since forecasting always involves predictions about the future values of a time series, some or all of the independent variables will probably be outside the region of observation on which the model was developed. It is important that the forecaster recognize the dangers of this type of prediction. If underlying conditions change drastically after the model is estimated (e.g., if federal price controls are imposed on the firm's products during the 36th year of operation), the forecasts and their confidence intervals are probably useless.

Problem 2

Although the straight-line model may adequately describe the secular trend of the sales, we have not attempted to build any cyclical effects into the model. Thus, the effect of inflationary and recessionary periods will be to increase the error of the forecasts because the model does not anticipate such periods.

Fortunately, the forecaster often has some degree of control over problem 2, as we demonstrate in the remainder of the chapter.

In forming the prediction intervals for the forecasts, we made the standard regression assumptions (Chapters 10 and 12) about the random error component of the model. We assumed the errors have mean 0, constant variance, normal probability distributions, and are *independent*. The latter assumption is dubious in time series models, especially in the presence of short-term trends. Often, if a year's value lies above the secular trend line, the next year's value has a tendency to be above the line also. That is, the errors tend to be correlated (see Figure 15.4).

We discuss how to deal with correlated errors in Section 15.6. For now, we can characterize the simple linear regression forecasting method as useful for discerning secular trends, but probably too simplistic for most time series. And, as with all forecasting methods, the simple linear regression forecasts should be applied only over the short term.

CASE STUDY 15.1 / Forecasting the Demand for Emergency Room Services

In Case Study 5.2, we described how queueing theory was used to model the emergency room in the Richmond Memorial Hospital in Richmond, Virginia. In this case study, we describe how a regression model was used to forecast the demand for emergency room services. In particular, we describe the procedure used by the hospital to forecast the average number of emergency room visits per day during the month of August 1970.

Data were collected on the emergency room's operations since its opening in October 1955 (month 1) through January 1970 (month 124). The data on pa-

TABLE 15.8 Emergency Room Data for the Month of August: 1959–1969

Month, t	Visits	Daily Average, Y_t	Month, t	Visits	Daily Average, Y_t
11	1,367	44.09	71	3,019	97.38
23	1,642	52.96	83	2,794	90.12
35	1,780	57.41	95	2,846	91.80
47	2,060	66.45	107	3,001	96.80
59	2,257	72.80	119	3,548	114.45

tient visits for each August since the emergency room opened are shown in Table 15.8.

A straight-line regression model was used to model the trend in Y_t, the average number of visits per day during August. With time t (measured in months) as the independent variable, the following least squares model was obtained: $\hat{Y}_t = 38.788 + .60990t$. This least squares line is plotted on a scattergram of the data in Figure 15.7. The plot reveals an upward trend in emergency room visits. Although some seasonal variation was present, a similar upward trend was observed for the other 11 months of the year. It was determined that the increase in the demand for emergency room services was greater than the growth in the Richmond area population. This finding substantiated management's belief that the emergency room was increasingly serving as a replacement for the family physician.

The model yielded a point forecast of $\hat{Y}_{131} = 38.788 + .60990(131) = 118.68$ for the average number of visits per day in August 1970. The associated 95% prediction interval was (100.20, 137.16). The actual demand in August turned out to be 119.20 and is indicated by a cross (×) in the plot of Figure 15.7. Thus, the actual demand fell close to the least squares trend line and well within the bounds of the prediction interval.

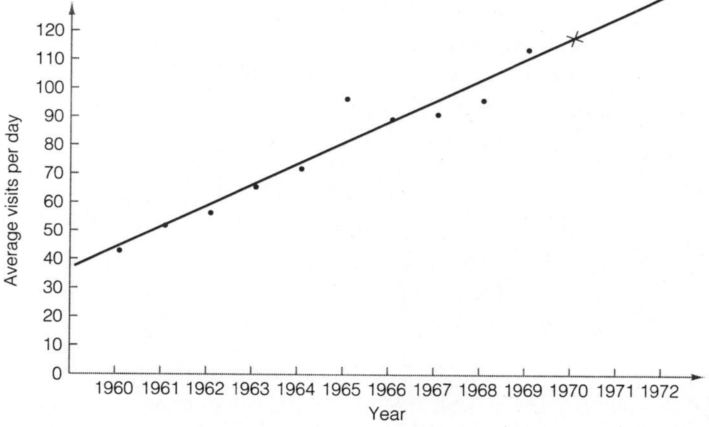

FIGURE 15.7 ▶ **Least squares trend line for average number of visits per day during August**

Even for projections 18 months into the future, the hospital found this least squares prediction procedure to be superior to previous methods employed by the hospital. As a result, it was adopted for regular use in budgeting for the emergency room and associated services (Bolling, 1972).

15.5 Seasonal Regression Models

Many time series have distinct seasonal patterns. Retail sales are usually highest around Christmas, spring, and fall, with lulls in the winter and summer periods. Energy usage is highest in summer and winter, and lowest in spring and fall. Teenage unemployment rises in summer months when schools are not in session and falls near Christmas when many businesses hire part-time help.

Multiple regression models can be used to forecast future values of a time series with strong seasonal components. To accomplish this, the mean value of the time series, $E(Y_t)$, is given a mathematical form that describes both the secular trend and seasonal components of the time series. Although the seasonal model can assume a wide variety of mathematical forms, the use of dummy variables to describe seasonal differences is common.

For example, consider the power load data for a southern utility company shown in Table 15.9. Data were obtained for each quarter from 1981 through 1992. A model that combines the expected growth in usage and the seasonal component is

TABLE 15.9 Quarterly Power Loads (Megawatts) for a Southern Utility Company, 1981–1992

Year	Quarter	Power Load	Year	Quarter	Power Load
1981	1	68.8	1987	1	130.6
	2	65.0		2	116.8
	3	88.4		3	144.2
	4	69.0		4	123.3
1982	1	83.6	1988	1	142.3
	2	69.7		2	124.0
	3	90.2		3	146.1
	4	72.5		4	135.5
1983	1	106.8	1989	1	147.1
	2	89.2		2	119.3
	3	110.7		3	138.2
	4	91.7		4	127.6
1984	1	108.6	1990	1	143.4
	2	98.9		2	134.0
	3	120.1		3	159.6
	4	102.1		4	135.1
1985	1	113.1	1991	1	149.5
	2	94.2		2	123.3
	3	120.5		3	154.4
	4	107.4		4	139.4
1986	1	116.2	1992	1	151.6
	2	104.4		2	133.7
	3	131.7		3	154.5
	4	117.9		4	135.1

$$E(Y_t) = \beta_0 + \beta_1 t + \beta_2 Q_1 + \beta_3 Q_2 + \beta_4 Q_3$$

where

t = Time period, ranging from $t = 1$ for quarter 1 of 1981 to $t = 48$ for quarter 4 of 1992

$$Q_1 = \begin{cases} 1 & \text{if quarter 1} \\ 0 & \text{if quarter 2, 3, or 4} \end{cases}$$

$$Q_2 = \begin{cases} 1 & \text{if quarter 2} \\ 0 & \text{if quarter 1, 3, or 4} \end{cases}$$

$$Q_3 = \begin{cases} 1 & \text{if quarter 3} \\ 0 & \text{if quarter 1, 2, or 4} \end{cases}$$

The printout in Figure 15.8 shows the least squares fit of this model to the data in Table 15.9.

```
DEPENDENT VARIABLE: POWER LOAD

SOURCE                 DF   SUM OF SQUARES      MEAN SQUARE    F VALUE

MODEL                   4   28374.99250583    7093.74812646     114.88
ERROR                  43    2655.13561917      61.74733998     PR > F
CORRECTED TOTAL        47   31030.12812500                      0.0001

R-SQUARE             C.V.         ROOT MSE           Y3 MEAN

0.914434           6.6766       7.85794757      117.69375000

                                 T FOR HO:       PR > |T|   STD ERROR OF
PARAMETER          ESTIMATE    PARAMETER=0                    ESTIMATE

INTERCEPT       70.50852273          22.63        0.0001     3.11552479
T                1.63621066          19.92        0.0001     0.08213932
QUARTER     1   13.65863199           4.25        0.0001     3.21744388
            2   -3.73591200          -1.16        0.2512     3.21219719
            3   18.46954400           5.76        0.0001     3.20904506
```

FIGURE 15.8 ▲
SAS printout of least squares fit to power load time series

Note that the model appears to fit well, with $R^2 = .91$, indicating that the model accounts for 91% of the sample variability in power loads over the 12-year period. The global $F = 114.88$ strongly supports the hypothesis that the model has predictive utility. The standard deviation **(ROOT MSE)** of 7.86 indicates that the model predictions will usually be accurate to within approximately $\pm 2(7.86)$, or about ± 16 megawatts. Furthermore, $\hat{\beta}_1 = 1.64$ indicates an estimated average growth in load of 1.64 megawatts per quarter. Finally the seasonal dummy variables have the following interpretations (refer to Chapter 12):*

*These interpretations assume a fixed value of t. In practical terms this is unrealistic, since each quarter is associated with a different value of t. Nevertheless, the coefficients of the seasonal dummy variables provide insight into the seasonality of these time series data.

$\hat{\beta}_2 = 13.66$ Quarter 1 loads average 13.66 megawatts more than quarter 4 loads.

$\hat{\beta}_3 = -3.74$ Quarter 2 loads average 3.74 megawatts less than quarter 4 loads.

$\hat{\beta}_4 = 18.47$ Quarter 3 loads average 18.47 megawatts more than quarter 4 loads.

Thus, as expected, winter and summer loads exceed spring and fall loads, with the peak occurring during the summer months.

In order to forecast the 1993 power loads, we calculate the predicted value $\hat{Y}$ for $k = 49$, 50, 51, and 52, at the same time substituting the dummy variable appropriate for each quarter. Thus, for 1993,

$$\hat{Y}_{\text{Quarter 1}} = \hat{\beta}_0 + \hat{\beta}_1(49) + \hat{\beta}_2 = 70.51 + 1.636(49) + 13.66 = 164.3$$
$$\hat{Y}_{\text{Quarter 2}} = \hat{\beta}_0 + \hat{\beta}_1(50) + \hat{\beta}_3 = 148.6$$
$$\hat{Y}_{\text{Quarter 3}} = \hat{\beta}_0 + \hat{\beta}_1(51) + \hat{\beta}_4 = 172.4$$
$$\hat{Y}_{\text{Quarter 4}} = \hat{\beta}_0 + \hat{\beta}_1(52) = 155.6$$

The predicted values and 95% prediction intervals are given in Table 15.10; the data and least squares predicted values are graphed in Figure 15.9. The color line on the graph connects the predicted values. Also shown in Table 15.10 and Figure 15.9 are the actual 1993 quarterly power loads. Notice that all 1993 power loads fall inside the forecast intervals.

TABLE 15.10 Predicted Power Loads, Confidence Intervals, and Actual Power Loads (Megawatts) for 1993

Quarter	Predicted Load	Lower 95% Confidence Limit	Upper 95% Confidence Limit	Actual Load
1	164.3	147.3	181.4	151.3
2	148.6	131.5	165.6	132.9
3	172.4	155.4	189.5	160.5
4	155.6	138.5	172.6	161.0

The seasonal model used to forecast the power loads is an **additive model** because the secular trend component ($\beta_1 t$) is added to the seasonal component ($\beta_2 Q_1 + \beta_3 Q_2 + \beta_4 Q_3$) to form the model. A **multiplicative model** would have the same form, except that the dependent variable would be the natural logarithm of power load; i.e.,

$$\ln Y_t = \beta_0 + \beta_1 t + \beta_2 Q_1 + \beta_3 Q_2 + \beta_4 Q_3 + \epsilon$$

To see the multiplicative nature of this model, we take the antilogarithm of both sides of the equation to get

FIGURE 15.9 ▶
Regression forecasting model for a southern utility company

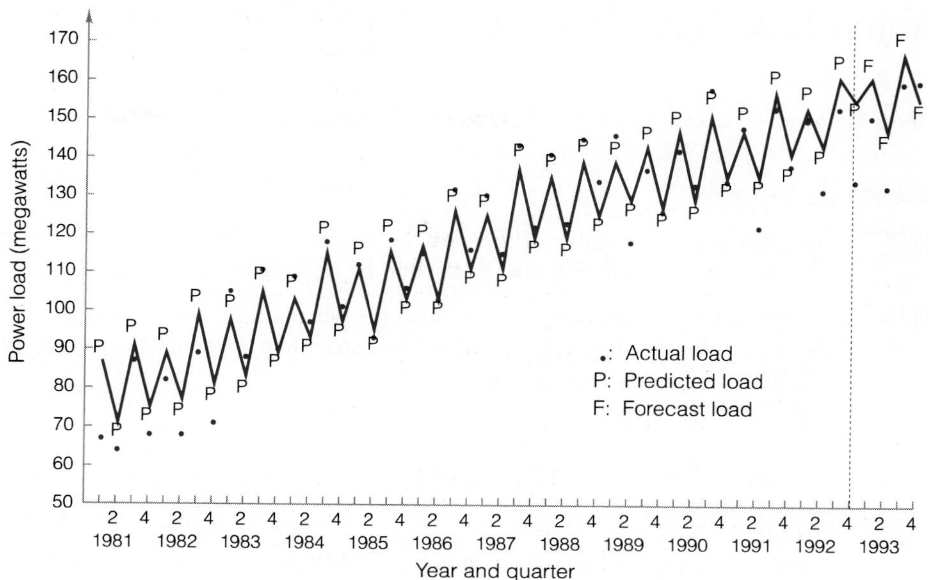

$$Y_t = \exp\{\beta_0 + \beta_1 t + \beta_2 Q_1 + \beta_3 Q_2 + \beta_4 Q_3 + \epsilon\}$$
$$= \underbrace{\exp\{\beta_0\}}_{\text{Constant}} \underbrace{\exp\{\beta_1 t\}}_{\substack{\text{Secular}\\\text{trend}}} \underbrace{\exp\{\beta_2 Q_1 + \beta_3 Q_2 + \beta_4 Q_3\}}_{\substack{\text{Seasonal}\\\text{component}}} \underbrace{\exp\{\epsilon\}}_{\substack{\text{Residual}\\\text{component}}}$$

The multiplicative model often provides a better forecasting model when the time series is changing at an increasing rate over time.

When time series data are observed monthly, a regression forecasting model needs 11 dummy variables to describe monthly seasonality; three dummy variables can be used (as in the previous models) if the seasonal changes are hypothesized to occur quarterly. In general, this approach to seasonal modeling requires one dummy variable fewer than the number of seasonal changes expected to occur.

There are approaches besides the regression dummy variable method for forecasting seasonal time series. Trigonometric (sine and cosine) terms can be used in regression models to model periodicity. Other time series models (the Holt-Winters exponential smoothing model, for example) do not use the regression approach at all, and there are various methods for adding seasonal components to these models. We have chosen to discuss the regression approach because it makes use of the important modeling concepts covered in Chapters 11 and 12, and because the regression forecasts are accompanied by prediction intervals that provide some measure of the forecast reliability. While most other methods do not have explicit measures of reliability, many have proved their merit by providing good forecasts for particular applications. Consult the references at the end of this chapter for details of other seasonal models.

Exercises 15.10–15.15

Note: Starred (*) exercises require the use of a computer.

Learning the Mechanics

15.10 What advantage do regression forecasts have over exponentially smoothed forecasts? Does this advantage assure that regression forecasts will prove to be more accurate? Explain.

15.11 The annual price of galvanized steel (in cents per pound) from 1971 to 1984 is given in the table, and the SAS printout for the simple linear regression model fit to these data is shown. The time variable t begins with $t = 1$ in 1971, and is incremented by 1 for each additional year:

$t = \text{Year} - 1970$

Year	*Price*	*Year*	*Price*	*Year*	*Price*
1971	9.61	1976	16.07	1981	26.88
1972	10.88	1977	18.10	1982	26.75
1973	10.59	1978	20.47	1983	28.43
1974	12.39	1979	22.32	1984	30.30
1975	14.80	1980	23.88		

Source: *Standard & Poor's Basic Statistics: Metals*, 1985.

a. Find and interpret the least squares estimates.
b. What are the forecasts for 1985 and 1986?
c. Find the 95% forecast intervals for 1985 and 1986.

***15.12** Retail sales in quarters 1–4 over a 10-year period for a department store are shown (in hundreds of thousands of dollars) in the table.

Year	*1*	*2*	*3*	*4*
1	8.3	10.3	8.7	13.5
2	9.8	12.1	10.1	15.4
3	12.1	14.5	12.7	17.1
4	13.7	16.0	14.2	19.2
5	17.4	19.7	18.0	23.1
6	18.2	20.5	18.6	24.0
7	20.0	22.2	20.5	25.1
8	22.3	25.1	22.9	27.7
9	24.7	26.9	25.1	29.8
10	25.8	28.7	26.0	32.2

a. Write a regression model that contains trend and seasonal components to describe the sales data.
b. Use a least squares regression program to fit the model. Evaluate the fit of the model and interpret the coefficients.

c. Use the regression model to forecast the quarterly sales during year 11. Give 95% confidence intervals for the forecasts.

Dep Variable: PRICE

Analysis of Variance

Source	DF	Sum of Squares	Mean Square	F Value	Prob>F
Model	1	660.72601	660.72601	936.327	0.0001
Error	12	8.46788	0.70566		
C Total	13	669.19389			

Root MSE	0.84003	R-Square	0.9873
Dep Mean	19.39071	Adj R-Sq	0.9863
C.V.	4.33215		

Parameter Estimates

Variable	DF	Parameter Estimate	Standard Error	T for H0: Parameter=0	Prob > |T|
INTERCEP	1	6.609231	0.47421483	13.937	0.0001
T	1	1.704198	0.05569371	30.599	0.0001

Obs	PRICE	Predict Value	Residual	Lower95% Predict	Upper95% Predict
1	9.6100	8.3134	1.2966	6.2613	10.3656
2	10.8800	10.0176	0.8624	8.0090	12.0263
3	10.5900	11.7218	-1.1318	9.7502	13.6935
4	12.3900	13.4260	-1.0360	11.4845	15.3676
5	14.8000	15.1302	-0.3302	13.2116	17.0489
6	16.0700	16.8344	-0.7644	14.9312	18.7377
7	18.1000	18.5386	-0.4386	16.6431	20.4341
8	20.4700	20.2428	0.2272	18.3473	22.1383
9	22.3200	21.9470	0.3730	20.0438	23.8503
10	23.8800	23.6512	0.2288	21.7326	25.5699
11	26.8800	25.3554	1.5246	23.4139	27.2969
12	26.7500	27.0596	-0.3096	25.0880	29.0312
13	28.4300	28.7638	-0.3338	26.7552	30.7724
14	30.3000	30.4680	-0.1680	28.4158	32.5202
15	.	32.1722	.	30.0704	34.2740
16	.	33.8764	.	31.7193	36.0335

Applying the Concepts

15.13 There was phenomenal growth in the transportation sector of the U.S. economy during the 1960s and 1970s. The personal consumption expenditure figures (in billions of dollars) for this sector are given in the table at the top of page 834 for the years 1960–1984.

Year	*Expenditure*	*Year*	*Expenditure*	*Year*	*Expenditure*	*Year*	*Expenditure*
1960	42.4	1967	63.3	1973	114.6	1979	219.4
1961	44.8	1968	69.3	1974	117.9	1980	236.6
1962	47.4	1969	75.7	1975	129.4	1981	261.5
1963	49.5	1970	80.6	1976	155.2	1982	267.3
1964	54.3	1971	92.3	1977	179.3	1983	291.9
1965	58.4	1972	105.4	1978	198.1	1984	319.5
1966	60.4						

Source: U.S. Bureau of the Census. *Statistical Abstract of the United States, 1986.*

a. Fit the simple regression model

$$E(Y_t) = \beta_0 + \beta_1 t$$

where t is the number of years since 1960 (i.e., $t = 0, 1, \ldots, 24$).

b. Forecast the personal consumption expenditure from 1985 to 1987. Calculate 95% confidence intervals for these forecasts.

15.14 The table presents the quarterly sales index for one brand of calculator at a campus bookstore. The quarters are based on an academic year, so the first quarter represents fall; the second, winter; the third, spring; and the fourth, summer.

Year	*First Quarter*	*Second Quarter*	*Third Quarter*	*Fourth Quarter*
1988	438	398	252	160
1989	464	429	376	216
1990	523	496	425	318
1991	593	576	456	398
1992	636	640	526	498

We defined the time variable as $t = 1$ for the first quarter of 1988, $t = 2$ for the second quarter of 1988, etc. The seasonal dummy variables are as follows:

$$Q_1 = \begin{cases} 1 & \text{if quarter 1} \\ 0 & \text{otherwise} \end{cases}$$

$$Q_2 = \begin{cases} 1 & \text{if quarter 2} \\ 0 & \text{otherwise} \end{cases}$$

$$Q_3 = \begin{cases} 1 & \text{if quarter 3} \\ 0 & \text{otherwise} \end{cases}$$

The SAS printout for the model

$$E(Y_t) = \beta_0 + \beta_1 t + \beta_2 Q_1 + \beta_3 Q_2 + \beta_4 Q_3$$

is reproduced on the facing page.

a. Interpret the least squares estimates, and evaluate the usefulness of the model.

b. Which of the assumptions about the random error component is in doubt when a regression model is fit to time series data?

c. Find the forecasts and the 95% prediction intervals for the 1993 quarterly sales.

Dep Variable: Y

Analysis of Variance

Source	DF	Sum of Squares	Mean Square	F Value	Prob>F
Model	4	318560.30000	79640.07500	117.817	0.0001
Error	15	10139.50000	675.96667		
C Total	19	328699.80000			

Root MSE	25.99936	R-Square	0.9692
Dep Mean	440.90000	Adj R-Sq	0.9609
C.V.	5.89688		

Parameter Estimates

Variable	DF	Parameter Estimate	Standard Error	T for H0: Parameter=0	Prob > ¦T¦
INTERCEP	1	119.850000	16.94950835	7.071	0.0001
T	1	16.512500	1.02771490	16.067	0.0001
Q1	1	262.337500	16.72998649	15.681	0.0001
Q2	1	222.825000	16.57140484	13.446	0.0001
Q3	1	105.512500	16.47552320	6.404	0.0001

Obs	Y	Predict Value	Residual	Lower95% Predict	Upper95% Predict
1	438.0	398.7	39.3000	335.5	461.9
2	398.0	375.7	22.3000	312.5	438.9
3	252.0	274.9	-22.9000	211.7	338.1
4	160.0	185.9	-25.9000	122.7	249.1
5	464.0	464.7	-0.7500	403.4	526.1
6	429.0	441.7	-12.7500	380.4	503.1
7	376.0	340.9	35.0500	279.6	402.3
8	216.0	251.9	-35.9500	190.6	313.3
9	523.0	530.8	-7.8000	470.1	591.5
10	496.0	507.8	-11.8000	447.1	568.5
11	425.0	407.0	18.0000	346.3	467.7
12	318.0	318.0	0	257.3	378.7
13	593.0	596.9	-3.8500	535.5	658.2
14	576.0	573.9	2.1500	512.5	635.2
15	456.0	473.1	-17.0500	411.7	534.4
16	398.0	384.0	13.9500	322.7	445.4
17	636.0	662.9	-26.9000	599.7	726.1
18	640.0	639.9	0.1000	576.7	703.1
19	526.0	539.1	-13.1000	475.9	602.3
20	498.0	450.1	47.9000	386.9	513.3
21	.	729.0	.	662.8	795.1
22	.	706.0	.	639.8	772.1
23	.	605.1	.	539.0	671.3
24	.	516.1	.	450.0	582.3

15.15 The U.S. labor force is comprised of the civilian labor force and resident military personnel. The table lists the size of the labor force (in thousands) over the period 1970–1991.

Year	*Labor Force*	*Year*	*Labor Force*	*Year*	*Labor Force*
1970	84,889	1978	103,882	1985	117,167
1971	86,355	1979	106,559	1986	119,540
1972	88,847	1980	108,544	1987	121,602
1973	91,203	1981	110,315	1988	123,378
1974	93,670	1982	111,872	1989	125,557
1975	95,453	1983	113,226	1990	126,424
1976	97,826	1984	115,241	1991	126,867
1977	100,665				

Source: U.S. Bureau of the Census. *Statistical Abstract of the United States, 1992.*

a. Use the method of least squares to fit a simple regression model to the data.
b. Forecast the labor force for 1992 and 1993.
c. Construct 95% confidence intervals for the forecasts of part **b**.
d. Check the accuracy of your forecasts by looking up the actual labor force size for 1992 and 1993 in either *Business Statistics* or the *Statistical Abstract of the United States*.

15.6 Autocorrelation and the Durbin-Watson Test

Recall that one of the assumptions we make when using a regression model for predictions is that the errors are independent. However, with time series data, this assumption is questionable. The cyclical component of a time series may result in deviations from the secular trend that tend to cluster alternately on the positive and negative sides of the trend, as shown in Figure 15.10.

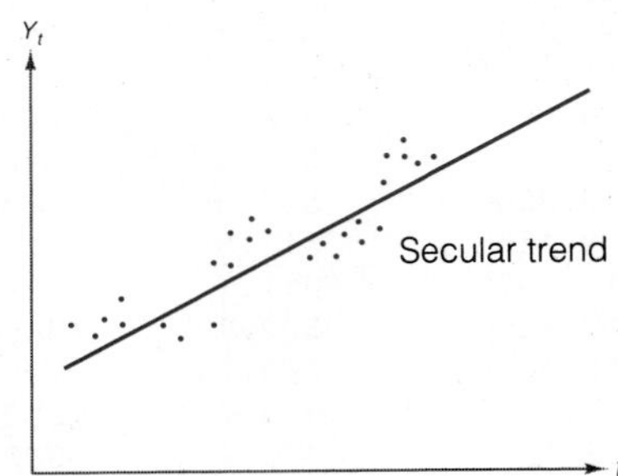

FIGURE 15.10 ▶ Illustration of cyclical errors

The observed errors between the time series and the regression model for the secular trend (and seasonal component, if present) are called **time series residuals.**

Thus, if the time series Y_t has an estimated trend of $\hat{Y}_t$, then the time series residual* is

$$\hat{R}_t = Y_t - \hat{Y}_t$$

Note that time series residuals are defined just as the residuals for any regression model. However, we will usually plot time series residuals versus time to determine whether a cyclical component is apparent.

For example, consider the sales forecasting data in Table 15.4 (page 815), to which we fit a simple straight-line regression model. The plot of the data and model are repeated in Figure 15.11, and a plot of the time series residuals is shown in Figure 15.12.

FIGURE 15.11 ▶ Plot of sales data

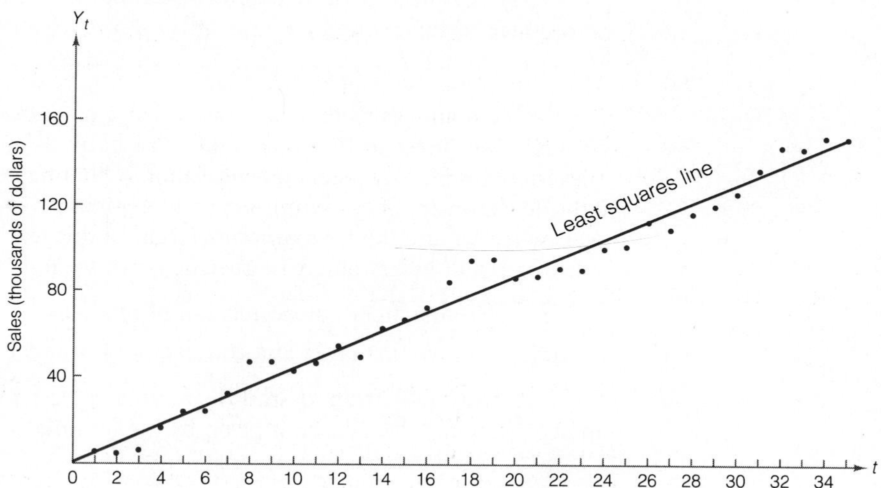

FIGURE 15.12 ▶ Plot of residuals vs. time for the sale data: Least squares model

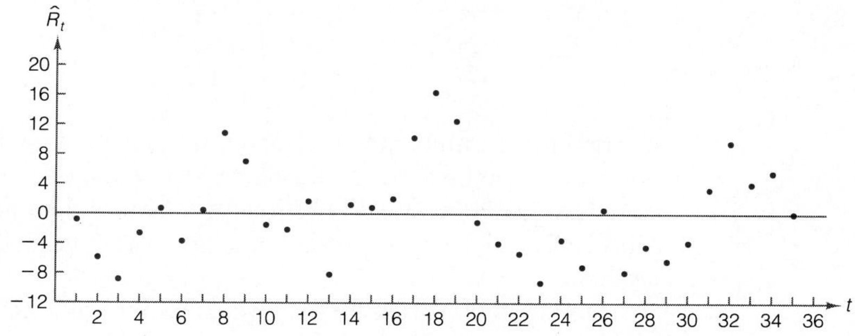

*We use $\hat{R}_t$ rather than $\hat{\epsilon}$ to denote a time series residual because, as we shall see, time series residuals often do not satisfy the regression assumptions associated with the random component ϵ.

Notice the tendency of the residuals to group alternately into positive and negative clusters. That is, if the residual for year t is positive, there is a tendency for the residual for year $(t + 1)$ to be positive. These cycles are indicative of possible positive correlation between neighboring residuals. The correlation between time series residuals at different points in time is called **autocorrelation**, and the autocorrelation of neighboring residuals (time periods t and $t + 1$) is called **first-order autocorrelation**.

Definition 15.1

The correlation between time series residuals at different points in time is called **autocorrelation**. Correlation between neighboring residuals (at times t and $t + 1$) is called **first-order autocorrelation**. In general, correlation between residuals at times t and $t + d$ is called dth-order autocorrelation.

Rather than speculate about the presence of autocorrelation among time series residuals, we prefer to test for it. For most business and economic time series, the relevant test is for first-order autocorrelation. Other higher-order autocorrelations may indicate seasonality, e.g., fourth-order autocorrelation in a quarterly time series. However, when we use the term *autocorrelation* in this text we are referring to first-order autocorrelation unless otherwise specified. So, we test

H_0: No first-order autocorrelation of residuals

H_a: Positive first-order autocorrelation of residuals

The **Durbin-Watson *d* statistic** is used to test for the presence of first-order autocorrelation. The statistic is given by the formula

$$d = \frac{\sum_{t=2}^{n} (\hat{R}_t - \hat{R}_{t-1})^2}{\sum_{t=1}^{n} \hat{R}_t^2}$$

where n is the number of observations (time periods) and $(\hat{R}_t - \hat{R}_{t-1})$ represents the difference between a pair of successive time series residuals. The value of d always falls in the interval from 0 to 4. The interpretations of the values of d are given in the box. Most statistical software packages include a routine that calculates d for time series residuals.

Durbin and Watson (1951) give tables for the lower-tail values of the d statistic, which we show in Tables XIV ($\alpha = .05$) and XV ($\alpha = .01$) of Appendix B. Part of Table XIV is reproduced in Table 15.11. For the sales example, we have $k = 1$ independent variable and $n = 35$ observations. Using $\alpha = .05$ for the one-tailed test for positive autocorrelation, we obtain the tabled values $d_L = 1.40$ and $d_U = 1.52$. The meaning of these values is illustrated in Figure 15.13. Because of the complexity of the

Interpretation of Durbin-Watson d Statistic

$$d = \frac{\sum_{t=2}^{n} (\hat{R}_t - \hat{R}_{t-1})^2}{\sum_{t=1}^{n} \hat{R}_t^2} \qquad \text{Range of } d\text{: } 0 \le d \le 4$$

1. If the residuals are uncorrelated, then $d \approx 2$.
2. If the residuals are positively autocorrelated, then $d < 2$, and if the autocorrelation is very strong, $d \approx 0$.
3. If the residuals are negatively autocorrelated, then $d > 2$, and if the autocorrelation is very strong, $d \approx 4$.

TABLE 15.11 Reproduction of Part of Table XIV of Appendix B: Critical Values for the Durbin-Watson d Statistic, $\alpha = .05$

n	$k = 1$		$k = 2$		$k = 3$		$k = 4$		$k = 5$	
	d_L	d_U	d_L	d_U	d_L	d_U	d_L	d_U	d_L	d_U
31	1.36	1.50	1.30	1.57	1.23	1.65	1.16	1.74	1.09	1.83
32	1.37	1.50	1.31	1.57	1.24	1.65	1.18	1.73	1.11	1.82
33	1.38	1.51	1.32	1.58	1.26	1.65	1.19	1.73	1.13	1.81
34	1.39	1.51	1.33	1.58	1.27	1.65	1.21	1.73	1.15	1.81
35	1.40	1.52	1.34	1.58	1.28	1.65	1.22	1.73	1.16	1.80
36	1.41	1.52	1.35	1.59	1.29	1.65	1.24	1.73	1.18	1.80
37	1.42	1.53	1.36	1.59	1.31	1.66	1.25	1.72	1.19	1.80
38	1.43	1.54	1.37	1.59	1.32	1.66	1.26	1.72	1.21	1.79
39	1.43	1.54	1.38	1.60	1.33	1.66	1.27	1.72	1.22	1.79
40	1.44	1.54	1.39	1.60	1.34	1.66	1.29	1.72	1.23	1.79

FIGURE 15.13 ► Rejection region for the Durbin-Watson d test: Sales example

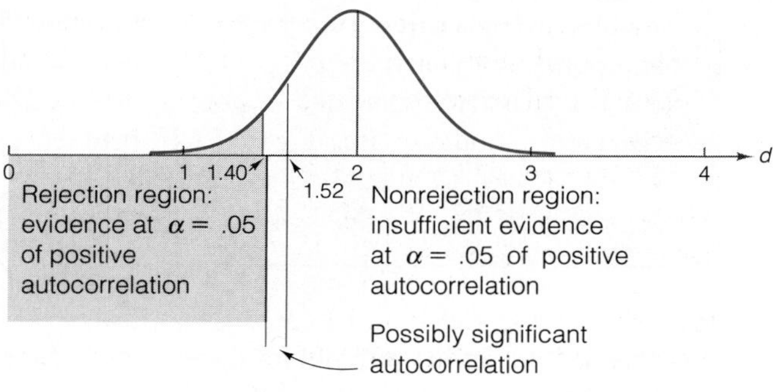

sampling distribution of d, it is not possible to specify a single point that acts as a boundary between the rejection and nonrejection regions, as we did for the z, t, F, and other test statistics. Instead, an upper (d_U) and lower (d_L) bound are specified. Thus a d value less than d_L *does* provide strong evidence of positive autocorrelation at $\alpha = .05$ (recall that small d values indicate positive autocorrelation); a d value greater than d_U does *not* provide evidence of positive autocorrelation at $\alpha = .05$; and a value of d between d_L and d_U might or might not be significant at the $\alpha = .05$ level. If $d_L < d < d_U$, more information is needed before we can reach any conclusion about the presence of autocorrelation.

Tests for negative autocorrelation and two-tailed tests can be conducted by making use of the symmetry of the sampling distribution of the d statistic about its mean, 2 (see Figure 15.13). The test procedure is summarized in the next box.

Durbin-Watson *d* Test

One-Tailed Test

H_0: No first-order autocorrelation

H_a: Positive first-order autocorrelation (or H_a: Negative first-order autocorrelation)

Test statistic:

$$d = \frac{\sum_{t=2}^{n} (\hat{R}_t - \hat{R}_{t-1})^2}{\sum_{t=1}^{n} \hat{R}_t^2}$$

Rejection region: $d < d_{L,\alpha}$ [or $(4 - d) < d_{L,\alpha}$ if H_a: Negative first-order autocorrelation]

where $d_{L,\alpha}$ is the lower tabled value corresponding to k independent variables and n observations. The corresponding upper value $d_{U,\alpha}$ defines a "possibly significant" region between $d_{L,\alpha}$ and $d_{U,\alpha}$ (see Figure 15.13).

Two-Tailed Test

H_0: No first-order autocorrelation

H_a: Positive or negative first-order autocorrelation

Test statistic:

$$d = \frac{\sum_{t=2}^{n} (\hat{R}_t - \hat{R}_{t-1})^2}{\sum_{t=1}^{n} \hat{R}_t^2}$$

Rejection region: $d < d_{L,\alpha/2}$ or $(4 - d) < d_{L,\alpha/2}$

where $d_{L,\alpha/2}$ is the lower tabled value corresponding to k independent variables and n observations. The corresponding upper value $d_{U,\alpha/2}$ defines a "possibly significant" region between $d_{L,\alpha/2}$ and $d_{U,\alpha/2}$ (see Figure 15.13).

Assumption: The residuals are normally distributed.

The SAS regression printout for the sales example is presented in Figure 15.14. It shows that the computed value of d is .82, which is less than the tabulated value of

DEPENDENT VARIABLE: Y SALES (THOUSANDS OF DOLLARS)

SOURCE	DF	SUM OF SQUARES	MEAN SQUARE	F VALUE
MODEL	1	65875.20816807	65875.20816807	1615.72
ERROR	33	1345.45354622	40.77131958	PR > F
CORRECTED TOTAL	34	67220.66171429		0.0001

R-SQUARE	C.V.	ROOT MSE	Y MEAN
0.979985	8.2154	6.38524233	77.72285714

<table>
<tr><th>PARAMETER</th><th>ESTIMATE</th><th>T FOR HO: PARAMETER=0</th><th>PR > |T|</th><th>STD ERROR OF ESTIMATE</th></tr>
<tr><td>INTERCEPT</td><td>0.40151261</td><td>0.18</td><td>0.8567</td><td>2.20570829</td></tr>
<tr><td>T</td><td>4.29563025</td><td>40.20</td><td>0.0001</td><td>0.10686692</td></tr>
</table>

OBSERVATION	OBSERVED VALUE	PREDICTED VALUE	RESIDUAL
1	4.80000000	4.69714286	0.10285714
2	4.00000000	8.99277311	-4.99277311
3	5.50000000	13.28840336	-7.78840336
4	15.60000000	17.58403361	-1.98403361
5	23.10000000	21.87966387	1.22033613
6	23.30000000	26.17529412	-2.87529412
7	31.40000000	30.47092437	0.92907563
8	46.00000000	34.76655462	11.23344538
9	46.10000000	39.06218487	7.03781513
10	41.90000000	43.35781513	-1.45781513
11	45.50000000	47.65344538	-2.15344538
12	53.50000000	51.94907563	1.55092437
13	48.40000000	56.24470588	-7.84470588
14	61.60000000	60.54033613	1.05966387
15	65.60000000	64.83596639	0.76403361
16	71.40000000	69.13159664	2.26840336
17	83.40000000	73.42722689	9.97277311
18	93.60000000	77.72285714	15.87714286
19	94.20000000	82.01848739	12.18151261
20	85.40000000	86.31411765	-0.91411765
21	86.20000000	90.60974790	-4.40974790
22	89.90000000	94.90537815	-5.00537815
23	89.20000000	99.20100840	-10.00100840
24	99.10000000	103.49663866	-4.39663866
25	100.30000000	107.79226891	-7.49226891
26	111.70000000	112.08789916	-0.38789916
27	108.20000000	116.38352941	-8.18352941
28	115.50000000	120.67915966	-5.17915966
29	119.20000000	124.97478992	-5.77478992
30	125.20000000	129.27042017	-4.07042017
31	136.30000000	133.56605042	2.73394958
32	146.80000000	137.86168067	8.93831933
33	146.10000000	142.15731092	3.94268908
34	151.40000000	146.45294118	4.94705882
35	150.90000000	150.74857143	0.15142857

SUM OF RESIDUALS	0.00000000
SUM OF SQUARED RESIDUALS	1345.45354622
SUM OF SQUARED RESIDUALS - ERROR SS	-0.00000000
FIRST ORDER AUTOCORRELATION	0.58962415
DURBIN-WATSON D	0.82072679

FIGURE 15.14 ▲ SAS printout for the regression analysis: Annual sales data

Thus, we conclude that the residuals of the straight-line model for sales are positively autocorrelated.

Once strong evidence of first-order autocorrelation has been established, as in the case of the sales example, doubt is cast on the least squares results and any inferences drawn from them. Under these circumstances, a time series model that accounts for the autocorrelation of the random errors is needed. A useful model is the first-order autoregressive model. Consult the references at the end of this chapter.

Exercises 15.16–15.20

Learning the Mechanics

15.16 Define autocorrelation. Explain why it is important in time series modeling and forecasting.

15.17 What do the following Durbin-Watson statistics suggest about the autocorrelation of the time series residuals from which each was calculated?

a. $d = 3.9$ b. $d = .2$ c. $d = 1.99$

15.18 For each case, indicate the decision regarding the test of the null hypothesis of no first-order autocorrelation against the alternative hypothesis of positive first-order autocorrelation.

a. $k = 2$, $n = 20$, $\alpha = .05$, $d = 1.1$
b. $k = 2$, $n = 20$, $\alpha = .01$, $d = 1.1$
c. $k = 5$, $n = 65$, $\alpha = .05$, $d = .95$
d. $k = 1$, $n = 31$, $\alpha = .01$, $d = 1.35$

Applying the Concepts

15.19 The decrease in the value of the dollar, Y_t, from 1960 to 1991 is illustrated by the data in the table. The buying power of the dollar (compared with 1982) is listed for each year. The first-order model

$$Y_t = \beta_0 + \beta_1 t + \epsilon$$

was fit to the data using the method of least squares. Then the least squares estimates of β_0 and β_1 were calculated as

$$\hat{\beta}_0 = 3.3177 \quad \text{and} \quad \hat{\beta}_1 = -.0884$$

a. Calculate and plot the regression residuals against t. Is there a tendency for the residuals to have long positive and negative runs? To what do you attribute this phenomenon?
b. Calculate the Durbin-Watson d statistic, and test the null hypothesis that the time series residuals are uncorrelated. Use $\alpha = .10$.

t	Y_t	t	Y_t	t	Y_t
1960	2.994	1971	2.469	1982	1.000
1961	2.994	1972	2.392	1983	0.984
1962	2.985	1973	2.193	1984	0.964
1963	2.994	1974	1.901	1985	0.955
1964	2.985	1975	1.718	1986	0.969
1965	2.933	1976	1.645	1987	0.949
1966	2.841	1977	1.546	1988	0.926
1967	2.809	1978	1.433	1989	0.880
1968	2.732	1979	1.289	1990	0.839
1969	2.632	1980	1.136	1991	0.822
1970	2.545	1981	1.041		

Source: U.S. Bureau of the Census. *Statistical Abstract of the United States, 1992.*

c. What assumption(s) must be satisfied in order for the test of part **b** to be valid?

15.20 The table gives the volume of retail sales (in millions of dollars) of passenger cars in the United States by motor vehicle dealers for the years 1984 and 1985.

Month	*Time, t*	*1984 Sales, Y_t*	*Time, t*	*1985 Sales, Y_t*
Jan.	1	\$20.76	13	\$22.70
Feb.	2	21.07	14	23.19
Mar.	3	19.97	15	22.94
Apr.	4	21.06	16	24.18
May	5	21.26	17	24.15
June	6	21.94	18	24.07
July	7	21.23	19	24.15
Aug.	8	20.40	20	25.30
Sept.	9	20.65	21	27.74
Oct.	10	21.83	22	23.02
Nov.	11	22.20	23	23.19
Dec.	12	22.16	24	24.12

Source: *Standard & Poor's Current Statistics*, Nov. 1986, p. 14.

We used the SAS System to fit the first-order time series model

$$Y_t = \beta_0 + \beta_1 t + \epsilon$$

to the data using the method of least squares, and obtained the printout shown on page 844.

a. Interpret the least squares estimates.

b. With what approximate precision do you expect this model to predict annual sales (assuming the necessary assumptions about the random component ϵ are satisfied)?

c. Is there evidence at the $\alpha = .10$ level of significance that the residuals are autocorrelated?

```
Dep Variable: Y

                      Analysis of Variance

                      Sum of          Mean
Source       DF      Squares        Square       F Value        Prob>F

Model         1     50.37956      50.37956        44.726        0.0001
Error        22     24.78078       1.12640
C Total      23     75.16033

    Root MSE         1.06132      R-Square        0.6703
    Dep Mean        22.63667      Adj R-Sq        0.6553
    C.V.             4.68850

                      Parameter Estimates

                   Parameter       Standard     T for H0:
Variable   DF       Estimate          Error   Parameter=0     Prob > |T|

INTERCEP    1      20.020362     0.44718746        44.770         0.0001
T           1       0.209304     0.03129660         6.688         0.0001

Durbin-Watson D               1.533
(For Number of Obs.)             24
1st Order Autocorrelation     0.211
```

Summary

Time series are often modeled as a combination of four components: **secular**, **seasonal**, **cyclical**, and **residual**. **Exponential smoothing** is an adaptive method for forecasting time series with little or no secular or seasonal trends. The **Holt-Winters model** provides an adaptive forecasting technique for a time series with a significant trend (secular) component. **Simple linear regression** can be used to forecast long-term trends of time series, and also allows the computation of prediction intervals to evaluate the forecast's reliability. **Multiple regression models** can be used to describe both long-term and seasonal components. Since many business and economic time series models exhibit cyclical behavior, the **Durbin-Watson *d* statistic** is important for testing **residual autocorrelation** and determining the appropriateness of regression modeling.

Forecasting is an especially difficult aspect of statistical inference, because, by definition, we are extrapolating out of the range of the time period containing the data. Forecasts should be confined to the short term and, when possible, some measure of forecast reliability should be calculated. The best measure of a forecasting technique's usefulness is obtained by comparing its forecasts to the future realizations of the time series. Such comparisons can be made using **MAD** or **RMSE**.

Supplementary Exercises 15.21 – 15.32

Note: Starred (*) exercises require the use of a computer.

15.21 Civilian employment is broadly classified by the federal government into two categories—agricultural and nonagricultural. Employment figures (in thousands of workers) for farm and nonfarm categories are given in the table for the years 1975–1991.

Year	*Farm*	*Nonfarm*	*Year*	*Farm*	*Nonfarm*
1975	3,408	82,438	1984	3,321	101,685
1976	3,331	85,421	1985	3,179	103,971
1977	3,283	88,734	1986	3,163	106,434
1978	3,387	92,661	1987	3,208	109,232
1979	3,347	95,477	1988	3,169	111,800
1980	3,364	95,938	1989	3,199	114,142
1981	3,368	97,030	1990	3,186	114,728
1982	3,401	96,125	1991	3,233	113,644
1983	3,383	97,450			

Source: U.S. Bureau of the Census. *Statistical Abstract of the United States*, 1992.

a. Use $w = .5$ to compute exponential smoothing forecasts for each of the two series for 1992.
b. Use the Holt-Winters model with $w = .5$ and $v = .5$ to compute 1992 forecasts for each of the series.

15.22 Refer to Exercise 15.13. Use $w = .3$ and $v = .7$ to compute the Holt-Winters forecasts for 1985–1987. Compare these to the linear regression forecasts obtained in Exercise 15.13.

15.23 The stock of Abbott Laboratories has had the yearly closing prices shown in the table.

Year	*Closing Price*	*Year*	*Closing Price*	*Year*	*Closing Price*
1973	$49.875	1978	$33.750	1983	$45.250
1974	50.750	1979	41.125	1984	41.750
1975	41.250	1980	56.500	1985	68.375
1976	49.125	1981	27.000	1986	45.625
1977	56.500	1982	38.750		

Source: *Daily Stock Price Record*, NYSE, 1973–1986; *Security Owners' Stock Guide*, Standard & Poor's, Jan. 1987.

a. Use exponential smoothing with $w = .8$ to forecast the 1987 and 1988 closing prices. If you buy at the end of 1986 and sell at the end of 1988, what is your expected gain (loss)?
b. Repeat part a using the Holt-Winters model with $w = .8$ and $v = .5$.
c. In which forecast do you have more confidence? Explain.

15.24 Refer to Exercise 15.23.

a. Fit a simple linear regression model to the stock price data.
b. Plot the fitted regression line on a scattergram of the data.
c. Forecast the 1987 and 1988 closing prices using the regression model.
d. Construct 95% prediction intervals for the forecasts of part **c**. Interpret the intervals in the context of the problem.

15.25 Refer to Exercise 15.24. Calculate the time series residuals for the simple linear model, and use the Durbin-Watson d statistic to test for the presence of autocorrelation.

15.26 The Gross Domestic Product (GDP) is the total U.S. output of goods and services valued at market prices. The quarterly GDP values (in billions of dollars) from 1976–1990 are given in the accompanying table.

Year	*1*	*2*	*3*	*4*
1976	1,717.8	1,746.4	1,779.9	1,829.6
1977	1,881.7	1,952.9	2,015.1	2,046.8
1978	2,090.2	2,213.9	2,274.7	2,352.0
1979	2,399.2	2,453.3	2,523.3	2,578.8
1980	2,650.1	2,643.9	2,705.3	2,832.9
1981	2,953.5	2,993.0	3,079.6	3,096.3
1982	3,092.9	3,146.2	3,164.2	3,195.1
1983	3,254.9	3,367.1	3,450.9	3,547.3
1984	3,666.9	3,754.6	3,818.2	3,869.1
1985	3,940.0	3,997.5	4,076.9	4,140.5
1986	4,215.7	4,232.0	4,290.2	4,336.6
1987	4,408.3	4,494.9	4,573.5	4,683.0
1988	4,752.4	4,857.2	4,947.3	5,044.6
1989	5,139.9	5,218.5	5,277.3	5,340.4
1990	5,422.4	5,504.7	5,570.5	5,557.5

Source: U.S. Bureau of Economic Analysis. *Business Statistics, 1963–1991,* 1992.

Use $w = .5$ and $v = .5$ to calculate Holt-Winters forecasts for 1991. Then complete the following table.

Quarter	*Actual GDP*	*Forecast*	*Forecast Error*
1	5,589.0		
2	5,652.6		
3	5,709.2		
4	5,739.7		

***15.27** Refer to Exercise 15.26.

a. Use the simple linear regression model to forecast the 1991 quarterly GDP. Place 95% confidence limits on the forecasts.
b. The GDP values given are **seasonally adjusted**, which means that an attempt to remove seasonality has been made prior to reporting the figures. Add quarterly dummy variables to the model. Use the partial F test (discussed in Section 12.4) to determine whether the data indicate the significance of the seasonal component. Does the test support the assertion that the GDP figures are seasonally adjusted?
c. Use the seasonal model to forecast the 1991 quarterly GDP values.

d. Calculate the time series residuals for the seasonal model, and use the Durbin-Watson test to determine whether the residuals are autocorrelated. Use $\alpha = .05$.

15.28 Refer to Exercises 15.26 and 15.27. For each of the forecasting models apply the MAD and RMSE criteria to evaluate the forecasts for the four quarters of 1991. Which of the forecasting models performs best according to each cirterion?

15.29 Commercial banks, finance companies, credit unions, and retail outlets are the principal issuers of consumer installment credit (CIC). Since commercial banks and credit unions typically offer lower interest rates, they are referred to as **primary lenders**. Finance companies and retail outlets are referred to as **secondary lenders**. From 1945 to 1970, CIC in the United States grew from \$2.5 billion to \$101 billion, a growth rate 4½ times greater than that of the Gross National Product. In the process, there was a dramatic increase in the yearly percentage market share of CIC held by primary lenders relative to secondary lenders. Regression analysis was used on the data given in the table to obtain projections of the market share of primary lenders beyond 1970.

Year	*Primary Lenders*		*Total*	*Secondary Lenders*		*Total*
	Commercial Banks	*Credit Unions*		*Finance Companies*	*Retail Outlets*	
1945	30.2%	4.1%	34.3%	36.9%	27.9%	64.8%
1950	39.4	4.0	43.4	36.1	19.7	55.8
1955	36.7	5.8	42.5	40.9	15.5	56.4
1960	38.9	9.2	48.1	35.4	14.7	50.1
1965	40.6	10.3	50.9	34.0	13.7	47.7
1966	40.4	10.6	51.0	33.6	13.9	47.5
1967	40.4	11.1	51.5	33.0	14.1	47.1
1968	41.1	11.3	52.4	32.4	13.8	46.2
1969	41.1	11.8	52.9	32.3	13.4	45.7
1970	41.4	12.4	53.8	30.8	13.9	44.7

Source: Dauten, J. J., Apilado, V. P., and Warner, D. C. "Consumer credit: Changing patterns in supply and demand variables, and their implications." *Journal of Consumer Affairs*, Winter 1973, pp. 7, 97.

a. Use the method of least squares to model the secular trend in the market share of primary lenders.
b. Plot the least squares model on a scattergram of the data.
c. Use your least squares model to find point forecasts for the market share of primary lenders for each of the years 1971 through 1975. Discuss some of the pitfalls of this forecasting technique.
d. The researchers concluded that if their ". . . projections are borne out over time, the relative position of primary and secondary lenders with respect to market share will have completely reversed in the 20-year period from 1955 to 1975." Do your projections yield the same conclusion?
e. Check the accuracy of your forecast for 1975 by determining the actual market share for primary lenders in 1975 from data supplied in the *Federal Reserve Bulletin*.

15.30 A major portion of total consumer credit is extended in the categories of automobile loans, mobile home loans, and revolving credit. Amounts outstanding (in billions of dollars) for the period 1980–1991 are given in the table on page 848.

a. Use simple linear regression models for each credit category to forecast the 1992 and 1993 values. Place 95% confidence bounds on each forecast.
b. Calculate the Holt-Winters forecasts for 1992 and 1993 using $w = .7$ and $v = .7$. Compare the results with the simple linear regression forecasts of part **a**.

Year	*Automobile*	*Mobile Home*	*Revolving*
1980	112.0	18.7	55.1
1981	119.0	20.1	61.1
1982	125.9	22.6	66.5
1983	143.6	23.6	79.1
1984	173.6	25.9	100.3
1985	210.2	26.8	121.8
1986	247.4	27.1	135.9
1987	265.9	25.9	153.1
1988	284.2	25.3	174.1
1989	290.7	22.5	199.1
1990	284.6	21.0	220.1
1991	267.9	19.1	234.5

Source: U.S. Bureau of the Census. *Statistical Abstract of the United States, 1992.*

15.31 Consider the monthly IBM stock prices from January 1990 to December 1992 shown in the table. Also shown are the exponentially smoothed values for $w = .5$.

1990	*Price*	*Smoothed*	*1991*	*Price*	*Smoothed*	*1992*	*Price*	*Smoothed*
Jan.	98.125	98.125	Jan.	107.500	108.594	Jan.	95.375	94.581
Feb.	103.750	100.938	Feb.	137.375	122.984	Feb.	89.750	92.165
Mar.	107.000	103.969	Mar.	126.625	124.805	Mar.	89.625	90.895
Apr.	107.125	105.547	Apr.	106.875	115.840	Apr.	88.500	89.698
May	115.625	110.586	May	102.750	109.295	May	91.625	90.661
June	120.375	115.480	June	100.375	104.835	June	94.125	92.393
July	121.000	118.240	July	99.250	102.042	July	99.250	95.822
Aug.	105.000	111.620	Aug.	98.375	100.209	Aug.	88.375	92.098
Sept.	104.375	107.998	Sept.	102.875	101.542	Sept.	85.500	88.799
Oct.	99.250	103.624	Oct.	104.250	102.896	Oct.	72.875	80.837
Nov.	112.625	108.124	Nov.	96.250	99.573	Nov.	64.875	72.856
Dec.	111.250	109.687	Dec.	88.000	93.786	Dec.	56.125	64.491

Source: U.S. Bureau of the Census, *Statistical Abstract of the United States, 1992.*

a. Plot both the stock price and the smoothed stock price on the same graph.
b. Use the exponentially smoothed series to forecast the monthly values of the IBM stock price from January 1993 to March 1993.

15.32 Refer to Exercise 15.31. The SAS printout for the simple linear regression model fit to the IBM stock prices is shown. The time t ranges from 1 to 36 over the 3-year period of the sample.
a. Interpret the least squares estimates.
b. With what approximate precision do you expect to be able to predict the IBM stock price using this model?
c. Give the forecasts and the 95% forecast intervals for the January–March 1993 prices. How does the precision of these forecasts agree with the approximation obtained in part **b**?

```
Model: MODEL1
Dependent Variable: PRICE

                          Analysis of Variance

                          Sum of          Mean
Source          DF       Squares        Square       F Value        Prob>F

Model            1    4530.81602    4530.81602        37.284        0.0001
Error           34    4131.78120     121.52298
C Total         35    8662.59722

    Root MSE       11.02375     R-square       0.5230
    Dep Mean       99.77778     Adj R-sq       0.5090
    C.V.           11.04830

                          Parameter Estimates

                   Parameter        Standard     T for H0:
Variable   DF       Estimate           Error   Parameter=0      Prob > |T|

INTERCEP    1     119.756349      3.75249701        31.914          0.0001
T           1      -1.079923      0.17686166        -6.106          0.0001

Durbin-Watson D                 0.653
(For Number of Obs.)               36
1st Order Autocorrelation       0.548

           Dep Var     Predict    Std Err   Lower95%   Upper95%
     Obs    PRICE        Value    Predict    Predict    Predict    Residual

       1   98.1250       118.7      3.599    95.1096      142.2    -20.5514
       2     103.8       117.6      3.448    94.1230      141.1    -13.8465
       3     107.0       116.5      3.300    93.1313      139.9     -9.5166
       4     107.1       115.4      3.155    92.1344      138.7     -8.3117
       5     115.6       114.4      3.013    91.1322      137.6      1.2683
       6     120.4       113.3      2.875    90.1247      136.4      7.0982
       7     121.0       112.2      2.741    89.1119      135.3      8.8031
       8     105.0       111.1      2.612    88.0936      134.1     -6.1170
       9     104.4       110.0      2.490    87.0698      133.0     -5.6620
      10   99.2500       109.0      2.374    86.0406      131.9     -9.7071
      11     112.6       107.9      2.266    85.0058      130.7      4.7478
      12     111.3       106.8      2.167    83.9655      129.6      4.4527
      13     107.5       105.7      2.079    82.9195      128.5      1.7826
      14     137.4       104.6      2.002    81.8679      127.4     32.7376
      15     126.6       103.6      1.939    80.8107      126.3     23.0675
      16     106.9       102.5      1.890    79.7478      125.2      4.3974
      17     102.8       101.4      1.856    78.6793      124.1      1.3523
      18     100.4       100.3      1.839    77.6051      123.0      0.0573
      19   99.2500     99.2378      1.839    76.5251      122.0      0.0122
      20   98.3750     98.1579      1.856    75.4395      120.9      0.2171
      21     102.9     97.0780      1.890    74.3482      119.8      5.7970
      22     104.3     95.9980      1.939    73.2513      118.7      8.2520
      23   96.2500     94.9181      2.002    72.1486      117.7      1.3319
      24   88.0000     93.8382      2.079    71.0404      116.6     -5.8382
      25   95.3750     92.7583      2.167    69.9265      115.6      2.6167
      26   89.7500     91.6784      2.266    68.8070      114.5     -1.9284
      27   89.6250     90.5984      2.374    67.6819      113.5     -0.9734
      28   88.5000     89.5185      2.490    66.5513      112.5     -1.0185
      29   91.6250     88.4386      2.612    65.4152      111.5      3.1864
      30   94.1250     87.3587      2.741    64.2736      110.4      6.7663
      31   99.2500     86.2787      2.875    63.1267      109.4     12.9713
      32   88.3750     85.1988      3.013    61.9743      108.4      3.1762
      33   85.5000     84.1189      3.155    60.8166      107.4      1.3811
      34   72.8750     83.0390      3.300    59.6537      106.4    -10.1640
      35   64.8750     81.9591      3.448    58.4856      105.4    -17.0841
      36   56.1250     80.8791      3.599    57.3123      104.4    -24.7541
      37         .     79.7992      3.752    56.1339      103.5           .
      38         .     78.7193      3.908    54.9505      102.5           .
      39         .     77.6394      4.065    53.7621      101.5           .

Sum of Residuals                        0
Sum of Squared Residuals        4131.7812
Predicted Resid SS (Press)      4752.3576
```

d. What assumptions does the random error component of the model have to satisfy in order to make the model inferences (such as the forecast intervals in part **c**) valid?

e. Test to determine whether there is evidence of first-order autocorrelation in the random error component. Use $\alpha = .05$. What can you infer about the validity of the model inferences?

On Your Own

Refer to **On Your Own** for Chapter 14. Use the same time series and simple index for this exercise. Hold out the last three values of the time series to evaluate the forecasting techniques utilized.

a. Using all but the last three values of the simple index, use the exponential smoothing forecasting technique with the same value of w you used for smoothing in part **b** of the Chapter 14 **On Your Own**.

b. Use the Holt-Winters technique to forecast the last three values of the simple index, selecting v consistent with the trend you found in part **c** of the Chapter 14 **On Your Own**, and using the same value of w used in part **a**.

c. Use a simple linear regression model to forecast the last three values of the simple index. Do not use these last three values in fitting the model.

d. Use the MAD and RMSE criteria to evaluate the forecasts of parts **a–c**. Which technique appears to provide the best forecasts for the last three index values? Plot all the forecasts and the actual values. Does the plot support your choice of the best technique?

e. Convert the best forecasts of the index values to forecasts of the actual time series. Compare the forecasts to the actual values using the MAD and RMSE criteria. Is your conclusion about the best forecasting technique for your series unchanged?

References

Abraham, B., and Ledholter, J. *Statistical Methods for Forecasting*. New York: Wiley, 1983.

Anderson, T. W. *The Statistical Analysis of Time Series*. New York: Wiley, 1971.

Bolling, W. B. "Queuing model of a hospital emergency room." *Industrial Engineering*, Sept. 1972, pp. 26–31.

Box, G. E. P., and Jenkins, G. M. *Time Series Analysis: Forecasting and Control*, 2nd ed. San Francisco: Holden-Day, 1977.

Durbin, J., and Watson, G. S. "Testing for serial correlation in least squares regression, I." *Biometrika*, 1950, Vol. 37, pp. 409–428.

Durbin, J., and Watson, G. S. "Testing for serial correlation in least squares regression, II." *Biometrika*, 1951, Vol. 38, pp. 159–178.

Durbin, J., and Watson, G. S. "Testing for serial correlation in least squares regression, III." *Biometrika*, 1971, Vol. 58, pp. 1–19.

Fuller, W. A. *Introduction to Statistical Time Series*. New York: Wiley, 1976.

Granger, C. W. J., and Newbold, P. *Forecasting Economic Time Series*. New York: Academic Press, 1977.

Nelson, C. R. *Applied Time Series Analysis for Managerial Forecasting*. San Francisco: Holden-Day, 1973.

CHAPTER SIXTEEN

Design of Experiments and Analysis of Variance

Contents

Case Study

Where We've Been

As we saw in preceding chapters, the solutions of many business problems are based on inferences about population means. Methods for estimating and testing hypotheses about a single mean and the comparison of two means were presented in Chapters 7–9 and Chapters 10–12 and dealt with methods for estimating the mean value of a response using regression models. Methods for monitoring the mean of a process were presented in Chapter 13.

Where We're Going

In this chapter we extend the methodology of Chapters 7–13 in two important ways. First, we discuss *designed* sampling experiments and their critical elements. Then we show how to *analyze* the experiment in order to compare more than two populations. We will present several of the more popular experimental designs and show how to use the computer to analyze designed experiments using both analysis of variance and regression programs.

Most of the data we analyzed in previous chapters were collected in *observational* sampling experiments rather than *designed* sampling experiments. In observational experiments the analyst has little or no control over the variables under study and merely observes their values. In contrast, designed experiments are those in which the analyst attempts to control the levels of one or more variables to determine their effect on a variable of interest. Although the opportunity for such control is not always present in real business settings, it is instructive, even for observational experiments, to have a working knowledge of the analysis and interpretation of data that result from designed experiments and to know at least the basics of how to design experiments when the opportunity arises.

In Section 16.1, we present the basic elements of an experimental design. We then discuss two simple but popular experimental designs in Section 16.2 and 16.3. In optional Section 16.4 we show how the statistical analysis of designed experiments can be accomplished using the regression methodology of Chapters 10–12.

16.1 Elements of a Designed Experiment

Many of the elements in a designed experiment are the same as or similar to those we introduced in regression analysis. For example, the **response** is the variable of interest in the experiment. The response might be the SAT score of a high school senior, the total sales of a firm last year, or the total income of a particular household this year. In regression analyses we referred to the response as the **dependent variable**, y. We use these terms interchangeably in this chapter.

> **Definition 16.1**
>
> The **response** is the variable of interest in the experiment. We also refer to the response as the **dependent variable**.

The intent of most statistical experiments is to determine the effect of one or more variables on the response. These variables, which we called the **independent variables** in regression analyses, are often referred to as the **factors** in a designed experiment. Like independent variables, factors are either **quantitative** or **qualitative** depending on whether the variable is measured on a numerical scale or not. For example, we might want to explore the effect of the qualitative factor "Gender" on the response "SAT score." In other words, we want to compare the SAT scores of male and female high school seniors. Or, we might wish to determine the effect of the quantitative factor "Number of salespeople" on the response "Total sales" for retail firms. Often two or more factors are of interest. For example, in studying a chemical process, we might want to determine the effect of the quantitative factors "Temperature" and "Pressure" on the response "Chemical produced."

Definition 16.2

Factors are those variables whose effect on the response is of interest to the experimenter. **Quantitative** factors are measured on a numerical scale, while **qualitative** factors are those that are not (naturally) measured on a numerical scale. We referred to factors as **independent variables** in regression analysis, and we use the terms synonymously.

Levels are the values of the factors that are utilized in the experiment. The levels of qualitative factors are usually nonnumerical. For example, the levels of Gender are Male and Female, and the levels of Location might be North, East, South, and West.* The levels of quantitative factors are numerical values. For example, the factor Number of salespeople may have levels 1, 3, 5, 7, and 9. The factor Pressure may have levels (in pounds per square inch) 10, 20, 30, and 40.

Definition 16.3

The **levels** of a factor are the values of the factor utilized in the experiment.

When a *single factor* is employed in an experiment, the **treatments** of the experiment are the levels of the factor. For example, if the effect of the factor Gender on the response SAT score is being investigated, the treatments of the experiment are the two levels of Gender: Female and Male. Or, if the effect of the Number of wage earners on Household income is the subject of the experiment, then the numerical values assumed by the quantitative factor Number of wage earners are the treatments. If *two or more factors* are utilized in an experiment, the treatments are the factor–level combinations employed. For example, if the effects of the factors Gender and GPA on the response SAT score are being investigated, then the treatments are the combinations of the levels of Gender and GPA employed; (Female, 2.61), (Male, 3.43), and (Female, 3.82) are all treatments if those particular factor–level combinations are utilized in the experiment.

Definition 16.4

The **treatments** of an experiment are the factor–level combinations utilized.

*The levels of a qualitative variable may bear numerical labels, e.g., the Locations could be numbered 1, 2, 3, and 4. However, such numerical labels are usually codes representing nonnumerical levels.

The objects on which the response variable and factors are observed are the **experimental units**. For example, SAT score, High school GPA, and Gender are all variables that can be observed on the same experimental unit—a high school senior. Or, Total sales, Earnings per share, and Number of salespeople can be measured on a particular firm in a particular year, and the firm–year combination is the experimental unit. As another example, Total income, Number of wage earners, and Location can be observed for a household at a particular point in time, and the household–time combination is the experimental unit. Every experiment, whether observational or designed, has experimental units on which the variables are measured. (Recall from Chapter 1 that the set of *all* experimental units is the *population*.)

Definition 16.5

An **experimental unit** is the object on which the response and factors are observed or measured.

When the specification of the treatments and the method of assigning the experimental units to each treatment is controlled by the analyst, the experiment is said to be **designed**. In contrast, if the analyst is just an observer of the treatments on a sample of experimental units, the experiment is **observational**. For example, if you give one randomly selected group of employees a training program and withhold it from another randomly selected group in order to evaluate the effect of the training on worker productivity, then you are designing an experiment. If, on the other hand, you compare the productivity of employees with college degrees with the productivity of employees without college degrees, the experiment is observational.

Definition 16.6

A **designed experiment** is one for which the analyst controls the specification of the treatments and the method of assigning the experimental units to each treatment. An **observational experiment** is one for which the analyst simply observes the treatments and the response on a sample of experimental units.

The diagram in Figure 16.1 provides an overview of the experimental process and a summary of the terminology we have introduced in this section. Note that the experimental unit is at the core of the process.

FIGURE 16.1 ▶
Experimental process and terminology

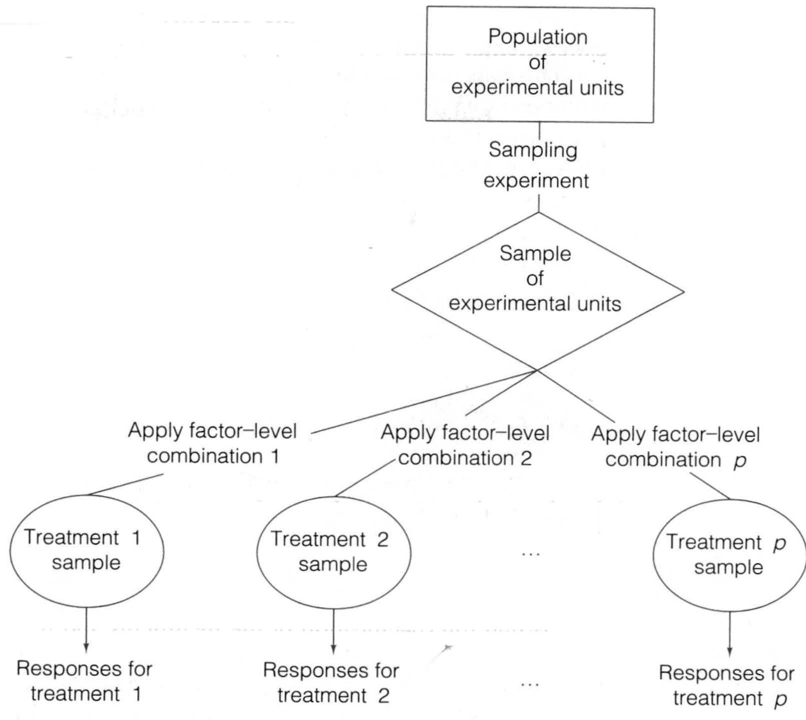

EXAMPLE 16.1

The United States Golf Association (USGA) regularly tests golf equipment to assure that it conforms to USGA standards. Suppose it wishes to compare the mean distance traveled by four brands of golf balls when struck by a driver (the club with least loft, used to maximize distance). The following experiment is conducted. Ten balls of each brand are randomly selected, each is struck by Iron Byron (a golf robot named for the famous golfer, Byron Nelson) using a driver, and the distance traveled is recorded. Identify each of the following elements in this experiment: response, factors, factor types, levels, treatments, and experimental units. Is the experiment designed or observational?

Solution

The response is the variable of interest, Distance traveled. The only factor being investigated is Brand of golf ball, which is nonnumerical and therefore qualitative. The four brands (say A, B, C, and D) represent the levels of this factor. Since only one factor is employed, the treatments are the four levels of this factor—that is, the four brands. The experimental unit is a golf ball; more specifically, it is a golf ball at a particular position in the striking sequence, since the distance traveled can be recorded only when the ball is struck, and we would expect the distance to be different (due to random factors such as wind resistance, landing place, and so forth) if the same ball were struck a second time. Note that 10 experimental units are sampled for each treatment, generating a total of 40 observations.

This experiment, like many real applications in business, is a blend of designed and observational: The analyst cannot control the assignment of the brand to each golf ball (observational), but he or she can control the assignment of each ball to the position in the striking sequence (designed).

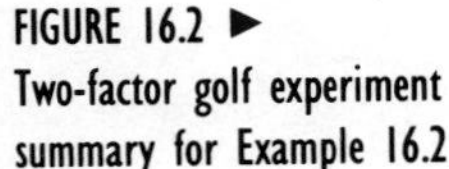

EXAMPLE 16.2 Refer to Example 16.1. Suppose the USGA is also interested in comparing the mean distances the four brands of golf balls travel when struck by a five-iron. Ten balls of each brand are randomly selected, five to be struck by the driver and five by the five-iron. Identify the elements of the experiment, and construct a schematic diagram similar to Figure 16.1 to provide an overview of this experiment.

Solution The response is unchanged: Distance traveled. The experiment now has two factors, Brand of golf ball and Club utilized. There are four levels of Brand and two of Club

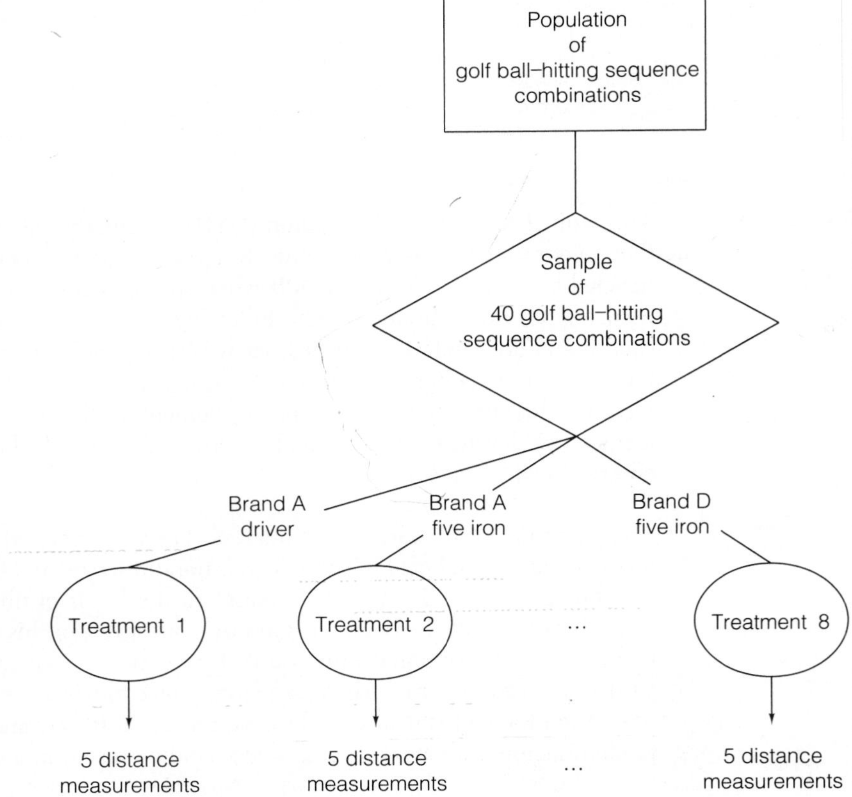

FIGURE 16.2 ► Two-factor golf experiment summary for Example 16.2

(driver and five-iron, or 1 and 5). Treatments are factor–level combinations, so there are $4 \times 2 = 8$ treatments in this experiment: (A, 1), (A, 5), (B, 1), (B, 5), (C, 1), (C, 5), (D, 1), and (D, 5). The experimental units are still the combinations of golf ball and hitting position. Note that five experimental units are sampled per treatment, generating 40 observations. The experiment is summarized in Figure 16.2

Our objective in designing an experiment is usually to maximize the amount of information obtained about the relationship between the treatments and the response. Of course, we are almost always subject to contraints on budget, time, and even the availability of experimental units. Nevertheless, designed experiments are generally preferred to observational experiments. Not only do we have better control of the amount and quality of the information collected, but also observational experiments are subject to biases in the selection of the experimental units representing each treatment.

Exercises 16.1–16.8

Learning the Mechanics

16.1 What are the treatments for a designed experiment that utilizes one qualitative factor with four levels (A, B, C, and D)?

16.2 What are the treatments for a designed experiment with two factors, one qualitative with two levels (A and B) and one quantitative with five levels (50, 60, 70, 80, and 90)?

16.3 What are the experimental units on which each of the following responses are observed?

a. College GPA
b. Statewide unemployment rate in December
c. Gasoline mileage rating for a model of automobile
d. Number of defective sectors on a computer diskette

16.4 What is the difference between an observational and a designed experiment?

Applying the Concepts

16.5 Suppose the mean debt-to-equity ratio is to be compared for four types of companies: insurance, publishing, electric utilities, and banking. If random samples of each type of company are to be taken to make the comparison, identify each of the following elements:

a. Response b. Factor(s) and factor type(s) c. Treatments d. Experimental units

16.6 Brockhaus (1980) compared the risk-taking propensity of three types of managers: entrepreneurs, newly hired managers, and newly promoted managers. Samples of individuals in each group were administered a test that measures one's propensity for taking risks. Identify each of the following elements:

a. Response b. Factor(s) and factor type(s) c. Treatments d. Experimental units

16.7 A quality control supervisor measures the quality of a steel ingot on a scale from 0 to 10. He designs an experiment in which three different temperatures (ranging from 1,100° to 1,200°F) and five different pressures (ranging from 500 to 600 psi) are utilized, with 20 ingots produced at each temperature–pressure combination. Identify the following elements of the experiment:

a. Response **b.** Factor(s) and factor type(s) **c.** Treatments **d.** Experimental units

16.8 For each experiment described, determine whether it is observational or designed, and explain your reasoning.

a. An economist obtains the unemployment rate and gross state product for a sample of states over the past 10 years, with the objective of examining the relationship between the unemployment rate and the gross state product by census region.

b. A manager in a paper production facility installs one of three incentive programs in each of nine plants to determine the effect of each program on productivity.

c. A marketer of microcomputers runs ads in each of four national publications for one quarter and keeps track of the number of sales that are attributable to each publication's ad.

d. An electric utility engages a consultant to monitor the discharge from its smokestack on a monthly basis over a 1-year period in order to relate the level of sulfur dioxide in the discharge to the load on the facility's generators.

e. Intrastate trucking rates are compared before and after governmental deregulation of prices charged, with the comparison also taking into account distance of haul, goods hauled, and the price of diesel fuel.

16.2 The Completely Randomized Design: Single Factor

The simplest experimental design, a **completely randomized design**, consists of the *independent random selection* of experimental units representing each treatment. For example, we could independently select random samples of 20 female and 15 male high school seniors to compare their mean SAT scores. Or, we could independently select random samples of 30 households from each of four census districts in order to compare the mean income per household among the districts. In both examples our objective is to compare treatment means by selecting random, independent samples for each treatment.

Definition 16.7

A **completely randomized design** is one for which independent random samples of experimental units are selected for each treatment.*

*We use *completely randomized design* to refer to both designed and observational experiments. Thus, the only requirement is that the experimental units to which treatments are applied (designed) or on which treatments are observed (observational) be independently selected for each treatment.

The objective of a completely randomized design is usually to compare the treatment means. If we denote the true, or population, means of the p treatments as μ_1, $\mu_2, \ldots, \mu_p$, then we will test the null hypothesis that the treatment means are all identical against the alternative that at least two of them differ:

H_0: $\mu_1 = \mu_2 = \cdots = \mu_p$

H_a: At least two of the p treatment means differ

The μ's might represent the means of *all* female and male high school seniors' SAT scores or the means of *all* households' income in each of four census regions.

In order to test these hypotheses, we will use the means of the independent random samples selected from the treatment populations using the completely randomized design. That is, we compare the p sample means $\bar{y}_1, \bar{y}_2, \ldots, \bar{y}_p$.

For example, suppose you select independent random samples of five female and five male high school seniors and obtain sample mean SAT scores of 550 and 590, respectively. Can we conclude that males score 40 points higher, on average, than females? To answer the question, we must consider the amount of sampling variability among the experimental units (students). If the scores are as depicted in the dot diagram shown in Figure 16.3, then the difference between the means is small relative to the sampling variability of the scores within the treatments, female and male. We would not be inclined to reject the null hypothesis of equal population means in this case.

FIGURE 16.3 ► Dot diagram of SAT scores: Difference between means dominated by sampling variability

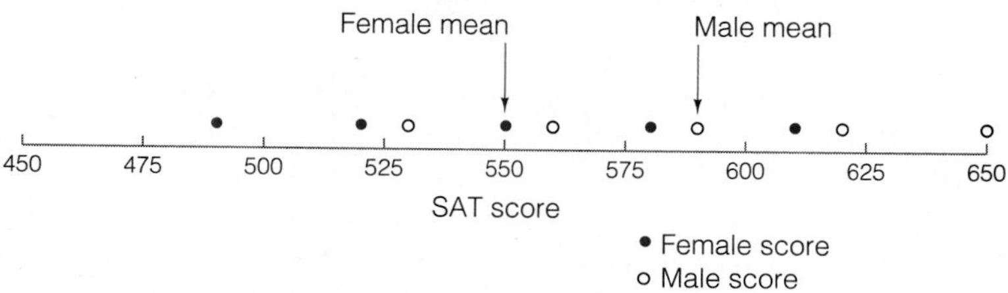

In contrast, if the data are as depicted in the dot diagram of Figure 16.4, then the sampling variability is small relative to the difference between the two means. We would be inclined to favor the alternative hypothesis that the population means differ in this case.

FIGURE 16.4 ► Dot diagram of SAT scores: Difference between means large relative to sampling variability

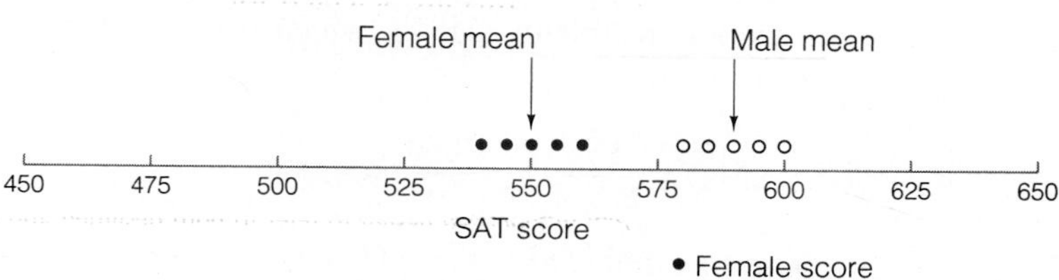

You can see that the key is to compare the difference between the treatment means to the amount of sampling variability. To conduct a formal statistical test of the hypotheses requires numerical measures of the difference between the treatment means and the sampling variability within each treatment. The variation between the treatment means is measured by the **sum of squares for treatment (SST)**, which is calculated by squaring the distance between each treatment mean and the overall mean of *all* sample measurements, multiplying each squared distance by the number of sample measurements for the treatment, and adding the results over all treatments:

$$\text{SST} = \sum_{i=1}^{p} n_i(\bar{y}_i - \bar{y})^2 = 5(550 - 570)^2 + 5(590 - 570)^2 = 4{,}000$$

where we use $\bar{y}$ to represent the overall mean response of all sample measurements—that is, the mean of the combined samples. The symbol n_i is used to denote the sample size for the ith treatment.

Next, we must measure the sampling variability within the treatments. We call this the **sum of squares for error (SSE)** because it measures the variability around the treatment means that is attributed to sampling error. Suppose the 10 measurements in the first dot diagram (Figure 16.3) are 490, 520, 550, 580, and 610 for the five females and 530, 560, 590, 620, and 650 for the five males. Then the value of SSE is computed by summing the squared distance between each response measurement and the corresponding treatment mean and then adding the squared differences over all measurements in the entire sample:

$$\text{SSE} = \sum_{j=1}^{n_1} (y_{1j} - \bar{y}_1)^2 + \sum_{j=1}^{n_2} (y_{2j} - \bar{y}_2)^2 + \cdots + \sum_{j=1}^{n_p} (y_{pj} - \bar{y}_p)^2$$

where the symbol y_{1j} is the jth measurement in sample 1, y_{2j} is the jth measurement in sample 2, and so on. This rather complex-looking formula is not difficult to interpret in practice. For our samples of SAT scores,

$$\begin{aligned}\text{SSE} &= [(490 - 550)^2 + (520 - 550)^2 + (550 - 550)^2 + (580 - 550)^2 + (610 - 550)^2] \\ &\quad + [(530 - 590)^2 + (560 - 590)^2 + (590 - 590)^2 + (620 - 590)^2 + (650 - 590)^2] \\ &= 18{,}000\end{aligned}$$

To make the two measures of variability comparable, we divide each by the degrees of freedom to convert the sum of squares to mean squares. First, the **mean square for treatments (MST)**, which measures the variability among the treatment means, is equal to

$$\text{MST} = \frac{\text{SST}}{p - 1} = \frac{4{,}000}{2 - 1} = 4{,}000$$

where the number of degrees of freedom for the p treatments is $(p - 1)$. Next, the **mean square for error (MSE)**, which measures the sampling variability within the treatments, is

$$\text{MSE} = \frac{\text{SSE}}{n - p} = \frac{18{,}000}{10 - 2} = 2{,}250$$

Finally, we calculate the ratio of MST to MSE, an ***F* statistic**:

$$F = \frac{\text{MST}}{\text{MSE}} = \frac{4{,}000}{2{,}250} = 1.78$$

Values of the F statistic near 1 indicate that the two sources of variation, that between treatment means and that within treatments, are approximately equal. In this case, the difference between the treatment means may well be attributable to sampling error, which provides little support for the alternative hypothesis that the population treatment means differ. Values of F well in excess of 1 indicate that the variation among treatment means well exceeds that within means and therefore support the alternative hypothesis that the population treatment means differ.

When does F exceed 1 by enough to reject the null hypothesis that the means are equal? This depends on the degrees of freedom for treatments and for error, and on the value of α selected for the test. We compare the calculated F value to a tabled F value (Tables VII–X of Appendix B) with $\nu_1 = (p - 1)$ degrees of freedom in the numerator and $\nu_2 = (n - p)$ degrees of freedom in the denominator, and corresponding to a Type I error probability of α. For the SAT example, the F statistic has $\nu_1 = (2 - 1) = 1$ numerator degree of freedom and $\nu_2 = (10 - 2) = 8$ denominator degrees of freedom. Thus, for $\alpha = .05$ we find (Table VIII of Appendix B) that

$$F_{.05} = 5.32 \quad \text{for } \nu_1 = 1 \text{ and } \nu_2 = 8$$

The implication is that MST would have to be 5.32 times greater than MSE before we could conclude at the .05 level of significance that the two population treatment means differ. Since the data yielded $F = 1.78$, our initial impressions for the dot diagram in Figure 16.3 are confirmed: There is insufficient information to conclude that the mean SAT scores differ for the populations of female and male high school seniors. The rejection region and the calculated F value are shown in Figure 16.5.

FIGURE 16.5 ▶
Rejection region and calculated F values for SAT score sample

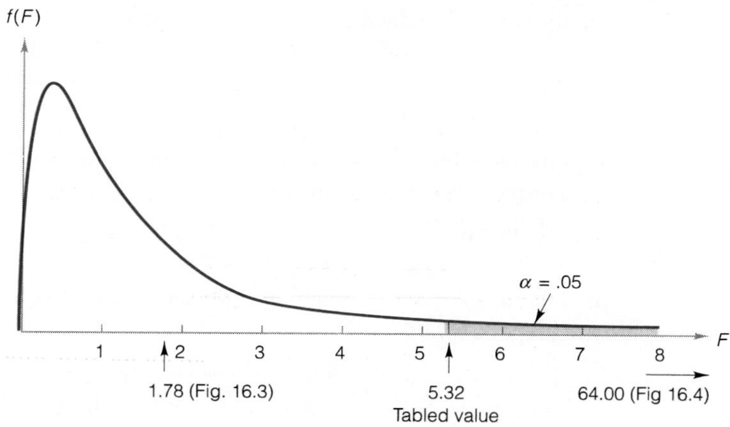

In contrast, consider the dot diagram in Figure 16.4. Since the means are the same as in the first example, 550 and 590, respectively, the variation between the means is the same, MST = 4,000. But the variation within the two treatments appears to be considerably smaller. The observed SAT scores are 540, 545, 550, 555, and 560 for females and 580, 585, 590, 595, and 600 for males. The variation within the treatments is measured by

$$\begin{aligned} \text{SSE} &= [(540 - 550)^2 + (545 - 550)^2 + (550 - 550)^2 + (555 - 550)^2 + (560 - 550)^2] \\ &\quad + [(580 - 590)^2 + (585 - 590)^2 + (590 - 590)^2 + (595 - 590)^2 + (600 - 590)^2] \\ &= 500 \end{aligned}$$

$$\text{MSE} = \frac{\text{SSE}}{n - p} = \frac{500}{8} = 62.5$$

Then the F ratio is

$$F = \frac{\text{MST}}{\text{MSE}} = \frac{4,000}{62.5} = 64.0$$

Again, our visual analysis of the dot diagram is confirmed statistically: $F = 64.0$ well exceeds the tabled F value, 5.32, corresponding to the .05 level of significance (see Figure 16.5). We would therefore reject the null hypothesis at that level and conclude that the SAT mean score of males differs from that of females.

Recall that we performed a hypothesis test for the difference between two means in Section 9.2 using a two-sample t statistic for two independent samples. When two independent samples are being compared, the t and F tests are equivalent. To see this, consider the second sample, depicted in Figure 16.4, and recall the formula

$$t = \frac{\bar{y}_1 - \bar{y}_2}{\sqrt{s_p^2\left(\frac{1}{n_1} + \frac{1}{n_2}\right)}} = \frac{590 - 550}{\sqrt{(62.5)\left(\frac{1}{5} + \frac{1}{5}\right)}} = \frac{40}{5} = 8$$

where we used the fact that $s_p^2 = \text{MSE}$, which you can verify by comparing the formulas. Note that the calculated F for these samples ($F = 64$) equals the square of the calculated t for the same samples ($t = 8$). Likewise, the tabled F value (5.32) equals the square of the tabled t value at the two-sided .05 level of significance ($t_{.025} = 2.306$ with 8 df). Since both the rejection region and the calculated values are related in the same way, the tests are equivalent. Moreover, the assumptions that must be met to assure the validity of the t and F test are the same:

1. The probability distributions of the populations of responses associated with each treatment must all be normal.
2. The probability distributions of the populations of responses associated with each treatment must have equal variances.
3. The samples of experimental units selected for the treatments must be random and independent.

In fact, the only real difference between the tests is that the F test can be used to compare *more than two* treatment means, whereas the t test is applicable to two samples only. The F test is summarized in the box.

Test to Compare p Treatment Means for a Completely Randomized Design

H_0: $\mu_1 = \mu_2 = \cdots = \mu_p$

H_a: At least two treatment means differ

Test statistic: $F = \dfrac{\text{MST}}{\text{MSE}}$

Assumptions:
1. All p population probability distributions are normal.
2. The p population variances are equal.
3. Samples are selected randomly and independently from the respective populations.

Rejection region: $F > F_\alpha$, where F_α is based on $(p - 1)$ numerator degrees of freedom (associated with MST) and $(n - p)$ denominator degrees of freedom (associated with MSE).

Calculation formulas for MST and MSE are given in Appendix D. We rely on some of the many computer programs available to compute the F statistic, concentrating on the interpretation of the results rather than their calculations.

EXAMPLE 16.3

Suppose the United States Golf Association (USGA) wants to compare the mean distances traveled by four brands of golf balls when struck with a driver. A completely randomized design is employed, with Iron Byron, the USGA's robotic golfer, using a driver to hit a random sample of 40 balls (10 balls of each brand) in a random sequence. The distance is recorded for each hit, and the results are shown in Table 16.1, organized by brand.

TABLE 16.1 Results of Completely Randomized Design: Iron Byron Driver

	Brand A	Brand B	Brand C	Brand D
	251.2	263.2	269.7	251.6
	245.1	262.9	263.2	248.6
	248.0	265.0	277.5	249.4
	251.1	254.5	267.4	242.0
	265.5	264.3	270.5	246.5
	250.0	257.0	265.5	251.3
	253.9	262.8	270.7	262.8
	244.6	264.4	272.9	249.0
	254.6	260.6	275.6	247.1
	248.8	255.9	266.5	245.9
Means	251.3	261.1	270.0	249.4

a. Set up the test to compare the mean distances for the four brands. Use $\alpha = .10$.
b. Use the SAS Analysis of Variance program to obtain the test statistic. Interpret the results.

Solution

a. To compare the mean distances of the four brands, we first specify the hypotheses to be tested. Denoting the population mean of the ith brand by μ_i, we test

H_0: $\mu_1 = \mu_2 = \mu_3 = \mu_4$

H_a: The mean distances differ for at least two of the brands

The test statistic compares the variation among the four treatment (brand) means to the sampling variability within each of the treatments.

Test statistic: $F = \dfrac{\text{MST}}{\text{MSE}}$

Rejection region: $F > F_\alpha = F_{.10}$
with $\nu_1 = (p - 1) = 3$ df and $\nu_2 = (n - p) = 36$ df

From Table VII of Appendix B, we find $F_{.10} \approx 2.25$ for 3 and 36 df. Thus, we reject H_0 if $F > 2.25$.

The assumptions necessary to assure the validity of the test are as follows: (1) The probability distributions of the distances for each brand are normal. (2) The variances of the distance probability distributions for the four brands are identical. (3) The samples of 10 golf balls for each brand are selected randomly and independently.

b. The SAS printout for the data in Table 16.1 resulting from this completely randomized design is given in Figure 16.6. Note that the top part of the printout is identical to that in a regression analysis. The Total Sum of Squares is designated the **Corrected Total**, and it is partitioned into the **Model** and **Error** Sums of Squares. The bottom part of the printout further partitions the Model component into the

FIGURE 16.6 ▶ SAS analysis of variance printout for golf ball distance data: Completely randomized design

```
Dependent Variable: DISTANCE
                                   Sum of          Mean
Source                    DF      Squares        Square      F Value      Pr > F

Model                      3   2709.19875     903.06625        35.81      0.0001

Error                     36    907.86100      25.21836

Corrected Total           39   3617.05975

                  R-Square          C.V.      Root MSE             DISTANCE Mean

                  0.749006     1.9469768       5.02179                257.927500

Source                    DF    Type I SS  Mean Square     F Value      Pr > F

BRAND                      3      2709.20       903.07       35.81      0.0001
```

factors that comprise the model. In this single-factor experiment, the Model and Brand sums of squares are the same. The Sum of Squares column is headed **Type I SS**, one of four types of sums of squares that SAS will calculate. The distinction becomes important only in multi-factor experiments with unequal numbers of observations per treatment; we will need to utilize only the Type I sum of squares. See the references at the end of this chapter for a more complete discussion of using SAS to analyze more complex experiments.

The values of the mean squares MST and MSE (shaded on the printout) are 903.07 and 25.22, respectively. The F ratio, 35.81, also shaded on the printout, exceeds the tabled value 2.25. We therefore reject the null hypothesis at the .10 level of significance, concluding that at least two of the brands differ with respect to mean distance traveled when struck by the driver.

The observed significance level of the F test is also given on the printout: .0001. This is the area to the right of the calculated F value and implies that we would reject the null hypothesis that the means are equal at any α level greater than .0001. The result of the test is summarized in Figure 16.7.

FIGURE 16.7 ► ***F* test for completely randomized design: Golf ball experiment**

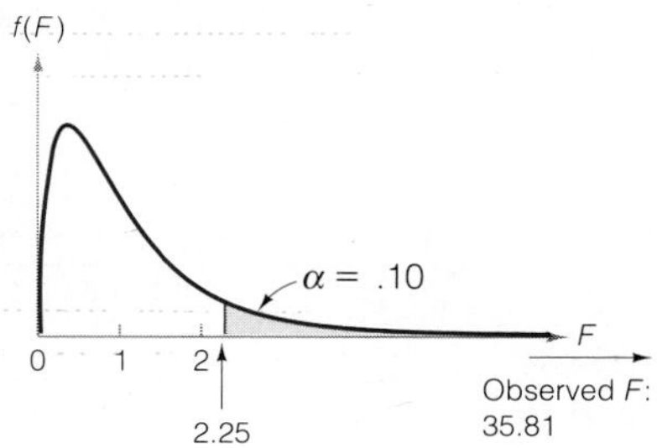

The results of an analysis of variance (ANOVA) can be summarized in a simple tabular format, similar to that obtained from the SAS program in Example 16.3. The general form of the table is shown in Table 16.2, where the symbols df, SS, and MS

TABLE 16.2 ANOVA Summary Table for a Completely Randomized Design

Source	df	SS	MS	F
Treatments	$p - 1$	SST	$\text{MST} = \dfrac{\text{SST}}{p - 1}$	$\dfrac{\text{MST}}{\text{MSE}}$
Error	$n - p$	SSE	$\text{MSE} = \dfrac{\text{SSE}}{n - p}$	
Total	$n - 1$	SS(Total)		

TABLE 16.3 ANOVA Summary Table for Example 16.3

Source	df	SS	MS	F
Brands	3	2,709.20	903.07	35.81
Error	36	907.86	25.22	
Total	39	3,617.06		

stand for degrees of freedom, Sum of Squares, and Mean Square, respectively. Note that the two sources of variation, Treatments and Error, add to the Total Sum of Squares, SS_{yy}, that we first introduced in simple linear regression analysis (Chapter 10). The ANOVA summary table for Example 16.3 is given in Table 16.3, and the partitioning of the Total Sum of Squares into its two components is illustrated in Figure 16.8.

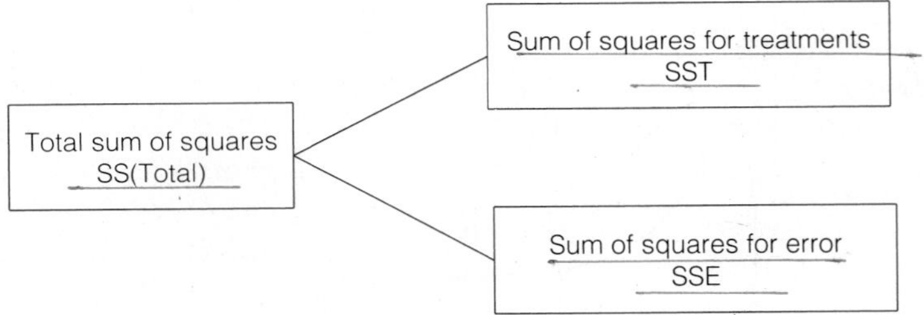

FIGURE 16.8 ► Partitioning of the total sum of squares for the completely randomized design

Suppose the F test results in a rejection of the null hypothesis that the treatment means are equal. Is the analysis complete? Usually, the conclusion that at least two of the treatment means differ leads to another question. Which of the means differ, and by how much? For example, the F test in Example 16.3 leads to the conclusion that at least two of the brands of golf balls are associated with different mean distances traveled when struck with a driver. Now the question is, which of the brands differ? How are the brands ranked with respect to mean distance?

We can place confidence intervals on the difference between the various pairs of treatment means in the experiment. If there are p treatment means, then there are $c = p(p - 1)/2$ pairs of means that can be compared. If we want to have $100(1 - \alpha)\%$ confidence that each of the c confidence intervals contains the true difference it is intended to estimate, then each individual confidence interval will have to be formed using a smaller value of α and, therefore, a higher confidence level, than would a single interval. For example, if we want to compare the $4(3)/2 = 6$ pairs of golf balls Brand means and we want 95% confidence that all six confidence intervals comparing the means contain the true differences between the Brand means, then each individual interval will need to be constructed using a smaller level of α than .05

in order to have 95% confidence that the six intervals collectively include the true differences.*

There are a number of procedures available for making **multiple comparisons** of a set of treatment means which, under various assumptions, assure that the overall confidence level associated with all the comparisons remains at or above the specified $100(1-\alpha)\%$ level. Perhaps the simplest of these to apply is the **Bonferroni procedure**, which calls for specifying the level of significance of each comparison at α/c, where α is the overall level of significance desired, and c is the number of pairs of means to be compared. The result is a set of confidence intervals in which we can be ***at least*** $100(1-\alpha)\%$ confident. The Bonferroni procedure is therefore conservative, since the confidence level is generally greater than that we specify. Of course, this means that the intervals are somewhat wider than they need to be in order to have the specified confidence level. Procedures that are more exact are also more complex than the Bonferroni procedure. The Bonferroni procedure is summarized in the next box. Other multiple comparison procedures can be found in the references at the end of the chapter.

Bonferroni Procedure for Multiple Comparisons: Completely Randomized Design

1. Specify the overall level of significance α or, equivalently, the overall confidence level $100(1-\alpha)\%$.
2. If there are c pairs of treatment means to be compared, specify the level of significance for each comparison at α/c.
3. Calculate the $100(1-\alpha/c)\%$ confidence interval for each of the c pairs of means using the following formula:

$$(\bar{y}_i - \bar{y}_j) \pm t_{\alpha/(2c)}\, s\sqrt{\left(\frac{1}{n_i} + \frac{1}{n_j}\right)}$$

   where $s = \sqrt{\text{MSE}}$ and $t_{\alpha/(2c)}$ is the tabulated value of t (Table VI of Appendix B) that locates area $\alpha/(2c)$ in the upper tail of the t distribution with $(n-p)$ degrees of freedom (the number of degrees of freedom associated with error in the ANOVA).
4. Summarize the results of the multiple comparisons by ranking the treatment means, showing which pairs are significantly different.

Assumptions: Same as for the ANOVA.

---

*The reason each interval must be formed at a higher confidence level than that specified for the collection of intervals can be demonstrated as follows:

$$\begin{aligned} &P\{\text{At least one of } c \text{ intervals fails to contain the true difference}\}\\ &\quad = 1 - P\{\text{All } c \text{ intervals contain the true differences}\} \geq 1 - (1-\alpha)^c \geq \alpha \end{aligned}$$

Thus, to make this probability of at least one failure equal to α, we must specify the individual levels of significance to be less than α.

EXAMPLE 16.4

Refer to Example 16.3, in which we concluded that at least two of the four brands of golf balls are associated with different mean distances traveled when struck with a driver. Use the Bonferroni procedure to compare the $4(3)/2 = 6$ pairs of treatment means. Use an overall confidence level of 90%.

Solution

The level of significance specified for the multiple comparisons is $\alpha = .10$. The Bonferroni procedure therefore specifies the level of significance for each comparison at $\alpha/c = .10/6 = .0167$. The t statistic to be used in the formation of the six confidence intervals splits this area into two tails, with $\alpha/(2c) = .0167/2 = .0083$ in each tail. The degrees of freedom associated with error are $(n - p) = (40 - 4) = 36$, which is more than the maximum (29) contained in the t table (Table VI of Appendix B). We use the standard normal z table (Table IV of Appendix B) to find $z_{.0083} \approx 2.4$. That is, as shown in Figure 16.9, we form six confidence intervals using plus and minus 2.4 standard deviations in order to have 90% confidence in the overall results of all six intervals. Contrast this with the fact that we would be using plus and minus $z = 1.645$ standard deviations if we were forming a single 90% confidence interval.*

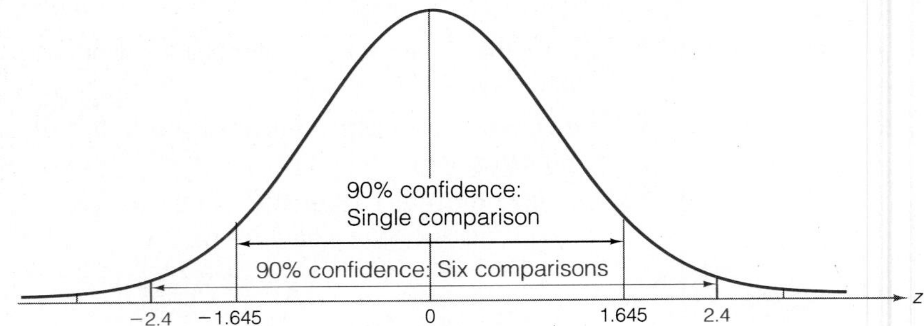

FIGURE 16.9 ► Confidence level for Bonferroni multiple comparisons for Example 16.4

We are now ready to form the confidence intervals. Calculating $s = \sqrt{\text{MSE}} = \sqrt{25.22} = 5.02$ (also given in the SAS printout of Figure 16.6 as **Root MSE**), we begin with a comparison of brands A and B:

$$(\bar{y}_A - \bar{y}_B) \pm t_{.0083}s\sqrt{(1/10 + 1/10)}$$

$$(251.3 - 261.1) \pm (2.4)(5.02)(.447)$$

$$-9.8 \pm 5.4 \quad \text{or} \quad (-15.2, -4.4)$$

*If the t table is used, we may have to round $\alpha/(2c)$ *down* to enable the use of Table VI. In Example 16.4, we would approximate $\alpha/(2c) = .0083 \approx .005$. This would provide slightly wider intervals than necessary but assures an overall confidence level of at least $100(1 - \alpha)\%$. Alternatively, many computer programs provide a more exact tabled t value corresponding to a given level of significance and a specified number of degrees of freedom.

Thus, we conclude that the brand B mean distance exceeds the brand A mean distance by between 4.4 and 15.2 yards. The remainder of the intervals are easier to calculate. The interval half-widths are all the same (5.4), since the sample sizes are the same for each brand. The rest of the comparisons are summarized as follows:

$$
\begin{array}{llll}
A - C: & -18.7 \pm 5.4 & \text{or} & (-24.1, -13.3) \\
A - D: & 1.9 \pm 5.4 & \text{or} & (-3.5, 7.3) \\
B - C: & -8.9 \pm 5.4 & \text{or} & (-14.3, -3.5) \\
B - D: & 11.7 \pm 5.4 & \text{or} & (6.3, 17.1) \\
C - D: & 20.6 \pm 5.4 & \text{or} & (15.2, 26.0)
\end{array}
$$

We are 90% confident that the intervals collectively contain all the differences between the true brand mean distances. Note that intervals that contain 0, such as the (brand A − brand D) interval from −3.5 to 7.3, do not support a conclusion that the true brand mean distances differ. If both endpoints of the interval are positive, as with the (brand B − brand D) interval, the implication is that the first brand (B) mean distance exceeds the second (D). Conversely, if both endpoints of the interval are negative, as with the (brand A − brand C) interval, the implication is that the second brand mean distance (C) exceeds the first (A).

A convenient summary of the results of the Bonferroni multiple comparisons is a listing of the brand means from highest to lowest, with a solid line connecting those that are *not* significantly different. This summary is shown in Figure 16.10. The interpretation is that brand C's mean distance exceeds all others; brand B's mean exceeds that of brands A and D; and the means of brands A and D do not differ significantly. All these inferences are made with 90% confidence, the overall confidence level of the Bonferroni multiple comparisons.

FIGURE 16.10 ▶
Summary of Bonferroni multiple comparisons for Example 16.4

Brand	*Mean*	
C	270.0	
B	261.1	
A	251.3	|
D	249.4	|

Since the samples were selected independently in a completely randomized design, we can also form a confidence interval for an individual treatment mean. We use the one-sample t confidence interval of Section 7.3 to form the interval, again using the standard deviation, $s = \sqrt{\text{MSE}}$, as the measure of sampling variability for the experiment. For example, if we wish to place a 90% confidence interval on the mean distance traveled by brand C (apparently the "longest ball" of those tested), we calculate

$$\bar{y}_C \pm t_{.05,36}\, s\sqrt{1/10}$$
$$270.0 \pm (1.645)(5.02)(.32)$$
$$270.0 \pm 2.6 \quad \text{or} \quad (267.4, 272.6)$$

Thus, we are 90% confident that the true mean distance traveled for brand C is between 267.4 and 272.6 yards, when hit with a driver by Iron Byron.

The procedure for conducting an analysis of variance for a completely randomized design is summarized in the box. Remember that the hallmark of this design is independent random samples of experimental units associated with each treatment. We discuss a design with dependent samples in the next section.

Steps for Conducting an ANOVA for a Completely Randomized Design

1. Be sure the design is truly completely randomized, with independent random samples for each treatment.
2. Create an ANOVA summary table that specifies the variability attributable to Treatments and Error and leads to the calculation of the F statistic to test the null hypothesis that the treatment means are equal in the population. Use either an ANOVA computer program or the calculation formulas in Appendix D to obtain the necessary numerical computations.
3. If the F test leads to the conclusion that the means differ:
   a. Conduct multiple comparisons of as many of the pairs of means as you wish to compare by using the Bonferroni (or some other) multiple comparisons procedure. Use the results to summarize the statistically significant differences among the treatment means.
   b. If desired, form confidence intervals for one or more individual treatment means.
4. If the F test leads to the nonrejection of the null hypothesis that the treatment means differ, several possibilities exist:
   a. The treatment means are equal—that is, the null hypothesis is true.
   b. The treatment means really differ, but other important factors affecting the response are not accounted for by the completely randomized design. These factors inflate the sampling variability, as measured by MSE, resulting in smaller values of the F statistic. Either increase the sample size for each treatment or use a different experimental design (as in Section 16.3) that accounts for the other factors affecting the response.

   Be careful not to automatically reach conclusion **a**, since the possibility of a Type II error must be considered if you accept H_0.

CASE STUDY 16.1 / Measuring Managers' Knowledge of Computers

In the past 10 years computers and computer training have become integral parts of the curricula of secondary schools and universities. As a result, younger business professionals tend to be more comfortable with computers than their more senior counterparts.

"The older the person is, the worse it is," said Arnold S. Kahn of the American Psychological Association. "The older they are and the longer they wait before learning, the more dissatisfied they will become." The computer's unrelenting

TABLE 16.4 Questionnaire Scores of Middle Managers

Manager	Expertise	Score	Manager	Expertise	Score	Manager	Expertise	Score
1	1	82	8	1	88	14	2	133
2	1	114	9	1	93	15	3	128
3	1	90	10	2	130	16	2	130
4	1	80	11	1	80	17	2	104
5	2	128	12	1	105	18	3	151
6	2	90	13	2	110	19	3	140
7	3	156						

Source: Personal communication from Gary W. Dickson, Jan. 1984.

march into offices and factories is often cited as a chief cause of work-related stress. As computers and robots come into the workplace, many workers fear they'll never master the new skills required (Aplin-Brownlee, 1984).

In 1984, Gary W. Dickson, Professor of Information and Decision Sciences at the University of Minnesota, investigated the computer literacy of middle managers with 10 years or more management experience. As part of his study, Dickson designed a questionnaire to measure a manager's technical knowledge of computers. If the questionnaire were properly designed, the scores received by managers could be used as predictors of their knowledge of computers, with higher scores indicating greater knowledge. To check the design of the questionnaire (i.e., its validity), 19 middle managers from the Minneapolis–St. Paul metropolitan area were randomly sampled and asked to complete the questionnaire. Their scores appear in Table 16.4. (The highest possible score on the questionnaire was 169.)

Prior to completing the questionnaire, the managers were asked to describe their knowledge of and experience with computers. This information was used to classify the managers as possessing a high (3), medium (2), or low (1) level of technical computer expertise. These data also appear in Table 16.4.

In order to evaluate the questionnaire's design, Dickson used analysis of variance to compare the mean scores of each of the three groups of managers. If the questionnaire were properly designed, the mean scores should differ. In particular, the mean score of managers with a high level of expertise should be greater than the mean score of managers with a medium level of expertise, etc. Dickson used Minitab to obtain the ANOVA table displayed in Figure 16.11.

```
ANALYSIS OF VARIANCE

DUE TO    DF        SS    MS=SS/DF    F-RATIO

FACTOR     2     7634.       3817.      19.19
ERROR     16     3182.        199.
TOTAL     18    10815.

LEVEL     N     MEAN     ST. DEV.

1         8     91.5       12.3
2         7    117.9       16.6
3         4    143.8       12.4

POOLED ST. DEV.=           14.1
```

FIGURE 16.11 ▲ ANOVA table for managers' scores

Since $F = 19.19$ is greater than $F_{.01} = 6.23$ ($\nu_1 = 2$, $\nu_2 = 16$), Dickson concluded that the mean scores differ for at least two of the three groups of managers. However, this result could be obtained even with a poorly designed questionnaire. For example, a significant F statistic could result even if the mean for the high group were lower than the mean for the low group. Accordingly, Dickson used confidence intervals to examine the differences between individual group means. Using the information on the Minitab printout along with the appropriate t value, he developed the following 95% confidence intervals:*

$$7.15 \le \mu_3 - \mu_2 \le 44.65$$
$$10.92 \le \mu_2 - \mu_1 \le 41.88$$

Since these confidence intervals indicate that $\mu_3 > \mu_2 > \mu_1$, Dickson concluded that the questionnaire could be used as a predictor of managers' technical knowledge of computers.

Exercises 16.9–16.27

Learning the Mechanics

16.9 Use Tables VII, VIII, IX, and X of Appendix B to find each of the following F values:

a. $F_{.05}$, $\nu_1 = 4$, $\nu_2 = 4$ b. $F_{.01}$, $\nu_1 = 4$, $\nu_2 = 4$
c. $F_{.10}$, $\nu_1 = 30$, $\nu_2 = 40$ d. $F_{.025}$, $\nu_1 = 15$, $\nu_2 = 12$

16.10 Find the following probabilities:

a. $P(F \le 2.88)$ for $\nu_1 = 20$, $\nu_2 = 21$ b. $P(F > 3.52)$ for $\nu_1 = 15$, $\nu_2 = 15$
c. $P(F > 2.40)$ for $\nu_1 = 15$, $\nu_2 = 15$ d. $P(F \le 1.69)$ for $\nu_1 = 40$, $\nu_2 = 40$

16.11 In which dot diagram is the difference between the sample means small relative to the variability within the sample observations? Justify your answer.

a.

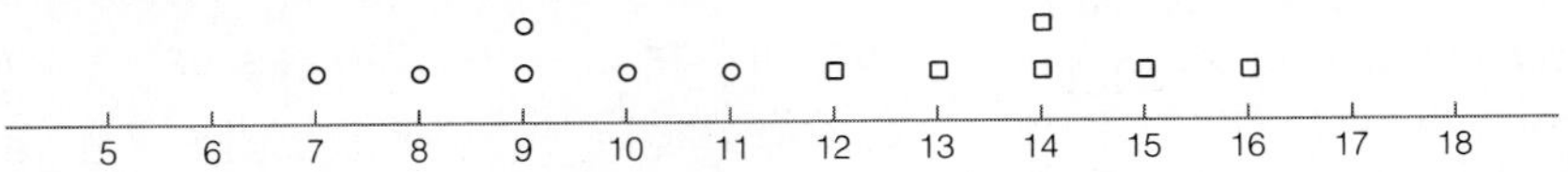

b.

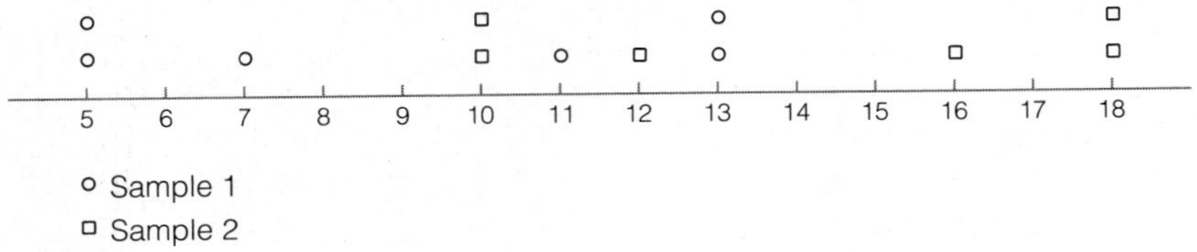

○ Sample 1
□ Sample 2

16.12 Refer to Exercise 16.11. Assume that the two samples represent independent, random samples corresponding to two treatments in a completely randomized design.

a. Calculate the treatment means—i.e., the means of samples 1 and 2—for both dot diagrams.
b. Use the means to calculate the sum of squares for treatments (SST) for each dot diagram.
c. Calculate the sum of squared differences between each sample measurement and the corresponding sample mean, and sum the squared differences over the two samples to obtain the sum of squares for error (SSE) for each dot diagram.

*Note that each confidence interval utilizes $\alpha = .05$. The Bonferroni technique utilizing an overall $\alpha = .05$ would result in slightly wider intervals.

d. Calculate the total sum of squares, SS(Total), for the two dot diagrams by adding the sums of squares for treatment and error. What percentage of SS(Total) is accounted for by the treatments—that is, what percentage of the total sum of squares is the sum of squares for treatment—in each case?
e. Convert the sums of squares for treatment and error to mean squares by dividing each by the appropriate number of degrees of freedom. Calculate the F ratio of the mean square for treatment (MST) to the mean square for error (MSE) for each dot diagram.
f. Use the F ratios to test the null hypothesis that the two samples are drawn from populations with identical means. Use $\alpha = .05$.
g. What assumptions must be made about the probability distributions corresponding to the responses for each treatment in order to assure the validity of the F tests conducted in part **f**?

16.13 Refer to Exercises 16.11 and 16.12. Conduct a two-sample t test (Section 9.2) of the null hypothesis that the two treatment means are equal for each dot diagram. Use $\alpha = .05$ and two-tailed tests. In the course of the test, compare each of the following with the F tests in Exercise 16.12:
a. The pooled variances and the MSEs
b. The t and the F test statistics
c. The tabled values of t and F that determine the rejection regions
d. The conclusions of the t and F tests
e. The assumptions that must be made in order to assure the validity of the t and F tests

16.14 Refer to Exercises 16.11 and 16.12. Complete the following ANOVA table for each of the two dot diagrams:

Source	df	SS	MS	F
Treatments				
Error				
Total				

16.15 Suppose that the total sum of squares for a completely randomized design with $p = 6$ treatments and $n = 36$ total measurements (six per treatment) is equal to 500. In each of the following cases, conduct an F test of the null hypothesis that the six treatment means are the same. Use $\alpha = .10$.
a. The SST is 20% of SS(Total). b. SST is 50% of SS(Total). c. SST is 80% of SS(Total).
d. What happens to the F ratio as the percentage of the total sum of squares attributable to treatments is increased?

16.16 A partially completed ANOVA summary for a completely randomized design is shown in the table.

Source	df	SS	MS	F
Treatments	6	17.5		
Error				
Total	41	46.5		

a. Complete the ANOVA table.
b. How many treatments are involved in the experiment?
c. Do the data provide sufficient evidence to indicate a difference among the population means? Test using $\alpha = .10$.
d. Find the approximate observed significance level for the test in part **c**, and interpret it.
e. Suppose that $\bar{x}_1 = 3.7$ and $\bar{x}_2 = 4.1$. Do the data provide sufficient evidence to indicate a difference between μ_1 and μ_2? Assume that there are six observations for each treatment. Test using $\alpha = .10$.
f. Refer to part **e**. Find a 90% confidence interval for $(\mu_1 - \mu_2)$.
g. Refer to part **e**. Find a 90% confidence interval for μ_1.

16.17 The Minitab printout for an experiment utilizing a completely randomized design is shown here. [*Note*: Minitab uses **FACTOR** instead of "Treatments."]

```
ANALYSIS OF VARIANCE
SOURCE    DF        SS        MS        F
FACTOR     3     57258     19086    14.80
ERROR     34     43836      1289
TOTAL     37    101094
```

a. How many treatments are involved in the experiment? What is the total sample size?
b. Conduct a test of the null hypothesis that the treatment means are equal. Use $\alpha = .01$.
c. What additional information is needed in order to be able to compare specific pairs of treatment means?

16.18 Refer to Exercise 16.17. Suppose the treatment means are 190.8, 260.1, 191.7, and 279.4, with sample sizes 8, 12, 10, and 8, respectively. Use the Bonferroni technique to compare all pairs of treatment means with an overall level of significance $\alpha = .10$.

16.19 The data in the table resulted from an experiment that utilized a completely randomized design.

Treatment 1	*Treatment 2*	*Treatment 3*
3.8	5.4	1.3
1.2	2.0	.7
4.1	4.8	2.2
5.5	3.8	
2.3		

a. Use the appropriate calculation formulas in Appendix D to complete the following ANOVA table:

Source	df	SS	MS	F
Treatments				
Error				
Total				

a. Test the null hypothesis that $\mu_1 = \mu_2 = \mu_3$, where μ_i represents the true mean for treatment i, against the alternative that at least two of the means differ. Use $\alpha = .01$.
c. Use the Bonferroni technique to compare all pairs of means. Use an $\alpha = .05$ overall level of significance.

Applying the Concepts

16.20 An accounting firm that specializes in auditing the financial records of large corporations is interested in evaluating the appropriateness of the fees it charges for it services. As part of its evaluation it wants to compare the costs it incurs in auditing corporations of different sizes. The accounting firm decided to measure the size of its client corporations in terms of their yearly sales. Accordingly, its population of client corporations was divided into three subpopulations:

A: Those with sales over $250 million

B: Those with sales between $100 million and $250 million

C: Those with sales under $100 million

The firm chose random samples of ten corporations from each of the subpopulations and determined the costs (in thousands of dollars) given in the table from its records.

A	B	C
250	100	80
150	150	125
275	75	20
100	200	186
475	55	52
600	80	92
150	110	88
800	160	141
325	132	76
230	233	200

a. Construct a dot diagram (refer to Figures 16.3 and 16.4) for the sample data using different types of dots for each of the three samples. Indicate the location of each of the sample means. Based on the information reflected in your dot diagram, do you believe that a significant difference exists among the subpopulation means? Explain.
b. SAS was used to conduct the analysis of variance calculations, resulting in the printout shown on page 876. Conduct a test to determine whether the three classes of firms have different mean costs incurred in audits. Use $\alpha = .05$.
c. What is the observed significance level for the test in part **b**? Interpret it.
d. Use the Bonferroni technique to compare all pairs of means. Use $\alpha = .05$ as the overall level of significance.
e. What assumptions must be met in order to assure the validity of the inferences you made in parts **b** and **d**?

```
                    General Linear Models Procedure

Dependent Variable: COST
                                      Sum of          Mean
Source                      DF       Squares        Square     F Value     Pr > F

Model                        2    318861.667    159430.833        8.44     0.0014

Error                       27    510163.000     18894.926

Corrected Total             29    829024.667

                  R-Square           C.V.       Root MSE                COST Mean

                  0.384623      72.220043        137.459               190.333333

Source                      DF     Type I SS   Mean Square     F Value     Pr > F

TREATMNT                     2     318861.67     159430.83        8.44     0.0014
```

16.21 R. H. Brockhaus (1980) of St. Louis University conducted a study to determine whether entrepreneurs, newly hired managers, and newly promoted managers differ in their risk-taking propensities. Entrepreneurs were defined as individuals who, within 3 months prior to the study, had ceased working for their employers in order to manage their own business ventures. Thirty-one individuals of each type were selected to participate. Each was asked to complete a questionnaire which required the respondent to choose between a safe alternative and a more attractive but risky one. Test scores were designed to measure risk-taking propensity. (Lower scores are associated with greater conservatism in risk-taking situations.) Summary statistics for the test scores of the three groups are given in the table.

Group	*Sample Size*	*Sample Mean*	*Standard Deviation*	*Group Totals*
Entrepreneurs	31	71.00	11.94	2,201
Newly hired managers	31	72.52	12.19	2,248
Promoted managers	31	66.97	10.84	2,076
	93			6,525

a. The following partial ANOVA table is calculated from the data. Complete the table.

Source	df	SS	MS	F
Treatments		509.87		
Error		12,259.96		
Total				

b. Do the data provide sufficient evidence to indicate differences in the mean risk-taking propensities among the three groups? Test using $\alpha = .05$.
c. What assumptions must be satisfied in order for the test of part **b** to be valid?
d. Would you advise conducting tests to compare the individual pairs of means? Explain.
e. Would you classify this experiment as observational or designed? Explain.

16.22 A study of the application of *management by objectives* (MBO), a method of performance appraisal (Shetty and Carlisle, 1974), dealt with the reactions of a university faculty to an MBO program. One hundred nine faculty members were asked to comment on whether they thought the MBO program was successful in improving their performance within their respective departments and the university. Each response was assigned a score from 1 (significant improvement) to 5 (significant decrease). The table shows the sample sizes, sample totals, mean scores, and sum of squares of deviations *within* each sample for samples of scores corresponding to the four academic ranks. Assume that the four samples in the table can be viewed as independent random samples of scores selected from among the four academic ranks.

	Instructor	*Assistant Professor*	*Associate Professor*	*Professor*
Sample size	15	41	29	24
Sample total	42.960	145.222	92.249	73.224
Sample mean	2.864	3.542	3.181	3.051
Within-sample sum of squared deviations	2.0859	14.0186	7.9247	5.6812

a. Given that SST = 6.816 and SSE = 29.710, construct an ANOVA table for this experiment.
b. Do the data provide sufficient evidence to conclude there is a difference in mean scores among the four academic ranks? Test using $\alpha = .05$.
c. Use the Bonferroni technique for comparing the pairs of means, using an overall significance level of .05.

16.23 Most new products are test marketed in several locations, frequently using different advertising techniques (Klompmaker, Hughes, and Haley, 1976). Suppose the table represents the number of sales for a new product at each of three locations during each of the last 4 months.

Location I	*Location II*	*Location III*
456	441	501
421	419	467
397	415	520
419	420	493

a. Treat this as a completely randomized design, and test to determine whether there is a difference among the mean sales at the three locations. Use Appendix D or a computer program to perform the calculations, and test at $\alpha = .05$.
b. Estimate the difference in the mean sales between locations I and III using a 90% confidence interval.

16.24 A company that employs a large sales staff is interested in learning which salespeople sell the most: those strictly on commission, those with a fixed salary, or those with a reduced fixed salary plus a commission. The previous month's records for a sample of salespeople are inspected and the amount of sales (in dollars) is recorded for each, as shown in the table on page 878.

Commissioned	*Fixed Salary*	*Commission Plus Salary*
\$425	\$420	\$430
507	448	492
450	437	470
483	432	501
466	444	
492		

a. Construct a dot diagram for these data. Indicate the location of each sample mean.
b. The Minitab ANOVA printout for these data is shown here. Is there sufficient evidence to indicate that the mean sales differ among the three types of compensation?

```
ANALYSIS OF VARIANCE
SOURCE    DF       SS      MS       F
FACTOR     2     4195    2098    3.17
ERROR     12     7945     662
TOTAL     14    12140
```

c. Use a 90% confidence interval to estimate the mean sales for salespeople who receive a commission plus salary.

16.25 How does *flextime*, which allows workers to set their individual work schedules, affect worker job satisfaction? Researchers recently conducted a study to compare a measure of job satisfaction for workers using three types of work scheduling: flextime, staggered starting hours, and fixed hours. Workers in each group worked according to their specified work scheduling system for 4 months. Although each worker filled out job satisfaction questionnaires both before and after the 4-month test period, we will examine only the post-test-period scores. The sample sizes, means, and standard deviations of the scores for the three groups are shown in the table.

	Flextime	*Staggered*	*Fixed*
Sample size	27	59	24
Mean	35.22	31.05	28.71
Standard deviation	10.22	7.22	9.28

a. Assume that the data were collected according to a completely randomized design. Identify the response, the factor, the factor type, the treatments, and the experimental units.
b. Use the sample means to calculate the sum of squares for treatments, SST.
c. Use the standard deviations to calculate the sum of squares for error, SSE. Remember that SSE is the sum of squared differences between each measurement in the experiment and the corresponding treatment mean.
d. Construct an ANOVA table for this experiment.
e. Do the data provide sufficient evidence that the three groups differ with respect to their mean job satisfaction? Test using $\alpha = .05$.
f. Use the Bonferroni technique to compare the pairs of means corresponding to the three treatments. Use $\alpha = .05$ as the overall significance level for the comparisons.

16.26 In Exercise 9.28 we compared the mean bond price changes over a 12-month period for two underwriters. Similar data providing a comparison of five underwriting firms (Logue and Rogalski, 1979) are shown in the table. Suppose the data represent independent random samples from the five populations.

	Firm 1	*Firm 2*	*Firm 3*	*Firm 4*	*Firm 5*
Sample size	27	20	23	11	15
Sample mean	−.0491	−.0479	−.0307	−.0438	−.0051
Sample variance	.009800	.006459	.002465	.001462	.002834

a. Given that SST = .023015 and SSE = .486047, construct an ANOVA table for this experiment.
b. Do the data provide sufficient evidence to indicate differences among the mean bond price changes over the 12-month period for the five underwriters? Use $\alpha = .05$.
c. Do the results of the test justify a comparison of the individual pairs of means? Explain.

16.27 Auditors may be called upon to perform compilations, reviews, or audits for their nonpublic clients. To do compilations and reviews of financial statements requires auditors an average of 25% and 50% fewer hours, respectively, than to perform an annual audit. Both auditors and clients fear that users of compilations and reviews might not recognize the limited nature of these reports and might assume that the auditor accepts the same degree of responsibility for them as for audited annual financial statements. To investigate these fears, Johnson, Pany, and White (1983) conducted an experiment designed to measure bankers' reactions to these three alternative forms of auditors' reports. Ninety-eight loan officers responded to a questionnaire in which they were asked to review a financial statement and background information for a commercial loan applicant. Thirty-one of the officers received a compilation, 25 received a review, 27 an audit, and 15 received a financial statement with no auditor association. One of the questions dealt with the loan officers' level of confidence in the financial statement's conformance to generally accepted accounting principles. They were asked to indicate their level of confidence on a scale from 0 to 10 (no confidence to extreme confidence). Johnson, Pany, and White hypothesized that the mean level of confidence would not be the same for all four types of financial statements. Further, they hypothesized that the mean level of confidence associated with the audit would be greater than with the next highest form of auditor association, the review. Some of the data obtained from their experiment are summarized in the tables.

Source	df	SS
Type of report	3	273
Error	94	494
Total	97	767

	Treatment Means
No auditor association	3.9
Compilation	5.5
Review	6.1
Audit	8.8

a. Do the data provide sufficient evidence to conclude that the mean level of confidence differs among the four forms of auditor association? Test using $\alpha = .05$.
b. Report the approximate p-value of your test.
c. Do the data provide sufficient evidence to conclude that the mean level of confidence associated with the audit is significantly higher than the mean for the review? Test using $\alpha = .05$.
d. Use the Bonferroni technique to compare the pairs of means using an overall significance level of $\alpha = .05$.
e. Relate your findings for parts **a**, **c**, and **d** to the expressed fear that users of financial reports might assume auditors were accepting the same degree of responsibility for each form of financial report.

16.3 Factorial Experiments

All the experiments discussed in Sections 16.2 were **single-factor experiments**. The treatments were levels of a single factor, with the sampling of experimental units performed using a completely randomized design. However, most responses are affected by more than one factor, and therefore we will often wish to design experiments involving more than one factor.

Consider an experiment in which the effects of two factors on the response are being investigated. Assume that factor A is to be investigated at a levels, and factor B at b levels. Recalling that treatments are factor–level combinations, you can see that the experiment has, potentially, ab treatments that could be included in the experiment. A **complete factorial experiment** is one in which all possible ab treatments are utilized.

Definition 16.8

A **complete factorial experiment** is one for which every factor–level combination is utilized. That is, the number of treatments in the experiment equals the total number of factor–level combinations.

For example, suppose the USGA wants to determine not only the relationship between distance and brand of golf ball, but also between distance and the club used to hit the ball. If it decides to use four brands and two clubs (say, driver and five-iron) in the experiment, then a complete factorial would call for utilizing all $4 \times 2 = 8$ brand–club combinations. This experiment is referred to more specifically as a **complete 4 × 2 factorial**. A layout for a two-factor factorial experiment (we are henceforth referring to a *complete factorial* when we use the term *factorial*) is given in Table 16.5. The factorial experiment is also referred to as a **two-way classification**, because it can be arranged in the row–column format exhibited in Table 16.5.

In order to complete the specification of the experimental design, the treatments must be assigned to experimental units. If the assignment of the ab treatments in the factorial experiment is random and independent, the design is completely randomized. For example, if the machine Iron Byron is used to hit 80 golf balls, 10 for each of the eight brand–club combinations, in a random sequence, the design would be completely randomized. In the remainder of this section, we confine our attention to factorial experiments utilizing completely randomized designs.

If we utilize a completely randomized design to conduct a factorial experiment with ab treatments, we can proceed with the analysis in exactly the same way we did in Section 16.2. That is, we calculate (or let the computer calculate) the measure of treatment mean variability (MST) and the measure of sampling variability (MSE) and use the F ratio of these two quantities to test the null hypothesis that the treatment means are equal. However, if this hypothesis is rejected, so that we conclude some differences exist among the treatment means, important questions remain. Are both

TABLE 16.5 Schematic Layout of Two-Factor Factorial Experiment

		Factor B at b Levels				
	Level	**1**	**2**	**3**	**. . .**	**b**
Factor A at a Levels	**1**	Trt. 1	Trt. 2	Trt. 3	. . .	Trt. b
	2	Trt. $b+1$	Trt. $b+2$	Trt. $b+3$	. . .	Trt. $2b$
	3	Trt. $2b+1$	Trt. $2b+2$	Trt. $2b+3$	. . .	Trt. $3b$
	⋮	⋮	⋮	⋮	. . .	⋮
	a	Trt. $(a-1)b+1$	Trt. $(a-1)b+2$	Trt. $(a-1)b+3$	. . .	Trt. ab

factors affecting the response, or only one? If both, do they affect the response independently, or do they interact to affect the response?

For example, suppose the distance data indicate that at least two of the eight treatment (brand–club combinations) means differ in the golf experiment. Does the Brand of ball (factor A) or the Club utilized (factor B) affect mean distance, or do both affect it? Several possibilities are shown in Figure 16.12 on page 882. In Figure 16.12(a), the Brand means (only three are shown for the purpose of illustration) are the same, but the distances differ for the two levels of factor B (club). Thus, there is no effect of Brand on distance, but a Club main effect is present. In Figure 16.12(b), the Brand means differ, but the Club means are identical for each brand. Here a Brand main effect is present, but no effect of Club is present.

Figures 16.12(c) and (d) illustrate cases in which both factors affect the response. In Figure 16.12(c) the mean distance between clubs does not change for the three brands, so the effect of Brand on distance is independent of Club. That is, the two factors Brand and Club *do not interact*. In contrast, Figure 16.12(d) shows that the difference between mean distances between clubs varies with brand. Thus, the effect of Brand on distance depends on Club, and therefore the two factors *do interact*.

In order to determine the nature of the treatment effect, if any, on the response in a factorial experiment, we need to break the treatment variability into three components: interaction between factors A and B, main effect of factor A, and main effect of factor B. The interaction component is used to test whether the factors combine to affect the response, while the main effect components are used to determine whether the factors separately affect the response.

The partitioning of the total sum of squares into its various components is illustrated in Figure 16.13 (page 882). Notice that at stage 1 the components are identical to those in the one-factor completely randomized designs of Section 16.2; the sums of

FIGURE 16.12 ▶
Illustration of possible treatment effects: Factorial experiment

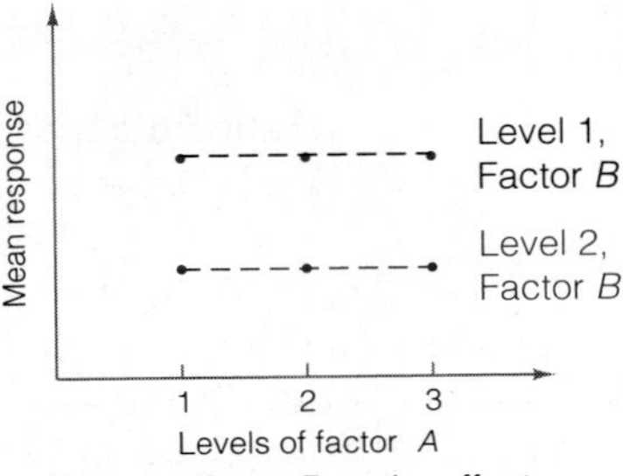

a. No A effect; B main effect

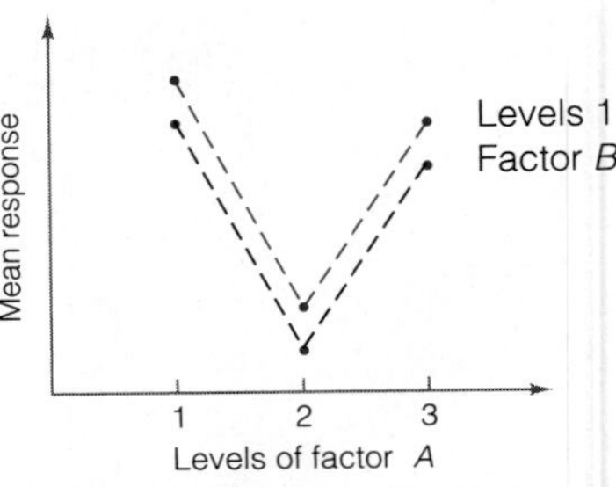

b. A main effect; no B effect

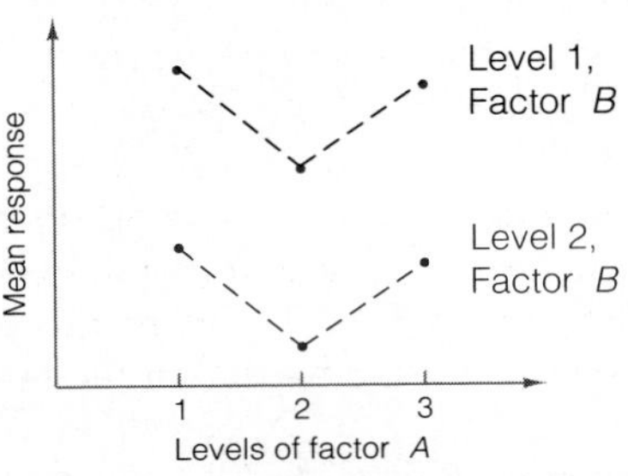

c. A and B main effects; no interaction

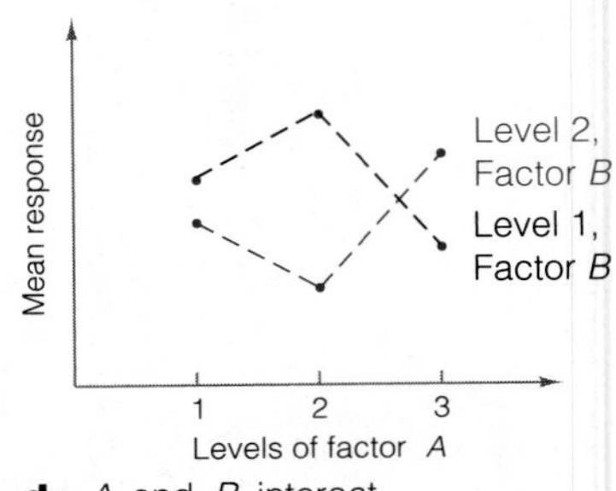

d. A and B interact

FIGURE 16.13 ▶
Partitioning the total sum of squares for a two-factor factorial

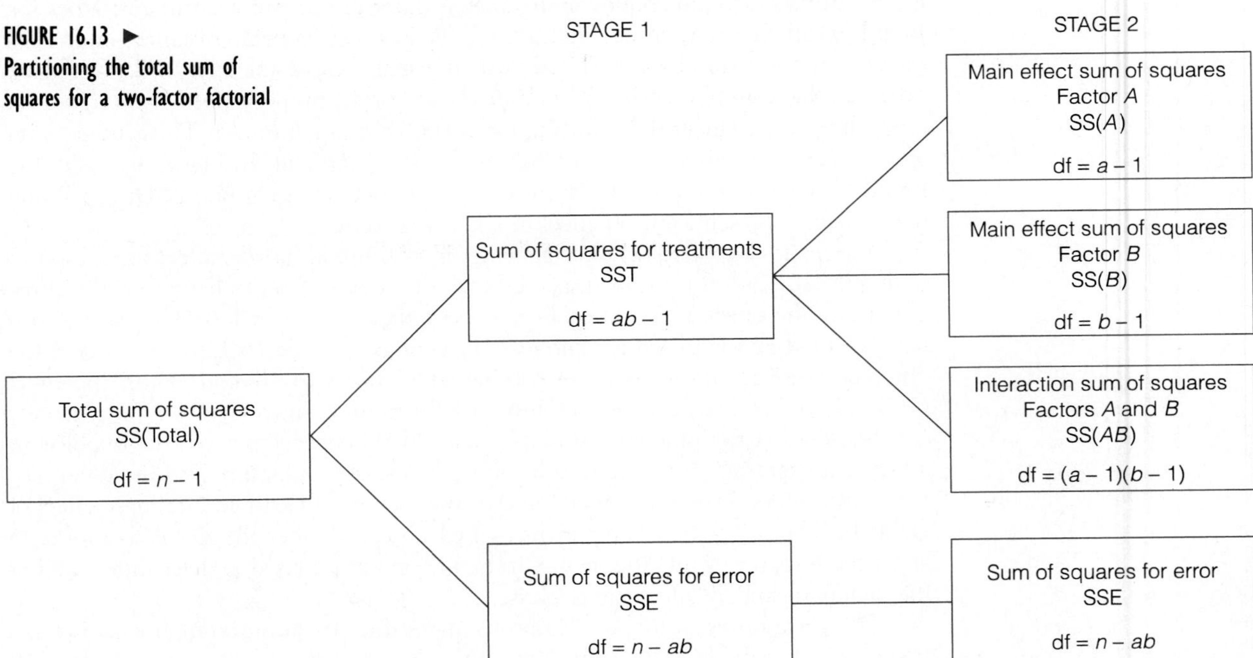

squares for treatment and error add to the total sum of squares. The degrees of freedom for treatments is equal to $(ab - 1)$, 1 less than the number of treatments. The degrees of freedom for error is equal to $(n - ab)$, the total sample size minus the number of treatments. Only at stage 2 of the partitioning is the factorial experiment differentiated from those previously discussed. Here we divide the treatment sum of squares into its three components: interaction and the two main effects. These components can then be utilized to test the nature of the differences, if any, among the treatment means.

There are a number of ways to proceed in the testing and estimation of factors in a factorial experiment. We present one approach in the box on page 884.

We assume the completely randomized design is **balanced**, meaning that the same number of observations are made for each treatment. That is, we assume that r experimental units are randomly and independently selected for each treatment. The numerical value of r must exceed 1 in order to have any degrees of freedom with which to measure the sampling variability. [Note that if $r = 1$, then $n = ab$, and the degrees of freedom associated with Error (Figure 16.13) is $\text{df} = n - ab = 0$.] The value of r is often referred to as the number of **replicates** of the factorial experiment, since we assume that all ab treatments are repeated, or replicated, r times. Whatever approach is adopted in the analysis of a factorial experiment, several tests of hypotheses are usually conducted. The tests are summarized in the box on page 885.

EXAMPLE 16.5

Suppose the USGA tests four different brands (A, B, C, D) of golf ball and two different clubs (driver, five-iron) in a completely randomized design. Each of the eight brand–club combinations (treatments) is randomly and independently assigned to four experimental units, each experimental unit consisting of a specific position in the sequence of hits by Iron Byron. The distance response is recorded for each of the 32 hits, and the results are shown in Table 16.6.

TABLE 16.6 Distance Data for 4 × 2 Factorial Golf Experiment

		Brand			
		A	B	C	D
Club	Driver	226.4 232.6 234.0 220.7	238.4 231.7 227.7 237.2	240.5 246.9 240.3 244.7	219.8 228.7 232.9 237.6
	Five-Iron	163.8 179.4 168.6 173.4	184.4 180.6 179.5 186.2	179.0 168.0 165.2 156.5	157.8 161.8 162.1 160.3

Procedure for Analysis of Two-Factor Factorial Experiment

1. Partition the total sum of squares into the treatment and error components (stage 1 of Figure 16.13). Use either a computer program or the calculation formulas in Appendix D to accomplish the partitioning.
2. Use the F ratio of mean square for treatments to mean square for error to test the null hypothesis that the treatment means are equal.*
   a. If the test results in nonrejection of the null hypothesis, consider refining the experiment by increasing the number of replications or introducing other factors. Also consider the possibility that the response is unrelated to the two factors.
   b. If the test results in rejection of the null hypothesis, proceed to step 3.
3. Partition the treatment sum of squares into the main effect and interaction sum of squares (stage 2 of Figure 16.13). Use either a computer program or the calculation formulas in Appendix D to accomplish the partitioning.
4. Test the null hypothesis that factors A and B do not interact to affect the response by computing the F ratio of the mean square for interaction to the mean square for error.
   a. If the test results in nonrejection of the null hypothesis, proceed to step 5.
   b. If the test results in rejection of the null hypothesis, conclude that the two factors interact to affect the mean response. Proceed to step 6a.
5. Conduct tests of two null hypotheses that the mean response is the same at each level of factor A and factor B. Compute two F ratios by comparing the mean square for each factor main effect to the mean square for error.
   a. If one or both tests result in rejection of the null hypothesis, conclude that the factor affects the mean response. Proceed to step 6b.
   b. If both tests result in nonrejection, an apparent contradiction has occurred. Although the treatment means apparently differ (step 2 test), the interaction (step 4) and main effect (step 5) tests have not supported that result. Further experimentation is advised.
6. a. If the test for interaction (step 4) is significant, use the Bonferroni multiple comparisons procedure to compare any or all pairs of the treatment means.
   b. If the test for one or both main effects (step 5) is significant, use the Bonferroni technique to compare the pairs of means corresponding to the levels of the significant factor(s).

*Some analysts prefer to proceed directly to test the interaction and main effect components, skipping the test of treatment means. We begin with this test to be consistent with our approach to regression analyses (global F test) and one-factor ANOVAs.

Tests Conducted in Analyses of Factorial Experiments: Completely Randomized Design, *r* Replicates per Treatment

Test for Treatment Means

H_0: No difference among the *ab* treatment means

H_a: At least two treatment means differ

Test statistic: $F = \dfrac{\text{MST}}{\text{MSE}}$

Rejection region: $F \geq F_\alpha$, based on $(ab - 1)$ numerator and $(n - ab)$ denominator degrees of freedom [*Note:* $n = abr$]

Test for Factor Interaction

H_0: Factors A and B do not interact to affect the response mean

H_a: Factors A and B do interact to affect the response mean

Test statistic: $F = \dfrac{\text{MS}(AB)}{\text{MSE}}$

Rejection region: $F \geq F_\alpha$, based on $(a - 1)(b - 1)$ numerator and $(n - ab)$ denominator degrees of freedom

Test for Main Effect of Factor *A*

H_0: No difference among the *a* mean levels of factor A

H_a: At least two factor A mean levels differ

Test statistic: $F = \dfrac{\text{MS(A)}}{\text{MSE}}$

Rejection region: $F \geq F_\alpha$, based on $(a - 1)$ numerator and $(n - ab)$ denominator degrees of freedom

Test for Main Effect of Factor *B*

H_0: No difference among the *b* mean levels of factor *B*

H_a: At least two factor *B* mean levels differ

Test statistic: $F = \dfrac{\text{MS}(B)}{\text{MSE}}$

Rejection region: $F \geq F_\alpha$, based on $(b - 1)$ numerator and $(n - ab)$ denominator degrees of freedom

Assumptions for All Tests

1. The response distribution for each factor-level combination (treatment) is normal.
2. The response variance is constant for all treatments.
3. Random and independent samples of experimental units are associated with each treatment.

a. Use a computer program to partition the total sum of squares, SS_{yy}, into the components necessary to analyze this 4×2 factorial experiment.

b. Follow the steps for analyzing a two-factor factorial experiment, and interpret the results of your analysis. Use $\alpha = .10$ for the tests you conduct.

Solution

a. The SAS printout that partitions the total sum of squares for this factorial experiment is given in Figure 16.14. As described previously, the partitioning takes place in two stages. First, the total sum of squares is partitioned into the treatment and error sums of squares at the top of the printout. Note that SST is 33,659.8 with 7 df, and SSE is 822.2 with 24 df, adding to 34,482.0 and 31 df. In the second stage of partitioning, the treatment sum of squares is further divided into the main effect and interaction sums of squares. At the bottom of the printout we see that SS(Club) is 32,093.1 with 1 df, SS(Brand) is 800.7 with 3 df, and SS(Club × Brand) is 766.0 with 3 df, adding to 33,659.8 and 7 df.

FIGURE 16.14 ▶ **SAS printout for factorial golf experiment**

Dependent Variable: DISTANCE

Source	DF	Sum of Squares	Mean Square	F Value	Pr > F
Model	7	33659.8087	4808.5441	140.35	0.0001
Error	24	822.2400	34.2600		
Corrected Total	31	34482.0487			

R-Square	C.V.	Root MSE	DISTANCE Mean
0.976155	2.8964608	5.85320	202.081250

Source	DF	Type I SS	Mean Square	F Value	Pr > F
CLUB	1	32093.11	32093.11	936.75	0.0001
BRAND	3	800.74	266.91	7.79	0.0008
CLUB*BRAND	3	765.96	255.32	7.45	0.0011

b. Once partitioning is accomplished, our first test is:

H_0: The eight treatment means are equal

H_a: At least two of the eight means differ

Test statistic: $F = \dfrac{\text{MST}}{\text{MSE}} = 140.35$ (top of printout)

Rejection region: $F > F_{.10} = 1.98$, with $(ab - 1) = 7$ numerator df, and $(n - ab) = (32 - 8) = 24$ denominator df.

We reject this null hypothesis and conclude that at least two of the brand–club combinations differ in mean distance.

After accepting the hypothesis that the treatment means differ, and therefore that the factors Brand and/or Club somehow affect the mean distance, we want to determine how the factors affect the mean response. We begin with a test of interaction between Brand and Club:

H_0: The factors Brand and Club do not interact to affect the mean response

H_a: Brand and Club interact to affect mean response

Test statistic: $F = \dfrac{\text{MS}(AB)}{\text{MSE}} = \dfrac{\text{MS(Brand} \times \text{Club)}}{\text{MSE}}$

$$= \frac{255.32}{34.26} = 7.45 \quad \text{(bottom of printout)}$$

Rejection region: $F > F_{.10} = 2.33$, with $(a - 1)(b - 1) = 3$ numerator df and 24 denominator df.

Since the test statistic falls in the rejection region, we conclude that the factors Brand and Club interact to affect mean distance.

Because the factors interact, we need not test the main effects for Brand and Club. Instead, we compare the treatment means in an attempt to learn the nature of the interaction. Rather than compare all $8(7)/2 = 28$ pairs of treatment means, we test for differences only between pairs of brands within each club. That differences exist between clubs can be assumed. Therefore, only $4(3)/2 = 6$ pairs of means need to be compared for each club, or a total of 12 comparisons for the two clubs. If the Bonferroni methodology is used with an overall $\alpha = .10$, each of the comparisons will use $\alpha = .10/12 = .0083$. Then the appropriate t value for the two-sided confidence intervals is based on $\alpha/2 = .0083/2 = .0042 \approx .005$. From Table VI of Appendix B, with 24 degrees of freedom, we find $t_{.005} = 2.797$.

Rather than producing confidence intervals for all 12 comparisons, we note that the half-width of each interval will be constant:

$$\text{Half-width} = t_{.005}\, s\sqrt{(1/4 + 1/4)}$$
$$= (2.797)(5.85)(.707) = 11.6$$

where we find $s = 5.85$ labeled **Root MSE** on the printout, and we have $n_i = 4$ observations for each club–brand combination. We can declare all pairs of treatment means that differ by more than 11.6 yards statistically significantly different at the $\alpha = .10$ level of significance. The means are listed in descending order in Figure 16.15, and those not significantly different are connected by a vertical line.

As shown in Figure 16.15, the picture is unclear with respect to Brand means. For the driver, the brand C mean significantly exceeds all except brand B. However, when hit with a five-iron, brand B's mean distance exceeds all except brand A's. Note the nontransitive nature of the multiple comparisons. For example, for the five-iron the brand B mean can be "the same" as the brand A mean, and the brand A mean "the same" as the brand C mean, and yet the brand B mean can

FIGURE 16.15 ▶
Comparison of treatment means for factorial golf experiment

Club	*Brand*	*Mean*
Driver	C	243.10
	B	233.72
	D	229.75
	A	228.42
Five-iron	B	182.68
	A	171.30
	C	167.18
	D	160.50

significantly exceed the brand C mean. The reason lies in the definition of "the same": we must be careful not to conclude that two means are equal simply because they are connected by a vertical line. The line indicates only that *the connected means are not significantly different*. You should conclude (at the overall α level of significance) only that means *not* connected are different, while withholding judgment on those that are connected. The picture of which means differ and by how much will become clearer as we increase the number of replicates of the factorial experiment.

The Club × Brand interaction can be seen in the plot of means in Figure 16.16. Note that the brand C mean drops considerably more than the others when going from driver to five-iron, whereas brand A gains relative to the others. Only brands B and D seem relatively consistent in their positions relative to the others, with brand B near the top and brand D near the bottom for both clubs.

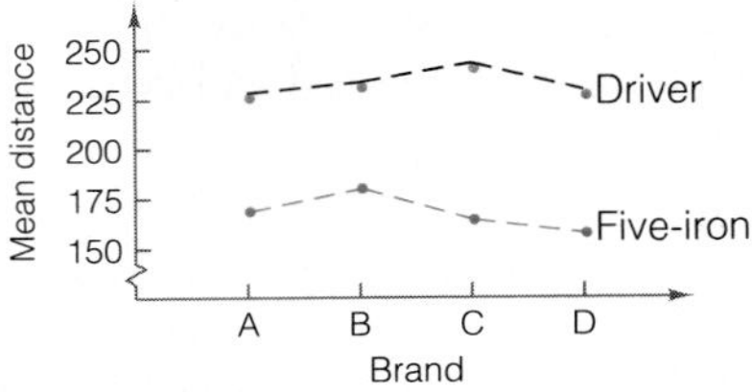

FIGURE 16.16 ▶
Sample mean plot for golf factorial experiment

EXAMPLE 16.6

Refer to Example 16.5. Suppose the same factorial experiment is performed on four other brands (E, F, G, and H), and the results are as shown in Table 16.7. Repeat the factorial analysis and interpret the results.

Solution

The printout for the second factorial experiment is shown in Figure 16.17. The F ratio for Treatments is $F = 290.1$, which exceeds the tabled value of $F_{.10} = 1.98$ for 7 numerator and 24 denominator degrees of freedom. (Note that the same rejection regions will apply in this example as in Example 16.5 since the factors, treatments, and

TABLE 16.7 Distance Data for Second Factorial Golf Experiment

		Brand			
		E	F	G	H
Club	Driver	238.6 241.9 236.6 244.9	261.4 261.3 254.0 259.9	264.7 262.9 253.5 255.6	235.4 239.8 236.2 237.5
	Five-Iron	165.2 156.9 172.2 163.2	179.2 171.0 178.0 182.7	189.0 191.2 191.3 180.5	171.4 159.3 156.6 157.4

FIGURE 16.17 ► SAS printout for second factorial golf experiment

```
Dependent Variable: DISTANCE
                                 Sum of          Mean
Source                   DF      Squares        Square     F Value     Pr > F

Model                     7   49959.3747     7137.0535      290.12     0.0001

Error                    24     590.4075       24.6003

Corrected Total          31   50549.7822

              R-Square          C.V.      Root MSE           DISTANCE Mean

              0.988320     2.3515897       4.95987              210.915625

Source                   DF   Type I SS  Mean Square     F Value     Pr > F

CLUB                      1    46443.90     46443.90     1887.94     0.0001
BRAND                     3     3410.32      1136.77       46.21     0.0001
CLUB*BRAND                3      105.16        35.05        1.42     0.2600
```

replicates are the same.) We conclude that at least two of the brand–club combinations are associated with different mean distances.

We next test for interaction between Brand and Club:

$$F = \frac{\text{MS(Brand} \times \text{Club)}}{\text{MSE}} = 1.42$$

Since this F ratio does not exceed the tabled value of $F_{.10} = 2.33$ with 3 and 24 df, we cannot conclude at the .10 level of significance that the factors interact. In fact, note that the observed significance level (on the SAS printout) for the test of interaction is .26. Thus, at any level of significance lower than $\alpha = .26$, we could not conclude that the factors interact. We therefore proceed to test the main effects for Brand and Club.

We first test the Brand main effect:

H_0: No difference exists among the true Brand mean distances

H_a: At least two Brand mean distances differ

Test statistic: $F = \dfrac{\text{MS(Brand)}}{\text{MSE}} = \dfrac{1{,}136.77}{24.60} = 46.21$

Rejection region: $F > F_{.10} = 2.33$, with $(a - 1) = (4 - 1) = 3$ numerator df and 24 denominator df.

Since 46.21 exceeds 2.33, we conclude that at least two of the Brand means differ. We will subsequently determine which Brand means differ using the Bonferroni technique. First, we also want to test the Club main effect:

H_0: No difference exists between the Club mean distances

H_a: The Club mean distances differ

Test statistic: $F = \dfrac{\text{MS(Club)}}{\text{MSE}} = \dfrac{46{,}443.9}{24.60} = 1{,}887.94$

Rejection region: $F > F_{.10} = 2.93$, with $(b - 1) = (2 - 1) = 1$ numerator df and 24 denominator df.

Since $F = 1{,}887.94$ exceeds 2.93, we conclude that the two clubs are associated with different mean distances. Since only two levels of Club were utilized in the experiment, this F test leads to the inference that the mean distance differs for the two clubs. It is no surprise (to golfers) that the mean distance for balls hit with the driver is significantly greater than that for those hit with the five-iron.

To determine which of the Brand's mean distances differ, we wish to compare the four Brand means. Since $4(3)/2 = 6$ pairs of means are to be compared, the Bonferroni multiple comparisons procedure requires that we specify $\alpha = .10/6 = .0167$ in order to have an overall significance level of (at least) .10. The two-tailed tabled t statistic has $\alpha/2 = .0167/2 = .0083 \approx .005$, with 24 df—that is, $t_{.005} = 2.797$. Each of the comparison intervals will have the same half-width:

$$\text{Half-width} = t_{.005}\, s\sqrt{(1/8 + 1/8)}$$
$$= (2.797)(4.96)(.5) = 6.9$$

Brand	*Mean*
G	223.59
F	218.44
E	202.44
H	199.20

FIGURE 16.18 ▲ **Comparison of mean distances by brand**

where we find $s = 4.96$ labeled **Root MSE** on the printout, and we have $n_i = 8$ observations for each brand. The Brand means are shown in descending order in Figure 16.18 with a vertical line connecting the means that are not significantly different. Brands F and G are apparently associated with significantly greater mean distances than brands E and H, but we cannot distinguish between brands F and G or between brands E and H utilizing these data. Since the interaction between Brand and Club was not significant, we conclude that this difference among brands applies to both clubs. The sample means for all club–brand combinations are shown in Figure 16.19 and appear to support the conclusions of the tests and comparisons. Note that the Brand means maintain their relative positions for each Club: Brands F and G dominate brands E and H for both driver and five-iron.

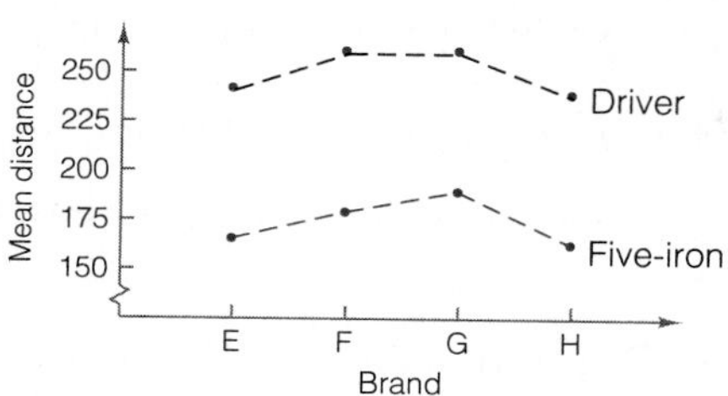

FIGURE 16.19 ▶ Sample mean plot for second factorial golf experiment

The analysis of factorial experiments can become complex if the number of factors is increased. Even the two-factor experiment becomes more difficult to analyze if some factor combinations have different numbers of observations than others. We have provided an introduction to these important experiments using two-factor factorials with equal numbers of observations for each treatment. Although similar principles apply to most factorial experiments, you should consult the references at the end of the chapter if you need to design and analyze more complex factorials.

Exercises 16.28–16.37

Learning the Mechanics

16.28 Suppose you conduct a 4×3 factorial experiment.

a. How many factors are utilized in the experiment?
b. Can you determine the factor type(s)—qualitative or quantitative—from the information given? Explain.
c. Can you determine the number of levels utilized for each factor? Explain.
d. Describe a treatment for this experiment, and determine the number of treatments employed.
e. What problem is caused by utilizing a single replicate of this experiment? How is the problem solved?

16.29 The partially completed ANOVA table for a 3×4 factorial experiment with two replications is shown here.

Source	df	SS	MS	F
A		.8		
B		5.3		
AB		9.6		
Error				
Total		17.0		

a. Complete the analysis of variance table.
b. Which sums of squares are combined to find the sum of squares for treatment? Do the data provide sufficient evidence to indicate that the treatment means differ? Use $\alpha = .05$.
c. Does the result of the test in part **b** warrant further testing? Explain.
d. What is meant by factor interaction, and what is the practical implication if it exists?
e. Test to determine whether these factors interact to affect the response mean. Use $\alpha = .05$, and interpret the results.
f. Does the result of the interaction test warrant further testing? Explain.

16.30 The partially completed ANOVA table given here is for a two-factor factorial experiment.

Source	df	SS	MS	F
A	3		.75	
B	1	.95		
AB			.30	
Error				
Total	23	6.5		

a. Give the number of levels for each factor.
b. How many observations were collected for each factor–level combination?
c. Complete the analysis of variance table.
d. Test to determine whether the treatment means differ. Use $\alpha = .10$.
e. Conduct the tests of factor interaction and main effects, each at the $\alpha = .10$ level of significance. Which of the tests are warranted as part of the factorial analysis? Explain.

16.31 The two-way table gives data for a 2 × 3 factorial experiment with two observations for each factor–level combination.

		Factor *B*		
	Level	1	2	3
Factor *A*	1	3.1, 4.0	4.6, 4.2	6.4, 7.1
	2	5.9, 5.3	2.9, 2.2	3.3, 2.5

a. Identify the treatments for this experiment. Calculate and plot the treatment means, using the response variable as the y-axis, and the levels of factor B as the x-axis. Use the levels of factor A as plotting symbols. Do the treatment means appear to differ? Do the factors appear to interact?
b. The Minitab ANOVA printout for this experiment is reproduced on the facing page. Test to determine whether the treatment means differ at the $\alpha = .05$ level of significance. Does the test support your visual interpretation from part **a**?
c. Does the result of the test in part **b** warrant a test for interaction between the two factors? If so, perform it using $\alpha = .05$.
d. Do the results of the previous tests warrant tests of the two factor main effects? If so, perform them using $\alpha = .05$.

```
ANALYSIS OF VARIANCE ON RESPONSE
SOURCE          DF        SS        MS
A                1     4.441     4.441
B                2     4.127     2.063
INTERACTION      2    18.007     9.003
ERROR            6     1.475     0.246
TOTAL           11    28.049
```

e. Interpret the results of the tests. Do they support your visual interpretation from part **a**?
f. Use the Bonferronni technique to compare all pairs of treatment means. Use an overall significance level of $\alpha = .10$.

16.32 The accompanying two-way table gives data for a 2×2 factorial experiment with two observations per factor–level combination.

	Factor *B* Level	1	2
Factor *A*	1	29.6 35.2	47.3 42.1
	2	12.9 17.6	28.4 22.7

a. Identify the treatments for this experiment. Calculate and plot the treatment means, using the response variable as the y-axis, and the levels of factor B as the x-axis. Use the levels of factor A as plotting symbols. Do the treatment means appear to differ? Do the factors appear to interact?
b. Use the computational formulas in Appendix D to create an ANOVA table for this experiment.
c. Test to determine whether the treatment means differ at the $\alpha = .05$ level of significance. Does the test support your visual interpretation from part **a**?
d. Does the result of the test in part **b** warrant a test for interaction between the two factors? If so, perform it using $\alpha = .05$.
e. Do the results of the previous tests warrant tests of the two factor main effects? If so, perform them using $\alpha = .05$.
f. Interpret the results of the tests. Do they support your visual interpretation from part **a**?
g. Given the results of your tests, which pairs of means, if any, should be compared? Use the Bonferroni technique to compare them, with an overall significance level of $\alpha = .05$. Interpret the results.

16.33 Suppose a 3×3 factorial experiment is conducted with three replications. Assume that SS(Total) = 1,000. For each of the following scenarios, form an ANOVA table, conduct the appropriate tests, and interpret the results.

a. The sum of squares of factor A main effect, SS(A), is 20% of SS(Total); the sum of squares for factor B main effect, SS(B), is 10% of SS(Total); and the sum of squares for interaction, SS(AB), is 10% of SS(Total).
b. SS(A) is 10%, SS(B) is 10%, and SS(AB) is 50% of SS(Total).
c. SS(A) is 40%, SS(B) is 10%, and SS(AB) is 20% of SS(Total).
d. SS(A) is 40%, SS(B) is 40%, and SS(AB) is 10% of SS(Total).

Applying the Concepts

16.34 Most short-run supermarket strategies such as price reductions, media advertising, and in-store promotions and displays are designed to increase unit sales of particular products temporarily. Factorial designs are often employed (Curhan, 1974) to evaluate the effectiveness of such strategies. Two of the factors examined by Wilkinson, Mason, and Paksoy (1982) were Price level (regular, reduced price, cost to supermarket) and Display level (normal display space, normal display space plus end-of-aisle display, twice the normal display space). A complete factorial design based on these two factors involves nine treatments. Suppose each treatment was applied three times to a particular product at a particular supermarket. Each application lasted a full week and the dependent variable (response) of interest was unit sales for the week. To minimize treatment carryover effects, each treatment was preceded and followed by a week in which the product was priced at its regular price and was displayed in its normal manner. The accompanying table reports the data collected, and the SAS ANOVA printout is also shown.

		Price		
		Regular	Reduced	Cost to Supermarket
Display	Normal	989 1,025 1,030	1,211 1,215 1,182	1,577 1,559 1,598
	Normal Plus	1,191 1,233 1,221	1,860 1,910 1,926	2,492 2,527 2,511
	Twice Normal	1,226 1,202 1,180	1,516 1,501 1,498	1,801 1,833 1,852

```
Dependent Variable: SALES
                                   Sum of           Mean
Source                   DF       Squares         Square      F Value     Pr > F

Model                     8    5291151.19      661393.90      1336.85     0.0001

Error                    18       8905.33         494.74

Corrected Total          26    5300056.52

                 R-Square          C.V.       Root MSE              SALES Mean

                 0.998320     1.4344689        22.2428              1550.59259


Source                   DF     Type I SS  Mean Square      F Value     Pr > F

DISPLAY                   2     1691392.5     845696.3      1709.37     0.0001
PRICE                     2     3089053.9    1544526.9      3121.89     0.0001
DISPLAY*PRICE             4      510704.8     127676.2       258.07     0.0001
```

a. Use the printout to complete the accompanying diagram partitioning the total sum of squares.

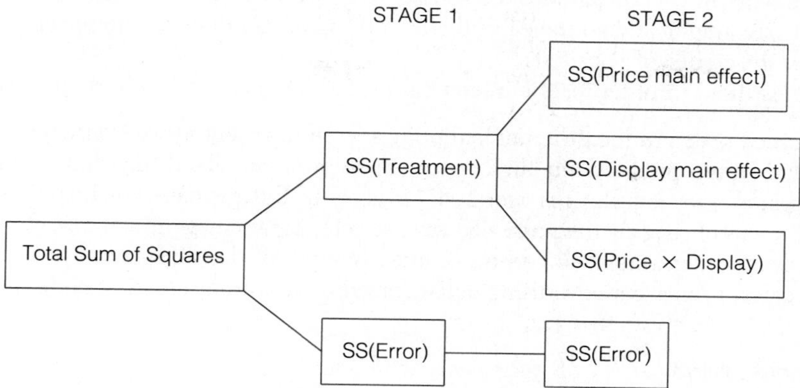

b. Do the data indicate that the mean sales differ among the nine treatments? Test using $\alpha = .10$.

c. Is the test of interaction between the factors Price and Display warranted as a result of the test in part **b**? If so, conduct the test using $\alpha = .10$.

d. Are the tests of the main effects for Price and Display warranted as a result of the previous tests? If so, conduct them using $\alpha = .10$.

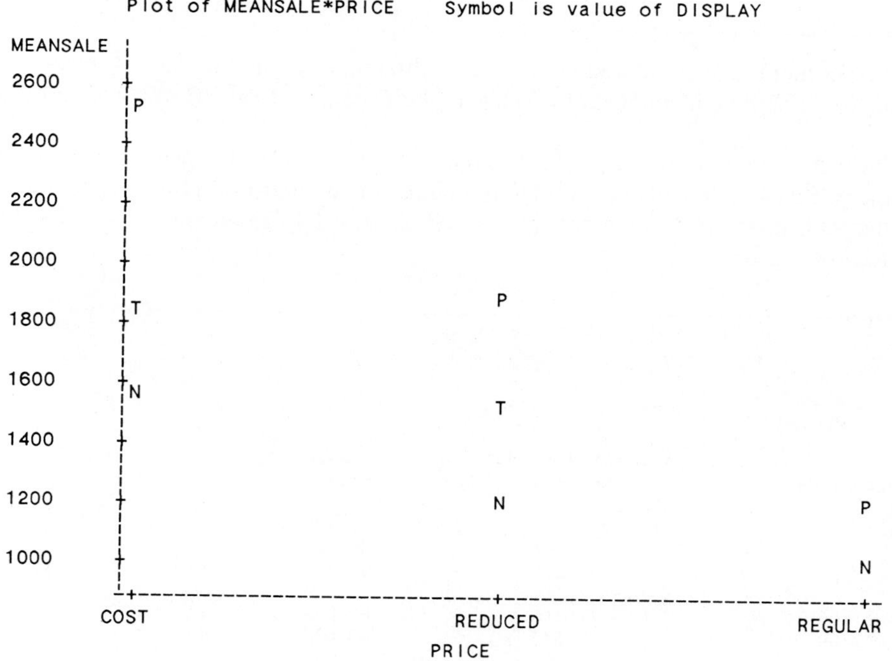

Note: The means for the Normal plus (P) and Twice normal (T) displays are nearly equal for Regular price. Both are represented by the symbol P at this level of price.

e. Which pairs of treatment means should be compared as a result of the tests in parts **b–d**? Use the Bonferroni technique with an overall significance level of $\alpha = .10$ to compare them.
f. The treatment means are graphically portrayed in the SAS ANOVA printout on the previous page. Use the graph to interpret the results of your tests.
g. What assumptions must hold in order for the inferential procedures you used to be appropriate?

16.35 A beverage distributor wanted to determine the combination of advertising agency (two levels) and advertising medium (three levels) that would produce the largest increase in sales per advertising dollar. Each of the advertising agencies prepared copy or film, as required for each of the media—newspaper, radio, and television. Twelve small towns of roughly the same size were selected for the experiment and two each were assigned to receive an advertisement prepared and transmitted by each of the six agency–medium combinations. The dollar increases in sales per advertising dollar, based on a 1-month sales period, are shown in the table.

		Advertising Medium		
		Newspaper	Radio	Television
Agency	1	15.3 12.7	20.1 17.4	12.7 16.2
	2	18.9 22.4	24.3 28.8	12.5 9.4

a. Describe the experiment. Include the identification of the response variable, factors, factor types, factor levels, treatments, and experimental units. What is the experiment called? What type of design was employed?
b. The SPSS ANOVA printout for this experiment is reproduced here. Use it to perform a complete analysis of the experiment. Be sure to conduct relevant tests, and use the Bonferroni technique to compare relevant pairs of means. Use $\alpha = .10$ throughout. [*Note*: SPSS uses **Explained** instead of "Treatment" and **Residual** instead of "Error."]

```
          * * *  A N A L Y S I S   O F   V A R I A N C E  * * *

                 SALES
          BY     AGENCY
                 MEDIUM

                                   Sum of            Mean              Signif
Source of Variation               Squares    DF     Square       F      of F

Main Effects                      238.299     3     79.433   13.934    .004
  AGENCY                           39.967     1     39.967    7.011    .038
  MEDIUM                          198.332     2     99.166   17.395    .003

2-way Interactions                 77.345     2     38.672    6.784    .029
  AGENCY    MEDIUM                 77.345     2     38.672    6.784    .029

Explained                         315.644     5     63.129   11.074    .005

Residual                           34.205     6      5.701

Total                             349.849    11     31.804
```

c. Graph the treatment means, placing advertisement medium on the horizontal axis, and using agency as a plotting symbol. Do the results of your analysis in part **b** appear reasonable? Use the graph to interpret the results of your analysis.

16.36 How do women compare with men in their ability to perform laborious tasks that require strength? Some information on this question is provided in a study (Phillips and Pepper, 1982) of the firefighting ability of men and women. A 2 × 2 factorial experiment investigated the effect of the factor Sex (male or female) and the factor Weight (light or heavy) on the length of time required for a person to perform a particular firefighting task. Eight persons were selected for each of the 2 × 2 = 4 Sex–Weight categories of the 2 × 2 factorial experiment and the length of time (in minutes) needed to complete the task was recorded for each of the 32 persons. The means and standard deviations of the four samples are shown in the table.

	Light		Heavy	
	Mean	Standard Deviation	Mean	Standard Deviation
Female	18.30	6.81	14.50	2.93
Male	13.00	5.04	12.25	5.70

a. Calculate the total of the $n = 8$ time measurements for each of the four categories of the 2 × 2 factorial experiment.
b. Calculate CM. (See Appendix D for computational formulas.)
c. Use the results of parts **a** and **b** to calculate the sums of squares for Sex, for Weight, and for the Sex × Weight interaction.
d. Calculate each sample variance. Then calculate the sum of squares of deviations *within* each sample for each of the four samples.
e. Calculate SSE. [*Hint*: SSE is the pooled sum of squares of the deviations calculated in part **d**.]
f. Now that you know SS(Sex), SS(Weight), SS(Sex × Weight), and SSE, find SS(Total).
g. Summarize the calculations in an analysis of variance table.
h. Conduct a complete analysis of these data. Use $\alpha = .05$ for any inferential techniques you employ. Interpret your conclusions graphically.
i. What assumptions are necessary to assure the validity of the inferential techniques you utilized? State them in terms of this experiment.

16.37 Refer to Exercise 16.36. Phillips and Pepper (1982) give data on another 2 × 2 factorial experiment utilizing 20 males and 20 females. The experiment involved the same treatments with 10 persons assigned to each Sex–Weight category. The response measured was the pulling force each person was able to exert on the starter cord of a P-250 fire pump. The means and standard deviations of the four samples (corresponding to the 2 × 2 = 4 categories of the experiment) are shown in the table.

	Light		Heavy	
	Mean	Standard Deviation	Mean	Standard Deviation
Females	46.26	14.23	62.72	13.97
Males	88.07	8.32	86.29	12.45

a. Use the procedures outlined in Exercise 16.36 to perform an analysis of variance for the experiment. Display your results in an analysis of variance table.
b. Conduct a complete analysis of these data. Use $\alpha = .05$ for any inferential techniques you employ. Interpret your conclusions graphically.
c. Note that the observed standard deviations for the four treatments vary from 8.32 to 14.23. Since the validity of the inferences requires that the treatment standard deviations be equal, how do you refute the criticism that the unequal standard deviations invalidate your analysis?

16.4 Using Regression Analysis for ANOVA (Optional)

We have seen in previous sections that analysis of variance involves the statistical analysis of the relationship between a response and one or more factors. In Chapters 10–12 we performed very similar analyses using regression techniques, so it is not surprising that the two methodologies are closely related. In fact, each of the ANOVAs performed in this chapter can be formulated as a regression model, and the same analysis can be performed in a regression context. We will illustrate the ANOVA–regression correspondence in this section.

A one-factor experiment utilizing a completely randomized design provides the simplest example of this correspondence. Suppose the factor has p levels—that is, there are p treatments in the experiment. Assuming the factor is qualitative, the corresponding regression model is

$$y = \beta_0 + \beta_1 x_1 + \beta_2 x_2 + \cdots + \beta_{p-1} x_{p-1} + \epsilon$$

where $x_1, x_2, \ldots, x_{p-1}$ are $(p - 1)$ dummy variables describing the p levels of the factor. If the factor is quantitative, then the x's become powers of the variable, i.e., $x_i = x^i$. In either case, to test the null hypothesis of interest in the ANOVA, we test

$$H_0\colon \quad \mu_1 = \mu_2 = \cdots = \mu_p$$

using the F ratio between mean square for treatments and mean square for error. This is precisely the same test as the global model test in the corresponding regression example:

$$H_0\colon \quad \beta_1 = \beta_2 = \cdots = \beta_{p-1} = 0$$

which we test using the F ratio of the mean square for (regression) model to the mean square for error. Both tests result in the same F ratio and have the same numerator and denominator degrees of freedom, $(p - 1)$ and $(n - p)$, respectively.

To better understand the equivalence of these two procedures, note that if all the β's are equal to 0 in the regression model, the mean response is equal to β_0 *no matter which level of the qualitative factor is utilized as the base level*. That is, the mean response is equal for all p treatments, which is exactly the ANOVA null hypothesis. For example, in Section 16.2 we used ANOVA to test the null hypothesis that four brands of golf balls travel the same distance, on average, when hit by Iron Byron with a driver. We tested

$$H_0: \quad \mu_A = \mu_B = \mu_C = \mu_D$$

where μ_i is the mean distance for brand i. Equivalently, we could have defined the regression model

$$E(y) = \beta_0 + \beta_1 x_1 + \beta_2 x_2 + \beta_3 x_3$$

with

$$x_1 = \begin{cases} 1 & \text{if brand A} \\ 0 & \text{otherwise} \end{cases} \qquad x_2 = \begin{cases} 1 & \text{if brand B} \\ 0 & \text{otherwise} \end{cases} \qquad x_3 = \begin{cases} 1 & \text{if brand C} \\ 0 & \text{otherwise} \end{cases}$$

Then β_0 represents the mean distance for brand D, β_1 the difference between the mean distances of brand A and brand D, etc. The null hypothesis

$$H_0: \quad \beta_1 = \beta_2 = \beta_3 = 0$$

is equivalent to the ANOVA null hypothesis that the Brand means are equal.

EXAMPLE 16.7

Refer to Example 16.3, in which we used ANOVA for a completely randomized design to test the hypothesis that the mean distances associated with four brands of golf balls were the same. Ten experimental units were randomly and independently assigned to each brand, and the distance was recorded. The data are repeated in Table 16.8. Use a regression program to fit the equivalent regression model, and conduct the global F test for the model's usefulness in predicting the response y. Compare the test to that conducted in Example 16.3, and interpret the results. Use $\alpha = .10$.

TABLE 16.8 Distance Data for Golf Experiment: Completely Randomized Design

	Brand A	Brand B	Brand C	Brand D
	251.2	263.2	269.7	251.6
	245.1	262.9	263.2	248.6
	248.0	265.0	277.5	249.4
	251.1	254.5	267.4	242.0
	265.5	264.3	270.5	246.5
	250.0	257.0	265.5	251.3
	253.9	262.8	270.7	262.8
	244.6	264.4	272.9	249.0
	254.6	260.6	275.6	247.1
	248.8	255.9	266.5	245.9
Means	251.28	261.06	269.95	249.42

Solution

The Minitab printout for the regression model using three dummy variables to describe Brand is given in Figure 16.20, page 900. To test the model's usefulness, we calculate the F ratio of the regression (model) mean square to the residual (error) mean square, both shown shaded in Figure 16.20. This is

FIGURE 16.20 ▶ Minitab printout for regression model of Example 16.7

```
THE REGRESSION EQUATION IS
DISTANCE = 249 + 1.86 X1 + 11.6 X2 + 20.5 X3

                                        ST. DEV.      T-RATIO =
COLUMN         COEFFICIENT              OF COEF.      COEF/S.D.
                  249.420               1.588         157.06
X1                  1.860               2.246           0.83
X2                 11.640               2.246           5.18
X3                 20.530               2.246           9.14

S = 5.022

R-SQUARED = 74.9 PERCENT
R-SQUARED = 72.8 PERCENT, ADJUSTED FOR D.F.

ANALYSIS OF VARIANCE

 DUE TO        DF           SS          MS=SS/DF
REGRESSION      3      2709.20            903.07
RESIDUAL       36       907.86             25.22
TOTAL          39      3617.06
```

$$F = \frac{\text{Regression mean square}}{\text{Residual mean square}} = \frac{903.07}{25.22} = 35.81$$

The tabled value $F_{.10} \approx 2.25$, with 3 model df (number of independent variables in the model) in the numerator and 36 error df (n minus the number of β's) in the denominator. We therefore reject the null hypothesis that the model is not useful and conclude at the .10 level of significance that brand is related to mean distance.

The regression F value is identical to that obtained in Example 16.3 using ANOVA. The conclusion may sound somewhat different, but on closer inspection they are also identical. In the ANOVA we concluded that the mean distances differ for at least two of the four brands, while here we conclude that the model is useful for predicting distance. But for the model to be useful, the factor Brand must have some association with mean distance, since Brand is the only factor in the regression model. That is, the mean distance must vary with Brand, which implies that at least two of the brands have different mean distances.

It is instructive to interpret the least squares estimates of the β parameters in Figure 16.20. The value of $\hat{\beta}_0 = 249.42$ is the sample mean distance for brand D (Table 16.8), since brand D is the base level at which all three dummy variables are equal to 0. The value of $\hat{\beta}_1 = 1.86$ is the difference between the sample mean distances of brand A and brand D $(251.28 - 249.42 = 1.86)$. Similarly, you can see that $\hat{\beta}_2$ and $\hat{\beta}_3$ are the sample mean differences between brand B and brand D and between brand C and brand D, respectively. Finally, note that the standard deviation of the regression model, $s = 5.022$, is the same as $\sqrt{\text{MSE}}$ in the ANOVA, and that both measure the variation in distance within a brand. In fact, we would expect to be able to predict the distance Iron Byron will hit a specified brand to within about ± 10 yards using this model. Or, to put it another way, we expect the actual distances for a specific brand to vary within a range of about 10 yards around the mean. Of course, this variability might be partially attributable to other factors that could be included in an expanded regression model.

The assumptions necessary to assure the validity of an ANOVA and a regression analysis are also identical, although usually phrased differently. Using the golf experiment as an example, we assumed that the distance probability distributions are normal with equal variances for each brand when we conducted the ANOVA. The regression assumption that the error component is normally distributed with constant variance for all settings of the independent variables is seen to be equivalent when we recognize that the error component of this particular model describes the random variability of distance within a particular brand. Thus, we are assuming that the distance measurements within each brand are normally distributed with the same variance for each brand. Similarly, the ANOVA assumption that the samples are random and independent is matched by the regression assumption that the error components are independent with mean 0.

The use of a regression model for a completely randomized design with a single factor can be extended to one with more than one factor.

EXAMPLE 16.8

Refer to Example 16.5 in which we employed a factorial experiment in a completely randomized design to evaluate the effects of two factors, Brand of golf ball and Club utilized, on mean distance. Four brands and two clubs were utilized, and the 4 × 2 factorial was replicated four times. The data are repeated in Table 16.9.

TABLE 16.9 Data for 4 × 2 Factorial Golf Experiment

		Brand			
		A	B	C	D
Club	Driver	226.4 232.6 234.0 220.7	238.3 231.7 227.7 237.2	240.5 246.9 240.3 244.7	219.8 228.7 232.9 237.6
	Five-Iron	163.8 179.4 168.6 173.4	184.4 180.6 179.5 186.2	179.0 168.0 165.2 156.5	157.8 161.8 162.1 160.3

a. Write a regression model to represent the factorial experiment.

b. Use a computer program to fit the regression model. Conduct an analysis similar to that employed in the ANOVA of Example 16.5.

Solution

a. The two factors Brand and Club are qualitative, so we employ dummy variables to describe them. Also, in contrast to the randomized block experiment, we are very interested in knowing whether the factors interact—that is, whether the difference in the Brand mean distances depends on the Club. Therefore, the model is

$$y = \overbrace{\beta_0 + \beta_1 x_1 + \beta_2 x_2 + \beta_3 x_3}^{\text{Brand main effect}} + \overbrace{\beta_4 x_4}^{\text{Club main effect}} + \overbrace{\beta_5 x_1 x_4 + \beta_6 x_2 x_4 + \beta_7 x_3 x_4}^{\text{Brand} \times \text{Club interaction}} + \epsilon$$

where x_1 to x_3 are dummy variables representing the three brands (using brand D as the base level at which all three dummy variables are equal to 0), and x_4 is the dummy variable representing the two clubs (using five-iron as the base level at which the dummy variable is equal to 0). Note that the interaction terms are just the cross-products of the Brand dummy variables with the Club dummy variable.

b. The SAS printout for the regression model is shown in Figure 16.21. The global F test for the model tests the null hypothesis

$$H_0\colon \ \beta_1 = \beta_2 = \cdots = \beta_7 = 0$$

In general, this hypothesis implies that the model has no predictive value. In this example, it implies that the mean distance is unrelated to Brand and Club; that is, the mean distance is the same for all brand–club combinations. This should be the same test as the ANOVA test that the $4 \times 2 = 8$ treatment means are all equal.

FIGURE 16.21 ▶
SAS printout for factorial golf experiment: Complete model including interaction

```
Model: MODEL1
Dep Variable: DISTANCE
                          Analysis of Variance

                          Sum of          Mean
Source        DF         Squares        Square        F Value         Prob>F

Model          7     33659.80875    4808.54411        140.354         0.0001
Error         24       822.24000      34.26000
C Total       31     34482.04875

     Root MSE            5.85320     R-Square          0.9762
     Dep Mean          202.08125     Adj R-Sq          0.9692
     C.V.                2.89646
                         Parameter Estimates

                   Parameter        Standard     T for H0:
Variable  DF        Estimate           Error   Parameter=0     Prob > |T|

INTERCEP   1      167.175000      2.92660213        57.123         0.0001
X1         1        4.125000      4.13884042         0.997         0.3289
X2         1       15.500000      4.13884042         3.745         0.0010
X3         1       -6.675000      4.13884042        -1.613         0.1199
X4         1       75.925000      4.13884042        18.345         0.0001
X1X4       1      -18.800000      5.85320425        -3.212         0.0037
X2X4       1      -24.875000      5.85320425        -4.250         0.0003
X3X4       1       -6.675000      5.85320425        -1.140         0.2654
```

We see that the value $F = 140.35$ for the SAS printout (Figure 16.21) is the same value we calculated using the ANOVA $F = \text{MST/MSE}$ in Example 16.5. The SAS printout gives the observed significance level of the test statistic as .0001 based on 7 numerator and 24 denominator df, implying that the null hypothesis is rejected at any level of significance greater than .0001. Therefore, using $\alpha = .10$ (as in Example 16.5), we reach the same conclusion as in the ANOVA analysis: There is sufficient evidence that the mean distance varies among two or more brand–club combinations.

The next test in a two-factor factorial experiment involves the interaction between the factors. In the regression context the null hypothesis that the factors Brand and Club do not interact is

$$H_0\colon \quad \beta_5 = \beta_6 = \beta_7 = 0$$

This test requires a comparison of the complete model in Figure 16.21 with the reduced model *without* the interaction terms. The SAS printout for the reduced model is shown in Figure 16.22. We then calculate

$$F = \frac{[\text{SSE(Reduced model)} - \text{SSE(Complete model)}]/3}{\text{MSE(Complete model)}}$$

$$= \frac{(1{,}588.201 - 822.240)/3}{34.260} = 7.45$$

FIGURE 16.22 ▶ **SAS printout for factorial golf experiment: Reduced model without interaction**

```
Model: MODEL2
Dep Variable: DISTANCE

                         Analysis of Variance

                        Sum of          Mean
Source       DF        Squares        Square      F Value      Prob>F

Model         4    32893.84750    8223.46187      139.802      0.0001
Error        27     1588.20125      58.82227
C Total      31    34482.04875

     Root MSE          7.66957      R-Square       0.9539
     Dep Mean        202.08125      Adj R-Sq       0.9471
     C.V.              3.79529

                         Parameter Estimates

                     Parameter       Standard     T for H0:
Variable   DF         Estimate          Error   Parameter=0    Prob > |T|

INTERCEP    1       173.468750     3.03166282        57.219        0.0001
X1          1        -5.275000     3.83478384        -1.376        0.1803
X2          1         3.062500     3.83478384         0.799        0.4315
X3          1       -10.012500     3.83478384        -2.611        0.0146
X4          1        63.337500     2.71160166        23.358        0.0001
```

This F ratio is identical to that obtained in the ANOVA in Example 16.5 and again results in rejection of the null hypothesis at the $\alpha = .10$ level of significance, since $F_{.10} = 2.33$ with 3 numerator and 24 denominator df. We therefore conclude that both Brand and Club affect the mean distance and that the differences among the Brand mean distances depend on the Club (i.e., the factors interact). The presence of interaction makes the tests of the factor main effects unnecessary.

We can use the least squares estimates of the β parameters in the complete model to estimate particular treatment means and differences between particular pairs of means. However, it is probably easier to use the Bonferroni procedure or some other multiple comparisons procedure, as we did in Example 16.5.

Note that the variability of the random error component is measured by the standard deviation of the regression model, $s = 5.85$. This measure is identical to that we obtained in the ANOVA, $\sqrt{\text{MSE}} = \sqrt{34.260} = 5.85$. The implication is that the model can be used to predict the distance Iron Byron will hit the ball to within approximately $\pm 2s = \pm 11.7$ yards if we specify the brand and club. Furthermore, the model accounts for $R^2 = .976$, or 97.6%, of the total sample variation in distance.

Most designed experiments can be analyzed using either ANOVA or regression techniques with identical results. The recommended method of analysis depends on a number of factors, including the ultimate objectives of the analysis (e.g., prediction of response or estimation of differences among treatment means), the type of computer software available, and—perhaps most important—the methodology with which the analyst is more familiar. Whichever analytical tools are employed, it is important to recognize that both ANOVA and regression relate the response mean to one or more factors and that both provide an estimate of sampling variability with which to make inferences.

Exercises 16.38–16.41

Learning the Mechanics

16.38 Suppose you conduct an experiment with one factor at five levels, utilizing a completely randomized design. Each level of the factor is randomly assigned three experimental units.

a. Write a regression model for this experiment. Interpret the β parameters of the model.

b. How many degrees of freedom will be available for estimating the standard deviation σ of the error component?

c. Write the null hypothesis that the treatment means are equal in terms of the β parameters of the model.

d. What is the rejection region for the test in part c using an $\alpha = .10$ level of significance?

16.39 The following regression model was proposed to describe a designed experiment:

$$y = \beta_0 + \beta_1 x_1 + \beta_2 x_2 + \beta_3 x_3 + \beta_4 x_1 x_3 + \beta_5 x_2 x_3 + \epsilon$$

where

$$x_1 = \begin{cases} 1 & \text{if factor A, level 1} \\ 0 & \text{otherwise} \end{cases}$$

$$x_2 = \begin{cases} 1 & \text{if factor A, level 2} \\ 0 & \text{otherwise} \end{cases}$$

$$x_3 = \begin{cases} 1 & \text{if factor } B\text{, level 1} \\ 0 & \text{if factor } B\text{, level 2} \end{cases}$$

a. What type of experiment does the model describe?
b. Describe how you would test the null hypothesis that the treatment means are equal. Include the hypotheses, test statistic, and rejection region. (Assume that three experimental units are randomly and independently assigned to each factor–level combination.)
c. Describe how you would test for interaction.

Applying the Concepts

16.40 Refer to Exercise 16.20, which used a completely randomized design to compare the mean costs associated with the audits of firms in three size groups. The data are repeated in the accompanying table.

Costs Incurred in Audits

A	B	C
250	100	80
150	150	125
275	75	20
100	200	186
475	55	52
600	80	92
150	110	88
800	160	141
325	132	76
230	233	200

The model

$$E(y) = \beta_0 + \beta_1 x_1 + \beta_2 x_2$$

where

$$x_1 = \begin{cases} 1 & \text{if group A} \\ 0 & \text{otherwise} \end{cases} \qquad x_2 = \begin{cases} 1 & \text{if group B} \\ 0 & \text{otherwise} \end{cases}$$

was fit using the SAS regression program. The result is shown in the printout on page 906.
a. Use the regression printout to conduct the test of the null hypothesis that the mean costs are the same for the three groups, and compare the result to that you obtained in Exercise 16.20.
b. Compare the observed significance levels of the tests.
c. Use the regression printout to partition the total sum of squares into the sums of squares for treatments and error. Compare them with those in Exercise 16.20.
d. Interpret the value of R^2 and the least squares estimates of the β parameters.

```
Model: MODEL1
Dep Variable: COST
                          Analysis of Variance

                         Sum of            Mean
Source        DF        Squares          Square       F Value        Prob>F

Model          2 318861.66667 159430.83333          8.438         0.0014
Error         27 510163.00000  18894.92593
C Total       29 829024.66667

      Root MSE      137.45882     R-Square          0.3846
      Dep Mean      190.33333     Adj R-Sq          0.3390
      C.V.           72.22004

                          Parameter Estimates

                     Parameter        Standard      T for H0:
Variable   DF         Estimate           Error    Parameter=0     Prob > |T|

INTERCEP    1       106.000000     43.46829411          2.439         0.0216
X1          1       229.500000     61.47345106          3.733         0.0009
X2          1        23.500000     61.47345106          0.382         0.7052
```

16.41 Refer to Exercise 16.34, which used a 3 × 3 factorial experiment to examine the effects of the factors Price and Display on the unit sales of a product. The data are repeated in the table.

		Price		
		Regular	**Reduced**	**Cost to Supermarket**
Display	**Normal**	989 1,025 1,030	1,211 1,215 1,182	1,577 1,559 1,598
	Normal Plus	1,191 1,233 1,221	1,860 1,910 1,926	2,492 2,527 2,511
	Twice Normal	1,226 1,202 1,180	1,516 1,501 1,498	1,801 1,833 1,852

The following regression model is proposed for the analysis of this experiment:

$$y = \beta_0 + \beta_1 x_1 + \beta_2 x_2 + \beta_3 x_3 + \beta_4 x_4 + \beta_5 x_1 x_3 + \beta_6 x_1 x_4 + \beta_7 x_2 x_3 + \beta_8 x_2 x_4 + \epsilon$$

where

$$x_1 = \begin{cases} 1 & \text{if Price = Regular} \\ 0 & \text{otherwise} \end{cases} \qquad x_2 = \begin{cases} 1 & \text{if Price = Reduced} \\ 0 & \text{otherwise} \end{cases}$$

$$x_3 = \begin{cases} 1 & \text{if Display = Normal} \\ 0 & \text{otherwise} \end{cases} \qquad x_4 = \begin{cases} 1 & \text{if Display = Normal Plus} \\ 0 & \text{otherwise} \end{cases}$$

The Minitab printout for this model is shown here.

```
THE REGRESSION EQUATION IS
SALES = 1829 - 626 X1 - 324 X2 - 251 X3 + 681 X4 + 62.7 X1X3 - 669 X1X4
          - 51.7 X2X3 - 288 X2X4

                                          ST. DEV.     T-RATIO =
COLUMN        COEFFICIENT                 OF COEF.     COEF/S.D.
                 1828.67                    12.84        142.40
X1               -626.00                    18.16        -34.47
X2               -323.67                    18.16        -17.82
X3               -250.67                    18.16        -13.80
X4                681.33                    18.16         37.52
X1X3               62.67                    25.68          2.44
X1X4             -669.00                    25.68        -26.05
X2X3              -51.67                    25.68         -2.01
X2X4             -287.67                    25.68        -11.20

S = 22.24

R-SQUARED = 99.8 PERCENT
R-SQUARED = 99.8 PERCENT, ADJUSTED FOR D.F.

ANALYSIS OF VARIANCE

 DUE TO       DF           SS        MS=SS/DF
REGRESSION     8      5291151          661394
RESIDUAL      18         8905             495
TOTAL         26      5300057
```

a. Interpret the least squares estimates by using them to estimate the mean response for each of the $3 \times 3 = 9$ treatments.

b. Interpret the standard deviation and R^2 for the model.

Printout for reduced model

```
THE REGRESSION EQUATION IS
SALES = 1934 - 828 X1 - 437 X2 - 247 X3 + 362 X4

                                          ST. DEV.     T-RATIO =
COLUMN        COEFFICIENT                 OF COEF.     COEF/S.D.
                 1933.74                    66.13         29.24
X1               -828.11                    72.45        -11.43
X2               -436.78                    72.45         -6.03
X3               -247.00                    72.45         -3.41
X4                362.44                    72.45          5.00

S = 153.7

R-SQUARED = 90.2 PERCENT
R-SQUARED = 88.4 PERCENT, ADJUSTED FOR D.F.

ANALYSIS OF VARIANCE

 DUE TO       DF           SS        MS=SS/DF
REGRESSION     4      4780446         1195112
RESIDUAL      22       519610           23619
TOTAL         26      5300057
```

c. Is there evidence that the mean unit sales differ for the nine treatments? Specify the null hypothesis in terms of the regression model and test it using $\alpha = .10$. Compare the result with that obtained in Exericse 16.34.
d. What is the appropriate null hypothesis to test for interaction between the factors. Price and Display? What are the test statistic and rejection region for this test at $\alpha = .10$?
e. A Minitab printout for the reduced model without interaction was shown on page 907. Conduct the test you set up in part **d**. Compare the result with that obtained for the interaction test in Exercise 16.34.
f. Do you recommend any further testing of the parameters in the model? Explain.

Summary

Designed experiments are those for which the analyst controls the assignment of treatments to experimental units. **Observational experiments** are those in which the analyst simply observes (i.e., does not influence) the treatments and the response of the experimental units. The **response**, **factors**, **levels**, **treatments**, and **experimental units** are elements of experiments with which you should now be familiar. A **completely randomized design** is one for which treatment assignments are made to experimental units in a random and independent manner. The analysis and interpretation of completely randomized designs are useful even when the assignment of treatments to experimental units is beyond the analyst's control—i.e., for observational as well as designed experiments.

Factorial experiments are those involving more than one factor, and **complete factorials** utilize all combinations of the factors. ***F* tests** can be conducted to determine whether treatment means differ, and, if so, whether the factors interact or independently affect the response mean. The **Bonferroni procedure** can be used to compare pairs of treatment means.

Most experiments can also be analyzed in a regression context. Both ANOVA and regression models relate the response mean to one or more factors, and provide an estimate of the sampling variability that can be used in making inferences.

Supplementary Exercises 16.42–16.58

Note: Starred (*) exercises require the use of a computer.

16.42 A direct-mail company assembles and stores paper products (envelopes, letters, brochures, order cards, etc.) for its customers. The company estimates the total number of pieces received in a shipment by estimating the weight per piece and then weighing the entire shipment. The company is unsure whether the sample of pieces used to estimate the mean weight per piece should be drawn from a single carton, or whether it is worth the extra time required to pull a few pieces from several cartons. To aid management in making a decision, eight brochures were pulled from each of five cartons of a typical shipment and weighed. The weights (in pounds) are shown in the table.

Carton 1	*Carton 2*	*Carton 3*	*Carton 4*	*Carton 5*
.01851	.01872	.01869	.01899	.01882
.01829	.01861	.01853	.01917	.01895
.01844	.01876	.01876	.01852	.01884
.01859	.01886	.01880	.01904	.01835
.01854	.01896	.01880	.01923	.01889
.01853	.01879	.01882	.01905	.01876
.01844	.01879	.01862	.01924	.01891
.01833	.01879	.01860	.01893	.01879

a. Identify the response, factor(s), treatments, and experimental units.
b. Do these data provide sufficient evidence to indicate differences in the mean weight per brochure among the five cartons?
c. What assumptions must be satisfied in order for the test of part **b** to be valid?
d. Use Bonferroni multiple comparisons to compare all pairs of means, with $\alpha = .05$ as the overall level of significance.
e. Given the results, make a recommendation to management about whether to sample from one carton or from many cartons.

16.43 Mowen et al. (1985) hypothesized that in evaluating salespeople, sales managers underutilize data on the salesperson's territory and instead rely on data associated with how much effort the salesperson expended. To investigate their hypothesis, they developed four scenarios based on the four factor–level combinations associated with the factors Salesperson's effort (high, low) and Sales territory difficulty (high, low). A random sample of 120 sales managers was selected, and 30 were randomly assigned to each of the four factor–level combinations. Each group was presented with the scenario corresponding to the appropriate factor–level combination, and then each sales manager was asked to evaluate the performance of a salesperson using a 7-point scale (1 = High, 7 = Low). The authors summarized their results in the accompanying partial ANOVA table.

Source	df	*F*
Sales territory difficulty	1	.39
Effort	1	53.27
Interaction	1	1.95

a. What type of experimental design was used in this study?
b. How many degrees of freedom are associated with SSE?
c. Do the data suggest that a sales manager's perceptions of sales performance are subject to an interaction effect between the difficulty of the territory and the level of effort? Test using $\alpha = .05$.
d. Do the data indicate that sales managers' performance ratings are influenced by sales territory data? Test using $\alpha = .05$.
e. Do the data indicate that the level of effort expended by a salesperson influences perceived performance? Test using $\alpha = .05$.
f. Do the results of parts **d** and **e** tend to confirm or to refute the hypothesis of Mowen et al.? Explain.
g. What assumptions must hold in order for the tests you conducted in parts **c**, **d**, and **e** to be valid?

16.44 The table shows the partially completed analysis of variance for a two-factor factorial experiment.

Source	df	SS	MS	F
A	3	2.6		
B	5	9.2		
AB			3.1	
Error		18.7		
Total	47			

a. Complete the analysis of variance table.
b. How many levels were utilized for each factor? How many treatments were employed? How many replications were performed?
c. Find the value of the sum of squares for treatments. Test to determine whether the data provide evidence that the treatment means differ. Use $\alpha = .05$.
d. Is further testing of the nature of the factor effects warranted? If so, test to determine whether the factors interact. Use $\alpha = .05$. Interpret the result.

***16.45** It has been hypothesized that treatment, after casting, of a plastic used in optic lenses will improve wear. Four treatments (A–D) are to be tested. To determine whether any differences in mean wear exist among treatments, 28 castings from a single formulation of the plastic were made, and seven castings were randomly assigned to each of the treatments. Wear was determined by measuring the increase in "haze" after 200 cycles of abrasion (better wear being indicated by smaller increases). The results are given in the table.

A	B	C	D
9.16	11.95	11.47	11.35
13.29	15.15	9.54	8.73
12.07	14.75	11.26	10.00
11.97	14.79	13.66	9.75
13.31	15.48	11.18	11.71
12.32	13.47	15.03	12.45
11.78	13.06	14.86	12.38

a. What type of experiment was utilized? Identify the response, factor(s), factor type(s), treatments, and experimental units.
b. Use either a computer program or the computational formulas in Appendix D to analyze the data. Is there evidence of a difference in mean wear among the treatments? Use $\alpha = .05$.
c. What is the observed significance level of the test? Interpret it.
d. Use the Bonferroni technique to compare all the pairs of treatment means with an overall significance level of $\alpha = .10$.
e. Use a 90% confidence interval to estimate the mean wear for lenses receiving treatment A.

***16.46** The data shown in the table (see facing page) are for a 4×3 factorial experiment with two replications.
a. Use a computer program or the calculation formulas in Appendix D to perform an analysis of variance for the data. Display the results in an ANOVA table.
b. Do the data indicate that the treatment means differ? Use $\alpha = .05$.

		Level of B		
		1	2	3
Level of A	1	2 4	5 6	1 3
	2	5 4	2 2	10 9
	3	7 10	1 0	5 3
	4	8 7	12 11	7 4

c. Is a test of factor interaction warranted? If so, perform it using $\alpha = .05$.
d. Is it necessary to perform tests of the main effects? If so, perform the tests using $\alpha = .05$.
e. Use the Bonferroni technique to compare all pairs of treatment means using an $\alpha = .10$ level of significance.

16.47 Several companies are experimenting with the concept of paying production workers (generally paid by the hour) on a salary basis. It is believed that absenteeism and tardiness will increase under this plan, yet some companies feel that the working environment and overall productivity will impove. Fifty production workers under the salary plan are monitored at company A, and 50 under the hourly plan are monitored at company B. The number of work-hours missed due to tardiness or absenteeism over a 1-year period is recorded for each worker. The results are partially summarized in the table.

Source	df	SS	MS	F
Companies		3,237.2		
Error		16,167.7		
Total	99			

a. Fill in the information missing from the table.
b. Is there evidence at $\alpha = .05$ that the mean number of hours missed differs for employees of the two companies?
c. Is there sufficient information given to form a confidence interval for the difference between the mean number of hours missed at the two companies?

16.48 A composite of a metal powder mixed with a plastic resin was invented to stop a radiation beam used for skin or oral cancer therapy. A 2×2 factorial experiment with two replications was conducted in a dental research lab to measure the degree of radiation traveling through the composite when the density of the metal powder was either a heavy alloy or a light alloy and when a second layer of plastic under the composite was present or absent. The table on page 912 shows the data collected. Use the techniques presented in this chapter to analyze these data, and make a recommendation regarding the preferred method of protecting patients from overexposure to potentially harmful radiation.

		Alloy Density	
		Heavy	Light
Second Plastic	Present	0.04, 0.02	0.46, 0.40
	Absent	0.38, 0.13	1.84, 2.29

Source: Personal communication from F. Eichmiller, Paffenbarger Research Center, Gaithersburg, MD, 1991.

16.49 Sixteen workers were randomly selected to participate in an experiment to determine the effects of work scheduling and method of payment on attitude toward the job. Two schedules were employed, the standard 8–5 workday and a modification whereby the worker could decide each day whether to start at 7 or 8 A.M.; in addition, the worker could choose between a ½-hour or 1-hour lunch period each day. The two methods of payment were a standard hourly rate and a reduced hourly rate with an added piece rate based on the worker's production. Four workers were randomly assigned to each of the four scheduling–payment combinations, and each completed an attitude test after 1 month on the job. The test scores are shown in the table.

		Payment	
		Hourly Rate	Hourly and Piece Rate
Scheduling	8–5	54, 68 55, 63	89, 75 71, 83
	Worker-Modified	79, 65 62, 74	83, 94 91, 86

a. What type of experiment was performed? Identify the response, factor(s), factor type(s), treatments, and experimental units.

```
                    General Linear Models Procedure

Dependent Variable: SCORE
                                 Sum of          Mean
Source                 DF       Squares        Square      F Value     Pr > F

Model                   3   1806.00000     602.00000        12.29     0.0006

Error                  12    588.00000      49.00000

Corrected Total        15   2394.00000

              R-Square         C.V.      Root MSE              SCORE Mean

              0.754386    9.3959732       7.00000              74.5000000

Source                 DF    Type I SS Mean Square     F Value     Pr > F

SCHEDULE                1     361.0000     361.0000        7.37     0.0188
PAYMENT                 1    1444.0000    1444.0000       29.47     0.0002
SCHEDULE*PAYMENT        1       1.0000       1.0000        0.02     0.8888
```

b. The SAS printout for this experiment is shown. Is there evidence that the treatment means differ? Use $\alpha = .05$.

c. If the test in part **b** warrants further analysis, conduct the appropriate tests of interaction and main effects. Interpret your results.

d. What assumptions are necessary to assure the validity of the inferences? State the assumptions in terms of this experiment.

16.50 A corporation that manages a large number of stores classifies them into three geographic divisions. Three stores are randomly selected from each division and a study is made to determine the mean inflation rate for the items in inventory. The inflation rate is recorded in the table as a percentage change in price over a year's time.

Division 1	*Division 2*	*Division 3*
1.1	1.4	.4
.9	1.6	.3
.8	1.0	.5

a. Is there sufficient evidence to indicate a difference in the mean inflation rates among the stores in different divisions?

b. In division 1, the numbers 1.1, .9, and .8 represent a random sample from what population?

16.51 One indicator of employee morale is the length of time employees stay with a company. A large corporation has three factories located in similar areas of the country. Although the corporation attempts to maintain uniformity in management, working conditions, and employee relations, it realizes that differences may exist among the various factories. To study this phenomenon, employee records are randomly selected at each of the three factories, and the length of service with the company is recorded. A summary of the data is shown. Is there evidence of a difference in mean length of service among the three factories? Use $\alpha = .05$.

Factory	1	2	3	
Number in Sample	15	21	17	

SST = 421.74
SSE = 3,574.06

16.52 England has experimented with different 40-hour work weeks to maximize production and minimize expenses. A factory tested a 5-day week (8 hours per day), a 4-day week (10 hours per day), and a 3⅓-day week (12 hours per day), with the weekly production results shown in the table (in thousands of dollars worth of items produced).

8-Hour Day	*10-Hour Day*	*12-Hour Day*
87	75	95
96	82	76
75	90	87
90	80	82
72	73	65
86		

a. What type of experiment was employed?

b. Use the SPSS printout (page 914) to test the null hypothesis that the mean productivity level is the same for the three lengths of workday. Use $\alpha = .10$.

c. Is further comparison of the pairs of means warranted? If so, use the Bonferroni technique to compare the pairs of mean productivity levels. Use an overall $\alpha = .10$.

```
  Variable    PRDCTION
By Variable   DAY
                                   Analysis of Variance

                              Sum of         Mean                 F        F
     Source          D.F.    Squares        Squares             Ratio    Prob.

Between Groups         2     57.6042        28.8021             .3375    .719

Within Groups         13   1109.3333        85.3333

Total                 15   1166.9375
```

16.53 To be able to provide its clients with comparative information on two large suburban residential communities, a realtor wants to know the average home value in each community. Eight homes are selected at random within each community and are appraised by the realtor. The appraisals are given in the table (in thousands of dollars). Can you conclude that the average home value is different in the two communities? You have three ways of analyzing this problem.

Suburb A		*Suburb B*	
43.5	57.5	73.5	44.5
49.5	32.0	62.0	56.0
38.0	67.5	47.5	68.0
66.5	71.5	36.5	63.5

a. Use the two-sample t statistic (Section 9.2) to test $H_0\colon \mu_A = \mu_B$.

b. Consider the regression model

$$y = \beta_0 + \beta_1 x + \epsilon$$

where

$$x = \begin{cases} 1 & \text{if community B} \\ 0 & \text{if community A} \end{cases} \qquad y = \text{Appraised price}$$

Since $\beta_1 = \mu_B - \mu_A$, testing $H_0\colon \beta_1 = 0$ is equivalent to testing $H_0\colon \mu_A = \mu_B$. Use the partial reproduction of the SAS printout shown here to test $H_0\colon \beta_1 = 0$. Use $\alpha = .05$.

```
SOURCE              DF    SUM OF SQUARES   MEAN SQUARE      F VALUE       PR > F

MODEL                1      40.64062500    40.64062500         0.21       0.6501
ERROR               14    2648.71875000   189.19419643                  ROOT MSE
CORRECTED TOTAL     15    2689.35937500                    R-SQUARE  13.75478813
                                                           0.015112

                                 T FOR H0:       PR > |T|    STD ERROR OF
PARAMETER        ESTIMATE     PARAMETER = 0                    ESTIMATE

INTERCEPT     53.25000000         10.95         0.0001       4.86305198
X              3.18750000          0.46         0.6501       6.87739406
```

c. The SAS ANOVA printout for this experiment is also given here. Is there evidence that the treatment means differ? Use $\alpha = .05$.

```
                     General Linear Models Procedure

Dependent Variable: PRICE
                                   Sum of          Mean
Source                     DF     Squares        Square    F Value      Pr > F

Model                       1  40.6406250    40.6406250       0.21      0.6501

Error                      14 2648.718750   189.1941964

Corrected Total            15 2689.359375

                 R-Square        C.V.      Root MSE              PRICE Mean

                 0.015112   25.079956       13.7548              54.8437500

Source                     DF    Type I SS  Mean Square   F Value      Pr > F

COMMUNTY                    1      40.6406      40.6406      0.21      0.6501
```

d. Using the results of the three tests in parts **a–c**, verify that the tests are the equivalent (for this special case, $p = 2$) of the completely randomized design in terms of the test statistic value and rejection region.

16.54 Random samples of size 36, 36, and 15 were drawn from stocks listed on the New York Stock Exchange (NYSE), the American Stock Exchange (ASE), and from those traded over-the-counter (OTC), respectively. The closing prices of all 87 stocks on December 15, 1986, are listed in the table.

NYSE			ASE			OTC		
14⅛	23	15½	8½	8½	14¾	27¼	11¾	8¾
15⅞	95	43¾	2⅝	3¼	16¾	9	32	11¾
24⅛	5⅜	80½	11¼	41⅞	⅜	3	9⅛	17
28	4⅛	11⅞	14⅛	12⅝	86	18	9	14½
77½	12¾	4½	5⅝	9½	6¼	8½	32½	21¼
30¾	35¼	16⅛	4⅞	5⅝	26½			
32	24¼	25	30¾	5	½			
11⅝	18	10¾	4⅜	24⅛	1½			
11⅛	21⅛	30⅜	6⅝	6¼	7⅞			
20⅛	14¾	83⅝	10¼	21⅝	13¼			
31⅞	19⅞	21¼	1⅝	17⅞	12¾			
13	8	6⅞	24⅛	16¼	35¼			

Source: *Wall Street Journal*, Dec. 16, 1986.

a. Is this experiment designed or observational? What type of experiment is it?
b. Given that the total sum of squares is 30,935.513 and the sum of squares for treatments is 2,759.905, complete the ANOVA summary table for the experiment.
c. Is there evidence that the mean closing price differed among the three markets? Use $\alpha = .10$.
d. Use the Bonferroni technique to compare the mean closing prices for all pairs of markets. Use an overall significance level of $\alpha = .10$.

16.55 Refer to Exercise 16.54.
a. Write a regression model to describe the experiment. Define all the variables you use.
b. The SAS regression printout for this experiment is shown here. Interpret the least squares estimates, and deduce the coding that was utilized for the dummy variables.

```
Model: MODEL1
Dep Variable: PRICE

                          Analysis of Variance

                         Sum of          Mean
Source         DF       Squares        Square      F Value       Prob>F

Model           2    2759.90461    1379.95230        4.114       0.0197
Error          84   28175.60868     335.42391
C Total        86   30935.51329

     Root MSE       18.31458      R-Square        0.0892
     Dep Mean       19.47270      Adj R-Sq        0.0675
     C.V.           94.05260

                          Parameter Estimates

                     Parameter       Standard      T for H0:
Variable   DF         Estimate          Error    Parameter=0    Prob > |T|

INTERCEP    1        15.558333     4.72880473          3.290        0.0015
X1          1        10.601389     5.62840342          1.884        0.0631
X2          1        -1.141667     5.62840342         -0.203        0.8398
```

c. Use the regression printout to test the null hypothesis that the three markets had the same mean closing prices on December 15, 1986. Use $\alpha = .10$. What is the observed significance level of the test?

16.56 In Case Study 16.1, Dickson used an analysis of variance to evaluate a questionnaire that he had designed to measure the computer expertise of middle managers. The data he employed in the analysis are displayed in Table 16.4. Use the data to replicate the analysis performed by Dickson—including the confidence intervals. Compare the results with those presented in Figure 16.11.

***16.57** In Example 12.6, we used a multiple regression analysis to analyze data for a 2 × 3 factorial experiment. The experiment measured worker productivity for each of two types of manufacturing plant (union and nonunion) and for each of three levels of bonus compensation. Three workers were randomly assigned to each of the six plant type–incentive factor level combinations. The objective of the experiment was to examine the relationship between worker productivity y and incentive level x for the two types of plants. Since the data in the table are the results of a 2 × 3 factorial experiment with the same number of observations per factor–level combination, we can analyze the data using an analysis of variance.

Type of Plant	*20¢/Casting Bonus*			*30¢/Casting Bonus*			*40¢/Casting Bonus*		
Union	1,435,	1,512,	1,491	1,583,	1,529,	1,610	1,601,	1,574,	1,636
Nonunion	1,575,	1,512,	1,488	1,635,	1,589,	1,661	1,645,	1,616,	1,689

a. Construct an analysis of variance table for the data. Use either a computer program or the formulas in Appendix D.

b. Do the data provide sufficient evidence to indicate an interaction between incentive level and type of plant? Test using $\alpha = .05$.

c. What are the practical implications of the test results of part **b**?

d. Compare your SSE with the one obtained in the multiple regression analysis, Figure 12.21. Why do they differ?

e. Compare the results of the analysis of Example 12.6 with those of your analysis of variance. Then consider the following: If one (or both) of the factors in a two-factor factorial experiment is a quantitative independent variable, a multiple regression analysis provides more practical information than an analysis of variance. Why? [*Note*: If both factors are qualitative variables, the two methods of analysis yield the same results.]

***16.58** Refer to Exercise 16.57.

a. Specify the regression model that could be utilized to yield results identical to those obtained in the ANOVA. [*Hint*: The model needs to have the same number of degrees of freedom for the main effects and the interaction as in the ANOVA summary table of Exercise 16.57, part **a**.]

b. Fit the model, and compare the value of SSE to that in Exercise 16.57.

c. Fit a reduced model and test to determine whether interaction terms contribute to the model. Test using $\alpha = .05$, and compare the result to part **b** of Exercise 16.57.

On Your Own

Design an experiment to compare the mean expenditure per customer for dinners in three fast-food restaurants. Your objective is to obtain a random sample of at least 25 customers' dinner expenditures in each of the restaurants. Write a statement of how you plan to proceed, listing all potential problems you might encounter and proposed solutions.

Conduct the experiment, and analyze the results. Carefully state the null and alternative hypotheses for each test you conduct, and specify the level of significance at which you conduct each test. Interpret the results carefully and completely.

Using the Computer

Do home values differ among the four census regions in the United States? Treat the 1,000 zip codes in the data set of Appendix C as if they were the population of all U.S. zip codes.

a. Use a completely randomized design with equal samples of size 50 to generate data to answer the above question. Construct a dot diagram and stem-and-leaf display for each of the four samples. What do your graphs suggest about the differences in mean home values among census regions?

b. Use ANOVA to answer the question of interest.

c. Use a multiple comparisons procedure to compare the four treatment means. Draw the appropriate conclusions.

d. What assumptions were necessary in order to employ the inferential procedures of parts **b** and **c**? Which of those assumptions are suspect in this problem?

References

Aplin-Brownlee, V. "Many workers facing terror of 'technophobia'." *Minneapolis Star and Tribune*, Jan. 10, 1984.

Box, G. E., Hunter, W. G., and Hunter, J. S. *Statistics for Experimenters*. New York: Wiley, 1978.

Brockhaus, R. H. "Risk-taking propensity of entrepreneurs." *Academy of Management Journal*, Sept. 1980, Vol. 23, pp. 509–520.

Curhan, R. C. "The effects of merchandising and temporary promotional activities on the sales of fresh fruit and vegetables in supermarkets." *Journal of Marketing Research*, Aug. 1974, Vol. 11, pp. 286–294.

Johnson, D. A., Pany, K., and White, R. "Audit reports and the loan decision: Actions and perceptions." *Auditing: A Journal of Practice and Theory*, Spring 1983, Vol. 2, pp. 38–51.

Klompmaker, J. E., Hughes, G. D., and Haley, R. I. "Test marketing in new products development." *Harvard Business Review*, May–June 1976, p. 128.

Logue, D., and Rogalski, R. *Harvard Business Review*, July–Aug. 1979.

Mendenhall, W. *Introduction to Linear Models and the Design and Analysis of Experiments*. Belmont, CA: Wadsworth, 1968. Chapter 8.

Miller, R. B., and Wichern, D. W. *Intermediate Business Statistics: Analysis of Variance, Regression, and Time Series*. New York: Holt, Rinehart and Winston, 1977. Chapter 4.

Mowen, J. C., Keith, J. E., Brown, S. W., and Jackson, D. W., Jr. "Utilizing effort and task difficulty information in evaluating salespeople." *Journal of Marketing Research*, May 1985, Vol. 22, pp. 185–191.

Neter, J., Wasserman, W., and Kutner, M. *Applied Linear Statistical Models*, 3rd ed. Homewood, Ill.: Richard D. Irwin, 1990.

Phillips, M. D., and Pepper, R. L. "Shipboard firefighting performance of females and males." *Human Factors*, 1982, Vol. 24, No. 3.

Scheffé, H. *The Analysis of Variance*. New York: Wiley, 1959.

Shetty, Y. K., and Carlisle, H. M. "Organizational correlates of a management by objectives program." *Academy of Management Journal*, 1974, Vol. 17, No. 1.

Wilkinson, J. B., Mason, J. B., and Paksoy, C. H. "Assessing the impact of short-term supermarket strategy variables." *Journal of Marketing Research*, Feb. 1982, Vol. 19, pp. 72–86.

CHAPTER SEVENTEEN
Nonparametric Statistics

Contents

Case Study

Where We've Been

Chapters 7–9 and 16 presented techniques for making inferences about the mean of a single population and for comparing the means of two or more populations. Chapters 10–12 treated simple and multiple regression—the problem of relating the mean of a population of y values to a set of independent variables $x_1, x_2, \ldots, x_k$. Most of the techniques discussed in Chapters 7–12 and 16 are based on the assumption that the sampled populations have probability distributions that are approximately normal with equal variances. But how can you analyze data from populations that do not satisfy these assumptions? And how can you make comparisons between populations when you cannot assign specific numerical values to your observations?

Where We're Going

In this chapter, we present inferential techniques that are based on an ordering of the sample measurements according to their relative magnitudes. These techniques, which require fewer or less stringent assumptions concerning the nature of the probability distributions of the populations, are called *nonparametric statistical methods*.

The t and F tests presented in Chapters 7–9 and 16 are unsuitable for some types of business data. These data fall into two categories: The first are data from populations that do not satisfy the assumptions upon which the t and F tests are based. For both tests, we assume that the random variables being measured have normal probability distributions with equal variances. Yet in practice, the observations from one population may exhibit much greater variability than those from another, or the probability distributions may be decidedly nonnormal. For example, the distribution might be very flat, peaked, or strongly skewed to the right or left. When any of the assumptions required for the t and F tests is seriously violated, the computed t and F statistics may not follow the standard t and F distributions. If this is true, the tabulated values of t and F (Table VI–X of Appendix B) are not applicable, the correct value of α for the test is unknown, and the t and F tests are of dubious value.

The second type of data for which t and F tests are inappropriate are responses that are not susceptible to numerical measurement but that can be *ranked in order of magnitude*. That is, in the terminology introduced in Chapter 2, the t and F tests are generally not appropriate for **ordinal** data. For example, if we want to compare the managerial ability of two executives based on subjective evaluations of trained observers, despite the fact that we cannot give an exact value to the managerial ability of a single executive, we may be able to decide that executive A has more ability than executive B. If executives A and B are evaluated by each of 10 observers, we have the standard problem of comparing the probability distributions for two populations of ratings, one for executive A and one for B. But the t test of Chapter 9 would be inappropriate because the only data that can be recorded are the ordinal preferences; i.e., each observer decides either that A is better than B or vice versa.

Consider another example of this type of data. Most firms that plan to market a new product nationally first test the product in a few cities or regions to determine its acceptability. For a food product this may entail taste tests in which consumers rank the new product in order of preference with respect to one or more currently popular brands. A consumer probably has a preference for each product, but the strength of the preference is difficult, if not impossible, to measure. Consequently, the best we can do is have each consumer examine the new product along with a few established products, and rank them according to preference: 1 for the most preferred, 2 for second, etc. The result is ordinal data, for which the t and F tests would be inappropriate.

The **nonparametric** counterparts of the t and F tests compare the probability distributions of the sampled populations, rather than specific parameters of those populations (such as the means or variances). For example, nonparametric tests can be used to compare the probability distribution of the strengths of preferences for a new product to the probability distributions of the strengths of preferences for the currently popular brands. If it can be inferred that the distribution for the new product lies above (to the right of) the others (see Figure 17.1), the implication is that the new product tends to be more preferred than the currently popular products. Such an inference might lead to a decision to market the product nationally.

Most nonparametric methods use the **relative ranks** of the sample observations rather than their actual numerical values. These tests are particularly valuable when we are unable to obtain numerical measurements of some phenomena but are able to

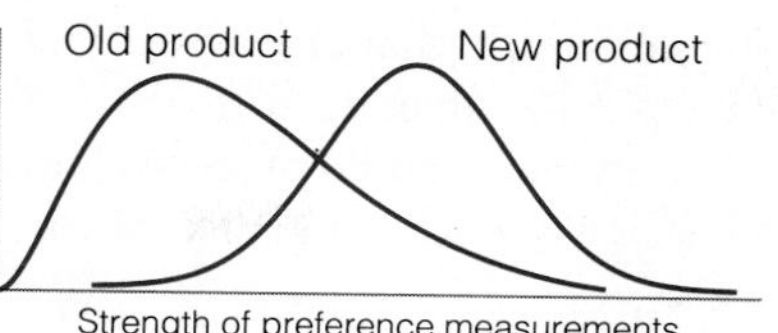

FIGURE 17.1 ► Probability distributions of strengths of preference measurements (new product is preferred)

rank them in comparison to each other. Statistics based on ranks of measurements are called **rank statistics**. In Section 17.1 we present a test to make inferences about the central tendency of a population. In Sections 17.2 and 17.4, we present rank statistics for comparing probability distributions using independent samples. In Section 17.3, paired samples are used to make nonparametric comparisons of populations. Finally, in Section 17.5 we present a nonparametric measure of correlation between two variables—*Spearman's rank correlation coefficient*. For a more complete discussion of tests based on rank statistics, see the references at the end of this chapter.

17.1 Single-Population Inferences: The Sign Test

In Chapter 8 we utilized the z and t statistics for testing hypotheses about a population mean. The z statistic is appropriate for large random samples selected from "general" populations—i.e., with few limitations on the probability distribution of the underlying population. The t statistic was developed for small-sample tests when the sample is selected at random from a *normal* distribution. The question is: How can we conduct a test of hypothesis when we have a small sample from a **nonnormal** distribution?

The **sign test** is a relatively simple nonparametric procedure to test hypotheses about the central tendency of a nonnormal probability distribution. Note that we used the phrase *central tendency* rather than *population mean*. This is because the sign test, like many nonparametric procedures, provides inferences about the population **median**, M, rather than the population mean, μ. Remember (Chapter 2) that the median is the 50th percentile of the distribution (Figure 17.2) and as such is less affected by the skewness of the distribution and the presence of outliers (extreme observations) than the mean. Since the nonparametric test must be suitable for all distributions, not just the normal, it is reasonable for nonparametric tests to focus on the more robust (less sensitive to extreme values) measure of central tendency, the median.

For example, suppose that a large metropolitan bank regularly samples several of its commercial loan accounts in order to audit their current financial stability. One of the important financial statistics calculated is the debt-to-equity ratio, and the bank's

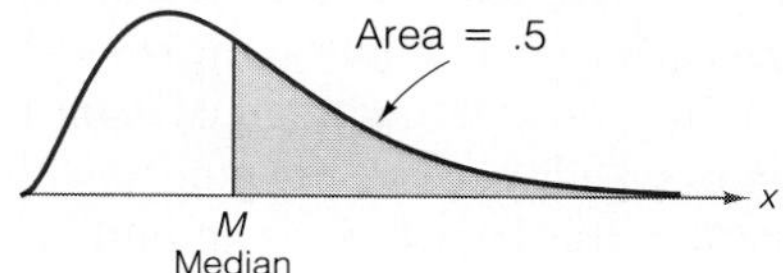

FIGURE 17.2 ► Location of the population median, M

board of directors wants to be sure that at least 50% of its commercial loan accounts have debt-to-equity ratios of less than 1.25 (that is, corporate debt is less than 1.25 times the company's equity). An audit of a random sample of eight commercial accounts produced the following debt-to-equity ratios:

1.18 .51 4.79 1.20 .77 1.23 1.06 1.04

If the objective is to determine whether the median is less than 1.25, we establish that as our alternative hypothesis and test

H_0: $M = 1.25$ H_a: $M < 1.25$

The one-tailed sign test is conducted by counting the number of sample measurements that "favor" the alternative hypothesis—in this case, the number that are less than 1.25. If the null hypothesis is true, we expect approximately half of the measurements to fall on each side of the hypothesized median; and if the alternative is true, we expect significantly more than half to favor the alternative—i.e., to be less than 1.25. Thus:

Test statistic: S = Number of measurements less than 1.25, the null hypothesized median

If we wish to conduct the test at the $\alpha = .05$ level of significance, then the rejection region can be expressed in terms of the observed significance level, or p-value, of the test:

Rejection region: $p\text{-value} \leq .05$

In this example, $S = 7$ of the 8 companies have debt-to-equity ratios less than 1.25. To determine the observed significance level associated with this outcome, we note that the number of measurements less than 1.25 is a binomial random variable (check the binomial characteristics presented in Chapter 4), and *if* H_0 *is true*, the binomial probability p that a measurement lies below (or above) the median 1.25 is equal to .5 (Figure 17.2). What is the probability of obtaining a result *as contrary or more contrary* to H_0 than the one observed? That is, what is the probability that 7 *or more* of 8 binomial measurements will result in success (be less than 1.25) if the probability of success is .5? Table II in Appendix B indicates that

$$
\begin{aligned}
P(x \geq 7) &= 1 - P(x \leq 6) \\
&= 1 - .965 \\
&= .035
\end{aligned}
$$

using $n = 8$ and $p = .5$. Thus, the probability that at least 7 of 8 companies would have ratios less than 1.25 *if the true median were* 1.25 is only .035. The p-value of the test is therefore .035, which is less than $\alpha = .05$. We therefore conclude that this sample provides sufficient evidence to reject the null hypothesis. The implication of this rejection is that the bank can conclude at the $\alpha = .05$ level of significance that the true median debt-to-equity ratio of its commercial loans is less than 1.25. However, we note that one of the loans greatly exceeds the others, with a ratio of 4.79, and deserves special attention. Note that this large ratio is an outlier that would make the use of a t test and its commensurate assumption of normality dubious.

Sign Test for a Population Median M

One-Tailed Test	*Two-Tailed Test*
H_0: $M = M_0$ H_a: $M > M_0$ [or H_a: $M < M_0$]	H_0: $M = M_0$ H_a: $M \neq M_0$
Test statistic: S = Number of sample measurements greater than M_0 [or S = Number of sample measurements less than M_0]	*Test statistic:* S = Larger of S_1 and S_2 where S_1 = Number of measurements less than M_0 and S_2 = Number of measurements greater than M_0
Observed significance level: p-value $= P(x \geq S)$	*Observed significance level:* p-value $= 2P(x \geq S)$

where x has a binomial distribution with parameters n and $p = .5$. (Use Table II, Appendix B.)

Rejection region: Reject H_0 if p-value $\leq .05$.

Assumption: The sample is selected randomly from a continuous probability distribution. [*Note:* No assumptions need to be made about the shape of the probability distribution.]

The use of the sign test for testing hypotheses about population medians is summarized in the box.

Recall that the normal probability distribution provides a good approximation for the binomial distribution when the sample size is large. For tests about the median of a distribution, the null hypothesis implies that $p = .5$, and the normal distribution will provide a good approximation if $n \geq 10$. (Samples with $n \geq 10$ satisfy the condition that $np \pm 3\sqrt{npq}$ be contained in the interval 0 to n.) Thus, we can use the standard normal z distribution to conduct the sign test for large samples. The large-sample sign test is summarized in the box at the top of page 924.

EXAMPLE 17.1

A manufacturer of compact disk (CD) players has established that the median time to failure for its players is 5,250 hours of utilization. A sample of 20 CD players from a competitor is obtained, and they are continuously tested until each fails. The 20 failure times range from 5 hours (a "defective" player) to 6,575 hours, and 14 of the 20 exceed 5,250 hours. Is there evidence that the median failure time of the competitor differs from 5,250 hours? Use $\alpha = .10$.

Large-Sample Sign Test for a Population Median M

One-Tailed Test

H_0: $M = M_0$
H_a: $M > M_0$
[or H_a: $M < M_0$]

Two-Tailed Test

H_0: $M = M_0$
H_a: $M \neq M_0$

Test statistic: $z = \dfrac{(S - .5) - .5n}{.5\sqrt{n}}$

[*Note:* S is calculated as shown in the previous box. We subtract .5 from S as the "correction for continuity." The null hypothesized mean value is $np = .5n$, and the standard deviation is

$$\sqrt{npq} = \sqrt{n(.5)(.5)} = .5\sqrt{n}$$

See Chapter 5 for details on the normal approximation to the binomial distribution.]

Rejection region: $z > z_\alpha$ *Rejection region:* $z > z_{\alpha/2}$

where tabulated z values can be found inside the front cover.

Solution

The null and alternative hypotheses of interest are

H_0: $M = 5{,}250$ hours H_a: $M \neq 5{,}250$ hours

Test statistic: Since $n \geq 10$, we use the standard normal z statistic:

$$z = \frac{(S - .5) - .5n}{.5\sqrt{n}}$$

where S is the maximum of S_1, the number of measurements greater than 5,250, and S_2, the number of measurements less than 5,250.

Rejection region: $z > 1.645$, where $z_{\alpha/2} = z_{.05} = 1.645$

Assumptions: The distribution of the failure times is continuous (time is a contintous variable), but nothing is assumed about the shape of its probability distribution.

Since the number of measurements exceeding 5,250 is $S_1 = 14$, and thus the number of measurements less than 5,250 is $S_2 = 6$, then S is 14, the greater of S_1 and S_2. The calculated z statistic is, therefore,

$$z = \frac{(S - .5) - .5n}{.5\sqrt{n}} = \frac{13.5 - 10}{.5\sqrt{20}} = \frac{3.5}{2.236} = 1.565$$

The value of z is not in the rejection region, so we cannot reject the null hypothesis at the $\alpha = .10$ level of significance. Thus, the CD manufacturer should not conclude, on the basis of this sample, that its competitor's CD players have a failure time median that differs from 5,250 hours.

The one-sample nonparametric sign test for a median provides an alternative to the t test for small samples from nonnormal distributions. However, if the distribution is approximately normal, the t test provides a more powerful test about the central tendency of the distribution.

Exercises 17.1 – 17.8

Learning the Mechanics

17.1 Under what circumstances is the sign test preferred to the t test for making inferences about the central tendency of a population?

17.2 What is the probability that a randomly selected observation exceeds each of the following?

a. Mean of a normal distribution
b. Median of a normal distribution
c. Mean of a nonnormal distribution
d. Median of a nonnormal distribution

17.3 Use Table II of Appendix B to calculate the following binomial probabilities:

a. $P(x \geq 7)$ when $n = 8$ and $p = .5$
b. $P(x \geq 5)$ when $n = 8$ and $p = .5$
c. $P(x \geq 8)$ when $n = 8$ and $p = .5$
d. $P(x \geq 10)$ when $n = 15$ and $p = .5$. Also use the normal approximation to calculate this probability, and compare the approximation with the exact value.
e. $P(x \geq 15)$ when $n = 25$ and $p = .5$. Also use the normal approximation to calculate this probability, and compare the approximation with the exact value.

17.4 Consider the following sample of ten measurements:

12.1 8.5 15.8 17.7 9.6 4.0 25.2 10.3 6.2 13.9

Use these data to conduct each of the following sign tests using the binomial tables (Table II, Appendix B) and $\alpha = .05$:

a. H_0: $M = 10$ versus H_a: $M > 10$
b. H_0: $M = 10$ versus H_a: $M \neq 10$
c. H_0: $M = 18$ versus H_a: $M < 18$
d. H_0: $M = 18$ versus H_a: $M \neq 18$
e. Repeat each of the preceding tests using the normal approximation to the binomial probabilities. Compare the results.
f. What assumptions are necessary to assure the validity of each test?

17.5 Suppose you wish to conduct a test of the research hypothesis that the median of a population is greater than 75. You randomly sample 25 measurements from the population and determine that 17 of them exceed 75. Set up and conduct the appropriate test of hypothesis at the .10 level of significance. Be sure to specify all necessary assumptions.

Applying the Concepts

17.6 In the early 1990s many firms found it necessary to reduce the size of their workforces in order to reduce costs. These reductions were referred to as *corporate downsizing* and *reductions in force* (RIF) by the business community and media. Subsequently, many companies were sued by former employees who alleged that the RIFs were discriminatory with regard to age. Federal law protects employees over 40 years of age against such discrimination.

Suppose one large company's employees have a median age of 37. Its RIF plan is to fire 15 employees aged 43, 32, 39, 28, 54, 41, 50, 62, 22, 45, 47, 54, 43, 33, and 59 years.

a. Calculate the median age of the employees who are being terminated.

b. What are the appropriate null and alternative hypotheses to test whether the population from which the terminated employees were selected has a median age that exceeds the entire company's median age?

c. Conduct the test of part **b**, and report the significance level of the test.

d. Assuming that courts generally require statistical evidence at the .10 level of significance before ruling that age discrimination laws were violated, what do you advise the company about its planned RIF? Explain.

17.7 Airline industry analysts expect that the Federal Aviation Administration (FAA) will increase the frequency and thoroughness of its review of aircraft maintenance procedures in response to the admission by Eastern Airlines in the summer of 1990 that it had not met some maintenance requirements. Suppose that the FAA samples the records of six aircraft currently utilized by one airline and determines the number of flights between the last two complete engine maintenances for each, with the following results: 24, 27, 25, 94, 29, 28.

The FAA requires that this maintenance be performed at least every 30 flights. Although it is obvious that not all aircraft are meeting the requirement, the FAA wishes to test whether the airline is meeting this particular maintenance requirement "on average."

a. Would you suggest the t test or sign test to conduct the test? Why?

b. Set up the null and alternative hypotheses such that the burden of proof is on the airline to show it is meeting the "on-average" requirement.

c. What are the test statistic and rejection region for this test if the level of significance is $\alpha = .01$? Why would the level of significance be set at such a low value?

d. Conduct the test, and state the conclusion in terms of this application.

17.8 A paper company requires that the median height of pine trees exceed 40 feet before they are harvested. A sample of 24 trees in one large plot is selected, and 17 of them are over 40 feet.

a. Test whether the company can conclude at the .05 level of significance that the median height of trees in the plot exceeds 40 feet.

b. Calculate and interpret the p-value of this test.

c. What assumptions must be made about the probability distribution of the tree heights in the plot in order to assure the validity of the test?

17.2 Comparing Two Populations: Wilcoxon Rank Sum Test for Independent Samples

Suppose two independent random samples are to be used to compare two populations, and the t test of Chapter 9 is inappropriate for making the comparison. Either we are unwilling to make assumptions about the form of the underlying probability distributions, or we are unable to obtain exact values of the sample measurements but can rank them in order of magnitude. For either of these situations, if the data can be ordered, the **Wilcoxon rank sum test** (developed by Frank Wilcoxon) can be used to test a hypothesis that the probability distributions associated with the two populations are equivalent.

For example, suppose six economists who work for the federal government and seven university economists are randomly selected, and each is asked to predict next year's percentage change in cost of living as compared with this year's figure. The objective of the study is to compare the government economists' predictions to those of the university economists. The data are shown in Table 17.1.

The two populations of predictions are those that would be obtained from *all* government and *all* university economists if they could all be questioned. To compare their probability distributions, we first *rank the sample observations as though they were all drawn from the same population*. That is, we pool the measurements from both samples and then rank the measurements from the smallest (a rank of 1) to the largest (a rank of 13). The ranks of the 13 economists' predictions are indicated in Table 17.1.

The test statistic for the Wilcoxon test is based on the totals of the ranks for each of the two samples—that is, on the **rank sums**. If the two rank sums are nearly equal, the implication is that there is no evidence that the probability distributions from which the samples were drawn are different. On the other hand, if the two rank sums are very different, the implication is that the two samples may have come from different populations.

TABLE 17.1 Percentage Cost of Living Change, as Predicted by Government and University Economists

Government Economist		University Economist	
Prediction	Rank	Prediction	Rank
3.1	4	4.4	6
4.8	7	5.8	9
2.3	2	3.9	5
5.6	8	8.7	11
0.0	1	6.3	10
2.9	3	10.5	12
		10.8	13

For the economists' predictions, we arbitrarily denote the rank sum for government economists by T_A and that for university economists by T_B. Then

$$T_A = 4 + 7 + 2 + 8 + 1 + 3 = 25$$
$$T_B = 6 + 9 + 5 + 11 + 10 + 12 + 13 = 66$$

The sum of T_A and T_B will always equal $n(n + 1)/2$, where $n = n_1 + n_2$. So, for this example, $n_1 = 6$, $n_2 = 7$, and

$$T_A + T_B = \frac{13(13 + 1)}{2} = 91$$

Since $T_A + T_B$ is fixed, a small value for T_A implies a large value for T_B (and vice versa) and a large difference between T_A and T_B. Therefore, the smaller the value of one of the rank sums, the greater is the evidence to indicate that the samples were selected from different populations.

Values that locate the rejection region for the rank sum associated with the smaller sample are given in Table XI of Appendix B. A partial reproduction of this table is shown in Table 17.2. The columns of the table represent n_1, the first sample size, and the rows represent n_2, the second sample size. *The T_L and T_U entries in the table are the boundaries of the lower and upper regions, respectively, for the rank sum associated with the sample that has fewer measurements.* If the sample sizes n_1 and n_2 are the same, either rank sum may be used as the test statistic. To illustrate, suppose $n_1 = 8$

Wilcoxon Rank Sum Test: Independent Samples*

One-Tailed Test

H_0: Two sampled populations have identical probability distributions

H_a: The probability distribution for population A is shifted to the right of that for B

Test statistic: The rank sum T associated with the sample with fewer measurements (for equal sample sizes, use either rank sum)

Rejection region: Assuming the smaller sample size is associated with distribution A (or, for equal sample sizes, use the rank sum T_A), we reject H_0 if $T_A \geq T_U$, where T_U is the upper value given by Table XI in Appendix B for the chosen *one-tailed* α value.

Two-Tailed Test

H_0: Two sampled populations have identical probability distributions

H_a: The probability distribution for population A is shifted to the left *or* to the right of that for B

Test statistic: The rank sum T associated with the sample with fewer measurements (for equal sample sizes, use either rank sum)

Rejection region: $T \leq T_L$ or $T \geq T_U$, where T_L is the lower value given by Table XI in Appendix B for the chosen *two-tailed* α value, and T_U is the upper value from Table XI.

[*Note*: If the one-sided alternative is that the probability distribution for A is shifted to the *left* of B (and T_A is the test statistic), we reject H_0 if $T_A \leq T_L$.]

Assumptions: 1. The two samples are random and independent.
2. The two probability distributions from which the samples are drawn are continuous.

Ties: Assign tied measurements the average of the ranks they would receive if they were unequal but occurred in successive order. For example, if the third-ranked and fourth-ranked measurements are tied, assign each a rank of $(3 + 4)/2 = 3.5$.

*Another statistic used for comparing two populations based on independent random samples is the **Mann-Whitney *U* statistic**. The U statistic is a simple function of the rank sums. It can be shown that the Wilcoxon rank sum test and the Mann-Whitney U test are equivalent.

and $n_2 = 10$. For a two-tailed test with $\alpha = .05$, we consult part **a** of the table and find that the null hypothesis will be rejected if the rank sum of sample 1 (the sample with fewer measurements), T, is less than or equal to $T_L = 54$ *or* greater than or equal to $T_U = 98$. The Wilcoxon rank sum test is summarized in the preceding box.

Note that the assumptions necessary for the validity of the Wilcoxon rank sum test do not specify the shape or type of probability distribution. However, the distributions are assumed to be continuous so that the probability of tied measurements is 0 (see Chapter 5), and each measurement can be assigned a unique rank. In practice, however, rounding of continuous measurements will sometimes produce ties. As long as the number of ties is small relative to the sample sizes, the Wilcoxon test procedure will still have approximate significance level α. The test is not recommended to compare discrete distributions for which many ties are expected.

TABLE 17.2 Reproduction of Part of Table XI of Appendix B: Critical Values of T_L and T_U for the Wilcoxon Rank Sum Test: Independent Samples

a. $\alpha = .025$ one-tailed; $\alpha = .05$ two tailed

n_2 \ n_1	3		4		5		6		7		8		9		10	
	T_L	T_U	T_L	T_U	T_L	T_U	T_L	T_U	T_L	T_U	T_L	T_U	T_L	T_U	T_L	T_U
3	5	16	6	18	6	21	7	23	7	26	8	28	8	31	9	33
4	6	18	11	25	12	28	12	32	13	35	14	38	15	41	16	44
5	6	21	12	28	18	37	19	41	20	45	21	49	22	53	24	56
6	7	23	12	32	19	41	26	52	28	56	29	61	31	65	32	70
7	7	26	13	35	20	45	28	56	37	68	39	73	41	78	43	83
8	8	28	14	38	21	49	29	61	39	73	49	87	51	93	54	98
9	8	31	15	41	22	53	31	65	41	78	51	93	63	108	66	114
10	9	33	16	44	24	56	32	70	43	83	54	98	66	114	79	131

EXAMPLE 17.2

Test the hypothesis that the university economists' predictions of next year's percentage change in cost of living tend to be higher than the government economists'. Conduct the test using the data in Table 17.1 and $\alpha = .05$.

Solution

H_0: The probability distributions corresponding to the government and university economists' predictions of inflation rate are identical

H_a: The probability distribution for the university economists' predictions lies above (to the right of) the probability distribution for the government economists' predictions*

*The alternative hypotheses in this chapter will be stated in terms of a difference in the *location* of the distributions. However, since the shapes of the distributions may also differ under H_a, some figures (e.g., Figure 17.3) depicting H_a will show probability distributions with different shapes.

Test statistic: Since fewer government economists ($n_1 = 6$) than university economists ($n_2 = 7$) were sampled, the test statistic is T_A, the rank sum of the government economists' predictions.

Rejection region: Since the test is one-sided, we consult part **b** of Table XI for the rejection region corresponding to $\alpha = .05$. We reject H_0 only for $T_A \leq T_L$, the lower value from Table XI, since we are specifically testing that the distribution of the government economists' predictions lies *below* the distribution of university economists' predictions, as shown in Figure 17.3. Thus, we reject H_0 if $T_A \leq 30$.

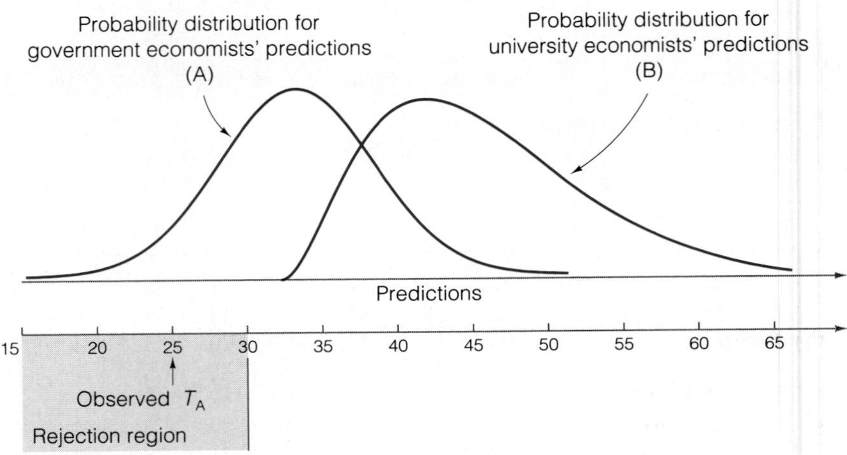

FIGURE 17.3 ► Alternative hypothesis and rejection region for Example 17.2

Since T_A, the rank sum of the government economists' predictions in Table 17.1, is 25, it is in the rejection region (see Figure 17.3). Therefore, we can conclude that the university economists' predictions tend, in general, to exceed the government economists' predictions.

Table XI of Appendix B gives values of T_L and T_U for sample sizes n_1 and n_2 less than or equal to 10. When both sample sizes are 10 or larger, the sampling distribution of T_A can be approximated by a normal distribution with mean and variance

$$E(T_A) = \frac{n_1(n_1 + n_2 + 1)}{2} \quad \text{and} \quad \sigma^2_{T_A} = \frac{n_1 n_2(n_1 + n_2 + 1)}{12}$$

Therefore, for $n_1 \geq 10$ and $n_2 \geq 10$, we can conduct the Wilcoxon rank sum test using the familiar z test of Chapters 7–9. The large-sample test is summarized in the next box.

Wilcoxon Rank Sum Test: Large Independent Samples

One-Tailed Test

H_0: Two sampled populations have identical probability distributions

H_a: The probability distribution for population A is shifted to the right of that for B

Test statistic: $$z = \frac{T_A - \frac{n_1(n_1 + n_2 + 1)}{2}}{\sqrt{\frac{n_1 n_2(n_1 + n_2 + 1)}{12}}}$$

Rejection region: $z > z_\alpha$

Two-Tailed Test

H_0: Two sampled populations have identical probability distributions

H_a: The probability distribution for population A is shifted left *or* right of that for B

Test statistic: $$z = \frac{T_A - \frac{n_1(n_1 + n_2 + 1)}{2}}{\sqrt{\frac{n_1 n_2(n_1 + n_2 + 1)}{12}}}$$

Rejection region: $z < -z_{\alpha/2}$ or $z > z_{\alpha/2}$

Assumptions: $n_1 \geq 10$ and $n_2 \geq 10$

Exercises 17.9–17.18

Learning the Mechanics

17.9 Specify the test statistic and the rejection region for the Wilcoxon rank sum test for independent samples in each of the following situations.

a. H_0: Two probability distributions, A and B, are identical
   H_a: Probability distribution for population A is shifted to the right or left of the probability distribution for population B
   $n_A = 10, \quad n_B = 6, \quad \alpha = .10$

b. H_0: Two probability distributions, A and B, are identical
   H_a: Probability distribution for population A is shifted to the right of the probability distribution for population B
   $n_A = 5, \quad n_B = 7, \quad \alpha = .05$

c. H_0: Two probability distributions, A and B, are identical
   H_a: Probability distribution for population A is shifted to the left of the probability distribution for population B
   $n_A = 9, \quad n_B = 8, \quad \alpha = .025$

d. H_0: Two probability distributions, A and B, are identical
   H_a: Probability distribution for population A is shifted to the right or left of the probability distribution for population B
   $n_A = 15, \quad n_B = 15, \quad \alpha = .05$

17.10 Suppose you want to compare two treatments, A and B. In particular, you wish to determine whether the distribution for population B is shifted to the right of the distribution for population A. You plan to use the Wilcoxon rank sum test.

a. Specify the null and alternative hypotheses you would test.
b. Suppose you obtained the following independent random samples of observations on experimental units subjected to the two treatments:

A: 36, 39, 33, 29, 42, 33, 35, 28, 34
B: 35, 48, 52, 66

Conduct a test of the hypotheses described in part **a**. Test using $\alpha = .05$.

17.11 Explain the difference between the one- and two-tailed versions of the Wilcoxon rank sum test for independent random samples.

17.12 Random samples of sizes $n_1 = 20$ and $n_2 = 15$ were drawn from populations 1 and 2, respectively. The measurements obtained are listed in the table.

Population 1				*Population 2*		
9.0	17.9	25.6	31.1	10.1	11.1	13.5
21.1	26.9	24.6	23.1	12.0	18.2	10.3
24.8	16.5	26.0	25.1	9.2	7.0	14.2
17.2	30.1	14.3	26.1	12.2	13.6	13.2
18.9	25.4	22.0	23.3	8.8	12.5	21.5

a. Conduct a hypothesis test to determine whether the probability distribution for population 2 is shifted to the left of the probability distribution for population 1. Use $\alpha = .05$.
b. What is the approximate p-value of the test of part **a**?

Applying the Concepts

17.13 A property's *assessment ratio* is calculated by dividing the property's assessed value (for tax purposes) by its market value (or a proxy for market value, such as a recent sale price; Freedman, 1985). Tax assessors' valuations are often evaluated by private real estate appraisers and government agencies using procedures such as the Wilcoxon rank sum test and the Kruskal-Wallis test (described in Section 17.4) to compare the distributions of the assessment ratios for different groups of properties. The table lists the assessment ratios for random samples of 10 properties in neighborhood A and eight properties in neighborhood B.

Neighborhood A		*Neighborhood B*	
.850	.880	.911	.835
1.060	.895	.770	.800
.910	.844	.815	.793
.813	.965	.748	.796
.737	.875		

a. Use the Wilcoxon rank sum test to investigate the fairness of the assessments between the two neighborhoods. Use $\alpha = .05$ and interpret your findings in the context of the problem.
b. Under what circumstances could the two-sample t test of Chapter 9 be used to investigate the fairness issue of part **a**?
c. What assumptions are necessary to assure the validity of the test you conducted in part **a**?

17.14 Recall that the variance of a binomial sample proportion, $\hat{p}$, depends on the value of the population parameter, p. As a consequence, the variance of a sample percentage, $(100\hat{p})\%$, also depends on p. Thus if you conduct an unpaired t test (Section 9.2) to compare the means of two populations of percentages, you may be violating the assumption that $\sigma_1^2 = \sigma_2^2$, upon which the t test is based. If the disparity in the variances is large, you will obtain more reliable test results using the Wilcoxon rank sum test for independent samples. In Exercises 9.29, we used a Student's t test to compare the mean annual percentages of labor turnover between U.S. and Japanese manufacturers of air conditioners. The annual percentage turnover rates for five U.S. and five Japanese plants are shown in the table. Do the data provide sufficient evidence to indicate that the mean annual percentage turnover for American plants exceeds the corresponding mean for Japanese plants? Test using the Wilcoxon rank sum test with $\alpha = .05$. Do your test conclusions agree with those of the t test in Exercise 9.29?

U.S. Plants	*Japanese Plants*
7.11%	3.52%
6.06	2.02
8.00	4.91
6.87	3.22
4.77	1.92

17.15 A state highway department has decided to investigate the increased severity of automobile accidents occurring at an urban intersection since the adoption of the right-turn-on-red law. From police records, it chooses a random sample of eight accidents that occurred at the intersection before the law was enacted and a random sample of eight accidents that occurred after the law was enacted. The total damage estimate for each accident was the measure of severity. The damage estimates are recorded in the table. Use the Wilcoxon rank sum test to determine whether the damages tended to increase after the enactment of the law. Test using $\alpha = .05$. Draw appropriate conclusions.

Damage per Accident			
Before Right-Turn Law		*After Right-Turn Law*	
$150	$242	$145	$ 899
500	435	390	1,250
250	100	680	290
301	402	560	963

17.16 A major razor blade manufacturer advertises that its twin-blade disposable razor "gets you lots more shaves" than any single-blade disposable razor on the market. A rival company that has been very successful in selling single-blade razors plans to test this claim. Independent random samples of eight single-blade users and eight twin-blade users are taken, and the number of shaves that each gets before indicating a preference to change blades is recorded. The results are shown in the table.

Twin Blades		*Single Blades*	
8	15	10	13
17	10	6	14
9	6	3	5
11	12	7	7

a. Do the data support the twin-blade manufacturer's claim? Use $\alpha = .05$.

b. Do you think this experiment was designed in the best possible way? If not, what design might have been better?

c. What assumptions are necessary for the validity of the test you performed in part **a**? Do the assumptions seem reasonable for this application?

17.17 A realtor wants to determine whether a difference exists between home prices in two subdivisions. Six homes from subdivision A and eight homes from subdivision B are sampled, and the prices (in thousands of dollars) are recorded in the table.

Subdivision A		*Subdivision B*	
43	39	57	88
48	47	39	46
42		55	41
60		52	64

a. Use the two-sample t test to compare the population mean prices per house in the two subdivisions. What assumptions are necessary for the validity of this procedure? Do you think they are reasonable in this case?

b. Use the Wilcoxon rank sum test to determine whether there is a shift in the locations of the probability distributions of house prices in the two subdivisions.

17.18 A *management information system* (MIS) is a computer-based information-processing system designed to support the operations, management, and decision functions of an organization. The development of an MIS involves three stages: definition, physical design, and implementation of the system (Davis and Olson, 1985). The successful implementation of an MIS is related to the quality of the entire development process. It could fail due to inadequate planning by and negotiating between the designers and the future users of the system prior to construction, or simply because the users were improperly trained (Alter and Ginzberg, 1978).

Thirty firms that recently implemented an MIS were surveyed: 16 were satisfied with the implementation results, 14 were not. Each firm was asked to rate the quality of the planning and negotiation stages of the development process, using a scale of 0 to 100, with higher numbers indicating better quality. (A score of 100 indicates that all the problems that occurred in the planning and negotiation stages were successfully resolved, while 0 indicates that none were resolved.) The results obtained are shown in the table.

Firms with Good MIS			*Firms with Poor MIS*		
52	59	95	60	40	90
70	60	90	50	55	85
40	90	86	55	65	80
80	75	95	70	55	90
82	80	93	41	70	
65					

a. Alter and Ginzberg used the Mann-Whitney U test (a procedure equivalent to the Wilcoxon rank sum test) on similar data to compare the quality of the development processes of successfully and unsuccessfully implemented MISs. Use the large-sample Wilcoxon rank sum test to determine whether the distribution of quality scores for successfully implemented systems lies above the distribution of scores for unsuccessfully implemented systems. Test using $\alpha = .05$.

b. Under what circumstances could you use the two-sample t test of Chapter 9 to conduct the same test?

17.3 Comparing Two Populations: Wilcoxon Signed Rank Test for the Paired Difference Experiment

Nonparametric techniques can also be used to compare two probability distributions when a paired difference design is used. For example, for some paper products, softness is an important consideration in determining consumer acceptance. One method of determining softness is to have judges give a sample of the products a softness rating. Suppose each of 10 judges is given a sample of two products that a company wants to compare. Each judge rates the softness of each product on a scale from 1 to 10, with higher ratings implying a softer product. The results are shown in Table 17.3

TABLE 17.3 Paper Softness Ratings

Judge	Product		Difference	Absolute Value of Difference	Rank of Absolute Value
	A	B	(A − B)		
1	6	4	2	2	5
2	8	5	3	3	7.5
3	4	5	−1	1	2
4	9	8	1	1	2
5	4	1	3	3	7.5
6	7	9	−2	2	5
7	6	2	4	4	9
8	5	3	2	2	5
9	6	7	−1	1	2
10	8	2	6	6	10

T_+ = Sum of positive ranks = 46
T_- = Sum of negative ranks = 9

Since this is a paired difference experiment, we analyze the differences between the measurements (see Section 9.4). However, the nonparametric approach requires that we calculate the ranks of the *absolute values* of the differences between the measurements—i.e., the ranks of the differences after removing any minus signs. Note that tied absolute differences are assigned the average of the ranks they would receive if they were unequal but successive measurements. After the absolute differences are ranked, the sum of the ranks of the positive differences, T_+, and the sum of the ranks of the negative differences, T_-, are computed.

We are now prepared to test the nonparametric hypotheses:

H_0: The probability distributions of the ratings for products A and B are identical

H_a: The probability distribution for product A is shifted to the right or left of the probability distribution of the ratings for product B

Test statistic: T = Smaller of the positive and negative rank sums T_+ and T_-

The smaller the value of T, the greater will be the evidence to indicate that the two probability distributions differ in location. The rejection region for T can be determined by consulting Table XII of Appendix B. A portion of that table is shown in Table 17.4. This table gives a value, T_0, for each value of n, the number of matched pairs.

TABLE 17.4 Reproduction of Part of Table XII of Appendix B: Critical Values of T_0 in the Wilcoxon Paired Difference Signed Rank Test

One-tailed	Two-tailed	$n = 5$	$n = 6$	$n = 7$	$n = 8$	$n = 9$	$n = 10$
$\alpha = .05$	$\alpha = .10$	1	2	4	6	8	11
$\alpha = .025$	$\alpha = .05$		1	2	4	6	8
$\alpha = .01$	$\alpha = .02$			0	2	3	5
$\alpha = .005$	$\alpha = .01$				0	2	3
		$n = 11$	$n = 12$	$n = 13$	$n = 14$	$n = 15$	$n = 16$
$\alpha = .05$	$\alpha = .10$	14	17	21	26	30	36
$\alpha = .025$	$\alpha = .05$	11	14	17	21	25	30
$\alpha = .01$	$\alpha = .02$	7	10	13	16	20	24
$\alpha = .005$	$\alpha = .01$	5	7	10	13	16	19
		$n = 17$	$n = 18$	$n = 19$	$n = 20$	$n = 21$	$n = 22$
$\alpha = .05$	$\alpha = .10$	41	47	54	60	68	75
$\alpha = .025$	$\alpha = .05$	35	40	46	52	59	66
$\alpha = .01$	$\alpha = .02$	28	33	38	43	49	56
$\alpha = .005$	$\alpha = .01$	23	28	32	37	43	49
		$n = 23$	$n = 24$	$n = 25$	$n = 26$	$n = 27$	$n = 28$
$\alpha = .05$	$\alpha = .10$	83	92	101	110	120	130
$\alpha = .025$	$\alpha = .05$	73	81	90	98	107	117
$\alpha = .01$	$\alpha = .02$	62	69	77	85	93	102
$\alpha = .005$	$\alpha = .01$	55	61	68	76	84	92

The values of T_0 are tabulated for both a one- and two-tailed test. For a two-tailed test with $\alpha = .05$, we will reject H_0 if $T \leq T_0$. The T_0 value that locates the rejection region for the judges' ratings in Table 17.3 is the value indicated for $n = 10$ pairs of observations. This value of T_0 is 8. Therefore, the rejection region for the test (see Figure 17.4) is

Rejection region: $T \leq 8$ for $\alpha = .05$

Since the smaller rank sum for the paper data, $T = T_- = 9$, does not fall within the rejection region, the experiment has not provided sufficient evidence to indicate that the two paper products differ with respect to their softness ratings at $\alpha = .05$.

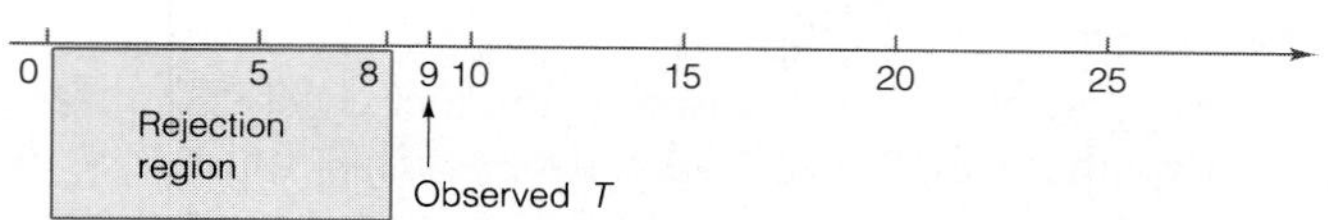

FIGURE 17.4 ► Rejection region for paired difference experiment

Note that, if a significance level of $\alpha = .10$ had been used, the rejection region would have been $T \le 11$, and we would have rejected H_0. In other words, the samples do provide evidence that the probability distributions of the softness ratings differ at $\alpha = .10$.

The **Wilcoxon signed rank test** is summarized in the next box. Note that the difference measurements are assumed to have a continuous probability distribution so that the absolute differences will have unique ranks. Although tied (absolute) differences can be assigned ranks by averaging, the number of ties should be small relative to the number of observations to assure the validity of the test.

Wilcoxon Signed Rank Test for a Paired Difference Experiment

One-Tailed Test

H_0: Two sampled populations have identical probability distributions

H_a: The probability distribution for population A is shifted to the right of that for population B

Test statistic: T_-, the negative rank sum (we assume the differences are computed by subtracting each paired B measurement from the corresponding A measurement)

Rejection region: $T_- \le T_0$, where T_0 is found in Table XII in Appendix B for the one-tailed significance level α and the number of untied pairs, n.

Two-Tailed Test

H_0: Two sampled populations have identical probability distributions

H_a: The probability distribution for population A is shifted to the right *or* to the left of that for population B

Test statistic: T, the smaller of the positive and negative rank sums, T_+ and T_-

Rejection region: $T \le T_0$, where T_0 is found in Table XII in Appendix B for the two-tailed significance level α and the number of untied pairs, n.

[*Note*: If the alternative hypothesis is that the probability distribution for A is shifted to the left of B, we use T_+ as the test statistic and reject H_0 if $T_+ \le T_0$.]

Assumptions:
1. The sample of differences is randomly selected from the population of differences.
2. The probability distribution from which the sample of paired differences is drawn is continuous.

Ties: Assign tied absolute differences the average of the ranks they would receive if they were unequal but occurred in successive order. For example, if the third-ranked and fourth-ranked differences are tied, assign both a rank of $(3 + 4)/2 = 3.5$.

EXAMPLE 17.3

Suppose the U.S. Consumer Product Safety Commission (CPSC) wants to test the hypothesis that New York City electrical contractors are more likely to install unsafe electrical outlets in urban homes than in suburban homes. A pair of homes, one urban and one suburban and both serviced by the same electrical contractor, is chosen for each of ten randomly selected electrical contractors. A CPSC inspector assigns each of the 20 homes a safety rating between 1 and 10, with higher numbers implying safer electrical conditions. The results are shown in Table 17.5. Use the Wilcoxon signed rank test to determine whether the CPSC hypothesis is supported at the $\alpha = .05$ level.

TABLE 17.5 Electrical Safety Ratings for 10 Pairs of New York City Homes

Contractor	Urban	Suburban	Contractor	Urban	Suburban
1	7	9	6	6	10
2	4	5	7	8	9
3	8	8	8	10	8
4	9	8	9	9	4
5	3	6	10	5	9

Solution

The null and alternative hypotheses are

H_0: The probability distributions of home electrical ratings are identical for urban and suburban homes

H_a: The electrical ratings for suburban homes tend to exceed the electrical ratings for urban homes

These hypotheses can be tested for the data in Table 17.5 using the Wilcoxon signed rank test. Since a paired difference design was used (the homes were selected in urban–suburban pairs so that the electrical contractor was the same for both), we first calculate the difference between the ratings for each pair of homes, and then rank the absolute values of the differences (see Table 17.6). Note that one pair of ratings was the same (both 8), and the resulting 0 difference contributes to neither the positive nor the negative rank sum. Thus, we eliminate this pair from the calculation of the test statistic.

Test statistic: T_+, the positive rank sum

In Table 17.6, we compute the urban minus suburban rating differences, and if the alternative hypothesis is true, we would expect most of these differences to be negative. Or, in other words, we would expect the *positive* rank sum T_+ to be small if the alternative hypothesis is true (see Figure 17.5).

TABLE 17.6 Differences in Ratings and the Ranks of Their Absolute Values

Rating		Difference	Rank of Absolute Difference
Urban	Suburban	(Urban − Suburban)	
7	9	−2	4.5
4	5	−1	2
8	8	0	(Eliminated)
9	8	1	2
3	6	−3	6
6	10	−4	7.5
8	9	−1	2
10	8	2	4.5
9	4	5	9
5	9	−4	7.5
		Positive rank sum = T_+ =	15.5

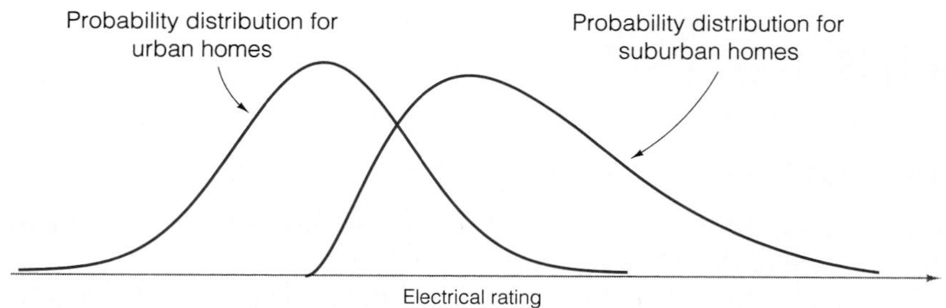

FIGURE 17.5 ► **The alternative hypothesis for Example 17.3: We expect T_+ to be small.**

Rejection region: For $\alpha = .05$, from Table XII of Appendix B, we use $n = 9$ (remember, one pair of observations was eliminated) to find the rejection region for this one-tailed test: $T_+ \leq 8$.

Since the computed value $T_+ = 15.5$ exceeds the critical value of 8, we conclude that this sample provides insufficient evidence at $\alpha = .05$ to support the alternative hypothesis. We *cannot* conclude on the basis of this sample information that suburban homes have safer electrical outlets than urban homes.

As in the case for the rank sum test for independent samples, the sampling distribution of the signed rank statistic can be approximated by a normal distribution when the number n of paired observations is large (say $n \geq 25$). The large-sample z test is summarized in the box.

Wilcoxon Signed Rank Test for a Paired Difference Experiment: Large Sample

One-Tailed Test

H_0: Two sampled populations have identical probability distributions

H_a: The probability distribution for population A is shifted to the right of that for population B

Test statistic: $z = \dfrac{T_+ - \dfrac{n(n+1)}{4}}{\sqrt{\dfrac{n(n+1)(2n+1)}{24}}}$

Rejection region: $z > z_\alpha$

Two-Tailed Test

H_0: Two sampled populations have identical probability distributions

H_a: The probability distribution for population A is shifted to the right *or* to the left of that for population B

Test statistic: $z = \dfrac{T_+ - \dfrac{n(n+1)}{4}}{\sqrt{\dfrac{n(n+1)(2n+1)}{24}}}$

Rejection region: $z < -z_{\alpha/2}$ or $z > z_{\alpha/2}$

Assumptions: $n \geq 25$

Exercises 17.19–17.29

Learning the Mechanics

17.19 Specify the test statistic and the rejection region for the Wilcoxon signed rank test for the paired difference design in each of the following situations.

a. H_0: Two probability distributions, A and B, are identical
H_a: Probability distribution for population A is shifted to the right or left of the probability distribution for population B
$n = 30, \quad \alpha = .10$

b. H_0: Two probability distributions, A and B, are identical.
H_a: Probability distribution for population A is shifted to the right of the probability distribution for population B
$n = 20, \quad \alpha = .05$

c. H_0: Two probability distributions, A and B, are identical
H_a: Probability distribution for population A is shifted to the left of the probability distribution for population B
$n = 8, \quad \alpha = .005$

17.20 Suppose you want to test a hypothesis that two treatments, A and B, are equivalent against the alternative hypothesis that the responses for A tend to be larger than those for B. You plan to use a paired difference experiment and to analyze the resulting data using the Wilcoxon signed rank test.

a. Specify the null and alternative hypotheses you would test.

b. Suppose the paired difference experiment yielded the data in the table. Conduct the test of part **a** using $\alpha = .025$.

Pair of Experimental Units	Treatment A	Treatment B	Pair of Experimental Units	Treatment A	Treatment B
1	56	42	6	76	75
2	54	45	7	74	63
3	98	87	8	29	35
4	45	25	9	63	59
5	82	71	10	80	81

17.21 Explain the difference between the one- and two-tailed versions of the Wilcoxon signed rank test for the paired difference experiment.

17.22 In order to conduct the Wilcoxon signed rank test, why do we need to assume that the probability distribution of differences is continuous?

17.23 A paired difference experiment with $n = 30$ pairs yielded $T_+ = 354$.

a. Specify the null and alternative hypotheses that should be used in conducting a hypothesis test to determine whether the probability distribution for population A is located to the right of that for population B.

b. Conduct the test of part **a** using $\alpha = .05$.

c. What is the approximate p-value of the test of part **b**?

d. What assumptions are necessary to assure the validity of the test you performed in part **b**?

Applying the Concepts

17.24 Which is the more effective means of dealing with complex group problem-solving tasks: face-to-face meetings or video teleconferencing? A report of an experiment (Rosetti and Surynt, 1985) concluded that video teleconferencing may be the more effective method. Ten groups of four people each were randomly assigned both to a specific communication setting (face-to-face or video teleconference) and to one of two specific complex problems. Upon completion of the problem-solving task, the same groups were placed in the alternative communication setting and asked to complete the second problem-solving task. The percentage of each problem task correctly completed was recorded for each group. Suppose the results were as given in the table.

Group	*Face-to-Face*	*Video Teleconferencing*	*Group*	*Face-to-Face*	*Video Teleconferencing*
1	65%	75%	6	85%	90%
2	82	80	7	98	98
3	54	60	8	35	40
4	69	65	9	85	89
5	40	55	10	70	80

a. What type of experimental design was used in this study?

b. Specify the null and alternative hypotheses that should be used in determining whether the data provide sufficient evidence to conclude that the problem-solving performance of video teleconferencing groups is superior to that of groups that interact face-to-face.

c. Conduct the hypothesis test of part **b**. Use $\alpha = .05$. Interpret the results of your test in the context of the problem.
d. What is the p-value of the test in part **c**?

17.25 Traditionally, workers in the United States have had a fixed 8-hour workday. A recent job-scheduling innovation that is helping managers to overcome the motivation and absenteeism problems associated with the fixed workday is a concept called *flextime*. This flexible working hours program permits employees to design their own 40-hour work week (Certo, 1980). The management of a large manufacturing firm may adopt a flextime program for its hourly employees, depending on the success or failure of a pilot program. Ten employees were randomly selected and given a questionnaire designed to measure their attitude toward their job. Each was then permitted to design and follow a flextime workday. After 6 months, attitudes toward their jobs were again measured. The resulting attitude scores are displayed in the table. The higher the score, the more favorable is the employee's attitude toward his or her work. Use a nonparametric test procedure to evaluate the success of the pilot flextime program. Test using $\alpha = .05$.

Employee	*Before*	*After*	*Employee*	*Before*	*After*
1	54	68	6	82	88
2	25	42	7	94	90
3	80	80	8	72	81
4	76	91	9	33	39
5	63	70	10	90	93

17.26 In Exercise 9.50, a paired difference test was used to compare the mean number of antitrust litigations per firm in the 1960s with the mean number per firm in the 1970s. The data are repeated in the table. Beckenstein, Gabel, and Roberts (1983) claimed that, on average, companies faced more antitrust litigations in the 1970s than in the 1960s. Do the data support their claim? Test using the Wilcoxon signed rank test with $\alpha = .05$. Be sure to specify your null and alternative hypotheses.

Firm	*1960s*	*1970s*	*Firm*	*1960s*	*1970s*
1	10	10	6	7	6
2	8	12	7	6	11
3	9	8	8	9	12
4	7	16	9	8	11
5	8	14	10	7	12

17.27 A manufacturer of household appliances is considering one of two chains of department stores to be the sales merchandiser for its product in a particular region of the United States. Before choosing one chain, the manufacturer wants to compare the product exposure that might be expected for the two chains. Eight locations are selected where both chains have stores, and on a specific day, the number of shoppers entering each store is recorded. The data are shown in the table. Is there sufficient evidence to indicate that one chain tends to have more customers per day than the other? Test using $\alpha = .05$.

Location	A	B	*Location*	A	B
1	879	1,085	5	2,326	2,778
2	445	325	6	857	992
3	692	848	7	1,250	1,303
4	1,565	1,421	8	773	1,215

17.28 A food vending company currently uses vending machines made by two different manufacturers. Before purchasing new machines, the company wants to compare the two types in terms of reliability. Records for 7 weeks are given in the table; the data indicate the number of breakdowns per week for each type of machine. The company has the same number of machines of each type. Is the probability distribution of the number of breakdowns for machine A shifted to the left or to the right of the probability distribution of the number of breakdowns for machine B? Use $\alpha = .05$.

Week	*Machine A*	*Machine B*
1	14	12
2	17	13
3	10	14
4	15	12
5	14	9
6	9	11
7	12	11

17.29 Economic indexes provide measures of economic change. The table lists the producer commodity price indexes for January 1985 and January 1986, for six product categories. By comparing these two sets of indexes, you can obtain information regarding changes in the economy that occurred during 1985.

Product Category	*January 1985*	*January 1986*
Processed poultry	198.8	192.4
Concrete ingredients	331.0	339.0
Lumber	343.0	329.6
Gas fuels	1,073.0	1,034.3
Drugs and pharmaceuticals	247.4	265.9
Synthetic fibers	157.6	151.1

Source: *Standard & Poor's Statistical Service, Current Statistics*, Jan. 1987, pp. 12–13.

a. Conduct a paired difference t test to compare the mean values of these indexes for January 1985 and January 1986. Use $\alpha = .05$. What assumptions are necessary for the validity of this procedure? Why might these assumptions be in doubt?

b. Use the Wilcoxon signed rank test to determine whether the data provide evidence that the probability distribution of the economic indexes has changed. Use $\alpha = .05$. What assumptions are necessary to assure the validity of this test?

17.4 The Kruskal-Wallis H Test for a Completely Randomized Design

Recall that a completely randomized design is one in which *independent* random samples are selected from each p populations (treatments) to be compared (Section 16.2). In Chapter 16, we used an analysis of variance and the F test to compare the means of the p populations (assuming the populations have normal probability distributions with equal variances). We now present a nonparametric technique that requires no assumptions concerning the population probability distributions to compare the p populations.

For example, suppose you want to compare the numbers of employees in companies representing each of three different business classifications: agriculture, manufacturing, and service. You sample 10 companies from each type and record the number of employees in each sampled business (see Table 17.7). You can see that the assumptions necessary for a parametric comparison of the means are doubtful for these data; the probability distributions are very likely to be skewed to the right, as indicated by the presence of some extremely large values. In addition, the variability in number of employees may not be constant for the different classifications. We therefore base our comparison on the rank sums for the classifications. The ranks are computed for each observation according to the relative magnitude of the measurements *when all p samples are combined* (see Table 17.7). Note that ties are handled in the usual manner, by assigning the average value of the ranks to each of the tied observations.

We test

H_0: All three populations have identical probability distributions

H_a: At least two of the three population probability distributions differ in location

TABLE 17.7 Number of Employees in 30 Companies

Agriculture	Rank	Manufacturing	Rank	Service	Rank
10	5	244	25	17	9.5
350	27	93	19	249	26
4	2	3,532	30	38	15
26	13	17	9.5	5	3
15	8	526	29	101	20
106	21	133	22	1	1
18	11	14	7	12	6
23	12	192	23	233	24
62	17	443	28	31	14
8	4	69	18	39	16
	$R_1 = 120$		$R_2 = 210.5$		$R_3 = 134.5$

If we denote the three sample rank sums by R_1, R_2, and R_3, the **test statistic** is given by

$$H = \frac{12}{n(n+1)} \sum_{j=1}^{p} \frac{R_j^2}{n_j} - 3(n+1)$$

where n_j is the number of measurements in the jth sample and n is the **total sample size** ($n = n_1 + n_2 + \cdots + n_p$). For the data in Table 17.7, we have $n_1 = n_2 = n_3 = 10$, and $n = 30$. The rank sums are $R_1 = 120$, $R_2 = 210.5$, and $R_3 = 134.5$. Thus,

$$H = \frac{12}{30(31)}\left[\frac{(120)^2}{10} + \frac{(210.5)^2}{10} + \frac{(134.5)^2}{10}\right] - 3(31)$$
$$= 99.097 - 93 = 6.097$$

The H statistic measures the extent to which the p samples differ with respect to their relative ranks. This is more easily seen by writing H in an alternative but equivalent form:

$$H = \frac{12}{n(n+1)}\sum_{j=1}^{p} n_j(\bar{R}_j - \bar{R})^2$$

where $\bar{R}_j$ is the mean rank corresponding to sample j, and $\bar{R}$ is the mean of all the ranks [i.e., $\bar{R} = (n+1)/2$]. Thus, the H statistic is 0 if all samples have the same mean rank, and becomes increasingly large as the distance between the sample mean ranks grows.

If the null hypothesis is true, the distribution of H in repeated sampling is approximately a χ^2 **(chi-square) distribution**. This approximation for the sampling distribution of H is adequate as long as each of the p sample sizes exceeds 5 (see the references for more detail). The χ^2 probability distribution is characterized by a single parameter, called the **degrees of freedom associated with the distribution**. Several χ^2 probability distributions with different degrees of freedom are shown in Figure 17.6.

FIGURE 17.6 ▶
Several χ^2 probability distributions

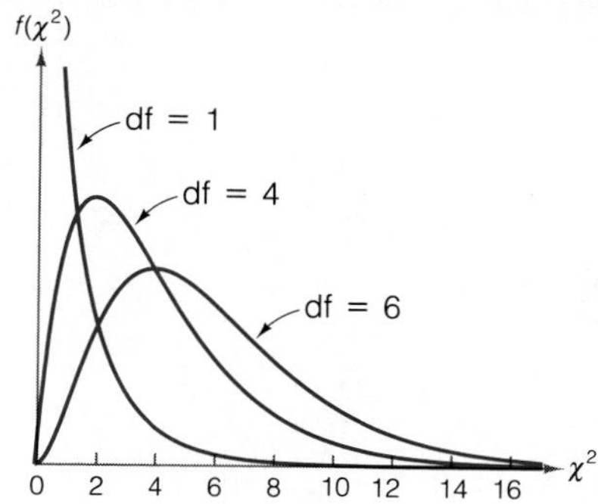

The degrees of freedom corresponding to the approximate sampling distribution of H will always be $(p - 1)$, one less than the number of probability distributions being compared. Because large values of H support the alternative hypothesis that at least two of the $(p - 1)$ population probability distributions differ in location, the rejection region for the test will be located in the upper tail of the χ^2 distribution, as shown in Figure 17.7 on page 946.

For the data of Table 17.7, the approximate distribution of the test statistic H is a χ^2 distribution with $(p - 1) = 2$ df. To determine how large H must be before we will reject the null hypothesis, we consult Table XIII in Appendix B; part of this table is

FIGURE 17.7 ▶ **Rejection region for the comparison of three probability distributions**

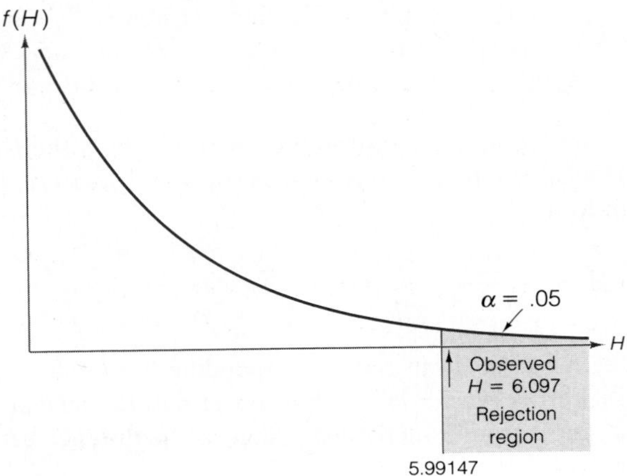

shown in Table 17.8. Entries in the table give an upper-tail value of χ^2, call it χ^2_α, such that $P(\chi^2 > \chi^2_\alpha) = \alpha$. The columns of the table identify the value of α associated with the tabulated value of χ^2_α, and the rows correspond to the degrees of freedom. Thus, for $\alpha = .05$ and df = 2, we can reject the null hypothesis that the three probability distributions are the same if $H > \chi^2_{.05}$ where $\chi^2_{.05} = 5.99147$.

TABLE 17.8 Reproduction of Part of Table XIII of Appendix B: Critical Values of χ^2

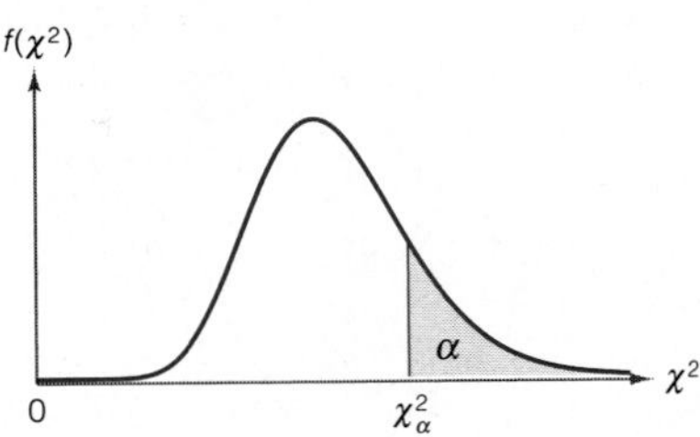

Degrees of Freedom	$\chi^2_{.100}$	$\chi^2_{.050}$	$\chi^2_{.025}$	$\chi^2_{.010}$	$\chi^2_{.005}$
1	2.70554	3.84146	5.02389	6.63490	7.87944
2	4.60517	5.99147	7.37776	9.21034	10.5966
3	6.25139	7.81473	9.34840	11.3449	12.8381
4	7.77944	9.48773	11.1433	13.2767	14.8602
5	9.23635	11.0705	12.8325	15.0863	16.7496
6	10.6446	12.5916	14.4494	16.8119	18.5476
7	12.0170	14.0671	16.0128	18.4753	20.2777
8	13.3616	15.5073	17.5346	20.0902	21.9550
9	14.6837	16.9190	19.0228	21.6660	23.5893
10	15.9871	18.3070	20.4831	23.2093	25.1882
11	17.2750	19.6751	21.9200	24.7250	26.7569

The rejection region is pictured in Figure 17.7. Since the calculated $H = 6.097$ exceeds the critical value of 5.99147, we conclude that at least two of the three probability distributions describing the number of employees for the three sampled business types differ in location.

The **Kruskal-Wallis H test** for comparing more than two probability distributions is summarized in the next box. We can use the Wilcoxon rank sum test to compare the separate pairs of populations if the Kruskal-Wallis H test supports the alternative hypothesis that at least two of the probability distributions differ. [*Note*: The Bonferroni procedure (Chapter 16) can be used to determine the appropriate significance level of each test such that the overall level of significance is less than or equal to a prespecified α.]

Kruskal-Wallis H Test for Comparing p Probability Distributions

H_0: The p probability distributions are identical

H_a: At least two of the p probability distributions differ in location

Test statistic: $$H = \frac{12}{n(n+1)} \sum_{j=1}^{p} \frac{R_j^2}{n_j} - 3(n+1)$$

where

n_j = Number of measurements in sample j

R_j = Rank sum for sample j, where the rank of each measurement is computed according to its relative magnitude in the totality of data for the p samples

n = Total sample size = $n_1 + n_2 + \cdots + n_p$

Rejection region: $H > \chi^2_\alpha$ with $(p - 1)$ df

Assumptions:
1. The p samples are random and independent.
2. There are 5 or more measurements in each sample.
3. The p probability distributions from which the samples are drawn are continuous.

Ties: Assign tied measurements the average of the ranks they would receive if they were unequal but occurred in successive order. For example, if the third- and fourth-ranked measurements tie, rank both $(3 + 4)/2 = 3.5$. The number of ties should be small relative to the total number of observations.

Exercises 17.30–17.39

Learning the Mechanics

17.30 Use Table XIII in Appendix B to find each of the following χ^2 values.

a. $\chi^2_{.05}$, df = 22 **b.** $\chi^2_{.025}$, df = 17 **c.** $\chi^2_{.01}$, df = 31

d. $\chi^2_{.10}$, df = 80 **e.** $\chi^2_{.05}$, df = 3 **f.** $\chi^2_{.005}$, df = 10

17.31 Use Table XIII in Appendix B to find each of the following probabilities.

a. $P(\chi^2 \geq 3.07382)$ where df = 12
b. $P(\chi^2 \leq 24.4331)$ where df = 40
c. $P(\chi^2 \geq 14.6837)$ where df = 9
d. $P(\chi^2 < 34.1696)$ where df = 20
e. $P(\chi^2 < 6.26214)$ where df = 15
f. $P(\chi^2 \leq .584375)$ where df = 3

17.32 Data were collected from three populations, A, B, and C, using a completely randomized design. The following describes the sample data:

$$n_A = n_B = n_C = 15$$
$$R_A = 230 \qquad R_B = 440 \qquad R_C = 365$$

a. Specify the null and alternative hypotheses that should be used in conducting a test of hypothesis to determine whether the probability distributions of populations A, B, and C differ in location.
b. Conduct the test of part **a**. Use $\alpha = .05$.
c. What is the approximate *p*-value of the test of part **b**?
d. Calculate the mean rank for each sample and compute H according to the formula on page 947 that utilizes these means. Verify that this formula yields the same value of H that you obtained in part **b**.

17.33 Suppose you want to use the Kruskal-Wallis H test to compare the probability distributions of three populations. The following are independent random samples selected from the three populations:

I	66	23	55	88	58	62	79	49
II	19	31	16	29	30	33	40	
III	75	96	102	75	98	78		

a. What type of experimental design was used?
b. Specify the null and alternative hypotheses you would test.
c. Specify the rejection region that would be used for your hypothesis test, at $\alpha = .01$.
d. Conduct the test using $\alpha = .01$.

17.34 Under what circumstances does the χ^2 distribution provide an appropriate characterization of the sampling distribution of the Kruskal-Wallis H statistic?

Applying the Concepts

17.35 In choosing a mutual fund, investors should compare their personal investment goals with those of the mutual fund. Among the hundreds of available mutual funds, three widely advertised goals are income, growth, and maximum growth. *Income funds* seek to maximize current income; *growth funds* seek long-term capital appreciation, with current income a secondary goal; and *maximum growth funds* seek greater capital appreciation by taking larger risks. The table lists the total rate of return to investors for samples of seven mutual funds in each of these three categories.

a. Do the data provide sufficient evidence to conclude that the rate-of-return distributions differ among the three types of mutual funds? Test using $\alpha = .05$.
b. What assumptions must hold for your test of part **a** to be valid?

Income Mutual Fund	*Return*	*Growth Mutual Fund*	*Return*	*Maximum Growth Mutual Fund*	*Return*
American National Income	8.3%	Babson Growth	20.2%	Fairfield	2.2%
Bull & Bear Equity–Income	19.0	Oppenheimer	1.9	44 Wall St. Equity	16.9
Oppenheimer Equity–Income	15.6	Midamerica Mutual	11.2	IDS Strategy–Aggressive Equity	23.3
T. Rowe Price Equity–Income	26.8	Investment Portfolios–Equity	4.5	Omega	12.1
Seligman Income	17.1	Franklin Equity	19.3	Pilot	10.4
Safeco Income	20.1	Fidelity Contrafund	13.3	Strong Opportunity	59.9
National Stock	14.9	Thomson McKinnon Growth	23.1	Lowry Market Timing	−9.3

Source: "Mutual fund scoreboard." *Business Week*, Feb. 23, 1987, pp. 70–103.

c. Describe the Type I and II errors associated with the test of part **a** in this context.
d. Under what circumstances could the F test of Section 16.2 help answer part **a**?

17.36 Random samples of seven lawyers employed by corporations were selected from each of three major cities. Their salaries were determined and are recorded in the table. You have been hired to determine whether differences exist among the salary distributions for corporate lawyers in the three cities.

Atlanta	*Los Angeles*	*Washington, D.C.*
\$45,500	\$52,000	\$41,500
47,900	72,000	40,100
43,100	41,000	39,000
42,000	54,000	56,500
49,000	33,000	37,000
52,000	42,000	49,000
39,000	50,000	43,500

Source: *American Almanac of Jobs and Salaries*, 1984, pp. 365–374.

a. Under what circumstances would it be appropriate to use the F test for a completely randomized design to perform the required analysis?
b. Which assumptions required by the F test are likely to be violated in this problem? Explain.
c. Use the Kruskal-Wallis H test to determine whether the salary distributions differ among the three cities. Specify your null and alternative hypotheses, and state your conclusions in the context of the problem. Use $\alpha = .05$. What assumptions are necessary to assure the validity of the nonparametric test?

17.37 An economist is interested in knowing whether property tax rates differ among three types of school districts—urban, suburban, and rural. A random sample of several districts of each type produced the data in the table (rate is in mills, where 1 mill = \$1/1,000). Do the data indicate a difference in the level of property taxes among the three types of school districts? Use $\alpha = .05$.

Urban	*Suburban*	*Rural*
4.3	5.9	5.1
5.2	6.7	4.8
6.2	7.6	3.9
5.6	4.9	6.2
3.8	5.2	4.2
5.8	6.8	4.3
4.7		

17.38 Three different brands (A, B, C) of magnetron tubes (the key components in microwave ovens) were subjected to stressful testing, and the number of hours each operated without repair was recorded. Although these times do not represent typical lifetimes, they do indicate how well the tubes can withstand extreme stress.

A	B	C
36	49	71
48	33	31
5	60	140
67	2	59
53	55	42

a. Use the F test for a completely randomized design (Chapter 16) to test the hypothesis that the mean length of life under stress is the same for the three brands. Use $\alpha = .05$. What assumptions are necessary for the validity of this procedure? Is there any reason to doubt these assumptions?

b. Use the Kruskal-Wallis H test to determine whether evidence exists to conclude that at least two of the probability distributions of length of life under stress differ in location. Use $\alpha = .05$.

17.39 In Exercise 12.48 regression analysis was applied to investigate whether the mean debt-to-equity ratio varies among four industries. The data are repeated in the table.

Insurance		Publishing	
Firm	Debt-to-equity	Firm	Debt-to-equity
Chubb	.24	Deluxe Check	.03
Kemper	.09	New York Times	.32
St. Paul Cos.	.12	Times Mirror	.51
Lincoln National	.09	Dow Jones	.15
USF & G	.00	Gannett	.73
Aetna Life & Cas.	.07		

Electric Utilities		Banking	
Firm	Debt-to-equity	Firm	Debt-to-equity
Pacific G & E	.84	U.S. Bancorp	.54
Houston Ind.	1.00	Sun Trust Banks	.29
Florida Progress	1.03	Mellon Bank	.82
Penn Power & Light	1.11	Michigan National	.13
North States Power	.83	Southeast Banking	.41
Ohio Edison	1.16		
Orange & Rockland	.70		

Source: "39th Annual Report on American Industry." *Forbes*, Jan. 12, 1987.

a. Compare the assumptions required by the regression analysis and the Kruskal-Wallis H test for comparing the mean debt-to-equity ratios among the four industries.

b. Use the Kruskal-Wallis H test to investigate whether debt-to-equity ratios differ among the four industries. Be sure to specify your null and alternative hypotheses and to state your conclusion in the context of the problem. Use $\alpha = .05$.

c. Assuming the Kruskal-Wallis H test indicates that differences exist among the four industries, which nonparametric procedure could be employed to compare the distribution of debt-to-equity ratios for the electric utility and banking industries?

17.5 Spearman's Rank Correlation Coefficient

When economic conditions are favorable, many banks advertise special loan rates for new cars, appliances, and other items to attract customers. Suppose a bank wants to determine whether to aim its advertising at a broad spectrum of potential borrowers or to concentrate on a specific income group. It randomly samples 10 noncommercial customers from recent files and ascertains their present incomes and the total amount each has borrowed over the past 3 years (excluding mortgages and business loans). The data are shown in Table 17.9.

TABLE 17.9 Income–Amount Borrowed Data

Customer	Income	Rank	Total Borrowed	Rank
1	\$14,800	5	\$4,300	7
2	8,900	1	4,800	8
3	83,600	10	500	2
4	22,100	8	3,300	5
5	18,200	7	5,500	9
6	13,700	4	3,700	6
7	41,800	9	0	1
8	9,300	2	3,200	4
9	12,700	3	6,100	10
10	16,100	6	1,800	3

One method of determining whether a correlation exists between income and amount borrowed is to calculate the Pearson product moment correlation, r (Section 10.6). However, to make an inference about the population correlation ρ (Greek rho), we must assume that the two random variables, income and amount borrowed, are normally distributed. This assumption is usually inappropriate for incomes, because they tend to have a relative frequency distribution that is heavily skewed to the right (Figure 17.8, page 952). Although the modal income (that with the largest relative frequency) may be relatively low, there are typically enough individuals with high incomes to make income distributions asymmetric.

FIGURE 17.8 ► Typical relative frequency distribution of incomes

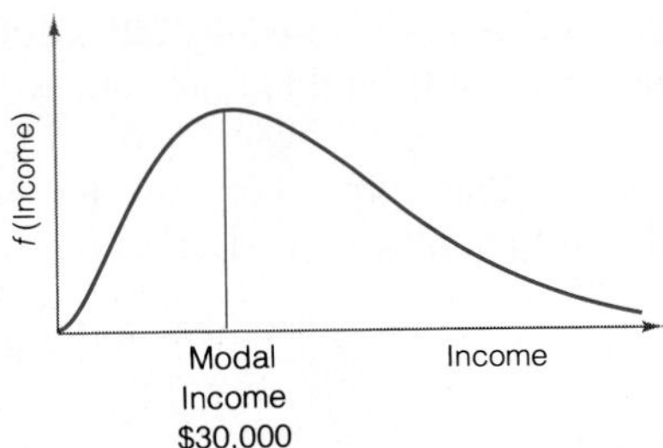

Thus, we turn to a nonparametric approach to correlation, which does not require underlying normal distributions. This nonparametric method, like those in the previous sections of this chapter, uses the ranks of the measurements to determine a measure of correlation. It is shown in the box.

Spearman's Rank Correlation Coefficient

$$r_s = \frac{SS_{uv}}{\sqrt{SS_{uu}SS_{vv}}}$$

where

$$SS_{uv} = \sum (u_i - \bar{u})(v_i - \bar{v}) = \sum u_i v_i - \frac{\left(\sum u_i\right)\left(\sum v_i\right)}{n}$$

$$SS_{uu} = \sum (u_i - \bar{u})^2 = \sum u_i^2 - \frac{\left(\sum u_i\right)^2}{n}$$

$$SS_{vv} = \sum (v_i - \bar{v})^2 = \sum v_i^2 - \frac{\left(\sum v_i\right)^2}{n}$$

u_i = Rank of the ith measurement in sample 1
v_i = Rank of the ith measurement in sample 2
n = Number of pairs of measurements (number of measurements in each sample)

Note that the definition of Spearman's rank correlation coefficient is identical to the definition of Pearson's r (see page 485), except that Spearman's r_s uses ranks. You can use the shortcut formula shown in the next box for calculating r_s if there are no ties. The shortcut formula will also provide a satisfactory approximation to r_s when the number of ties is small relative to the number of pairs.

Shortcut Formula for r_s

$$r_s = 1 - \frac{6 \sum d_i^2}{n(n^2 - 1)}$$

where $d_i = u_i - v_i$

(difference in the ranks of the ith measurement for sample 1 and sample 2)

EXAMPLE 17.4

Calculate Spearman's rank correlation coefficient, r_s, for the bank customer data given in Table 17.9.

Solution

The ranks are reproduced in Table 17.10, along with the difference, d_i, for each pair of measurements. Then

$$r_s = 1 - \frac{6 \sum d_i^2}{n(n^2 - 1)} = 1 - \frac{6(260)}{10(100 - 1)} = 1 - 1.576 = -.576$$

The sign of r_s indicates the nature of the relationship between the two variables. Positive values indicate a tendency for the variables to increase together, and negative values indicate a tendency for one variable to increase while the other decreases.

TABLE 17.10 Calculation of Spearman's Rank Correlation Coefficient

Customer	Income Rank	Total Borrowed Rank	Difference	Difference Squared
	u_i	v_i	$d_i = u_i - v_i$	d_i^2
1	5	7	−2	4
2	1	8	−7	49
3	10	2	8	64
4	8	5	3	9
5	7	9	−2	4
6	4	6	−2	4
7	9	1	8	64
8	2	4	−2	4
9	3	10	−7	49
10	6	3	3	9
			Total	260

The strength of the relationship between the ranks is indicated by the numerical size of r_s. Since r_s is really a Pearson product moment correlation of the ranks, it must lie between -1 and $+1$. Recall that a correlation of 0 implies no linear relationship, while correlations of -1 and $+1$ imply perfect negative and positive relationships, respectively. In Example 17.4, we obtained $r_s = -.576$. Is this different enough from 0 to conclude that the variables are related in the population?

If we define ρ_s as the population Spearman rank correlation coefficient, this question can be answered by conducting the test

H_0: $\rho_s = 0$ (There is no population correlation between ranks)

H_a: $\rho_s \neq 0$ (There is a population correlation between ranks)

Test statistic: r_s, the sample Spearman rank correlation coefficient

To determine a rejection region, we consult Table XVI in Appendix B, which is partially reproduced in Table 17.11. Note that the left-hand column gives values of n, the number of pairs of measurements. The entries in the table are values for an upper-tail rejection region, since only positive values are given. Thus, for $n = 10$ and $\alpha = .05$, the value .564 is the boundary of the upper-tailed rejection region, so that $P(r_s > .564) = .05$ if in fact H_0 is true. That is, we would expect to see r_s exceed .564 only 5% of the time if in fact there is no relationship between the variables. The two-tailed rejection region is $r_s > .564$ *or* $r_s < -.564$, and the α value is therefore double the table value: $\alpha = 2(.05) = .10$. Thus, we have

Rejection region: $r_s < -.564$ or $r_s > .564$ for $\alpha = .10$

TABLE 17.11 Reproduction of Part of Table XVI of Appendix B: Critical Values of Spearman's Rank Correlation Coefficient

n	$\alpha = .05$	$\alpha = .025$	$\alpha = .01$	$\alpha = .005$
5	.900	—	—	—
6	.829	.886	.943	—
7	.714	.786	.893	—
8	.643	.738	.833	.881
9	.600	.683	.783	.833
10	.564	.648	.745	.794
11	.523	.623	.736	.818
12	.497	.591	.703	.780
13	.475	.566	.673	.745
14	.457	.545	.646	.716
15	.441	.525	.623	.689
16	.425	.507	.601	.666
17	.412	.490	.582	.645
18	.399	.476	.564	.625
19	.388	.462	.549	.608
20	.377	.450	.534	.591

Since the calculated value from Example 17.4 is $r_s = -.576$, which is less than $-.564$, we reject H_0 at $\alpha = .10$ and conclude that the population rank correlation coefficient, ρ_s, differs from 0. In fact, it appears that bank customers at lower income levels tend to borrow more than those at higher income levels. Therefore, the bank would be wise to aim its loan advertising at those in the middle- and lower-income groups, unless it wants to attempt to entice those who rarely borrow to become customers, in which case higher-income groups should be the target.

A summary of Spearman's nonparametric test for correlation is shown in the next box.

Spearman's Nonparametric Test for Rank Correlation

One-Tailed Test

H_0: $\rho_s = 0$
H_a: $\rho_s > 0$
(or H_a: $\rho_s < 0$)

Test statistic: r_s, the sample rank correlation (formulas for r_s are given on pages 952 and 953)

Rejection region: $r_s > r_{s,\alpha}$
(or $r_s < -r_{s,\alpha}$ when H_a: $\rho_s < 0$)

where $r_{s,\alpha}$ is the value from Table XVI corresponding to the upper-tail area α and n pairs of observations

Two-Tailed Test

H_0: $\rho_s = 0$
H_a: $\rho_s \neq 0$

Test statistic: r_s, the sample rank correlation (formulas for r_s are given on pages 952 and 953)

Rejection region: $r_s < -r_{s,\alpha/2}$
or $r_s > r_{s,\alpha/2}$

where $r_{s,\alpha/2}$ is the value from Table XVI corresponding to the upper-tail area $\alpha/2$ and n pairs of observations

Assumptions:
1. The sample of experimental units on which the two variables are measured is randomly selected.
2. The probability distributions of the two variables are continuous.

Ties: Assign tied measurements the average of the ranks they would receive if they were unequal but occurred in successive order. E.g., if the third- and fourth-ranked measurements tie, rank both $(3 + 4)/2 = 3.5$. The number of ties should be small relative to the total number of observations.

EXAMPLE 17.5

Manufacturers of perishable foods often use preservatives to retard spoilage. One concern is that too much preservative will change the flavor of the food. Suppose an experiment is conducted using samples of a food product with varying amounts of preservative added. Both length of time until the food shows signs of spoiling and a taste rating are recorded for each sample. The taste rating is the average rating for three tasters, each of whom rates each sample on a scale from 1 (good) to 5 (bad). Twelve sample measurements are shown in Table 17.12 (page 956). Use a nonparametric test to find out whether the spoilage times and taste ratings are negatively correlated. Use $\alpha = .05$.

TABLE 17.12 Data for Example 17.5

Sample	Days Until Spoilage	Rank	Taste Rating	Rank
1	30	2	4.3	11
2	47	5	3.6	7.5
3	26	1	4.5	12
4	94	11	2.8	3
5	67	7	3.3	6
6	83	10	2.7	2
7	36	3	4.2	10
8	77	9	3.9	9
9	43	4	3.6	7.5
10	109	12	2.2	1
11	56	6	3.1	5
12	70	8	2.9	4

Note: Tied measurements are assigned the average of the ranks they would be given if they were different but consecutive.

Solution

The test is one-tailed, with

$$H_0\colon\ \rho_s = 0 \qquad H_a\colon\ \rho_s < 0$$

*Test statistic:** $r_s = 1 - \dfrac{6\sum d_i^2}{n(n^2 - 1)}$

Rejection region: Reject H_0 if $r_s < -r_{s,.05}$, where from Table XVI, for $\alpha = .05$ and $n = 12$, $-r_{s,.05} = -.497$. [*Note*: The value of α need not be doubled, since the test is one-tailed.]

The first step in the computation of r_s is to sum the squares of the differences between ranks:

$$\sum d_i^2 = (2 - 11)^2 + (5 - 7.5)^2 + \cdots + (8 - 4)^2 = 536.5$$

Then

$$r_s = 1 - \frac{6(536.5)}{12(144 - 1)} = -.876$$

Since $-.876 < -.497$, we reject H_0 and conclude that the preservative does affect the taste of this food adversely.

*The shortcut formula is not exact when there are tied measurements, but it is a good approximation when the total number of ties is not large relative to n.

CASE STUDY 17.1 / The Problem of Nonresponse Bias in Mail Surveys

Researchers who collect their sample data via mail questionnaires often run the risk that their respondents will not be a representative sample of the entire population. The reason for this, according to Rosenthal and Rosnow (1975), is the tendency for respondents to be (1) better educated, (2) of higher social-class status, (3) more intelligent, (4) in need of social approval, (5) more social, and (6) more interested in the research topic than nonrespondents.

Researchers sometimes attempt to verify the representativeness of their sample by obtaining information about the demographic characteristics of the nonrespondents and comparing them to those of the sample of respondents. Finding similarities gives researchers confidence that their sample is representative. Another approach is to contact a small sample of the nonrespondents and obtain responses to the questionnaire. These responses can be compared with those of the original respondents; the more similar the patterns of responses, the more confident the researcher can be that the original sample of returned questionnaires is representative of the population from which the sample came.

In one marketing research study (Finn, Wang, and Lamb, 1983), a random sample of 20 nonrespondents to a mail questionnaire was contacted by telephone and asked a question from the questionnaire: "In general, what is your willingness to buy products made in each of the following countries?" The replies used a 5-point scale that ranged from "extremely willing" to "extremely unwilling." The mean willingness score was computed for each country and the countries were ranked accordingly. Similarly, a rank ordering was developed for the 273 respondents to the mail questionnaire. Both sets of rankings are displayed in Table 17.13.

TABLE 17.13 Ranking of Consumer Willingness to Buy Products Made in Indicated Countries

Country	Rank Order	
	Respondent	Nonrespondent
United Kingdom	1	1
Japan	2	3
France	3	2
Taiwan	4	4
Brazil	5	5
India	6	6
Iran	7	7
Angola	8	8
USSR	9	9
Cuba	10	10

Finn, Wang, and Lamb compared the rank orderings for the respondents and nonrespondents using Spearman's rank correlation coefficient. They obtained $r_s = .9879$ (p-value $< .01$) and concluded that "respondents and nonrespondents in this study did not differ attitudinally." Further, they noted that even if respondents and nonrespondents to a mail survey differ demographically, as was the case in their study, the Spearman rank correlation result indicates that such differences do not automatically signal the existence of nonresponse bias. The sample of opinions obtained from the respondents may, in fact, be representative of the population of opinions even though respondents and nonrespondents differ demographically.

Exercises 17.40–17.47

Learning the Mechanics

17.40 Use Table XVI of Appendix B to find each of the following probabilities:

a. $P(r_s > .496)$ when $n = 23$ **b.** $P(r_s > .496)$ when $n = 28$

c. $P(r_s \leq .600)$ when $n = 9$ **d.** $P(r_s < -.377 \text{ or } r_s > .377)$ when $n = 20$

17.41 Compute Spearman's rank correlation coefficient for each of the following pairs of sample observations.

a.

x	30	55	60	19	40
y	26	36	65	25	35

b.

x	90	100	120	137	41
y	81	95	75	52	136

c.

x	1	15	4	10
y	11	26	15	21

d.

x	5	20	15	10	3
y	80	83	91	82	87

17.42 The sample data shown in the table were collected on variables x and y.

x	−1	2	−5	−1	3	0	4
y	−1	1	−4	1	3	1	2

a. Specify the null and alternative hypotheses that should be used in conducting a hypothesis test to determine whether the variables x and y are correlated.

b. Conduct the test of part **a** using $\alpha = .05$.

c. What is the approximate p-value of the test of part **b**?

d. What assumptions are necessary to assure the validity of the test of part **b**?

Applying the Concepts

17.43 Because public safety and millions of tax dollars are at stake in highway design decisions, designers should consider both the physical and mental skills of drivers. A researcher investigated the correlation between the magnitude of brake pressure applied by drivers and various physiological variables such as heart rate and electrodermal response (EDR). The table lists 14 traffic events to which 60 test drivers were exposed. The events are listed in rank order according to the average magnitude of brake pressure associated with each (1 is highest, 14 is lowest). The events were also ranked according to the drivers' average electrodermal response (1, highest; 14, lowest). High electrodermal responses are associated with mental stress. Based on a one-sided hypothesis test employing Spearman's rank correlation coefficient, it was concluded that events involving the use of the brake are perceived as stressful. Therefore, highway designs that minimize braking should be preferred.

Traffic Event	*Brake Pressure Rank*	*EDR Rank*
Cyclist or pedestrian and oncoming car	1	1
Other car merges in front of own car	2	2
Multiple events	3	3
Leading car diverges	4	4
Cyclist or pedestrian	5	6
Own car passes other car with car following	6	5
Cyclist or pedestrian and car following	7	9
Car following and meeting other car	8	11
Meeting other car	9	8
Car following	10	10
Parked car	11	14
No event	12	12
Other car passes own car	13	7
Parked car and car following	14	13

Source: Helander, M. "Applicability of drivers' electrodermal responses to the design of the traffic environment." *Journal of Applied Psychology*, 1978, Vol. 63, pp. 481–488.

a. Calculate Spearman's rank correlation coefficient and test its significance using $\alpha = .05$. Does your analysis confirm the reported findings? Explain.
b. Describe the Type I and Type II errors associated with the hypothesis test.
c. What is the approximate p-value of the test of part **a**?
d. What assumptions are necessary to assure the validity of the test?

17.44 A *negotiable certificate of deposit* is a marketable receipt for funds deposited in a bank for a specified period of time at a specified rate of interest (Cook, 1977). The table lists the end-of-quarter interest rate for 3-month certificates of deposit during the period January 1979 through June 1986. The table also lists end-of-quarter values of Standard & Poor's 500 Stock Composite Average (an indicator of stock market activity) for the same time period.

Year	*Quarter*	*Interest Rate*	*S&P 500*
1979	I	10.13%	101.59
	II	9.95	102.91
	III	11.89	109.32
	IV	13.43	107.94
1980	I	17.57	104.69
	II	8.49	114.24
	III	11.29	125.46
	IV	18.65	135.76
1981	I	14.43	136.00
	II	16.90	131.21
	III	16.84	116.18
	IV	12.49	122.55
1982	I	14.21	111.96
	II	14.46	109.61
	III	10.66	120.42
	IV	8.66	135.28
1983	I	8.69	152.96
	II	9.20	168.11
	III	9.39	166.07
	IV	9.69	164.93
1984	I	10.08	159.18
	II	11.34	153.18
	III	11.29	166.10
	IV	8.60	167.24
1985	I	9.02	180.66
	II	7.44	191.85
	III	7.93	182.08
	IV	7.80	211.28
1986	I	7.24	238.90
	II	6.73	250.84

Source: *Standard & Poor's Statistical Service, Current Statistics*, 1986.

a. Compute Spearman's rank correlation coefficient to measure the strength of the relationship between the interest rate on certificates of deposit and the S&P 500.

b. Test the null hypothesis that the interest rate on certificates of deposit and the S&P 500 are not correlated against the alternative hypothesis that these variables are correlated. Use $\alpha = .10$.

c. Repeat parts **a** and **b** using data from 1986 through 1992, which can be obtained at your library in *Standard & Poor's Current Statistics*. Compare your results for the newer data with your results for the earlier period.

17.45 It has been conjectured that income is a primary determinant of job satisfaction. To investigate this theory, 15 employees of a particular firm are chosen at random and their gross salaries are noted. Each employee is then asked to complete a questionnaire designed to measure job satisfaction. The resulting scores (higher scores mean greater satisfaction) and gross incomes (in thousands of dollars) are given in the table.

Employee	*Job Score*	*Income*	*Employee*	*Job Score*	*Income*
1	92	29.9	9	45	16.0
2	51	18.7	10	72	25.0
3	88	32.0	11	53	17.2
4	65	15.0	12	43	9.7
5	80	26.0	13	87	20.1
6	31	9.0	14	30	15.5
7	38	11.3	15	74	16.5
8	75	22.1			

a. Compute Spearman's rank correlation coefficient for these data.

b. Is there evidence that job satisfaction and income are positively correlated? Use $\alpha = .05$.

17.46 Many large businesses send representatives to college campuses to conduct job interviews. To aid the interviewer, one company decides to study the correlation between the strength of an applicant's references (the company requires three references) and the performance of the applicant on the job. Eight recently hired employees are sampled, and independent evaluations of both references and job performance are made on a scale from 1 to 20. The scores are given in the table.

Employee	*References*	*Performance*	*Employee*	*References*	*Performance*
1	18	20	5	16	14
2	14	13	6	11	18
3	19	16	7	20	15
4	13	9	8	9	12

a. Compute Spearman's rank correlation coefficient for these data.

b. Is there evidence that strength of references and job performance are positively correlated? Use $\alpha = .05$.

17.47 The decision to build a new plant or to move an existing plant to a new location involves long-term commitment of both human and monetary resources. Accordingly, such decisions should be made only after carefully considering the relevant factors at alternative sites. G. Michael Epping (1982) examined the relationship between the location factors deemed important by businesses that located in Arkansas and those that considered Arkansas but located elsewhere. A questionnaire that asked manufacturers to rate the importance of 13 general location factors on a 9-point scale was completed by 118 firms that had moved a plant to Arkansas in the period 1955–1977 and by 73 firms that had considered Arkansas but went elsewhere.

Epping averaged the importance ratings and arrived at the rankings shown in the table. Calculate Spearman's rank correlation coefficient and carefully interpret its value in the context of the problem.

Location Factor	*Firms Choosing Arkansas: Rank*	*Firms Rejecting Arkansas: Rank*
Labor	1	1
Taxes	2	2
Industrial site	3	4
Information sources, special inducements	4	5
Legislative laws and structure	5	3
Utilities and resources	6	7
Transportation facilities	7	8
Raw material supplies	8	10
Community	9	6
Industrial financing	10	9
Markets	11	12
Business services	12	11
Personal preferences	13	13

Summary

We have presented several useful **nonparametric techniques** for testing hypotheses about the median of a single population and for comparing two or more populations. Nonparametric techniques are useful when the underlying assumptions for their parametric counterparts are not justified or when it is impossible to assign specific numerical values to the observations. The **sign test** is useful for testing hypotheses about the central tendency of a population, particularly when the assumptions necessary to conduct a t test are not satisfied.

Rank sums are the primary tools of nonparametric statistics. The **Wilcoxon rank sum statistic** and the **Wilcoxon signed rank statistic** can be used to compare two populations for either an **independent sampling experiment** or a **paired difference experiment**. The **Kruskal-Wallis H test** is applied when comparing p populations using a **completely randomized design**. **Spearman's rank correlation coefficient** is a nonparametric alternative to Pearson's product moment correlation coefficient.

The strength of nonparametric statistics lies in their general applicability. Few restrictive assumptions are required, and they may be used for observations that can be ranked but not measured exactly. Therefore, nonparametric methods provide useful alternatives to the parametric tests of Chapters 7–9 and 16.

Supplementary Exercises 17.48–17.67

17.48 When is it inappropriate to use the t and F tests of Chapters 9 and 16 for testing hypotheses about population means?

17.49 A study was conducted to determine whether the installation of a traffic light was effective in reducing the number of accidents at a busy intersection. Samples of 6 months prior to installation and 5 months after installation of the light yielded the numbers of accidents per month listed in the table.

Before	*After*
12	4
5	2
10	7
9	3
14	8
6	

a. Is there sufficient evidence to conclude that the traffic light aided in reducing the number of accidents? Test using $\alpha = .025$.

b. Explain why this type of data might or might not be suitable for analysis using the t test of Chapter 9.

17.50 The Wilcoxon signed rank test was used to compare the performance of organizations using an innovative decentralized organization structure, the M-form, and their principal competitors. Teece (1981) identified the first firm in each of 20 industries that adopted the M-form structure. The principal competitor of each of these firms was identified, and if it also had adopted the M-form structure, but at a later date, it qualified for the sample. The difference in performance within the 14 qualifying pairs of firms was measured over two 3-year time periods: "before," when only the M-form originator in each pair used the M-form, and "after," when both had M-form structures. Differential performance in each time period was measured as the average difference in yearly return on stockholders' equity, where the differences were formed by subtracting the competitor's return on equity from the M-form originator's. The performance data appear in the accompanying table.

Industry	*Average Difference in Yearly Return on Equity*		*Industry*	*Average Difference in Yearly Return on Equity*	
	Before	*After*		*Before*	*After*
Grocery	25.70%	10.78%	Grain milling	3.86	4.81
Chemicals	6.26	−.39	Petroleum	3.29	2.38
Textiles	5.84	1.94	Tires	−3.36	−2.46
Aluminum	4.16	.56	Autos	13.48	14.38
Meat packing	−2.43	−5.80	Electrical equipment	−10.85	−10.40
Packaged foods	.77	−1.18	Tobacco	1.50	1.90
Can manufacturing	3.79	2.49	Retail department stores	12.32	12.37

a. Teece concluded that "the M-form innovation has been shown to display a statistically significant impact on firm performance." Do you agree? Test using $\alpha = .05$.

b. What is the approximate p-value of the test of part **a**?

c. What assumptions are necessary to assure the validity of the test procedure you used in part **a**?

17.51 Exercise 10.47 described a calibration study (Wright, Owen, and Pena, 1983) undertaken by the Minnesota Department of Transportation to evaluate a new weigh-in-motion scale. Pearson's product moment correlation coefficient was used to measure the strength of the relationship between the static weight of a truck and the truck's weight as measured by the weigh-in-motion equipment. The data (in thousands of pounds) are repeated in the table.

Truck	Static Weight x	Weigh-in-Motion Reading Prior to Calibration Adjustment y_1	Weigh-in-Motion Reading After Calibration Adjustment y_2
1	27.9	26.0	27.8
2	29.1	29.9	29.1
3	38.0	39.5	37.8
4	27.0	25.1	27.1
5	30.3	31.6	30.6
6	34.5	36.2	34.3
7	27.8	25.1	26.9
8	29.6	31.0	29.6
9	33.1	35.6	33.0
10	35.5	40.2	35.0

a. Calculate Spearman's rank correlation coefficient for x and y_1 and for x and y_2. Interpret, in the context of the problem, the values you obtain. Compare your results with those of Exercise 10.47, part **c**.

b. In the context of this problem, what circumstances would result in Spearman's rank correlation coefficient being exactly 1? Being exactly 0?

17.52 An experiment was conducted to compare two print types, A and B, to determine whether type A is easier to read. Ten subjects were randomly divided into two groups of five. Each subject was given the same material to read, one group receiving the material printed with type A, the other group receiving print type B. The times necessary for each subject to read the material (in seconds) were

Type A: 95, 122, 101, 99, 108

Type B: 110, 102, 115, 112, 120

Do the data provide sufficient evidence to indicate that print type A is easier to read? Test using $\alpha = .05$.

17.53 Refer to Exericse 17.52. Test the research hypothesis that the median of the type A probability distribution exceeds 100 seconds. Repeat the test for the type B probability distribution. Use $\alpha = .05$ for both tests.

17.54 Suppose a company wants to study how personality relates to leadership. Four supervisors (1–4) with different types of personalities are selected. Several employees are then selected from the group supervised by each and asked to rate the leader of their group on a scale from 1 to 20 (20 signifies highly favorable). The resulting data are shown in the table.

1	2	3	4
20	17	16	8
19	11	15	12
20	13	13	10
18	15	18	14
17	14	11	9
	16		10

a. What type of experimental design was employed? Identify the key elements of the experiment: response, factor(s), factor type(s), treatments, and experimental units.
b. Test to determine whether evidence exists that the probability distributions of ratings differ for at least two of the four supervisors. Use $\alpha = .05$.
c. What assumptions are necessary to assure the validity of the test?
d. Do the results of the test warrant further comparisons of the pairs of supervisors? If so, compare all pairs of probability distributions using $\alpha = .05$ for each comparison. Does one supervisor appear to be most popular?

17.55 A national clothing store franchise operates two stores in one city—one urban and one suburban. To stock the stores with clothing suited to the customers' needs, a survey is conducted to determine the incomes of the customers. Ten customers in each store are offered significant discounts if they will reveal the annual income of their household. The results are listed in the table (in thousands of dollars). Is there evidence that the probability distributions of the customers' incomes differ in location for the two stores? Use $\alpha = .05$.

Store 1		*Store 2*	
18.8	29.5	12.3	10.3
27.9	16.3	19.2	15.6
12.2	22.1	6.3	9.8
85.3	15.7	24.5	8.6
13.1	24.0	11.0	19.3

17.56 A state highway patrol was interested in knowing whether frequent patrolling of highways substantially reduced the number of speeders. Two similar interstate highways were selected for the study—one heavily patrolled and the other only occasionally patrolled. After 1 month, random samples of 100 cars were chosen on each highway and the number of cars exceeding the speed limit was recorded. This process was repeated on 5 randomly selected days. The data are shown in the table.

Day	*Well Patrolled Road*	*Seldom Patrolled Road*
1	35	60
2	40	36
3	25	48
4	38	54
5	47	63

a. Use the paired t test with $\alpha = .10$ to compare the population mean number of speeders per 100 cars for the two highways. What assumptions are necessary for the validity of this procedure? Do you think the assumptions are reasonable in this situation?
b. Use a nonparametric procedure to determine whether the data provide evidence to indicate that heavy patrolling tends to reduce the number of speeders. Test using $\alpha = .10$.

17.57 A hotel had a problem with people reserving rooms for a weekend and then not honoring their reservations (no-shows). As a result, the hotel developed a new reservation and deposit plan that it hoped would reduce the number of no-shows. One year after the policy was initiated, management evaluated its effect in comparison with the old policy. Compare the records given in the table on the number of no-shows for the ten nonholiday weekends preceding the institution of the new policy and the ten nonholiday weekends preceding the evaluation time. Has the situation improved under the new policy? Test at $\alpha = .05$.

Before		*After*	
10	11	4	4
5	8	3	2
3	9	8	5
6	6	5	7
7	5	6	1

17.58 In recent years, many magazines have been forced to raise their prices because of increased postage, printing, and paper costs. Because magazines are now more expensive, some households may subscribe to fewer magazines than they did 3 years ago. Ten households are selected at random, and the number of magazines subscribed to 3 years ago and now is determined, as shown in the table. Does this sample provide sufficient evidence to indicate that households tend to subscribe to fewer magazines now than they did 3 years ago? Use $\alpha = .05$.

Household	*3 Years Ago*	*Now*	*Household*	*3 Years Ago*	*Now*
1	8	4	6	6	5
2	3	5	7	4	3
3	6	4	8	2	2
4	3	3	9	9	6
5	10	5	10	8	2

17.59 A savings and loan association is considering three locations in a large city as potential office sites. The company has hired a marketing firm to compare the incomes of people living in the area surrounding each site. The market researchers interview 10 households chosen at random in each area to determine the type of job, length of employment, etc., of those in the households who work. This information enables them to estimate the annual income of each household, as shown in the table (in thousands of dollars).

Site 1		*Site 2*		*Site 3*	
34.3	36.2	39.3	42.2	34.5	38.3
35.5	43.5	45.5	103.5	29.3	43.3
32.1	34.7	50.2	47.9	37.2	36.7
28.3	38.0	72.1	41.2	33.2	40.0
40.5	35.1	48.6	44.0	32.6	35.2

a. What type of design was utilized for this experiment?

b. Use the appropriate nonparametric test to compare the treatments. Specify the hypotheses and interpret the results in terms of this experiment. Use $\alpha = .05$.

c. Does the result of your test warrant further comparison of the pairs of treatments? If so, compare the pairs using the appropriate nonparametric technique and $\alpha = .05$ for each comparison. Interpret the results in terms of this experiment.

d. What assumptions are necessary to assure the validity of the nonparametric procedures you employed? How do they compare with the assumptions that would have to be satisfied in order to use the appropriate parametric technique to analyze this experiment?

17.60 Twelve samples of variously priced carpeting were selected and tested for wearability. The cost per square yard and the number of months of wear for each of the 12 samples of carpeting are listed in the table. Do the data provide sufficient evidence to indicate that wearability increases as the price increases? Test using $\alpha = .05$.

Cost	*Months of Wear*	*Cost*	*Months of Wear*
$ 6.95	32.5	$ 8.45	25.2
4.25	24.8	17.95	35.3
10.85	25.6	12.95	34.6
7.99	18.4	9.99	29.7
15.25	28.3	14.85	29.9
20.50	20.4	6.25	26.3

17.61 For many years, the Girl Scouts of America have sold cookies. One troop experimented with several sales techniques and reported the number of boxes sold per scout as listed in the table. Is there evidence that the probability distributions of number of boxes sold per scout differ in location for at least two of the four techniques? Use $\alpha = .10$.

Door-to-Door	*Telephone*	*Grocery Store Stand*	*Department Store Stand*
47	63	113	25
93	19	50	36
58	29	68	21
37	24	37	27
62	33	39	18
		77	31

17.62 Refer to Exercise 17.61. Compare the locations of the probability distributions of the number of sales for the door-to-door and grocery store stand techniques. Use the Bonferroni technique (Chapter 16) to assure that the overall α is .05.

17.63 The personnel director of a company interviewed six potential job applicants without knowing anything about their backgrounds and then rated them on a scale from 1 to 10. Independently, the director's supervisor made an evaluation of the background qualifications of each candidate on the same scale. The results are shown in the table. Is there evidence that candidates' qualification scores are related to their interview performance? Use $\alpha = .10$.

Candidate	*Qualifications*	*Interview Performance*
1	10	8
2	8	9
3	9	10
4	4	5
5	5	3
6	6	6

17.64 Two car-rental companies have long waged an advertising war. An independent testing agency is hired to compare the number of rentals at one major airport. After 10 days, the agency has the data listed in the table. At this point, can either car-rental firm claim to be number one at this airport? Use $\alpha = .05$.

Day	*Firm A*	*Firm B*	*Day*	*Firm A*	*Firm B*
1	29	22	6	16	20
2	26	29	7	35	30
3	19	30	8	43	45
4	28	25	9	29	38
5	27	26	10	32	40

17.65 David K. Campbell, James Gaertner, and Robert P. Vecchio (1983) investigated the perceptions of accounting professors with respect to the present and desired importance of various factors considered in promotion and tenure decisions at major universities. One hundred fifteen professors at universities with accredited doctoral programs responded to a mailed questionnaire. The questionnaire asked the professors to rate (1) the actual importance placed on 20 factors in the promotion and tenure decisions at their universities and (2) how they believe the factors *should* be weighted. Responses were obtained on a 5-point scale ranging from "no importance" to "extreme importance." The resulting ratings were averaged and converted to the rankings shown in the table. Calculate Spearman's rank correlation coefficient for the data and carefully interpret its value in the context of the problem.

Factor	*Actual*	*Ideal*
I. Teaching (and related items):		
Teaching performance	6	1
Advising and counseling students	19	15
Students' complaints/praise	14	17
II. Research:		
Number of journal articles	1	6.5
Quality of journal articles	4	2
Refereed publications:		
a. Applied studies	5	4
b. Theoretical empirical studies	2	3
c. Educationally oriented	11	8
Papers at professional meetings	10	12
Journal editor or reviewer	9	10
Other (textbooks, etc.)	7.5	11
III. Service and professional interaction:		
Service to profession	15	9
Professional/academic awards	7.5	6.5
Community service	18	19
University service	16	16
Collegiality/cooperativeness	12	13
IV. Other:		
Academic degrees attained	3	5
Professional certification	17	14
Consulting activities	20	20
Grantsmanship	13	18

Source: Campbell, D. K., Gaertner, J., and Vecchio, R. P. "Perceptions of promotion and tenure criteria: A survey of accounting educators." *Journal of Accounting Education*, Spring 1983, Vol. 1, pp. 83–92.

17.66 Increasing numbers of private and public agencies require employee testing for substance abuse. One laboratory has developed a system with a normalized measurement scale, in which values less than 1.00 indicate normal ranges while values of or greater than 1.00 indicate potential substance abuse. The lab reports a normal result as long as the median level for an individual is less than 1.00. Nine independent measurements of each individual's sample are made; suppose that one individual's results were as follows:

.78 .51 4.32 .23 .77 .98 .96 .89 1.11

a. Set up the appropriate null and alternative hypotheses if the employer wants "proof" beyond a reasonable doubt that the individual is in the normal range.
b. Test to determine whether the laboratory can conclude that the individual's median level is in the normal range using a .05 level of significance.
c. What assumptions are necessary to assure the validity of this test?

17.67 Many water treatment facilities add hydrofluorosilicic acid to the water to supplement the natural fluoride concentration in order to reach a target concentration of fluoride in the drinking water. Certain levels are thought to enhance dental health, but very high concentrations can be dangerous. Suppose that one such treatment plant targets .75 milligram per liter (mg/L) for its water. The plant tests 25 samples each day to determine whether the median level differs from the target.
a. Set up the null and alternative hypotheses.
b. Set up the test statistic and rejection region using $\alpha = .10$.
c. Explain the implication of a Type I error in the context of this application. A Type II error.
d. Suppose that one day's samples result in 18 values that exceed .75 mg/L. Conduct the test and state the appropriate conclusion in the context of this application.
e. When it was suggested to the plant's supervisor that a t test should be used to conduct the daily test, she replied that the probability distribution of the fluoride concentrations was "heavily skewed to the right." Show graphically what she meant by this, and explain why this is a reason to prefer the sign test to the t test.

On Your Own

In the Chapter 16 **On Your Own** you used a completely randomized design to compare the mean dinner expenditure per customer at three fast-food restaurants. Analyze these data again using nonparametric procedures. How do the necessary assumptions differ? Compare the results of the two analyses. Do your conclusions differ? Explain.

Using the Computer

A large department store chain has decided to locate a new store in either Mississippi or Florida. As part of the information to be used in deciding between the two states, the firm would like confirmation of its belief that the Overall Purchasing Potential Index tends to be higher in Florida's zip codes than in Mississippi's. Assume that the Florida zip codes in the data set of Appendix C are a random sample of all of Florida's zip codes. Make a similar assumption for Mississippi. Use the Wilcoxon rank sum test to provide the desired information.

References

Alter, S., and Ginzberg, M. "Managing uncertainty in MIS implementation." *Sloan Management Review*, Fall 1978, Vol. 20, pp. 23–31.

Beckenstein, A. R., Gabel, H. L., and Roberts, K. "An executive's guide to antitrust compliance." *Harvard Business Review*, Sept.–Oct. 1983, pp. 94–102.

Certo, S. C. *Principles of Modern Management*. Dubuque, IA: Wm. C. Brown, 1980. Chapter 14.

Conover, W. J. *Practical Nonparametric Statistics*, 2nd ed. New York: Wiley, 1980.

Cook, T. Q., ed. *Instruments of the Money Market*, 4th ed. Richmond, VA.: Federal Reserve Bank of Richmond, 1977.

Davis, G. B., and Olson, M. H. *Management Information Systems*, 2nd ed. New York: McGraw-Hill, 1985.

Epping, G. M. "Importance factors in plant location in 1980." *Growth and Change*, Apr. 1982, Vol. 13, pp. 47–51.

Finn, D. W., Wang, C.-K., and Lamb, C. W. "An examination of the effects of sample composition bias in a mail survey." *Journal of the Market Research Society*, Oct. 1983, Vol. 25, pp. 331–338.

Freedman, D. A. "The mean versus the median: A case study in 4-R Act litigation." *Journal of Business and Economic Statistics*, Jan. 1985, Vol. 3, pp. 1–13.

Gibbons, J. D. *Nonparametric Statistical Inference*, 2nd ed. New York: McGraw-Hill, 1985.

Hollander, M., and Wolfe, D. A. *Nonparametric Statistical Methods*. New York: Wiley, 1973.

Lehmann, E. L. *Nonparametrics: Statistical Methods Based on Ranks*. San Francisco: Holden-Day, 1975.

Rosenthal, R., and Rosnow, R. L. *The Volunteer Subject*. New York: Wiley, 1975.

Rosetti, D. K., and Surynt, T. J. "Video teleconferencing and performance." *Journal of Business Communication*, Fall 1985, Vol. 22, pp. 25–31.

Siegel, S. *Nonparametric Statistics for the Behavioral Sciences*. New York: McGraw-Hill, 1956.

Teece, D. J. "Internal organization and economic performance: An empirical analysis of the profitability of principal firms." *The Journal of Industrial Economics*, Dec. 1981, Vol. 30, pp. 173–199.

Winkler, R. L., and Hays, W. L. *Statistics: Probability, Inference, and Decision*, 2nd ed. New York: Holt, Rinehart and Winston, 1975. Chapter 12.

Wright, J. L., Owen, F., and Pena, D. "Status of MN/DOT's weigh-in-motion program." St. Paul: Minnesota Department of Transportation, Jan. 1983.

CHAPTER EIGHTEEN

The Chi-Square Test and the Analysis of Contingency Tables

Contents

Case Studies

Where We've Been

The preceding chapters have presented statistical methods for analyzing many types of business data. Chapters 7–12 and 16 were appropriate for populations of data generated by quantitative random variables that were independent and had (at least approximately) normal probability distributions with a common variance. Nonparametric statistical procedures were presented in Chapter 17. They included methods to compare two or more populations when the assumptions of normality or common variance were likely to be violated or when the responses could be ranked only according to their relative magnitudes.

Where We're Going

The methods of this chapter are appropriate for a type of data known as *count*, or *classificatory*, data. For example, a brokerage company might want to investigate the relationship between its customers' investment preferences (stocks, bonds, mutual funds, etc.) and its customers' occupations. To do this, the company would sample its customers and count the number in each preference–occupation category. Then, the data would be used to make inferences about the actual proportions of their population of customers in each category. Problems of this type, as well as others that involve count data, are the topic of Chapter 18.

Many business analyses involve enumerating the occurrences of some event. For example, we may count the number of defectives during one shift at a manufacturing plant, or the number of consumers who choose each of three brands of coffee, or the number of sales made by each of five automobile salespeople during the month of June.

In some instances, the objective of collecting the count data is to analyze the distribution of the counts in the various **classes**, or **cells**. For example, we may want to estimate the proportion of smokers who prefer each of three different brands of cigarettes by counting the number in a sample of smokers who buy each brand. We will say that count data classified on a single scale have a **one-dimensional classification**. The analysis of one-dimensional count data is discussed in Section 18.1.

More often the objective of collecting the count data is to determine the relationship between two different methods of classifying the data. For example, we may be interested in knowing whether the size and model of the automobile purchased by new car buyers are related. Or the relationship between the shift and the number of defectives produced in a plant could be of interest. When count data are classified in a **two-dimensional** table, we call the result a **contingency table**. The analysis of general contingency tables is discussed in Section 18.2

18.1 One-Dimensional Count Data: Multinomial Distribution

Consumer preference surveys can be valuable aids in making marketing decisions. Suppose a large supermarket chain conducts a consumer preference survey by recording the brand of bread purchased by customers in its stores. Assume the chain carries three brands of bread—two major brands (A and B) and its own store brand. The brand preferences of a random sample of 150 buyers are observed, and the resulting count data appear in Table 18.1. Do these data indicate that a preference exists for any of the brands?

TABLE 18.1 Consumer Preference Survey

A	B	Store Brand
61	53	36

To answer this question, we have to know the underlying probability distribution of these count data. This distribution, called the **multinomial probability distribution**, is an extension of the binomial distribution (Section 4.4). The properties of the multinomial distribution are shown in the box.

Properties of the Multinomial Probability Distribution

1. The experiment consists of n identical trials.
2. There are k possible outcomes to each trial.
3. The probabilities of the k outcomes, denoted by $p_1, p_2, \ldots, p_k$, remain the same from trial to trial, where $p_1 + p_2 + \cdots + p_k = 1$.
4. The trials are independent.
5. The random variables of interest are the counts $n_1, n_2, \ldots, n_k$ in each of the k cells.

You can see that the properties of the multinomial experiment closely resemble those of the binomial experiment and that, in fact, a binomial experiment is a multinomial experiment for the special case where $k = 2$.

In most practical applications involving a multinomial experiment, the true values of the k outcome probabilities, $p_1, p_2, \ldots, p_k$, will be unknown. The objective is therefore to make inferences about these probabilities.

Note that the consumer preference survey given in Table 18.1 satisfies the multinomial conditions. Suppose we want to test the null hypothesis that there is no preference for any of the three brands versus the alternative hypothesis that a preference exists for one or more of the brands. Then, letting

p_1 = Proportion of all customers who prefer brand A
p_2 = Proportion of all customers who prefer brand B
p_3 = Proportion of all customers who prefer the store brand

we want to test

H_0: $p_1 = p_2 = p_3 = 1/3$ (No preference)
H_a: At least one of the proportions exceeds $1/3$ (A preference exists)

If the null hypothesis is true (i.e., if $p_1 = p_2 = p_3 = 1/3$), then we would expect to see approximately $1/3$ of the customers in the sample purchase each brand. Or, more formally, the expected value (mean value) of the number of customers purchasing brand A is given by

$$E(n_1) = np_1 = n(1/3) = 150(1/3) = 50$$

Similarly, $E(n_2) = E(n_3) = 50$ if no preference exists.

The following test statistic measures the degree of disagreement between the data and the null hypothesis:

$$X^2 = \frac{[n_1 - E(n_1)]^2}{E(n_1)} + \frac{[n_2 - E(n_2)]^2}{E(n_2)} + \frac{[n_3 - E(n_3)]^2}{E(n_3)}$$
$$= \frac{(n_1 - 50)^2}{50} + \frac{(n_2 - 50)^2}{50} + \frac{(n_3 - 50)^2}{50}$$

Note that the farther the observed numbers n_1, n_2, and n_3 are from their expected value (50), the larger X^2 will become. That is, large values of X^2 cast doubt on the null hypothesis and suggest that it is false.

We have to know the distribution of X^2 in repeated sampling before we can decide whether the data indicate that a preference exists. If in fact H_0 is true, X^2 can be shown to have approximately a χ^2 distribution with $(k - 1)$ degrees of freedom.* The χ^2 distribution was first introduced in Section 17.4, and the critical values are given in Table XIII of Appendix B. For the consumer preference survey in Table 18.1, with $\alpha = .05$ and $k - 1 = 3 - 1 = 2$ df, we will reject H_0 if $X^2 > \chi^2_{.05}$. This value of χ^2

*Deriving the df involves the number of linear restrictions imposed on the count data, so we will simply give the df for each X^2. See the references at the end of the chapter for more detail.

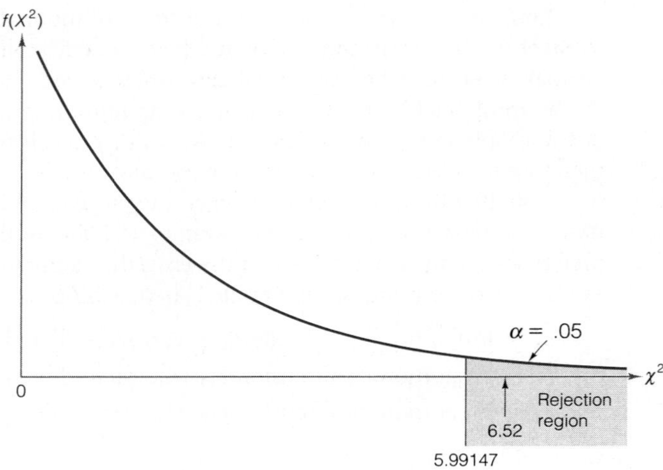

FIGURE 18.1 ▶ Rejection region for consumer preference survey

(found in Table XIII) is 5.99147 (see Figure 18.1). The computed value of the test statistic is

$$X^2 = \frac{(n_1 - 50)^2}{50} + \frac{(n_2 - 50)^2}{50} + \frac{(n_3 - 50)^2}{50}$$
$$= \frac{(61 - 50)^2}{50} + \frac{(53 - 50)^2}{50} + \frac{(36 - 50)^2}{50} = 6.52$$

Since the computed $X^2 = 6.52$ exceeds the critical value of 5.99147, we conclude at $\alpha = .05$ that there is a customer preference for one or more of the brands of bread.

The general form for a test of hypothesis concerning multinomial probabilities is shown in the next box.

A Test of Hypothesis About Multinomial Probabilities

H_0: $p_1 = p_{1,0}, p_2 = p_{2,0}, \ldots, p_k = p_{k,0}$, where $p_{1,0}, p_{2,0}, \ldots, p_{k,0}$ represent the hypothesized values of the multinomial probabilities

H_a: At least one of the multinomial probabilities does not equal its hypothesized value

Test statistic: $X^2 = \sum_{i=1}^{k} \frac{[n_i - E(n_i)]^2}{E(n_i)}$

where $E(n_i) = np_{i,0}$, the expected number of outcomes of type i assuming H_0 is true. The total sample size is n.

Rejection region: $X^2 > \chi^2_\alpha$ where df $= k - 1$

Assumptions: The sample size n will be large enough so that, for every cell, the expected cell count, $E(n_i)$, will be equal to 5 or more.

EXAMPLE 18.1

TABLE 18.2 Distribution of Pay Increases

None	Standard	Merit
42	365	193

A large firm has established what it hopes is an objective system of deciding on annual pay increases for its employees. The system is based on a series of evaluation scores determined by the supervisors of each employee. Employees with scores above 80 receive a merit pay increase, those with scores between 50 and 80 receive the standard increase, and those below 50 receive no increase. The firm designed the plan with the objective that, on the average, 25% of its employees would receive merit increases, 65% would receive standard increases, and 10% would receive no increase.

After 1 year of operation using the new plan, the distribution of pay increases for the 600 company employees was as shown in Table 18.2. Test at the $\alpha = .01$ level to determine whether these data indicate that the distribution of pay increases differs significantly from the proportions established by the firm.

Solution

Define

p_1 = Proportion of employees who receive no pay increase
p_2 = Proportion of emloyees who receive a standard increase
p_3 = Proportion of employees who receive a merit increase

Then the null and alternative hypotheses representing the intent are

H_0: $p_1 = .10,\ p_2 = .65,\ p_3 = .25$

H_a: At least two of the proportions differ from the firm's proposed plan

Test statistic: $X^2 = \sum \frac{[n_i - E(n_i)]^2}{E(n_i)}$

where

$$E(n_1) = np_{1,0} = 600(.10) = 60$$
$$E(n_2) = np_{2,0} = 600(.65) = 390$$
$$E(n_3) = np_{3,0} = 600(.25) = 150$$

Rejection region: For $\alpha = .01$ and df $= k - 1 = 2$, reject H_0 if $X^2 > \chi^2_{.01}$, where (from Table XIII of Appendix B) $\chi^2_{.01} = 9.21034$.

We now calculate the test statistic:

$$X^2 = \frac{(42 - 60)^2}{60} + \frac{(365 - 390)^2}{390} + \frac{(193 - 150)^2}{150} = 19.33$$

Since this value exceeds the table value of χ^2 (9.21034), the data provide strong evidence ($\alpha = .01$) that the company's pay plan is not working as planned.

By focusing on one particular outcome of a multinomial experiment, we can use the methods developed in Section 7.4 for a binomial proportion to establish a confidence interval for any one of the multinomial probabilities.* For example, if we want a 95% confidence interval for the proportion of the company's employees who will receive merit increases under the new system, we calculate

$$\hat{p}_3 \pm 1.96\sigma_{\hat{p}_3} \approx \hat{p}_3 \pm 1.96\sqrt{\frac{\hat{p}_3(1-\hat{p}_3)}{n}} \qquad \text{where } \hat{p}_3 = \frac{n_3}{n} = \frac{193}{600} = .32$$

$$= .32 \pm 1.96\sqrt{\frac{(.32)(1-.32)}{600}} = .32 \pm .04$$

Thus, we estimate that between 28% and 36% of the firm's employees will qualify for merit increases under the new plan. It appears that the firm will have to raise the requirements for merit increases in order to achieve the stated goal of a 25% employee qualification rate.

CASE STUDY 18.1 / Investigating Response Bias in a Diary Survey

Marketing researchers sometimes collect data from consumers by asking them to keep diaries of their purchases or product usage. Such *diary methods* generally provide more accurate data than collection methods (such as the telephone interview) that require the consumer to recall from memory the details of purchases and activities. However, diary methods are not problem-free. Frequently, response rates by consumers to requests for diary data are lower than for other types of market research surveys such as mailed questionnaires. In addition, recording biases (such as not recording some events or recording inappropriate events) may be present in diary data.

John McKenzie (1983) studied the accuracy of telephone-call data collected by diary methods in Great Britain. This required comparing the demographic profile of a sample of 1,802 telephone users who responded to a request for diary data to the demographic profile of the population from which the sample of all those who were asked to keep diaries was selected. The population data were available from telephone company records for the study population of 29,507 households.

TABLE 18.3 Terminal Education Age of Head of Household

	Population		Sample	
	Frequency	Relative Frequency	Frequency	Relative Frequency
Up to 15	13,137	$p_1 = .445$	791	.439
16–18	7,021	$p_2 = .238$	531	.295
19+	3,074	$p_3 = .104$	202	.112
Not Known	6,275	$p_4 = .213$	278	.154

Table 18.3 shows the education profiles of the population and the sample. These data can be used to determine if the distribution of the age at which schooling ended for the sample differs significantly from the distribution for the population. If a significant difference exists, then it can be inferred that the sample is not representative of the population and that the survey results have been affected by **response bias**.

The sample data should lead to rejection of the following null hypothesis if the population and sample profiles differ:

*Note that focusing on one outcome has the effect of combining the other $(k-1)$ outcomes into a single group. Thus, we obtain, in effect, two outcomes—or a binomial experiment.

H_0: $p_1 = .445$, $p_2 = .238$, $p_3 = .104$, $p_4 = .213$
H_a: At least one of the relative frequencies differs from its hypothesized value

The sample data yield

$$X^2 = \frac{(791 - 801.89)^2}{801.89} + \frac{(531 - 428.88)^2}{428.88} + \frac{(202 - 187.41)^2}{187.41} + \frac{(278 - 383.83)^2}{383.83} = 54.78$$

Since $X^2 = 54.78 > \chi^2_{.005} = 12.8381$ (df = 3), the null hypothesis is rejected and the existence of response bias is confirmed. McKenzie found similar discrepancies between the population and sample with respect to the age, sex, and social class of the head of household.

When response biases such as these can be identified or are suspected, they can generally be overcome by increasing the response rates of the survey. For a discussion of ways to increase response rates, see Sudman and Ferber (1971, 1974).

Exercises 18.1 – 18.14

Learning the Mechanics

18.1 Use Table XIII of Appendix B to find each of the following χ^2 values.
a. $\chi^2_{.05}$ for df = 17 b. $\chi^2_{.990}$ for df = 100
c. $\chi^2_{.10}$ for df = 15 d. $\chi^2_{.005}$ for df = 3

18.2 Find the following probabilities.
a. $P(\chi^2 \leq .872085)$ for df = 6 b. $P(\chi^2 > 30.5779)$ for df = 15
c. $P(\chi^2 \geq 82.3581)$ for df = 100 d. $P(\chi^2 < 13.7867)$ for df = 30

18.3 Find the rejection region for a one-dimensional χ^2 test of a null hypothesis concerning $p_1, p_2, \ldots, p_k$ if:
a. $k = 4$ and $\alpha = .10$ b. $k = 6$ and $\alpha = .01$ c. $k = 8$ and $\alpha = .05$

18.4 What are the characteristics of a multinomial experiment? Compare the characteristics to those of a binomial experiment.

18.5 What conditions must n satisfy to make the χ^2 test valid?

18.6 A multinomial experiment with $k = 5$ cells and $n = 300$ produced the data shown in the table.

Cell	n_i
1	48
2	69
3	83
4	61
5	39

a. Do these data provide sufficient evidence to contradict the null hypothesis that $p_1 = .15$, $p_2 = .25$, $p_3 = .30$, $p_4 = .20$, and $p_5 = .10$? Test using $\alpha = .05$.
b. Find the approximate observed significance level for the test in part **a**.

18.7 A multinomial experiment with $k = 4$ cells and $n = 200$ produced the data shown in the table.

Cell	1	2	3	4
n_i	48	56	63	53

a. Do these data provide sufficient evidence to conclude that the multinomial probabilities differ? Test using $\alpha = .05$.
b. What are the Type I and Type II errors associated with the test of part **a**?

18.8 Refer to Exercise 18.7. Construct a 95% confidence interval for the multinomial probability associated with cell 3.

Applying the Concepts

18.9 A 1985 Gallup survey portrayed U.S. entrepreneurs as "the mavericks, dreamers, and loners whose rough edges and uncompromising need to do it their own way set them in sharp contrast to senior executives in major American corporations" (Graham, 1985). One of the many questions put to a sample of $n = 100$ entrepreneurs about their job characteristics, work habits, social activities, etc. concerned the origin of the car they usually drive. The responses are given in the table.

United States	*Europe*	*Japan*
45	46	9

a. Do these data provide evidence of a difference in the preference of entrepreneurs for the cars of the United States, Europe, and Japan? Test using $\alpha = .05$.
b. Do these data provide evidence of a difference in the preference of entrepreneurs for domestic versus foreign cars? Test using $\alpha = .05$.
c. What assumptions must be made in order for your inferences of parts **a** and **b** to be valid?

18.10 Refer to Exercise 18.9. Use a 90% confidence interval to estimate the proportion of U.S. entrepreneurs who drive foreign cars.

18.11 Overweight trucks are responsible for much of the damage sustained by our local, state, and federal highway systems. Although illegal, overloading is common. Truckers may avoid weigh stations run by enforcement officers by taking back roads or by traveling when weigh stations are likely to be closed. A state highway planning agency monitored the movements of overweight trucks on an interstate highway using an unmanned, computerized scale that is built into the highway (Dahlin and Owen, 1984). Unknown to the truckers, the scale weighed their vehicles as they passed over it. Each day's proportion of one week's total truck traffic (5-axle tractor truck semitrailers) is shown in the table.

Monday	*Tuesday*	*Wednesday*	*Thursday*	*Friday*	*Saturday*	*Sunday*
.191	.198	.187	.180	.155	.043	.046

During the same week, the number of overweight trucks per day was as follows:

Monday	*Tuesday*	*Wednesday*	*Thursday*	*Friday*	*Saturday*	*Sunday*
90	82	72	70	51	18	31

a. The planning agency would like to know whether the number of overweight trucks per week is distributed over the 7 days of the week in direct proportion to the volume of truck traffic. Test using $\alpha = .05$.
b. Find the approximate p-value for the test of part **a**.

18.12 The manager of a firm's data processing division believes that the distribution of the number of errors per invoice has changed dramatically since the physical layout of the fields (areas of information) on the invoices was changed 6 months ago. The table describes the distribution of the number of errors per invoice using the previous format.

Errors per Invoice	0	1	2	3	4	4+
Proportion of Finished Invoices	.90	.04	.03	.02	.005	.005

A random sample of 300 printed invoices was selected from those that printed last week. Each invoice was examined for errors. The following data resulted:

Errors per Invoice	0	1	2	3	4	4+
Number of Invoices	150	120	15	7	4	4

a. Do the data provide sufficient evidence to indicate that the proportions of printed invoices in the six error categories differ from the proportions using the previous format?
b. Find the approximate observed significance level for the test in part **a**.

18.13 A company that manufactures dice for gambling casinos in Nevada and New Jersey regularly inspects its product to be sure that only "fair" (balanced) dice are supplied to the casinos. One die was randomly chosen from a production lot and rolled 120 times. Counts of the numbers showing face up are recorded in the table. Do the data provide sufficient evidence to indicate that the die is unbalanced? Test using $\alpha = .10$.

Numbers Face Up	1	2	3	4	5	6
Frequency	28	27	20	18	15	12

18.14 The threat of earthquakes is a part of life in California, where scientists have warned about "the big one" for decades. A recent article (Palm and Hodgson, 1992) investigated what influences homeowners in purchasing earthquake insurance. One factor investigated was the proximity to a major fault. The researchers hypothesized that the nearer a county is to a major fault, the more likely residents are to own earthquake insurance. Suppose that a random sample of 700 earthquake-insured residents from four California counties is selected, and the number in each county is counted and recorded in the table:

	Contra Costa	*Santa Clara*	*Los Angeles*	*San Bernardino*
Number insured:	103	213	241	143

a. What are the appropriate null and alternative hypotheses to test whether the proportions of all earthquake-insured residents in the four counties differ?
b. Do the data provide sufficient evidence that the proportions of all earthquake-insured residents differ among the four counties? Test using $\alpha = .05$.
c. Los Angeles County is closest to a major earthquake fault. Construct a 95% confidence interval for the proportion of all earthquake-insured residents in the four counties that reside in Los Angeles County.
d. Does the confidence interval you formed in part **c** support the conclusion of the test conducted in part **b**? Explain.

18.2 Contingency Tables

Rising fuel prices have made consumers more aware of what size cars they purchase. An automobile manufacturer who is interested in determining the relationship between the size and make of newly purchased automobiles randomly samples 1,000 recent buyers of American-made cars. The manufacturer classifies each purchase with respect to the size and manufacturer of the purchased automobile. The data are shown in Table 18.4, which is an example of a **contingency table**. Contingency tables consist of **multinomial count data classified on two scales,** or **dimensions.**

Let the probabilities for the multinomial experiment in Table 18.4 be those shown in Table 18.5. Thus, p_{11} is the probability that a new-car buyer purchases a small car of manufacturer A. Note the probability totals, called **marginal probabilities,** for each row and column. The marginal probability p_1 is the probability that a small car is purchased, and the marginal probability p_A is the probability that a car of manufacturer A is purchased.

TABLE 18.4 Contingency Table for Automobile Size Example

		Manufacturer				Totals
		A	B	C	D	
	Small	157	65	181	10	413
Size	Intermediate	126	82	142	46	396
	Large	58	45	60	28	191
	Totals	341	192	383	84	1,000

TABLE 18.5 Probabilities for Contingency Table 18.4

		Manufacturer				Totals
		A	B	C	D	
	Small	p_{11}	p_{12}	p_{13}	p_{14}	p_1
Size	Intermediate	p_{21}	p_{22}	p_{23}	p_{24}	p_2
	Large	p_{31}	p_{32}	p_{33}	p_{34}	p_3
	Totals	p_A	p_B	p_C	p_D	1

Suppose we want to know whether the two classifications, manufacturer and size, are dependent. That is, if we know which size car a buyer will choose, does that information give us a clue about which manufacturer the buyer will choose? In a probabilistic sense we know (Chapter 3) that independence of events A and B implies that $P(A \cap B) = P(A)P(B)$. Similarly, in the contingency table analysis, if the classifications are independent, the probability that an item is classified in any particular cell of the table is the product of the corresponding marginal probabilities. Thus, under the hypothesis of independence, in Table 18.5 we must have

$$p_{11} = p_1 p_A \qquad p_{12} = p_1 p_B$$

and so forth.

To test the hypothesis of independence, we use the same reasoning employed in the one-dimensional tests of Section 18.1. First, we calculate the expected (or mean) count in each cell assuming the null hypothesis of independence is true. We do this by noting that the expected count in the upper left-hand corner of the table, for example, is just the total number of multinomial trials, n, times the probability, p_{11}. Then $E(n_{11}) = np_{11}$ and, if the classifications are independent, $E(n_{11}) = np_1 p_A$.

We can estimate p_1 and p_A by the sample proportions $\hat{p}_1 = n_1/n$ and $\hat{p}_A = n_A/n$. Thus, the estimate of the expected value $E(n_{11})$ is

$$\hat{E}(n_{11}) = n\left(\frac{n_1}{n}\right)\left(\frac{n_A}{n}\right) = \frac{n_1 n_A}{n}$$

Similarly,

$$\hat{E}(n_{12}) = \frac{n_1 n_B}{n}$$

$$\vdots$$

$$\hat{E}(n_{34}) = \frac{n_3 n_D}{n}$$

Using the data in Table 18.4, we find

$$\hat{E}(n_{11}) = \frac{n_1 n_A}{n} = \frac{(413)(341)}{1{,}000} = 140.833$$

$$\hat{E}(n_{12}) = \frac{n_1 n_B}{n} = \frac{(413)(192)}{1{,}000} = 79.296$$

$$\vdots$$

$$\hat{E}(n_{34}) = \frac{n_3 n_D}{n} = \frac{(191)(84)}{1{,}000} = 16.044$$

The observed data and the estimated expected values are shown in Table 18.6 on page 982.

We now use the X^2 statistic to compare the observed and expected (estimated) counts in each cell of the contingency table:

TABLE 18.6 Observed and Estimated Expected (in Parentheses) Counts

		Manufacturer				Totals
		A	B	C	D	
Size	Small	157 (140.833)	65 (79.296)	181 (158.179)	10 (34.692)	413
	Intermediate	126 (135.036)	82 (76.032)	142 (151.668)	46 (33.264)	396
	Large	58 (65.131)	45 (36.672)	60 (73.153)	28 (16.044)	191
	Totals	341	192	383	84	1,000

$$X^2 = \frac{[n_{11} - \hat{E}(n_{11})]^2}{\hat{E}(n_{11})} + \frac{[n_{12} - \hat{E}(n_{12})]^2}{\hat{E}(n_{12})} + \cdots + \frac{[n_{34} - \hat{E}(n_{34})]^2}{\hat{E}(n_{34})}$$

$$= \sum_{i=1}^{3} \sum_{j=1}^{4} \frac{[n_{ij} - \hat{E}(n_{ij})]^2}{\hat{E}(n_{ij})}$$

Substituting the data of Table 18.6 into this expression yields

$$X^2 = \frac{(157 - 140.833)^2}{140.833} + \frac{(65 - 79.296)^2}{79.296} + \cdots + \frac{(28 - 16.044)^2}{16.044} = 45.81$$

Large values of X^2 imply that the observed and expected counts do not closely agree and therefore that the hypothesis of independence is false. To determine how large X^2 must be before it is too large to be attributed to chance, we make use of the fact that the sampling distribution of X^2 is approximately a χ^2 probability distribution when the classifications are independent. The number of degrees of freedom for the approximating χ^2 distribution will be $(r - 1)(c - 1)$, where r is the number of rows and c is the number of columns in the table.

For the size and manufacturer of automobiles example, the number of degrees of freedom for χ^2 is $(r - 1)(c - 1) = (3 - 1)(4 - 1) = 6$. Then, for $\alpha = .05$, we reject the hypothesis of independence if

$$X^2 > \chi^2_{.05} = 12.5916$$

Since the computed $X^2 = 45.81$ exceeds the value 12.5916, we conclude that the size and manufacturer of a car selected by a purchaser are dependent events.

What is the observed significance level of X^2 for this test? While we cannot use Table XIII to find the exact significance level, we can use it to place bounds on the p-value. Note that the value of X^2 exceeds even the χ^2 value corresponding to a significance level of .005 with 6 degrees of freedom, 18.5476. Thus, we know that the observed significance level corresponding to the test statistic, 45.81, is smaller than .005, i.e., p-value $< .005$. The evidence in favor of the alternative hypothesis that the manufacturer and size of car are dependent is therefore quite strong.

The pattern of dependence can be seen more clearly by expressing the data as percentages. We first select one of the two classifications to be used as the base variable. In the automobile size preference example, suppose we select manufacturer as the classificatory variable to be the base. Next, we represent the responses for each level of the second categorical variable (car size) as a percentage of the subtotal for the base variable. For example, from Table 18.6 we convert the response for small car sales for manufacturer A (157) to a percentage of the total sales for manufacturer A (341). That is,

$$\left(\frac{157}{341}\right)100\% = 46\%$$

The conversion of all Table 18.6 entries is accomplished in the same way, and the values are shown in Table 18.7. The value shown at the right of each row is the row's total expressed as a percentage of the total number of responses in the entire table. Thus, the small car percentage is $^{413}\!/_{1,000}(100)\% = 41\%$ (rounded to the nearest percent).

TABLE 18.7 Percentages of Car Sizes by Manufacturer

		Manufacturer				All
		A	B	C	D	
Size	Small	46	34	47	12	41
	Intermediate	37	43	37	55	40
	Large	17	23	16	33	19
	Totals	100	100	100	100	100

If the size and manufacturer variables are independent, then the percentages in the cells of the table are expected to be approximately equal to the row percentages. Thus, we would expect the small car percentages for each of the four manufacturers to be approximately 41% if size and manufacturer were independent. The extent to which each manufacturer's percentage departs from this value determines the dependence of the two classifications, with greater variability of the row percentages meaning a greater degree of dependence. A plot of the percentages helps summarize the observed pattern. In Figure 18.2 (page 984) we show the manufacturer (the base variable) on the horizontal axis and the size percentages on the vertical axis. The "expected" percentages under the assumption of independence are shown as horizontal lines, and each observed value is represented by a symbol indicating the size category.

FIGURE 18.2 ► Size as a percentage of manufacturer subtotals

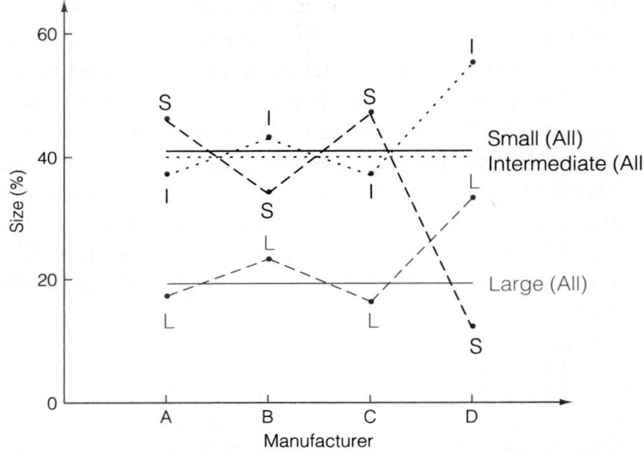

Figure 18.2 clearly indicates why the test resulted in the conclusion that the two classifications in the contingency table are dependent. Note that the sales of manufacturers A, B, and C fall relatively close to the expected percentages under the assumption of independence. However, the sales of manufacturer D deviate significantly from the expected values, with much higher percentages for large and intermediate cars and a much smaller percentage for small cars than expected under independence. Also, manufacturer B deviates slightly from the expected pattern, with a greater percentage of intermediate than small car sales. Statistical measures of the degree of dependence and procedures for making comparisons of pairs of levels for classifications are available. They are beyond the scope of this text but can be found in the references at the end of the chapter. We do, however, utilize descriptive

TABLE 18.8 General $r \times c$ Contingency Table

		Column				Row Totals
		1	2	. . .	c	
Row	1	n_{11}	n_{12}	. . .	n_{1c}	r_1
	2	n_{21}	n_{22}	. . .	n_{2c}	r_2
	⋮	⋮	⋮		⋮	⋮
	r	n_{r1}	n_{r2}	. . .	n_{rc}	r_r
Column Totals		c_1	c_2	. . .	c_c	n

summaries such as Figure 18.2 to examine the degree of dependence exhibited by the sample data.

The general form of a contingency table is shown in Table 18.8. Note that the observed count in the cell located in the ith row and the jth column is denoted by n_{ij}, the ith row total is r_i, the jth column total is c_j, and the total sample size is n. Using this notation, we give the general form of the contingency table test for independent classifications in the box.

General Form of a Contingency Table Analysis: A Test for Independence

H_0: The two classifications are independent

H_a: The two classifications are dependent

Test statistic: $$X^2 = \sum_{i=1}^{r}\sum_{j=1}^{c}\frac{[n_{ij} - \hat{E}(n_{ij})]^2}{\hat{E}(n_{ij})}$$

where $\hat{E}(n_{ij}) = \dfrac{r_i c_j}{n}$

Rejection region: $X^2 > \chi^2_\alpha$, where χ^2_α is based on $(r - 1)(c - 1)$ df

Assumption: The sample size, n, is large enough so that, for every cell, the expected cell count, $E(n_{ij})$, is equal to 5 or more.

EXAMPLE 18.2

A large brokerage firm wants to determine whether the service it provides to affluent customers differs from the service it provides to lower-income customers. A sample of 500 customers is selected, and each customer is asked to rate his or her broker. The results are shown in Table 18.9.

TABLE 18.9 Observed and Estimated Expected (in Parentheses) Counts for Example 18.2

		Customer's Income			Totals
		Under \$20,000	\$20,000–\$50,000	Over \$50,000	
Broker Rating	Outstanding	48 (53.856)	64 (66.402)	41 (32.742)	153
	Average	98 (94.336)	120 (116.312)	50 (57.352)	268
	Poor	30 (27.808)	33 (34.286)	16 (16.906)	79
Totals		176	217	107	500

a. Test to determine whether there is evidence that broker rating and customer income are independent. Use $\alpha = .10$.

b. Plot the data and describe the patterns revealed. Is the result of the test supported by the plot?

Solution

a. The first step is to calculate estimated expected cell frequencies under the assumption that the classifications are independent. Thus,

$$\hat{E}(n_{11}) = \frac{r_1 c_1}{n} = \frac{(153)(176)}{500} = 53.856$$

$$\hat{E}(n_{12}) = \frac{r_1 c_2}{n} = \frac{(153)(217)}{500} = 66.402$$

and so forth. All the estimated expected counts are shown in Table 18.9.

We are now ready to conduct the test for independence:

H_0: The rating a customer gives his or her broker is independent of the customer's income

H_a: Broker rating and customer income are dependent

Test statistic: $$X^2 = \sum_{i=1}^{3}\sum_{j=1}^{3}\frac{[n_{ij} - \hat{E}(n_{ij})]^2}{\hat{E}(n_{ij})}$$

Rejection region: For $\alpha = .10$ and $(r - 1)(c - 1) = (2)(2) = 4$ df, reject H_0 if $X^2 > \chi^2_{.10}$, where $\chi^2_{.10} = 7.77944$.

The calculated value of X^2 is

$$X^2 = \frac{(48 - 53.856)^2}{53.856} + \frac{(64 - 66.402)^2}{66.402} + \cdots + \frac{(16 - 16.906)^2}{16.906}$$
$$= 4.28$$

Since $X^2 = 4.28$ does not exceed the critical value, 7.77944, there is insufficient evidence at $\alpha = .10$ to conclude that broker rating and customer income are dependent. This survey does not support the firm's alternative hypothesis that affluent customers get broker service different from lower-income customers.

b. The broker rating frequencies are expressed as percentages of income category frequencies in Table 18.10. The expected percentages under the assumption of independence are shown at the right of each row. The plot of the percentage data is shown in Figure 18.3, where horizontal lines represent the expected percentages assuming independence. Note that the response percentages deviate only slightly from those expected under the assumption of independence, supporting the result of the test in part **a**. That is, neither the descriptive plot nor the statistical test provides evidence that the rating given the broker services depends on (varies with) the customer's income.

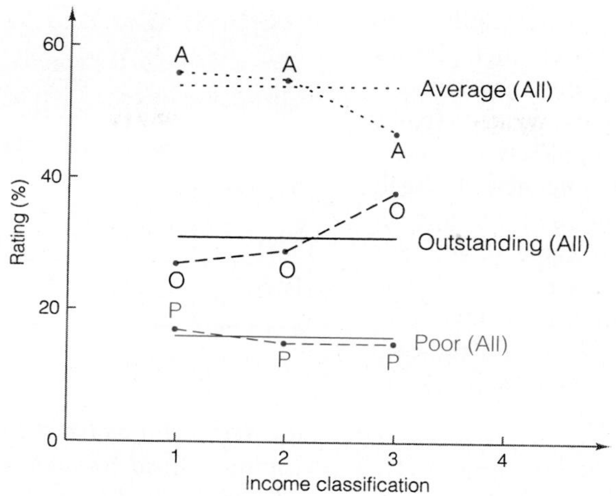

FIGURE 18.3 ► Plot of broker rating–customer income contingency table

TABLE 18.10 Broker Ratings as Percentages of Income Class

		Customer's Income			All
		Under $20,000	$20,000–$50,000	Over $50,000	
Broker Rating	Outstanding	27	29	38	31
	Average	56	55	47	54
	Poor	17	15	15	16
Totals		100	99[a]	100	101[a]

[a]Percentages do not add to 100 because of rounding.

CASE STUDY 18.2 / Deceived Survey Respondents: Once Bitten, Twice Shy

An article by Sheets et al. (1974) explores a situation sometimes encountered by marketing research personnel:

For some time, people engaged in marketing and other field-based research have had to contend with the consequences of a fairly widely used ploy in the direct selling field: gaining a potential customer's attention and interest by requesting cooperation in some sort of false survey. Despite the efforts of the American Association for Public Opinion Research, the American Marketing Association, and other groups, and regardless of Federal Trade Commission orders, this gambit is still in use, although perhaps somewhat modified.

The authors hypothesized that a previous exposure to a false survey increases the probability that a person will refuse to respond to a legitimate survey. They conducted an experiment in which 104 individuals were asked to cooperate in a marketing research study. The 54 people who agreed to participate were then given a low-key sales presentation for a fictitious encyclopedia. Two to 4 days later, 49 of the original 54 participants (five were not available) and 70 completely new individuals (the control group) were interviewed. Each group was asked the same opening question. The results of this survey are presented in Table 18.11.

A χ^2 test was used to analyze these count data. The χ^2 test statistic is 8.691, significant at $\alpha = .005$. This indicates dependence of the refusal rate on previous exposure to false surveys. The interpretation given to these data by Sheets et al. (1974) is:

The findings indicate support for the hypothesis: false market surveys have a deleterious effect upon respondent willingness to cooperate in subsequent market research studies. By inference, households that have been previously exposed to false research are half again as likely to refuse to cooperate in legitimate field research as those who have not. The implication for field researchers is either to stay away from areas that have had recent, heavy, direct, sales efforts or to plan for higher refusal rates in such areas.

TABLE 18.11 Experimental and Control Group Willingness to Participate in True Market Research

	Experimental	Control	Totals
Consented	12	36	48
Refused	37	34	71
Totals	49	70	119

Exercises 18.15–18.28

Learning the Mechanics

18.15 Find the rejection region for a test of independence of two classifications where the contingency table contains r rows and c columns and

a. $r = 5,\quad c = 4,\quad \alpha = .05$ b. $r = 4,\quad c = 6,\quad \alpha = .10$ c. $r = 3,\quad c = 3,\quad \alpha = .01$

18.16 Consider the accompanying 2×3 (i.e., $r = 2$ and $c = 3$) contingency table:

		Column 1	Column 2	Column 3
Row	1	12	32	45
	2	16	29	25

a. Specify the null and alternative hypotheses that should be used in testing the independence of the row and column classifications.
b. Specify the test statistic and the rejection region that should be used in conducting the hypothesis test of part **a**. Use $\alpha = .01$.
c. Assuming the row classification and the column classification are independent, find estimates for the expected cell counts.
d. Conduct the hypothesis test of part **a**. Interpret your result.

18.17 Refer to Exercise 18.16.

a. Convert the frequency responses to percentages by calculating the percentage of each column total falling in each row. Also convert the row totals to percentages of the total number of responses. Display the percentages in a table.

b. Create a graph with percentage on the vertical axis and the column number on the horizontal axis. Show the row total percentages as horizontal lines on the plot, and plot the cell percentages from part **a** using the row number as a plotting symbol.

c. What pattern do you expect to see if the rows and columns are independent? Does the plot support the results of the test of independence in Exercise 18.16?

18.18 Test the null hypothesis of independence of the two classifications, A and B, of the 3×3 contingency table shown. Test using $\alpha = .05$.

		B		
		B_1	B_2	B_3
	A_1	39	70	42
A	A_2	67	51	70
	A_3	30	38	26

18.19 Refer to Exercise 18.18. Convert the responses to percentages by calculating the percentage of each B class total falling into each A classification. Also, calculate the percentage of the total number of responses that constitute each of the A classification totals. Create a graph with percentage on the vertical axis and the B classification on the horizontal axis. Show the A classification total percentages as horizontal lines on the graph, and plot the individual cell percentages using the A class number as a plotting symbol. Does the graph support the result of the test of hypothesis in Exercise 18.18? Explain.

18.20 Test the null hypothesis of independence of the two classifications, A and B, of the 3×4 contingency table shown below. Test using $\alpha = .05$.

		B			
		B_1	B_2	B_3	B_4
	A_1	22	38	29	47
A	A_2	42	24	68	53
	A_3	29	85	105	68

18.21 A chi-square test was applied to each of the following contingency tables and the null hypothesis of independence was rejected. Construct a graph for each table that enables you to interpret the pattern of dependence in each case.

a.

		B	
		B_1	B_2
A	A_1	50	10
	A_2	10	40

b.

		D	
		D_1	D_2
C	C_1	15	70
	C_2	80	15

c.

		F		
		F_1	F_2	F_3
E	E_1	10	30	15
	E_2	30	10	15

18.22 In a contingency table test for independence, explain why the null hypothesis is independence and the alternative hypothesis is dependence (instead of vice versa).

Applying the Concepts

18.23 Are all employees equally likely to have accidents? Are certain employee groups—say, younger employees—more prone to particular kinds of accidents? The effective implementation of hiring policies, training programs, and safety programs requires such knowledge. A study was recently conducted to address these questions for a particular manufacturing company. A portion of the research results is summarized in the accompanying contingency table.

		Kind of Accident		
		Sprain	Burn	Cut
Age	Under 25	9	17	5
	25 and Over	61	13	12

Source: Parry, A. E. "Changing assumptions about loss frequency." *Professional Safety*, Oct. 1985, pp. 39–43.

a. From the contingency table, the researcher concluded that there is a relationship between an employee's age and the kind of accident that the employee may have. Do you agree? Test using $\alpha = .05$.
b. According to the frequencies of the contingency table, which is the most frequent type of accident? Are younger or older employees more likely to have sprains? Burns? Justify your answers.
c. What assumptions must hold in order for your test of part **a** to be valid?
d. Plot the percentage of employees under 25 who are injured based on the total injuries for each kind of accident. Compare this to the percentage of the total number of employees under 25 who are injured based on the total of all kinds of accidents. What does the plot indicate about the pattern of dependence in the data?

18.24 Over the years, pollsters have found that the public's confidence in big business is closely tied to the economic climate. When businesses are growing and employment is increasing, public confidence is high. When the opposite occurs, public confidence is low. Harvey Kahalas (1981) hypothesized that there is a relationship between the level of confidence in business and job satisfaction, and that this is true for both union and nonunion workers. He analyzed sample data collected by the National Opinion Research Center and shown in the tables.

		Job Satisfaction			
		Very Satisfied	Moderately Satisfied	A Little Dissatisfied	Very Dissatisfied
Union Member Confidence in Major Corporations	A Great Deal	26	15	2	1
	Only Some	95	73	16	5
	Hardly Any	34	28	10	9

		Job Satisfaction			
		Very Satisfied	Moderately Satisfied	A Little Dissatisfied	Very Dissatisfied
Nonunion Confidence in Major Corporations	A Great Deal	111	52	12	4
	Only Some	246	142	37	18
	Hardly Any	73	51	19	9

a. Kahalas concluded that his hypothesis was not supported by the data. Do you agree? Conduct the appropriate tests using $\alpha = .05$. Be sure to specify your null and alternative hypotheses.

b. Find and interpret the approximate p-values of the tests you conducted in part **a**.

18.25 A study was conducted to determine the effects of television viewing on the purchase of products advertised on television. A sample of 2,452 women from the Davenport, Iowa, metropolitan area was interviewed, first in February and then again in May. Each time, the women were asked whether they watch a specific program and whether they purchase the product advertised during the program. Their responses are categorized in the table.

		Buying, Feb./May				Totals
		Yes/Yes	Yes/No	No/Yes	No/No	
Viewing Feb./May	Yes/Yes	460	173	191	351	1,175
	Yes/No	76	59	44	113	292
	No/Yes	86	27	53	80	246
	No/No	175	104	113	347	739
Totals		797	363	401	891	2,452

Source: *Journal of Marketing Research*, Feb. 1966, pp. 13–24.

a. If a χ^2 test of independence were conducted for the table, what would be the null and alternative hypotheses?

b. Conduct the test referred to in part **a**. Test using $\alpha = .05$.
c. What assumptions must you make so that the χ^2 test in part **b** will be valid?

18.26 In recent years, corporate boards of directors have been pressed to improve their monitoring of corporate economic performance and to become more careful overseers of the activities of management. In addition, boards are being asked to guide the long-term responsiveness of their respective organizations to the economic and social climate. These pressures have forced many boards of directors to better articulate the mission and strategies of their firms. To study the extent and nature of strategic planning being undertaken by boards of directors, Ahmed Tashakori and William Boulton (1983) questioned a sample of 119 chief executive officers of major U.S. corporations. One objective was to determine if a relationship exists between the composition of a board—i.e., a majority of outside directors versus a majority of in-house directors—and its level of participation in the strategic planning process. To this end, the questionnaire data were used to classify the responding corporations according to the level of their board's participation in the strategic planning process:

Level 1: Board participates in formulation or implementation or evaluation of strategy
Level 2: Board participates in formulation and implementation, formulation and evaluation, or implementation and evaluation of strategy
Level 3: Board participates in formulation, implementation, and evaluation of strategy

The results obtained are shown in the left-hand table. Of these 119 firms, 100 had boards where outside directors constituted a majority. Their levels of participation in strategic planning are shown at the right.

All Firms

Level	1	2	3
Number of Firms	22	37	60

Outside Directed

Level	1	2	3
Number of Firms	20	27	53

a. Tashakori and Boulton concluded that a relationship exists between a board's level of participation in the strategic planning process and the composition of the board. Do you agree? Construct the appropriate contingency table, and test using $\alpha = .10$.
b. In the context of the problem, specify the Type I and Type II errors associated with the test of part **a**.
c. Find the approximate p-value for the test in part **a**. Based on the p-value, would the null hypothesis have been rejected at $\alpha = .05$? Explain.
d. Construct a graph that helps to interpret the result of the test in part **a**.

18.27 An insurance company that sells hospitalization policies wants to know whether there is a relationship between the amount of hospitalization coverage a person has and the length of stay in the hospital. Records are selected at random at a large hospital by hospital personnel, and the information on length of stay and hospitalization coverage is given to the insurance company. The results are summarized in the table. Can you conclude that there is a relationship between length of stay and hospitalization coverage? Use $\alpha = .01$.

		Length of Stay (days)			
		5 or Under	6–10	11–15	Over 15
Coverage of Costs	Under 25%	26	30	6	5
	25–50%	21	30	11	7
	51–75%	25	25	45	9
	Over 75%	11	32	17	11

18.28 One criterion used to evaluate employees in the assembly section of a large factory is the number of defective pieces per 1,000 parts produced. The quality control department wants to find out whether there is a relationship between years of experience and defect rate. Since the job is repetitious, after the initial training period any improvement due to a learning effect might be offset by a loss of motiviation. A defect rate is calculated for each worker in a yearly evaluation. The results for 100 workers are given in the table.

		Years Since Training Period		
		<1	1–4	5–9
	High	6	9	9
Defect Rate	Average	9	19	23
	Low	7	8	10

a. Is there evidence of a relationship between defect rate and years of experience? Use $\alpha = .05$.
b. Find the approximate observed significance level for the test in part **a**.

18.3 Caution

Because the X^2 statistic for testing hypotheses about multinomial probabilities is one of the most widely applied statistical tools, it is also one of the most abused statistical procedures. Always be certain that the experiment satisfies the properties of the multinomial experiment given in Section 18.1. Furthermore, be certain that the sample is drawn from the correct population—that is, from the population about which the inference is to be made.

The use of the χ^2 probability distribution as an approximation to the sampling distribution for X^2 should be avoided when the expected counts are very small. The approximation can be very poor when these expected counts are small, and thus the actual value of α may be very different from the tabled value. As a rule of thumb, an expected cell count of at least 5 will mean that the χ^2 probability distribution can be used to determine an approximate critical value.

Finally, if the X^2 value does not exceed the established value of χ^2, *do not accept* the hypothesis of independence. You would be risking a Type II error (accepting H_0 if in fact it is false), and the probability, β, of committing such an error is unknown. The usual alternative hypothesis is that the classifications are dependent. Because there is literally an infinite number of ways two classifications can be dependent, it is difficult to calculate one or even several values of β to represent such a broad alternative hypothesis. Therefore, we avoid concluding that two classifications are independent, even when X^2 is small.

Summary

The use of **count data** to test hypotheses about **multinomial probabilities** represents a very useful statistical technique. In a **one-dimensional table** we can use count data to test the hypothesis that the multinomial probabilities are equal to specified values. In a **two-dimensional contingency table**, we can test the independence of the two classifications. And these by no means exhaust the uses of the X^2 statistic. Many other applications can be found in the references at the end of this chapter.

Caution should be exercised to avoid misuse of the χ^2 procedure. The experiment must be multinomial,* and the expected counts should not be too small so that the χ^2 critical value may be used. Also, the X^2 statistic should not always be viewed as the final answer. If two classifications are found to be dependent, many measures of association exist for quantifying the nature and strength of their dependence (see the references).

Supplementary Exercises 18.29–18.48

18.29 Many investors believe that the stock market's directional change in January signals the market's direction for the remainder of the year. This so-called January indicator is frequently cited in the popular press. But is this indicator valid? If so, the well-known *random walk and efficient markets* theories (basically postulating that market movements are unpredictable) of stock-price behavior would be called into question. The accompanying table summarizes the relevant changes in the Dow Jones Industrial Average for the period December 31, 1927, through January 31, 1981. Joseph S. Martinich (1984) applied the chi-square test of independence to these data to investigate the January indicator.

		Next 11-Month Change	
		Up	Down
January Change	Up	25	10
	Down	9	9

a. Examine the contingency table. Based solely on your visual inspection, do the data appear to confirm the validity of the January indicator? Explain.

b. Construct a plot of the percentage of years for which the 11-month movement is up based on the January change. Compare these two percentages to the percentage of times the market moved up during the last 11 months over all years in the sample. What do you think of the January indicator now?

*When the row (or column) totals are fixed, each row (or column) represents a separate multinomial experiment.

c. If a chi-square test of independence is to be used to investigate the January indicator, what are the appropriate null and alternative hypotheses?
d. Conduct the test of part **c**. Use $\alpha = .05$. Interpret your results in the context of the problem.
e. Would you get the same result in part **d** if $\alpha = .10$ were used? Explain.

18.30 The classification of solder joints as acceptable or not is a difficult inspection task due to its subjective nature. Westinghouse Electric Company has experimented with different means of evaluating the performance of solder inspectors. One approach involves comparing an individual inspector's classifications with those of the group of experts that comprise Westinghouse's Work Standards Committee. In an experiment 153 solder connections were evaluated by the committee and 111 were classified as acceptable. An inspector evaluated the same 153 connections and classified 124 as acceptable. Of the items rejected by the inspector, the committee agreed with 19 (Meagher and Scazzero, 1985).
a. Construct a contingency table that summarizes the classifications of the committee and the inspectors.
b. Based on a visual examination of the table you constructed in part **a**, does it appear that there is a relationship between the inspector's classifications and the committee's? Explain. (A plot of the percentage rejected by committee and inspector will aid your examination.)
c. Conduct a chi-square test of independence for these data. Use $\alpha = .05$. Carefully interpret the results of your test in the context of the problem.

18.31 Consumers have traditionally viewed products with warranties more favorably than products without warranties. In fact, when given the choice between two similar products, one of which is warranted, consumers prefer the warranted product, even at a higher price. Thus, consumers generally perceive warranties as a kind of value added to the product. However, a substantial number of firms perceive their warranties primarily as legal disclaimers of responsibility and nothing more. As a result of the differences in perceptions by consumers and businesses, Congress passed the Magnuson-Moss Warranty Act, which took effect in 1977. According to this act, all warranties must be designated as "full" or "limited," must be clearly written in readily understood language, and must contain specific information.

McDaniel and Rao (1982) investigated consumer satisfaction with warranty practices since the advent of the Magnuson-Moss Warranty Act. Using a mailed questionnaire, they sampled 237 midwestern consumers who had purchased a major appliance within the past 6 to 18 months. One question was, "Do most retailers and dealers make a conscientious effort to satisfy their customers' warranty claims?" One hundred fifty-six answered yes, 61 were uncertain, and 20 said no.

The population of consumers from which this sample was drawn had also been investigated 2 years prior to the Magnuson-Moss Warranty Act. At that time, 37.0% of the population answered yes to the same question, 53.3% were uncertain, and 9.7% said no.
a. As reflected in the answers to the preceding question, have consumer attitudes toward warranties changed since the earlier study? Test using $\alpha = .05$.
b. Compare the pre- and post-Magnuson-Moss Warranty Act responses, and describe any changes.

18.32 When a buyer charges a purchase, the seller records the sale in his or her record books under a category called *accounts receivable*. Some retailers monitor the status of their accounts receivable by regularly classifying each in one of the following categories: current, 1–30 days late, 31–60 days late, more than 60 days late, or uncollectable. Historical data indicate that the status of a particular retailer's accounts receivable can be described as follows:

Current:	65%	1–30 days late:	15%
31–60 days late:	10%	Over 60 days late:	7%
Uncollectable:	3%		

Six months after the interest rate charged to late accounts was increased, the status of the retailer's 200 accounts receivable was as follows:

Current:	78%
1–30 days late:	12%
31–60 days late:	5%
Over 60 days late:	2%
Uncollectable:	3%

a. Is there evidence to indicate that the increase in interest rates affected the timing of buyers' payments? Test using $\alpha = .10$.

b. Find the approximate observed significance level for the test.

18.33 In Case Study 9.2, we described part of a statistical analysis (Kaufman and Wolf, 1982) examining the preferences and anxieties of men and women with respect to attending job interviews in hotel rooms. The table gives additional data from this study. A random sample of 302 students (95 men and 207 women) was asked whether they would be uncomfortable about interviewing in a hotel room. The responses are shown in the table. Do these data provide sufficient evidence to conclude that anxiety over hotel room interviewing is related to the interviewee's sex? Test using $\alpha = .05$.

	Yes	No
Men	63.16%	36.84%
Women	77.29%	22.71%

18.34 Organizations that loan money to businesses are in effect gambling that the business will become (or remain) successful long enough so that the loan will not be defaulted. Loan applicants are carefully screened to weed out those firms with a high probability of defaulting. Characteristics of an applicant firm that might be examined by a loan officer include such things as the firm's age and legal structure. The loan officer's evaluation will also consider the amount of the requested loan, its duration, and its type (e.g., for an existing business, to buy an existing business, or to start a new business).

Albert L. Page et al. (1977) conducted an empirical investigation of the past loan performance of the Greater Cleveland Growth Corporation, an affiliate of the Office of Minority Business Enterprise. Part of the study involved identifying demographic and firm characteristic variables that relate to the status of the loans made. Loans were classified as paid off, current, or defaulted. Paid off and current loans are regarded as good loans.

In a sample of 64 loan histories, the frequencies for six loan status and legal structure categories were as shown in the table.

Loan Status	*Firm's Legal Structure*	*Frequency*
Defaulted	Sole proprietorship	14
Paid off or current	Sole proprietorship	13
Defaulted	Partnership	10
Paid off or current	Partnership	1
Defaulted	Corporation	12
Paid off or current	Corporation	14

a. From a contingency table analysis of the data, Page et al. reported a relationship between the legal structure of an applicant firm and the success or failure of the loan. Test their conclusion using $\alpha = .05$.
b. Find the approximate observed significance level for the test in part **a**.
c. What assumptions must you make so that the test conclusions in part **a** will be valid?
d. Use a 95% confidence interval to estimate the proportion of loans made by the Growth Corporation that will be defaulted.

18.35 A computer used by a 24-hour banking service is supposed to randomly assign each transaction to one of five memory locations. A check at the end of a day's transactions gave the counts shown in the table to each of the five memory locations. Is there evidence to indicate a difference in the proportions of transactions assigned to the five memory locations? Test using $\alpha = .025$.

Memory Location	*Number of Transactions*
1	90
2	78
3	100
4	72
5	85

18.36 Refer to Case Study 18.2.
a. Verify that the value of the test statistic is 8.691.
b. Conduct the χ^2 test described in the case study.
c. Is the p-value of the test you conducted in part **b** less than or greater than .005? Explain.
d. In the context of the problem, describe the Type I and Type II errors associated with the test of part **b**.

18.37 A restaurateur who owns restaurants in four cities is considering the possibility of building separate dining rooms for nonsmokers to accommodate customers who wish to dine in a smoke-free environment. Since this would be costly, the restaurateur plans to survey the customers at each restaurant and ask, "Would you be more comfortable dining here if there were a separate dining room for nonsmokers only?" Suppose 75 people were randomly selected and surveyed at each restaurant with the results shown in the table. Is there sufficient evidence to indicate that customer preferences are different for the four restaurants (i.e., that customer preference and restaurant are dependent)? Use $\alpha = .10$.

		Answer to Question		
		Yes	No	Indifferent
Restaurant	1	38	32	5
	2	42	26	7
	3	35	34	6
	4	37	30	8

18.38 It is commonly assumed that the more experience a job applicant has, the better that person will perform the necessary duties. Other factors, such as whether the person has a college degree or is male or female, may also be indicative of future performance. H. M. Greenberg and J. Greenberg (1980) argue that for sales jobs, the most important factor is matching the job requirement with an applicant's personal characteristics. This, they claim, will result in better retention rates and higher levels of job performance. To validate this claim they studied two groups of recently hired sales personnel. In the first group, which numbered 1,980, all were job-matched; the 3,961 members of the second group were not. After 6 months they were evaluated, and the aggregate data are shown in the table (1, highest performance; 4, lowest). The tabulated values are the percentages of total sales personnel contained in the respective samples.

	Performance					Totals
	1	2	3	4	Quit or Fired	
Job-Matched	9%	40%	32%	14%	5%	100%
Not Job-Matched	2%	17%	25%	31%	25%	100%

a. Use both the percentages given in the table and the sample sizes to construct a contingency table that shows the numbers of sales personnel falling in each category of the table.

b. Do the data provide sufficient evidence to indicate that the proportions of sales personnel falling in the performance categories depend on whether the people are job-matched? Test using $\alpha = .05$.

c. Do the data provide sufficient evidence to indicate the proportion of sales personnel receiving the highest rating (1) is larger if job-matched than if not? Test using $\alpha = .05$. [*Note*: This will require a one-sided test.]

18.39 Product or service quality is often defined as *fitness for use*. This means the product or service meets the customer's needs. Generally speaking, fitness for use is based on five quality characteristics: technological (e.g., strength, hardness), psychological (taste, beauty), time-oriented (reliability), contractual (guarantee provisions), and ethical (courtesy, honesty). The quality of a service may involve all these characteristics, while the quality of a manufactured product generally depends on technological and time-oriented characteristics (Schroeder, 1993). After a barrage of customer complaints about poor quality, a manufacturer of gasoline filters for cars had its quality inspectors sample 600 filters—200 per work shift—and check for defects. The data in the table resulted.

Shift	*Defectives Produced*
First	25
Second	35
Third	80

a. Do the data indicate that the quality of the filters being produced may be related to the shift producing the filter? Test using $\alpha = .05$.

b. Estimate the proportion of defective filters produced by the first shift. Use a 95% confidence interval.

18.40 A national survey was conducted to determine the general public's view of the federal government's involvement in the regulation of private enterprise. Two hundred people from each of three income levels were asked if they thought the government is too involved, not involved enough, or involved just enough. A summary of their responses is shown in the table. Do the data provide sufficient information to indicate a relationship between income and view on government regulation of private enterprise? Test using $\alpha = .05$.

		Involvement			Totals
		Too Little	Just Enough	Too Much	
Income	Low	125	48	27	200
	Medium	103	58	39	200
	High	72	69	59	200
Totals		300	175	125	600

18.41 An industrial security firm wants to conduct a study of criminal cases involving stolen company money in which employees were found guilty. Among the data recorded are the employee's salary (wages) and the amount of money stolen from the company for 400 recent cases. The results are given here.

		Amount Stolen ($)			
		Under 5,000	5,000–9,999	10,000–19,999	20,000 or More
Income of Employee ($1,000s)	Under 15	46	39	17	5
	15–25	78	79	61	19
	Over 25	5	14	25	12

a. Does this information provide evidence of a relationship between employee income and amount stolen? Use $\alpha = .05$.

b. Convert the responses to percentages by calculating the percentage in each income category based on the total number of cases in each amount stolen category. Also calculate the percentage of all 400 cases in each income category. Plot percentage on the vertical axis and amount stolen on the horizontal axis, showing the overall income category percentages as horizontal lines on the plot. Plot the income percentages for each cell using 1, 2, and 3 as the plotting symbols for the three categories. Does the plot support the result of the test in part **a**?

18.42 A local bank plans to offer a special service to its young customers. To determine their economic interests, a survey of 100 people under 30 years of age is conducted. Each person is asked his or her top two financial priorities among the six choices shown in the table. Use the χ^2 test to determine whether the proportions of responses differ for the six pairs of priorities. Test at $\alpha = .10$.

First Priority	*Second Priority*	*Number of Respones*
Buy a car	Go on a trip	15
Car	Save money	14
Save	Car	22
Save	Trip	23
Trip	Car	10
Trip	Save	16

18.43 A corporation owns several convenience stores that are open 24 hours a day. It is interested in knowing whether there is a relationship between time of day and size of purchase. One of its stores is selected at random to be involved in a study. Store records are collected over a period of several weeks and then 300 purchases are randomly selected. Since the register also prints the time of the purchase, this random selection procedure yields both amount and time of purchase. The information is summarized in the table. Is there a relationship between time and size of purchase? Use $\alpha = .05$.

		Size of Purchase		
		$2 or Less	$2.01–$7	Over $7
	8 A.M.–3.59 P.M.	65	38	14
Time of Purchase	4 P.M.–11:59 P.M.	61	49	10
	12 Midnight–7:59 A.M.	29	27	7

18.44 Refer to Exercise 18.43. Use a 90% confidence interval to estimate the difference between the proportions of customers who spend $2 or less for the periods 8 A.M.–3:59 P.M. and 12 midnight–7:59 A.M.

18.45 Five candidates have just entered the race for mayor of a large city. To determine whether any of them has an early lead in popularity, 2,000 voters are polled and each is asked to indicate the candidate he or she prefers. A summary of their responses is shown in the table.

Candidate	I	II	III	IV	V
Voters Who Prefer Candidate	385	493	628	235	259

a. Do the data provide sufficient evidence to indicate a difference in preference for the five candidates? Test using $\alpha = .01$.
b. Find the approximate observed significance level for the test in part **a**.

18.46 A city has three television stations, each with its own evening news program from 6:00 to 6:30 P.M. every weekday. An advertising firm wants to know whether there is an unequal breakdown of the evening news audience among the three stations. One hundred people are selected at random from the evening news audiences of these three stations. Each is asked which news program he or she watches. Do the results in the table provide sufficient evidence to indicate that the three stations do not have equal shares of the evening news audience? Use $\alpha = .05$.

Station	1	2	3
Number of Viewers	35	43	22

18.47 Several life insurance firms have policies geared to college students. To get more information about this group, a major insurance firm interviewed college students to find out the type of life insurance they preferred, if any. The accompanying table was produced after surveying 1,600 students.

	Term Policy	*Whole-Life Policy*	*No Preference*
Females	116	27	676
Males	215	33	533

a. Is there evidence that the life insurance preference of students depends on their sex?
b. Find the approximate observed significance level for the test in part **a**.
c. Construct a plot of percentage responses that will help to interpret the result of the test in part **a**.

18.48 Refer to Exercise 18.47. Estimate the difference in the proportions of female and male college students who have no preference about life insurance.

On Your Own

Market researchers rely on surveys to estimate the proportions of the consumer market that prefer various brands of a product. Choose a product with which you are familiar, and guess the proportion of consumers you think favor the major brands of the product. (Choose a product for which there are at least three major brands sold in the same store.)

Now go to a store that carries these brands, and observe how many consumers purchase each brand. Be sure to observe long enough so that at least five (and preferably at least ten) purchases of each brand have been made. Stop sampling after a predetermined length of time or after a predetermined number of total purchases, rather than at some arbitrary time, which could bias your results.

Use the count data to test the null hypothesis that the true proportions of consumers who favor each brand equal your presampling guesses of the proportions. Would failure to reject this null hypothesis imply that your guesses are correct?

Using the Computer

Are monthly homeowner costs related to the region of the country in which a homeowner lives? Use the zip code data of Appendix C to investigate this question.

a. Conduct a χ^2 test of independence to answer the question. Treat the variable median monthly homeowner cost as if it had four classes: \$0–\$200, \$200.01–\$400, \$400.01–\$600, and \$601.01–\$800.
b. Construct a table similar to Table 18.10 and a graph similar to Figure 18.3 to help explain the results of the test you conducted in part **a**. Interpret your results.

References

Conover, W. J. *Practical Nonparametric Statistics*, 2nd ed. New York: Wiley, 1980.

Dahlin, C., and Owen, F. *An Analysis of Data Collected at the I-494 Weighing-in-Motion Site*. St. Paul: Minnesota Department of Transportation, 1984.

Graham, E. "The entrepreneurial mystique." *Wall Street Journal*, May 20, 1985.

Greenberg, H. M., and Greenberg, J. "Job-matching for better sales performance." *Harvard Business Review*, Sept.–Oct. 1980.

Hollander, M., and Wolfe, D. A. *Nonparametric Statistical Methods*. New York: Wiley, 1973.

Kahalas, H. "The relationship between confidence in business and job satisfaction for union and nonunion members." *Baylor Business Studies*, Feb.–Apr. 1981, Vol. 127, pp. 45–53.

Kaufman, L., and Wolf, J. "Hotel room interviewing—Anxiety and suspicion." *Sloan Management Review*, Spring 1982, Vol. 23, pp. 57–64.

Martinich, J. S. "The January indicator: A nonrandom but unprofitable walk." *Mid-South Business Journal*, 1984, Vol. 4, No. 4.

McDaniel, S. W., and Rao, C. P. "Consumer attitudes toward and satisfaction with warranties and warranty performance—Before and after Magnuson-Moss." *Baylor Business Studies*. Nov.–Dec. 1982, Vol. 130, pp. 47–61.

McKenzie, J. "The accuracy of telephone call data collected by diary methods." *Journal of Marketing Research*, Nov. 1983, Vol. 20, pp. 417–427.

Meagher, J. J., and Scazzero, J. A. "Measuring inspector variability." *1985 ASQC Quality Congress Transaction*, Baltimore, May 1985, pp. 75–81.

Neter, J., Wasserman, W., and Whitmore, G. A. *Applied Statistics*, 2nd ed. Boston: Allyn & Bacon, 1982. Chapter 17.

Page, A. L., Trombetta, W. L., Werner, C., and Kulifay, M. "Identifying successful versus unsuccessful loans held by the minority small business clients of an OMBE affiliate." *Journal of Business Research*, June 1977, Vol. 5, pp. 139–153.

Palm, R., and Hodgson, M. "Earthquake insurance: Mandated disclosure and homeowner response in California." *Annals of the Association of American Geographers*, June 1992, Vol. 82, No. 2.

Schroeder, R. G. *Operations Management*, 4th ed. New York: McGraw-Hill, 1993.

Sheets, T., Radlinski, A., Kohne, J., and Brunner, G. A. "Deceived respondents: Once bitten, twice shy." *Public Opinion Quarterly*, 1974, Vol. 18, pp. 261–263.

Siegel, S. *Nonparametric Statistics for the Behavioral Sciences*. New York: McGraw-Hill, 1956. Chapter 9.

Sudman, S., and Ferber, R. "A comparison of alternative procedures for collecting consumer expenditure data for frequently purchased products." *Journal of Marketing Research*, May 1974, Vol. 11, pp. 129–135.

Sudman, S., and Ferber, R. "Experiments in obtaining consumer expenditures by diary methods." *Journal of the American Statistical Association*, Dec. 1971, Vol. 66, pp. 725–735.

Tashakori, A., and Boulton, W. "A look at the board's role in planning." *Journal of Business Strategy*, Winter 1983, Vol. 3, pp. 64–70.

Winkler, R. L., and Hays, W. L. *Statistics: Probability, Inference and Decision*, 2nd ed. New York: Holt, Rinehart and Winston, 1975. Chapter 12.

CHAPTER NINETEEN

Decision Analysis

Contents

Case Studies

Where We've Been

In previous chapters we used a decision procedure to test hypotheses about population parameters. Using sample information, we decided to reject or accept the null hypothesis based on the calculated probabilities of making incorrect decisions—namely, the probability (α) of rejecting the null hypothesis if it was in fact true, and the probability (β) of accepting the null hypothesis if it was actually false. In this simplistic process, we assumed that a manager would be able to assess the gains or losses associated with each type of error and choose a test with acceptable values of α and β.

Where We're Going

In this chapter we present the basic concepts of a general theory for making decisions that explicitly accounts for the gains or losses associated with alternative decisions and the probabilities of the occurrence of these gains or losses. First we describe how to handle decision problems using only information that is currently available about the problem. Then we extend the analysis to include the case in which additional information can be obtained by sampling.

19

Suppose you have been given the responsibility of determining whether your firm should expand its sales region to include the southwestern part of the United States. Before making your decision, you would probably want answers to many questions. How large would the yearly demand for the product be? How many salespeople would be assigned to the new territory? How much and what types of advertising would be used? Are adequate warehousing facilities available? Who would the company's principal competitors be, and how would they react to new competition? Even if it were possible to obtain accurate answers (*perfect information*) to these and other pertinent questions, your decision problem would be extremely complex. Realistically, however, you cannot expect to receive perfect information. Thus, in making your decision, you will face the more complex problem of having to deal with answers about which you are uncertain.

How would you tackle such a decision problem? The most natural first step—and one that is used by most decision analysts—is to reduce the problem to a manageable size by considering only questions that bear significantly on the objective of your decision. In this case, your objective may be to increase corporate profit. It may turn out that the profitability of the decision to expand the sales region depends primarily on the extent of the demand for the product in the new territory during the first year after it has been introduced. However, since this demand is unknown, even with this simplified decision problem you still must make your decision in the face of uncertainty. Given your uncertainty about the demand, how much information about demand in the new region would you want prior to making your decision, and how would you process such information? Answers to these questions are provided by the methodology referred to as *decision analysis*, which is the subject of this chapter.

Decision analysis is a systematic approach to solving decision problems under conditions of uncertainty. It does not *describe* how or why an individual makes a decision; rather, it *prescribes* a decision for the individual that is consistent with his or her *preferences and attitudes toward risk*. You might be asking yourself: "Why do I need to study decision-making as though it were a science, when I know most decisions (business or otherwise) are made on an intuitive level?" The answer is fourfold:

1. Yes, the vast majority of decisions made in business do not require, and are made without, formal analysis. But for that one crucial decision upon which "everything depends," it is important to have a systematic, logical decision procedure to follow.
2. Most of us have had little experience intuitively processing the probabilistic and sample information that may confront us in a complex decision-making problem. Consequently, it is frequently better to rely on the systematically generated information of decision analysis to guide decision-making than on the less reliable information-processing capabilities of our intuition.
3. Decision analysis forces us to consider carefully and logically all possible courses of action and the outcomes that could result from each. By so doing, we may disclose a new aspect of the problem or even find we have been addressing the wrong problem. Thus, the information obtained from decision analysis may more than compensate for the effort expended.

4. Another reason business and economics majors should study decision analysis is that many firms and government agencies use it regularly. Consequently, you may very well be required to use decision analysis in your future employment.

One of the alternatives we face in making a decision is whether the decision should be made *now*—utilizing information we currently possess about the problem (we will refer to this as **prior information**)—or *postponed* until we have gathered additional information. We first study decision-making under uncertainty and assume that only prior information is available. Next, we expand our study of decision-making to include situations in which additional information is available. We discuss how to determine the value of additional information as well as when and how to use additional information in decision-making.

19.1 Three Types of Decision Problems

Although all decision problems involve the selection of a course of action from among two or more alternatives, we can classify them into one of three categories:

1. Decision-making under certainty
2. Decision-making under uncertainty
3. Decision-making under conflict

Decision-making under certainty entails the selection of a course of action when we *know* the result each alternative action will yield. If the number of alternatives being considered is small, such decisions may be easy to make. However, if the number of alternatives is large, the optimal decision may be difficult—if not impossible—to obtain. It may take too long or cost too much to evaluate so many alternatives individually and select the one with the most favorable results. Decision problems of this type are not addressed in this text, but here is a simple example of this type of problem:

A furniture company constructs and finishes tables and chairs. Each table produced by the company nets a profit of $100 and each chair a profit of $60. During 1 week the company has 305 work-hours available for assembly operations and 355 work-hours available for finishing. From past experience it is known that each chair requires 3 hours to be assembled and 1½ hours of finishing, while each table requires 4 hours for assembly and 2 hours for finishing. How many tables and how many chairs should the company produce over the week in order to maximize profits?

Note that a unique solution to this problem exists that will maximize the firm's profit. Many problems involving decision-making under certainty, including this one, are solved using a technique known as **linear programming**.

Decision-making under uncertainty entails the selection of a course of action when we *do not know* with certainty the results that each alternative action will yield.* Furthermore, we assume that the outcome of whatever course of action we select is affected only by chance and not by an opponent or competitor. We discuss decision-making under uncertainty in detail in this chapter. Our introductory example concerning the decision of whether to expand the sales region to the southwestern United States demonstrates decision-making under uncertainty.

Decision-making under conflict is similar to decision-making under uncertainty in that we do not know with certainty the result each available alternative course of action will yield. However, the reason for this uncertainty is different in the case of decision-making under conflict. In such cases, we are in effect "playing against" one or more opponents or competitors. The outcome of our chosen course of action depends on decisions made by our competitors. Decision problems of this type require the approach known as **game theory**. Game theory is not discussed in this text; however, an example of this type of decision problem follows:

An entrepreneur is interested in purchasing real estate near Dallas for the purpose of building a fast-food restaurant. He has narrowed his alternatives to five suburban neighborhoods. He is certain his venture will be profitable as long as no major fast-food chain decides to locate nearby. Thus, the decision problem involves choosing a parcel of land while knowing that the results of this decision depend on the expansion plans of potential competitors. Furthermore, the location decision facing other fast-food chains interested in the area depends to some extent on the decision of the Dallas entrepreneur. They, too, would prefer not to locate near another resaturant of the same type.

Since our objective throughout this text has been to make inferences when we have only partial (or *imperfect*) information, we concentrate on decision-making under uncertainty in the remaining sections.

Exercises 19.1 – 19.4

Applying the Concepts

19.1 Compare and contrast decision-making under certainty, uncertainty, and conflict.

19.2 Describe two decision problems that you face every day. Categorize each as being a problem requiring a decision made under certainty, uncertainty, or conflict. Justify your categorization.

19.3 Categorize the following decision problems as decision-making under certainty, uncertainty, or conflict. Justify your categorization.

---

*Some texts further distinguish "decision-making under risk" if probabilities are available to describe the degree of uncertainty. Since probablities can always be obtained through judgment and reasoning (subjective probabilities; Section 19.4), we make no such distinction.

a. A bank manager is considering an application for a commercial loan. If she decides to make the loan but the customer defaults, the bank will lose the amount of the loan plus the lost profits. On the other hand, if the bank fails to grant the loan and the customer would have repaid it, the bank will lose the interest on the loan.
b. A manufacturer must decide about the price of an electric lawn mower it makes. If the company sets the price too high, potential customers will purchase competitors' mowers. If, on the other hand, the price is set too low, the competitors will also drop their prices, thereby reducing everyone's profits.
c. A computer hardware company has two contracts for producing electronic components for the space program. Since the contracts are for a fixed number of components at a fixed price, the decision problem involves how to allocate fixed production resources to maximize profit.

19.4 Categorize the following decision problems as decision-making under certainty, uncertainty, or conflict. Justify your categorization.
a. A plant manager wants to replace an obsolete piece of machinery with a new model. The two brands on the market from which to choose are of equal quality, have the same guarantee, and give the same yield per hour. However, brand B is $500 cheaper than brand A.
b. A company is faced with the decision of whether or not to increase its production capacity by adding a new building to the existing facilities. If the company decides not to expand, it expects to make a profit of $550,000 for each of the next 3 years regardless of the state of the economy. If it builds the addition and the economy continues to expand, the addition is expected to increase company profits to at least $650,000 a year for the next 3 years. If it builds the addition and the economy remains stable or experiences a downward trend over the next 3 years, the company would incur a reduction in profits to $475,000 or less per year for the next 3 years.

19.2 Decision-Making Under Uncertainty: Basic Concepts

We will use the following example to introduce some basic concepts: A profit-motivated entrepreneur is committed to producing a concert that will feature a major rock star sometime in late June next year. The promoter has been unable to finalize plans, however, due to indecision over whether to take a chance on rain and hold the concert in Memorial Stadium (40,000 seats outdoors) or play it safe and hold the concert in the Civic Center (15,000 seats indoors). A sellout is expected at least a month in advance, whichever facility is chosen. If the stadium is chosen and the weather cooperates, the promoter will make a net profit of about $350,000 (ticket proceeds less costs, taxes, and other expenses). If it is raining at concert time, the rock star may choose not to perform (according to the contract) and the promoter will lose about $40,000 (stadium rental, commitment to the rock star, administrative costs, salaries of security personnel, advertising). However, if the Civic Center is chosen, the entrepreneur would make a net profit of about $150,000 regardless of the weather. Which option would you choose? We will use decision analysis to make our selection later in the chapter.

We can identify three specific elements of this decision problem: First, a choice must be made between two possible courses of action—rent the stadium or rent the

Civic Center. We refer to these alternatives as **actions**. Second, it is uncertain which event will occur—rain or no rain. We refer to these events as **states of nature**. Third, depending on which action is chosen and which state of nature occurs on the evening of the concert, the decision-maker* will receive either a financial reward or a penalty for the chosen action. The consequences of the decision problem are referred to as *outcomes*; they may be either positive or negative. For example, if the action chosen by the promoter is Rent the stadium and the state of nature that occurs in Rain, the outcomes that result is −$40,000. That is, the action/state of nature combination Rent the stadium/Rain will *cost* the promoter $40,000. The combination Rent the stadium/No rain will yield a *profit* of $350,000. The reward (or penalty) corresponding to each action/state of nature combination is called the **outcome**, or **payoff**.

We can conveniently summarize all three elements of a decision problem in a **payoff table** (Table 19.1). Each of the set of possible actions the decision-maker has

TABLE 19.1 Payoff Table for the Rock Concert Decision Problem

		State of Nature	
		Rain	No Rain
Action	Rent Stadium	−$40,000	$350,000
	Rent Civic Center	$150,000	$150,000

chosen to consider is associated with a *row* of the payoff table. Each state of nature is associated with a *column*. The numbers in the table are the outcomes of the decision problem. For example, $350,000 is the outcome that would result from the implementation of the action associated with the top row of the table (Rent stadium) and the occurrence of the state of nature associated with the right-hand column of the table (No rain).

A decision problem can also be illustrated by a **decision tree**. A payoff table and a decision tree may display the same information, but as we will see in the next chapter, it is sometimes more convenient to use a decision tree. The decision tree in Figure 19.1 corresponds to Table 19.1. Conceptually, the promoter's movement through time toward the outcome of the decision problem is repesented by movement from left to right through the decision tree. The ■ denotes a **decision fork** and signals that a decision must be made. At this position on the tree, the decision-maker must choose between the two actions, Rent stadium and Rent Civic Center. If Rent stadium is chosen, then from the decision fork we move along the upper branch of the tree. The

*The term *decision-maker* is used to refer not only to an individual but also to a corporation, a community, or, in general, any entity faced with a decision problem.

● denotes a **chance fork** and signals that the next branch of the tree the promoter will follow will be determined by the chance occurrence of a state of nature. If the decision-maker is positioned at the upper chance fork of Figure 19.1 and it rains on the day of the concert, then the upper branch of the chance fork (labeled Rain) will lead the promoter to the consequence (−$40,000) of the action/state combination Rent stadium/Rain.

FIGURE 19.1 ▶
Decision tree for the rock concert decision problem

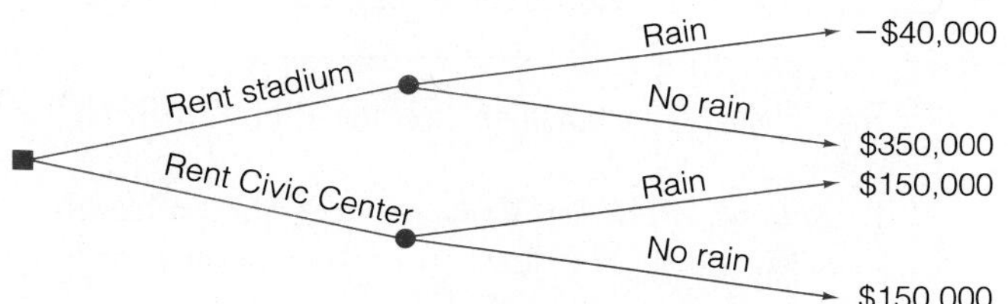

Both the selection of actions to be considered in a decision problem and the choice of an action to implement are under the control of the decision-maker, but the state of nature is not. The decision-maker must choose a course of action *prior* to knowing which state of nature will occur and *without* being able to influence the random process generating the states of nature. *Note also that the states of nature considered in any decision problem must be mutually exclusive and collectively exhaustive.* That is, the state of nature that occurs must be clearly identifiable as one and only one of the listed states, and the list of states considered must include all possible states that can occur. The former constraint precludes any overlapping of, or vagueness in, state definitions; the latter precludes the possibility of any state occurrence not anticipated by the decision-maker. For practical purposes, the states of nature in the above example are mutually exclusive and collectively exhaustive. They are mutually exclusive because the weather on a June evening can be classified as either rainy or not rainy, but not both. They are collectively exhaustive because the list of possible weather patterns can be narrowed to include only rainy or not rainy.

All outcomes in a decision problem should be stated in terms of the same numerical quantity, and that quantity should be chosen to rank the outcomes relative to the decision-maker's overall objective. We refer to this measure of the outcomes of a decision problem as the **objective variable**. In the preceding example, the promoter's motive—or objective—for producing the rock concert was net profit. Accordingly, the decision outcomes are expressed in terms of net profit (dollars). If the objective were to give as many people as possible an opportunity to see a live performance by the rock star, we would measure the outcomes of potential actions in terms of the number of people attending the concert.

We conclude by summarizing the concept of decision-making under uncertainty, the four elements common to this type of decision problem, and the methods for displaying these elements.

Decision-Making Under Uncertainty

If a decision-maker is faced with choosing one action from among two or more alternative actions, and at least one of these actions has possible outcomes that depend on the chance occurrence of one of a set of mutually exclusive and collectively exhaustive states of nature, the decision-maker is said to be faced with **decision-making under uncertainty**.

Four Elements Common to Decision Problems Involving Uncertainty

1. *Actions*: The set of two or more alternatives the decision-maker has chosen to consider. The decision-maker's problem is to choose one action from this set.
2. *States of nature*: The set of two or more mutually exclusive and collectively exhaustive chance events upon which the outcome of the decision-maker's chosen action depends.
3. *Outcomes*: The set of consequences resulting from all possible action/state of nature combinations.
4. *Objective variable*: The quantity used to measure and express the outcomes of a decision problem.

Table 19.2 and Figure 19.2 depict the general format for the payoff table and the decision tree, respectively. If the ith action is denoted by a_i and the jth state of nature by S_j, then the outcome resulting from the combination of the ith action with the jth state of nature is O_{ij}.

TABLE 19.2 General Form of a Payoff Table

		State of Nature			
		S_1	S_2	. . .	S_m
Action	a_1	O_{11}	O_{12}	. . .	O_{1m}
	a_2	O_{21}	O_{22}	. . .	O_{2m}
	⋮	⋮	⋮		⋮
	a_n	O_{n1}	O_{n2}	. . .	O_{nm}

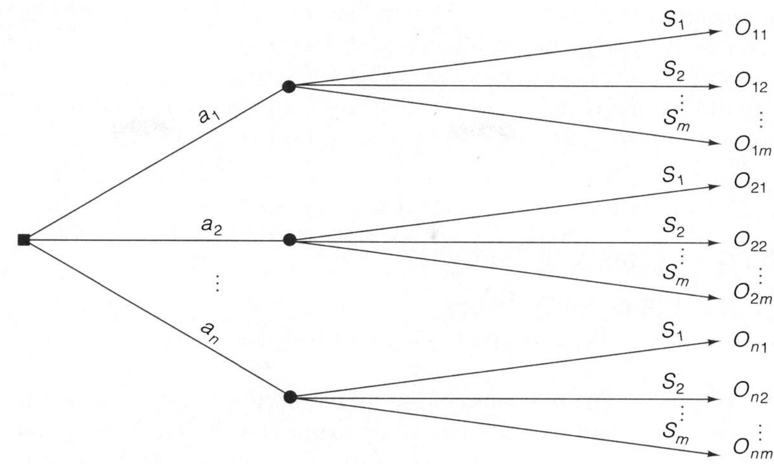

FIGURE 19.2 ▶
General form of a decision tree

Exercises 19.5–19.10

Applying the Concepts

19.5 List and define the four primary elements of a decision problem under uncertainty.

19.6 The states of nature that are defined in a decision problem must be mutually exclusive and collectively exhaustive. In this context, what is meant by the phrase *mutually exclusive and collectively exhaustive?*

19.7 Describe a decision you make every day that is done in the face of uncertainty. Identify the actions and states of nature of your decision problem. Construct a decision tree to illustrate your decision problem.

19.8 For each of these decision problems, identify the actions, states of nature, outcomes, and objective variable:

a. A winery is considering introducing an inexpensive dinner wine. The introduction of the wine will cost \$3 million in promotional and fixed costs per year. Each bottle sold will contribute \$.30 to profits. The management believes that sales could range from 5 to 25 million bottles per year.

b. An administrator in the Environmental Protection Agency (EPA) must determine how much to fine companies that discharge a certain effluent into U.S. waterways. If the fine is set too low, all companies will simply pay the fine because it will be less costly than installing pollution-control equipment. On the other hand, if the fine is set too high, small companies may be driven out of business because they cannot afford the capital expenditures necessary to meet the new EPA standard. Thus, the problem is to set the fine to minimize the total social cost—that is, the cost to society of (1) firms being driven out of business and (2) waterways being polluted.

19.9 Refer to Exercise 19.4, part **b**, in which a company was deciding whether to increase its production capacity. Identify the actions, states of nature, outcomes, and objective variable.

19.10 A company that manufactures a well-known line of designer jeans is contemplating whether to increase its advertising budget by \$1 million for next year. If the expanded advertising campaign is successful, the company expects sales to increase by $1.6 million next year. If the advertising campaign fails, the company expects sales to increase by only $400,000 next year. If the company decides not to increase its advertising expenditures, it expects sales to increase by $200,000 next year.

a. Identify the actions, states of nature, outcomes, and objective variable for this decision problem.

b. Construct a decision tree that illustrates the jeans manufacturer's decision problem.

19.3 Two Ways of Expressing Outcomes: Payoffs and Opportunity Losses

The outcomes of the rock concert example discussed in the previous section were in terms of the net profit that would be realized by the promoter depending on the decision (action) concerning the location of the concert and on the weather (state of nature) on the day of the concert. That is, net profit was the objective variable. Recall that the objective variable can assume both positive and negative values. In general, we refer to outcomes that reflect the *actual* reward to the decision-maker in terms of the objective variable as **payoffs**.

Alternatively, outcomes can be expressed in terms of *opportunities* for higher profits that the decision-maker has *lost* as a result of the action selected. For example, in the rock concert example, if the weather is not rainy, a decision to hold the concert in the Civic Center will bring in a profit of $150,000 and a decision to use the stadium will bring in a profit of $350,000. If, in fact, the Civic Center was chosen, then by not choosing to rent the stadium, the promoter will have *lost the opportunity* to net an additional $200,000. We refer to this $200,000 as the *opportunity loss* (or *regret*) associated with the action/state combination Rent Civic Center/No rain. An opportunity loss may be determined in a similar fashion for each action/state combination of a decision problem, and an **opportunity loss table** may be constructed as in Table 19.3. Notice that none of the opportunity losses of Table 19.3 is less than 0. A little thought should convince you that this is true in general for any opportunity loss. In Section 19.5, we show that decision problems may be solved using outcomes expressed either as payoffs or as opportunity losses.

TABLE 19.3 Opportunity Loss Table for the Rock Concert Decision Problem

		State of Nature	
		Rain	No Rain
Action	Rent Stadium	$190,000	0
	Rent Civic Center	0	$200,000

Definition 19.1

The **opportunity loss** is the difference between the payoff a decision-maker receives for a chosen action and the maximum that the decision-maker could have received for choosing the action yielding the highest payoff for the state of nature that occurred.

Opportunity Loss Determination

Repeat the following procedure for each state of nature in a decision problem—i.e., each column of a payoff table:

1. Find the maximum payoff in a column. The opportunity loss associated with this payoff is 0.
2. The opportunity loss associated with any other payoff in this column is found by subtracting that payoff from the maximum payoff in the column.

EXAMPLE 19.1

A beer producer with breweries located in the western part of the United States and a distribution network that extends only as far east as the Mississippi River is considering expanding its sales region to include the northeastern part of the country. The producer would need to build a new brewery in the Northeast to overcome refrigeration problems in transporting its beer. The problem is to determine how large a brewery to construct.

It has been decided that the size should be based on the projected gross profits (profit before taxes) for the fifth year of operation for each of the four sizes of breweries under consideration. The firm's marketing department recognizes that the company cannot possibly obtain more than a 15% market share during the fifth year of operation and has put together a payoff table for the firm's planning committee (Table 19.4). Construct the corresponding opportunity loss table.

TABLE 19.4 Payoff Table for the Brewer's Decision Problem

		State of Nature: Market Share in Year 5		
		S_1: 0% < 5%	S_2: 5% < 10%	S_3: 10%–15%
Action: Brewery Size	a_1: Small	$300,000	$350,000	$450,000
	a_2: Medium	$250,000	$700,000	$800,000
	a_3: Large	$200,000	$600,000	$1,000,000
	a_4: Very Large	−$100,000	$100,000	$500,000

Solution

For each column of the payoff table (Table 19.4) find the maximum payoff:

	Column 1	*Column 2*	*Column 3*
Maximum payoff:	\$300,000	\$700,000	\$1,000,000

TABLE 19.5 Calculation of Opportunity Losses for the Brewer's Decision Problem

		State of Nature: Market Share in Year 5		
		S_1: 0% < 5%	S_2: 5% <10%	S_3: 10%–15%
Action: Brewery Size	a_1: Small	0	\$700,000 − \$350,000 = \$350,000	\$1,000,000 − \$450,000 = \$550,000
	a_2: Medium	\$300,000 − \$250,000 = \$50,000	0	\$1,000,000 − \$800,000 = \$200,000
	a_3: Large	\$300,000 − \$200,000 = \$100,000	\$700,000 − \$600,000 = \$100,000	0
	a_4: Very Large	\$300,000 − (−\$100,000) = \$400,000	\$700,000 − \$100,000 = \$600,000	\$1,000,000 − \$500,000 = \$500,000

TABLE 19.6 Opportunity Loss Table for the Brewer's Decision Problem

		State of Nature: Market Share in Year 5		
		S_1: 0% < 5%	S_2: 5% < 10%	S_3: 10%–15%
Action: Brewery Size	a_1: Small	0	\$350,000	\$550,000
	a_2: Medium	\$50,000	0	\$200,000
	a_3: Large	\$100,000	\$100,000	0
	a_4: Very Large	\$400,000	\$600,000	\$500,000

The opportunity loss associated with each column maximum is 0. The opportunity associated with, for example, any other payoff in column 1 is found by subtracting that payoff from the column's maximum payoff, \$300,000, as shown in Table 19.5. The resulting opportunity loss table is shown in Table 19.6.

After formulating and displaying a decision problem as just described, we may notice that an action has been included that should never be selected *no matter which state of nature occurs*. In Example 19.1, a_4 is such an action. Inspection of the payoff table (Table 19.4) or the opportunity loss table (Table 19.6) reveals that both actions a_2 and a_3 result in higher payoffs (and lower opportunity losses) than a_4 for each possible state of nature. Thus, a_4 is said to be *dominated* by both actions a_2 and a_3 and should never be chosen by the decision-maker. Accordingly, dominated actions are dropped from consideration. Becauase dominated actions should not be admitted for consideration by the decision-maker, they are sometimes referred to as being *inadmissible actions*. We will see later that eliminating inadmissible actions eases the computation burdens in solving a decision problem.

Definition 19.2

Action a_i is said to **dominate** action a_j, thereby making a_j **inadmissible**, if each of the following is true:

1. For each state of nature, the payoff for action a_i is greater than or equal to the payoff for a_j.
2. For at least one state of nature, the payoff for action a_i is greater than the payoff for a_j.

The decision-maker should eliminate any dominated actions (inadmissible actions) from all decision problems.

To use decision analysis to prescribe a decision in the face of uncertainty, we must characterize our uncertainty concerning the states of nature of the problem with a probability distribution. We discuss the determination of such distributions in the next section.

Exercises 19.11–19.22

Learning the Mechanics

19.11 Explain the difference between payoffs and opportunity losses.

19.12 Why should a dominated action be eliminated from a decision problem?

19.13 Eliminate any inadmissible actions from the payoff table (shown on page 1016) and convert it to an opportunity loss table.

		State of Nature			
		S_1	S_2	S_3	S_4
Action	a_1	−58	−2	0	50
	a_2	−69	−10	40	71
	a_3	−70	−15	40	70
	a_4	−100	−40	−10	80

19.14 Identify any inadmissible actions in the decision tree shown here, and redraw the tree without the inadmissible actions. The objective variable is total sales (in thousands of dollars).

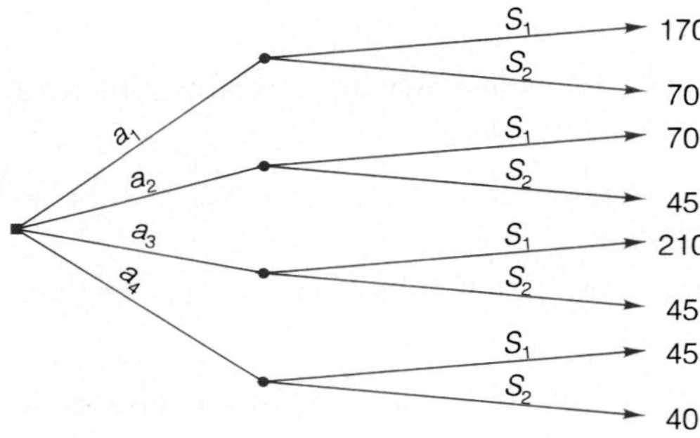

19.15 An opportunity loss table that has been derived from a payoff table is shown.

	S_1	S_2	S_3
a_1	15	40	0
a_2	0	10	8
a_3	43	0	15
a_4	20	20	10

a. Which action(s) in the table is (are) inadmissible? Justify your answer.
b. Why is it not possible to convert this opportunity loss table to the original payoff table?

19.16 The outcomes displayed on this decision tree (see facing page) are in terms of payoffs. Convert the outcomes to opportunity losses.

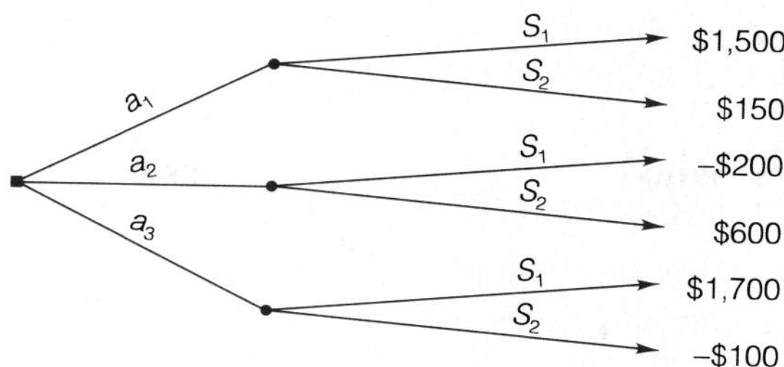

Applying the Concepts

19.17 A U.S. manufacturer of minicomputers is considering marketing its product in Europe. If it does so, the company expects to attain a European market share of no more than 3% and no less than 1% over the next 3 years. The company's accounting department has determined that the payoff for this expansion should be an increase in total company profits for the next 3 years of $3 million if the product is able to attain a 3% market share, an increase of $1 million for a 2% market share, and a decrease of $1 million for a 1% market share. Total domestic profits for the next 3 years are expected to be $15 million regardless of whether or not the minicomputer is marketed in Europe.

a. Formulate the payoff table for this decision problem.

b. Convert the payoff table to an opportunity loss table.

19.18 A common problem in business is the management of perishable inventories—items that lose most of their economic value after a given date due to obsolescence or spoilage. For example, if a corner newsstand orders too many papers and cannot sell them all, the excess papers have little value. Hence, these types of problems are called *newsboy problems*. Newsboy problems are common in retailing situations such as the following: A buyer for a large department store must decide how many of a new style of dress to order. Because of rapid changes in fashion, she does not want to order too many dresses, but if she orders too few she will lose profits for her department. Each dress purchased costs the store $30 and sells for $50. Any dresses not sold at the end of the season are sold at the annual half-price sale. The buyer believes the department will sell three, four, five, six, seven, or eight dozen dresses. Dresses must be purchased by the department store in lots of one dozen.

a. Formulate the payoff table for this decision problem.

b. Convert the payoff table to an opportunity loss table.

19.19 Refer to Exercise 19.10, in which a company that manufactures designer jeans must decide whether to increase its advertising budget for next year.

a. Formulate the payoff table for this decision problem.

b. Are either of the actions inadmissible? Explain.

c. Draw the decision tree that corresponds to your payoff table.

19.20 The purchasing agent for an automobile manufacturer is concluding a purchase agreement with a supplier of preformed body pieces. The pieces will be purchased in lots of 500, and the cost of a lot is $15,000. The body stamping has a frequent defect called a *burr*, which can be filed off by the automobile manufacturer at a cost of $10 per piece. Past experience with this supplier has shown that the proportion of defects tends to be 5%, 10%, or 50%. The supplier now offers a guarantee that it will assume the costs for all defective body moldings greater than 50 in a production lot of 500. This guarantee may be purchased for a cost of $1,000 before the lot begins production. The purchasing agent is interested in determining whether the guarantee should be purchased.

a. Formulate the payoff table for this problem.
b. Draw the decision tree that corresponds to your payoff table.

19.21 Computer products company A has sued computer products company B for patent infringements. The management of B is deciding whether to settle the suit out of court. If they do so, the accounting and marketing departments calculate that company B would lose $50 million over the next 5 years in royalty payments. If they go to court and win, they will have $1 million in court costs. However, if they lose they will owe company A $100 million in royalties.

a. Formulate the payoff table for this decision problem.
b. Draw the decision tree that corresponds to your payoff table.

19.22 A winery can introduce an inexpensive dinner wine for $3 million in fixed and promotional costs per year. Each bottle sold will contribute $.30 to profits. The management believes that sales will be 5 million, 10 million, 15 million, 20 million, or 25 million bottles per year.

a. Formulate the payoff table for this problem.
b. Construct the opportunity loss table.

19.4 Characterizing the Uncertainty in Decision Problems

In Chapter 3 we explained that probability distributions are used to characterize an individual's uncertainty about the outcomes of an experiment. In a decision problem, the observation of a state of nature associated with the problem may be regarded as an experiment and the various states of nature as experimental outcomes. Thus, it follows that a decision-maker can use a probability distribution to characterize the uncertainties associated with the states of nature in a decision problem. These probabilities indicate the likelihood of the occurrence of the various states of nature and may be unknown. Consequently, we assess these values using one of the following:

1. Information about the relative frequencies of the states
2. Judgmental (subjective) information about the states
3. A combination of relative frequency information and subjective information

In other words, we use any type of information that is available to assign probabilities to the states of nature.

For example, in the rock concert example of Section 19.2, the concert promoter could use historical weather data (relative frequency data) along with personal knowledge of the rock star's tendency to declare a day too rainy to perform (subjective information) to assess judgmentally the probabilities of states for the decision problem. Or, if you wanted to decide whether to open your own business next year, you would be interested in knowing whether the economy will move at a healthy pace for the next 3 years or whether it will fall into a recession. Since the observation of this experiment—observing the health of the economy over the next 3 years—can never be repeated, you must rely on your own experience or the advice of economic experts to assess the probabilities associated with these states of nature.

Caution Great care must be taken in determining probabilities for the states of nature. Decision analyses based on poorly chosen probabilities may lead to inappropriate actions.

CASE STUDY 19.1 / Evaluating Uncertainty in Research and Development

Sandoz, a Swiss pharmaceutical company, uses subjective probabilities as a basis for its research and development (R&D) planning and decisions:

Although all management functions must cope with uncertainty, R&D is generally agreed to be the function involving the largest number and the widest range of uncertainties. Thus, the R&D manager faces huge problems not only in selecting the most promising avenues for R&D effort and expenditures but also in attempting to insure a steady flow of technically successful projects (Balthasar, Boschi, and Menke, 1978).

Twice a year a small group of experts is asked to assess the probability of technical success for each of Sandoz's R&D projects. The group of experts includes R&D line managers and other technical experts familiar with particular requirements for a project's success. When the program was begun, these probability assessments were obtained through interviews. Two basic methods of eliciting subjective probabilities were used—direct and indirect. A *direct method* is one that requires the expert to state the probability, or odds, of a project's success explicitly in numerical terms. An *indirect method* is one in which the expert's responses are not probabilities per se.

One indirect method used by Sandoz in the early stages of the program is the *probability wheel*, a disk divided into a blue section and an orange one. The relative size of each section is adjustable. In the center is a pointer to spin that will stop in one of the two sections. The expert is asked which event is more likely: (1) the spinner will stop in the orange section, or (2) project A will succeed. If the answer is (1), the wheel is adjusted to reduce the orange section. If the answer is (2), the wheel is adjusted to enlarge the orange section. This procedure is repeated until the expert says the two events are equally likely. The relative size of the orange section is the expert's subjective probability of the success of project A.

Initially, all probabilities were elicited indirectly. As individual forecasters became more familiar with the process, probabilities were assigned directly to each project. When all the experts were sufficiently familiar with the technique, Sandoz replaced interviews by questionnaires. After individual probability assessments have been obtained from each expert, the entire group meets to discuss their assessments and arrive at one consensus probability for each R&D project. The advantage of the group assessment is that it brings together the opinions of individuals with different experience and information.

The consensus probabilities are then used in making decisions concerning R&D projects. There are many models and techniques for planning and controlling R&D that require input regarding the probability of a project's success. Obtaining subjective probability assessments allows management to use these models as an aid in decision-making.

Sandoz is pleased with the results of using subjective probabilities in managing R&D. By comparing the experts' consensus success probabilities with the actual relative frequency of project successes over time, Balthasar, Boschi, and Menke (1978) found that the probabilistic predictions are reliable indicators of future results. In line with its goal of "a steady flow of technically successful projects," Sandoz has found that decision-making based on subjective probabilities has reduced the variability in its expected success rate for projects in the early stage of development. The R&D managers at Sandoz consider subjective probabilities a useful input to assist in management planning and control in a highly uncertain environment.

19.5 Solving the Decision Problem Using the Expected Payoff Criterion

Now that we have shown you how to structure a decision problem by constructing a payoff table or an opportunity loss table and how to characterize the uncertainty associated with the states of nature in a decision problem, we have to choose a rule for reaching a decision. Numerous rules have been proposed, but the one most commonly used in decision analyses uses a payoff table and chooses the action that produces the **maximum expected payoff**. This is called the **expected payoff criterion**. Equivalently, you could use an opportunity loss table and choose the action that produces the **minimum expected opportunity loss**. This is called the **expected opportunity loss criterion**. It can be shown that both criteria lead to the same solution.

To understand how the expected payoff criterion is used to reach a decision, recall the definition of the expected value of a discrete random variable (Chapter 4). If x is a discrete random variable with probability distribution $p(x)$, then the expected (or mean) value of x is

$$E(x) = \sum_{\text{All } x} xp(x)$$

As we will illustrate, the random variable x in a decision problem is the payoff, and the probabilities associated with x are the same as those that describe the likelihood of occurrence of states of nature.

For example, consider the brewery decision problem in Example 19.1. The payoff table is reproduced in Table 19.7. Note that we have eliminated the inadmissible action, a_4: Very large brewery.

TABLE 19.7 Payoff Table for the Brewery Example

		State of Nature: Market Share in Year 5		
		S_1: 0% < 5%	S_2: 5% < 10%	S_3: 10%–15%
Action: Brewery Size	a_1: Small	\$300,000	\$350,000	\$450,000
	a_2: Medium	\$250,000	\$700,000	\$800,000
	a_3: Large	\$200,000	\$600,000	\$1,000,000

Suppose the brewery has assessed the following probability distribution for the states of nature:

State	0% < 5%	5% < 10%	10%–15%
P(State will occur)	.4	.5	.1

Then if you choose action a_1 (see row 1 of Table 19.7), the payoffs can be \$300,000, \$350,000, or \$450,000 with probabilities .4, .5, and .1, respectively. Thus, the probability distribution for the payoff x if you choose action a_1 is

Payoff x	\$300,000	\$350,000	\$450,000
$p(x)$	.4	.5	.1

The expected payoff of action a_1, denoted by the symbol $EP(a_1)$, is

$$\begin{aligned} EP(a_1) &= \sum xp(x) \\ &= (\$300{,}000)(.4) + (\$350{,}000)(.5) + (\$450{,}000)(.1) = \$340{,}000 \end{aligned}$$

This tells us that, if we were faced with this decision problem a very large number of times and chose action a_1 each time, the mean or expected payoff would be \$340,000—assuming that the probabilities accurately reflect the likelihood of occurrence for the states of nature.

Similarly, we can write the probability distributions and expected payoffs for actions a_2 and a_3. For action a_2:

Payoff x	\$250,000	\$700,000	\$800,000
$p(x)$	.4	.5	.1

$$\begin{aligned} EP(a_2) &= \sum xp(x) \\ &= (\$250{,}000)(.4) + (\$700{,}000)(.5) + (\$800{,}000)(.1) = \$530{,}000 \end{aligned}$$

For action a_3:

Payoff x	\$200,000	\$600,000	\$1,000,000
$p(x)$	.4	.5	.1

$$\begin{aligned} EP(a_3) &= \sum xp(x) \\ &= (\$200{,}000)(.4) + (\$600{,}000)(.5) + (\$1{,}000{,}000)(.1) = \$480{,}000 \end{aligned}$$

Now examine the expected (mean) payoffs for each action, as summarized in Table 19.8, page 1022. Which action would you choose? We think you would choose action a_2; i.e., you would recommend that the brewer construct a medium-sized brewery because this strategy will produce the largest expected payoff—namely, \$530,000.

The Expected Payoff Criterion

Choose the action that produces the largest expected payoff.

TABLE 19.8 Expected Payoff Table for the Brewer's Decision Problem

Action a_i	Expected Payoff for Action a_i $EP(a_i)$
a_1	\$340,000
a_2	\$530,000
a_3	\$480,000

As noted at the beginning of this section, we arrive at the same solution to the brewer's decision problem if we use an opportunity loss table, find the expected opportunity loss for each action, and then choose the action that produces the minimum expected opportunity loss. The solution is again action a_2.

The Expected Opportunity Loss Criterion

Choose the action that produces the smallest expected opportunity loss. (This always leads to the same decision as the expected payoff criterion.)

EXAMPLE 19.2

A well-known cosmetics firm has been approached by a television producer to determine whether the firm is interested in sponsoring a new television series next fall. The firm is faced with the choice of one of two actions. It can continue to sponsor a popular television show it has sponsored in the past, or it can shift to the producer's new prime-time show. The states of nature are the projected averages of the biweekly Nielsen ratings for the entire season. A Nielsen rating for a show is the percentage of homes in a sample taken by the A. C. Nielsen Co. that have watched the show for at least 6 minutes during the rating period. The objective variable is the firm's projected market share at the end of the television season next year. According to the expected payoff criterion, which show should the cosmetics firm sponsor? The firm's advertising agency and marketing department have come up with the respresentation of the firm's decision problem shown in Table 19.9.

TABLE 19.9 Cosmetics Firm's Decision Problem in Example 19.2

		State of Nature: Projected Average Nielsen Rating for the New Show (Probabilities in Parentheses)			
		S_1: <12 (.1)	S_2: 12–17 (.2)	S_3: 18–29 (.4)	S_4: ≥30 (.3)
Action: Sponsorship	a_1: Old Show	.12	.12	.12	.12
	a_2: New Show	.06	.10	.14	.17

Solution

Let x denote the payoff (market share) for a particular action and $p(x)$ its probability distribution. We now compute $E(x) = \sum_{\text{All } x} xp(x)$ for each action that produces the larger expected payoff. Thus, for action a_1:

$$EP(a_1) = .12(.1) + .12(.2) + .12(.4) + .12(.3) = .12$$

and for action a_2:

$$EP(a_2) = .06(.1) + .10(.2) + .14(.4) + .17(.3) = .133$$

Since the expected payoff for action a_2 is larger than that for a_1, the firm should sponsor the new show rather than the old show. Notice that in this example the objective variable was market share and not profit. The expected payoff and expected opportunity loss criteria can be used, as stated above, for any objective variable we want to maximize. For some situations, however, we may want to minimize the expected value of the objective variable (or, equivalently, maximize the expected opportunity loss). For example, with an objective variable such as time to process an order, we would prefer less time to more time. Then the expected payoff and expected opportunity loss criteria would have to be redefined, and we would seek the *minimum expected payoff*.

CASE STUDY 19.2 / Hurricanes: To Seed or Not to Seed?

The seeding of hurricanes as a means of lessening their destructive power was suggested by R. H. Simpson in the early 1960s. Even though the results of experimental hurricane seeding were encouraging, government policy through the early 1970s prohibited the seeding of hurricanes that threatened coastal areas. Howard, Matheson, and North (1972) used decision analysis to determine (1) whether existing government policy should be modified and (2) whether further experiments on hurricane seeding should be carried out. We will discuss the second question in Case Study 19.4.

To answer the first question, Howard, Matheson, and North had to define the states of nature, assign probabilities to each of the possible states, and assess the consequences of each action/state combination. The actions specified were Seed a threatening hurricane and Do not seed a threatening hurricane. The states of nature were defined in terms of the percentage change in maximum wind speed over a 12-hour period. This measure was chosen because it is related to the primary cause of the destruction inflicted by most hurricanes, and it is this characteristic that seeding is expected to influence. Based on historical data, a probability distribution was assessed for changes in wind speed for a "representative hurricane" over a 12-hour period. Using Simpson's theoretical work on hurricane seeding and early experimental work on seeding, Howard, Matheson, and North also assessed a probability distribution for changes in wind speed of a seeded hurricane. These probability distributions are given in Table 19.10 on page 1024.

The consequences of the action/state combinations considered were property damage resulting from the hurricane and government liability in the case of seeded hurricanes. Using a least squares regression analysis (see Chapter 10) and past records of hurricane damage, the authors modeled the relationship between changes in wind speed and property damage (in millions of dollars) for a "representative hurricane." Results of their analysis appear in Table 19.11.

TABLE 19.10 Probability Distributions for States of Nature

State	State Probability	State Probability
Change in Wind Speed	Unseeded Hurricanes	Seeded Hurricanes
S_1: +25% or more	.054	.038
S_2: +10% to +25%	.206	.143
S_3: −10% to +10%	.480	.392
S_4: −25% to −10%	.206	.255
S_5: −25% or more	.054	.172

TABLE 19.11 Predicted Property Damage for Hurricane

Change in Wind Speed	Predicted Damage
+25% or more	335.8
+10% to +25%	191.1
−10% to +10%	100.0
−25% to −10%	46.7
−25% or more	16.3

It was assumed that if a hurricane did not slow after seeding, the government would bear increased legal and social costs. A percentage of the property damage resulting from the states S_1, S_2, and S_3 was used to estimate these costs. The total costs associated with a particular state was found by adding these government responsibility costs to the predicted property damages. Estimates were 50% for S_1, 30% for S_2, and 5% for S_3. Thus, the outcomes for each state under the decision to seed would be (including the $.25 million cost of seeding) as shown in Table 19.12. These quantities are expressed (in millions of dollars) in terms of negative gains—i.e., losses.

TABLE 19.12 Outcomes Associated with States of Nature

State	Outcome
S_1: +25% or more	−503.95
S_2: +10% to +25%	−248.65
S_3: −10% to +10%	−105.25
S_4: −25% to −10%	−46.95
S_5: −25% or more	−16.55

The expected payoff criterion can now be applied as usual. Based on the given probabilities and outcomes, the expected payoffs are

$$EP(\text{No seeding}) = -\$116.0 \text{ million}$$
$$EP(\text{Seeding}) = -\$110.78 \text{ million}$$

As a result of their analysis, the researchers recommended that the government change its policy and allow seeding of hurricanes that threaten coastal areas. They noted that each decision to seed a hurricane, however, should be based on a decision analysis that uses all the meteorological and geographic factors relevant to the particular case.

Exercises 19.23–19.33

Learning the Mechanics

19.23 Consider the payoff table shown here. Find the action that would be prescribed by the expected payoff criterion. Probabilities are in parentheses in the table.

		State of Nature			
		S_1 (.1)	S_2 (.3)	S_3 (.4)	S_4 (.2)
	a_1	−150	−500	10	300
Action	a_2	−200	−100	300	100
	a_3	−150	−40	−10	85

19.24 In the decision tree shown here, state probabilities are displayed in parentheses to the right of the state symbols, S_1 and S_2.

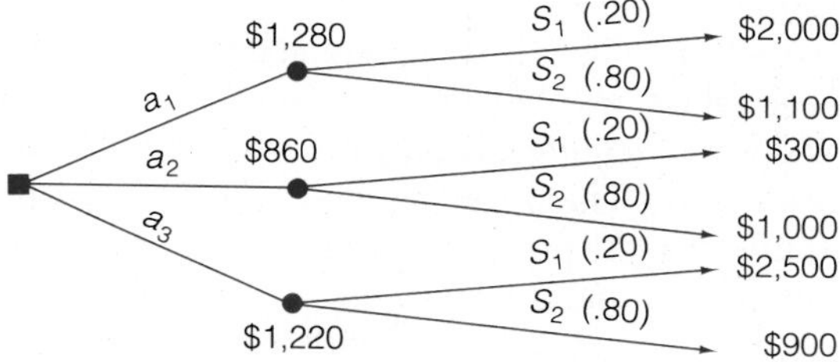

a. Identify any inadmissible actions, and explain why they are inadmissible.
b. The expected payoff for each action has been computed and appears to the left of the chance fork associated with each action in the decision tree. Verify that these expected payoffs are correct.
c. Identify the action that is prescribed by the expected payoff criterion.

19.25 Construct the decision tree for the decision problem described in Exercise 19.23. Include on your decision tree the state probabilities and the expected payoffs associated with each action. (For an example of such a tree, see Exercise 19.24.)

19.26 Consider the accompanying payoff table (probabilities are in parentheses).

		State of Nature		
		S_1 (.20)	S_2 (.30)	S_3 (.50)
Action	a_1	150	55	130
	a_2	−35	0	210

a. Use the expected payoff criterion to select the better action.
b. Convert the payoff table to an opportunity loss table and use the expected opportunity loss criterion to select the better action.

19.27 In the decision tree shown here, state probabilities are displayed in parentheses to the right of the state symbols, S_1 and S_2. The outcomes on the decision tree are expressed in terms of opportunity losses. According to the expected opportunity loss criterion, which action should be selected?

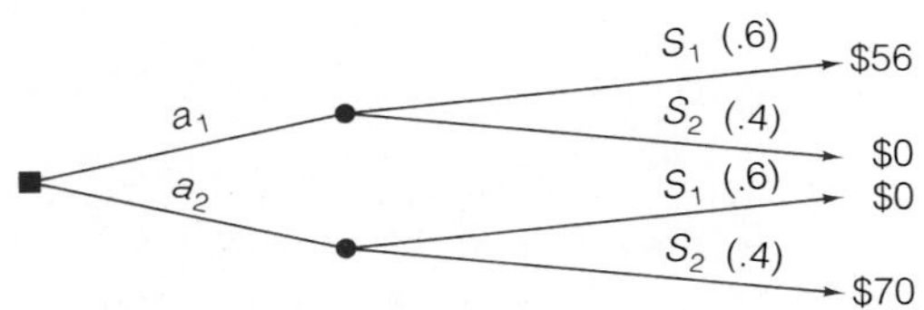

Applying the Concepts

19.28 *Materials requirements planning* (MRP) systems are computerized planning and control systems for manufacturing operations. They are used to manage raw materials and work-in-progress inventories. Since their introduction in the mid-1960s, MRP systems have made it possible to simultaneously reduce inventories and improve customer service (Schroeder, 1993). The operations manager for a company that produces a defense product wants an MRP system. He has narrowed his choices to three vendors, X, Y, and Z, whose systems cost $5,000, $4,000, and $3,000, respectively. In addition, each vendor charges a fee for implementing the system (e.g., adapting the system to fit the purchaser's need, installing the system, and debugging the system); the size of the fee depends primarily on the extent to which the system must be modified to meet the purchaser's needs. Systems X, Y, and Z are similar and would require basically the same modifications, but the implementation fees charged by the vendors vary widely. These fees are described in the table. The manager, in consultation with the vendors, has determined that the probability of a major modification being required is .3, the probability of a moderate modification is .5, and the probability of a minor modification is .2. This uncertainty concerning the extent of the modification required exists because the operations manager has not finalized the specifications of the system he desires so the vendors are not yet sure how much modification will be required.

Modification	X	Y	Z
Major	$3,000	$4,000	$6,000
Moderate	1,000	1,400	2,000
Minor	200	400	500

a. The objective variable for this decision problem is the initial cost plus the implementation fee for a system. Since we want to minimize (rather than maximize) the value of this objective variable, we need to convert our problem to one of maximization. This can be done by multiplying each value of system cost (initial cost plus implementation fee) by -1. The action that maximizes the value of this new objective variable will be the one that minimizes the system cost. Eliminate any inadmissible actions, and formulate the payoff table for this decision problem.

b. According to the expected payoff criterion, which MRP system should the manager purchase? Explain.

19.29 A company that manufactures hair dryers wants you to help choose the travel case design for its new portable model. It has constructed the accompanying payoff table to characterize the decision problem. The objective variable is the contribution to the company's profits over the next year that results from marketing the new dryer. Use the expected payoff criterion to answer the questions.

		State of Nature	
		S_1: Consumers Like Design	S_2: Consumers Dislike Design
Action	a_1: Design A	$800,000	−$200,000
	a_2: Design B	$660,000	$10,000

a. If $P(S_1) = P(S_2) = .5$, which design should the company choose?
b. If $P(S_1) = 2/3$ and $P(S_2) = 1/3$, which design should it choose?

19.30 Reconsider Exercise 19.17, in which a minicomputer company is considering marketing its product in Europe. The company's marketing department has assessed the probability distribution in the table for the share of the European minicomputer market that the company will attain over the next 3 years. According to the expected payoff criterion, should the company enter the European market? Explain.

Market Share	1%	2%	3%
Probability	.1	.3	.6

19.31 Reconsider Exercise 19.20, which concerned the purchase of body pieces by an automobile manufacturer. A lot of 500 pieces is purchased for $15,000, and the proportion of defects in each lot is .05, .10, or .50. The supplier offers a guarantee that it will assume the costs for all defective body moldings greater than 50 in a production lot of 500. This guarantee may be purchased for a cost of $1,000 before the lot begins production. The data in the table have been gathered on 100 past production lots of preformed body pieces. Given only this information, should the guarantee be purchased? Explain. [*Hint*: Use the sample of 100 lots to assess the probabilities of the states of nature.]

Proportion of Defects	.05	.10	.50
Times Observed	55	24	21

19.32 Refer to the payoff table for the decision problem in Exercise 19.18. The dress buyer has assessed the probability distribution given in the table for the number of dresses (in dozens) she can sell. How many dozen dresses should the buyer order according to the expected payoff criterion?

Sales	3	4	5	6	7	8
Probability	.15	.25	.30	.15	.10	.05

19.33 Reconsider Exercises 19.10 and 19.19, in which a company that manufactures designer jeans is concerned about whether to increase its advertising budget for the year. The company has determined that, if the advertising budget is increased, the expanded advertising campaign will succeed with probability 2/3 and will fail with probability 1/3.

a. According to the expected payoff criterion, should the jeans manufacturer increase its advertising budget? Explain.
b. Convert the payoff table you developed in part **a** to an opportunity loss table.
c. According to the expected opportunity loss criterion, should the jeans manufacturer increase its advertising budget? Explain.

19.6 The Expected Utility Criterion

The expected payoff (or opportunity loss) criterion sometimes fails to provide decisions that are consistent with the decision-maker's attitude toward risk. For example, suppose you were required to choose one of these two actions:

a_1: Deposit your $100,000 inheritance in the bank for 1 year at 7% interest

a_2: Invest your $100,000 inheritance for 1 year with a .5 probability of having a total of $250,000 at the end of the year and a .5 probability of losing all your inheritance

Which would you choose? Why? Most of us would probably choose action a_1, basing our decision on a "safety first" strategy. The usual argument goes, "Why pass up a chance for a sure \$107,000 at the end of the year for a 50–50 chance to end up with nothing?" Notice that if you choose a_1, your decision is not in accord with the action prescribed by the expected payoff criterion, because

$$EP(a_1) = \$107{,}000(1) = \$107{,}000$$
$$EP(a_2) = \$250{,}000(.5) + 0(.5) = \$125{,}000$$

Thus, if we use the expected payoff criterion, we would select action a_2.

The expected payoff criterion can be adapted to reflect our attitude toward risk if we can express the outcomes of the decision problem in terms of an objective variable that better reflects the true relative values of outcomes. An objective variable that reflects the decision-maker's attitude toward risk is called a **utility function**, and the values assigned to the outcomes are referred to as **utility values**, or **utilities** (or sometimes **utiles**).

Definition 19.3

A **utility function** is a rule that assigns numerical values to the potential outcomes of a decision problem in such a way that

1. The values rank the outcomes in accordance with the decision-maker's preferences.
2. The function itself describes the decision-maker's attitude toward risk.

We now illustrate how to assign utility values to the monetary outcomes of the inheritance example. The payoff table for this example is shown in Table 19.13. As noted previously, many of us would be reluctant to gamble on the outcome of the investment if the chance of success were only .5, even though the expected payoff is higher for action a_2 (\$125,000) than for action a_1 (\$107,000). *The key to assigning utility values to the action/state of nature combinations is to answer the following question: What would the probability of success for the investment have to be for you to value actions a_1 and a_2 equally?* The answer to this question is a probability, p, such that if the probability of the investment's success exceeds p, you would prefer to invest the money (a_2), but if the success probability is less than p, you would prefer to deposit the money in the bank and take the safe return. The probability p is called the **utility of the outcome** \$107,000, and we write $U(107{,}000) = p$. The minimum payoff (\$0) and the maximum payoff (\$250,000) are assigned utility values of 0 and 1, respectively, so that $U(0) = 0$, and $U(250{,}000) = 1$.

TABLE 19.13 Payoff Table for the Inheritance Example

		State of Nature	
		S_1: Investment Fails	S_2: Investment Succeeds
Action	a_1: Deposit \$100,000	\$107,000	\$107,000
	a_2: Invest \$100,000	0	\$250,000

Suppose you decide that the probability of the investment's success would have to be .7 before the two actions were equally appealing. Then the utility value assigned to \$107,000 is .7, and the payoff table with the utility values as outcomes is shown in Table 19.14 on page 1030.

The general rule for assigning utility values to monetary outcomes* is given in the box.

Assigning Utility Values to Monetary Outcomes†

1. Identify the maximum and minimum payoffs in the decision table. Call them O_M and O_L, respectively.
2. Set $U(O_M) = 1$ and $U(O_L) = 0$, where $U(O)$ represents the utility value of outcome O.
3. To determine the utility value for any other outcome O_{ij} in the payoff table, determine the value of p that makes you indifferent between the following:
   a. Receiving O_{ij} with certainty
   b. Participating in a gamble in which you can win O_M with probability p or O_L with probability $(1 - p)$

   Then, $U(O_{ij}) = p$.

*Utility functions may be assessed for the outcomes corresponding to any objective variable, but because the objective variable in business decision-making is typically monetary (or can be converted to a monetary equivalent), we have restricted our discussion to utility functions for money.

†The choice of 0 and 1 as the minimum and maximum utility values is arbitrary. The same result is obtained if the utility values are all multiplied by the same constant or if a constant is added to each value—but then the utility values lose their probabilistic interpretation. For more detail on the choice of scale, its implications for interpersonal and intrapersonal comparisons of utility values, and other methods for assessing utility functions, see the references at the end of this chapter.

TABLE 19.14 Payoff Table for the Inheritance Example with Utility Values as Payoffs

		State of Nature (Probabilities in Parentheses)	
		S_1: Investment Fails (.5)	S_2: Investment Succeeds (.5)
Action	a_1: Deposit \$100,000	.7	.7
	a_2: Invest \$100,000	0	1

We are now prepared to make a decision based on the **expected utility criterion**. We first find the expected utility for each action using the formula

$$EU(a_i) = \sum_{\substack{\text{All states}\\ \text{of nature}}} \begin{pmatrix} \text{Utility of the} \\ \text{action } a_i\text{/state of nature} \\ \text{combination} \end{pmatrix} \begin{pmatrix} \text{Probability} \\ \text{of the} \\ \text{state of nature} \end{pmatrix}$$

Thus, for the inheritance example,

$$EU(a_1) = .7(.5) + .7(.5) = .7$$
$$EU(a_2) = 0(.5) + 1(.5) = .5$$

We now choose the action with the higher expected utility, which is action a_1. You can see that this decision differs from that yielded by the expected payoff criterion and that it more accurately reflects a safety-first attitude toward risk.

Expected Utility Criterion

Choose the action that has the greatest expected utility, where the expected utility of action a_i is given by

$$EU(a_i) = \sum_{\substack{\text{All states}\\ \text{of nature}}} \begin{pmatrix} \text{Utility of the} \\ \text{action } a_i\text{/state of nature} \\ \text{combination} \end{pmatrix} \begin{pmatrix} \text{Probability} \\ \text{of the} \\ \text{state of nature} \end{pmatrix}$$

EXAMPLE 19.3

Recall the brewery size decision problem of Example 19.1. The payoff table is repeated in Table 19.15 (omitting the inadmissible action a_4). Suppose we assign the utility values shown in Table 19.16 to the monetary payoffs.

TABLE 19.15 Payoff Table for the Brewery Size Decision Problem

		State of Nature: Market Share in Year 5		
		S_1: 0% < 5%	S_2: 5% < 10%	S_3: 10%–15%
Action: Brewery Size	a_1: Small	\$300,000	\$350,000	\$450,000
	a_2: Medium	\$250,000	\$700,000	\$800,000
	a_3: Large	\$200,000	\$600,000	\$1,000,000

TABLE 19.16 Utility Values for the Brewery Decision Problem

		State of Nature: Market Share in Year 5 (Probabilities in Parentheses)		
		S_1: 0% < 5% (.4)	S_2: 5% < 10% (.5)	S_3: 10%–15% (.1)
Action: Brewery Size	a_1: Small	.35	.50	.65
	a_2: Medium	.20	.90	.95
	a_3: Large	0	.85	1.00

a. Interpret the utility value assigned to the \$700,000 payoff, $U(700{,}000) = .9$.

b. Determine which action should be taken according to the expected utility criterion.

Solution

a. Referring to the rules for assigning utility values, we see that the utility value of .9 represents the probability that makes us have no preference between receiving \$700,000 with certainty and participating in a gamble in which we can gain \$1,000,000 (the maximum payoff) with probability .9 or \$200,000 (the minimum payoff) with probability $(1 - .9) = .1$. You can see that we have again adopted a conservative strategy, because the expected payoff for the gamble is $1{,}000{,}000(.9) + 200{,}000(.1) = \$920{,}000$, which exceeds the fixed payoff of \$700,000. In other words, the assignment of $U(700{,}000) = .9$ reflects our desire to receive the fixed payoff unless the odds are very high that we will win the gamble. The rest of the utility values can be similarly interpreted, and they all reflect this conservative attitude toward risk.

b. The expected utilities are calculated as follows:

$$EU(a_1) = .35(.4) + .50(.5) + .65(.1) = .455$$
$$EU(a_2) = .20(.4) + .90(.5) + .95(.1) = .625$$
$$EU(a_3) = 0(.4) + .85(.5) + 1.0(.1) = .525$$

Thus, according to our assignment of utility values to the monetary outcomes, the expected utility criterion indicates that we should select action a_2 and build a medium-sized brewery.

You can see that the assignment of utility values is personal and subjective. If careful thought is given to this task, the result will be a utility function that reflects your preferences for outcomes and your attitude toward risk. Applying the expected utility criterion will then yield a decision consistent with your preferences and risk attitude.

Exercises 19.34–19.40

Learning the Mechanics

19.34 Suppose you are indifferent (have no preference) between receiving \$150 with certainty and participating in a gamble that offers you a probability of .3 of receiving \$1,000 and a probability of .7 of receiving nothing. You are also indifferent between receiving \$400 with certainty and participating in a gamble offering you a probability of .6 of receiving \$1,000 and a probability of .4 of receiving nothing. Finally, you are also indifferent between receiving \$800 with certainty and participating in a gamble offering you a probability of .95 of receiving \$1,000 and a probability of .05 of receiving nothing.

a. Let $U(\$1{,}000) = 1$ and $U(\$0) = 0$. Find $U(\$150)$, $U(\$400)$, and $U(\$800)$.

b. Use these utility values to plot your utility function for money over the range from \$0 to \$1,000.

19.35 Consider the utility function* for money,

$$U(x) = x^2 \qquad 0 \le x \le 100$$

and the payoff table shown here (probabilities in parentheses).

		State of Nature		
		S_1 (.25)	S_2 (.30)	S_3 (.45)
Action	a_1	\$100	\$75	\$25
	a_2	\$70	\$95	0

a. Use the utility function to convert the outcomes in the payoff table from monetary values to utility values (utiles).

b. Which action would be prescribed by the expected utility criterion?

*Notice that the minimum and maximum values of this utility function are not 0 and 1, respectively. Refer to the second footnote on page 1029.

19.36 Consider the utility function for money,

$$U(x) = \frac{\ln(x + 51)}{5.525} \qquad -50 < x < 200$$

and the given payoff table (probabilities in parentheses).

		State of Nature		
		S_1 (.50)	S_2 (.40)	S_3 (.10)
Action	a_1	−\$10	\$50	\$90
	a_2	0	\$20	\$20
	a_3	−\$10	\$60	\$50
	a_4	−\$10	−\$50	\$200

a. Eliminate any inadmissible actions.
b. Use the utility function to convert the outcomes in the payoff table from monetary values to utility values (utiles).
c. Which action is prescribed by the expected utility criterion?
d. Does this decision differ from the decision prescribed by the expected payoff criterion?

19.37 Consider the utility function for money,

$$U(x) = -1 + .01x \qquad 100 \le x \le 200$$

and the payoff table shown (probabilities are in parentheses).

		State of Nature			
		S_1 (.30)	S_2 (.05)	S_3 (.45)	S_4 (.20)
Action	a_1	\$150	\$180	\$100	\$80
	a_2	\$130	\$165	\$190	\$140
	a_3	\$115	\$120	\$160	\$210

a. Use the utility function to convert the outcomes in the payoff table from monetary values to utility values (utiles).
b. Which action is prescribed by the expected utility criterion?

Applying the Concepts

19.38 An investor is trying to decide whether to invest in a wildcat oil well. If the well is drilled and it is dry, she will lose \$500,000. On the other hand, if the well is a gusher, she will make \$1.5 million. It is also possible for the well to yield less oil than a gusher, in which case she will make \$600,000.

a. A decision analyst asks her the following questions:

At what value of p would you be indifferent between the two situations: receive \$600,000 with certainty, and receive \$1.5 million with probability p or lose \$500,000 with probability $(1 - p)$? She replies, at $p = .90$. What is the investor's utility value for \$600,000?

At what value of p would you be indifferent between receiving \$0 with certainty, and receiving \$1.5 million with probability p or losing \$500,000 with probability $(1 - p)$? She replies, at $p = .8$. Find her utility value for \$0.

b. Graph the utility function for this investor for dollar values between −\$500,000 and \$1,500,000.

19.39 Refer to Exercise 19.38. The investor now must decide whether to keep the \$500,000 or invest it in the oil well. She assesses the probabilities of the states to as

$$P(\text{Dry well}) = .5 \qquad P(\text{Moderate success}) = .3 \qquad P(\text{Gusher}) = .2$$

a. Set up the payoff table for this decision problem.
b. Use the expected payoff criterion to determine which action the investor should select.
c. Substitute the utility values for the monetary outcomes in the payoff table of part **a** and use the expected utility criterion to select the action.

19.40 Refer to Exercise 19.32, in which you selected an action for the dress purchaser using the expected payoff criterion. Now suppose the buyer's utility function for money is

$$U(x) = \frac{\sqrt{x + 2{,}000}}{110} \qquad -2{,}000 \leq x \leq 10{,}000$$

How many dresses should she order? Should she be using her own utility function to make this decision? Why or why not?

19.7 Classifying Decision-Makers by Their Utility Functions

The attitude of a decision-maker toward risk is revealed by the shape of his or her utility function. Recall the inheritance decision problem of Section 19.6, with the payoff table given in Table 19.13 and a table of utility values given in Table 19.14. These tables are repeated for convenience as Tables 19.17 and 19.18.

TABLE 19.17 Payoff Table for the Inheritance Example

		State of Nature (Probabilities in Parentheses)	
		Investment Fails (.5)	Investment Succeeds (.5)
Action	a_1: Deposit \$100,000	\$107,000	\$107,000
	a_2: Invest \$100,000	0	\$250,000

TABLE 19.18 Utility Value Table for the Inheritance Example

		State of Nature (Probabilities in Parentheses)	
		Investment Fails (.5)	Investment Succeeds (.5)
Action	a_1: Deposit $100,000	.7	.7
	a_2: Invest $100,000	0	1

We know that these utility values reflect a conservative attitude toward risk, since the expected payoff of the investment must exceed 0(.3) + 250,000(.7) = $175,000 before the decision-maker prefers this gamble to the fixed $107,000 return. This attitude may be graphically portrayed by plotting the utility values on a vertical axis against the payoffs on a horizontal axis, as shown in Figure 19.3. The points are connected with a smooth curve. Utility functions with this shape are called **concave utility functions**. This shape is characteristic of a conservative attitude toward risk, and decision-makers whose utility functions are concave are called **risk-avoiders**.

Now, suppose we decided to assign a utility value of .2 to the $107,000 payoff in Table 19.17. This value reflects a liberal, or gambling, attitude toward risk, since the interpretation is that we have no preference between a sure $107,000 and a .2 probability of receiving $250,000 (with a .8 probability of nothing). The expected payoff of the gamble is 0(.8) + $250,000(.2) = $50,000, which is less than the fixed payoff if we deposit the money in the bank. The graph of this utility function is shown in Figure 19.4. Utility functions with this shape are called **convex utility functions**. This shape is characteristic of an aggressive attitude toward risk, and decision-makers who have convex utility functions are called **risk-takers**.

If the utility value of .428 is assigned to the $107,000 outcome, the implication is that we would have no preference between accepting $107,000 and gambling on receiving $250,000 with probability .428, a gamble that has an expected payoff of

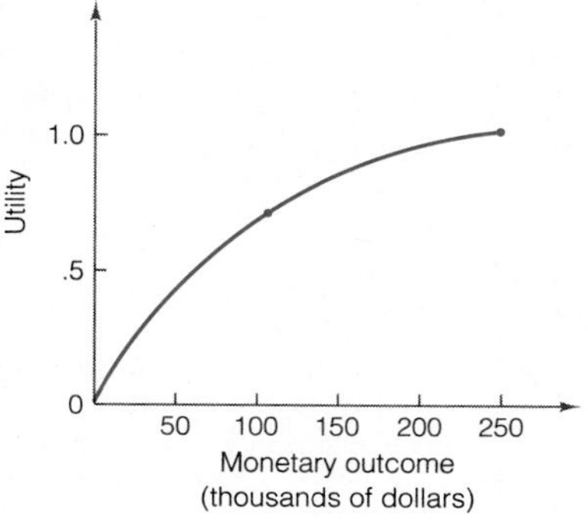

FIGURE 19.3 ▲
Utility function for the inheritance example: A risk-avoiding attitude

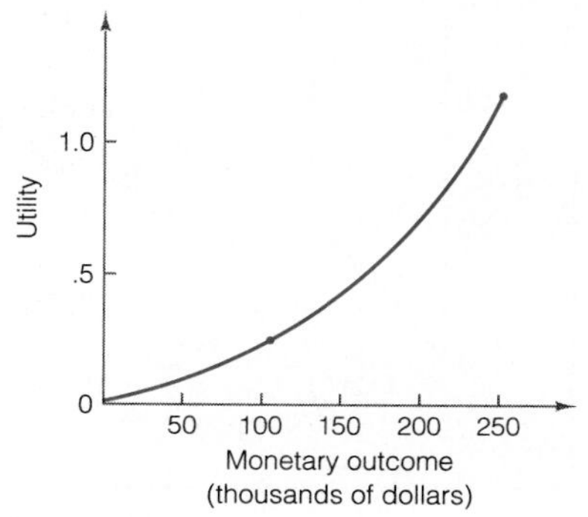

FIGURE 19.4 ▲
Utility function for the inheritance example: A risk-taking attitude

0(.572) + 250,000(.428) = \$107,000. The fact that both courses of action have the same expected payoff reflects a neutral attitude toward risk. The graph of the utility function in this case and, for **risk-neutral** decision-makers in general, is a straight line, as shown in Figure 19.5. Since the expected utility of action a_i is equal to the utility of the expected payoff for action a_i for straight-line utility functions, *the expected payoff criterion and the expected utility criterion are equivalent for risk-neutral decision-makers.*

FIGURE 19.5 ▶
Utility function for the inheritance example: A risk-neutral attitude

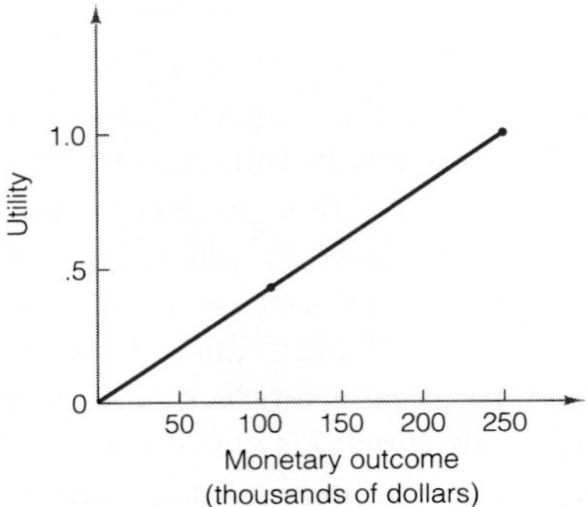

The three types of utility functions and the risk attitudes they characterize are summarized in the box.

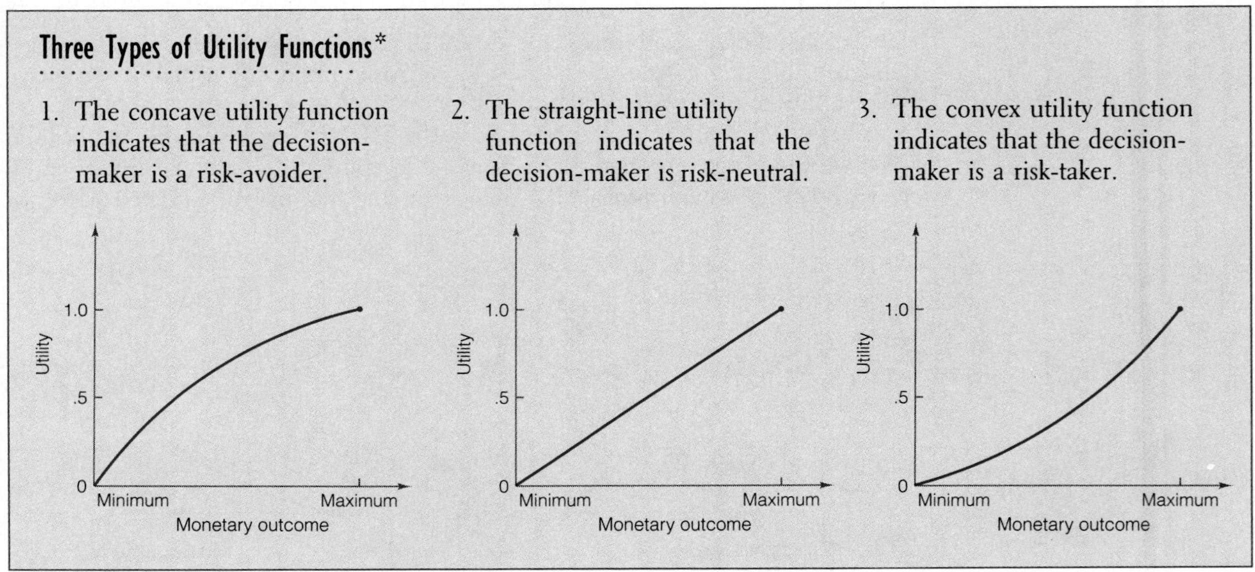

Three Types of Utility Functions*

1. The concave utility function indicates that the decision-maker is a risk-avoider.
2. The straight-line utility function indicates that the decision-maker is risk-neutral.
3. The convex utility function indicates that the decision-maker is a risk-taker.

*Other types of utility functions may be obtained by combining these. For example, a decision-maker may be risk-neutral for relatively small monetary outcomes and a risk-avoider for relatively large monetary outcomes. Thus, the utility function might appear S-shaped.

EXAMPLE 19.4

For the brewery decision problem (Examples 19.1 and 19.3), consider the two utility functions reflected in Table 19.19.

TABLE 19.19 Two Utility Functions for the Brewery Decision Problem

		State of Nature: Year 5 Market Share (Probabilities in Parentheses)					
		Situation A			Situation B		
		S_1: 0% < 5% (.4)	S_2: 5% < 10% (.5)	S_3: 10%–15% (.1)	S_1: 0% < 5% (.4)	S_2: 5% < 10% (.5)	S_3: 10%–15% (.1)
Action: Brewery Size	a_1: Small	.35	.50	.65	.04	.07	.11
	a_2: Medium	.20	.90	.95	.02	.30	.40
	a_3: Large	0	.85	1.00	0	.20	1.00

a. Graph the utility functions, and identify the attitude toward risk that each characterizes.

b. Determine the action that should be taken in each situation.

Solution

a. The utility functions are shown in Figure 19.6. Situation A is the same one we used and characterized as conservative in Example 19.3. You can see that the utility function in Figure 19.6(a) is *concave*, indicating a *risk-avoiding* attitude. However, the utility function corresponding to situation B is *convex*, which characterizes a *risk-taking* situation.

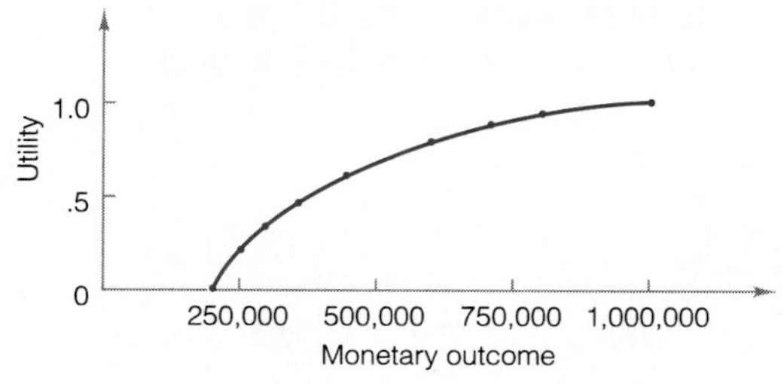

a. Situation A

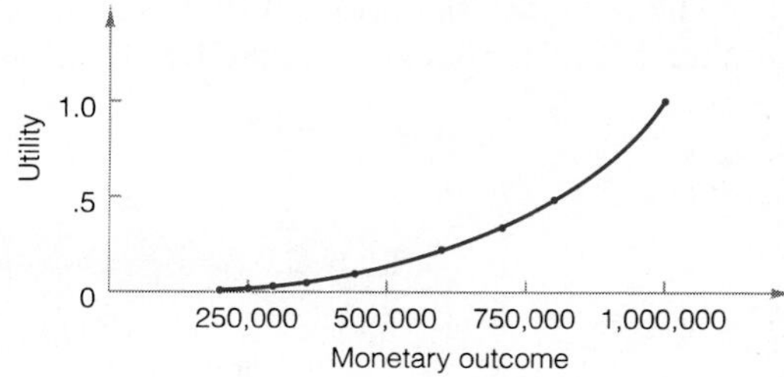

b. Situation B

FIGURE 19.6 ▲
Utility functions for the brewery decision problem

b. We showed in Example 19.3 that for situation A,

$$EU(a_1) = .455 \qquad EU(a_2) = .625 \qquad EU(a_3) = .525$$

so the expected utility criterion selects action a_2: Build a medium-sized brewery. For situation B, we find

$$EU(a_1) = .04(.4) + .07(.5) + .11(.1) = .062$$
$$EU(a_2) = .02(.4) + .30(.5) + .40(.1) = .198$$
$$EU(a_3) = 0(.4) + .20(.5) + 1.00(.1) = .200$$

Thus, the expected utility criterion selects action a_3: Build a large brewery. Note that the risk-taking utility function leads to the selection of a riskier action, a_3, than the risk-avoiding and risk-neutral functions, which both select a_2 (see Section 19.5 for the risk-neutral—that is, expected payoff—solution).

CASE STUDY 19.3 / An Airport: To Expand or Not to Expand?

The Mexican government received two conflicting recommendations on airport development for Mexico City. One study recommended expanding the present airport at Texcoco, and the other urged rapidly transferring all operations to a new airport to be built at Zumpango. As a result of this disagreement, the Mexican Ministry of Public Works employed analysts to evaluate alternatives for airport development and recommend the most effective strategy (Keeney, 1973).

The basic decision problem involved which types of aircraft (international, I; domestic, D; general, G; and military, M) should operate at each of the two locations (Texcoco, T, and Zumpango, Z) over the next 30 years. To simplify the problem, it was assumed that changes in operations from one site to another could occur at only three times: 1975, 1985, and 1995. Each airport development strategy (action) indicated which types of aircraft would operate at each location for each time period. Some examples of possible strategies are given in Table 19.20, where 1 and 2 are the recommendations of the previous studies, while 3 is one possible intermediate strategy.

TABLE 19.20 Possible Strategies for Airport Expansion

		1975		1985		1995	
		T	Z	T	Z	T	Z
Strategy	1	IDGM		IDGM		IDGM	
	2		IDGM		IDGM		IDGM
	3	DGM	I	MG	ID		IDGM

The outcomes of 100 possible strategies were evaluated in terms of six attributes:

1. Cost
2. Capacity (number of aircraft operating)
3. Average access time to airport
4. Number of people killed or injured per aircraft accident
5. Number of people displaced by airport development
6. Number of people subjected to a high noise level

One step in the decision analysis was to evaluate utility functions for each of these attributes. The following description of the utility assessment for access time is representative of the other utility assessments. Questioning of the clients determined that the bounds on average access time over all strategies were 12 minutes and 90 minutes. The best possible time, 12 minutes, was assigned a utility of 1; the worst possible time, 90 minutes, was assigned a utility of 0. The clients were then presented with a series of choices of the following type: 62 minutes access time for certain, or a lottery with p chance at 12 minutes and $(1 - p)$ chance at 90 minutes. The size of p was varied until the client indicated indifference between these two choices. In the case of 62 minutes, the client was indifferent for $p = .5$. Thus, $U(62 \text{ minutes}) = .5$.

This process was repeated for several other access times, and the results were plotted as shown in Figure 19.7. You can see that the clients are risk-avoiders with respect to access time.

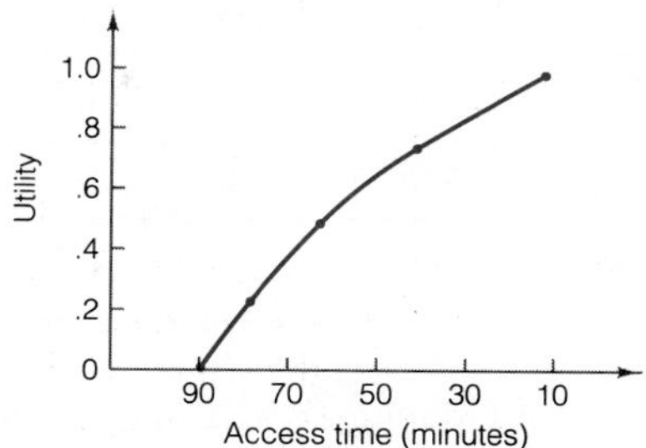

FIGURE 19.7 ▲
Utility function for Case Study 19.3

The results of the utility analysis for each attribute were combined to develop a set of utility functions. Probabilities for the various outcomes were also assessed. The utilities and probabilities were combined to determine an expected utility for each of the airport development strategies. The results of the decision analysis indicated that a strategy of gradual shift of operations from Texcoco to Zumpango over the next 30 years had the highest expected utility.

Exercises 19.41 – 19.46

Learning the Mechanics

19.41 Graph each of the following utility functions* for money and classify each as the utility function of a risk-avoiding, risk-taking, or risk-neutral decision-maker.

a. $U(x) = 2x^3, \quad 0 \le x \le 100$
b. $U(x) = (1/100)x, \quad 0 \le x \le 100$
c. $U(x) = \log(x + 100), \quad 10 \le x \le 100$
d. $U(x) = -1 + .01x, \quad 100 \le x \le 200$

19.42 Consider the following utility function for money:

$$U(x) = (x + 1{,}000)^{1/2} \qquad -1{,}000 \le x \le 1{,}000$$

a. Plot the function.

*Notice that the minimum and maximum values of these utility functions are not 0 and 1, respectively. Refer to the second footnote on page 1029.

b. Is this the utility function of a risk-avoiding, risk-taking, or risk-neutral decision-maker?
c. Use this utility function and the expected utility criterion to identify the optimal action for the decision problem characterized by the payoff table shown here (probabilities in parentheses).

		State of Nature		
		S_1 (.1)	S_2 (.5)	S_3 (.4)
Action	a_1	−\$600	\$1,000	\$300
	a_2	−800	00	750
	a_3	60	−175	610

19.43 Rework all parts of Exercise 19.42 using the utility function for money,

$$U(x) = (x + 1{,}000)^2 \qquad -1{,}000 \le x \le 1{,}000$$

Is the action prescribed by the expected utility criterion the same as the action prescribed in Exercise 19.42? If not, explain the reason for this disagreement.

19.44 Consider the utility function for money,

$$U(x) = .001x \qquad 0 \le x \le 1{,}000$$

and the payoff table shown here (probabilities in parentheses).

		State of Nature			
		S_1 (.05)	S_2 (.30)	S_3 (.45)	S_4 (.20)
Action	a_1	0	50	310	900
	a_2	1,000	400	200	80
	a_3	100	500	1,000	100

a. What does this utility function indicate about the decision-maker's attitude toward risk?
b. Which action is prescribed by the expected utility criterion? Should it differ from the action prescribed by the expected payoff criterion? Explain.

Applying The Concepts

19.45 Refer to Exercise 19.38, in which an investor considers whether to invest in a wildcat oil well.
a. Plot the utility function you constructed for the investor.
b. Categorize her as a risk-avoider, risk-taker, or risk-neutral.

19.46 Refer to Exercise 19.40, in which the dress purchaser used the utility function

$$U(x) = \frac{\sqrt{x + 2{,}000}}{110} \qquad -2{,}000 \le x \le 10{,}000$$

a. Plot the purchaser's utility function.
b. Is she a risk-taker, risk-avoider, or risk-neutral?

19.8 Revising State of Nature Probabilities: Bayes' Rule

Up until now in this chapter, we have presented solutions to decision problems using only currently available information. Now, we incorporate **sample information** into the decision process.

Suppose you are trying to decide whether to purchase 100 shares of common stock in a company. You have decided that you want to make the purchase only if the probability is at least .75 that the stock's price will be higher a month from now. Based on your *prior information* about the company and stock market conditions, you assign a *prior probability* of only .6 to this event. Therefore, if you base your decision on your prior information, you will not buy the stock.

However, you may want to obtain sample information about the stock before making your decision. Your sample information might be an opinion from a reputable stock analyst whose predictions have the following reliability. Among the stocks that increase in price over a 1-month period, the analyst is able to predict the increase in 80% of the cases. But for stocks that will be stable or decrease in price, she predicts an increase in 40% of the cases. The analyst advises you that, in her opinion, the price of the stock in which you are interested will be higher 1 month from now. How should you revise your prior probability to incorporate this sample information?

Define the events

S_1: {Stock price will be higher 1 month from now}

S_2: {Stock price will be the same or lower 1 month from now}

I: {Analyst predicts the price will be higher 1 month from now}

We are interested in the conditional probability that the stock price will increase *given that* the analyst says it will increase—i.e., $P(S_1 \mid I)$. Recall that the definition of conditional probability stipulates

$$P(S_1 \mid I) = \frac{P(I \cap S_1)}{P(I)}$$

where $P(I \cap S_1)$ is the probability that the analyst says it will increase *and* it does in fact increase. We will find $P(S_1 \mid I)$ in two steps.

Step 1 Find $P(I \cap S_1)$. Recall that $P(I \cap S_1) = P(I \mid S_1)P(S_1)$. We know that $P(I \mid S_1)$—the probability that the analyst predicts a stock price will be higher in a month given that it will in fact be higher—is .8. Furthermore, $P(S_1)$ is the probability that the price will be higher a month from now, given no sample information; i.e., this is the prior probability of S_1, which we have assessed to be .6. Then

$$P(I \cap S_1) = P(I \mid S_1)P(S_1) = (.8)(.6) = .48$$

Step 2 Find $P(I)$. Note that $P(I)$ is the probability that the analyst predicts an increase in the stock's price. This will occur simultaneously with one of the two mutually exclusive, collectively exhaustive (i.e., no other states can occur) states of nature, S_1 and S_2. That is, either the stock will in fact cost more 1 month from now, or its price will be the same or lower. Thus,

$$P(I) = P(I \cap S_1) + P(I \cap S_2)$$

We have already found that $P(I \cap S_1) = .48$. In the same way,

$$P(I \cap S_2) = P(I \mid S_2)P(S_2)$$

We know that $P(I \mid S_2)$—the probability that the analyst predicts a stock will increase in price when in fact it will not—is .4. Also,

$$P(S_2) = 1 - P(S_1) = 1 - .6 = .4$$

Then

$$P(I \cap S_2) = P(I \mid S_2)P(S_2) = (.4)(.4) = .16$$

and

$$P(I) = P(I \cap S_1) + P(I \cap S_2) = .48 + .16 = .64$$

Finally, we combine steps 1 and 2 to find

$$P(S_1 \mid I) = \frac{P(I \cap S_1)}{P(I)} = \frac{.48}{.64} = .75$$

Thus, the probablity that the stock will increase in price, given the sample information that the analyst predicts its rise, is .75. We refer to probablities revised on the basis of sample information as *posterior probabilities.** Since this posterior probability meets our previously established criterion that the probability of increase be at least .75, we now decide to purchase the stock.

Definition 19.4

A probability $P(S_i)$ of the state of nature S_i that does not incorporate sample information is called a **prior probability** of S_i.

A probability $P(S_i \mid I)$ of the state of nature S_i, given the sample information I, is called a **posterior probability** of S_i.

*We use the term *posterior* because the probablities are determined *after* the sample information has been obtained, whereas *prior* probabilities are determined *before* sample information is obtained.

The process of revising prior probabilities to incorporate sample information is known as **Bayes' rule**. It is named for Thomas Bayes, an English Presbyterian minister and mathematician who lived from 1702 to 1761. He was one of the first to develop methods of calculating posterior probabilities. Bayes' rule for two states of nature and for k states of nature is given in the accompanying box. The states of nature are assumed to be mutually exclusive, collectively exhaustive events.

Bayes' Rule for Calculating Posterior Probabilities

Two states of nature: S_1 and S_2

$$P(S_1 \mid I) = \frac{P(I \cap S_1)}{P(I)}$$

$$= \frac{P(I \mid S_1)P(S_1)}{P(I \mid S_1)P(S_1) + P(I \mid S_2)P(S_2)}$$

where I is the sample information.

k states of nature: $S_1, S_2, \ldots, S_k$

$$P(S_i \mid I) = \frac{P(I \cap S_i)}{P(I)} = \frac{P(I \mid S_i)P(S_i)}{\sum_{j=1}^{k} P(I \mid S_j)P(S_j)}$$

A **probability revision table** organizes the calculation of posterior probabilities. In Table 19.21 we show the probability revision table for the stock price example, and the general form of the table is given in Table 19.22 (page 1044).

TABLE 19.21 Probability Revision Table for the Stock Price Example

(1)		(2)	(3)	(4)	(5)
State of Nature		Prior Probability	Conditional Probability of Sample Information	Probability of Intersection of State and Sample Information (2) × (3)	Posterior Probability (4) ÷ Total of (4)
S_1		.6	.8	.48	.75
S_2		.4	.4	.16	.25
	Total	1.0		.64	1.00

TABLE 19.22 Probability Revision Table (General)

(1) State of Nature	(2) Prior Probability	(3) Conditional Probability of Sample Information	(4) Probability of Intersection of State and Sample Information (2) × (3)	(5) Posterior Probability (4) ÷ Total of (4)
S_1	$P(S_1)$	$P(I \mid S_1)$	$P(I \cap S_1)$	$P(S_1 \mid I)$
S_2	$P(S_2)$	$P(I \mid S_2)$	$P(I \cap S_2)$	$P(S_2 \mid I)$
⋮	⋮	⋮	⋮	⋮
S_k	$P(S_k)$	$P(I \mid S_k)$	$P(I \cap S_k)$	$P(S_k \mid I)$
	Total 1		$P(I)$	1

In Example 19.5, we show how to apply Bayes' rule to a decision problem with monetary outcomes.

EXAMPLE 19.5

A company has developed a home smoke detector that is considerably more reliable than those currently on the market, but it is also more expensive to produce. To market the detector competitively, the company will have to price it so low that the profit margin per unit will be quite small. Accordingly, the sales volume, and therefore the market share, will have to be high for the product to be worth marketing.

The company's accounting and marketing departments prepared Table 19.23 using the objective variable Net contribution to profit by the detector during its first 2 years on the market. The marketing department assessed the prior state probabilities (shown in parentheses in the table) using industry sales data for other detectors.

TABLE 19.23 Payoff Table for Example 19.5

		State of Nature: Market Share by Year 3 (Prior Probabilities in Parentheses)			
		S_1: .01 (.10)	S_2: .05 (.40)	S_3: .10 (.40)	S_4: .15 (.10)
Action	a_1: Market Detector	−\$1,500,000	−\$200,000	\$300,000	\$1,000,000
	a_2: Do Not Market Detector	0	0	0	0

Now suppose the company wants to incorporate *sample information* into the decision analysis. Twenty prospective purchasers are randomly sampled and asked their opinion of the new detector. Two of the 20 say they will purchase the detector if it is marketed. Use this sample information to *revise* the prior state probabilities by calculating the *posterior probabilities* associated with the states of nature.

Solution

This decision problem has four states of nature, and the prior probabilities are known (see Table 19.23). The next step is to calculate the probability of the sample information given each state of nature (column 3 of the probability revision table). Note that the sample consists of a consumer preference survey and that we can regard its outcome as a binomial random variable (see Section 4.4). The number of trials is $n = 20$, and the probability, p, represents the true market share the company will obtain upon marketing the detector. Thus, each state of nature represents a different value of p. The binomial random variable, x, is the number of the sampled consumers who will purchase the new smoke detector, and the survey result is $x = 2$. We now calculate the conditional probabilities of this sample result (using Table II in Appendix B):

$$P(I \mid S_1) = P(x = 2 \mid p = .01) = \binom{20}{2}(.01)^2(.99)^{18} = .016$$

$$P(I \mid S_2) = P(x = 2 \mid p = .05) = \binom{20}{2}(.05)^2(.95)^{18} = .189$$

$$P(I \mid S_3) = P(x = 2 \mid p = .10) = \binom{20}{2}(.10)^2(.90)^{18} = .285$$

$$P(I \mid S_4) = P(x = 2 \mid p = .15) = \binom{20}{2}(.15)^2(.85)^{18} = .229$$

TABLE 19.24 Probability Revision Table for Example 19.5

(1)	(2)	(3)	(4)	(5)
State of Nature	Prior Probability	Conditional Probability of Sample Information	Probability of Intersection of State and Sample Information (2) × (3)	Posterior Probability (4) ÷ Total of (4)
S_1: $p = .01$	.10	.016	.0016	.0075
S_2: $p = .05$	.40	.189	.0756	.3531
S_3: $p = .10$	.40	.285	.1140	.5325
S_4: $p = .15$	.10	.229	.0229	.1070
Total	1.00		.2141	1.0001[a]

[a]Rounding error

We are now prepared to use the probability revision table to calculate the posterior probabilities for the states of nature. The calculations are shown in Table 19.24. Note that the sample information revised the probabilities of states S_3 and S_4 *upward* and the probabilities of states S_1 and S_2 *downward*. The posterior probabilities indicate that the company can be more optimistic about the market share of the smoke detector than it was prior to obtaining the sample information.

The extent to which the prior probabilities are revised depends on the amount of information contained in the sample. For example, if 100 consumers had been surveyed in Example 19.5 and if the same *proportion* had responded favorably to the new smoke detector (i.e., $x = 10$), the four posterior state probabilities would be 0, .104, .826, and .069, respectively. Thus, the prior probability of state S_3: $p = .10$ is revised from .4 to .826. These probabilities are shown in Table 19.25 along with the posterior probabilities for a survey of $n = 20$ consumers (Example 19.5). As the sample size is increased, the weight given to the prior probabilities is diminished and more importance is attached to the sample information.

TABLE 19.25 Prior and Posterior Probabilities for the Smoke Detector Example: Two Different Sample Sizes

State	Prior Probability	Posterior Probability	
		$n = 20$ $(x = 2)$	$n = 100$ $(x = 10)$
S_1: $p = .01$	.10	.008	.000
S_2: $p = .05$	.40	.353	.104
S_3: $p = .10$	.40	.532	.826
S_4: $p = .15$	.10	.107	.069

Exercises 19.47–19.56

Learning the Mechanics

19.47 Give Bayes' rule for the case in which there are three states of nature, S_1, S_2, and S_3, and information I has been received by sampling.

19.48 Use the formula obtained in Exercise 19.47 to complete the following probability revision table.

(1) State of Nature	(2) Prior Probability	(3) Conditional Probability of Sample Information	(4) (2) × (3)	(5) Posterior Probability (4) ÷ Total of (4)
S_1	.30	.60		
S_2	.50	.80		
S_3	.20	.10		
Total	1.00			

19.49 Complete the following probability revision table.

(1) State of Nature	(2) Prior Probability	(3) Conditional Probability of Sample Information	(4) (2) × (3)	(5) Posterior Probability (4) ÷ Total of (4)
S_1	.10	.90		
S_2	.25	.10		
S_3	.45	.65		
S_4	.20	.30		
Total	1.00			

19.50 What role does sampling play in decision analysis?

19.51 Compare and contrast prior and posterior information.

Applying the Concepts

19.52 A systems analyst is concerned about the proportion, p, of records that are in error in the inventory control system he designed. The probability distribution shown characterizes his beliefs regarding p.

Proportion of Errors	*Prior Probability*
.09	.40
.10	.25
.11	.15
.12	.10
.13	.10

a. Find the mean and variance of the analyst's prior probability distribution for p.
b. A random sample of 25 records yielded three incorrect records. Find the analyst's posterior probability distribution for p.
c. Find the mean and variance of the analyst's posterior probability distribution for p.
d. The posterior probability distribution you computed in part **b** combines the analyst's beliefs about p *and* the information about p contained in the sample. Using your results from parts **a** and **c** and graphs of the prior and posterior probability distributions, describe how the posterior distribution differs from the prior distribution. In so doing, you are explaining what the analyst learned about p by sampling.

19.53 A press produces masks used in the manufacture of television tubes. If the press is correctly adjusted, it produces masks with a scrap rate of 5%. If it is not adjusted correctly, it produces scrap at a 50% rate. From past company records, the machine is known to be correctly adjusted 90% of the time. A quality control inspector randomly selects one mask from those recently produced by the press and discovers it is defective. What is the probability that the machine is incorrectly adjusted?

19.54 Refer to Exercise 19.53. The quality control inspector observes a second mask and finds that it is also defective. Using the posterior probabilities from Exercise 19.52 as prior probabilities, calculate the posterior probability that the machine is incorrectly adjusted. What assumptions are required to make this calculation?

19.55 Suppose the proportion of defectives produced by a certain production process is .01, .05, or .10. The prior probability that it is .05 is twice the probability that it is .10. The prior probability that it is .01 is equal to the probability that it is .10.

a. Determine the prior probabilities associated with the proportion of defectives produced by the production process.
b. A sample of five units drawn at random yields three defectives. Revise the prior distribution you obtained in part a.

19.56 [*Note: This exercise refers to optional Section 4.5.*] A replacement parts inventory manager for a computer manufacturer believes the demand for a certain component is distributed as a Poisson random variable with a mean monthly demand of two, three, or four units. Experience suggests that a mean monthly demand for three units is twice as likely as a mean monthly demand of two units, and that a mean monthly demand of two units is as likely as a mean monthly demand of four units.
a. If the demand for this component last month was four units, find the posterior distribution for the mean monthly demand for next month.
b. If the demand in the following month was two units, find the posterior distribution for the mean monthly demand.
c. Comment on the ability of Bayes' rule to help a decision-maker "learn" about the shifting values of a parameter over time.

19.9 Solving Decision Problems Using Posterior Probabilities

Recall that the expected payoff criterion (Section 19.5) selects the action with the highest expected payoff. The expected payoff for each action is computed using either the *prior probabilities* or the *posterior probabilities* of the states of nature, depending on whether sample information is available. For example, for the smoke detector decision problem (Example 19.5), the payoff table prior to obtaining sample information is given in Table 19.26.

TABLE 19.26 Payoff Table for the Smoke Detector Decision Problem

		State of Nature (Prior Probabilities in Parentheses)			
		S_1: $p = .01$ (.10)	S_2: $p = .05$ (.40)	S_3: $p = .10$ (.40)	S_4: $p = .15$ (.10)
Action	a_1: Market Detector	−\$1,500,000	−\$200,000	\$300,000	\$1,000,000
	a_2: Do Not Market Detector	0	0	0	0

The expected payoffs for the two actions are

$$
\begin{aligned}
EP(a_1) &= \sum_{\text{All states}} (\text{Payoff})(\text{Prior probability of state}) \\
&= (-1{,}500{,}000)(.10) + (-200{,}000)(.40) + (300{,}000)(.40) + (1{,}000{,}000)(.10) \\
&= -\$10{,}000 \\
EP(a_2) &= 0(.10) + 0(.40) + 0(.40) + 0(.10) = 0
\end{aligned}
$$

Thus, the expected payoff criterion, using prior probabilities, selects action a_2 and indicates that the firm should not market the new smoke detector.

In Example 19.5 we revised these prior probabilities based on the sample information that 2 out of 20 randomly sampled consumers would purchase the new detector if it were marketed. The payoff table with the posterior state probabilities is shown in Table 19.27.

TABLE 19.27 Payoff Table for the Smoke Detector Decision Problem

		State of Nature (Posterior Probabilities in Parentheses)			
		S_1: $p = .01$ (.008)	S_2: $p = .05$ (.353)	S_3: $p = .10$ (.532)	S_4: $p = .15$ (.107)
Action	a_1: Market Detector	−$1,500,000	−$200,000	$300,000	$1,000,000
	a_2: Do Not Market Detector	0	0	0	0

The expected payoffs for the actions using the posterior probabilities are

$$\begin{aligned} EP(a_1) &= \sum_{\text{All states}} (\text{Payoff})(\text{Posterior probability of state}) \\ &= (-1,500,000)(.008) + (-200,000)(.353) + (300,000)(.532) + (1,000,000)(.107) \\ &= +\$184,000 \\ EP(a_2) &= 0 \end{aligned}$$

Thus, the expected payoff criterion using posterior probabilities selects action a_1, indicating that the company should market the new detector. The sample information has altered the selection of the expected payoff criterion, changing it from a_2: Do not market to a_1: Market. Since the **posterior decision analysis** incorporates both prior and sample information, whereas the **prior decision analysis** incorporates only prior information, *we prefer the posterior decision to the prior decision.*

The expected payoff criterion using posterior probabilities is summarized in the box.

Expected Payoff Criterion Using Posterior Probabilities

Choose the action with the maximum expected payoff, where the expected payoffs are calculated using the posterior state probabilities.

$$EP(a_i) = \sum_{\text{All states}} \begin{pmatrix} \text{Payoff for} \\ \text{action } a_i\text{/state} \\ \text{combination} \end{pmatrix} \begin{pmatrix} \text{Posterior} \\ \text{probability} \\ \text{of state} \end{pmatrix}$$

EXAMPLE 19.6

Suppose the company that is considering marketing the new smoke detector randomly samples 100 consumers and finds that 10 would purchase the new detector if it were marketed. The payoff table with the posterior probabilities is shown in Table 19.28. Which action should the company take if the expected payoff criterion is used?

TABLE 19.28 Payoff Table for the Smoke Detector Example

		State of Nature (Posterior Probabilities in Parentheses)			
		S_1: $p = .01$ (.000)	S_2: $p = .05$ (.104)	S_3: $p = .10$ (.826)	S_4: $p = .15$ (.069)
Action	a_1: Market Detector	−\$1,500,000	−\$200,000	\$300,000	\$1,000,000
	a_2: Do Not Market Detector	0	0	0	0

Solution

We calculate the expected payoffs using the posterior probabilities. Thus,

$$\begin{aligned} EP(a_1) &= \sum_{\text{All states}} \begin{pmatrix} \text{Payoff for} \\ \text{action } a_1\text{/state} \\ \text{combination} \end{pmatrix} \begin{pmatrix} \text{Posterior} \\ \text{probability} \\ \text{of state} \end{pmatrix} \\ &= (-1{,}500{,}000)(.000) + (-200{,}000)(.104) + (300{,}000)(.826) + (1{,}000{,}000)(.069) \\ &= \$296{,}000 \\ EP(a_2) &= 0(.000) + 0(.104) + 0(.826) + 0(.069) = 0 \end{aligned}$$

The expected payoff criterion selects action a_1. This selection criterion incorporates both prior information and the information contained in the sample of 100 consumers.

The expected utility criterion presented in Section 19.6 may also be extended to incorporate sample information. Simply replace the prior probabilities by the posterior probabilities. This criterion will then prescribe an action that is consistent with the decision-maker's preferences for outcomes and risk attitude, as well as with the sample information. Thus, no matter which probabilistic criterion you use—expected payoff or expected utility—the incorporation of sample information requires only the substitution of posterior probabilities for prior probabilities.

Exercises 19.57–19.61

Learning the Mechanics

19.57 Use the posterior probabilities of Exercise 19.49 to select the action with the highest expected payoff in the following payoff table (outcomes are in thousands of dollars).

		State of Nature			
		S_1	S_2	S_3	S_4
	a_1	30	20	0	15
Action	a_2	40	20	5	0
	a_3	9	15	−21	25

19.58 Consider the payoff table given here (with prior probabilites in parentheses).

		State of Nature		
		S_1 (.4)	S_2 (.1)	S_3 (.5)
	a_1	9	25	−14
Action	a_2	−16	18	15
	a_3	25	12	9
	a_4	6	30	−20

a. Use the expected payoff criterion to choose an action.
b. Sampling information I has been obtained that suggests that S_1 is the true state of nature. The reliability of this information is reflected in the following probabilities:

$$P(I \mid S_1) = .80 \qquad P(I \mid S_2) = .30 \qquad P(I \mid S_3) = .10$$

Find the posterior state probabilities.
c. Compute the expected payoff for each action using the posterior probabilities.
d. Use your results of part **c** to determine which action is prescribed by the expected payoff criterion. Compare this action with the action selected in part **a** using prior probabilities.

19.59 Consider the accompanying payoff table.

		State of Nature		
		S_1 (.5)	S_2 (.3)	S_3 (.2)
	a_1	−150	225	120
Action	a_2	200	−172	100
	a_3	300	80	−100

a. Use the expected payoff criterion to choose an action.

b. Sampling information I has been obtained that suggests that S_2 is the true state of nature. The reliability of this information is reflected by the following probabilities:

$$P(I \mid S_1) = .20$$
$$P(I \mid S_2) = .60$$
$$P(I \mid S_3) = .20$$

Find the posterior state probabilities.

c. Given the sample information I in part **b**, which action does the expected payoff criterion imply?

Applying the Concepts

19.60 A large hospital is considering purchasing 100 new color television sets under one of two different purchase agreements. Under one agreement the TV sets would cost $460 each and all sets that are seriously defective would be replaced at no cost. Under the other agreement, the TV sets would cost $400 each and any seriously defective sets would have to be replaced by the hospital at $400 each. (Assume all replacement sets are nondefective.) The hospital's purchasing agent believes the probabilities shown in the table appropriately characterize the proportion of seriously defective sets in a shipment of 100 sets from the manufacturer.

Proportion Defective	.00	.05	.10	.15	.20
Probability	.4	.3	.1	.1	.1

a. Construct the payoff table for this decision problem. [*Note*: Costs represent negative payoffs.]
b. By the expected payoff criterion, which purchase agreement should the hospital choose?
c. Suppose the hospital was able to randomly sample one TV set from the incoming shipment of 100 sets before deciding which purchase agreement to choose. According to the expected payoff criterion, which agreement should it select if the sampled set was defective? Nondefective?

19.61 A small company that produces auto parts for American-made cars expects to lose $1.5 million next year unless Congress passes a bill to limit the number of foreign-made cars that can be imported into the United States. If the bill passes, the company expects to make a profit of $2.5 million next year. The company's lobbyists in Washington have reported that the probability of the bill passing this year is .6. The auto parts company must decide whether to stay in business and risk a huge loss next year or to lease its facility for a year to a Japanese firm for $500,000. The company has just received new information from "a source close to the White House" that indicates Congress will pass the bill. The company assesses the reliability of the source as follows:

$$P(\text{Source says bill will pass} \mid \text{Bill will pass}) = .8$$
$$P(\text{Source says bill will pass} \mid \text{Bill will not pass}) = .1$$

a. Construct the payoff table for the company's decision problem. Include the company's prior state probabilities.
b. Compute the posterior state probabilities.
c. Describe the effect of the sample information on the company's prior beliefs regarding the passage of the bill by comparing the company's prior and posterior state probabilities.
d. Given the sample information, which action is prescribed by the expected payoff criterion?

19.10 The Expected Value of Sample Information: Preposterior Analysis (Optional)

We now want to make a realistic assessment of the expected worth of sample information. Of course, for the information to be of any use to us in deciding whether to obtain sample information, we must make the assessment *before* the sample is taken. Accordingly, the analysis is referred to as **preposterior analysis**.

For example, consider the smoke detector example of the previous sections. We repeat the payoff table with the prior state probabilities in Table 19.29. The expected payoff criterion using the prior probabilities yields $EP(a_1) = \$10,000$ and $EP(a_2) = 0$, so we would not market the detector (action a_2) based on the prior information.

TABLE 19.29 Payoff Table for the Smoke Detector Example

		State of Nature (Prior Probabilities in Parentheses)			
		S_1: $p = .01$ (.10)	S_2: $p = .05$ (.40)	S_3: $p = .10$ (.40)	S_4: $p = .15$ (.10)
Action	a_1: Market Detector	−\$1,500,000	−\$200,000	\$300,000	\$1,000,000
	a_2: Do Not Market Detector	0	0	0	0

Now suppose the company is considering taking a sample of two consumers to obtain their opinions about the detector. (We choose a small sample size to reduce the computational difficulty.) The value of x, the number of the sampled consumers who state they will purchase the detector, will be either 0, 1, or 2. *Preposterior analysis involves examining the decision problem for each possible sample outcome.* Thus, in Table 19.30 (see page 1054) we use a probability revision table to derive the posterior probabilities for each of the three possible sample outcomes. The first two columns show the states and prior probabilities, respectively. Column 3 presents the conditional probability of observing the sample outcome given a particular state of nature. For the smoke detector example, these are binomial probabilities with $n = 2$ and a value of p corresponding to the market share of each state. Thus, for the first two entries in column 3 under the heading $x = 0$, we find

$$P(x = 0 \mid p = .01) = \binom{2}{x} p^x (1 - p)^{2-x} = \binom{2}{0}(.01)^0(.99)^2 = (.99)^2 = .9801$$

$$P(x = 0 \mid p = .05) = \binom{2}{0}(.05)^0(.95)^2 = (.95)^2 = .9025$$

and so on. In column 4 we find the probability of the intersection of the states and the sample outcome using the formula

$$P(x \cap S_i) = P(S_i)P(x \mid S_i)$$

TABLE 19.30 Probability Revision Table for Each Sample Outcome in the Smoke Detector Example

	(1) State	(2) Prior Probability	(3) Conditional Probability of Sample Outcome Given State	(4) Intersection of States and Sample Outcome (2) × (3)	(5) Posterior Probability (4) ÷ Total of (4)
$x = 0$	S_1: $p = .01$	.10	.9801	.0980	.1146
	S_2: $p = .05$	.40	.9025	.3610	.4221
	S_3: $p = .10$	.40	.8100	.3240	.3789
	S_4: $p = .15$	.10	.7225	.0722	.0844
	Total	1.00		.8552	1.0000
$x = 1$	S_1: $p = .01$	.10	.0198	.0020	.0145
	S_2: $p = .05$	.40	.0950	.0380	.2764
	S_3: $p = .10$	.40	.1800	.0720	.5236
	S_4: $p = .15$	.10	.2550	.0255	.1855
	Total	1.00		.1375	1.0000
$x = 2$	S_1: $p = .01$	.10	.0001	.0000	.0000
	S_2: $p = .05$	.40	.0025	.0010	.1389
	S_3: $p = .10$	.40	.0100	.0040	.5556
	S_4: $p = .15$	.10	.0225	.0022	.3056
	Total	1.00		.0072	1.0001[a]

[a]Rounding error

which is the product of the respective entries in columns 2 and 3. Finally, we find the revised state probabilities—the posterior probabilities—in column 5 by applying Bayes' rule:

$$P(S_i \mid x) = \frac{P(x \cap S_i)}{\sum_{\text{All states}} P(x \cap S_j)} = \frac{P(S_i)P(x \mid S_i)}{\sum_{\text{All states}} P(S_j)P(x \mid S_j)}$$

Then each element in column 5 is equal to the corresponding element in column 4 divided by the column 4 total.

We now determine which action is dictated by each posterior distribution. That is, if none of the two sampled consumers will buy the detector (i.e., $x = 0$), which action should be selected? We use the expected payoff criterion with the posterior probabilities corresponding to $x = 0$:

$$\begin{aligned} EP(a_1 \mid x = 0) &= \sum_{\text{All states}} (\text{Payoff})(\text{Posterior probability of state for } x = 0) \\ &= (-1{,}500{,}000)(.1146) + (-200{,}000)(.4221) + (300{,}000)(.3789) + (1{,}000{,}000)(.0844) \\ &= -\$58{,}250 \\ EP(a_2 \mid x = 0) &= 0(.1146) + 0(.4221) + 0(.3789) + 0(.0844) = 0 \end{aligned}$$

Thus, the expected payoff criterion selects action a_2: Do not market the detector when $x = 0$ is the sample outcome.

Now suppose we observe $x = 1$ when the sample information is collected:

$$EP(a_1 \mid x = 1) = \sum_{\text{All states}} (\text{Payoff})(\text{Posterior probability of state for } x = 1)$$

$$= (-1{,}500{,}000)(.0145) + (-200{,}000)(.2764) + (300{,}000)(.5236) + (1{,}000{,}000)(.1855)$$

$$= \$265{,}550$$

$$EP(a_2 \mid x = 1) = 0$$

So we choose a_1: Market the detector if $x = 1$.

Finally.

$$EP(a_1 \mid x = 2) = (-1{,}500{,}000)(.0000) + (-200{,}000)(.1389) + (300{,}000)(.5556) + (1{,}000{,}000)(.3056)$$

$$= \$444{,}500$$

$$EP(a_2 \mid x = 2) = 0$$

Therefore, the expected payoff criterion selects action a_1: Market the detector when $x = 2$.

TABLE 19.31 Actions Selected and Expected Payoffs for the Smoke Detector Example

Sample Outcome	Action Selected	Expected Payoff	Marginal Probability of Sample Outcome
$x = 0$	a_2	0	.8552
$x = 1$	a_1	$265,550	.1375
$x = 2$	a_1	$444,500	.0072
			.9999[a] ≈ 1.0

[a]Rounding error

Table 19.31 summarizes our work to this point. Note that we included the probability of observing each of the sample outcomes in the last column of Table 19.31. These probabilities are the sums of the column 4 entries for each sample outcome in the probability revision table (Table 19.30). Thus,

$$P(x = 0) = \sum_{\text{All states}} P[(x = 0) \cap S_i]$$

$$= .0980 + .3610 + .3240 + .0722 = .8552$$

$P(x = 1)$ and $P(x = 2)$ can be computed in a similar manner. They form the **marginal** (or **predictive**) **probability distribution** for the random variable x. Note that the marginal probabilities are not dependent on the state, and they will sum to 1 if no rounding errors are present.

Now recall our ultimate objective. We are trying to determine how much is to be gained from sampling. Since we now know the expected payoff and the marginal probability of each sample outcome (Table 19.31), we can calculate the *expected payoff of sampling* (EPS):

$$\text{EPS} = \sum_x (EP \mid x)p(x) = \sum_{\substack{\text{All sample}\\ \text{outcomes}}} \begin{pmatrix}\text{Maximum expected}\\ \text{payoff for}\\ \text{sample outcome}\end{pmatrix}\begin{pmatrix}\text{Marginal}\\ \text{probability of}\\ \text{sample outcome}\end{pmatrix}$$

$$= (0)(.8552) + (265{,}550)(.1375) + (444{,}500)(.0072) = \$39{,}713.53$$

or \$39,700, rounding to the nearest hundred dollars. Thus, the mean payoff is \$39,700 when $n = 2$ consumers are sampled.

To determine the expected gain attributed to sampling, we compare the EPS to the *expected payoff for no sampling* (EPNS). But EPNS is the expected payoff of the action selected by the expected payoff criterion using prior probabilities. We showed in previous sections that action a_2 is chosen using prior decision analysis, so that

$$\text{EPNS} = EP(a_2) = 0$$

Finally, the *expected value of sample information* (EVSI) is the expected payoff of sampling less the expected payoff with no sampling; i.e.,

$$\text{EVSI} = \text{EPS} - \text{EPNS} = \$39{,}700 - \$0 = \$39{,}700$$

This figure represents the mean dollar amount the company will gain from sampling and therefore provides a figure it will be reluctant to exceed in the cost of obtaining the sample information.

The steps for computing the expected value of sample information are summarized in the box. The calculation of EVSI becomes very tedious when the number of possible sample outcomes is large. However, because the method is the same no matter how many possible outcomes exist, computers can be useful aids in determining EVSI.

Computing the Expected Value of Sample Information (EVSI)

Step 1 Obtain the posterior state probabilities for each possible sample outcome using a probability revision table.

Step 2 Use the expected payoff criterion with the posterior probabilities to determine the action with the maximum expected payoff *for each sample outcome*.

Step 3 Find the marginal probability distribution for the sample outcomes by using

$$P(x) = \sum_{\text{All states}} P(x \cap S_i)$$

For each sample outcome, the marginal probability will be the sum of column 4 in the probability revision table (step 1).

Step 4 Find the expected payoff of sampling (EPS) by combining the results of steps 2 and 3:

$$\text{EPS} = \sum_{\substack{\text{All sample}\\ \text{outcomes}}} \begin{pmatrix}\text{Maximum}\\ \text{expected payoff}\\ \text{for sample outcome}\end{pmatrix}\begin{pmatrix}\text{Marginal}\\ \text{probability of}\\ \text{sample outcome}\end{pmatrix}$$

Step 5 Calculate the EVSI by

$$\text{EVSI} = \text{EPS} - \text{EPNS}$$

where EPNS is the expected payoff of no sampling, computed using the prior probabilities.

After the EVSI has been determined, it should be compared with the *cost of sampling* (CS) by computing the *expected net gain for sampling* (ENGS). For example, if the smoke detector company determines that a total cost of \$500 will be incurred during the sampling of two consumers, the expected net gain of sampling is

$$\text{ENGS} = \text{EVSI} - \text{CS} = \$39{,}700 - \$500 = \$39{,}200$$

As long as the net gain is positive, the company can expect to gain by obtaining the sample information. We thus have the preposterior expected gain decision rule given in the next box.

Preposterior Expected Gain Decision Rule

If the expected net gain of sampling (ENGS) is greater than 0, the decision analyst should obtain the sample information before making a decision.

The ENGS calculation formula is ENGS = EVSI − CS, where CS is the cost of sampling.

Decision trees (Section 19.2) are often used to conduct a preposterior analysis or to summarize the results of a preposterior analysis. The decision tree for the smoke detector example is shown in Figure 19.8 on page 1058. The first decision fork (always reading from left to right) represents the decision of whether to sample. Along the upper branch (Do not sample) we place a decision fork representing the two possible actions of the decision problem. Beside this fork we write the expected payoff of no sampling (EPNS), which is 0 for this example. At the ends of the action branches, we place chance forks and branch into the four states of nature. Beside each chance fork we write the expected payoff for the associated action, where the expected payoffs are

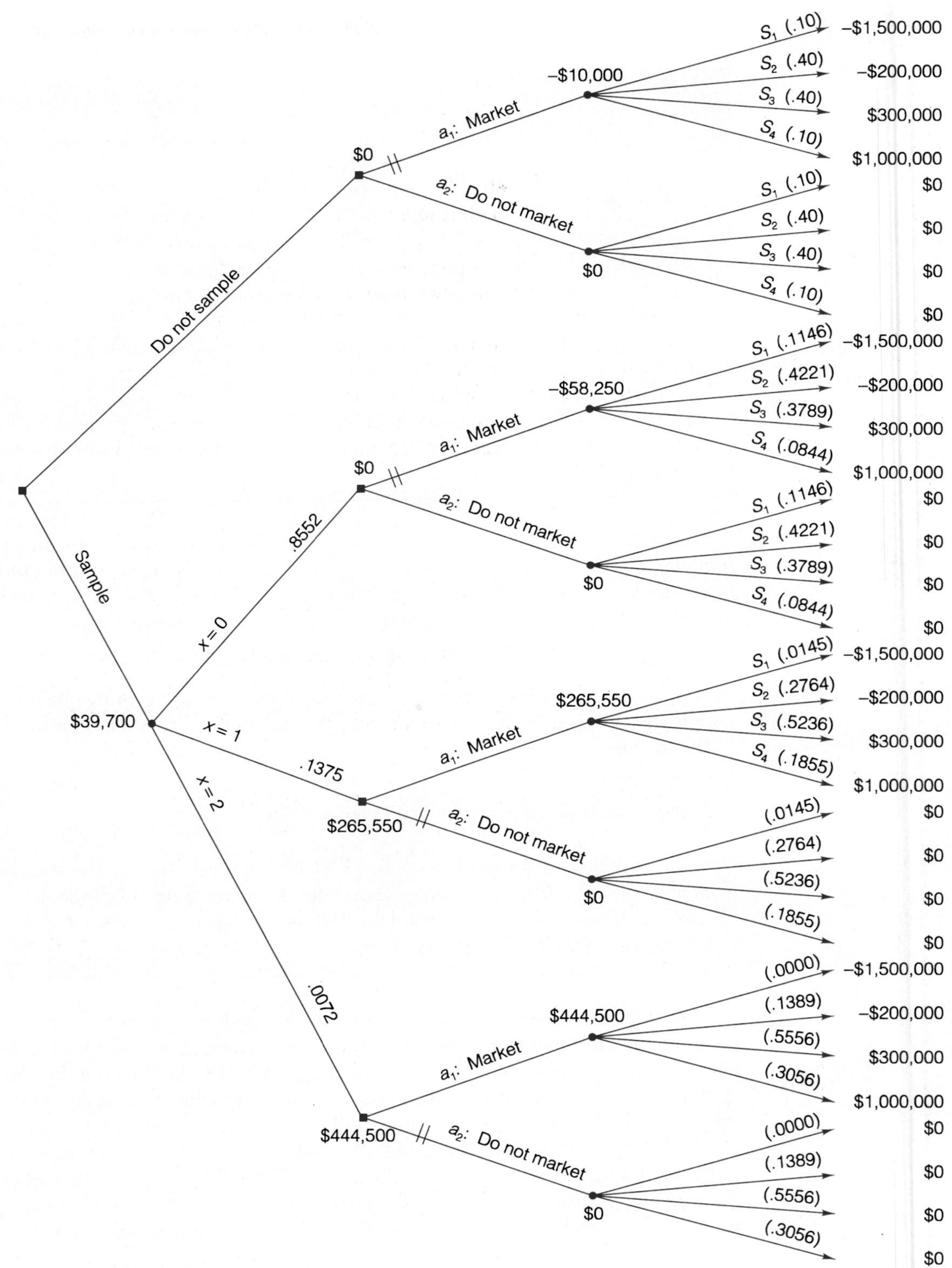

FIGURE 19.8 ▲ Decision tree for the smoke detector example

computed using the prior probabilities. The expected payoff is $-\$10,000$ for action a_1 and \$0 for action a_2 in our example. On the state branches, we write the prior probabilities, and at the ends of the branches are the outcomes corresponding to the action/state combinations.

The lower branch of the decision tree in Figure 19.8, the Sample branch, has an extra chance fork corresponding to the possible sample outcomes. The expected payoff of sampling (EPS), which is \$39,700 for this example, is written beside the sampling chance fork. The marginal probability for each sample outcome is written on the appropriate sample outcome branch, and the expected payoff for the optimal action corresponding to each sample outcome is recorded beside the decision fork at the end of each sample outcome branch. The remainder of the branching is into actions and then into states, just as for the Do not sample branch. The only difference is that posterior rather than prior probabilities are recorded on each state of nature branch. Note that two vertical lines are used to block the branches corresponding to actions with expected payoffs that are not maximum. This reminds us which action should be chosen at each decision fork.*

EXAMPLE 19.7

In Section 19.2, we introduced the decision problem facing the promoter of a rock concert. The payoff table characterizing this decision problem is reproduced in Table 19.32.

The promoter can purchase a long-range weather forecast for the night of the concert from an expert meteorologist for \$15,000. The meteorologist's track record in terms of the percentage of times her past predictions have or have not been accurate is shown in Table 19.33, page 1060.

TABLE 19.32 Payoff Table for the Rock Concert Example

		State of Nature (Prior Probabilities in Parentheses)	
		Rain, S_1(⅓)	No Rain, S_2(⅔)
Action	a_1: Rent Stadium	−\$40,000	\$350,000
	a_2: Rent Civic Center	\$150,000	\$150,000

*The computations for EVSI are often performed directly from a decision tree such as Figure 19.8. Both EPNS and EPS can be determined by starting at the far right-hand side of the tree and taking expectations backward through the tree.

TABLE 19.33 Long-Range Prediction Record for Meteorologist

		Actual Weather	
		Rain	No Rain
Meteorologist's Prediction	Rain	85%	30%
	No Rain	15%	70%

The meteorologist has correctly predicted 85% of the rainy days and 70% of the days without rain. Should the promoter purchase this sample information (the meteorologist's opinion) or make a decision concerning which facility to rent utilizing just the prior information?

a. Calculate the EVSI.

b. Calculate the ENGS and make the decision about whether to obtain the meteorologist's prediction.

c. Summarize the results of this preposterior analysis using a decision tree.

Solution

a. We follow the five-step approach for finding the EVSI.

Step 1 The first step is to construct a probability revision table to obtain the posterior probability distribution corresponding to each sample outcome. The two possible sample outcomes in this decision problem are that the meteorologist will predict no rain and that she will predict rain. The probability revision tables corresponding to these outcomes are shown in Table 19.34. Thus, the posterior probability of rain is .586 if the meteorologist predicts rain, but it is only .097 if the meteorologist predicts no rain.

Step 2 We now use the expected payoff criterion and the posterior probabilities to determine the preferred action for each sample outcome.

$$\begin{aligned} EP(a_1 \mid \text{Predicts rain}) &= (-40{,}000)(.586) + (350{,}000)(.414) \\ &= \$121{,}460 \\ EP(a_2 \mid \text{Predicts rain}) &= (150{,}000)(.586) + (150{,}000)(.414) \\ &= \$150{,}000 \\ EP(a_1 \mid \text{Predicts no rain}) &= (-40{,}000)(.097) + (350{,}000)(.903) \\ &= \$312{,}170 \\ EP(a_2 \mid \text{Predicts no rain}) &= (150{,}000)(.097) + (150{,}000)(.903) \\ &= \$150{,}000 \end{aligned}$$

Thus, if the meteorologist predicts rain, the expected payoff criterion selects action a_2: Rent Civic Center. But if the meteorologist predicts no rain, the criterion selects a_1: Rent stadium.

TABLE 19.34 Probability Revision Tables for the Rock Concert Example

	(1) State	(2) Prior Probability	(3) Conditional Probability of Sample Outcome Given State	(4) Intersection of Sample Outcome and State (2) × (3)	(5) Posterior Probability (4) ÷ Total of (4)
Predicts Rain	S_1: Rain	.333	.85	.283	.586
	S_2: No rain	.667	.30	.200	.414
	Total	1.000		.483	1.000
Predicts No Rain	S_1: Rain	.333	.15	.050	.097
	S_2: No rain	.667	.70	.466	.903
	Total	1.000		.516	1.000

Step 3 We now find the marginal probabilities for each sample outcome by summing the probabilities of the combinations of sample outcome and state over all states. These are the column 4 sums in the probability revision table (Table 19.34). We find

$$P(\text{Predicts rain}) = .483 \qquad P(\text{Predicts no rain}) = .516$$

Step 4 We summarize the results of steps 2 and 3 in Table 19.35. We now want to obtain the expected payoff of sampling:

$$\begin{aligned} \text{EPS} &= \sum_{\substack{\text{All sample}\\ \text{outcomes}}} \begin{pmatrix}\text{Maximum expected payoff}\\ \text{for sample outcome}\end{pmatrix}\begin{pmatrix}\text{Marginal probability}\\ \text{of sample outcome}\end{pmatrix} \\ &= (150{,}000)(.483) + (312{,}170)(.516) \\ &= \$233{,}530 \text{ (rounding to the nearest dollar)} \end{aligned}$$

Step 5 Finally, the EVSI is the difference between the EPS and the EPNS. Using the prior probabilities from Table 19.32, we find

$$EP(a_1) = (-40{,}000)(1/3) + (350{,}000)(2/3) = \$220{,}000$$

$$EP(a_2) = (150{,}000)(1/3) + (150{,}000)(2/3) = \$150{,}000$$

TABLE 19.35 Summary for the Rock Concert Example

Sample Outcome	Action Selected	Expected Payoff	Marginal Probability of Sample Outcome
Predicts rain	a_2	\$150,000	.483
Predicts no rain	a_1	\$312,170	.516

so that based on prior information we select action a_1, and the EPNS is \$220,000. Thus,

$$\begin{aligned} \text{EVSI} &= \text{EPS} - \text{EPNS} \\ &= \$233{,}530 - \$220{,}000 = \$13{,}530 \end{aligned}$$

The expected gain from sampling is \$13,530.

b. We now want to decide whether the promoter should hire the meteorologist. The ENGS = EVSI − CS and the meteorologist charges \$15,000 for her prediction, so

$$\text{ENGS} = \$13{,}530 - \$15{,}000 = -\$1{,}470$$

Since the ENGS is negative, the promoter should not pay for the sample information.

c. The decision tree summarizing the results of the preposterior analysis is shown in Figure 19.9. Note that the upper branch represents the Do not sample decision. The action and state branches emanate from the Do not sample fork. The expected payoff of no sampling, the expected payoffs for each action, the prior probabilities, and finally, the payoffs are also shown. The fact that a_1 is the preferred action is indicated by the vertical lines through the a_2 branch.

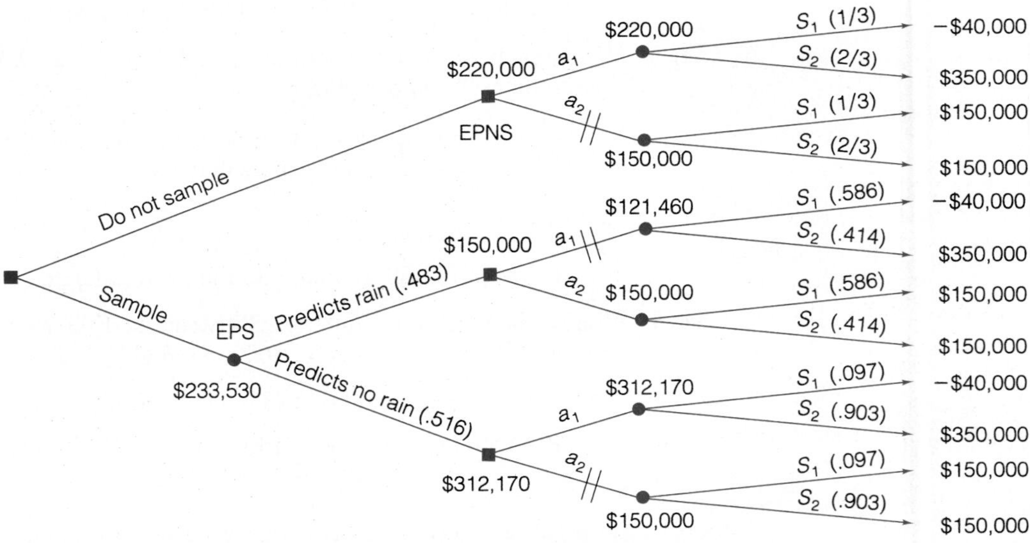

FIGURE 19.9 ▶ **Decision tree for the rock concert example**

In the lower (Sample) branch, the first branching consists of the two possible sample outcomes, Predicts rain and Predicts no rain. The action and state branches emanate from the sample outcome branches. The expected payoff of sampling, the maximum expected payoff for each sample outcome, the expected payoff for each action, the posterior probabilities, and the actual payoffs are also recorded. We have again drawn vertical lines through action branches with expected payoffs that are exceeded by other actions emanating from the same decision fork.

In the context of decision analysis, the term *sampling* means any procedure or process for gathering information. This includes statistical sampling, such as random sampling, as well as less technical methods, such as obtaining an expert opinion, as in the preceding example. If we are interested in random sampling, the *optimal sample size* can be determined by conducting a preposterior analysis for each potential sample size and choosing the sample size with the maximum ENGS. If this optimal sample size is 0 (or negative), the decision should be based on presently available information. In the former case, the decision-maker should select a sample of the optimal size and then repeat the analysis to determine whether further sampling is expected to yield a positive net gain.

Considerable effort is required to conduct a preposterior analysis when the sample size is moderate or large. Because the methodology remains the same for any sample size, computers are typically used to conduct preposterior analyses.

CASE STUDY 19.4 / An Example of the Benefits of Additional Information

In addition to studying whether hurricanes should be seeded (see Case Study 19.2), Howard, Matheson, and North (1972) addressed the question of whether more evidence on the effects of hurricane seeding should be gathered before the government reverses its policy against seeding. Would the expected loss from a representative hurricane be reduced by carrying out an additional seeding experiment?

First, the authors looked at the cost and possible outcomes of a seeding experiment. The cost to seed a hurricane and observe the results is $250,000. The possible outcomes are the same as those given in Case Study 19.2 for a seeded hurricane; the prior probabilities for these outcomes are also given in Case Study 19.2. Using the probabilities of each experimental outcome conditioned on each of the states of nature, the prior probabilities of the outcomes of seeding can be revised using Bayes' rule. The resulting posterior probabilities can then be used to evaluate the *expected loss* (EL) for seeding and not seeding. The decision to experiment or not can be made by comparing the expected loss of the "best" strategy *with* an experiment to the expected loss of the "best" strategy *without* an experiment. A condensed decision tree is shown in Figure 19.10 on page 1064.

The tree indicates that the expected loss with the experiment is $2.83 million lower than without the experiment. Since the net gain from the experiment is greater than the cost ($.25 million), the experiment should be conducted. As a result of their analysis, Howard, Matheson, and North recommended that further experiments with hurricane seeding be conducted prior to altering government policy against seeding hurricanes.

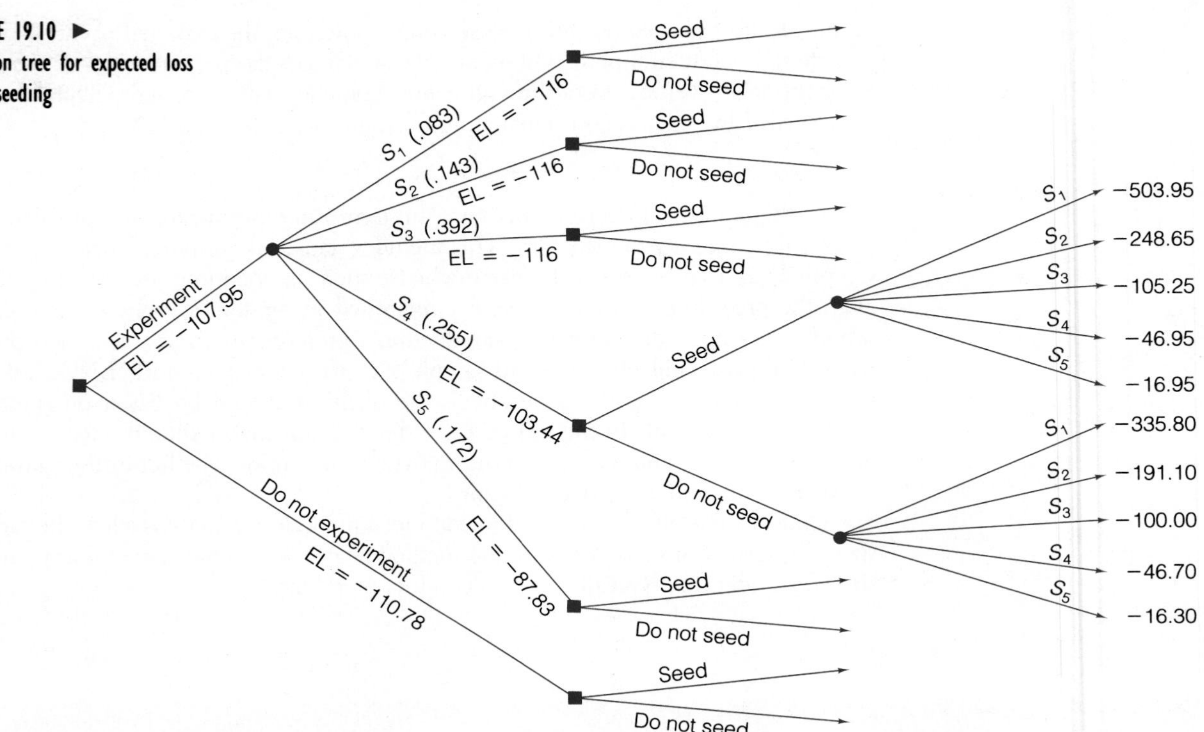

FIGURE 19.10 ▶ Decision tree for expected loss from seeding

Exercises 19.62–19.67

Learning the Mechanics

19.62 Consider the following payoff table (prior probabilities are shown in parentheses).

		State of Nature	
		S_1 (.40)	S_2 (.60)
Action	a_1	\$600	−\$100
	a_2	−\$200	\$300

The decision-maker must decide whether to purchase sample information about the true state of nature prior to choosing an action. The sample information would cost \$100. The reliability of the sample information is described by the following conditional probabilities:

P(Sample information indicates S_1 true | S_1 is true) = .8
P(Sample information indicates S_2 true | S_1 is true) = .2
P(Sample information indicates S_1 true | S_2 is true) = .1
P(Sample information indicates S_2 true | S_2 is true) = .9

a. Find the expected payoff of sampling (EPS) and the expected payoff of no sampling (EPNS).
b. Use the results of part **a** to find the expected value of sample information (EVSI).
c. Find the expected net gain of sampling (ENGS).
d. According to the ENGS, should the decision-maker purchase the sample information prior to making a decision? Explain.

19.63 Refer to Exercise 19.62. A second source of sample information is also available to the decision-maker. This information costs only $10, but it is much less reliable than the first source, as indicated by the following conditional probabilities:

P(Sample information indicates S_1 true | S_1 is true) = .4
P(Sample information indicates S_2 true | S_1 is true) = .6
P(Sample information indicates S_1 true | S_2 is true) = .3
P(Sample information indicates S_2 true | S_2 is true) = .7

a. Find the expected net gain of sampling (ENGS) for the second source of sample information.
b. From which source should the decision-maker purchase sample information? Explain.

19.64 Consider the following decision tree, with outcomes expressed as payoffs and prior probabilities shown in parentheses. The decision-maker is considering purchasing sample information about the true state of nature

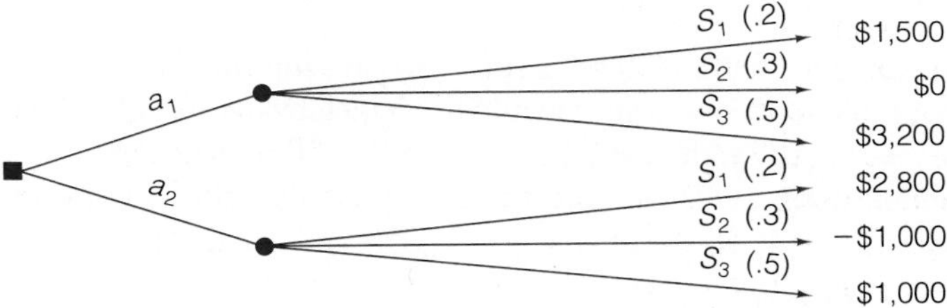

prior to choosing an action. The sample information costs $250. The reliability of the sample information is described by the following conditional probabilities:

P(Sample information indicates S_1 true | S_1 true) = .7
P(Sample information indicates S_2 true | S_1 true) = .2
P(Sample information indicates S_3 true | S_1 true) = .1
P(Sample information indicates S_1 true | S_2 true) = .05
P(Sample information indicates S_2 true | S_2 true) = .9
P(Sample information indicates S_3 true | S_2 true) = .05
P(Sample information indicates S_1 true | S_3 true) = .1
P(Sample information indicates S_2 true | S_3 true) = .1
P(Sample information indicates S_3 true | S_3 true) = .8

a. What are the prior probabilities for S_1, S_2, and S_3?
b. Find the posterior probabilities for S_1, S_2, and S_3 when the sample information indicates S_1 is true.
c. Repeat part **b** for the case when the sample information indicates S_2 is true.
d. Repeat part **b** for the case when the sample information indicates S_3 is true.
e. Find the predictive (marginal) probabilities for the three sample results.
f. Expand the decision tree to include the decision-maker's sampling decision. Enter the payoffs on the tree along with all the probabilities you found in parts **a–e**. (For an example of such a tree, see Figure 19.9.)
g. Find EPNS and EPS, and enter them on your tree.
h. Find EVSI.
i. Use the preposterior expected gain decision rule to determine whether the decision-maker should purchase the sample information.

Applying the Concepts

19.65 The decision problem for a company that may market a new product is characterized in the following payoff table (prior probabilities are in parentheses).

	Product's Status After 1 Year on the Market		
	Failure (.6)	**Successful (.3)**	**Very Successful (.1)**
Market	−$200,000	$300,000	$600,000
Do Not Market	0	0	0

The company is considering whether to conduct a $30,000 market survey to gain information concerning the product's potential for success. One of four different conclusions would be yielded by the survey: "Product will be very successful," "Product will be successful," "Product will be a failure," or "Product's status after 1 year is uncertain." The reliability of the conclusions yielded by the survey can be described by the following conditional probabilities:

P(Survey concludes "very successful" | Product is very successful) = .6
P(Survey concludes "successful" | Product is very successful) = .2
P(Survey concludes "failure" | Product is very successful) = .1
P(Survey concludes "uncertain" | Product is very successful) = .1
P(Survey concludes "very successful" | Product is successful) = .2
P(Survey concludes "successful" | Product is successful) = .4
P(Survey concludes "failure" | Product is successful) = .2
P(Survey concludes "uncertain" | Product is successful) = .2
P(Survey concludes "very successful" | Product is a failure) = .1
P(Survey concludes "successful" | Product is a failure) = .1
P(Survey concludes "failure" | Product is a failure) = .5
P(Survey concludes "uncertain" | Product is a failure) = .3

a. Find the expected value of sample information (EVSI), and interpret your result in the context of the problem.

b. Find the expected net gain of sampling (ENGS), and use it to determine whether the company should undertake the proposed market survey.

19.66 Reconsider Exercises 19.18 and 19.32. The dress buyer expects to sell between three and eight dozen dresses with the prior probabilities shown in the table at the left. A market research firm can be hired to forecast the demand for the new style of dress. Past records indicate that the research firm's conditional probabilities of forecasts, given the various states of nature (sales), are as shown in the table at the right, sales numbers are in dozens. Perform a preposterior analysis and indicate how much the buyer should be willing to spend for the sample information.

Sales	Probability
3	.15
4	.25
5	.30
6	.15
7	.10
8	.05

		Actual Sales					
		3	4	5	6	7	8
Forecast Sales	3	.70	.10	.06	.04	0	0
	4	.20	.50	.10	.05	.05	0
	5	.10	.20	.50	.11	.05	.05
	6	0	.10	.20	.40	.20	.10
	7	0	.10	.04	.30	.60	.15
	8	0	0	.10	.10	.10	.70

19.67 Refer to Exercise 19.60, in which a hospital was considering purchasing 100 color television sets. On the basis of its prior information, what is it worth to the hospital to be able to test one of the 100 television sets before selecting a purchase agreement?

Summary

Decision analysis can be used to prescribe a course of action when the decision-maker is uncertain about the outcome that will result from a chosen action. For the prescribed action to reflect the decision-maker's preferences regarding the outcomes *and* his or her attitude toward risk, we use the **expected utility criterion**. To use this criterion, it is necessary for the decision-maker to assess a **utility function** by which the decision-maker can quantify his or her preferences regarding the potential outcomes of the decision problem, as well as his or her attitudes toward risk. In fact, a decision-maker can be classified as a **risk-avoider**, a **risk-taker**, or **risk-neutral** by examining the shape of his or her utility function.

If a decision-maker is risk-neutral, then both the expected utility criterion and the **expected payoff criterion** prescribe the same action. The expected payoff criterion has the advantage of not requiring the assessment of a utility function. Thus, many decision-makers assume risk-neutrality and use the expected payoff criterion as an approximation to the expected utility criterion.

To obtain meaningful results from decision analysis, the decision-maker must take great care in developing the inputs to the analysis. The **actions**, **states of nature**, **outcomes**, and **objective variable** must be properly defined. Much careful, logical thought should go into the assessment of the utility function. The **probability distribution of the states of nature** should be assessed using as much information about the states of nature as it is possible to obtain. Frequently, we may wish to collect additional information—a sample—and revise these probabilities. The state probabilities assessed before sampling are called **prior probabilities**. Their revised values, obtained using **sample information** and the procedures of this chapter, are called **posterior probabilities**. The decision analyst can compute the **expected value of sample information** *before* actually obtaining the sample; this is called a **preposterior analysis**. This quantity can then be compared to the **cost of sampling** to determine whether the sample information is worth its cost. If we decide to sample, we observe the sample outcome and use the appropriate posterior probabilities to make a decision.

We showed that the procedures for making decisions based on prior analysis or posterior analysis are identical once the probabilities of the states of nature have been assigned to the payoff table. In either case, use of the **expected utility criterion** will lead to decisions that are consistent with the decision-maker's preferences for outcomes and attitudes toward risk.

If conducted properly, decision analysis can provide valuable inputs into the decision-making process, even if the decision it prescribes is not implemented. The detailed analysis and logical thinking it requires may uncover courses of action or states of nature not previously recognized. It may even be discovered that the wrong decision problem is being addressed.

Supplementary Exercises 19.68–19.88

Note: Starred (*) exercises refer to the optional section in this chapter.

19.68 A small company that produces auto parts for American-made cars expects to lose \$1.5 million next year unless Congress passes a bill to limit the number of foreign-made cars that can be imported into the United States. If the bill passes, the company expects to make a profit of \$2.5 million next year. The company's lobbyists in Washington believe that the probability that the bill will pass this year is .6. The auto parts company must decide whether to stay in business and risk a huge loss next year, or to lease its facility for a year to a Japanese firm for \$500,000.

a. Identify the objective variable, actions, and states of nature of the company's decision problem.
b. Construct the payoff table for the decision problem.
c. Convert the payoff table to an opportunity loss table.
d. Verify that the expected payoff and expected opportunity loss criteria prescribe the same action.

19.69 Refer to Exercise 19.68. Assume that the auto parts manufacturer's utility function for money is

$$U(x) = .5 + .20x \qquad -2.5 \text{ million} \leq x \leq 2.5 \text{ million}$$

a. Plot the utility function.
b. Is the utility function that of a risk-avoiding, risk-taking, or risk-neutral decision-maker?
c. According to the expected utility criterion, which action should the parts manufacturer select?

19.70 Consider the following payoff table, with state probabilities shown in parentheses:

	S_1 (.5)	S_2 (.3)	S_3 (.2)
a_1	12	77	−27
a_2	24	2	12
a_3	−20	41	80
a_4	25	5	70
a_5	40	50	10

a. Eliminate any inadmissible actions.
b. Construct an opportunity loss table for this decision problem.
c. Calculate both the expected payoffs and the expected opportunity losses for the actions, and verify that both criteria select the same action.
d. Construct a decision tree for this problem that includes both state probabilities and expected payoffs. [*Hint*: Exercise 19.24 describes where to locate these quantities on your tree.]

19.71 Indicate what type of information (subjective, relative frequency, or both) you would use to assign probabilities to the events described here. Explain the reasoning behind your choice. How would you obtain the necessary information to assign the probabilities?
a. Your firm is in the automobile insurance business, and you want to assess a probability distribution to characterize the total dollar value of damage to a certain model car when an 18- to 21-year-old driver is in an accident.
b. Your company is introducing a new product, and you are assigned the task of assessing a probability distribution for the market share the product will obtain.
c. A drug company is attempting to develop a new type of antihistamine. The research is still in the early stages, and you want to assess the probability that the research will be successful.
d. You are a banker and are considering giving a 30-year mortgage of $75,000 to a family with a yearly income of $25,000. You want to assess the probability of a default on the loan.

19.72 A bank is trying to decide whether to make a 1-year commercial loan of $75,000 to an automobile repair shop. Past experience has shown that one of three outcomes will result if the loan is made:

Outcome 1: The customer repays the loan plus the 10% interest with no complications.
Outcome 2: The customer has difficulty repaying the loan. It is eventually repaid with the 10% interest, but a $1,000 collection cost is incurred by the bank.
Outcome 3: The customer declares bankruptcy, and the bank recovers only 60% of the loan.

If the bank decides not to make the loan, the money will earn 8% for the year.
a. Construct the payoff table for this decision problem.
b. Convert the payoff table to an opportunity loss table.
c. Draw a decision tree for this problem.

19.73 Refer to Exercise 19.72. Suppose past records yield the frequency distribution for commercial loans shown in the table.

Outcome	*Frequency*
1. Repaid	1,104
2. Repaid with difficulty	120
3. Defaulted	56

a. Use this information to assess the probabilities of the three outcomes.
b. Using the assessed probabilities and the expected payoff criterion, decide whether the bank should make the loan.

19.74 Refer to Exercises 19.72 and 19.73. Suppose the bank uses the following utility function for money:

$$U(x) = \frac{\sqrt{x + 100{,}000}}{1{,}000} \qquad -100{,}000 \leq x \leq 900{,}000$$

a. Graph this function over the given range of x, and classify the bank as risk-avoiding, risk-neutral, or risk-taking.
b. Use this function to assign utility values to the monetary payoffs in the decision table of Exercise 19.72. Based on the expected utility criterion, should the loan be made?

19.75 A hosiery company must make a pricing decision regarding its new line of stockings. Two economists have been hired to develop forecasting equations that relate the quantity demanded to the price of the stockings. The forecasting equations the economists derive are

$$\textit{Economist 1:}\ q = 10 - 2p \qquad \textit{Economist 2:}\ q = 16 - 4p$$

where q is the quantity demanded in units of 100,000 and p is the price in dollars. Four prices are being considered: \$.99, \$1.98, \$2.75, and \$3.50. Assume that one of the economists' forecasting equations will be correct, but you do not know which one.
a. Construct the payoff table for this problem.
b. Construct the opportunity loss table for this problem.
c. If the company believes the forecasting equations have an equal probability of being correct, which price should be charged according to the expected payoff criterion?

19.76 A medical doctor is involved in a \$1 million malpractice suit. He can either settle out of court for \$250,000 or go to court. If he goes to court and loses, he must pay the \$925,000 plus \$75,000 in court costs. If he wins in court, the plaintiffs pay the court costs.
a. Construct a payoff table for this decision problem.
b. Draw a decision tree for this problem.
c. The doctor's lawyer estimates the probability of winning to be .2. Use the expected payoff criterion to decide whether the doctor should settle or go to court. Enter the lawyer's assessed probabilities and the expected payoffs associated with each action on your decision tree. (See Exercise 19.24 for a description of where to locate these quantities.)

19.77 Refer to Exercise 19.76. Suppose the doctor's utility function for money is given by

$$U(x) = \frac{\sqrt{1{,}000{,}000 + x}}{1{,}400} \qquad -1{,}000{,}000 \leq x \leq 1{,}000{,}000$$

a. Convert the payoffs in Exercise 19.76 to utility values.
b. Use the expected utility criterion to decide whether the doctor should settle or go to court.

19.78 A common problem with management information systems (MIS) is that managers fail to use them, even when they are technically sound. A large computer firm is considering an MIS that will cost \$1,000,000 to build and operate over a 5-year period. If the company's managers use the system, a savings of \$750,000 per year will be realized.
a. Construct the payoff table for this decision problem.
b. Convert the payoff table to an opportunity loss table.
c. Suppose the probability that the MIS will be used is assessed to be only .05. Use the expected opportunity loss criterion to decide whether to install the MIS.

19.79 Graph each of the following utility functions and classify the decision-maker's attitude toward risk.

a. $U(x) = \dfrac{100 + .5x}{250} \quad -200 \le x \le 300$ b. $U(x) = \dfrac{\sqrt{x + 100}}{20} \quad -100 \le x \le 300$

c. $U(x) = \dfrac{x^3 - 2x^2 - x + 10}{800} \quad 0 \le x \le 10$ d. $U(x) = \dfrac{.1x^2 + 10x}{110} \quad 0 \le x \le 10$

19.80 Consider the payoff table shown. The decision-maker is indifferent between the following pairs of events:

Receive \$0 with certainty, and win \$80 with probability .15 or lose \$20 with probability .85
Receive \$20 with certainty, and win \$80 with probability .3 or lose \$20 with probability .7
Receive \$40 with certainty, and win \$80 with probability .55 or lose \$20 with probability .45
Receive \$60 with certainty, and win \$80 with probability .8 or lose \$20 with probability .2

	S_1	S_2	S_3
a_1	12	77	−10
a_2	−20	41	80
a_3	25	5	70
a_4	40	50	10

a. Plot the utility values versus monetary values between −\$20 and \$80 (assign these endpoints utility values of 0 and 1, respectively).
b. Draw a smooth curve through the points in part **a**, and use this curve to assign utility values to all the payoffs in the table.
c. If the state probabilities are $P(S_1) = .4$, $P(S_2) = .35$, and $P(S_3) = .25$, which action should be selected according to the expected utility criterion? How does this compare with the action selected by the expected payoff criterion?

19.81 The jury you are sitting on hears the following evidence: A woman's body was found in a ditch following a violent argument with her boyfriend the previous evening. The murder weapon shows a palm print that matches the boyfriend's print, but this evidence is not conclusive, a fingerprint expert asserts, because such palm prints are possessed by 1 person in 1,000 (Finkelstein and Fairley, 1977).

a. What is the posterior probability of the boyfriend's guilt if the prior probability of guilt is .10? Repeat this procedure for each of the following prior probabilities: .2, .3, .4, .5, .6, .7, .8, and .9.

b. Graph the posterior probabilities you computed in part **a** versus their respective prior probabilities. Such graphs can help show the sensitivity of a particular posterior probability to different prior probabilities. Comment on the importance of such sensitivity analyses for decision analysis.

19.82 The management of a bank must decide whether to install a commercial loan decision-support system (an on-line management information system) to aid its analysts. Experience suggests that each correct loan decision (accepting good loan applications and rejecting those that will eventually be defaulted) adds, on the average, approximately $25,000 per decision to the bank's profit. Further, it is estimated that the additional number, x, of correct loan decisions (per year) that could be attributed to the decision-support system has the probability distribution given in the table.

x	0	10	20	30	40	50	60	70	80	90
$p(x)$	.01	.04	.10	.15	.20	.15	.10	.10	.10	.05

a. If the decision-support system is estimated to have a useful life of 5 years, what would be the expected increase in profits that could be attributed to it?

b. The increase in profits will accrue only if the system is used by the analysts. Past experience has shown that for various behavioral and political reasons the system was not used by analysts in 80% of the installations. Given this information and that the system costs $1,500,000 to purchase, install, and maintain over a 5-year period, should the bank purchase the system?

*c. The bank could hire a consulting firm for $50,000 to interview its loan analysts and then predict whether this group of analysts will use the decision-support system. The reliability of the firm's predictions is measured by the probabilities given in the table. Should the bank purchase the survey? Explain.

		Actual Outcome	
		Used System	Did Not Use System
Forecast	Will Use System	.7	.1
	Will Not Use System	.3	.9

19.83 *Acceptance sampling* is commonly used by manufacturers to screen incoming lots of material for an excessive number of defective units. A sample is selected from each incoming lot of units, the number of defective units is counted, and the lot is either rejected or accepted depending on whether the number of defectives, x, exceeds a predetermined acceptance number (Montgomery, 1991). As an example of the decision analysis approach to acceptance sampling, assume the proportion of the number of defectives in an incoming lot is either 5% or 10%. The prior probability that the proportion of defectives will be 5% is .80. Suppose the cost of accepting a lot with 10% defectives is $4,100, and the cost of rejecting a lot with 5% defectives is $1,000. There are no costs for making a correct decision (i.e., accepting a lot with 5% defectives or rejecting a lot with 10% defectives).

a. Formulate the payoff table for this problem.

b. If no sample is drawn, should the lot be accepted or rejected?

c. Assume a unit is drawn at random from the lot and is found to be defective. Should the lot be accepted or rejected?
d. If the unit drawn in part **d** is not defective, should the lot be accepted or rejected?
*e. Calculate the EVSI when the sample size is 1. Draw the corresponding decision tree.
*f. It will cost management $10 in fixed costs plus $10 for every item inspected. Should management use a sample size equal to 1?

***19.84** Assume the same basic facts given in Exercise 19.83, but now consider a sample size of 3.
a. Draw a decision tree for a preposterior analysis and calculate EVSI.
b. What is the ENGS?

***19.85** Repeat Exericse 19.83 with a sample size of 5. Then, using these results and the results of Exercises 19.83 and 19.84, plot ENGS versus sample size.

19.86 The chief forester of a large midwestern city must plan for the identification and removal next year of trees infected with Dutch elm disease. State law requires that all diseased elms must be removed and disposed of by October 30. For each diseased tree left standing after October 30, the city loses $300 in state funds budgeted for reforestation. Since the forester's staff can cut only 15,000 trees per season, private contractors must be hired to cut the trees in excess of 15,000. Unfortunately, the private contractors must be hired at the beginning of the disease season (May) before the spread of the epidemic is known. There is a fixed cost of $2,000 per contract signed, and each contractor is paid $250 for each tree removed. Assume each contractor can perform 6,000 removals per season. The forester has hired a consultant to help determine how many contractors to hire. The first action taken by the consultant is to assess a probability distribution that characterizes local experts' prior opinions regarding the number of elms that will be infected in the coming season. This distribution is given in the table.

Number of Trees	*Probability*
0– 5,000	.05
5,001–10,000	.20
10,001–20,000	.30
20,001–30,000	.20
30,001–40,000	.10
40,001–50,000	.08
50,001–60,000	.07

a. Using the midpoints of the intervals in the probability distribution as the states of nature, formulate the payoff table for this decision problem.
b. From your payoff table, construct a decision tree for the forester's decision problem.
c. Based on the prior probabilities, how many contractors should be hired to maximize the expected payoff?

***19.87** Refer to Exercise 19.86. The consultant reports to the forester that a statistical model can be developed to predict the number of trees that will be infected. The conditional probabilities of the forecasts given the various states of nature are shown in the table on page 1074. It will cost $5 million to develop the model. Conduct a preposterior analysis to determine whether the model should be developed.

		Actual Number of Trees Infected						
		0–5,000	5,001–10,000	10,001–20,000	20,001–30,000	30,001–40,000	40,001–50,000	50,001–60,000
	0– 5,000	.5	.3	.1	0	0	0	0
	5,001–10,000	.35	.4	.2	.1	0	0	0
	10,001–20,000	.15	.2	.5	.2	.1	0	0
Forecast	20,001–30,000	0	.1	.2	.5	.2	0	0
	30,001–40,000	0	0	0	.2	.5	.1	.1
	40,001–50,000	0	0	0	0	.2	.5	.4
	50,001–60,000	0	0	0	0	0	.4	.5

19.88 Certified Public Accounting (CPA) firms are hired by private corporations to audit or certify their accounting records. The auditors check two main problem areas—possible fraud by company employees and acceptability of record-keeping practices. Upon completing its audit, the CPA firm either will certify that the client firm's records and financial statements are in order or will fail to certify them and will report on existing irregularities. If the auditor certifies the records when in fact errors or irregularities exist, the CPA firm may be sued by the client firm's stockholders for malpractice. If the CPA firm refuses to certify the records when in fact the records contain no errors or irregularities, it may be sued by the client.

Suppose you have audited a company's financial records and now face a certification decision. In your judgment, you have assessed the following costs. If you report incorrectly that the books are not certifiable when in fact they are, then you will lose the account and be sued by the client; you estimate the total cost for this error to be $2 million. If, on the other hand, you certify the accounts when irregularities exist, you will be sued for $10 million by the stockholders.

a. Formulate the payoff table for this problem.
b. Describe how you would assess the state probabilities for this decision problem.
c. Assume you have assessed that the probability that the records are in order is .9. Which action would decision analysis prescribe?
d. What should you be willing to pay for perfect information about the records?

On Your Own

a. Suppose you are trying to decide whether to promote an outdoor event scheduled for this spring. Choose an event that can be held only in good weather (lawn concert, baseball game, festival, etc.) and for which your city has an available facility. Based on the expected attendance and a reasonable ticket price, determine the payoff if the event is held. Remember to subtract your costs for advertising, facility rental, etc. These costs represent the (negative) payoff in case you decide to promote the event and it must be cancelled. Construct the payoff table corresponding to the two-action/two-state decision problem.

Now, based on *your own* knowledge of the climate in the area at the proposed time of the concert, assign probabilities to the states of nature: Rain and No rain. Based on these probabilities, which action (Promote or Do not promote) is selected by the expected payoff criterion?

b. Ask a local meteorologist for a long-range forecast based on a careful study of all pertinent data. You need to know the meteorologist's prediction 60 days in advance of the scheduled date of the event. Before you can assess the worth of the meteorologist's prediction, you have to ask for two conditional probabilities: the probability that the meteorologist will predict rain 60 days in advance *given that* it actually will rain, and the probability that he or she will predict no rain *given that* it will not rain. Conduct a complete preposterior analysis using the meteorologist's conditional probabilities and your own prior probabilities from part **a**. Will the meteorologist's prediction affect the action selected by the expected payoff criterion? How much would you be willing to pay the meteorologist for the prediction? Explain.

References

Baird, B. F. *Introduction to Decision Analysis*. North Scituate, MA: Duxbury, 1978.

Balthasar, H. U., Boschi, R. A. A., and Menke, M. M. "Calling the shots in R&D." *Harvard Business Review*, May–June 1978, Vol. 56, pp. 151–160.

Brown, R. V., Kahr, A. S., and Peterson, C. *Decision Analysis for the Manager*. New York: Holt, Rinehart and Winston, 1974.

Bunn, D. *Applied Decision Analysis*. New York: McGraw-Hill, 1984.

Clemen, R. *Making Hard Decisions*. Boston: PWS-Kent, 1991.

Dawes, R. M. *Rational Choice in an Uncertain World*. San Diego: Harcourt Brace Jovanovich, 1988.

Farquhar, P. H. "Utility assessment methods." *Management Science*, Nov. 1984, Vol. 30, No. 11, pp. 1283–1300.

Finkelstein, M. O., and Fairley, W. "A comment on 'Trial by mathematics.' " In *Statistics and Public Policy*, W. Fairley and F. Mosteller, eds. Reading, MA: Addison-Wesley, 1977.

Howard, R. A., Matheson, J. E., and North, D. W. "The decision to seed hurricanes." *Science*, June 1971, Vol. 176, pp. 1191–1202.

Keeney, R. L. "A decision analysis with multiple objectives: The Mexico City airport." *Bell Journal of Economics and Management*, 1973, Vol. 4, pp. 101–117.

Keeney, R. L., and Raiffa, H. *Decisions with Multiple Objectives: Preferences and Value Tradeoffs*. New York: Wiley, 1976.

LaValle, I. *Fundamentals of Decision Analysis*. New York: Holt, Rinehart and Winston, 1978.

Lindley, D. V. *Making Decisions*, 2nd ed. London: Wiley, 1985.

Luce, R. D., and Raiffa, H. *Games and Decisions*. New York: Wiley, 1957.

Montgomery, D. C. *Introduction to Statistical Quality Control*. 2nd ed. New York: Wiley, 1991.

Raiffa, H. *Decision Analysis. Introductory Lectures on Choices Under Uncertainty*. Reading, MA: Addison-Wesley, 1968.

Raiffa, H., and Schlaifer, R. *Applied Statistical Decision Theory*. Cambridge, MA: MIT Press, 1961.

Samson, D. *Managerial Decision Analysis*. Homewood, IL: Irwin, 1988.

Schroeder, R. G. *Operations Management: Decision Making in the Operations Function*, 4th ed. New York: McGraw-Hill, 1993.

Smith, J. Q. *Decision Analysis: A Bayesian Approach*. London: Chapman and Hull, 1988.

von Winterfeldt, D., and Edwards, W. *Decision Analysis and Behavioral Research*. Cambridge: Cambridge University Press, 1986.

Watson, S. R., and Buede, D. M. *Decision Synthesis*. Cambridge: Cambridge University Press, 1987.

Winkler, R. L. *An Introduction to Bayesian Inference and Decision*. New York: Holt, Rinehart and Winston, 1972.

Winkler, R. L., and Hays, W. L. *Statistics: Probability, Inference, and Decision*, 2nd ed. New York: Holt, Rinehart and Winston, 1975. Chapters 2 and 9.

CHAPTER TWENTY

Survey Sampling

Contents

Case Studies

Where We've Been

Although many methods are available for selecting a sample, the statistical methods described in the preceding chapters were based primarily on simple random sampling from populations of measurements that were large in relation to the sample size. Two exceptions to this method of data collection—the paired difference experiment (Chapter 9) and the factorial design (Chapter 16)—demonstrated the power of experimental design to increase the amount of information in sample data.

Where We're Going

The term *sample survey* is usually used in conjunction with sampling of people, households, businesses, etc. The Current Population Survey and the Gallup Poll are examples of such surveys. Special problems arise in survey sampling that may require more elaborate sampling designs than simple random sampling. This chapter introduces some of the problems encountered in survey sampling and the sampling designs and methods that have been developed to handle them.

20

Almost all the statistical methods we have covered were based on simple random sampling (Chapter 3). Three exceptions to this method of data collection—the paired difference experiment, the randomized block design, and the factorial design—demonstrated that sampling designs other than simple random sampling can be used to increase the amount of information obtained in a sample. In this chapter, we present sampling designs and estimation procedures of a specific type, those used in *sample surveys*.

The term **sample survey** is used in conjuncton with the sampling of populations, i.e., collections of people, households, businesses, etc. A consumer preference poll is an example of a sample survey. Samplings conducted to estimate the general level of business inventories or to estimate the proportion of households that watched a particular television program are also examples of sample surveys.

Most sample surveys are conducted to estimate one or more of three population parameters. For example, suppose we are interested in the market for seafood. One population parameter we might want to estimate is the mean amount, μ, of money spent monthly per household on seafood in a given market. A second population parameter of interest would be the total money, τ, spent on seafood per month in the market (i.e., the sum of the expenditures for all households in the market). Third, we might be interested in the proportion, p, of households that consume some seafood each month. Procedures for estimating μ and p for simple random samples were discussed in Chapter 7. We will discuss the estimation of the population total, τ, in this chapter. We present a summary of these estimation objectives in the box.

Common Objectives of Sample Surveys

1. Estimation of the population mean μ
2. Estimation of population total τ
3. Estimation of population proportion p

Sample surveys cost time and money, and sometimes they are almost impossible to conduct. For example, suppose we want to obtain an estimate of the proportion of households in the United States that plan to purchase new television sets next year, and we plan to base our estimate on the intentions of a random sample of 3,000 households. What are the problems associated with collecting these data? In order to use a random number table (Chapter 3) to select the sample, we would need a list of all the households in the United States. Obtaining such a list would be a monumental obstacle. After we obtain a list of households, we need to contact each of the 3,000 selected for the sample. Will all be at home when the surveyor reaches the household? And will all answer the surveyor's question? You can see that collecting a random sample is easier said than done.

The large body of knowledge called **survey sampling** or **sample survey design** was developed to help solve some of the problems we have noted. It includes sample survey designs that help reduce the cost and time involved in conducting a sample survey, and it includes the statistical estimation procedures associated with those designs. Since survey sampling is a course in itself (or several courses), we will present only a few of the most widely used sample survey designs and address only a few of the problems you might encounter. Further information on this important subject can be found in the references at the end of the chapter.

CASE STUDY 20.1 / Who Does Sample Surveys?

We all know of the public opinion polls that are reported in the news media. The Gallup Poll and the Harris Survey issue reports periodically, describing national public opinion on a wide range of current issues. State polls and metropolitan area polls, often supported by a local newspaper or television station, are reported regularly in many localities. The major broadcasting networks and national news magazines also conduct polls and report their findings.

But the great majority of surveys are not exposed to public view. The reason is that, unlike the public opinion polls, most surveys are directed to a specific administrative or commercial purpose. The wide variety of issues with which surveys deal is illustrated by the following listing of actual uses (Ferber et al., 1980):

1. The U.S. Department of Agriculture conducted a survey to find out how poor people use food stamps.
2. Major television networks rely on surveys to tell them how many and what types of people are watching their programs.
3. Auto manufacturers use surveys to find out if people are satisfied with their cars.
4. The U.S. Bureau of the Census compiles a survey every month—the Current Population Survey (see Case Study 1.3 for more details)—to obtain information on employment and unemployment in the nation.
5. The National Center for Health Statistics sponsors a survey every year to determine how much money people are spending for different types of medical care.
6. Local housing authorities conduct surveys to ascertain satisfaction of people in public housing with their living accommodations.
7. The Illinois Board of Higher Education surveys the interest of Illinois residents in adult education.
8. Local, state, and national transportation authorities conduct surveys to acquire information on residents' commuting and travel habits.
9. Magazines and trade journals utilize surveys to find out what their subscribers are reading.
10. Surveys are used to ascertain the characteristics of people who use our national parks and other recreation facilities.
11. Bank auditors survey a bank's savings account customers to verify the accuracy of the bank's records.

Sample surveys are also heavily used by marketing researchers to uncover new uses for products already on the market. Such information is helpful in redirecting existing advertising campaigns or creating new ones, as illustrated in the following examples (Cox, 1979):

a. The producer of Ben-Gay, a topical analgesic, believed that consumers use Ben-Gay primarily for the relief of simple muscle aches. Data collected from a large sample of consumers revealed, however, that more than 50% of all consumers use Ben-Gay for arthritis relief.

b. More than 4,000 consumers are surveyed weekly by Lever Brothers. The sample information obtained regarding Wisk detergent indicated that many consumers used Wisk to "pretreat" shirt collars. This information spawned the familiar "ring-around-the-collar" advertising campaign.

20.1 Terminology

In survey sampling, the unit or entity upon which a measurement is made is called an **element**. Thus, a **population** is a collection of elements—usually people, objects, or events—about which we wish to make an inference.

> **Definition 20.1**
>
> The entity upon which a measurement is made is called an **element.**

Sometimes we may want to reduce the cost of sampling by taking measurements on collections of elements that are physically near one another or bear some other relationship that makes them more easily observed as a group. For example, if we plan to sample the opinions of all adults regarding some particular product, we might wish to randomly select households and then interview all the adults in the households. When nonoverlapping sets of elements are randomly selected and each element in the set is measured, the sets are called **sampling units**. For the product preference survey just described, a household would be a sampling unit and the adults in the household would be the elements. Note that each element in a sampling unit (a household) is measured and that the elements in one sampling unit do not occur in the elements of another.

In the preceding chapters, we randomly selected units or experimental units and made a single observation on each. Thus, the earlier chapters were restricted to the special case where each sampling unit contained only one element.

> **Definition 20.2**
>
> A **sampling unit** is a collection of elements. The elements must satisfy the condition that those in any one sampling unit do not overlap with the elements in other sampling units.

In order to select a sample of sampling units from the total of those available, we must have a listing of them. Such a listing, which must include *all* the sampling units in the population of interest (to enable us to draw a sample that is representative of the population), is called a **frame**. Then, a *sample* is a subset of sampling units selected from a frame. The plan that specifies which sampling units will be included in a sample is called a **sampling design** or, for sample surveys, a **sample survey design**.

Definition 20.3

A **frame** is a list of sampling units.

Definition 20.4

A **sample** is a collection of sampling units selected from a frame.

We summarize the terminology in the next box.

Term	*Definition*	*Example: Product Preference Survey*
Element	Entity on which a measurement is made	Individual consumer
Sampling unit	Collection of elements	Household
Frame	List of sampling units	List of all households in relevant population
Sample	Collection of sampling units selected from frame	Set of households from which product preferences are obtained

20.2 Sample Survey Designs

分層的.

Several useful sampling designs are available for sample surveys. One of the most common (besides random sampling), called *stratified random sampling,* is described in this section, and the appropriate procedures for estimating the population mean, μ, a proportion, p, and the total, τ, of all measurements in the population is presented in Section 20.6. Two other sampling designs, *systematic sampling* and *randomized response sampling,* are discussed briefly in this section, but we refer you to the references at the end of the chapter for the associated estimation procedures.

Stratified random sampling is used when the sampling units associated with the population are physically separated into two or more groups of sampling units (called **strata**) where the within-strata response variation is less than the variation within the entire population. For example, if y is the rent paid for a two-bedroom apartment in a city, we might want to divide the city into regions (strata) where the rents within each stratum are relatively homogeneous. Then we would estimate a population mean, proportion, or total by selecting random samples from within each stratum and combining the strata estimates as explained in Section 20.6. Stratified random sampling often produces estimators with smaller standard errors than those achieved using simple random sampling. Furthermore, by sampling from each stratum we are more likely to obtain a sample representative of the entire population. In addition, the administrative and labor costs of selecting the strata samples are often less than those for simple random sampling.

Sometimes it is difficult or too costly to select random samples. For example, it would be easier to obtain a sample of student opinions at a large university by selecting every hundredth name from the student directory, with the first name selected randomly from the first 100 names in the directory. Although **systematic samples** are usually easier to select than other types of samples, one difficulty is the possibility of a systematic sampling bias. For example, if every fifth item in an assembly line is selected for quality control inspection, and if five different machines are sequentially producing the items, all the items sampled may have been manufactured by the same machine. If we use systematic sampling we must be certain that no cycles (like every fifth item manufactured by the same machine) exist in the list of the sampling units.

Randomized response sampling is particularly useful when the questions of the pollsters are likely to elicit false answers. For example, suppose each person in a sample of wage earners is asked whether he or she cheated on an income tax return. A person who has not cheated most likely would give an honest answer to this question. A cheater might lie, thus biasing an estimate of what proportion of persons cheats on their income tax return.

One method of coping with the false responses produced by sensitive questions is randomized response sampling. Each person is presented *two* questions; one question is the object of the survey and the other is an innocuous question to which the interviewee will give an honest answer. For example, each person might be asked these two questions:

1. Did you cheat on your income tax return?
2. Did you drink coffee this morning?

Then a procedure is used to select randomly which of the two questions the person is to answer. For example, the interviewee might be asked to flip a coin. If the coin shows a head, the interviewee answers the sensitive question, 1. If the coin shows a tail, the interviewee answers the innocuous question, 2. Since the interviewer never has the opportunity to see the coin, the interviewee can answer the question and feel assured that his or her guilt (if guilty) will not be exposed. Consequently, the random response procedure can elicit an honest response to a sensitive question.

Four of the most important sampling designs are discussed in this chapter. In each case, we will present the methodology for selecting the sample, calculating the estimates of population parameters, and measuring the standard error of the estimates.

The size of the standard error will serve as a measure of the amount of information on a parameter that is provided by a specific sampling design.

CASE STUDY 20.2 / Methods of Data Collection in Sample Surveys

The survey designs described in this chapter prescribe methods for selecting elements from a frame. Once selected, the attribute of interest must be measured for each of the elements. That is, the data must be collected. Surveys that involve human populations are classified by their method of data collection, such as **mail surveys**, **telephone surveys**, or **personal interview surveys** (Feber et al., 1980).

Mail surveys require the development of questionnaires that respondents complete on their own (*self-enumeration*). Mail surveys are seldom used to collect information from the general public because names and addresses are not often available and response rates tend to be low. However, this method may be effective with particular groups, such as subscribers to specialized magazines or members of a professional organization.

Telephone interviewing is an efficient and popular method of collecting some types of data. Random samples of telephone numbers may be randomly or systematically selected from telephone directories, or a methodology called **random-digit dialing** may be employed. This approach involves using a random number generator to mechanically create the sample of phone numbers to be called. Random-digit dialing was developed to help overcome the sampling biases introduced into survey results by sampling from telephone directories (Glasser and Metzger, 1972).

Personal interviews are generally conducted in a respondent's home or office. They cost much more than mail or telephone surveys but may be necessary when complex information is being collected.

New methods of data collection can enter information directly into computers. They include the A. C. Nielsen Company's measurement of TV audiences using electronic devices—called **audimeters**—attached to a sample of TV sets. Nielsen places an audimeter in each of a sample of about 1,700 homes across the United States. The audimeter, usually located in a closet or in the basement, is wired to every TV set in the home and records when the sets are on or off and which channels are tuned in. The audimeter is connected via special telephone lines to Nielsen's computer. When you read in your newspaper that a particular TV show received, say, a 20 rating for the week by A. C. Nielsen, it means that 20% of the sample of Nielsen families tuned in to that show for at least 6 minutes (Chagall, 1978).

Some surveys combine various methods. Survey workers may use the telephone to screen eligible respondents (say, women of a particular age group) and then make appointments for a personal interview. The U.S. Bureau of the Census' monthly Current Population Survey (see Case Study 1.3) uses both telephone and personal interviews.

Because changes in attitude or behavior cannot be reliably ascertained from a single interview, some surveys use a **panel** of respondents who are interviewed two or more times. Such surveys are often used during election campaigns, or to chart a family's health or purchasing pattern over a period of time. The Nielsen families, for example, constitute a panel of respondents whose TV watching patterns are monitored over time. Panels are also used to trace changes in behavior over time, as with social experiments that study changes in the work behavior of low-income families in response to an income maintenance plan.

CASE STUDY 20.3 / The *Literary Digest* Poll: FDR Versus Alf Landon

Regardless of the survey design and data collection method employed, great care must be exercised in implementing the survey. Poorly implemented surveys may yield disastrous results, as this case study illustrates.

In 1936, the *Literary Digest*, a popular magazine, mailed 10 million questionnaires to voters in the United States. The questionnaire asked which presidential candidate was preferred, the Democratic incumbent, F. D. Roosevelt, or the Republican governor of Kansas, Alfred Landon. The *Digest* had previously predicted the winner of the presidency in every election since 1916. Prior to receiving the responses to its questionnaire, the *Digest* boasted, "When the last figure has been totted and checked, if past experience is a criterion, the country will know to within a fraction of 1% the actual popular vote of forty million" (Aug. 22, 1936, p. 3). The *Digest* received 2.4 million responses—a sample size approximately 800 times larger than is currently used by the Gallup Poll. The sample results indicated Landon would win by a landslide: Landon 57% and FDR 43%. Unfortunately for Landon and the *Literary Digest*, the actual election results yielded a landslide for FDR: FDR 62% and Landon 38%.

What went wrong? How could such a large sample generate such misleading results? Part of the answer lies in the *Digest*'s choice of a sampling frame. The frame was constructed from sources such as telephone directories, club membership lists, magazine subscriptions, and car registrations. Although use of such lists might not yield such misleading results today, the country was split politically along economic lines in 1936—Republicans were generally wealthier than Democrats. As a result, the majority of people listed in the *Digest*'s frame were probably Republicans. Accordingly, the sample was probably not representative of the population of voters in the United States; it was biased in favor of Republican voters.

Another reason for the lack of representativeness of the sample was the *Literary Digest*'s reliance on *voluntary response* to its poll. Respondents to mail questionnaires typically represent only that portion of the population will relatively intense interests in the subject matter of the questionnaire. The anti-Roosevelt voters—although a minority—apparently felt more strongly about the election than did the pro-Roosevelt majority. As a result, the *Literary Digest*'s poll was apparently affected by *nonresponse bias* (see Section 20.5).

If you have never heard of the *Literary Digest*, there is a reason: It is now defunct—thanks in part to the credibility lost as a result of its 1936 presidential poll (Bryson, 1976).

20.3 Estimation in Survey Sampling: Bounds on the Error of Estimation

Estimation procedures developed for the various sample survey designs may differ from those presented in earlier chapters for two reasons. The standard errors of estimators presented in earlier chapters were based on the assumption that the number of sampling units, N, in the population is large relative to the sample size, n. This assumption may not hold in survey sampling and thus will necessitate a modification of the formulas given for the standard errors of the estimators.

The second difference is that the sampling distributions of estimators are often unknown. For this reason, it is difficult to construct exact confidence intervals for population parameters. The usual procedure (see Scheaffer, Mendenhall, and Ott, 1979) is to give an estimate along with an approximate upper limit on the error of estimation—i.e., on the difference that might occur between the estimate and the unknown value of the population parameter. This upper limit, which we call a **bound on the error of estimation**, is calculated using the Empirical Rule of Section 2.6. The logic is that, according to the Empirical Rule, most (approximately 95%) of the estimates produced by an unbiased estimator should lie within 2 standard errors of the estimated population parameter. Or, we could form an approximate large-sample confidence interval for a parameter using the logic of Section 7.1; that is, we will find the endpoints of the confidence interval by adding and subtracting 2 standard errors to the estimate. Consequently, we will present the formulas for estimators, the estimated bounds on the error of estimation, and approximate confidence intervals for each sample survey design using the procedure shown in the box.

General Procedures for Estimating Population Parameters Based on Sample Surveys

1. Present a formula for calculating the estimate.
2. Give a bound on the error of estimation equal to 2 standard errors (or the sample estimate thereof) of the estimator.
3. Calculate an approximate 95% confidence interval for the parameter by forming the interval given by

   Estimate ± (Bound on error)

   that is, Estimate ± (2 estimated standard errors).

CASE STUDY 20.4 / Sampling Error Versus Nonsampling Error

In Chapter 6, we learned that the behavior of the sample mean, $\bar{x}$, in repeated sampling can be described by its sampling distribution. We described the difference between a particular value of the estimator, $\bar{x}$, and the true value of the population parameter, μ, as **estimation error**. This difference is also known as **sampling error**. It is not error in the sense that anyone or anything is at fault or deserves blame; it is simply due to the fact that $\bar{x}$ is computed from a subset of the population rather than from the entire population. The standard error of the sampling distribution of $\bar{x}$ is a measure of the magnitude of the sampling error (estimation error) that may be present in the results of a survey that has been conducted to estimate μ. Accordingly, the standard error of $\bar{x}$ is used to place a bound on the sampling error associated with $\bar{x}$. As we will see in Section 20.7, this bound can be tightened simply by increasing the sample size of the survey.

Unfortunately, the other types of errors that plague surveys—known as **nonsampling errors**—are not so easily measured or controlled. Nonsampling errors are any phenomena other than sampling errors that cause a difference between an estimate and the true value of the population parameter. Nonsampling errors can be classified into two groups: **random errors**, whose effects approximately cancel out if large samples are used, and **biases** that tend to create errors in the same direction and thus do not cancel out over the entire sample.

Biases can arise from any aspect of the survey operation. Some of the main contributing causes are the following (Ferber et al., 1980):

1. *Sampling operations* Mistakes are made in drawing the sample, or part of the population is omitted from the sampling frame. (This was part of the problem with the *Literary Digest* poll in Case Study 20.3.)
2. *Noninterviews* Information is obtained for only part of the sample due to, for example, "not-at-homes" or nonresponse to mail questionnaires. This causes a problem because, typically, there are differences between the noninterviewed part of the sample and the part that is interviewed.
3. *Adequacy of respondent* Sometimes respondents cannot be interviewed, so information is obtained about them from others; the proxy respondent is not always as knowledgeable about the facts.
4. *Understanding the concepts* Some respondents do not understand what is wanted.
5. *Lack of knowledge* Respondents do not always know the information requested or do not try to obtain the correct information.
6. *Concealment of the truth* Out of fear or suspicion, respondents conceal the truth. This concealment may reflect a desire to answer in a way that is socially acceptable, such as claiming to follow an energy conservation program when this is not actually so.
7. *Loaded questions* The question as worded influences the respondents to answer in a specific (not necessarily correct) way.
8. *Processing errors* These include coding errors, data keying, computer programming errors, etc.
9. *Conceptual problems* What is desired may differ from what the survey actually covers. For example, the population or the time period may not be the one for which information is needed, but may have to be used to meet a deadline.
10. *Interviewer errors* Interviewers who misread the question or recast the answers in their own words thereby introduce bias.

Although not every survey will be subject to all these biases, a good survey statistician is aware of them and will attempt to control as many as possible.

In the case of the U.S. Bureau of the Census' Current Population Survey (see Case Study 1.3), many safeguards have been built into the survey process to protect against biases due to interviewer errors (Taeuber, 1978):

1. The survey's 1,100 interviewers are continuously trained and retrained.
2. Each interviewer's work is reviewed each month.
3. Periodically, interviewers are accompanied by supervisory personnel.
4. Approximately twice each year, a sample of the addresses assigned to an interviewer is reinterviewed by a supervisor. The interviewers have no way of knowing when their work will be checked or which addresses will be reinterviewed.

These precautions not only protect against interviewer error but also provide a measure of quality of the Current Population Survey.

20.4 Estimation for Simple Random Sampling

We discussed the estimation of a population mean μ and a proportion p based on simple random sampling in Chapter 7. The confidence intervals for these parameters were based on the assumption that the sample size n is sufficiently large and, although we did not state it, that the number N of sampling units in the population is large relative to the sample size n.

In some sample surveys, the sample size n may represent 5% or perhaps 10% of the total number N of sampling units in the population. When the sample size is large relative to the number of measurements in the population, the standard errors of the estimators of μ and p (given in Chapter 7) should be multiplied by a **finite population correction factor**.

The form of the finite population correction factor depends on how the population variance σ^2 is defined. In order to simplify the formulas of the standard errors that are used in sample surveys, it is common to define σ^2 as division of the sum of squares of deviations by $N - 1$ rather than by N (analogous to the way we defined the sample variance). If we adopt this convention, the finite population correction factor becomes $\sqrt{(N - n)/N}$. Then the point estimators and the estimated bounds on the errors of estimation for μ and p are as shown in the boxes (below and on page 1088).*

Estimation of the Population Mean, μ: Simple Random Sampling

Estimator of μ: $\bar{x} = \dfrac{\sum x_i}{n}$

Estimated bound on the error of estimation:

$$2\hat{\sigma}_{\bar{x}} = 2\frac{s}{\sqrt{n}}\sqrt{\frac{N - n}{N}}$$

where

$$s = \sqrt{\frac{\sum (x_i - \bar{x})^2}{n - 1}}$$

N = Number of sampling units in the population

n = Number of sampling units in the sample

[*Note*: In simple random sampling, each sampling unit contains only one element.]

Approximate 95% confidence interval: $\bar{x} \pm 2\hat{\sigma}_{\bar{x}}$

*For most sample surveys, the finite population correction factor is approximately equal to 1 and, if desired, can be safely ignored. However, if $n/N > .05$, the finite population correction factor should be included in the calculation of the standard error and the bound on the error of estimation.

Estimation of the Population Proportion, p: Simple Random Sampling

Estimator of p: $\hat{p} = \frac{x}{n}$

where x is the number of sampling units that possess a specific attribute (in terms of the binomial distribution, x is the number of "successes").

Estimated bound on the error of estimation:

$$2\hat{\sigma}_{\hat{p}} = 2\sqrt{\frac{\hat{p}(1-\hat{p})}{n}}\sqrt{\frac{N-n}{N}}$$

where N = Number of sampling units in the population
n = Number of sampling units in the sample

Approximate 95% confidence interval: $\hat{p} \pm 2\hat{\sigma}_{\hat{p}}$

The point estimator and the estimated bound on the error for estimating a population total, τ, were not presented in Chapter 7. Their formulas are shown in the next box.

Estimation of the Population Total, τ: Simple Random Sampling

Estimator of τ: $\hat{\tau} = N\bar{x}$

where N = Number of sampling units in the population
n = Number of sampling units in the sample
$\bar{x}$ = Sample mean

Estimated bound on the error of estimation: $2\hat{\sigma}_{\hat{\tau}} = 2\sqrt{N^2 \frac{s^2}{n}\left(\frac{N-n}{N}\right)}$

where s^2 is the sample variance: $s^2 = \frac{\sum (x_i - \bar{x})^2}{n-1}$

Approximate 95% confidence interval: $\hat{\tau} \pm 2\hat{\sigma}_{\hat{\tau}}$

EXAMPLE 20.1

A specialty manufacturer wants to purchase remnants of sheet aluminum foil. The foil, all of which is the same thickness, is stored on 7,462 rolls, each containing a varying amount of foil. To obtain an estimate of the total number of square feet of foil on all the rolls, the manufacturer randomly sampled 100 rolls and measured the number of square feet on each roll. The sample mean was 47.4, and the sample variance was 153.1. Find an approximate 95% confidence interval for the total amount of foil on the 7,462 rolls.

Solution

Each roll of foil is a sampling unit, and there are $N = 7{,}462$ units in the population and $n = 100$ in the sample. Further, $\bar{x} = 47.4$ and $s^2 = 153.1$. Substituting these quantities into the formula for the confidence interval, we obtain (for $\hat{\tau} = N\bar{x}$):

$$\hat{\tau} \pm 2\sqrt{N^2 \frac{s^2}{n}\left(\frac{N-n}{N}\right)} = (7{,}462)(47.4) \pm 2\sqrt{(7{,}462)^2 \frac{153.1}{100}\left(\frac{7{,}462 - 100}{7{,}462}\right)}$$

or, the approximate 95% confidence interval is $353{,}698.8 \pm 18{,}341.8$.

Consequently, the manufacturer estimates the total amount of foil to be in the interval 335,357.0 square feet to 372,040.6 square feet. If the manufacturer wants to adopt a conservative approach, the bid for the foil will be based on the lower confidence limit, 335,357 square feet of foil.

Examples of the estimation of a population mean μ and sample proportion p are not presented in this section because the examples would be identical to those presented in Chapter 7, except for the use of the finite population correction factor. We include exercises of this type at the end of this exercise set.

Exercises 20.1 – 20.18

Learning the Mechanics

20.1 Distinguish among the following terms: element, sampling unit, and sample.

20.2 List three different methods of data collection used in surveys and describe an advantage associated with each.

20.3 What went wrong with the *Literary Digest*'s 1936 election poll?

20.4 Distinguish between sampling error and nonsampling error, and give an example of each.

20.5 Calculate the percentage of the population sampled and the finite population correction factor for each of the following situations.

a. $n = 1{,}000$, $N = 2{,}500$ b. $n = 1{,}000$, $N = 5{,}000$

c. $n = 1{,}000$, $N = 10{,}000$ d. $n = 1{,}000$, $N = 100{,}000$

20.6 Suppose the standard deviation of the population is known to be $\sigma = 200$. Calculate the standard error of $\bar{x}$ for each of the situations described in Exercise 20.5.

20.7 Suppose $N = 5{,}000$, $n = 64$, and $s = 24$.

a. Compare the size of the standard error of $\bar{x}$ computed with and without the finite population correction factor.

b. Repeat part **a**, but this time assume $n = 400$.

c. Theoretically, when sampling from a finite population, the finite population correction factor should always be used in computing the standard error of $\bar{x}$. However, when n is small relative to N, the finite population correction factor is close to 1 and can safely be ignored. Explain how parts **a** and **b** illustrate this point.

20.8 Suppose N = 10,000, n = 2,000, and s = 50.

a. Compute the standard error of $\bar{x}$ using the finite population correction factor.
b. Repeat part **a** assuming n = 4,000.
c. Repeat part **a** assuming n = 10,000.
d. Compare parts **a**, **b**, and **c**, and describe what happens to the standard error of $\bar{x}$ as n is increased.
e. The answer to part **c** is 0. This indicates that there is no sampling error in this case. Explain.

20.9 Suppose you want to estimate a population mean, μ, and $\bar{x} = 422$, $s = 14$, $N = 375$, and $n = 40$. Find an approximate 95% confidence interval for μ.

20.10 Suppose you want to estimate a population proportion, p, and $\hat{p} = .42$, $N = 6{,}000$, and $n = 1{,}600$. Find an approximate 95% confidence interval for p.

20.11 Suppose you want to estimate a population total, τ, and $\bar{x} = 39.4$, $s = 4.0$, $N = 3{,}500$, and $n = 100$. Find an approximate 95% confidence interval for τ.

20.12 A random sample of size $n = 30$ was drawn from a population of size $N = 3{,}000$. The following measurements were obtained:

21	33	19	29	22	38	58	29	52	36	37	30	53	37	29
18	35	42	36	41	35	36	33	38	29	38	39	54	42	42

a. Estimate τ and place a bound on the error of estimation.
b. Estimate μ and place a bound on the error of estimation.
c. Estimate p, the proportion of measurements in the population that are greater than 30. Place a bound on the error of estimation.

Applying the Concepts

20.13 Because external audits by CPA firms can become quite expensive, many firms are creating or increasing the size of their existing internal audit departments. Internal auditors can lower the cost of an external audit by improving the company's accounting controls, performing financial examinations that support the outside auditors' activities, and providing general support for the outside auditors. As part of a study designed to determine the effect of internal audit activities on the cost of external audits, Wallace (1984) questioned a sample of 32 large, diverse companies concerning expenditures for external audits. The mean external audit fee paid by the 32 companies in 1981 was \$779,030; the standard deviation was \$1,083,162.

a. Construct an approximate 95% confidence interval for the mean external audit fee in 1981 for the population of firms from which Wallace drew her sample. Assume that N = 1,500.
b. Construct an approximate 95% confidence interval for the total amount spent by all firms in the population on external audits in 1981. Again, assume N = 1,500.
c. Carefully interpret your confidence interval of part **b** in the context of the problem.
d. What assumption must hold in order to assure the validity of the confidence intervals in parts **a** and **b**?

20.14 Organizations hire independent public accountants to perform audit examinations of their financial statements and to judge the fairness with which the statements characterize their financial position. The audit examination includes reviews and tests that are designed to provide the auditor with evidence from which an opinion can be developed. In addition, this evidence gives the auditor a basis for deciding whether the financial statements were prepared according to "generally accepted accounting principles." Since the early

1950's, auditors have relied to a great extent on sampling techniques, rather than 100% audits, to help them test and evaluate financial records. For example, sampling is frequently used to obtain an estimate of the total dollar value of an account—the account balance. The estimate can be used to check the account balance reported in the organization's financial statements. Such an examination of an account balance is known as a *substantive test* (Arkin, 1982). In order to evaluate the reasonableness of a firm's stated total value of its parts inventory, an auditor randomly samples 100 of the total of 5,000 parts in stock, prices each part, and reports the results shown in the accompanying table.

Part Number	*Part Price*	*Sample Size*	*Part Number*	*Part Price*	*Sample Size*
002	$108	3	271	$ 50	9
101	55	2	399	125	12
832	500	1	761	1,000	2
077	73	10	093	62	8
688	300	1	505	205	7
910	54	4	597	88	11
839	92	6	830	100	19
121	833	5			

a. Find a point estimate of the total value of the parts inventory.

b. Estimate the bound on the error of estimation associated with your point estimate of part **a**. [*Hint:* $s =$ \$209.10]

c. Construct an approximate 95% confidence interval for the total value of the parts inventory.

d. The firm reported a total parts inventory value of \$1,500,000. What does your confidence interval of part **c** suggest about the reasonableness of the firm's reported figure? Explain.

20.15 In 1984 the U.S. Environmental Protection Agency (EPA) announced a ban on further use of the cancer-causing pesticide ethylene dibromide (EDB) as a fumigant for grain and flour-milling equipment. EDB was used to protect against infestation by microscopic roundworms called nematodes. Maximum safe levels were set for EDB presence in raw grain, flour, cake mixes, cereals, bread, and other grain products already on supermarket shelves and in warehouses. Because the federal government lacked the authority to regulate the amount of chemicals in foods, these safe levels were intended as guidelines for state governments. It was estimated that, if state governments followed the EPA guidelines, approximately 7% of the existing corn products would have to be removed from supermarket and warehouse shelves. Following the announcement, state agriculture agencies began sampling the grain products sold in their respective states and testing for the presence of unsafe levels of EDB (Berg, Klauda, and Feyder, 1984). Of the 3,000 corn-related products sold in one state, tests indicated that 15 of a random sample of 175 had EDB residues above the safe level.

a. In the context of the problem, describe the population parameter p for which $\hat{p} = 15/175$ is a point estimate.

b. Estimate the bound on the error of estimation associated with $\hat{p}$ in part **a**. Interpret this bound in the context of the problem.

c. Construct an approximate 95% confidence interval for p.

d. Do the data provide sufficient evidence to indicate that more than 7% of the corn-related products in this state would have to be removed from shelves and warehouses? Test using $\alpha = .05$, and interpret your test results.

20.16 A sample survey is undertaken to determine the proportion of voters in a county who favor a proposal to create urban enterprise zones that would seek to attract new business and job opportunities in declining areas of the county's cities. A random sample of 1,000 voters is selected from 50,840 eligible voters in the county. Of the

1,000 voters, 620 said they would favor the proposal. Use the techniques outlined in this section to find an approximate 95% confidence interval for the true proportion of the county's voters who favor the creation of urban enterprise zones.

20.17 A small grocery chain, which stocks 410 items, conducted an audit to compare the dollar value of the inventory shown on its books with the actual value of the inventory on hand. Sixty items were randomly selected from the 410, each of the 60 items was inventoried, and the difference between the book and actual values of the inventory was recorded. The difference between the book and actual inventories for the 60 items had a mean equal to \$330 and a standard deviation equal to \$546.

a. Estimate the mean difference per item between the book and actual inventories using a 95% confidence interval.

b. Estimate the total difference between the book and actual inventories for the chain. Use a 95% confidence interval.

20.18 In an urban industrial community, 70,500 persons are classified as potential members of the work force. An economist who wishes to investigate the unemployment rate in the community interviews 6,150 potential members of the work force and finds that 572 are currently jobless. Estimate the current unemployment rate in the community, and place a bound on the error of estimation.

20.5 Simple Random Sampling: Nonresponse

We explained in Section 3.7 how to draw a simple random sample, but we did not comment on the physical problem of actually doing it. For example, as we mentioned early in this chapter, it would be extremely difficult to select a random sample of 3,000 households from all the households in the United States. And, even if we had a frame, it would be costly to contact the selected households.

Two methods for reducing the cost of random sampling are to use a telephone survey or a mailed survey. This type of sampling eliminates transportation costs and reduces labor costs, but it introduces a serious difficulty, the problem of **nonresponse**. By this, we mean that sampling units contained in a sample do not produce sample observations. For example, an individual may not be at home when telephoned or may refuse to complete and mail back a questionnaire.

Nonresponse is a serious problem because it may lead to very biased results. There may be a high correlation between the type of response and whether or not a person responds. For example, most citizens in a community might have an opinion on a school bond issue, but the respondents in a mail survey might very well be those with vested interests in the outcome of the survey—say, parents with children of school age, or school teachers, or those whose taxes might be substantially affected. Others with no vested interests might have opinions on the issue but might not take the time to respond. For this example, the absence of the nonrespondents' data could lead to a larger estimate of the percentage in favor of the issue than was actually the case. In other words, the absence of the nonrespondents' data could lead to a biased estimate. This was one of the problems in the *Literary Digest* poll discussed in Case Study 20.3.

The problem of nonresponse identifies a very important sampling problem. If your sampling plan calls for a specific collection of sampling units, failure to acquire the responses from those units may violate your sampling plan and lead to biased estimates. If you intend to select a random sample and you cannot obtain the responses from some of the sampling units, then your sampling procedure is *no longer random*, and the methodology based on it and the product of the methodology are suspect.

There are ways for coping with nonresponse. Most involve tracking down all or part of the nonrespondents and using the additional information to adjust for the missing nonrespondent data. For mailed surveys, however, it has been found that the inclusion of a monetary incentive with the questionnaire—even as little as 25¢—will substantially increase the response rate of the survey (Armstrong, 1975).

20.6 Stratified Random Sampling

Suppose you were in the wholesale seafood business in a city that had three distinctly different market areas. To plan your purchasing, you wish to obtain an estimate of the mean monthly seafood consumption per household in the city.

If you base your estimate of the mean monthly consumption, μ, of seafood per household on the mean, $\bar{x}$, of a random sample of n households selected within the city, the standard error that measures the variation associated with your estimate is

$$\sigma_{\bar{x}} = \frac{\sigma}{\sqrt{n}}\sqrt{\frac{N-n}{N}}$$

Notation for Stratified Random Sampling

k = Number of strata

N_i = Number of sampling units in stratum i

N = Number of sampling units in the population

$= N_1 + N_2 + \cdots + N_k$

n_i = Number of sampling units selected from stratum i

n = Total number of sampling units in the sample

$= n_1 + n_2 + \cdots + n_k$

$$\bar{x} = \text{Mean of the sample for stratum } i = \frac{\sum_{j=1}^{n_i} x_{ij}}{n_i}$$

where x_{ij} is the jth measurement obtained from stratum i. Also,

$$s_i^2 = \text{Sample variance for stratum } i = \frac{\sum_{j=1}^{n_i} (x_{ij} - \bar{x}_i)^2}{n_i - 1}$$

Estimation of the Population Mean, μ: Stratified Random Sampling

Estimator of μ: $\bar{x}_{st} = \frac{1}{N}(N_1\bar{x}_1 + N_2\bar{x}_2 + \cdots + N_k\bar{x}_k)$

Estimated bound on the error of estimation: $2\hat{\sigma}_{\bar{x}_{st}} = 2\sqrt{\frac{1}{N^2}\sum_{i=1}^{k} N_i^2\left(\frac{N_i - n_i}{N_i}\right)\frac{s_i^2}{n_i}}$

Approximate 95% confidence interval: $\bar{x}_{st} \pm 2\hat{\sigma}_{\bar{x}_{st}}$

Estimation of the Population Total, τ: Stratified Random Sampling

Estimator of τ: $\hat{\tau}_{st} = N\bar{x}_{st} = N_1\bar{x}_1 + N_2\bar{x}_2 + \cdots + N_k\bar{x}_k$

Estimated bound on the error of estimation: $2\hat{\sigma}_{\hat{\tau}_{st}} = 2\sqrt{\sum_{i=1}^{k} N_i^2\left(\frac{N_i - n_i}{N_i}\right)\frac{s_i^2}{n_i}}$

Approximate 95% confidence interval: $\hat{\tau} \pm 2\hat{\sigma}_{\hat{\tau}_{st}}$

Estimation of a Population Proportion, p: Stratified Random Sampling

Estimator of p: $\hat{p}_{st} = \frac{1}{N}(N_1\hat{p}_1 + N_2\hat{p}_2 + \cdots + N_k\hat{p}_k)$

where $\hat{p}_i$ is the sample proportion for stratum i ($i = 1, 2, \ldots, k$). Also,

Estimated bound on the error of estimation: $2\hat{\sigma}_{\hat{p}_{st}} = 2\sqrt{\frac{1}{N^2}\sum_{i=1}^{k} N_i^2\left(\frac{N_i - n_i}{N_i}\right)\frac{\hat{p}_i(1-\hat{p}_i)}{n_i - 1}}$

Approximate 95% confidence interval: $\hat{p}_{st} \pm 2\hat{\sigma}_{\hat{p}_{st}}$

One way to reduce $\sigma_{\bar{x}}$ and reduce the costs of collecting the sample is to select samples within the three markets. The seafood consumption per household is likely to be less in some neighborhoods than in others. Consequently, there will be a substantial amount of variability in the household consumption, x, within the city. In contrast, the variation in consumption within one of the relatively homogeneous (socially and economically) neighborhoods is likely to be less, as is also the variation in consumption within each of the other neighborhoods. This suggests an alternative to simple random sampling. We select a random sample from within each of the three relatively homogeneous marketing areas (called **strata**), estimate the mean consumption within each, and then combine these estimates to obtain an estimate of the mean monthly consumption per household for the whole city. This type of sampling plan, called **stratified random sampling**, has three advantages:

1. Stratified sampling provides additional information; that is, it gives estimates of the mean for *each* stratum as well as of the mean for the entire population.

2. Stratified sampling usually provides more accurate estimates of the population mean than does a simple random sample of the same size because the variability within the strata is usually less than the variability over the entire population.
3. The transportation and administrative costs of sampling within strata are usually less than the costs of sampling within the entire population. This is because the sampling units are frequently geographically closer when selected within the strata than when they are selected randomly from within the entire population.

To summarize, a stratified random sampling plan consists of partitioning the population into a group of k strata, each of which is more homogeneous than the population itself. This sampling plan usually results in more precise estimates (lower variability) at a lower cost. To implement a stratified sampling plan, select a random sample of n_1 sampling units from stratum 1, n_2 from stratum 2, . . ., and n_k from stratum k. Then, the total sample size selected from the population is $n = n_1 + n_2 + \cdots + n_k$. The notation and the formulas for parameter estimators are given in the preceding boxes.

EXAMPLE 20.2

The seafood wholesaler described earlier selected random samples of $n_1 = n_2 = n_3 = 400$ households from within each of the three markets (strata) and obtained from each household an estimate of the dollar amount spent per month on seafood. The number of households in each market along with the sample means and variances are shown in the table.

Neighborhood	N_i	$\bar{x}_i$	s_i^2
1	20,800	\$5.31	16.83
2	6,400	\$9.49	15.10
3	12,600	\$6.75	23.78
	N = 39,800		

a. Estimate the total amount τ spend per month on seafood in the city.

b. Place bounds on the error of estimation.

Solution

a. Substituting the values of N_i and $\bar{x}_i$ into the formula for $\hat{\tau}$, we obtain

$$\hat{\tau} = N\bar{x}_{st} = N_1\bar{x}_1 + N_2\bar{x}_2 + N_3\bar{x}_3$$
$$= (20{,}800)(5.31) + (6{,}400)(9.49) + (12{,}600)(6.75) = \$256{,}234$$

b. The bound on the error of estimation is

$$2\hat{\sigma}_{\hat{\tau}} = 2\sqrt{\sum_{i=1}^{k} N_i^2\left(\frac{N_i - n_i}{N_i}\right)\frac{s_i^2}{n_i}}$$
$$= 2\sqrt{(20{,}800)^2\left(\frac{20{,}800 - 400}{20{,}800}\right)\left(\frac{16.83}{400}\right) + (6{,}400)^2\left(\frac{6{,}400 - 400}{6{,}400}\right)\left(\frac{15.10}{400}\right) + (12{,}600)^2\left(\frac{12{,}600 - 400}{12{,}600}\right)\left(\frac{23.78}{400}\right)}$$
$$= \$10{,}666$$

Thus, we estimate the total monthly expenditure for seafood in the city (for the month sampled) to be \$256,234, and an approximate 95% confidence interval is \$256,234 ± \$10,666, or \$245,568 to \$266,900.

EXAMPLE 20.3

Refer to Example 20.2, and estimate the mean monthly expenditure for seafood per household in neighborhood 2.

Solution

Estimates of the mean expenditure per month per household for seafood for the three neighborhoods might play an important role in deciding how to allocate sales effort and in deciding where to locate retail markets. An estimate of the mean monthly expenditure per household for neighborhood 2 is $\bar{x}_2 = \$9.49$. The estimated bound on the error of estimation is

$$2\hat{\sigma}_{\bar{x}_2} = 2\frac{s_2}{\sqrt{n_2}}\sqrt{\frac{N_2 - n_2}{N_2}} = 2\frac{\sqrt{15.10}}{\sqrt{400}}\sqrt{\frac{6{,}400 - 400}{6{,}400}} = \$.38$$

Thus, we estimate the mean monthly expenditure per household in neighborhood 2 for the sampled month to be \$9.49. We are reasonably certain that the true mean monthly expenditure per household in neighborhood 2 is between \$9.11 and \$9.87.

Examples 20.2 and 20.3 illustrate the methods for estimating parameters based on the stratified random sampling of $n_1, n_2, \ldots, n_k$ sampling units from the k strata. Without being specific, we know that the standard errors of the estimators will decrease as the total sample size, $n = n_1 + n_2 + \cdots + n_k$ increases, but we have not commented on the relative magnitudes of $n_1, n_2, \ldots, n_k$. As a general rule, we select larger samples from strata with greater variability. More precise determination of the sample size requires numerical estimates of the strata variances. Also, the cost of sampling for each stratum will usually play a role in determining strata sample sizes because the total cost of sampling must be kept within the budget for the project. An example of sample size determination is given in Section 20.7.

Exercises 20.19–20.26

Learning the Mechanics

20.19 A survey based on a stratified random sample produced the data shown in the table.

Stratum	*Stratum Size*	*Measurements*
1	25,000	40, 70, 85, 63, 75, 82, 56, 49, 85, 98, 79, 90, 96, 88, 72, 66, 71, 79, 90, 79
2	10,000	26, 55, 42, 47, 58, 51, 62, 55, 45, 49, 65, 72, 33, 55, 61
3	5,000	10, 32, 30, 21, 40, 19, 23, 36, 30, 27

a. Find k, N_1, N_2, N_3, and N. b. Find n_1, n_2, n_3, and n.
c. Find $\bar{x}_1$, $\bar{x}_2$, and $\bar{x}_3$. d. Find s_1^2, s_2^2, and s_3^2.
e. Estimate the population mean and place a bound on the error of estimation.
f. Estimate the population total and place a bound on the error of estimation.
g. Estimate the proportion of measurements in the population that are over 50 and place a bound on the error of estimation.

20.20 A survey based on a stratified random sample produced the data shown in the table.

Stratum	*Sampling Units in Stratum*	*Measurements*
1	4,000	10, 15, 5, 30, 25, 26, 38, 50, 10, 28
2	6,000	5, 33, 15, 45, 47, 36, 25, 40, 17, 31, 62, 28, 33, 45, 68
3	10,000	28, 75, 62, 43, 31, 48, 35, 26, 5, 81, 66, 18, 33, 38, 40, 45, 46, 18, 62, 40
4	15,000	45, 43, 15, 78, 92, 105, 38, 45, 49, 10, 36, 48, 17, 82, 76, 51, 39, 46, 40, 52, 88, 20, 40, 41, 50

a. Estimate the population mean, μ, and place a bound on the error of estimation.
b. Estimate the population total, τ, and place a bound on the error of estimation.
c. Estimate the proportion of the measurements in the population, p, that are between 35 and 55, inclusive, and place a bound on the error of estimation.

20.21 Refer to Exercise 20.20. Estimate the mean of stratum 4 and place a bound on the estimation error.

Applying the Concepts

20.22 How much does corporate America spend on employee training? This and many other training-related questions were addressed in a survey of U.S. organizations (Gordon, 1986). A stratified sample of 15,210 companies was drawn from Dun and Bradstreet's directory of U.S. businesses. The strata and the number of usable responses received are described in the accompanying table. The table also presents data (in thousands of dollars) on the outside training expenditures (e.g., expenditures for seminars, conferences, audiovisual equipment, computer courseware, books, and films) budgeted for 1986 by the survey respondents; the standard deviations are fictitious.

Employees in Organization	**Organizations in Stratum**	**Usable Responses**	**Mean Outside Training Budget**	**Standard Deviation of Outside Training Budget**
50–99	114,464	87	11.7	2.0
100–499	91,754	444	26.6	4.3
500–999	11,011	357	42.3	7.1
1,000–2,499	7,340	553	89.6	10.8
2,500–9,999	3,670	575	142.5	15.3
≥10,000	1,147	534	604.5	100.5
Total	229,386	2,550		

a. The use of these data to estimate the mean or the total budgeted outside training expenditure for the population of firms with 50 or more employees could yield a biased estimate. Explain.
b. Use an approximate 95% confidence interval to estimate the mean outside training expenditure for the population.
c. Use an approximate 95% confidence interval to estimate the total outside training expenditure for the population.
d. What assumptions would have to hold in order for your inferences of parts **b** and **c** to be valid?
e. Would the estimated bound on the error of estimation in part **b** change substantially if the finite population correction factor were ignored? Explain.

20.23 In 1980, the U.S. Department of Labor classified approximately 185,000 persons as health care administrators and estimated that another 105,000 positions would be created during the 1980s. Health care administrators include hospital administrators, managers of nursing homes, and managers of health maintenance organizations. In order to estimate the mean 1980 income of the head administrators of the 6,965 hospitals in the United States, a labor economist used stratified random sampling to select 30 administrators to be questioned about their incomes. The population was stratified according to the number of beds in each administrator's hospital. The income results (in thousands of dollars) are shown for each stratum.

<100 Beds ($N_1 = 3{,}210$)	*100–299 Beds* ($N_2 = 2{,}015$)		*≥300 Beds* ($N_3 = 1{,}740$)		
32.0	39.2	48.9	69.2	60.0	68.1
39.1	55.4	46.1	65.0	54.8	62.4
35.6	51.6	44.6	58.9	68.8	45.0
36.2	48.0	45.2	49.3	57.3	56.7
38.7	37.5	27.3	70.5	71.1	59.5

Source: Wright, J. W. *The American Almanac of Jobs and Salaries*. New York: Avon, 1982. P. 608.

a. Find a point estimate for the mean 1980 income of hospital administrators.
b. Place bounds on the error of estimation associated with your point estimate in part **a**, and interpret the bounds in the context of the problem.
c. Find an approximate 95% confidence interval for the mean income of administrators of hospitals with 300 or more beds.
d. Examine the sample data and suggest a reason why the labor economist chose to allocate the sample size unevenly across the strata.

20.24 Since 1978, Internal Revenue Service (IRS) agents used sampling procedures to facilitate the auditing of tax returns of individuals and businesses. For example, sampling is used to estimate the total value of the error associated with a particular account balance reported on a tax return. Generally, agents use 95% confidence intervals to estimate such quantities. If substantial error is found, adjustments to the tax return will be suggested (Brown, 1982; Hull and Everett, 1982).

In auditing the investment credit of a particular corporation, the IRS stratified the firm's population of 1,000 invoices containing the appropriate investment credit information into four strata according to the size of the expenditure involved: \$0 to \$999, \$1,000 to \$2,999, \$3,000 to \$9,999, or \$10,000 and over. Random samples of invoices of sizes $n_1 = 6$, $n_2 = 8$, $n_3 = 10$, and $n_4 = 15$ were drawn from each of the respective strata. Each sampled invoice was examined to determine whether it was properly treated by the firm in determining the firm's investment credit. The table describes the error associated with each sampled invoice as identified by the IRS. Positive errors reflect an overstatement by the firm of its investment credit and negative errors reflect an understatement.

Investment Credit Errors

\$0–\$999 ($N_1 = 100$)	\$1,000–\$2,999 ($N_2 = 400$)		\$3,000–\$9,999 ($N_3 = 300$)		≥\$10,000 ($N_4 = 200$)		
\$ 10	\$ 0	\$550	\$ 750	\$ 0	\$ 0	\$1,800	\$ 0
0	0	0	0	1,500	0	0	0
0	100		0	0	5,000	0	500
−15	0		1,000	2,000	0	0	
25	−50		0		0	0	
20	0		0		0	0	

a. Find a point estimate for the total value of the error in the investment credit claimed by the firm.
b. Place bounds on the error of estimation associated with your point estimate of part **a**, and interpret the bounds in the context of the problem.
c. Find an approximate 95% confidence interval for the total value of the error.
d. The firm claimed an investment credit of \$500,000. Based on your answers to parts **a–c**, approximately how much investment credit should the firm have claimed? Explain.

20.25 An economist wants to estimate the mean annual income of families in a mainly industrial community. Since one section of the city houses primarily factory workers, one mostly company executives, and the remaining area mostly farmers, the economist decides to use the three relatively homogeneous areas as strata. The economist selected random samples of 30 homes from within each of the three strata and gathered information on the annual income for each family. The number of households in each section of the city along with the sample means and variances are given in the table. Estimate the mean annual income for households in the community, and place a bound on the error of estimation.

City Section	N_i	$\bar{x}_i$	s_i^2
Factory workers	360	\$14,900	9,150,500
Executives	74	39,250	25,003,000
Farmers	95	23,800	16,801,100

20.26 The owners of a chain of department stores wish to estimate the proportion of customer accounts for which payments are 6 or more weeks overdue. A random sample of customers is taken in each of the chain's four stores, and $\hat{p}$, the sample proportion of overdue accounts in each store, is determined. These data and the total number of customer accounts are given in the table. Using the stores as strata, give an estimate of the true proportion of customer accounts that are 6 or more weeks overdue in this chain of department stores. Place a bound on the error of estimation.

Store	N_i	n_i	$\hat{p}$
1	1,572	100	.28
2	2,369	100	.31
3	3,007	120	.35
4	2,981	120	.10

20.7 Determining the Sample Size

To determine the sample size for sample surveys, you use essentially the same procedures as explained in Sections 7.2 and 7.5. Generally speaking, the bound on the error of estimation will be approximately inversely proportional to the number of sampling units. This relationship does not hold exactly because of the effect of the finite population correction factor and because, for stratified random sampling, you are really selecting k random samples, one corresponding to each stratum. Nevertheless, doubling the sample size n will decrease, approximately, the bound on the error of estimation to $1/\sqrt{2}$ times its original value (even for stratified random sampling if you keep $n_1, n_2, \ldots, n_k$ in the same proportions). If you quadruple the sample size n, you will cut the bound on the error estimation in half.

To select the sample size to estimate a population parameter based on a specific sample survey design, first decide on the accuracy you desire in your estimate: what bound on the error of estimation are you willing to tolerate? Set this number equal to the estimated bound on the error of estimation and solve the resulting equation for n. We will illustrate with an example.

EXAMPLE 20.4

Suppose the seafood wholesaler in Example 20.2 wanted to reduce the bound on the error of estimating the total, τ, of monthly seafood expenditure in the city to \$5,000. That is, the wholesaler wants to estimate the total monthly expenditure to within \$5,000 with approximate 95% confidence. If the wholesaler plans to use equal sample sizes, approximately how many households must be selected from within each stratum to estimate τ with a bound on the error of estimation of \$5,000?

Solution

We found in Example 20.2 that the bound on the error of estimation for $n_1 = n_2 = n_3 = 400$ was \$10,666. Since \$5,000 is slightly less than half of this value, we know (without calculating) that it will require approximately four times as many households to reduce the bound on the error of estimation to $1/\sqrt{4} = ½$ its original size.

To solve the problem formally, let

$$2\hat{\sigma}_{\hat{\tau}} = 2\sqrt{\sum_{i=1}^{k} N_i^2\left(\frac{N_i - n_i}{N_i}\right)\frac{s_i^2}{n_i}} = \$5{,}000$$

where (since we want equal sample sizes) we will let $n_1 = n_2 = n_3 = n_s$. We will also assume, for a first approximation to n_s, that $(N_i - n_i)/N_i \approx 1$. Substituting these values, along with $N_1 = 20{,}800$, $N_2 = 6{,}400$, $N_3 = 12{,}600$, and the sample variances from the earlier samples into the formula for the bound on the error of estimation yields

$$2\sqrt{\frac{(20{,}800)^2(16.83)}{n_s} + \frac{(6{,}400)^2(15.10)}{n_s} + \frac{(12{,}600)^2(23.78)}{n_s}} = 5{,}000$$

and solving for n_s yields $n_s = 1{,}868$.

This solution will be larger than the actual sample sizes required to achieve a \$5,000 bound on the error of estimation because the finite population correction factors will not equal 1. We now substitute $n_s = 1,868$ into the equation and resolve it for n_s:

$$2\sqrt{(20,800)^2\left(\frac{20,800 - 1,868}{20,800}\right)\left(\frac{16.83}{n_s}\right) + (6,400)^2\left(\frac{6,400 - 1,868}{6,400}\right)\left(\frac{15.10}{n_s}\right) + (12,600)^2\left(\frac{12,600 - 1,868}{12,600}\right)\left(\frac{23.78}{n_s}\right)}$$
$$= 5,000$$

or $n_s = 1,645$.

This solution will be too small because we used $n_s = 1,868$ in the finite population correction factor. If we use this new value to calculate the finite population correction factors and re-solve the equation for n_s, we obtain $n_s = 1,672$. Thus, we would select approximately 1,672 households from each stratum in order to reduce the bound on the error of estimation to \$5,000.

Example 20.4 illustrates that although the finite population correction factor does affect the sample size required to obtain a specified bound on the error of estimation, the effect is not large. The solution, assuming the finite population correction factor equals 1, was 1,868—a value not much larger than the solution obtained using the correction factor.

Often previous sample data will not be available from which to obtain estimates of the strata variances. These must then be more crudely estimated, possibly by using each stratum range divided by 4, or by calculating s_i^2 from small pilot samples drawn from each stratum. Although this will produce only crude estimates of the required sample sizes, the procedure is still worthwhile, since the general magnitude of the sample size can be determined. This, in turn, allows the researcher to determine an approximate relationship between sampling costs and the bound on the error of estimation.

Exercises 20.27–20.31

Applying the Concepts

20.27 Refer to Exercise 20.13. In estimating the mean external audit fee, suppose the estimated bound on the error of estimation is to be no more than \$200,000. How many additional firms should be sampled?

20.28 Refer to Exercise 20.14. Suppose the auditor wants to estimate the total value of the parts inventory to within \$100,000 of the true value with approximately 95% confidence. In order to obtain this result, how many of the firm's 5,000 inventory items should be sampled?

20.29 Refer to Exercise 20.23. Suppose the labor economist wants to estimate the mean 1980 income of the head administrators to within \$1,000 with approximately 95% confidence.

a. If an equal number of administrators is to be sampled from each stratum, what is the total number of administrators that should be sampled?

b. If 30 administrators are to be sampled from each of the first two strata, "<100 beds" and "100–299 beds," approximately how many should be sampled from the third stratum?

20.30 Refer to Exercise 20.25. The economist now desires to reduce the bound on the error of estimation to $600. Assuming equal sample sizes, how many homes from each city section should be sampled so that the economist estimates the mean annual income for households in the community to within $600 of the true value (with approximately 95% confidence)?

20.31 Refer to Exercise 20.16. It is desired to estimate the proportion of the county's voters who favor the creation of urban enterprise zones to within .01 of the true value with approximate 95% confidence. In order to obtain this accuracy, how many of the county's 50,840 eligible voters should be sampled?

Summary

This chapter introduced the important topic of **survey sampling**. We presented several sampling designs for reducing the cost of conducting a sample survey and also presented the associated methods of estimation.

The objective of a sample survey is usually to estimate one or more of three population parameters: a **population mean**, μ, a **population total**, τ, and a **population proportion**, p. We introduced three sample survey designs for collecting a sample: simple random sampling, stratified random sampling, and cluster sampling.

Simple random sampling is conceptually easy to understand, but it is often difficult to conduct, and it may be more costly than other sample survey designs. This is because it may be difficult and costly to construct the frame, and the cost of collecting the sample may be relatively large due to geographic separation of the sampling units.

Stratified random sampling is used when the population can be subdivided into groups (strata) of sampling units that possess smaller variability within each stratum than between strata. Random samples are selected from within each stratum, and the information contained in these samples is then pooled to obtain an estimate of the desired population parameter. Stratified sampling has the advantage that it enables the sampler to obtain estimates of the individual stratum parameters. In addition, it may be less costly than simple random sampling because the strata frames are often easier to construct and the sampling units are often geographically closer to one another in comparison to simple random sampling within the entire population. Finally, stratified samples often result in more precise estimates because the variance within strata is less than the variance of the entire population.

There are many other sampling designs available to a sample surveyor; some are variations on stratified random sampling, and other designs are completely different. In addition, different types of estimators can be used with these designs. In this introduction to survey sampling, our intent was to present only the basic elements of a sample survey and several of the most important sample survey designs. More thorough presentations are given in textbooks devoted to this topic (see the references at the end of this chapter).

Supplementary Exercises 20.32–20.39

20.32 How do the makers of Kleenex facial tissue determine how many tissues to include in a box or personal pack? In the case of the personal pack, the marketing experts at Kimberly-Clark Corporation have "little doubt that the company should put 60 tissues in each pack" (Koten, 1984). By having hundreds of customers keep count of their Kleenex use in diaries, researchers determined that 60 is the average number of times a person uses a tissue during a cold. Suppose a 1987 study of a random sample of 250 people from a midwestern town of 5,000 people yielded these summary data on the number of times they used a tissue when they had a cold: $\bar{x} = 57$ and $s = 26$.

a. Estimate the mean number of times people in the midwestern town use a tissue during a cold. Place a bound on the estimation error.

b. Assume each person in the town had one cold during 1987. Use an approximate 95% confidence interval to estimate the total number of tissues used in the town in 1987 during colds. How might such information be used by Kimberly-Clark?

20.33 When a buyer charges a purchase, the seller records the sale in his or her accounting records under the category accounts receivable. At any point in time, a large business may have many thousands of accounts receivable. A department store in Boston is interested in estimating (1) the mean age of its accounts receivable, (2) the total monetary value of its accounts receivable, and (3) the proportion of accounts receivable that are more than 90 days old. On September 10, the store's accounting department randomly selected 100 of its 15,887 accounts receivable and recorded the age, x, and the value, y, of each account. Define in the context of the problem (in words, not formulas):

a. n and N **b.** $\bar{x}$ and μ **c.** $\hat{\tau}$ and τ **d.** $\hat{p}$ and p **e.** $\hat{\tau} \pm 2\hat{\sigma}_{\hat{\tau}}$

20.34 Refer to Exercise 20.33. Besides Boston, the chain has department stores in three other cities. It was decided to estimate the characteristics of the accounts receivable for the entire chain. Each store was treated as a stratum and the stratified sample design shown in the table was used to sample accounts receivable.

Stratum	*Accounts Receivable on September 10*	*Sample Size*
1. Boston	15,887	75
2. New York	25,010	100
3. Atlanta	10,982	50
4. Dallas	18,931	80

The information collected from each account and the parameters to be estimated are described in Exercise 20.33. Define in the context of the problem (in words, not formulas):

a. N **b.** n_2 and n_4 **c.** $\bar{x}_3$ and $\bar{x}_{st}$ and μ **d.** s_1 and σ_1 **e.** $\hat{p}_2$ and $\hat{p}_{st}$ and p

20.35 Corporations, labor unions, and trade and medical associations can set up organizations known as political action committees (PACs) to raise money from their members for support of political candidates believed to represent their best interests. Approximately 3,500 PACs currently exist. Federal law limits the amount a PAC can contribute to a presidential candidate to $5,000 per election. However, a loophole in the law permits committees to spend as much as they want to elect or defect a candidate as long as there is no cooperation or contact with the candidate or the candidate's authorized agents ("ABC's of how America chooses a president," *U.S. News and World Report*, Feb. 20, 1984, pp. 39–46). In early 1984, in order to estimate the amount of money that would be spent in support of Ronald Reagan in the 1984 presidential election, a political economist hired by the Democratic party randomly sampled 30 PACs and asked them how much they expected to spend in support of Reagan. The following results were obtained (in thousands of dollars):

10	0	5	18	0	5	22	0	50	60
35	0	0	18	35	0	40	10	50	20
150	15	0	0	30	15	0	20	15	10

a. Estimate the total amount that PACs expected to spend in support of Reagan in 1984. Place a bound on the error of estimation.
b. Use an approximate 95% confidence interval to estimate the proportion of PACs that planned to support Reagan.
c. In addition to sampling error, what might cause your estimate for part **a** to be inaccurate?

20.36 When a poll reports, for example, that 61% of the public supports a program of national health insurance, it usually also reports the sampling error. For example, a poll might report that the estimate is accurate to within plus or minus 3%. An essay in *Time* magazine ("How not to read the polls," Apr. 28, 1980, pp. 72–73) points out:

Readers consistently misinterpret the meaning of this 'warning label.' . . . [The sampling error warning] says nothing about errors that might be caused by a sloppily worded question or a biased one or a single question that evokes complex feelings. Example: 'Are you satisfied with your job?' Most important of all, warning labels about sampling error say nothing about whether or not the public is conflict-ridden or has given a subject much thought. This is the most serious source of opinion poll misinterpretation.

Carefully explain the difference between sampling error and nonsampling error, both in general and in the context of the above quote.

20.37 Publishers of a weekly nationwide business magazine believe that a large proportion of their Florida subscribers invest in the stock market. They would like to be able to use this information to persuade brokerage firms in Florida to advertise in their magazine. The publishers send each of the 500,000 subscribers in Florida a questionnaire about stock investments. A total of 10,000 of the questionnaires are returned, and of these, 9,296 subscribers respond that they do currently have stock market investments.
a. Use this information to estimate the proportion of Florida subscribers who invest in the stock market, and place a bound on the error of estimation.
b. Should a brokerage firm in Florida consider the estimate in part **a** to be reliable? Explain.

20.38 Refer to Exercise 20.37. The publishers of the magazine have decided to alter their sampling scheme. Instead of sending out questionnaires, they will personally interview a random selection of their Florida subscribers. In order to estimate, with aproximate 95% confidence, the true proportion investing in the stock market with a bound on the error of estimation of .05, how many of the magazine's 500,000 Florida subscribers should be included in the sample?

20.39 In its next advertising campaign, a tobacco company will use an estimate of the average number of cigarettes smoked per day by its employees. The company expects a difference in amounts smoked by men and women, so it has decided to stratify on sex. From the company's 535 male employees, 50 are selected, and of the 366 female employees, 40 are sampled. An estimate of the number of cigarettes smoked per day by each is recorded. The accompanying table gives the respective means and variances of the samples. Estimate the average number of cigarettes smoked per day by the company employees, and place bounds on the error of estimation.

	N_i	n_i	$\bar{x}_i$	s_i^2
Men	535	50	8.5	16.8
Women	366	40	5.2	21.2

Using the Computer

Treat the 1,000 zip codes in the data set of Appendix C as if they were the population of all U.S. zip codes. The four census regions geographically stratify this population.

a. Use stratified random sampling with $n_1 = 30$, $n_2 = 20$, $n_3 = 35$, and $n_4 = 15$ to estimate the mean number of households per zip code in the United States. Place a bound on the error of estimation. Since your statistical software package probably does not have a command to compute an estimator from a stratified sample, you will need to either (1) write a short program to compute the estimator and its bound, or (2) use the computer to find $\bar{x}_i$ and s_i^2 $(i = 1, 2, 3, 4)$ and then calculate $\bar{x}_{st}$ and $\hat{\sigma}_{\bar{x}_{st}}$ by hand.

b. Use the sample data from part **a** to construct a 95% confidence interval for the mean number of households per zip code in the Northwest census region.

References

Arkin, H. *Sampling Methods for the Auditor.* New York: McGraw-Hill, 1982. Chapters 1, 3, and 4.

Armstrong, J. S. "Monetary incentives in mail surveys." *Public Opinion Quarterly,* 1975, Vol. 39, pp. 111–116.

Berg, A., Klauda, P., and Feyder, S. "EDB banned as grain pesticide." *Minneapolis Star and Tribune,* Feb. 4, 1984.

Brown, D. B. "Statistical sampling in the IRS examination of large cases." *The Tax Executive,* Apr. 1982, Vol. 34, pp. 175–179.

Bryson, M. C. "The *Literary Digest* poll: Making of a statistical myth." *American Statistician,* Nov. 1976.

Chagall, D. "Can you believe the ratings?" *TV Guide,* June 24, 1978, 3.

Cochran, W. G. *Sampling Techniques,* 3d ed. New York: Wiley, 1977.

Cox, E. P. III. *Marketing Research, Information for Decision-Making.* New York: Harper & Row, 1979, P. 7.

Ferber, R., Sheatsley, P., Turner, A., and Waksberg, J. *What Is a Survey?* Washington, D.C.: American Statistical Association, 1980.

Freedman, D., Pisani, R., and Purves, R. *Statistics.* New York: Norton, 1978. Chapter 19.

Glasser, G. J., and Metzger, G. D. "Random-digit dialing as a method of telephone sampling." *Journal of Marketing Research,* Feb. 1972, Vol. 9, pp. 59–64.

Gordon, J. "*Training* magazine's industry report 1986." *Training,* Oct. 1986, Vol. 23, pp. 26–42.

Greenberg, B. G., Kuebler, R. T., Abernathy, J. R., and Horvitz, D. G. "Application of randomized response technique in obtaining quantitative data." *Journal of the American Statistical Association,* 1971, 66.

Hansen, M. H., Hurwitz, W. N., and Madow, W. G. *Sampling Survey Methods and Theory,* Vol. 1, New York: Wiley, 1953.

Huff, D. *How to Lie with Statistics.* New York: Norton, 1954, p. 20.

Hull, R. P., and Everett, J. O. "On the use of statistical sampling in tax audits." *Tax Executive*, Oct. 1982, Vol. 35, pp. 51–54.

Kish, L. *Survey Sampling*. New York: Wiley, 1965.

Koten, J. "Why do hot dogs come in packs of 10 and buns in 8's or 12's?" *Wall Street Journal*, Sept. 21, 1984.

Scheaffer, R., Mendenhall, W., and Ott, R. L. *Elementary Survey Sampling*, 2nd ed. North Scituate, MA: Duxbury, 1979.

Taeuber, C. "Information for the nation from a sample survey." In Tanur et al., eds., *Statistics: A Guide to the Unknown*, 2nd ed. San Francisco: Holden-Day, 1978.

Wallace, W. A. "Internal auditors can cut outside CPA costs." *Harvard Business Review*, Mar.–Apr. 1984, pp. 16–20.

APPENDIX A

Basic Counting Rules

Simple events associated with many experiments have identical characteristics. If you can develop a counting rule to count the number of simple events, it can be used to aid in the solution of many probability problems. For example, many experiments involve sampling n elements from a population of N. Then, as explained in Section 3.1, we can use the formula

$$\binom{N}{n} = \frac{N!}{n!(N-n)!}$$

to find the number of different samples of n elements that could be selected from the total of N elements. This gives the number of simple events for the experiment.

Here, we give you a few useful counting rules. You should learn the characteristics of the situation to which each rule applies. Then, when working a probability problem, carefully examine the experiment to see whether you can use one of the rules.

Learning how to decide whether a particular counting rule applies to an experiment takes patience and practice. If you want to develop this skill, try to use the rules to solve some of the exercises in Chapter 3. You will also find large numbers of exercises in the texts listed in the references at the end of Chapter 3. Proofs of the rules below can be found in the text by W. Feller listed in the references to Chapter 3.

1. **Multiplicative rule:** You have k *sets* of different elements, n_1 in the first set, n_2 in the second set, . . . , and n_k in the kth set. Suppose you want to form a sample of k elements *by taking one element from each of the k sets*. The number of different samples that can be formed is the product

$$n_1 \cdot n_2 \cdot n_3 \cdot \cdots \cdot n_k$$

EXAMPLE A.1

If a product can be shipped by four different airlines and each airline can ship via three different routes, how many ways can you ship the product?

Solution

A method of shipment corresponds to a pairing of one airline and one route. Therefore, $k = 2$, the number of airlines is $n_1 = 4$, the number of routes is $n_2 = 3$, and the number of ways to ship the product is $n_1 \cdot n_2 = (4)(3) = 12$.

How the multiplicative rule works can be seen by using a **decision tree**. The airline choice is shown by three branching lines in Figure A.1.

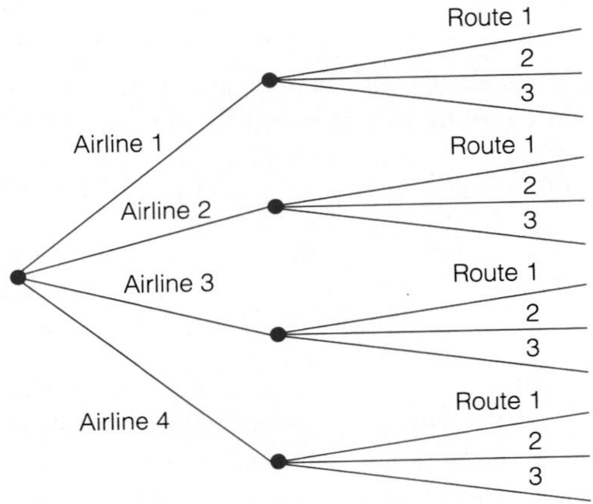

FIGURE A.1 ▶
Decision tree for Example A.1

EXAMPLE A.2

You have twenty candidates for three different executive positions, E_1, E_2, and E_3. How many different ways could you fill the positions?

Solution

For this example, there are $k = 3$ sets of elements:

Set 1: The candidates available to fill position E_1

Set 2: The candidates remaining (after filling E_1) that are available to fill E_2

Set 3: The candidates remaining (after filling E_1 and E_2) that are available to fill E_3

The numbers of elements in the sets are $n_1 = 20$, $n_2 = 19$, $n_3 = 18$. Thus, the number of different ways to fill the three positions is $n_1 \cdot n_2 \cdot n_3 = (20)(19)(18) = 6{,}840$.

2. **Partitions rule:** You have a *single set* of N distinctly different elements, and you want to partition them into k sets, the first set containing n_1 elements, the second containing n_2 elements, . . . , and kth containing n_k elements. The number of different partitions is

$$\frac{N!}{n_1!n_2! \cdot \cdots \cdot n_k!} \quad \text{where} \quad n_1 + n_2 + n_3 + \cdots + n_k = N$$

EXAMPLE A.3

You have twelve construction workers and you want to assign three to job 1, four to job 2, and five to job 3. How many different ways could you make this assignment?

Solution

For this example, $k = 3$ (corresponding to the $k = 3$ job sites), $N = 12$, and $n_1 = 3$, $n_2 = 4$, $n_3 = 5$. Then, the number of different ways to assign the workers to the job sites is

$$\frac{N!}{n_1!n_2!n_3!} = \frac{12!}{3!4!5!} = \frac{12 \cdot 11 \cdot 10 \cdot \cdots \cdot 3 \cdot 2 \cdot 1}{(3 \cdot 2 \cdot 1)(4 \cdot 3 \cdot 2 \cdot 1)(5 \cdot 4 \cdot 3 \cdot 2 \cdot 1)} = 27{,}720$$

3. **Combinations rule:** The combinations rule given in Chapter 3 is a special case ($k = 2$) of the partitions rule. That is, sampling is equivalent to partitioning a set of N elements into $k = 2$ groups: elements that appear in the sample and those that do not. Let $n_1 = n$, the number of elements in the sample, and $n_2 = N - n$, the number of elements remaining. Then the number of different samples of n elements that can be selected from N is

$$\frac{N!}{n_1!n_2!} = \frac{N!}{n!(N-n)!} = \binom{N}{n}$$

This formula was given in Section 3.1.

EXAMPLE A.4

How many samples of four fire fighters can be selected from a group of ten?

Solution

We have $N = 10$ and $n = 4$; then,

$$\binom{N}{n} = \binom{10}{4} = \frac{10!}{4!6!} = \frac{10 \cdot 9 \cdot 8 \cdot \cdots \cdot 3 \cdot 2 \cdot 1}{(4 \cdot 3 \cdot 2 \cdot 1)(6 \cdot 5 \cdot \cdots \cdot 2 \cdot 1)} = 210$$

APPENDIX B

Tables

Contents

TABLE I Random Numbers

Row \ Column	1	2	3	4	5	6	7	8	9	10	11	12	13	14
1	10480	15011	01536	02011	81647	91646	69179	14194	62590	36207	20969	99570	91291	90700
2	22368	46573	25595	85393	30995	89198	27982	53402	93965	34095	52666	19174	39615	99505
3	24130	48360	22527	97265	76393	64809	15179	24830	49340	32081	30680	19655	63348	58629
4	42167	93093	06243	61680	07856	16376	39440	53537	71341	57004	00849	74917	97758	16379
5	37570	39975	81837	16656	06121	91782	60468	81305	49684	60672	14110	06927	01263	54613
6	77921	06907	11008	42751	27756	53498	18602	70659	90655	15053	21916	81825	44394	42880
7	99562	72905	56420	69994	98872	31016	71194	18738	44013	48840	63213	21069	10634	12952
8	96301	91977	05463	07972	18876	20922	94595	56869	69014	60045	18425	84903	42508	32307
9	89579	14342	63661	10281	17453	18103	57740	84378	25331	12566	58678	44947	05585	56941
10	85475	36857	53342	53988	53060	59533	38867	62300	08158	17983	16439	11458	18593	64952
11	28918	69578	88231	33276	70997	79936	56865	05859	90106	31595	01547	85590	91610	78188
12	63553	40961	48235	03427	49626	69445	18663	72695	52180	20847	12234	90511	33703	90322
13	09429	93969	52636	92737	88974	33488	36320	17617	30015	08272	84115	27156	30613	74952
14	10365	61129	87529	85689	48237	52267	67689	93394	01511	26358	85104	20285	29975	89868
15	07119	97336	71048	08178	77233	13916	47564	81056	97735	85977	29372	74461	28551	90707
16	51085	12765	51821	51259	77452	16308	60756	92144	49442	53900	70960	63990	75601	40719
17	02368	21382	52404	60268	89368	19885	55322	44819	01188	65255	64835	44919	05944	55157
18	01011	54092	33362	94904	31273	04146	18594	29852	71585	85030	51132	01915	92747	64951
19	52162	53916	46369	58586	23216	14513	83149	98736	23495	64350	94738	17752	35156	35749
20	07056	97628	33787	09998	42698	06691	76988	13602	51851	46104	88916	19509	25625	58104
21	48663	91245	85828	14346	09172	30168	90229	04734	59193	22178	30421	61666	99904	32812
22	54164	58492	22421	74103	47070	25306	76468	26384	58151	06646	21524	15227	96909	44592
23	32639	32363	05597	24200	13363	38005	94342	28728	35806	06912	17012	64161	18296	22851
24	29334	27001	87637	87308	58731	00256	45834	15398	46557	41135	10367	07684	36188	18510
25	02488	33062	28834	07351	19731	92420	60952	61280	50001	67658	32586	86679	50720	94953

(*continued*)

TABLE I Continued

Row \ Column	1	2	3	4	5	6	7	8	9	10	11	12	13	14
26	81525	72295	04839	96423	24878	82651	66566	14778	76797	14780	13300	87074	79666	95725
27	29676	20591	68086	26432	46901	20849	89768	81536	86645	12659	92259	57102	80428	25280
28	00742	57392	39064	66432	84673	40027	32832	61362	98947	96067	64760	64584	96096	98253
29	05366	04213	25669	26422	44407	44048	37937	63904	45766	66134	75470	66520	34693	90449
30	91921	26418	64117	94305	26766	25940	39972	22209	71500	64568	91402	42416	07844	69618
31	00582	04711	87917	77341	42206	35126	74087	99547	81817	42607	43808	76655	62028	76630
32	00725	69884	62797	56170	86324	88072	76222	36086	84637	93161	76038	65855	77919	88006
33	69011	65795	95876	55293	18988	27354	26575	08625	40801	59920	29841	80150	12777	48501
34	25976	57948	29888	88604	67917	48708	18912	82271	65424	69774	33611	54262	85963	03547
35	09763	83473	73577	12908	30883	18317	28290	35797	05998	41688	34952	37888	38917	88050
36	91576	42595	27958	30134	04024	86385	29880	99730	55536	84855	29080	09250	79656	73211
37	17955	56349	90999	49127	20044	59931	06115	20542	18059	02008	73708	83517	36103	42791
38	46503	18584	18845	49618	02304	51038	20655	58727	28168	15475	56942	53389	20562	87338
39	92157	89634	94824	78171	84610	82834	09922	25417	44137	48413	25555	21246	35509	20468
40	14577	62765	35605	81263	39667	47358	56873	56307	61607	49518	89656	20103	77490	18062
41	98427	07523	33362	64270	01638	92477	66969	98420	04880	45585	46565	04102	46880	45709
42	34914	63976	88720	82765	34476	17032	87589	40836	32427	70002	70663	88863	77775	69348
43	70060	28277	39475	46473	23219	53416	94970	25832	69975	94884	19661	72828	00102	66794
44	53976	54914	06990	67245	68350	82948	11398	42878	80287	88267	47363	46634	06541	97809
45	76072	29515	40980	07391	58745	25774	22987	80059	39911	96189	41151	14222	60697	59583
46	90725	52210	83974	29992	65831	38857	50490	83765	55657	14361	31720	57375	56228	41546
47	64364	67412	33339	31926	14883	24413	59744	92351	97473	89286	35931	04110	23726	51900
48	08962	00358	31662	25388	61642	34072	81249	35648	56891	69352	48373	45578	78547	81788
49	95012	68379	93526	70765	10592	04542	76463	54328	02349	17247	28865	14777	62730	92277
50	15664	10493	20492	38391	91132	21999	59516	81652	27195	48223	46751	22923	32261	85653
51	16408	81899	04153	53381	79401	21438	83035	92350	36693	31238	59649	91754	72772	02338
52	18629	81953	05520	91962	04739	13092	97662	24822	94730	06496	35090	04822	86774	98289
53	73115	35101	47498	87637	99016	71060	88824	71013	18735	20286	23153	72924	35165	43040
54	57491	16703	23167	49323	45021	33132	12544	41035	80780	45393	44812	12512	98931	91202
55	30405	83946	23792	14422	15059	45799	22716	19792	09983	74353	68668	30429	70735	25499
56	16631	35006	85900	98275	32388	52390	16815	69290	82732	38480	73817	32523	41961	44437
57	96773	20206	42559	78985	05300	22164	24369	54224	35083	19687	11052	91491	60383	19746
58	38935	64202	14349	82674	66523	44133	00697	35552	35970	19124	63318	29686	03387	59846
59	31624	76384	17403	53363	44167	64486	64758	75366	76554	31601	12614	33072	60332	92325
60	78919	19474	23632	27889	47914	02584	37680	20801	72152	39339	34806	08930	85001	87820
61	03931	33309	57047	74211	63445	17361	62825	39908	05607	91284	68833	25570	38818	46920
62	74426	33278	43972	10110	89917	15665	52872	73823	73144	88662	88970	74492	51805	99378

Row \ Column	1	2	3	4	5	6	7	8	9	10	11	12	13	14
63	09066	00903	20795	95452	92648	45454	09552	88815	16553	51125	79375	97596	16296	66092
64	42238	12426	87025	14267	20979	04508	64535	31355	86064	29472	47689	05974	52468	16834
65	16153	08002	26504	41744	81959	65642	74240	56302	00033	67107	77510	70625	28725	34191
66	21457	40742	29820	96783	29400	21840	15035	34537	33310	06116	95240	15957	16572	06004
67	21581	57802	02050	89728	17937	37621	47075	42080	97403	48626	68995	43805	33386	21597
68	55612	78095	83197	33732	05810	24813	86902	60397	16489	03264	88525	42786	05269	92532
69	44657	66999	99324	51281	84463	60563	79312	93454	68876	25471	93911	25650	12682	73572
70	91340	84979	46949	81973	37949	61023	43997	15263	80644	43942	89203	71795	99533	50501
71	91227	21199	31935	27022	84067	05462	35216	14486	29891	68607	41867	14951	91696	85065
72	50001	38140	66321	19924	72163	09538	12151	06878	91903	18749	34405	56087	82790	70925
73	65390	05224	72958	28609	81406	39147	25549	48542	42627	45233	57202	94617	23772	07896
74	27504	96131	83944	41575	10573	08619	64482	73923	36152	05184	94142	25299	84387	34925
75	37169	94851	39117	89632	00959	16487	65536	49071	39782	17095	02330	74301	00275	48280
76	11508	70225	51111	38351	19444	66499	71945	05422	13442	78675	84081	66938	93654	59894
77	37449	30362	06694	54690	04052	53115	62757	95348	78662	11163	81651	50245	34971	52924
78	46515	70331	85922	38329	57015	15765	97161	17869	45349	61796	66345	81073	49106	79860
79	30986	81223	42416	58353	21532	30502	32305	86482	05174	07901	54339	58861	74818	46942
80	63798	64995	46583	09785	44160	78128	83991	42865	92520	83531	80377	35909	81250	54238
81	82486	84846	99254	67632	43218	50076	21361	64816	51202	88124	41870	52689	51275	83556
82	21885	32906	92431	09060	64297	51674	64126	62570	26123	05155	59194	52799	28225	85762
83	60336	98782	07408	53458	13564	59089	26445	29789	85205	41001	12535	12133	14645	23541
84	43937	46891	24010	25560	86355	33941	25786	54990	71899	15475	95434	98227	21824	19585
85	97656	63175	89303	16275	07100	92063	21942	18611	47348	20203	18534	03862	78095	50136
86	03299	01221	05418	38982	55758	92237	26759	86367	21216	98442	08303	56613	91511	75928
87	79626	06486	03574	17668	07785	76020	79924	25651	83325	88428	85076	72811	22717	50585
88	85636	68335	47539	03129	65651	11977	02510	26113	99447	68645	34327	15152	55230	93448
89	18039	14367	61337	06177	12143	46609	32989	74014	64708	00533	35398	58408	13261	47908
90	08362	15656	60627	36478	65648	16764	53412	09013	07832	41574	17639	82163	60859	75567
91	79556	29068	04142	16268	15387	12856	66227	38358	22478	73373	88732	09443	82558	05250
92	92608	82674	27072	32534	17075	27698	98204	63863	11951	34648	88022	56148	34925	57031
93	23982	25835	40055	67006	12293	02753	14827	23235	35071	99704	37543	11601	35503	85171
94	09915	96306	05908	97901	28395	14186	00821	80703	70426	75647	76310	88717	37890	40129
95	59037	33300	26695	62247	69927	76123	50842	43834	86654	70959	79725	93872	28117	19233
96	42488	78077	69882	61657	34136	79180	97526	43092	04098	73571	80799	76536	71255	64239
97	46764	86273	63003	93017	31204	36692	40202	35275	57306	55543	53203	18098	47625	88684
98	03237	45430	55417	63282	90816	17349	88298	90183	36600	78406	06216	95787	42579	90730
99	86591	81482	52667	61582	14972	90053	89534	76036	49199	43716	97548	04379	46370	28672
100	38534	01715	94964	87288	65680	43772	39560	12918	86537	62738	19636	51132	25739	56947

Source: Abridged from W. H. Beyer (ed.). *CRC Standard Mathematical Tables*, 24th edition. (Cleveland: The Chemical Rubber Company), 1976. Reproduced by permission of the publisher.

TABLE II Binomial Probabilities

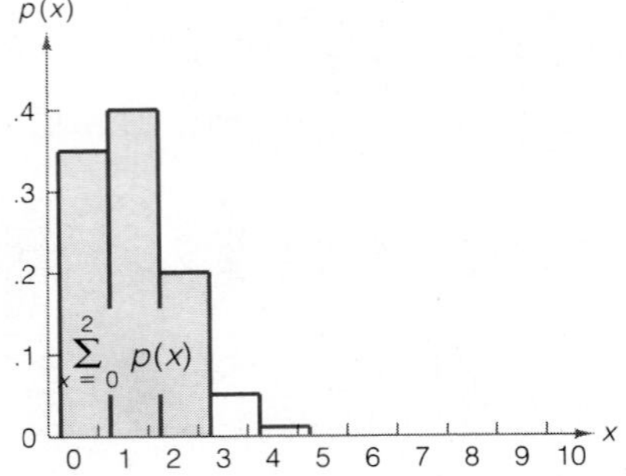

Tabulated values are $\sum_{x=0}^{k} p(x)$. *(Computations are rounded at the third decimal place.)*

a. $n = 5$

k \ p	.01	.05	.10	.20	.30	.40	.50	.60	.70	.80	.90	.95	.99
0	.951	.774	.590	.328	.168	.078	.031	.010	.002	.000	.000	.000	.000
1	.999	.977	.919	.737	.528	.337	.188	.087	.031	.007	.000	.000	.000
2	1.000	.999	.991	.942	.837	.683	.500	.317	.163	.058	.009	.001	.000
3	1.000	1.000	1.000	.993	.969	.913	.812	.663	.472	.263	.081	.023	.001
4	1.000	1.000	1.000	1.000	.998	.990	.969	.922	.832	.672	.410	.226	.049

b. $n = 6$

k \ p	.01	.05	.10	.20	.30	.40	.50	.60	.70	.80	.90	.95	.99
0	.941	.735	.531	.262	.118	.047	.016	.004	.001	.000	.000	.000	.000
1	.999	.967	.886	.655	.420	.233	.109	.041	.011	.002	.000	.000	.000
2	1.000	.998	.984	.901	.744	.544	.344	.179	.070	.017	.001	.000	.000
3	1.000	1.000	.999	.983	.930	.821	.656	.456	.256	.099	.016	.002	.000
4	1.000	1.000	1.000	.998	.989	.959	.891	.767	.580	.345	.114	.033	.001
5	1.000	1.000	1.000	1.000	.999	.996	.984	.953	.882	.738	.469	.265	.059

c. $n = 7$

k \ p	.01	.05	.10	.20	.30	.40	.50	.60	.70	.80	.90	.95	.99
0	.932	.698	.478	.210	.082	.028	.008	.002	.000	.000	.000	.000	.000
1	.998	.956	.850	.577	.329	.159	.063	.019	.004	.000	.000	.000	.000
2	1.000	.996	.974	.852	.647	.420	.227	.096	.029	.005	.000	.000	.000
3	1.000	1.000	.997	.967	.874	.710	.500	.290	.126	.033	.003	.000	.000
4	1.000	1.000	1.000	.995	.971	.904	.773	.580	.353	.148	.026	.004	.000
5	1.000	1.000	1.000	1.000	.996	.981	.937	.841	.671	.423	.150	.044	.002
6	1.000	1.000	1.000	1.000	1.000	.998	.992	.972	.918	.790	.522	.302	.068

TABLE II Continued

d. $n = 8$

k \ p	.01	.05	.10	.20	.30	.40	.50	.60	.70	.80	.90	.95	.99
0	.923	.663	.430	.168	.058	.017	.004	.001	.000	.000	.000	.000	.000
1	.997	.943	.813	.503	.255	.106	.035	.009	.001	.000	.000	.000	.000
2	1.000	.994	.962	.797	.552	.315	.145	.050	.011	.001	.000	.000	.000
3	1.000	1.000	.995	.944	.806	.594	.363	.174	.058	.010	.000	.000	.000
4	1.000	1.000	1.000	.990	.942	.826	.637	.406	.194	.056	.005	.000	.000
5	1.000	1.000	1.000	.999	.989	.950	.855	.685	.448	.203	.038	.006	.000
6	1.000	1.000	1.000	1.000	.999	.991	.965	.894	.745	.497	.187	.057	.003
7	1.000	1.000	1.000	1.000	1.000	.999	.996	.983	.942	.832	.570	.337	.077

e. $n = 9$

k \ p	.01	.05	.10	.20	.30	.40	.50	.60	.70	.80	.90	.95	.99
0	.914	.630	.387	.134	.040	.010	.002	.000	.000	.000	.000	.000	.000
1	.997	.929	.775	.436	.196	.071	.020	.004	.000	.000	.000	.000	.000
2	1.000	.992	.947	.738	.463	.232	.090	.025	.004	.000	.000	.000	.000
3	1.000	.999	.992	.914	.730	.483	.254	.099	.025	.003	.000	.000	.000
4	1.000	1.000	.999	.980	.901	.733	.500	.267	.099	.020	.001	.000	.000
5	1.000	1.000	1.000	.997	.975	.901	.746	.517	.270	.086	.008	.001	.000
6	1.000	1.000	1.000	1.000	.996	.975	.910	.768	.537	.262	.053	.008	.000
7	1.000	1.000	1.000	1.000	1.000	.996	.980	.929	.804	.564	.225	.071	.003
8	1.000	1.000	1.000	1.000	1.000	1.000	.998	.990	.960	.866	.613	.370	.086

f. $n = 10$

k \ p	.01	.05	.10	.20	.30	.40	.50	.60	.70	.80	.90	.95	.99
0	.904	.599	.349	.107	.028	.006	.001	.000	.000	.000	.000	.000	.000
1	.996	.914	.736	.376	.149	.046	.011	.002	.000	.000	.000	.000	.000
2	1.000	.988	.930	.678	.383	.167	.055	.012	.002	.000	.000	.000	.000
3	1.000	.999	.987	.879	.650	.382	.172	.055	.011	.001	.000	.000	.000
4	1.000	1.000	.998	.967	.850	.633	.377	.166	.047	.006	.000	.000	.000
5	1.000	1.000	1.000	.994	.953	.834	.623	.367	.150	.033	.002	.000	.000
6	1.000	1.000	1.000	.999	.989	.945	.828	.618	.350	.121	.013	.001	.000
7	1.000	1.000	1.000	1.000	.998	.988	.945	.833	.617	.322	.070	.012	.000
8	1.000	1.000	1.000	1.000	1.000	.998	.989	.954	.851	.624	.264	.086	.004
9	1.000	1.000	1.000	1.000	1.000	1.000	.999	.994	.972	.893	.651	.401	.096

(continued)

TABLE II Continued

g. $n = 15$

k \ p	.01	.05	.10	.20	.30	.40	.50	.60	.70	.80	.90	.95	.99
0	.860	.463	.206	.035	.005	.000	.000	.000	.000	.000	.000	.000	.000
1	.990	.829	.549	.167	.035	.005	.000	.000	.000	.000	.000	.000	.000
2	1.000	.964	.816	.398	.127	.027	.004	.000	.000	.000	.000	.000	.000
3	1.000	.995	.944	.648	.297	.091	.018	.002	.000	.000	.000	.000	.000
4	1.000	.999	.987	.838	.515	.217	.059	.009	.001	.000	.000	.000	.000
5	1.000	1.000	.998	.939	.722	.403	.151	.034	.004	.000	.000	.000	.000
6	1.000	1.000	1.000	.982	.869	.610	.304	.095	.015	.001	.000	.000	.000
7	1.000	1.000	1.000	.996	.950	.787	.500	.213	.050	.004	.000	.000	.000
8	1.000	1.000	1.000	.999	.985	.905	.696	.390	.131	.018	.000	.000	.000
9	1.000	1.000	1.000	1.000	.996	.966	.849	.597	.278	.061	.002	.000	.000
10	1.000	1.000	1.000	1.000	.999	.991	.941	.783	.485	.164	.013	.001	.000
11	1.000	1.000	1.000	1.000	1.000	.998	.982	.909	.703	.352	.056	.005	.000
12	1.000	1.000	1.000	1.000	1.000	1.000	.996	.973	.873	.602	.184	.036	.000
13	1.000	1.000	1.000	1.000	1.000	1.000	1.000	.995	.965	.833	.451	.171	.010
14	1.000	1.000	1.000	1.000	1.000	1.000	1.000	1.000	.995	.965	.794	.537	.140

h. $n = 20$

k \ p	.01	.05	.10	.20	.30	.40	.50	.60	.70	.80	.90	.95	.99
0	.818	.358	.122	.012	.001	.000	.000	.000	.000	.000	.000	.000	.000
1	.983	.736	.392	.069	.008	.001	.000	.000	.000	.000	.000	.000	.000
2	.999	.925	.677	.206	.035	.004	.000	.000	.000	.000	.000	.000	.000
3	1.000	.984	.867	.411	.107	.016	.001	.000	.000	.000	.000	.000	.000
4	1.000	.997	.957	.630	.238	.051	.006	.000	.000	.000	.000	.000	.000
5	1.000	1.000	.989	.804	.416	.126	.021	.002	.000	.000	.000	.000	.000
6	1.000	1.000	.998	.913	.608	.250	.058	.006	.000	.000	.000	.000	.000
7	1.000	1.000	1.000	.968	.772	.416	.132	.021	.001	.000	.000	.000	.000
8	1.000	1.000	1.000	.990	.887	.596	.252	.057	.005	.000	.000	.000	.000
9	1.000	1.000	1.000	.997	.952	.755	.412	.128	.017	.001	.000	.000	.000
10	1.000	1.000	1.000	.999	.983	.872	.588	.245	.048	.003	.000	.000	.000
11	1.000	1.000	1.000	1.000	.995	.943	.748	.404	.113	.010	.000	.000	.000
12	1.000	1.000	1.000	1.000	.999	.979	.868	.584	.228	.032	.000	.000	.000
13	1.000	1.000	1.000	1.000	1.000	.994	.942	.750	.392	.087	.002	.000	.000
14	1.000	1.000	1.000	1.000	1.000	.998	.979	.874	.584	.196	.011	.000	.000
15	1.000	1.000	1.000	1.000	1.000	1.000	.994	.949	.762	.370	.043	.003	.000
16	1.000	1.000	1.000	1.000	1.000	1.000	.999	.984	.893	.589	.133	.016	.000
17	1.000	1.000	1.000	1.000	1.000	1.000	1.000	.996	.965	.794	.323	.075	.001
18	1.000	1.000	1.000	1.000	1.000	1.000	1.000	.999	.992	.931	.608	.264	.017
19	1.000	1.000	1.000	1.000	1.000	1.000	1.000	1.000	.999	.988	.878	.642	.182

TABLE II Continued

i. $n = 25$

k \ p	.01	.05	.10	.20	.30	.40	.50	.60	.70	.80	.90	.95	.99
0	.778	.277	.072	.004	.000	.000	.000	.000	.000	.000	.000	.000	.000
1	.974	.642	.271	.027	.002	.000	.000	.000	.000	.000	.000	.000	.000
2	.998	.873	.537	.098	.009	.000	.000	.000	.000	.000	.000	.000	.000
3	1.000	.966	.764	.234	.033	.002	.000	.000	.000	.000	.000	.000	.000
4	1.000	.993	.902	.421	.090	.009	.000	.000	.000	.000	.000	.000	.000
5	1.000	.999	.967	.617	.193	.029	.002	.000	.000	.000	.000	.000	.000
6	1.000	1.000	.991	.780	.341	.074	.007	.000	.000	.000	.000	.000	.000
7	1.000	1.000	.998	.891	.512	.154	.022	.001	.000	.000	.000	.000	.000
8	1.000	1.000	1.000	.953	.677	.274	.054	.004	.000	.000	.000	.000	.000
9	1.000	1.000	1.000	.983	.811	.425	.115	.013	.000	.000	.000	.000	.000
10	1.000	1.000	1.000	.994	.902	.586	.212	.034	.002	.000	.000	.000	.000
11	1.000	1.000	1.000	.998	.956	.732	.345	.078	.006	.000	.000	.000	.000
12	1.000	1.000	1.000	1.000	.983	.846	.500	.154	.017	.000	.000	.000	.000
13	1.000	1.000	1.000	1.000	.994	.922	.655	.268	.044	.002	.000	.000	.000
14	1.000	1.000	1.000	1.000	.998	.966	.788	.414	.098	.006	.000	.000	.000
15	1.000	1.000	1.000	1.000	1.000	.987	.885	.575	.189	.017	.000	.000	.000
16	1.000	1.000	1.000	1.000	1.000	.996	.946	.726	.323	.047	.000	.000	.000
17	1.000	1.000	1.000	1.000	1.000	.999	.978	.846	.488	.109	.002	.000	.000
18	1.000	1.000	1.000	1.000	1.000	1.000	.993	.926	.659	.220	.009	.000	.000
19	1.000	1.000	1.000	1.000	1.000	1.000	.998	.971	.807	.383	.033	.001	.000
20	1.000	1.000	1.000	1.000	1.000	1.000	1.000	.991	.910	.579	.098	.007	.000
21	1.000	1.000	1.000	1.000	1.000	1.000	1.000	.998	.967	.766	.236	.034	.000
22	1.000	1.000	1.000	1.000	1.000	1.000	1.000	1.000	.991	.902	.463	.127	.002
23	1.000	1.000	1.000	1.000	1.000	1.000	1.000	1.000	.998	.973	.729	.358	.026
24	1.000	1.000	1.000	1.000	1.000	1.000	1.000	1.000	1.000	.996	.928	.723	.222

TABLE III Poisson Probabilities

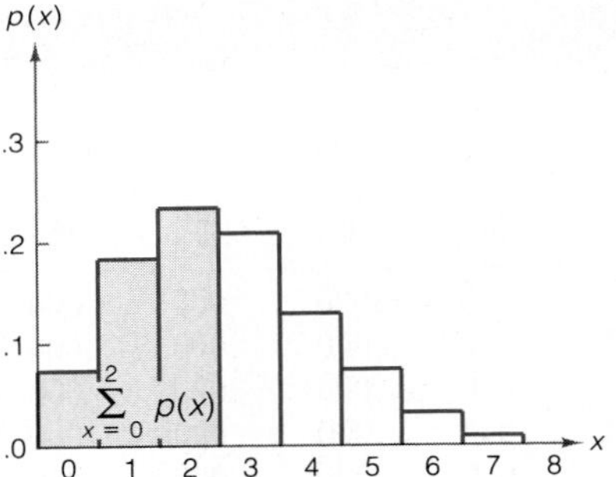

Tabulated values are $\sum_{x=0}^{k} p(x)$. *(Computations are rounded at the third decimal place.)*

λ \ x	0	1	2	3	4	5	6	7	8	9
.02	.980	1.000								
.04	.961	.999	1.000							
.06	.942	.998	1.000							
.08	.923	.997	1.000							
.10	.905	.995	1.000							
.15	.861	.990	.999	1.000						
.20	.819	.982	.999	1.000						
.25	.779	.974	.998	1.000						
.30	.741	.963	.996	1.000						
.35	.705	.951	.994	1.000						
.40	.670	.938	.992	.999	1.000					
.45	.638	.925	.989	.999	1.000					
.50	.607	.910	.986	.998	1.000					
.55	.577	.894	.982	.998	1.000					
.60	.549	.878	.977	.997	1.000					
.65	.522	.861	.972	.996	.999	1.000				
.70	.497	.844	.966	.994	.999	1.000				
.75	.472	.827	.959	.993	.999	1.000				
.80	.449	.809	.953	.991	.999	1.000				
.85	.427	.791	.945	.989	.998	1.000				
.90	.407	.772	.937	.987	.998	1.000				
.95	.387	.754	.929	.981	.997	1.000				
1.00	.368	.736	.920	.981	.996	.999	1.000			
1.1	.333	.699	.900	.974	.995	.999	1.000			
1.2	.301	.663	.879	.966	.992	.998	1.000			
1.3	.273	.627	.857	.957	.989	.998	1.000			
1.4	.247	.592	.833	.946	.986	.997	.999	1.000		
1.5	.223	.558	.809	.934	.981	.996	.999	1.000		

TABLE III Continued

λ \ x	0	1	2	3	4	5	6	7	8	9
1.6	.202	.525	.783	.921	.976	.994	.999	1.000		
1.7	.183	.493	.757	.907	.970	.992	.998	1.000		
1.8	.165	.463	.731	.891	.964	.990	.997	.999	1.000	
1.9	.150	.434	.704	.875	.956	.987	.997	.999	1.000	
2.0	.135	.406	.677	.857	.947	.983	.995	.999	1.000	
2.2	.111	.355	.623	.819	.928	.975	.993	.998	1.000	
2.4	.091	.308	.570	.779	.904	.964	.988	.997	.999	1.000
2.6	.074	.267	.518	.736	.877	.951	.983	.995	.999	1.000
2.8	.061	.231	.469	.692	.848	.935	.976	.992	.998	.999
3.0	.050	.199	.423	.647	.815	.916	.966	.988	.996	.999
3.2	.041	.171	.380	.603	.781	.895	.955	.983	.994	.998
3.4	.033	.147	.340	.558	.744	.871	.942	.977	.992	.997
3.6	.027	.126	.303	.515	.706	.844	.927	.969	.988	.996
3.8	.022	.107	.269	.473	.668	.816	.909	.960	.984	.994
4.0	.018	.092	.238	.433	.629	.785	.889	.949	.979	.992
4.2	.015	.078	.210	.395	.590	.753	.867	.936	.972	.989
4.4	.012	.066	.185	.359	.551	.720	.844	.921	.964	.985
4.6	.010	.056	.163	.326	.513	.686	.818	.905	.955	.980
4.8	.008	.048	.143	.294	.476	.651	.791	.887	.944	.975
5.0	.007	.040	.125	.265	.440	.616	.762	.867	.932	.968
5.2	.006	.034	.109	.238	.406	.581	.732	.845	.918	.960
5.4	.005	.029	.095	.213	.373	.546	.702	.822	.903	.951
5.6	.004	.024	.082	.191	.342	.512	.670	.797	.886	.941
5.8	.003	.021	.072	.170	.313	.478	.638	.771	.867	.929
6.0	.002	.017	.062	.151	.285	.446	.606	.744	.847	.916

λ \ x	10	11	12	13	14	15	16
2.8	1.000						
3.0	1.000						
3.2	1.000						
3.4	.999	1.000					
3.6	.999	1.000					
3.8	.998	.999	1.000				
4.0	.997	.999	1.000				
4.2	.996	.999	1.000				
4.4	.994	.998	.999	1.000			
4.6	.992	.997	.999	1.000			
4.8	.990	.996	.999	1.000			
5.0	.986	.995	.998	.999	1.000		
5.2	.982	.993	.997	.999	1.000		
5.4	.977	.990	.996	.999	1.000		
5.6	.972	.988	.995	.998	.999	1.000	
5.8	.965	.984	.993	.997	.999	1.000	
6.0	.957	.980	.991	.996	.999	.999	1.000

(continued)

TABLE III Continued

λ \ x	0	1	2	3	4	5	6	7	8	9
6.2	.002	.015	.054	.134	.259	.414	.574	.716	.826	.902
6.4	.002	.012	.046	.119	.235	.384	.542	.687	.803	.886
6.6	.001	.010	.040	.105	.213	.355	.511	.658	.780	.869
6.8	.001	.009	.034	.093	.192	.327	.480	.628	.755	.850
7.0	.001	.007	.030	.082	.173	.301	.450	.599	.729	.830
7.2	.001	.006	.025	.072	.156	.276	.420	.569	.703	.810
7.4	.001	.005	.022	.063	.140	.253	.392	.539	.676	.788
7.6	.001	.004	.019	.055	.125	.231	.365	.510	.648	.765
7.8	.000	.004	.016	.048	.112	.210	.338	.481	.620	.741
8.0	.000	.003	.014	.042	.100	.191	.313	.453	.593	.717
8.5	.000	.002	.009	.030	.074	.150	.256	.386	.523	.653
9.0	.000	.001	.006	.021	.055	.116	.207	.324	.456	.587
9.5	.000	.001	.004	.015	.040	.089	.165	.269	.392	.522
10.0	.000	.000	.003	.010	.029	.067	.130	.220	.333	.458

λ \ x	10	11	12	13	14	15	16	17	18	19
6.2	.949	.975	.989	.995	.998	.999	1.000			
6.4	.939	.969	.986	.994	.997	.999	1.000			
6.6	.927	.963	.982	.992	.997	.999	.999	1.000		
6.8	.915	.955	.978	.990	.996	.998	.999	1.000		
7.0	.901	.947	.973	.987	.994	.998	.999	1.000		
7.2	.887	.937	.967	.984	.993	.997	.999	.999	1.000	
7.4	.871	.926	.961	.980	.991	.996	.998	.999	1.000	
7.6	.854	.915	.954	.976	.989	.995	.998	.999	1.000	
7.8	.835	.902	.945	.971	.986	.993	.997	.999	1.000	
8.0	.816	.888	.936	.966	.983	.992	.996	.998	.999	1.000
8.5	.763	.849	.909	.949	.973	.986	.993	.997	.999	.999
9.0	.706	.803	.876	.926	.959	.978	.989	.995	.998	.999
9.5	.645	.752	.836	.898	.940	.967	.982	.991	.996	.998
10.0	.583	.697	.792	.864	.917	.951	.973	.986	.993	.997

λ \ x	20	21	22
8.5	1.000		
9.0	1.000		
9.5	.999	1.000	
10.0	.998	.999	1.000

TABLE III Continued

λ \ x	0	1	2	3	4	5	6	7	8	9
10.5	.000	.000	.002	.007	.021	.050	.102	.179	.279	.397
11.0	.000	.000	.001	.005	.015	.038	.079	.143	.232	.341
11.5	.000	.000	.001	.003	.011	.028	.060	.114	.191	.289
12.0	.000	.000	.001	.002	.008	.020	.046	.090	.155	.242
12.5	.000	.000	.000	.002	.005	.015	.035	.070	.125	.201
13.0	.000	.000	.000	.001	.004	.011	.026	.054	.100	.166
13.5	.000	.000	.000	.001	.003	.008	.019	.041	.079	.135
14.0	.000	.000	.000	.000	.002	.006	.014	.032	.062	.109
14.5	.000	.000	.000	.000	.001	.004	.010	.024	.048	.088
15.0	.000	.000	.000	.000	.001	.003	.008	.018	.037	.070

λ \ x	10	11	12	13	14	15	16	17	18	19
10.5	.521	.639	.742	.825	.888	.932	.960	.978	.988	.994
11.0	.460	.579	.689	.781	.854	.907	.944	.968	.982	.991
11.5	.402	.520	.633	.733	.815	.878	.924	.954	.974	.986
12.0	.347	.462	.576	.682	.772	.844	.899	.937	.963	.979
12.5	.297	.406	.519	.628	.725	.806	.869	.916	.948	.969
13.0	.252	.353	.463	.573	.675	.764	.835	.890	.930	.957
13.5	.211	.304	.409	.518	.623	.718	.798	.861	.908	.942
14.0	.176	.260	.358	.464	.570	.669	.756	.827	.883	.923
14.5	.145	.220	.311	.413	.518	.619	.711	.790	.853	.901
15.0	.118	.185	.268	.363	.466	.568	.664	.749	.819	.875

λ \ x	20	21	22	23	24	25	26	27	28	29
10.5	.997	.999	.999	1.000						
11.0	.995	.998	.999	1.000						
11.5	.992	.996	.998	.999	1.000					
12.0	.988	.994	.997	.999	.999	1.000				
12.5	.983	.991	.995	.998	.999	.999	1.000			
13.0	.975	.986	.992	.996	.998	.999	1.000			
13.5	.965	.980	.989	.994	.997	.998	.999	1.000		
14.0	.952	.971	.983	.991	.995	.997	.999	.999	1.000	
14.5	.936	.960	.976	.986	.992	.996	.998	.999	.999	1.000
15.0	.917	.947	.967	.981	.989	.994	.997	.998	.999	1.000

(continued)

TABLE III Continued

λ \ x	4	5	6	7	8	9	10	11	12	13
16	.000	.001	.004	.010	.022	.043	.077	.127	.193	.275
17	.000	.001	.002	.005	.013	.026	.049	.085	.135	.201
18	.000	.000	.001	.003	.007	.015	.030	.055	.092	.143
19	.000	.000	.001	.002	.004	.009	.018	.035	.061	.098
20	.000	.000	.000	.001	.002	.005	.011	.021	.039	.066
21	.000	.000	.000	.000	.001	.003	.006	.013	.025	.043
22	.000	.000	.000	.000	.001	.002	.004	.008	.015	.028
23	.000	.000	.000	.000	.000	.001	.002	.004	.009	.017
24	.000	.000	.000	.000	.000	.000	.001	.003	.005	.011
25	.000	.000	.000	.000	.000	.000	.001	.001	.003	.006

λ \ x	14	15	16	17	18	19	20	21	22	23
16	.368	.467	.566	.659	.742	.812	.868	.911	.942	.963
17	.281	.371	.468	.564	.655	.736	.805	.861	.905	.937
18	.208	.287	.375	.469	.562	.651	.731	.799	.855	.899
19	.150	.215	.292	.378	.469	.561	.647	.725	.793	.849
20	.105	.157	.221	.297	.381	.470	.559	.644	.721	.787
21	.072	.111	.163	.227	.302	.384	.471	.558	.640	.716
22	.048	.077	.117	.169	.232	.306	.387	.472	.556	.637
23	.031	.052	.082	.123	.175	.238	.310	.389	.472	.555
24	.020	.034	.056	.087	.128	.180	.243	.314	.392	.473
25	.012	.022	.038	.060	.092	.134	.185	.247	.318	.394

λ \ x	24	25	26	27	28	29	30	31	32	33
16	.978	.987	.993	.996	.998	.999	.999	1.000		
17	.959	.975	.985	.991	.995	.997	.999	.999	1.000	
18	.932	.955	.972	.983	.990	.994	.997	.998	.999	1.000
19	.893	.927	.951	.969	.980	.988	.993	.996	.998	.999
20	.843	.888	.922	.948	.966	.978	.987	.992	.995	.997
21	.782	.838	.883	.917	.944	.963	.976	.985	.991	.994
22	.712	.777	.832	.877	.913	.940	.959	.973	.983	.989
23	.635	.708	.772	.827	.873	.908	.936	.956	.971	.981
24	.554	.632	.704	.768	.823	.868	.904	.932	.953	.969
25	.473	.553	.629	.700	.763	.818	.863	.900	.929	.950

λ \ x	34	35	36	37	38	39	40	41	42	43
19	.999	1.000								
20	.999	.999	1.000							
21	.997	.998	.999	.999	1.000					
22	.994	.996	.998	.999	.999	1.000				
23	.988	.993	.996	.997	.999	.999	1.000			
24	.979	.987	.992	.995	.997	.998	.999	.999	1.000	
25	.966	.978	.985	.991	.991	.997	.998	.999	.999	1.000

TABLE IV Normal Curve Areas

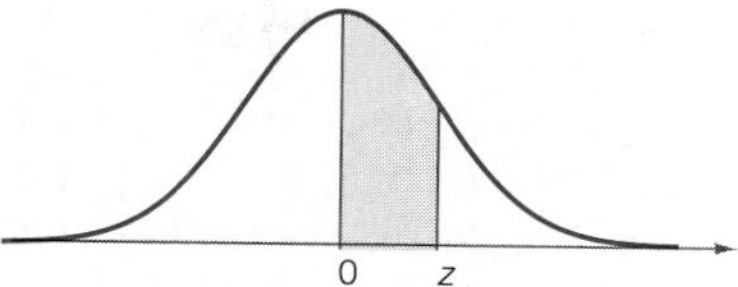

z	.00	.01	.02	.03	.04	.05	.06	.07	.08	.09
.0	.0000	.0040	.0080	.0120	.0160	.0199	.0239	.0279	.0319	.0359
.1	.0398	.0438	.0478	.0517	.0557	.0596	.0636	.0675	.0714	.0753
.2	.0793	.0832	.0871	.0910	.0948	.0987	.1026	.1064	.1103	.1141
.3	.1179	.1217	.1255	.1293	.1331	.1368	.1406	.1443	.1480	.1517
.4	.1554	.1591	.1628	.1664	.1700	.1736	.1772	.1808	.1844	.1879
.5	.1915	.1950	.1985	.2019	.2054	.2088	.2123	.2157	.2190	.2224
.6	.2257	.2291	.2324	.2357	.2389	.2422	.2454	.2486	.2517	.2549
.7	.2580	.2611	.2642	.2673	.2704	.2734	.2764	.2794	.2823	.2852
.8	.2881	.2910	.2939	.2967	.2995	.3023	.3051	.3078	.3106	.3133
.9	.3159	.3186	.3212	.3238	.3264	.3289	.3315	.3340	.3365	.3389
1.0	.3413	.3438	.3461	.3485	.3508	.3531	.3554	.3577	.3599	.3621
1.1	.3643	.3665	.3686	.3708	.3729	.3749	.3770	.3790	.3810	.3830
1.2	.3849	.3869	.3888	.3907	.3925	.3944	.3962	.3980	.3997	.4015
1.3	.4032	.4049	.4066	.4082	.4099	.4115	.4131	.4147	.4162	.4177
1.4	.4192	.4207	.4222	.4236	.4251	.4265	.4279	.4292	.4306	.4319
1.5	.4332	.4345	.4357	.4370	.4382	.4394	.4406	.4418	.4429	.4441
1.6	.4452	.4463	.4474	.4484	.4495	.4505	.4515	.4525	.4535	.4545
1.7	.4554	.4564	.4573	.4582	.4591	.4599	.4608	.4616	.4625	.4633
1.8	.4641	.4649	.4656	.4664	.4671	.4678	.4686	.4693	.4699	.4706
1.9	.4713	.4719	.4726	.4732	.4738	.4744	.4750	.4756	.4761	.4767
2.0	.4772	.4778	.4783	.4788	.4793	.4798	.4803	.4808	.4812	.4817
2.1	.4821	.4826	.4830	.4834	.4838	.4842	.4846	.4850	.4854	.4857
2.2	.4861	.4864	.4868	.4871	.4875	.4878	.4881	.4884	.4887	.4890
2.3	.4893	.4896	.4898	.4901	.4904	.4906	.4909	.4911	.4913	.4916
2.4	.4918	.4920	.4922	.4925	.4927	.4929	.4931	.4932	.4934	.4936
2.5	.4938	.4940	.4941	.4943	.4945	.4946	.4948	.4949	.4951	.4952
2.6	.4953	.4955	.4956	.4957	.4959	.4960	.4961	.4962	.4963	.4964
2.7	.4965	.4966	.4967	.4968	.4969	.4970	.4971	.4972	.4973	.4974
2.8	.4974	.4975	.4976	.4977	.4977	.4978	.4979	.4979	.4980	.4981
2.9	.4981	.4982	.4982	.4983	.4984	.4984	.4985	.4985	.4986	.4986
3.0	.4987	.4987	.4987	.4988	.4988	.4989	.4989	.4989	.4990	.4990

Source: Abridged from Table I of A. Hald, *Statistical Tables and Formulas* (New York: Wiley), 1952. Reproduced by permission of A. Hald and the publisher, John Wiley & Sons, Inc.

TABLE V Exponentials

λ	$e^{-\lambda}$	λ	$e^{-\lambda}$	λ	$e^{-\lambda}$	λ	$e^{-\lambda}$	λ	$e^{-\lambda}$
.00	1.000000	2.05	.128735	4.05	.017422	6.05	.002358	8.05	.000319
.05	.951229	2.10	.122456	4.10	.016573	6.10	.002243	8.10	.000304
.10	.904837	2.15	.116484	4.15	.015764	6.15	.002133	8.15	.000289
.15	.860708	2.20	.110803	4.20	.014996	6.20	.002029	8.20	.000275
.20	.818731	2.25	.105399	4.25	.014264	6.25	.001930	8.25	.000261
.25	.778801	2.30	.100259	4.30	.013569	6.30	.001836	8.30	.000249
.30	.740818	2.35	.095369	4.35	.012907	6.35	.001747	8.35	.000236
.35	.704688	2.40	.090718	4.40	.012277	6.40	.001661	8.40	.000225
.40	.670320	2.45	.086294	4.45	.011679	6.45	.001581	8.45	.000214
.45	.637628	2.50	.082085	4.50	.011109	6.50	.001503	8.50	.000204
.50	.606531	2.55	.078082	4.55	.010567	6.55	.001430	8.55	.000194
.55	.576950	2.60	.074274	4.60	.010052	6.60	.001360	8.60	.000184
.60	.548812	2.65	.070651	4.65	.009562	6.65	.001294	8.65	.000175
.65	.522046	2.70	.067206	4.70	.009095	6.70	.001231	8.70	.000167
.70	.496585	2.75	.063928	4.75	.008652	6.75	.001171	8.75	.000158
.75	.472367	2.80	.060810	4.80	.008230	6.80	.001114	8.80	.000151
.80	.449329	2.85	.057844	4.85	.007828	6.85	.001059	8.85	.000143
.85	.427415	2.90	.055023	4.90	.007447	6.90	.001008	8.90	.000136
.90	.406570	2.95	.052340	4.95	.007083	6.95	.000959	8.95	.000130
.95	.386741	3.00	.049787	5.00	.006738	7.00	.000912	9.00	.000123
1.00	.367879	3.05	.047359	5.05	.006409	7.05	.000867	9.05	.000117
1.05	.349938	3.10	.045049	5.10	.006097	7.10	.000825	9.10	.000112
1.10	.332871	3.15	.042852	5.15	.005799	7.15	.000785	9.15	.000106
1.15	.316637	3.20	.040762	5.20	.005517	7.20	.000747	9.20	.000101
1.20	.301194	3.25	.038774	5.25	.005248	7.25	.000710	9.25	.000096
1.25	.286505	3.30	.036883	5.30	.004992	7.30	.000676	9.30	.000091
1.30	.272532	3.35	.035084	5.35	.004748	7.35	.000643	9.35	.000087
1.35	.259240	3.40	.033373	5.40	.004517	7.40	.000611	9.40	.000083
1.40	.246597	3.45	.031746	5.45	.004296	7.45	.000581	9.45	.000079
1.45	.234570	3.50	.030197	5.50	.004087	7.50	.000553	9.50	.000075
1.50	.223130	3.55	.028725	5.55	.003887	7.55	.000526	9.55	.000071
1.55	.212248	3.60	.027324	5.60	.003698	7.60	.000501	9.60	.000068
1.60	.201897	3.65	.025991	5.65	.003518	7.65	.000476	9.65	.000064
1.65	.192050	3.70	.024724	5.70	.003346	7.70	.000453	9.70	.000061
1.70	.182684	3.75	.023518	5.75	.003183	7.75	.000431	9.75	.000058
1.75	.173774	3.80	.022371	5.80	.003028	7.80	.000410	9.80	.000056
1.80	.165299	3.85	.021280	5.85	.002880	7.85	.000390	9.85	.000053
1.85	.157237	3.90	.020242	5.90	.002739	7.90	.000371	9.90	.000050
1.90	.149569	3.95	.019255	5.95	.002606	7.95	.000353	9.95	.000048
1.95	.142274	4.00	.018316	6.00	.002479	8.00	.000336	10.00	.000045
2.00	.135335								

TABLE VI Critical Values of t

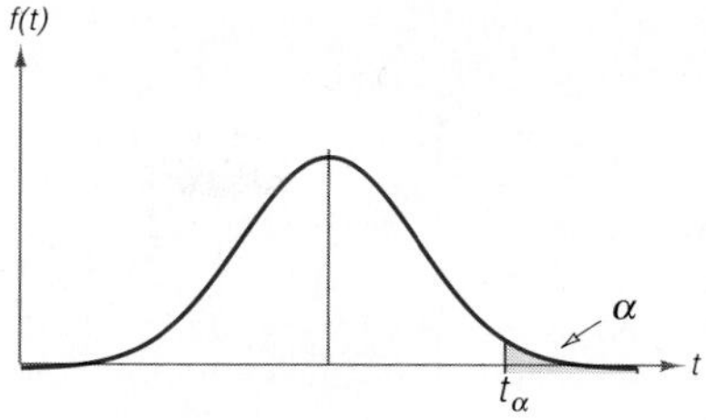

ν	$t_{.100}$	$t_{.050}$	$t_{.025}$	$t_{.010}$	$t_{.005}$	$t_{.001}$	$t_{.0005}$
1	3.078	6.314	12.706	31.821	63.657	318.31	636.62
2	1.886	2.920	4.303	6.965	9.925	22.326	31.598
3	1.638	2.353	3.182	4.541	5.841	10.213	12.924
4	1.533	2.132	2.776	3.747	4.604	7.173	8.610
5	1.476	2.015	2.571	3.365	4.032	5.893	6.869
6	1.440	1.943	2.447	3.143	3.707	5.208	5.959
7	1.415	1.895	2.365	2.998	3.499	4.785	5.408
8	1.397	1.860	2.306	2.896	3.355	4.501	5.041
9	1.383	1.833	2.262	2.821	3.250	4.297	4.781
10	1.372	1.812	2.228	2.764	3.169	4.144	4.587
11	1.363	1.796	2.201	2.718	3.106	4.025	4.437
12	1.356	1.782	2.179	2.681	3.055	3.930	4.318
13	1.350	1.771	2.160	2.650	3.012	3.852	4.221
14	1.345	1.761	2.145	2.624	2.977	3.787	4.140
15	1.341	1.753	2.131	2.602	2.947	3.733	4.073
16	1.337	1.746	2.120	2.583	2.921	3.686	4.015
17	1.333	1.740	2.110	2.567	2.898	3.646	3.965
18	1.330	1.734	2.101	2.552	2.878	3.610	3.922
19	1.328	1.729	2.093	2.539	2.861	3.579	3.883
20	1.325	1.725	2.086	2.528	2.845	3.552	3.850
21	1.323	1.721	2.080	2.518	2.831	3.527	3.819
22	1.321	1.717	2.074	2.508	2.819	3.505	3.792
23	1.319	1.714	2.069	2.500	2.807	3.485	3.767
24	1.318	1.711	2.064	2.492	2.797	3.467	3.745
25	1.316	1.708	2.060	2.485	2.787	3.450	3.725
26	1.315	1.706	2.056	2.479	2.779	3.435	3.707
27	1.314	1.703	2.052	2.473	2.771	3.421	3.690
28	1.313	1.701	2.048	2.467	2.763	3.408	3.674
29	1.311	1.699	2.045	2.462	2.756	3.396	3.659
30	1.310	1.697	2.042	2.457	2.750	3.385	3.646
40	1.303	1.684	2.021	2.423	2.704	3.307	3.551
60	1.296	1.671	2.000	2.390	2.660	3.232	3.460
120	1.289	1.658	1.980	2.358	2.617	3.160	3.373
∞	1.282	1.645	1.960	2.326	2.576	3.090	3.291

Source: This table is reproduced with the kind permission of the Trustees of Biometrika from E. S. Pearson and H. O. Hartley (eds.), *The Biometrika Tables for Statisticians*, Vol. 1, 3d ed., *Biometrika*, 1966.

TABLE VII Percentage Points of the F Distribution, $\alpha = .10$

ν_2 \ ν_1 (Denominator Degrees of Freedom \ Numerator Degrees of Freedom)	1	2	3	4	5	6	7	8	9
1	39.86	49.50	53.59	55.83	57.24	58.20	58.91	59.44	59.86
2	8.53	9.00	9.16	9.24	9.29	9.33	9.35	9.37	9.38
3	5.54	5.46	5.39	5.34	5.31	5.28	5.27	5.25	5.24
4	4.54	4.32	4.19	4.11	4.05	4.01	3.98	3.95	3.94
5	4.06	3.78	3.62	3.52	3.45	3.40	3.37	3.34	3.32
6	3.78	3.46	3.29	3.18	3.11	3.05	3.01	2.98	2.96
7	3.59	3.26	3.07	2.96	2.88	2.83	2.78	2.75	2.72
8	3.46	3.11	2.92	2.81	2.73	2.67	2.62	2.59	2.56
9	3.36	3.01	2.81	2.69	2.61	2.55	2.51	2.47	2.44
10	3.29	2.92	2.73	2.61	2.52	2.46	2.41	2.38	2.35
11	3.23	2.86	2.66	2.54	2.45	2.39	2.34	2.30	2.27
12	3.18	2.81	2.61	2.48	2.39	2.33	2.28	2.24	2.21
13	3.14	2.76	2.56	2.43	2.35	2.28	2.23	2.20	2.16
14	3.10	2.73	2.52	2.39	2.31	2.24	2.19	2.15	2.12
15	3.07	2.70	2.49	2.36	2.27	2.21	2.16	2.12	2.09
16	3.05	2.67	2.46	2.33	2.24	2.18	2.13	2.09	2.06
17	3.03	2.64	2.44	2.31	2.22	2.15	2.10	2.06	2.03
18	3.01	2.62	2.42	2.29	2.20	2.13	2.08	2.04	2.00
19	2.99	2.61	2.40	2.27	2.18	2.11	2.06	2.02	1.98
20	2.97	2.59	2.38	2.25	2.16	2.09	2.04	2.00	1.96
21	2.96	2.57	2.36	2.23	2.14	2.08	2.02	1.98	1.95
22	2.95	2.56	2.35	2.22	2.13	2.06	2.01	1.97	1.93
23	2.94	2.55	2.34	2.21	2.11	2.05	1.99	1.95	1.92
24	2.93	2.54	2.33	2.19	2.10	2.04	1.98	1.94	1.91
25	2.92	2.53	2.32	2.18	2.09	2.02	1.97	1.93	1.89
26	2.91	2.52	2.31	2.17	2.08	2.01	1.96	1.92	1.88
27	2.90	2.51	2.30	2.17	2.07	2.00	1.95	1.91	1.87
28	2.89	2.50	2.29	2.16	2.06	2.00	1.94	1.90	1.87
29	2.89	2.50	2.28	2.15	2.06	1.99	1.93	1.89	1.86
30	2.88	2.49	2.28	2.14	2.05	1.98	1.93	1.88	1.85
40	2.84	2.44	2.23	2.09	2.00	1.93	1.87	1.83	1.79
60	2.79	2.39	2.18	2.04	1.95	1.87	1.82	1.77	1.74
120	2.75	2.35	2.13	1.99	1.90	1.82	1.77	1.72	1.68
∞	2.71	2.30	2.08	1.94	1.85	1.77	1.72	1.67	1.63

Source: From M. Merrington and C. M. Thompson, "Tables of Percentage Points of the Inverted Beta (*F*)-Distribution," *Biometrika*, 1943, 33, 73–88. Reproduced by permission of the *Biometrika* Trustees.

ν_2 \ ν_1	Numerator Degrees of Freedom									
Denominator Degrees of Freedom	10	12	15	20	24	30	40	60	120	∞
1	60.19	60.71	61.22	61.74	62.00	62.26	62.53	62.79	63.06	63.33
2	9.39	9.41	9.42	9.44	9.45	9.46	9.47	9.47	9.48	9.49
3	5.23	5.22	5.20	5.18	5.18	5.17	5.16	5.15	5.14	5.13
4	3.92	3.90	3.87	3.84	3.83	3.82	3.80	3.79	3.78	3.76
5	3.30	3.27	3.24	3.21	3.19	3.17	3.16	3.14	3.12	3.10
6	2.94	2.90	2.87	2.84	2.82	2.80	2.78	2.76	2.74	2.72
7	2.70	2.67	2.63	2.59	2.58	2.56	2.54	2.51	2.49	2.47
8	2.54	2.50	2.46	2.42	2.40	2.38	2.36	2.34	2.32	2.29
9	2.42	2.38	2.34	2.30	2.28	2.25	2.23	2.21	2.18	2.16
10	2.32	2.28	2.24	2.20	2.18	2.16	2.13	2.11	2.08	2.06
11	2.25	2.21	2.17	2.12	2.10	2.08	2.05	2.03	2.00	1.97
12	2.19	2.15	2.10	2.06	2.04	2.01	1.99	1.96	1.93	1.90
13	2.14	2.10	2.05	2.01	1.98	1.96	1.93	1.90	1.88	1.85
14	2.10	2.05	2.01	1.96	1.94	1.91	1.89	1.86	1.83	1.80
15	2.06	2.02	1.97	1.92	1.90	1.87	1.85	1.82	1.79	1.76
16	2.03	1.99	1.94	1.89	1.87	1.84	1.81	1.78	1.75	1.72
17	2.00	1.96	1.91	1.86	1.84	1.81	1.78	1.75	1.72	1.69
18	1.98	1.93	1.89	1.84	1.81	1.78	1.75	1.72	1.69	1.66
19	1.96	1.91	1.86	1.81	1.79	1.76	1.73	1.70	1.67	1.63
20	1.94	1.89	1.84	1.79	1.77	1.74	1.71	1.68	1.64	1.61
21	1.92	1.87	1.83	1.78	1.75	1.72	1.69	1.66	1.62	1.59
22	1.90	1.86	1.81	1.76	1.73	1.70	1.67	1.64	1.60	1.57
23	1.89	1.84	1.80	1.74	1.72	1.69	1.66	1.62	1.59	1.55
24	1.88	1.83	1.78	1.73	1.70	1.67	1.64	1.61	1.57	1.53
25	1.87	1.82	1.77	1.72	1.69	1.66	1.63	1.59	1.56	1.52
26	1.86	1.81	1.76	1.71	1.68	1.65	1.61	1.58	1.54	1.50
27	1.85	1.80	1.75	1.70	1.67	1.64	1.60	1.57	1.53	1.49
28	1.84	1.79	1.74	1.69	1.66	1.63	1.59	1.56	1.52	1.48
29	1.83	1.78	1.73	1.68	1.65	1.62	1.58	1.55	1.51	1.47
30	1.82	1.77	1.72	1.67	1.64	1.61	1.57	1.54	1.50	1.46
40	1.76	1.71	1.66	1.61	1.57	1.54	1.51	1.47	1.42	1.38
60	1.71	1.66	1.60	1.54	1.51	1.48	1.44	1.40	1.35	1.29
120	1.65	1.60	1.55	1.48	1.45	1.41	1.37	1.32	1.26	1.19
∞	1.60	1.55	1.49	1.42	1.38	1.34	1.30	1.24	1.17	1.00

TABLE VIII Percentage Points of the F Distribution, $\alpha = .05$

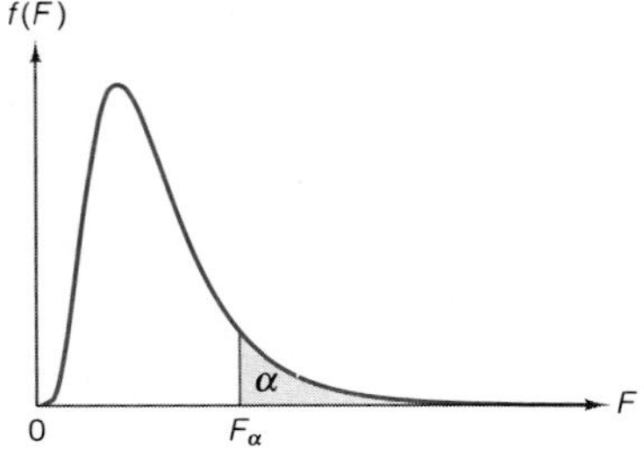

ν_2 \ ν_1	Numerator Degrees of Freedom								
	1	2	3	4	5	6	7	8	9
1	161.4	199.5	215.7	224.6	230.2	234.0	236.8	238.9	240.5
2	18.51	19.00	19.16	19.25	19.30	19.33	19.35	19.37	19.38
3	10.13	9.55	9.28	9.12	9.01	8.94	8.89	8.85	8.81
4	7.71	6.94	6.59	6.39	6.26	6.16	6.09	6.04	6.00
5	6.61	5.79	5.41	5.19	5.05	4.95	4.88	4.82	4.77
6	5.99	5.14	4.76	4.53	4.39	4.28	4.21	4.15	4.10
7	5.59	4.74	4.35	4.12	3.97	3.87	3.79	3.73	3.68
8	5.32	4.46	4.07	3.84	3.69	3.58	3.50	3.44	3.39
9	5.12	4.26	3.86	3.63	3.48	3.37	3.29	3.23	3.18
10	4.96	4.10	3.71	3.48	3.33	3.22	3.14	3.07	3.02
11	4.84	3.98	3.59	3.36	3.20	3.09	3.01	2.95	2.90
12	4.75	3.89	3.49	3.26	3.11	3.00	2.91	2.85	2.80
13	4.67	3.81	3.41	3.18	3.03	2.92	2.83	2.77	2.71
14	4.60	3.74	3.34	3.11	2.96	2.85	2.76	2.70	2.65
15	4.54	3.68	3.29	3.06	2.90	2.79	2.71	2.64	2.59
16	4.49	3.63	3.24	3.01	2.85	2.74	2.66	2.59	2.54
17	4.45	3.59	3.20	2.96	2.81	2.70	2.61	2.55	2.49
18	4.41	3.55	3.16	2.93	2.77	2.66	2.58	2.51	2.46
19	4.38	3.52	3.13	2.90	2.74	2.63	2.54	2.48	2.42
20	4.35	3.49	3.10	2.87	2.71	2.60	2.51	2.45	2.39
21	4.32	3.47	3.07	2.84	2.68	2.57	2.49	2.42	2.37
22	4.30	3.44	3.05	2.82	2.66	2.55	2.46	2.40	2.34
23	4.28	3.42	3.03	2.80	2.64	2.53	2.44	2.37	2.32
24	4.26	3.40	3.01	2.78	2.62	2.51	2.42	2.36	2.30
25	4.24	3.39	2.99	2.76	2.60	2.49	2.40	2.34	2.28
26	4.23	3.37	2.98	2.74	2.59	2.47	2.39	2.32	2.27
27	4.21	3.35	2.96	2.73	2.57	2.46	2.37	2.31	2.25
28	4.20	3.34	2.95	2.71	2.56	2.45	2.36	2.29	2.24
29	4.18	3.33	2.93	2.70	2.55	2.43	2.35	2.28	2.22
30	4.17	3.32	2.92	2.69	2.53	2.42	2.33	2.27	2.21
40	4.08	3.23	2.84	2.61	2.45	2.34	2.25	2.18	2.12
60	4.00	3.15	2.76	2.53	2.37	2.25	2.17	2.10	2.04
120	3.92	3.07	2.68	2.45	2.29	2.17	2.09	2.02	1.96
∞	3.84	3.00	2.60	2.37	2.21	2.10	2.01	1.94	1.88

Denominator Degrees of Freedom (ν_2)

Source: From M. Merrington and C. M. Thompson, "Tables of Percentage Points of the Inverted Beta (F)-Distribution." *Biometrika*, 1943, 33, 73–88. Reproduced by permission of the *Biometrika* Trustees.

ν_2 \ ν_1 (Denominator Degrees of Freedom \ Numerator Degrees of Freedom)	10	12	15	20	24	30	40	60	120	∞
1	241.9	243.9	245.9	248.0	249.1	250.1	251.1	252.2	253.3	254.3
2	19.40	19.41	19.43	19.45	19.45	19.46	19.47	19.48	19.49	19.50
3	8.79	8.74	8.70	8.66	8.64	8.62	8.59	8.57	8.55	8.53
4	5.96	5.91	5.86	5.80	5.77	5.75	5.72	5.69	5.66	5.63
5	4.74	4.68	4.62	4.56	4.53	4.50	4.46	4.43	4.40	4.36
6	4.06	4.00	3.94	3.87	3.84	3.81	3.77	3.74	3.70	3.67
7	3.64	3.57	3.51	3.44	3.41	3.38	3.34	3.30	3.27	3.23
8	3.35	3.28	3.22	3.15	3.12	3.08	3.04	3.01	2.97	2.93
9	3.14	3.07	3.01	2.94	2.90	2.86	2.83	2.79	2.75	2.71
10	2.98	2.91	2.85	2.77	2.74	2.70	2.66	2.62	2.58	2.54
11	2.85	2.79	2.72	2.65	2.61	2.57	2.53	2.49	2.45	2.40
12	2.75	2.69	2.62	2.54	2.51	2.47	2.43	2.38	2.34	2.30
13	2.67	2.60	2.53	2.46	2.42	2.38	2.34	2.30	2.25	2.21
14	2.60	2.53	2.46	2.39	2.35	2.31	2.27	2.22	2.18	2.13
15	2.54	2.48	2.40	2.33	2.29	2.25	2.20	2.16	2.11	2.07
16	2.49	2.42	2.35	2.28	2.24	2.19	2.15	2.11	2.06	2.01
17	2.45	2.38	2.31	2.23	2.19	2.15	2.10	2.06	2.01	1.96
18	2.41	2.34	2.27	2.19	2.15	2.11	2.06	2.02	1.97	1.92
19	2.38	2.31	2.23	2.16	2.11	2.07	2.03	1.98	1.93	1.88
20	2.35	2.28	2.20	2.12	2.08	2.04	1.99	1.95	1.90	1.84
21	2.32	2.25	2.18	2.10	2.05	2.01	1.96	1.92	1.87	1.81
22	2.30	2.23	2.15	2.07	2.03	1.98	1.94	1.89	1.84	1.78
23	2.27	2.20	2.13	2.05	2.01	1.96	1.91	1.86	1.81	1.76
24	2.25	2.18	2.11	2.03	1.98	1.94	1.89	1.84	1.79	1.73
25	2.24	2.16	2.09	2.01	1.96	1.92	1.87	1.82	1.77	1.71
26	2.22	2.15	2.07	1.99	1.95	1.90	1.85	1.80	1.75	1.69
27	2.20	2.13	2.06	1.97	1.93	1.88	1.84	1.79	1.73	1.67
28	2.19	2.12	2.04	1.96	1.91	1.87	1.82	1.77	1.71	1.65
29	2.18	2.10	2.03	1.94	1.90	1.85	1.81	1.75	1.70	1.64
30	2.16	2.09	2.01	1.93	1.89	1.84	1.79	1.74	1.68	1.62
40	2.08	2.00	1.92	1.84	1.79	1.74	1.69	1.64	1.58	1.51
60	1.99	1.92	1.84	1.75	1.70	1.65	1.59	1.53	1.47	1.39
120	1.91	1.83	1.75	1.66	1.61	1.55	1.50	1.43	1.35	1.25
∞	1.83	1.75	1.67	1.57	1.52	1.46	1.39	1.32	1.22	1.00

TABLE IX Percentage Points of the F Distribution, $\alpha = .025$

ν_2 \ ν_1 (Numerator Degrees of Freedom)	1	2	3	4	5	6	7	8	9
1	647.8	799.5	864.2	899.6	921.8	937.1	948.2	956.7	963.3
2	38.51	39.00	39.17	39.25	39.30	39.33	39.36	39.37	39.39
3	17.44	16.04	15.44	15.10	14.88	14.73	14.62	14.54	14.47
4	12.22	10.65	9.98	9.60	9.36	9.20	9.07	8.98	8.90
5	10.01	8.43	7.76	7.39	7.15	6.98	6.85	6.76	6.68
6	8.81	7.26	6.60	6.23	5.99	5.82	5.70	5.60	5.52
7	8.07	6.54	5.89	5.52	5.29	5.12	4.99	4.90	4.82
8	7.57	6.06	5.42	5.05	4.82	4.65	4.53	4.43	4.36
9	7.21	5.71	5.08	4.72	4.48	4.32	4.20	4.10	4.03
10	6.94	5.46	4.83	4.47	4.24	4.07	3.95	3.85	3.78
11	6.72	5.26	4.63	4.28	4.04	3.88	3.76	3.66	3.59
12	6.55	5.10	4.47	4.12	3.89	3.73	3.61	3.51	3.44
13	6.41	4.97	4.35	4.00	3.77	3.60	3.48	3.39	3.31
14	6.30	4.86	4.24	3.89	3.66	3.50	3.38	3.29	3.21
15	6.20	4.77	4.15	3.80	3.58	3.41	3.29	3.20	3.12
16	6.12	4.69	4.08	3.73	3.50	3.34	3.22	3.12	3.05
17	6.04	4.62	4.01	3.66	3.44	3.28	3.16	3.06	2.98
18	5.98	4.56	3.95	3.61	3.38	3.22	3.10	3.01	2.93
19	5.92	4.51	3.90	3.56	3.33	3.17	3.05	2.96	2.88
20	5.87	4.46	3.86	3.51	3.29	3.13	3.01	2.91	2.84
21	5.83	4.42	3.82	3.48	3.25	3.09	2.97	2.87	2.80
22	5.79	4.38	3.78	3.44	3.22	3.05	2.93	2.84	2.76
23	5.75	4.35	3.75	3.41	3.18	3.02	2.90	2.81	2.73
24	5.72	4.32	3.72	3.38	3.15	2.99	2.87	2.78	2.70
25	5.69	4.29	3.69	3.35	3.13	2.97	2.85	2.75	2.68
26	5.66	4.27	3.67	3.33	3.10	2.94	2.82	2.73	2.65
27	5.63	4.24	3.65	3.31	3.08	2.92	2.80	2.71	2.63
28	5.61	4.22	3.63	3.29	3.06	2.90	2.78	2.69	2.61
29	5.59	4.20	3.61	3.27	3.04	2.88	2.76	2.67	2.59
30	5.57	4.18	3.59	3.25	3.03	2.87	2.75	2.65	2.57
40	5.42	4.05	3.46	3.13	2.90	2.74	2.62	2.53	2.45
60	5.29	3.93	3.34	3.01	2.79	2.63	2.51	2.41	2.33
120	5.15	3.80	3.23	2.89	2.67	2.52	2.39	2.30	2.22
∞	5.02	3.69	3.12	2.79	2.57	2.41	2.29	2.19	2.11

Rows: Denominator Degrees of Freedom (ν_2).

Source: From M. Merrington and C. M. Thompson, "Tables of Percentage Points of the Inverted Beta (F)-Distribution," *Biometrika*, 1943, 33, 73–88. Reproduced by permission of the *Biometrika* Trustees.

ν_2 \ ν_1 (Denominator Degrees of Freedom \ Numerator Degrees of Freedom)	10	12	15	20	24	30	40	60	120	∞
1	968.6	976.7	984.9	993.1	997.2	1,001	1,006	1,010	1,014	1,018
2	39.40	39.41	39.43	39.45	39.46	39.46	39.47	39.48	39.49	39.50
3	14.42	14.34	14.25	14.17	14.12	14.08	14.04	13.99	13.95	13.90
4	8.84	8.75	8.66	8.56	8.51	8.46	8.41	8.36	8.31	8.26
5	6.62	6.52	6.43	6.33	6.28	6.23	6.18	6.12	6.07	6.02
6	5.46	5.37	5.27	5.17	5.12	5.07	5.01	4.96	4.90	4.85
7	4.76	4.67	4.57	4.47	4.42	4.36	4.31	4.25	4.20	4.14
8	4.30	4.20	4.10	4.00	3.95	3.89	3.84	3.78	3.73	3.67
9	3.96	3.87	3.77	3.67	3.61	3.56	3.51	3.45	3.39	3.33
10	3.72	3.62	3.52	3.42	3.37	3.31	3.26	3.20	3.14	3.08
11	3.53	3.43	3.33	3.23	3.17	3.12	3.06	3.00	2.94	2.88
12	3.37	3.28	3.18	3.07	3.02	2.96	2.91	2.85	2.79	2.72
13	3.25	3.15	3.05	2.95	2.89	2.84	2.78	2.72	2.66	2.60
14	3.15	3.05	2.95	2.84	2.79	2.73	2.67	2.61	2.55	2.49
15	3.06	2.96	2.86	2.76	2.70	2.64	2.59	2.52	2.46	2.40
16	2.99	2.89	2.79	2.68	2.63	2.57	2.51	2.45	2.38	2.32
17	2.92	2.82	2.72	2.62	2.56	2.50	2.44	2.38	2.32	2.25
18	2.87	2.77	2.67	2.56	2.50	2.44	2.38	2.32	2.26	2.19
19	2.82	2.72	2.62	2.51	2.45	2.39	2.33	2.27	2.20	2.13
20	2.77	2.68	2.57	2.46	2.41	2.35	2.29	2.22	2.16	2.09
21	2.73	2.64	2.53	2.42	2.37	2.31	2.25	2.18	2.11	2.04
22	2.70	2.60	2.50	2.39	2.33	2.27	2.21	2.14	2.08	2.00
23	2.67	2.57	2.47	2.36	2.30	2.24	2.18	2.11	2.04	1.97
24	2.64	2.54	2.44	2.33	2.27	2.21	2.15	2.08	2.01	1.94
25	2.61	2.51	2.41	2.30	2.24	2.18	2.12	2.05	1.98	1.91
26	2.59	2.49	2.39	2.28	2.22	2.16	2.09	2.03	1.95	1.88
27	2.57	2.47	2.36	2.25	2.19	2.13	2.07	2.00	1.93	1.85
28	2.55	2.45	2.34	2.23	2.17	2.11	2.05	1.98	1.91	1.83
29	2.53	2.43	2.32	2.21	2.15	2.09	2.03	1.96	1.89	1.81
30	2.51	2.41	2.31	2.20	2.14	2.07	2.01	1.94	1.87	1.79
40	2.39	2.29	2.18	2.07	2.01	1.94	1.88	1.80	1.72	1.64
60	2.27	2.17	2.06	1.94	1.88	1.82	1.74	1.67	1.58	1.48
120	2.16	2.05	1.94	1.82	1.76	1.69	1.61	1.53	1.43	1.31
∞	2.05	1.94	1.83	1.71	1.64	1.57	1.48	1.39	1.27	1.00

TABLE X Percentage Points of the F Distribution, $\alpha = .01$

Denominator Degrees of Freedom ν_2 \ Numerator Degrees of Freedom ν_1	1	2	3	4	5	6	7	8	9
1	4,052	4,999.5	5,403	5,625	5,764	5,859	5,928	5,982	6,022
2	98.50	99.00	99.17	99.25	99.30	99.33	99.36	99.37	99.39
3	34.12	30.82	29.46	28.71	28.24	27.91	27.67	27.49	27.35
4	21.20	18.00	16.69	15.98	15.52	15.21	14.98	14.80	14.66
5	16.26	13.27	12.06	11.39	10.97	10.67	10.46	10.29	10.16
6	13.75	10.92	9.78	9.15	8.75	8.47	8.26	8.10	7.98
7	12.25	9.55	8.45	7.85	7.46	7.19	6.99	6.84	6.72
8	11.26	8.65	7.59	7.01	6.63	6.37	6.18	6.03	5.91
9	10.56	8.02	6.99	6.42	6.06	5.80	5.61	5.47	5.35
10	10.04	7.56	6.55	5.99	5.64	5.39	5.20	5.06	4.94
11	9.65	7.21	6.22	5.67	5.32	5.07	4.89	4.74	4.63
12	9.33	6.93	5.95	5.41	5.06	4.82	4.64	4.50	4.39
13	9.07	6.70	5.74	5.21	4.86	4.62	4.44	4.30	4.19
14	8.86	6.51	5.56	5.04	4.69	4.46	4.28	4.14	4.03
15	8.68	6.36	5.42	4.89	4.56	4.32	4.14	4.00	3.89
16	8.53	6.23	5.29	4.77	4.44	4.20	4.03	3.89	3.78
17	8.40	6.11	5.18	4.67	4.34	4.10	3.93	3.79	3.68
18	8.29	6.01	5.09	4.58	4.25	4.01	3.84	3.71	3.60
19	8.18	5.93	5.01	4.50	4.17	3.94	3.77	3.63	3.52
20	8.10	5.85	4.94	4.43	4.10	3.87	3.70	3.56	3.46
21	8.02	5.78	4.87	4.37	4.04	3.81	3.64	3.51	3.40
22	7.95	5.72	4.82	4.31	3.99	3.76	3.59	3.45	3.35
23	7.88	5.66	4.76	4.26	3.94	3.71	3.54	3.41	3.30
24	7.82	5.61	4.72	4.22	3.90	3.67	3.50	3.36	3.26
25	7.77	5.57	4.68	4.18	3.85	3.63	3.46	3.32	3.22
26	7.72	5.53	4.64	4.14	3.82	3.59	3.42	3.29	3.18
27	7.68	5.49	4.60	4.11	3.78	3.56	3.39	3.26	3.15
28	7.64	5.45	4.57	4.07	3.75	3.53	3.36	3.23	3.12
29	7.60	5.42	4.54	4.04	3.73	3.50	3.33	3.20	3.09
30	7.56	5.39	4.51	4.02	3.70	3.47	3.30	3.17	3.07
40	7.31	5.18	4.31	3.83	3.51	3.29	3.12	2.99	2.89
60	7.08	4.98	4.13	3.65	3.34	3.12	2.95	2.82	2.72
120	6.85	4.79	3.95	3.48	3.17	2.96	2.79	2.66	2.56
∞	6.63	4.61	3.78	3.32	3.02	2.80	2.64	2.51	2.41

Source: From M. Merrington and C. M. Thompson, "Tables of Percentage Points of the Inverted Beta (*F*)-Distribution," *Biometrika*, 1943, 33, 73–88. Reproduced by permission of the *Biometrika* Trustees.

ν_2 \ ν_1	Numerator Degrees of Freedom									
Denominator Degrees of Freedom	10	12	15	20	24	30	40	60	120	∞
1	6,056	6,106	6,157	6,209	6,235	6,261	6,287	6,313	6,339	6,366
2	99.40	99.42	99.43	99.45	99.46	99.47	99.47	99.48	99.49	99.50
3	27.23	27.05	26.87	26.69	26.60	26.50	26.41	26.32	26.22	26.13
4	14.55	14.37	14.20	14.02	13.93	13.84	13.75	13.65	13.56	13.46
5	10.05	9.89	9.72	9.55	9.47	9.38	9.29	9.20	9.11	9.02
6	7.87	7.72	7.56	7.40	7.31	7.23	7.14	7.06	6.97	6.88
7	6.62	6.47	6.31	6.16	6.07	5.99	5.91	5.82	5.74	5.65
8	5.81	5.67	5.52	5.36	5.28	5.20	5.12	5.03	4.95	4.86
9	5.26	5.11	4.96	4.81	4.73	4.65	4.57	4.48	4.40	4.31
10	4.85	4.71	4.56	4.41	4.33	4.25	4.17	4.08	4.00	3.91
11	4.54	4.40	4.25	4.10	4.02	3.94	3.86	3.78	3.69	3.60
12	4.30	4.16	4.01	3.86	3.78	3.70	3.62	3.54	3.45	3.36
13	4.10	3.96	3.82	3.66	3.59	3.51	3.43	3.34	3.25	3.17
14	3.94	3.80	3.66	3.51	3.43	3.35	3.27	3.18	3.09	3.00
15	3.80	3.67	3.52	3.37	3.29	3.21	3.13	3.05	2.96	2.87
16	3.69	3.55	3.41	3.26	3.18	3.10	3.02	2.93	2.84	2.75
17	3.59	3.46	3.31	3.16	3.08	3.00	2.92	2.83	2.75	2.65
18	3.51	3.37	3.23	3.08	3.00	2.92	2.84	2.75	2.66	2.57
19	3.43	3.30	3.15	3.00	2.92	2.84	2.76	2.67	2.58	2.49
20	3.37	3.23	3.09	2.94	2.86	2.78	2.69	2.61	2.52	2.42
21	3.31	3.17	3.03	2.88	2.80	2.72	2.64	2.55	2.46	2.36
22	3.26	3.12	2.98	2.83	2.75	2.67	2.58	2.50	2.40	2.31
23	3.21	3.07	2.93	2.78	2.70	2.62	2.54	2.45	2.35	2.26
24	3.17	3.03	2.89	2.74	2.66	2.58	2.49	2.40	2.31	2.21
25	3.13	2.99	2.85	2.70	2.62	2.54	2.45	2.36	2.27	2.17
26	3.09	2.96	2.81	2.66	2.58	2.50	2.42	2.33	2.23	2.13
27	3.06	2.93	2.78	2.63	2.55	2.47	2.38	2.29	2.20	2.10
28	3.03	2.90	2.75	2.60	2.52	2.44	2.35	2.26	2.17	2.06
29	3.00	2.87	2.73	2.57	2.49	2.41	2.33	2.23	2.14	2.03
30	2.98	2.84	2.70	2.55	2.47	2.39	2.30	2.21	2.11	2.01
40	2.80	2.66	2.52	2.37	2.29	2.20	2.11	2.02	1.92	1.80
60	2.63	2.50	2.35	2.20	2.12	2.03	1.94	1.84	1.73	1.60
120	2.47	2.34	2.19	2.03	1.95	1.86	1.76	1.66	1.53	1.38
∞	2.32	2.18	2.04	1.88	1.79	1.70	1.59	1.47	1.32	1.00

TABLE XI Critical Values of T_L and T_U for the Wilcoxon Rank Sum Test: Independent Samples

Test statistic is the rank sum associated with the smaller sample (if equal sample sizes, either rank sum can be used).

a. $\alpha = .025$ one-tailed; $\alpha = .05$ two-tailed

n_2 \ n_1	3		4		5		6		7		8		9		10	
	T_L	T_U	T_L	T_U	T_L	T_U	T_L	T_U	T_L	T_U	T_L	T_U	T_L	T_U	T_L	T_U
3	5	16	6	18	6	21	7	23	7	26	8	28	8	31	9	33
4	6	18	11	25	12	28	12	32	13	35	14	38	15	41	16	44
5	6	21	12	28	18	37	19	41	20	45	21	49	22	53	24	56
6	7	23	12	32	19	41	26	52	28	56	29	61	31	65	32	70
7	7	26	13	35	20	45	28	56	37	68	39	73	41	78	43	83
8	8	28	14	38	21	49	29	61	39	73	49	87	51	93	54	98
9	8	31	15	41	22	53	31	65	41	78	51	93	63	108	66	114
10	9	33	16	44	24	56	32	70	43	83	54	98	66	114	79	131

b. $\alpha = .05$ one-tailed; $\alpha = .10$ two-tailed

n_2 \ n_1	3		4		5		6		7		8		9		10	
	T_L	T_U	T_L	T_U	T_L	T_U	T_L	T_U	T_L	T_U	T_L	T_U	T_L	T_U	T_L	T_U
3	6	15	7	17	7	20	8	22	9	24	9	27	10	29	11	31
4	7	17	12	24	13	27	14	30	15	33	16	36	17	39	18	42
5	7	20	13	27	19	36	20	40	22	43	24	46	25	50	26	54
6	8	22	14	30	20	40	28	50	30	54	32	58	33	63	35	67
7	9	24	15	33	22	43	30	54	39	66	41	71	43	76	46	80
8	9	27	16	36	24	46	32	58	41	71	52	84	54	90	57	95
9	10	29	17	39	25	50	33	63	43	76	54	90	66	105	69	111
10	11	31	18	42	26	54	35	67	46	80	57	95	69	111	83	127

Source: From F. Wilcoxon and R. A. Wilcox, "Some Rapid Approximate Statistical Procedures," 1964, 20–23. Reproduced with the permission of American Cyanamid Company.

TABLE XII Critical Values of T_0 in the Wilcoxon Paired Difference Signed Rank Test

One-Tailed	Two-Tailed						
		$n = 5$	$n = 6$	$n = 7$	$n = 8$	$n = 9$	$n = 10$
$\alpha = .05$	$\alpha = .10$	1	2	4	6	8	11
$\alpha = .025$	$\alpha = .05$		1	2	4	6	8
$\alpha = .01$	$\alpha = .02$			0	2	3	5
$\alpha = .005$	$\alpha = .01$				0	2	3
		$n = 11$	$n = 12$	$n = 13$	$n = 14$	$n = 15$	$n = 16$
$\alpha = .05$	$\alpha = .10$	14	17	21	26	30	36
$\alpha = .025$	$\alpha = .05$	11	14	17	21	25	30
$\alpha = .01$	$\alpha = .02$	7	10	13	16	20	24
$\alpha = .005$	$\alpha = .01$	5	7	10	13	16	19
		$n = 17$	$n = 18$	$n = 19$	$n = 20$	$n = 21$	$n = 22$
$\alpha = .05$	$\alpha = .10$	41	47	54	60	68	75
$\alpha = .025$	$\alpha = .05$	35	40	46	52	59	66
$\alpha = .01$	$\alpha = .02$	28	33	38	43	49	56
$\alpha = .005$	$\alpha = .01$	23	28	32	37	43	49
		$n = 23$	$n = 24$	$n = 25$	$n = 26$	$n = 27$	$n = 28$
$\alpha = .05$	$\alpha = .10$	83	92	101	110	120	130
$\alpha = .025$	$\alpha = .05$	73	81	90	98	107	117
$\alpha = .01$	$\alpha = .02$	62	69	77	85	93	102
$\alpha = .005$	$\alpha = .01$	55	61	68	76	84	92
		$n = 29$	$n = 30$	$n = 31$	$n = 32$	$n = 33$	$n = 34$
$\alpha = .05$	$\alpha = .10$	141	152	163	175	188	201
$\alpha = .025$	$\alpha = .05$	127	137	148	159	171	183
$\alpha = .01$	$\alpha = .02$	111	120	130	141	151	162
$\alpha = .005$	$\alpha = .01$	100	109	118	128	138	149
		$n = 35$	$n = 36$	$n = 37$	$n = 38$	$n = 39$	
$\alpha = .05$	$\alpha = .10$	214	228	242	256	271	
$\alpha = .025$	$\alpha = .05$	195	208	222	235	250	
$\alpha = .01$	$\alpha = .02$	174	186	198	211	224	
$\alpha = .005$	$\alpha = .01$	160	171	183	195	208	
		$n = 40$	$n = 41$	$n = 42$	$n = 43$	$n = 44$	$n = 45$
$\alpha = .05$	$\alpha = .10$	287	303	319	336	353	371
$\alpha = .025$	$\alpha = .05$	264	279	295	311	327	344
$\alpha = .01$	$\alpha = .02$	238	252	267	281	297	313
$\alpha = .005$	$\alpha = .01$	221	234	248	262	277	292
		$n = 46$	$n = 47$	$n = 48$	$n = 49$	$n = 50$	
$\alpha = .05$	$\alpha = .10$	389	408	427	446	466	
$\alpha = .025$	$\alpha = .05$	361	379	397	415	434	
$\alpha = .01$	$\alpha = .02$	329	345	362	380	398	
$\alpha = .005$	$\alpha = .01$	307	323	339	356	373	

Source: From F. Wilcoxon and R. A. Wilcox, "Some Rapid Approximate Statistical Procedures," 1964, p. 28. Reproduced with the permission of American Cyanamid Company.

TABLE XIII Critical Values of χ^2

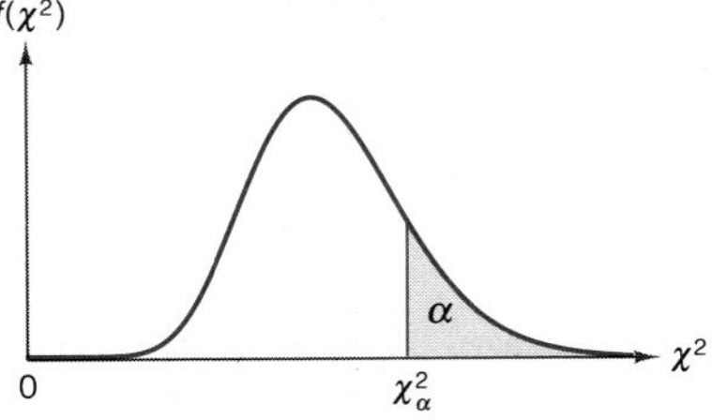

Degrees of Freedom	$\chi^2_{.995}$	$\chi^2_{.990}$	$\chi^2_{.975}$	$\chi^2_{.950}$	$\chi^2_{.900}$
1	.0000393	.0001571	.0009821	.0039321	.0157908
2	.0100251	.0201007	.0506356	.102587	.210720
3	.0717212	.114832	.215795	.351846	.584375
4	.206990	.297110	.484419	.710721	1.063623
5	.411740	.554300	.831211	1.145476	1.61031
6	.675727	.872085	1.237347	1.63539	2.20413
7	.989265	1.239043	1.68987	2.16735	2.83311
8	1.344419	1.646482	2.17973	2.73264	3.48954
9	1.734926	2.087912	2.70039	3.32511	4.16816
10	2.15585	2.55821	3.24697	3.94030	4.86518
11	2.60321	3.05347	3.81575	4.57481	5.57779
12	3.07382	3.57056	4.40379	5.22603	6.30380
13	3.56503	4.10691	5.00874	5.89186	7.04150
14	4.07468	4.66043	5.62872	6.57063	7.78953
15	4.60094	5.22935	6.26214	7.26094	8.54675
16	5.14224	5.81221	6.90766	7.96164	9.31223
17	5.69724	6.40776	7.56418	8.67176	10.0852
18	6.26481	7.01491	8.23075	9.39046	10.8649
19	6.84398	7.63273	8.90655	10.1170	11.6509
20	7.43386	8.26040	9.59083	10.8508	12.4426
21	8.03366	8.89720	10.28293	11.5913	13.2396
22	8.64272	9.54249	10.9823	12.3380	14.0415
23	9.26042	10.19567	11.6885	13.0905	14.8479
24	9.88623	10.8564	12.4011	13.8484	15.6587
25	10.5197	11.5240	13.1197	14.6114	16.4734
26	11.1603	12.1981	13.8439	15.3791	17.2919
27	11.8076	12.8786	14.5733	16.1513	18.1138
28	12.4613	13.5648	15.3079	16.9279	18.9392
29	13.1211	14.2565	16.0471	17.7083	19.7677
30	13.7867	14.9535	16.7908	18.4926	20.5992
40	20.7065	22.1643	24.4331	26.5093	29.0505
50	27.9907	29.7067	32.3574	34.7642	37.6886
60	35.5346	37.4848	40.4817	43.1879	46.4589
70	43.2752	45.4418	48.7576	51.7393	55.3290
80	51.1720	53.5400	57.1532	60.3915	64.2778
90	59.1963	61.7541	65.6466	69.1260	73.2912
100	67.3276	70.0648	74.2219	77.9295	82.3581

Source: From C. M. Thompson, "Tables of the Percentage Points of the χ^2-Distribution," *Biometrika*, 1941, 32, 188–189. Reproduced by permission of the *Biometrika* Trustees.

Degrees of Freedom	$\chi^2_{.100}$	$\chi^2_{.050}$	$\chi^2_{.025}$	$\chi^2_{.010}$	$\chi^2_{.005}$
1	2.70554	3.84146	5.02389	6.63490	7.87944
2	4.60517	5.99147	7.37776	9.21034	10.5966
3	6.25139	7.81473	9.34840	11.3449	12.8381
4	7.77944	9.48773	11.1433	13.2767	14.8602
5	9.23635	11.0705	12.8325	15.0863	16.7496
6	10.6446	12.5916	14.4494	16.8119	18.5476
7	12.0170	14.0671	16.0128	18.4753	20.2777
8	13.3616	15.5073	17.5346	20.0902	21.9550
9	14.6837	16.9190	19.0228	21.6660	23.5893
10	15.9871	18.3070	20.4831	23.2093	25.1882
11	17.2750	19.6751	21.9200	24.7250	26.7569
12	18.5494	21.0261	23.3367	26.2170	28.2995
13	19.8119	22.3621	24.7356	27.6883	29.8194
14	21.0642	23.6848	26.1190	29.1413	31.3193
15	22.3072	24.9958	27.4884	30.5779	32.8013
16	23.5418	26.2962	28.8454	31.9999	34.2672
17	24.7690	27.5871	30.1910	33.4087	35.7185
18	25.9894	28.8693	31.5264	34.8053	37.1564
19	27.2036	30.1435	32.8523	36.1908	38.5822
20	28.4120	31.4104	34.1696	37.5662	39.9968
21	29.6151	32.6705	35.4789	38.9321	41.4010
22	30.8133	33.9244	36.7807	40.2894	42.7956
23	32.0069	35.1725	38.0757	41.6384	44.1813
24	33.1963	36.4151	39.3641	42.9798	45.5585
25	34.3816	37.6525	40.6465	44.3141	46.9278
26	35.5631	38.8852	41.9232	45.6417	48.2899
27	36.7412	40.1133	43.1944	46.9630	49.6449
28	37.9159	41.3372	44.4607	48.2782	50.9933
29	39.0875	42.5569	45.7222	49.5879	52.3356
30	40.2560	43.7729	46.9792	50.8922	53.6720
40	51.8050	55.7585	59.3417	63.6907	66.7659
50	63.1671	67.5048	71.4202	76.1539	79.4900
60	74.3970	79.0819	83.2976	88.3794	91.9517
70	85.5271	90.5312	95.0231	100.425	104.215
80	96.5782	101.879	106.629	112.329	116.321
90	107.565	113.145	118.136	124.116	128.299
100	118.498	124.342	129.561	135.807	140.169

TABLE XIV Critical Values for the Durbin-Watson d Statistic, $\alpha = .05$										
n	$k = 1$		$k = 2$		$k = 3$		$k = 4$		$k = 5$	
	d_L	d_U	d_L	d_U	d_L	d_U	d_L	d_U	d_L	d_U
15	1.08	1.36	.95	1.54	.82	1.75	.69	1.97	.56	2.21
16	1.10	1.37	.98	1.54	.86	1.73	.74	1.93	.62	2.15
17	1.13	1.38	1.02	1.54	.90	1.71	.78	1.90	.67	2.10
18	1.16	1.39	1.05	1.53	.93	1.69	.92	1.87	.71	2.06
19	1.18	1.40	1.08	1.53	.97	1.68	.86	1.85	.75	2.02
20	1.20	1.41	1.10	1.54	1.00	1.68	.90	1.83	.79	1.99
21	1.22	1.42	1.13	1.54	1.03	1.67	.93	1.81	.83	1.96
22	1.24	1.43	1.15	1.54	1.05	1.66	.96	1.80	.96	1.94
23	1.26	1.44	1.17	1.54	1.08	1.66	.99	1.79	.90	1.92
24	1.27	1.45	1.19	1.55	1.10	1.66	1.01	1.78	.93	1.90
25	1.29	1.45	1.21	1.55	1.12	1.66	1.04	1.77	.95	1.89
26	1.30	1.46	1.22	1.55	1.14	1.65	1.06	1.76	.98	1.88
27	1.32	1.47	1.24	1.56	1.16	1.65	1.08	1.76	1.01	1.86
28	1.33	1.48	1.26	1.56	1.18	1.65	1.10	1.75	1.03	1.85
29	1.34	1.48	1.27	1.56	1.20	1.65	1.12	1.74	1.05	1.84
30	1.35	1.49	1.28	1.57	1.21	1.65	1.14	1.74	1.07	1.83
31	1.36	1.50	1.30	1.57	1.23	1.65	1.16	1.74	1.09	1.83
32	1.37	1.50	1.31	1.57	1.24	1.65	1.18	1.73	1.11	1.82
33	1.38	1.51	1.32	1.58	1.26	1.65	1.19	1.73	1.13	1.81
34	1.39	1.51	1.33	1.58	1.27	1.65	1.21	1.73	1.15	1.81
35	1.40	1.52	1.34	1.58	1.28	1.65	1.22	1.73	1.16	1.80
36	1.41	1.52	1.35	1.59	1.29	1.65	1.24	1.73	1.18	1.80
37	1.42	1.53	1.36	1.59	1.31	1.66	1.25	1.72	1.19	1.80
38	1.43	1.54	1.37	1.59	1.32	1.66	1.26	1.72	1.21	1.79
39	1.43	1.54	1.38	1.60	1.33	1.66	1.27	1.72	1.22	1.79
40	1.44	1.54	1.39	1.60	1.34	1.66	1.29	1.72	1.23	1.79
45	1.48	1.57	1.43	1.62	1.38	1.67	1.34	1.72	1.29	1.78
50	1.50	1.59	1.46	1.63	1.42	1.67	1.38	1.72	1.34	1.77
55	1.53	1.60	1.49	1.64	1.45	1.68	1.41	1.72	1.38	1.77
60	1.55	1.62	1.51	1.65	1.48	1.69	1.44	1.73	1.41	1.77
65	1.57	1.63	1.54	1.66	1.50	1.70	1.47	1.73	1.44	1.77
70	1.58	1.64	1.55	1.67	1.52	1.70	1.49	1.74	1.46	1.77
75	1.60	1.65	1.57	1.68	1.54	1.71	1.51	1.74	1.49	1.77
80	1.61	1.66	1.59	1.69	1.56	1.72	1.53	1.74	1.51	1.77
85	1.62	1.67	1.60	1.70	1.57	1.72	1.55	1.75	1.52	1.77
90	1.63	1.68	1.61	1.70	1.59	1.73	1.57	1.75	1.4	1.78
95	1.64	1.69	1.62	1.71	1.60	1.73	1.58	1.75	1.56	1.78
100	1.65	1.69	1.63	1.72	1.61	1.74	1.59	1.76	1.57	1.78

Source: From J. Durbin and G. S. Watson, "Testing for Serial Correlation in Least Squares Regression, II." *Biometrika*, 1951, 30, 159–178. Reproduced by permission of the *Biometrika* Trustees.

TABLE XV Critical Values for the Durbin-Watson d Statistic, $\alpha = .01$

n	$k = 1$		$k = 2$		$k = 3$		$k = 4$		$k = 5$	
	d_L	d_U	d_L	d_U	d_L	d_U	d_L	d_U	d_L	d_U
15	.81	1.07	.70	1.25	.59	1.46	.49	1.70	.39	1.96
16	.84	1.09	.74	1.25	.63	1.44	.53	1.66	.44	1.90
17	.87	1.10	.77	1.25	.67	1.43	.57	1.3	.48	1.85
18	.90	1.12	.80	1.26	.71	1.42	.61	1.60	.52	1.80
19	.93	1.13	.83	1.26	.74	1.41	.65	1.58	.56	1.77
20	.95	1.15	.86	1.27	.77	1.41	.68	1.57	.60	1.74
21	.97	1.16	.89	1.27	.80	1.41	.72	1.55	.63	1.71
22	1.00	1.17	.91	1.28	.83	1.40	.75	1.54	.66	1.69
23	1.02	1.19	.94	1.29	.86	1.40	.77	1.53	.70	1.67
24	1.04	1.20	.96	1.30	.88	1.41	.80	1.53	.72	1.66
25	1.05	1.21	.98	1.30	.90	1.41	.83	1.52	.75	1.65
26	1.07	1.22	1.00	1.31	.93	1.41	.85	1.52	.78	1.64
27	1.09	1.23	1.02	1.32	.95	1.41	.88	1.51	.81	1.63
28	1.10	1.24	1.04	1.32	.97	1.41	.90	1.51	.83	1.62
29	1.12	1.25	1.05	1.33	.99	1.42	.92	1.51	.85	1.61
30	1.13	1.26	1.07	1.34	1.01	1.42	.94	1.51	.88	1.61
31	1.15	1.27	1.08	1.34	1.02	1.42	.96	1.51	.90	1.60
32	1.16	1.28	1.10	1.35	1.04	1.43	.98	1.51	.92	1.60
33	1.17	1.29	1.11	1.36	1.05	1.43	1.00	1.51	.94	1.59
34	1.18	1.30	1.13	1.36	1.07	1.43	1.01	1.51	.95	1.59
35	1.19	1.31	1.14	1.27	1.08	1.44	1.03	1.51	.97	1.59
36	1.21	1.32	1.15	1.38	1.10	1.44	1.04	1.51	.99	1.59
37	1.22	1.32	1.16	1.38	1.11	1.45	1.06	1.51	1.00	1.59
38	1.23	1.33	1.18	1.39	1.12	1.45	1.07	1.52	1.02	1.58
39	1.24	1.34	1.19	1.39	1.14	1.45	1.09	1.52	1.03	1.58
40	1.25	1.34	1.20	1.40	1.15	1.46	1.10	1.52	1.05	1.58
45	1.29	1.38	1.24	1.42	1.20	1.48	1.16	1.53	1.11	1.58
50	1.32	1.40	1.28	1.45	1.24	1.49	1.20	1.54	1.16	1.59
55	1.36	1.43	1.32	1.47	1.28	1.51	1.25	1.55	1.21	1.59
60	1.38	1.45	1.35	1.48	1.32	1.52	1.28	1.56	1.25	1.60
65	1.41	1.47	1.38	1.50	1.35	1.53	1.31	1.57	1.28	1.61
70	1.43	1.49	1.40	1.52	1.37	1.55	1.34	1.58	1.31	1.61
75	1.45	1.50	1.42	1.53	1.39	1.56	1.37	1.59	1.34	1.62
80	1.47	1.52	1.44	1.54	1.42	1.57	1.39	1.60	1.36	1.62
85	1.48	1.53	1.46	1.55	1.43	1.58	1.41	1.60	1.39	1.63
90	1.50	1.54	1.47	1.56	1.45	1.59	1.43	1.61	1.41	1.64
95	1.51	1.55	1.49	1.57	1.47	1.60	1.45	1.62	1.42	1.64
100	1.52	1.56	1.50	1.58	1.48	1.60	1.46	1.63	1.44	1.65

Source: From J. Durbin and G. S. Watson, "Testing for Serial Correlation in Least Squares Regression, II." *Biometrika*, 1951, 30, 159–78. Reproduced by permission of the *Biometrika* Trustees.

TABLE XVI Critical Values of Spearman's Rank Correlation Coefficient

The α values correspond to a one-tailed test of H_0: $\rho_s = 0$. The value should be doubled for two-tailed tests.

n	$\alpha = .05$	$\alpha = .025$	$\alpha = .01$	$\alpha = .005$	n	$\alpha = .05$	$\alpha = .025$	$\alpha = .01$	$\alpha = .005$
5	.900	—	—	—	18	.399	.476	.564	.625
6	.829	.886	.943	—	19	.388	.462	.549	.608
7	.714	.786	.893	—	20	.377	.450	.534	.591
8	.643	.738	.833	.881	21	.368	.438	.521	.576
9	.600	.683	.783	.833	22	.359	.428	.508	.562
10	.564	.648	.745	.794	23	.351	.418	.496	.549
11	.523	.623	.736	.818	24	.343	.409	.485	.537
12	.497	.591	.703	.780	25	.336	.400	.475	.526
13	.475	.566	.673	.745	26	.329	.392	.465	.515
14	.457	.545	.646	.716	27	.323	.385	.456	.505
15	.441	.525	.623	.689	28	.317	.377	.448	.496
16	.425	.507	.601	.666	29	.311	.370	.440	.487
17	.412	.490	.582	.645	30	.305	.364	.432	.478

Source: From E. G. Olds, "Distribution of Sums of Squares of Rank Differences for Small Samples," *Annals of Mathematical Statistics*, 1938, 9. Reproduced with the permission of the Editor, *Annals of Mathematical Statistics*.

TABLE XVII Control Chart Constants

Number of Observations in Subgroup, n	A_2	d_2	d_3	D_3	D_4
2	1.880	1.128	.853	.000	3.267
3	1.023	1.693	.888	.000	2.574
4	.729	2.059	.880	.000	2.282
5	.577	2.326	.864	.000	2.114
6	.483	2.534	.848	.000	2.004
7	.419	2.704	.833	.076	1.924
8	.373	2.847	.820	.136	1.864
9	.337	2.970	.808	.184	1.816
10	.308	3.078	.797	.223	1.777
11	.285	3.173	.787	.256	1.744
12	.266	3.258	.778	.283	1.717
13	.249	3.336	.770	.307	1.693
14	.235	3.407	.762	.328	1.672
15	.223	3.472	.755	.347	1.653
16	.212	3.532	.749	.363	1.637
17	.203	3.588	.743	.378	1.622
18	.194	3.640	.738	.391	1.608
19	.187	3.689	.733	.403	1.597
20	.180	3.735	.729	.415	1.585
21	.173	3.778	.724	.425	1.575
22	.167	3.819	.720	.434	1.566
23	.162	3.858	.716	.443	1.557
24	.157	3.895	.712	.451	1.548
25	.153	3.931	.709	.459	1.541
More than 25	$3/\sqrt{n}$				

Source: *ASTM Manual on the Presentation of Data and Control Chart Analysis,* Philadelphia, PA: American Society for Testing Materials, pp. 134–136, 1976.

APPENDIX C

Demographic Data Set

A demographic data set was assembled based on a systematic random sample of 1,000 United States zip codes. To obtain the sample, the more than 30,000 zip codes were sorted, and approximately every 30th was selected. The map in Figure C.1 shows the number of zip codes selected in each state. Note that each state is classified according to its census region: North Central, Northeast, South, and West.

Demographic data for each zip code area selected were supplied by CACI, an international demographic and market information firm, and are reproduced with its permission. CACI produces interim estimates of many of the demographic variables measured decennially by the Bureau of the Census. CACI also produces market information based on the Bureau of the Census Consumer Expenditure Survey (such

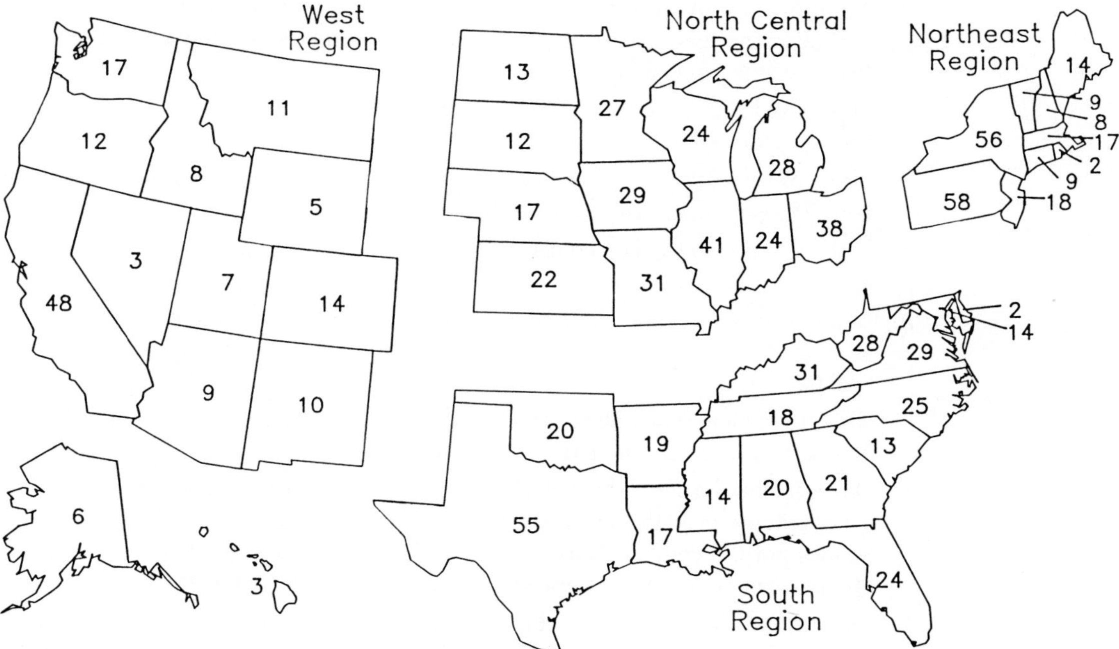

FIGURE C.1 ▲ Number of zip code areas selected by state and census region

as the four purchasing indexes given below for each zip code), which measure the relative propensity of a given market (zip code) to purchase the goods, a higher index value indicating a higher propensity to buy such goods as compared to other similar zip codes in the same census region. In general, the 1980 measurements are official U.S. Bureau of the Census estimates, while the 1986 measurements are CACI estimates.

Fifteen demographic measurements are presented for each zip code area. Portions of the data are referenced at the end of each chapter in "Using the Computer." The objectives are to enable the student to analyze real data in a relatively large sample using the computer, and to gain experience using statistical techniques and concepts on real data. Of course, neither the student nor the instructor need be bound by the suggestions in "Using the Computer"; the data are rich enough to support many more analyses than could be listed (or imagined) by the authors.

The following 15 measurements are reported for each zip code in the sample:

1. Population (1986): Total population for the zip code, 1986.
2. Number of Households (1986): Number of households for the zip code in 1986.
3. Age (Median, 1986): Median age for the zip code in 1986.
4. Household Income (Median, 1986): Median household income for the zip code in 1986.
5. Home Value (Median, 1980): Median residential owner-occupied home value for the zip code in 1980.
6. Monthly Cost of Housing (Median, 1980): Median monthly homeowner cost, including mortgage payments, real estate taxes, property insurance, utilities, and fuels, for the zip code in 1980.
7. Household Size (Average, 1980): Average number of persons per household in the zip code, 1980.
8. Years of Education (Median, 1980): Median years of education for adults (persons 25 years of age and over) in the zip code, 1980.
9. College Education (Percentage, 1980): Percentage of adults in the zip code having a college education.
10. Women in Labor Force (Percentage, 1980): Percentage of women age 16 and over in the zip code who are in the labor force (either having jobs or actively seeking one).
11. Unemployment (Percentage, 1980): Percentage of the labor force in the zip code that is unemployed and actively seeking work.
12. Average Purchasing Potential (1986): Purchasing potential index for all consumer items over the zip code area, based on the Bureau of Labor Statistics Consumer Expenditure Survey. The index compares a zip code to similar urban or rural zip codes in the same census region, with 100 set as the average value.
13. Sporting Goods Purchasing Potential (1986): Purchasing potential index for sporting goods over the zip code area.
14. Groceries Purchasing Potential (1986): Purchasing potential index for groceries over the zip code area.

15. Home Improvement Purchasing Potential (1986): Purchasing potential index for home improvements over the zip code area.

Numerical codes have been assigned to each state and census region to facilitate computer utilization of these data. The codes were assigned alphabetically as shown in Tables C.1 and C.2.

TABLE C.1 Census Region Codes

Code	Census Region
1	North Central region
2	Northeast region
3	South region
4	West region

TABLE C.2 State Codes

Code	State		Code	State	
01	AK:	Alaska	26	MT:	Montana
02	AL:	Alabama	27	NC:	North Carolina
03	AR:	Arkansas	28	ND:	North Dakota
04	AZ:	Arizona	29	NE:	Nebraska
05	CA:	California	30	NH:	New Hampshire
06	CO:	Colorado	31	NJ:	New Jersey
07	CT:	Connecticut	32	NM:	New Mexico
08	DE:	Delaware	33	NV:	Nevada
09	FL:	Florida	34	NY:	New York
10	GA:	Georgia	35	OH:	Ohio
11	HI:	Hawaii	36	OK:	Oklahoma
12	IA:	Iowa	37	OR:	Oregon
13	ID:	Idaho	38	PA:	Pennsylvania
14	IL:	Illinois	39	RI:	Rhode Island
15	IN:	Indiana	40	SC:	South Carolina
16	KS:	Kansas	41	SD:	South Dakota
17	KY:	Kentucky	42	TN:	Tennessee
18	LA:	Louisiana	43	TX:	Texas
19	MA:	Massachusetts	44	UT:	Utah
20	MD:	Maryland	45	VA:	Virginia
21	ME:	Maine	46	VT:	Vermont
22	MI:	Michigan	47	WA:	Washington
23	MN:	Minnesota	48	WI:	Wisconsin
24	MO:	Missouri	49	WV:	West Virginia
25	MS:	Mississippi	50	WY:	Wyoming

The demographic data base is contained in two ASCII files, which are both sorted by census region and state within census region. The file names and corresponding layouts are given in Tables C.3 and C.4. The data files are available on magnetic tape or diskette from the publisher.

TABLE C.3 ZIPCOD01.DAT

Columns	Description
1–4	Observation number
7	Census region number
9–10	State number
12–16	Zip code
18–22	Population in '86
24–28	Number of households in '86
30–33	Median age in '86
35–39	Median household income in '86
41–46	Median home value in '80
48–50	Median monthly homeowner cost in '80
52–54	Median household size in '80
56–59	Median years of education in '80

TABLE C.4 ZIPCOD02.DAT

Columns	Description
1–4	Observation number
7	Census region number
9–10	State number
12–16	Zip code
18–21	Percent college education in '80
23–26	Percent women in work force in '80
28–31	Percent unemployed in '80
33–37	Purchasing potential index '86: Overall average
39–43	Purchasing potential index '86: Sporting goods
45–49	Purchasing potential index '86: Grocery
51–55	Purchasing potential index '86: Home improvement

APPENDIX D

Calculation Formulas for Analysis of Variance

Contents

D.1 Formulas for the Calculations in the Completely Randomized Design

$$\begin{aligned}
\text{CM} &= \text{Correction for mean} \\
&= \frac{(\text{Total of all observations})^2}{\text{Total number of observations}} = \frac{\left(\sum y_i\right)^2}{n} \\
\text{SS(Total)} &= \text{Total sum of squares} \\
&= (\text{Sum of squares of all observations}) - \text{CM} = \sum y_i^2 - \text{CM} \\
\text{SST} &= \text{Sum of squares for treatments} \\
&= \begin{pmatrix} \text{Sum of squares of treatment totals with} \\ \text{each square divided by the number of} \\ \text{observations for that treatment} \end{pmatrix} - \text{CM} \\
&= \frac{T_1^2}{n_1} + \frac{T_2^2}{n_2} + \cdots + \frac{T_p^2}{n_p} - \text{CM} \\
\text{SSE} &= \text{Sum of squares for error} = \text{SS(Total)} - \text{SST} \\
\text{MST} &= \text{Mean square for treatments} = \frac{\text{SST}}{p-1}
\end{aligned}$$

$$\text{MSE} = \text{Mean square for error} = \frac{\text{SSE}}{n - p}$$

$$F = \text{Test statistic} = \frac{\text{MST}}{\text{MSE}}$$

where

$$n = \text{Total number of observations}$$
$$p = \text{Number of treatments}$$
$$T_i = \text{Total for treatment } i \ (i = 1, 2, \ldots, p)$$

D.2 Formulas for the Calculations for a Two-Factor Factorial Experiment

$$\text{CM} = \text{Correction for the mean}$$

$$= \frac{(\text{Total of all } n \text{ measurements})^2}{n} = \frac{\left(\sum_{i=1}^{n} y_i\right)^2}{n}$$

$$\text{SS(Total)} = \text{Total sum of squares}$$

$$= \text{Sum of squares of all } n \text{ measurements} - \text{CM} = \sum_{i=1}^{n} y_i^2 - \text{CM}$$

$$\text{SS(A)} = \text{Sum of squares for main effects, factor A}$$

$$= \left(\begin{array}{c}\text{Sum of squares of the totals } A_1, A_2, \ldots, A_a \\ \text{divided by the number of measurements} \\ \text{in a single total, namely } br\end{array}\right) - \text{CM}$$

$$= \frac{\sum_{i=1}^{a} A_i^2}{br} - \text{CM}$$

$$\text{SS}(B) = \text{Sum of squares for main effects, factor } B$$

$$= \left(\begin{array}{c}\text{Sum of squares of the totals } B_1, B_2, \ldots, B_b \\ \text{divided by the number of measurements} \\ \text{in a single total, namely } ar\end{array}\right) - \text{CM}$$

$$= \frac{\sum_{i=1}^{b} B_i^2}{ar} - \text{CM}$$

$$\text{SS}(AB) = \text{Sum of squares for } AB \text{ interaction}$$

$$= \begin{pmatrix} \text{Sum of squares of the cell} \\ \text{totals } AB_{11}, AB_{12}, \ldots, AB_{ab} \\ \text{divided by the number of} \\ \text{measurements in a single} \\ \text{total, namely } r \end{pmatrix} - \text{SS(A)} - \text{SS}(B) - \text{CM}$$

$$= \frac{\sum_{j=1}^{b} \sum_{i=1}^{a} AB_{ij}^2}{r} - \text{SS(A)} - \text{SS}(B) - \text{CM}$$

where

a = Number of levels of factor A
b = Number of levels of factor B
r = Number of replicates (observations per treatment)
A_i = Total for level i of factor A ($i = 1, 2, \ldots, a$)
B_i = Total for level i of factor B ($i = 1, 2, \ldots, b$)
AB_{ij} = Total for treatment (i, j), i.e., for ith level of factor A and jth level of factor B

APPENDIX E

ASP Tutorial

This appendix provides an overview of the ASP program. It gives the minimal hardware requirements and start-up procedures necessary to begin an ASP session on a personal computer (PC). This tutorial is not intended to replace any of the ASP documentation manuals available from the publisher or DMC Software, Inc.

Hardware Requirements

ASP must be run on an IBM-compatible PC with at least 512K of memory, two disk drives (either one hard drive and one floppy drive, or two floppy drives), and DOS 2.0 or higher. A blank formatted floppy disk is also required for data storage, unless your PC has a hard drive (i.e., fixed disk) available for storing data.

Getting Started

To use the ASP program, you must first load it into the memory of the computer. To accomplish this when starting ASP from a floppy disk:

1. Insert your copy of ASP into either of your two disk drives, drive A or drive B. (Assume drive A.)
2. Type **A:** and press **ENTER** to make drive A the current drive:

   A: ⟨ENTER⟩

3. Type **ASP** and press **ENTER** to load the ASP program into memory.

   ASP ⟨ENTER⟩

The ASP disk must remain in drive A for as long as you are using the program.

To start ASP from a fixed disk or hard drive (e.g., drive C), it is first necessary to install ASP on the fixed disk. This is accomplished by placing your copy of the ASP disk into floppy drive A and entering the following commands at the DOS prompt:

C: ⟨ENTER⟩
MD \ASP ⟨ENTER⟩
CD \ASP ⟨ENTER⟩
COPY A:*.* ⟨ENTER⟩

(This sequence of DOS commands assumes the drive letter of the fixed disk is **C** and that the subdirectory in which the ASP program resides is **\ASP**.) Once ASP has been installed on the fixed disk, it need not be installed again. The ASP program can then be started at any point in the future by entering the following commands at the DOS prompt:

C: **⟨ENTER⟩**
CD \ASP **⟨ENTER⟩**
ASP **⟨ENTER⟩**

The Main Menu

The initial screen to appear as the ASP program is loaded into memory displays copyright and licensing information. After reading this information, press any key to obtain the MAIN MENU shown in Figure E.1.

The MAIN MENU is a typical ASP "bounce bar" menu. The highlighted bar can be moved from option to option by pressing the SPACE BAR, the cursor control keys (→ ← ↑ ↓), or the TAB key. Once your selection is made, press **ENTER** to display submenus associated with the option. (You can also make a selection by pressing the letter of the desired option.)

Table E.1 gives a brief description of each of the MAIN MENU options and the corresponding chapters in the text. Several of these options contain statistical procedures that are beyond the scope of the text. Only the statistical routines covered in the text are described in the table.

FIGURE E.1 ►
The ASP Main Menu

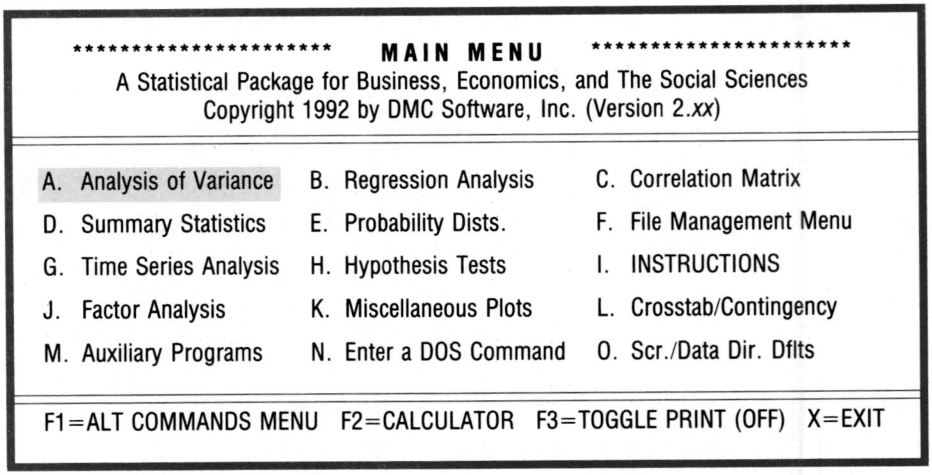

```
***********************     MAIN MENU     ***********************
   A Statistical Package for Business, Economics, and The Social Sciences
           Copyright 1992 by DMC Software, Inc. (Version 2.xx)

A. Analysis of Variance    B. Regression Analysis     C. Correlation Matrix
D. Summary Statistics      E. Probability Dists.      F. File Management Menu
G. Time Series Analysis    H. Hypothesis Tests        I. INSTRUCTIONS
J. Factor Analysis         K. Miscellaneous Plots     L. Crosstab/Contingency
M. Auxiliary Programs      N. Enter a DOS Command     O. Scr./Data Dir. Dflts

F1=ALT COMMANDS MENU  F2=CALCULATOR  F3=TOGGLE PRINT (OFF)  X=EXIT
```

TABLE E.1 Options on the Main Menu

Option	Description	Chapter(s)
A. Analysis of Variance	One-way and two-way ANOVAs	16
B. Regression Analysis	Simple and multiple regression; residual analysis	10, 11, 12
C. Correlation Matrix	Bivariate correlations	10
D. Summary Statistics	Mean, median, standard deviation, etc.	2
E. Probability Dists.	Binomial and normal distributions	4, 5
F. File Management Menu	Creating, saving, editing data	—
G. Time Series Analysis	(Beyond the scope of this text)	—
H. Hypothesis Tests	Confidence intervals and hypothesis tests for means, proportions, and variances; one-way table χ^2 test; nonparametric tests	7, 8, 9, 17, 18
I. INSTRUCTIONS	A short tutorial on the use of ASP	—
J. Factor Analysis Menu	(Beyond the scope of this text)	—
K. Miscellaneous Plots	Stem-and-leaf display, box plot, normal probability plot, scatter plot	2, 5, 10
L. Crosstab/Contingency	Two-way (contingency) table χ^2 test	18
M. Auxiliary Programs	(Beyond the scope of this text)	—
N. Enter a DOS Command	Enter and execute DOS commands within an ASP session	—
O. Scr./Data Dir. Dflts.	Set the color scheme on the monitor; set the default directory and printer port	—

Alternate Commands Menu

All of the statistical routines in ASP are accessible through the MAIN MENU. However, additional commands can be executed through the ALT COMMANDS MENU. The ALT COMMANDS MENU is called by pressing the **F1** function key from the main menu or from any menu one level below the main menu. The ALT COMMANDS MENU appears as shown in Figure E.2 on page 1156.

FIGURE E.2 ► The Alt Commands Menu

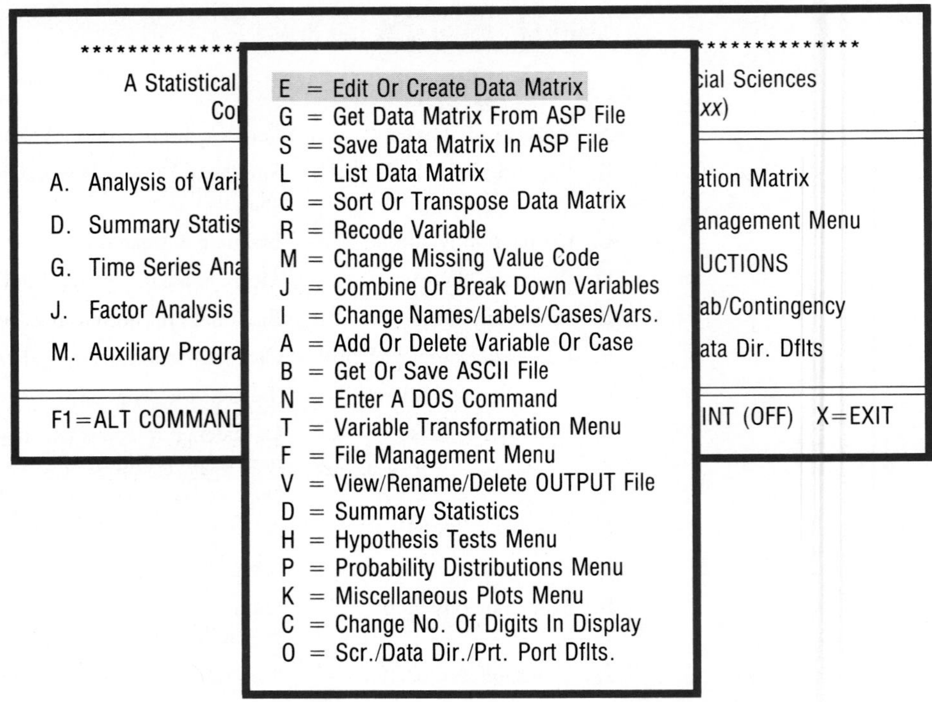

You can execute the commands on this menu either by moving the cursor to the desired option and pressing **ENTER** or by pressing the letter associated with the option. You will find this menu most useful for:

- creating or editing data sets (option **E**)
- listing data (option **L**)
- getting data from an already created ASP data set (option **G**)
- creating new variables for a data set (option **T**)
- adding or deleting variables and/or cases (option **A**)
- changing the names of variables (option **I**)
- getting or saving data in an external ASCII file (option **B**)
- saving an ASP data set (option **S**)

Creating a Data Matrix

Typically, you will use ASP to analyze a data set. To do this, you must first create an ASP "data matrix." Select **E = Edit or Create Data Matrix** on the ALT COM-

MANDS MENU, and ASP responds with a series of questions and prompts. The first question is:

EDIT or CREATE? E

Note that the ASP default answer is **E** for EDIT. This is used when you want to edit an existing ASP data matrix. To create a data matrix, press the letter **C** (for CREATE). You are now prompted with the question:

Number of Variables? 1

Change the default to the correct number and press **ENTER.** ASP creates names for the variables using the convention Var1, Var2, Var3, etc., then asks:

Are Names OK? Y

To change the name, Press **N** (for No). ASP will then ask you to enter the new name of each variable. Once this is completed, ASP will prompt you, one case (i.e., one observation) and one variable at a time, to enter the data into the data matrix.

Important: The ASP data editor will not accept letters or special characters (e.g., dollar sign, comma) as data. Only whole numbers or numbers with decimals should be entered into the data matrix.

When data entry is complete, press **X** to exit the numerical data editor. Several questions will be asked, the most important being:

Do You Wish to Save the Data Matrix? Y

Answer "yes" by pressing **ENTER**. ASP will then ask for the drive letter (e.g., drive A) and directory of the disk where you want to save the data:

DATA DIRECTORY: A:

If the default is correct, press **ENTER**. Otherwise, enter the correct drive/path. You will be asked to name the ASP data file, provide a file label (optional), and whether you want to save all variables and all cases.

Suppose you enter the following file name:

File Name: MYDATA

ASP will save your data matrix in the ASP file named MYDATA.ASP in the directory specified earlier. In future ASP sessions, you can access this data set by first selecting the option **G = Get Data Matrix from ASP File** from the ALT COMMANDS MENU, and then selecting MYDATA from the resulting menu-list of ASP data files.

Analyzing a Data Matrix

To analyze an ASP data matrix that you have just created or accessed, return to the MAIN MENU by pressing **X** or **ESC**. From the MAIN MENU, select the desired statistical routine. Each choice will result in a series of submenus, prompts, and/or questions similar to those shown previously. After making your selections, ASP will perform the analysis and display the results immediately on the monitor screen. ASP menu selections at the bottom of the screen permit you to send the output directly to a printer or to save the output in a file for future use.

Available Documentation

ASP User's Manual (by DMC Software, Inc.)—available free to adopters of the text from the publisher of the text, Macmillan/Dellen Publishing.

ASP Tutorial and Student Guide (by George Blackford)—can be purchased directly from DMC Software, Inc., or from your campus bookstore.

Answers to Selected Exercises

Chapter 2

2.3. Qualitative; qualitative **2.5a.** Nominal **b.** Ratio **c.** Ratio **d.** Ratio **e.** Interval **2.7.** I: Nominal, II: Ratio, III: Ordinal, IV: Nominal, V: Ordinal, VI: Ratio **2.11a.** Frequency histogram **b.** 14 **c.** 49 **2.13b.** 24; .8 **2.23.** 14.545; 15; 15 **2.29a.** 180; 162 **b.** Skewed to the right **c.** No **2.31a.** 70.01; 68.75; two modes: 64.6 and 76.8 **b.** 68.83; 69; two modes: 64.6 and 76.8 **c.** 69.0125 **2.33.** \$.35 **2.37a.** 2; 37.5; 6.124 **b.** 1.25; 1.835; 1.355 **2.39a.** 6.2; 77.2; 8.786 **b.** 28.25; 2,304.92; 48.010 **2.43.** No; yes **2.45a.** A: 33.14; B: 34.57 **b.** A: 3.98; B: .98 **d.** A: .82; B: 2.12 **2.47.** Any data set; mound-shaped data sets **2.49a.** At least 0% **b.** At least 75% **c.** At least 88.9% **2.51.** No **2.53a.** .1488; .0273; .1652 **b.** Cannot say anything; at least 84%; at least 93.75% **c.** 68%; 96%; 100% **d.** 0% **2.55.** At most 13 **2.57.** At least .889 **2.59a.** At least .889 **b.** $\approx$.16 **c.** Yes **2.61.** Do not buy **2.63a.** 1; sample **b.** -1.5; sample **c.** 1.765; population **d.** $-.6$; population **2.65a.** 3.7; 2.2; 2.95; 1.45 **b.** 1.9 **c.** Cum laude: $z > 1$, GPA > 3.2; summa cum laude: $z > 2$, GPA > 3.7; mound-shaped distribution **2.67a.** Brazil: -3.825; Egypt: 1.745; France: -2.787; Italy: -5.510; Japan: -56.326; Mexico: -5.689; Panama: .387; Soviet Union: 1.055; Sweden: -2.864; Turkey: .622 **b.** -2.81; .48 **2.69.** 1989: 69.4th percentile; 1985: 85.8th percentile **2.71a.** $\approx$68%; $\approx$100% **b.** 31 **c.** Original **2.73a.** At most 25%; cannot be determined; at most 11.1% **b.** 1.5; 1.5 **2.77a.** 39 **b.** 45; 31.5 **c.** 13.5 **d.** Skewed to the left **e.** 50%; 75% **2.81a.** 3.03; -2.68; 0 **2.83b.** Customers 238, 264, 268, and 269 **c.** Customer 238: 1.92; Customer 264: 3.14; Customer 268: 2.06; Customer 269: 2.13 **2.85a.** Ratio **b.** Nominal **c.** Interval **d.** Ordinal **e.** (a) Quantitative; (b) Qualitative; (c) Quantitative; (d) Qualitative **2.87a.** 239; 29; 841 **b.** 634; 50; 2,500 **c.** 254; 2; 4 **d.** 10,404; 102; 10,404 **2.93a.** 11.65 **b.** 6.6441 **c.** At least 75%; 96.7% **2.95a.** Skewed to the right **c.** Empirical rule: $\approx$49; Chebyshev: at least 38 **d.** 3.33; no **2.97f.** -2.5 **2.99a.** Frequency bar chart **2.111a.** At least 0% **b.** At most 25% **2.115a.** \$7,500 **b.** Over 120,000 **c.** Yes, but very unlikely **2.117a.** None; V = 0 for all data sets **2.121a.** At least .75 **b.** At least .889 **c.** At least .56 **d.** At least .87 **2.123a.** U.S.: $\bar{x} = 41.08$, $s^2 = 330.08$; Canada: $\bar{x} = 70.83$, $s^2 = 529.24$; United Kingdom: $\bar{x} = 60.42$, $s^2 = 634.27$; Sweden: $\bar{x} = 83.42$, $s^2 = 511.17$; France: $\bar{x} = 66.33$, $s^2 = 1{,}051.88$; West Germany: $\bar{x} = 56.08$, $s^2 = 907.17$ **b.** Brewing: $\bar{x} = 54.17$, $s^2 = 640.97$; cigarettes: $\bar{x} = 91.00$, $s^2 = 142.00$; fabric weaving: $\bar{x} = 35.67$, $s^2 = 365.07$; paints: $\bar{x} = 40.67$, $s^2 = 727.47$; petroleum refining: $\bar{x} = 62.50$, $s^2 = 666.70$; shoes: $\bar{x} = 20.33$, $s^2 = 71.87$; glass bottles: $\bar{x} = 85.83$, $s^2 = 210.97$; cement: $\bar{x} = 67.67$, $s^2 = 805.07$; ordinary steel: $\bar{x} = 60.67$, $s^2 = 352.67$; antifriction bearings: $\bar{x} = 80.67$, $s^2 = 390.27$; refrigerators: $\bar{x} = 77.50$, $s^2 = 202.70$; storage batteries: $\bar{x} = 79.67$, $s^2 = 269.87$ **c.** U.S., Sweden **d.** Shoes, fabric weaving, and paints; cigarettes, glass bottles, and antifriction bearings

Chapter 3

3.3a. .3 **b.** .25 **c.** .6 **3.5a.** {under 2,000, 2,000–4,999, 5,000–7,999, 8,000 or more} **b.** .184 **c.** .216 **3.7.** $\frac{1}{20}$ **3.11a.** {Visa or MasterCard, Diners Club, Carte Blanche, Choice} **b.** *P*(Visa or MasterCard) = .632; *P*(Diners Club) = .232; *P*(Carte Blanche) = .032; *P*(Choice) = .105 **c.** .368 **3.13b.** $\frac{1}{16}$ **c.** $\frac{5}{16}$ **3.15a.** 1 to 2 **b.** $\frac{1}{2}$ **c.** .4 **3.19a.** $\frac{7}{8}$ **b.** $\frac{3}{8}$ **c.** $\frac{3}{8}$ **d.** 0 **e.** Yes **f.** $\frac{3}{4}$ **3.21a.** $A \cap F$ **b.** $B \cup C$ **c.** A^c **3.23a.** $B \cap C$ **b.** A^c **c.** $C \cup B$ **d.** $A \cap C^c$ **3.25a.** {(R, 1), (R, 2), (R, 3), (S, 1), (S, 2), (S, 3), (E, 1), (E, 2), (E, 3)} **b.** Sample space **c.** .24 **d.** .10 **e.** .47 **f.** .62 **g.** .28 **3.27b.** .184; yes **c.** .008 **d.** .556 **e.** .326 **3.29a.** .81 **b.** .65 **c.** .41 **d.** .22 **e.** *B* and *C*; *B* and *D*; *C* and *D* **3.31a.** .052 **b.** .259 **c.** .157 **d.** .692 **3.33a.** $P(A) = .8$; $P(B) = .70$; $P(A \cap B) = .60$ **b.** $P(E_1 \mid A) = .25$; $P(E_2 \mid A) = .375$; $P(E_3 \mid A) = .375$ **c.** $P(B \mid A) = .75$ **d.** No **3.35a.** No **b.** No **c.** No **3.37a.** No **b.** Yes

c. No **d.** No **3.39.** No **3.41a.** .550 **b.** .450 **c.** .272 **d.** .040 **e.** .182 **f.** .857 **g.** .182 **3.43.** .005 **3.45a.** .18 **b.** .82 **c.** .31 **d.** .69 **e.** .26 **3.47a.** .3; .6 **b.** Dependent **c.** Independent **3.51a.** {(0001, 0002), (0001, 0003), (0001, 0004), (0001, 0005), (0002, 0003), (0002, 0004), (0002, 0005), (0003, 0004), (0003, 0005), (0004, 0005)} **b.** $1/10$ **c.** $1/10$ **d.** $3/10$ **3.57.** .5 **3.59a.** .8145 **b.** .0135 **c.** Evidence to indicate claim is incorrect **d.** No **3.61a.** .75 **b.** .2875 **c.** .6 **d.** .06 **3.63.** .801 **3.65b.** .95 **c.** .25 **d.** .5 **3.69a.** .7127 **b.** .2873 **3.71.** No **3.73a.** $1/15$ **b.** $6/15$ **c.** $14/15$ **d.** $6/14$ **e.** $5/14$ **3.75.** .79 **3.79.** False

Chapter 4

4.3a. Continuous **b.** Discrete **c.** Continuous **d.** Discrete **4.11a.** {(HHH), (HHT), (HTH), (THH), (HTT), (THT), (TTH), (TTT)} **b.** $p(0) = 1/8$; $p(1) = 3/8$; $p(2) = 3/8$; $p(3) = 1/8$ **4.13a.** Not valid **b.** Valid **c.** Not valid **d.** Not valid **4.15b.** .56; .32 **4.17a.** .729; .001 **b.** .243; .027 **c.** Yes **4.19a.** 3.85 **4.21a.** 60; 250; 15.8113 **c.** (23.38, 91.62); 1 **4.23a.** Expected total loss is $2,450 for both firms **b.** A: $\sigma = 661.44$; B: $\sigma = 701.78$; greater risk is B **4.25a.** .8 **b.** No

4.27a.

Disease	*Hepatitis*	*Cirrhosis*	*Gallstone*	*Pancreatic cancer*
Cost	$700	$1,110	$3,320	$16,450
p(cost)	.4	.1	.45	.05

b. $2,707.50

c.

Disease	*Hepatitis*	*Cirrhosis*
Cost	$700	$1,110
p(cost)	.8	.2

d. $782.00

4.29a.

Cost	$1,000	$2,000	$3,000
p(cost)	.25	.25	.50

b. .25 **c.** $2,250 **d.** $2,250 **4.31a.** Discrete **b.** Binomial **d.** 2.4; 1.2 **4.33a.** .21875 **b.** .0250 **c.** .1296 **4.35a.** .09767 **b.** .99869 **c.** .22884 **4.37a.** .5 **b.** $p < .5$ **c.** $p > .5$ **d.** .5; $p < .5$; $p > .5$ **4.39a.** .006; .046 **b.** .0293; .0003 **c.** .0662; .0019 **d.** .8992 **4.41b.** 2.4; 1.470 **c.** Binomial with $p = .9$, $q = .1$, and $n = 24$; $E(y) = 21.6$; $\sigma = 1.470$ **4.43a.** 1 **b.** .998 **c.** .537 **d.** .009 **e.** 0 **f.** 0 **g.** 0 **h.** 1 **4.45a.** 520; 13.491 **b.** No **c.** No **4.47a.** .265 **b.** .647 **c.** .981 **d.** Increases **4.49b.** 5; 2.236; (.5278, 9.4722) **c.** .961 **4.51a.** 2 **b.** No **4.53** .193; .660 **4.55.** .224; .050; $.05^8$; hours are independent **4.57.** .632 **4.59a.** .2734 **b.** .4096 **c.** .3432 **4.61a.** .084 **b.** .073 **c.** .028 **4.63a.** Discrete **b.** Continuous **c.** Continuous **d.** Continuous **4.65a.** $497,000 **b.** 6.606×10^{11} **4.67a.** A: 4.6; B: 3.7 **b.** A: $46,000; B: $55,550 **c.** A: $\sigma^2 = 1.34$, $\sigma = 1.16$; B: $\sigma^2 = 1.21$, $\sigma = 1.10$ **d.** A: .95; B: .95 **4.69.** .3028 **4.71.** .036 **4.73a.** .34272 **b.** .004 **4.75a.** 5; 4 **b.** .617 **c.** .006

Chapter 5

5.1a. $f(x) = 1/25$; $20 \le x \le 45$ **b.** 32.5; 52.0833 **c.** 1 **5.3a.** $f(x) = 1/4$, $3 \le x \le 7$ **b.** 5; 1.333 **c.** .577 **5.5a.** 0 **b.** 1 **c.** 1 **5.9b.** .5; .2; .2; 0 **5.11.** .4444 **5.13a.** .2789 **b.** .9544 **c.** .0819 **d.** .9974 **e.** .4878 **f.** .4878 **5.15a.** 0 **b.** .8413 **c.** .8413 **d.** .1587 **5.17a.** .6826 **b.** .95 **c.** .90 **d.** .9544 **5.19a.** −.81 **b.** .55 **c.** 1.43 **d.** .21 **e.** −2.05 **f.** .50 **5.21a.** 1.25 below **b.** 1.875 above **c.** 0 **d.** 1.5 above **5.23a.** .0456; .0026 **b.** .6826; .9544 **c.** 1,008.4; 987.2 **5.25.** 182 **5.27a.** .2843 **b.** .0228 **c.** 71.64 **5.29a.** .1151 **b.** .6554 **c.** $10.50 **5.31a.** XYZ **b.** ABC: $105; XYZ: $107 **c.** ABC **5.33.** 5.068 **5.35a.** .367879 **b.** .082085 **c.** .000553 **d.** .223130 **5.37a.** .999447 **b.** .999955 **c.** .981684 **d.** .632121 **5.39a.** .018316 **b.** .950213 **c.** .383401 **5.41.** .223130 **5.43a.** .367879; .212248 **b.** .049787 **c.** Less **d.** 40.6 **5.45a.** $R(x) = e^{-.5x}$ **b.** .135335 **c.** .367879 **d.** No **e.** 820.85; 3,934.69 **f.** 37 days **5.49a.** Yes **b.** 17.5; 5.25 **c.** .902 **d.** .9049 **5.51a.** .4880 **b.** .2334 **c.** 0 **5.53a.** .0559 **c.** No **5.55b.** ≈0 **c.** No **d.** Yes **5.57a.** $f(x) = 1/80$, $10 \le x \le 90$ **b.** 50; 23.094 **d.** .625 **e.** 0 **f.** .875 **g.** .577 **h.** .1875 **5.59a.** .3300 **b.** .0918 **c.** .9245 **d.** .0255 **5.61a.** −.13 **b.** .02 **c.** 1.04 **d.** −.69 **5.63a.** .451188 **b.** .406570 **c.** 0 **d.** .877544 **e.** .273870 **f.** 0 **5.65a.** .0918 **b.** 0 **c.** Lower by 4.87 decibels **5.67.** $9.6582 **5.69a.** 13.33 weeks **b.** .860708; .637628 **c.** .593430 **5.71.** From high to low: **a.** Bank 3, bank 1, bank 2 **b.** Bank 3, bank 1, bank 2 **c.** Bank 1, bank 3, bank 2 **5.73a.** .0548 **b.** .6006 **c.** .3446 **d.** $6,503.80 **5.75a.** .8264 **b.** 17 **c.** .6217 **d.** 0; −157

Chapter 6

6.3a. 8; .75 **b.** 200; 3.873 **c.** 50; .4 **d.** −10; 1.621 **6.5a.** .3085 **b.** .6915 **c.** .0606 **d.** .3830 **e.** .0495 **6.7a.** 2.5; .1904 **6.11a.** 6: (chip 1, chip 2), (chip 1, chip 3), (chip 1, chip 4), (chip 2, chip 3), (chip 2, chip 4), (chip 3, chip 4) **b.** 1/6 **c.** 1.5; 1.5; 2; 2; 2.5; 2.5 **d.**

$\bar{x}$	1.5	2	2.5
$p(\bar{x})$	1/3	1/3	1/3

6.13.

n	1	5	10	20	30	40	50
$\sigma/\sqrt{n}$	10	4.472	3.162	2.236	1.826	1.581	1.414

6.15a. 36.5148 **b.** $n = 120$ **c.** $n = 54$ **6.17a.** $\mu_{\bar{x}} = 10$; $\sigma_{\bar{x}} = 2.0486$ **6.19a.** 95.25% **b.** 100% **6.21a.** $\mu_{\bar{x}} = 840$; $\sigma_{\bar{x}} = 2.1213$ **b.** ≈0 **d.** $\mu_{\bar{x}} = 840$; $\sigma_{\bar{x}} = 6.3640$; .0582 **6.23.** .4514 **6.25a.** $\sigma_{\bar{x}} = .0717$ **b.** $\sigma_{\bar{x}} = .0536$ **c.** 502 **6.27a.** 1.5% **b.** .0026 **c.** .1587 **6.29a.** .05 **b.** 1 **c.** .000625 **6.31.** .9332

Chapter 7

7.1a. 1.645 **b.** 2.575 **c.** 1.96 **d.** 1.28 **7.3a.** (14.28, 19.72) **7.5a.** (24.929, 27.471) **b.** Wider, narrower **7.11b.** μ **c.** (57.907, 67.213) **7.13c.** (−47.84, 49.84) **7.15a.** (21.613, 25.247) **7.17.** 1,230 **7.19a.** 68 **b.** 31 **7.21a.** 656 **b.** Wider **c.** 38.3% **7.23.** 55 **7.25.** 271 **7.27a.** $z = 1.28$; $t = 1.533$ **b.** $z = 1.645$; $t = 2.132$ **c.** $z = 1.96$; $t = 2.776$ **d.** $z = 2.33$; $t = 3.747$ **e.** $z = 2.575$; $t = 4.604$ **7.29a.** 2.228 **b.** 2.228 **c.** −1.812 **d.** 2.086 **e.** 4.032 **7.31a.** (96.284, 103.716) **b.** (94.171, 105.829) **7.33a.** (73.956, 82.772) **d.** Same, narrower **7.35b.** (1,688.132, 2,115.618) **c.** 427.486; wider **7.37a.** (121.161, 148.839) **b.** Population must be normal **7.41a.** Yes **b.** (.831, .929) **7.43a.** (.217, .383) **7.45.** (.445, .685) **7.47b.** Yes **c.** (.340, .460) **7.49a.** 482 **b.** 752 **7.51.** 34 **7.53.** No; would need $n = 980$ **7.55.** 322 **7.57a.** t; 2.074 **b.** z; 1.96 **c.** z; 1.96 **d.** z; 1.96 **e.** Neither **7.63a.** (11,738, 13,306) **7.65.** (.562, .664) **7.67a.** 42,250 **b.** (41,743.43, 42,756.57) **c.** Interval **7.69.** 818 **7.71a.** 853

Chapter 8

8.1. Null; alternative **8.7.** No **8.9a.** H_0: Drug unsafe; H_a: Drug safe **c.** α **8.11g.** (a) .025; (b) .05; (c) .005; (d) .0985; (e) .10; (f) .01 **8.13a.** Reject H_0 if $z > 2.14$ **b.** .0162 **8.15a.** .1056 **b.** .0384 **c.** $z = 3.88$; reject H_0; yes **8.17a.** H_a: $\mu > 35$; H_0: $\mu = 35$ **b.** $z = 2.16$; reject H_0; yes **8.19a.** H_0: $\mu = 1{,}502.5$ vs H_a: $\mu > 1{,}502.5$ **b.** $z = 7.34$; reject H_0; yes **8.21a.** No **b.** Yes **c.** Yes **d.** No **e.** No **8.23.** Just greater than .083 **8.25.** .0768 **8.27.** p-value = .0119; evidence to reject H_0 for $\alpha > .0119$ **8.29a.** $z = -1.29$; do not reject H_0; no **b.** p-value = .0985 **c.** Small **8.31a.** H_0: $\mu = \$9{,}083$ vs H_a: $\mu > \$9{,}083$ **b.** $z = 1.86$; p-value = .0314; evidence to reject H_0 for $\alpha > .0314$ **8.33a.** $z = 1.37$; p-value =.0853; no evidence to reject H_0 for $\alpha = .05$; evidence to reject H_0 for $\alpha > .0853$ **b.** p-value = .1706; no evidence to reject H_0 for $\alpha \le .10$ **8.37a.** .10 **b.** .05 **c.** .05 **8.39a.** .05 < p-value < .10 **b.** .10 < p-value < .20 **8.41a.** Sampled population normal **b.** $t = 1.894$; p-value = .0382; evidence to reject H_0 for $\alpha > .0382$ **c.** p-value = .0764; no evidence to reject H_0 for $\alpha = .05$; evidence to reject H_0 for $\alpha > .0764$ **8.43.** $t = -4.193$; reject H_0; yes **8.45a.** $t = 1.113$; do not reject H_0; no **b.** Population of differences normal **8.47a.** Yes **b.** No **c.** Yes **d.** No **e.** No **8.49a.** $z = -2.33$ **c.** Reject H_0 **d.** p-value = .0099 **8.51a.** $z = 1.13$; do not reject H_0 **b.** p-value = .1292 **8.53a.** $z = -4.83$; reject H_0 **b.** Yes **c.** $z = -.48$; do not reject H_0; no **8.55a.** $z = 2.66$; reject H_0; yes **b.** p-value = .0039 **8.57.** $z = 2.80$; reject H_0; yes **8.61a.** .0174 **b.** .9826 **8.63.** Power = .3632 **8.65a.** Approximately normal, $\mu_{\bar{x}} = 30$, $\sigma_{\bar{x}} = .1091$ **b.** Approximately normal, $\mu_{\bar{x}} = 29.8$, $\sigma_{\bar{x}} = .1091$ **c.** .5359 **d.** .0409 **8.67.** .1075 **8.69.** Power becomes larger in each case. Power curve shifts upward. **8.71.** Alternative **8.73a.** H_0: patient does not have disease vs H_a: patient has disease **b.** Type I: false positive; Type II: false negative **8.75.** Large **8.77b.** .8212 **c.** .1788 **d.** $\beta = .3121$; power = .6879; increases

8.79a. $t = .93$; do not reject H_0; no **b.** Sampled population is normal **8.81.** p-value $> .10$ **8.83.** p-value $= .0012$; evidence to reject H_0 for $\alpha > .0012$ **8.85.** $z = 3.79$; evidence to conclude claim is true **8.87a.** p-value $= .0304$; evidence to reject H_0 for $\alpha > .0304$ **b.** p-value $= .0152$ **8.89.** p-value $= .0793$ **8.91a.** H_0: $p = .5$ vs H_a: $p > .5$ **b.** p-value $= .0329$; evidence to reject H_0 for $\alpha > .0329$

Chapter 9

9.1a. (144, 156) **b.** (142, 158) **c.** $\mu = 0$; $\sigma = 5$ **d.** $(-10, 10)$ **e.** Larger **9.3a.** $z = .4$ **b.** .3446 **c.** .6554 **d.** .9726 **9.5.** $(-12.08, -10.12)$ **9.7a.** p-value $= .1150$; no evidence to reject H_0 for $\alpha \le .10$ **b.** p-value $= .0575$; no evidence to reject H_0 for $\alpha = .05$; evidence to reject H_0 for $\alpha > .0575$ **9.9a.** $z = -2.78$; reject H_0; yes **b.** p-value $= .0027$ **9.11a.** $z = 12.71$; reject H_0; yes; p-value ≈ 0 **9.13b.** $z = 4.27$; reject H_0; yes **c.** p-value ≈ 0 **9.15c.** $z = 2.30$; reject H_0; yes **d.** ($67,187.8, $150,496.2) **9.17a.** No **b.** No **c.** No **d.** Yes **e.** No **9.19a.** 110 **b.** 14.5714 **c.** .1821 **d.** 2,741.94 **9.21a.** $(-1.622, -.172)$ **9.23.** $(-11.077, -4.923)$ **9.25a.** H_0: $\mu_1 - \mu_2 = 0$ vs H_a: $\mu_1 - \mu_2 > 0$ **b.** $t = 2.616$; reject H_0; $.01 < p\text{-value} < .025$ **9.27a.** $t = -1.96$; reject H_0; yes **c.** $.05 < p\text{-value} < .10$ **d.** $(-13.772, -1.028)$ **9.29a.** $t = 4.46$; reject H_0; yes **b.** $.001 < p\text{-value} < .005$ **9.33a.** .01 **b.** .95 **c.** .95 **d.** .01 **9.35a.** $F > 2.19$ **b.** $F > 2.75$ **c.** $F > 3.37$ **d.** $F > 4.30$ **9.37a.** $F = 4.287$; do not reject H_0; no **b.** $.10 < p\text{-value} < .20$ **9.39.** $F = 2.626$; reject H_0; yes; choose line 1 **9.41a.** $F = 5.03$; reject H_0; not appropriate **d.** $.02 < p\text{-value} < .05$ **9.43a.** -8.2; 3.8987 **c.** $t = -4.70$; reject H_0; yes **d.** $.002 < p\text{-value} < .01$ **9.45a.** $t < -1.333$ **b.** $t = -3.24$; reject H_0 **d.** $(-5.379, -1.621)$ **e.** Confidence interval **9.47a.** $t = -1.14$; do not reject H_0 **b.** p-value $> .20$ **9.49a.** $t = -1.92$; reject H_0; yes **9.51a.** $t = 7.68$; reject H_0; yes **b.** p-value $< .01$ **d.** (.2772, .5562) **9.53a.** H_0: $\mu_D = 0$ vs H_a: $\mu_D \neq 0$ **b.** $t = 5.76$; p-value $= .000$; strong evidence to reject H_0 **9.57a.** $z < -2.33$ **b.** $z < -1.96$ **c.** $z < -1.645$ **d.** $z < -1.28$ **9.59.** $(-.1935, .0095)$ **9.61.** $(-.0796, -.0264)$ **9.63a.** H_0: $p_1 - p_2 = 0$ vs H_a: $p_1 - p_2 \neq 0$ **b.** Yes **c.** $z = -4.30$; reject H_0; p-value ≈ 0 **9.65a.** Yes **b.** $z = 5.03$; reject H_0; yes **9.67a.** $z = -5.21$; reject H_0 **b.** $(-.324, -.106)$ **c.** Confidence interval **9.69a.** (.1656, .1808) (95% confidence) **b.** (.0692, .0820) (95% confidence) **c.** (.0877, .1075) **9.71.** Less likely **9.73.** 3,383 each **9.75.** No; $n_1 = n_2 = 136$ required **9.77a.** $n_1 = n_2 = 911$ **b.** No; $n = 1{,}692$ required **9.79.** $n_1 = n_2 = 542$ **9.81.** $n_1 = n_2 = 293$ **9.83.** $n_1 = n_2 = 1{,}905$ **9.85.** $(-5.081, 5.681)$ **9.87a.** $z = 2.60$; reject H_0; yes **b.** p-value $= .0047$ **9.89.** $n_1 = n_2 = 542$ **9.91a.** (.407, .633) **9.93.** $z = 3.62$; reject H_0; yes **9.95a.** H_0: $\mu_1 - \mu_2 = 0$ vs H_a: $\mu_1 - \mu_2 \neq 0$ **b.** $z = -2.69$; p-value $= .0072$; evidence to reject H_0 for $\alpha > .0072$ **9.97a.** Reject H_0 **b.** Do not reject H_0 **c.** Reject H_0 **d.** Reject H_0 **e.** Do not reject H_0 **f.** Do not reject H_0 **9.99a.** $n_1 = n_2 = 546$ **b.** $z = 2.30$; reject H_0; p-value $= .0107$ **9.101.** $(-.1306, .0372)$ **9.103a.** H_0: $\mu_1 - \mu_2 = 180$ vs H_a: $\mu_1 - \mu_2 \neq 180$ **b.** $z = -2.07$; p-value $= .0384$; evidence to reject H_0 for $\alpha > .0384$ **d.** (148.856, 179.144); no

Chapter 10

10.3a. $y = -10/3 + 5/3x$ **b.** $y = 3/4x$ **c.** $y = -2 + 8/7x$ **d.** $y = -8/5 + 7/5x$ **10.5a.** 1; 4 **b.** 1; -4 **c.** 2; 4 **d.** -2; 0 **e.** 1; 0 **f.** .75; .50 **10.9b.** -26.2857 **c.** 33.7143 **d.** $-.7797$ **e.** 3.4286; 4.4286 **f.** 7.102 **g.** $\hat{y} = 7.102 - .7797x$ **10.11b.** $y = 1 + x$ **d.** $y = 1 + x$ **e.** $\hat{y} = 1 + x$ **10.13a.** $E(y) = \beta_0 + \beta_1 x$ **b.** $-30{,}000$; 70 **e.** $180,000; yes **f.** $320,000; no **10.15a.** $\hat{y} = 36.3511 + 2.8442x$ **c.** 79.0141 **10.17a.** Positive **c.** $\hat{y} = -37.435 + 455.271x$ **10.19.** .3475 **10.21.** .24407, .4940; 1.1427, 1.069 **10.23a.** $\hat{y} = 7.3805 + .3726x$ **b.** 603.5405 billion dollars **c.** 2,225.629819; 741.8766 **10.25a.** $\hat{y} = 74.7129 - 1.2883x$ **c.** 6,341.0887; 352.2827; 18.7692 **10.27a.** (29.83, 32.17); (30.06, 31.94) **b.** (58.92, 69.08); (59.84, 68.16) **c.** $(-9.15, -7.65)$; $(-9.02, -7.78)$ **10.29.** (.49, 1.15); (.06, 1.58) **10.31a.** p-value $< .0010$ **10.33a.** $\hat{y} = 12.71 + 1.50x$ **b.** $t = 18.26$; reject H_0 **10.35.** $t = -.96$; do not reject H_0; yes **10.37.** $t = -4.43$; reject H_0; yes **10.41a.** .985; .971 **b.** $-.993$; .987 **c.** 0; 0 **d.** 0; 0 **10.43.** .9046; .8182 **10.45.** $r_1 = -.967$; $r_2 = -.110$; x_1 **10.47c.** Prior: $r = .965$; After: $r = .996$ **10.49a.** $\hat{y} = -235.1 + 1.273x_2$ **b.** $t = 18.30$; reject H_0; yes **c.** .977 **d.** $\hat{y} = -235.1 + 1.273x_2$ **e.** x_2 **10.51a.** $\hat{y} = 44.17 - .0255x$ **c.** $t = -.03$; do not reject H_0 **d.** No **e.** .0003 **10.53c.** (3.526, 5.762) **d.** (1.655, 2.913) **e.** $(-1.585, .757)$ **10.55a.** $\hat{y} = 1.375 + .875x$ **c.** 1.5 **d.** .1875 **e.** (3.235, 3.891) **f.** (3.813, 5.937) **10.57b.** $\hat{y} = 4.861 - .3466x$ **c.** $t = -5.91$; reject H_0; yes **d.** (.65, 2.83) **e.** (2.23, 2.99) **10.59a.** $\hat{y} = 5.325 + .5861x$ **c.** $t = 15.35$; reject H_0 **d.** (26.91, 29.456); (23.554, 32.812) **10.61.** Gets wider

10.63a. $\hat{y} = 44.130 + .2366x$ **b.** 19.40375 **c.** .0865 **d.** $t = 1.269$; do not reject H_0; p-value $= .2216$ **10.65c.** $\hat{y} = x$ **d.** SSE $= 0$ **10.67b.** $\hat{y} = 12.594 + .10936x$ **c.** $t = 3.49$; reject H_0; yes **d.** (16.50, 41.50) **10.69a.** $-.4431$; .1963 **b.** -1.85; do not reject H_0; no **10.71a.** .8429 **b.** $t = 2.71$; do not reject H_0; no **c.** No **10.73b.** $\hat{y} = -15{,}124 + 76.1745x$ **c.** .9185 **d.** $t = 15.743$; reject H_0; yes **e.** (66.1395, 86.2095); yes **f.** p-value $= .0001$ **g.** (146,713, 158,206) **10.75a.** $\hat{y} = 40.657 - .7902x$ **b.** $r = -.0707$; $r_2 = .0050$; $s = 32.6326$ **c.** $t = -.17$; do not reject H_0

Chapter 11

11.1a. 506.346; -941.900; -429.060 **b.** $\hat{y} = 506.346 - 941.9x_1 - 429.06x_2$ **c.** 151,015.72376; 8,883.27787; 94.25114 **d.** $t = -3.424$; p-value $= .0032$; reject H_0 **e.** $(-1{,}230.4922, 372.3722)$; yes **11.3a.** $t = 3.13$; reject H_0 **b.** $t = 3.13$; reject H_0 if $t > 1.717$; reject H_0; yes **11.5a.** $\hat{y} = 1.4326 + .0100x_1 + .3793x_2$ **b.** $t = 3.15$; reject H_0; yes **11.7a.** $\hat{y} = 20.091 - .6705x + .0095x^2$ **c.** $t = 1.51$; p-value $= .1576$; do not reject H_0; no **d.** $\hat{y} = 19.2791 - .4449x$ **e.** $(-.5176, -.3722)$ **11.9.** $t = 2.11$; reject H_0; yes **11.11a.** .8911 **b.** $F = 65.478$; p-value $\le .0001$; reject H_0 **c.** p-value $\le .0001$ **d.** $t = -6.803$; p-value $\le .0001$; reject H_0

11.13a.

Source	df	SS	MS	F
Model	2	11.38	5.6900	7.82
Error	17	12.37	.7276	
Total	19	23.75		

$R^2 = .4792$ **b.** $F = 7.82$; reject H_0

11.15a. $E(y) = \beta_0 + \beta_1x_1 + \beta_2x_2 + \beta_3x_3 + \beta_4x_4 + \beta_5x_5$ **b.** $\hat{y} = 15.491 + 12.774x_1 + .713x_2 + 1.519x_3 + .320x_4 + .205x_5$ **c.** $F = 11.68$; reject H_0 **d.** p-value $< .025$; reject H_0 **11.17a.** $\hat{y} = 131.924 + 2.726x_1 + .04721x_2 - 2.587x_3$ **b.** $F = 17.87$; p-value $\le .0001$; reject H_0 **c.** $t = .51$; p-value $= .6199$; do not reject H_0 **d.** .7701 **e.** $-\$19.4025$ **f.** $\pm\$19.62$ **11.19b.** $F = 58.56$; reject H_0; yes **c.** $t = -50.0$; reject H_0; yes **11.21a.** $\hat{y} = 2.41 + 1.43x_1 - .366x_2$ **b.** $F = 54.807$; reject H_0 **c.** .9400 **d.** $\hat{y} = -.349 + 2.07x_1 + .0215x_2 - .0919x_1x_2$ **e.** .9865 **g.** $t = -4.544$; reject H_0; yes **h.** No **11.23a.** $\hat{y} = 506.35 - 941.9x_1 - 429.1x_2$ **b.** $s = 94.25$ **c.** $F = 7.22$; reject H_0; yes **d.** $(-1{,}089.95, 231.75)$ **e.** .459 **11.25.** Need for quadratic term; yes **11.27a.** 25 observations: $\hat{\beta}_0 = 1.4326$, $\hat{\beta}_1 = .00999$, $\hat{\beta}_2 = .3793$; 26 observations: $\hat{\beta}_0 = 1.1554$, $\hat{\beta}_1 = .0187$, $\hat{\beta}_2 = .4058$ **b.** 25 observations: $s = .2769$; 26 observations: $s = .5581$ **c.** 25 observations: $F = 100.80$, reject H_0; 26 observations: $F = 33.88$, reject H_0 **d.** 25 observations: (.322, .436); 26 observations: (.291, .521) **11.33a.** $F = 6.38$; reject H_0; yes **c.** p-value $= .002$; reject H_0 **11.35a.** $\hat{y} = .6013 + .595x_1 - 3.725x_2 - 16.232x_3 + .235x_1x_2 + .308x_1x_3$ **b.** $R^2 = .928$; $F = 139.42$; reject H_0 **11.39a.** $\hat{y} = .0562 + .273x_1 + .0006x_2$ **b.** $F = 164.74$; reject H_0 **c.** $t = 4.34$; reject H_0 **d.** 14.25 **11.41.** $F = 15.32$; reject H_0 **11.43a.** 13.6812 **b.** Yes **11.45b.** $F = 52.21$; reject H_0; yes **c.** $t = -3.33$; reject H_0; yes **d.** $t = 4$; reject H_0; yes **11.47b.** $F = 16.10$; reject H_0; yes **c.** $t = 2.5$; reject H_0; yes **d.** 945 **11.51b.** $\hat{y}_2 = 238.006 - .442x_1 - 19.379x_4 + 17.930x_5$ **c.** $R^2 = .6485$ **d.** H_0: $\beta_1 = \beta_2 = \beta_3 = 0$ vs H_a: At least one $\beta_i \neq 0$ $(i = 1, 2, 3)$ **e.** $F = 8.609$; reject H_0; p-value $= .0017$ **f.** $t = -2.086$; reject H_0 **g.** $(-3.569, 2.685)$ **h.** 192.613

Chapter 12

12.1a. Quantitative **b.** Qualitative **c.** Quantitative **d.** Quantitative **e.** Qualitative **12.3.** Normally distributed **12.5a.** Second; $\beta_0 = 4$; $\beta_2 < 0$ **b.** Second; $\beta_0 = 8$; $\beta_2 > 0$ **12.7.** $E(y) = \beta_0 + \beta_1x + \beta_2x^2$ **12.9a.** $E(y) = \beta_0 + \beta_1x + \beta_2x^2$ **b.** $\beta_0 > 0$; $\beta_2 > 0$ **12.11.** $E(y) = \beta_0 + \beta_1x + \beta_2x^2$ **12.13a.** $E(y) = \beta_0 + \beta_1x + \beta_2x^2$ **12.15a.** $\hat{y} = 32.964 - .0282x$; $R^2 = .9430$ **d.** $\hat{y} = -22.008 + .102x - .00008x^2$; $R^2 = .9481$ **f.** $t = -1.443$; do not reject H_0 **12.17a.** $E(y) = \beta_0 + \beta_1x_1 + \beta_2x_2$ **b.** $E(y) = \beta_0 + \beta_1x_1 + \beta_2x_2 + \beta_3x_1x_2$ **c.** $E(y) = \beta_0 + \beta_1x_1 + \beta_2x_2 + \beta_3x_1x_2 + \beta_4x_1^2 + \beta_5x_2^2$ **12.19a.** Second **f.** -4 **12.21a.** $\hat{y} = -2.55 + 3.82x_1 + 2.63x_2 - 1.29x_1x_2$ **e.** H_0: $\beta_3 = 0$ vs H_a: $\beta_3 \neq 0$ **f.** $t = -8.06$; reject H_0

12.23a. Both quantitative **b.** $E(y) = \beta_0 + \beta_1 x_1 + \beta_2 x_2$ **c.** $E(y) = \beta_0 + \beta_1 x_1 + \beta_2 x_2 + \beta_3 x_1 x_2$ **d.** $E(y) = \beta_0 + \beta_1 x_1 + \beta_2 x_2 + \beta_3 x_1 x_2 + \beta_4 x_1^2 + \beta_5 x_2^2$ **12.25a.** $\hat{y} = 149 + .472x_1 - .0993x_2 - .0005x_1^2 + .000015x_2^2$ **c.** $F = 31.66$; reject H_0; yes **d.** Yes **12.27a.** H_a: At least one $\beta_i \neq 0$ ($i = 3, 4, 5$) **c.** Numerator df = 3; denominator df = 24 **12.31a.** H_0: $\beta_1 = \beta_2 = \beta_3 = \beta_4 = \beta_5 = 0$ vs H_a: At least one $\beta_i \neq 0$ ($i = 1, 2, 3, 4, 5$) **b.** H_0: $\beta_3 = \beta_4 = \beta_5 = 0$ vs H_a: At least one $\beta_i \neq 0$ ($i = 3, 4, 5$) **c.** $F = 18.24$; reject H_0 **d.** $F = 8.46$; reject H_0 **12.33.** $F = 24.19$; reject H_0; yes **12.35a.** $E(y) = \beta_0 + \beta_1 x_1 + \beta_2 x_2 + \beta_3 x_3 + \beta_4 x_1 x_2 + \beta_5 x_1 x_3 + \beta_6 x_2 x_3 + \beta_7 x_1^2 + \beta_8 x_2^2 + \beta_9 x_3^2$ **b.** $\hat{y} = 655.81 - 57.327x_1 - 3.390x_2 - 28.271x_3 + .224x_1x_2 + 2.201x_1x_3 + .089x_2x_3 + .453x_1^2 + .004x_2^2 + .208x_3^2$ **c.** $F = 10.25$; reject H_0; yes **d.** $F = 2.25$; do not reject H_0 **12.37.** $E(y) = \beta_0 + \beta_1 x_1$, $x_1 = 1$ if level 2; $x_1 = 0$ if not **12.39a.** 10.2 **b.** 6.2 **c.** 22.2 **d.** 12.2 **12.41a.** Large-sample confidence interval **b.** $E(y) = \beta_0 + \beta_1 x_1$, $x_1 = 1$ if public college, $x_1 = 0$ if not **12.43a.** $F = 4.80$; reject H_0; yes **b.** \$11,400 **c.** \$20,000 **12.45a.** Qualitative **b.** $E(y) = \beta_0 + \beta_1 x_1 + \beta_2 x_2$; $x_1 = 1$ if brand B_2, $x_1 = 0$ if not; $x_2 = 1$ if brand B_3, $x_2 = 0$ if not **d.** $\beta_0 + \beta_2$ **12.47a.** $E(y) = \beta_0 + \beta_1 x_1 + \beta_2 x_2 + \beta_3 x_3 + \beta_4 x_4$ **b.** $\hat{y} = 20.35 + 4.75x_1 + 8.60x_2 + 11.3x_3 + 2.1x_4$ **c.** H_0: $\beta_1 = \beta_2 = \beta_3 = \beta_4 = 0$ vs H_a: At least one $\beta_i \neq 0$ ($i = 1, 2, 3, 4$) **d.** $F = 23.966$; reject H_0 **e.** $(-.746, 4.946)$ **12.49a.** $E(y) = \beta_0 + \beta_1 x_1$ **b.** $E(y) = \beta_0 + \beta_1 x_1 + \beta_2 x_2 + \beta_3 x_3$; $x_2 = 1$ if qualitative variable at level 2, $x_2 = 0$ if not; $x_3 = 1$ if qualitative variable at level 3, $x_3 = 0$ if not **c.** $E(y) = \beta_0 + \beta_1 x_1 + \beta_2 x_2 + \beta_3 x_3 + \beta_4 x_1 x_2 + \beta_5 x_1 x_3$ **d.** $\beta_4 = \beta_5 = 0$ **e.** $\beta_2 = \beta_3 = \beta_4 = \beta_5 = 0$ **12.51a.** Extractor is qualitative; diameter is quantitative **b.** $E(y) = \beta_0 + \beta_1 x_1 + \beta_2 x_2$; $x_2 = 1$ if brand B, $x_2 = 0$ if not **c.** $E(y) = \beta_0 + \beta_1 x_1 + \beta_2 x_2 + \beta_3 x_1 x_2$ **e.** H_0: $\beta_3 = 0$ vs H_a: $\beta_3 \neq 0$ **12.53a.** $E(y) = \beta_0 + \beta_1 x_1 + \beta_2 x_2 + \beta_3 x_3$; $x_2 = 1$ if condition E, $x_2 = 0$ if not; $x_3 = 1$ if condition G, $x_3 = 0$ if not **c.** $\hat{y} = 36{,}388 + 15{,}617x_1 + 152{,}487x_2 + 49{,}441x_3$; F: $\hat{y} = 36{,}388 + 15{,}617x_1$; G: $\hat{y} = 85{,}829 + 15{,}617x_1$; E: $\hat{y} = 188{,}875 + 15{,}617x_1$ **e.** $F = 8.43$; reject H_0; yes **12.55.** $F = 2.60$; do not reject H_0 **12.57b.** $E(y) = \beta_0 + \beta_1 x_1 + \beta_2 x_2 + \beta_3 x_1 x_2$; $x_2 = 1$ if after 1973, $x_2 = 0$ if not **c.** $\hat{y} = 884.8416 - 612.5777x_1 - 409.361x_2 + 502.2018x_1x_2$ **e.** $F = 54.32$; reject H_0; yes **f.** $t = 5.957$; reject H_0; yes **12.59a.** $\beta_5 = \beta_6 = \beta_7 = \beta_8 = 0$ **b.** $\beta_2 = \beta_5 = \beta_6 = \beta_7 = \beta_8 = 0$ **c.** $\beta_3 = \beta_4 = \beta_5 = \beta_6 = \beta_7 = \beta_8 = 0$ **12.61a.** $\hat{y} = 48.8 - 3.36x_1 + .0749x_1^2 - 2.36x_2 - 7.60x_3 + 3.71x_1x_2 + 2.66x_1x_3 - .0183x_1^2x_2 - .0372x_1^2x_3$ **b.** $\hat{y} = 48.8 - 3.36x_1 + .0749x_1^2$; $\hat{y} = 46.44 + .35x_1 + .0566x_1^2$; $\hat{y} = 41.2 - .7x_1 + .0377x_1^2$ **d.** H_0: $\beta_3 = \beta_4 = \beta_5 = \beta_6 = \beta_7 = \beta_8 = 0$ vs H_a: At least one $\beta_i \neq 0$ ($i = 3, 4, 5, 6, 7, 8$) **e.** $F = 242.65$; reject H_0 **12.63a.** H_0: $\beta_4 = \beta_5 = 0$ vs H_a: At least one $\beta_i \neq 0$ ($i = 4, 5$) **b.** H_0: $\beta_3 = \beta_4 = \beta_5 = 0$ vs H_a: At least one $\beta_i \neq 0$ ($i = 3, 4, 5$) **12.65.** $F = 15.78$; reject H_0; yes **12.67a.** $E(y) = \beta_0 + \beta_1 x_1 + \beta_2 x_2 + \beta_3 x_1 x_2$; $x_2 = 1$ if bear market, $x_2 = 0$ if not **c.** H_0: $\beta_3 = 0$ vs H_a: $\beta_3 \neq 0$ **d.** H_0: $\beta_2 = \beta_3 = 0$ vs H_a: At least one $\beta_i \neq 0$ ($i = 2, 3$) **12.69a.** Quantitative: $x_1, x_2, x_3, x_4, x_5, x_6, x_{11}$; qualitative: x_7, x_8, x_9, x_{10} **b.** 5 **c.** $R^2 = .1836$ **d.** $F = 11.96$; reject H_0 **e.** H_0: $\beta_7 = \beta_8 = \beta_9 = \beta_{10} = 0$ vs H_a: At least one $\beta_i \neq 0$ ($i = 7, 8, 9, 10$) **12.71a.** H_0: $\beta_2 = \beta_5 = 0$ vs H_a: At least one $\beta_i \neq 0$ ($i = 2, 5$) **b.** H_0: $\beta_3 = \beta_4 = \beta_5 = 0$ vs H_a: At least one $\beta_i \neq 0$ ($i = 3, 4, 5$) **12.73a.** $\hat{y} = 10.2333 + .5x_1 + 2.0167x_2 + .6833x_3$ **b.** $F = 63.089$; reject H_0; yes **12.75a.** $E(y) = \beta_0 + \beta_1 x_1 + \beta_2 x_1^2 + \beta_3 x_2 + \beta_4 x_3 + \beta_5 x_1 x_2 + \beta_6 x_1 x_3 + \beta_7 x_1^2 x_2 + \beta_8 x_1^2 x_3$; $x_2 = 1$ if paint B, $x_2 = 0$ if not; $x_3 = 1$ if paint C, $x_3 = 0$ if not **12.81a.** $F = 3.74$; do not reject H_0; yes **b.** $F = 209$; reject H_0

Chapter 13

13.7. Rule 6 indicates process out of control **13.9a.** 1.023 **b.** .308 **c.** .167 **13.11b.** $\bar{\bar{x}} = 20.11625$, $\bar{R} = 3.31$ **c.** Centerline = 20.116; UCL = 22.529; LCL = 17.703 **d.** UA − B = 21.725; LA − B = 18.507; UB − C = 20.920; LB − C = 19.312 **e.** In control **13.13a.** $\bar{\bar{x}} = 23.9971$; $\bar{R} = .1815$ **b.** Process is in control **c.** No **13.15b.** Mean **c.** $\bar{\bar{x}} = 5.0423$; $\bar{R} = .9685$ **d.** Rule 6 indicates process out of control **e.** No **13.19a.** Centerline = 7.948; UCL = 16.802 **b.** UA − B = 13.853; LA − B = 2.043; UB − C = 10.900; LB − C = 4.996; process is in control **13.21.** R-chart: centerline = 4.03; UCL = 7.754; LCL = .306; UA − B = 6.513; LA − B = 1.547; UB − C = 5.271; LB − C = 2.789; process is in control; $\bar{x}$-chart: centerline = 21.728; UCL = 23.417; LCL = 20.039; UA − B = 22.854; LA − B = 20.602; UB − C = 22.291; LB − C = 21.165; process is out of control **13.23a.** Yes **b.** Centerline = .0796; UCL = .168; UA − B = .139; LA − B = .020; UB − C = .109; LB − C = .050 **c.** Out of control **d.** No **e.** No **13.25a.** Centerline = .1868; UCL = .426; UA − B = .346; LA − B = .027; UB − C = .267; LB − C = .107 **b.** In control **c.** Yes **13.27.** 104 **13.29a.** Centerline = .0575; UCL = .1145; LCL = .0005 **b.** UA − B = .0955; LA − B = .0195; UB − C = .0765; LB − C = .0385 **d.** Out of control **e.** No **13.31a.** Yes **b.** UCL = .02013;

LCL = .00081 **c.** UA − B = .01691; LA − B = .00403; UB − C = .01369; LB − C = .00725; in control **13.33a.** Centerline = .0455; UCL = .085; LCL = .006; UA − B = .072; LA − B = .019; UB − C = .059; LB − C = .032 **b.** Out of control **c.** No **13.53.** 1.645 standard deviations **13.55a.** Centerline = 7.529 **b.** Level shift **13.57a.** Yes **b.** Centerline = .0614; UCL = .112; LCL = .010; UA − B = .095; LA − B = .027; UB − C = .078; LB − C = .044 **c.** Out of control; no **13.59a.** Centerline = 31.468; UCL = 38.205; LCL = 24.731; UA − B = 35.960; LA − B = 26.976; UB − C = 33.714; LB − C = 29.222; assume variation stable **b.** In control **c.** Yes **13.61a.** Centerline = .063; UCL = .123; LCL = .003; UA − B = .103; LA − B = .023; UB − C = .083; LB − C = .043 **c.** Out of control **e.** No

Chapter 14

14.3.

1970	1971	1972	1973	1974	1975	1976	1977	1978	1979
78.06	80.59	82.87	87.16	91.61	94.19	96.01	100.00	105.04	108.04

1980	1981	1982	1983	1984	1985	1986	1987	1988	1989
113.84	113.61	115.07	114.60	112.73	113.96	114.02	114.90	115.78	116.01

14.5.

1970	1971	1972	1973	1974	1975	1976	1977	1978	1979
68.57	70.79	72.80	76.56	80.47	82.74	84.34	87.84	92.27	94.90

1980	1981	1982	1983	1984	1985	1986	1987	1988	1989
100.00	99.79	101.08	100.67	99.02	100.10	100.15	100.93	101.70	101.91

14.7

1982	1983	1984	1985	1986	1987	1988	1989	1990
100.00	104.55	107.97	101.58	78.67	67.60	68.59	68.72	69.94

14.17a.

Month	Jan.	Feb.	Mar.	Apr.	May	Jun.	Jul.	Aug.	Sep.	Oct.	Nov.	Dec.
Paasche	100.0	100.3	101.0	101.3	100.6	100.3	100.0	99.9	99.4	99.1	99.9	99.2

14.19a.

1977	1978	1979	1980	1981	1982	1983	1984	1985
2.133	2.241	2.237	1.989	1.716	1.351	1.320	1.388	3.135

14.21a.

1975	1976	1977	1978	1979	1980	1981	1982	1983	1984
161.00	132.20	144.84	184.17	283.23	547.05	477.41	396.28	418.46	372.49

1985	1986	1987	1988	1989	1990
328.90	360.18	430.44	436.49	393.70	388.34

14.23a.

Year	$w = .2$ Exp. Smoothed Value	$w = .8$ Exp. Smoothed Value	Year	$w = .2$ Exp. Smoothed Value	$w = .8$ Exp. Smoothed Value
1960	42.4	42.4	1973	84.7	112.1
1961	42.9	44.3	1974	91.3	116.7
1962	43.8	46.8	1975	98.9	126.9
1963	44.9	49.0	1976	110.2	149.5
1964	46.8	53.2	1977	124.0	173.3
1965	49.1	57.4	1978	138.8	193.1
1966	51.4	59.8	1979	154.9	214.1
1967	53.8	62.6	1980	171.3	232.1
1968	56.9	68.0	1981	189.3	255.6
1969	60.6	74.2	1982	204.9	265.0
1970	64.6	79.3	1983	222.3	286.5
1971	70.2	89.7	1984	241.8	312.9
1972	77.2	102.3			

14.25a.

1976	1977	1978	1979	1980	1981	1982	1983	1984
66.58	75.22	84.72	92.79	100.00	110.40	112.37	119.88	127.94

1985	1986	1987	1988	1989
127.94	120.00	123.75	125.60	125.60

b. Price

14.27

1960	1961	1962	1963	1964	1965	1966	1967	1968	1969
32.8	34.6	36.6	38.3	42.0	45.1	46.7	48.9	53.6	58.5

1970	1971	1972	1973	1974	1975	1976	1977	1978	1979
62.3	71.3	81.5	88.6	91.1	100.0	119.9	138.6	153.1	169.6

1980	1981	1982	1983	1984
182.8	202.1	206.6	225.6	246.9

14.29a. Using simple composite index for 1980:

1980	1981	1982	1983	1984	1985	1986	1987	1988	1989	1990	1991
100.0	102.3	106.3	114.2	128.4	147.8	169.7	190.6	211.5	230.8	246.4	256.7

14.31a.

1985	1986	1987	1988	1989	1990	1991
64.5	73.0	77.7	86.3	94.4	100.0	102.6

14.33a.

	1970	*1971*	*1972*	*1973*	*1974*	*1975*	*1976*	*1977*	*1978*
$w = .2$	106.4	106.8	108.6	109.9	109.8	109.2	110.5	113.0	116.1
$w = .8$	106.4	108.1	114.3	114.8	110.3	107.7	114.0	121.4	126.8

	1979	*1980*	*1981*	*1982*	*1983*	*1984*
$w = .2$	118.5	119.3	119.1	118.0	117.3	117.3
$w = .8$	128.0	123.5	119.6	114.5	114.7	116.6

Chapter 15

15.3a.

Year	Forecast $w = .3$	Error	Forecast $w = .7$	Error
1987	192.29	3.61	194.27	1.63
1988	192.29	5.11	194.27	3.13
1989	192.29	5.51	194.27	3.53

b.

Year	Forecast $w = .7, v = .3$	Error	Forecast $w = .3, v = .7$	Error
1987	195.52	.38	194.18	1.72
1988	196.21	1.19	191.87	5.53
1989	196.90	.90	189.56	8.24

15.5.

1992, Quarter	a. Forecast	b. Error
I	386.79	20.57
II	386.79	21.48
III	386.79	31.69
IV	386.79	48.85

c.

1992, Quarter	Forecast	Error
I	370.91	36.45
II	370.91	37.36
III	370.91	47.57
IV	370.91	64.73

d.

1992, Quarter	Forecast $w = .7, v = .5$	Error	Forecast $w = .3, v = .5$	Error
I	401.86	5.50	395.09	12.27
II	411.10	−2.83	403.28	4.99
III	420.35	−1.87	411.47	7.01
IV	429.60	6.04	419.66	15.98

15.7.

		a. $w = .3$	$w = .7$	b. $w = .3, v = .5$	$w = .7, v = .5$
1993	I	409.57	429.38	441.23	444.06
	II	409.57	429.38	452.26	455.19
	III	409.57	429.38	463.29	466.32
	IV	409.57	429.38	474.32	477.45

15.9a.

Model	*MAD*	*RMSE*
$w = .5$	22.68	23.55
$w = .5, v = .5$	49.07	52.85

b.

Model	*MAD*	*RMSE*
$w = .5$	6.23	7.35
$w = .5, v = .5$	7.71	8.68

15.11a. $\hat{\beta}_0 = 6.609231$; $\hat{\beta}_1 = 1.704198$ **b.** $\hat{Y}_{1985} = 32.1722$; $\hat{Y}_{1986} = 33.8764$

c.

Year	*Prediction Interval*
1985	(30.0704, 34.2740)
1986	(31.7193, 36.0335)

15.13a. $\hat{Y}_t = -2.7138 + 11.3412t$ **b.**

Year	*Forecast*	*Prediction Interval*
1985	280.8162	(218.7324, 342.9000)
1986	292.1574	(229.5250, 354.7898)
1987	303.4986	(240.2822, 366.7150)

15.15a. $\hat{Y}_t = 83{,}542 + 2{,}099.462451t$ **b.** $\hat{Y}_{1992} = 131{,}829$; $\hat{Y}_{1993} = 133{,}929$ **c.** 1992: (129,120, 134,539); 1993: (131,189, 136,669) **15.17a.** Very strong negative autocorrelation **b.** Very strong positive autocorrelation **c.** Residuals are probably uncorrelated **15.19a.** Yes **b.** $d = .122$; reject H_0

15.21.

Year	Variable	a. Forecast $w = .5$	b. Forecast $w = .5, v = .5$
1992	Agricultural	3,211.6	3,214.9
1992	Nonagricultural	113,422.8	116,970.9

15.23.

Year	a. Forecast	Expected Gain/Loss	b. Forecast	Expected Gain/Loss
1987	49.126		50.802	
1988	49.126	+3.501/share	50.245	+4.62/share

15.25. $d = 2.367$; do not reject H_0

15.27a.

Quarter	Forecast	95% Lower Limit	95% Upper Limit
1	5,570.0	5,406.2	5,733.9
2	5,636.7	5,472.6	5,800.8
3	5,703.4	5,539.0	5,867.8
4	5,770.1	5,605.4	5,934.8

b. $F = .039$; do not reject H_0; no **c.**

Quarter	Forecast
1	5,573.5
2	5,639.5
3	5,704.2
4	5,764.5

d. $d = .155$; reject H_0

15.29a. $\hat{Y}_t = 36.6666 + .691722t$ **c.**

1971	1972	1973	1974	1975
54.6514	55.3431	56.0348	56.7265	57.4182

d. Projections are in agreement

Year	Primary	Secondary
1955	42.5	56.4
1975	57.42	40.91

15.31b.

Jan.	Feb.	Mar.
64.491	64.491	64.491

Chapter 16

16.9a. 6.39 **b.** 15.98 **c.** 1.54 **d.** 3.18 **16.11.** Diagram b **16.13a.** $s_p^2 = 2 = \text{MSE}$; $s_p^2 = 14.4 = \text{MSE}$ **b.** $t_1^2 = (-6.124)^2 = F_1$; $t_2^2 = (-2.282)^2 = F_2$ **c.** $t_{.025}^2$ with 10 df $= (2.228)^2 = F_{.05}$ **d.** Reject H_0 **16.15a.** $F = 1.5$; do not reject H_0 **b.** $F = 6.0$; reject H_0 **c.** $F = 24.0$; reject H_0 **d.** Increases **16.17a.** 4; 38 **b.** $F = 14.80$; reject H_0

16.19a.

Source	df	SS	MS	F
Treatment	2	12.301	6.1505	2.931
Error	9	18.888	2.0987	
Total	11	31.189		

b. $F = 2.931$; do not reject H_0

c. $T_1 - T_2$: $(-3.778, 2.538)$; $T_1 - T_3$: $(-1.458, 5.418)$; $T_2 - T_3$: $(-.996, 6.196)$

16.21a.

Source	df	SS	MS	F
Treatment	2	509.87	254.935	1.871
Error	90	12,259.96	136.2218	
Total	92	12,769.83		

b. $F = 1.871$; do not reject H_0; no **d.** No

16.23a. $F = 16.95$; reject H_0 **c.** $(-98.081, -45.919)$ **16.25b.** 571.9662 **c.** 7,719.8288

d.

Source	df	SS	MS	F
Treatment	2	571.9662	285.9831	3.96
Error	107	7,719.8288	72.1479	
Total	109	8,291.7950		

e. $F = 3.96$; reject H_0; yes

f. 1 − 2: (−.567, 8.907); 1 − 3: (.791, 12.229); 2 − 3: (−2.595, 7.275) **16.27a.** $F = 17.32$; reject H_0; yes **b.** p-value < .01 **c.** $t = 4.24$; reject H_0; yes **d.** 1 − 2: (−3.5, .3); 1 − 3: (−4.173, −.227); 1 − 4: (−6.845, −2.955); 2 − 3: (−2.224, 1.024); 2 − 4: (−4.890, −1.710); 3 − 4: (−4.377, −1.023)

16.29a.

Source	df	SS	MS	F
A	2	.8	.4	3.69
B	3	5.3	1.7667	16.31
AB	6	9.6	1.6	14.77
Error	12	1.3	.1083	
Total	23	17.0		

b. $F = 13.18$; reject H_0; yes **c.** Yes

e. $F = 14.77$; reject H_0 **f.** No **16.31b.** $F = 21.61$; reject H_0; yes **c.** Yes; $F = 36.60$, reject H_0 **d.** No **f.** Half-width: 1.838; pairs of treatments that differ: (A_1B_1, A_1B_3), (A_1B_1, A_2B_1), (A_1B_2, A_1B_3), (A_1B_2, A_2B_2), (A_1B_3, A_2B_2), (A_1B_3, A_2B_3), (A_2B_1, A_2B_2), (A_2B_1, A_2B_3) **16.33a.** F(Treatment) = 1.5; do not reject H_0 **b.** F(Treatment) = 5.25, reject H_0; $F(AB) = 7.5$, reject H_0 **c.** F(Treatment) = 5.25, reject H_0; $F(AB) = 3.0$, reject H_0 **d.** F(Treatment) = 20.25, reject H_0; $F(AB) = 4.5$, reject H_0 **16.35b.** F(Treatment) = 11.074, reject H_0; F(Interaction) = 6.784, reject H_0; half-width: 8.85; pairs of treatments that differ: (1N, 2R), (1T, 2R), (2T, 2N), (2T, 2R)

16.37a.

Source	df	SS	MS	F
Sex	1	10,686.361	10,686.361	68.74
Weight	1	538.756	538.756	3.47
Sex × Weight	1	831.744	831.744	5.35
Error	36	5,596.9083	155.4697	
Total	39	17,653.7693		

b. F(Treatment) = 25.85, reject H_0; F(Interaction) = 5.35, reject H_0; compare sexes at each weight; half-width: 13.701, pairs that differ are (FH − MH), (FL − ML) **16.39a.** 3 × 2 factorial **b.** H_0: $\beta_1 = \beta_2 = \beta_3 = \beta_4 = \beta_5 = 0$ vs H_a: At least one $\beta_i \neq 0$ (i = 1, 2, 3, 4, 5); $F = \dfrac{\text{MS(Model)}}{\text{MSE}}$; reject H_0 if $F > 3.11$ ($\alpha = .05$) **c.** H_0: $\beta_4 = \beta_5 = 0$ vs H_a: At least one $\beta_i \neq 0$ (i = 4, 5) **16.41c.** H_0: $\beta_1 = \beta_2 = \beta_3 = \beta_4 = \beta_5 = \beta_6 = \beta_7 = \beta_8 = 0$ vs H_a: At least one $\beta_i \neq 0$ (i = 1, 2, 3, 4, 5, 6, 7, 8); $F = 1{,}336.15$; reject H_0; yes **d.** H_0: $\beta_5 = \beta_6 = \beta_7 = \beta_8 = 0$ vs H_a: At least one $\beta_i \neq 0$ (i = 5, 6, 7, 8); reject H_0 if $F > 2.29$ **e.** $F = 258.05$; reject H_0 **f.** No **16.43a.** Completely randomized design **b.** 116 **c.** $F = 1.95$; do not reject H_0; no **d.** $F = .39$; do not reject H_0; no **e.** $F = 53.27$; reject H_0; yes **16.45a.** Completely randomized design **b.** $F = 4.88$; reject H_0; yes **c.** p-value < .01 **d.** Half-width: 2.369; pair of treatments that differ: (B − D) **e.** (10.961, 13.011)

16.47a.

Source	df	SS	MS	F
Companies	1	3,237.2	3,237.2	19.62
Error	98	16,167.7	164.9765	
Total	99	19,404.9		

b. $F = 19.62$; reject H_0; yes **c.** No

16.49b. $F = 12.29$; reject H_0; yes **c.** F(Interaction) = .02, do not reject H_0; F(Schedule) = 7.37, reject H_0; F(Payment) = 29.47, reject H_0 **16.51.** $F = 2.95$; do not reject H_0; no **16.53a.** $t = -.46$; do not reject H_0 **b.** $t = .46$; do not reject H_0 **c.** $F = .21$; do not reject H_0 **d.** $t^2 = (\pm.46)^2 = .21 = F$ **16.55a.** $E(y) = \beta_0 + \beta_1x_1 + \beta_2x_2$ where $x_1 = 1$ if NYSE, $x_1 = 0$ if not; $x_2 = 1$ if ASE, $x_2 = 0$ if not **c.** $F = 4.114$; reject H_0; p-value = .0197

16.57a.

Source	df	SS	MS	F
Type	1	10,706.7222	10,706.7222	7.18
Incentive	2	52,003.1167	26,001.5584	17.43
Type × Inc.	2	69.7611	34.8806	.023
Error	12	17,902.01	1,491.8342	
Total	17	80,681.61		

b. $F = .023$; do not reject H_0; no **d.** Different models fit

Chapter 17

17.3a. .035 **b.** .363 **c.** .004 **d.** .151; .1515 **e.** .212; .2119 **17.5.** $S = 17$; p-value = .054; reject H_0 **17.7a.** Sign test **b.** H_0: $M = 30$ vs H_a: $M < 30$ **c.** $S = 5$; reject H_0 if p-value < .01 **d.** p-value = .109; do not reject H_0 **17.9a.** $T_B \leq 35$ or $T_B \geq 67$ **b.** $T_A \geq 43$ **c.** $T_B \geq 93$ **d.** $z < -1.96$ or $z > 1.96$ **17.13a.** $T_A = 53$; reject H_0 **17.15.** $T_{After} = 86$; reject H_0 **17.17a.** $t = -1.26$; do not reject H_0 ($\alpha = .05$) **b.** $T_A = 37.5$; do not reject H_0 ($\alpha = .05$) **17.19a.** $\text{Min}(T_-, T_+) \leq 152$ **b.** $T_- \leq 60$ **c.** $T_+ \leq 0$ **17.23a.** H_0: Two sampled populations have identical distributions vs H_a: The probability distribution for population A is shifted to the right of that for population B **b.** $z = 2.499$; reject H_0 **c.** p-value = .0062 **17.25.** $T_+ = 2$; reject H_0 **17.27.** $T_+ = 6$; do not reject H_0 **17.29a.** $t = .802$; do not reject H_0 **b.** $T_- = 8$; do not reject H_0 **17.31a.** .995 **b.** .025 **c.** .10 **d.** .975 **e.** .025 **f.** .10 **17.33a.** Completely randomized design **b.** H_0: The three population distributions are identical vs H_a: At least two of the three population distributions differ in location **c.** Reject H_0 if $H > 9.21034$ **d.** $H = 13.85$; reject H_0 **17.35a.** $H = .809$; do not reject H_0 **17.37.** $H = 5.85$; do not reject H_0; no **17.39b.** $H = 16.505$; reject H_0 **c.** Wilcoxon rank sum test **17.41a.** 1 **b.** −.9 **c.** 1 **d.** .2 **17.43a.** $r_s = .864$; reject H_0 **c.** p-value < .005 **17.45a.** .861 **b.** Reject H_0; yes **17.47.** $r_s = .9341$ **17.49a.** $T_B = 19$; reject H_0; yes **17.51a.** $r_{s_1} = .9848$; $r_{s_2} = .9879$ **17.53.** A: $S = 3$, p-value = .5, do not reject H_0; B: $S = 5$, p-value = .031, reject H_0 **17.55.** $T_A = 135$; reject H_0; yes **17.57.** $T_{Before} = 132.5$; reject H_0; yes **17.59a.** Completely randomized design **b.** $H = 16.392$; reject H_0; yes **c.** Yes; 1 vs 2: $T_A = 59$, reject H_0; 1 vs 3: $T_A = 103$, do not reject H_0; 2 vs 3: $T_B = 151$, reject H_0 **17.61.** $H = 12.793$; reject H_0 **17.63.** $r_s = .7714$; do not reject H_0 **17.65.** $r_s = .8574$ **17.67a.** H_0: $M = .75$ vs H_a: $M \neq .75$ **b.** S, reject H_0 if p-value < .10 **d.** $S = 18$; p-value = .044; reject H_0

Chapter 18

18.1a. 27.5871 **b.** 70.0648 **c.** 22.3072 **d.** 12.8381 **18.3a.** $X^2 > 6.25139$ **b.** $X^2 > 15.0863$ **c.** $X^2 > 14.0671$ **18.7a.** $X^2 = 2.145$; do not reject H_0; no **18.9a.** $X^2 = 26.66$; reject H_0; yes **b.** $X^2 = 1$; do not reject H_0; no **18.11a.** $X^2 = 12.374$; do not reject H_0 **b.** $.05 < p\text{-value} < .10$ **18.13.** $X^2 = 10.3$; reject H_0; yes **18.15a.** $X^2 > 21.0261$ **b.** $X^2 > 22.3072$ **c.** $X^2 > 13.2767$

18.17a.

	1	2	3	
1	42.9%	52.5%	64.3%	56.0%
2	57.1%	47.5%	35.7%	44.0%

18.19.

		B			
		B_1	B_2	B_3	
	A_1	28.7%	44.0%	30.4%	34.9%
A	A_2	49.3%	32.1%	50.7%	43.4%
	A_3	22.1%	23.9%	18.8%	21.7%

18.23a. $X^2 = 20.781$; reject H_0; yes **b.** Sprain; older; younger **18.25b.** $X^2 = 89.09$; reject H_0 **18.27.** $X^2 = 40.697$; reject H_0; yes **18.29d.** $X^2 = 2.373$; do not reject H_0 **e.** Yes **18.31a.** $X^2 = 87.379$; reject H_0; yes **18.33.** $X^2 = 6.579$; reject H_0; yes **18.35.** $X^2 = 5.506$; do not reject H_0; no **18.37.** $X^2 = 2.601$; do not reject H_0; no **18.39a.** $X^2 = 47.981$; reject H_0; yes **b.** (.079, .171) **18.41a.** $X^2 = 38.68$; reject H_0; yes **18.43.** $X^2 = 3.132$; do not reject H_0; no **18.45a.** $X^2 = 269.91$; reject H_0; yes **b.** p-value < .005 **18.47a.** $X^2 = 46.24$; reject H_0; yes **b.** p-value < .005

Chapter 19

19.3a. Uncertainty **b.** Conflict **c.** Certainty **19.5.** Actions, states of nature, outcomes, objective variables

19.13a. a_3 is inadmissible **b.**

	State of Nature			
	S_1	S_2	S_3	S_4
a_1	0	0	40	30
a_2	11	8	0	9
a_3	42	38	50	0

19.15a. a_4 is inadmissible

19.17a. Payoffs in millions of dollars

	State of Nature		
	1%	2%	3%
Market	14	16	18
Don't market	15	15	15

b. Opportunity losses in millions of dollars

	State of Nature		
	1%	2%	3%
Market	1	0	0
Don't market	0	1	3

19.19a. Payoffs in dollars

	State of Nature	
	Successful	Unsuccessful
Increase advertising	600,000	−600,000
Don't increase advertising	200,000	200,000

b. No

19.21a. Payoffs in millions of dollars

	State of Nature	
	Win	Lose
Court	−1	−101
No court	−50	−50

19.23. a_2 **19.27.** a_2

19.29a. a_2 **b.** a_1 **19.31.** Do not purchase **19.33a.** Both actions have same expected payoff

b.

	State of Nature	
	(⅔) Successful	(⅓) Unsuccessful
Increase advertising	0	800,000
Don't increase advertising	400,000	0

c. Both actions have same expected opportunity loss

19.35a.

	S_1	S_2	S_3
a_1	10,000	5,625	625
a_2	4,900	9,025	0

b. a_1

19.37a.

	S_1	S_2	S_3	S_4
a_1	.5	.8	0	−.2
a_2	.3	.65	.9	.4
a_3	.15	.2	.6	1.1

b. a_2

19.39a.

	Dry	Moderate	Gusher
Invest	−500,000	600,000	1,500,000
Don't Invest	0	0	0

b. Invest **c.** Do not invest

19.41a. Risk-taking **b.** Risk-neutral **c.** Risk-avoiding **d.** Risk-neutral **19.43b.** Risk-taking **c.** a_1

19.45b. Risk-avoiding **19.49a.**

	(4)	(5)
S_1	.0900	.1925
S_2	.0250	.0535
S_3	.2925	.6257
S_4	.0600	.1283
	.4675	1.0000

19.53. .526

19.55a. $P(S_1) = .25$; $P(S_2) = .50$; $P(S_3) = .25$ **b.**

	(3)	(4)	(5)
S_1	.0000098	.0000025	.00096
S_2	.0011281	.0005641	.21766
S_3	.0081000	.0020250	.78137
		.0025916	1.00000

19.57. a_2 **19.59a.** a_3 **b.**

	(4)	(5)
S_1	.10	.3125
S_2	.18	.5625
S_3	.04	.1250
	.32	1.0000

c. a_3

19.61a. Payoffs in millions of dollars

	State of Nature	
	(.6) Pass bill	(.4) Don't pass bill
Stay	2.5	−1.5
Lease	.5	.5

b.

	(4)	(5)
S_1	.48	.923
S_2	.04	.077
	.52	1.000

d. a_1: Stay in business **19.63a.** −10.014 **b.** First source **19.65a.** \$48,000 **b.** \$18,000; purchase survey **19.67a.** \$2.68 **19.69b.** Risk-neutral **c.** a_1 **19.71a.** Relative frequency **b.** Subjective **c.** Subjective **d.** Relative frequency **19.73a.** P(Repaid) = .8625; P(Repaid with difficulty) = .09375; P(Defaulted) = .04375

b. a_2: Do not make loan **19.75a.** Payoffs in \$100,000

	State of Nature	
	Economist 1	Economist 2
.99	7.94	11.92
1.98	11.96	16.00
2.75	12.38	13.75
3.50	10.50	7.00

b. Opportunity losses in \$100,000

	State of Nature	
	Economist 1	Economist 2
.99	4.44	4.08
1.98	.42	0
2.75	0	2.25
3.50	1.88	9.00

c. a_2: Set price at \$1.98

19.77a. Utilities in thousands of dollars

	Win	Lose
Court	.7143	0
Settle	.6186	.6186

b. a_2: Settle out of court

19.79a. Risk-neutral **b.** Risk-avoiding **c.** Risk-taking **d.** Risk-taking **19.81a.** .9911, .9960, .9977, .9985, .9990, .9993, .9996, .9998, .9999

19.83a.

	State of Nature	
	5%	10%
Accept	0	−4,100
Reject	−1,000	0

b. Reject **c.** Reject **d.** Accept **e.** 21.957 **f.** Yes

19.87. ENGS = −\$4,993,900; do not develop model

Chapter 20

20.5a. 40%; .7746 **b.** 20%; .8944 **c.** 10%; .9487 **d.** 1%; .9950 **20.7a.** 2.981; 3 **b.** 1.151; 1.2 **20.9.** (417.816, 426.184) **20.11.** (135,140.29, 140,659.71) **20.13a.** (400,181.28, 1,157,878.72) **b.** (600,271,925.2, 1,736,818,074.8) **20.15b.** .041 **c.** (.045, .127) **d.** t = .85; do not reject H_0; no **20.17a.** (199.75, 460.25) **b.** (81,896.10, 188,703.90) **20.19a.** $k = 3$; $N_1 = 25{,}000$; $N_2 = 10{,}000$; $N_3 = 5{,}000$; $N = 40{,}000$ **b.** $n_1 = 20$; $n_2 = 15$; $n_3 = 10$; $n = 45$ **c.** $\bar{x}_1 = 75.65$; $\bar{x}_2 = 51.7333$; $\bar{x}_3 = 26.8$ **d.** $s_1^2 = 232.3447$; $s_2^2 = 145.2095$; $s_3^2 = 77.5111$ **e.** 63.5646; ±4.5867 **f.** 2,542,583; ±183,466.23 **g.** .7125; ±.1080 **20.21.** 49.84; ±9.7889 **20.23a.** \$44,844 **b.** ±\$2,128 **c.** (57,077.5, 65,135.9) **20.25.** 19,904.54; ±778.57 **20.27.** 77 **20.29a.** 129 **b.** 203 **20.31.** 7,951 **20.35a.** 73,850; ±37,845.89 **b.** .7; ±.167 **20.37a.** .9296; ±.0051 **20.39.** 7.1595; ±.8609

Index of Exercise Data Sets

Most of the exercise data sets that contain 30 or more measurements are listed below and are available on a computer disk (ASCII format). Instructors who adopt this text may obtain a copy of the disk by writing Dellen Publishing Company, 400 Pacific Avenue, San Francisco, California 94133, or by calling 415/433-9900.

Exercise Number	File Name	Description	Page
9.15	CHAP09.015	Prices of single-family homes	400
9.24	CHAP09.024	Data for Learning the Mechanics	408
9.27	CHAP09.027	Ages of toothpaste purchasers	409
9.53	CHAP09.053	Pupillometer readings	432
10.14	CHAP10.014	Retaliation index and salaries of whistle blowers	469
10.15	CHAP10.015	Gasoline and crude oil prices: 1975–1990	470
10.24	CHAP10.024	Cutting tool data	476
10.25	CHAP10.025	Nielsen data: Channels available and number watched	476
10.32	CHAP10.032	Managerial success data	481
10.33	CHAP10.033	Bicycle data: Production and overhead	482
10.44	CHAP10.044	GDP and new housing starts data	492
10.46	CHAP10.046	3-month CD rates and S&P 500 Index: 1980–1991	493
10.47	CHAP10.047	Calibration data	493
10.48	CHAP10.048	Florida counties: 1982 population density and crime	494
10.57	CHAP10.057	Industry quit rates and average wages, 1986	502
10.59	CHAP10.059	Manager demand	503
10.66	CHAP10.066	Exhaustion index and concentration	514
10.69	CHAP10.069	Air and rail passenger data	515
10.70	CHAP10.070	Flour shipments data	516
10.72	CHAP10.072	Evaluation points and salaries	517
10.73	CHAP10.073	Single-family homes: Living area and selling price	519
10.74	CHAP10.074	Fortune 500 companies: Sales and number of employees	521
11.5	CHAP11.005	1990 Washington, D.C., household data	540
11.7	CHAP11.007	Production times and experience	542
11.10	CHAP11.010	1990 Minneapolis apartment data	544
11.17	CHAP11.017	Shipping department data	554
11.20	CHAP11.020	Survival size of nonprofit hospitals	557
11.21	CHAP11.021	Service call data	558
11.51	CHAP11.051	Sales data and Producer Price Index: 1967–1984	600
11.52	CHAP11.052	Predicting GPA of college freshmen	602
11.56	CHAP11.056	Fuel, population, and price data: 1965–1983	605
12.14	CHAP12.014	Mill production data	619
12.15	CHAP12.015	Interest rates and money supply data	619

Exercise Number	File Name	Description	Page
15.29	CHAP15.029	Market share of primary and secondary lenders: 1945–1970	847
15.30	CHAP15.030	Consumer credit: 1980–1991	847
15.31	CHAP15.031	Monthly IBM stock prices: 1990–1992	848
16.20	CHAP16.020	Client data for accounting firm	875
16.42	CHAP16.042	Shipping weights of cartons	908
16.45	CHAP16.045	Wear data for lenses	910
16.54	CHAP16.054	Closing prices for stocks	915
17.12	CHAP17.012	Learning the Mechanics data	932
17.18	CHAP17.018	Success of MIS	934
17.44	CHAP17.044	Quarterly S&P 500 index: 1979–1986	959
17.45	CHAP17.045	Job satisfaction and income	959
17.51	CHAP17.051	Calibration data	962
17.59	CHAP17.059	Potential office sites data	965
20.12	CHAP20.012	Learning the Mechanics data	1090
20.14	CHAP20.014	Parts inventory audit	1090
20.23	CHAP20.023	Hospital bed data	1098
20.35	CHAP20.035	PAC money expenditures	1103

Index

Symbol	Description
$\hat{p}_1 - \hat{p}_2$	Difference between sample proportions of success in two independent binomial samples 433
$\hat{p}_i$	Sample proportion for stratum i in a stratified random sample 1094
$\hat{p}_{st}$	Estimator of population proportion computed from a stratified random sample 1094
$\bar{p}$	(1) The overall proportion of defective units in k samples of size n 753 (2) Centerline of p-chart 754
P(A)	Probability of event A 132
P(A|B)	Probability of event A given that event B occurs 154
q	Probability of Failure for the binomial and geometric distributions 198
r	Pearson product moment coefficient of correlation between two samples 485
r^2	Coefficient of determination in simple regression 489
r_s	Spearman's rank correlation coefficient between two samples 952
R	Sample range 727
$\overline{R}$	(1) Mean of a set of sample ranges 727 (2) Centerline of R-chart 743
R_i	(1) Range of ith sample 743 (2) Rank sum for the ith sample in a nonparametric analysis of variance 974
R^2	Coefficient of determination in multiple regression 546
R_t	Residual effect in a time series 810
RMSE	Root mean squared error of a set of forecasts 818
ρ (rho)	Pearson product moment coefficient of correlation between two populations 487
ρ_s	Spearman's rank correlation coefficient between two populations 954
s	Sample standard deviation 67
s^2	Sample variance 67
s_D	Sample standard deviation of differences in a paired-difference experiment 423
s_i^2	Sample variance for stratum i 1093
s_p^2	Pooled sample variance in two-sample t test 402
S	Sample space 131
S_i	State of nature i in a decision analysis 1010
S_t	Seasonal effect in a time series 809
SS(Total)	Total sum of squares in an analysis of variance 882
SS_{xx}	Sum of squares of the distances between x measurements and their mean 465
SS_{xy}	Sum of products of distances of x and y measurements from their means 465
SS_{yy}	Sum of squares of the distances between y measurements and their mean 474
SSE	(1) Sum of squared distances between observed and predicted values in a regression model 474 (2) Sum of squares for error in an analysis of variance 860
SST	Sum of squares for treatments in an analysis of variance 860

Symbol	Description
σ (lowercase sigma)	Population standard deviation 68
σ^2	Population variance 67
$\sigma_{\hat{\beta}_1}$	Standard deviation of the sampling distribution of β_1 478
$\sigma_{\hat{p}}$	Standard deviation of the sampling distribution of $\hat{p}$ 326
$\sigma_{(\hat{p}_1-\hat{p}_2)}$	Standard deviation of the sampling distribution of $(\hat{p}_1 - \hat{p}_2)$ 434
σ_R	The standard deviation of the sampling distribution of R 743
$\sigma_{\bar{x}}$	Standard deviation of the sampling distribution of $\bar{x}$ 280
$\sigma_{(\bar{x}_1-\bar{x}_2)}$	Standard deviation of the sampling distribution of $(\bar{x}_1 - \bar{x}_2)$ 392
$\sigma_{\hat{y}}$	Standard deviation of the sampling distribution of $\hat{y}$ 497
$\sigma_{(y-\hat{y})}$	Standard deviation of prediction error when $\hat{y}$ is used to predict a particular value of y 497
$\sum$ (uppercase sigma)	Symbol for summation 51
t	Statistic used for small-sample tests of hypotheses 317
t_α	Value of t distribution with an area α to its right 318
T_A, T_B	Rank sums corresponding to the two samples in a Wilcoxon rank sum test 928
T_L, T_U	Upper and lower rejection region values in a Wilcoxon rank sum test 928
T_t	Secular trend of a time series 808
T_0	Rejection region value in a Wilcoxon signed rank test 936
T_+, T_-	Sum of the ranks of the positive and negative differences in a Wilcoxon signed rank test 935
τ (tau)	Population total 1078
$\hat{\tau}$	Estimator of population total τ 1088
$U(O_{ij})$	Utility value corresponding to action i and state of nature j outcome in a decision analysis 1029
$\bar{x}$	Sample mean 51
$\bar{\bar{x}}$	(1) Mean of a set of sample means 727 (2) Centerline of $\bar{x}$-chart 727
x_i	ith sample measurement 50
$\bar{x}_i$	Sample mean for stratum i in a stratified random sample 1093
$\bar{x}_D$	Sample mean of differences in a paired difference experiment 423
$\bar{x}_{st}$	Estimator of population mean μ computed from a stratified random sample 1094
X^2	Test statistic for multinomial and contingency table tests 973
Y_t	Value of a time series Y at time t 779
$\bar{y}$	Mean of all observations in an analysis of variance 860
$\bar{y}_i$	Sample mean for ith treatment in an analysis of variance 859
$\hat{y}$	Least squares prediction of y using a regression model 464
z	z-score 86
$z_{\alpha/2}$	Value of standard normal random variable with an area $\alpha/2$ to its right 307